GRAPHS

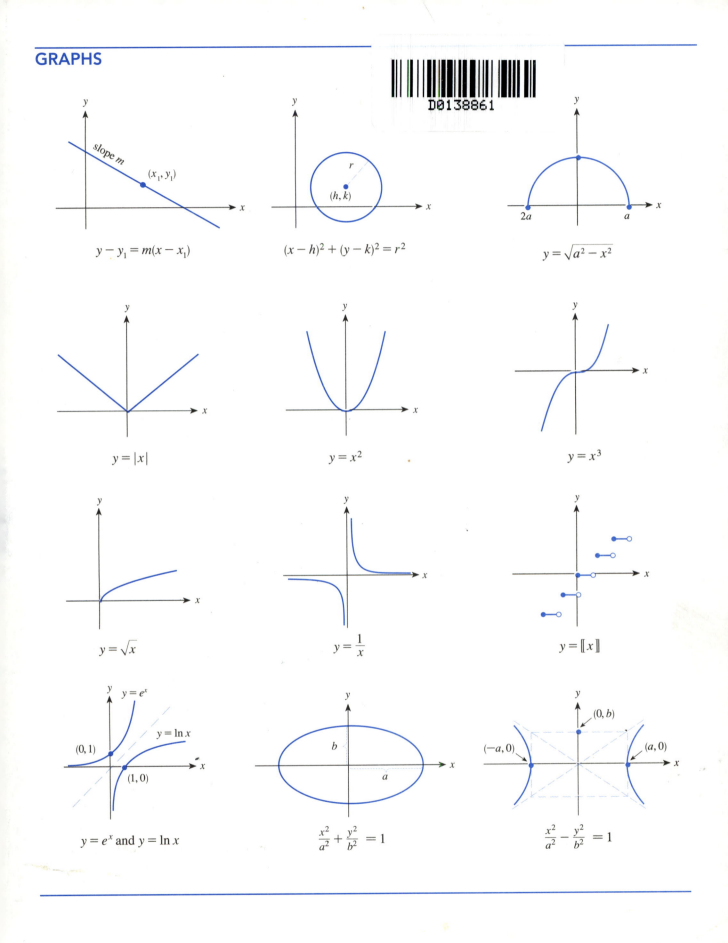

$y - y_1 = m(x - x_1)$

$(x-h)^2 + (y-k)^2 = r^2$

$y = \sqrt{a^2 - x^2}$

$y = |x|$

$y = x^2$

$y = x^3$

$y = \sqrt{x}$

$y = \dfrac{1}{x}$

$y = [\![\, x \,]\!]$

$y = e^x$ and $y = \ln x$

$\dfrac{x^2}{a^2} + \dfrac{y^2}{b^2} = 1$

$\dfrac{x^2}{a^2} - \dfrac{y^2}{b^2} = 1$

College Algebra

Fifth Edition

David Cohen
University of California
Los Angeles

With contributions by

Ross Rueger
College of the Sequoias

THOMSON
BROOKS/COLE

Australia • Canada • Mexico • Singapore • Spain
United Kingdom • United States

THOMSON

BROOKS/COLE

Mathematics Editors: Jennifer Huber/John-Paul Ramin
Assistant Editors: Rachael Sturgeon/Molly Nance
Editorial Assistant: Lisa Chow
Technology Project Manager: Star MacKenzie
Marketing Manager: Karin Sandberg
Marketing Assistant: Stephanie Taylor
Advertising Project Manager: Bryan Vann
Project Manager, Editorial Production: Janet Hill
Print/Media Buyer: Vena Dyer
Permissions Editor: Joohee Lee

Production Service: Helen Walden
Text Designer: Rokusek Design
Photo Researcher: Helen Walden
Illustrator: TECHarts
Cover Designer: Cheryl Carrington
Cover Image: Peter Samuels/Getty Images
Cover Printer: Lehigh Press (NJ)
Compositor: G & S Typesetters, Inc.
Printer: R. R. Donnelley & Sons, Willard

Printed in the United States of America

2 3 4 5 6 7 06 05 04 03

For more information about our products, contact us at:
Thomson Learning Academic Resource Center
1-800-423-0563

For permission to use material from this text, contact us by:
Phone: 1-800-730-2214   **Fax:** 1-800-730-2215
Web: http://www.thomsonrights.com

Library of Congress Control Number: 2002110403

Student Edition with InfoTrac College Edition:
ISBN 0-534-35771-7

Student Edition without InfoTrac College Edition:
ISBN 0-534-45318-X

Instructor's Edition: ISBN 0-534-39709-3

Brooks/Cole–Thomson Learning
511 Forest Lodge Road
Pacific Grove, CA 93950
USA

Asia
Thomson Learning
5 Shenton Way #01-01
UIC Building
Singapore 068808

Australia
Nelson Thomson Learning
102 Dodds Street
South Melbourne, Victoria 3205
Australia

Canada
Nelson Thomson Learning
1120 Birchmount Road
Toronto, Ontario M1K 5G4
Canada

Europe/Middle East/Africa
Thomson Learning
High Holborn House
50/51 Bedford Row
London WC1R 4LR
United Kingdom

Latin America
Thomson Learning
Seneca, 53
Colonia Polanco
11560 Mexico D.F.
Mexico

Spain
Paraninfo Thomson Learning
Calle/Magallanes, 25
28015 Madrid, Spain

With deep gratitude to these three colleagues and mentors for affording me the privilege to teach.
Newman Fisher (San Francisco)
Phil Curtis and Ted Gamelin (Los Angeles)

Tribute to David Cohen

Because he loved the part in "The Myth of Sisyphus," where Camus re-envisions Sisyphus eternally condemned to the task of pushing a giant boulder to the summit of a mountain only to watch it roll back down again—descending the hill and smiling, I like to remember my dad in similar moments of inbetweenness. When the unknown and hope would combine to make anything seem attainable. When becoming conscious of a situation did not necessarily signify limitation by it, but instead, liberation from it.

And so I imagine my dad, after hours of working at the computer, between a thoughtful sip of coffee and suddenly realizing the clearest way to word a problem. After years of early morning commutes to avoid 405 traffic, solitary interludes that began with waking to a still-dark sky and still-dreaming family, and ended with walking, transformed into a professor, into a UCLA lecture hall. After a lifetime of doing his best for those he cared about, intervals of good and poor health simply challenged him to give a little more. My dad endured his task bravely, from his decision 15 years ago to fight leukemia until last May, when the illness reminded us that even fathers and teachers are mortal, regardless of scheduled office hours or books left unwritten.

A couple months ago, my dad told me the five most important things in life to him; teaching and writing these math books were both on that list. I was never a "registered" student of his, but to anyone using this book, especially students, I pass on what I know he would have told me: Your best is always good enough. Enjoy.

Emily Cohen

• • •

My dad was gifted at teaching. He had an innate ability to explain things simply and relevantly. He excelled at combining his explanations with a patient ear and his self-defined nerdy sense of humor.

I remember when my dad taught me to read. I had become upset one night while we were out to dinner because I couldn't read the menu. Frustrated, I expressed my dismay from my booster chair. My dad responded with the first of my reading lessons. He began by pointing out something I already knew, the letter "a." He carefully explained that this letter was actually a word . . . every day after that he would write new words on one of his ubiquitous lined yellow notepads and teach them to me. Eventually we began to form simple sentences. He always kept my attention because he made up sentences that made me laugh. The frogs go jog, the cat loves the dog, so on and so forth. My dad's sense of humor was quite captivating.

I was lucky in high school to have someone to assist me with all of my mathematical questions. My math teachers seemed to have a special aptitude for making formulas, theories, and problems both complicated and boring. My dad was good at simplifying these matters for me. He was so good at explaining math to me that I would often remark afterward, with the hindsight of an enlightened one, "Oh, that's all? Why didn't they just say that in the first place?" I took comfort in knowing that math didn't have to be complicated when it was explained well.

Hemingway said that the key to immortality was to write a book. I feel lucky because my dad left behind several. Try reading one—you might find that you actually like it!

Jennifer Cohen

• • •

I first met Dave Cohen in spring 1981, when I signed up for an algebra course he was teaching as part of my graduate studies at UCLA. I quickly took a liking to his conversational style of teaching and the clear explanations he would use to illustrate an idea. Dave and I began an informal, weekly meeting to discuss solutions to various problems he would pose. Usually he would buy me a cup of coffee and present me with some new trigonometric identity he had discovered, or some archaic conic section property he had come across in an old 1886 algebra book. We would discuss the problem ("Did it have a solution? Was it really an identity? Could we prove it?"), then go about our business.

In summer 1981, I had the opportunity to TA a course called Precalculus, which used Dave's notes (sometimes written by hand), rather than a traditional textbook. I had no idea what the term "precalculus" meant, and soon learned that these notes Dave had prepared were beginning to define, or at the least greatly expand, the subject. It was the beginning of a textbook.

Two years later, after graduating UCLA, I was pleased to receive a copy of *Precalculus,* by David Cohen. Dave had scribbled a note to me saying: "Thanks for the great coffee breaks." I was pleased to see that some of our weekly problems appeared in that book (others showed up years later). More importantly, that book, and every other text Dave has ever written, talked to you. Even now, when I read this book, I can hear Dave talking, and sometimes even listening. Lecturing was never a part of Dave Cohen's vocabulary.

Since that time Dave has written books entitled *College Algebra, Trigonometry,* and *Algebra and Trigonometry.* Most have gone on to multiple editions, including this one. In each edition Dave has sought the better explanation, the "cooler" problem, the more interesting data set. I've had the honor of working with him on all of these books, and we had developed a mathematical friendship. Perhaps some of you reading this have a great study partner; one who intuitively knows what the other is thinking. That was the relationship Dave and I had.

I last saw Dave in November 2001, when we met at UCLA to (naturally) have a cup of coffee and discuss this textbook. He was excited about the quality of problems and interesting data sets he had found to use in this book. He felt this was going to be the best book he had ever written.

In May 2002, just after completing this manuscript, Dave Cohen passed away from complications caused by his leukemia. His legacy of fine textbooks (and this is, perhaps, his finest) and great teaching has influenced a generation of students and teachers. Everyone who had contact with Dave feels a bit richer from the experience. His total lack of ego, curiosity about the world, and respect for what others think, made him one of the finest human beings I have ever met.

Ross Rueger

Contents

Preface

This text develops the elements of college algebra in a straightforward manner. As in the earlier editions, my goal has been to create a book that is *accessible* to the student. The presentation is student-oriented in three specific ways. First, I've tried to talk to, rather than lecture at, the student. Second, examples are consistently used to introduce, to explain, and to motivate concepts. And third, many of the initial exercises for each section are carefully coordinated with the worked examples in that section.

AUDIENCE

In writing *College Algebra,* I have assumed that the students have been exposed to intermediate algebra, but that they have not necessarily mastered that subject. Also, for many college algebra students, there may be a gap of several years between their last mathematics course and the present one. Appendix B consists of review sections for such students, reviewing topics on integer exponents, nth roots, rational exponents, factoring, and fractional expressions. In Chapter 1, many references refer the reader to Appendix B for further practice.

CURRICULUM REFORM

This new edition of *College Algebra* reflects several of the major themes that have developed in the curriculum reform movement of the past decade. Graphs, visualization of data, and functions are now introduced much earlier and receive greater emphasis. Many sections now contain more examples and exercises involving applications and real-life data. In addition to the Writing Mathematics sections from the previous edition, there are now mini projects or projects at the ends of many sections. These give the students additional opportunities to discuss, explore, learn, and write mathematics, often using real-life data.

TECHNOLOGY

In the following discussion and throughout this text, the term *graphing utility* refers to either a graphing calculator or a computer with software for graphing and analyzing functions. Over the past decade, all of us in the mathematics teaching community have become increasingly aware of the *graphing utility* and its potential for making a positive impact on our students' learning. We are also aware of the limitations of the graphing utility as a *sole* analysis device.

The role of the graphing utility is expanded in this new edition: The existence of the graphing utility is taken for granted and some examples do make use of this technology. However, just as with the previous edition, this remains a text in which the central focus is on mathematics and its applications, and if the instructor chooses, the text can be used without reference to the graphing utility. As a minimum, though, the students will need to read the page in Section 1.5 explaining how to specify the dimensions of a viewing rectangle, since the notation accompanies some figures in the text. Overall, the quality and relevance of the graphing utility exercises is vastly improved over the previous edition. Graphing utility exercises (identified by the symbol ⓖ) are now integrated into the regular exercise sets.

CHANGES IN THIS EDITION

Comments and suggestions from students, instructors, and reviewers have helped me to revise this text a number of ways that I believe will make the book more useful to the instructor and more accessible to the student. As previously mentioned, graphing utility exercises now appear in virtually every exercise set as well as in the text examples. Some of the major changes occur in the following areas.

Chapter 1   Some of the previous edition's Chapter 1 review material that is developed at length in intermediate algebra courses has been shifted to two appendices in the new edition. There are then notes to students at appropriate places reminding them to consult one of these appendices if the use of one of these topics seems unclear. Additionally, the material on long division of polynomials that was in the first chapter of the previous edition now appears in Chapter 7 where it more naturally belongs, preceding synthetic division. Section 1.4 introduces graphs and data visualization, and Section 1.5 includes an introduction to graphing utilities.

Chapter 2   This is essentially Chapter 3 of the former text, although graphing utilities are utilized more extensively in examples and exercises.

Chapter 3   Section 3.1 includes an expanded introduction to the function concept. Functions are introduced at length using algebraic, verbal, tabular, and graphical forms. Functions as models are introduced and used in examples and exercises. Implicit functions are introduced in the section project. In Section 3.3, material on the average rate of change of a function has been expanded, and there is an increased emphasis on applications.

Chapter 4   Section 4.1 includes additional examples on applications of the regression line. There is also a new discussion on the use of spreadsheets in generating scatter plots and regression lines. In Section 4.2, there is new material on modeling real-life data with quadratic functions. Also, first and second differences are introduced as a tool for determining whether a set of data points may be generated by a linear or quadratic function. In Section 4.3, this section on iteration and population dynamics is now marked "optional," and some exercises that don't refer directly to the text exposition have been deleted. In Section 4.5, in order to better focus on problem solving rather than algebraic manipulation in the applied max/min problems, the vertex formula $x = -b/2a$ is introduced as a simple tool for determining the vertex of a parabola. Instructors who require the use of completing the square rather than the vertex formula should find their students well prepared in view of the previous work on that topic in Sections 1.7, 2.1, and 4.2.

Chapter 5   Section 5.2 contains new material explaining in a careful, but student-friendly, manner exactly why the use of e as the base for an exponential function indeed simplifies matters. In Section 5.3 the Richter magnitude scale is discussed as an example of a logarithmic scale. Historical background motivating the use of the logarithmic scale is provided. The Section 5.6 project explains how loan payments in an amortization schedule are calculated under compound interest. In Section 5.7 the concept of an average relative growth rate is explained in the text and a follow-up exercise gives an empirical introduction to the instantaneous relative growth rate. These ideas are used to supply insight into the interpretation and use of the exponential growth constant k.

Chapter 6   In Section 6.1 the work on simple linear systems is applied to supply and demand models in determining market equilibrium. The Section 6.2–6.4 projects and mini projects provide a detailed introduction to the Leontief input-output model in economics. Section 6.3 discusses the use of graphing calculators and spreadsheets in computing matrix products. Students can apply this technology in an extended project on the use of matrices in the study of communication networks. Section 6.4 discusses the use of graphing calculators and spreadsheets in computing matrix inverses. The section includes new examples and exercises on using matrices to code and decode messages. In an extended project, students will need to use a graphing utility to solve a classic input-output problem involving a seven-sector model for the U.S. economy.

Chapter 7   In Section 7.2 long division for polynomials is discussed just prior to synthetic division. (In the previous edition, long division was included in Chapter 1.) Two methods for solving certain types of cubic equations are explained and applied in a project for this section. This complements the discussion in the text (from previous editions) on the history of polynomial equations.

Chapter 8   Section 8.1 defines focal length and focal ratio in preparation for an example and exercises on radio telescopes and parabolic reflectors. Mini project 2 at the end of the section discusses the classic string and T-square construction of the parabola. Section 8.2 includes new material on the perihelion and aphelion for an elliptical orbit. In Section 8.4 the use of hyperbolas in determining a location appears in the section project.

The Accompanying CD   *Interactive Video Skillbuilder CD* **(0-534-38618-0)** Think of it as portable office hours! The Interactive Video Skillbuilder CD-ROM contains more than eight hours of video instruction. The problems worked during each video lesson are shown next to the viewing screen so that students can try working them before watching the solution. To help students evaluate their progress, each section contains a 10-question Web quiz (the results of which can be emailed to the instructor) and each chapter contains a chapter test, with answers to each problem on each test. The CD also includes MathCue Tutorial software. This dual-platform software presents and scores problems and tutors students by displaying annotated, step-by-step solutions. Problem sets can be customized as desired.

SUPPLEMENTARY MATERIALS

For the Instructor

Annotated Instructor's Edition **(0-534-39709-3)**   This special version of the complete student text contains a Resource Integration Guide as well as complete answers at the back of the book.

Test Bank **(0-534-36901-4)**   The text bank includes 8 tests per chapter as well as 3 final exams. The tests are made up of a combination of multiple-choice, free-response, true/false, and fill-in-the-blank questions.

Complete Solutions Manual **(0-534-36898-0)**   The complete solutions manual provides worked-out solutions to all of the problems in the text.

BCA Testing **(0-534-36902-2)**   With a balance of efficiency and high-performance, simplicity and versatility, *Brooks/Cole Assessment* gives you the power to transform the learning and teaching experience. This revolutionary, Internet-ready testing suite is text-specific and allows instructors to customize exams and track student progress in an accessible, browser-based format. *BCA Testing* offers full algorithmic generation of problems and free-response mathematics. No longer are you limited to multiple choice or true/false test questions. The complete integration of the testing and course management components simplifies your routine tasks. Test results flow automatically to your gradebook, and you can easily communicate to individuals, sections, or entire courses.

Text-Specific Videotapes **(0-534-36900-6)**   This set of videotapes is available free upon adoption of the text. Each tape offers one chapter of the text and is broken down into 10–20 minute problem-solving lessons that cover each section of the chapter.

Available to qualified adopters—please consult your local sales representative for details.

For the Student

Student Solutions Manual **(0-534-36899-9)**   The student solutions manual provides worked-out solutions to the odd-numbered problems in the text.

Web Site **www.brookscole.com/mathematics**   The Brooks/Cole Mathematics Resource Center offers book-specific student and instructor resources, discipline-specific links, and a complete catalog of Brooks/Cole mathematics products.

BCA Tutorial Instructor and Student Versions **(0-534-38829-9-Instructor/ 0-534-37820-X-Student)**   This text-specific, interactive tutorial software is delivered via the Web (at http://bca.brookscole.com) and is offered in both student and instructor versions. Like *BCA Testing*, it is browser-based, making it an intuitive mathematical guide even for students with little technological proficiency. So sophisticated, it's simple, *BCA Tutorial* allows students to work with real math notation in real time, providing instant analysis and feedback. The tracking program built into the instructor version of the software enables instructors to carefully monitor student progress.

ACKNOWLEDGMENTS

Many students and teachers from both colleges and high schools have made useful constructive suggestions about the text and exercises, and I thank them for that. Special thanks go to Ian Crewe and Ignacio Alarcon for their careful work in checking the text for accuracy, and to Matt Bourez for checking exercise answers. In preparing and revising the manuscript, I received valuable suggestions and comments from the following reviewers.

Donna J. Bailey
Truman State University

Laurence Small
Los Angeles Pierce College

Josephine Battaglia
Pennsylvania State University

Mary Jo Steig
Huntingdon College

Marianne Bedee
Bowling Green State University—
Firelands College

Ping Wang
Pennsylvania State University

I would also like to thank Ross Rueger, who wrote the supplementary manuals and created the answer section for the text. Finally, to Jennifer Huber, Jean-Paul Ramin, Bob Pirtle, Molly Nance, Rachael Sturgeon, Janet Hill, Helen Walden, and the staff at Brooks/Cole–Thomson Learning, thank you for all your work and help in bringing this manuscript into print.

David Cohen

CHAPTER

1

Fundamentals

Natural numbers have been used since time immemorial; fractions were employed by the ancient Egyptians as early as 1700 B.C.; and the Pythagoreans, in ancient Greece, about 400 B.C., discovered numbers, like $\sqrt{2}$, which cannot be fractions. —Stefan Drobot in *Real Numbers* (Englewood Cliffs, N.J.: Prentice-Hall, Inc., 1964)

What secrets lie hidden in decimals? —Stephan P. Richards in *A Number for Your Thoughts* (New Providence, N.J.: S. P. Richards, 1982)

Real numbers, equations, graphs—these topics set the stage for our work in college algebra. How much from previous algebra courses should you remember about solving equations? Section 1.3 provides a review of the fundamentals. The rest of the chapter reviews and begins to extend what you've learned in previous courses about graphs and graphing. For example, we use graphs to visualize trends in

- Spending by the television networks to broadcast the Olympic Games (Exercises 23 and 24, page 31)
- Internet usage (Exercise 25, page 31)
- Carbon dioxide levels in the atmosphere (Example 5, page 28)
- U.S. population growth (Exercises 7 and 8 on page 58)

1.1 SETS OF REAL NUMBERS

Here, as in your previous mathematics courses, most of the numbers we deal with are real numbers. These are the numbers used in everyday life, in the sciences, in industry, and in business. Perhaps the simplest way to define a real number is this: A **real number** is any number that can be expressed in decimal form. Some examples of real numbers are

$$7 (= 7.000\ldots)$$
$$\sqrt{2} (= 1.4142\ldots)$$
$$-2/3 (= -0.\overline{6})$$

(Recall that the bar above the 6 in the decimal $-0.\overline{6}$ indicates that the 6 repeats indefinitely.)

Certain sets of real numbers are referred to often enough to be given special names. These are summarized in the box that follows.

As you've seen in previous courses, the real numbers can be represented as points on a *number line,* as shown in Figure 1. As indicated in Figure 1, the point associated with the number zero is referred to as the **origin.**

The fundamental fact here is that there is a **one-to-one correspondence** between the set of real numbers and the set of points on the line. This means that each real number is identified with exactly one point on the line; conversely, with each point on the line we identify exactly one real number. The real number associated with a given point is called the **coordinate** of the point. As a practical matter, we're

Figure 1

▌PROPERTY SUMMARY   Sets of Real Numbers

Name	Definition and Comments	Examples
Natural numbers	These are the ordinary counting numbers: 1, 2, 3, and so on.	1, 4, 29, 1066
Integers	These are the natural numbers along with their negatives and zero.	$-26, 0, 1, 1776$
Rational numbers	As the name suggests, these are the real numbers that are *ratios* of two integers (with nonzero denominators). It can be proved that a real number is rational if and only if its decimal expansion *terminates* (e.g., 3.15) or *repeats* (e.g., $2.\overline{43}$).	$4 (= \frac{4}{1}), -\frac{2}{3},$ $1.7 (= \frac{17}{10}), 4.\overline{3},$ $4.1\overline{73}$
Irrational numbers	These are the real numbers that are not rational. Section A.4 of the Appendix contains a proof of the fact that the number $\sqrt{2}$ is irrational. The proof that π is irrational is more difficult. The first person to prove that π is irrational was the Swiss mathematician J. H. Lambert (1728–1777).	$\sqrt{2}, 3 + \sqrt{2},$ $3\sqrt{2}, \pi, 4 + \pi,$ 4π

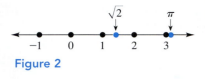

Figure 2

usually more interested in relative locations than precise locations on a number line. For instance, since π is approximately 3.1, we show π slightly to the right of 3 in Figure 2. Similarly, since $\sqrt{2}$ is approximately 1.4, we show $\sqrt{2}$ slightly less than halfway from 1 to 2 in Figure 2.

It is often convenient to use number lines that show reference points other than the integers used in Figure 2. For instance, Figure 3(a) displays a number line with reference points that are multiples of π. In this case it is the integers that we then locate approximately. For example, in Figure 3(b) we show the approximate location of the number 1 on such a line.

Figure 3
(a)   (b)

Two of the most basic relations for real numbers are **less than** and **greater than,** symbolized by $<$ and $>$, respectively. For ease of reference, we review these and two related symbols in the box on page 3.

In general, relationships involving real numbers and any of the four symbols $<$, $\leq$, $>$, and $\geq$ are called **inequalities.** One of the simplest uses of inequalities is in defining certain sets of real numbers called *intervals.* Roughly speaking, any uninterrupted portion of the number line is referred to as an **interval.** In the definitions that follow, you'll see notations such as $a < x < b$. This means that *both* of the inequalities $a < x$ and $x < b$ hold; in other words, the number x is between a and b.

DEFINITION │ Open Intervals and Closed Intervals

The **open interval** (a, b) consists of all real numbers x such that $a < x < b$. See Figure 4(a).

The **closed interval** $[a, b]$ consists of all real numbers x such that $a \leq x \leq b$. See Figure 4(b).

PROPERTY SUMMARY   Notation for Less Than and Greater Than

Notation	Definition	Examples
$a < b$	a is less than b. On a number line, oriented as in Figure 1, 2, or 3, the point a lies to the left of b.	$2 < 3; -4 < 1$
$a \leq b$	a is less than or equal to b.	$2 \leq 3; 3 \leq 3$
$b > a$	b is greater than a. On a number line oriented as in Figure 1, 2, or 3, the point b lies to the right of a. [$b > a$ is equivalent to $a < b$.]	$3 > 2; 0 > -1$
$b \geq a$	b is greater than or equal to a.	$3 \geq 2; 3 \geq 3$

Note that the brackets in Figure 4(b) are used to indicate that the numbers a and b are included in the interval $[a, b]$, whereas the parentheses in Figure 4(a) indicate that a and b are excluded from the interval (a, b). At times you'll see notation such as $[a, b)$. This stands for the set of all real numbers x such that $a \leq x < b$. Similarly, $(a, b]$ denotes the set of all numbers x such that $a < x \leq b$.

(a) The open interval (a, b) contains all real numbers from a to b, excluding a and b.

(b) The closed interval $[a, b]$ contains all real numbers from a to b, including a and b.

Figure 4

EXAMPLE 1   Understanding interval notation

Show each interval on a number line, and specify inequalities describing the numbers x in each interval.

$$[-1, 2] \qquad (-1, 2) \qquad (-1, 2] \qquad [-1, 2)$$

SOLUTION
See Figure 5.

$$[-1, 2] \qquad\qquad (-1, 2) \qquad\qquad (-1, 2] \qquad\qquad [-1, 2)$$
$$-1 \leq x \leq 2 \qquad -1 < x < 2 \qquad -1 < x \leq 2 \qquad -1 \leq x < 2$$

Figure 5

In addition to the four types of intervals shown in Figure 5, we can also consider **unbounded intervals.** These are intervals that extend indefinitely in one direction or the other, as shown, for example, in Figure 6. We also have a convenient notation for unbounded intervals; for example, we indicate the unbounded interval in Figure 6 with the notation $(2, \infty)$.

Figure 6

COMMENT AND CAUTION   The symbol ∞ is read *infinity*. It is not a real number, and its use in the context $(2, \infty)$ is only to indicate that the interval has no right-hand boundary. In the box that follows we define the five types of unbounded intervals. Note that the last interval, $(-\infty, \infty)$, is actually the entire real-number line.

PROPERTY SUMMARY   Unbounded Intervals

Notation	Defining Inequality	Example
(a, ∞)	$x > a$	$(2, \infty)$
$[a, \infty)$	$x \geq a$	$[2, \infty)$
$(-\infty, a)$	$x < a$	$(-\infty, 2)$
$(-\infty, a]$	$x \leq a$	$(-\infty, 2]$
$(-\infty, \infty)$		$(-\infty, \infty)$

EXAMPLE   2   Understanding notation for unbounded intervals

Indicate each set of real numbers on a number line:
(a) $(-\infty, 4]$;   **(b)** $(-3, \infty)$.

SOLUTION
(a) The interval $(-\infty, 4]$ consists of all real numbers that are less than or equal to 4. See Figure 7.
(b) The interval $(-3, \infty)$ consists of all real numbers that are greater than -3. See Figure 8.

Figure 7

Figure 8

We conclude this section by mentioning that our treatment of the real-number system has been rather informal, and we have not derived any of the rules of arithmetic and algebra using the most basic properties of the real numbers. However, we do list those basic properties and derive some of their consequences in Section A.2 of the Appendix. (For example, in the Appendix we prove that the product of two negative numbers is positive.)

EXERCISE SET 1.1

A

In Exercises 1–12, determine whether the number is a natural number, an integer, a rational number, or an irrational number. (Some numbers fit in more than one category.) The following facts will be helpful in some cases: Any number of the form $\sqrt{n}$, where n is a natural number that is not a perfect square, is irrational. Also, the sum, difference, product, and quotient of an irrational number and a nonzero rational are all irrational. (For example, the following four numbers are irrational: $\sqrt{6}$, $\sqrt{10} - 2$, $3\sqrt{15}$, and $-5\sqrt{3}/2$.)

1. (a) 7
   (b) -7

2. (a) -203
   (b) 204/2

3. (a) 27/4
   (b) $\sqrt{27/4}$

4. (a) $\sqrt{7}$
   (b) $4\sqrt{7}$

5. (a) 10^6
   (b) $10^6/10^7$

6. (a) 8.7
   (b) $8.\overline{7}$

7. (a) 8.74
   (b) $8.\overline{74}$

8. (a) $\sqrt{99}$
   (b) $\sqrt{99} + 1$

9. $3\sqrt{101} + 1$

10. $(3 - \sqrt{2}) + (3 + \sqrt{2})$

11. $(\sqrt{5} + 1)/4$

12. $(0.1234)/(0.5677)$

In each of Exercises 13–22, draw a number line similar to the one shown in Figure 1. Then indicate the approximate location of the given number. Where necessary, make use of the approximations $\sqrt{2} \approx 1.4$ and $\sqrt{3} \approx 1.7$. (The symbol $\approx$ means is approximately equal to.)

13. $11/4$ **14.** $-7/8$ **15.** $1 + \sqrt{2}$

16. $1 - \sqrt{2}$ **17.** $\sqrt{2} - 1$ **18.** $-\sqrt{2} - 1$

19. $\sqrt{2} + \sqrt{3}$ **20.** $\sqrt{2} - \sqrt{3}$ **21.** $(1 + \sqrt{2})/2$

22. $(2\sqrt{3} + 1)/2$

In Exercises 23–32, draw a number line similar to the one shown in Figure 3(a). Then indicate the approximate location of the given number.

23. $\pi/2$ **24.** $3\pi/2$ **25.** $\pi/6$ **26.** $7\pi/4$

27. -1 **28.** 3 **29.** $\pi/3$ **30.** $3/2$

31. $2\pi + 1$ **32.** $2\pi - 1$

33. Use a calculator to show the approximate location of the irrational number $(\sqrt{139} - 5)/3$ on a number line.

34. Use a calculator to show the approximate location of the irrational number $\sqrt{\dfrac{3 - \sqrt{5}}{2}}$ on a number line.

In Exercises 35–44, say whether the statement is TRUE *or* FALSE. *(In Exercises 41–44, do not use a calculator or tables; use instead the approximations $\sqrt{2} \approx 1.4$ and $\pi \approx 3.1$.)*

35. $-5 < -50$ **36.** $0 < -1$ **37.** $-2 \leq -2$

38. $\sqrt{7} - 2 \geq 0$ **39.** $\frac{13}{14} > \frac{15}{16}$ **40.** $0.\overline{7} > 0.7$

41. $2\pi < 6$ **42.** $2 \leq (\pi + 1)/2$ **43.** $2\sqrt{2} \geq 2$

44. $\pi^2 < 12$

In Exercises 45–58, show the given interval on a number line.

45. $(2, 5)$ **46.** $(-2, 2)$ **47.** $[1, 4]$

48. $[-\frac{3}{2}, \frac{1}{2}]$ **49.** $[0, 3)$ **50.** $(-4, 0]$

51. $(-3, \infty)$ **52.** $(\sqrt{2}, \infty)$ **53.** $[-1, \infty)$

54. $[0, \infty)$ **55.** $(-\infty, 1)$ **56.** $(-\infty, -2)$

57. $(-\infty, \pi]$ **58.** $(-\infty, \infty)$

B

From grade school on, we all acquire a good deal of experience in working with rational numbers. Certainly we can tell when two rational numbers are equal, although it may require some calculation, as, for example, in the case of $\frac{129}{31} = \frac{2193}{527}$. In Exercises 59 and 60, you'll see that in working with irrational numbers, it's sometimes hard to recognize when two numbers are equal. For each equation in Exercises 59 and 60, use a calculator to verify that the quantities on both sides agree, to six decimal places. (In each case it can be shown that the two quantities are indeed equal.)

59. (a) $\sqrt{6} + \sqrt{2} = 2\sqrt{2 + \sqrt{3}}$

    (b) $\sqrt{3 + \sqrt{5}} + \sqrt{3 - \sqrt{5}} = \sqrt{10}$

    (c) $\sqrt{\sqrt{6 + 4\sqrt{2}}} = \sqrt{2 + \sqrt{2}}$

60. (a) $\sqrt{7} - \sqrt{8 - 2\sqrt{7}} = 1$

    (b) $\sqrt[3]{2 + \sqrt{5}} + \sqrt[3]{2 - \sqrt{5}} = 1$

    (c) $\sqrt[5]{176 + 80\sqrt{5}} = 1 + \sqrt{5}$

61. The value of the irrational number π, correct to ten decimal places (without rounding off), is 3.1415926535. By using a calculator, determine to how many decimal places each of the following quantities agrees with π.

    (a) $(4/3)^4$: This is the value used for π in the Rhind papyrus, an ancient Babylonian text written about 1650 B.C.

    (b) $22/7$: Archimedes (287–212 B.C.) showed that $223/71 < \pi < 22/7$. The use of the approximation $22/7$ for π was introduced to the Western world through the writings of Boethius (ca 480–520), a Roman philosopher, mathematician, and statesman. Among all fractions with numerators and denominators less than 100, the fraction $22/7$ is the best approximation to π.

    (c) $355/113$: This approximation of π was obtained in fifth-century China by Zu Chong-Zhi (430–501) and his son. According to David Wells in *The Penguin Dictionary of Curious and Interesting Numbers* (Harmondsworth, Middlesex, England: Viking Penguin, Ltd., 1986), "This is the best approximation of any fraction below 103993/33102."

    (d) $\dfrac{63}{25} \left(\dfrac{17 + 15\sqrt{5}}{7 + 15\sqrt{5}} \right)$: This approximation for π was obtained by the Indian mathematician Scrinivasa Ramanujan (1887–1920).

Remark: A simple approximation that agrees with π through the first 14 decimal places is $\dfrac{355}{113} \left(1 - \dfrac{0.0003}{3533} \right)$. This approximation was also discovered by Ramanujan. For a fascinating account of the history of π, see the book by Petr Beckmann, *A History of π*, 4th ed. (Boulder, Colo.: Golem Press, 1977), and for a more modern look at π, see Richard Preston's article, "The Mountains of Pi," in *The New Yorker* (March 2, 1992; pp. 36–67).

C

In Exercises 62–64, give an example of irrational numbers a and b such that the indicated expression is **(a)** *rational and* **(b)** *irrational.*

62. $a + b$ **63.** ab **64.** a/b

65. (a) Give an example in which the result of raising a rational number to a rational power is an irrational number.

    (b) Give an example in which the result of raising an irrational number to a rational power is a rational number.

66. Can an irrational number raised to an irrational power yield an answer that is rational? This problem shows that the answer is "yes." (However, if you study the following solution very carefully, you'll see that even though we've answered the question in the affirmative, we've not pinpointed the specific case in which an irrational number raised to an irrational power is rational.)

(a) Let $A = (\sqrt{2})^{\sqrt{2}}$. Now, either A is rational or A is irrational. If A is rational, we are done. Why?

(b) If A is irrational, we are done. Why?
    Hint: Consider $A^{\sqrt{2}}$.

Remark: For more about this problem and related questions, see the article "Irrational Numbers," by J. P. Jones and S. Toporowski in *American Mathematical Monthly*, vol. 80 (1973), pp. 423–424.

| | 1.2   ABSOLUTE VALUE |

There has been a real need in analysis for a convenient symbolism for "absolute value" . . . and the two vertical bars introduced in 1841 by Weierstrass, as in $|z|$, have met with wide adoption; . . .—Florian Cajori in A History of Mathematical Notations, Vol. 1 (La Salle, Ill.: The Open Court Publishing Co., 1928)

As an aid in measuring distances on the number line, we review the concept of *absolute value*. We begin with a definition of absolute value that is geometric in nature. Then, after you have developed some familiarity with the concept, we explain a more algebraic approach that is often useful in analytical work.

DEFINITION | **Absolute Value (geometric version)**

The **absolute value** of a real number x, denoted by $|x|$, is the distance from x to the origin.

For instance, because the numbers 5 and -5 are both five units from the origin, we have $|5| = 5$ and $|-5| = 5$. Here are three more examples:

$$|17| = 17 \qquad |-2/3| = 2/3 \qquad |0| = 0$$

In dealing with an expression such as $|-5 + 3|$, the convention is to compute the quantity $-5 + 3$ first and then to take the absolute value. We therefore have, in this case,

$$|-5 + 3| = |-2| = 2$$

EXAMPLE **1** **Evaluating expressions containing absolute values**

Evaluate each expression:
(a) $5 - |6 - 7|$;   **(b)** $||-2| - |-3||$.

SOLUTION

(a) $5 - |6 - 7| = 5 - |-1|$
$= 5 - 1 = 4$

(b) $||-2| - |-3|| = |2 - 3|$
$= |-1| = 1$

As we said at the beginning of this section, there is an equivalent, more algebraic way to define absolute value. According to this equivalent definition, the value of $|x|$ is x itself when $x \geq 0$, and the value of $|x|$ is $-x$ when $x < 0$. We can write this symbolically as follows:

DEFINITION	Absolute Value (algebraic version)

$$|x| = \begin{cases} x & \text{when } x \geq 0 \\ -x & \text{when } x < 0 \end{cases}$$

EXAMPLE
$$|-7| = -(-7) = 7$$

By looking at examples with specific numbers, you should be able to convince yourself that both definitions yield the same result. We use the algebraic definition of absolute value in Examples 2 and 3.

EXAMPLE 2 **Rewriting expressions to eliminate absolute value**

Rewrite each expression in a form that does not contain absolute values:
(a) $|\pi - 4| + 1$;
(b) $|x - 5|$, given that $x \geq 5$;
(c) $|t - 5|$, given that $t < 5$.

SOLUTION
(a) The quantity $\pi - 4$ is negative (since $\pi \approx 3.14$), and therefore its absolute value is equal to $-(\pi - 4)$. In view of this, we have

$$|\pi - 4| + 1 = -(\pi - 4) + 1 = -\pi + 5$$

(b) Since $x \geq 5$, the quantity $x - 5$ is nonnegative, and therefore its absolute value is equal to $x - 5$ itself. Thus, we have

$$|x - 5| = x - 5 \qquad \text{when } x \geq 5$$

(c) Since $t < 5$, the quantity $t - 5$ is negative. Therefore its absolute value is equal to $-(t - 5)$, which in turn is equal to $5 - t$. In view of this, we have

$$|t - 5| = 5 - t \qquad \text{when } t < 5$$

EXAMPLE 3 **Simplifying an expression containing absolute values**

Simplify the expression $|x - 1| + |x - 2|$, given that x is in the open interval $(1, 2)$.

SOLUTION
Since x is greater than 1, the quantity $x - 1$ is positive and, consequently,

$$|x - 1| = x - 1$$

On the other hand, we are also given that x is less than 2. Therefore the quantity $x - 2$ is negative, and we have

$$|x - 2| = -(x - 2) = -x + 2$$

Putting things together now, we can write

$$|x - 1| + |x - 2| = (x - 1) + (-x + 2)$$
$$= -1 + 2 = 1$$

The solutions in Examples 2 and 3 use the algebraic version of the definition of absolute value. In the error box that follows, the first two items indicate possible misuses of this definition. *Suggestion*: Before reading further, cover up the columns headed "Correction" and "Comment" in the box and try to decide for yourself where each error lies.

ERRORS TO AVOID

Error	Correction	Comment
$\lvert 3 - \sqrt{2} \rvert \neq -(3 - \sqrt{2}) = -3 + \sqrt{2}$	$\lvert 3 - \sqrt{2} \rvert = 3 - \sqrt{2}$	Because the quantity $3 - \sqrt{2}$ is positive, the appropriate formula is $\lvert x \rvert = x$, not $\lvert x \rvert = -x$.
$\lvert x - y \rvert \neq -(x - y) = -x + y$	$\lvert x - y \rvert = \begin{cases} x - y & \text{if } x - y \geq 0 \\ -(x - y) & \text{if } x - y < 0 \end{cases}$	The error arises in assuming that the quantity $x - y$ is negative. Unless additional information is available, $\lvert x - y \rvert$ cannot be simplified.
$\lvert a + b \rvert \neq \lvert a \rvert + \lvert b \rvert$	$\lvert a + b \rvert = \begin{cases} a + b & \text{if } a + b \geq 0 \\ -(a + b) & \text{if } a + b < 0 \end{cases}$	In general, the absolute value of a sum is *not* equal to the sum of the individual absolute values. For example, if $a = -1$ and $b = 2$, then $\lvert a + b \rvert = \lvert -1 + 2 \rvert = \lvert 1 \rvert = 1$, but $\lvert a \rvert + \lvert b \rvert = \lvert -1 \rvert + \lvert 2 \rvert = 3$.

In the box that follows, we list several basic properties of the absolute value. Each of these properties can be derived from the definitions. (With the exception of the *triangle inequality,* we shall omit the derivations. For a proof of the triangle inequality, see Exercise 71.)

PROPERTY SUMMARY   **Properties of Absolute Value**

1. For all real numbers x, we have
   (a) $\lvert x \rvert \geq 0$;
   (b) $x \leq \lvert x \rvert$ and $-x \leq \lvert x \rvert$.
   (c) $\lvert x \rvert^2 = x^2$.

2. For all real numbers a and b, we have
   (a) $\lvert ab \rvert = \lvert a \rvert \lvert b \rvert$ and $\lvert a/b \rvert = \lvert a \rvert / \lvert b \rvert$ $(b \neq 0)$;
   (b) $\lvert a + b \rvert \leq \lvert a \rvert + \lvert b \rvert$ (the triangle inequality).

EXAMPLE **4**   **Rewriting an expression to eliminate absolute value**

Write the expression $\lvert -2 - x^2 \rvert$ in an equivalent form that does not contain absolute values.

SOLUTION

We have

$$|-2 - x^2| = |-1(2 + x^2)|$$
$$= |-1||2 + x^2| \qquad \text{using Property 2(a)}$$
$$= 2 + x^2$$

The last equality follows from the fact that x^2 is nonnegative for any real number x and, consequently, the quantity $2 + x^2$ is positive.

Distance $= |5 - 7| = |7 - 5| = 2$

Figure 1

If we think of the real numbers as points on a number line, the distance between two numbers a and b is given by the absolute value of their difference. For instance, as indicated in Figure 1, the distance between 5 and 7—namely, 2—is given by either $|5 - 7|$ or $|7 - 5|$. For reference, we summarize this simple but important fact as follows.

PROPERTY SUMMARY   Distance on a Number Line

The **distance** between a and b is $|a - b| = |b - a|$.

EXAMPLE 5 **Using absolute value to rewrite statements regarding distance**

Rewrite each of the following statements using absolute value notation:
(a) The distance between 12 and -5 is 17.
(b) The distance between x and 2 is 4.
(c) The distance between x and 2 is less than 4.
(d) The point t is more than five units from the origin.

SOLUTION
(a) $|12 - (-5)| = 17$ or $|-5 - 12| = 17$
(b) $|x - 2| = 4$ or $|2 - x| = 4$
(c) $|x - 2| < 4$ or $|2 - x| < 4$
(d) $|t| > 5$

EXAMPLE 6 **Displaying intervals defined by absolute-value inequalities**

In each case, the set of real numbers satisfying the given inequality is one or more intervals on the number line. Show the interval(s) on a number line.
(a) $|x| < 2$ **(b)** $|x| > 2$ **(c)** $|x - 3| < 1$ **(d)** $|x - 3| \geq 1$

Figure 2
$|x| < 2$

SOLUTION
(a) The given inequality tells us that the distance from x to the origin is less than two units. So, as indicated in Figure 2, the point x must lie in the open interval $(-2, 2)$.

Figure 3
$|x| > 2$

(b) The condition $|x| > 2$ means that x is more than two units from the origin. Thus, as indicated in Figure 3, the point x lies either to the right of 2 or to the left of -2.

(c) The given inequality tells us that x must be less than one unit away from 3 on the number line. Looking one unit to either side of 3, then, we see that x must lie between 2 and 4 and x cannot equal 2 or 4. See Figure 4.

Figure 4
$|x - 3| < 1$

(d) The inequality $|x - 3| \geq 1$ says that x is at least one unit away from 3 on the number line. This means that either $x \geq 4$ or $x \leq 2$, as shown in Figure 5. [Here's an alternative way of thinking about this: The numbers satisfying the given inequality are precisely those numbers that do *not* satisfy the inequality in part (c). So for part (d), we need to shade that portion of the number line that was not shaded in part (c).]

Figure 5
$|x - 3| \geq 1$

EXERCISE SET 1.2

A

In Exercises 1–16, evaluate each expression.

1. $|3|$ **2.** $3 + |-3|$

3. $|-6|$ **4.** $-6 - |-6|$

5. $|-1 + 3|$ **6.** $|-6 + 3|$

7. $\left|-\frac{4}{5}\right| - \frac{4}{5}$ **8.** $\left|\frac{4}{5}\right| - \frac{4}{5}$

9. $|-6 + 2| - |4|$ **10.** $|-3 - 4| - |-4|$

11. $||-8| + |-9||$ **12.** $||-8| - |-9||$

13. $\left|\dfrac{27 - 5}{5 - 27}\right|$ **14.** $\dfrac{|27 - 5|}{|5 - 27|}$

15. $|7(-8)| - |7| \cdot |-8|$

16. $|(-7)^2| + |-7|^2 - (-|-3|)^3$

In Exercises 17–24, evaluate each expression, given that $a = -2$, $b = 3$, and $c = -4$.

17. $|a - b|^2$ **18.** $a^2 - |bc|$

19. $|c| - |b| - |a|$ **20.** $|b + c| - |b| - |c|$

21. $|a + b|^2 - |b + c|^2$ **22.** $\dfrac{|a| + |b| + |c|}{|a + b + c|}$

23. $\dfrac{a + b + |a - b|}{2}$ **24.** $\dfrac{a + b - |a - b|}{2}$

In Exercises 25–38, rewrite each expression without using absolute value notation.

25. $|\sqrt{2} - 1| - 1$ **26.** $|1 - \sqrt{2}| + 1$

27. $|x - 3|$ given that $x \geq 3$ **28.** $|x - 3|$ given that $x < 3$

29. $|t^2 + 1|$ **30.** $|x^4 + 1|$

31. $|-\sqrt{3} - 4|$ **32.** $|-\sqrt{3} - \sqrt{5}|$

33. $|x - 3| + |x - 4|$ given that $x < 3$

34. $|x - 3| + |x - 4|$ given that $x > 4$

35. $|x - 3| + |x - 4|$ given that $3 < x < 4$

36. $|x - 3| + |x - 4|$ given that $x = 4$

37. $|x + 1| + 4|x + 3|$ given that $-\frac{5}{2} < x < -\frac{3}{2}$

38. $|x + 1| + 4|x + 3|$ given that $x < -3$

In Exercises 39–50, rewrite each statement using absolute value notation, as in Example 5.

39. The distance between x and 4 is 8.

40. The distance between x and 4 is at least 8.

41. The distance between x and 1 is $\frac{1}{2}$.

42. The distance between x and 1 is less than $\frac{1}{2}$.

43. The distance between x and 1 is at least $\frac{1}{2}$.

44. The distance between x and 1 exceeds $\frac{1}{2}$.

45. The distance between y and -4 is less than 1.

46. The distance between x^3 and -1 is at most 0.001.

47. The number y is less than three units from the origin.

48. The number y is less than one unit from the number t.

49. The distance between x^2 and a^2 is less than M.

50. The sum of the distances of a and b from the origin is greater than or equal to the distance of $a + b$ from the origin.

In Exercises 51–62, the set of real numbers satisfying the given inequality is one or more intervals on the number line. Show the interval(s) on a number line.

51. $|x| < 4$ **52.** $|x| < 2$

53. $|x| > 1$ **54.** $|x| > 0$

55. $|x - 5| < 3$ **56.** $|x - 4| < 4$

57. $|x - 3| \leq 4$ **58.** $|x - 1| \leq 1/2$

59. $\left|x + \frac{1}{3}\right| < \frac{3}{2}$ **60.** $\left|x + \frac{\pi}{2}\right| > 1$

61. $|x - 5| \geq 2$ **62.** $|x + 5| \geq 2$

B

63. In parts (a) and (b), sketch the interval or intervals corresponding to the given inequality:

(a) $|x - 2| < 1$;

(b) $0 < |x - 2| < 1$.

(c) In what way do your answers in (a) and (b) differ? (The distinction is important in the study of *limits* in calculus.)

64. Given two real numbers a and b, the notation **max(a, b)** denotes the larger of the two numbers. For instance, if $a = 1$ and $b = 2$, then $\max(1, 2) = 2$. In cases where $a = b$, then $\max(a, b)$ denotes the common value of a and b. It can be shown (see Exercise 69) that $\max(a, b)$ can be expressed in terms of absolute value as follows:

$$\max(a, b) = \frac{a + b + |a - b|}{2}$$

Verify this equation in each of the following cases:
(a) $a = 1$ and $b = 2$;
(b) $a = -1$ and $b = -5$;
(c) $a = b = 10$.

65. Given two real numbers a and b, the notation **min(a, b)** denotes the smaller of the two numbers. In cases where $a = b$, then $\min(a, b)$ denotes the common value of a and b. It can be shown (see Exercise 70) that $\min(a, b)$ can be expressed in terms of absolute value as follows:

$$\min(a, b) = \frac{a + b - |a - b|}{2}$$

Verify this equation in each of the following cases:
(a) $a = 6$ and $b = 1$;
(b) $a = 1$ and $b = -6$;
(c) $a = b = -6$.

66. Show that for all real numbers a and b, we have

$$|a| - |b| \le |a - b|$$

Hint: Beginning with the identity $a = (a - b) + b$, take the absolute value of each side and then use the triangle inequality.

67. Show that

$$|a + b + c| \le |a| + |b| + |c|$$

for all real numbers a, b, and c. *Hint:* The left-hand side can be written $|a + (b + c)|$. Now use the triangle inequality.

68. Explain why there are no real numbers that satisfy the equation $|x^2 + 4x| = -12$.

C

69. (As background for this exercise, you'll need to have worked Exercise 64.) Prove that

$$\max(a, b) = \frac{a + b + |a - b|}{2}$$

Hint: Consider three separate cases: $a = b$; $a > b$; and $b > a$.

70. (As background for this exercise, you'll need to have worked Exercise 65.) Prove that

$$\min(a, b) = \frac{a + b - |a - b|}{2}$$

71. Complete the following steps to prove the triangle inequality.
(a) Let a and b be real numbers. Which property in the summary box on page 8 tells us that $a \le |a|$ and $b \le |b|$?
(b) Add the two inequalities in part (a) to obtain $a + b \le |a| + |b|$.
(c) In a similar fashion, add the two inequalities $-a \le |a|$ and $-b \le |b|$ and deduce that $-(a + b) \le |a| + |b|$.
(d) Why do the results in parts (b) and (c) imply that $|a + b| \le |a| + |b|$?

1.3 SOLVING EQUATIONS (REVIEW AND PREVIEW)

I learned algebra fortunately by not learning it at school, and knowing that the whole idea was to find out what x was, and it didn't make any difference how you did it. —Physicist Richard Feynman (1918–1988) in Jagdish Mehra's *The Beat of a Different Drum* (New York: Oxford University Press, 1994)

The title of al-Khwarizmi's second and most important book, Hisab al-jabr w'al muqabala *[830] . . . has given us the word* algebra. Al-jabr *means transposing a quantity from one side of an equation to the other, while* muqabala *signifies the simplification of the resulting equation.* —Stuart Hollingdale in *Makers of Mathematics* (Harmondsworth, Middlesex, England: Penguin Books, Ltd., 1989)

"Algebra is a merry science," Uncle Jakob would say. "We go hunting for a little animal whose name we don't know, so we call it x. When we bag our game we pounce on it and give it its right name." —Physicist Albert Einstein (1879–1955)

As background for our review of solving equations, we first recall the terms *constant* and *variable*. By way of example, consider the familiar expression for the area of a circle of radius r—namely, πr^2. Here π is a constant; its value never changes throughout the discussion. On the other hand, r is a variable; we can substitute any positive number for r to obtain the area of a particular circle. More generally, by a **constant** we mean either a particular number (such as π, or -17, or $\sqrt{2}$) or a letter with a value that remains fixed (although perhaps unspecified) throughout a given discussion. In contrast, a **variable** is a letter for which we can substitute any number selected from a given set of numbers. The given set of numbers is called the **domain** of the variable.

Some expressions will make sense only for certain values of the variable. For instance, $1/(x - 3)$ will be undefined when x is 3 (for then the denominator is zero). So in this case we would agree that the domain of the variable x consists of all real numbers except $x = 3$. Similarly, throughout this chapter, we adopt the following convention:

The Domain Convention

The domain of a variable in a given expression is the set of all real-number values of the variable for which the expression is defined.

In algebra it's customary (but certainly not mandatory) to use letters near the end of the alphabet for variables; letters from the beginning of the alphabet are generally used for constants. For example, in the expression $ax + b$, the letter x is the variable and a and b are constants.

EXAMPLE 1   Specifying variables, constants, and the domain in an expression

Specify the variable, the constants, and the domain of the variable for each of the following expressions:

(a) $3x + 4$;   **(b)** ;   **(c)** $ay^2 + by + c$;   **(d)** $4x + 3x^{-1}$.

SOLUTION

	VARIABLE	CONSTANTS	DOMAIN
(a) $3x + 4$	x	$3, 4$	The set of all real numbers.
(b) $\dfrac{1}{(t - 1)(t + 3)}$	t	$1, -1, 3$	The set of all real numbers except $t = 1$ and $t = -3$.
(c) $ay^2 + by + c$	y	a, b, c	The set of all real numbers.
(d) $4x + 3x^{-1}$	x	$4, 3$	The set of all real numbers except $x = 0$. (Remember, $x^{-1} = 1/x$).

We are ready now to review the terminology and skills used in solving two basic types of equations, *linear equations* and *quadratic equations*.

DEFINITION   Linear Equation in One Variable

A **linear** or **first-degree equation in one variable** is an equation that can be written in the form

$$ax + b = 0 \quad \text{with } a \text{ and } b \text{ real numbers and } a \neq 0$$

As examples of linear equations in one variable, we cite the following:

$$2x - 10 = 0 \qquad 3m + 1 = 2 \qquad \frac{y}{2} = \frac{y}{3} + 1$$

As with any equation involving a variable, each of these equations is neither true nor false *until* we replace the variable with a number. By a **solution** or a **root** of an equation in one variable, we mean a value for x (or m or y in the foregoing examples) that makes the equation a true statement. For example, the value $x = 5$ is a solution of the equation $2x = 10$, since, with $x = 5$, the equation becomes $2(5) = 10$, which is certainly true. We also say in this case that the value $x = 5$ **satisfies** the equation.

Equations that become true statements for all values in the domain of the variable are called **identities.** Two examples of identities are

$$x^2 - 9 = (x - 3)(x + 3) \qquad \text{and} \qquad \frac{4x^2}{x} = 4x$$

In contrast to this, a **conditional equation** is true only for some (or perhaps none) of the values of the variable. Two examples of conditional equations are $2x = 10$ and $x = x + 1$. The first of these is true only when $x = 5$. The second equation has no solution (because, intuitively at least, no number can be one more than itself).

We say that two equations are **equivalent** when they have exactly the same solutions. In this section, and throughout the text, the basic method for solving an equation in one variable involves writing a sequence of equivalent equations until we finally reach an equation of the form

$$\text{variable} = \text{a number}$$

In generating equivalent equations, we rely on the following three principles. (These can be justified using the properties of real numbers discussed in Appendix A.2.)

Procedures That Yield Equivalent Equations

1. Adding or subtracting the same quantity on both sides of an equation produces an equivalent equation.
2. Multiplying or dividing both sides of an equation by the same nonzero quantity produces an equivalent equation.
3. Simplifying an expression on either side of an equation produces an equivalent equation.

The examples that follow show how these principles are applied in solving various equations. *Note:* Beginning in Example 5, we use some basic factoring techniques from elementary or intermediate algebra. If you find that you need a quick reminder of factoring, see the inside back cover of this book. For a detailed review (with many examples and practice exercises) see Appendix B.4.

EXAMPLE **2** **Solving and checking linear equations**

(a) Solve: $4x + 5 = 33$.

(b) Solve: $ax + b = c$; where a, b, and c are constants, $a \neq 0$.

SOLUTION

(a) $4x + 5 = 33$

$$4x = 28 \qquad \text{subtracting 5 from both sides}$$
$$x = 7 \qquad \text{dividing both sides by 4}$$

CHECK   Replacing x with 7 in the original equation yields

$$4(7) + 5 = 33$$
$$28 + 5 = 33 \qquad \text{True}$$

(b) $ax + b = c$

$$ax = c - b \qquad \text{subtracting } b \text{ from both sides}$$
$$x = \frac{c - b}{a} \qquad \text{dividing both sides by } a \text{ (recall } a \neq 0\text{)}$$

CHECK   Replacing x with $\dfrac{c - b}{a}$ in the original equation yields

$$a\left(\frac{c - b}{a}\right) + b = c$$
$$c - b + b = c$$
$$c = c \qquad \text{True}$$

EXAMPLE 3   Solving a slightly more complicated linear equation

Solve: $3[1 - 2(x + 1)] = 2 - x$.

SOLUTION

$$3[1 - 2(x + 1)] = 2 - x$$
$$3[1 - 2x - 2] = 2 - x \qquad \text{simplifying the left-hand side}$$
$$3(-1 - 2x) = 2 - x$$
$$-3 - 6x = 2 - x$$
$$-3 - 5x = 2 \qquad \text{adding } x \text{ to both sides}$$
$$-5x = 5 \qquad \text{adding 3 to both sides}$$
$$x = -1 \qquad \text{dividing both sides by } -5$$

CHECK   Replacing x with -1 in the original equation yields

$$3[1 - 2(0)] = 2 - (-1)$$
$$3(1) = 2 + 1 \qquad \text{True}$$

In the next two examples the equations involve fractions. A common strategy in such cases is to multiply through by the least common denominator; that eliminates the need to work with fractions.

EXAMPLE 4   Solving equations involving fractions

Solve:

(a) $\dfrac{x}{2} - \dfrac{3x}{4} = 1$;   **(b)** $\dfrac{1}{x + 5} = \dfrac{2}{x - 3} + \dfrac{2x + 2}{x^2 + 2x - 15}$.

SOLUTION

(a)
$$\frac{x}{2} - \frac{3x}{4} = 1$$

$$4\left(\frac{x}{2}\right) - 4\left(\frac{3x}{4}\right) = 4(1) \qquad \text{multiplying both sides by the least common}$$
$$\text{denominator}$$

$$2x - 3x = 4 \qquad \text{simplifying}$$
$$-x = 4 \qquad \text{simplifying}$$
$$x = -4 \qquad \text{multiplying both sides by } -1$$

CHECK Replacing x with -4 in the original equation yields

$$\frac{-4}{2} - \frac{3(-4)}{4} = 1$$
$$-2 + 3 = 1 \qquad \text{True}$$

(b) By factoring the denominator $x^2 + 2x - 15$, we obtain

$$\frac{1}{x + 5} = \frac{2}{x - 3} + \frac{2x + 2}{(x + 5)(x - 3)}$$

From this we see that the least common denominator for the three fractions is $(x + 5)(x - 3)$. Now, multiplying both sides by this least common denominator, we have

$$\frac{(x + 5)(x - 3)}{x + 5} = \frac{2(x + 5)(x - 3)}{x - 3} + \frac{(2x + 2)(x + 5)(x - 3)}{(x + 5)(x - 3)}$$

$$x - 3 = 2(x + 5) + 2x + 2$$
$$x - 3 = 2x + 10 + 2x + 2 \qquad \text{simplifying}$$
$$x - 3 = 4x + 12 \qquad \text{simplifying}$$
$$-3x - 3 = 12 \qquad \text{subtracting } 4x \text{ from both sides}$$
$$-3x = 15 \qquad \text{adding } 3 \text{ to both sides}$$
$$x = -5 \qquad \text{dividing both sides by } -3$$

CHECK The preceding steps show that *if* the equation has a solution, then the solution is $x = -5$. With $x = -5$, however, the left-hand side of the original equation becomes $1/(-5 + 5)$, or $1/0$, which is undefined. We conclude therefore that the given equation has no solution.

In Example 4(b), the value $x = -5$, which does not check in the original equation, is called an **extraneous root** or **extraneous solution.** How is it that an extraneous solution was generated in Example 4(b), but not in Example 4(a)? (After all, both solutions involve the same general procedure.) In Example 4(a), we multiplied both sides of the original equation by 4 to produce an equivalent equation. In Example 4(b), however, we multiplied both sides of the original equation by $(x + 5)(x - 3)$. Since we didn't know at that point whether the quantity $(x + 5)(x - 3)$ was nonzero, we could not be certain that the resulting equation was actually an equivalent equation. [Indeed, as it turns out with $x = -5$, the quantity $(x + 5)(x - 3)$ *is* equal to zero.] For this reason, it is always necessary to check in the original equation any solutions you obtain as a result of multiplying both sides of an equation by an expression involving the variable. We restate this advice in the box that follows. (We'll update this box in later chapters when extraneous solutions occur with other types of equations.)

> **PROPERTY SUMMARY** Extraneous Solutions
>
> Multiplying both sides of an equation by an expression involving the variable may introduce extraneous solutions that do not check in the original equation. Therefore, it is always necessary to check any candidates for solutions that you obtain in this manner.

EXAMPLE **5** Solving equations where the unknown is the denominator

Solve each equation for x in terms of the other letters:

(a) $\dfrac{1}{x} = \dfrac{1}{a} + \dfrac{1}{b}$, where $a + b \neq 0$;

(b) $y = \dfrac{ax + b}{cx + d}$, where $cx + d \neq 0$, $yc - a \neq 0$.

SOLUTION

(a)
$$\frac{1}{x} = \frac{1}{a} + \frac{1}{b}$$

$$(xab) \cdot \frac{1}{x} = (xab) \cdot \frac{1}{a} + (xab) \cdot \frac{1}{b} \qquad \text{\color{red}{multiplying both sides by the least common denominator, } } \textit{\color{red}{xab}}$$

$$ab = xb + xa \qquad \text{\color{red}{simplifying}}$$

$$ab = x(b + a) \qquad \text{\color{red}{factoring}}$$

$$\frac{ab}{b + a} = x \qquad \text{\color{red}{dividing both sides by } } b + a \neq 0$$

CHECK As Exercise 88 asks you to verify, the value $x = ab/(a + b)$ indeed satisfies the original equation.

(b) Multiplying both sides of the given equation by the nonzero quantity $cx + d$ yields

$$y(cx + d) = ax + b$$
$$ycx + yd = ax + b$$
$$ycx - ax = b - yd$$
$$x(yc - a) = b - yd$$
$$x = \frac{b - yd}{yc - a}$$

You should check for yourself now that the expression for x on the right-hand side of this last equation indeed satisfies the original equation.

In the example just concluded, we used a basic factoring technique from elementary algebra to solve the equations. Factoring is also useful in solving *quadratic equations*. (For a review of factoring, see Appendix B.4.)

DEFINITION	Quadratic Equation

A **quadratic equation** is an equation that can be written in the form
$$ax^2 + bx + c = 0 \qquad \text{with } a, b, \text{ and } c \text{ real numbers and } a \neq 0$$

Some examples of quadratic equations are
$$x^2 - 8x - 9 = 0 \qquad 2y^2 = 5 \qquad 3m^2 = 1 - m$$

To solve a quadratic equation by factoring, we rely on the following familiar and important property of the real-number system.

PROPERTY SUMMARY
Zero-Product Property of Real Numbers

$$pq = 0 \qquad \text{if and only if} \qquad p = 0 \text{ or } q = 0 \quad \text{(or both)}$$

For example, to solve the equation $x^2 - 2x - 3 = 0$ by factoring, we have
$$x^2 - 2x - 3 = 0$$
$$(x - 3)(x + 1) = 0$$

$x - 3 = 0$	$x + 1 = 0$
$x = 3$	$x = -1$

As you can easily check, the values $x = 3$ and $x = -1$ both satisfy the given equation.

EXAMPLE 6	Applying the zero-product property to solve quadratic equations

Solve:

(a) $8x^2 - 3 = 10x$; **(b)** $x^2 - 9 = 0$.

SOLUTION

(a) In preparation for factoring, we first rewrite the equation so that the right-hand side is zero: $8x^2 - 10x - 3 = 0$. Then we have
$$8x^2 - 10x - 3 = 0$$
$$(2x - 3)(4x + 1) = 0 \qquad \textcolor{red}{\text{Check the factoring.}}$$

$2x - 3 = 0$	$4x + 1 = 0$
$x = 3/2$	$x = -1/4$

As you can now check, the values $x = 3/2$ and $x = -1/4$ both satisfy the given equation.

(b) Using difference-of-squares factoring, we have
$$(x - 3)(x + 3) = 0$$

$x - 3 = 0$	$x + 3 = 0$
$x = 3$	$x = -3$

As you can easily check, the values $x = 3$ and $x = -3$ both satisfy the given equation.

Here's another perspective on Example 6(b). Instead of using factoring to solve the equation $x^2 - 9 = 0$, we can instead rewrite it as $x^2 = 9$. Taking the principal square root of both sides then yields

$$\sqrt{x^2} = \sqrt{9}$$

and therefore

$$|x| = 3 \qquad \text{(Why?)}$$

By looking at this last equation, we can see that there are two solutions, $x = 3$ and $x = -3$. (Those are the only two numbers with absolute values of 3.) We abbreviate these two solutions by writing $x = \pm 3$. In practice, we usually omit showing the step involving the absolute value. For example, to solve the equation $9x^2 - 2 = 0$, just rewrite it as $x^2 = 2/9$. Then, "taking square roots" immediately yields the two solutions $x = \pm\sqrt{2/9} = \pm\sqrt{2}/3$.

Not all quadratic equations can be solved by factoring. Consider, for example, the equation $x^2 - 2x - 4 = 0$. There are only three possible factorizations with integer coefficients, but none yields the appropriate middle term, $-2x$, when multiplied out:

$$(x - 4)(x + 1), \qquad (x + 4)(x - 1), \qquad (x - 2)(x + 2)$$

Moreover, it is not always obvious whether a particular equation, such as $x^2 + 156x + 5963 = 0$, can be solved by factoring. [See Exercise 58(a).] In cases such as these, we can use the *quadratic formula*, given in the box that follows. (In Section 2.1, we'll derive this formula and look at some of its implications. For now, though, the focus is simply on using this formula to calculate solutions.)

The Quadratic Formula

The solutions of the quadratic equation $ax^2 + bx + c = 0$, where $a \neq 0$, are given by

$$x = \frac{-b \pm \sqrt{b^2 - 4ac}}{2a}$$

EXAMPLE 7 Using the quadratic formula to solve a quadratic equation

Use the quadratic formula to solve the equation $2x^2 = 3 - 4x$.

SOLUTION
We first rewrite the given equation as $2x^2 + 4x - 3 = 0$, so that it has the form $ax^2 + bx + c = 0$. By comparing these last two equations, we see that $a = 2$, $b = 4$, and $c = -3$. Therefore

$$x = \frac{-b \pm \sqrt{b^2 - 4ac}}{2a} = \frac{-4 \pm \sqrt{4^2 - 4(2)(-3)}}{2(2)}$$

$$= \frac{-4 \pm \sqrt{40}}{4} = \frac{-4 \pm 2\sqrt{10}}{4} = \frac{-2 \pm \sqrt{10}}{2}$$

Thus, the two solutions are $\dfrac{-2 + \sqrt{10}}{2}$ and $\dfrac{-2 - \sqrt{10}}{2}$.

The techniques that we've reviewed in this section for solving linear and quadratic equations will be used throughout this book; you'll see applications in analyzing graphs and functions and in solving many types of applied problems. Linear and quadratic equations both fall under the general heading of *polynomial equations.*

DEFINITION | **Polynomial Equation**

A **polynomial equation** is an equation of the form

$$a_n x^n + a_{n-1} x^{n-1} + \cdots + a_1 x + a_0 = 0$$

where the subscripted letter a's represent constants and the exponents on the variable are nonnegative integers.

EXAMPLES

(a) $4x^2 - 5x - 1 = 0$

(b) $x^3 - 2x^2 - 3x = 0$

(c) $2x^4 - \frac{4}{3}x^3 + x^2 - 3x + \sqrt{2} = 0$

If a_n is not zero, the **degree** of the polynomial equation is the largest exponent of the variable that appears in the equation. For example, the degrees of equations (a), (b), and (c) in the box above are 2, 3, and 4, respectively.

As we've seen in this section, polynomial equations of degree 1 (linear equations) and polynomial equations of degree 2 (quadratic equations) can be solved by using fairly basic algebra. So too can some higher degree equations. For instance, we can use factoring and the zero-product property to solve equation (b) in the box above. We have

$$x^3 - 2x^2 - 3x = 0$$

$$x(x^2 - 2x - 3) = 0 \qquad \text{factoring out the common term } x$$

$$x(x - 3)(x + 1) = 0 \qquad \text{factoring the quadratic}$$

Therefore

$$x = 0 \quad \text{or} \quad x - 3 = 0 \quad \text{or} \quad x + 1 = 0 \qquad \text{using the zero-product property}$$

From these last three equations we conclude that the solutions of the third-degree polynomial equation $x^3 - 2x^2 - 3x = 0$ are $x = 0, 3,$ and -1. (You should check for yourself that each of these numbers indeed satisfies the equation.)

Unfortunately, not all polynomial equations are as easy to solve as this last one. Here are three examples to show you some of the possibilities that can occur.

EQUATION	COMMENT	TECHNIQUES REQUIRED
$x^3 - 2x + 1 = 0$	Although this polynomial can be factored, none of the basic factoring techniques (reviewed in Appendix B.4) is applicable.	The remainder theorem from Section 7.3 and the rational roots theorem from Section 7.5
$x^3 + 2x - 4 = 0$	This polynomial cannot be factored using integer or rational coefficients.	The location theorem and the method of successive approximations from Section 7.5
$x^2 - 2x - 4 = 0$	There are no real-number solutions.	The quadratic formula is applicable, but the solutions involve the complex-number system, which is discussed in Section 7.1

Chapter 7 contains a complete discussion of these ideas and also an answer to the following question: Is there a general formula, similar to the quadratic formula, for solving *any* polynomial equation?

EXERCISE SET 1.3

A

In Exercises 1–5, determine whether the given value is a solution of the equation.

1. $4x - 5 = -13; x = -2$

2. $\dfrac{1}{x} = \dfrac{3}{x} - 1; x = 2$

3. $\dfrac{2}{y - 1} - \dfrac{3}{y} = \dfrac{7}{y^2 - y}; y = -3$

4. $(y - 1)(y + 5) = 0; y = 5$

5. $m^2 + m - \frac{5}{16} = 0; m = \frac{1}{4}$

6. Verify that the numbers $1 + \sqrt{5}$ and $1 - \sqrt{5}$ both satisfy the equation $x^2 - 2x - 4 = 0$.

Solve each equation in Exercises 7–32.

7. $2x - 3 = -5$

8. $x + 4 = 0$

9. $-6x + 1 = 49$

10. $3 - 4y = 8y + 3$

11. $2m - 1 + 3m + 5 = 6m - 8$

12. $1 - (2m + 5) = -3m$

13. $(x + 2)(x + 1) = x^2 + 11$

14. $t - \{4 - [t - (4 + t)]\} = 6$

15. $\dfrac{x}{3} + \dfrac{2x}{5} = \dfrac{-11}{5}$

16. $\dfrac{x}{2} + 1 = 0$

17. $1 - \dfrac{y}{3} = 6$

18. $y - \dfrac{y}{2} - \dfrac{y}{3} = 2$

19. $\dfrac{x - 1}{4} + \dfrac{2x + 3}{-1} = 0$

20. $\dfrac{x - 1}{2} + \dfrac{x - 2}{4} = 1 + \dfrac{x}{3}$

21. $\dfrac{1}{x - 3} - \dfrac{2}{x + 3} = \dfrac{1}{x^2 - 9}$

22. $\dfrac{1}{x - 5} + \dfrac{1}{x + 5} = \dfrac{6}{x^2 - 25}$

23. $\dfrac{1}{x - 5} + \dfrac{1}{x + 5} = \dfrac{2x + 1}{x^2 - 25}$

24. $\dfrac{4}{x + 2} + \dfrac{1}{x - 2} = \dfrac{4}{x^2 - 4}$

25. $\dfrac{3}{2x + 1} - \dfrac{4}{x + 1} = \dfrac{2}{2x^2 + 3x + 1}$

26. $\dfrac{5}{x - 4} - \dfrac{3}{2x^2 - 5x - 12} = \dfrac{1}{2x + 3}$

27. $\dfrac{2(x + 1)}{x - 1} - 3 = \dfrac{5x - 1}{x - 1}$

28. $\dfrac{x - 1}{x + 1} = \dfrac{x + 1}{x - 1}$

29. $\dfrac{1}{x} = \dfrac{4}{x} - 1$

30. $\dfrac{1}{y} + 1 = \dfrac{3}{y} - \dfrac{1}{2y}$

31. **(a)** $\dfrac{2}{3x} = \dfrac{3}{x}$

(b) $\dfrac{2}{3x} = \dfrac{3}{x + 1}$

(c) $\dfrac{2}{3x} = \dfrac{3}{x} + 1$

32. **(a)** $\dfrac{3}{x - 2} = \dfrac{5}{9x}$

(b) $\dfrac{3}{x - 2} = \dfrac{5}{9x - 2}$

(c) $\dfrac{3}{x - 2} = \dfrac{5}{\frac{5}{3}x - 2}$

In Exercises 33–42, solve each equation by factoring.

33. $x^2 - 5x - 6 = 0$

34. $x^2 - 5x = -6$

35. $10z^2 - 13z - 3 = 0$

36. $3t^2 - t - 4 = 0$

37. $(x + 1)^2 - 4 = 0$

38. $x^2 + 3x - 40 = 0$

39. $x(2x - 13) = -6$

40. $x(3x - 23) = 8$

41. $x(x + 1) = 156$

42. $x^2 + (2\sqrt{5})x + 5 = 0$

In Exercises 43–52, use the quadratic formula to solve each equation. In Exercises 43–50, give two forms for each solution: an expression containing a radical, and a calculator approximation rounded off to two decimal places.

43. $x^2 - x - 5 = 0$

44. $x^2 - x + 5 = 0$

45. $2x^2 + 3x - 4 = 0$

46. $4x^2 - 3x - 9 = 0$

47. $x(x + 6) = -2$

48. $x(3x + 8) = -2$

49. $2x^2 - 10 = -\sqrt{2}x$

50. $\sqrt{3}x^2 + \sqrt{3} = 6x$

51. $12x^2 - 25x = -12$

52. $24x^2 + 23x = -5$

In Exercises 53–58, solve the equations using any method you choose.

53. $x^2 = 24$

54. $2y^2 - 50 = 0$

55. $\frac{1}{8} - t^2 = 0$

56. $x^2 - \sqrt{5} = 0$

57. **(a)** $u(u + 18) = -81$

(b) $u(u + 18) = 81$

58. **(a)** $x^2 + 156x + 5963 = 0$

(b) $144y^2 - 54y = 13$

For Exercises 59–70, solve each equation for x in terms of the other letters.

59. $3ax - 2b = b + 3$

60. $ax + b = bx - a$

61. $ax + b = bx + a$

62. $\dfrac{x}{a} + \dfrac{x}{b} = 1$

63. $\dfrac{1}{x} = a + b$

64. $\dfrac{1}{ax} = \dfrac{1}{bx} - \dfrac{1}{c}$

65. $\dfrac{1}{a} - \dfrac{1}{x} = \dfrac{1}{x} - \dfrac{1}{b}$

66. **(a)** $y = mx + b$, where $m \neq 0$

(b) $y - y_1 = m(x - x_1)$, where $m \neq 0$

(c) $\dfrac{x}{a} + \dfrac{y}{b} = 1$

(d) $Ax + By + C = 0$, where $A \neq 0$

67. $2y^2x^2 - 3yx + 1 = 0$, where $y > 0$

68. $(ax + b)^2 - (bx + a)^2 = 0$, where $a \neq \pm b$

69. $(x - p)^2 + (x - q)^2 = p^2 + q^2$

70. $21x^2 - 2kx - 3k^2 = 0$, where $k > 0$

B

In Exercises 71–76, solve each equation for x in terms of the other letters.

71. $a^2(a - x) = b^2(b + x) - 2abx$, where $a \neq b$

72. $\dfrac{b}{ax - 1} - \dfrac{a}{bx - 1} = 0$, where $a \neq b$

73. $\dfrac{a - x}{a - b} - 2 = \dfrac{c - x}{b - c}$

74. $\dfrac{x + 2p}{2q - x} + \dfrac{x - 2p}{2q + x} - \dfrac{4pq}{4q^2 - x^2} = 0$

75. $\dfrac{x - a}{x - b} = \dfrac{b - x}{a - x}$, where $a \neq b$

76. $1 - \dfrac{a}{b}\left(1 - \dfrac{a}{x}\right) - \dfrac{b}{a}\left(1 - \dfrac{b}{x}\right) = 0$

In Exercises 77–80, solve each equation for the indicated variable.

77. $d = \dfrac{r}{1 + rt}$; for r

78. $\dfrac{x_1 x}{a^2} + \dfrac{y_1 y}{b^2} = 1$; for y

79. $S = 2\pi r^2 + 2\pi rh$; for h

80. $S = \dfrac{rl - a}{r - 1}$; for r

Solve the equations in Exercises 81–86. (In these exercises, you'll need to multiply both sides of the equations by expressions involving the variable. Remember to check your answers in these cases.)

81. $\dfrac{3}{x + 5} + \dfrac{4}{x} = 2$

82. $\dfrac{5}{x + 2} - \dfrac{2x - 1}{5} = 0$

83. $1 - x - \dfrac{2}{6x + 1} = 0$

84. $\dfrac{x^2 - 3x}{x + 1} = \dfrac{4}{x + 1}$

85. $\dfrac{x}{x - 2} + \dfrac{x}{x + 2} = \dfrac{8}{x^2 - 4}$

86. $\dfrac{2x}{x^2 - 1} - \dfrac{1}{x + 3} = 0$

87. Solve each of the following equations for x. *Hint*: As in the text, begin by factoring out a common term.
(a) $x^3 - 13x^2 + 42x = 0$
(b) $x^3 - 6x^2 + x = 0$

88. Verify that $x = ab/(a + b)$ is a solution of the equation $\dfrac{1}{x} = \dfrac{1}{a} + \dfrac{1}{b}$. [This is to check the answer obtained in Example 5(a).]

**1.4   RECTANGULAR COORDINATES.
VISUALIZING DATA**

The name coordinate does not appear in the work of Descartes. This term is due to Leibniz and so are abscissa *and* ordinate *(1692).* —David M. Burton in *The History of Mathematics, an Introduction,* 2nd ed. (Dubuque, Iowa: Wm. C. Brown Publishers, 1991)

With equal passion I have sought knowledge. I have wished to understand the hearts of men. I have wished to know why the stars shine. And I have tried to apprehend the Pythagorean power by which number holds sway about the flux. —Bertrand Russell (1872–1970)

There are three kinds of lies: lies, damned lies, and statistics. —Benjamin Disraeli (1804–1881)

In previous courses you learned to work with a rectangular coordinate system such as that shown in Figure 1. In this section we review some of the most basic formulas and techniques that are useful here.

The point of intersection of the two perpendicular number lines, or **axes,** is called the **origin** and is denoted by the letter O. The horizontal and vertical axes are often labeled the *x*-**axis** and the *y*-**axis,** respectively, but any other variables will do just as well for labeling the axes. See Figure 2 for examples of this. (We'll discuss curves or graphs like the ones in Figure 2 in later sections.)

Note that in Figures 1 and 2 the axes divide the plane into four regions, or **quadrants,** labeled I through IV, as shown in Figure 1. Unless indicated otherwise, we assume that the same unit of length is used on both axes. In Figure 1, the same scales are used on both axes; not so in Figure 2.

Figure 1

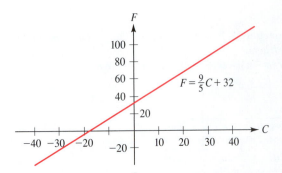

(a) A graph of the formula $s = 16t^2$ in a t-s coordinate system. [The formula relates the distance s (in feet) and the time t (in seconds) for an object falling in a vacuum.]

(b) A graph of the equation $F = \frac{9}{5}C + 32$ in a C-F coordinate system. (The equation gives the relatationship between the temperature C on the Celsius scale and F on the Fahrenheit scale.)

Figure 2

Now look at the point P in Figure 3(a). Starting from the origin O, one way to reach P is to move three units in the positive x-direction and then two units in the positive y-direction. That is, the location of P relative to the origin and the axes is "right 3, up 2." We say that the **coordinates** of P are $(3, 2)$. The first number within the parentheses conveys the information "right 3," and the second number conveys the information "up 2." We say that the **x-coordinate** of P is 3 and the **y-coordinate** of P is 2. Likewise, the coordinates of point Q in Figure 3(a) are $(-2, 4)$. With this coordinate notation in mind, observe in Figure 3(b) that $(3, 2)$ and $(2, 3)$ represent different points; that is, the order in which the two numbers appear within the parentheses affects the location of the point. Figure 3(c) displays various points with given coordinates; you should check for yourself that the coordinates correspond correctly to the location of each point.

Some terminology and notation: The x-y coordinate system that we have described is often called a **Cartesian coordinate system.** The term *Cartesian* is used in honor of René Descartes, the seventeenth-century French philosopher and mathematician. The coordinates (x, y) of a point P are referred to as an **ordered pair.** Recall, for example, that $(3, 2)$ and $(2, 3)$ represent different points; that is, the order of the numbers matters. The x-coordinate of a point is sometimes referred to as the **abscissa** of the point; the y-coordinate is the **ordinate.** The notation $P(x, y)$ means that P is a point that has coordinates (x, y). At times, we abbreviate the phrase *the point whose coordinates are* (x, y) to simply *the point* (x, y).

The next part of our work in this section depends on a key result from elementary geometry, the Pythagorean theorem. For reference, we state this theorem and

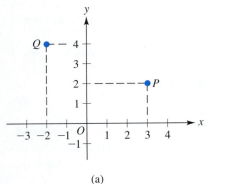

(a)

(b)

(c)

Figure 3

its converse in the box that follows. (For proofs of the Pythagorean theorem, see Exercises 34 and 35 at the end of this section or Exercise 100 in the Chapter Review Exercises.)

The Pythagorean Theorem and Its Converse

1. Pythagorean Theorem
   (See Figure 4.) In a right triangle, the lengths of the sides are related by the equation

   $$a^2 + b^2 = c^2$$

   where a and b are the lengths of the sides forming the right angle and c is the length of the hypotenuse (the side opposite the right angle).

2. Converse
   If the lengths of the sides of a triangle are related by an equation of the form $a^2 + b^2 = c^2$, then the triangle is a right triangle, and c is the length of the hypotenuse.

Figure 4

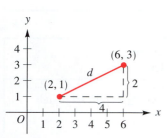

Figure 5

EXAMPLE 1 Using the Pythagorean theorem to find a distance

Use the Pythagorean theorem to calculate the distance d between the points $(2, 1)$ and $(6, 3)$.

SOLUTION
We plot the two given points and draw a line connecting them, as shown in Figure 5. Then we draw the broken lines as shown, parallel to the axes, and apply the Pythagorean theorem to the right triangle that is formed. The base of the triangle is four units long. You can see this by simply counting spaces or by using absolute value, as discussed in Section 1.2: $|6 - 2| = 4$. The height of the triangle is found to be two units, either by counting spaces or by computing the absolute value: $|3 - 1| = 2$. Thus, we have

$$d^2 = 4^2 + 2^2 = 20$$
$$d = \sqrt{20} = \sqrt{4}\sqrt{5} = 2\sqrt{5}$$

Note: Since d is a distance, we disregard the solution $-\sqrt{20}$ of the equation $d^2 = 20$.

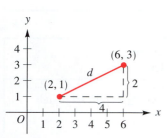

The method we used in Example 1 can be applied to derive a general formula for the distance d between any two points (x_1, y_1) and (x_2, y_2) (see Figure 6). Just as before, we draw in the right triangle and apply the Pythagorean theorem. We have

$$d^2 = |x_2 - x_1|^2 + |y_2 - y_1|^2$$
$$= (x_2 - x_1)^2 + (y_2 - y_1)^2 \quad \text{(Why?)}$$

and therefore

$$d = \sqrt{(x_2 - x_1)^2 + (y_2 - y_1)^2}$$

This last equation is referred to as the **distance formula.** For reference, we restate it in the box that follows.

Figure 6

The Distance Formula

The distance d between the points (x_1, y_1) and (x_2, y_2) is given by

$$d = \sqrt{(x_2 - x_1)^2 + (y_2 - y_1)^2}$$

Examples 2–4 demonstrate some simple calculations involving the distance formula.

NOTE   In computing the distance between two given points, it does not matter which one you treat as (x_1, y_1) and which as (x_2, y_2), because quantities such as $x_2 - x_1$ and $x_1 - x_2$ are negatives of each other and so have equal squares.

EXAMPLE	2	Using the distance formula

Calculate the distance between the points $(2, -6)$ and $(5, 3)$.

SOLUTION
Substituting $(2, -6)$ for (x_1, y_1) and $(5, 3)$ for (x_2, y_2) in the distance formula, we have

$$\begin{aligned}
d &= \sqrt{(5 - 2)^2 + [3 - (-6)]^2} \\
&= \sqrt{3^2 + 9^2} = \sqrt{90} \\
&= \sqrt{9}\sqrt{10} = 3\sqrt{10}
\end{aligned}$$

You should check for yourself that the same answer is obtained using $(2, -6)$ as (x_2, y_2) and $(5, 3)$ as (x_1, y_1).

EXAMPLE	3	Using the distance formula and the converse of the Pythagorean theorem

Is the triangle with vertices $D(-2, -1)$, $E(4, 1)$, and $F(3, 4)$ a right triangle?

SOLUTION
First we sketch the triangle in question (see Figure 7). From the sketch it appears that angle E could be a right angle, but certainly this is not a proof. We

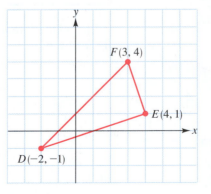

Figure 7

need to use the distance formula to calculate the lengths of the three sides and then check whether any relation of the form $a^2 + b^2 = c^2$ holds. The calculations are as follows:

$$DE = \sqrt{[4 - (-2)]^2 + [1 - (-1)]^2} = \sqrt{36 + 4} = \sqrt{40}$$
$$EF = \sqrt{(4 - 3)^2 + (1 - 4)^2} = \sqrt{1 + 9} = \sqrt{10}$$
$$DF = \sqrt{[3 - (-2)]^2 + [4 - (-1)]^2} = \sqrt{25 + 25} = \sqrt{50}$$

Because $(\sqrt{40})^2 + (\sqrt{10})^2 = (\sqrt{50})^2$, we are indeed guaranteed that $\triangle DEF$ is a right triangle. (In Section 1.6, you'll see that this result can be obtained more efficiently using the concept of "slope.")

Figure 8

EXAMPLE   4   Using the distance formula to find a radius

(a) Find the radius r of the circle in Figure 8. (Assume that the center of the circle is located at the origin.)

(b) Compute the area and the circumference of the circle. For each answer, give exact expressions and also calculator approximations rounded to one decimal place.

SOLUTION

(a) The radius r is the distance from center $(0, 0)$ to the given point $(-3, 2)$ on the circle. Using the distance formula, we have

$$r = \sqrt{(-3 - 0)^2 + (2 - 0)^2}$$
$$= \sqrt{9 + 4} = \sqrt{13} \text{ units}$$

(b) Recall the formulas for the area A and the circumference C of a circle: $A = \pi r^2$ and $C = 2\pi r$. Using the value for r from part (a), we have

$$A = \pi r^2 = \pi(\sqrt{13})^2 \qquad\qquad C = 2\pi r = 2\pi\sqrt{13} \text{ units}$$
$$= 13\pi \text{ square units} \qquad\qquad \approx 22.7 \text{ units}$$
$$\approx 40.8 \text{ square units}$$

One of the important applications of rectangular coordinates is in displaying quantitative data. You see instances of this every day in newspapers, in magazines, and in textbooks as diverse as astronomy to biology. We show some examples in the figures and discussion that follow.

Table 1 provides world population data for the period 1965–1995. In Figure 9(a) the familiar *bar graph* (or *column chart*) format is used to display the data from the table. In Figure 9(b) we've plotted the data in a rectangular coordinate system. On the horizontal axis, the variable t represents years; the variable P represents population in units of one billion. The data in the first row of the table (which state that the population in 1965 was 3.345 billion) is plotted in Figure 9(b) as the point (1965, 3.345). Likewise, the second row of data in the table gives us the point (1975, 4.086), and so on. Sometimes it is more convenient to work with smaller numbers on the horizontal axis than those used in Figure 9(b). One very common way to do this is indicated in Figure 9(c), where we are now letting the variable t

TABLE 1	World Population 1965–1995
Year	**Population (billions)**
1965	3.345
1975	4.086
1985	4.850
1995	5.687

Source: U.S. Census Bureau (International Data Base)

(a) Bar graph or column chart

(b) Rectangular coordinates with the variable t representing the year

(c) Rectangular coordinates with the variable t representing the number of years since 1965

Figure 9
World population 1965–1995

represent *years since 1965*. In other words, the year 1965 is $t = 0$, 1966 is $t = 1$, 1967 is $t = 2$, and so on. The data in the first row of Table 1 are then plotted in Figure 9(c) as the point (0, 3.345), rather than (1965, 3.345). Likewise, the second row of data in Table 1 is plotted in Figure 9(c) as (10, 4.086), and so on.

Both the bar graph and the rectangular plots in Figure 9 make it immediately clear that the world population is increasing. Is it increasing at a steady rate? Is it increasing rapidly? In fact, one needs to exercise caution in using graphs to draw conclusions about *how fast* the quantity being graphed (in this case, population) is increasing or decreasing. For instance, Figure 10 shows another graph of world population, this time covering the period 1800–1995. Compare the graph to the one shown in Figure 9(b). Certainly Figure 10 leads to different ideas, different interpretations, about the nature of world population growth. [In Figure 10, the four blue dots are the data points that appear in Figure 9(b).]

As another example about the need for caution in interpreting graphs, look at Figure 11, which shows two very different interpretations of the data for SAT mathe-

Figure 10
World population 1800–1995.
Source: U.S. Census Bureau

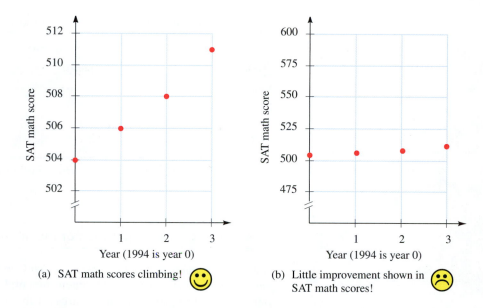

Figure 11

Two visualizations and interpretations of the data in Table 2

(a) SAT math scores climbing! 🙂

(b) Little improvement shown in SAT math scores! 😦

TABLE 2   SAT Math Scores: National Averages 1994–1997

Year	SAT math score
0 (1994)	504
1 (1995)	506
2 (1996)	508
3 (1997)	511

Source: The College Board

matics scores in Table 2. The bottom line is that graphs are useful, even indispensible, in giving us an easy way to see general trends in data, but exercise care in drawing further conclusions, especially regarding rates of increase or decrease. A complete analysis of how fast a quantity is increasing or decreasing may require topics from calculus or the field of statistics. For straight-line graphs, however, the concept of *slope* (reviewed in Section 1.6) tells us definitively about rates of increase or decrease. Also, when we study functions in Chapter 3, we'll make a first step toward answering general questions about rates of change.

In Example 5 we make use of a simple result that you may recall from previous courses: the *midpoint formula*. This result is summarized in the box that follows. (For a proof of the formula, see Exercise 33.)

THE MIDPOINT FORMULA

The midpoint of the line segment joining the points $P(x_1, y_1)$ and $Q(x_2, y_2)$ is

$$\left(\frac{x_1 + x_2}{2}, \frac{y_1 + y_2}{2} \right)$$

Example

The midpoint of the line segment joining $(2, -15)$ and $(4, 5)$ is

$$\left(\frac{2 + 4}{2}, \frac{-15 + 5}{2} \right) = (3, -5)$$

TABLE 3 Atmospheric Carbon Dioxide	
Year	Carbon dioxide in atmosphere (ppm)
1990	354.0
1992	356.3
1994	358.9
1996	362.6

Source: C. D. Keeling and T. P. Whorf, Scripps Institution of Oceanography

EXAMPLE 5 An application of the midpoint formula

Data concerning the amount of carbon dioxide in the atmosphere (measured in *parts per million* or *ppm*) is used by environmental scientists in the study of global warming. Table 3 provides some figures for the period 1990–1996.

(a) Plot the data in a rectangular coordinate system. Use the variable t on the horizontal axis, with $t = 0$ corresponding to the year 1990. Use the variable c (to denote carbon dioxide levels, in ppm) on the vertical axis.

(b) Use the midpoint formula and the data for 1992 and 1994 to estimate the amount of carbon dioxide for the year 1993.

(c) Compute the percentage error in the estimation in part (b), given that the actual 1993 value is 357.0 ppm. The general formula for percentage error in an estimation or approximation is

$$\text{percentage error} = \frac{|(\text{actual value}) - (\text{approximate value})|}{\text{actual value}} \times 100$$

SOLUTION

(a) See Figure 12. Note that if $t = 0$ corresponds to 1990, then $t = 2$ corresponds to 1992, $t = 4$ corresponds to 1994, and $t = 6$ corresponds to 1996.

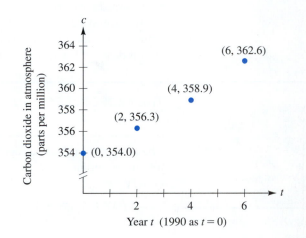

Figure 12

(b) The midpoint of the line segment joining the points $(2, 356.3)$ and $(4, 358.9)$ is

$$\left(\frac{2 + 4}{2}, \frac{356.3 + 358.9}{2} \right)$$

which, as you can verify, works out to

$$(3, 357.6)$$

Thus, our approximation for the carbon dioxide level in 1993 ($t = 3$) is 357.6 ppm.

(c) We have

$$\text{percentage error} = \frac{|(\text{actual value}) - (\text{approximate value})|}{\text{actual value}} \times 100$$

$$= \frac{|357.0 - 357.6|}{357.0} \times 100$$

$$\approx 0.2\%$$ *using a calculator and round-ing to one decimal place*

In Figure 13 we show the given data for 1992 and 1994 along with the estimated and the actual value for 1993.

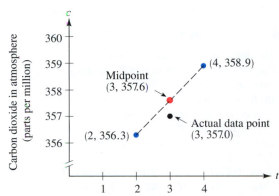

Figure 13
The midpoint of the line segment is close to the actual data point.

CAUTION Do not assume on the basis of this one example that the midpoint approximation always works as well as it does here. In this regard, be sure to work Exercise 25 at the end of this section.

EXERCISE SET 1.4

A

1. Plot the points $(5, 2)$, $(-4, 5)$, $(-4, 0)$, $(-1, -1)$, and $(5, -2)$.

2. Draw the square $ABCD$ with vertices (corners) $A(1, 0)$, $B(0, 1)$, $C(-1, 0)$, and $D(0, -1)$.

3. **(a)** Draw the right triangle PQR with vertices $P(1, 0)$, $Q(5, 0)$, and $R(5, 3)$.
   (b) Use the formula for the area of a triangle, $A = \frac{1}{2}bh$, to find the area of triangle PQR in part (a).

4. **(a)** Draw the trapezoid $ABCD$ with vertices $A(0, 0)$, $B(7, 0)$, $C(6, 4)$, and $D(4, 4)$.
   (b) Compute the area of the trapezoid. (See the inside front cover of this book for the appropriate formula.)

In Exercises 5–10, calculate the distance between the given points.

5. **(a)** $(0, 0)$ and $(-3, 4)$
   (b) $(2, 1)$ and $(7, 13)$

6. **(a)** $(-1, -3)$ and $(-5, 4)$
   (b) $(6, -2)$ and $(-1, 1)$

7. **(a)** $(-5, 0)$ and $(5, 0)$
   (b) $(0, -8)$ and $(0, 1)$
8. **(a)** $(-5, -3)$ and $(-9, -6)$
   (b) $(\frac{9}{2}, 3)$ and $(-2\frac{1}{2}, -1)$
9. $(1, \sqrt{3})$ and $(-1, -\sqrt{3})$ 10. $(-3, 1)$ and $(374, -335)$
11. Which point is farther from the origin?
    (a) $(3, -2)$ or $(4, \frac{1}{2})$
    (b) $(-6, 7)$ or $(9, 0)$
12. Use the distance formula to show that, in each case, the triangle with given vertices is an isosceles triangle.
    (a) $(0, 2), (7, 4), (2, -5)$
    (b) $(-1, -8), (0, -1), (-4, -4)$
    (c) $(-7, 4), (-3, 10), (1, 3)$
13. In each case, determine whether the triangle with the given vertices is a right triangle.
    (a) $(7, -1), (-3, 5), (-12, -10)$
    (b) $(4, 5), (-3, 9), (1, 3)$
    (c) $(-8, -2), (1, -1), (10, 19)$
14. **(a)** Two of the three triangles specified in Exercise 13 are right triangles. Find their areas.
    (b) Calculate the area of the remaining triangle in Exercise 13 by using the following formula for the area A of a triangle with vertices (x_1, y_1), (x_2, y_2), and (x_3, y_3):

    $$A = \tfrac{1}{2}|x_1 y_2 - x_2 y_1 + x_2 y_3 - x_3 y_2 + x_3 y_1 - x_1 y_3|$$

    The derivation of this formula is given in Exercise 36.
    (c) Use the formula given in part (b) to check your answers in part (a).
15. Use the formula given in Exercise 14(b) to calculate the area of the triangle with vertices $(1, -4)$, $(5, 3)$, and $(13, 17)$. What do you conclude?
16. The coordinates of points A, B, and C are $A(-4, 6)$, $B(-1, 2)$, and $C(2, -2)$.
    (a) Show that $AB = BC$ by using the distance formula.
    (b) Show that $AB + BC = AC$ by using the distance formula.
    (c) What can you conclude from parts (a) and (b)?

In Exercises 17 and 18, find the midpoint of the line segment joining points P and Q.
17. **(a)** $P(3, 2)$ and $Q(9, 8)$
    (b) $P(-4, 0)$ and $Q(5, -3)$
    (c) $P(3, -6)$ and $Q(-1, -2)$
18. **(a)** $P(12, 0)$ and $Q(12, 8)$
    (b) $P(\frac{3}{5}, -\frac{2}{3})$ and $Q(0, 0)$
    (c) $P(1, \pi)$ and $Q(3, 3\pi)$

*In Exercises 19 and 20, the given points P and Q are the endpoints of a diameter of a circle. Find **(a)** the center of the circle; **(b)** the radius of the circle.*
19. $P(-4, -2)$ and $Q(6, 4)$
20. $P(1, -3)$ and $Q(-5, -5)$

The number of new car sales in the United States each year has two components: domestic car sales and import car sales. For

Exercises 21 and 22 refer to Table 4, which gives the percentage of U.S. car sales from imports over the period 1990–1996. (The percentages are of the total number of retail car sales, not the dollar amounts.)

TABLE 4 Percentage of Retail New Car Sales in United States due to Imports, 1990–1996

Year	1990	1992	1994	1996
Percentage due to imports	25.8	23.6	19.3	14.9

Source: American Automobile Manufacturers Assn.

21. **(a)** Copy or photocopy the following coordinate system and use it to plot the data in Table 4.

(b) What are the coordinates of the point in your graph that corresponds to the data for 1990? For 1992?
(c) Use the midpoint formula with the two points that you listed in part (b) to estimate the percentage of import sales for the year 1991. (Round the answer to one decimal place.)
(d) Compute the percentage error in the estimate in part (c), given that the actual 1991 value is 24.9%. [The formula for percentage error is given in Example 5(c).]
22. (Requires Exercise 21 as background)
(a) What are the coordinates of the point in your graph [in Exercise 21(a)] that corresponds to the data for 1994? For 1996?

(b) Use the midpoint formula with the two points that you listed in part (a) to estimate the percentage of import sales for the year 1995. (Round the answer to one decimal place.)

(c) Compute the percentage error in the estimate in part (b), given that the actual 1995 value is 17.4%.

For Exercises 23 and 24, refer to Table 5, which indicates how much the television networks paid (per hour) to televise the Winter Olympics over the period 1980–1998.

Table 5 How Much the Networks Paid (per TV hour) to Televise the Winter Olympics, 1980–1998

Year (network)	1980 (ABC)	1984 (ABC)	1988 (ABC)	1992 (CBS)	1994 (CBS)	1998 (CBS)
Millions of dollars per TV hour	0.3	1.5	3.3	2.1	2.5	2.9

Source: World Features Syndicate

23. (a) Using a coordinate system similar to the one shown in the following figure (or a photocopy), plot the two points corresponding to the data for the years 1984 and 1992. (Ignore the information about the name of the network.)

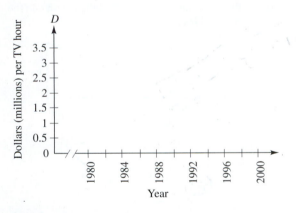

(b) Use the midpoint formula with the two points that you plotted in part (a) to obtain an approximation for the dollar amount paid per TV hour for 1988. (Round the answer to one decimal place.)

(c) Compute the percentage error in the approximation in part (b). The actual 1988 value is given in the table.

24. (a) Using a coordinate system similar to the one shown in the figure accompanying Exercise 23(a), plot the two points corresponding to the data for the years 1980 and 1988.

(b) Use the midpoint formula with the two points that you plotted in part (a) to obtain an approximation for the dollar amount paid per TV hour for 1984. (Round the answer to one decimal place.)

(c) Compute the percentage error in the approximation in part (b). The actual 1984 value is given in the table.

25. Over the past two decades the Internet has grown very rapidly. Figures A and B provide estimates for the number n of Internet host computers, worldwide, for the years $t = 1995$–1997 and 1985–1987. Source: *Network Wizards* <http//www.nw.com>

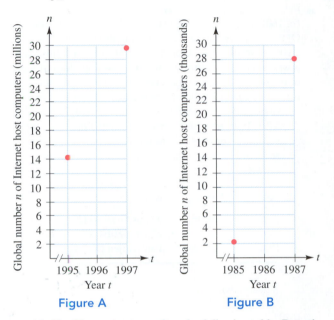

Figure A	Figure B

(a) Use Figure A to complete the following table. Round the values of n to the nearest two million. Then use the midpoint formula and the numbers in your table to estimate the global number of Internet host computers for the year 1996.

t	1995	1997
n		

(b) Use Figure B to complete the following table. Round the values of n to the nearest two thousand. Then use the midpoint formula and the numbers in your table to estimate the global number of Internet host computers for the year 1986.

t	1985	1987
n		

(c) Compute the percentage errors to determine which estimate, the one for 1986 or the one for 1996, is more accurate. Use the following data from *Network Wizards* <http//www.nw.com> in computing the percentage errors: The number of host computers for 1986 and 1996 were 5089 and 21,819,000, respectively. (Round each answer to the nearest one percent.)

(continues)

Note: You'll find out in part (c) that one estimate is very good, the other is way off. The point here is that without more initial information, it's hard to say whether the midpoint formula will produce a useful estimate. In subsequent chapters, we'll use *functions* and larger data sets to obtain more reliable estimates.

26. Have you or a friend ever run in a 10K (10,000 meter) race? When the author polled his precalculus class at UCLA Fall 1997, he found that there were five students in the class (of 160) who said they had run a 10K in under 50 minutes. Of those five, two (one male, one female) said they had run a 10K in under 40 minutes. The world record for this event is well under 30 minutes. In this exercise you'll look at some of the world records in this event over the past decade.

(a) The table that follows lists the world records in the (men's) 10,000 meter race as of the end of the years 1993, 1995, and 1997. After converting the times into seconds, plot the three points corresponding to these records in a coordinate system similar to the one shown.

Year	Time	Runner
1993	26:58.38	Yobes Ondieki (Kenya)
1995	26:43.53	Haile Gebrselassie (Kenya)
1997	26:27.85	Paul Tergat (Kenya)

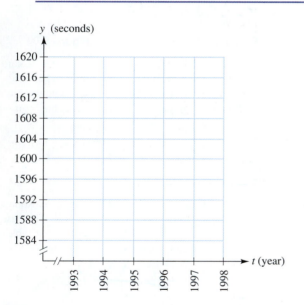

(b) Use the midpoint formula and the data for 1993 and 1995 to compute an estimate for what the world record might have been by the end of 1994. Then compute the percentage error (rounded to two decimal places), given that the record at the end of 1994 was 26:52.23 (set by William Seigei of Kenya). Was your estimate too high or too low?

(c) Use the midpoint formula and the data for 1995 and 1997 to compute an estimate for what the world record might have been by the end of 1996. Then com-

pute the percentage error given that the record at the end of 1996 was 26:38.08 (set by Salah Hissou of Morocco). Was your estimate too high or too low? Is the percentage error more or less than that obtained in part (b)?

(d) Using a coordinate system similar to the one shown in part (a), or using a photocopy, plot the points corresponding to the (actual, not estimated) world records for the years 1993, 1994, 1995, 1996, 1997, and 1998. Except for 1998, all the records have been given above. The world record at the end of 1998 was 26:22.75 (set by Haile Gebrselassie of Kenya). Use the picture you obtain to say whether or not the record times seem to be leveling off.

27. (a) Sketch the parallelogram with vertices $A(-7, -1)$, $B(4, 3)$, $C(7, 8)$, and $D(-4, 4)$.

(b) Compute the midpoints of the diagonals $\overline{AC}$ and $\overline{BD}$.

(c) What conclusion can you draw from part (b)?

28. The vertices of $\triangle ABC$ are $A(1, 1)$, $B(9, 3)$, and $C(3, 5)$.

(a) Find the perimeter of $\triangle ABC$.

(b) Find the perimeter of the triangle that is formed by joining the midpoints of the three sides of $\triangle ABC$.

(c) Compute the ratio of the perimeter in part (a) to the perimeter in part (b).

(d) What theorem from geometry provides the answer for part (c) without using the results in (a) and (b)?

29. Use the Pythagorean theorem to find the length a in the figure. Then find $b, c, d, e, f,$ and g.

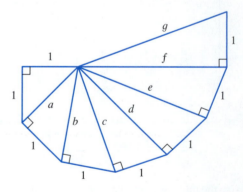

Note: This figure provides a geometric construction for the irrational numbers $\sqrt{2}, \sqrt{3}, \ldots, \sqrt{n}$, where n is a nonsquare natural number. According to Boyer's *A History of Mathematics*, 2nd ed. (New York: John Wiley & Sons, Inc., 1991), "... Plato ... says that his teacher Theodorus of Cyrene ... was the first to prove the irrationality of the square roots of the nonsquare integers from 3 to 17, inclusive. It is not known how he did this, nor why he stopped with $\sqrt{17}$." One plausible reason for Theodorus's stopping with $\sqrt{17}$ may have to do with the figure shown here. Theodorus may have known that the figure begins to overlap itself at the stage where $\sqrt{18}$ would be constructed.

30. (A numerologist's delight) Using the Pythagorean theorem and your calculator, compute the area of a right triangle in which the lengths of the hypotenuse and one leg are 2045 and 693, respectively.

B

31. The diagonals of a parallelogram bisect each other. Steps (a), (b), and (c) outline a proof of this theorem.
- **(a)** In the parallelogram $OABC$ shown in the figure, check that the coordinates of B must be $(a + b, c)$.
- **(b)** Use the midpoint formula to calculate the midpoints of diagonals $\overline{OB}$ and $\overline{AC}$.
- **(c)** The two answers in part (b) are identical. This shows that the two diagonals do indeed bisect each other, as we wished to prove.

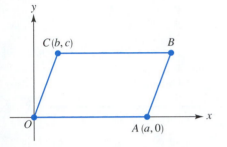

32. Prove that in a parallelogram, the sum of the squares of the lengths of the diagonals equals the sum of the squares of the lengths of the four sides. (Use the figure in Exercise 31.)

33. Suppose that the coordinates of points P, Q, and M are

$$P(x_1, y_1) \qquad Q(x_2, y_2)$$
$$M\left(\frac{x_1 + x_2}{2}, \frac{y_1 + y_2}{2}\right)$$

Follow steps (a) and (b) to prove that M is the midpoint of the line segment from P to Q.
- **(a)** By computing both of the distances PM and MQ, show that $PM = MQ$. (This shows that M lies on the perpendicular bisector of line segment $\overline{PQ}$, but it does not show that M actually lies on $\overline{PQ}$.)
- **(b)** Show that $PM + MQ = PQ$. (This shows that M does lie on $\overline{PQ}$.)

34. This problem outlines one of the shortest proofs of the Pythagorean theorem. The proof was discovered by the Hindu mathematician Bhāskara (1114–ca 1185). (For other proofs, see the next exercise and also Exercise 100 on page 83.) In the figure we are given a right triangle ACB with the right angle at C, and we want to prove that $a^2 + b^2 = c^2$. In the figure, $\overline{CD}$ is drawn perpendicular to $\overline{AB}$.
- **(a)** Check that $\angle CAD = \angle DCB$ and that $\triangle BCD$ and $\triangle BAC$ are similar.
- **(b)** Use the result in part (a) to obtain the equation $a/y = c/a$, and conclude that $a^2 = cy$.
- **(c)** Show that $\triangle ACD$ is similar to $\triangle ABC$, and use this to deduce that $b^2 = c^2 - cy$.

(d) Combine the two equations deduced in parts (b) and (c) to arrive at $a^2 + b^2 = c^2$.

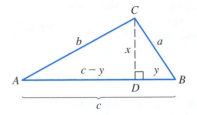

35. One of the oldest and simplest proofs of the Pythagorean theorem is found in the ancient Chinese text *Chou Pei Suan Ching*. This text was written during the Han period (206 B.C.–A.D. 222), but portions of it may date back to 600 B.C. The proof in *Chou Pei Suan Ching* is based on this diagram from the text. In this exercise we explain the details of the proof.

A diagram accompanying a proof of the "Pythagorean" theorem in the ancient Chinese text *Chou Pei Suan Ching* [from *Science and Civilisation in China*, Vol. 3, by Joseph Needham (Cambridge: Cambridge University Press, 1959)].

- **(a)** Starting with the right triangle in Figure A, we make four replicas of this triangle and arrange them to form the pattern shown in Figure B. Explain why the outer quadrilateral in Figure B is a square.

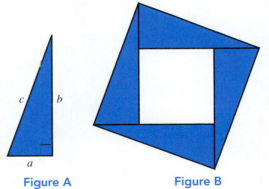

Figure A   **Figure B**

- **(b)** The unshaded region in the center of Figure B is a square. What is the length of each side?

(c) The area of the outer square in Figure B is (side)$^2 = c^2$. This area can also be computed by adding up the areas of the four right triangles and the inner square. Compute the area in this fashion. After simplifying, you should obtain $a^2 + b^2$. Now conclude that $a^2 + b^2 = c^2$, since both expressions represent the same area.

C

36. This problem indicates a method for calculating the area of a triangle when the coordinates of the three vertices are given.

(a) Calculate the area of $\triangle ABC$ in the figure.

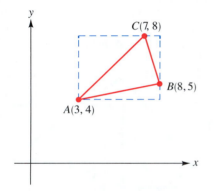

Hint: First calculate the area of the rectangle enclosing $\triangle ABC$, and then subtract the areas of the three right triangles.

(b) Calculate the area of the triangle with vertices $(1, 3)$, $(4, 1)$, and $(10, 4)$. *Hint:* Work with an enclosing rectangle and three right triangles, as in part (a).

(c) Using the same technique that you used in parts (a) and (b), show that the area of the triangle in the following figure is given by

$$A = \tfrac{1}{2}(x_1 y_2 - x_2 y_1 + x_2 y_3 - x_3 y_2 + x_3 y_1 - x_1 y_3)$$

Remark: If we use absolute value signs instead of the parentheses, then the formula will hold regardless of the relative positions or quadrants of the three vertices. Thus, the area of a triangle with vertices (x_1, y_1), (x_2, y_2), (x_3, y_3) is given by

$$A = \tfrac{1}{2}|x_1 y_2 - x_2 y_1 + x_2 y_3 - x_3 y_2 + x_3 y_1 - x_1 y_3|$$

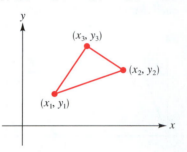

MINI PROJECT | DISCUSS, COMPUTE, REASSESS

As background for this discussion you should review Example 5. In that example, we used the midpoint formula to estimate the amount of carbon dioxide in the atmosphere for the year 1993 using data from the surrounding years 1992 and 1994. Consider now the following three figures (independent of Example 5). Each figure shows four data points, and there is a question mark regarding the location of a fifth. Note that the data points in each figure appear to be rising in a straight-line or near straight-line fashion.

(a) In your opinion, which of the following scenarios might lead to the most reliable estimate for the actual location of the fifth data point? The least reliable? What are the reasons or intuitions behind your replies?

(i) Figure A: Using the midpoint formula and the data for 1994 and 1996 to obtain a possible value for the global average temperature in 1995;

(ii) Figure B: Using the midpoint formula with the points $(10, 50)$ and $(30, 86)$ to obtain a possible value for the Fahrenheit temperature corresponding to 20 degrees on the Celsius scale;

(iii) Figure C: Using the midpoint formula and the data for 1965 and 1985 to obtain a possible value for the percentage of the earth's population living in urban areas in 1975.

(b) Carry out the three calculations described in part (a). Then use the following information to compute the percentage error for each of your estimates: **(i)** The average global temperature for 1995 was 57.9°F; **(ii)** The temperature 20°C corresponds to 68°F; **(iii)** In 1975, 37.8% of the world's population lived in urban areas.

(c) Use the percentage errors you've computed as a check on your part (a) answers. Were you correct about which estimate was the most reliable and which was the least? If you missed either one, how might you adjust the reasons or intuitions that you expressed in part (a)?

Figure A

Global average temperature at Earth's surface.

Source: Adapted from data compiled by Worldwatch Institute from Goddard Institute for Space Studies

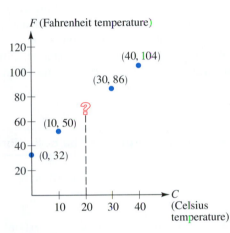

Figure B

Some correspondences between temperatures on the Celsius and Fahrenheit scales. [For example, the rightmost point (40, 104) indicates that a temperature of 40 degrees on the Celsius scale corresponds to 104 degrees on the Fahrenheit scale.]

Figure C

Percentage of Earth's human population living in urban areas.

Source: Compiled by Worldwatch Institute from United Nations, World Population Prospects: The 1996 Revision

1.5 GRAPHS AND GRAPHING UTILITIES

In this section we look at the connection between two basic skills that you've worked on in previous courses:

- graphing equations
- solving equations

We also look at these topics in terms of *graphing utilities*. By a **graphing utility** we mean either a graphing calculator or a computer with software for graphing and analyzing equations. Whether we are working "by hand" or with technology, the following definition underlies all.

> **DEFINITION** │ **The Graph of an Equation**
>
> The **graph** of an equation in two variables is the set of all points with coordinates satisfying the equation.

TABLE 1   y = 3x − 2

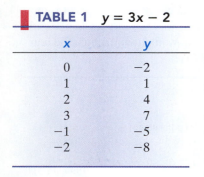

x	y
0	−2
1	1
2	4
3	7
−1	−5
−2	−8

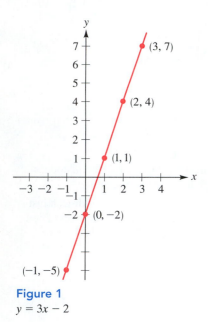

Figure 1
$y = 3x - 2$

Suppose, for example, that we want to graph the equation $y = 3x - 2$. We begin by noting that the domain of the variable x in the expression $3x - 2$ is the set of all real numbers. Now we choose values for x and in each case compute the corresponding y-value from the equation $y = 3x - 2$. For example, if x is zero, then $y = 3(0) - 2 = -2$. Table 1 summarizes the results of some of these calculations. The first line in Table 1 tells us that the point with coordinates $(0, -2)$ is on the graph of $y = 3x - 2$. Reading down the table, we see that some other points on the graph are $(1, 1)$, $(2, 4)$, $(3, 7)$, $(-1, -5)$, and $(-2, -8)$.

We now plot (locate) the points we have determined and we note, in this example, that they all appear to lie on a straight line. We draw the line indicated; this is the graph of $y = 3x - 2$ (see Figure 1). In Example 1 we ask a simple question that will test your understanding of the process we've just described and of graphing in general.

EXAMPLE 1 Applying the definition of the "graph of an equation"

Does the point $(1\frac{2}{3}, 2\frac{2}{3})$ lie on the graph of $y = 3x - 2$?

SOLUTION
Looking at the graph in Figure 1, we certainly see that the point $(1\frac{2}{3}, 2\frac{2}{3})$ is very close to the line, but from this visual inspection alone, we cannot be certain that this point actually lies on the line. To settle the question, then, we check to see whether the values $x = 1\frac{2}{3} = \frac{5}{3}$ and $y = 2\frac{2}{3} = \frac{8}{3}$ together satisfy the equation $y = 3x - 2$. We have

$$\frac{8}{3} \stackrel{?}{=} 3\left(\frac{5}{3}\right) - 2$$

$$\frac{8}{3} = 3 \qquad \text{No}$$

Since the coordinates $(1\frac{2}{3}, 2\frac{2}{3})$ evidently do not satisfy the equation $y = 3x - 2$, we conclude that the point $(1\frac{2}{3}, 2\frac{2}{3})$ is not on the graph. [The calculation further tells us that the point $(1\frac{2}{3}, 3)$ *is* on the graph.]

A graphing utility produces graphs of equations in essentially the same way that we did in Table 1 and Figure 1. The graphing utility uses the given equation to compute pairs of points (x, y), and it plots these points one at a time on the viewing screen. Of course, the machine is able to quickly compute and plot many more points than we used in Table 1—that's an advantage. On the other hand, the graphing utility is not "smart." For example, it does not know ahead of time that the equation $y = 3x - 2$ will be a straight line. (*You'll* know this by the end of the next section; or maybe you already remember from a previous course that the graph of any equation of the form $y = mx + b$ is always a straight line.) Later in this section

GRAPHICAL PERSPECTIVE

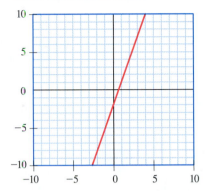

Figure 2

A graph of $y = 3x - 2$

and in the exercises, we'll say more about some of the limitations of a graphing utility, and how some thought or experimentation on your part is often necessary.

In Figure 2 we show a graph of the equation $y = 3x - 2$ obtained with a graphing utility. Actually, since there are so many different types of graphing utilities, the picture we've chosen to show in this text is a kind of amalgam containing features from several types of utilities. If you have a graphing utility you should now produce a graph of $y = 3x - 2$ for yourself and compare it to Figure 2. (If necessary refer to the owner's manual that came with your calculator or software.) In this text, the label GRAPHICAL PERSPECTIVE (as you see above Figure 2) is a signal for those students with graphing utilities: Use your graphing utility, making adjustments as necessary, to obtain a view that is similar to the one shown here in the textbook. For many graphing utilities, the default view or **standard viewing rectangle** is $-10 \leq x \leq 10$ and $-10 \leq y \leq 10$. This is the view shown in Figure 2. In the next example we graph an equation first by hand, then with a graphing utility. In the graphing utility portion, you'll see a case where the standard viewing rectangle is inappropriate and needs to be modified.

EXAMPLE	2	An application of linear equations and graphs to temperature scales

The equation $5F - 9C = 160$ relates temperature F on the Fahrenheit scale to temperature C on the Celsius scale.

(a) Solve the given equation for F in terms of C. Then set up a table and graph the equation in a C-F coordinate system (C on the horizontal axis, F on the vertical).

(b) Use a graphing utility to obtain a graph of $5F - 9C = 160$.

SOLUTION:

(a) We have

$$5F - 9C = 160$$
$$5F = 9C + 160 \qquad \text{adding } 9C \text{ to both sides}$$
$$F = \tfrac{9}{5}C + 32 \qquad \text{dividing both sides by 5} \qquad (1)$$

Using equation (1) now, we pick values for C and compute the corresponding values for F. To avoid working with fractions, we take values for C that are multiples of 5, as indicated in Table 2. (Check for yourself that the entries in the right-hand column of the table are correct.)

TABLE 2 $F = \tfrac{9}{5}C + 32$

C	F
0	32
5	41
10	50
15	59
−5	23
−10	14

In view of the data in Table 2, we'll mark off the axes as indicated in Figure 3(a), with tick marks 5 units apart on the C-axis and 10 units apart on the F-axis. [Other markings are feasible. For example, it would be reasonable to use markings of 20 instead of 10 on the vertical axis (but it would be very clumsy to use markings at 1-unit intervals).] In Figure 3(b) we've plotted the data pairs (C, F) given by Table 2. These points appear to lie on a straight line, and we've drawn the line in Figure 3(c). This is the required graph.

Note: In elementary graphing, there is always the question of how many points must be plotted before the essential features of a graph are clear. As you'll see throughout this text, there are a number of techniques and concepts that make it unnecessary to plot a large number of points. For instance, if you knew ahead of time that the graph of $F = \frac{9}{5}C + 32$ were a straight line, then you would only need to plot two points in order to draw the line.

(b) With a graphing utility, just as in graphing by hand, we first need to solve the given equation for one variable in terms of the other. From part (a) we have $F = \frac{9}{5}C + 32$. Because most graphing calculators (and some types of computer software) require that we use x and y for the variables, we'll rewrite the equation as $y = \frac{9}{5}x + 32$. Then we can enter it in the graphing utility as

$$y = (9/5) * x + 32 \qquad \text{The symbol * denotes multiplication.}$$

or better yet (with fewer keystrokes and less chance of typing error)

$$y = 9x/5 + 32 \tag{2}$$

Figure 4(a) shows the "graph" of equation (2) in the standard viewing rectangle. Yes, nothing's there; evidently no point of the required graph passes through the standard viewing rectangle. We need to find a more appropriate viewing rectangle. After some experimenting with the dimensions of the viewing rectangle, we obtain the views shown in Figure 4(b), then Figure 4(c). Figure 4(c) is an appropriate viewing rectangle; the essential features of the graph are clear, and there is not a lot of wasted space in the viewing screen as there is in Figure 4(b).

(a) Scales on axes are chosen to accommodate the range of numbers in Table 2.

(b) Plot of data points (C, F) from Table 2.

(c) Graph of $F = \frac{9}{5}C + 32$ (or $5F - 9C = 160$)

Figure 3

GRAPHICAL PERSPECTIVE

(a) The graph of $y = \frac{9}{5}x + 32$ does not pass through the standard viewing rectangle.

(b) Better, but still not an appropriate viewing rectangle.

(c) An appropriate viewing rectangle. The essential features of the graph are clear.

Figure 4

Finding an appropriate viewing rectangle for $y = \frac{9}{5}x + 32$ (or $F = \frac{9}{5}C + 32$)

We can use Figure 4 to introduce some common notation that is often used in describing viewing rectangles. (Even if you are not working with a graphing utility, you are apt to see this notation in other math books or on a friend's math paper.) In Figure 4(a), x runs from -10 to 10 in increments of 5. A notation for this is

$$\texttt{Xmin = -10} \qquad \texttt{Xmax = 10} \qquad \texttt{Xscl = 5}$$

In the rightmost equation, "Xscl" refers to the *scale* on the x-axis, and the equation Xscl = 5 tells you that it is 5 units between adjacent tick marks or grid lines. In Figure 4(a), y also runs from -10 to 10 in increments of 5, so we write Ymin = -10, Ymax = 10, and Yscl = 5. As another example of this notation, for Figure 4(c) we have (as you should check)

$$\begin{array}{lll} \texttt{Xmin = -30} & & \texttt{Ymin = -20} \\ \texttt{Xmax = 30} & \text{and} & \texttt{Ymax = 80} \\ \texttt{Xscl = 10} & & \texttt{Yscl = 20} \end{array}$$

Another type of shorthand that you'll see is based on the closed interval notation $[a, b]$ described in Section 1.1. With this notation, we specify the size of the viewing rectangle in Figure 4(c) as

$$[-30, 30] \text{ by } [-20, 80]$$

In this notation, the x-interval precedes the y-interval.

The only drawback to this notation is that the scale or number of units between tick marks is not specified. We remedy this by inserting a third number, denoting scale, in each bracket. The viewing rectangle in Figure 4(c) is then completely specified by writing

$$[-30, 30, 10] \text{ by } [-20, 80, 20]$$

EXAMPLE **3** **Specifying viewing rectangles on a graphing utility**

Figure 5 shows two views of the graph of the equation $y = 2x^3 - x^2$. Describe each viewing rectangle using the two types of notation discussed above (i.e., the min-max-scl equations and the bracket notation).

Figure 5
Two views of $y = 2x^3 - x^2$

SOLUTION:
For Figure 5(a) we have

$$
\begin{array}{ll}
\texttt{Xmin = -4} & \texttt{Ymin = -40} \\
\texttt{Xmax = 4} \qquad \text{and} \qquad & \texttt{Ymax = 60} \\
\texttt{Xscl = 2} & \texttt{Yscl = 20}
\end{array}
$$

The bracket notation describing the viewing rectangle in Figure 5(a) is

$$[-4, 4, 2] \text{ by } [-40, 60, 20]$$

For Figure 5(b) we have

$$
\begin{array}{ll}
\texttt{Xmin = -0.2} & \texttt{Ymin = -0.06} \\
\texttt{Xmax = 0.6} \qquad \text{and} \qquad & \texttt{Ymax = 0.02} \\
\texttt{Xscl = 0.2} & \texttt{Yscl = 0.02}
\end{array}
$$

The bracket notation describing the viewing rectangle in Figure 5(b) is

$$[-0.2, 0.6, 0.2] \text{ by } [-0.06, 0.02, 0.02]$$

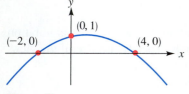

Figure 6
x-intercepts: -2 and 4
y-intercept: 1

When we graph an equation, it's helpful to know where the curve or line intersects the x- or y-axis. By an ***x-intercept*** of a graph we mean the x-coordinate of a point where the graph intersects the x-axis. For instance, in Figure 6 there are two x-intercepts: -2 and 4. We define y-intercepts in a similar manner: A ***y-intercept*** of a graph is the y-coordinate of a point where the graph intersects the y-axis. Thus, the y-intercept of the graph in Figure 6 is 1.

EXAMPLE **4**   **Computing x- and y-intercepts**

Figure 7 shows the graph of the equation $3x + 5y = 15$. Find the x- and y-intercepts.

SOLUTION
At the point where the graph crosses the x-axis, the y-coordinate is zero. So, to find the x-intercept, we set $y = 0$ in the given equation $3x + 5y = 15$ and solve for x:

$$
\begin{aligned}
3x + 5(0) &= 15 \\
3x &= 15 \\
x &= 5
\end{aligned}
$$

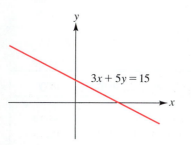

Figure 7

Thus, the x-intercept is 5. Similarly, to find the y-intercept, we set $x = 0$ in the given equation $3x + 5y = 15$ and solve for y:

$$3(0) + 5y = 15$$
$$5y = 15$$
$$y = 3$$

The y-intercept is 3.

In the next example we show two different ways to determine x-intercepts. First we use a graphing utility to obtain approximate values for the intercepts; then we use algebra to find exact values.

EXAMPLE 5   Finding approximate and exact values for x-intercepts

Figure 8 shows a graph of the equation $y = x^3 - 2x^2 - 2x$ obtained with a graphing utility.

GRAPHICAL PERSPECTIVE

Figure 8
$y = x^3 - 2x^2 - 2x$
viewing rectangle: $[-4, 4, 2]$
by $[-6, 4, 2]$

(a) Use a graphing utility to estimate (to one decimal place) the x-intercepts of the graph.
(b) Use algebra to determine the exact values for the x-intercepts. Then use a calculator to check that the answers are consistent with the estimates obtained in part (a).

SOLUTION:
(a) According to Figure 8, there appear to be three x-intercepts: one between -1 and 0, one that is either zero or very close to zero, and one between 2 and 3. To estimate the intercept that is between -1 and 0 as well as the intercept near (or possibly at) zero, we'll change the viewing rectangle shown in Figure 8 so that x runs from -1 to 1.1 in increments of 0.1. After some experimenting with the y-values, we obtain the graph in Figure 9(a), which shows that the leftmost intercept is between -0.8 and -0.7. Is it closer to -0.8 or -0.7? On a graphing calculator screen (as opposed to a larger picture on a computer monitor) it may not be clear that the intercept is closer to -0.7 than to -0.8. In Figure 9(b) we've adjusted the viewing rectangle for a closer view, which shows that the intercept is indeed closer to -0.7 than to -0.8. So, to one decimal place, we can say that the leftmost intercept is -0.7. [Actually, Figure 9(b) tells us more than this. The intercept is between -0.74 and -0.73.]

GRAPHICAL PERSPECTIVE

Figure 9
Adjusting the viewing rectangle to
estimate an x-intercept for
$y = x^3 - 2x^2 - 2x$

(a) $[-1.0, 0.1, 0.1]$ by $[-0.5, 0.5, 0.5]$

(b) $[-0.80, -0.70, 0.01]$ by
$[-0.2, 0.1, 0.1]$

Regarding the intercept that appears in Figure 8 to be at or near zero, Figure 9(a) tells us that to one decimal place, the intercept is 0.0. In part (b) we'll show that this intercept is indeed exactly zero. (You cannot conclude this fact from Figure 9(a); see Exercise 35 for an example along these lines.)

There is one more x-intercept to estimate, the one in Figure 8 that is between 2 and 3. Exercise 36 asks you to use the methods we've just demonstrated to show that this rightmost x-intercept is approximately 2.7.

(b) To compute the exact values for the x-intercepts of the graph, we set y equal to zero in the given equation $y = x^3 - 2x^2 - 2x$ to obtain $0 = x^3 - 2x^2 - 2x$. We then have

$$x^3 - 2x^2 - 2x = 0 \tag{3}$$
$$x(x^2 - 2x - 2) = 0 \qquad \text{factoring out the common term } x$$

The zero product property from Section 1.3 now tells us that $x = 0$ or $x^2 - 2x - 2 = 0$. We can use the quadratic formula (from Section 1.3) to solve the equation $x^2 - 2x - 2 = 0$:

$$x = \frac{-b \pm \sqrt{b^2 - 4ac}}{2a}$$

$$= \frac{-(-2) \pm \sqrt{(-2)^2 - 4(1)(-2)}}{2(1)} \qquad \text{using } a = 1, b = -2, c = -2$$

$$= \frac{2 \pm \sqrt{12}}{2}$$

$$= \frac{2 \pm 2\sqrt{3}}{2} = 1 \pm \sqrt{3} \qquad \text{See Appendix B.2 if you need to review working with radicals.}$$

Putting things together now, we have found three roots for equation (3): 0, $1 - \sqrt{3}$, and $1 + \sqrt{3}$. The largest (and the only positive root) of these three is $1 + \sqrt{3}$; that is the rightmost x-intercept for the graph in Figure 8. The smallest (and the only negative root) is $1 - \sqrt{3}$; that is the leftmost x-intercept for the graph in Figure 8. Finally, the root $x = 0$ tells us that the middle x-intercept is indeed exactly zero. As you should check for yourself using a calculator, we have $1 - \sqrt{3} \approx -0.73$ and $1 + \sqrt{3} \approx 2.73$. Note that these values are indeed consistent with the estimates obtained in part (a).

In the graphing utility portion of the example just concluded, we started with the viewing rectangle in Figure 8 and then adjusted it to obtain a closer look at the x-intercepts. Most graphing utilities also allow you to obtain a closer look in another way, by using a ZOOM feature and a TRACE feature. For details, see the owner's manual for your graphing utility. Another technology note: In part (b) of the example just concluded we used algebra to determine the exact roots of the equation $x^3 - 2x^2 - 2x = 0$. Most graphing utilities have a SOLVE feature that can be used to obtain roots that, while only approximate, are accurate to a good number of decimal places. Again, for details consult your owner's manual.

As mentioned previously in this section, when we first graph a given equation (either by hand or with a graphing utility), we try to present a view that shows all the essential features of the graph. The next example deals with one aspect of this issue.

EXAMPLE 6 | **Obtaining information about a graph through its equation**

Figure 10 shows a graph of the equation $x^2y - x^3 + 3 = 0$.

Figure 10
Does the graph ever cross
the y-axis?

$$x^2y - x^3 + 3 = 0$$

(a) From the portion of the graph shown in Figure 10, it appears that there is no y-intercept. Use the given equation to show that this is indeed the case.
(b) As indicated in Figure 10, there is an x-intercept between 1 and 2. Find this intercept. Give two forms for the answer: an exact expression and a calculator approximation rounded to two decimal places.

SOLUTION
(a) If there were a y-intercept (perhaps somewhere below the portion of the graph displayed in Figure 10), we could find it by setting x equal to zero in the given equation. But with $x = 0$, the given equation becomes $3 = 0$, which is clearly impossible. We thus conclude that there is no y-intercept; the graph never crosses the y-axis.
(b) The x-intercept of the graph is determined by setting y equal to zero in the given equation. That gives us $x^2(0) - x^3 + 3 = 0$, and therefore $x^3 = 3$. To solve this equation for x, just take the cube root of both sides to obtain

$$x = \sqrt[3]{3}$$
$$\approx 1.44 \qquad \text{using a calculator}$$

GRAPHICAL PERSPECTIVE

Figure 11
To graph $x^2y - x^3 + 3 = 0$, first solve for y:
$$x^2y = x^3 - 3$$
$$y = (x^3 - 3)/x^2$$

(a) Zoom-in view of $y = (x^3 - 3)/x^2$ shows that the graph has an x-intercept between 1.4 and 1.5.

(b) Further magnification shows that the x-intercept is between 1.44 and 1.45 (closer to 1.44 than 1.45).

As you can see, the value $x = 1.44$ is consistent with the graph shown in Figure 10. Also, see the Graphical Perspective in Figure 11. Note that in the caption for Figure 11 we've solved the given equation for y in terms of x. Most graphing utilities require that you enter the equation in this form.

EXERCISE SET 1.5

A

In Exercises 1–6, determine whether the given point lies on the graph of the equation, as in Example 1. Note: You are not asked to draw the graph.

1. $(8, 6)$; $y = \frac{1}{2}x + 3$
2. $(\frac{3}{5}, -\frac{17}{5})$; $y = -\frac{2}{3}x - 3$
3. $(4, 3)$; $3x^2 + y^2 = 52$
4. $(4, -2)$; $3x^2 + y^2 = 52$
5. $(a, 4a)$; $y = 4x$
6. $(a - 1, a + 1)$; $y = x + 2$
7. (a) Solve the equation $2x - 3y = -3$ for y and then complete the following table.

x	−6	−3	0	3	6
y					

   (b) Use your table from part (a) to graph the equation $2x - 3y = -3$.
8. (a) Solve the equation $3x + 2y = 6$ for y and then complete the following table.

x	−4	−2	0	2	4
y					

   (b) Use your table from part (a) to graph the equation $3x + 2y = 6$.

In Exercises 9–14, the graph of each equation is a straight line. Graph the equation after finding the x- and the y-intercepts. (Since you are given that the graph is a line, you need only plot two points before drawing the line.)

9. $3x + 4y = 12$
10. $3x - 4y = 12$
11. $y = 2x - 4$
12. $x = 2y - 4$
13. $x + y = 1$
14. $2x - 3y = 6$

For Exercises 15 and 16: As in Example 5, describe each viewing rectangle using the two types of notation discussed in the text (i.e., the min-max-scl equations and the bracket notation).

15. Two views of $y = x^5 - x^4$

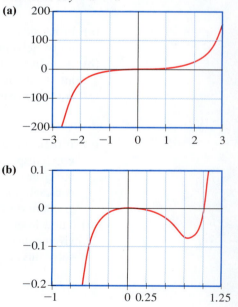

16. Two views of $y = \sqrt{1 - x^2}$

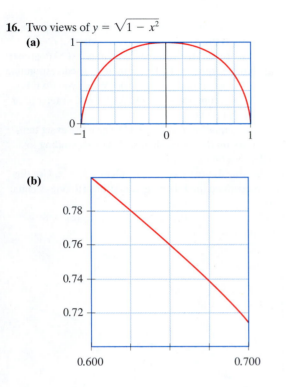

In Exercises 17–20, determine any x- or y-intercepts for the graph of the equation. Note: You're not asked to draw the graph.

17. (a) $y = x^2 + 3x + 2$
   (b) $y = x^2 + 2x + 3$
18. (a) $y = x^2 - 4x - 12$
   (b) $y = x^2 - 4x + 12$
19. (a) $y = x^2 + x - 1$
   (b) $y = x^2 + x + 1$
20. (a) $y = 6x^3 + 9x^2 + x$
   (b) $y = 9x^3 + 6x^2 + x$

In Exercises 21–24, each figure shows the graph of an equation. Find the x- and y-intercepts of the graph. If an intercept involves a radical, give both the radical form of the answer and a calculator approximation rounded to two decimal places. (Check to see that your answer is consistent with the given figure.)

21.

22.

$y = x^3 + x^2 - 3x$

23.

$y^2 - 4y - 8 = 3x$

24.

$x^3 + x^2 y = 12$

For Exercises 25–30:

Ⓖ (a) Use a graphing utility to graph the equation.
Ⓖ (b) Use a graphing utility, as in Example 5, to estimate to one decimal place the x-intercepts.
   (c) Use algebra to determine the exact values for the x-intercepts. Then use a calculator to check that the answers are consistent with the estimates obtained in part (b).

25. $y = x^2 - 2x - 2$ **26.** $y = 2x^2 + x - 5$
27. $y = 2x^3 - 5x$ **28.** $y = 3x^3 + 5x^2 + x$
29. $2xy - x^3 - 5 = 0$
   Hints: See the caption for Figure 11 and Example 6(b).
30. $xy - 12y = x^2 - x - 1$
   Hints for part (a). To solve the equation for y, first factor out the common term y on the left-hand side. Regarding the essential features of the graph, you should find that there are are two distinct pieces, only one of which is visible in the standard viewing rectangle.

Ⓖ *In Exercises 31–34, use a graphing utility to graph the equations and to approximate the x-intercepts. In approximating the x-intercepts, use a "solve" key or a sufficiently magnified view to ensure that the values you give are correct in the first three decimal places. Remark: None of the x-intercepts for these four equations can be obtained using factoring techniques.)*

31. $y = x^3 - 3x + 1$
32. $y = 8x^3 - 6x - 1$
33. $y = x^5 - 6x^4 + 3$
34. $y = 2x^5 - 5x^4 + 5$

Ⓖ **35. (a)** Graph the equation $y = x^3 + 10x + 2$ in the standard viewing rectangle. Based upon this view, what can you say about the location of any x-intercepts?

(b) The following viewing rectangle shows that $y = x^3 + 10x + 2$ has an x-intercept either at -0.2 or very close to -0.2.

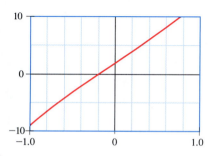

(c) Find a viewing rectangle to demonstrate that -0.2 is *not* an x-intercept.

(d) Looking at the picture in part (b) and your result in part (c), what would you say is the moral of the story?

Ⓖ **36.** This exercise refers to Example 5 and Figure 8.

(a) According to Figure 8, the graph of $y = x^3 - 2x^2 - 2x$ has an x-intercept between 2 and 3, slightly closer to 3 than to 2. Graph the equation in the viewing rectangle $[2.5, 3, 0.1]$ by $[-2, 3, 1]$ to see that the intercept is between 2.7 and 2.8, closer to 2.7 than to 2.8. Thus, to one decimal place, the intercept is 2.7, as stated in Example 5.

(b) Estimate the x-intercept to three decimal places by finding a suitable viewing rectangle.

B

For Exercises 37 and 38, refer to Figure 4(c) on page 39, which shows a graph of the equation $F = \frac{9}{5}C + 32$. The equation relates temperature C on the Celsius scale to temperature F on the Fahrenheit scale. [In Figure 4(c), C is the variable for the horizontal axis and F is the variable for the vertical axis.]

37. Use Figure 4(c) to estimate (to the nearest 5 degrees) the Celsius temperature corresponding to 0 degrees on the

Fahrenheit scale. When you do this, which intercept are you looking at, the C-intercept or the F-intercept?

38. (a) Use Figure 4(c) to estimate (to the nearest 10 degrees) the temperature on the Fahrenheit scale corresponding to 0 degrees on the Celsius scale. When you do this, which intercept are you looking at, the C-intercept or the F-intercept?

(b) Use the equation $F = \frac{9}{5}C + 32$ to give the exact temperature on the Fahrenheit scale corresponding to 0 degrees on the Celsius scale.

39. The following figure shows the graph of $y = \sqrt{x}$. Use this graph to estimate the following quantities (to one decimal place).

(a) $\sqrt{2}$ **(b)** $\sqrt{3}$ **(c)** $\sqrt{6}$

Hint: $\sqrt{ab} = \sqrt{a}\sqrt{b}$

40. The following figure shows the graph of $y = \sqrt[3]{x}$; use it to estimate the following quantities. **(a)** $\sqrt[3]{2}$; **(b)** $\sqrt[3]{5}$; **(c)** $\sqrt[3]{6.5}$. In each case, compare your estimate with a value obtained from a calculator.

41. In a certain biology experiment, the number N of bacteria increases with time t in hours as indicated in the following figure.

(a) How many bacteria are initially present when $t = 0$?
(b) Approximately how long does it take for the original colony to double in size?
(c) For which value of t is the population approximately 2500?
(d) During which time interval does the population increase more rapidly, between $t = 0$ and $t = 1$ or between $t = 3$ and $t = 4$?

In Exercises 42 and 43, determine the x-coordinates of the points A and B. (Each dashed line is parallel to a coordinate axis.)

42

$y = \frac{1}{2}x + 1$

43.

$y = \frac{1}{2}x^2$

MINI PROJECT   DRAWING CONCLUSIONS FROM VISUAL EVIDENCE

This discussion concerns x-intercepts and requires that at least one person in the group have a graphing utility. How much graphical evidence do you feel is necessary before you are "reasonably certain" about the exact value of an x-intercept? Or would graphical evidence alone never leave you reasonably certain? To focus the discussion, begin by considering the following two views of the graph of the equation $y = x^3 + 550x + 11$.

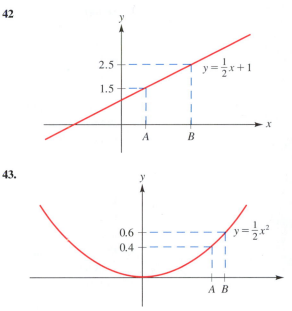

Figure A
Close-up view of $y = x^3 + 550x + 11$ near an x-intercept. The intercept is either -0.02 or something very close to that.

Figure B
Still closer view of the x-intercept in Figure A (zooming in by a factor of 2). Would you say now that it is quite likely that the intercept is exactly -0.02?

1.6   EQUATIONS OF LINES

He [Pierre de Fermat (1601–1665)] *introduced perpendicular axes and found the general equations of straight lines and circles and the simplest equations of parabolas, ellipses, and hyperbolas; and he further showed in a fairly complete and systematic way that every first- or second-degree equation can be reduced to one of these types.* —George F. Simmons in *Calculus Gems, Brief Lives and Memorable Moments* (New York: McGraw-Hill Book Company, 1992)

Consider a straight line, for example, one suspended in midair and dutifully propagating itself throughout all of space. What can one say of this line except something on the order of there she goes? *But* [now] *imagine it, that line, passing through a Cartesian coordinate system and ascending upward for parts unknown.* —David Berlinski, *A Tour of the Calculus* (New York: Vintage Books, 1995)

For sufficiently short periods of time, the graphs of many real-world quantities can be closely approximated by a straight-line graph. In this section we take a systematic look at equations of lines and their graphs. (In Section 4.1 we'll see applications that use the material developed here.) We begin by recalling the concept of *slope*, which you've seen in previous courses. As defined in the box that follows, the slope of a nonvertical line is a number that measures the slant or direction of the line.

DEFINITION | **Slope**

The slope of a nonvertical line passing through the two points (x_1, y_1) and (x_2, y_2) is the number m defined by

$$m = \frac{y_2 - y_1}{x_2 - x_1}$$

EXAMPLE

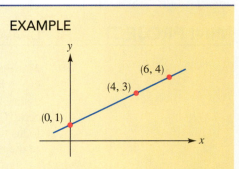

Calculating m using $(0, 1)$ and $(4, 3)$:

$$m = \frac{3 - 1}{4 - 0} = \frac{2}{4} = \frac{1}{2}$$

or calculating m using $(4, 3)$ and $(6, 4)$:

$$m = \frac{4 - 3}{6 - 4} = \frac{1}{2}$$

In the example in the box, note that in both calculations the same number is obtained for the slope. Later in this section we'll explain why it is always the case that the slope of a given line comes out the same no matter which two points on the line are used in the calculation.

The two quantities $x_2 - x_1$ and $y_2 - y_1$ that appear in the definition of slope can be interpreted geometrically. As indicated in Figure 1, $x_2 - x_1$ is the amount by which x changes as we move from (x_1, y_1) to (x_2, y_2) along the line. We denote this

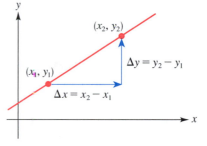

Figure 1
The delta notation

change in x by Δx (read: *delta x*). Thus, $\Delta x = x_2 - x_1$. Similarly, Δy is defined to mean the change in y: $\Delta y = y_2 - y_1$. Using this notation, we can rewrite our definition of slope as $m = \Delta y/\Delta x$.

EXAMPLE **1** **Using the delta notation in computing slope**

Suppose a cardiologist asks a patient to walk on a treadmill at a slow steady pace, and among the data collected are the figures in Table 1. In Table 1, x represents time, in seconds, with $x = 0$ corresponding to the instant that the patient starts walking. The variable y stands for the total distance covered by the patient after x seconds. Figure 2 shows a graph obtained from these data. Compute Δx, Δy, and the slope m for each of the following intervals: **(a)** from $x = 5$ seconds to $x = 10$ seconds; **(b)** from $x = 10$ seconds to $x = 30$ seconds.

TABLE 1

x (time in seconds)	0	5	10	15	20	25	30
y (distance in feet)	0	15	30	45	60	75	90

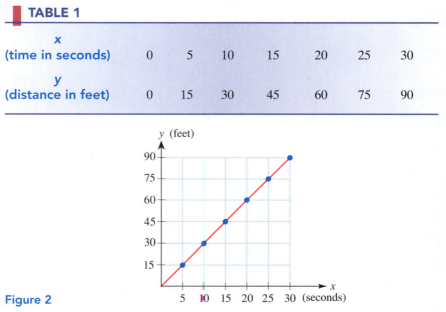

Figure 2

SOLUTION
(a) From $x = 5$ seconds to $x = 10$ seconds:

$$\Delta x = x_2 - x_1 = 10 \text{ sec} - 5 \text{ sec} = 5 \text{ sec}$$
$$\Delta y = y_2 - y_1 = 30 \text{ ft} - 15 \text{ ft} = 15 \text{ ft}$$
$$\frac{\Delta y}{\Delta x} = \frac{15 \text{ ft}}{5 \text{ sec}} = 3 \text{ ft/sec}$$

(b) From $x = 10$ seconds to $x = 30$ seconds:

$$\Delta x = x_2 - x_1 = 30 \text{ sec} - 10 \text{ sec} = 20 \text{ sec}$$
$$\Delta y = y_2 - y_1 = 90 \text{ ft} - 30 \text{ ft} = 60 \text{ ft}$$
$$\frac{\Delta y}{\Delta x} = \frac{60 \text{ ft}}{20 \text{ sec}} = 3 \text{ ft/sec}$$

Note that both parts of this exercise lead to the same value for slope. Exercises 5 and 6 ask you to check other instances of this, and in the next paragraph we prove that for a given line, the same value is obtained for slope no matter which pair of points on the line is used. Another point to observe: Slope is a rate of change. In this example, it is the rate of change of distance with respect to time, or *velocity*. We'll talk more about slope as a rate of change in Section 4.1.

To see why slope of a line does not depend on which two particular points on the line are used in the calculation, consider Figure 3. The two right triangles are similar (because the corresponding angles are equal). This implies that the corresponding sides of the two triangles are proportional, and so we have

$$\frac{a}{d} = \frac{b}{c}$$

Now notice that the left-hand side of this equation represents the slope $\Delta y/\Delta x$ calculated using the points A and D, and the right-hand side represents the slope calculated using the points B and C. Thus the values we obtain for the slope are indeed equal.

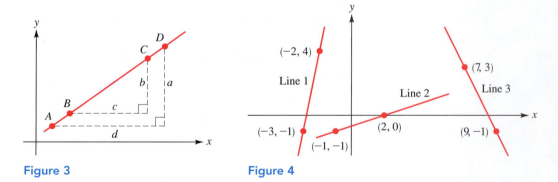

Figure 3

Figure 4

EXAMPLE **2** **Comparing slopes**

Compute and compare the slopes of the three lines shown in Figure 4.

SOLUTION
First, we will calculate the slope of line 1, using the formula $m = \dfrac{y_2 - y_1}{x_2 - x_1}$. Which point should we use as (x_1, y_1) and which as (x_2, y_2)? In fact, it doesn't matter how we label our points. Using $(-3, -1)$ for (x_1, y_1) and $(-2, 4)$ for (x_2, y_2), we have

$$m = \frac{y_2 - y_1}{x_2 - x_1} = \frac{4 - (-1)}{-2 - (-3)} = 5$$

So the slope of line 1 is 5. If, instead, we had used $(-2, 4)$ for (x_1, y_1) and $(-3, -1)$ for (x_2, y_2), then we'd have $m = (-1 - 4)/(-3 + 2) = 5$, the same result. This is not accidental because, in general,

$$\frac{y_2 - y_1}{x_2 - x_1} = \frac{y_1 - y_2}{x_1 - x_2}$$

(Exercise 55 asks you to verify this identity). Next, we calculate the slopes of lines 2 and 3.

Slope of line 2: $\dfrac{0 - (-1)}{2 - (-1)} = \dfrac{1}{3}$ Slope of line 3: $\dfrac{3 - (-1)}{7 - 9} = -2$

Lines 1 and 2 both have positive slopes and slant upward to the right. Note that line 1 is steeper than line 2 and correspondingly has the larger slope, 5. Line 3 has a negative slope, -2, and slants downward to the right.

◼

The observations made in Example 2 are true in general. Lines with a positive slope slant upward to the right, the steeper line having the larger slope. Likewise, lines with a negative slope slant downward to the right. See Figure 5.

We have yet to mention slopes for horizontal or vertical lines. In Figure 6, line 1 is horizontal; it passes through the points (a, b) and (c, b). Note that the two y-coordinates must be the same for the line to be horizontal. Line 2 in Figure 6 is vertical; it passes through (d, e) and (d, f). Note that the two x-coordinates must be the same for the line to be vertical. For the slope of line 1 we have $m = \dfrac{b - b}{c - a} = \dfrac{0}{c - a} = 0$ (provided $a \neq c$). Thus the slope of line 1, a horizontal line, is zero. For the vertical line in Figure 6, the calculation of slope begins with writing $\dfrac{e - f}{d - d}$, but the denominator is zero, and since division by zero is undefined, we conclude that slope is undefined for vertical lines. We summarize these results in the box that follows. Figure 7 shows for comparison some values of m for various lines.

Figure 5

Figure 6

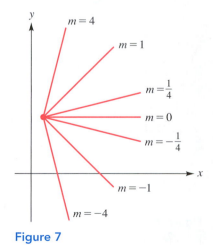

Figure 7

Horizontal and Vertical Lines

1. The slope of a horizontal line is zero.
2. Slope is not defined for vertical lines.

We can use the concept of slope to find the equation of a line. Suppose we have a line with slope m and passing through the point (x_1, y_1), as shown in Figure 8(a). Let (x, y) be any other point on the line, as in Figure 8(b). Then the slope of the line is given by $m = (y - y_1)/(x - x_1)$, and therefore, $y - y_1 = m(x - x_1)$. Note that the given point (x_1, y_1) also satisfies this last equation, because in that case we have $y_1 - y_1 = m(x_1 - x_1)$, or $0 = 0$. The equation $y - y_1 = m(x - x_1)$ is called the **point–slope formula.** We have shown that any point on the line satisfies this equation. (Conversely, it can be shown that if a point satisfies this equation, then the point does lie on the given line.)

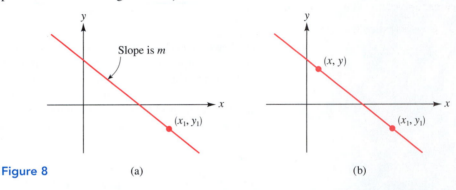

Figure 8   (a)   (b)

The Point–Slope Formula

The equation of a line with slope m passing through the point (x_1, y_1) is

$$y - y_1 = m(x - x_1)$$

EXAMPLE **3**   **Using the point–slope formula**

Write the equation of the line passing through $(-3, 1)$ with a slope of -2. Sketch a graph of the line.

SOLUTION
Since the slope and a point are given, we use the point–slope formula:

$$y - y_1 = m(x - x_1)$$
$$y - 1 = -2[x - (-3)]$$
$$y - 1 = -2x - 6$$
$$y = -2x - 5$$

This is the required equation. One way to graph this line is to pick values for x and then compute the corresponding y-values, as we did in Section 1.5. The table and graph are displayed in Figure 9. Note that if we know ahead of time that the graph is a line, a table as brief as the one in Figure 9 is sufficient.

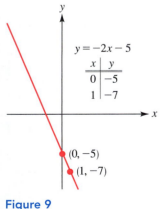

$y = -2x - 5$

x	y
0	-5
1	-7

$(0, -5)$

$(1, -7)$

Figure 9
$y = -2x - 5$

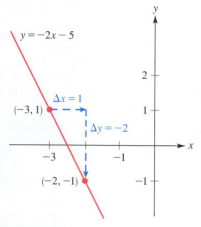

$y = -2x - 5$

$\Delta x = 1$

$(-3, 1)$

$\Delta y = -2$

$(-2, -1)$

Figure 10

There is another way to go about graphing the line $y = -2x - 5$ in Example 3. Since slope is (change in y)/(change in x), a slope of -2 (or $-2/1$) can be interpreted as telling us that if we start at $(-3, 1)$ and let x increase by one unit, then y must decrease by two units to bring us back to the line. Following this path in Figure 10 takes us from $(-3, 1)$ to $(-2, -1)$. We now draw the line through these two points, as shown in Figure 10.

EXAMPLE 4 | **Finding the equation of a line through two points**

Find the equation of the line passing through the points $(-2, -3)$ and $(2, 5)$.

SOLUTION
The slope of the line is $m = \dfrac{y_2 - y_1}{x_2 - x_1} = \dfrac{5 - (-3)}{2 - (-2)} = \dfrac{8}{4} = 2$

Knowing the slope, we can apply the point–slope formula, making use of either $(-2, -3)$ or $(2, 5)$. Using the point $(2, 5)$ as (x_1, y_1), we have

$$y - y_1 = m(x - x_1)$$
$$y - 5 = 2(x - 2)$$
$$y - 5 = 2x - 4$$
$$y = 2x + 1$$

Thus the required equation is $y = 2x + 1$. You should check for yourself that the same answer is obtained using the point $(-2, -3)$ instead of $(2, 5)$ in the last set of calculations.

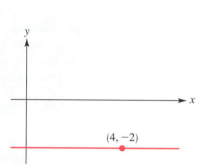

$(4, -2)$

Figure 11

EXAMPLE 5 | **Finding the equation of a horizontal line**

Find the equation of the horizontal line passing through the point $(4, -2)$. See Figure 11.

SOLUTION
Since the slope of a horizontal line is zero, we have

$$y - y_1 = m(x - x_1)$$
$$y - (-2) = 0(x - 4)$$
$$y = -2$$

Thus the equation of the horizontal line passing through $(4, -2)$ is $y = -2$.

(a, b)

$y = b$

Figure 12

$x = a$

(a, b)

Figure 13

By using the point–slope formula exactly as we did in Example 5, we can show more generally that the equation of the horizontal line in Figure 12 is $y = b$. What about the equation of the vertical line in Figure 13 passing through the point (a, b)? Because slope is not defined for vertical lines, the point–slope formula is not applicable this time. However, note that as we move along the vertical line, the x-coordinate is always a; it is only the y-coordinate that varies. The equation $x = a$ expresses exactly these two facts; it says that x must always be a and it places no restrictions on y.

In the box that follows, we summarize our results concerning horizontal and vertical lines.

Equations of Horizontal and Vertical Lines

1. The equation of a horizontal line through the point (a, b) is $y = b$. (See Fig. 12.)
2. The equation of a vertical line through the point (a, b) is $x = a$. (See Fig. 13.)

Figure 14

Another basic form for the equation of a line is the *slope–intercept form*. We are given the slope m and the y-intercept b, as shown in Figure 14, and we want to find the equation of the line. To say that the line has a y-intercept of b is the same as saying that the line passes through $(0, b)$. The point–slope formula is applicable now, using the slope m and the point $(0, b)$. We have

$$y - y_1 = m(x - x_1)$$
$$y - b = m(x - 0)$$
$$y = mx + b$$

This last equation is called the **slope–intercept formula.**

The Slope-Intercept Formula

The equation of a line with slope m and y-intercept b is

$$y = mx + b$$

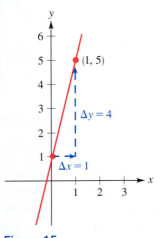

Figure 15
$y = 4x + 1$

EXAMPLE	6	Using the slope–intercept formula

Write the equation of a line with slope 4 and y-intercept 1. Graph the line.

SOLUTION
Substituting $m = 4$ and $b = 1$ in the equation $y = mx + b$ yields

$$y = 4x + 1$$

This is the required equation. We could draw the graph by first setting up a simple table, but for purposes of emphasis and review we proceed as we did just after Example 3. Starting from the point $(0, 1)$, we interpret the slope of 4 as saying that if x increases by 1, then y increases by 4. This takes us from $(0, 1)$ to the point $(1, 5)$, and the line can now be sketched as in Figure 15.

EXAMPLE 7   Determining the slope and y-intercept

Find the slope and y-intercept of the line $3x - 5y = 15$.

SOLUTION
First we solve for y so that the equation is in the form $y = mx + b$:

$$3x - 5y = 15$$
$$-5y = -3x + 15$$

and therefore

$$y = \frac{3}{5}x - 3$$

The slope m and the y-intercept b can now be read directly from the equation: $m = 3/5$ and $b = -3$.

As a consequence of our work up to this point, we can say that the graph of any **linear equation** $Ax + By + C = 0$, where A and B are not both zero, is a line, since an equation of this type can always be rewritten in one of the following three forms: $y = mx + b$, $x = a$, or $y = b$. In the box that follows, we summarize our basic results on equations of lines.

PROPERTY SUMMARY   Equations of Lines

Equation	Comment
$y - y_1 = m(x - x_1)$ (the point–slope formula)	This is the equation of a line with slope m, passing through the point (x_1, y_1).
$y = mx + b$ (the slope–intercept formula)	This is the equation of a line with slope m and y-intercept b.
$Ax + By + C = 0$	The graph of any equation of this form (where A and B are not both zero) is a line. Special cases: If $A = 0$, then the equation can be written $y = -C/B$, which is the equation of a horizontal line with y-intercept $-C/B$. If $B = 0$, then the equation can be written $x = -C/A$, which represents a vertical line with x-intercept $-C/A$.

GRAPHICAL PERSPECTIVE

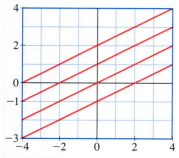

Figure 16
Parallel lines have the same slope.
Each line here has a slope of 0.5.
The equations of the lines, from top to bottom, are $y = 0.5x + 2$, $y = 0.5x + 1$, $y = 0.5x$, and $y = 0.5x - 1$.

We conclude this section by discussing two useful relationships regarding the slopes of parallel lines and the slopes of perpendicular lines. First, nonvertical parallel lines have the same slope. (See Figure 16.) This should seem reasonable if you recall that slope is a number indicating the direction or slant of a line. The relationship concerning slopes of perpendicular lines is not so obvious. The slopes of two nonvertical perpendicular lines are opposite reciprocals of each other. That is, if m_1 and m_2 denote the slopes of the two perpendicular lines, then $m_1 = -1/m_2$ or, equivalently, $m_1 m_2 = -1$. For example, if a line has a slope of $2/3$, then any line perpendicular to it must have a slope of $-3/2$. For reference we summarize these facts in the box that follows. (Proofs of these facts are outlined in detail in Exercises 53 and 54.)

Parallel and Perpendicular Lines

Let m_1 and m_2 denote the slopes of two nonvertical lines. Then:
1. The lines are parallel if and only if $m_1 = m_2$;
2. The lines are perpendicular if and only if $m_1 = -1/m_2$.

EXAMPLE 8   Determining whether two lines are parallel

Determine whether the two lines $3x - 6y - 8 = 0$ and $2y = x + 1$ are parallel.

SOLUTION

By solving each equation for y, we can see what the slopes are:

$$3x - 6y - 8 = 0 \qquad\qquad 2y = x + 1$$
$$-6y = -3x + 8 \qquad\qquad y = \frac{1}{2}x + \frac{1}{2}$$
$$y = \frac{-3}{-6}x + \frac{8}{-6}$$
$$y = \frac{1}{2}x - \frac{4}{3}$$

From this we see that both lines have the same slope—namely, $m = 1/2$. It follows therefore that the lines are parallel.

EXAMPLE 9   Finding parallel lines and perpendicular lines

In each case, find an equation for the line satisfying the given conditions, and then use a graphing utility to check that the result appears reasonable.
(a) The line passes through the point $(2, 0)$ and is parallel to the line $5x + 6y = 30$;
(b) The line passes through the point $(2, 0)$ and is perpendicular to $5x + 6y = 30$.

SOLUTION

(a) The line we are looking for is parallel to $5x + 6y = 30$ and therefore has the same slope. To find that slope, we solve the equation $5x + 6y = 30$ for y as follows:

$$6y = -5x + 30$$
$$y = -\frac{5}{6}x + 5 \qquad \text{dividing by 6}$$

The slope is the x-coefficient in this last equation, so $m = -5/6$. Knowing now that the slope of the required line is $-5/6$ and that it passes through the given point $(2, 0)$, we can use the point–slope formula $y - y_1 = m(x - x_1)$ to obtain $y - 0 = -\frac{5}{6}(x - 2)$, and consequently

$$y = -\frac{5}{6}x + \frac{5}{3}$$

GRAPHICAL PERSPECTIVE

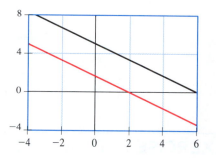

Figure 17

The line $y = -\frac{5}{6}x + \frac{5}{3}$ (in red) passes through the point $(2, 0)$ and is parallel to $5x + 6y = 30$.

This is the required equation of the line passing through the point $(2, 0)$ and parallel to $5x + 6y = 30$. See the graphical perspective in Figure 17.

(b) In part (a) we found that the slope of the line $5x + 6y = 30$ is $-5/6$. The slope of a perpendicular line is the opposite reciprocal of this, which is $6/5$. Knowing now that the slope of the required line is $6/5$ and that it passes through the point $(2, 0)$, we use the point–slope formula to obtain

$$y - 0 = \frac{6}{5}(x - 2)$$

$$y = \frac{6}{5}x - \frac{12}{5}$$

This is the required equation of the line that passes through the point $(2, 0)$ and that is perpendicular to $5x + 6y = 30$. For a quick visual check to see whether our result is reasonable, we've used a graphing utility in Figure 18(a) to plot the given line $5x + 6y = 30$ (entering it as $y = -\frac{5}{6}x + 5$) and the perpendicular through $(2, 0)$ that we found to be $y = \frac{6}{5}x - \frac{12}{5}$. The latter line, shown in red in Figure 18(a), does appear to pass through $(2, 0)$. It does not, however, appear to be perpendicular to $5x + 6y = 30$. Assuming that our graphing utility work is correct, does this mean we've made an error in our algebra? The answer is "No, not necessarily." The problem with the graph in Figure 18(a) is that it does not show true proportions. The actual length (or number of pixels) for one unit on the x-axis is not the same as that on the y-axis. In Figure 18(b) we show a view using true proportions, and the two lines do indeed appear to be perpendicular. (If you need help

GRAPHICAL PERSPECTIVE

(a) This view does not show true proportions.

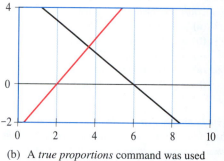

(b) A *true proportions* command was used to generate this view. The two lines do appear to be perpendicular.

Figure 18

Two views of the lines $y = \frac{6}{5}x - \frac{12}{5}$ (in red) and $y = -\frac{5}{6}x + 5$ (in black).

obtaining a view showing true proportions, see the owner's manual that came with your graphing utility. Example: On the Texas Instruments graphing calculators, the zoom choice "ZSquare" yields a viewing screen showing true proportions, albeit often with awkward numbers for Xmin and Xmax.)

EXERCISE SET 1.6

A

In Exercises 1–3, compute the slope of the line passing through the two given points. In Exercise 3, include a sketch with your answers.

1. **(a)** $(-3, 2), (1, -6)$
   (b) $(2, -5), (4, 1)$
   (c) $(-2, 7), (1, 0)$
   (d) $(4, 5), (5, 8)$

2. **(a)** $(-3, 0), (4, 9)$
   (b) $(-1, 2), (3, 0)$
   (c) $(\frac{1}{2}, -\frac{3}{5}), (\frac{3}{2}, \frac{3}{4})$
   (d) $(\frac{17}{3}, -\frac{1}{2}), (-\frac{1}{2}, \frac{17}{3})$

3. **(a)** $(1, 1), (-1, -1)$
   (b) $(0, 5), (-8, 5)$

4. Compute the slope of the line in the following figure using each pair of points indicated.
   (a) A and B　　**(b)** B and C　　**(c)** A and C
   The principle involved here is that no matter which pair of points you choose, the slope is the same.

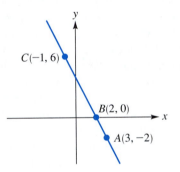

For Exercises 5 and 6, refer to Figure 2 on page 49 and find Δx, Δy, and $m = \Delta y / \Delta x$ for each of the indicated intervals. (Be sure to include units with your answers.)

5. **(a)** $x = 10$ sec to $x = 15$ sec
   (b) $x = 10$ sec to $x = 25$ sec
   (c) $x = 5$ sec to $x = 30$ sec

6. **(a)** $x = 0$ sec to $x = 5$ sec
   (b) $x = 0$ sec to $x = 30$ sec

For Exercises 7 and 8, refer to Figure A, which follows. At the beginning of this section it was mentioned that for sufficiently short periods of time, the graphs of many real-world quantities can be approximated quite accurately by a straight line. For instance, the following graph provides a close approximation to U.S. population data for the years 1970–1990.

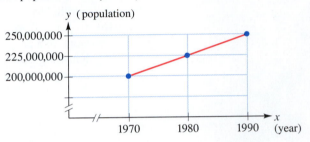

Figure A

U.S. population (as of July 1) for years 1970–1990

Adapted from U.S. Bureau of the Census data

7. Specify Δx and Δy for each of the following intervals.
   (a) $x = 1970$ to $x = 1980$　　**(c)** $x = 1980$ to $x = 1990$
   (b) $x = 1970$ to $x = 1990$

8. What is the slope of the population graph in Figure A? (Be sure to include units with your answer.)

9. The following straight-line graph summarizes the data for world cigarette production over the years 1990–1993. Specify Δt and ΔN for the period 1990–1993. (Include units with each answer.)

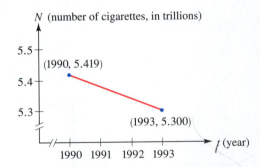

World cigarette production, 1990–1993.

Adapted from data compiled in *Vital Signs, 1998*, Lester R. Brown et al. (New York: W. W. Norton & Company, 1998)

10. Compute the slope $\dfrac{\Delta N}{\Delta t}$ for the straight-line graph shown in Exercise 9. (Be sure to include units as part of your answer.)

11. The slopes of four lines are indicated in the figure. List the slopes $m_1, m_2, m_3,$ and m_4 in order of increasing value.

12. Refer to the accompanying figure.
   (a) List the slopes $m_1, m_2,$ and m_3 in order of increasing size.
   (b) List the numbers $b_1, b_2,$ and b_3 in order of increasing size.

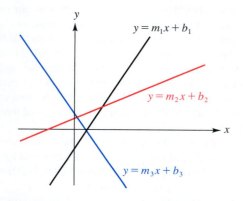

*In Exercises 13 and 14(a), three points A, B, and C are specified. Determine whether A, B, and C are **collinear** (lie on the same line) by checking to see whether the slope of $\overline{AB}$ equals the slope of $\overline{BC}$.*

13. $A(-8, -2); B(2, \frac{1}{2}); C(11, -1)$
14. **(a)** $A(0, -5); B(3, 4); C(-1, -8)$
   (b) If the area of the "triangle" formed by three points is zero, then the points must in fact be collinear. Use this observation, along with the formula in Exercise 14(b) of Exercise Set 1.4, to rework part (a).

In Exercises 15 and 16, find the equation for the line having the given slope and passing through the given point. Write your answers in the form $y = mx + b$.

15. **(a)** $m = -5$; through $(-2, 1)$
   (b) $m = \frac{1}{3}$; through $(-6, -\frac{2}{3})$
16. **(a)** $m = 22$; through $(0, 0)$
   (b) $m = -222$; through $(0, 0)$

In Exercises 17 and 18, find the equation for the line passing through the two given points. Write your answer in the form $y = mx + b$.

17. **(a)** $(4, 8)$ and $(-3, -6)$
   (b) $(-2, 0)$ and $(3, -10)$
   (c) $(-3, -2)$ and $(4, -1)$
18. **(a)** $(7, 9)$ and $(-11, 9)$
   (b) $(5/4, 2)$ and $(3/4, 3)$
   (c) $(12, 13)$ and $(13, 12)$

In Exercises 19 and 20, write the equation of:
(a) *a vertical line passing through the given point;*
(b) *a horizontal line passing through the given point.*
19. $(-3, 4)$ **20.** $(5, 8)$
21. Is the graph of the line $x = 0$ the x-axis or the y-axis?
22. Is the graph of the line $y = 0$ the x-axis or the y-axis?

In Exercises 23 and 24, find the equation of the line with the given slope and y-intercept.
23. **(a)** slope -4; y-intercept 7
   (b) slope 2; y-intercept 3/2
24. **(a)** slope 0; y-intercept 14
   (b) slope 14; y-intercept 0

In Exercises 25–28, find an equation for the line that is described, and sketch the graph. For Exercises 25 and 26, write the final answer in the form $y = mx + b$; for Exercises 27 and 28, write the answer in the form $Ax + By + C = 0$.
25. **(a)** Passes through $(-3, -1)$ and has slope 4
   (b) Passes through $(5/2, 0)$ and has slope 1/2
   (c) Has x-intercept 6 and y-intercept 5
   (d) Has x-intercept -2 and slope 3/4
   (e) Passes through $(1, 2)$ and $(2, 6)$
26. **(a)** Passes through $(-7, -2)$ and $(0, 0)$
   (b) Passes through $(6, -3)$ and has y-intercept 8
   (c) Passes through $(0, -1)$ and has the same slope as the line $3x + 4y = 12$
   (d) Passes through $(6, 2)$ and has the same x-intercept as the line $-2x + y = 1$
   (e) Has x-intercept -6 and y-intercept $\sqrt{2}$
27. Passes through $(-3, 4)$ and is parallel to the x-axis
28. Passes through $(-3, 4)$ and is parallel to the y-axis

In Exercises 29 and 30, find the x- and y-intercepts of the line, and find the area and the perimeter of the triangle formed by the line and the axes.
29. **(a)** $3x + 5y = 15$ **30.** **(a)** $5x + 4y = 40$
   (b) $3x - 5y = 15$ **(b)** $2x + 4y = \sqrt{2}$
31. Determine whether each pair of lines is parallel, perpendicular, or neither.
   (a) $3x - 4y = 12; 4x - 3y = 12$
   (b) $y = 5x - 16; y = 5x + 2$
   (c) $5x - 6y = 25; 6x + 5y = 0$
   (d) $y = -\frac{2}{3}x - 1; y = \frac{3}{2}x - 1$
32. Are the lines $y = x + 1$ and $y = 1 - x$ parallel, perpendicular, or neither?

In Exercises 33–38, find an equation for the line that is described. Write the answer in the two forms $y = mx + b$ and $Ax + By + C = 0$.

33. Is parallel to $2x - 5y = 10$ and passes through $(-1, 2)$

34. Is parallel to $4x + 5y = 20$ and passes through $(0, 0)$

35. Is perpendicular to $4y - 3x = 1$ and passes through $(4, 0)$

36. Is perpendicular to $x - y + 2 = 0$ and passes through $(3, 1)$

Ⓖ **37. (a)** Use a graphing utility to graph the following three parallel lines in the standard viewing rectangle:
$$y + 4 = -0.5(x - 2); \quad y - 3 = -0.5(x + 2); \quad y = -0.5x.$$

  (b) Experiment with different settings for `Xmin`, `Xmax`, `Ymin`, and `Ymax`. In each case, do the three lines still appear to be parallel?

Ⓖ **38. (a)** The lines $y = 4x$ and $y = -0.25x$ are perpendicular because their slopes are negative reciprocals. Use a graphing utility to graph these two lines in the standard viewing rectangle. Unless your graphing utility automatically shows true proportions, the lines will not appear to be perpendicular.

  (b) If necessary, modify the viewing rectangle in part (a) so that true proportions are used and the two lines indeed appear perpendicular.

39. (a) Find the equation of the line that passes through the origin and is perpendicular to the line $3x + 4y = 12$.

Ⓖ **(b)** Use a graphing utility to check that your answer in part (a) is reasonable. (That is, graph the two lines using true proportions; the line you found should appear to pass through the origin and be perpendicular to $3x + 4y = 12$.)

Ⓖ **40.** In each of parts (a) through (d), first solve the equation for y so that you can enter it in your graphing utility. Then use the graphing utility to graph the equation in an appropriate viewing rectangle. In each case, the graph is a line. Given that the x- and y-intercepts are (in every case here) integers, read their values off the screen and write them down for easy reference when you get to part (e).

  (a) $\dfrac{x}{2} + \dfrac{y}{3} = 1$ **(c)** $\dfrac{x}{6} + \dfrac{y}{5} = 1$

  (b) $\dfrac{x}{-2} + \dfrac{y}{-3} = 1$ **(d)** $\dfrac{x}{-6} + \dfrac{y}{-5} = 1$

  (e) Based on your results in parts (a) through (d), describe, in general, the graph of the equation $\dfrac{x}{a} + \dfrac{y}{b} = 1$, where a and b are nonzero constants.

B

41. After analyzing sales figures for a particular model of CD player, the accountant for College Sound Company has produced the following graph relating the selling price P (in dollars) and the number y of units that can be sold each month at that price. For instance, as the graph shows, setting the selling price at $225 yields sales of 260 units per month.

y (number of units sold each month)

(225, 260)   (270, 230)

P (price per unit)

  (a) Find the equation of the line. (Remember to use the letter P instead of the usual x.)

  (b) Use the equation that you found in part (a) to determine how many units can be sold in a month when the price is $303 per unit.

  (c) What should the price be to sell 288 units per month?

42. Imagine that you own a grove of orange trees, and suppose that from past experience you know that when 100 trees are planted, each tree will yield approximately 240 oranges per year. Furthermore, you've noticed that when additional trees are planted in the grove, the yield per tree decreases. Specifically, you have noted that the yield per tree decreases by about 20 oranges for each additional tree planted.

  (a) Let y denote the yield per tree when x trees are planted. Find a linear equation relating x and y. *Hint:* You are given that the point $(100, 240)$ is on the line. What is given about $\Delta y / \Delta x$?

  (b) Use the equation in part (a) to determine how many trees should be planted to obtain a yield of 400 oranges per tree.

  (c) If the grove contains 95 trees, what yield can you expect from each tree?

43. Show that the slope of the line passing through the two points (a, a^2) and (x, x^2) is $x + a$. *Hint:* You'll need to use difference of squares factoring from intermediate algebra. If you need a review, see Appendix B.4.

44. Show that the slope of the line passing through the two points $(3, 9)$ and $(3 + h, (3 + h)^2)$ is $6 + h$.

45. Show that the slope of the line passing through the two points (a, a^3) and (x, x^3) is $x^2 + ax + a^2$. *Hint:* You'll need to use difference of cubes factoring from intermediate algebra. If you need a review, see Appendix B.4.

46. Write down, and then simplify as much as possible, an expression for the slope of the line passing through the two points $(a, 1/a)$ and $(x, 1/x)$.

47. A line with a slope of -5 passes through the point $(3, 6)$. Find the area of the triangle in the first quadrant formed by this line and the coordinate axes.

48. The y-intercept of the line in the figure on the next page is 6. Find the slope of the line if the area of the shaded triangle is 72 square units.

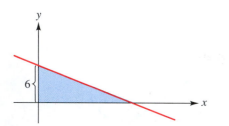

49. (a) Sketch the line $y = \frac{1}{2}x - 5$ and the point $P(1, 3)$. Follow parts (b)–(d) to calculate the perpendicular distance from point $P(1, 3)$ to the line.
(b) Find the equation of the line that passes through $P(1, 3)$ and is perpendicular to the line $y = \frac{1}{2}x - 5$.
(c) Find the coordinates of the point where these two lines intersect. *Hint:* From intermediate algebra, to find where two lines $y = mx + b$ and $y = Mx + B$ intersect, set the expressions $mx + b$ and $Mx + B$ equal to each other, and solve for x.
(d) Use the distance formula to find the perpendicular distance from $P(1, 3)$ to the line $y = \frac{1}{2}x - 5$.
50. The following figure shows a circle centered at the origin and a line that is tangent to the circle at the point $(3, -4)$.

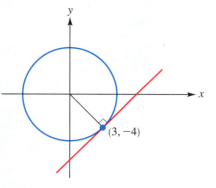

(a) Find the equation of the tangent line. *Hint:* Make use of the theorem from elementary geometry stating that the tangent line is perpendicular to the radius drawn to the point of contact.
(b) Find the intercepts of the tangent line.
(c) Find the length of the portion of the tangent line in quadrant IV.

*For Exercises 51 and 52, you'll need to recall the following definitions and results from elementary geometry. In a triangle, a line segment drawn from a vertex to the midpoint of the opposite side is called a **median**. The three medians of a triangle are **concurrent**; that is, they intersect in a single point. This point of intersection is called the **centroid** of the triangle. A line segment drawn from a vertex perpendicular to the opposite side is an **altitude**. The three altitudes of a triangle are concurrent; the point where the altitudes intersect is the **orthocenter** of the triangle.*

ⓖ 51. This exercise provides an example of the fact that the medians of a triangle are concurrent.
(a) The vertices of $\triangle ABC$ are as follows:

$$A(-4, 0) \qquad B(2, 0) \qquad C(0, 6)$$

Use a graphing utility to draw $\triangle ABC$. (Since $\overline{AB}$ coincides with the x-axis, you won't need to draw a line segment for this side.) *Note:* If the graphing utility you use does not have a provision for drawing line segments, you will need to determine an equation for the line in each case and then graph the line.
(b) Find the coordinates of the midpoint of each side of the triangle, then add the three medians to your picture in part (a). Note that the three medians do appear to intersect in a single point. Use the graphing utility to estimate the coordinates of the centroid.
(c) Using paper and pencil, find the equation of the medians from A to $\overline{BC}$ and from B to $\overline{AC}$. Then (using simultaneous equations from intermediate algebra), determine the exact coordinates of the centroid. How do these numbers compare with your estimates in part (b)?
ⓖ 52. This exercise illustrates the fact that the altitudes of a triangle are concurrent. Again, we'll be using $\triangle ABC$ with vertices $A(-4, 0)$, $B(2, 0)$, and $C(0, 6)$. Note that one of the altitudes of this triangle is just the portion of the y-axis extending from $y = 0$ to $y = 6$; thus, you won't need to graph this altitude; it will already be in the picture.
(a) Using paper and pencil, find the equations for the three altitudes. (Actually, you are finding equations for the lines that coincide with the altitude segments.)
(b) Use a graphing utility to draw $\triangle ABC$ along with the three altitude lines that you determined in part (a). Note that the altitudes appear to intersect in a single point. Use the graphing utility to estimate the coordinates of this point.
(c) Using simultaneous equations (from intermediate algebra), find the exact coordinates of the orthocenter. Are your estimates in part (b) close to these values?
53. This exercise outlines a proof of the fact that two nonvertical lines are parallel if and only if their slopes are equal. The proof relies on the following observation for the given figure: The lines $y = m_1x + b_1$ and $y = m_2x + b_2$ will be parallel if and only if the two vertical distances AB and CD are equal. (In the figure, the points C and D both have x-coordinate 1.)

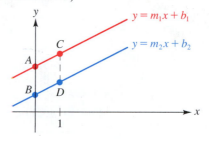

(a) Verify that the coordinates of A, B, C, and D are

$$A(0, b_1) \quad B(0, b_2) \quad C(1, m_1 + b_1) \quad D(1, m_2 + b_2)$$

(b) Using the coordinates in part (a), check that

$$AB = b_1 - b_2 \quad \text{and} \quad CD = (m_1 + b_1) - (m_2 + b_2).$$

(c) Use part (b) to show that the equation $AB = CD$ is equivalent to $m_1 = m_2$.

54. This exercise outlines a proof of the fact that two nonvertical lines with slopes m_1 and m_2 are perpendicular if and only if $m_1 m_2 = -1$. In the following figure, we've assumed that our two nonvertical lines $y = m_1 x$ and $y = m_2 x$ intersect at the origin. [If they did not intersect there, we could just as well work with lines parallel to these that do intersect at $(0, 0)$, recalling that parallel lines have the same slope.] The proof relies on the following geometric fact:

$$\overline{OA} \perp \overline{OB} \quad \text{if and only if} \quad (OA)^2 + (OB)^2 = (AB)^2$$

(a) Verify that the coordinates of A and B are $A(1, m_1)$ and $B(1, m_2)$.

(b) Show that

$$OA^2 = 1 + m_1^2$$
$$OB^2 = 1 + m_2^2$$
$$AB^2 = m_1^2 - 2m_1 m_2 + m_2^2$$

(c) Use part (b) to show that the equation

$$OA^2 + OB^2 = AB^2$$

is equivalent to $m_1 m_2 = -1$.

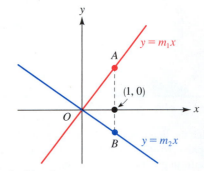

55. Verify the identity

$$(y_2 - y_1)/(x_2 - x_1) = (y_1 - y_2)/(x_1 - x_2)$$

What does this identity tell you about calculating slope?

C

56. Find the slope m of the line in the following figure.

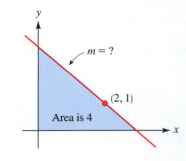

MINI PROJECT | THINKING ABOUT SLOPE

Working within a group or with the class at large, carry out the two activities indicated below. Then, on your own, write a summary of what you have learned.

1. What if the definition of slope that we gave in this section had instead been

$$m = \frac{y_2 + y_1}{x_2 + x_1}?$$ Try reworking Example 1 using this version of "slope" and describe what happens. Why is this unsatisfactory?

2. Why, where, and when did the use of the letter m for slope originate? See what you can find out by reading comments from mathematicians and math historians at the following Internet Web site:

    http://mathforum.org/epigone/math-history-list

    On the right-hand side of the resulting Web page, click on "search this discussion," and type in the word *slope*. As with many historical issues in many fields, don't expect one simple definitive answer. Summarize what you find out.

1.7 SYMMETRY AND GRAPHS. CIRCLES

Symmetry is a working concept. If all the object is symmetrical, then the parts must be halves (or some other rational fraction) and the amount of information necessary to describe the object is halved (etc.). —Alan L. Macay, Department of Crystallography, University of London

As you will see throughout this text, there are some basic techniques that help us to understand the essential features of a graph. In the box that follows, we introduce three types of *symmetry* that are useful in analyzing graphs.

THREE TYPES OF SYMMETRY

Type of Symmetry	Example	Definition
1. Symmetry about the x-axis	**Figure 1** Symmetry about the x-axis	For each point (x, y) on the graph, the point $(x, -y)$ is also on the graph. The points (x, y) and $(x, -y)$ are *reflections of one another about* (or *in*) *the x-axis*. In Figure 1, the portions of the graph above and below the x-axis also are said to be reflections of one another about the x-axis.
2. Symmetry about the y-axis	**Figure 2** Symmetry about the y-axis	For each point (x, y) on the graph, the point $(-x, y)$ is also on the graph. The points (x, y) and $(-x, y)$ are *reflections of one another about* (or *in*) *the y-axis*. In Figure 2, the portions of the graph to the right and left of the y-axis also are said to be reflections of one another about the y-axis.
3. Symmetry about the origin	**Figure 3** Symmetry about the origin	For each point (x, y) on the graph, the point $(-x, -y)$ is also on the graph. The points (x, y) and $(-x, -y)$ are *reflections of one another about the origin*. In Figure 3, the first- and third-quadrant portions of the curve also are said to be reflections of one another about the origin. In terms of the two previous symmetries, the point $(-x, -y)$ can be obtained from (x, y) as follows: First reflect (x, y) about the y-axis to obtain $(-x, y)$, then reflect $(-x, y)$ about the x-axis to obtain $(-x, -y)$.

EXAMPLE 1   Sketching reflections

A line segment $\mathscr{L}$ has endpoints $(1, 2)$ and $(5, 3)$. Sketch the reflection of $\mathscr{L}$ about: **(a)** the x-axis; **(b)** the y-axis; and **(c)** the origin.

SOLUTION

(a) First reflect the endpoints $(1, 2)$ and $(5, 3)$ about the x-axis to obtain the new endpoints $(1, -2)$ and $(5, -3)$, respectively. Then join these two points as indicated in Figure 4(a), in quadrant IV.

(b) Reflect the given endpoints about the y-axis to obtain the new endpoints $(-1, 2)$ and $(-5, 3)$; then join these two points as shown in Figure 4(a), in quadrant II.

Figure 4   (a)   (b)

(c) As described in the box preceding this example, reflection about the origin can be carried out in two steps: First reflect about the y-axis, then reflect about the x-axis. In part (b) we obtained the reflection of $\mathscr{L}$ in the y-axis, so now we need only reflect the line segment obtained in part (b) about the x-axis. See Figure 4(b).

In the box that follows, we list three rules for testing whether the graph of an equation possesses any of the types of symmetry we've been discussing. (The validity of each rule follows directly from the definitions of symmetry.)

Three Tests for Symmetry

1. The graph of an equation is symmetric about the y-axis if replacing x with $-x$ yields an equivalent equation.
2. The graph of an equation is symmetric about the x-axis if replacing y with $-y$ yields an equivalent equation.
3. The graph of an equation is symmetric about the origin if replacing x and y with $-x$ and $-y$, respectively, yields an equivalent equation.

EXAMPLE 2   Testing for symmetry about the x-axis and the y-axis

(a) In Figure 5(a), obtained with a graphing utility, the graph of $y = x^4 - 3x^2 + 1$ appears to be symmetric about the y-axis. Use an appropriate symmetry test to find out whether the graph indeed possesses this type of symmetry.

(b) In Figure 5(b), obtained with a graphing utility, the graph of $x = y^2 - \frac{1}{10}y - 15$ appears to be symmetric about the x-axis. Use an appropriate symmetry test to find out whether the graph indeed possesses this type of symmetry. *Remark*: Most graphing utilities require that you solve the equation for y before you enter it. See Exercise 61 for details.

GRAPHICAL PERSPECTIVE

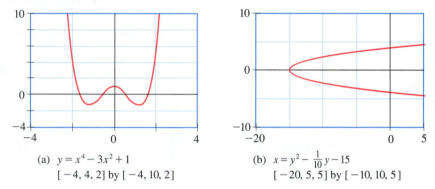

(a) $y = x^4 - 3x^2 + 1$
$[-4, 4, 2]$ by $[-4, 10, 2]$

(b) $x = y^2 - \frac{1}{10}y - 15$
$[-20, 5, 5]$ by $[-10, 10, 5]$

Figure 5

SOLUTION

(a) To test for symmetry about the y-axis, we replace x with $-x$ in the given equation, to obtain

$$y = (-x)^4 - 3(-x)^2 + 1 \qquad \text{or} \qquad y = x^4 - 3x^2 + 1$$

Since this last equation is the same as the original equation, we conclude that the graph is indeed symmetric about the y-axis.

(b) To test for symmetry about the x-axis, we replace y with $-y$ in the given equation, to obtain

$$x = (-y)^2 - \tfrac{1}{10}(-y) - 15 \qquad \text{or} \qquad x = y^2 + \tfrac{1}{10}y - 15$$

This last equation is not equivalent to the given equation. This tells us that the graph is not symmetric about the x-axis. (Exercise 61(c), a graphing utility exercise, asks for a viewing rectangle that clearly shows this lack of symmetry.)

EXAMPLE   3   Testing for symmetry about the origin

In Figure 6, it appears that the graph of $y = 2x^3 - 3x$ is symmetric about the origin. Use a symmetry test to find out whether this is indeed the case.

GRAPHICAL PERSPECTIVE

Figure 6
$y = 2x^3 - 3x$,
$[-2, 2, 1]$ by $[-10, 10, 5]$
The graph appears to be symmetric about the origin.
Is this really the case?

SOLUTION
Replacing x and y with $-x$ and $-y$, respectively, gives us

$$-y = 2(-x)^3 - 3(-x) = -2x^3 + 3x$$
$$y = 2x^3 - 3x \qquad \text{multiplying through by } -1$$

This last equation is identical to the given equation, and we conclude that, in this case, appearances are not deceiving; the graph is indeed symmetric about the origin.

In the next two examples, we use the notions of symmetry and intercepts as guides for drawing graphs. True, the graphs could be obtained more quickly using a graphing utility. The idea here, however, is to *understand* why the graphs look as they do.

EXAMPLE **4** **Using symmetry and intercepts as aids in graphing**

Graph the equation: $y = -x^2 + 5$.

SOLUTION
The domain of the variable x in the expression $-x^2 + 5$ is the set of all real numbers. However, since the graph must be symmetric about the y-axis (why?), we need only sketch the graph to the right of the y-axis; the portion to the left will then be the mirror image of this (with the y-axis as the mirror). The x- and y-intercepts (if any) are computed as follows:

x-INTERCEPTS
$$-x^2 + 5 = 0$$
$$x^2 = 5$$
$$x = \pm\sqrt{5} \ (\approx \pm 2.2)$$

y-INTERCEPTS
$$y = -(0)^2 + 5$$
$$y = 5$$

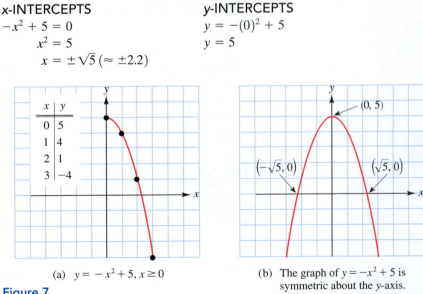

(a) $y = -x^2 + 5, x \geq 0$

(b) The graph of $y = -x^2 + 5$ is symmetric about the y-axis.

Figure 7

In Figure 7(a) we've set up a short table of values and sketched the graph for $x \geq 0$. The complete graph is then obtained by reflection about the y-axis, as shown in Figure 7(b). [The curve in Figure 7(b) is a *parabola*. This type of curve will be considered in detail in later chapters.]

EXAMPLE 5 **Using symmetry and domain as aids in graphing**

Graph: $y = -4/x$.

SOLUTION
Before doing any calculations, note that the domain of the variable x consists of all real numbers other than zero. So, however the resulting graph may look, it cannot contain a point with an x-coordinate of zero. In other words, the graph cannot cross the y-axis. (Check for yourself that the graph cannot cross the x-axis either.) Now, since the graph is symmetric about the origin (why?), we need only sketch the graph for $x > 0$; the portion corresponding to $x < 0$ can then be obtained by reflection about the origin. In Figure 8(a) we've set up a table of values and sketched the graph for $x > 0$. (Note the fractional x-values at the end of the table: since y is undefined when x is zero, it's informative to pick x-values near zero and look at the corresponding y-values.) In Figure 8(b) we have reflected the first-quadrant portion of the graph first about the y-axis and then about the x-axis to obtain the required reflection about the origin. Figure 9 shows our final graph of $y = -4/x$.

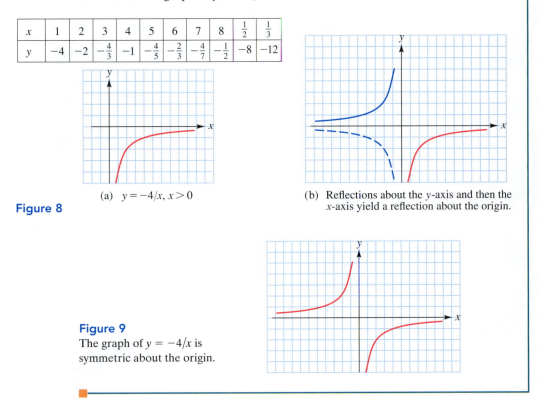

x	1	2	3	4	5	6	7	8	$\frac{1}{2}$	$\frac{1}{3}$
y	-4	-2	$-\frac{4}{3}$	-1	$-\frac{4}{5}$	$-\frac{2}{3}$	$-\frac{4}{7}$	$-\frac{1}{2}$	-8	-12

(a) $y = -4/x$, $x > 0$

Figure 8

(b) Reflections about the y-axis and then the x-axis yield a reflection about the origin.

Figure 9
The graph of $y = -4/x$ is symmetric about the origin.

We can use the graph in Figure 8(a) to introduce the idea of an *asymptote*. A line is an **asymptote** for a curve if the distance between the line and the curve approaches zero as we move farther and farther out along the line. So, for the curve in Figure 8(a), both the x-axis and the y-axis are asymptotes. (We'll return to this idea several times later in the text. For other pictures of asymptotes, see Figure 1 in Section 4.7.)

Figure 10

Since we've been discussing symmetry in this section, it seems appropriate to conclude with a discussion of the circle, because in some sense, this is the most symmetric curve. We can use the distance formula to obtain *the equation of a circle*. Figure 10 shows a circle with center (h, k) and radius r. By definition, a point (x, y) is on this circle if and only if the distance from (x, y) to (h, k) is r. Thus we have

$$\sqrt{(x - h)^2 + (y - k)^2} = r$$

or, equivalently,

$$(x - h)^2 + (y - k)^2 = r^2 \qquad (1)$$

(These last two equations are equivalent because two nonnegative quantities are equal if and only if their squares are equal.)

The work in the previous paragraph tells us two things. First, if a point (x, y) lies on the circle in Figure 10, then x and y together satisfy equation (1). Second, if a pair of numbers x and y satisfies equation (1), then the point (x, y) lies on the circle in Figure 10. (Does it sound to you as if the previous two sentences say the same thing? They don't! Think about it.) Equation (1) is called the **standard form for the equation of a circle.** For reference, we record the result in the box that follows.

The Equation of a Circle in Standard Form

The equation of a circle with center (h, k) and radius r is

$$(x - h)^2 + (y - k)^2 = r^2$$

EXAMPLE 6   Finding the equation of a circle

(a) Write the equation of the circle with center $(-2, 1)$ and radius 3.
(b) Does the point $(-4, 3)$ lie on this circle?

SOLUTION
(a) In the equation $(x - h)^2 + (y - k)^2 = r^2$, we substitute the given values $h = -2, k = 1$, and $r = 3$. This yields

$$[x - (-2)]^2 + (y - 1)^2 = 3^2$$

or

$$(x + 2)^2 + (y - 1)^2 = 9 \qquad (2)$$

This is the standard form for the equation of the given circle.
(b) To find out whether the point $(-4, 3)$ lies on the circle, we check to see if the coordinates $x = -4$ and $y = 3$ satisfy equation (2). We have

$$(-4 + 2)^2 + (3 - 1)^2 \stackrel{?}{=} 9$$
$$(-2)^2 + 2^2 \stackrel{?}{=} 9$$
$$8 = 9 \qquad \text{False}$$

This shows that the values $x = -4$ and $y = 3$ do not satisfy equation (2). Consequently, the point $(-4, 3)$ does not lie on the circle.

In the example just concluded we found that an equation for the circle with center $(-2, 1)$ and radius 3 is $(x + 2)^2 + (y - 1)^2 = 9$. There is another way to write this equation that is useful for work with graphing utilities. It involves solving the equation for y in terms of x, as follows:

$$(y - 1)^2 = 9 - (x + 2)^2$$

$$y - 1 = \pm\sqrt{9 - (x + 2)^2}$$

$$y = 1 \pm \sqrt{9 - (x + 2)^2}$$

Thus, the equation of the circle is equivalent to the *pair* of equations $y = 1 + \sqrt{9 - (x + 2)^2}$ and $y = 1 - \sqrt{9 - (x + 2)^2}$. In working with a graphics calculator or a computer, both of these equations are graphed to yield the required circle. See Figure 11 and the GRAPHICAL PERSPECTIVE in Figure 12. In Figure 11, note that the equation with the positive square root represents the upper semicircle, while the equation with the negative square root represents the lower semicircle.

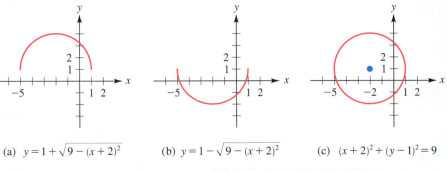

Figure 11
(a) $y = 1 + \sqrt{9 - (x + 2)^2}$ (b) $y = 1 - \sqrt{9 - (x + 2)^2}$ (c) $(x + 2)^2 + (y - 1)^2 = 9$

GRAPHICAL PERSPECTIVE

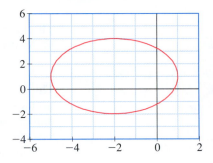

Figure 12
A distorted view of the circle $(x + 2)^2 + (y - 1)^2 = 9$ results when a *true proportions* option is not selected on the graphing utility.

Again, let us go back to the equation obtained in Example 6 for the circle with center $(-2, 1)$ and radius 3: $(x + 2)^2 + (y - 1)^2 = 9$. An alternative form for this equation is found by carrying out the indicated algebra:

$$(x + 2)^2 + (y - 1)^2 = 9$$

$$x^2 + 4x + 4 + y^2 - 2y + 1 = 9$$

$$x^2 + 4x + y^2 - 2y - 4 = 0 \qquad (3)$$

The disadvantage of this last equation is that the center and radius are no longer readily apparent. For this reason, it is useful to have a systematic procedure for converting equations such as equation (3) back into standard form. The algebraic technique from intermediate algebra known as *completing the square* allows us to accomplish this. We'll review (or reintroduce) this technique for you with two examples. (We'll have several other uses for completing the square later in this text. It's also useful at times in calculus.)

As a first example in completing the square, we convert the equation $x^2 + y^2 + 10x - 2y - 4 = 0$ into the standard form for a circle. We begin by grouping the x-terms, grouping the y-terms, and moving the constant to the other side (i.e., adding 4 to both sides):

$$(x^2 + 10x \quad) + (y^2 - 2y \quad) = 4 \tag{4}$$

We've left the extra space within each set of parentheses because we are going to add something there. (You'll see in a moment why we want to do this.) To determine the number that we want to add in the first set of parentheses, we follow the completing-the-square procedure:

Take half of the coefficient of x and square it.

From equation (4), the coefficient of x is 10. Taking half of 10 and then squaring it gives us 5^2, or 25. That's the number we want to add in the first set of parentheses. Of course, to keep the equation in balance, we have to add 25 on the other side too. So equation (4) becomes

$$(x^2 + 10x + 25) + (y^2 - 2y \quad) = 4 + 25 \tag{5}$$

In the same way now, to determine the number that we want to add in the second set of parentheses, we

Take half of the coefficient of y and square it.

From equation (5), the coefficient of y is -2. Taking half of -2 and squaring gives us $(-1)^2$, or 1. That's the number we want to add in the second set of parentheses. To keep the equation in balance, we also add 1 to the right side to obtain

$$(x^2 + 10x + 25) + (y^2 - 2y + 1) = 4 + 25 + 1 = 30 \tag{6}$$

The whole point in completing the square is that the expressions within the parentheses are now perfect squares, and we can rewrite equation (6) as

$$(x + 5)^2 + (y - 1)^2 = 30 \qquad \textcolor{red}{\text{Verify this factoring for yourself.}}$$

This last equation is the standard form for the equation of a circle. From the equation we see that the center of the circle is $(-5, 1)$. What about the radius? In standard form, the right side of the equation is r^2. So we have $r^2 = 30$, and consequently $r = \sqrt{30}$.

The process that we've just described for completing the square requires that the coefficients of x^2 and y^2 both be 1. If this is not the case at the start, then you need to divide both sides of the equation by an appropriate constant, as indicated in the next example.

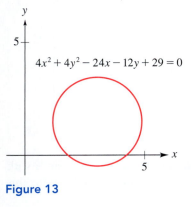

$4x^2 + 4y^2 - 24x - 12y + 29 = 0$

Figure 13

| **EXAMPLE** | **7** | **Completing the square and determining x-intercepts** |

Refer to Figure 13, which shows a graph of the circle

$$4x^2 + 4y^2 - 24x - 12y + 29 = 0$$

(a) Find the center and radius of the circle.
(b) Use a graphing utility to estimate (to two decimal places) the x-intercepts of the circle.
(c) Use algebra to find exact values for the x-intercepts.

SOLUTION

(a) To obtain coefficients of 1 in front of both x and y, we divide both sides of the equation by 4:

$$x^2 + y^2 - 6x - 3y + \frac{29}{4} = 0$$

Next, we group the x-terms, group the y-terms, and subtract the constant 29/4 from both sides. This gives us

$$(x^2 - 6x \quad) + (y^2 - 3y \quad) = -\frac{29}{4}$$

To complete the square within the first set of parentheses we need to add $\left(\frac{-6}{2}\right)^2$, which is 9. Likewise in the second set of parentheses, we want to add $\left(\frac{-3}{2}\right)^2$ or $\frac{9}{4}$. So we obtain

$$(x^2 - 6x + 9) + \left(y^2 - 3y + \frac{9}{4}\right) = -\frac{29}{4} + 9 + \frac{9}{4}$$

or equivalently

$$(x - 3)^2 + \left(y - \frac{3}{2}\right)^2 = 4 \qquad \text{\color{red}Check the arithmetic.} \qquad (7)$$

Looking at this last equation, we see that the center of the circle is $(3, \frac{3}{2})$ and the radius is 2.

(b) To use a graphing utility to graph the circle, we solve equation (7) for y in terms of x as follows.

$$\left(y - \frac{3}{2}\right)^2 = 4 - (x - 3)^2$$

$$y - \frac{3}{2} = \pm\sqrt{4 - (x - 3)^2}$$

$$y = \frac{3}{2} + \sqrt{4 - (x - 3)^2} \qquad \text{or} \qquad y = \frac{3}{2} - \sqrt{4 - (x - 3)^2}$$

Figure 14(a) shows a graph of this pair of equations (using true proportions). The view in Figure 14(a) indicates that the smaller x-intercept is between 1 and 2, closer to 2 than to 1; the larger x-intercept is between 4 and 5, closer to 4 than to 5. Figure 14(b) shows a close-up view of the smaller x-intercept.

GRAPHICAL PERSPECTIVE

Figure 14

A graph of the circle $(x - 3)^2 + (y - \frac{3}{2})^2 = 4$ along with a magnified view near the smaller of the two x-intercepts

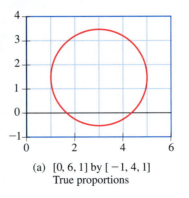

(a) $[0, 6, 1]$ by $[-1, 4, 1]$
True proportions

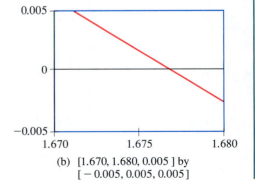

(b) $[1.670, 1.680, 0.005]$ by $[-0.005, 0.005, 0.005]$

Evidently it is closer to 1.680 than to 1.670, and so our estimate to two decimal places is 1.68. Exercise 62 asks you to use a graphing utility in a similar manner to show that the larger x-intercept is approximately 4.32.

(c) The easiest way to find exact expressions for the x-intercepts is to use equation (7). Setting $y = 0$ in equation (7) yields

$$(x - 3)^2 + \frac{9}{4} = 4$$

$$(x - 3)^2 = \frac{7}{4} \qquad \textcolor{red}{\text{Check the arithmetic.}}$$

$$x - 3 = \pm\sqrt{\frac{7}{4}} = \pm\frac{\sqrt{7}}{2}$$

$$x = 3 \pm \frac{\sqrt{7}}{2}$$

In summary, the two x-intercepts are $3 - \sqrt{7}/2$ and $3 + \sqrt{7}/2$. As you can check for yourself with a calculator, these last two expressions are approximately 1.68 and 4.32, respectively. These values are consistent with the numbers that we obtained graphically in part (b).

EXERCISE SET 1.7

A

In Exercises 1–6, the endpoints of a line segment $\overline{AB}$ are given. Sketch the reflection of $\overline{AB}$ about **(a)** *the x-axis;* **(b)** *the y-axis; and* **(c)** *the origin.*

1. $A(1, 4)$ and $B(3, 1)$
2. $A(-1, -2)$ and $B(-5, -2)$
3. $A(-2, -3)$ and $B(2, -1)$
4. $A(-3, -3)$ and $B(-3, -1)$
5. $A(0, 1)$ and $B(3, 1)$
6. $A(-2, -2)$ and $B(0, 0)$

In Exercises 7–24, graph the equation after determining the x- and y-intercepts and whether the graph possesses any of the three types of symmetry described on page 63.

7. $y = 4 - x^2$
8. $y = -x^3$
9. $y = -1/x$
10. $x = y^2 - 1$
11. $y = -x^2$
12. $y = 1/x^2$
13. $y = -1/x^3$
14. $y = |x| - 2$
15. $y = \sqrt{x^2}$
16. $y = x + 1$
17. $y = x^2 - 2x + 1$
18. $x = y^3 - 1$
19. $y^2 = 2x - 4$
20. $|y| = 2x - 4$
21. $y = 2x^2 + x - 4$
22. $y = 2x^2$
23. (a) $x + y = 2$
    (b) $|x| + y = 2$
24. (a) $x + |y| = 2$
    (b) $|x + y| = 2$

Ⓖ *In Exercises 25–38:*
   (a) *Use a graphing utility to graph each equation in the standard viewing rectangle.*
   (b) *Does the graph in part (a) appear to possess any of the three types of symmetry defined on page 63?*
   (c) *In cases where your answer to part (b) is Yes, adjust your viewing rectangle for a second, more careful inspection. (In Exercises 25–31, suggestions for the second look are provided.)*
   (d) *In those cases where, after the second look, it still appears that the graph possesses symmetry, use an appropriate symmetry test from page 64 to settle the matter.*

25. $y = x^2 - 3x$ (second view: x from -2 to 5; y from -4 to 10)
26. $y = x^3 - 3x$ (second view: x from -2 to 5; y from -4 to 10)
27. $y = 2^x$ (second view: x from -3 to 3; y from 0 to 8)
28. $y = 2^{|x|}$ (second view: x from -3 to 3; y from 0 to 8)
29. $y = 1/(x^2 - x)$ (second view: x from -2 to 2; y unchanged)
30. $y = 1/(x^3 - x)$ (second view: x from -2 to 2; y unchanged)
31. $y = x^2 - 0.2x - 15$ *Hint*: Look closely at the x-intercepts.
32. $y = x^2 - 2x^3$
33. $y = \sqrt{|x|}$
34. $y = \sqrt{|x|^3}$
35. $y = 2x - x^3 - x^5 + x^7$
36. $y = |2x - x^3 - x^5 + x^7|$
37. $y = x^4 - 10x^2 + \frac{1}{4}x$ *Suggestion*: For the second view, try zooming out in the y-direction.
38. $y = x^4 - 10x^2 + \frac{1}{4}$

In Exercises 39–42, specify the center and radius of each circle. Also, determine whether the given point lies on the circle.

39. $(x - 1)^2 + (y - 5)^2 = 169$; $(6, -7)$

40. $(x + 4)^2 + (y + 2)^2 = 20$; $(0, 1)$

41. $(x + 8)^2 + (y - 5)^2 = 13$; $(-5, 2)$

42. $x^2 + y^2 = 1$; $(1/2, \sqrt{3}/2)$

In Exercises 43–48, determine the center and the radius for the circle. Also, find the y-coordinates of the points (if any) where the circle intersects the y-axis.

43. $x^2 + y^2 = \sqrt{2}$

44. $x^2 + y^2 - 10x + 2y + 17 = 0$

45. $x^2 + y^2 + 8x - 6y = -24$

46. $4x^2 - 4x + 4y^2 - 63 = 0$

47. $9x^2 + 54x + 9y^2 - 6y + 64 = 0$

48. $3x^2 + 3y^2 + 5x - 4y = 1$

In Exercises 49 and 50, use the techniques shown in Example 7 to carry out the following procedures.

(a) *Find the center and radius of the circle.*

Ⓖ **(b)** *Use a graphing utility to graph the circle and to estimate (to two decimal places) the x-intercepts.*

(c) *Use algebra to find exact values for the x-intercepts, and then use a calculator to check that the results are consistent with the estimates in part (b).*

49. $16x^2 - 64x + 16y^2 + 48y - 69 = 0$

50. $3x^2 + x + 3y^2 + 3y - 1 = 0$

Ⓖ **51.** In the text we said that a line is an *asymptote* for a curve if the distance between the line and the curve approaches zero as we move farther and farther out along the line. In terms of graphing, this means that as we zoom out, the curve and the line eventually appear indistinguishable. In this exercise, we'll demonstrate this using the curve $y = -4/x$ (which we graphed in Figure 9). As indicated in the text, both the x- and y-axes are asymptotes for this curve. First, graph $y = -4/x$ using a viewing rectangle that extends from -5 to 5 in both the x- and the y-directions. Then take a second look using a viewing rectangle that extends from -30 to 30 in both the x- and y-directions. At this scale, you'll see that the curve is virtually indistinguishable from an asymptote when either $|x| > 8$ or $|y| > 8$.

Ⓖ **52. (a)** Graph the equation $y = 20/x$ using a standard viewing rectangle.

(b) Although both the x- and the y-axes are asymptotes for this curve, the graph in part (a) does not show this clearly. Take a second look using a viewing rectangle that extends from -100 to 100 in both the x- and the y-directions. Note that the curve indeed appears indistinguishable from an asymptote when either $|x|$ or $|y|$ is sufficiently large.

B

53. The center of a circle is the point $(3, 2)$. If the point $(-2, -10)$ lies on this circle, find the equation of the circle.

54. Find the equation of the circle tangent to the x-axis and with center $(3, 5)$. *Hint:* First draw a sketch.

55. Find the equation of the circle tangent to the y-axis and with center $(3, 5)$.

56. Find the equation of the circle passing through the origin and with center $(3, 5)$.

57. The points $A(-1, 6)$ and $B(3, -2)$ are the endpoints of a diameter of a circle, as indicated in the accompanying figure. Find the y-intercepts of the circle. *Hint:* Could you do the problem if you had the equation of the circle?

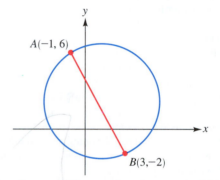

$\overline{AB}$ is a diameter. What are the y-intercepts of the circle?

58. (a) Verify that the point $(3, 7)$ is on the circle
$$x^2 + y^2 - 2x - 6y - 10 = 0$$

(b) Find the equation of the line tangent to this circle at the point $(3, 7)$. *Hint:* A result from elementary geometry says that the tangent to a circle is perpendicular to the radius drawn to the point of contact.

59. The accompanying figure shows the graphs of $y = \frac{3}{4}x - 2$ and $y = |\frac{3}{4}x - 2|$.

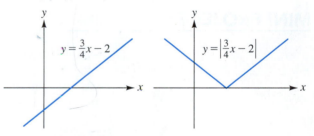

(a) Determine the x- and y-intercepts for each graph.

(b) Which portions of the two graphs are identical? (Give your answer in terms of an interval along the x-axis.)

(c) Explain how the graph of $y = |\frac{3}{4}x - 2|$ can be obtained from that of $y = \frac{3}{4}x - 2$ by means of reflection.

60. The accompanying figure shows the graphs of the two equations $y = x^2 - 6x + 8$ and $y = |x^2 - 6x + 8|$.

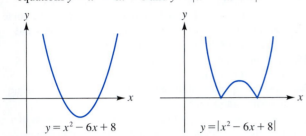

$y = x^2 - 6x + 8$ $y = |x^2 - 6x + 8|$

(a) Determine the x- and y-intercepts for each graph.

(b) Which portions of the two graphs are identical? (Give your answer in terms of intervals along the x-axis.)

(c) Explain how the graph of $y = |x^2 - 6x + 8|$ can be obtained from that of $y = x^2 - 6x + 8$ by means of reflection.

Ⓖ **61.** This exercise relates to Figure 5(b) on page 65. That figure shows a graph of the equation $x = y^2 - 0.1y - 15$. Note that this equation is solved for x rather than y. In the standard Cartesian graphing mode, however, most graphing utilities require that the equation be solved for y. This exercise shows one method for obtaining the graph. (A remark at the end of this exercise mentions a second method.)

(a) Rewrite the equation as

$$y^2 - 0.1y - (15 + x) = 0$$

Then solve for y in terms of x by using the quadratic formula with $a = 1$, $b = -0.1$, and $c = -(15 + x)$. After simplifying, you should obtain

$$y = \frac{0.1 \pm \sqrt{60.01 + 4x}}{2}$$

(b) Enter and graph the pair of equations

$$y = \frac{0.1 - \sqrt{60.01 + 4x}}{2} \quad \text{and} \quad y = \frac{0.1 + \sqrt{60.01 + 4x}}{2}$$

using the viewing rectangle $[-20, 5, 5]$ by $[-10, 10, 5]$. Check that your graph agrees with that shown in Figure 5(b).

(c) Experiment to find a viewing rectangle in which it is clear that the graph of the equation $x = y^2 - 0.1y - 15$ is *not* symmetric about the x-axis.

Remark: A second method for graphing $x = y^2 - 0.1y - 15$ requires the *parametric mode* on your graphing utility. Consult the owner's manual for your graphing utility to see how to access and operate this mode. Entering the following pair of equations in the parametric mode will yield the graph (or a portion of the graph): $x = t^2 - 0.1t - 15$; $y = t$.

Ⓖ **62.** Use a graphing utility and the method shown in the solution of Example 7(b) to show that the larger of the two x-intercepts of the circle $4x^2 + 4y^2 - 24x - 12y + 29 = 0$ is (to two decimal places) 4.32.

C

63. Suppose that the circle $x^2 + 2Ax + y^2 + 2By = C$ has two x-intercepts, x_1 and x_2, and two y-intercepts, y_1 and y_2. Prove each statement.

(a) $\dfrac{x_1 + x_2}{y_1 + y_2} = \dfrac{A}{B}$

(b) $x_1 x_2 - y_1 y_2 = 0$

(c) $x_1 x_2 + y_1 y_2 = -2C$

MINI PROJECT | THINKING ABOUT SYMMETRY

This mini project works well in the context of a small group discussion but, of course, it can also be carried out by individuals. Begin by studying and/or discussing the following example involving a regular hexagon. Then work on the exercise that follows. For each part of the exercise, label each figure that you create (e.g., "symmetric about the y- but not the x-axis").

Example: The following figure shows a regular hexagon.

The next four figures show some of the (infinitely many) ways to locate the regular hexagon with respect to an x-y coordinate system.

(a) Symmetric about y-axis, but not about x-axis

(b) Symmetric about x-axis, but not about y-axis

(c) Symmetric about origin and about both axes

(d) Symmetric about origin and not about either axis

In each case below, you are given a geometric figure. Try to find a way to draw the figure with an x-y coordinate system so that the figure is

(a) symmetric about the y-axis, but not about the x-axis;

(b) symmetric about the x-axis, but not about the y-axis;

(c) symmetric about the origin and both axes;

(d) symmetric about the origin but not about either axis.

(In some cases there are many ways, in others none.)

1. equilateral triangle:

2. rectangle:

3. regular pentagon:

4. semicircle:

5. connected semicircles:

Chapter 1

Summary of Principal Terms and Formulas

Terms or notation	Page reference	Comments
1. Natural numbers, integers, rational numbers, and irrational numbers	2	The box on page 2 provides both definitions and examples. Also note the theorem in the box, which explains how to distinguish between rationals and irrationals in terms of their decimal representations.
2. $a < b$ $b > a$	3	a is less than b. b is greater than a.
3. (a, b) $[a, b]$	2	The open interval (a, b) consists of all real numbers between a and b, excluding a and b. The closed interval $[a, b]$ consists of all real numbers between a and b, including a and b.
4. (a, ∞)	4	The unbounded interval (a, ∞) consists of all real numbers x such that $x > a$. The infinity symbol, ∞, does not denote a real number. It is used in the context (a, ∞) to indicate that the interval has no right-hand boundary. For the definitions of the other types of unbounded intervals, see the box on page 4.
5. $|x|$	6	The absolute value of a real number x is the distance of x from the origin. This is equivalent to the following algebraic definition: $$|x| = \begin{cases} x & \text{if } x \geq 0 \\ -x & \text{if } x < 0 \end{cases}$$
6. Constant, variable, and the domain convention	12	By a constant we mean either a particular number (such as -8 or π) or a letter that remains fixed (although perhaps unspecified) throughout a given discussion. In contrast, a variable is a letter for which we can substitute any number from a given set of numbers. The given set is called the domain of the variable. According to the domain convention, the domain of a variable in a given expression consists of all real numbers for which the expression is defined.
7. Linear (or first-degree) equation in one variable	12	These are equations that can be written in the form $ax + b = 0$, where $a \neq 0$.
8. Solution (or root) of an equation	13	A solution or root is a number that, when substituted for the variable in an equation, yields a true statement.
9. Equivalent equations	13	Two equations are equivalent if they have the same set of solutions.
10. Extraneous solution (or extraneous root)	15	In solving equations, certain processes can lead to answers that do not check in the original equation. These numbers are called extraneous solutions (or extraneous roots).

Terms or notation	Page reference	Comments
11. Quadratic (or second-degree) equation	17	A quadratic equation is one that can be written in the form $ax^2 + bx + c = 0$, where $a \neq 0$.
12. Zero-product property of real numbers	17	If $pq = 0$ then $p = 0$ or $q = 0$; conversely, if $p = 0$ or $q = 0$, then $pq = 0$. (We used this property in solving quadratic equations by factoring.)
13. Quadratic formula $$x = \frac{-b \pm \sqrt{b^2 - 4ac}}{2a}$$	18	This is the quadratic formula; it provides the solutions of the quadratic equation $ax^2 + bx + c = 0$, where $a \neq 0$. The formula is derived in Section 2.1.
14. Polynomial equation	19	A polynomial equation is an equation of the form $$a_n x^n + a_{n-1} x^{n-1} + \cdots + a_1 x + a_0 = 0$$ where the subscripted letter a's represent constants and the exponents on the variable are nonnegative integers. The degree of the polynomial equation is the largest exponent of the variable that appears in the equation.
15. Pythagorean theorem $a^2 + b^2 = c^2$	23	In a right triangle, the lengths of the sides are related by this equation, where c is the length of the hypotenuse. Conversely, if the lengths of the sides of a triangle are related by an equation of the form $a^2 + b^2 = c^2$, then the triangle is a right triangle, and c is the length of the hypotenuse.
16. Distance formula $$d = \sqrt{(x_2 - x_1)^2 + (y_2 - y_1)^2}$$	24	d is the distance between the points (x_1, y_1) and (x_2, y_2).
17. Midpoint formula $$\left(\frac{x_1 + x_2}{2}, \frac{y_1 + y_2}{2} \right)$$	27	This is the midpoint of the line segment joining (x_1, y_1) and (x_2, y_2).
18. Graph of an equation	36	The graph of an equation in the variables x and y is the set of all points (x, y) with coordinates that satisfy the equation.
19. Graphing utility and standard viewing rectangle	37	By a graphing utility we mean either a graphing calculator or a computer with software for graphing and analyzing equations or data. The standard viewing rectangle on a graphing utility is the default view screen and coordinate system for graphing. For many graphing utilities, the standard viewing rectangle extends from -10 to 10 in both the x- and the y-directions.
20. x-intercept and y-intercept	40	An x-intercept of a graph is the x-coordinate of a point where the graph intersects the x-axis. Similarly, a y-intercept is the y-coordinate of a point where the graph intersects the y-axis.

Terms or notation	Page reference	Comments
21. Slope $$m = \frac{y_2 - y_1}{x_2 - x_1}$$	48	m is the slope of a nonvertical line passing through the points (x_1, y_1) and (x_2, y_2).
22. Δx, Δy	49	If the value of a variable x changes from x_1 to x_2, then Δx denotes the amount by which x changes: $\Delta x = x_2 - x_1$. Similarly, $\Delta y = y_2 - y_1$. With this notation, the slope formula becomes $m = \Delta y / \Delta x$.
23. Point–slope formula $y - y_1 = m(x - x_1)$	52	This is the point–slope form for the equation of a line passing through the point (x_1, y_1) with slope m.
24. Slope–intercept formula $y = mx + b$	54	This is the slope–intercept form for the equation of a line with slope m and y-intercept b.
25. Parallel lines $m_1 = m_2$	56	Nonvertical parallel lines have the same slope.
26. Perpendicular lines $m_1 = -1/m_2$	56	Two nonvertical lines are perpendicular if and only if the slopes are negative reciprocals of one another.
27. Symmetry about the x-axis	63	A graph is symmetric about the x-axis if, for each point (x, y) on the graph, the point $(x, -y)$ is also on the graph. The points (x, y) and $(x, -y)$ are *reflections* of each other about (or in) the x-axis.
28. Symmetry about the y-axis	63	A graph is symmetric about the y-axis if, for each point (x, y) on the graph, the point $(-x, y)$ is also on the graph. The points (x, y) and $(-x, y)$ are *reflections* of each other about (or in) the y-axis.
29. Symmetry about the origin	63	A graph is symmetric about the origin if, for each point (x, y) on the graph, the point $(-x, -y)$ is also on the graph. The points (x, y) and $(-x, -y)$ are *reflections* of each other about the origin.
30. Symmetry tests	64	(i) The graph of an equation is symmetric about the y-axis if replacing x with $-x$ yields an equivalent equation. (ii) The graph of an equation is symmetric about the x-axis if replacing y with $-y$ yields an equivalent equation. (iii) The graph of an equation is symmetric about the origin if replacing x and y with $-x$ and $-y$, respectively, yields an equivalent equation.
31. Equation of a circle $(x - h)^2 + (y - k)^2 = r^2$	68	This is the standard form for the equation of a circle with center (h, k) and radius r.

Writing Mathematics

1. The following geometric method for solving quadratic equations of the form $x^2 - ax + b = 0$ is due to the British writer Thomas Carlyle (1795–1881):

   To solve the equation $x^2 - ax + b = 0$, plot the points $A(0, 1)$ and $B(a, b)$. Then draw the circle with $\overline{AB}$ as diameter. The x-intercepts of this circle are the roots of the equation.

   Use the techniques of this chapter to verify for yourself that this method indeed yields the roots for the equation $x^2 - 6x + 5 = 0$. (You need to find the equation of the circle, determine the x-intercepts, and then check that the numbers you obtain are the roots of the given equation.) After you have done this, carefully write out your verification in detail (using complete sentences), as if you were explaining the method and why it works to a classmate. Be sure to make explicit reference to any formulas or equations that you use from the chapter.

2. In Section 1.6 we developed the point–slope formula, $y - y_1 = m(x - x_1)$, and the slope–intercept formula, $y = mx + b$. Although these two formulas look quite different from one another, both are merely restatements of the definition of slope. Write a paragraph (or two, at the most) to explain this last sentence. Make use of the following two figures.

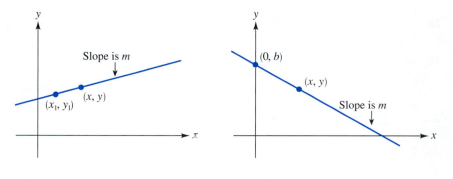

Chapter 1   Review Exercises

For Exercises 1–6 rewrite the statements using absolute values and inequalities or equalities.

1. The distance between x and 6 is 2.
2. The distance between x and a is less than $1/2$.
3. The distance between a and b is 3.
4. The distance between x and -1 is 5.
5. The distance between x and 0 exceeds 10.
6. What can you say about x if $|x - 5| = 0$?

Rewrite each of the expressions in Exercises 7–12 in a form that does not contain absolute values.

7. $|\sqrt{6} - 2|$
8. $|2 - \sqrt{6}|$
9. $|x^4 + x^2 + 1|$
10. (a) $|x - 3|$, if $x < 3$
    (b) $|x - 3|$, if $x > 3$
11. $|x - 2| + |x - 3|$, if:
    (a) $x < 2$
    (b) $2 < x < 3$
    (c) $x > 3$
12. $|x + 2| + |x - 1|$, if:
    (a) $x < -2$
    (b) $-2 < x < 1$
    (c) $x > 1$

In Exercises 13–18, sketch each interval on a number line.
13. $(3, 5)$
14. $(3, 5]$

15. The set of all negative real numbers that are in the interval $[-5, 2]$

16. $(-\infty, 4)$ 　　　　　　　**17.** $[-1, \infty)$

18. The set of real numbers that belongs to either of the intervals $(0, \sqrt{2}]$ or $(\sqrt{2}, 2]$

In Exercises 19–23, sketch the intervals described by the given inequalities.

19. $|x - 6| < 3$ 　　　　　　　**20.** $|x - \frac{1}{2}| < 1$

21. $|x + 1| \geq 1$ 　　　　　　　**22.** $|x| \geq 5$

23. **(a)** $0 < |x - 4| < 5$ 　　　　**(b)** $|x - 4| < 5$

24. Determine whether each of the following is true or false. (You should be able to do these without a calculator.)

　　(a) $\pi \leq \sqrt{3}$ 　　　　　　　**(b)** $\dfrac{1}{5} > -\sqrt{2}$

In Exercises 25–40, find all the real solutions of each equation.

25. $5 - 9x = 2$ 　　　　　　**26.** $\dfrac{x}{3} - \dfrac{3x}{5} = \dfrac{x}{6} - 13$

27. $(t - 4)(t + 3) = (t + 5)^2$

28. $\dfrac{1 - x}{1 + x} = 1$ 　　　　　**29.** $\dfrac{2t - 1}{t + 2} = 5$

30. $\dfrac{1}{1 + \dfrac{1}{x + 1}} = \dfrac{5}{6}$ 　　　**31.** $\dfrac{2y - 5}{4y + 1} = \dfrac{y - 1}{2y + 5}$

32. $\dfrac{2x - 3}{x - 2} = \dfrac{1}{x - 2}$ 　　　**33.** $12x^2 + 2x - 2 = 0$

34. $4y^2 - 21y = 18$ 　　　　**35.** $\frac{1}{2}x^2 + x - 12 = 0$

36. $x^2 + \frac{13}{2}x + 10 = 0$ 　　**37.** $\dfrac{x}{5 - x} = \dfrac{-2}{11 - x}$

38. $4x^2 + x - 2 = 0$ 　　　　**39.** $t^2 + t - \frac{1}{2} = 0$

40. $x^3 - 8x^2 - 9x = 0$

In Exercises 41–57, find an equation for the line satisfying the given conditions. For Exercises 41–49, write the answer in the form $y = mx + b$; in Exercises 50–57, write the answer in the form $Ax + By + C = 0$.

41. Passes through $(-4, 2)$ and $(-6, 6)$

42. $m = -2$; y-intercept 5

43. $m = 1/4$; passes through $(-2, -3)$

44. $m = 1/3$; x-intercept -1

45. x-intercept -4; y-intercept 8

46. $m = -10$; y-intercept 0

47. y-intercept -2; parallel to the x-axis

48. Passes through $(0, 0)$ and is parallel to $6x - 3y = 5$

49. Passes through $(1, 2)$ and is perpendicular to the line $x + y + 1 = 0$

50. Passes through $(1, 1)$ and through the center of the circle $(x - 2)^2 + (y - 4)^2 = 20$

51. Passes through the centers of the circles $(x + 2)^2 + (y + 1)^2 = 5$ and $(x - 2)^2 + (y - 8)^2 = 68$

52. $m = 3$ and has the same x-intercept as the line $3x - 8y = 12$

53. Passes through the origin and the midpoint of the line segment joining the points $(-2, -3)$ and $(6, -5)$

54. Is tangent to the circle $x^2 + y^2 = 20$ at the point $(-2, 4)$

55. Is tangent to the circle $(x - 3)^2 + (y + 4)^2 = 25$ at the point $(0, 0)$

56. Passes through $(2, 4)$; the y-intercept is twice the x-intercept

57. Passes through $(2, -1)$; the sum of the x- and y-intercepts is 2. (There are two answers.)

58. **(a)** Find the perimeter of the triangle with vertices $A(3, 1)$, $B(7, 4)$, and $C(-2, 13)$.

　　(b) Find the perimeter of the triangle formed by joining the midpoints of the sides of the triangle in part (a).

In Exercises 59–62, test each equation for symmetry about the x-axis, the y-axis, and the origin.

59. $y = x^4 - 2x^2$ 　　　　　**60.** $y = x^3 + 5x + 1$

61. $y = 2^x + 2^{-x}$ 　　　　　**62.** $y = 2^x - 2^{-x}$

For Exercises 63–68, tell whether each graph appears to be symmetric about the x-axis, the y-axis, or the origin.

63. 　　　　　　　　　　**64.**

65. 　　　　　　　　　　**66.**

67. 　　　　　　　　　　**68.**

Graph the equations in Exercises 69–74.

69. $x = 9 - y^2$

70. $(x + 2)^2 + y^2 = 1$

71. $y = 1 - |x|$

72. $y = |x - 2| + 2$

73. $(4x - y + 4)(4x + y - 4) = 0$

    Hint: Use the zero-product property.

74. $(y - x)(y - x + 2)(y - x - 2) = 0$

You know from the text that the graph of the equation $x^2 + y^2 = 9$ is a circle with center (0, 0) and radius 3. In Exercises 75–81, use a graphing utility to investigate what happens when the equation is changed slightly. (As in graphing circles, you'll first need to solve the given equations for y.)

75. (a) $x^2 + 2y^2 = 9$ (The resulting curve is an *ellipse.* We'll study this curve in a later chapter.)

    (b) On the same set of axes, graph the ellipse $x^2 + 2y^2 = 9$ and the circle $x^2 + y^2 = 9$. At which points do the two curves appear to intersect?

    (c) Carry out the calculations to verify that the points indicated in part (b) are indeed the exact intersection points.

76. $2x^2 + y^2 = 9$

77. (a) $x^2 - y^2 = 9$ (The resulting curve is a *hyperbola.* We'll study this curve in a later chapter.)

    (b) $x^2 - y^2 = 0$

    (c) On the same set of axes, graph the equations given in parts (a) and (b). Are there any intersection points?

78. $x^2 - y^3 = 9$

79. $x^3 - y^2 = 9$

80. (a) $x^3 + y^3 = 9$

    (b) $x^5 + y^5 = 9$

    (c) On the same set of axes, graph the equations given in parts (a) and (b). Based on the pattern you see, what do you think the graph of the equation $x^7 + y^7 = 9$ will look like? Draw a sketch (by hand) to show your prediction. Then use a graphing utility to see how accurate your prediction is.

81. (a) $x^4 + y^4 = 9$

    (b) $x^6 + y^6 = 9$

    (c) On the same set of axes, graph the equations given in parts (a) and (b). Based on the pattern you see, what do you think the graph of the equation $x^8 + y^8 = 9$ will look like? Draw a sketch (by hand) to show your prediction. Then use a graphing utility to see how accurate your prediction is.

82. Find a value for t such that the slope of the line passing through (2, 1) and (5, t) is 6.

83. The vertices of a right triangle are $A(0, 0)$, $B(0, 2b)$, and $C(2c, 0)$. Let M be the midpoint of the hypotenuse. Compute the three distances MA, MB, and MC. What do you observe?

84. The vertices of parallelogram $ABCD$ are $A(-4, -1)$, $B(2, 1)$, $C(3, 3)$, and $D(-3, 1)$.

    (a) Compute the sum of the squares of the two diagonals.

    (b) Compute the sum of the squares of the four sides. What do you observe?

In Exercises 85 and 86, two points are given. In each case compute **(a)** *the distance between the two points,* **(b)** *the slope of the line segment joining the two points, and* **(c)** *the midpoint of the line segment joining the two points.*

85. (2, 5) and (5, −6)

86. $(\sqrt{3}/2, 1/2)$ and $(-\sqrt{3}/2, -1/2)$

87. A line passes through the points (1, 2) and (4, 1). Find the area of the triangle bounded by this line and the coordinate axes.

88. Let $P_1(x_1, y_1)$ and $P_2(x_2, y_2)$ be two given points. Let Q be the point $(\frac{1}{3}x_1 + \frac{2}{3}x_2, \frac{1}{3}y_1 + \frac{2}{3}y_2)$.

    (a) Show that the points P_1, Q, and P_2 are collinear. *Hint:* Compute the slope of $\overline{P_1Q}$ and the slope of $\overline{QP_2}$.

    (b) Show that $P_1Q = \frac{2}{3}P_1P_2$. (In other words, Q is on the line segment $\overline{P_1P_2}$ and two-thirds of the way from P_1 to P_2.)

89. (a) Let the vertices of $\triangle ABC$ be $A(-5, 3)$, $B(7, 7)$, and $C(3, 1)$. Find the point on each median that is two-thirds of the way from the vertex to the midpoint of the opposite side. (Recall that a median of a triangle is a line segment drawn from a vertex to the midpoint of the opposite side.) What do you observe? *Hint:* Use the result in Exercise 88.

    (b) Follow part (a) but take the vertices to be $A(0, 0)$, $B(2a, 0)$, and $C(2b, 2c)$. What do you observe? What does this prove?

90. In the accompanying figure, points P and Q trisect the hypotenuse in $\triangle ABC$. Prove that the square of the hypotenuse is equal to 9/5 the sum of the squares of the distances from the trisection points to the vertex A of the right angle.

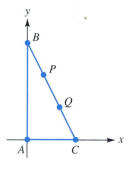

Hint: Let the coordinates of B and C be $(0, 3b)$ and $(3c, 0)$, respectively. Then the coordinates of P and Q are $(c, 2b)$ and $(2c, b)$, respectively.

91. Figure A shows a triangle with sides of lengths s, t, and u and a median of length m. Prove that

$$m^2 = \frac{1}{2}(s^2 + t^2) - \frac{1}{4}u^2$$

Hint: Set up a coordinate system as indicated in Figure B. Then each of the quantities m^2, s^2, t^2, and u^2 can be computed in terms of a, b, and c.

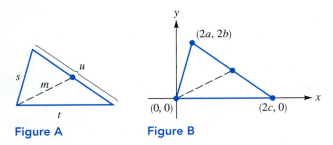

Figure A   **Figure B**

In Exercises 92 and 93, the endpoints of a line segment $\overline{AB}$ are given. Sketch the reflection of $\overline{AB}$ about **(a)** the origin; **(b)** the x-axis; and **(c)** the y-axis.

92. $A(-2, -3)$ and $B(0, 2)$

93. $A(3, 1)$ and $B(3, -2)$

94. In this exercise you'll derive a useful formula for the (perpendicular) distance d from the point (x_0, y_0) to the line $y = mx + b$. The formula is

$$d = \frac{|y_0 - mx_0 - b|}{\sqrt{1 + m^2}}$$

(a) Refer to the figure. Use similar triangles to show that

$$\frac{d}{AB} = \frac{1}{\sqrt{1 + m^2}}$$

Therefore, $d = AB/\sqrt{1 + m^2}$.

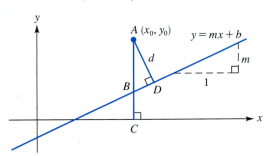

(b) Check that $AB = AC - BC = y_0 - mx_0 - b$.

(c) Conclude from parts (a) and (b) that

$$d = \frac{y_0 - mx_0 - b}{\sqrt{1 + m^2}}$$

For the general case (in which the point and line may not be situated as in our figure), we need to use the absolute value of the quantity in the numerator to ensure that AB and d are nonnegative.

95. Use the formula given in Exercise 94 to find the distance from the point $(1, 2)$ to the line $y = \frac{1}{2}x - 5$.

96. Use the formula given in Exercise 94 to demonstrate that the distance from the point (x_0, y_0) to the line $Ax + By + C = 0$ is

$$d = \frac{|Ax_0 + By_0 + C|}{\sqrt{A^2 + B^2}}$$

97. Use the formula given in Exercise 96 to find the distance from the point $(-1, -3)$ to the line $2x + 3y - 6 = 0$.

98. Find the equation of the circle that has center $(2, 3)$ and is tangent to the line $x + y - 1 = 0$.
Hint: Use the formula given in Exercise 96 to find the radius.

99. In the figure, the circle is tangent to the x-axis, to the y-axis, and to the line $3x + 4y = 12$.

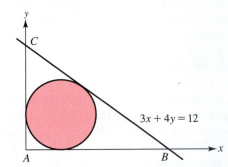

(a) Find the equation of the circle. *Suggestion:* First decide what the relationships must be between h, k, and r in the equation $(x - h)^2 + (y - k)^2 = r^2$. Then find a way to apply the formula given in Exercise 94.

(b) Let S, T, and U denote the points where the circle touches the x-axis, the line $3x + 4y = 12$, and the y-axis, respectively. Find the equation of the line through A and T; through B and U; through S and C.

(c) Where do the line segments $\overline{AT}$ and $\overline{CS}$ intersect? Where do $\overline{AT}$ and $\overline{BU}$ intersect? What do you observe? *Remark:* The point determined here is called the *Gergonne point* of triangle

ABC, so named in honor of its discoverer, French mathematician Joseph Diaz Gergonne (1771–1859).

100. This exercise outlines a proof of the Pythagorean theorem that was discovered by James A. Garfield, the twentieth President of the United States. Garfield published the proof in 1876, when he was the Republican leader in the House of Representatives. We start with a right triangle with legs of length a and b and hypotenuse of length c. We want to prove that $a^2 + b^2 = c^2$.

(a) Take two copies of the given triangle and arrange them as shown in Figure A. Explain why the angle marked θ is a right angle. (θ is the Greek letter "theta.")

(b) Draw the line segment indicated in Figure B. Notice that the outer quadrilateral in Figure B

is a trapezoid. (Two sides are parallel.) The area of this trapezoid can be computed in two distinct ways: using the formula for the area of a trapezoid (given on the inside front cover of this book) or adding the areas of the three right triangles in Figure B. Compute the area of the trapezoid in each of these two ways. When you equate the two answers and simplify, you should obtain $a^2 + b^2 = c^2$.

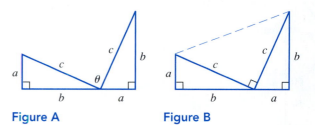

Figure A   **Figure B**

Chapter 1   Test

1. Simplify $|x - 6| + |x - 7|$, where $6 < x < 7$.
2. Specify the intervals that are described by the inequalities.
   (a) $|x - 4| < \frac{1}{10}$ (b) $x \geq 2$

In Problems 3–5, find all the real solutions of each equation.

3. $\dfrac{2}{x + 4} - \dfrac{1}{x - 4} = \dfrac{-7}{x^2 - 16}$

4. $\dfrac{1}{1 - x} + \dfrac{4}{2 - x} = \dfrac{11}{6}$

5. (a) $x^2 + 4x = 5$
   (b) $x^2 + 4x = 1$

6. Solve the following equation for x in terms of the other letters:
$$\frac{ax + b}{cx + d} = e, \text{ where } a \neq ce$$

7. Refer to the following graph. As indicated in the graph, the number N of U.S. radio stations on the air increased in (approximately) a straight-line fashion from 1990 to 1994. Source of data: *The Universal Almanac 1997,* John W. Wright (ed.) (Kansas City: Andrews and McMeel, 1997)

(a) As t increases from 0 to 4, we have $\Delta t = 4$ years. What is the corresponding value of ΔN?
(b) Find the slope $\Delta N/\Delta t$ for the line. Round the answer to the nearest integer. (Include units with your answer.)
(c) Use your result in part (b) to answer the following question. If the number of radio stations had continued to grow in a straight-line fashion, how many stations would there have been in 1995?

8. The following data are from a classic experiment in biology that measured the average height of a group of sunflower plants at 7-day intervals.

Growth in (Mean) Height of Sunflower Plants

Day t	Mean Height h (cm)
7	17.93
21	67.76

Source: H.S. Reed and R.H. Holland in *Proceedings of the National Academy of Sciences,* vol. 5 (1919), p. 140.

(a) Plot the two data points in the following coordinate system.

(b) Use the midpoint formula and the data for $t = 7$ days and $t = 21$ days to estimate what the average height of the sunflowers might have been at $t = 14$ days. (Round the answer to one decimal place.)

(c) Compute the percentage error in the estimation in part (b), given that the actual average height at $t = 14$ days was 36.36 cm. Round the answer to the nearest whole percent. *Check:* You should obtain a substantial percentage error. In Chapter 5 we'll study some types of equations

that do a much better job than the midpoint formula in modeling biological growth.)

In Problems 9 and 10, find an equation for the line satisfying the given conditions. Write the answer in the form $y = mx + b$.

9. Passes through $(1, -2)$ and the y-intercept is 6

10. Passes through $(2, -1)$ and is perpendicular to the line $5x + 6y = 30$

11. Which point is farther from the origin, $(3, 9)$ or $(5, 8)$?

12. Test each equation for symmetry about the x-axis, the y-axis, and the origin:
  (a) $y = x^3 + 5x$
  (b) $y = 3^x + 3^{-x}$
  (c) $y^2 = 5x^2 + x$

In Problems 13 and 14, graph the equations and specify all x- and y-intercepts.

13. $3x - 5y = 15$

14. $(x - 1)^2 + (y + 2)^2 = 9$

Ⓖ **15. (a)** Use a graphing utility to graph the equation $y = x^3 - 2x^2 - 9x$ in an appropriate viewing rectangle.

Ⓖ   **(b)** Use a graphing utility to estimate to one decimal place the x-intercepts of the graph.

    (c) Use algebra to find exact values for the x-intercepts.

16. Does the point $(-1/2, 5)$ lie on the graph of the equation $y = 4x^2 - 8x$?

17. The endpoints of line segment $\overline{AB}$ are $A(3, -1)$ and $B(-1, 2)$. Sketch the reflection of $\overline{AB}$ about **(a)** the x-axis; **(b)** the y-axis; and **(c)** the origin.

18. A line passes through the points $(6, 2)$ and $(1, 4)$. Find the area of the triangle bounded by this line and the x- and y-axes.

CHAPTER 2

Equations and Inequalities

We continue the work begun in Section 1.3 on solving equations, and then we adapt those skills in solving inequalities. Some of the applications that you'll see in this chapter are:

- Tracking population growth in the United States (Example 1 in Section 2.1)
- Predicting world records for the men's and women's 10,000-meter race (Exercises 21 and 22 in Section 2.1)
- Computing interest in savings accounts (Example 4 in Section 2.3)
- Predicting sulfur dioxide emissions for the United States and Asia (Exercises 47 and 48 in Section 2.5) (Sulfur dioxide emissions are the main contributor to acid rain.)

By 1900 B.C. the Babylonians had a well-established algebra. They could solve quadratic equations (positive coefficients and positive solutions only) and some types of higher degree equations as well. —Stuart Hollingdale in *Makers of Mathematics* (London: Penguin Books, 1989)

['Abd-al-Hamid ibn-Turk, a ninth-century Persian mathematician] . . . gives geometric figures to prove if the discriminant is negative, a quadratic equation has no [real] solution. —Carl B. Boyer in *A History of Mathematics,* 2nd ed. (Uta C. Merzbach, revision editor) (New York: John Wiley & Sons, 1991)

2.1 QUADRATIC EQUATIONS: THEORY AND EXAMPLES

A wealth of quadratic equations applied to genetic problems can be found in . . . [the textbook by Albert Jacquard, *The Genetic Structure of Populations*]. —Edward Batschelet, *Introduction to Mathematics for Life Scientists,* 3rd ed. (New York: Springer-Verlag, 1979)

Recall (from Section 1.3, or from a previous course) that a **quadratic equation** is an equation that can be written in the form $ax^2 + bx + c = 0$, where a, b, and c are real numbers and a is not zero. In Section 1.3 we solved quadratic equations in two ways: by factoring and by the quadratic formula. The factoring method is the simpler of the two, but it can't be used in every instance. (Example: Try solving $x^2 - 2x - 4 = 0$ by factoring.) The quadratic formula, on the other hand, can be used to solve *any* quadratic equation. In this section we take a more careful look at the quadratic formula: We derive the formula and we use it, not only to solve equations, but also to *analyze* them. The technique that we'll use to derive the quadratic formula is completing the square (introduced in Section 1.7).

We'll review the technique of completing the square by solving the equation $x^2 - 2x - 4 = 0$. First, we rewrite the equation in the form

$$x^2 - 2x = 4 \qquad (1)$$

with the x-terms isolated on the left-hand side of the equation. To complete the square, we follow these two steps:

Step 1 Take half of the coefficient of x and square it.

Step 2 Add the number obtained in Step 1 to both sides of the equation.

For equation (1), the coefficient of x is -2. Taking half of -2 and then squaring it gives us $(-1)^2$, or 1. Now, as directed in Step 2, we add 1 to both sides of equation (1). This yields

$$x^2 - 2x + 1 = 4 + 1$$
$$(x - 1)^2 = 5$$
$$x - 1 = \pm\sqrt{5}$$
$$x = 1 \pm \sqrt{5}$$

We have now obtained the two solutions, $1 + \sqrt{5}$ and $1 - \sqrt{5}$. Let's check that these numbers are indeed solutions of $x^2 - 2x - 4 = 0$. (We'll show the work for $x = 1 + \sqrt{5}$; you should carry out the calculations for $1 - \sqrt{5}$ on your own.) We have

$$(1 + \sqrt{5})^2 - 2(1 + \sqrt{5}) - 4 = (1 + 2\sqrt{5} + 5) - 2 - 2\sqrt{5} - 4$$
$$= 6 - 2 - 4$$
$$= 0 \quad \text{as required}$$

In the box that follows, we summarize the process of completing the square and we indicate how it got its name.

Algebraic Procedure for Completing the Square in the Expression $x^2 + bx$

Add the square of half of the x-coefficient:

$$(x^2 + bx) + \left(\frac{b}{2}\right)^2 = \left(x + \frac{b}{2}\right)^2$$

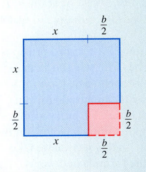

Geometric Interpretation of Completing the Square for $x^2 + bx$

The blue region in the figure represents the quantity $x^2 + bx$ [since the area is

$$x^2 + (b/2)x + (b/2)x]$$

By adding the red square to the blue region, we fill out or "complete" the larger square. The area of the red region that completes the square is

$$\frac{b}{2} \cdot \frac{b}{2} = \frac{b^2}{4}$$

The technique of completing the square can be used to derive the quadratic formula. We start with the general quadratic equation $ax^2 + bx + c = 0$ $(a \neq 0)$ and divide both sides by a (so that the coefficient of x^2 will be 1, as in the box above). This yields

$$x^2 + \frac{b}{a}x + \frac{c}{a} = 0$$

Subtracting c/a from both sides yields

$$x^2 + \frac{b}{a}x = -\frac{c}{a}$$

Now, to complete the square, we add $[\frac{1}{2}(b/a)]^2$, or $b^2/4a^2$, to both sides. That gives us

$$x^2 + \frac{b}{a}x + \frac{b^2}{4a^2} = \frac{b^2}{4a^2} - \frac{c}{a}$$

$$\left(x + \frac{b}{2a}\right)^2 = \frac{b^2 - 4ac}{4a^2}$$

$$x + \frac{b}{2a} = \pm\sqrt{\frac{b^2 - 4ac}{4a^2}} = \pm\frac{\sqrt{b^2 - 4ac}}{2|a|}$$

$$= \pm\frac{\sqrt{b^2 - 4ac}}{2a}$$

This last equality follows from the fact that for any real number $a \neq 0$, the expressions $\pm 2|a|$ and $\pm 2a$ both represent the same two numbers. We now conclude that the solutions are

$$x = -\frac{b}{2a} + \frac{\sqrt{b^2 - 4ac}}{2a} \quad \text{and} \quad x = -\frac{b}{2a} - \frac{\sqrt{b^2 - 4ac}}{2a}$$

For reference, we summarize our work in the box that follows.

The Quadratic Formula

The solutions of the equation $ax^2 + bx + c = 0$ $(a \neq 0)$ are given by

$$x = \frac{-b \pm \sqrt{b^2 - 4ac}}{2a}$$

EXAMPLE 1 Using the quadratic formula

For the years 1850 through 1990, the population of the United States can be closely approximated using the equation

$$y = 0.006609x^2 - 23.771x + 21{,}382 \tag{2}$$

where y is the population in millions, and x is the year.

(a) To get a feeling for working with equation (2), use the equation to estimate the U.S. population in 1900. (Round the answer to one decimal place.)

(b) Use equation (2) and the quadratic formula to estimate the year in which the U.S. population reached 200 million.

SOLUTION

(a) Substituting $x = 1900$ in equation (2) yields

$$y = 0.006609(1900)^2 - 23.771(1900) + 21{,}382$$
$$= 75.6 \text{ million} \qquad \text{using a calculator and rounding to one decimal place}$$

Thus, our estimate for the U.S. population in 1900 is 76.6 million. *Remark*: This estimate is very good, for according to the U.S. Bureau of the Census, the population in 1900 was approximately 76.2 million.

(b) Using equation (2), we want to find the year x in which the population y reached 200 million. Substituting $y = 200$ in equation (2) yields

$$0.006609x^2 - 23.771x + 21{,}382 = 200$$

This is a quadratic equation, and to put it in the form $ax^2 + bx + c = 0$ (so that the quadratic formula can be used) we subtract 200 from both sides to obtain

$$0.006609x^2 - 23.771x + 21{,}182 = 0$$

From this last equation we see that

$$a = 0.006609 \qquad b = -23.771 \qquad c = 21{,}182$$

and consequently

$$x = \frac{-b \pm \sqrt{b^2 - 4ac}}{2a}$$

$$= \frac{-(-23.771) \pm \sqrt{(-23.771)^2 - 4(0.006609)(21{,}182)}}{2(0.006609)}$$

Using a calculator now and rounding to the nearest integer, we obtain the two values $x \approx 1969$ and $x \approx 1628$. We choose the value in the required range between 1850 and 1990. Thus, our estimate is that 1969 was the year in which the U.S. population reached 200 million. *Remark*: This is a reasonable estimate, for according to Census Bureau data, the 1960 population was approximately 179 million, whereas the 1970 population was approximately 203 million.

EXAMPLE 2   Using the quadratic formula

Consider the quadratic equation $x^2 = 5 - 3x$.
(a) Use the quadratic formula to find the roots of the equation.
(b) Compute the product of the roots.
(c) Compute the sum of the roots.

SOLUTION
(a) Just as we did in the previous example, we first write the equation in the form $ax^2 + bx + c = 0$:

$$x^2 + 3x - 5 = 0$$

From this last equation, we see that $a = 1$, $b = 3$, and $c = -5$. The quadratic formula then yields

$$x = \frac{-b \pm \sqrt{b^2 - 4ac}}{2a} = \frac{-3 \pm \sqrt{3^2 - 4(1)(-5)}}{2(1)}$$

$$= \frac{-3 \pm \sqrt{9 + 20}}{2} = \frac{-3 \pm \sqrt{29}}{2}$$

(b) Computing the product of the roots, we have

$$\left(\frac{-3 + \sqrt{29}}{2}\right)\left(\frac{-3 - \sqrt{29}}{2}\right) = \frac{9 - 29}{4} = \frac{-20}{4} = -5$$

So, the product of the two roots is -5.

(c) For the sum of the roots, we have

$$\frac{-3 + \sqrt{29}}{2} + \frac{-3 - \sqrt{29}}{2} = \frac{-3 + \sqrt{29} + (-3) - \sqrt{29}}{2}$$

$$= \frac{-6}{2} = -3$$

Therefore, the sum of the roots is -3.

If you look over the results in Example 2, you'll see that there appears to be something of a coincidence. We found that the product of the roots of the equation $x^2 + 3x - 5 = 0$ is -5, which happens to be the "c" term in the equation. Furthermore, the sum of the roots is -3, which happens to be the negative of the "b" term in the equation. In fact, however, these are not coincidences. Rather, they are examples of the following general result.

THEOREM   The Product and the Sum of the Roots of $x^2 + bx + c = 0$

Let r_1 and r_2 be the roots of the quadratic equation $x^2 + bx + c = 0$. Then

$$r_1 r_2 = c \qquad \text{and} \qquad r_1 + r_2 = -b$$

In words: The product of the roots of $x^2 + bx + c = 0$ is the constant term in the equation, and the sum of the roots is the negative of the coefficient of the x-term.

The proof of this theorem follows exactly the same steps that we used in Example 2. For the equation $x^2 + bx + c = 0$, we have $a = 1$ and therefore the quadratic formula gives us

$$x = \frac{-b \pm \sqrt{b^2 - 4(1)c}}{2(1)} = \frac{-b \pm \sqrt{b^2 - 4c}}{2}$$

So the two roots are $r_1 = \dfrac{-b + \sqrt{b^2 - 4c}}{2}$ and $r_2 = \dfrac{-b - \sqrt{b^2 - 4c}}{2}$. Then, for the product $r_1 r_2$, we have

$$r_1 r_2 = \frac{-b + \sqrt{b^2 - 4c}}{2} \cdot \frac{-b - \sqrt{b^2 - 4c}}{2}$$

$$= \frac{b^2 - (b^2 - 4c)}{4} = \frac{4c}{4} = c \qquad \text{as required}$$

For the sum of the roots, we have

$$r_1 + r_2 = \frac{-b + \sqrt{b^2 - 4c}}{2} + \frac{-b - \sqrt{b^2 - 4c}}{2}$$

$$= \frac{-b + \sqrt{b^2 - 4c} + (-b) - \sqrt{b^2 - 4c}}{2} = \frac{-2b}{2} = -b$$

as we wished to show. The next two examples show how these results are applied.

EXAMPLE 3   Finding the product and sum of the roots of a quadratic equation

Find the product and the sum of the roots of the quadratic equation $2x^2 + 6x - 7 = 0$.

SOLUTION
To apply the theorem on the product and sum of roots, the coefficient of x^2 in the equation must be 1. To arrange this, we divide both sides of the given equation by 2. This yields the equivalent equation

$$x^2 + 3x - \frac{7}{2} = 0$$

In this equation we have $b = 3$ and $c = -7/2$. Therefore

$$r_1 r_2 = c = -\frac{7}{2}$$

and

$$r_1 + r_2 = -b = -3$$

That is, the product of the roots is $-7/3$ and the sum of the roots is -3.

EXAMPLE 4   Finding a quadratic equation with given roots

Find a quadratic equation with roots

$$r_1 = \frac{1 + \sqrt{5}}{4} \qquad \text{and} \qquad r_2 = \frac{1 - \sqrt{5}}{4}$$

Write the equation in a form involving integer coefficients only.

SOLUTION
If r_1 and r_2 are the roots of the equation $x^2 + bx + c = 0$, we have

$$c = r_1 r_2 = \frac{1 + \sqrt{5}}{4} \cdot \frac{1 - \sqrt{5}}{4} = \frac{1 - 5}{16} = -\frac{1}{4}$$

and

$$b = -(r_1 + r_2) = -\left(\frac{1 + \sqrt{5}}{4} + \frac{1 - \sqrt{5}}{4}\right) = -\frac{2}{4} = -\frac{1}{2}$$

With these values for b and c, the quadratic equation is $x^2 - \frac{1}{2}x - \frac{1}{4} = 0$. To obtain an equation with integer coefficients (but still the same roots), we multiply both sides of this last equation by 4 to obtain

$$4x^2 - 2x - 1 = 0$$

This is the equation with the required roots and with integer coefficients.

SUGGESTION   For practice in using the quadratic formula, solve this last equation to verify that the roots are indeed $(1 \pm \sqrt{5})/4$.

In Examples 1 and 2 we found that each quadratic equation had two real roots (although in Example 1 we were interested only in one root). As the next example indicates, however, it is also possible for quadratic equations to have only one real root, or even no real roots.

EXAMPLE 5   Using the quadratic formula

Use the quadratic formula to solve each equation:
(a) $4x^2 - 12x + 9 = 0$;   **(b)** $x^2 + x + 2 = 0$.

SOLUTION
(a) Here $a = 4$, $b = -12$, and $c = 9$, so

$$x = \frac{12 \pm \sqrt{(-12)^2 - 4(4)(9)}}{2(4)} = \frac{12 \pm \sqrt{144 - 144}}{8}$$
$$= \frac{12 \pm 0}{8} = \frac{12}{8} = \frac{3}{2}$$

In this case our only solution is $3/2$. We refer to $3/2$ as a **double root.** *Note:* We've used the quadratic formula here only for an illustration; in this case we can solve the original equation more efficiently by factoring.

(b) Since $a = 1$, $b = 1$, and $c = 2$, we have

$$x = \frac{-1 \pm \sqrt{(1)^2 - 4(1)(2)}}{2(1)} = \frac{-1 \pm \sqrt{-7}}{2}$$

From this we conclude that the given equation has no real-number solutions because the expression $\sqrt{-7}$ is undefined within the real-number system. [However, if we take a broader point of view and work within the *complex-number system,* then there are indeed two solutions, albeit not real-number solutions. As you'll see in Section 7.1, or as you perhaps already know from a previous course, these two complex-number solutions can be written $(-1 \pm i\sqrt{7})/2$.]

The results in Example 5 can be interpreted in terms of graphs. For as we know from Section 1.5, the x-intercepts of the graph of $y = ax^2 + bx + c$ are found by solving the equation $ax^2 + bx + c = 0$. As indicated in Figure 1(a), the graph of $y = 4x^2 - 12x + 9$ has only one x-intercept; this intercept is the root of the equation $4x^2 - 12x + 9 = 0$ that we solved in Example 5(a). Figure 1(b) shows that the graph of $y = x^2 + x + 2$ has no x-intercepts; this reflects the fact that the equation $x^2 + x + 2 = 0$ has no real solutions. For completeness, we've also shown a graph [Figure 1(c)] that has two x-intercepts. We computed these two x-intercepts in Example 2(a) when we solved the equation $x^2 + 3x - 5 = 0$.

GRAPHICAL PERSPECTIVE

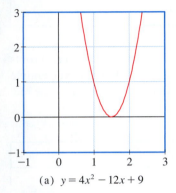

(a) $y = 4x^2 - 12x + 9$

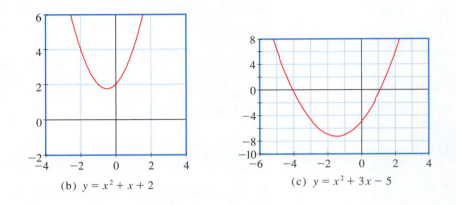

(b) $y = x^2 + x + 2$

(c) $y = x^2 + 3x - 5$

Figure 1
In Figure 1(a), the graph of $y = 4x^2 - 12x + 9$ has only one x-intercept; the equation $4x^2 - 12x + 9 = 0$ has only one root. In Figure 1(b), the graph of $y = x^2 + x + 2$ has no x-intercepts; the equation $x^2 + x + 2 = 0$ has no real roots. In Figure 1(c), the graph of $y = x^2 + 3x - 5$ has two x-intercepts; the equation $x^2 + 3x - 5 = 0$ has two real roots.

The quantity $b^2 - 4ac$ that appears under the radical sign in the quadratic formula is called the *discriminant*. (The term "discriminant" was coined by the nineteenth-century British mathematician James Joseph Sylvester.) If you look back at Example 2(a), you'll see that the discriminant is positive and that this leads to two real solutions. In Example 5(a), the discriminant is zero, which leads to only one real solution. Finally, in Example 5(b), the discriminant is negative and, consequently, there are no real solutions. These observations are generalized in the box that follows.

The Discriminant $b^2 - 4ac$

Consider the quadratic equation $ax^2 + bx + c = 0$, where a, b, and c are real numbers and $a \neq 0$. The expression $b^2 - 4ac$ is called the **discriminant.**

1. If $b^2 - 4ac > 0$, then the equation has two distinct real roots.
2. If $b^2 - 4ac = 0$, then the equation has exactly one real root, referred to as a **double root** or a root of **multiplicity two.**
3. If $b^2 - 4ac < 0$, then the equation has no real root. [There are two (nonreal) complex-number roots. This is discussed in Chapter 7.]

EXAMPLE	6	Using the discriminant

(a) Compute the discriminant to determine how many real solutions there are for the equation $x^2 + x - 1 = 0$.

(b) Find a value for k such that the quadratic equation $x^2 + \sqrt{2}x + k = 0$ has exactly one real solution.

SOLUTION
(a) Here $a = 1$, $b = 1$, and $c = -1$. Therefore,

$$b^2 - 4ac = 1^2 - 4(1)(-1) = 5$$

Since the discriminant is positive, the equation has two distinct real solutions.

(b) For the equation to have exactly one real solution, the discriminant must be zero; that is,

$$b^2 - 4ac = 0$$
$$(\sqrt{2})^2 - 4(1)(k) = 0$$
$$2 - 4k = 0$$
$$k = 1/2$$

The required value for k is $1/2$.

EXERCISE SET 2.1

A

In Exercises 1–18, solve the quadratic equations. If an equation has no real roots, state this. In cases where the solutions involve radicals, give both the radical form of the answer and a calculator approximation rounded to two decimal places.

1. $x^2 + 8x - 2 = 0$
2. $x^2 - 6x - 2 = 0$
3. $x^2 + 4x + 1 = 0$
4. $x^2 + 12x + 18 = 0$
5. $2y^2 - 5y - 2 = 0$
6. $3y^2 - 3y - 4 = 0$
7. $4y^2 + 8y + 5 = 0$
8. $y^2 + y + 1 = 0$
9. $4s^2 - 20s + 25 = 0$
10. $16s^2 + 8s + 1 = 0$
11. $x^2 = 8x - 6$
12. $x^2 = 8x + 9$
13. $-3x^2 + x = -3$
14. $(x - 5)(x + 3) = 1$
15. $-y^2 - 8y = 1$
16. $5y(y - 2) = 2$
17. $t^2 = -3t - 4$
18. $t^2 = -3t + 4$

19. **(a)** Use the formula given in Example 1 to estimate the year that the U.S. population reached 100 million.
    (b) Use the following U.S. Census Bureau data to say whether your estimate in part (a) is in the ballpark. In 1910 the population was approximately 92.2 million; in 1920 it was approximately 106.0 million.

20. **(a)** Use the formula given in Example 1 to estimate the year that the U.S. population reached 225 million.
    (b) According to the U.S. Census Bureau, the U.S. population in 1980 was 226.5 million. Is your estimate in part (a) in the ballpark?

21. The chart that follows shows the world records for the men's 10,000 meter run in the years 1972 and 1998.

Year	Time (minutes and seconds)	Runner
1972	27:38.4	Lasse Viren (Finland)
1998	26:22.75	Haile Gebrselassie (Ethiopia)

Over the years between 1972 and 1998, the world record can be approximated by the equation

$$y = -0.09781t^2 + 385.8336t - 378,850.4046$$
$$(1972 \le t \le 1998)$$

where y is the world-record time (in seconds) in the year t.

(a) Use the given equation, the quadratic formula, and your calculator to estimate the year in which the record might have been 27 minutes ($= 1620$ seconds). (You'll get two solutions for the quadratic; be sure to pick the appropriate one.) Then say by how many years your prediction is off, given the following information: On July 5, 1993, Richard Chelimo of Kenya ran a record time of 27:07.91; five days after that, Yobes Ondieki, also of Kenya, ran 26:58.38.

(b) Estimate the year in which the record might have been 26:30 ($= 1590$ seconds). Then say by how many years your prediction is off, given the following information: On July 4, 1997, Haile Gebrselassie ran a record time of 27:07.91; eighteen days later, Paul Tergat of Kenya ran 26:27.85.

22. The chart below shows the world records for the women's 10,000 meter run in the years 1970 and 1993.

Year	Time (minutes and seconds)	Runner
1970	35:30.5	Paola Pigni (Italy)
1993	29:31.78	Wang Junxia (China)

Over the years between 1970 and 1993, the world record can be approximated by the equation

$$y = 0.37553t^2 - 1503.7154t + 1,507,042.699$$
$$(1970 \le t \le 1993)$$

where y is the world-record time (in seconds) in the year t. Use the given equation, the quadratic formula, and your calculator to estimate the year in which the record might have been 32 minutes ($= 1920$ seconds). (As in Exercise 21, you'll get two solutions for the quadratic; pick the appropriate one.) Then say by how many years your prediction is off, given the following information: On September 19, 1981, Yekena Sipatova of the former Soviet Union ran a record time of 32:17.2; in the following year on July 16, Mary Decker of the United States ran 31:35.3.

Ⓖ *In each of Exercises 23–28, you are given an equation of the form* $y = ax^2 + bx + c$. **(a)** *Use a graphing utility to graph the equation and to estimate the x-intercepts. (Use a zoom-in process to obtain the estimates; keep zooming in until the first three decimal places of the estimate remain the same as you progress to the next step.)* **(b)** *Determine the exact values of the intercepts by using the quadratic formula. Then use a calculator to evaluate the expressions that you obtain. Round off the results to four decimal places. [Check to see that your results are consistent with the estimates in part (a).]*

23. $y = x^2 - 4x + 1$ **24.** $y = x^2 - 10x + 15$
25. $y = 0.5x^2 + 8x - 3$ **26.** $y = 2x^2 + 2x - 1$
27. $y = 2x^2 + 2\sqrt{26}\,x + 13$ **28.** $y = 3x^2 - 12\sqrt{3}\,x + 36$

In Exercises 29–32, find the sum and the product of the roots of each quadratic equation.

29. $x^2 + 8x - 20 = 0$ **30.** $x^2 - 3x + 12 = 0$
31. $4y^2 - 28y + 9 = 0$ **32.** $\frac{1}{2}y^2 = 4y - 5$

In Exercises 33–38, find a quadratic equation with the given roots r_1 and r_2. Write each answer in the form $ax^2 + bx + c = 0$, where a, b, and c are integers and $a > 0$.

33. $r_1 = 3$ and $r_2 = 11$ **34.** $r_1 = -4$ and $r_2 = -9$
35. $r_1 = 1 - \sqrt{2}$ and $r_2 = 1 + \sqrt{2}$
36. $r_1 = 2 - \sqrt{5}$ and $r_2 = 2 + \sqrt{5}$
37. $r_1 = \frac{1}{2}(2 + \sqrt{5})$ and $r_2 = \frac{1}{2}(2 - \sqrt{5})$
38. $r_1 = \frac{1}{3}(4 - \sqrt{5})$ and $r_2 = \frac{1}{3}(4 + \sqrt{5})$

In Exercises 39 and 40, solve the equations.
Hint: Look before you leap.

39. $x^2 - (\sqrt{2} + \sqrt{5})x + \sqrt{10} = 0$
40. $x^2 + (\sqrt{2} - 1)x = \sqrt{2}$

41. A ball is thrown straight upward. Suppose that the height of the ball at time t is given by the formula $h = -16t^2 + 96t$, where h is in feet and t is in seconds, with $t = 0$ corresponding to the instant that the ball is first tossed.
  (a) How long does it take before the ball lands?
  (b) At what time is the height 80 ft? Why does this question have two answers?

42. During a flu epidemic in a small town, a public health official finds that the total number of people P who have caught the flu after t days is closely approximated by the formula $P = -t^2 + 26t + 106$, where $1 \le t \le 13$.
  (a) How many have caught the flu after 10 days?
  (b) After approximately how many days will 250 people have caught the flu?

In Exercises 43–50, use the discriminant to determine how many real roots each equation has.

43. $x^2 - 12x + 16 = 0$ **44.** $2x^2 - 6x + 5 = 0$
45. $4x^2 - 5x - \frac{1}{2} = 0$ **46.** $4x^2 - 28x + 49 = 0$
47. $x^2 + \sqrt{3}\,x + \frac{3}{4} = 0$ **48.** $\sqrt{2}\,x^2 + \sqrt{3}\,x + 1 = 0$
49. $y^2 - \sqrt{5}\,y = -1$ **50.** $\dfrac{m^2}{4} - \dfrac{4m}{3} + \dfrac{16}{9} = 0$

In each of Exercises 51–54, find the value(s) of k such that the equation has exactly one real root.

51. $x^2 + 12x + k = 0$ **52.** $3x^2 + (\sqrt{2k})x + 6 = 0$
53. $x^2 + kx + 5 = 0$ **54.** $kx^2 + kx + 1 = 0$

B

In Exercises 55–58, solve for the indicated letter.

55. $2\pi r^2 + 2\pi rh = 20\pi$; for r
  Hint: Rewrite the equation as $(2\pi)r^2 + (2\pi h)r - 20\pi = 0$ and use the quadratic formula with $a = 2\pi$, $b = 2\pi h$, and $c = -20\pi$.

56. $2\pi y^2 + \pi yx = 12$; for y
57. $-16t^2 + v_0 t = 0$; for t
58. $-\frac{1}{2}gt^2 + v_0 t + h_0 = 0$; for t

Ⓖ **59. (a)** On the same set of axes, graph the equations $y = x^2 + 8x + 16$ and $y = x^2 - 8x + 16$.
  (b) Use the graphs to estimate the roots of the two equations $x^2 + 8x + 16 = 0$ and $x^2 - 8x + 16 = 0$. How do the roots appear to be related?
  (c) Solve the two equations in part (b) to determine the exact values of the roots. Do your results support the response you gave to the question at the end of part (b)?

Ⓖ **60. (a)** Figure 1(c) in the text shows a graph of the equation $y = x^2 + 3x - 5$. Use a graphing utility to reproduce the graph. [Use the same viewing rectangle that is used in Figure 1(c).]
  (b) Add the graph of the equation $y = x^2 - 3x - 5$ to the picture that you obtained in part (a). (This new equation is the same as the one in part (a) except that the sign of the coefficient of x has been reversed.) Note that the x-intercepts of the two graphs appear to be negatives of one another.
  (c) Use the quadratic formula to determine exact expressions for the roots of the two equations $x^2 + 3x - 5 = 0$ and $x^2 - 3x - 5 = 0$. You'll find that the roots of one equation are the opposites of the roots of the other equation. [In general, the graphs of the two equations $y = ax^2 + bx + c$ and $y = ax^2 - bx + c$ are symmetric about the y-axis. Thus, the x-intercepts (when they exist) will always be opposites of one another.]

61. (a) Use the quadratic formula to show that the roots of the equation $x^2 + 3x + 1 = 0$ are $\frac{1}{2}(-3 \pm \sqrt{5})$.
  (b) Show that $\frac{1}{2}(-3 + \sqrt{5}) = 1/[\frac{1}{2}(-3 - \sqrt{5})]$.
  Hint: Rationalize the denominator on the right-hand side of the equation.
  (c) The result in part (b) shows that the roots of the equation $x^2 + 3x + 1 = 0$ are reciprocals. Can you find another, much simpler way to establish this fact?

62. If r_1 and r_2 are the roots of the quadratic equation $ax^2 + bx + c = 0$, show that $r_1 + r_2 = -b/a$ and $r_1 r_2 = c/a$.

63. Show that the quadratic equation

$$ax^2 + bx - a = 0 \qquad (a \ne 0)$$

has two distinct real roots.

64. Show that the quadratic equation

$$(x - p)(x - q) = r^2 \qquad (p \neq q)$$

has two distinct real roots.

In Exercises 65 and 66, determine the value(s) of the constant k for which the equation has equal roots (that is, only one distinct root).

65. $x^2 = 2x(3k + 1) - 7(2k + 3)$

66. $x^2 + 2(k + 1)x + k^2 = 0$

67. Here is an outline for a slightly different derivation of the quadratic formula. The advantage of this method is that fractions are avoided until the very last step. Fill in the details.

(a) Beginning with $ax^2 + bx = -c$, multiply both sides by $4a$. Then add b^2 to both sides.

(b) Now factor the resulting left-hand side and take square roots.

68. In this section and in Section 1.3, we solved quadratic equations by factoring and by using the quadratic formula. This exercise shows how to solve a quadratic equation by the **method of substitution.** As an example, we use the equation

$$x^2 + x - 1 = 0 \qquad (1)$$

(a) In equation (1), make the substitution $x = y + k$. Show that the resulting equation can be written

$$y^2 + (2k + 1)y = 1 - k - k^2 \qquad (2)$$

(b) Find a value for k so that the coefficient of y in equation (2) is 0. Then, using this value of k, show that equation (2) becomes $y^2 = 5/4$.

(c) Solve the equation $y^2 = 5/4$. Then use the equation $x = y + k$ to obtain the solutions of equation (1).

69. Use the substitution method (explained in Exercise 68) to solve the quadratic equation $2x^2 - 3x + 1 = 0$.

70. Assume that a and b are the roots of the equation $x^2 + px + q = 0$.

(a) Find the value of $a^2b + ab^2$ in terms of p and q.
Hint: Factor the expression $a^2b + ab^2$.

(b) Find the value of $a^3b + ab^3$ in terms of p and q.
Hint: Factor. Then use the fact that $a^2 + b^2 = (a + b)^2 - 2ab$.

Ⓖ **71.** In this exercise we investigate the effect of the constant c upon the roots of the quadratic equation $x^2 - 6x + c = 0$. We do this by looking at the x-intercepts of the graphs of the corresponding equations $y = x^2 - 6x + c$.

(a) Set a viewing rectangle that extends from 0 to 5 in the x-direction and from -2 to 3 in the y-direction. Then (on the same set of axes) graph the equations $y = x^2 - 6x + c$ with c running from 8 to 10 at increments of 0.25. In other words, graph the equations $y = x^2 - 6x + 8, y = x^2 - 6x + 8.25, y = x^2 - 6x + 8.50$, and so on, up through $y = x^2 - 6x + 10$.

(b) Note from the graphs in part (a) that, initially, as c increases, the x-intercepts draw closer and closer together. For which value of c do the two x-intercepts seem to merge into one?

(c) Use algebra as follows to check your observation in part (b). Using that value of c for which there appears to be only one intercept, solve the quadratic equation $x^2 - 6x + c = 0$. How many roots do you obtain?

(d) Some of the graphs in part (a) have no x-intercepts. What are the corresponding values of c in these cases? Pick any one of these values of c and use the quadratic formula to solve the equation $x^2 - 6x + c = 0$. What happens?

C

72. Find nonzero real numbers A and B so that the roots of the equation $x^2 + Ax + B = 0$ are A and B.

PROJECT ▎ PUT THE QUADRATIC EQUATION IN ITS PLACE!

This is easier as a group project (divide and conquer), but of course, is open to individuals too.

Despite the critical influence of science and mathematics in today's culture, many high school history books tell very little about the history of these fields. In mathematics, for example, it's easy to develop the false impression that mathematics is strictly a "Western" achievement, developed rather exclusively in Greece (Pythagoras and Euclid), in France (René Descartes) and in England (Isaac Newton). A mere glance, however, at the following list of a few mathematicians (from the period 0 to A.D. 1400) will tell you otherwise. (All right, one Greek name does appear in the list!) The common element linking the names is that, among other mathematical achievements, each person solved quadratic equations. By checking math history books or encyclopedias in your college library, locate these names on the time line that follows. The World Wide Web is also a useful source here. *Some hints:* The listing here is in no particular order. In some cases, you'll find that only very approximate dates are known. Also, be prepared to see numerous alternate spellings.

Omar Khayyam   Abraham bar Hiyya (or Chiya)
Mahāvira   Jia Xian
Bhāskara   Muhammad al-Khwārizmi
Diophantus   Āryabhata
Brahmagupta   Joranus de Nemore

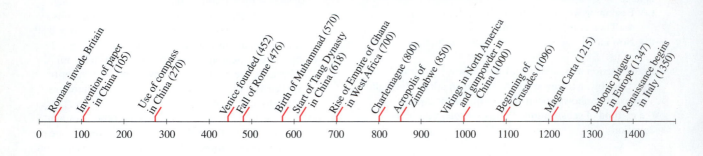

2.2 OTHER TYPES OF EQUATIONS

In this chapter we propose to consider some miscellaneous equations; it will be seen that many of these can be solved by the ordinary rules for quadratic equations. . . .—H. S. Hall and S. R. Knight, *Higher Algebra* [London: Macmillan and Co., 1946 (first edition published in 1887)]

In Sections 1.3 and 2.1, we developed techniques for solving linear and quadratic equations. In this section we consider some other types of equations that can be solved by applying those techniques. As in the previous section, we will consider only solutions that are real numbers. (Solutions involving the complex-number system are discussed in Chapter 7.) We begin with two examples involving absolute values.

| EXAMPLE | 1 | Solving an equation where the variable is inside absolute value signs |

Solve: $|2x - 1| = 7$.

SOLUTION
There are two cases to consider.

If $2x - 1 \geq 0$, the equation becomes

$$2x - 1 = 7$$
$$2x = 8$$
$$x = 4$$

If $2x - 1 < 0$, the equation becomes

$$-(2x - 1) = 7$$
$$-2x + 1 = 7$$
$$-2x = 6$$
$$x = -3$$

We've now obtained the values $x = 4$ and $x = -3$. You should check for yourself that both of these numbers indeed satisfy the original equation.

The solution in Example 1 proceeds by considering two separate cases. There is another way to solve the equation in Example 1 that avoids this two-case technique. This alternate method relies on the following basic property of the absolute value (from Section 1.2):

For any real number a, we have $|a|^2 = a^2$.

To apply this property in solving the equation $|2x - 1| = 7$, we square both sides of the equation to obtain

$$|2x - 1|^2 = (2x - 1)^2 = 49$$
$$4x^2 - 4x - 48 = 0$$
$$x^2 - x - 12 = 0$$

This last equation is a quadratic equation that can be solved by factoring (as in Section 1.3). We have

$$(x - 4)(x + 3) = 0$$

$$
\begin{array}{c|c}
x - 4 = 0 & x + 3 = 0 \\
x = 4 & x = -3
\end{array}
$$

This yields the two roots $x = 4$ and $x = -3$, in agreement with the result in Example 1.

CAUTION   We obtained these two roots by squaring both sides of the original equation. However, it is not always the case that squaring both sides yields roots that check in the original equation. Consider, for example, the equation $|x| = -1$. This equation has no solution (why?) yet squaring both sides yields $x^2 = 1$, with roots $x = \pm 1$. However, neither 1 nor -1 checks in the original equation $|x| = -1$. The roots 1 and -1 are extraneous roots. We'll return to this point again, later in this section.

The solutions in Example 1 can be checked graphically. One way to do this is first to rewrite the given equation $|2x - 1| = 7$ in the equivalent form $|2x - 1| - 7 = 0$. Then, as we've seen in previous sections, the roots of this last equation are the x-intercepts of the graph of

$$y = |2x - 1| - 7$$

As indicated in Figure 1(a), the x-intercepts of the graph appear to be -3 and 4, which are indeed the two roots obtained in Example 1. A different way to check the solutions of the equation $|2x - 1| = 7$ involves graphing the *two* equations $y = |2x - 1|$ and $y = 7$ on the same set of axes. As indicated in Figure 1(b), the graphs intersect in two points. The x-coordinates of those two points appear to be -3 and 4, which again are the two roots determined in Example 1.

CAUTION   Some care needs to be taken in drawing conclusions from a graph, even a graph obtained with a computer or graphing calculator. For instance, taken

GRAPHICAL PERSPECTIVE

Figure 1

Two ways to check the solutions in Example 1. In Figure 1(a), the x-intercepts of the graph of $y = |2x - 1| - 7$ are the solutions of the equation $|2x - 1| = 7$. In Figure 1(b), the x-coordinates of the intersection points of the two graphs $y = |2x - 1|$ and $y = 7$ are the solutions of the equation $|2x - 1| = 7$.

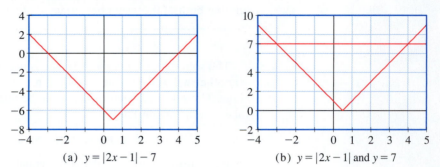

(a) $y = |2x - 1| - 7$ (b) $y = |2x - 1|$ and $y = 7$

in isolation, the graphs in Figure 1 do not *prove* that the roots we are looking for are -3 and 4. Indeed, the only honest conclusion from the graphs themselves is that the two roots we are looking for must be very close to -3 and 4. The only way to tell if the roots are exactly -3 and 4 is to use an algebraic method, as in Example 1, or to substitute each purported root in the given equation to see if it checks. Figure 2 serves to emphasize this point. In Figure 2(a) we are looking at a portion of the curve $y = x^2 - 2.2x$ and the line $y = 0.8$ in order to find a root of the equation $x^2 - 2.2x = 0.8$. Based on the evidence in Figure 2(a), it appears that $x = 2.5$ may be a root. But, if you replace x with 2.5 in the equation $x^2 - 2.2x = 0.8$, you'll see that it does *not* check; 2.5 is not a root. Indeed, as indicated in the magnified view in Figure 2(b), the graphs of $y = x^2 - 2.2x$ and $y = 0.8$ do not intersect at $x = 2.5$. That is, 2.5 is not a root. *Note*: Figure 2(b) does more than just tell us that 2.5 is not a root. It tells us that the root lies in the open interval $(2.5, 2.55)$, so certainly the digit in the first decimal place of the root is a 5. More zooming in with a graphing calculator or a computer would yield additional decimal places.

When n is a natural number, equations of the form $x^n = a$ are solved simply by rewriting the equation in terms of the appropriate nth root or roots. For example, if $x^3 = 34$, then $x = \sqrt[3]{34}$. If the exponent n is even, however, we have to remember that there are *two* real nth roots for each positive number. For instance, if $x^4 = 81$, then $x = \pm\sqrt[4]{81} = \pm 3$. *Note*: If you feel that you need to review the subject of nth roots from intermediate algebra, see Appendix B.2 at the back of this book.

GRAPHICAL PERSPECTIVE

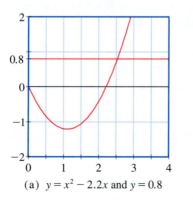

(a) $y = x^2 - 2.2x$ and $y = 0.8$

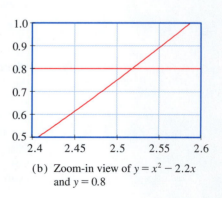

(b) Zoom-in view of $y = x^2 - 2.2x$ and $y = 0.8$

Figure 2

Looking for a root of the equation $x^2 - 2.2x = 0.8$.

EXAMPLE 2   Solving equations using nth roots

Solve:
(a) $(x - 1)^4 = 15$;   **(b)** $3x^4 = -48$;   **(c)** $3x^5 = -48$.

SOLUTION

(a) $(x - 1)^4 = 15$

$$x - 1 = \pm\sqrt[4]{15}$$

$$x = 1 \pm \sqrt[4]{15}$$

As you can check for yourself now, both of the values, $1 + \sqrt[4]{15}$ and $1 - \sqrt[4]{15}$, satisfy the given equation.

Remark: We can estimate these values without a calculator. For instance, for $x = 1 + \sqrt[4]{15}$ we have

$$1 + \sqrt[4]{15} \approx 1 + \sqrt[4]{16} = 1 + 2 = 3$$

Therefore, $1 + \sqrt[4]{15}$ is a little less than 3. (In fact, a calculator shows that $1 + \sqrt[4]{15} \approx 2.97$.)

(b) For any real number x, the quantity x^4 is nonnegative. Therefore, the left-hand side of the equation is nonnegative, whereas the right-hand side is negative. Consequently, there are no real numbers satisfying the given equation.

(c) Dividing both sides of the given equation by 3 yields $x^5 = -16$, and therefore

$$x = \sqrt[5]{-16} \quad \text{or equivalently} \quad x = -\sqrt[5]{16}$$

Calculator note: To approximate $-\sqrt[5]{16}$ using a calculator, first recall that $\sqrt[5]{16}$ can be rewritten in the equivalent form $16^{1/5}$. (See Appendix B.3 if you need a review of fraction exponents.) We then have

$$x = -\sqrt[5]{16} = -16^{1/5} \approx -1.74 \qquad \text{using a calculator and rounding to two decimal places}$$

In Example 3 we solve two equations by factoring. As with quadratic equations, the factoring method is justified by the zero-product property of real numbers (from Section 1.3).

EXAMPLE 3   Using factoring to solve equations

Find the real-number solutions for each equation:
(a) $3x^3 - 12x^2 - 15x = 0$;   **(b)** $x^4 + x^2 - 6 = 0$.

SOLUTION

(a) First we factor the equation

$$3x^3 - 12x^2 - 15x = 0$$

$$3x(x^2 - 4x - 5) = 0$$

$$3x(x - 5)(x + 1) = 0$$

Now, by setting each factor equal to zero, we obtain the three values $x = 0$, $x = 5$, and $x = -1$. As you can check, each of these numbers indeed satisfies the original equation. So we have three solutions: 0, 5, and -1.

NOTE If initially we had divided both sides of the given equation by $3x$ to obtain $x^2 - 4x - 5 = 0$, then the solution $x = 0$ would have been overlooked.

(b) Again we factor the equation

$$x^4 + x^2 - 6 = 0$$
$$(x^2 - 2)(x^2 + 3) = 0$$

$x^2 - 2 = 0$	$x^2 + 3 = 0$
$x^2 = 2$	$x^2 = -3$
$x = \pm\sqrt{2}$	

By setting the first factor equal to zero, we've obtained the two values $x = \pm\sqrt{2}$. As you can check, these values do satisfy the given equation. When we set the second factor equal to zero, however, we obtained $x^2 = -3$, which has no real solutions (because the square of a real number must be nonnegative). We conclude therefore that the given equation in this case has only two real solutions, $\sqrt{2}$ and $-\sqrt{2}$.

The equation $x^4 + x^2 - 6 = 0$ that we considered in Example 3(b) is called **an equation of quadratic type.** This means that with a suitable substitution, the resulting equation becomes quadratic. In particular here, suppose that we let

$$x^2 = t \qquad \text{and therefore} \qquad x^4 = t^2$$

Then the equation $x^4 + x^2 - 6 = 0$ becomes $t^2 + t - 6 = 0$, which is quadratic. This idea is exploited in the next three examples.

EXAMPLE **4** **Solving an equation of quadratic type**

As shown in Figure 3, the graph of $y = x^4 - 6x^2 + 4$ has four x-intercepts. Determine these x-intercepts. Give two forms for each answer: an exact expression involving radicals and a calculator approximation rounded to two decimal places.

SOLUTION
As explained in Section 1.5, the x-intercepts are obtained by setting y equal to zero in the given equation and then solving for x. So, we need to solve the equation

$$x^4 - 6x^2 + 4 = 0 \tag{1}$$

This equation cannot readily be solved by factoring; however, it is of quadratic type. We let $x^2 = t$; then $x^4 = t^2$, and the equation becomes

$$t^2 - 6t + 4 = 0$$

The quadratic formula then yields

$$t = \frac{6 \pm \sqrt{36 - 4(4)}}{2} = \frac{6 \pm \sqrt{20}}{2}$$

$$= \frac{6 \pm 2\sqrt{5}}{2} = 3 \pm \sqrt{5}$$

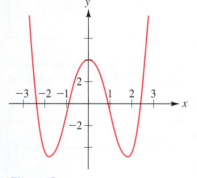

Figure 3
$y = x^4 - 6x^2 + 4$

Consequently, we have (in view of the definition of t)

$$x^2 = 3 \pm \sqrt{5}$$

and therefore

$$x = \pm\sqrt{3 \pm \sqrt{5}}$$

We've now found four values for x:

$$\sqrt{3 + \sqrt{5}} \qquad -\sqrt{3 + \sqrt{5}} \qquad \sqrt{3 - \sqrt{5}} \qquad -\sqrt{3 - \sqrt{5}}$$

These are the required x-intercepts for the graph in Figure 3. [Exercise 93 asks you to check that these values indeed satisfy equation (1).] Which is which, though? Without using a calculator, but with a little thought, you should be able to list these four quantities in order of increasing size and thereby assign them to the appropriate locations in Figure 3. (Try it before reading on!) Using a calculator now, and referring to Figure 3, we obtain the following results.

the x-intercept between -3 and -2: $-\sqrt{3 + \sqrt{5}} \approx -2.29$

the x-intercept between -1 and 0: $-\sqrt{3 - \sqrt{5}} \approx -0.87$

the x-intercept between 0 and 1: $\sqrt{3 - \sqrt{5}} \approx 0.87$

the x-intercept between 2 and 3: $\sqrt{3 + \sqrt{5}} \approx 2.29.$

EXAMPLE 5   Solving an equation of quadratic type

Solve: $6x^{-2} - x^{-1} - 2 = 0$.

SOLUTION
We'll show two methods. The first method depends on recognizing the equation as one of quadratic type.

FIRST METHOD
Let $x^{-1} = t$. Then $x^{-2} = t^2$ and the equation becomes

$$6t^2 - t - 2 = 0$$
$$(3t - 2)(2t + 1) = 0$$

$3t - 2 = 0 \qquad\mid\qquad 2t + 1 = 0$

$t = 2/3 \qquad\quad\mid\qquad\quad t = -1/2$

If $t = 2/3$, then $x^{-1} = 2/3$. Therefore, $\dfrac{1}{x} = \dfrac{2}{3}$ and, consequently, $x = 3/2$.

On the other hand, if $t = -1/2$, then $\dfrac{1}{x} = -\dfrac{1}{2}$ and, consequently, $x = -2$.

In summary, the two solutions are $3/2$ and -2.

ALTERNATIVE METHOD
Multiplying both sides of the given equation by x^2 yields

$$6 - x - 2x^2 = 0$$
$$2x^2 + x - 6 = 0$$
$$(2x - 3)(x + 2) = 0$$

$2x - 3 = 0 \qquad\mid\qquad x + 2 = 0$

$x = 3/2 \qquad\quad\mid\qquad\quad x = -2$

Thus we have $x = 3/2$ or $x = -2$, as we obtained using the first method.

In the next example we first use a graphing utility to approximate the solutions of an equation, then we use algebra to determine the solutions exactly. Note, however, that even in the graphing utility portion of the problem, algebra is useful. It turns out that the graph is symmetric about the y-axis, and that cuts the work in half.

EXAMPLE **6** **Solving an equation with fraction exponents**

Consider the equation $2x^{4/3} - x^{2/3} - 6 = 0$.
(a) Use a graphing utility to approximate the roots to the nearest hundredth.
(b) Use algebra to determine the exact solutions. Then check that the answers are consistent with the approximations obtained in part (a).

SOLUTION
(a) The roots of the given equation are the x-intercepts of the graph of $y = 2x^{4/3} - x^{2/3} - 6$. After experimenting with different viewing rectangles, we find that the standard viewing rectangle in Figure 4(a) does show the essential features of the graph.

GRAPHICAL PERSPECTIVE

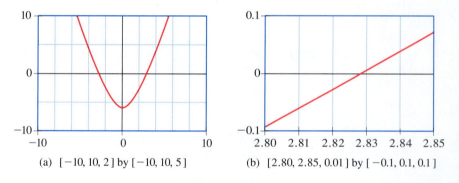

Figure 4
Two views of $y = 2x^{4/3} - x^{2/3} - 6$

(a) $[-10, 10, 2]$ by $[-10, 10, 5]$

(b) $[2.80, 2.85, 0.01]$ by $[-0.1, 0.1, 0.1]$

Looking at Figure 4(a), it appears that there are two roots, one between 2 and 4, the other between -2 and -4. It also appears that the graph may be symmetric about the y-axis. If so, once we find the first root between 2 and 4, then we'll automatically know the second root: it will just be the opposite of the first. To show that the graph is indeed symmetric about the y-axis, we can rewrite the equation $y = 2x^{4/3} - x^{2/3} - 6$ as $y = 2(x^4)^{1/3} - (x^2)^{1/3} - 6$. In this form, because of the even integer exponents on x, replacing x with $-x$ will make no difference. So, according to the second symmetry test on page 64, the graph is symmetric about the y-axis. In Figure 4(b), we've zoomed in on the positive root. Since we are supposed to approximate the root to the nearest hundredth, we've zoomed in far enough to allow us to mark off the x-scale in units of 0.01. Figure 4(b) shows that the positive root is 2.83, to the nearest hundredth. By symmetry, the other root then is -2.83, also to the nearest hundredth. (We note in passing that many graphing utilities also have "solve" keys for approximating roots of equations. For details, consult your owner's manual.)

(b) In the given equation $2x^{4/3} - x^{2/3} - 6 = 0$, we let $x^{2/3} = t$. Then $x^{4/3} = t^2$, and the equation becomes

$$2t^2 - t - 6 = 0$$
$$(2t + 3)(t - 2) = 0$$

$$2t + 3 = 0 \qquad\mid\qquad t - 2 = 0$$
$$t = -3/2 \qquad\mid\qquad t = 2$$

We now have two cases to consider: $t = -3/2$ and $t = 2$. With $t = -3/2$, we have (in view of the definition of t)

$$x^{2/3} = -3/2 \qquad \text{or equivalently} \qquad (x^{1/3})^2 = -3/2$$

This last equation has no real-number solutions, because the square of a real number is never negative. Turning our attention then to the other case where $t = 2$, we have

$$x^{2/3} = 2$$
$$x^2 = 8 \qquad\qquad \color{red}{\text{Two real numbers are equal if}}$$
$$\color{red}{\text{and only if their cubes are equal.}}$$
$$x = \pm\sqrt{8} = \pm 2\sqrt{2}$$

As Exercise 94 asks you to verify, each of the values $2\sqrt{2}$ and $-2\sqrt{2}$ satisfies the original equation. These are the required roots. As you can check now by picking up a calculator, we have $2\sqrt{2} \approx 2.828$, which is indeed consistent with the approximations in part (a).

Some equations can be solved by raising both sides to the same power. In fact, if you review the solution given for Example 6, you'll see that we've already made use of this idea in cubing both sides of one of the equations there. As another example, consider the equation

$$\sqrt{x - 3} = 5$$

By squaring both sides of this equation we obtain $x - 3 = 25$ and, consequently, $x = 28$. As you can easily check, the value $x = 28$ does satisfy the original equation.

There is a complication, however, that can arise in raising both sides of an equation to the same power. Consider, for example, the equation

$$x - 2 = \sqrt{x}$$

By squaring both sides, we obtain

$$x^2 - 4x + 4 = x$$
$$x^2 - 5x + 4 = 0$$
$$(x - 4)(x - 1) = 0$$

Therefore

$$x = 4 \qquad \text{or} \qquad x = 1$$

The value $x = 4$ checks in the original equation, but the value $x = 1$ does not. (Verify this.) Therefore, $x = 1$ is an extraneous solution, and the only solution of the given equation is $x = 4$.

The extraneous solution in this case arises from the fact that if $a^2 = b^2$, then it need not be true that $a = b$. For instance, $(-3)^2 = 3^2$, but certainly $-3 \neq 3$. In the box that follows, we summarize this remark about extraneous solutions. For reference, we also include the results from Section 1.3.

▌ PROPERTY SUMMARY   Extraneous Solutions

1. Squaring both sides of an equation (or raising both sides to an even integral power) may introduce extraneous solutions that do not check in the original equation.
2. Multiplying sides of an equation by an expression involving the variable may introduce extraneous solutions.

Therefore it is always necessary to check any candidates for solutions that you obtain in either of these ways.

In the next example we solve equations involving radicals. In each case, notice that *before* squaring, we first isolate one of the radicals on one side of the equation. Otherwise things become more complicated, rather than less. For instance, in Example 7(a) we're given the equation $2 + \sqrt{10 - x} = -x$. If we square both sides without first isolating the radical, we obtain

$$(2 + \sqrt{10 - x})^2 = (-x)^2$$

and consequently

$$4 + 4\sqrt{10 - x} + 10 - x = x^2 \qquad \text{or} \qquad 14 + 4\sqrt{10 - x} - x = x^2$$

As you can see, this last equation is indeed more complicated than the original.

 EXAMPLE | **7** | **Solving equations with radicals**

Find all real-number solutions of each equation.

(a) $2 + \sqrt{10 - x} = -x$ **(b)** $\sqrt{2x + 3} - 2\sqrt{x - 2} = 1$

SOLUTION

(a) First isolate the radical by rewriting the given equation as

$$\sqrt{10 - x} = -x - 2$$

Squaring both sides now yields

$$(\sqrt{10 - x})^2 = (-x - 2)^2$$
$$10 - x = x^2 + 4x + 4$$
$$0 = x^2 + 5x - 6$$
$$0 = (x + 6)(x - 1)$$

$$x + 6 = 0 \qquad \Big| \qquad x - 1 = 0$$
$$x = -6 \qquad \Big| \qquad x = 1$$

Check for extraneous roots: With $x = -6$, the *original* equation becomes

$$2 + \sqrt{10 - (-6)} = -(-6)$$
$$2 + 4 = 6 \qquad \text{True}$$

With $x = 1$, the original equation becomes

$$2 + \sqrt{10 - 1} = -1$$
$$2 + 3 = -1 \qquad \text{False}$$

In summary, the given equation has but one root: $x = -6$.

(b) The given equation contains two radicals, so initially the best that we can do is to isolate one of them before squaring. (It does not matter which.) Rewriting the given equation as $\sqrt{2x + 3} = 2\sqrt{x - 2} + 1$ and then squaring yields

$$(\sqrt{2x + 3})^2 = (2\sqrt{x - 2} + 1)^2$$
$$2x + 3 = 4(x - 2) + 4\sqrt{x - 2} + 1$$

You should check now for yourself that if we combine like terms in this last equation, then it can be rewritten as

$$-2x + 10 = 4\sqrt{x - 2}$$

Next, don't square both sides quite yet! It won't be wrong, but you can work with smaller numbers if you first divide both sides by 2 to obtain

$$-x + 5 = 2\sqrt{x - 2}$$

Now square both sides:

$$(-x + 5)^2 = (2\sqrt{x - 2})^2$$
$$x^2 - 10x + 25 = 4x - 8$$
$$x^2 - 14x + 33 = 0$$
$$(x - 3)(x - 11) = 0$$

From this last equation we conclude that $x = 3$ or $x = 11$.
Check for extraneous roots: With $x = 3$, the *original* equation becomes

$$\sqrt{2(3) + 3} - 2\sqrt{3 - 2} = 1$$
$$3 - 2 = 1 \qquad \text{True}$$

Next, with $x = 11$, the original equation becomes

$$\sqrt{2(11) + 3} - 2\sqrt{11 - 2} = 1$$
$$5 - 2(3) = 1 \qquad \text{False}$$

So the only solution of the given equation is $x = 3$. For a graphical perspective, see Figure 5.

GRAPHICAL PERSPECTIVE

Figure 5
The curve $y = \sqrt{2x + 3} - 2\sqrt{x - 3}$ meets the horizontal line $y = 1$ at $x = 3$. That is, $x = 3$ is a solution of the equation $\sqrt{2x + 3} - 2\sqrt{x - 3} = 1$.

EXERCISE SET 2.2

A

In Exercises 1–6, solve each equation.

1. $|x - 5| = 1$
2. $|x - 4| - 5 = 2$
3. $|x + 6| = 1/2$
4. $|x + 6| + 1/2 = 0$
5. $|6x - 5| = 25$
6. $|5 - 6x| = 0$

In Exercises 7–10, solve each equation by using the squaring technique that is discussed following Example 1 in the text.

7. $|x + 3| = 2x - 2$
8. $4|x - 2| = 3x - 4$
9. $|2x - 1| = 1 - \frac{1}{2}x$
10. $|x + 1| - |3x - 2| = 0$

G **11.** **(a)** On the same set of axes, graph the equations $y = x^2 - 2.2x$ and $y = 0.8$. Use the same viewing rectangle that is shown in Figure 2(a) in the text. Check that your graphs are consistent with those drawn in Figure 2(a). The graphs show that the line $y = 0.8$ intersects the curve $y = x^2 - 2.2x$ at a point with an x-coordinate very close to 2.5. In other words, one of the roots of the equation $x^2 - 2.2x = 0.8$ is approximately 2.5.

(b) Take a closer look at the intersection point using the same viewing rectangle that is shown in Figure 2(b) in the text. As pointed out in the text, this view shows that the root we are looking for is actually slightly larger than 2.5; it lies in the open interval (2.5, 2.55). Thus, the first decimal place of the root must indeed be 5, and the second decimal place must be a digit between 0 and 4, inclusive. Continue to zoom in on this intersection point until you are sure of the first three decimal places of the root.

(c) Use the quadratic formula to find an expression for the exact value of the root in part (b). Then evaluate the expression and round to four decimal places. Check that your answer is consistent with the result obtained graphically in part (b).

G **12.** **(a)** Graph the equation $y = 3x^3 - 12x^2 - 15x$.

(b) Based on the graph, how many roots are there for the equation $3x^3 - 12x^2 - 15x = 0$? Use the graph to estimate these roots. Then compare your estimates to the actual values obtained in Example 3(a).

G **13.** **(a)** Graph the equation $y = x^4 + x^2 - 6$.

(b) Based on the graph, how many roots are there for the equation $x^4 + x^2 - 6 = 0$? Use the graph to estimate these roots. Zoom in until you are sure about the first three decimal places of each root. Then compare your estimates to the actual values obtained in Example 3(b).

G **14.** **(a)** Using a viewing rectangle that extends from -5 to 5 in both the x- and y-directions, graph the equation $y = 6x^{-2} - x^{-1} - 2$. Use the graph to estimate the roots of the equation $6x^{-2} - x^{-1} - 2 = 0$. How do your estimates compare to the actual values determined in Example 5?

(b) Using the viewing rectangle specified in part (a), graph the equation $y = 6x^{-2} - x^{-1} + 2$. Based on the graph, make a statement about the roots of the equation $6x^{-2} - x^{-1} + 2 = 0$. Then solve the equation algebraically to confirm your statement.

In Exercises 15–64, find all real solutions of each equation. For Exercises 31–36, give two forms for each answer: an exact answer (involving a radical) and a calculator approximation rounded to two decimal places.

15. $3x^2 - 48x = 0$
16. $3x^3 - 48x = 0$
17. $t^3 - 125 = 0$
18. $t - t^3 = 0$
19. $7x^4 - 28x^2 = 0$
20. $y^4 - 81 = 0$
21. $t^4 + 2t^3 - 3t^2 = 0$
22. $2t^5 + 5t^4 - 12t^3 = 0$
23. $6x = 23x^2 + 4x^3$
24. $x^5 = 36x$
25. $x^4 - x^2 = 6$
26. $x^4 - 5x^2 = -6$
27. $4y^2 = 5 - y^4$
28. $6y^2 = -5 - y^4$
29. $3t^2 + 2 = 9t^4$
30. $5t^2 - 1 = 4t^4$
31. $(x - 2)^3 - 5 = 0$
32. $(x + 4)^3 + 2 = 0$
33. $(x + 4)^5 + 16 = 0$
34. $(1 - x)^5 - 40 = 0$
35. **(a)** $(x - 3)^4 - 30 = 0$
    (b) $(x - 3)^4 + 30 = 0$
36. **(a)** $(x^2 - 1)^4 - 81 = 0$
    (b) $(x^2 - 1)^4 + 81 = 0$
37. $y^4 + 4y^2 - 5 = 0$
38. $y^4 + 6y^2 + 5 = 0$
39. $9t^4 - 3t^2 - 2 = 0$
40. $4t^4 - 5t^2 + 1 = 0$
41. $x^6 - 10x^4 + 24x^2 = 0$
42. $2x^5 - 15x^3 - 27x = 0$
43. $x^4 + x^2 - 1 = 0$
44. $x^4 - x^2 + 1 = 0$
45. $x^4 + 3x^2 - 2 = 0$
46. $2x^4 + x^2 + 2 = 0$
47. $x^6 + 7x^3 = 8$
48. $x^6 + 27 = -28x^3$
49. $t^{-2} - 7t^{-1} + 12 = 0$
50. $8t^{-2} - 17t^{-1} + 2 = 0$
51. $12y^{-2} - 23y^{-1} = -5$
52. $y^{-2} - y^{-1} = 0$
53. $4x^{-4} - 33x^{-2} - 27 = 0$
54. $x^{-4} + x^{-2} + 1 = 0$
55. $t^{2/3} = 9$
56. $t^{3/2} = 8$
57. $x^4 = 81$
58. $x^4 = -81$
59. $(y - 1)^3 = 7$
60. $(2y + 3)^4 = 5$
61. $(t + 1)^5 = -243$
62. $(t + 3)^4 = 625$
63. $9x^{4/3} - 10x^{2/3} + 1 = 0$
64. $x^{4/3} + 3x^{2/3} - 28 = 0$

In Exercises 65 and 66, find the x-intercepts of the graphs. Round each answer to two decimal places. Check to see that your answers are consistent with the graphs.

65.

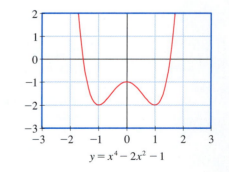

$y = x^4 - 2x^2 - 1$

66.

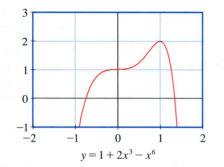

$$y = 1 + 2x^3 - x^6$$

In Exercises 67–82, determine all of the real-number solutions for each equation. (Remember to check for extraneous solutions.)

67. $\sqrt{1 - 3x} = 2$

68. $\sqrt{x^2 + 5x - 2} = 2$

69. $\sqrt{x + 6} = x$

70. $x - \sqrt{x} = 20$

71. $x - \sqrt{3 - x} = -3$

72. $\sqrt{2 - x} - 10 = x$

73. $4x + \sqrt{2x + 5} = 0$

74. $\sqrt{7 - 3x} - 6(x + 1) = 1$

75. $\sqrt{x^4 - 13x^2 + 37} = 1$

76. $\sqrt{y + 2} = y - 4$

77. $\sqrt{1 - 2x} + \sqrt{x + 5} = 4$

78. $\sqrt{x - 5} - \sqrt{x + 4} + 1 = 0$

79. $\sqrt{3 + 2t} + \sqrt{-1 + 4t} = 1$

80. $\sqrt{2t + 5} - \sqrt{8t + 25} + \sqrt{2t + 8} = 0$

81. $\sqrt{2y - 3} - \sqrt{3y + 3} + \sqrt{3y - 2} = 0$

82. $\sqrt{a - x} + \sqrt{b - x} = \sqrt{a + b - 2x}$, where $b > a > 0$

Ⓖ In Exercises 83–92, **(a)** use zoom-in techniques to estimate the roots of each equation to the nearest hundredth, as in Example 6; and **(b)** use algebraic techniques to determine an exact expression

for each root, then evaluate the expression and round to four decimal places. Check to see that your answers are consistent with the graphical results obtained in part (a).

83. (a) $x^4 - 5x^2 = -6.2$

    (b) $x^4 - 5x^2 = -6.3$

84. (a) $x^4 - 5x^2 = -25/4$

    (b) $x^4 - 5x^2 = -13/2$

85. $x - 5\sqrt{x} = -3$

86. $x^{-2} - 5x^{-1} = -3$

87. $\sqrt{2x - 1} - \sqrt{x - 2} = 2$

88. $\sqrt{2x - 1} - \sqrt{x - 2} = 1$

89. $\dfrac{\sqrt{x} - 4}{\sqrt{x} + 3} + \dfrac{1}{2} = 0$

90. $\dfrac{\sqrt[4]{x} - 4}{\sqrt[4]{x} + 3} + \dfrac{1}{2} = 0$

91. $x^{2/3} - x^{1/3} = 1$

92. $x^{-2/3} - x^{-1/3} = 1$

93. Verify that the four solutions obtained in Example 4 indeed satisfy the original equation.

94. Verify that the values $x = \pm 2\sqrt{2}$ obtained in Example 6(b) satisfy the equation $2x^{4/3} - x^{2/3} - 6 = 0$.

B

In Exercises 95–98, solve each equation.

95. $\sqrt{\sqrt{x} + \sqrt{a}} + \sqrt{\sqrt{x} - \sqrt{a}} = \sqrt{2\sqrt{x} + 2\sqrt{b}}$

96. $\sqrt{x^2 - x - 1} - \dfrac{2}{\sqrt{x^2 - x - 1}} = 1$

    Hint: Let $t = x^2 - x - 1$.

97. $\sqrt{\dfrac{x - a}{x}} + 4\sqrt{\dfrac{x}{x - a}} = 5$, where $a \neq 0$

    Hint: Let $t = \dfrac{x - a}{x}$; then $\dfrac{1}{t} = \dfrac{x}{x - a}$.

98. $\sqrt{x^2 + 3x - 4} - \sqrt{x^2 - 5x + 4} = x - 1$, where $x > 4$

    Hint: Factor the expressions beneath the radicals. Then note that $\sqrt{x - 1}$ is a factor of both sides of the equation.

MINI PROJECT | FLYING THE FLAG

According to Clint Brookhart in his book *Go Figure!* (Lincolnwood, Illinois: Contemporary Books, 1998),

> *A flag the area of A in square feet, furled in a wind with a velocity of v in miles per hour, exerts a force of F in pounds on the flagpole equal to:* $F = 0.0003Av^{1.9}$.

(a) Solve the given equation for the velocity v in terms of the force F.

(b) According to the *Encyclopedia Americana,* "A 5-foot by 6-foot 9-inch U.S. flag flown in bad weather [at a U.S. military post] is known as a storm flag." Suppose that the flagpole for a storm flag can withstand a force of 40 pounds before it breaks away from the wall to which it is attached. Use the result in part (a) to determine the wind speed that produces a 40-pound force on the flagpole. Round your answer to the nearest integer. (Actually, in a more realistic model, the length of the flagpole, not just the force on it, would have to be taken into account.)

PROJECT

SPECIFIC OR GENERAL? WHATEVER WORKS!

Sometimes in math and science, when we're stuck on a problem, we can make progress by turning from the specific to the general. This approach sometimes surprises students because of a mistaken belief that the general or more abstract case is always tougher than the specific. The specific equation that you'll work to solve in this project is

$$\sqrt{4 - \sqrt{4 + x}} - x = 0 \tag{1}$$

(a) Use a graphing utility to graph the equation $y = \sqrt{4 - \sqrt{4 + x}} - x$. Then, by zooming in on the x-intercept, approximate the root of equation (1) to the nearest hundredth.

(b) Now let's try to find the exact value for the root. By using the process of isolating the radical and then squaring, show that equation (1) eventually becomes

$$x^4 - 8x^2 - x + 12 = 0$$

(c) At this point we are stuck; the equation obtained in part (b) can't be solved using any of the elementary factoring techniques from intermediate algebra. Here is where we shift our focus to a slightly more general problem. Instead of equation (1), we consider

$$\sqrt{k - \sqrt{k + x}} - x = 0 \tag{2}$$

where k is a constant. Use the process of isolating the radical and then squaring to show that equation (2) eventually becomes

$$x^4 - 2kx^2 - x + k^2 - k = 0 \tag{3}$$

(d) Although equation (3) is a fourth-degree equation in x, it is only a quadratic equation in k. That is, equation (3) can be written

$$k^2 - (2x^2 + 1)k + (x^4 - x) = 0 \qquad \text{Verify that this is equivalent to equation (3).}$$

Use the quadratic formula to solve this last equation for k in terms of x. You should obtain (after simplifying) two solutions:

$$k = x^2 + x + 1 \qquad \text{or} \qquad k = x^2 - 1$$

(e) Finally now, in both of the equations obtained in part (d), replace k with 4 [that's what it is in equation (1)], and solve each equation for x.

(f) Three of the four values that you obtain in part (e) are extraneous solutions for equation (1). Use your approximation in part (a) and a calculator to decide which value is the actual root of equation (1).

2.3   APPLIED PROBLEMS (A WARM-UP)

Algebra is a language which does not consist of words but of symbols. If we are familiar with it we can translate into it appropriate sentences of everyday life. — George Polya in *Mathematics of Discovery* (New York: John Wiley & Sons, 1981)

In this section and the next, we practice setting up and solving equations that describe some simple situations. In solving the equations that arise, we'll be using many of the techniques developed in Sections 1.3, 2.1, and 2.2.

EXAMPLE	1	A number problem

Follow parts (a) through (e) to solve the following problem:

> The difference of two positive numbers is 2, and the sum of their squares is 6. What are the two numbers?

(a) What is the problem asking for? Assign variables to denote the required quantities.
(b) Restate the problem using algebraic notation.
(c) Find an equation that can be solved for one of the variables (or "unknowns") introduced in part (a).
(d) Solve the equation obtained in part (c).
(e) Use the root or roots obtained in part (d) to answer the original question. Check that your answer does satisfy the conditions of the problem.

SOLUTION

(a) We first read and reread the problem until we can say just what the problem requires us to find. (By focusing on this, we can initially, at least, skim over many of the details that sometimes seem so overwhelming in an applied problem. Often these details are best handled *after* the variables have been assigned.)

The problem asks us to find **two numbers.** Let x and y denote the two numbers.

(b) See Table 1. [The format in Table 1 is due to the mathematician and master teacher George Polya (1887–1985).]

TABLE 1   Restate the Problem

In English	In algebraic notation
Find two numbers	x and y
Their difference is 2	$x - y = 2$
The sum of their squares is 6	$x^2 + y^2 = 6$

(c) From part (b) we have the equations

$$x - y = 2 \tag{1}$$
$$x^2 + y^2 = 6 \tag{2}$$

Each of these equations involves both x and y. To obtain an equation involving only one variable, we first solve equation (1), the simpler of the two equations, for y in terms of x. [Alternately, we could solve equation (1) for x in terms of y.] From equation (1) we have

$$y = x - 2 \qquad\qquad (3)$$

and therefore

$$x^2 + (x - 2)^2 = 6 \qquad \text{replacing } y \text{ with } x - 2 \text{ in equation (2)}$$

This is an equation involving only one of the variables, as required. However, looking ahead to solving the equation, we carry out the indicated multiplication and then collect like terms. As you can readily check, this yields

$$2x^2 - 4x - 2 = 0$$

and therefore

$$x^2 - 2x - 1 = 0 \qquad\qquad (4)$$

(d) Equation (4) is a quadratic equation. Exercise 25 asks you to solve it (using the quadratic formula) to obtain the two roots $x = 1 \pm \sqrt{2}$.

(e) From part (d) we have two roots, $1 - \sqrt{2}$ and $1 + \sqrt{2}$; these are the two possibilities for x. However, we can discard the value $x = 1 - \sqrt{2}$ because the original problem asks for positive numbers. Using the positive root $x = 1 + \sqrt{2}$ then gives us

$$y = (1 + \sqrt{2}) - 2 \qquad \text{substituting } x = 1 + \sqrt{2} \text{ in equation (3)}$$

$$= -1 + \sqrt{2}$$

Now we have two numbers, $x = 1 + \sqrt{2}$ and $y = -1 + \sqrt{2}$. Do they check in the original problem? For the difference $x - y$ we have

$$x - y = (1 + \sqrt{2}) - (-1 + \sqrt{2}) = 2, \text{ as required}$$

For the sum of the squares, we have

$$x^2 + y^2 = (1 + \sqrt{2})^2 + (-1 + \sqrt{2})^2 = (1 + 2\sqrt{2} + 2) + (1 - 2\sqrt{2} + 2)$$
$$= 6, \text{ as required}$$

In summary, the two numbers are $1 + \sqrt{2}$ and $-1 + \sqrt{2}$.

Comments on the Solution in Example 1.   We were given that the two numbers have a difference of 2, and initially we denoted the numbers by x and y. However, in view of equation (3), we could have begun by calling the two numbers x and $x - 2$, thereby eliminating the need to introduce a second variable. Or, we could have denoted the two numbers by x and $x + 2$. Exercise 26 asks you to check that these approaches both lead to the same results we obtained in Example 1. *Question*: Why would it be incorrect to denote the two numbers by x and $2 - x$?

EXAMPLE   2   A graphing utility approach for the previous example

Use a graphing utility to rework Example 1, approximating the solutions to the nearest tenth.

SOLUTION

Even with a graphing utility at our disposal, parts (a) and (b) are the same here as in Example 1: Calling the required numbers x and y, we obtain the two equations $x - y = 2$ and $x^2 + y^2 = 6$. Now we're ready for the geometric or graphical approach.

Figure 1(a) shows the graphs of these equations, a line and a circle (as we knew ahead of time from our work in Sections 1.6 and 1.7). Recall that before typing in the equations in the graphing utility, we solve for y in terms of x. Thus, we enter the equation $x - y = 2$ in its equivalent form $y = x - 2$; the equation $x^2 + y^2 = 6$ is entered as the pair of equations $y = \pm\sqrt{6 - x^2}$. (See Section 1.7 if this seems unfamiliar to you.)

GRAPHICAL PERSPECTIVE

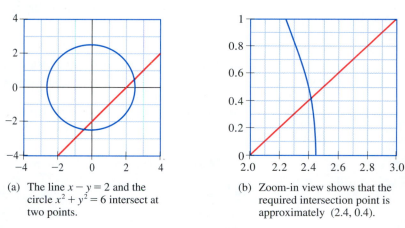

(a) The line $x - y = 2$ and the circle $x^2 + y^2 = 6$ intersect at two points.

(b) Zoom-in view shows that the required intersection point is approximately $(2.4, 0.4)$.

Figure 1
A graphical approach to the problem posed in Example 1: Find two positive numbers x and y with a difference of 2 and such that the sum of their squares is 6.

Figure 1(a) shows that there are two points where the graphs intersect. We want to focus on the intersection point (x, y) in quadrant I; at this point the coordinates x and y satisfy *both* of the required equations. (The coordinates of the intersection point in quadrant III also satisfy the two equations, but the initial problem asks for *positive*, not negative, numbers.) In Figure 1(b), the zoom-in view indicates that, to the nearest tenth, the intersection point is

$$x \approx 2.4, \qquad y \approx 0.4$$

Subsequent zoom-ins could be used to obtain more decimal places in these approximations, but they would still be just that, *approximations*. (On graphing calculators, the TRACE key could also be used with the ZOOM key in obtaining closer approximations; see the owner's manual for your calculator.)

Before concluding this example, let's check that the approximations we've obtained are consistent with the actual solutions in Example 1. Using a calculator to evaluate the results from Example 1, we have

$$x = 1 + \sqrt{2} \approx 2.414 \qquad \text{and} \qquad y = -1 + \sqrt{2} \approx 0.414$$

So, yes, our approximations $x \approx 2.4$ and $y \approx 0.4$ are consistent with these values; they do agree to the nearest tenth.

EXAMPLE **3**   **A problem about a rectangle**

Follow parts (a) through (e) to solve the following problem:

The perimeter of a rectangle is 26 cm and the area is 40 cm². What are the dimensions of the rectangle?

(a) What is the problem asking for? Draw a sketch of the situation, and assign variables to denote the required quantities.
(b) Restate the problem using algebraic notation.
(c) Find an equation that can be solved for one of the variables introduced in part (a).
(d) Solve the equation obtained in part (c).
(e) Use the root or roots obtained in part (d) to answer the original question. Check that your answer does in fact satisfy the conditions of the problem.

SOLUTION
(a) The problem asks for the dimensions (i.e., the length and the width) of a rectangle. In Figure 2 we've sketched a rectangle and labeled the length l and the width w.
(b) See Table 2. Before going on to part (c), note that the equation $2w + 2l = 26$ in Table 2 will be simpler if we divide both sides by 2 to obtain $w + l = 13$.

w

l

area = 40 cm²
perimeter = 26 cm

Figure 2

TABLE 2   Restate the Problem

In English	In algebraic notation
Find the width and length of a rectangle	w and l
The perimeter is 26 cm	$2w + 2l = 26$
The area is 40 cm²	$lw = 40$

(c) From part (b) we have two equations:

$$w + l = 13 \qquad (5)$$
$$lw = 40 \qquad (6)$$

Each of these equations involves both w and l. We want an equation involving only one of the variables. From equation (5), we have

$$l = 13 - w \qquad (7)$$

Using this last equation to substitute for l in equation (6) gives us

$$(13 - w)w = 40$$

or

$$w^2 - 13w + 40 = 0 \qquad \text{Check the algebra.} \qquad (8)$$

This equation involves only one variable, as required. *Question:* Suppose instead that we had solved equation (6) for l and then substituted in equation (5). Would this lead to the same final equation? (Try it.)

(d) Equation (8) can be solved by factoring:

$$(w - 5)(w - 8) = 0$$

$$\begin{array}{c|c} w - 5 = 0 & w - 8 = 0 \\ w = 5 & w = 8 \end{array}$$

(e) From part (d) we have two roots, 5 and 8. If $w = 5$ cm, then equation (7) gives us $l = 13 - 5 = 8$ cm. On the other hand, using $w = 8$ cm, equation (7) tells us $l = 13 - 8 = 5$ cm. Thus, in both cases, we reach the same conclusion: The dimensions of the rectangle are

$$5 \text{ cm by } 8 \text{ cm}$$

CHECK   The perimeter is 2(5 cm) + 2(8 cm) = 26 cm, as required. The area is (5 cm)(8 cm) = 40 cm^2, also as required.

The next example involves *simple interest* earned on money in a savings account. As an illustration of simple interest, suppose that you place $100 in a savings account for one year at 5% simple interest. This means that at the end of the year, the bank contributes 5% of $100, or $5, to your account. Furthermore, assuming that you make no subsequent withdrawals or deposits, the bank continues to contribute $5 to your account at the end of each year. The following general formula can be derived for computing simple interest.

Simple Interest Formula $I = Prt$

where I is the simple interest earned on an initial deposit P (the *principal*) after t years at an annual rate r.

To use this formula, we need to convert the annual rate r from a percent to a decimal. For instance, to compute the simple interest earned on a principal of $1000 invested for six years at 5%, we write

$$I = Prt = (\$1000)(0.05)(6) = \$300$$

EXAMPLE   4   Money in the bank: simple interest

Follow parts (a) through (d) to solve the following problem.

Suppose that you open two savings accounts. In the first account you deposit $5000 at an annual simple interest rate of 4%. In the second you deposit $10,000 at 6% (also simple interest). After how many years will the total interest amount to $5600?

(a) What is the problem asking for? Assign a variable to denote the required quantity.
(b) Find an equation that can be solved for the variable introduced in part (a). (Don't simplify or solve the equation yet.)
(c) Solve the equation obtained in part (b).
(d) Use the root or roots obtained in part (c) to answer the original question. Check your answer.

SOLUTION

(a) The problem asks for the number of years required until the interest from both accounts adds up to $5600. Let t denote the required number of years.

(b) We apply the formula $I = Prt$ to each account. After t years, the interest on the first account is $(5000)(0.04)t$, while that on the second is $(10,000)(0.06)t$. Since we want the sum of these two amounts to be $5600, the required equation is

$$(5000)(0.04)t + (10,000)(0.06)t = 5600$$

(c) To solve this last equation, we start by carrying out the indicated multiplication on the left-hand side. This gives us

$$200t + 600t = 5600$$
$$800t = 5600$$
$$t = \frac{5600}{800} = 7$$

(d) After seven years, the total interest will amount to $5600.

CHECK After seven years, the interest from the first account will be ($5000)(0.04)(7), which is $1400. The interest from the second account after seven years will be ($10,000)(0.06)(7), which is $4200. So the total interest will be the sum of $1400 and $4200, which is $5600, as required. For a graphical check of the solution, see Figure 3. In the figure, we've plotted the (linear) interest equation

$$I = (5000)(0.04)t + (10,000)(0.06)t$$
$$= 800t$$

in a t-I coordinate system. The graph illustrates our result that, after seven years, the interest will be $5600.

GRAPHICAL PERSPECTIVE

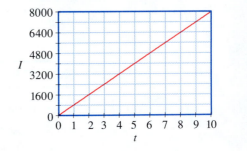

Figure 3
A graph of the interest equation $I = 800t$ indicates that after seven years, the interest is $5600.

In the next example we solve a problem in coordinate geometry. As background, you need to recall the following two facts about coordinates and graphs from Chapter 1:

1. If a point P in the x-y plane lies on the graph of an equation, then the coordinates of P satisfy that equation.
2. The distance d between two points (x_1, y_1) and (x_2, y_2) is given by

$$d = \sqrt{(x_2 - x_1)^2 + (y_2 - y_1)^2}$$

EXAMPLE	5	Setting up and solving a problem in coordinate geometry

Follow parts (a) through (e) to solve the following problem.

The point (a, b) lies in the third quadrant on the graph of the line $y = \frac{1}{2}x + 1$. Find a and b given that the distance from the point (a, b) to the origin is 13 units.

(a) What is the problem asking for?
(b) Restate the problem using algebraic notation.
(c) Find an equation that can be solved for one of the unknowns.
(d) Solve the equation obtained in part (c).
(e) Use the root or roots obtained in part (d) to answer the original question. Check that your answer does satisfy the conditions of the problem.

SOLUTION
(a) The problem asks for two numbers a and b that are the coordinates of a certain point.
(b) See Table 3.

TABLE 3   Restate the Problem

In English	In algebraic notation
Find two numbers	a and b
(a, b) is in the third quadrant	$a < 0$ and $b < 0$
(a, b) is on the graph of $y = \frac{1}{2}x + 1$	$b = \frac{1}{2}a + 1$
The distance from (a, b) to $(0, 0)$ is 13	$\sqrt{(a - 0)^2 + (b - 0)^2} = 13$

(c) From part (b) we have the two equations

$$b = \tfrac{1}{2}a + 1 \tag{9}$$

and

$$\sqrt{a^2 + b^2} = 13 \tag{10}$$

One way to obtain an equation involving only one unknown is to go to equation (10) and substitute for b using the result from equation (9). This yields

$$\sqrt{a^2 + (\tfrac{1}{2}a + 1)^2} = 13 \tag{11}$$

or

$$a^2 + (\tfrac{1}{2}a + 1)^2 = 169 \qquad \text{squaring both sides of} \tag{12}$$
$$\text{the previous equation}$$

(d) Equation (12) is a quadratic equation that can be solved for a. Exercise 27 asks you to do three things: first, to simplify the equation and to show that it becomes $5a^2 + 4a - 672 = 0$; second, to use the quadratic formula to obtain the roots $a = -12$ or $a = 56/5$; and third, to show that these values are not extraneous roots, that they both check in equation (11).

(e) As indicated in part (d), we have two roots, $a = -12$ and $a = 56/5$. Which do we choose? The point (a, b) is in the third quadrant, and therefore a must be negative. Consequently, we choose $a = -12$. To find b now, we can substitute in equation (9) to obtain $b = \frac{1}{2}(-12) + 1 = -5$. In summary, then, the values for a and b are

$$a = -12 \qquad \text{and} \qquad b = -5$$

As Exercise 28 asks you to check, these values for a and b do satisfy all of the required conditions: $(-12, -5)$ lies in the third quadrant on the graph of $y = \frac{1}{2}x + 1$, and the distance from $(-12, -5)$ to the origin is 13 units.

EXERCISE SET 2.3

A

1. Follow parts (a) through (e) to solve the following problem.

> The sum of two numbers is -9, and the sum of their squares is 153. What are the two numbers?

(a) What is the problem asking for? Assign variables to denote the required quantities.

(b) Complete the following table, as in Example 1.

In English	In algebraic notation
Find two numbers	_____
Their sum is -9	_____
The sum of their squares is 153	_____

(c) Find and simplify an equation that can be solved for one of the variables or "unknowns" introduced in part (a).

(d) Solve the equation obtained in part (c).

(e) Use the root or roots obtained in part (d) to answer the original question. Check that your answer does satisfy the conditions of the problem.

2. Follow parts (a) through (e) to solve the following problem.

> Twice the sum of two numbers is 4, and the sum of their squares is 52. What are the two numbers?

(a) What is the problem asking for? Assign variables to denote the required quantities.

(b) Complete the following table, as in Example 1.

In English	In algebraic notation
Find two numbers	_____
Twice their sum is 4	_____
The sum of their squares is 52	_____

(c) Find and simplify an equation that can be solved for one of the variables or "unknowns" introduced in part (a).

(d) Solve the equation obtained in part (c).

(e) Use the root or roots obtained in part (d) to answer the original question. Check that your answer does satisfy the conditions of the problem.

3. Follow parts (a) through (e) to solve the following problem.

> Find two numbers with a sum of 17 and a product of 52.

(a) What is the problem asking for? Assign variables to denote the required quantities.

(b) Complete the following table, as in Example 1.

In English	In algebraic notation
Find two numbers	_____
Their sum is 17	_____
Their product is 52	_____

(c) Find and simplify an equation that can be solved for one of the variables or "unknowns" introduced in part (a).

(d) Solve the equation obtained in part (c).

(e) Use the root or roots obtained in part (d) to answer the original question. Check that your answer does satisfy the conditions of the problem.

4. Follow parts (a) through (e) to solve the following problem.

> The difference of two numbers is 20 and their product is -75. Find the two numbers.

(a) What is the problem asking for? Assign variables to denote the required quantities.

(b) Complete the following table, as in Example 1.

In English	In algebraic notation
Find two numbers	_____
Their difference is 20	_____
Their product is -75	_____

(c) Find and simplify an equation that can be solved for the larger of the two numbers. *Hint*: If x and y are two numbers and $x - y$ is positive, then x is the larger of the two numbers.

(d) Solve the equation obtained in part (c).

(e) Use the root or roots obtained in part (d) to answer the original question. Check that your answer does satisfy the conditions of the problem.

In Exercises 5 and 6, you are asked to find two consecutive integers (for example, 17 and 18, or −3 and −2) satisfying certain conditions. A pair of consecutive integers can be denoted by x and $x + 1$; two variables are not required.

5. Follow parts (a) through (e) to solve the following problem.

The sum of the squares of two consecutive positive integers is 365. Find the two integers.

(a) What is the problem asking for? Assign variables or expressions to denote the required quantities.

(b) Complete the following table.

In English	In algebraic notation
Find two consecutive positive integers	
The sum of their squares is 365	

(c) Find and simplify an equation that can be solved for the smaller of the two integers.

(d) Solve the equation obtained in part (c).

(e) Use the root or roots obtained in part (d) to answer the original question. Check that your answer does satisfy the conditions of the problem.

6. Follow parts (a) through (e) to solve the following problem.

Find two consecutive positive integers such that the sum of the square of the smaller and three times the larger is 111.

(a) What is the problem asking for? Assign variables or expressions to denote the required quantities.

(b) Complete the following table.

In English	In algebraic notation
Find two consecutive positive integers	
The square of the smaller one	
Three times the larger one	
The sum of the square of the smaller and three times the larger is 111	

(c) Find and simplify an equation that can be solved for one of the required integers.

(d) Solve the equation obtained in part (c).

(e) Use the root or roots obtained in part (d) to answer the original question. Check that your answer does satisfy the conditions of the problem.

7. Follow parts (a) through (e) to solve the following problem.

The average of two numbers is 4 and the average of their squares is 80. Find the numbers.

(a) What is the problem asking for? Assign variables to denote the required quantities.

(b) Complete the following table, as in Example 1.

In English	In algebraic notation
Find two numbers	
Their average is 4	
The average of their squares is 80	

(c) Find and simplify an equation that can be solved for one of the variables introduced in part (a).

(d) Solve the equation obtained in part (c).

(e) Use the root or roots obtained in part (d) to answer the original question. Check that your answer does satisfy the conditions of the problem.

8. Follow parts (a) through (e) to solve the following problem.

Find two numbers that have an average of 11 and a product of 72.

(a) What is the problem asking for? Assign variables to denote the required quantities.

(b) Complete the following table, as in Example 1.

In English	In algebraic notation
Find two numbers	
The average of the two numbers is 11	
The product of the two numbers is 72	

(c) Find and simplify an equation that can be solved for one of the variables introduced in part (a).

(d) Solve the equation obtained in part (c).

(e) Use the root or roots obtained in part (d) to answer the original question. Check that your answer does satisfy the conditions of the problem.

9. Follow parts (a) through (e) to solve the following problem.

The perimeter of a rectangle is 40 cm and the area is 64 cm². What are the dimensions of the rectangle?

(a) What is the problem asking for? Draw a sketch of the situation, and assign variables to denote the required quantities.

(b) Complete the following table, as in Example 3.

In English	In algebraic notation
Find the width and length of a rectangle	_____
The perimeter is 40 cm	_____
The area is 64 cm²	_____

(c) Find and simplify an equation that can be solved for the width of the rectangle.
(d) Solve the equation obtained in part (c).
(e) Use the root or roots obtained in part (d) to answer the original question. Check that your answer does satisfy the conditions of the problem.

10. Follow parts (a) through (e) to solve the following problem.

> Determine the base and the height of a triangle given that the area is 96 m² and that the base is three-fourths of the height.

(a) What is the problem asking for? Draw a sketch of the situation, and assign variables to denote the required quantities.
(b) Complete the following table, as in Example 3.

In English	In algebraic notation
Find the base and height of a triangle	_____
The area is 96 m²	_____
The base is three-fourths of the height.	_____

(c) Find and simplify an equation that can be solved for the height of the triangle.
(d) Solve the equation obtained in part (c).
(e) Use the root or roots obtained in part (d) to answer the original question. Check that your answer does satisfy the conditions of the problem.

11. Follow parts (a) through (d) to solve the following problem.

> Suppose that you open three savings accounts (each earning simple interest). In the first account you deposit $3000 at an annual simple interest rate of 7%. In the second you deposit $5000 at 8%, and in the third you deposit $10,000 at 8.5%. After how many years will the total interest amount to $8760?

(a) What is the problem asking for? Assign a variable to denote the required quantity.
(b) Find an equation that can be solved for the variable introduced in part (a). (Don't simplify the equation.)
(c) Solve the equation obtained in part (b).
(d) Use the root or roots obtained in part (c) to answer the original question. Check that your answer does satisfy the conditions of the problem.

12. Follow parts (a) through (d) to solve the following problem.

> Suppose that you open three savings accounts (each earning simple interest). In the first account you deposit $2500 at an annual simple interest rate of 6%, in the second you deposit $3500 at 6.5%, and in the third you deposit $10,000 at 8.5%. After how many years will the total interest amount to $9820?

(a) What is the problem asking for? Assign a variable to denote the required quantity.
(b) Find an equation that can be solved for the variable introduced in part (a). (Don't simplify the equation.)
(c) Solve the equation obtained in part (b).
(d) Use the root or roots obtained in part (c) to answer the original question. Check that your answer does satisfy the conditions of the problem.

13. Follow parts (a) through (c) to solve the following problem.

> Two savings accounts are opened with initial deposits of $7400 and $12,600. Both accounts earn simple interest, and the rate for the larger account is 1% higher than that for the smaller. After ten years the total value of the two accounts is $31,260. Find the interest rate (expressed as a percent) for the account in which the initial deposit was $7400.

(a) Find an equation that can be solved for the required interest rate. (Don't simplify the equation.)
(b) Solve the equation obtained in part (a).
(c) Use the root or roots obtained in part (b) to answer the original question. Check that your answer does satisfy the conditions of the problem.

14. Follow parts (a) through (c) to solve the following problem.

> Two savings accounts are opened with initial deposits of $14,000 and $25,000. Both accounts earn simple interest, and the rate for the smaller account is 2% less than that for the larger. After five years the total value of the two accounts is $54,165. Find the interest rate (expressed as a percent) of the $14,000 account.

(a) Find an equation that can be solved for the required interest rate. (Don't simplify the equation.)
(b) Solve the equation obtained in part (a).
(c) Use the root or roots obtained in part (b) to answer the original question. Check that your answer does satisfy the conditions of the problem.

15. Follow parts (a) through (e) to solve the following problem.

> The point (a, b) lies in the first quadrant on the graph of the line $y = 2x + 3$. Find a and b given that the distance from the point (a, b) to the origin is 5 units.

(a) What is the problem asking for?
(b) Complete the following table, as in Example 5.

In English	In algebraic notation
Find two numbers	_____
(a, b) is in the first quadrant	_____
(a, b) is on the line $y = 2x + 3$	_____
The distance from (a, b) to $(0, 0)$ is 5	_____

(c) Find an equation that can be solved for one of the unknowns.
(d) Solve the equation obtained in part (c).
(e) Use the root or roots obtained in part (d) to answer the original question. Check that your answer does satisfy the conditions of the problem.

16. Follow parts (a) through (e) to solve the following problem.

The point (a, b) lies in the second quadrant on the graph of $y = x^2$. Find a and b given that the distance from the point (a, b) to the origin is 3 units.

(a) What is the problem asking for?
(b) Complete the following table, as in Example 5.

In English	In algebraic notation
Find two numbers	_____
(a, b) is in the second quadrant	_____
(a, b) is on the graph of $y = x^2$	_____
The distance from (a, b) to $(0, 0)$ is 3	_____

(c) Find an equation that can be solved for one of the unknowns.
(d) Solve the equation obtained in part (c).
(e) Use the root or roots obtained in part (d) to answer the original question. Check that your answer does satisfy the conditions of the problem.

17. Follow parts (a) through (e) to solve the following problem.

The point (a, b) lies in the third quadrant on the graph of the equation $y = 1/x$. Find a and b given that the distance from the point (a, b) to the origin is $\sqrt{257}/4$.

(a) What is the problem asking for?
(b) Complete the table as in Example 5.
(c) Find an equation that can be solved for one of the unknowns.
(d) Solve the equation obtained in part (c).
(e) Use the root or roots obtained in part (d) to answer the original question. Check that your answer does satisfy the conditions of the problem.

In English	In algebraic notation
Find two numbers	_____
(a, b) is in the third quadrant	_____
(a, b) is on the graph of $y = 1/x$	_____
The distance from (a, b) to $(0, 0)$ is $\sqrt{257}/4$.	_____

18. Follow parts (a) through (e) to solve the following problem.

The point (a, b) lies on the graph of the equation $y = \sqrt{x}$, and the slope of the line joining the points (a, b) and $(-1, -3)$ is 3/5. Find a and b.

(a) What is the problem asking for?
(b) Complete the following table, as in Example 5.

In English	In algebraic notation
Find two numbers	_____
(a, b) lies on the graph of $y = \sqrt{x}$	_____
The slope of the line joining (a, b) and $(-1, -3)$ is 3/5.	_____

(c) Find an equation that can be solved for one of the unknowns.
(d) Solve the equation obtained in part (c).
(e) Use the root or roots obtained in part (d) to answer the original question. Check that your answer does satisfy the conditions of the problem.

(G) **19.** This exercise provides a graphical look at Example 3 in the text. (As background, you need to reread Example 3.)

(a) In Example 3 we found that the width w and the length l of the rectangle were related by the two equations $w + l = 13$ and $lw = 40$. On the same set of axes, graph these two equations, using w for the horizontal axis and l for the vertical axis. In other words, graph the line $x + y = 13$ and the curve $yx = 40$. (Begin with a viewing rectangle extending from -20 to 20 in both the horizontal and vertical directions. Then, for a better look, change to a viewing rectangle that extends from 0 to 14 in both directions.)

(b) The graph in part (a) shows that the curve and the line intersect twice in the first quadrant. Zoom in on each intersection point and estimate the coordinates. Check that the estimates for w are consistent with the values calculated in part (d) of Example 3.

(c) In the standard viewing rectangle, graph the equation $l = w^2 - 13w + 40$. (We are still using w for the horizontal axis and l for the vertical axis.) Note the w-intercepts. What does this have to do with the work in Example 3(d)?

Ⓖ **20.** In Example 5 in the text, we determined a point (a, b) satisfying the following three conditions: (a, b) is in the third quadrant; (a, b) lies on the line $y = \frac{1}{2}x + 1$; and the distance from (a, b) to the origin is 13 units. This exercise outlines a graphical approach to the problem.

(a) In the standard viewing rectangle, graph the line $y = \frac{1}{2}x + 1$.

(b) The points that are 13 units from the origin lie on the circle $x^2 + y^2 = 13^2$. On the same set of axes, graph this circle and the line in part (a). Use a viewing rectangle that is large enough to show the entire circle.

(c) Your graph in part (b) should show that the line intersects the circle at two points, one in the first quadrant and one in the third quadrant. The third-quadrant point is the point that was to be determined in Example 5. (The point meets all three conditions of the problem.) Zoom in on this point and estimate its coordinates. Check that your results are consistent with the values that were calculated algebraically in Example 5.

In Exercises 21–24:

Ⓖ (a) *Use a graphing utility (as in the previous exercise and in Example 2 of the text) to approximate to the nearest tenth each of the required numbers.*

(b) *Find exact expressions for the required numbers (as in Example 5).*

21. Find a point P in the third quadrant on the line $y = 3x - 1$ such that the distance from P to the origin is 9 units.

22. Find a point P satisfying the following conditions: P lies on the first-quadrant portion of the curve $y = x^2$, and the distance from P to the origin is 12 units.

23. Find two numbers whose sum and product are both equal to 8.

24. Two positive numbers p and q (with $p > q$) are reciprocals of one another, and the point with coordinates (p, q) lies on the circle with radius 2 and center $(0, 0)$. Find p and q.

In Exercises 25–28, you are asked to solve equations or to fill in details relating to the examples in the text.

25. Use the quadratic formula to show that the roots of the equation $x^2 - 2x - 1 = 0$ are $1 \pm \sqrt{2}$. [This equation was obtained in part (c) of Example 1.]

26. In this exercise we verify statements made in the comments following Example 1.

(a) If the two numbers are denoted by x and $x - 2$, then the fact that the sum of their squares is 6 can be written

$$x^2 + (x - 2)^2 = 6$$

Solve this equation. Show that it leads to the same final answers as those in Example 1.

(b) If the two numbers are denoted by x and $x + 2$, then the fact that the sum of their squares is 6 can be written $x^2 + (x + 2)^2 = 6$. Solve this equation and show that it too leads to the same final answers as those obtained in Example 1.

27. Consider the quadratic equation

$$a^2 + (\tfrac{1}{2}a + 1)^2 = 169$$

[This was obtained in part (c) of Example 5.]

(a) On the left-hand side of the equation, carry out the indicated multiplication and combine like terms. Then multiply both sides of the resulting equation by 4. Show that the equation becomes $5a^2 + 4a - 672 = 0$.

(b) Use the quadratic formula and your calculator to solve the equation $5a^2 + 4a - 672 = 0$.

(c) Verify that the two roots obtained in part (b) satisfy the equation $\sqrt{a^2 + (\tfrac{1}{2}a + 1)^2} = 13$.

28. In this exercise, we check that the values $a = -12$ and $b = -5$ obtained in Example 5 indeed satisfy the conditions of the problem.

(a) In which quadrant is the point $(-12, -5)$?

(b) Show that the point $(-12, -5)$ lies on the line $y = \frac{1}{2}x + 1$.

(c) Show that the distance from $(-12, -5)$ to the origin is 13.

MINI PROJECT ANCIENT HISTORY/MODERN TECHNOLOGY

Figure A shows a sketch of the contents of an ancient Babylonian clay tablet discovered in southern Iraq. The tablet, dating back to 1700 B.C., contains the statement and solution of a math problem about a rectangle. The problem translates as follows:

Length, width. I have multiplied length and width, thus obtaining the area. Then I added to the area the excess of the length over the width: 183. Furthermore, I have added length and width: 27. Required: length, width, and area. —Source: B. L. van der Waerden, *Science Awakening,* English translation by Arnold Dresden (Groningen, Holland: F. Noordhoff Ltd., 1954)

(a) In Example 3 we solved a problem that is similar to this ancient Babylonian problem. By adapting parts (a) through (c) in Example 3, show that the

Figure A

Sketch by Otto Neugebauer. The notation *AO 8862* is a museum cataloging number in the Louvre Museum, Paris, where the actual tablet is housed. Source: O. Neugebauer, *Mathematische Keilschrift-texte*

(Berlin: Julius Springer, 1935)

width w satisfies the quadratic equation $w^2 - 25w + 156 = 0$. Include with your work a table similar to Table 2 in the text.

(b) Use a graphing utility to graph the equation $y = w^2 - 25w + 156$. Zoom in on the w-intercepts to estimate the two roots of the equation in part (a).

(c) Now, use the quadratic formula to solve the equation in part (a). How close were your estimates?

(d) Using one of your roots from part (c), verify that a solution for the given problem is $w = 12$, $l = 15$, area $= 180$. These are the values obtained in the Babylonian tablet, 3700 years ago.

(e) Use the other root from part (c) to obtain another solution to the given problem.

Remark: For the Babylonian method of solving quadratics, see the Writing Mathematics section at the end of this chapter.

2.4   APPLICATIONS

The steps that we followed in the previous section can be used in solving a wide variety of applied problems. For reference, we summarize those steps in the box that follows. As you approach each example and exercise in this section for the first time, you may find it useful to refer to this list of suggestions. Eventually, of course, you'll probably want to modify the list to fit your own style.

Solving problems is a practical art, like swimming, or skiing, or playing the piano: you can learn it only by imitation and practice. This book cannot offer you a magic key that opens all doors and solves all the problems, but it offers you good examples for imitation and many opportunities for practice. . . .
—George Polya in *Mathematical Discovery* (New York: John Wiley & Sons, 1981)

Steps for Problem Solving

1. Read the problem carefully. When appropriate, draw a picture or a chart to summarize the situation.
2. State in your own words, as specifically as you can, what the problem is asking for. (This usually requires rereading the problem.) Assign variables or expressions to denote the required quantities. Then, as necessary, assign auxiliary variables to any other quantities that appear relevant. Are there equations relating these auxiliary variables?
3. Set up an equation (or equations) involving the required variables. Where necessary, use the results from Step 2 to obtain an equation involving but one required unknown.
4. Solve the equation that you obtained in Step 3.
5. Check your answer. Relate it to the original problem. Does it satisfy the conditions of the *original* problem?

EXAMPLE   1   Raising your test average

If you have three exam scores of 91, 76, and 84, what score do you need on the fourth exam to raise your average to 85?

SOLUTION
We let

$$x = \text{the score on the fourth exam}$$

The average of the four scores 91, 76, 84, and x is $(91 + 76 + 84 + x) \div 4$. Since this average is to be 85, we write

$$\frac{91 + 76 + 84 + x}{4} = 85$$
$$91 + 76 + 84 + x = 340 \qquad \text{multiplying by 4}$$
$$251 + x = 340$$
$$x = 340 - 251 = 89$$

We've now found that $x = 89$. This is the required score because the average of the four numbers 91, 76, 84, and 89 is, in fact, 85, as required.

In the next example, we use the Pythagorean theorem to answer a question about right triangles. [For a statement of the Pythagorean theorem (and its converse), see

Section 1.4.] One of the simplest examples of a right triangle in which the lengths of the sides are integers is the 3-4-5 right triangle. Note that $3^2 + 4^2 = 5^2$, so the Pythagorean theorem (or, really, its converse) guarantees that such a triangle is indeed a right triangle. This brings up a question: Are there any other right triangles in which the lengths of the sides are consecutive integers? We answer this question in the next example.

| EXAMPLE | 2 | A question about right triangles |

Are there right triangles other than the 3-4-5 right triangle in which the lengths of the sides are three consecutive integers?

SOLUTION
Any three consecutive integers can be denoted by x, $x + 1$, and $x + 2$ (see Figure 1). If these three quantities are to be the lengths of the sides of a right triangle, then according to the Pythagorean theorem we must have

$$x^2 + (x + 1)^2 = (x + 2)^2$$
$$x^2 + x^2 + 2x + 1 = x^2 + 4x + 4$$
$$x^2 - 2x - 3 = 0 \qquad \text{simplifying}$$
$$(x - 3)(x + 1) = 0$$
$$x - 3 = 0 \quad | \quad x + 1 = 0$$
$$x = 3 \quad | \quad x = -1$$

Figure 1

Thus, $x = 3$ or $x = -1$. We can immediately discard the value $x = -1$, because x is supposed to represent a length. On the other hand, with $x = 3$ we find that

$$x + 1 = 3 + 1 = 4 \qquad \text{and} \qquad x + 2 = 3 + 2 = 5$$

From this we conclude that if the lengths of the three sides in the right triangle are consecutive integers, then those integers must be 3, 4, and 5. Or, to answer the original question, there are no other right triangles in which the lengths of the sides are three consecutive integers.

In some problems it's helpful to set up a chart or table to see the given data at a glance. This is done in the next two examples.

| EXAMPLE | 3 | Mixtures in the chemistry lab |

Suppose that a chemistry student can obtain two acid solutions from the stockroom. The first solution is 20% acid and the second is 45% acid. (The percentages are by volume.) How many cubic centimeters (cm^3) of each solution should the student mix together to obtain 100 cm^3 of a 30% acid solution?

SOLUTION
We begin by assigning letters or expressions to denote the required quantities. We let x denote the number of cm^3 of the 20% solution to be used. Then

$$100 - x$$

denotes the number of cm^3 of the 45% solution to be used. All the data can now be summarized in Table 1.

TABLE 1

Type of solution	Number of cm^3	Percent acid	Total acid (cm^3)
First solution (20% acid)	x	20	$(0.20)x$
Second solution (45% acid)	$100 - x$	45	$(0.45)(100 - x)$
Mixture	100	30	$(0.30)(100)$

Looking at the data in the right-hand column of the table, we can write

$$0.20x \quad + \quad 0.45(100 - x) \quad = \quad (0.30)(100)$$

Amount of acid in x cm^3 of the 20% solution   Amount of acid in $(100 - x)$ cm^3 of the 45% solution   Amount of acid in the final mixture

So we have

$$0.20x + 0.45(100 - x) = 30$$
$$20x + 4500 - 45x = 3000 \qquad \text{multiplying by 100}$$
$$-25x = -1500$$
$$x = 60$$

Thus $x = 60$, and therefore $100 - x = 100 - 60 = 40$. Consequently, the student should use 60 cm^3 of the first solution and 40 cm^3 of the second solution.

CHECK   The amount of acid in 60 cm^3 of the first solution is $(60 \text{ cm}^3)(0.20)$, which is 12 cm^3; and the amount of acid in 40 cm^3 of the second solution is $(40 \text{ cm}^3)(0.45)$, which is 18 cm^3. So the total amount of acid in the mixture is the sum of 12 cm^3 and 18 cm^3, which is 30 cm^3. Since the volume of the final mixture is 100 cm^3, the amount of acid there is $\frac{30}{100}(100)$ cm^3; this is 30%, as required.

Our next example is a motion problem. For travel at a constant speed or *rate,* the basic relationship is

$$\text{distance} = \text{rate} \times \text{time}$$

or, more concisely,

$$d = rt$$

For example, if you drive 55 mph for three hours, then the distance you cover is $d = (55)(3) = 165$ miles. If a trip involves several parts, each at a different constant speed, then the *average speed* for the trip is defined by

$$\text{average speed} = \frac{\text{total distance}}{\text{total time}}$$

TABLE 2(a)

	d (miles)	r (miles/ min)	t (min)
Andy		$\frac{1}{8}$	
Jerry		$\frac{1}{7}$	

TABLE 2(b)

	d (miles)	r (miles/ min)	t (min)
Andy	$\frac{1}{8}T$	$\frac{1}{8}$	T
Jerry	$\frac{1}{7}T$	$\frac{1}{7}$	T

EXAMPLE	4	Running on the track: distance, rate, and time

Andy and Jerry are going to run on a quarter-mile track. Andy can run at a pace of 8 minutes per mile ($= \frac{1}{8}$ mile/min), while Jerry can run at a pace of 7 minutes per mile ($= \frac{1}{7}$ mile/min). If they start together at the same time (running in the same direction), how long will it take for Jerry to lap Andy?

SOLUTION
As in Example 3, we set up a table to organize our data. Table 2(a) indicates the format. (Notice that the top row is just taken from the formula $d = rt$.) In the r column of Table 2(a), we've filled in the given rates for Andy and Jerry.

Now we complete the table, as shown in Table 2(b). In the t column, we've used T to represent the (unknown) number of minutes for Jerry to lap Andy; the d column was filled in using the fact that $d = rt$. Thus, Andy's and Jerry's distances are represented by $\frac{1}{8}T$ and $\frac{1}{7}T$, respectively. Consequently, the condition that Jerry be exactly $\frac{1}{4}$ mile ahead of Andy is

$$\frac{1}{7}T = \frac{1}{8}T + \frac{1}{4}$$
$$8T = 7T + 14 \qquad \text{\textcolor{red}{multiplying by 56}}$$
$$T = 14$$

Thus, after 14 minutes Jerry will have lapped Andy. (Check for yourself that this answer indeed satisfies the conditions of the problem.)

EXAMPLE	5	Upgrading a cooling pipe at a power plant

A safety engineer at a power plant decides that one of the backup systems requires a cooling pipe with a larger capacity. As indicated in Figure 2, the cross section of the existing pipe is a circle of radius 20 cm. The engineer has calculated that the cross-sectional area for the new pipe must be 200 cm^2 greater than the existing cross-sectional area. By how many centimeters should the radius be increased to accomplish this? On-site measurements indicate that there is only enough space to increase the radius by a mere 2 cm. Would this be sufficient?

Figure 2
The radius of the original pipe is 20 cm.

— 20 cm

SOLUTION
We let

$x =$ the number of cm by which the radius should be increased

Original circle

20

Area $= \pi r^2$
$= \pi(20^2) = 400\pi$

New circle

20   x

$20 + x$

Area $= \pi r^2$
$= \pi(20 + x)^2$

Figure 3

Then, as the calculations accompanying Figure 3 indicate, the existing cross-sectional area is 400π cm^2, while the new cross-sectional area is $\pi(20 + x)^2$ cm^2. Now, according to the statement of the problem, we want the two areas to differ by 200 cm^2. Consequently, we have

$$\pi(20 + x)^2 - 400\pi = 200$$

and therefore

$$(20 + x)^2 = \frac{200 + 400\pi}{\pi}$$

or

$$20 + x = \sqrt{\frac{200 + 400\pi}{\pi}} \qquad \text{disregarding the negative square root (Why?)}$$

Finally, subtracting 20 from both sides of this last equation and then using a calculator, we have

$$x = \sqrt{\frac{200 + 400\pi}{\pi}} - 20 \approx 1.533 \text{ cm}$$

This shows that the required increase is approximately 1.533 cm. Thus, there is sufficient space to make the change (since there is room for a 2-cm increase).

EXERCISE SET 2.4

A

1. Find two numbers with sum 7 and difference 17.
2. The larger of two numbers is one less than twice the smaller. The sum of the two numbers is 29. Find the two numbers.
3. Find two numbers with sum 17 and product 52.
4. The sum of two numbers is 6 and the sum of their squares is 90. What are the two numbers?
5. The sum of a number and its double is 63. What is the number?
6. Find three consecutive integers such that twice the first plus half of the second is nine more than twice the third.
7. The product of a certain positive number and four more than that number is 96. What is the number?
8. The sum of the squares of three consecutive positive integers is 365. Find the integers.
9. What number must be added to both the numerator and the denominator of the fraction 3/7 to yield a fraction that is equal to 2/3?
10. Find two consecutive even integers such that when they are divided by 8 and 9, respectively, the sum of the quotients is 4.

11. The difference of two positive numbers is one-third of their sum. The sum of the squares of the two numbers is 180. Find the two numbers.
12. (a) Is there a positive real number such that the sum of the number and its square is 1? If so, find the number.
    (b) Is there a positive real number such that the difference between the number and its square is 1? If so, find the number.
13. If you have exam scores of 70, 77, and 75, what score do you need on the fourth exam to raise your average to 75?
14. The average of three numbers is 42. The first two numbers differ by 1, and the average of the second and third numbers is 47. What are the three numbers?
15. The average of four numbers is 52. Find the fourth number, given that the first three are 80, 22, and 62.
16. The average of the reciprocals of two consecutive integers is 9/40. Find the smaller of the two integers.
17. In a certain right triangle, the hypotenuse is 8 cm longer than the first leg and 1 cm longer than the second. Find the lengths of the sides of the triangle.
18. In a certain right triangle, the lengths of the two shorter sides are consecutive integers (measured in cm). Find the

area of the triangle given that the sum of the lengths of the hypotenuse and longer leg is 50 cm.

19. When the sides of a square are each increased by 2 cm, the area increases by 14 cm^2. Find the length of a side in the original square.

20. In a certain rectangle, the length exceeds the width by 2 m. If the length is increased by 4 m and the width is decreased by 2 m, the area of the new rectangle is the same as that of the original rectangle. Find the dimensions of the original rectangle.

21. The perimeter and area of a rectangle are 104 cm and 640 cm^2, respectively. Find the length of the shorter side of the rectangle.

22. A rectangular picture with dimensions 5 in. by 8 in. is surrounded by a frame of uniform width. If the area of the frame (not including the space for the picture) is 114 in.^{2} greater than the area of the picture, find the width of the frame.

23. A piece of wire 12 in. long is bent to form a right triangle in which the shortest side is 2 in. Find the lengths of the other two sides.

24. A piece of wire 24 cm long is cut into two pieces. Each piece is then formed into a square. If the sum of the areas of the two squares is 20 cm^2, find the lengths of the two pieces of wire.

25. In a certain isosceles triangle, each of the two equal legs is three times as long as the base. If the perimeter of the triangle is 70 cm, find the lengths of the sides.

26. In a certain triangle, the second angle is four times the first, while the third angle is 60° more than the first angle. Find the three angles. (Use the fact that the sum of the angles in any triangle is 180°.)

27. Suppose that you open two savings accounts. In the first you deposit $4000 at an annual simple interest rate of 7%. In the second you deposit $5000 at 9% (also simple interest). After how many years will the total interest amount to $5840?

28. You invest $600 in stock paying a 6% annual dividend and $800 in a stock paying an 8% annual dividend. How many years will it take for the total earnings to equal the total initial investment? (Use the formula $I = Prt$.)

29. If you deposit $10,000 at 11% (simple interest), how long will it be until the total amount in the account is $21,000?

30. Mr. X has $5000 to deposit. He puts part of this in an account earning 6%. The remainder he puts in an account earning 8.5%. (Both rates are simple interest.) If the total interest earned each year is $387.50, how much was initially deposited in each account?

31. You deposit $7500 at 7.5% and $4000 at 5%, both simple interest. What is the **effective** interest rate? (In other words, what interest rate on $11,500 would yield the same total interest at year's end?) Express your answer as a percentage, rounded to the nearest tenth of one percent.

32. You deposit $2000 at 8% and $3000 at 9%, both simple interest. What additional amount would you need to

invest at 12% to yield an overall return of 10% after one year?

33. Ms. X deposits $8000 in one account and $9000 in another. The interest rate on the $8000 account is one percent less than that on the $9000 account. If the $9000 account earns $165 more per year than the $8000 account, find the interest rates on each account.

34. Jenny inherits $42,900 and invests it for a year in two stocks. Stock A earns 12.5% annually and stock B earns 8.5% annually. How much did she invest in each stock if stock B earned three-quarters as much as stock A?

35. A student in a chemistry laboratory has two acid solutions available to her. The first solution is 10% acid and the second is 35% acid. (The percentages are by volume.) How many cm^3 of each solution should she mix to obtain 200 cm^3 of a 25% acid solution?

36. One salt solution is 15% salt and another is 20% salt. How many cm^3 of each solution must be mixed to obtain 50 cm^3 of a 16% salt solution?

37. A shopkeeper has two types of coffee beans. The first type sells for $5.20/lb, the second for $5.80/lb. How many pounds of the first type must be mixed with 5 lb of the second to produce a blend selling for $5.35/lb?

38. How much pure copper must be melted with 300 lb of an alloy that is 20% copper to yield an alloy that is 37.5% copper? (The percentages are by weight.)

39. Two sources of iron ore are available to a steel company. The first source of ore contains 20% iron (by weight). The second source contains 35% iron. How many tons of the second ore must be added to 16 tons of the first ore to produce a mixture that is 25% iron?

40. How much water must be evaporated from 400 lb of a 10% salt solution to yield a solution that is 16% salt?

41. The two shorter sides of a right triangle differ by 7 cm. If the area is 60 cm^2, find the hypotenuse.

42. If the sides of a square are doubled, the area increases by 147 cm^2. Find the sides of the original square.

43. A ball is thrown straight up. Suppose that the height of the ball at time t is given by $h = -16t^2 + 96t$, where h is in feet and t is in seconds, with $t = 0$ corresponding to the instant that the ball is first tossed.

(a) How long does it take before the ball lands?

(b) At what time is the height 80 ft? Why are there two answers here?

44. During a flu epidemic in a small town, a public health official finds that the total number P of people who have caught the flu after t days is closely approximated by the formula

$$P = -t^2 + 26t + 106, \text{ where } 1 \le t \le 13$$

(a) How many have caught the flu after 10 days?

(b) After approximately how many days will 250 people have caught the flu?

45. Emily and Ernie are going to jog on a quarter-mile track. Emily can jog at a pace of 9 min/mile (= 1/9 mile/min), while Ernie can jog at a pace of 11 min/mile (= 1/11 mile/min). If they start at the same time (running in the same direction), how long will it take Emily to lap Ernie?

46. Consider the following motion problem.

Suppose that you drive from town A to town B at 40 mph. At what (constant) speed should you make the return trip so as to average 50 mph for the entire round trip? (The answer is not 60 mph.)

Solve this problem by completing the steps that follow.

(a) Let D denote the distance (in miles) from A to B, and let R denote the speed (in mph) on the return trip. Complete the table.

	d (miles)	r (mph)	t (h)
Going	D		
Returning		R	

(b) Explain why the following equation is valid.

$$\frac{2D}{(D/40) + (D/R)} = 50$$

(c) Solve the equation in part (b) for R, and round your result to the nearest mph.

47. The speed of an eastbound train is 5 mph greater than that of a westbound train. If the trains leave at the same time from stations 630 miles apart and pass each other after 6 hours, find the speed of the eastbound train.

48. Jim and Juan are going to drive their motorcycles from San Francisco to Los Angeles. Jim leaves at 8:00 A.M., but Juan can't leave until 8:30 A.M. If Jim drives at 50 mph and Juan at 55 mph, at what time will Juan catch up with Jim?

49. Two airplanes take off from the airport at the same time, heading in opposite directions. After 2 hours the planes are 1500 miles apart. If the speeds of the planes differ by 50 mph, find the speed of the slower plane.

50. Annie drives from town A to town B in 3 hours (at a fixed speed). The traffic is heavier on the return trip, so Annie's speed is 16 mph less and the return trip takes 4 hours. What were her speeds on both parts of the trip?

51. A charter boat captain has 2 hours in which to leave the dock, head out to sea as far as possible, and then return to the dock. The boat's speeds on the outward and return journeys are 20 mph and 12 mph, respectively. How far out to sea can the boat travel?

52. You invest $5200 in a stock paying a 6.75% annual dividend and $8400 in a stock paying a 7.85% annual dividend. How many years will it take for the total earnings to be $2500? (Use a calculator and round your answer to the nearest one-half year.)

53. You deposit $2650 at 8.1% and $3300 at 9.2% simple interest. What additional amount would you need to deposit at

11.5% to yield an overall annual return of 10%? (Use a calculator and round your answer to the nearest dollar.)

54. In a certain right triangle, the hypotenuse is 205 cm and the two legs differ by 23 cm. Find the length of the shorter leg.

B

55. At noon Larry and Linda start walking in the same direction. Larry's speed is 1/15 mile/min; Linda's is less than that. However, when Larry has walked 2 miles, Linda doubles her pace and catches up with Larry at 1:15 P.M. How fast was Linda walking initially?

56. The length of a rectangular sheet of metal is 4 inches more than the width. As indicated in the figure, 3-in. squares are cut out from each corner of the rectangular sheet, and the sides are then folded up to form a box with no top. If the volume of the box is 180 in.^{3}, determine the dimensions of the original rectangular sheet.

57. Suppose that you travel from town A to town B at a speed of v_1 mph and then return at a speed of v_2 mph. What is your average speed for the round trip (in terms of v_1 and v_2)?

58. If you travel for a hours at b mph and then c hours at d mph, what is your average speed for the trip?

59. In one savings account you deposit d dollars at an annual simple interest rate of r%. At the same time you also deposit D dollars at R% in a second account. After how many years will the total interest amount to $d + D$ dollars?

60. A chemistry student has two acid solutions available to him. The first is a% acid and the second is b% acid. (The percentages are by volume.) How many cm^3 of each solution should be mixed to obtain d cm^3 of a solution that is c% acid? (Assume that a, b, c, and d are positive and that $a < c < b$.)

61. A piece of wire a inches long is bent to form a right triangle of which the shortest side is b in. Find the length of the hypotenuse.

62. How much water must be evaporated from b lb of an a% salt solution to obtain a solution that is c% salt? (Assume that the percentages are by weight and that $c > a$.)

63. If the length of each side of a square is multiplied by a, the area increases by b^2 square units. Find the sides of the original square.

64. A piece of wire L inches long is cut into two pieces. Each piece is then bent to form a square. If the sum of the areas of the two squares is $5L^2/128$, how long are the two pieces of wire?

65. A piece of wire 16 cm long is cut into two pieces, and each piece is then bent to form a circle. If the sum of the areas of the two circles is 12 cm^2, how long is the shorter piece of wire? Round your answer to two decimal places.

66. The height of a right circular cylinder is 1 ft. Find the radius, given that the total surface area is 660 ft^2. Round your answer to two decimal places.

MINI PROJECT | MORE ON ANCIENT HISTORY/MODERN TECHNOLOGY

Figure A shows a page from the ancient Chinese text *Nine Chapters on the Mathematical Art (Jiuzhang suanshu or Chiu Chang Suan Shu)*. In parts (b) and (c) of this project you're asked to solve a problem from this text. Your instructor will tell you whether this is to be an individual or group project.

Figure A

(a) Historical studies indicate that in its present form, *Nine Chapters* dates back to the first century, but that it was compiled or assembled from earlier sources. Among the many reasons why some of these earlier sources and authors are unknown, one contributing event in Chinese history stands out. It occurred in 213 B.C. By consulting an encyclopedia or book on the history of China, find out what this singular event was and why it occurred. (A concise answer can be given in two or three sentences or at most one paragraph; but do, of course, write in complete sentences.)

(b) The following is a translation of a problem appearing in the final chapter of *Nine Chapters*:

> A square, walled village of unknown sides is crossed by a street joining the centers of the north and south sides. There is a gate at each end of the street. A man walks 14 paces southward from the south gate; then he turns west and walks 1775 paces. At this point, a tree that is 20 paces northward of the north gate just becomes visible. Find the length of a side of the village.

Let x denote the length of a side of the square village. Draw a sketch of the situation, and then find an equation that can be solved for x. *Hint*: Use similar triangles.

(c) Use a graphing utility and its zoom feature to solve the equation in part (b). You'll find that there are two roots; choose the appropriate one.

(d) As a check on your answer in part (c), use the quadratic formula to solve the equation in part (b).

2.5   INEQUALITIES

The fundamental results of mathematics are often inequalities rather than equalities. —E. Beckenbach and R. Bellman in *An Introduction to Inequalities* (New York: Random House, 1961)

If we replace the equal sign in an equation with any one of the four symbols $<$, $\leq$, $>$, or $\geq$, we obtain an **inequality.** As with equations in one variable, a real number is a **solution** of an inequality if we obtain a true statement when the variable is replaced by the real number. For example, the value $x = 5$ is a solution of the inequality $2x - 3 < 8$, because when $x = 5$ we have

$$2(5) - 3 < 8$$

$$7 < 8 \qquad \text{which is true}$$

We also say in this case that the value $x = 5$ **satisfies** the inequality. To **solve** an inequality means to find all of the solutions. The set of all solutions of an inequality is called (naturally enough) the **solution set.**

Recall that two equations are said to be equivalent if they have exactly the same solutions. Similarly, two inequalities are **equivalent** if they have the same solution set. Most of the procedures used for solving inequalities are similar to those for equalities. For example, adding or subtracting the same number on both sides of an inequality produces an equivalent inequality. We need to be careful, however, in multiplying or dividing both sides of an inequality by the same nonzero number. For instance, suppose that we start with the inequality $2 < 3$ and multiply both sides by 5. That yields $10 < 15$, which is certainly true. But if we multiply both sides of the inequality $2 < 3$ by -5, we obtain $-10 < -15$, which is false. Multiplying both sides of an inequality by the same *positive* number preserves the inequality, whereas multiplying by a *negative* number reverses the inequality. In the following box, we list some of the principal properties of inequalities. In general, whenever we use Property 1 or 2 in solving an inequality, we obtain an equivalent inequality. Also, note

▌PROPERTY SUMMARY   Properties of Inequalities

Property	Example
1. If $a < b$, then $a + c < b + c$ and $a - c < b - c$.	If $x - 3 < 0$, then $(x - 3) + 3 < 0 + 3$ and, consequently, $x < 3$.
2. (a) If $a < b$ and c is positive, then $ac < bc$ and $a/c < b/c$.	If $\frac{1}{2}x < 4$, then $2(\frac{1}{2}x) < 2(4)$ and, consequently, $x < 8$.
(b) If $a < b$ and c is negative, then $ac > bc$ and $a/c > b/c$.	If $-\dfrac{x}{5} < 6$, then $(-5)\left(-\dfrac{x}{5}\right) > (-5)(6)$ and, consequently, $x > -30$.
3. The transitive property: If $a < b$ and $b < c$, then $a < c$.	If $a < x$ and $x < 2$, then $a < 2$.

that each property can be rewritten to reflect the fact that $a < b$ is equivalent to $b > a$. For example, Property 3 can just as well be written this way: If $b > a$ and $c > b$, then $c > a$.

EXAMPLE **1**   **Solving two simple inequalities**

Solve each of the following inequalities:
(a) $2x - 3 < 5$;   **(b)** $5t + 8 \le 7(1 + t)$.

SOLUTION
(a) Our work follows the same pattern that we would use to solve the equation $2x - 3 = 5$. We have

$$2x - 3 < 5$$
$$2x < 8 \qquad \text{adding 3 to both sides}$$
$$x < 4 \qquad \text{dividing both sides by 2}$$

Figure 1

The solution set is therefore $(-\infty, 4)$; see Figure 1.
(b) Again, we follow the pattern that we would use to solve the corresponding equation.

$$5t + 8 \le 7 + 7t$$
$$-2t \le -1 \qquad \text{subtracting } 7t \text{ and 8 from both sides}$$
$$t \ge \tfrac{1}{2} \qquad \text{dividing both sides by } -2, \text{ which reverses the inequality}$$

Figure 2

The solution set is therefore $[1/2, \infty)$; see Figure 2.

In Section 1.5 we saw that there is a close connection between solving equations and graphing. (Example: The roots of the equation $x^2 - 4 = 0$ are the x-intercepts of the graph of $y = x^2 - 4$.) It's also true that solving inequalities and graphing are closely related. Consider, for instance, the inequality $2x - 3 < 5$ that we solved in Example 1(a). To interpret this inequality geometrically, we graph the two equations $y = 2x - 3$ and $y = 5$. Then, as indicated in Figure 3(a), the graph of $y = 2x - 3$

Figure 3

(a) To the left of $x = 4$, the line $y = 2x - 3$ is below the line $y = 5$.

(b) The graph indicates that if $t \geq 1/2$ then $5t + 8 \leq 7(1 + t)$.

is below the line $y = 5$ as long as x is to the left of 4. The previous sentence translates algebraically into

$$2x - 3 < 5 \qquad \text{provided} \qquad x < 4$$

This confirms the result in Example 1(a).

Similarly, for a geometric interpretation of the inequality in Example 1(b), we graph the two lines $y = 5t + 8$ and $y = 7(1 + t)$ in a t-y coordinate system. As indicated in Figure 3(b), the line $y = 5t + 8$ is below the line $y = 7(1 + t)$ provided t is to the right of $1/2$. Furthermore, at $t = 1/2$, the lines intersect. That is, for $t = 1/2$, the quantities $5t + 8$ and $7(1 + t)$ are equal. Algebraically, we can restate this information by saying

$$5t + 8 \leq 7(1 + t) \qquad \text{provided} \qquad t \geq 1/2$$

This confirms the result in Example 1(b).

In the next example, we solve the inequality

$$-\frac{1}{2} < \frac{3 - x}{-4} < \frac{1}{2}$$

By definition, this is equivalent to the pair of inequalities

$$-\frac{1}{2} < \frac{3 - x}{-4} \qquad \text{and} \qquad \frac{3 - x}{-4} < \frac{1}{2}$$

One way to proceed here would be first to determine the solution set for each inequality. Then the set of real numbers common to both solution sets would be the solution set for the original inequality. However, the method shown in Example 2 is more efficient.

EXAMPLE 2 **Solving a compound inequality**

Solve: $-\dfrac{1}{2} < \dfrac{3 - x}{-4} < \dfrac{1}{2}$.

SOLUTION
We begin by multiplying through by -4. Remember that this will reverse the inequalities:

$$2 > 3 - x > -2$$

Next, with a view toward isolating x, we first subtract 3 to obtain

$$-1 > -x > -5$$

Finally, multiplying through by -1, we have

$$1 < x < 5$$

The solution set is therefore the open interval $(1, 5)$.

The next example refers to the Celsius and Fahrenheit scales for measuring temperature.* The formula relating the temperature readings on the two scales is

$$F = \frac{9}{5}C + 32$$

EXAMPLE	3	Converting a Fahrenheit temperature range into Celsius

Over the temperature range $32° \leq F \leq 39.2°$ on the Fahrenheit scale, water contracts (rather than expands) with increasing temperature. What is the corresponding temperature range on the Celsius scale?

SOLUTION

$$32 \leq F \leq 39.2 \qquad \text{given}$$

$$32 \leq \frac{9}{5}C + 32 \leq 39.2 \qquad \text{substituting } \frac{9}{5}C + 32 \text{ for } F$$

$$0 \leq \frac{9}{5}C \leq 7.2 \qquad \text{subtracting 32}$$

$$0 \leq C \leq \frac{5}{9}(7.2) \qquad \text{multiplying by } \frac{5}{9}$$

$$0 \leq C \leq 4$$

Thus, a range of 32°F to 39.2°F on the Fahrenheit scale corresponds to 0°C to 4°C on the Celsius scale.

In the next three examples, we solve inequalities that involve absolute values. The following theorem is very useful in this context.

THEOREM   Absolute Value and Inequalities

If $a > 0$, then

$$|u| < a \qquad \text{if and only if} \qquad -a < u < a$$

and

$$|u| > a \qquad \text{if and only if} \qquad u < -a \quad \text{or} \quad u > a$$

---

*The Celsius scale was devised in 1742 by the Swedish astronomer Anders Celsius. The Fahrenheit scale was first used by the German physicist Gabriel Fahrenheit in 1724.

Figure 4

You can see why this theorem is valid if you think in terms of distance and position on a number line. The condition $-a < u < a$ means that u lies between $-a$ and a, as indicated in Figure 4. But this is the same as saying that the distance from u to zero is less than a, which in turn can be written $|u| < a$. (The second part of the theorem can be justified in a similar manner.)

EXAMPLE 4   Solving inequalities containing absolute values

Solve: **(a)** $|x| < 1$;   **(b)** $|x| \geq 1$.

SOLUTION

(a) By the theorem we just discussed, the condition $|x| < 1$ is equivalent to

$$-1 < x < 1$$

The solution set is therefore the open interval $(-1, 1)$.

(b) In view of the second part of the theorem, the inequality $|x| \geq 1$ is satisfied when x satisfies either of the inequalities

$$x \leq -1 \qquad \text{or} \qquad x \geq 1$$

The solution sets for these last two inequalities are $(-\infty, -1]$ and $[1, \infty)$, respectively. Consequently, the solution set for the given inequality $|x| \geq 1$ consists of the two intervals $(-\infty, -1]$ and $[1, \infty)$.

In Example 4(b), we found that the solution set consisted of two intervals on the number line. We have a convenient notation for describing such sets. Given any two sets A and B, we define the set $A \cup B$ (read **A union B**) to be the set of all elements that are in A or in B (or in both). For example, if $A = \{1, 2, 3\}$ and $B = \{4, 5\}$, then $A \cup B = \{1, 2, 3, 4, 5\}$. As another example, the union of the two closed intervals $[3, 5]$ and $[4, 7]$ is given by

$$[3, 5] \cup [4, 7] = [3, 7]$$

because the numbers in the interval $[3, 7]$ are precisely those numbers that are in $[3, 5]$ or $[4, 7]$ (or in both). Using this notation, we can write the solution set for Example 4(b) as

$$(-\infty, -1] \cup [1, \infty)$$

EXAMPLE 5   Two different methods for solving an inequality containing absolute values

Solve: $|x - 3| < 1$.

SOLUTION
We'll show two methods.

FIRST METHOD
We use the theorem preceding Example 4. With $u = x - 3$ and $a = 1$, the theorem tells us that the given inequality is equivalent to

$$-1 < x - 3 < 1$$

or (by adding 3)

$$2 < x < 4$$

The solution set is therefore the open interval $(2, 4)$.

ALTERNATIVE METHOD
The given inequality tells us that x must be less than one unit away from 3 on the number line. Looking one unit to either side of 3, then, we see that x must lie strictly between 2 and 4. The solution set is therefore $(2, 4)$, as obtained using the first method.

EXAMPLE 6 **Another inequality with absolute values**

Solve: $\left| 1 - \dfrac{t}{2} \right| > 5$.

SOLUTION
Referring again to the theorem, we use the fact that $|u| > a$ means that either $u < -a$ or $u > a$. So, in the present example, the given inequality means that either

$$1 - \frac{t}{2} < -5 \qquad \text{or} \qquad 1 - \frac{t}{2} > 5$$
$$-\frac{t}{2} < -6 \qquad\qquad\qquad -\frac{t}{2} > 4$$
$$t > 12 \qquad\qquad\qquad\quad t < -8$$

This tells us that the given inequality is satisfied when t is in either of the intervals $(12, \infty)$ or $(-\infty, -8)$. Thus the solution set is $(-\infty, -8) \cup (12, \infty)$.

Earlier in this section we saw that there is a close connection between inequalities and graphs. The next example emphasizes this point again.

EXAMPLE 7 **A graphical approach to solving an inequality**

As indicated in Figure 5, the graph of the equation $y = |7x - 10| - 9$ has two x-intercepts. Let a denote the x-intercept between 0 and 1, and let b denote the x-intercept between 2 and 3.

(a) Use the graph in Figure 5 to specify the general form of the solution set for the inequality

$$|7x - 10| - 9 \leq 0$$

Use interval notation. (You can assume that Figure 5 shows all of the essential features of the graph.)

GRAPHICAL PERSPECTIVE

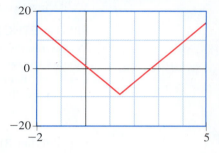

Figure 5
$y = |7x - 10| - 9$
$[-2, 5, 1]$ by $[-20, 20, 10]$

(b) Use a graphing utility and the ZOOM feature to estimate to the nearest hundredth the x-intercepts a and b. Then state the corresponding estimate for the solution set of the inequality $|7x - 10| - 9 \leq 0$.

(c) Use algebra to solve the inequality $|7x - 10| - 9 \leq 0$. Check that the solution is consistent with the graphical results in parts (a) and (b).

SOLUTION

(a) The graph has two x-intercepts, which we are calling a and b (with $a < b$). At each of these x-intercepts, the quantity $|7x - 10| - 9$ is zero, while for x between a and b, the quantity $|7x - 10| - 9$ is negative (because the graph is *below* the x-axis). In other words, the inequality $|7x - 10| - 9 \leq 0$ is satisfied by all values of x in the closed interval $[a, b]$. Furthermore, the graph shows that these are the only values of x satisfying the required inequality (because the graph is *above* the x-axis when x is outside of this closed interval). In summary then, the solution set for the inequality $|7x - 10| - 9 \leq 0$ has the form

$$[a, b] \quad \text{where } a \text{ is between 0 and 1 and } b \text{ is between 2 and 3}$$

(b) After some experimenting with viewing rectangles of various sizes, we obtain the views displayed in Figure 6. Figure 6(a) shows that the smaller of the two intercepts is between 0.14 and 0.15, closer to 0.14 than to 0.15, so $a \approx 0.14$ to the nearest hundredth. Similarly, for the larger intercept b, Figure 6(b) indicates that $b \approx 2.71$, to the nearest hundredth. Our corresponding estimate for the solution set of the given inequality is then the closed interval $[0.14, 2.71]$.

GRAPHICAL PERSPECTIVE

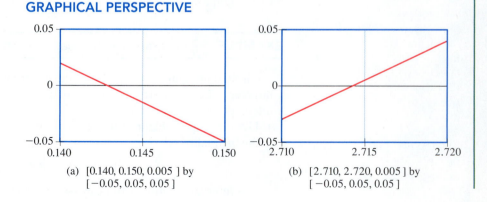

Figure 6
Zoom-in views of the x-intercepts
for $y = |7x - 10| - 9$

(a) $[0.140, 0.150, 0.005]$ by
     $[-0.05, 0.05, 0.05]$

(b) $[2.710, 2.720, 0.005]$ by
     $[-0.05, 0.05, 0.05]$

(c) So that we can apply the theorem on page 133, we first rewrite the given inequality in the equivalent form $|7x - 10| \le 9$. Then we have

$$-9 \le 7x - 10 \le 9 \qquad \text{using the first part of the theorem on page 133}$$

$$1 \le 7x \le 19 \qquad \text{adding 10}$$

$$\frac{1}{7} \le x \le \frac{19}{7} \qquad \text{dividing by 7}$$

The solution set is therefore the closed interval $[\frac{1}{7}, \frac{19}{7}.]$ The fact that it's a closed interval agrees with our graphical work in part (a). Furthermore, if you now use a calculator, you'll find that $\frac{1}{7} \approx 0.143$ and $\frac{19}{7} \approx 2.714$, values that are indeed consistent with the estimates obtained in part (b).

EXERCISE SET 2.5

A

In Exercises 1–34, solve the inequality and specify the answer using interval notation.

1. $x + 5 < 4$

2. $2x - 7 < 11$

3. $1 - 3x \le 0$

4. $6 - 4x \le 22$

5. $4x + 6 < 3(x - 1) - x$

6. $2(t - 1) - 3(t + 1) \le -5$

7. $1 - 2(t + 3) - t \le 1 - 2t$

8. $t - 4[1 - (t - 1)] > 7 + 10t$

9. $\dfrac{3x}{5} - \dfrac{x - 1}{3} < 1$

10. $\dfrac{2x + 1}{2} + \dfrac{x - 1}{3} < x + \dfrac{1}{2}$

11. $\dfrac{x - 1}{4} - \dfrac{2x + 3}{5} \le x$

12. $\dfrac{x}{2} - \dfrac{8x}{3} + \dfrac{x}{4} > \dfrac{23}{6}$

13. $-2 \le x - 6 \le 0$

14. $-3 \le 2x + 1 \le 5$

15. $-1 \le \dfrac{1 - 4t}{3} \le 1$

16. $\dfrac{2}{3} \le \dfrac{5 - 3t}{-2} \le \dfrac{3}{4}$

17. $0.99 < \dfrac{x}{2} - 1 < 0.999$

18. $\dfrac{9}{10} < \dfrac{3x - 1}{-2} < \dfrac{91}{100}$

19. (a) $|x| \le 1/2$
(b) $|x| > 1/2$

20. (a) $|x| > 2$
(b) $|x| \le 2$

21. (a) $|x| > 0$
(b) $|x| < 0$

22. (a) $|t| \ge 0$
(b) $|t| \le 0$

23. (a) $x - 2 < 1$
(b) $|x - 2| < 1$
(c) $|x - 2| > 1$

24. (a) $x - 4 \ge 4$
(b) $|x - 4| \ge 4$
(c) $|x - 4| \le 4$

25. (a) $1 - x \le 5$
(b) $|1 - x| \le 5$
(c) $|1 - x| > 5$

26. (a) $3x + 5 < 17$
(b) $|3x + 5| < 17$
(c) $|3x + 5| > 17$

27. (a) $a - x < c$
(b) $|a - x| < c$
(c) $|a - x| \ge c$

28. (Assume $b < c$ throughout this exercise.)
(a) $|x - a| + b < c$
(c) $|x + a| + b < c$
(b) $|x + a| + b > c$

29. $\left| \dfrac{x - 2}{3} \right| < 4$

30. $\left| \dfrac{4 - 5x}{2} \right| > 1$

31. $\left| \dfrac{x + 1}{2} - \dfrac{x - 1}{3} \right| < 1$ *Hint:* Combine the fractions.

32. $\left| \dfrac{3(x - 2)}{4} + \dfrac{4(x - 1)}{3} \right| \le 2$

33. (a) $|(x + h)^2 - x^2| < 3h^2$, where $h > 0$
(b) $|(x + h)^2 - x^2| < 3h^2$, where $h < 0$

34. (a) $|3(x + 2)^2 - 3x^2| < 1/10$
(b) $|3(x + 2)^2 - 3x^2| < \varepsilon$, where $\varepsilon > 0$

For Exercises 35–38:

Ⓖ **(a)** *Use a graph [as in Example 7(a)] to determine which of the following general forms describes the solution set of the given inequality:*

$$[a, b] \qquad (-\infty, a) \cup (b, \infty) \qquad [a, \infty) \qquad (a, \infty)$$

Ⓖ **(b)** *Use a graphing utility [as in Example 7(b)] to estimate to the nearest hundredth the value of a, and where appropriate, b.*

(c) *Solve the inequality algebraically and write the solution set using interval notation. Check that your answers are consistent with the graphical results in parts (a) and (b).*

35. $7x - 2 \ge 0$

36. $6 - 13x < 0$

37. $|8x - 3| - 2 \le 0$

38. $1 - |15x - 3| < 0$

39. Data from the *Apollo 11* moon mission in July 1969 showed temperature readings on the lunar surface varying over the interval $-183° \le C \le 112°$ on the Celsius scale. What is the corresponding interval on the Fahrenheit scale? (Round the numbers you obtain to the nearest integers.)

40. Measurements sent back to Earth from the *Viking I* spacecraft in June 1976 indicated temperature readings on the surface of Mars in the range $-139° \le C \le -28°$ on the Celsius scale. What is the corresponding range on the Fahrenheit scale? (Round the numbers you obtain to the nearest integers.)

41. The temperature of the variable star Delta Cephei varies over the interval $5100° \leq C \leq 6500°$ on the Celsius scale. What is the corresponding range on the Fahrenheit scale? (Round the numbers in your answer to the nearest 100°F.)

42. Data from the *Mariner 10* spacecraft (launched November 3, 1973) indicate that the surface temperature on the planet Mercury varies over the interval $-170° \leq C \leq 430°$ on the Celsius scale. What is the corresponding interval on the Fahrenheit scale? (Round the values that you obtain to the nearest 10°F.)

B

43. Solve the inequality $|x - 1| + |x - 2| < 3$.
Hint: Begin by considering three cases: $x < 1$; $1 \leq x < 2$; $x \geq 2$.

44. Solve the inequality $6 - |x + 3| - |x - 2| < 0$.
Hint: Adapt the hint in Exercise 43.

45. Given two positive numbers a and b, we define the **geometric mean** and the **arithmetic mean** as follows:

$$\text{G.M.} = \sqrt{ab} \qquad \text{A.M.} = \frac{a + b}{2}$$

(a) Complete the table, using a calculator as necessary so that the entries in the third and fourth columns are in decimal form.

a	b	$\sqrt{ab}$ (G.M.)	$(a + b)/2$ (A.M.)	Which is larger, G.M. or A.M.?
1	2			
1	3			
1	4			
2	3			
3	4			
5	10			
9	10			
99	100			
999	1000			

(b) Prove that for all nonnegative numbers a and b we have

$$\sqrt{ab} \leq \frac{a + b}{2} \qquad (1)$$

Hint: Use the following property of inequalities: If x and y are nonnegative, then the inequality $x \leq y$ is equivalent to $x^2 \leq y^2$.

(c) Assuming that $a = b$ (and that a and b are nonnegative), show that inequality (1) becomes an equality.

(d) Assuming that a and b are nonnegative and that $\sqrt{ab} = \frac{a + b}{2}$, show that $a = b$.

Remark: Parts (b) through (d) can be summarized as follows. For all nonnegative numbers a and b, we have

$$\sqrt{ab} \leq \frac{a + b}{2}, \text{ with equality holding if and only if } a = b.$$

This result is known as the *arithmetic-geometric mean inequality* for two numbers. The mini project at the end of this section shows an application of this result.

46. Use the result in Exercise 45(b) to prove each of the following inequalities. Assume that p, q, r, and s all are nonnegative.

(a) $\sqrt{pqr} \leq (pq + r)/2$ **(b)** $\sqrt{pqrs} \leq (pq + rs)/2$

*Exercises 47 and 48 concern the environmental problem known as **acid rain.** The principal source contributing to acid rain is sulfur dioxide, emitted into the atmosphere through the burning of fossil fuels. Once in the clouds, the sulfur dioxide combines with water vapor to form one of the nasty ingredients of acid rain—sulfuric acid.*

47. In the United States over the years 1980–2000, sulfur dioxide emissions due to the burning of fossil fuels can be approximated by the equation

$$y = -0.4743t + 24.086$$

where y represents the sulfur dioxide emissions (in millions of tons) for the year t, with $t = 0$ corresponding to 1980. *Source:* This equation (and the equation in Exercise 48) were computed using data from the book *Vital Signs 1999,* Lester Brown et al. (New York: W. W. Norton & Co., 1999).

(a) Use a graphing utility to graph the equation $y = -0.4743t + 24.086$ in the viewing rectangle $[0, 25, 5]$ by $[0, 30, 5]$. According to the graph, sulfur dioxide emissions are decreasing. What piece of information in the equation $y = -0.4743t + 24.086$ tells you this even before looking at the graph?

(b) Assuming this equation remains valid, estimate the year in which sulfur dioxide emissions in the United States might fall below 10 million tons per year. (You need to solve the inequality $-0.4743t + 24.086 \leq 10$.)

48. In Asia over the years 1980–2000, sulfur dioxide emissions due to the burning of fossil fuels can be approximated by the equation

$$y = 1.84x + 14.8$$

where y represents the sulfur dioxide emissions (in millions of tons) for the year t, with $t = 0$ corresponding to 1980.

(a) Use a graphing utility to graph the equation $y = 1.84x + 14.8$ in the viewing rectangle $[0, 25, 5]$ by $[0, 60, 20]$. According to the graph, sulfur dioxide emissions are increasing. What piece of information in the equation $y = 1.84x + 14.8$ tells you this even before looking at the graph?

(b) Assuming this equation remains valid, estimate the year in which sulfur dioxide emissions in Asia might exceed 65 million tons per year.

C

49. Let a, b, and c be nonnegative numbers. Follow steps (a) through (e) to show that

$$\sqrt[3]{abc} \le \frac{a + b + c}{3}$$

with equality holding if and only if $a = b = c$.

This result is known as the *arithmetic-geometric mean inequality* for three numbers. (Applications are developed in the projects at the ends of Sections 4.6 and 4.7.)

(a) By multiplying out the right-hand side, show that the following equation holds for all real numbers A, B, and C.

$$3ABC = A^3 + B^3 + C^3 - \frac{1}{2}(A + B + C) \quad (1)$$
$$\times [(A - B)^2 + (B - C)^2 + (C - A)^2]$$

(b) Now assume for the remainder of this exercise that A, B, and C are nonnegative numbers. Use equation (1) to explain why

$$3ABC \le A^3 + B^3 + C^3 \quad (2)$$

(c) Make the following substitutions in inequality (2): $A^3 = a$, $B^3 = b$, and $C^3 = c$. Show that the result can be written

$$\sqrt[3]{abc} \le \frac{a + b + c}{3} \quad (3)$$

(d) Assuming $a = b = c$, show that inequality (3) becomes an equality.

(e) Assuming $\sqrt[3]{abc} = \dfrac{a + b + c}{3}$, show that $a = b = c$.

Hint: In terms of A, B, and C, the assumption becomes $ABC = \dfrac{A^3 + B^3 + C^3}{3}$. Use this to substitute for ABC on the right side of equation (1). Then use the resulting equation to deduce that $A = B = C$, and consequently $a = b = c$.

MINI PROJECT | AN INEQUALITY FOR THE GARDEN

Part of a landscape architect's plans call for a rectangular garden with a perimeter of 150 ft. The architect wants to find the width w and length l that maximize the area A of the garden. Two preliminary sketches are shown in the figure that follows. (In both cases, note that the perimeter is 150 ft, as required.) Of the two sketches, the one on the right yields the larger area. In this project you'll use the inequality given in Exercise 45(b) to find exactly which combination of w and l yields the largest possible area.

(a) As background, first review or rework Exercise 45(b). If you are working in a group, assign one person to do this and then present the result to the others.

$A = 25 \text{ ft} \times 50 \text{ ft} = 1250 \text{ ft}^2 \qquad A = 30 \text{ ft} \times 45 \text{ ft} = 1350 \text{ ft}^2$

(b) Letting w denote the width of the rectangular garden, and assuming that the perimeter is 150 ft, show or explain why the area A is given by $A = w(75 - w)$.

(c) Use the inequality in Exercise 45(b) to show that $A \le 5625/4 = 1406.25$. In words: Given a perimeter of 150 ft, no matter how we pick w and l, the area can never exceed 1406.25 ft^2. *Hint*: Apply the inequality to the product $w(75 - w)$.

(G) **(d)** From part (c) we know that the area can't exceed 1406.25 ft², so now the job is to find a combination of w and l yielding that area. To accomplish this, take the equation obtained in part (b), replace A with 1406.25, and then use a graphing utility to solve for w.

(e) As a check on the work with the graphing utility (and to stay in shape with the algebra), use the quadratic formula to solve the same equation.

(f) Now that you know w, find l and describe the proportions of the required rectangle.

2.6   MORE ON INEQUALITIES

The symbols of inequality $>$ and $<$ were introduced by [Thomas] Harriot [1560–1621]. The signs $\geq$ and $\leq$ were first used about a century later by the Parisian hydrographer Pierre Bouguer. —Florian Cajori in *A History of Mathematics,* 4th ed. (New York: Chelsea Publishing Co., 1985)

It is rather surprising to think that the man who surveyed and mapped Virginia was one of the founders of algebra as we know the subject today. Such however is the case, for Thomas Harriot was sent by Sir Walter Raleigh to accompany Sir Richard Grenville (1585) to the New World, where he made the survey of that portion of American territory. —David Eugene Smith in *History of Mathematics,* vol. I (New York: Ginn and Co., 1923)

The last example in Section 2.5 emphasized the connection between inequalities and graphs. This is our starting point for the present section. We begin with an example in which polynomial inequalities are solved by relying on a graph. Then we'll use the ideas there to explain a method for solving inequalities even when a graph is not given.

EXAMPLE **1**   **A graphical approach to solving inequalities**

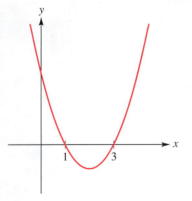

Figure 1
$y = x^2 - 4x + 3$

Figure 1 shows the graph of the equation $y = x^2 - 4x + 3$. Use the graph to solve each of the following inequalities:
(a) $x^2 - 4x + 3 < 0$;   **(b)** $x^2 - 4x + 3 \geq 0$.

SOLUTION
(a) As indicated in Figure 1, the graph of $y = x^2 - 4x + 3$ is below the x-axis when (and only when) the x-values are in the open interval $(1, 3)$. This is the same as saying that the quantity $x^2 - 4x + 3$ is negative when (and only when) the x-values are in the open interval $(1, 3)$. Or, rewording once more, the solution set for the inequality $x^2 - 4x + 3 < 0$ is the open interval $(1, 3)$.

(b) The graph in Figure 1 is above the x-axis when (and only when) the x-values are in either of the two intervals $(-\infty, 1)$ or $(3, \infty)$. This is the same as saying that the quantity $x^2 - 4x + 3$ is positive when (and only when) the x-values are in the set $(-\infty, 1) \cup (3, \infty)$. Furthermore, the graph shows that the quantity $x^2 - 4x + 3$ is equal to 0 when $x = 1$ and when $x = 3$. In summary, the solution set for the inequality $x^2 - 4x + 3 \geq 0$ is the set $(-\infty, 1] \cup [3, \infty)$.

We can summarize the essential features of our work in Example 1 as follows. The x-intercepts of the graph in Figure 1 divide the number line into three intervals—namely, $(-\infty, 1)$, $(1, 3)$, and $(3, \infty)$. Within each of these intervals, the algebraic sign of $x^2 - 4x + 3$ stays the same. [For example, to the left of $x = 1$—that is, on the interval $(-\infty, 1)$—the quantity $x^2 - 4x + 3$ is always positive.] More generally, it can be shown (using calculus) that this same type of behavior—regarding *persistence of sign*—occurs with all polynomials and, indeed, with quotients of polynomials as well. This important fact, along with the definition of a *key number*, is presented in the box that follows.

PROPERTY SUMMARY **Key Numbers and Persistence of Sign**

Let P and Q be polynomials with no common factors (other than, possibly, constants), and consider the following four inequalities:

$$\frac{P}{Q} < 0 \qquad \frac{P}{Q} \le 0 \qquad \frac{P}{Q} > 0 \qquad \frac{P}{Q} \ge 0$$

The **key numbers** for each of these inequalities are the real numbers for which $P = 0$ or $Q = 0$. (Geometrically speaking, the key numbers are the x-intercepts for the graphs of the equations $y = P$ and $y = Q$.) It can be proved (using calculus) that the algebraic sign of P/Q does not change within each of the intervals determined by the key numbers.

EXAMPLES
1. The key numbers for the inequality $(x - 7)/(x - 8) \ge 0$ are 7 and 8.
2. The key numbers for the inequality $(x - 3)(x + 3) < 0$ are ± 3. (In this case, the polynomial Q is the constant 1.)

We'll show how this result is applied by solving the polynomial inequality

$$x^3 - 2x^2 - 3x > 0$$

First, using factoring techniques from basic algebra, we rewrite the inequality in the equivalent form

$$x(x + 1)(x - 3) > 0$$

(If you need a review of factoring, see Appendix B.4.) The key numbers, then, are the solutions of the equation $x(x + 1)(x - 3) = 0$. That is, the key numbers are $x = 0$, $x = -1$, and $x = 3$. Next, we locate these numbers on a number line. As indicated in Figure 2, this divides the number line into four distinct intervals.

Figure 2

Now, according to the result stated in the box just prior to this example, no matter what x-value we choose in the interval $(-\infty, -1)$, the resulting sign of $x^3 - 2x^2 - 3x$ [$= x(x + 1)(x - 3)$] will always be the same. Thus, to see what that sign is, we first choose any convenient *test number* in the interval $(-\infty, -1)$—say, $x = -2$. Then, using $x = -2$, we determine the sign of $x(x + 1)(x - 3)$ simply by considering the sign of each factor, as indicated in Table 1.

Looking at Table 1, we conclude that the values of $x^3 - 2x^2 - 3x$ are negative *throughout* the interval $(-\infty, -1)$ and, consequently, no number in this interval satisfies the given inequality. Next, we carry out similar analyses for the remaining

▌ **TABLE 1**

Interval	Test number	x	$x + 1$	$x - 3$	$x(x + 1)(x - 3)$
$(-\infty, -1)$	-2	neg.	neg.	neg.	neg.

On the interval $(-\infty, -1)$, the sign of $x^3 - 2x^2 - 3x\ [= x(x + 1)(x - 3)]$ is negative because it is the product of three negative factors.

▌ **TABLE 2**

Interval	Test number	x	$x + 1$	$x - 3$	$x(x + 1)(x - 3)$
$(-1, 0)$	$-\frac{1}{2}$	neg.	pos.	neg.	pos.
$(0, 3)$	1	pos.	pos.	neg.	neg.
$(3, \infty)$	4	pos.	pos.	pos.	pos.

On the interval $(-1, 0)$, the product $x(x + 1)(x - 3)$ is positive because it has two negative factors and one positive factor. On $(0, 3)$, the product is negative because it has two positive factors and one negative factor. And on $(3, \infty)$, the product is positive because all the factors are positive.

three intervals, as shown in Table 2. (You should verify for yourself that the entries in the table are correct.)

Looking at the two tables now, we conclude that the sign of $x(x + 1)(x - 3)$ is positive throughout both of the intervals $(-1, 0)$ and $(3, \infty)$ and, consequently, all the numbers in these intervals satisfy the given inequality. Moreover, our work also shows that the other two intervals that we considered are not part of the solution set. As you can readily check, the key numbers themselves do not satisfy the given inequality for this example. In summary, then, the solution set for the inequality $x^3 - 2x^2 - 3x > 0$ is $(-1, 0) \cup (3, \infty)$. Furthermore, it is important to note that the work we just carried out also provides us with three additional pieces of information.

The solution set for $x^3 - 2x^2 - 3x \geq 0$ is $[-1, 0] \cup [3, \infty)$.

The solution set for $x^3 - 2x^2 - 3x < 0$ is $(-\infty, -1) \cup (0, 3)$.

The solution set for $x^3 - 2x^2 - 3x \leq 0$ is $(-\infty, -1] \cup [0, 3]$.

In the box that follows, we summarize the steps for solving polynomial inequalities.

Steps for Solving Polynomial Inequalities

1. If necessary, rewrite the inequality so that the polynomial is on the left-hand side and zero is on the right-hand side.
2. Find the key numbers for the inequality and locate them on a number line.
3. List the intervals determined by the key numbers.
4. From each interval, choose a convenient test number. Then use the test number to determine the sign of the polynomial throughout the interval.
5. Use the information obtained in the previous step to specify the required solution set. [Don't forget to take into account whether the original inequality is strict ($<$ or $>$) or nonstrict ($\leq$ or $\geq$).]

| EXAMPLE | 2 | **Using factoring and key numbers to solve an inequality** |

Solve $x^4 \leq 14x^3 - 48x^2$.

SOLUTION

First we rewrite the inequality so that zero is on the right-hand side. Then we factor the left-hand side as follows:

$$x^4 - 14x^3 + 48x^2 \leq 0$$
$$x^2(x^2 - 14x + 48) \leq 0$$
$$x^2(x - 6)(x - 8) \leq 0$$

From this last line, we see that the key numbers are $x = 0, 6$, and 8. As indicated in Figure 3, these numbers divide the number line into four distinct intervals.

$(-\infty, 0) \qquad (0, 6) \qquad (6, 8) \quad (8, \infty)$

Figure 3

We need to choose a test number from each interval and see whether the polynomial is positive or negative on the interval. This work is carried out in the following table.

Interval	Test number	x^2	$x - 6$	$x - 8$	$x^2(x - 6)(x - 8)$
$(-\infty, 0)$	-1	pos.	neg.	neg.	pos.
$(0, 6)$	1	pos.	neg.	neg.	pos.
$(6, 8)$	7	pos.	pos.	neg.	neg.
$(8, \infty)$	9	pos.	pos.	pos.	pos.

The table shows that the quantity $x^2(x - 6)(x - 8)$ is negative only for x-values in the interval $(6, 8)$. Also, as noted at the start, the quantity is equal to zero when $x = 0, 6$, or 8. Thus, the solution set consists of the numbers in the closed interval $[6, 8]$, along with the number 0. We can write this set

$$[6, 8] \cup \{0\}$$

where $\{0\}$ denotes the set that has zero as its only member.

| EXAMPLE | 3 | **An inequality with no key numbers** |

Solve: **(a)** $x^2 - 4x + 5 > 0$;   **(b)** $x^2 - 4x + 5 < 0$.

SOLUTION

(a) The equation $x^2 - 4x + 5 = 0$ has no real solution (because the discriminant of the quadratic is -4, which is negative), so there is no key number. Consequently, the polynomial $x^2 - 4x + 5$ never changes sign. To see what that sign is, choose the most convenient test number—namely, $x = 0$—and evaluate the polynomial: $0^2 - 4(0) + 5 > 0$. So the polynomial is positive for every value of x, and the solution set is $(-\infty, \infty)$, the set of all real numbers.

GRAPHICAL PERSPECTIVE

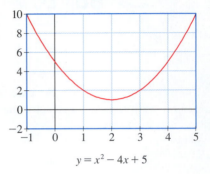

Figure 4
The graph of $y = x^2 - 4x + 5$ lies entirely above the x-axis. Algebraically this means that the quantity $x^2 - 4x + 5$ is positive for all values of x. Consequently, the solution set for $x^2 - 4x + 5 > 0$ is the set of all real numbers. Equivalently, there are no real numbers satisfying $x^2 - 4x + 5 < 0$.

$y = x^2 - 4x + 5$

(b) Our work in part (a) shows that no real number satisfies the inequality $x^2 - 4x + 5 < 0$. See Figure 4 for a graphical perspective on this and on the result in part (a).

The technique used in the previous examples can also be used to solve inequalities involving quotients of polynomials. For these cases recall that the definition of a key number includes the x-values for which the denominator is zero. For example, the key numbers for the inequality $\dfrac{x + 3}{x - 4} \geq 0$ are -3 and 4.

EXAMPLE **4** **An inequality involving a quotient of polynomials**

Solve: $\dfrac{x + 3}{x - 4} \geq 0$.

SOLUTION
The key numbers are -3 and 4. As indicated in Figure 5, these numbers divide the number line into three intervals. In the table that follows, we've chosen a test number from each interval and determined the sign of the quotient $(x + 3)/(x - 4)$ for each interval.

Figure 5

Interval	Test number	$x + 3$	$x - 4$	$\dfrac{x + 3}{x - 4}$
$(-\infty, -3)$	-4	neg.	neg.	pos.
$(-3, 4)$	0	pos.	neg.	neg.
$(4, \infty)$	5	pos.	pos.	pos.

From these results, we conclude that the solution set for $\dfrac{x + 3}{x - 4} \geq 0$ contains the two intervals $(-\infty, -3)$ and $(4, \infty)$. However, we still need to consider the two endpoints, -3 and 4. As you can easily check, the value $x = -3$ does satisfy the given inequality, but $x = 4$ does not. In summary, then, the solution set is $(-\infty, -3] \cup (4, \infty)$. See Figure 6 for a graphical interpretation of this result. [In

GRAPHICAL PERSPECTIVE

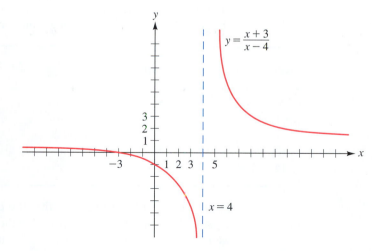

Figure 6
The graph of $y = (x + 3)/(x - 4)$ is above the x-axis when x is to the left of -3 and also when x is to the right of 4. Furthermore, the graph intersects the x-axis (where $y = 0$) when $x = 3$. Consequently, the solution set for the inequality $(x + 3)/(x - 4) \geq 0$ is the set $(-\infty, -3] \cup (4, \infty)$.

Figure 6, the graph of the equation $y = (x + 3)/(x - 4)$ has two distinct pieces, or *branches,* separated by the vertical line $x = 4$. The points on the vertical line itself are not a part of the graph of $y = (x + 3)/(x - 4)$. We'll study graphs like this in Section 5.7, *Rational Functions.*]

| EXAMPLE | 5 | Another inequality involving quotients of polynomials |

Solve: $\dfrac{2x + 1}{x - 1} - \dfrac{2}{x - 3} < 1$.

SOLUTION
Our first inclination here might be to multiply through by $(x - 1)(x - 3)$ to eliminate fractions. This strategy is faulty, however, since we don't know whether the quantity $(x - 1)(x - 3)$ is positive or negative. Thus, we begin by rewriting the inequality in an equivalent form, with zero on the right-hand side and a single fraction on the left-hand side.

$$\frac{2x + 1}{x - 1} - \frac{2}{x - 3} - 1 < 0$$

$$\frac{(2x + 1)(x - 3) - 2(x - 1) - 1(x - 1)(x - 3)}{(x - 1)(x - 3)} < 0$$

$$\frac{x^2 - 3x - 4}{(x - 1)(x - 3)} < 0 \qquad \text{Check the algebra!}$$

$$\frac{(x + 1)(x - 4)}{(x - 1)(x - 3)} < 0$$

(1)

The key numbers are those x-values for which the denominator or the numerator is zero. By inspection, then, we see that these numbers are $-1, 4, 1$, and 3. As Figure 7 indicates, these numbers divide the number line into five distinct intervals.

Figure 7

Now, just as in the previous examples, we choose a test number from each interval and determine the sign of the quotient for that interval. (You should check each entry in the following table for yourself.)

Interval	Test number	$(x + 1)(x - 4)$	$(x - 1)(x - 3)$	$\dfrac{(x + 1)(x - 4)}{(x - 1)(x - 3)}$
$(-\infty, -1)$	-2	pos.	pos.	pos.
$(-1, 1)$	0	neg.	pos.	neg.
$(1, 3)$	2	neg.	neg.	pos.
$(3, 4)$	$7/2$	neg.	pos.	neg.
$(4, \infty)$	5	pos.	pos.	pos.

From these results, we can see that the quotient on the left-hand side of inequality (1) is negative (as required) on the two intervals $(-1, 1)$ and $(3, 4)$. Now we need to check the endpoints of these intervals. When $x = -1$ or $x = 4$, the quotient is zero, and so, in view of the original inequality, we exclude these two x-values from the solution set. Furthermore, the quotient is undefined when $x = 1$ or $x = 3$, so we must also exclude those two values from the solution set. In summary, then, the solution set is $(-1, 1) \cup (3, 4)$.

EXERCISE SET 2.6

A

In Exercises 1–6, use the graph to solve each inequality. (Assume that each figure shows all of the essential features of the graph of the equation; that is, that there are no surprises "out of camera range.")

1. (a) $x^2 - 3x - 4 \geq 0$
   (b) $x^2 - 3x - 4 \leq 0$

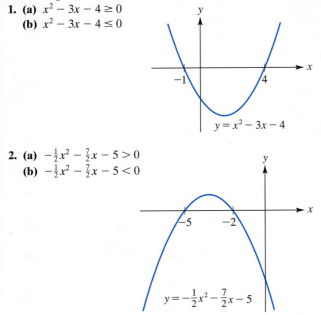

$y = x^2 - 3x - 4$

2. (a) $-\frac{1}{2}x^2 - \frac{7}{2}x - 5 > 0$
   (b) $-\frac{1}{2}x^2 - \frac{7}{2}x - 5 < 0$

$y = -\frac{1}{2}x^2 - \frac{7}{2}x - 5$

3. (a) $x^4 - 4x^3 + 6x^2 - 4x + 2 < 0$
   (b) $x^4 - 4x^3 + 6x^2 - 4x + 2 > 0$

$y = x^4 - 4x^3 + 6x^2 - 4x + 2$

4. (a) $-3(x + 3)^2 \leq 0$
   (b) $-3(x + 3)^2 < 0$

$y = -3(x + 3)^2$

5. (a) $x^3 - 3x^2 - x + 3 \geq 0$
   (b) $x^3 - 3x^2 - x + 3 < 0$

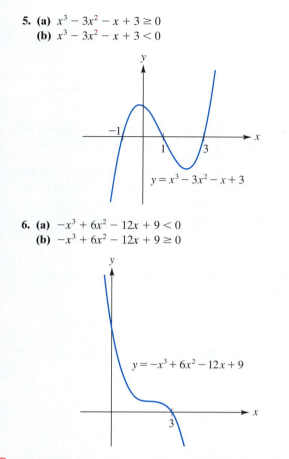

$y = x^3 - 3x^2 - x + 3$

6. (a) $-x^3 + 6x^2 - 12x + 9 < 0$
   (b) $-x^3 + 6x^2 - 12x + 9 \geq 0$

$y = -x^3 + 6x^2 - 12x + 9$

Ⓖ 7. In the text (pages 141–142) we solved the inequality $x^3 - 2x^2 - 3x > 0$. Graph the equation $y = x^3 - 2x^2 - 3x$ and explain or describe (in complete sentences) the relationship between the graph and the solution set of the given inequality.

Ⓖ 8. In Example 2 in the text, we solved the inequality $x^4 \leq 14x^3 - 48x^2$. Graph the equation $y = x^4 - 14x^3 + 48x^2$ and explain or describe (in complete sentences) the relationship between the graph and the solution set of the given inequality.

Solve the inequalities in Exercises 9–58. Suggestion: A calculator is useful for approximating the key numbers in Exercises 25, 26, 48, and 50.

9. $x^2 + x - 6 < 0$
10. $x^2 + 4x - 32 < 0$
11. $x^2 - 11x + 18 > 0$
12. $2x^2 + 7x + 5 > 0$
13. $9x - x^2 \leq 20$
14. $3x^2 + x \leq 4$
15. $x^2 - 16 \geq 0$
16. $24 - x^2 \geq 0$
17. $16x^2 + 24x < -9$
18. $x^4 - 16 < 0$
19. $x^3 + 13x^2 + 42x > 0$
20. $2x^3 - 9x^2 + 4x \geq 0$
21. $2x^2 + 1 \geq 0$
22. $1 + x^2 < 0$
23. $12x^3 + 17x^2 + 6x < 0$
24. $8x^4 < x^2 - 2x^3$
25. $x^2 + x - 1 > 0$
26. $2x^2 + 9x - 1 > 0$
27. $x^2 - 8x + 2 \leq 0$
28. $3x^2 - x + 5 \leq 0$

29. $(x - 1)(x + 3)(x + 4) \geq 0$
30. $x^4(x - 2)(x - 16) \geq 0$
31. $(x + 4)(x + 5)(x + 6) < 0$
32. $(x - \frac{1}{2})(x + \frac{1}{2})(x + \frac{3}{2}) < 0$
33. $(x - 2)^2(3x + 1)^3(3x - 1) > 0$
34. $(2x - 1)^3(2x - 3)^5(2x - 5) > 0$
35. $(x - 3)^2(x + 1)^4(2x + 1)^4(3x + 2) \leq 0$
36. $x^4 - 25x^2 + 144 \leq 0$
37. $20 \geq x^2(9 - x^2)$
38. $x^2(3x^2 + 11) \geq 4$
39. $9(x - 4) - x^2(x - 4) < 0$
40. $(x + 1)^2 - 5(x + 1) > 14$
41. $4(x^2 - 9) - (x^2 - 9)^2 > -5$
42. $x(1 - x^2)^4 + (x + 3)(1 - x^2)^4 \geq 0$

43. $\dfrac{x - 1}{x + 1} \leq 0$
44. $\dfrac{x + 4}{2x - 5} \leq 0$

45. $\dfrac{2 - x}{3 - 2x} \geq 0$
46. $\dfrac{x^2 - 1}{x^2 + 8x + 15} \geq 0$

47. $\dfrac{x^2 - 8x - 9}{x} < 0$
48. $\dfrac{x^2 - 3x + 1}{1 - x} < 0$

49. $\dfrac{2x^3 + 5x^2 - 7x}{3x^2 + 7x + 4} > 0$
50. $\dfrac{x^2 - x - 1}{x^2 + x - 1} > 0$

51. $\dfrac{x}{x + 1} > 1$
52. $\dfrac{2x}{x - 2} < 3$

53. $\dfrac{1}{x} \leq \dfrac{1}{x + 1}$
54. $\dfrac{2}{x} < \dfrac{x}{2}$

55. $\dfrac{1}{x - 2} - \dfrac{1}{x - 1} \geq \dfrac{1}{6}$
56. $\dfrac{2x}{x + 5} + \dfrac{x - 1}{x - 5} < \dfrac{1}{5}$

57. $\dfrac{1 + x}{1 - x} - \dfrac{1 - x}{1 + x} < -1$
58. $\dfrac{x + 1}{x + 2} > \dfrac{x - 3}{x + 4}$

For Exercises 59–74:

Ⓖ (a) *Use a graph to estimate the solution set for each inequality. Zoom in far enough so that you can estimate the relevant endpoints to the nearest thousandth.*

(b) *Exercises 59–68 can be solved algebraically using the techniques presented in this section. Carry out the algebra to obtain exact expressions for the endpoints that you estimated in part (a). Then use a calculator to check that your results are consistent with the previous estimates.*

59. $x^2 - 5x + 3 \leq 0$
60. $x^2 + x - 4 \leq 0$
61. $0.25x^2 - 6x - 2 < 0$
62. $0.25x^3 - 6x^2 - 2x < 0$
63. $x^4 - 2x^2 - 1 > 0$
64. $x^4 - 2x^2 + 1 > 0$
65. $(x^2 - 5)/(x^2 + 1) \leq 0$
66. $(x^3 - 5)/(x^2 + 1) \leq 0$
67. $(x^2 + 1)/(x^2 - 5) \leq 0$
68. $(x^2 + 1)/(x^3 - 5) \leq 0$
69. $x^3 + 2x \geq -1$
70. $x^3 - 2x \geq -1$
71. $x^4 - 2x > -1$
72. $x^5 - 2x > -1$

73. $x - \dfrac{x^3}{3!} + \dfrac{x^5}{5!} - \dfrac{x^7}{7!} < 0$ *Note: For natural numbers n, the symbol $n!$ (read n factorial) denotes the product of the first n natural numbers. For instance, $3! = 1 \times 2 \times 3 = 6$.*

74. $1 - \dfrac{x^2}{2!} + \dfrac{x^4}{4!} - \dfrac{x^6}{6!} + \dfrac{x^8}{8!} < 0$

B

(G) **75.** Suppose that after studying a corporation's records, a business analyst predicts that the corporation's monthly revenues R for the near future can be closely approximated by the equation

$$R = -0.0217x^5 + 0.626x^4 - 6.071x^3 + 25.216x^2 \\ -57.703x + 159.955 \quad (1 \le x \le 12)$$

where R is the revenue (in thousands of dollars) for the month x, with $x = 1$ denoting January, $x = 2$ denoting February, and so on.

(a) According to this model, for which months will the monthly revenue be no more than $80,000? *Hint*: You need to solve the inequality $R \le 80$. Round each key number to the nearest integer. (Why?)

(b) For which months, if any, will the monthly revenue be at least $120,000?

(G) **76.** (Continuation of Exercise 75)

(a) Solve the inequality $R > 165$ to determine the months, if any, that the revenue will exceed $165,000.

(b) Are there any months when the revenue will fall below $45,000?

77. For which values of b will the equation $x^2 + bx + 1 = 0$ have real solutions?

78. The sum of the first n natural numbers is given by

$$1 + 2 + 3 + \cdots + n = \frac{n(n + 1)}{2}$$

For which values of n will the sum be less than 1225?

79. For which values of a is $x = 1$ a solution of the inequality

$$\frac{2a + x}{x - 2a} < 1?$$

80. Solve: $\dfrac{ax + b}{\sqrt{x}} > 2\sqrt{ab}$, where a and b are positive constants.

81. The two shorter sides in a right triangle have lengths x and $1 - x$, where $x > 0$. For which values of x will the hypotenuse be less than $\sqrt{17}/5$?

82. A piece of wire 12 cm long is cut into two pieces. Denote the lengths of the two pieces by x and $12 - x$. Both pieces are then bent into squares. For which values of x will the combined areas of the squares exceed 5 cm²?

C

83. Find a nonzero value for c so that the solution set for the inequality

$$x^2 + 2cx - 6c < 0$$

is the open interval $(-3c, c)$.

84. Solve: $(x - a)^2 - (x - b)^2 > (a - b)^2/4$, where a and b are constants and $a > b$.

PROJECT | WIND POWER

The relatively recent [in terms of human history] *transition to coal that began in Europe in the seventeenth century marked a major shift to dependence on a finite stock of fossilized fuels whose remaining energy is now equivalent to less than 11 days of sunshine. From a millennial perspective, today's hydrocarbon-based civilization is but a brief interlude in human history.* —Christopher Flavin and Seth Dunn, "Reinventing the Energy System" in *Vital Signs, 1999,* Lester R. Brown et al. (New York: W.W. Norton & Co., 1999)

As the world's nonrenewable energy resources such as oil and coal start to become more scarce, alternative sources such wind and solar become increasingly important. In this project you'll calculate some estimates concerning wind power, an industry and technology that has grown rapidly over the past two decades. (Later, in Chapter 5, after studying exponential and logarithmic functions, we'll be able to make some rough estimates for possible depletion dates regarding resources such as oil and coal.) In working this project, you'll be reviewing math from the following sections of this book:

- 1.3 Solving Equations
- 1.6 Equations of Lines
- 2.5 Inequalities
- 2.6 More on Inequalities

Wind turbines in Altamont, California. Photo by Glen Allison, PhotoDisc.

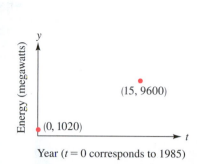

Year ($t = 0$ corresponds to 1985)

World wind-energy generating capacity, 1985 ($t = 0$) and 1998 ($t = 15$).

Source: Data from *BTM Consult,* published in *Vital Signs, 1999,* p. 53, Lester R. Brown et al. (New York: W. W. Norton & Co., 1999)

(a) The two data points in the above figure show world wind-energy generating capacity for the years 1985 and 1998. What is the slope of the line joining the two points? What are the units for the slope?

(b) Find the equation of the line joining the two given points. Write the answer in the form $y = mt + b$.

(c) Use the result in part (b) to estimate the years in which world wind-energy generating capacity was between 3000 megawatts and 7500 megawatts (inclusive). *Hint:* You need to solve the inequality $3000 \leq y \leq 7500$; round the answers to the nearest integers.

(d) Your estimates in part (c) are based upon the assumption that world wind-energy generating capacity increased in a straight-line fashion from 1985 to 1998. By using techniques from statistics (and more detailed data), it can be shown that a more accurate summary of world wind-energy generating capacity over these years is given by the equation

$$y = 6.977t^3 - 64.786t^2 + 325.4t + 1012$$

where again, $t = 0$ corresponds to 1985. Using this expression for y, estimate the years in which world wind-energy generating capacity was between 3000 megawatts and 7500 megawatts (inclusive). (Round the key numbers to the nearest integers.) Compare your results to those in part (c).

(e) Although increasing rapidly, wind-energy generation is currently but a small portion of the world energy picture. For instance, whereas world wind-energy generating capacity in 1998 was 9600 megawatts, the corresponding figure for nuclear energy in 1998 was 343,000 megawatts. When might wind-energy reach this level? Use the equation in part (d) to obtain one possible answer to this question. Round your answer to the nearest five years. *Hint:* In an appropriate viewing rectangle, you want to look for the point where the curve $y = 6.977t^3 - 64.786t^2 + 325.4t + 1012$ meets the horizontal line $y = 343,000$.

(continues)

Ⓖ **(f)** Rework part (e), but instead of using the cubic equation from part (d), use the straight-line equation obtained in part (b). Compare the results; which scenario do you think is more realistic?

Chapter 2

Summary of Principal Terms and Formulas

Terms or notation	Page reference	Comments
1. Quadratic equation	85	A quadratic equation is an equation that can be written in the form $ax^2 + bx + c = 0$, where $a \neq 0$.
2. Quadratic formula $$x = \frac{-b \pm \sqrt{b^2 - 4ac}}{2a}$$	87	This is the quadratic formula; it provides the solutions of the quadratic equation $ax^2 + bx + c = 0$. The formula is derived on page 87 by means of the useful technique of completing the square.
3. Product-and-sum-of-roots theorem	89	The product and the sum of the roots of the quadratic equation $x^2 + bx + c = 0$ are c and $-b$, respectively.
4. Discriminant	92	The discriminant of the quadratic equation $ax^2 + bx + c = 0$ is the number $b^2 - 4ac$. As indicated in the box on page 92, the discriminant provides information about the roots of the equation.
5. Extraneous solution (or extraneous root)	104	In solving equations, certain processes (such as squaring both sides) can lead to answers that do not check in the original equation. These numbers are called extraneous solutions (or extraneous roots).
6. Solution set of an inequality	130	This is the set of numbers that satisfy the inequality.
7. $A \cup B$	134	The set $A \cup B$ consists of all elements that are in at least one of the two sets A and B.
8. Key numbers of an inequality	141	Let P and Q denote polynomials and consider the following four inequalities: $$\frac{P}{Q} < 0 \qquad \frac{P}{Q} \leq 0 \qquad \frac{P}{Q} > 0 \qquad \frac{P}{Q} \geq 0$$ The key numbers for each of these inequalities are those real numbers for which $P = 0$ or $Q = 0$. It can be proved that the algebraic sign of P/Q is constant within each of the intervals determined by these key numbers.

Writing Mathematics

1. More than 3000 years ago, the ancient Babylonian mathematicians solved quadratic equations. A method they used is demonstrated (using modern notation) in the following example.

   To find the positive root of $x^2 + 8x = 84$:

   Rewrite the equation as $x(x + 8) = 84$, and let $y = x + 8$. Then the equation to be solved becomes $xy = 84$. Now take half of the coefficient of x in the original equation, which is 4, and define another variable t by $t = x + 4$. Then we have $x = t - 4$ and $y = t + 4$. Therefore

$$(t - 4)(t + 4) = 84$$
$$t^2 = 100$$
$$t = 10$$

   With $t = 10$ we get $x = 10 - 4 = 6$, the required positive root.

   On your own or with a group of classmates, work through the above example, filling in any missing details if necessary. Then (strictly on your own), use the Babylonian method to find the positive root of each of the following equations. Write out your solutions in detail, as if you were explaining the method to another student who had not seen it before. This will involve a combination of English composition and algebra. Also, in part (b), check your answer by using the quadratic formula.
   (a) $x^2 + 14x = 72$ **(b)** $x^2 + 2Ax = B$, where $B > 0$

2. The ancient Greek mathematicians (2500 years ago) used geometric methods to solve quadratic equations or, rather, to construct line segments whose lengths were the roots of the equations. According to historian Howard Eves in *An Introduction to the History of Mathematics,* 6th ed. (Philadelphia: Saunders College Publishing, 1990):

   Imbued with the representation of a number by a length and completely lacking any adequate algebraic notation, the early Greeks devised ingenious geometrical processes for carrying out algebraic operations.

   One of the methods used by the ancient Greeks to solve a quadratic equation is described in the following example (using modern algebraic notation).
   To construct a line segment whose length is equal to the (positive) root of the equation $x^2 + 8x = 84$:

   Begin with a line segment $\overline{AB}$ of length 8. At B, construct a line segment $\overline{BC} \perp \overline{AB}$ such that the length of $\overline{BC}$ is $\sqrt{84}$. Next, let M be the midpoint of $\overline{AB}$. With M as center, draw a circular arc of radius $\overline{MC}$ intersecting $\overline{AB}$ (extended) at P. Then the length of $\overline{AP}$ is the required root.

   On your own or with a group of classmates, work through the preceding construction. That is, sketch the appropriate figure and verify that the construction indeed yields the positive root of the equation. Then, on your own, use the Greek method to determine the positive root of each of the following equations. Write out your work in detail, as if you were explaining it to a student who had not seen it before. Be sure to include the appropriate geometric figures and an explanation of why the method works.

(a) $x^2 + 14x = 72$ **(b)** $x^2 + 2cx = d$ $(c, d > 0)$

3. The example in Exercise 1 demonstrates a Babylonian method for solving a quadratic equation. According to Professor Victor J. Katz in *A History of Mathematics* (New York: HarperCollins College Publishers, 1993), "... whatever the ultimate origin of this method, a close reading of the wording of the [ancient clay] tablets seems to indicate that the scribe had in mind a geometric procedure [completing-the-square]. ..."

One of the earliest *explicit* uses of the completing-the-square technique to solve a quadratic is due to the ninth-century Persian mathematician and astronomer Muhammed al-Khwārizmī. The following example demonstrates al-Khwārizmī's use of completing the square to solve the equation $x^2 + 8x = 84$:

To find the positive root of $x^2 + 8x = 84$:

Begin with a square of side x, as in Figure A. Take half of the coefficient of x: this is one-half of 8, or 4. Now, form two rectangles, each with dimensions 4-by-x, and adjoin them to the square, as indicated in Figure B. Then, as in Figure C, draw the dashed lines to complete the (outer) square.

In Figure B, the combined area of the square and the two rectangles is $x^2 + 8x$. But, from the equation we wish to solve, $x^2 + 8x$ is equal to 84. In Figure C, the area of the small square in the lower right-hand corner is 16. Thus the area of the entire outer square is $16 + 84$, or 100. But the area of the outer square is also $(x + 4)^2$. Therefore,

$$(x + 4)^2 = 100 \quad \text{and consequently} \quad x + 4 = 10$$

The positive root we are looking for is therefore $x = 6$.

On your own or with a group of classmates, work through the preceding example, filling in the missing details as necessary. Then, on your own, use this completing-the-square process to find the positive root of each of the following equations. Write out your solutions in detail, as if you were explaining the method to another student who had not seen it before. Be sure to include the appropriate geometric figures.

(a) $x^2 + 14x = 72$ **(b)** $x^2 + 2Ax = B$, where $B > 0$

Figure A

Figure B

Figure C

Chapter 2   Review Exercises

In Exercises 1–12, answer TRUE *if the statement is true without exception. Otherwise, answer* FALSE.

1. If $x < 3$, then $x + 7 < 10$.

2. If $x < y$, then $x^2 < y^2$.

3. If $x \geq -4$, then $x \leq 4$.

4. If $-x > y$, then $x < -y$.

5. If $x < 2$, then $\dfrac{1}{x} < \dfrac{1}{2}$.   **6.** $x \leq x^2$

7. If $\sqrt{x + 1} = 3$, then $x + 1 = 9$.

8. If $0 < x < 1$, then $\dfrac{1}{x} > x$.

9. If $x < 2$ and $y < 3$, then $x < y$.

10. If $x < 2$ and $y > 3$, then $x < y$.

11. If $a - b \leq a^2 - b^2$, then $1 \leq a + b$.

12. If $0 < a - b \leq a^2 - b^2$, then $1 \leq a + b$.

In Exercises 13–32, find all of the real solutions of each equation.

13. $12x^2 + 2x - 2 = 0$

14. $4y^2 - 21y = 18$

15. $\frac{1}{2}x^2 + x - 12 = 0$

16. $x^2 + \frac{13}{2}x + 10 = 0$

17. $\dfrac{x}{5 - x} = \dfrac{-2}{11 - x}$

18. $\dfrac{x^2}{(x - 1)(x + 1)} = \dfrac{4}{x + 1} + \dfrac{4}{(x - 1)(x + 1)}$

19. $\dfrac{1}{3x - 7} - \dfrac{2}{5x - 5} - \dfrac{3}{3x + 1} = 0$

20. $4x^2 + x - 2 = 0$

21. $t^2 + t - \frac{1}{2} = 0$

22. $x^3 - 6x^2 + 7x = 0$

23. $1 + 14x^{-1} + 48x^{-2} = 0$

24. $x^{-2} - x^{-1} - 1 = 0$

25. $x^{1/2} - 13x^{1/4} + 36 = 0$

26. $y^{2/3} - 21y^{1/3} + 80 = 0$

27. $\sqrt{4x + 3} = \sqrt{11 - 8x} - 1$

28. $2 - \sqrt{3}\sqrt{2x - 1} + x = 0$

29. $\dfrac{2}{\sqrt{x^2 - 36}} + \dfrac{1}{\sqrt{x + 6}} - \dfrac{1}{\sqrt{x - 6}} = 0$

30. $\sqrt{x} - \sqrt{1 - x} + \sqrt{x - 1} = 0$

31. $\sqrt{x + 7} - \sqrt{x + 2} = \sqrt{x - 1} - \sqrt{x - 2}$

32. $\dfrac{\sqrt{x} - 4}{\sqrt{x} - 18} = 3$

In Exercises 33–38, solve each equation for x in terms of the other letters.

33. $4x^2y^2 - 4xy = -1$, where $y \neq 0$

34. $4x^4y^4 + 2x^2y^2 + \frac{1}{4} = 0$, where $y \neq 0$

35. $x + \dfrac{1}{a} - \dfrac{1}{b} = \dfrac{2}{a^2x} + \dfrac{2}{abx}$

36. $\dfrac{a}{x - b} + \dfrac{b}{x - a} = \dfrac{x - a}{b} + \dfrac{x - b}{a}$, where $a \neq -b$

Suggestion: Before clearing fractions, carry out the indicated additions on each side of the equation.

37. $\dfrac{1}{x + a + b} = \dfrac{1}{x} + \dfrac{1}{a} + \dfrac{1}{b}$, where $a + b \neq 0$

38. $\dfrac{1}{x} + \dfrac{1}{a - x} + \dfrac{1}{x + 3a} = 0$, where $a \neq 0$

Solve the inequalities in Exercises 39–55.

39. $-1 < \dfrac{1 - 2(1 + x)}{3} < 1$

40. $3 < \dfrac{x - 1}{-2} < 4$

41. $|x| \leq 1/2$

42. $|3 - 5x| < 2$

43. $|2x - 1| \geq 5$

44. $x^2 - 21x + 108 \leq 0$

45. $x^2 + 3x - 40 < 0$

46. $x^2 \geq 15x$

47. $x^2 - 6x - 1 < 0$

48. $(x + 12)(x - 1)(x - 8) < 0$

49. $x^4 - 34x^2 + 225 < 0$

50. $\dfrac{x + 12}{x - 5} > 0$

51. $\dfrac{(x - 7)^2}{(x + 2)^3} \geq 0$

52. $\dfrac{(x - 6)^2(x - 8)(x + 3)}{(x - 3)^2} \leq 0$

53. $\dfrac{3x + 1}{x - 4} < 1$

54. $\dfrac{1 - 2x}{1 + 2x} \leq \dfrac{1}{2}$

55. $x^2 + \dfrac{1}{x^2} > 3$ *Suggestion:* Use a calculator to evaluate the key numbers.

For Exercises 56 and 57, find the values of k for which the roots of the equations are real numbers.

56. $kx^2 - 6x + 5 = 0$

57. $x^2 + (k + 1)x + 2k = 0$

In Exercises 58 and 59, find a quadratic equation that has integer coefficients and the given values as roots.

58. $r_1 = (1 + \sqrt{5})/3$, $r_2 = (1 - \sqrt{5})/3$

59. $r_1 = -3$, $r_2 = -4$

60. Solve for x:

$$5\sqrt[3]{b^2 - x^2} = \sqrt[3]{(b + x)^2} + 4\sqrt[3]{(b - x)^2},$$

where $a \neq 0$

Hint: Divide through by $\sqrt[3]{(b - x)^2}$. The resulting equation can be solved by using the substitution

$$t = \sqrt[3]{\dfrac{b + x}{b - x}}.$$

61. The sum of the cubes of two numbers is 2071, while the sum of the two numbers themselves is 19. Find the two numbers.

62. How many pounds of an alloy that is 12% copper must be added to 1000 lb of an alloy that is 9% copper to yield a new alloy that is 11.2% copper?

63. The sum of the digits in a certain two-digit number is 11. If the order of the digits is reversed, the number is increased by 27. Find the original number.

64. Robin deposits $6000, part at 10% and the remainder at 12%. If the net interest rate for the entire $6000 is $11\frac{1}{3}$%, how much was deposited at each rate? (Assume simple interest.)

65. The four corners of a square $ABCD$ have been cut off to form a regular octagon, as shown in the figure. If each side of the square is 1 cm long, how long is each side of the octagon?

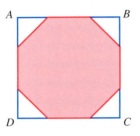

66. Determine p if the larger root of the equation $x^2 + px - 2 = 0$ is p.

67. A piece of wire x cm long is bent into a square. For which values of x will the area be (numerically) greater than the perimeter?

68. A piece of wire $6x$ cm long is bent into an equilateral triangle. For which values of x will the area be (numerically) less than the perimeter?

69. Find three consecutive positive integers such that the sum of their squares is 1454.

70. The sum of two numbers is 1 and the sum of their cubes is 91. What are the numbers?

71. In a 10-mile race, Joan covers the first 4 miles at a constant rate. Then she speeds up a bit and runs the last 6 miles at a rate that is $\frac{1}{2}$ mph faster. If the entire race had been run at this faster pace, Joan's overall time would have been 2 min better. What was Joan's speed for the first 4 miles, and what was her time for the race?

72. The length of a rectangular piece of tin exceeds the width by 8 cm. A 1 cm square is cut from each corner of the piece of tin, and then the resulting flaps are turned up to form a box with no top. What are the dimensions of the box if its volume is 48 cm³?

73. The length and width of a rectangular flower garden are a and b, respectively. The garden is bordered on all four sides by a gravel path of uniform width. Find the width of the path, given that the area of the garden equals the area of the path.

If an object is thrown vertically upward from a height of h_0 ft with an initial speed of v_0 ft/sec, then its height h (in feet) after t seconds is given by

$$h = -16t^2 + v_0 t + h_0$$

Make use of this formula in working Exercises 74–76.

74. A ball is thrown vertically upward from ground level with an initial speed of 64 ft/sec.
   (a) At what time will the height of the ball be 15 ft? (Two answers.)
   (b) For how long an interval of time will the height exceed 63 ft?

75. One ball is thrown vertically upward from a height of 50 ft with an initial speed of 40 ft/sec. At the same instant, another ball is thrown vertically upward from a height of 100 ft with an initial speed of 5 ft/sec. Which ball hits the ground first?

76. An object is projected vertically upward. Suppose that its height is H ft at t_1 sec and again at t_2 sec. Express the initial speed in terms of t_1 and t_2.
   Answer: $16(t_1 + t_2)$

77. A rectangle is inscribed in a semicircle of radius 1 cm, as shown. For which value of x is the area of the rectangle 1 cm²? *Note:* x is defined in the figure.

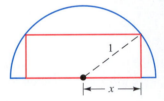

78. The height of an isosceles triangle is $a + b$ units, where $a > b > 0$, and the area does not exceed $a^2 - b^2$ square units. What is the range of possible values for the base of the triangle?

79. A circle is inscribed in a quadrant of a larger circle of radius r (as shown in the figure). Find the radius of the inscribed circle.

Chapter 2 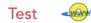 Test

In Problems 1–4, find all of the real solutions of each equation.

1. (a) $x^2 + 4x = 5$
   (b) $x^2 + 4x = 1$
2. $x^2(x^2 - 7) + 12 = 0$
3. $1 + \sqrt{2 - x} - \sqrt{5 - 2x} = 0$
4. $|3x - 1| = 2$
5. Find the sum of the roots of the equation $2x^2 + 8x - 9 = 0$.
6. Find a quadratic equation with integer coefficients and with roots $r_1 = 2 + 3\sqrt{7}$ and $r_2 = 2 - 3\sqrt{7}$.
7. As indicated in the accompanying figure, the graph of $y = x^4 - 3x^2 - 1$ has two x-intercepts, one between 1.5 and 2 and one between -1.5 and -2. Find exact expressions (containing radicals) for these intercepts and also calculator approximations rounded to three decimal places. Check that your calculator values are consistent with the graph.

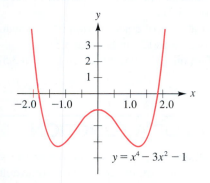

$$y = x^4 - 3x^2 - 1$$

8. Consider the equation $2x^{4/3} - x^{2/3} - 6 = 0$.
   (a) Use a graphing utility to approximate the roots to the nearest hundredth.

(b) Use algebra to determine the exact solutions. Then check that the answers are consistent with the approximations obtained in part (a).

In Problems 9–13, solve the inequalities. Write the answers using interval notation.

9. $4(1 + x) - 3(2x - 1) \geq 1$
10. $\dfrac{3}{5} < \dfrac{3 - 2x}{-4} < \dfrac{4}{5}$ 11. $|3x - 8| \leq 1$
12. $(x - 4)^2(x + 8)^3 \geq 0$
13. $\dfrac{1}{x} + \dfrac{1}{x + 1} + \dfrac{1}{x + 2} \geq 0$
14. Kona coffee beans sell for \$6/lb and Colombian beans for \$5/lb. How many pounds of each bean must be mixed to yield 280 lb of a blend selling for \$5.60/lb?
15. At 6 A.M. a train traveling at 50 mph leaves Los Angeles for San Francisco. Three hours later a train traveling at 40 mph leaves San Francisco for Los Angeles. What time will the trains pass each other? Assume that the distance between the two cities is 450 miles.
16. Find the values of k for which the roots of the equation $x^2 + 3x + k^2 = 0$ are real numbers.
17. The point (a, b) lies in the third quadrant on the graph of the line $y = 2x + 1$. Find a and b given that the distance from (a, b) to the origin is $\sqrt{65}$.
18. After solving the equation $1 + \sqrt{3x + 7} = x$, give an example of what is meant by an extraneous root, and explain (in complete sentences) how, in this case, the extraneous root was generated.

CHAPTER 3

Functions

The concept of a function is one of the most useful and broad-ranging ideas in all of mathematics. After introducing the definition of a function and function notation in Section 3.1, we use graphs in the next three sections to understand and to analyze functions. In Sections 3.5 and 3.6 we find that two functions can be combined to produce new functions. One of these ways of combining two functions is known as *composition of functions.* As you'll see, this is the unifying theme between Sections 3.5 and 3.6. Finally, in Section 3.7 we study certain types of functions and formulas that occur so frequently in applications that a special terminology exists to describe them. In this chapter, for example, we use functions to

- Estimate how many units of a particular item a store can expect to sell at a given price (Example 7 in Section 3.1)
- Compute the average rate of increase or decrease of a patient's fever over given time periods (Example 3 in Section 3.3)
- Analyze and compare changing patterns of consumption of red meat in China and in the United States (Exercises 19 and 20 in Section 3.3)
- Determine the rate of spread of an oil spill (Example 5 in Section 3.5)

The word "function" was introduced into mathematics by Leibniz, who used the term primarily to refer to certain kinds of mathematical formulas. It was later realized that Leibniz's idea of function was much too limited in scope, and the meaning of the word has since undergone many steps of generalization. —Tom M. Apostol in *Calculus,* 2nd ed. (New York: John Wiley & Sons, 1967)

To make a start out of particulars and make them general . . . —William Carlos Williams (1883–1963) in his poem *Paterson*

3.1 THE DEFINITION OF A FUNCTION

There are numerous instances in mathematics, business, and the sciences in which one quantity corresponds to or depends on another according to some definite *rule.* Consider, for example, the equation $y = 2x + 1$. Each time that we select an x-value, a corresponding y-value is determined, in this case according to the rule *multiply by 2, then add 1.* In this sense, the equation $y = 2x + 1$ represents a *function.* It is a rule specifying a y-value corresponding to each x-value. It is useful to think of the x-values as inputs and the corresponding y-values as outputs. The function or rule then tells us what output results from a given input, as indicated schematically in Figure 1.

In general, a function can be represented in one or more of the following forms:

- Algebraic, using an equation or formula (as in the previous paragraph)
- Verbal (also as in the previous paragraph)
- Tabular, using a two-column table
- Graphical, using a coordinate system and graph

After looking at some examples of these ways of representing functions, we'll give an "official" definition of the term *function* on page 160.

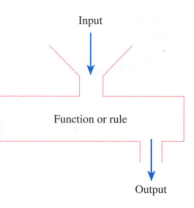

Figure 1

Algebraic. Many of the familiar formulas from geometry represent functions. The area A of a circle depends on the radius r according to the formula $A = \pi r^2$. For each value of r, a corresponding value for A is determined by this formula or rule. The values for r are the inputs, and the corresponding values for A are the outputs. The field of statistics provides another rich source of examples of functions. For instance, over the years 1850–1900, the population of the United States can be closely approximated using the equation

$$y = 0.006609x^2 - 23.777x + 21{,}382$$

where y is the population in millions and x is the year. (We worked with this equation in Example 1 in Section 2.1.) For this rule or function, the inputs are the numbers 1850, 1851, . . . 1900, and the corresponding outputs are the population estimates (in millions).

Verbal. If you count the number of times that a cricket chirps in one minute, you can estimate the air temperature (Fahrenheit) using the following rule or function: *Divide the number of chirps by 4, then add 40*; the answer will be a good estimate for the air temperature. (We're assuming here that at a given temperature, all crickets chirp at the same rate.) For instance, suppose that you count 80 chirps in one minute. This is our input. Applying the rule, we divide 80 by 4, which is 20, and then add 40 to obtain 60. Our estimate for the air temperature is 60°F; that's the output corresponding to an input of 80 chirps/min.

The terminology of proportion and variation introduced in intermediate algebra is sometimes used in the sciences to represent functions verbally. (Don't worry now if you don't recall this terminology; in Section 3.7 it will be reintroduced from scratch.) Here's an example from physics. If you drop a ball off the roof of a building, then before it hits the ground (and neglecting air resistance), *the distance the ball falls is directly proportional to the square of the elapsed time*. The equivalent algebraic representation for this verbal rule or function is

$$d = kt^2$$

where d is the distance that the ball falls in time t, and k is a constant. (If the distance d is measured in feet and the time t in seconds, then the numerical value for k turns out to be 16, and the algebraic form of the rule or function becomes $d = 16t^2$.)

TABLE 1 Federal Minimum Wage

Year	Minimum wage (dollars)
1975	2.00
1980	3.10
1985	3.35
1990	3.80
1995	4.25
2000	5.15

Source: U.S. Bureau of Labor Statistics

TABLE 2 Number of Moons Each Planet in the Solar System Is Known to Have as of 2000

Planet	Number of moons
Mercury	0
Venus	0
Earth	1
Mars	2
Jupiter	16
Saturn	18
Uranus	18
Neptune	8
Pluto	1

Tabular. In many applications, a function is specified by means of a two-column table. Unless stated otherwise, we usually think of the entries in the left-hand column as the inputs and the entries in the right-hand column as the outputs. In Table 1, for example, the input 1975 yields the output $2.00. The inputs and outputs for a function needn't always be numbers. Notice that in Table 2, the inputs aren't numbers, but rather names of the planets in our solar system. In Table 3, neither the inputs nor the outputs are numbers. Another item to note: as indicated in the last row of Table 3, it is possible for an input and output to be the same item. For an algebraic example of this, consider the rule defined by $y = x^2$. For the input $x = 1$, the corresponding output is $y = 1^2 = 1$. That is, the input and output are both 1.

TABLE 3 Original Names of Some Movie Stars

Movie star	Original name
Kate Capshaw	Kathleen Sue Nail
Tom Cruise	Thomas Cruise Mapother, IV
Demi Moore	Demetria Guynes
Kevin Spacey	Kevin Fowler
Arnold Schwarzenegger	Arnold Schwarzenegger

Graphical. In newspapers, business, and the sciences, functions are often presented graphically. Sometimes, but certainly not always, this is in conjunction with an explicit table of values or an equation. For example, consider the minimum wage information given in Table 1. As you know from Chapter 1 (or from previous courses), each row of the table gives us an ordered pair of numbers that we can plot in a coordinate system. Looking at Table 1, we see that the first row gives us the point with coordinates (1975, 2.00), the second row gives us the point (1980, 3.10), and so on. The resulting graph is shown in Figure 2.

In the first paragraph of this section we used the equation $y = 2x + 1$ as an example of a function represented algebraically. A graphical representation of this function is shown in Figure 3(a). (We've used the techniques of Section 1.6 to graph the equation $y = 2x + 1$.) Given this graphical representation of the function (and ignoring, for the moment, the fact that we have the equation $y = 2x + 1$), what is

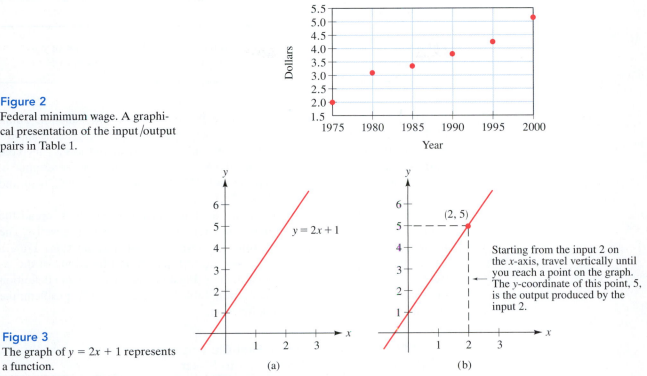

Figure 2

Federal minimum wage. A graphical presentation of the input/output pairs in Table 1.

Figure 3

The graph of $y = 2x + 1$ represents a function.

(a)   (b)

Starting from the input 2 on the x-axis, travel vertically until you reach a point on the graph. The y-coordinate of this point, 5, is the output produced by the input 2.

the underlying rule here for obtaining the output y from a given input x? One way to state the rule is this:

> Given an input on the x-axis, for example $x = 2$ as in Figure 3(b), travel vertically until you reach a point on the graph. The y-coordinate of this point is the required output. Thus, according to Figure 3(b), the input $x = 2$ produces the output $y = 5$.

A less dynamic but more succinct (and more general) way to state the rule is this:

> Suppose we have a graph of a function in an x-y coordinate system. Then a point (x, y) is on the graph if and only if the function assigns the output y to the input x.

(Actually, when we take a more careful look at graphs of functions in Section 3.2, we'll adapt this last version for the very definition of the graph of a function.)

Now that we've looked at a few introductory examples, it's time to state the definition of the term *function* as it is used in mathematics. We need this definition before we can answer the following two questions, which may have already occurred to you:

> Does every rule or table represent a function?
> Does every graph represent a function?

As you'll see later in this section, the answer to the first question is *no*. For the second question, the answer is also negative, but we postpone that discussion until the next section. The following definition of the term *function* is broad enough to encompass all of the examples that we've looked at.

DEFINITION | Function

Let A and B be two nonempty sets. A function from A to B is a rule of correspondence that assigns to each element in set A exactly one element in B.

The set A in the definition just given is called the **domain** of the function. Think of the domain as the set of all inputs. For each input, the function gives us an output (from the set B mentioned in the definition). The set of all outputs is called the **range** of the function. For example, for the minimum wage function represented in Table 1 (on page 158), the domain is the set {1975, 1980, 1985, 1990, 1995, 2000}, and the range is {\$2.00, \$3.10, \$3.35, \$3.80, \$4.25, \$5.15}.

In a moment we'll talk more about domain and range, but first reread the definition of function in the box above and note the use of the word "exactly." The inclusion of that word implies that some rules will *not* qualify as functions. For example, the rule represented in Table 4(a) doesn't qualify as a function; for the input year 1981 there are two outputs (namely, 18¢ and 20¢), whereas the definition of a function requires exactly one output. Table 4(b) remedies this problem; the rule in Table 4(b) does represent a function.

TABLE 4(a) Price of U.S. First-Class Stamp and the Year the Price Went into Effect

Year (inputs)	1975	1978	1981	1981	1985	1988	1991	1995	1999
Price in cents (outputs)	13	15	18	20	22	25	29	32	33

TABLE 4(b) Price of U.S. First-Class Stamp and the Date the Price Went into Effect

Date (inputs)	12/31/75	5/29/78	3/22/81	11/1/81	2/17/85	4/3/88	2/3/91	1/1/95	1/10/99
Price in cents (outputs)	13	15	18	20	22	25	29	32	33

For an algebraic example of this same issue, consider square roots and the symbol $\sqrt{\ }$. (*Note*: The following discussion is intended to be self contained. If after reading it, however, things don't seem clear, you should refer to the detailed review of roots in Appendix B.2 at the back of this book.) Recall from basic algebra that each positive number has two square roots. For instance, the square roots of 16 are 4 and -4. Thus, the rule in Table 5(a) does not represent a function because for each input there are two outputs. To remedy this, the symbol $\sqrt{\ }$ is defined in algebra to mean the *positive* square root only. Thus $\sqrt{16} = 4$, and if you are thinking of the other root, you write $-\sqrt{16} = -4$. Consequently, the correspondences in Table 5(b) do represent a function; for each input there is exactly one output.

TABLE 5(a) Not a Function	
x	**Square roots of x**
4	2, −2
9	3, −3
16	4, −4

TABLE 5(b) A Function	
x	**$\sqrt{x}$**
4	2
9	3
16	4

EXAMPLE 1 Using the definition of a function

Let $A = \{b, g\}$ and $B = \{s, t, u, z\}$. Which of the four correspondences in Figure 4 represent functions from A to B? For those correspondences that do represent functions, specify the range in each case.

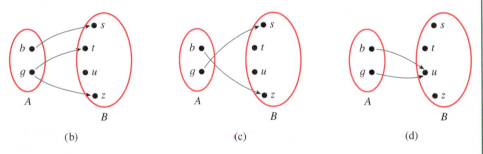

(a)   (b)   (c)   (d)

Figure 4

SOLUTION

(a) This is not a function. The definition requires that *each* element in A be assigned an element in B. The element g in this case has no assignment.

(b) This is not a function. The definition requires *exactly one* output for a given input. In this case there are two outputs for the input g.

(c) and **(d)** Both of these rules qualify as functions from A to B. For each input there is exactly one output. (Regarding the function in (d) in particular, notice that nothing in the definition of a function prohibits two different inputs from producing the same output.) For the function in (c), the outputs are s and z, and so the range is the set $\{s, z\}$. For the function in (d), the only output is u, consequently the range is the set $\{u\}$.

For functions defined by equations, we'll agree on the following convention regarding the domain: Unless otherwise indicated, the domain is assumed to be the set of all real numbers that lead to unique real-number outputs. (This is essentially the domain convention described in Section 1.3.) Thus, the domain of the function defined by $y = 3x - 2$ is the set of all real numbers, whereas the domain of the function defined by $y = 1/(x - 5)$ is the set of all real numbers except 5. [The expression $1/(x - 5)$ is undefined when $x = 5$ because the denominator is then zero.] In general, the letter representing elements from the domain (that is, the inputs) is called the **independent variable.** For example, in the equation $y = 3x - 2$, the independent variable is x. The letter representing elements from the range (that is, the outputs) is called the **dependent variable.** In the equation $y = 3x - 2$, the dependent variable is y; its value *depends* on x. This is also expressed by saying that *y is a function of x.*

EXAMPLE **2** **Determining the domain of a function**

Find the domain of the function defined by each equation:
(a) $y = \sqrt{2x + 6}$;   **(b)** $s = 1/(t^2 - 6t - 7)$.

SOLUTION

(a) The quantity under the radical sign must be nonnegative, so we have

$$2x + 6 \geq 0$$
$$2x \geq -6$$
$$x \geq -3$$

The domain is therefore the interval $[-3, \infty)$.

(b) Since division by zero is undefined, the domain of this function consists of all real numbers t except those for which the denominator is zero. Thus to find out which values of t to exclude, we solve the equation $t^2 - 6t - 7 = 0$. We have

$$t^2 - 6t - 7 = 0$$
$$(t - 7)(t + 1) = 0$$
$$t - 7 = 0 \quad \mid \quad t + 1 = 0$$
$$t = 7 \quad \mid \quad t = -1$$

It follows now that the domain of the function defined by $s = 1/(t^2 - 6t - 7)$ is the set of all real numbers except $t = 7$ and $t = -1$.

EXAMPLE **3** **Determining the domain of a function**

Find the domain of the function defined by each equation:

(a) $y = \sqrt{\dfrac{x + 3}{x - 4}}$;   **(b)** $y = \sqrt[3]{\dfrac{x + 3}{x - 4}}$.

SOLUTION

(a) The quantity underneath the square root sign must be nonnegative, so we require

$$\frac{x + 3}{x - 4} \geq 0$$

This inequality can be solved using the techniques of Section 2.6. In fact, we did solve exactly this inequality in Example 4 of Section 2.6. As you can check (preferably on your own, without looking back), the solution set is $(-\infty, -3] \cup (4, \infty)$. This is the domain of the given function.

(b) As opposed to the situation with square roots, cube roots are defined for all real numbers. Thus, the only trouble spot for $y = \sqrt[3]{(x + 3)/(x - 4)}$ occurs when the denominator is zero—that is, when x is 4. Consequently, the domain of the given function consists of all real numbers except $x = 4$. Using interval notation, this set can be written $(-\infty, 4) \cup (4, \infty)$.

EXAMPLE **4** **Determining the range of a function**

Find the range of the function defined by $y = \dfrac{x + 2}{x - 3}$.

SOLUTION
The range of this function is the set of all outputs y. One way to see what restrictions the given equation imposes on y is to solve the equation for x as follows:

$$y(x - 3) = x + 2 \qquad \textcolor{red}{\text{multiplying by } x - 3 \text{ (for } x \neq 3)}$$
$$xy - 3y = x + 2$$
$$xy - x = 3y + 2$$
$$x(y - 1) = 3y + 2$$
$$x = \frac{3y + 2}{y - 1}$$

From this last equation we see that the value of y cannot be 1. (The denominator is zero when $y = 1$.) The range therefore consists of all real numbers except $y = 1$. Using interval notation, this is the set $(-\infty, 1) \cup (1, \infty)$.

We often use single letters to name functions. If f is a function and x is an input for the function, then the resulting output is denoted by $f(x)$. This is read *f of x* or *the value of f at x*. As an example of this notation, suppose that f is the function defined by

$$f(x) = x^2 - 3x + 1 \tag{1}$$

Then $f(-2)$ denotes the output that results when the input is -2. To calculate this output, replace x with -2 throughout equation (1). This yields

$$f(-2) = (-2)^2 - 3(-2) + 1$$
$$= 4 + 6 + 1 = 11$$

That is, $f(-2) = 11$. Figure 5 summarizes this result and the notation.

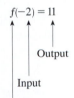

$$f(-2) = 11$$

Output

Input

Name of function

Figure 5

EXAMPLE **5** **Using function notation**

Let g denote the minimum wage function defined by Table 1 (on page 158).
(a) Find $g(1975)$.
(b) Find $g(1990) - g(1980)$ and interpret the result.

SOLUTION
(a) The notation $g(1975)$ stands for the output corresponding to the input 1975. According to Table 1, that output is $2.00, and so $g(1975) = \$2.00$.
(b) We have

$$g(1990) - g(1980) = \$3.80 - \$3.10$$
$$= \$0.70$$

This represents the increase in the minimum wage from 1980 to 1990.

EXAMPLE **6** **Using function notation**

Let $f(x) = \dfrac{x}{x - 4}$.

(a) Compute each of the following: $f(2)$, $f(t)$, and $f(a^2)$.

(b) Find an input x such that $f(x) = 3$.

SOLUTION

(a) For $f(2)$, replace each occurrence of x in the given equation by 2 to obtain

$$f(2) = \frac{2}{2 - 4} = -1$$

Similarly, for $f(t)$ and for $f(a^2)$ we have

$$f(t) = \frac{t}{t - 4} \quad \text{and} \quad f(a^2) = \frac{a^2}{a^2 - 4}$$

(b) To solve the equation $f(x) = 3$, replace $f(x)$ by the given expression $x/(x - 4)$. This yields

$$\frac{x}{x - 4} = 3$$

$$x = 3(x - 4) \qquad \textcolor{red}{\text{multiplying both sides}}$$
$$\textcolor{red}{\text{by the quantity } x - 4 \text{ (for } x \neq 4)}$$
$$x = 3x - 12$$
$$-2x = -12$$
$$x = 6$$

CHECK Using the input $x = 6$ we have

$$f(6) = \frac{6}{6 - 4} = 3, \quad \text{as required}$$

So $x = 6$ is indeed the input that we were looking for.

As background for the next example (and for work in later sections), we introduce the concept of a *demand function* from the field of economics. As you well know, the price of an item that you're interested in has a strong influence on whether or not you actually buy that item. A function relating the selling price of an item to the number of units sold at that price is called a **demand function.** Such a function might be obtained after analyzing either a company's sales records or the results of a marketing survey. As a simple example, suppose that an auto manufacturer invites 12 people to participate in a focus group. At some point in the session, the manufacturer shows them pictures and specifications for a proposed new model. Then each of the 12 people fills out a questionnaire asking whether he or she would purchase the car at various price levels. The results, shown in Table 6, represent a demand function. Figure 6 provides a graphical representation of this same function.

TABLE 6   Tabular Representation of a Demand Function

Price (1000's of dollars)	Number of cars that can be sold
12	12
14	10
16	6
18	5
20	2
22	1

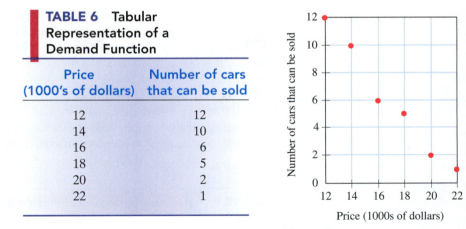

Figure 6
Graphical representation of the demand function

The demand function represented in Table 6 and Figure 6 involves only 6 data points, whereas an actual marketing survey might well produce 600 or even 6000 points. In order to detect patterns in the data or to make predictions, it's useful to have an equation with a relatively simple graph that somehow "fits" or summarizes the data. In Figure 7(a), for instance, you can see that the blue line does seem to fit the data points from Figure 6 quite well.

Assuming that we are using x on the horizontal axis for price and y on the vertical axis for the number of cars, the equation for the line in Figure 7(a) is $y = -1.143x + 25.43$. The equation was obtained using statistical techniques that are built into many types of graphing utilities and spreadsheets. (More about that in Section 4.1.) The line is known variously as a *trend line, regression line,* or *least squares line.* We say in this case that the function defined by $y = -1.143x + 25.43$ is a *mathematical model* for the demand data displayed in Table 6 and in Figure 6.

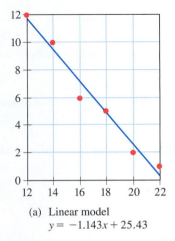

(a) Linear model
$y = -1.143x + 25.43$

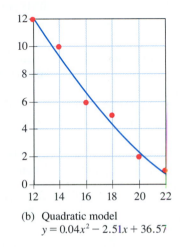

(b) Quadratic model
$y = 0.04x^2 - 2.51x + 36.57$

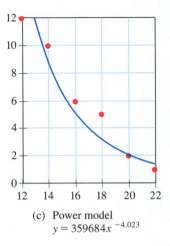

(c) Power model
$y = 359684x^{-4.023}$

Figure 7
Three models for the demand data in Table 6

More generally, any function used to describe or summarize a real-world situation is called a **mathematical model,** and the process of creating or developing that function is called **mathematical modeling.** In Figures 7(b) and 7(c) we show two other functions that can be used to model the demand data. We'll study these and other types of functions used in mathematical modeling in subsequent chapters.

For the demand function represented in Figure 6, prices are represented on the horizontal axis, and the numbers of items sold are on the vertical axis. However, this is not a standard convention. Economists sometimes put prices on the vertical axis and numbers of items on the horizontal axis. Example 7 shows one instance of this.

EXAMPLE   7   Using a demand function and function notation

Suppose that an economist for a factory outlet store determines that the demand function p for a certain type of tee shirt is given by

$$p(n) = -0.012n + 20.49 \qquad (100 \le n \le 1500)$$

where n is the number of tee shirts that can be sold per month at a price of $p(n)$ dollars per shirt. Solve the equation $p(n) = 16$ and interpret the result.

SOLUTION

In the equation $p(n) = 16$, we replace $p(n)$ by the given expression $-0.012n + 20.49$. This yields

$$-0.012n + 20.49 = 16$$
$$-0.012n = 16 - 20.49$$
$$n = \frac{16 - 20.49}{-0.012} = 374.166\ldots \qquad \text{\color{red}using a calculator}$$

Now since n represents the number of tee shirts, we round to the nearest integer to obtain $n = 374$. Interpretation: When the price is set at $16 per shirt, the store can expect to sell 374 tee shirts per month.

Before going on to the next example, we offer a word of caution to help you avoid a common error in using function notation.

NOTE OF CAUTION REGARDING FUNCTION NOTATION   It is not in general true that $f(a + b) = f(a) + f(b)$.

As an example, suppose we take f to be the function defined by $f(x) = x^2$ and use the two inputs $a = 1$ and $b = 2$. Then we have

$$f(1 + 2) = (1 + 2)^2 = 3^2 = 9$$

whereas

$$f(1) = 1^2 = 1, f(2) = 2^2 = 4, \quad \text{and consequently} \quad f(1) + f(2) = 1 + 4 = 5$$

In summary, we have $f(1 + 2) = 9$ and $f(1) + f(2) = 5$, so certainly

$$f(1 + 2) \ne f(1) + f(2)$$

EXAMPLE | **8** | **Using function notation**

Compute $g(x + 1)$ for the function g defined by $g(x) = 1 - x - x^2$.

SOLUTION
We're given $g(x)$. To find $g(x + 1)$, we go to the given equation $g(x) = 1 - x - x^2$ and replace each occurrence of x with the quantity $x + 1$. This yields

$$g(x + 1) = 1 - (x + 1) - (x + 1)^2$$
$$= 1 - x - 1 - (x^2 + 2x + 1)$$
$$= -x^2 - 3x - 1 \qquad \text{(Check the algebra.)}$$

Here's a slightly different perspective that we can apply in Example 8 and in similar situations. Instead of writing $g(x) = 1 - x - x^2$, we can write

$$g(\ ) = 1 - (\ ) - (\ )^2 \qquad (1)$$

with the understanding that whatever quantity goes into the parentheses on the left-hand side of equation (1) must also be placed into each set of parentheses on the right-hand side. In particular then, placing the quantity $x + 1$ into all three sets of parentheses in equation (1) yields $g(x + 1) = 1 - (x + 1) - (x + 1)^2$. From here on, the algebra is the same as in Example 8.

EXERCISE SET 3.1

A

1. As in Example 5, let g denote the minimum wage function represented in Table 1.
   (a) Find $g(1975)$.
   (b) Find $g(1995) - g(1975)$ and interpret the result.
2. Let f denote the function represented in Table 3.
   (a) Find f(Kate Capshaw).
   (b) For which input x are x and $f(x)$ identical?
3. Let h denote the function represented in Table 2.
   (a) Is it the domain or the range of h that consists of real numbers?
   (b) Find h(Mars).
   (c) Which is larger: h(Neptune) or h(Pluto)? Interpret the result (using a complete sentence).
4. Again let h denote the function represented in Table 2.
   (a) List those inputs x for which $h(x) = 1$.
   (b) For which real number t will it be true that h(Jupiter) $+ t = h$(Saturn)?

For Exercises 5 and 6, two sets A and B are defined as follows:
$A = \{x, y, z\}; B = \{1, 2, 3\}.$
(a) *Which of the rules displayed in the figures represent functions from A to B?*

(b) *For each rule that does represent a function, specify the range.*

5.

6.

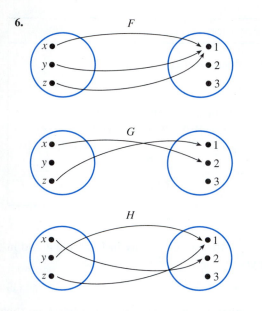

For Exercises 7 and 8, two sets D and C are defined as follows: $D = \{a, b\}$; $C = \{i, j, k\}$.

(a) Which of the rules displayed in the figures represent functions from D to C?

(b) For each rule that represents a function, specify the range.

7.

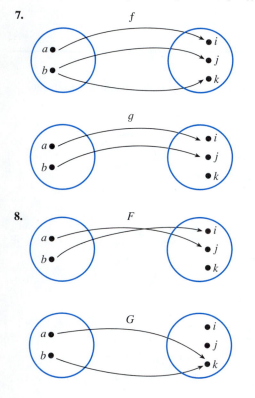

8.

In Exercises 9–16, determine the domain of each function.

9. (a) $y = -5x + 1$
(b) $y = 1/(-5x + 1)$
(c) $y = \sqrt{-5x + 1}$
(d) $y = \sqrt[3]{-5x + 1}$

10. (a) $s = 3t + 12$
(b) $s = 1/(3t + 12)$
(c) $s = \sqrt{3t + 12}$
(d) $s = \sqrt[3]{3t + 12}$

11. (a) $f(x) = x^2 - 9$
(b) $g(x) = 1/(x^2 - 9)$
(c) $h(x) = \sqrt{x^2 - 9}$
(d) $k(x) = \sqrt[3]{x^2 - 9}$

12. (a) $F(t) = t^2 + 4t$
(b) $G(t) = 1/(t^2 + 4t)$
(c) $H(t) = \sqrt{t^2 + 4t}$
(d) $K(t) = \sqrt[3]{t^2 + 4t}$

13. (a) $f(t) = t^2 - 8t + 15$
(b) $g(t) = 1/(t^2 - 8t + 15)$
(c) $h(t) = \sqrt{t^2 - 8t + 15}$
(d) $k(t) = \sqrt[3]{t^2 - 8t + 15}$

14. (a) $F(x) = 2x^2 + x - 6$
(b) $G(x) = 1/(2x^2 + x - 6)$
(c) $H(x) = \sqrt{2x^2 + x - 6}$
(d) $K(x) = \sqrt[3]{2x^2 + x - 6}$

15. (a) $f(x) = (x - 2)/(2x + 6)$
(b) $g(x) = \sqrt{(x - 2)/(2x + 6)}$
(c) $h(x) = \sqrt[3]{(x - 2)/(2x + 6)}$

16. (a) $F(t) = (3t - 4)/(7 - 2t)$
(b) $G(t) = \sqrt{(3t - 4)/(7 - 2t)}$
(c) $H(t) = \sqrt[3]{(3t - 4)/(7 - 2t)}$

In Exercises 17–26, determine the domain and the range of each function.

17. $y = 4x - 5$

18. $y = 125 - 12x$

19. $y = 4x^3 - 5$

20. $y = 125 - 12x^3$

21. $g(x) = \dfrac{4x - 20}{3x - 18}$

22. $f(x) = \dfrac{1 - x}{x}$

23. (a) $f(x) = \dfrac{x + 3}{x - 5}$

(b) $F(x) = \dfrac{x^3 + 3}{x^3 - 5}$

24. (a) $g(x) = \dfrac{2x - 7}{3x + 24}$

(b) $G(x) = \dfrac{2x^3 - 7}{3x^3 + 24}$

25. $s = t^2 + 4$

26. $s = 2t^2 - 10$

27. Each of the following rules defines a function with domain the set of all real numbers. Express each rule in the form of an equation.

EXAMPLE
The rule *For each real number, compute its square* can be written $y = x^2$.

(a) For each real number, subtract 3 and then square the result.
(b) For each real number, compute its square and then subtract 3 from the result.
(c) For each real number, multiply it by 3 and then square the result.
(d) For each real number, compute its square and then multiply the result by 3.

28. Each of the following rules defines a function with domain equal to the set of all real numbers. Express each rule in words.

(a) $y = 2x^3 + 1$
(b) $y = 2(x + 1)^3$
(c) $y = (2x + 1)^3$
(d) $y = (2x)^3 + 1$

29. Let $f(x) = x^2 - 3x + 1$. Compute the following.
    (a) $f(1)$ (e) $f(z)$ (i) $|f(1)|$
    (b) $f(0)$ (f) $f(x + 1)$ (j) $f(\sqrt{3})$
    (c) $f(-1)$ (g) $f(a + 1)$ (k) $f(1 + \sqrt{2})$
    (d) $f(3/2)$ (h) $f(-x)$ (l) $|1 - f(2)|$

30. Let $H(x) = 1 - x + x^2 - x^3$.
    (a) Which number is larger, $H(0)$ or $H(1)$?
    (b) Find $H(\frac{1}{2})$. Does $H(\frac{1}{2}) + H(\frac{1}{2}) = H(1)$?

31. Let $f(x) = 3x^2$. Find the following.
    (a) $f(2x)$ (c) $f(x^2)$ (e) $f(x/2)$
    (b) $2f(x)$ (d) $[f(x)]^2$ (f) $f(x)/2$
    For checking: No two answers are the same.

32. Let $f(x) = 4 - 3x$. Find the following.
    (a) $f(2)$ (f) $2f(x)$ (k) $1/f(x)$
    (b) $f(-3)$ (g) $f(x^2)$ (l) $f(-x)$
    (c) $f(2) + f(-3)$ (h) $f(1/x)$ (m) $-f(x)$
    (d) $f(2 + 3)$ (i) $f[f(x)]$ (n) $-f(-x)$
    (e) $f(2x)$ (j) $x^2 f(x)$

33. Let $H(x) = 1 - 2x^2$. Find the following.
    (a) $H(\sqrt{2})$ (c) $H(x + 1)$
    (b) $H(5/6)$ (d) $H(x + h)$

34. (a) If $f(x) = 2x + 1$, does $f(3 + 1) = f(3) + f(1)$?
    (b) If $f(x) = 2x$, does $f(3 + 1) = f(3) + f(1)$?
    (c) If $f(x) = \sqrt{x}$, does $f(3 + 1) = f(3) + f(1)$?

35. Let $g(x) = 2$, for all x. Find each output.
    (a) $g(0)$ (b) $g(5)$ (c) $g(x + h)$

36. Let $g(t) = |t - 4|$. Find $g(3)$. Find $g(x + 4)$.

37. Let $f(x) = x^2 - 6x$. In each case, find all real numbers x (if any) that satisfy the given equation.
    (a) $f(x) = 16$ (b) $f(x) = -10$ (c) $f(x) = -9$

38. Let $g(t) = (4t - 6)/(t - 4)$. In each case, find all the real number solutions (if any) for the given equation.
    (a) $g(t) = 14$ (b) $g(t) = 4$ (c) $g(t) = 0$

For Exercises 39 and 40, refer to the demand function given in Example 7. In each case, solve the indicated equation for n. Round each answer to the nearest integer and interpret the result.
39. $p(n) = 8$ 40. $p(n) = 18$

In Exercises 41 and 42, p refers to the demand function given in Example 7. Now assume that a second economist proposes an alternative model f for the demand function:

$$f(n) = 4 + \frac{3000}{n + 100} \qquad (100 \le n \le 1500)$$

where n is the number of tee shirts that can be sold per month at a price of f(n) dollars per shirt.

41. At a price level of $19 per shirt, compare the predictions for monthly sales obtained using each model. *Hint:* You need to solve each of the equations $p(n) = 19$ and $f(n) = 19$ and interpret the results.

42. Show that at a price level of $10 per shirt, the model p predicts more than twice as many sales per month as does the model f. *Hint:* Solve each of the equations $p(n) = 10$ and $f(n) = 10$, and interpret the results.

In Exercises 43 and 44, let $L(x) = 6 - 3x$. Find (and simplify) each expression.
43. $L(2x - 1)$ 44. $L(2 + x)$

In Exercises 45 and 46, let $T(x) = 2x^2 - 3x$. Find (and simplify) each expression.
45. (a) $T(x + 2)$ 46. (a) $T(x + h)$
    (b) $T(x - 2)$ (b) $T(x - h)$
    (c) $T(x + 2) - T(x - 2)$ (c) $T(x + h) - T(x - h)$

In Exercises 47–50, refer to the following table. The left-hand column of the table lists four errors to avoid in working with function notation. In each case, use the function $f(x) = x^2 - 1$ and give a numerical example showing that the expressions on each side of the equation are not equal.

Errors to avoid	Numerical example showing that the equation is not, in general, valid
47. $f(a + b) = f(a) + f(b)$	
48. $f(ab) = f(a) \cdot f(b)$	
49. $f\left(\dfrac{1}{a}\right) = \dfrac{1}{f(a)}$	
50. $\dfrac{f(a)}{f(b)} = \dfrac{a}{b}$	

B

51. Let $f(x) = (x - a)/(x + a)$.
    (a) Find $f(a), f(2a)$, and $f(3a)$. Is it true that
        $f(3a) = f(a) + f(2a)$?
    (b) Show that $f(5a) = 2f(2a)$.

52. Let $k(x) = 5x^3 + \dfrac{5}{x^3} - x - \dfrac{1}{x}$. Show that $k(x) = k(1/x)$.

53. Let $f(x) = 2x + 3$. Find values for a and b such that the equation $f(ax + b) = x$ is true for all values of x. *Hint:* Use the fact that if two polynomials (in the variable x) are equal for all values of x, then the corresponding coefficients are equal.

54. If $p(x) = 2^x$, verify each identity:
    (a) $2p(x) = p(x + 1)$ (c) $p(x) \cdot p(-x) - 1 = 0$
    (b) $p(a + b) = p(a) \cdot p(b)$

55. Let $f(t) = (t - x)/(t + y)$. Show that
    $$f(x + y) + f(x - y) = \frac{-2y^2}{x^2 + 2xy}$$

56. Let $f(z) = \dfrac{3z - 4}{5z - 3}$. Find $f\left(\dfrac{3z - 4}{5z - 3}\right)$.

57. Let $F(x) = \dfrac{ax + b}{cx - a}$. Show that $F\left(\dfrac{ax + b}{cx - a}\right) = x$, where $a^2 + bc \ne 0$.

58. If $f(x) = -2x^2 + 6x + k$ and $f(0) = -1$, find k.

59. If $g(x) = x^2 - 3xk - 4$ and $g(1) = -2$, find k.

60. Let $h(x) = x^2 - 4x - c$. Find a nonzero value for c such that $h(c) = c$.

61. Let the function L be defined by the following rule: $L(x)$ *is the exponent to which 2 must be raised to yield x.* (For the moment, we won't concern ourselves with the domain and range.) Then $L(8) = 3$, for example, since the exponent to which 2 must be raised to yield 8 is 3 (that is, $8 = 2^3$). Find the following outputs.

(a) $L(1)$ **(e)** $L(1/2)$
(b) $L(2)$ **(f)** $L(1/4)$
(c) $L(4)$ **(g)** $L(1/64)$
(d) $L(64)$ **(h)** $L(\sqrt{2})$

The function L is called a *logarithm function.* The usual notation for $L(x)$ in this example is $\log_2 x$. Logarithm functions will be studied in Chapter 5.

62. Let $q(x) = ax^2 + bx + c$. Evaluate

$$q\left(\frac{-b + \sqrt{b^2 - 4ac}}{2a}\right)$$

63. By definition, a **fixed point** for the function f is a number x_0 such that $f(x_0) = x_0$. For instance, to find any fixed points for the function $f(x) = 3x - 2$, we write $3x_0 - 2 = x_0$. On solving this last equation, we find that $x_0 = 1$. Thus, 1 is a fixed point for f. Calculate the fixed points (if any) for each function.

(a) $f(x) = 6x + 10$ **(c)** $S(t) = t^2$
(b) $g(x) = x^2 - 2x - 4$ **(d)** $R(z) = (z + 1)/(z - 1)$

64. Consider the following two rules, F and G, where F is the rule that assigns to each person his or her mother, and G is the rule that assigns to each person his or her aunt. Explain why F is a function but G is not.

*In Exercises 65–67, use this definition: **A prime number** is a positive whole number with no factors other than itself and 1. For example, 2, 13, and 37 are primes, but 24 and 39 are not. By convention, 1 is not considered prime, so the list of the first few primes is as follows:*

$$2, 3, 5, 7, 11, 13, 17, 19, 23, 29, \ldots$$

65. Let G be the rule that assigns to each positive integer the nearest prime. For example, $G(8) = 7$, since 7 is the prime

nearest 8. Explain why G is not a function. How could you alter the definition of G to make it a function?

66. Let f be the function that assigns to each natural number x the number of primes that are less than or equal to x. For example, $f(12) = 5$ because, as you can easily check, five primes are less than or equal to 12. Similarly, $f(3) = 2$, because two primes are less than or equal to 3. Find $f(8)$, $f(10)$, and $f(50)$.

67. (a) If $P(x) = x^2 - x + 17$, find $P(1)$, $P(2)$, $P(3)$, and $P(4)$. Can you find a natural number x for which $P(x)$ is not prime?
     (b) If $Q(x) = x^2 - x + 41$, find $Q(1)$, $Q(2)$, $Q(3)$, and $Q(4)$. Can you find a natural number x for which $Q(x)$ is not prime?

68. $\pi = 3.141592653589793 \ldots$ *and so on!*
For each natural number n, let $G(n)$ be the digit in the nth decimal place of π. For instance, according to the expression for π given above, we have $G(1) = 1$, $G(2) = 4$, and $G(5) = 9$.
     (a) Use the expression for π given above to evaluate $G(10)$ and $G(14)$.
     (b) Use the Internet to help you evaluate $G(100)$, $G(750)$, and $G(1000)$. *Suggestions:* Using any of the common search engines on the World Wide Web, under the categories of mathematics or science, search for "pi." Here, for example, are two sites that contain the information you need. (They were accessible at the time of this writing, March 2002.)

(University of Exeter)   http://www.ex.ac.uk/cimt/general/pi10000.htm

(UCLA)   http://cad.ucla.edu/repository/useful/PI.txt

C

69. If $f(x) = mx + b$, show that $\dfrac{f(x_1) + f(x_2)}{2} = f\left(\dfrac{x_1 + x_2}{2}\right)$.

70. Let $f(x) = ax^2 + bx + c$, where $a < 0$. Show that $\dfrac{f(x_1) + f(x_2)}{2} \le f\left(\dfrac{x_1 + x_2}{2}\right)$.

PROJECT   A PRIME FUNCTION

Primes fuel cryptography, the technological field behind the secret coding of computer communication as well as communication between governments.—The Oregonian, June 26, 1999

. . . the security of large international data networks nowadays relies on the inability of mathematicians to find an efficient method of factoring large numbers (while at the same

time being able to produce large primes with ease). —Keith Devlin, *Mathematics: The New Golden Age* (New York: Columbia University Press, 1999)

. . . why does a large supercomputer manufacturer like Cray Research invest so much money . . . [in searching for huge prime numbers]? The answer is that the computation required . . . is a heavy one, stretching over days or weeks, and so it provides an excellent way to test the efficiency and accuracy of a new computer system. —Keith Devlin, *op. cit.*

As indicated by the quotes above, prime numbers do have real-world applications. (For the definition of a prime number, see the instructions preceding Exercise 65 on page 170. For a readable explanation of how prime numbers are used in secret codes, refer to Chapter One in the book by Keith Devlin cited above.)

In this project you'll work with a function F involving prime numbers: *For each natural number n, let F(n) be the nth prime number.* For instance, since the first, second, third, and fourth primes are 2, 3, 5, and 7, respectively, we have

$$F(1) = 2 \qquad F(2) = 3 \qquad F(3) = 5 \qquad F(4) = 7$$

At the outset, we remark that there is no simple algebraic formula for F.

(a) After creating a list of the first 40 prime numbers, complete the following table. (Though certainly not conceptually difficult, creating the list takes some time, and each error may throw off your ordering. Therefore, this is a good small-group project: some people should work as checkers.)

n	5	10	20	30	40
F(n)					

(b) Make use of the Internet or a library to complete the following table.

n	100	250	500	750	1000
F(n)					

Suggestion: Using any of the common search engines on the World Wide Web and typing *prime numbers* can lead to Web sites that have the information you need here. Here are two that were accessible at the time of this writing, March 2002:

http://www.utm.edu/research/primes
(University of Tennessee at Martin Web site)

and

http://www.svobodat.com/primes

An advantage to this later site is that a list of primes is available in spreadsheet format, in which it's immediately clear which is the nth prime.

(c) Although there is no simple algebraic formula for F, there are simple functions that do a good job of approximating F along relatively small portions of its domain. For example, look at the graphs in Figures A and B. Figure A shows the graph of F plotted from actual prime number data. Figure B shows the graph of a function g used to model these data:

$$g(n) = 0.0014n^2 + 6.712n - 141.85$$

As you can see, at this scale at least, it's hard to discern any difference between the two functions.

Figure A
$F(n) = n$th prime, $1 \leq n \leq 1000$

Figure B
$g(n) = 0.0014n^2 + 6.7121n - 141.85$,
$1 \leq n \leq 1000$

Complete the following table to see some examples of how well the function g approximates the nth prime. Round the values for $g(n)$ to the nearest integer. Round the percentage error to one decimal place. Recall that percentage error is defined by

$$\left| \frac{(\text{actual value}) - (\text{approximate value})}{\text{actual value}} \right| \times 100$$

n			100	250	500	750	1000
$g(n)$							
Percentage error in approximation $g(n) \approx F(n)$							

3.2 THE GRAPH OF A FUNCTION

In my own case, I got along fine without knowing the name of the distributive law until my sophomore year in college; meanwhile I had drawn lots of graphs. —Professor Donald E. Knuth (one of the world's preeminent computer scientists) in *Mathematical People* (Boston: Birkhäuser, 1985)

When the domain and range of a function are sets of real numbers, we can graph the function in the same way that we graphed equations in Chapters 1 and 2. In graphing functions, the usual practice is to reserve the horizontal axis for the independent variable (the inputs) and the vertical axis for the dependent variable (the outputs). In terms of the familiar x-y coordinate system, for each x-coordinate or input in the domain of the function, the function or rule tells you the corresponding y-coordinate. These ideas are summarized in the following definition.

DEFINITION | **Graph of a Function**

The graph of a function f in the x-y plane consists of those points (x, y) such that x is in the domain of f and $y = f(x)$. See Figure 1.

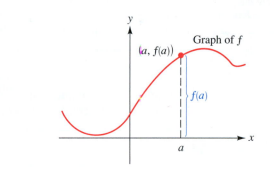

Figure 1

EXAMPLE | **1** | **Using the definition of the graph of a function**

In Figures 2(a) and 2(b), specify the y-coordinates of the points P and Q, respectively. In each case, give an exact expression and also a calculator approximation rounded to two decimal places. (We'll look at the graphs of these two functions in more detail later in this section.)

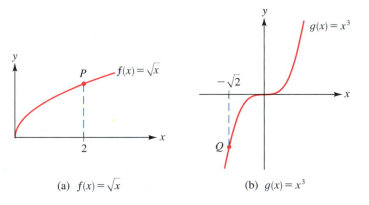

Figure 2
What are the y-coordinates of the points P and Q?

(a) $f(x) = \sqrt{x}$ \qquad (b) $g(x) = x^3$

SOLUTION
In Figure 2(a), which shows a graph of the function f defined by $f(x) = \sqrt{x}$, the x-coordinate of P is 2. The y-coordinate is therefore $f(2) = \sqrt{2}$. Using a calculator, we find $\sqrt{2} \approx 1.41$. In Figure 2(b), which shows a graph of the function g defined by $g(x) = x^3$, the x-coordinate of Q is $-\sqrt{2}$. The y-coordinate is therefore

$$g(-\sqrt{2}) = (-\sqrt{2})^3 = (-\sqrt{2})^2(-\sqrt{2})^1$$
$$= 2(-\sqrt{2}) = -2\sqrt{2}$$
$$\approx -2.28 \qquad \text{using a calculator}$$

Two y-values are assigned to one x-value. Thus this is not the graph of a function.

Figure 3

All of the graphs that we looked at in Section 3.1 are graphs of functions. So too are most of the graphs in Chapter 1. However, it's important to understand that not every graph represents a function. Consider, for example, the graph in Figure 3 (the red curve). Figure 3 shows a vertical line intersecting the graph in two distinct points. The specific coordinates of the two points are unimportant. What the vertical line helps us to see is that two different y-values (outputs) have been assigned to the same x-value (an input), and therefore the graph does not represent a function. These remarks are generalized in the box that follows.

Vertical Line Test

A graph in the x-y plane represents a function $y = f(x)$ provided that any vertical line intersects the graph in at most one point.

EXAMPLE 2 **Applying the vertical line test**

The vertical line test implies that the graph in Figure 4(a) represents a function and that the graph in Figure 4(b) does not.

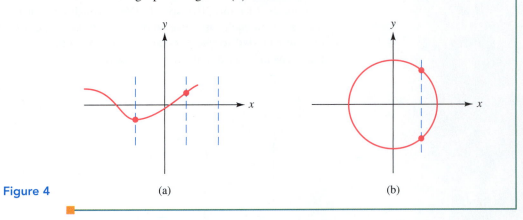

Figure 4   (a)   (b)

In Examples 2 through 4 in the previous section, we used algebra to determine the domain and range of functions. If the graph of a function is given, however, we can read off that information directly. The next two examples show instances of this.

EXAMPLE 3 **Determining the domain and range of a function from its graph**

Specify the domain and the range of the function g graphed in Figure 5(a).

SOLUTION

The domain of g is just that portion of the x-axis (the inputs) utilized in graphing g. As Figure 5(b) indicates, this amounts to all real numbers x from 1 to 5, inclusive: $1 \leq x \leq 5$. In interval notation, this set of numbers is denoted by $[1, 5]$. To find the range of g, we need to check which part of the y-axis is utilized in graphing g. As Figure 5(b) indicates, this is the set of all real numbers y be-

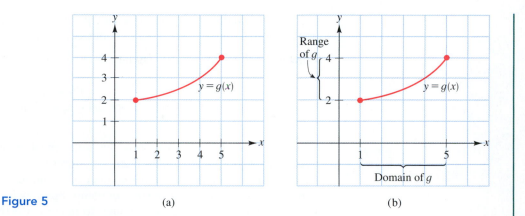

Figure 5   (a)   (b)

tween 2 and 4, inclusive: $2 \leq y \leq 4$. Interval notation for this set of numbers is $[2, 4]$.

EXAMPLE 4   **Determining domain, range, and function values from a graph**

The graph of a function h is shown in Figure 6. The open circle in the figure is used to indicate that the point $(3, 3)$ does not belong to the graph of h.

(a) Specify the domain and the range of the function h.

(b) Determine each value: **(i)** $h(-2)$;   **(ii)** $h(3)$;   **(iii)** $|h(-4)|$.

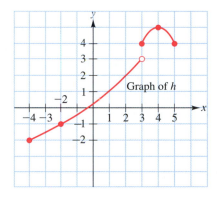

Figure 6

SOLUTION

(a) The domain of h is the interval $[-4, 5]$. The range is the set $[-2, 3) \cup [4, 5]$.

(b) **(i)**   The function notation $h(-2)$ stands for the y-coordinate of that point on the graph of h whose x-coordinate is -2. Since the point $(-2, -1)$ is on the graph of h, we conclude that $h(-2) = -1$.

   (ii)   We have $h(3) = 4$ because the point $(3, 4)$ lies on the graph of h. Note that h would not be considered a function if the point $(3, 3)$ were also part of the graph. (Why?)

   (iii)   Since the point $(-4, -2)$ lies on the graph of h, we write $h(-4) = -2$. Thus $|h(-4)| = |-2| = 2$.

Given a function represented by an equation $y = f(x)$, how do we obtain the graph? From Chapter 1 (or from previous courses) we have these three options:

1. Set up a table, plot points, and then draw the line or curve that the points seem to describe.
2. If the equation has the form $y = mx + b$, the graph is a line (as explained in Section 1.6), and we can draw the graph using slope-intercept information.
3. Use a graphing utility.

As a practical matter, option 1 above has a drawback. How many points must we plot before we're sure about the essential features of the graph? In later sections of this text, you'll see that there are a number of techniques and concepts that make it unnecessary to plot a large number of points. For now, however, we point out that some functions arise frequently enough to make it worth memorizing the basic shapes and features of their graphs. True, you can always reach for a graphing calculator, but sometimes it helps to be able to think on your feet, so to speak. Also, if you've memorized the graph of a function, then you automatically know the domain and the range. Figure 7 displays the graphs of six basic functions. Exercises 21

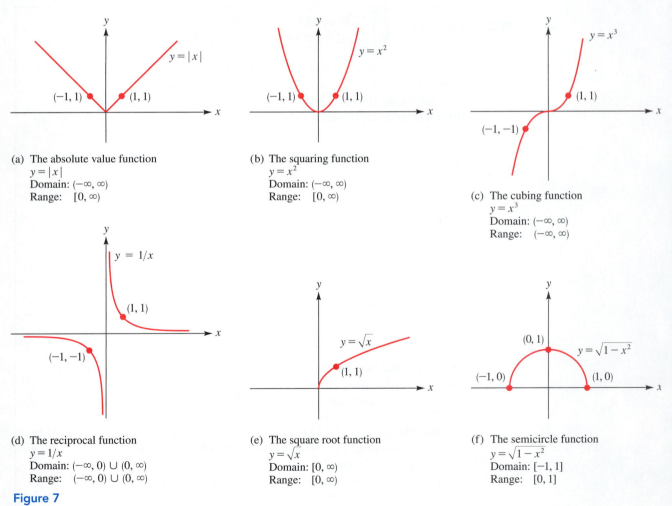

(a) The absolute value function
$y = |x|$
Domain: $(-\infty, \infty)$
Range: $[0, \infty)$

(b) The squaring function
$y = x^2$
Domain: $(-\infty, \infty)$
Range: $[0, \infty)$

(c) The cubing function
$y = x^3$
Domain: $(-\infty, \infty)$
Range: $(-\infty, \infty)$

(d) The reciprocal function
$y = 1/x$
Domain: $(-\infty, 0) \cup (0, \infty)$
Range: $(-\infty, 0) \cup (0, \infty)$

(e) The square root function
$y = \sqrt{x}$
Domain: $[0, \infty)$
Range: $[0, \infty)$

(f) The semicircle function
$y = \sqrt{1 - x^2}$
Domain: $[-1, 1]$
Range: $[0, 1]$

Figure 7
The graphs of six basic functions

and 22 at the end of this section ask you to set up tables and verify for yourself that the graphs in Figure 7 are indeed correct. From now on (with the exception of Exercises 21 and 22), if you need to sketch by hand a graph of one of these basic six functions, you should do so from memory.

All the graphs in Figure 7 represent functions defined by rather simple equations. In some instances, however, functions may be defined by combinations of equations. This is illustrated in the next two examples, where we graph *piecewise-defined functions*.

EXAMPLE **5**   **Graphing a piecewise-defined function**

A function g is defined by $g(x) = \begin{cases} x^2 & \text{if } x < 2 \\ 1/x & \text{if } x \geq 2 \end{cases}$

(a) Find $g(3/2)$, $g(2)$, and $g(3)$.

(b) Sketch the graph of the function g.

SOLUTION

(a) To find $g(3/2)$, which of the two equation do we use: $g(x) = x^2$ or $g(x) = 1/x$? According to the given inequalities, we should use $g(x) = x^2$ whenever the x-values are less than 2. So for $x = 3/2$ (which is surely less than 2), we have $g(3/2) = (3/2)^2 = 9/4$. On the other hand, the inequalities tell us to use the equation $g(x) = 1/x$ whenever the x-values are greater than or equal to 2. Thus we have $g(2) = 1/2$ and $g(3) = 1/3$.

(b) For the graph, we look back at Figures 7(b) and (d) and choose the appropriate portion of each. The result is displayed in Figure 8. The open circle in the figure is used to indicate that the point $(2, 4)$ does not belong to the graph. The filled-in circle, on the other hand, is used to indicate that the point $(2, 1/2)$ does belong to the graph.

Figure 8

A graph of $g(x) = \begin{cases} x^2, & x < 2 \\ \dfrac{1}{x}, & x \geq 2 \end{cases}$

EXAMPLE **6**

Sketch a graph of each function:

(a) $y = \begin{cases} |x| & \text{if } -1 \leq x \leq 1, \\ \sqrt{x} & \text{if } 1 \leq x \leq 4; \end{cases}$

(b) $y = \begin{cases} |x| & \text{if } -1 \leq x < 1, \\ \sqrt{x} & \text{if } 1 < x \leq 4. \end{cases}$

SOLUTION

(a) We want to "paste together" portions of the graphs of $y = |x|$ and $y = \sqrt{x}$. The appropriate portions of each graph are indicated in color in Figures 9(a) and 9(b). Using Figures 9(a) and 9(b) as a guide, we sketch the required graph in Figure 9(c).

(b) Comparing the given equations and inequalities with those in part (a), we see that there is only one difference: in part (b), the number 1 is excluded from the domain of the variable x. To graph the equations in part (b), we

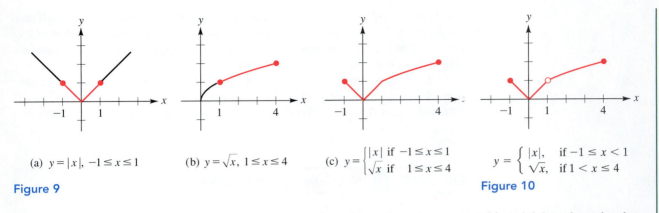

(a) $y = |x|$, $-1 \le x \le 1$

(b) $y = \sqrt{x}$, $1 \le x \le 4$

(c) $y = \begin{cases} |x| & \text{if } -1 \le x \le 1 \\ \sqrt{x} & \text{if } 1 \le x \le 4 \end{cases}$

$y = \begin{cases} |x|, & \text{if } -1 \le x < 1 \\ \sqrt{x}, & \text{if } 1 < x \le 4 \end{cases}$

Figure 9

Figure 10

need only take the graph obtained in Figure 9(c) and delete the point that corresponds to $x = 1$. That is, we delete the point $(1, 1)$; see Figure 10.

It is useful to compare the graphs in Figures 8, 9(c), and 10. Each graph was obtained by combining portions of the basic graphs in Figure 7. In Figures 8 and 10 the graphs have a break or gap in them. We say that the graph (or more precisely, the *function*) in Figure 8 is *discontinuous* when $x = 2$. (The two pieces do not meet at $x = 2$.) The function in Figure 10 is discontinuous when $x = 1$. (There is a gap or missing point when $x = 1$.) In Figure 9(c), however, the two portions of the graph form a graph with no break or gap. We say that the function in Figure 9(c) is *continuous* when $x = 1$. A rigorous definition of continuity is properly a subject for calculus. However, even at the intuitive level at which we've presented the idea here, you'll find that the concept is useful in helping you to organize your thoughts about graphs.

The graph of a piecewise-defined function can also be obtained using a graphing utility. For details, consult the owner's manual that accompanies your calculator or software. We will, however, point out one problem to watch for. In Figure 11(a), we've used a graphing utility to graph the function g from Example 5. Note the extraneous vertical line segment that was generated, connecting what should be two disjoint curves. Not all graphing utilities make this error; but if yours does, it can be corrected by switching the curve-drawing mode from *connected* to *dotted*. See Figure 11(b).

GRAPHICAL PERSPECTIVE

Figure 11

$$g(x) = \begin{cases} x^2 & \text{if } x < 2 \\ 1/x & \text{if } x \ge 2 \end{cases}$$

On a TI-83 graphing calculator, for example, this function g can be entered with the keystrokes

$(x < 2)(x^2) + (x \ge 2)(x^{-1})$.

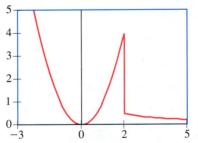

(a) Graphing utility produces an extraneous vertical segment in the graph of g.

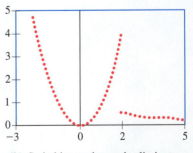

(b) Switching to dot mode eliminates the extraneous segment.

We began this section with a definition of what is meant by the *graph of a function*. The next example uses one of the six basic functions along with the line $y = x$ to help you develop some additional insight into that definition.

EXAMPLE	7	Applying the definition of the graph of a function

Figure 12 shows portions of the graphs of the functions $f(x) = \sqrt{x}$ and $y = x$. What are the coordinates of the points P, Q, and R in Figure 12? (Each dashed line is parallel to one of the coordinate axes.) For those coordinates involving radicals, also supply a calculator approximation rounded to two decimal places.

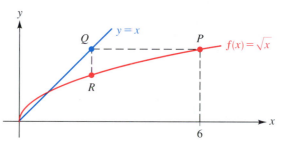

Figure 12
What are the coordinates of P, Q, and R?

SOLUTION
In Figure 12, the point P lies on the graph of $f(x) = \sqrt{x}$. Since the x-coordinate of P is 6, the y-coordinate is $f(6)$, which, according to the rule for f, is $\sqrt{6}$. Thus, the coordinates of P are $(6, \sqrt{6})$. Next, we want the coordinates of Q. Figure 12 shows that the points P and Q have the same y-coordinate, so the y-coordinate of Q must also be $\sqrt{6}$. What about the x-coordinate of Q? Since Q lies on the graph of $y = x$, the x-coordinate of Q must be the same as its y-coordinate— namely, $\sqrt{6}$. Thus, the coordinates of Q are $(\sqrt{6}, \sqrt{6})$.

Finally, let us determine the coordinates of the point R. According to Figure 12, the x-coordinate of R is the same as that of Q, which we know to be $\sqrt{6}$. The y-coordinate of R is then found by taking the square root of this x-coordinate, to obtain $\sqrt{\sqrt{6}}$. (We've used the fact that R lies on the graph of $f(x) = \sqrt{x}$.)

Summarizing now, the required coordinates are as follows. (You should check for yourself that the calculator values given are correct.)

P: $(6, \sqrt{6}) \approx (6, 2.45)$

Q: $(\sqrt{6}, \sqrt{6}) \approx (2.45, 2.45)$

R: $(\sqrt{6}, \sqrt{\sqrt{6}}) \approx (2.45, 1.57)$

EXERCISE SET 3.2

A

In Exercises 1–4, specify the y-coordinate of the point P on the given function. In each case, give an exact expression and also a calculator approximation rounded to three decimal places.

1.

$f(x) = \sqrt{x}$

2.

$g(x) = x^3$

3.

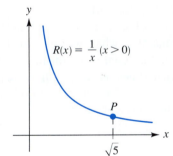

$R(x) = \dfrac{1}{x} \ (x > 0)$

4.

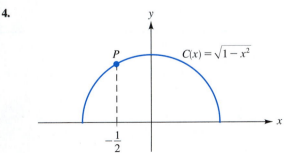

$C(x) = \sqrt{1 - x^2}$

In Exercises 5 and 6, use the vertical line test to determine whether each graph represents a function $y = f(x)$.

5.

(a)

(b)

(c)

(d)

6.

(a)

(b)

(c)

(d)

In Exercises 7–14, the graph of a function is given. In each case, specify the domain and the range of the function. (The axes are marked off in one-unit intervals.)

7.

8.

9.

10.

11.

12.

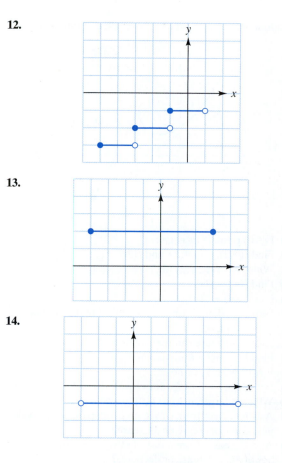

13.

14.

In Exercises 15 and 16, refer to the graph of the function F in the figure. (Assume that the axes are marked off in one-unit intervals.)

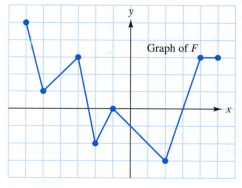

Graph of *F*

15. (a) Find $F(-5)$
   (b) Find $F(2)$.
   (c) Is $F(1)$ positive?
   (d) For which value of x is $F(x) = -3$?
   (e) Find $F(2) - F(-2)$.

16. (a) Find $F(4)$.
   (b) Find $F(-1)$.
   (c) Is $F(-4)$ positive?
   (d) For which value of x is $F(x) = 5$?
   (e) Find $F(5) - F(-3)$.

17. The following figure displays the graph of a function f.

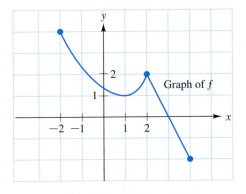

(a) Is $f(0)$ positive or negative?
(b) Find $f(-2), f(1), f(2),$ and $f(3)$.
(c) Which is larger, $f(2)$ or $f(4)$?
(d) Find $f(4) - f(1)$.
(e) Find $|f(4) - f(1)|$.
(f) Write the domain and range of f using the interval notation $[a, b]$.

18. The following figure shows the graph of a function h.
(a) Find $h(a), h(b), h(c),$ and $h(d)$.
(b) Is $h(0)$ positive or negative?
(c) For which values of x does $h(x) = 0$?
(d) Which is larger, $h(b)$ or $h(0)$?
(e) As x increases from c to d, do the corresponding values of $h(x)$ increase or decrease?
(f) As x increases from a to b, do the corresponding values of $h(x)$ increase or decrease?

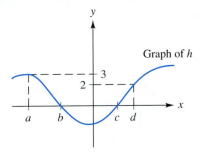

Graph of h

In Exercises 19 and 20, refer to the graphs of the functions f and g in the figure. Assume that the domain of each function is $[-3, 3]$ and that the axes are marked off in one-unit intervals.

19. (a) Which is larger, $f(-2)$ or $g(-2)$?
(b) Compute $f(0) - g(0)$.
(c) Which among the following three quantities is the smallest?

$$f(1) - g(1) \qquad f(2) - g(2) \qquad f(3) - g(3)$$

(d) For which value(s) of x does $g(x) = f(1)$?
(e) Is the number 4 in the range of f or in the range of g?

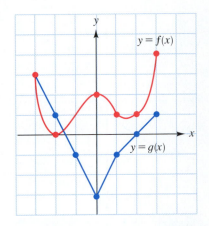

20. (a) For the interval $[0, 3]$, is the quantity $g(x) - f(x)$ positive or negative?
(b) For the interval $(-3, -2)$, is the quantity $g(x) - f(x)$ positive or negative?

In Exercises 21 and 22, set up a table and graph each function. [In Exercises 22(b) and (c), use a calculator to compute the square roots.]

21. (a) $y = |x|$
(b) $y = x^2$
(c) $y = x^3$

22. (a) $y = 1/x$
(b) $y = \sqrt{x}$
(c) $y = \sqrt{1 - x^2}$

In Exercises 23–30, sketch the graphs, as in Examples 5 and 6.

23. $A(x) = \begin{cases} x^3 & \text{if } -2 \le x \le -1 \\ x^2 & \text{if } x > -1 \end{cases}$

24. $B(x) = \begin{cases} \sqrt{1 - x^2} & \text{if } -1 \le x < 1 \\ 1/x & \text{if } x \ge 1 \end{cases}$

25. $C(x) = \begin{cases} x^3 & \text{if } x < 1 \\ \sqrt{x} & \text{if } x > 1 \end{cases}$

26. $y = \begin{cases} |x| & \text{if } x \le 0 \\ x^2 & \text{if } x > 0 \end{cases}$

27. (a) $y = \begin{cases} \sqrt{x} & \text{if } 0 \le x \le 1 \\ 1/x & \text{if } 1 < x < 2 \end{cases}$

(b) $y = \begin{cases} \sqrt{x} & \text{if } 0 \le x < 1 \\ 1/x & \text{if } 1 < x < 2 \end{cases}$

28. (a) $y = \begin{cases} x^3 & \text{if } x \le 0 \\ |x| & \text{if } x > 0 \end{cases}$

(b) $y = \begin{cases} |x| & \text{if } x \le 0 \\ x^3 & \text{if } x > 0 \end{cases}$

29. (a) $y = \begin{cases} \sqrt{1 - x^2} & \text{if } -1 \le x \le 0 \\ x^2 & \text{if } 0 < x \le 2 \end{cases}$

(b) $y = \begin{cases} \sqrt{1 - x^2} & \text{if } -1 \le x < 0 \\ x^2 & \text{if } 0 < x \le 2 \end{cases}$

30. $y = \begin{cases} x^3 & \text{if } -1 \le x < 0 \\ \sqrt{x} & \text{if } 0 \le x < 1 \\ 1/x & \text{if } 1 \le x \le 3 \end{cases}$

Ⓖ *In Exercises 31 and 32, use a graphing utility to obtain the graphs of the piecewise-defined functions. If there are cases where the graphing utility generates extraneous segments, use the dot mode to correct the situation.*

31. (a) $f(x) = \begin{cases} x^3 - x & \text{if } -1 \le x < 0 \\ x^3 + x & \text{if } 0 \le x \le 1 \end{cases}$

   (b) $g(x) = \begin{cases} x^3 - x & \text{if } -1 \le x < 0 \\ x^3 + x + 1 & \text{if } 0 \le x \le 1 \end{cases}$

32. (a) $h(x) = \begin{cases} \sqrt{1 - x^6} & \text{if } -1 \le x < 1 \\ \sqrt{1 - (x-2)^6} & \text{if } 1 \le x \le 3 \end{cases}$

   (b) $k(x) = \begin{cases} \sqrt{1 - x^6} & \text{if } -1 \le x < 1 \\ \sqrt{1 - (x-2)^6} + 3 & \text{if } 1 \le x \le 3 \end{cases}$

In Exercises 33–36, determine the coordinates of the points P, Q, and R in each figure; give an exact expression and also a calculator approximation rounded to three decimal places. Assume that each dashed line is parallel to one of the coordinate axes. (In Exercise 36, note that the line is $y = x/2$ rather than $y = x$.)

33.

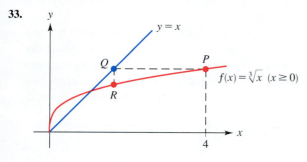

$f(x) = \sqrt[3]{x} \; (x \ge 0)$

34.

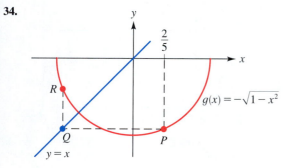

$g(x) = -\sqrt{1 - x^2}$

35.

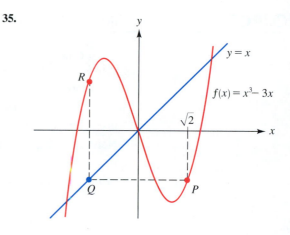

$f(x) = x^3 - 3x$

36.

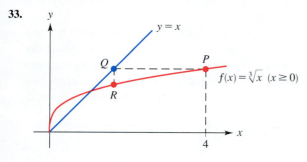

$G(x) = \dfrac{\sqrt{2}}{1 + x^2}$

$y = \dfrac{x}{2}$

B

Ⓖ **37. (a)** Graph each of the following functions on the same screen, using the viewing rectangle $[0, 1, 0.2] \times [0, 1, 0.2]$.

   $y = x \qquad y = x^2 \qquad y = x^3 \qquad y = x^4 \qquad y = x^5$

   (b) Describe, in complete sentences, the pattern you see in your results in part (a). Based on your results in part (a), draw (by hand) a sketch of what you think the graph of $y = x^{100}$ must look like in this interval. Then use a graphing utility to see how accurate your sketch was.

38. (Although this is not a graphing-utility exercise, the previous exercise, which does use a graphing utility, provides some useful background here.)

   (a) Suppose that the thickness of the line drawn by your pencil or pen is 0.05 cm, and suppose that the common unit of length marked off along both axes is 1 cm. Under these drawing conditions, where is the graph of $y = x^2$ indistinguishable from the x-axis? That is, for which x-values will the corresponding y-values be less than 0.05?

   (b) Follow part (a), using $y = x^4$ instead of $y = x^2$.

PROJECT

IMPLICIT FUNCTIONS: BATTERIES REQUIRED?

From Chapter 1, we know that the graph of the equation $x^2 + y^2 = 1$ is a circle, as shown in Figure A. From our work in this section, we know the graph does not represent a function. However, certain pieces of the graph do represent functions. In particular, solving the equation for y yields $y = \pm\sqrt{1 - x^2}$. Figures B and C show the graphs of $y = \sqrt{1 - x^2}$ and $y = -\sqrt{1 - x^2}$, respectively. Note that the graphs in Figures B and C indeed represent functions. We say that the initial equation $x^2 + y^2 = 1$ defines these functions *implicitly,* whereas each of the equations $y = \sqrt{1 - x^2}$ and $y = -\sqrt{1 - x^2}$ defines a function *explicitly.*

Figure A
$x^2 + y^2 = 1$

Figure B
$y = \sqrt{1 - x^2}$

Figure C
$y = -\sqrt{1 - x^2}$

Here's another example of this terminology. One of the basic functions in Figure 7 on page 176 is defined by the equation $y = 1/x$. In this form, the function is defined explicitly. If the equation is given in the equivalent form $xy = 1$, then the function is defined implicitly.

The reason we ask "Batteries Required?" in the title of this project is that a graphing calculator or a computer is often the option of choice in graphing implicitly defined functions. In some cases, it's the only option (assuming the tools of calculus are not at our disposal). As examples, consider the two equations graphed in Figures D and E. We'll talk about how these graphs were obtained.

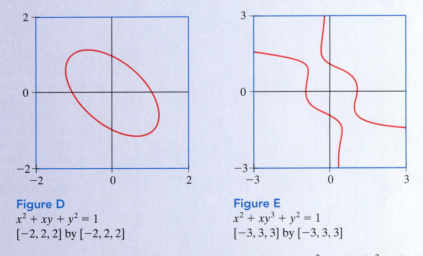

Figure D
$x^2 + xy + y^2 = 1$
$[-2, 2, 2]$ by $[-2, 2, 2]$

Figure E
$x^2 + xy^3 + y^2 = 1$
$[-3, 3, 3]$ by $[-3, 3, 3]$

For Figure D, we can rewrite the equation in the form $y^2 + xy + (x^2 - 1) = 0$. In this form, the quadratic formula can be used to solve for y in terms of x. As you should check for yourself (using $a = 1$, $b = x$, and $c = x^2 - 1$), the result is $y = \frac{1}{2}(-x \pm \sqrt{4 - 3x^2})$. The graph of $x^2 + xy + y^2 = 1$ is now obtained by using a graphing utility to graph the *two* explicit functions,

$$y_1 = \frac{-x + \sqrt{4 - 3x^2}}{2} \quad \text{and} \quad y_1 = \frac{-x - \sqrt{4 - 3x^2}}{2}$$

In Figure E, the equation cannot be solved for y in terms of x by elementary means. (Try it!) Furthermore, as of this writing, most graphing calculators require that functions be entered in explicit rather than implicit form. The graph in Figure E was obtained using a computer with commercial software that doesn't necessarily require the function to be in explicit form. (Three examples of such software are *Maple, Mathematica,* and *LiveMath Maker.*)

For Exercises 1–10: Each equation defines one or more functions implicitly. Solve the equation for y to determine explicit equations for the functions. Then use a graphing calculator or a computer to obtain the graph of the original equation.

For Exercises 11–14, if you have access to a computer with appropriate software, graph the given equation. Otherwise, you can look up the general form of the graph at the University of St. Andrews MacTutor Web site:

http://www-groups.dcs.st-and.ac.uk/~history/Curves/Curves.html

(Actually, in Exercises 11–13, the equations can be solved for y explicitly, using elementary means, and then a graphing calculator could be used. But we are not asking for that here.)

1. $xy = 0.1x^3 - 2x^2 - 3x - 4$ *(trident of Newton)*
2. $x^4 - 10(x^2 - y^2) = 0$ *(figure eight curve)*
3. $x^2y + 4y - 36x = 0$ *(serpentine)*
4. $9x^2 + 16y^2 = 144$ *(ellipse)*
5. $y^2(x^2 + y^2) = 9x^2$ *(kappa curve)*
6. $9y^2 = x^3(6 - x)$ *(pear-shaped quartic)*
7. $3y^2 = x(x - 1)^2$ *(Tschirnhaus' cubic)*
8. $(x - 3)^2(x^2 + y^2) = 8x^2$ *(conchoid)*
9. $4y^2 = x(x^2 - 2x - 10)$ *(Newton's diverging parabolas)*
10. $y^2(7 - x) = x^3$ *(cissoid of Diocles)*
11. $y^2(1 - x^2) = (x^2 + 2y - 1)^2$ *(bicorn)*
12. $(x^2 + y^2 - 2x)^2 = x^2 + y^2$ *(limaçon of Pascal)*
13. $(x^2 + y^2)^2 - 8x(x^2 - 3y^2) + 18(x^2 + y^2) = 27$ *(deltoid)*
14. $x^3 + y^3 = 3xy$ *(folium of Descartes)*

3.3 SHAPES OF GRAPHS. AVERAGE RATE OF CHANGE

In many applications, one of the most basic questions regarding a function is this: What are the highest and lowest points (if any) on the graph? We'll use the graph of the function G in Figure 1 to introduce some terminology that is useful here.

Figure 1
The graph of the function G is a fever graph: $G(t)$ is a patient's temperature t hours after midnight, where $0 \leq t \leq 12$.

In the figure, each of the points, P, Q, and R is called a *turning point*. At a **turning point,** a graph changes from rising to falling, or vice versa. In Figure 1, the point R, with coordinates $(5, 103)$, is the highest point on the graph of the function G. We say that the *maximum value* of the function G is 103, and this maximum occurs at $t = 5$. In terms of the temperature interpretation (described in the caption for Figure 1), the patient's temperature reaches a maximum of 103°F, and this maximum occurs 5 hours after midnight. More generally, we say that a number $f(x_0)$ is the **maximum value** of a function f if the inequality $f(x_0) \geq f(x)$ holds for every x in the domain of f. Minimum values are defined similarly: $f(x_0)$ is the minimum value for a function f if the inequality $f(x_0) \leq f(x)$ holds for every x in the domain of f. Assuming that the domain of the function in Figure 1 is $[0, 12]$, the minimum value of the function G is 99 and it occurs at $t = 12$. (How would you express this in terms of the temperature interpretation?) *Caution*: Not every function has a maximum or minimum value. See Figure 2 for an example.

Returning to Figure 1, there are two time intervals when the patient's temperature is rising: from midnight to 1 A.M., and from 3 A.M. to 5 A.M. We say that the function G is *increasing* on each of the intervals $[0, 1]$ and $[3, 5]$. Similarly (again looking at the graph), the patient's temperature is falling from 1 A.M. to 3 A.M. and again from 5 A.M. to 12 noon. We say that the function G is decreasing on each of the intervals $[1, 3]$ and $[5, 12]$. For another example of this terminology, look at the graph of $f(x) = x^2$ in Figure 2. The function is decreasing on the interval $(-\infty, 0]$ and increasing on the interval $[0, \infty)$. For analytical work involving functions, the terms increasing and decreasing are defined in terms of inequalities. A function f is **increasing** on an interval provided the following condition holds: For all pairs of numbers a and b in the interval, if $a < b$, then $f(a) < f(b)$. Similarly, a function f is **decreasing** on an interval if the following condition holds: For all pairs of numbers a and b in the interval, if $a < b$, then $f(a) > f(b)$.

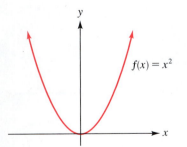

Figure 2
There is no maximum value for the function f defined by $f(x) = x^2$, assuming that the domain is $(-\infty, \infty)$. The minimum value of the function is 0, and it occurs at $x = 0$.

In general, for functions defined by equations, the techniques of calculus are required to compute exact coordinates for turning points, maximum or minimum values, and intervals where the function is increasing or decreasing. However, these quantities can be easily approximated by using a graphing utility to zoom in on the turning points. We do this in Example 1. *Technology note*: On a graphing calculator, the TRACE key can be used in conjunction with a ZOOM key to facilitate this process. Alternatively, graphing calculators have MIN and MAX operations that produce very accurate approximations for the minimum or maximum value of a function within any interval that the user requests. For details refer to the owner's manual that came with your calculator. *Algebra note*: In Sections 4.2 and 4.5, you'll see that in certain cases we can use the methods of algebra to compute *exact* maximum and minimum values.

| EXAMPLE | 1 | Approximating the coordinates of a turning point |

In Figure 3, a graphing utility was used to generate a graph of the function f defined by $f(x) = x^4 - 6x$. Assume that the chosen viewing rectangle shows all the essential features of the graph. Use a graphing utility to estimate to the nearest hundredth the coordinates of the turning point. What are the corresponding estimates for the minimum value of the function and the intervals where the function is increasing or decreasing?

GRAPHICAL PERSPECTIVE

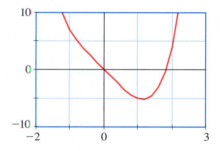

Figure 3
$f(x) = x^4 - 6x$
$[-2, 3, 1]$ by $[-10, 10, 5]$

SOLUTION
According to Figure 3, the turning point is located in quadrant IV, quite close to the point $(1, -5)$. Our goal is to obtain views that allow us to estimate the coordinates of the turning point to the nearest hundredth. We'll do this by repeatedly zooming in and, when appropriate, fine-tuning the viewing rectangle. At the outset, we remark that some trial and error is usually required here and that many different approaches are possible.

The view obtained in Figure 4(a) (after some experimenting) shows that the y-coordinate of the turning point is -5.15, to the nearest hundredth. The x-coordinate appears to be something between 1.14 and 1.15, but evidently we need a closer look to say whether it's nearer to 1.14 or 1.15. With that in mind, we try adjusting the x-specifications of the viewing rectangle so that x runs from 1.140 to 1.150.

As indicated in Figure 4(b), however, adjusting only the x-specifications isn't very helpful; much more magnification in the y-direction is required. [Are you

GRAPHICAL PERSPECTIVE

(a) [1.10, 1.20, 0.01] by
[−5.16, −5.12, 0.01]

(b) [1.140, 1.50, 0.002] by
[−5.16, −5.12, 0.01]

(c) [1.142, 1.48, 0.005] by
[−5.15125, −5.15115, 0.00005]

Figure 4
Three views of $f(x) = x^4 − 6x$

surprised by the seemingly horizontal line obtained in Figure 4(b)? We'll say more about it after this example.] In Figure 4(c), after more zooming and then fine-tuning of the viewing rectangle, we finally are able to see that the x-coordinate of the turning point is closer to 1.14 than to 1.15 (because it is to the *left* of 1.145).

In summary now, to the nearest hundredth, the coordinates of the turning point are $(1.14, −5.15)$. Thus, our estimates for the minimum value and intervals of increase or decrease are as follows:

$$\begin{aligned} \text{minimum value of } f\colon &\quad −5.15 \\ f \text{ is decreasing on:} &\quad (−\infty, 1.14] \\ f \text{ is increasing on:} &\quad [1.14, \infty) \end{aligned}$$

AFTERWORD ON ZOOMING FOR TURNING POINTS   In the process of trying to estimate the coordinates of the turning point in Example 1, we produced the view in Figure 4(b), which seems to show a horizontal line segment rather than a curve. This is actually a common occurrence when zooming in on turning points. There are two factors at work here, one having to do with the actual shape of the graph, the other with the limitations of the display screen and technology. For many types of functions, including those defined by polynomial expressions (as in Example 1), the graph does very closely resemble a horizontal line in the immediate vicinity of a turning point. We say "resemble" here because the "true" graph, as opposed to the approximation displayed on your calculator or computer screen, is slightly curved. Remember that the calculator display or the computer screen has only a finite number of dots or *pixels* with which to display a graph. Thus, in general, the display on the screen must necessarily be an approximation, because the true graph contains infinitely many points (one point for each x-value in the domain). In Figure 4(b), the calculator display shows a horizontal segment because, evidently, the actual curve differs from the horizontal segment by less than one pixel.

We can use the temperature graph back in Figure 1 (on page 186) to introduce another important concept relating to functions. According to Figure 1, the patient's temperature is increasing from midnight to 1 A.M. and again from 3 A.M. to 5 A.M. During which of these two time intervals is the temperature increasing faster? That is, how can we measure the *rate* at which the temperature is rising during a given time interval? Similarly, how can we measure and thereby compare the rates at which the temperature is falling during the time intervals from 1 A.M. to

3 A.M. and from 5 A.M. to 12 noon? One way to measure this is to use the *average rate of change,* which we define in the box that follows. (In Example 3 we'll answer the questions raised in this paragraph regarding the temperature graph.)

DEFINITION | **The Average Rate of Change of a Function**

Refer to Figure 5. The **average rate of change** of a function f on the interval $[a, b]$ is the slope of the line joining the two points $(a, f(a))$ and $(b, f(b))$. If $y = f(x)$, the average rate of change is denoted by $\Delta y/\Delta x$ or $\Delta f/\Delta x$ and we have

$$\frac{\Delta y}{\Delta x} = \frac{f(b) - f(a)}{b - a}$$

Figure 5

EXAMPLE **2** **Computing and comparing average rates of change**

Compute and compare the average rates of change for the function $f(x) = x^2$ on the following intervals: **(a)** $[-2, 0]$;   **(b)** $[0, 1]$;   **(c)** $[1, 2]$.

SOLUTION

(a) On $[-2, 0]$: $\dfrac{\Delta f}{\Delta x} = \dfrac{f(0) - f(-2)}{0 - (-2)} = \dfrac{0^2 - (-2)^2}{2} = -2$

(b) On $[0, 1]$: $\dfrac{\Delta f}{\Delta x} = \dfrac{f(1) - f(0)}{1 - 0} = \dfrac{1^2 - 0^2}{1} = 1$

(c) On $[1, 2]$: $\dfrac{\Delta f}{\Delta x} = \dfrac{f(2) - f(1)}{2 - 1} = \dfrac{2^2 - 1^2}{1} = 3$

(Refer to Figure 6.) On the interval $[-2, 0]$, the function is decreasing and the average rate of change is negative. On both of the intervals $[0, 1]$ and $[1, 2]$, the function is increasing and the average rates of change are positive. The graph rises more steeply on the interval $[1, 2]$ than on $[0, 1]$ and, correspondingly, the average rate of change is greater for the interval $[1, 2]$ than for $[0, 1]$.

Figure 6
$f(x) = x^2$

A WORD OF CAUTION   Like other types of averages occurring in arithmetic and statistics, the average rate of change can mask or hide some details. For instance, the fact that the average rate of change is positive on an interval does not guarantee that the function is increasing throughout that interval. For example, look back at Figure 5; for the function shown there, the average rate of change on $[a, b]$ is positive (because the slope of the solid blue line is positive). The function, however, is increasing only on a portion of $[a, b]$, not throughout the entire interval.

EXAMPLE **3** **Computing and interpreting average rates of change**

Compute and compare the average rates of change for the patient's temperature function G in Figure 1 (page 186) on each of the following pairs of intervals. Keep track of the units in the computations.

(a) $t = 0$ to $t = 1$ hr and $t = 3$ to $t = 5$ hr;

(b) $t = 1$ to $t = 3$ hr and $t = 5$ to $t = 12$ hr.

SOLUTION

(a) On the interval $t = 0$ to $t = 1$ hr:

$$\frac{\Delta G}{\Delta t} = \frac{G(1) - G(0)}{1 - 0}$$

$$= \frac{102° - 101°}{1 \text{ hr}} \qquad \text{reading the coordinates from the graph in Figure 1}$$

$$= \frac{1°}{1 \text{ hr}} = 1 \text{ degree/hr}$$

Similarly, on the interval $t = 3$ to $t = 5$ hr, we have

$$\frac{\Delta G}{\Delta t} = \frac{G(5) - G(3)}{5 - 3}$$

$$= \frac{103° - 100°}{2 \text{ hr}} = \frac{3°}{2 \text{ hr}} = 1.5 \text{ degrees/hr}$$

The calculations we've just made show that, on average, the temperature is increasing slightly faster from 3 A.M. to 5 A.M. than it is from midnight to 1 A.M. This is consistent with the graph in Figure 1: The curve looks slightly steeper from 3 A.M. to 5 A.M. than from midnight to 1 A.M.

(b) Before computing, let us note that both answers will be negative here because the graph shows the function is decreasing throughout each of the given intervals.

On the interval $t = 1$ to $t = 3$ hr:

$$\frac{\Delta G}{\Delta t} = \frac{G(3) - G(1)}{3 - 1}$$

$$= \frac{100° - 102°}{2 \text{ hr}} = \frac{-2°}{2 \text{ hr}} = -1 \text{ degree/hr}$$

On the interval $t = 5$ to $t = 12$ hr:

$$\frac{\Delta G}{\Delta t} = \frac{G(12) - G(5)}{12 - 5}$$

$$= \frac{99° - 103°}{7 \text{ hr}} = \frac{-4°}{7 \text{ hr}} = -\frac{4}{7} \text{ degree/hr} \approx -0.6 \text{ degree/hr}$$

These answers tell us that, on average, the patient's temperature is dropping faster between 1 A.M. and 3 A.M. than it is between 5 A.M. and noon. Again, this is consistent with what we see in Figure 1: The portion of the curve between points P and Q appears to be steeper, in general, than that between R and S.

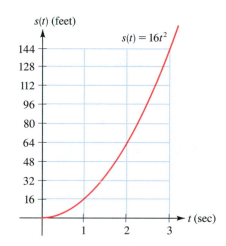

Figure 7

For an example using the average rate of change in a different context, consider the function s represented in Figure 7, with defining equation $s(t) = 16t^2$, and $t \geq 0$. This function relates the distance $s(t)$ and the time t for an object falling in a vacuum. Here t is measured in seconds, $s(t)$ is in feet, and $t = 0$ corresponds to the instant that the object begins to fall. For instance, after 1 sec, the object falls a distance of $s(1) = 16(1)^2 = 16$ ft. After 2 sec, the total distance will be $s(2) = 16(2)^2 = 64$ ft. Let's calculate the average rate of change of this function from, say, $t = 1$ to $t = 3$ sec. As in the temperature example, we'll keep track of the units. We have

$$\frac{\Delta s}{\Delta t} = \frac{s(3) - s(1)}{3 - 1}$$

$$= \frac{16(3)^2 \text{ ft} - 16(1)^2 \text{ ft}}{2 \text{ sec}} = \frac{128 \text{ ft}}{2 \text{ sec}} = 64 \text{ ft/sec}$$

Notice that the units here have the form *distance per unit time*. So in this case $\Delta s/\Delta t$ gives us the *average velocity* of the object. More generally, whenever we have a function expressing distance $s(t)$ in terms of time t, the **average velocity** over an interval is defined to be the average rate of change $\Delta s/\Delta t$ over that interval. We'll return to this idea in Example 6 at the end of this section.

Let's talk for a moment about notation. We defined the average rate of change of a function f on the interval $[a, b]$ to be the quantity

$$\frac{f(b) - f(a)}{b - a} \tag{1}$$

If, instead of $[a, b]$, we have the interval $[a, x]$, then, by definition, the average rate of change is

$$\frac{f(x) - f(a)}{x - a} \tag{2}$$

Figure 8
The closed interval $[x, x + h]$.

Finally, if the interval is $[x, x + h]$ as displayed in Figure 8, then the average rate of change is $\dfrac{f(x + h) - f(x)}{(x + h) - x}$, which simplifies to

$$\frac{f(x + h) - f(x)}{h} \tag{3}$$

Each of the algebraic expressions given in (1), (2), or (3) is known as a **difference quotient.** (But whatever the name, remember, it represents an average rate of change.) Our last three examples for this section show how difference quotients can be simplified once a specific function is given. (This skill is required in calculus when studying rates of change.)

EXAMPLE 4 Simplifying difference quotients of the form $\dfrac{f(x) - f(a)}{x - a}$

(a) If $f(x) = x^2$, find $\dfrac{f(x) - f(5)}{x - 5}$.

(b) If $g(x) = x^2 + 3x$, find $\dfrac{g(x) - g(a)}{x - a}$.

SOLUTION

PRELIMINARY NOTE   The solutions that follow make use of factoring techniques from intermediate algebra. If you find that you need a review of this topic, either now or in working the exercises, refer to Appendix B.4 at the back of this book.

(a)
$$\frac{f(x) - f(5)}{x - 5} = \frac{x^2 - 5^2}{x - 5}$$
$$= \frac{(x - 5)(x + 5)}{x - 5} \qquad \text{using difference of squares factoring}$$
$$= x + 5 \qquad \text{simplifying}$$

(b) Since $g(x) = x^2 + 3x$ and $g(a) = a^2 + 3a$, we have

$$\frac{g(x) - g(a)}{x - a} = \frac{(x^2 + 3x) - (a^2 + 3a)}{x - a} = \frac{x^2 + 3x - a^2 - 3a}{x - a}$$
$$= \frac{x^2 - a^2 + 3x - 3a}{x - a} \qquad \text{rearranging}$$
$$= \frac{(x - a)(x + a) + 3(x - a)}{x - a}$$
$$= \frac{(x - a)[(x + a) + 3]}{x - a} \qquad \text{factoring by grouping}$$
$$= x + a + 3 \qquad \text{simplifying}$$

EXAMPLE 5 Simplifying a difference quotient of the form $\dfrac{f(x + h) - f(x)}{h}$

Let $f(x) = 2/x$. Find $\dfrac{f(x + h) - f(x)}{h}$.

SOLUTION

We have $\dfrac{f(x + h) - f(x)}{h} = \dfrac{\dfrac{2}{x + h} - \dfrac{2}{x}}{h}$. This last expression needs to be simplified because the numerator itself contains fractions. An easy way to simplify here is to multiply by $\dfrac{x(x + h)}{x(x + h)}$, which equals 1. This yields

$$
\begin{aligned}
\frac{f(x + h) - f(x)}{h} &= \frac{x(x + h)}{x(x + h)} \cdot \frac{\dfrac{2}{x + h} - \dfrac{2}{x}}{h} \\
&= \frac{2x - 2(x + h)}{xh(x + h)} \\
&= \frac{2x - 2x - 2h}{xh(x + h)} = \frac{-2h}{xh(x + h)} \\
&= \frac{-2}{x(x + h)}
\end{aligned}
$$

EXAMPLE 6   **Computing average velocity**

Refer to the distance function s in Figure 7, $s = 16t^2$.
(a) Find a general expression for the average velocity $\Delta s/\Delta t$ over the interval $[1, 1 + h]$.
(b) Use the result in part (a) to complete the following table.

h (seconds)	Average velocity $\Delta s/\Delta t$ on the interval $[1, 1 + h]$ (feet/second)
1.0	
0.5	
0.1	

SOLUTION
(a) On the interval $[1, 1 + h]$ we have

$$
\begin{aligned}
\frac{\Delta s}{\Delta t} &= \frac{16(1 + h)^2 - 16(1)^2}{(1 + h) - 1} \\
&= \frac{16 + 32h + 16h^2 - 16}{h} = \frac{32h + 16h^2}{h} \\
&= \frac{16h(2 + h)}{h} = 16(2 + h) = 32 + 16h
\end{aligned}
$$

(b) From part (a) we have $\Delta s/\Delta t = 32 + 16h$. Using this result, we complete the required table as follows.

h (seconds)	Average velocity $\Delta s/\Delta t$ on the interval $[1, 1 + h]$ (feet/second)
1.0	$32 + 16(1) = 48$
0.5	$32 + 16(0.5) = 40$
0.1	$32 + 16(0.1) = 33.6$

NOTE   Exercise 36 asks you to carry out these calculations several steps further, using shorter and shorter time intervals. In calculus, this procedure leads to the concept of *instantaneous* (as opposed to *average*) *velocity*.

EXERCISE SET 3.3

A

1. Using the functions in Figure 7 on page 176 for example, explain what is meant by each of the following. Your answers should be in complete sentences.
   (a) turning point
   (b) maximum value
   (c) minimum value
   (d) f is increasing on an interval.
   (e) f is decreasing on an interval.
2. What is the definition of the average rate of change of a function g on an interval $[c, d]$? What does this have to do with slope?

In Exercises 3–6, you are given functions with domain $[0, 4]$. Specify:
(a) *the range of each function;*
(b) *the maximum value of the function;*
(c) *the minimum value of the function;*
(d) *interval(s) where the function is increasing; and*
(e) *interval(s) where the function is decreasing.*

3.

4.

5.

6.

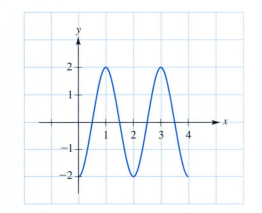

G **7.** Assume that the accompanying viewing rectangle shows the essential features of the graph of $f(x) = x^3 - 4x - 2$. Use a graphing utility to estimate to the nearest hundredth the coordinates of the turning points. What are the corresponding estimates for the intervals where the function is increasing or decreasing?

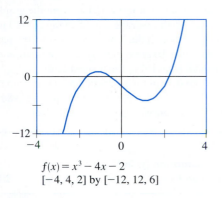

$$f(x) = x^3 - 4x - 2$$
$$[-4, 4, 2] \text{ by } [-12, 12, 6]$$

G **8.** Assume that the accompanying viewing rectangle shows the essential features of the graph of $f(x) = -x^4 + 3x + 5$. Use a graphing utility to estimate to the nearest hundredth the coordinates of the turning point. What is the corresponding estimate for the maximum value of the function? Estimate the intervals where the function is increasing or decreasing.

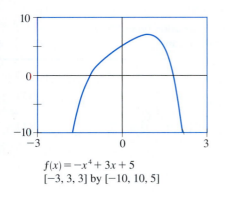

$$f(x) = -x^4 + 3x + 5$$
$$[-3, 3, 3] \text{ by } [-10, 10, 5]$$

In Exercises 9–14, compute the average rate of change of the function on the given interval.

9. $f(x) = x^2 + 2x$ on $[3, 5]$ **10.** $f(x) = \sqrt{x}$ on $[4, 9]$

11. $g(x) = 2x^2 - 4x$ on $[-1, 3]$ **12.** $g(x) = x^3 - x$ on $[1, 2]$

13. $h(t) = 2t - 6$ on $[5, 12]$

14. $h(t) = 16 - 7t$ on $[-\sqrt{2}, 2\sqrt{2}]$

15. The following graph shows the temperature $G(t)$ of a solution during the first 8 minutes of a chemistry experiment.

Compute the average rate of change of temperature, $\Delta G/\Delta t$, over the following intervals. (Be sure to specify the units as part of each answer.)

(a) $t = 0$ min to $t = 3$ min

(b) $t = 3$ min to $t = 6$ min

(c) $t = 6$ min to $t = 8$ min

16. Iodine-131 is a radioactive substance. The accompanying graph shows how an initial one-gram sample decays over a 32-day period; $f(t)$ represents the number of grams present after t days.

(a) Compute $\Delta f/\Delta t$ over each of the following intervals: $t = 0$ days to $t = 8$ days; $t = 8$ days to $t = 16$ days; $t = 16$ days to $t = 24$ days; $t = 24$ days to $t = 32$ days. (Be sure to specify the units with each answer.)

(b) Find the average of the four answers in part (a).

(c) Compute $\Delta f/\Delta t$ over the interval from $t = 0$ to $t = 32$. Is the answer the same as that obtained in part (b)?

For Exercises 17 and 18, let y = P(t) denote the percentage of U.S. households in the year t with at least one VCR (video cassette recorder). The following figure shows a graph of this function P over the period 1978 ≤ t ≤ 1997.

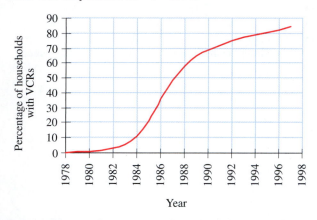

Source of data used to create graph: A. C. Nielsen and Nielsen Media Research

17. (a) According to the graph, is $\Delta P/\Delta t$ greater over the period 1978–1984 or over the period 1984–1990? After answering, use rough numerical values obtained from the graph to calculate estimates for these two rates of change. Include units with answers.

  (b) Using the following data, compute $\Delta P/\Delta t$ for the periods mentioned in part (a). Round the answers to one decimal place. P(1978) = 0.3%; P(1984) = 10.6%; P(1990) = 68.6%.

18. (a) In the given graph (or in a photocopy), use a ruler to draw two line segments, one connecting the two points (1978, P(1978)) and (1988, P(1988)), the other connecting the two points (1984, P(1984)) and (1997, P(1997)). You'll find that the two line segments appear to be parallel (or nearly so). Now complete the following sentence (concerning $\Delta P/\Delta t$ on the two intervals 1978 ≤ t ≤ 1988 and 1984 ≤ t ≤ 1997). Because parallel lines have equal slopes, we can estimate that _____.

  (b) Using the following data, compute $\Delta P/\Delta t$ for each of the two periods mentioned in part (a). Round the answers to one decimal place. Do your answers support the estimate you made in part (a)?

  P(1978) = 0.3%; P(1988) = 58.0%;
  P(1984) = 10.6%; P(1997) = 84.2%

For Exercises 19 and 20, refer to the following figure, which shows consumption of red meat in the United States and in China over the period 1980–1996. Let y = f(t) denote the consumption function for the United States, and let y = g(t) denote the consumption function for China.

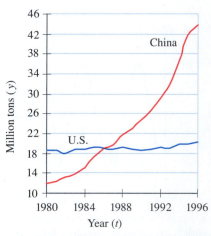

Consumption of red meat in United States and in China, 1980–1996

Source of data used to create graph: Worldwatch Institute

19. (a) Using values that you estimate from the graph, calculate $\Delta f/\Delta t$, the average rate of change of red meat consumption in the United States over the period 1980–1996. Express the answer as a fraction and as a decimal rounded to the nearest tenth.

  (b) Use the data in the following table to obtain a more accurate value for $\Delta f/\Delta t$ over the period 1980–1996. Round the answer to two decimal places.

Year (t)	1980	1996
Consumption (y) in U.S. (million tons)	18.68	20.32

20. (a) Using values that you estimate from the graph, calculate $\Delta g/\Delta t$, the average rate of change of red meat consumption in China over the period 1980–1996.

  (b) Use the data in the following table to obtain a more accurate value for $\Delta g/\Delta t$ over the period 1980–1996. Round the answer to two decimal places.

Year (t)	1980	1996
Consumption (y) in China (million tons)	11.90	43.83

21. Let $f(x) = 2x^2$.

  (a) Find $\dfrac{f(x) - f(3)}{x - 3}$.

  (b) Find $\dfrac{f(x) - f(a)}{x - a}$.

22. Let $f(x) = 4x^2$.

  (a) Find $\dfrac{f(x) - f(-2)}{x + 2}$.

  (b) Find $\dfrac{f(x) - f(a)}{x - a}$.

For Exercises 23 and 24, let $f(x) = x^2$. *Find and simplify the indicated difference quotient.*

23. (a) $\dfrac{f(2 + h) - f(2)}{h}$

24. (a) $\dfrac{f(1 + h) - f(1)}{h}$

(b) $\dfrac{f(x + h) - f(x)}{h}$

(b) $\dfrac{f(t + h) - f(t)}{h}$

In Exercises 25–32:

(a) *Find the difference quotient* $\dfrac{f(x) - f(a)}{x - a}$ *for each function, as in Example 4.*

(b) *Find the difference quotient* $\dfrac{f(x + h) - f(x)}{h}$ *for each function, as in Example 5.*

25. $f(x) = 8x - 3$

26. $f(x) = -2x + 5$

27. $f(x) = x^2 - 2x + 4$

28. $f(x) = 2x^2 - x + 1$

29. $f(x) = 1/x$

30. $f(x) = -3/x^2$

31. $f(x) = 2x^3$

32. $f(x) = 1 - x^3$

Hint: For Exercises 31 and 32, you'll need to use difference-of-cubes factoring from intermediate algebra. See the inside back cover for the relevant formula. For a more detailed review of the topic, refer to Appendix B.4.

For Exercises 33–36, use the distance function $s(t) = 16t^2$ *discussed on page 191 and in Example 6. Recall that this function relates the distance* $s(t)$ *and the time t for a freely falling object (neglecting air resistance). The time t is measured in seconds, with* $t = 0$ *corresponding to the instant that the object begins to fall; the distance* $s(t)$ *is in feet.*

33. Find the average velocity $\Delta s/\Delta t$ over the time interval $1 \le t \le 2$.

34. (a) Find the average velocity over each of the following time intervals: [2, 3], [3, 4], and [2, 4].

(b) Let a, b, and c denote the three average velocities that you computed in part (a), in the order given. Is it true that the arithmetical average of a and b is c?

35. (a) Follow the method of Example 6(a) to find a general expression for the average velocity $\Delta s/\Delta t$ over the interval $[2, 2 + h]$.

(b) Complete a table similar to the one shown in the solution of Example 6(b); for the h-values in the left-hand column use 0.1, 0.01, 0.001, 0.0001, and 0.00001.

(c) Looking at your results in part (b), answer the following question. As h approaches zero, what value does the average velocity in the right-hand column seem to be approaching? This target value or limit is the *instantaneous* (as opposed to average) *velocity* of the object when $t = 2$ sec.

36. (a) After rereading Example 6, extend the results in part (b) of the example by completing the following table. Don't round your answers.

h (seconds)	0.01	0.001	0.0001	0.00001
Average velocity $\Delta s/\Delta t$ on interval [1, 1 + h] (ft/second)				

(b) As h approaches zero, what value does the average velocity $\Delta s/\Delta t$ seem to be approaching? This target value or limit is called the *instantaneous velocity* of the object when $t = 1$ sec.

37. Suppose that the demand function for a certain item is given by $p(x) = \dfrac{24}{2^{x/100}}$, where x is the number of items that can be sold when the price of each item is $p(x)$ dollars. Compute $\Delta p/\Delta x$ over each of the intervals $0 \le x \le 100$ and $300 \le x \le 400$. Include units in your answers. Why does it make sense (economically) that the answers are negative?

38. Suppose that during the first few hours of a laboratory experiment, the temperature of a certain substance is closely approximated by the function $f(t) = t^3 - 6t^2 + 9t$, where t is measured in hours, with $t = 0$ corresponding to the instant the experiment begins, and $f(t)$ is the temperature (°F) of the substance after t hours.

(a) Find an expression for $\Delta f/\Delta t$. What are the units for $\Delta f/\Delta t$?

(b) Use the result in part (a) to complete the following table. Round the answers to four decimal places.

Interval	[0, 0.1]	[0, 0.01]	[0, 0.001]	[0, 0.0001]
$\Delta f/\Delta t$				

(c) In part (b), as the right-hand endpoint gets closer and closer to 0, what value does $\Delta f/\Delta t$ seem to be approaching? This target value tells us the rate at which the temperature is changing at the *instant* the experiment begins.

39. Complete the following table.

Function	$|x|$	x^2	x^3
Domain			
Range			
Turning point			
Maximum value			
Minimum value			
Interval(s) where increasing			
Interval(s) where decreasing			

40. Set up and complete a table like the one in Exercise 39 for the three functions $1/x$, $\sqrt{x}$, and $\sqrt{1 - x^2}$.

B

41. Let $f(x) = 1/x$. Find a number b so that the average rate of change of f on the interval $[1, b]$ is $-1/5$.

42. Let $f(x) = \sqrt{x}$. Find a number b so that the average rate of change of f on the interval $[1, b]$ is $1/7$.

43. Let $f(x) = ax^2 + bx + c$. Show that $$\dfrac{f(x + h) - f(x)}{h} = 2ax + ah + b.$$

44. Find a number a between 0 and 1 so that the average rate of change of $f(x) = x^2$ on the interval $[a, 1/a]$ is $10a$.

Ⓖ *In Exercises 45 and 46 (as opposed to Exercises 7 and 8), the viewing rectangles do not show all of the essential features of the graphs. In each case, use a graphing utility and experiment with different viewing rectangles to determine approximate x- and y-coordinates for all turning points. What are the corresponding intervals of increase and decrease for the function?*

45.

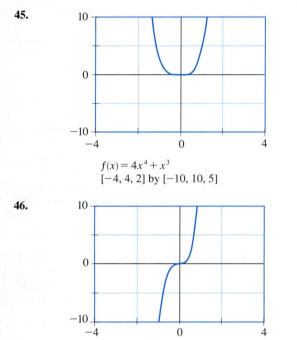

$f(x) = 4x^4 + x^3$
$[-4, 4, 2]$ by $[-10, 10, 5]$

46.

$g(x) = x^4 + 12x^3 + x^2$
$[-4, 4, 2]$ by $[-10, 10, 5]$

47. Let $h(x) = \sqrt{x}$. Two functions f and g are defined in terms of h as follows:

$$f(x) = \frac{h(x) - h(2)}{x - 2} \qquad g(x) = \frac{1}{h(x) + h(2)}$$

Ⓖ **(a)** Using a graphing utility, graph the two functions f and g in the same viewing screen. What do you observe?

(b) Use algebra to explain the result in part (a). That is, either derive or verify the identity $f(x) = g(x)$. Are there any positive values of x for which this equation does not hold?

48. Consider the function f defined by

$$f(x) = x^2 + \frac{2}{x^2} \quad (x > 0)$$

(a) Complete the table. (Round the results to four decimal places.)

x	1	1.05	1.10	1.15	1.20	1.25
f(x)						

(b) Which $f(x)$-value in the table is the smallest? What is the corresponding input?

Ⓖ **(c)** Use a graphing utility to graph the function f. Then use the zoom and/or trace features to find an output $f(x)$ that is less than the output you indicated in part (b).

(d) It can be shown using calculus that the input x yielding the minimum value for the function f is $\sqrt[4]{2}$. Compute $f(\sqrt[4]{2})$. Which value of $f(x)$ in the table is closest to $f(\sqrt[4]{2})$? To how many decimal places does your estimate in part (c) agree with $f(\sqrt[4]{2})$?

3.4 TECHNIQUES IN GRAPHING

. . . geometrical figures are graphic formulas. —David Hilbert (1862–1943)

The simple geometric concepts of reflection and translation can be used to great advantage in graphing. We discussed the idea of reflection in the x-axis and in the y-axis in Section 1.7. By a **translation** of a graph, we mean a shift in its location such that every point of the graph is moved the same distance in the same direction. The size and the shape of a graph are unchanged by a translation.

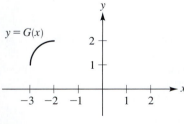

Figure 1

EXAMPLE **1** **Sketching combinations of translations and reflections**

The graph of a function G is the quarter-circle shown in Figure 1. In each case, sketch the resulting graph after the following operations are carried out on the graph of G:

(a) a translation of four units to the right followed by a translation of one unit vertically downward;

(b) a translation of three units to the right followed by a reflection in the *y*-axis;

(c) a reflection in the *y*-axis followed by a translation of three units to the right. [Note that these are the same two operations used in part (b), but here the order is reversed.]

SOLUTION

(a) Figure 2 shows the result of translating the graph of *G* four units to the right and then one unit down. (As you can check by drawing a sketch, the same end result is obtained if we first translate one unit down and then four units to the right.) *Fact:* If several translations are to be carried out in succession, the end result will be the same, no matter in what order the individual translations are carried out.

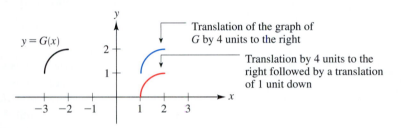

Figure 2
The red curve is the end result when the graph of *G* is translated four units to the right and then one unit down.

(b) Figure 3(a) shows the graph of *G* after a translation of three units to the right. When this is followed by a reflection in the *y*-axis, we obtain the result shown in Figure 3(b).

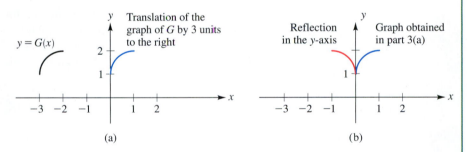

Figure 3
Figure (a) shows the result of translating the graph of *G* three units to the right. When this is followed by a reflection in the *y*-axis, we obtain the red curve in Figure (b).

(c) Figure 4 shows what happens to the graph of *G* if we use the same two operations that were used in part (b) but reverse the order in which they are carried out. Note that the end result is quite different from that obtained in part (b).

Figure 4
The reflection of the graph of *G* in the *y*-axis followed by a translation of three units to the right.

Given the graph of a function f, it is useful to know how to sketch efficiently the graphs of the following closely related functions. (In the following equations, c denotes a positive constant.)

$$y = f(x) + c \qquad y = f(x + c) \qquad y = -f(x)$$
$$y = f(x) - c \qquad y = f(x - c) \qquad y = f(-x)$$

Each of these can be graphed by translating or reflecting the graph of $y = f(x)$. In the box that follows, we summarize the techniques that are involved. (Exercises 47–50 will help you to see why these techniques are valid, and Section 8.1 will explain these techniques from a more general point of view.)

█ PROPERTY SUMMARY   Translations and Reflections

(In Items 1–4, the letter c denotes a positive constant.)

Equation	How to Obtain the Graph from That of $y = f(x)$
1. $y = f(x) + c$	Translate c units vertically upward
2. $y = f(x) - c$	Translate c units vertically downward
3. $y = f(x + c)$	Translate c units to the left
4. $y = f(x - c)$	Translate c units to the right
5. $y = -f(x)$	Reflect in the x-axis
6. $y = f(-x)$	Reflect in the y-axis

In Figure 5 we show examples of the first four techniques in the box, the techniques involving translation. Figure 6 displays examples of the remaining two techniques, the ones involving reflection. You should look over these figures carefully before going on to Examples 2 through 4, in which the various graphing techniques are combined.

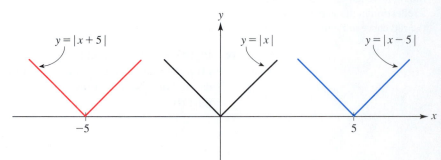

(a) To graph $y = |x| + 2$, translate $y = |x|$ up 2 units. To graph $y = |x| - 2$, translate $y = |x|$ down 2 units.

(b) To graph $y = |x + 5|$, translate $y = |x|$ to the left 5 units. To graph $y = |x - 5|$, translate $y = |x|$ to the right 5 units.

Figure 5

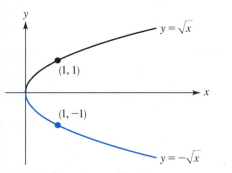

(a) To graph $y = -\sqrt{x}$, reflect $y = \sqrt{x}$ in the x-axis. More generally, to graph $y = -f(x)$, reflect the graph of $y = f(x)$ in the x-axis.

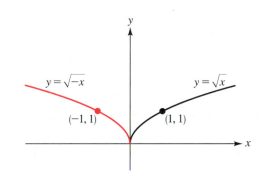

(b) To graph $y = \sqrt{-x}$, reflect $y = \sqrt{x}$ in the y-axis. More generally, to graph $y = f(-x)$, reflect the graph of $y = f(x)$ in the y-axis.

Figure 6

EXAMPLE 2 **Using translations of $y = x^2$**

Graph $y = (x - 2)^2 + 1$.

SOLUTION
Begin with the graph of $y = x^2$ in Figure 7(a). Move the curve two units to the right to obtain the graph of $y = (x - 2)^2$; see Figure 7(b). Then move the curve in Figure 7(b) up one unit to obtain the graph of $y = (x - 2)^2 + 1$; see Figure 7(c).

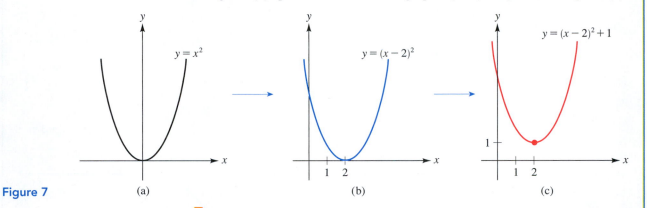

Figure 7   (a)   (b)   (c)

EXAMPLE 3 **Using translations of $y = 1/x$**

Graph the functions $y = \dfrac{1}{x - 1}$ and $y = \dfrac{1}{x - 1} + 1$.

SOLUTION
Begin with the graph of $y = 1/x$ in Figure 8(a). The x- and y-axes are asymptotes for this graph. Moving this graph to the right one unit yields the graph of $y = \dfrac{1}{x - 1}$, shown in Figure 8(b). Note that the vertical asymptote moves one unit to the right also, but the horizontal asymptote is unchanged. Next, we move

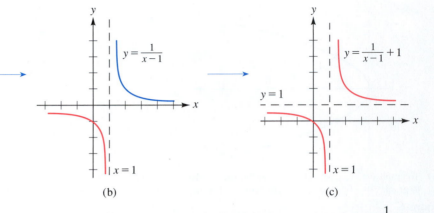

(a)   (b)   (c)

Figure 8

the graph in Figure 8(b) up one unit (why?) to obtain the graph of $y = \dfrac{1}{x-1} + 1$, as shown in Figure 8(c). (You should verify for yourself that the y-intercepts in Figures 8(b) and 8(c) are -1 and 0, respectively.)

| EXAMPLE | 4 | Applying reflection and translations to $y = |x|$ |

Graph each equation:
(a) $y = -|x|$;   **(b)** $y = -|x-2| + 3$.

SOLUTION
See Figure 9.

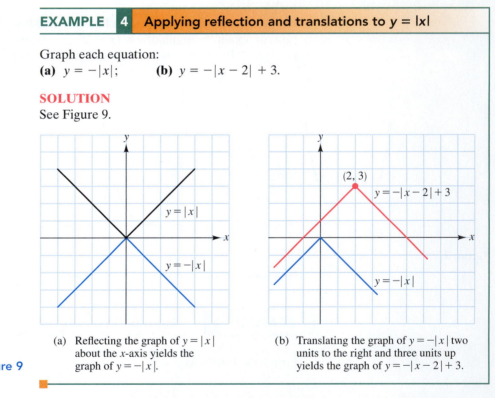

(a) Reflecting the graph of $y = |x|$ about the x-axis yields the graph of $y = -|x|$.

(b) Translating the graph of $y = -|x|$ two units to the right and three units up yields the graph of $y = -|x-2|+3$.

Figure 9

| EXAMPLE | 5 | Using algebra to determine how two graphs are related |

Figure 10 shows a graph of the function k defined by $k(x) = x^4 - 3x + 1$. In each case, say how the graph of the given function can be obtained from the graph of k. Then use a graphing utility to verify your statement:
(a) $f(x) = -x^4 + 3x - 1$;   **(b)** $g(x) = x^4 + 3x + 1$;   **(c)** $h(x) = x^4 - 3x + 3$.

GRAPHICAL PERSPECTIVE

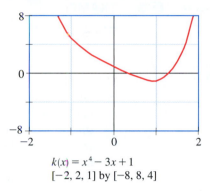

$$k(x) = x^4 - 3x + 1$$
$$[-2, 2, 1] \text{ by } [-8, 8, 4]$$

Figure 10

SOLUTION

(a) Note that the given expression for $f(x)$ is just the negative of $k(x)$. That is,

$$f(x) = -x^4 + 3x - 1 = -(x^4 - 3x + 1) = -k(x)$$

Therefore, the graph of f can be obtained by reflecting the graph of k in the x-axis. See Figure 11(a).

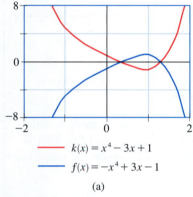

$$k(x) = x^4 - 3x + 1$$
$$f(x) = -x^4 + 3x - 1$$

(a)

$$k(x) = x^4 - 3x + 1$$
$$g(x) = x^4 + 3x + 1$$

(b)

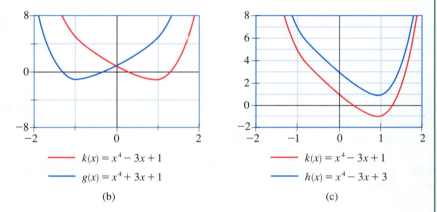

$$k(x) = x^4 - 3x + 1$$
$$h(x) = x^4 - 3x + 3$$

(c)

Figure 11
Some graphs obtained by reflecting or translating the graph of
$k(x) = x^4 - 3x + 1$

(b) The expressions for $g(x)$ and $k(x)$ are the same except for the sign in front of the $3x$-term. After some thought (or perhaps some trial and error) we find that

$$k(-x) = (-x)^4 - 3(-x) + 1$$
$$= x^4 + 3x + 1$$
$$= g(x)$$

That is, we have $k(-x) = g(x)$, and therefore the graph of g can be obtained by reflecting the graph of k in the y-axis. See Figure 11(b).

(c) Comparing the two equations $k(x) = x^4 - 3x + 1$ and $h(x) = x^4 - 3x + 3$, we see that

$$h(x) = k(x) + 2$$

Thus, the graph of h is obtained by translating the graph of k up two units in the y-direction. See Figure 11(c).

EXAMPLE | **6** | **A case where the translation rule for $y = f(x + c)$ is not directly applicable**

(a) Graph $y = \sqrt{-x + 2}$. **(b)** Graph $y = |1 - x|$.

SOLUTION

(a) Our translation rule for $y = f(x + c)$ is not directly applicable here because in the generic equation $y = f(x + c)$, the sign of the x-term is positive; the rule provides no information about $y = f(-x + c)$. Rather than introducing yet another rule, we'll explain how to proceed using what you've already learned. To graph $y = \sqrt{-x + 2}$, first ignore the negative sign and graph the simpler equation $y = \sqrt{x + 2}$, as shown in Figure 12(a). Then for the graph of $y = \sqrt{-x + 2}$, just reflect the graph of $y = \sqrt{x + 2}$ in the y-axis, as shown in Figure 12(b). (We are using the fact that replacing x with $-x$ throughout an equation reflects the graph in the y-axis.)

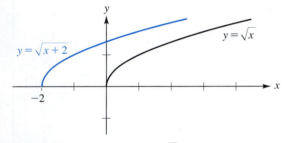

(a) Translate the graph of $y = \sqrt{x}$ to the left 2 units to obtain the graph of $y = \sqrt{x + 2}$.

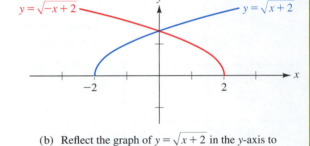

(b) Reflect the graph of $y = \sqrt{x + 2}$ in the y-axis to obtain the graph of $y = \sqrt{-x + 2}$.

Figure 12
Steps in graphing $y = \sqrt{-x + 2}$: first graph $y = \sqrt{x + 2}$; then reflect in the y-axis to obtain the graph of $y = \sqrt{-x + 2}$.

(b) Write the given equation in the equivalent but somewhat more familiar form $y = |-x + 1|$. Now apply the strategy explained in part (a). That is, first graph $y = |x + 1|$, and then reflect in the y-axis to obtain the graph of $y = |-x + 1|$. See Figure 13.

NOTE We can use one of the properties of absolute value to obtain the graph in a different way and thus check the result in Figure 13. We have

$$y = |-x + 1| = |-1(x - 1)| = |-1||x - 1|$$
$$= |x - 1|$$

This shows that the required graph can be obtained by graphing $y = |x - 1|$ —that is, by translating $y = |x|$ one unit to the right. As you can see, this agrees with the end result in Figure 13.

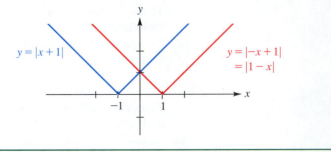

Figure 13
Steps in graphing $y = |-x + 1|$:
First graph $y = |x + 1|$, then reflect in the y-axis.

The following example makes use of all the graphing techniques presented in this section. When you reach the point where you can work each part of this example on your own, then you know the material in this section.

EXAMPLE | **7** | **A comprehensive example**

The graph of a function f is a line segment joining the points $(-3, 1)$ and $(2, 4)$. Graph each of the following functions:

(a) $y = f(-x)$; **(b)** $y = -f(x)$; **(c)** $y = -f(-x)$;

(d) $y = f(x + 1)$; **(e)** $y = -f(x + 1)$; **(b)** $y = f(1 - x)$.

SOLUTION
See Figure 14.

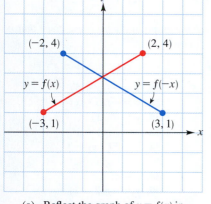

(a) Reflect the graph of $y = f(x)$ in the y-axis to obtain the graph of $y = f(-x)$.

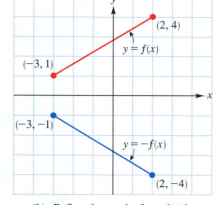

(b) Reflect the graph of $y = f(x)$ in the x-axis to obtain the graph of $y = -f(x)$.

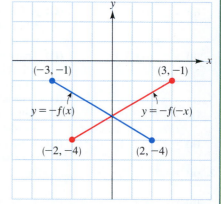

(c) Reflect the graph of $y = -f(x)$ in the y-axis to obtain the graph of $y = -f(-x)$. [Or, reflect the graph of $y = f(-x)$ in the x-axis.]

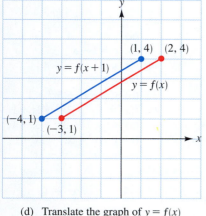

(d) Translate the graph of $y = f(x)$ one unit to the left to obtain the graph of $y = f(x + 1)$.

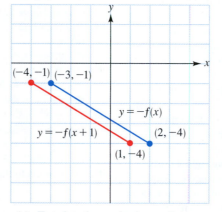

(e) Translate the graph of $y = -f(x)$ one unit to the left to obtain the graph of $y = -f(x + 1)$. [Or, reflect the graph of $y = f(x + 1)$ in the x-axis.]

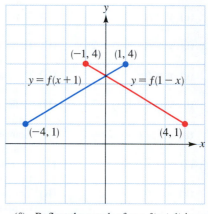

(f) Reflect the graph of $y = f(x + 1)$ in Figure 14(d) in the y-axis to obtain the graph of $y = f(1 - x)$. [Reason: Replacing x with $-x$ in $y = f(x + 1)$ yields $y = f(1 - x)$.]

Figure 14

EXERCISE SET 3.4

A

In Exercises 1 and 2, the right-hand column contains instructions for translating and/or reflecting the graph of $y = f(x)$. Match each equation in the left-hand column with an appropriate set of instructions in the right-hand column.

1. (a) $y = f(x - 1)$ **(A)** Translate left 1 unit.

    (b) $y = f(x) - 1$ **(B)** Reflect in the x-axis, then translate left 1 unit.

    (c) $y = f(x) + 1$ **(C)** Translate right 1 unit.

    (d) $y = f(x + 1)$ **(D)** Reflect in the x-axis, then translate up 1 unit.

    (e) $y = f(-x) + 1$ **(E)** Reflect in the x-axis, then translate down 1 unit.

    (f) $y = f(-x) - 1$ **(F)** Translate down 1 unit.

    (g) $y = -f(x) + 1$ **(G)** Reflect in the x-axis, reflect in the y-axis, then translate up 1 unit.

    (h) $y = -f(x + 1)$ **(H)** Translate left 1 unit, then reflect in the y-axis, then translate up 1 unit.

    (i) $y = -f(x) - 1$ **(I)** Translate up 1 unit.

    (j) $y = f(1 - x) + 1$ **(J)** Reflect in the y-axis, then translate up 1 unit.

      Hint: See Example 6.

    (k) $y = -f(-x) + 1$ **(K)** Reflect in the y-axis, then translate down 1 unit.

2. (a) $y = f(x + 2) + 3$ **(A)** Translate left 2 units, then translate down 3 units.

    (b) $y = f(x + 3) + 2$ **(B)** Translate left 3 units, then translate up 2 units.

    (c) $y = f(x - 2) + 3$ **(C)** Translate right 3 units, then translate up 2 units.

    (d) $y = f(x - 2) - 3$ **(D)** Translate left 3 units, then translate down 2 units.

    (e) $y = f(x + 2) - 3$ **(E)** Translate right 3 units and down 2 units.

    (f) $y = f(x - 3) + 2$ **(F)** Reflect in the y-axis, then translate up 2 units.

    (g) $y = f(x - 3) - 2$ **(G)** Reflect in the x-axis, then translate right 2 units.

    (h) $y = f(x + 3) - 2$ **(H)** Reflect in the x-axis, then translate left 2 units.

    (i) $y = -f(x + 2)$ **(I)** Translate left 2 units, then reflect in the y-axis.

    (j) $y = -f(x - 2)$ **(J)** Translate right 2 units, then translate up 3 units.

    (k) $y = f(2 - x)$ **(K)** Translate left 2 units, then translate up 3 units.

    (l) $y = f(-x) + 2$ **(L)** Translate right 2 units and down 3 units.

In Exercises 3–24, sketch the graph of the function.

3. $y = x^3 - 3$ **4.** $y = x^2 + 3$

5. $y = (x + 4)^2$ **6.** $y = (x + 4)^2 - 3$

7. $y = (x - 4)^2$ **8.** $y = (x - 4)^2 + 1$

9. $y = -x^2$ **10.** $y = -x^2 - 3$

11. $y = -(x - 3)^2$ **12.** $y = -(x - 3)^2 - 3$

13. $y = \sqrt{x - 3}$ **14.** $y = \sqrt{x - 3} + 1$

15. $y = \sqrt{-x + 3}$ **16.** $y = \sqrt{-x + 3} + 2$

17. $y = \dfrac{1}{x + 2} + 2$ **18.** $y = \dfrac{1}{x - 3} - 1$

19. $y = (x - 2)^3$ **20.** $y = (x - 2)^3 + 1$

21. $y = -x^3 + 4$ **22.** $y = -(x - 1)^3 + 4$

23. (a) $y = |x + 4|$ **24.** (a) $y = \sqrt{x + 2}$

     (b) $y = |4 - x|$ (b) $y = \sqrt{2 - x}$

     (c) $y = -|4 - x| + 1$ (c) $y = -\sqrt{2 - x}$

In Exercises 25–40, sketch the graph of the function, given that f, F, and g are defined as follows:

$$f(x) = |x| \qquad F(x) = 1/x \qquad g(x) = \sqrt{1 - x^2}$$

25. $y = f(x - 5)$ **26.** $y = -f(x - 5)$

27. $y = f(5 - x)$ **28.** $y = -f(5 - x)$

29. $y = 1 - f(x - 5)$ **30.** $y = f(-x)$

31. $y = F(x + 3)$ **32.** $y = F(x) + 3$

33. $y = -F(x + 3)$ **34.** $y = F(-x) + 3$

35. $y = g(x - 2)$ **36.** $y = -g(x - 2)$

37. $y = 1 - g(x - 2)$ **38.** $y = g(-x)$

39. $y = g(2 - x)$ **40.** $y = -g(2 - x)$

Ⓖ As background for Exercises 41–46, you should review Example 5. In Exercises 41–46, a function f is given. Say how the graph of each of the related functions can be obtained from the graph of f, and then use a graphing utility to verify your statement (as in Figure 11).

41. $f(x) = 3x - 2$ **42.** $f(x) = 4 - x^2$

     (a) $y = -3x - 2$ (a) $y = x^2 - 4$

     (b) $y = -3x + 2$ (b) $y = 1 - x^2$

43. $f(x) = x^2 + 4x + 2$ **44.** $f(x) = x^3 + 3x^2 - 4$

     (a) $y = x^2 - 4x + 2$ (a) $y = -x^3 + 3x^2 - 4$

     (b) $y = x^2 + 4x$ (b) $y = -x^3 + 3x^2 + 1$

45. $f(x) = x^4 - 3x + 3$ **46.** $f(x) = -x^3 + 3x^2 - 3x + 1$

     (a) $y = x^4 + 3x + 3$ (a) $y = -x^3 + 3x^2 - 3x - 1$

     (b) $y = -x^4 + 3x - 3$ (b) $y = x^3 + 3x^2 + 3x + 1$

     (c) $y = -x^4 + 3x$ (c) $y = x^3 - 3x^2 + 3x - 1$

47. (a) Complete the following table.

x	x^2	$x^2 - 1$	$x^2 + 1$
0			
±1			
±2			
±3			

(b) Using the results in the table, graph the functions $y = x^2$, $y = x^2 - 1$, and $y = x^2 + 1$ on the same set of axes. How are the graphs related?

48. (a) Complete the given table.

x	x^2	$(x - 1)^2$	$(x + 1)^2$
0			
1			
2			
3			
-1			
-2			
-3			

(b) Using the results in the table, graph the functions $y = x^2$, $y = (x - 1)^2$, and $y = (x + 1)^2$ on the same set of axes. How are the graphs related?

49. (a) Complete the following table. (Use a calculator where necessary.)

x	$\sqrt{x}$	$-\sqrt{x}$
0		
1		
2		
3		
4		
5		

(b) Using the results in the table, graph the functions $y = \sqrt{x}$ and $y = -\sqrt{x}$ on the same set of axes. How are the two graphs related?

50. (a) Complete the given tables. (Use a calculator where necessary.)

x	0	1	2	3	4	5
$\sqrt{x}$						

x	0	-1	-2	-3	-4	-5
$\sqrt{-x}$						

(b) Using the tables, graph the functions $y = \sqrt{x}$ and $y = \sqrt{-x}$ on the same set of axes. How are the graphs related?

In Exercises 51–53 you'll use a graphing utility to provide examples of the six graphing techniques listed in the box on page 200.

Ⓖ **51. (a)** Graph the two functions $f(x) = \sqrt{1 + x^2}$ and $g(x) = \sqrt{1 + x^2} + 3$. Observe that the graph of g is obtained by translating the graph of f *up* three units. This illustrates Property 1 on page 200.

(b) Graph the two functions $f(x) = \sqrt{1 + x^2}$ and $h(x) = \sqrt{1 + x^2} - 3$. Observe that the graph of h is

obtained by translating the graph of f *down* three units. This illustrates Property 2 on page 200.

Ⓖ **52.** Using the viewing rectangle $[-5, 3]$ by $[0, 5]$, graph the three functions

$$F(x) = \sqrt{1 - x^3}$$
$$L(x) = \sqrt{1 - (x + 2)^3}$$
$$R(x) = \sqrt{1 - (x - 2)^3}$$

Observe that the graph of L is obtained by translating the graph of F to the *left* 2 units. Also, the graph of R is obtained by translating the graph of F to the *right* 2 units. This illustrates Properties 3 and 4 on page 200.

Ⓖ **53.** For this exercise, use the viewing rectangle $[-4, 4]$ by $[-4, 4]$.

(a) On the same set of axes, graph the two functions $f(x) = 2^x$ and $g(x) = -2^x$. [*Algebra reminder*: -2^x means $-(2^x)$, not $(-2)^x$]. Observe that the graph of g is obtained by reflecting the graph of f in the x-axis. This illustrates Property 5 on page 200.

(b) On the same set of axes graph the two functions $f(x) = 2^x$ and $h(x) = 2^{-x}$. Observe that the graph of h is obtained by reflecting the graph of f in the y-axis. This illustrates Property 6 on page 200.

B

54. Reflect the graph of $y = \sqrt{x}$ in the y-axis and then translate that 2 units to the left. What is the equation of the resulting graph? *Hint*: The answer is not $y = \sqrt{-x + 2}$.

55. Let P be a point with coordinates (a, b), and assume that c and d are positive numbers. (The condition that c and d are positive isn't really necessary in this problem, but it will help you to visualize things.)

(a) Translate the point P by c units in the x-direction to obtain a point Q, then translate Q by d units in the y-direction to obtain a point R. What are the coordinates of the point R?

(b) Translate the point P by d units in the y-direction to obtain a point S, then translate S by c units in the x-direction to obtain a point T. What are the coordinates of the point T?

(c) Compare your answers for parts (a) and (b). What have you demonstrated? (Answer in complete sentences.)

56. Let P be a point with coordinates (a, b).

(a) Reflect P in the x-axis to obtain a point Q, then reflect Q in the y-axis to obtain a point R. What are the coordinates of the point R?

(b) Reflect P in the y-axis to obtain a point S, then reflect S in the x-axis to obtain a point T. What are the coordinates of the point T?

(c) Compare your answers for parts (a) and (b). What have you demonstrated? (Answer in complete sentences.)

Ⓖ **57.** **(a)** Use a graphing utility to graph the function
$y = x/(x - 1)$.

(b) From the screen display, it appears that the graph may be a translation of the basic $y = 1/x$ graph. Prove that this is indeed the case by using algebra to show that

$$\frac{x}{x - 1} = \frac{1}{x - 1} + 1. \quad \textit{Hint}: \text{Combine the quantities}$$

on the right side of the equation.

(c) Beginning with the graph of $y = 1/x$, what are the translations used to obtain the graph of $y = x/(x - 1)$?

For Exercises 58 and 59, assume that (a, b) is a point on the graph of $y = f(x)$, and specify the corresponding point on the graph of each equation. [For example, the point that corresponds to (a, b) on the graph of $y = f(x - 1)$ is $(a + 1, b)$.]

58. **(a)** $y = f(x - 3)$ **(e)** $y = f(-x)$
  (b) $y = f(x) - 3$ **(f)** $y = -f(-x)$
  (c) $y = f(x - 3) - 3$ **(g)** $y = f(3 - x)$
  (d) $y = -f(x)$ **(h)** $y = -f(3 - x) + 1$

59. **(a)** $y = f(-x) + 2$ **(d)** $y = 1 - f(x + 1)$
  (b) $y = -f(-x) + 2$ **(e)** $y = f(1 - x)$
  (c) $y = -f(x - 3)$ **(f)** $y = -f(1 - x) + 1$

60. **(a)** A function f is said to be **even** if the equation $f(-x) = f(x)$ is satisfied by all values of x in the domain of f. Explain why the graph of an even function must be symmetric about the y-axis.

(b) Show that each function is even by computing $f(-x)$ and then noting that $f(x)$ and $f(-x)$ are equal.

(i) $f(x) = x^2$ **(iii)** $f(x) = 3x^6 - \dfrac{4}{x^2} + 1$
(ii) $f(x) = 2x^4 - 6$

61. **(a)** A function f is said to be **odd** if the equation $f(-x) = -f(x)$ is satisfied by all values of x in the domain of f. Show that if (x, y) is a point on the graph of an odd function f, then the point $(-x, -y)$ is also on the graph. (This implies that the graph of an odd function must be symmetric about the origin.)

(b) Show that each function is odd by computing $f(-x)$ as well as $-f(x)$ and then noting that the two expressions obtained are equal.

(i) $f(x) = x^3$ **(iii)** $f(x) = |x|/(x + x^7)$
(ii) $f(x) = -2x^5 + 4x^3 - x$

62. Is each function odd, even, or neither? (See Exercises 60 and 61 for definitions.)

(a) $f(x) = \dfrac{1 - x^2}{2 + x^2}$

(b) $g(x) = \dfrac{x - x^3}{2x + x^3}$

(c) $h(x) = x^2 + x$

(d) $F(x) = (x^2 + x)^2$

(e) $G(x) = \begin{cases} 1 & \text{if } x > 0 \\ 0 & \text{if } x = 0 \\ -1 & \text{if } x < 0 \end{cases}$

Suggestion for part (e): Look at the graph.

MINI PROJECT | CORRECTING A GRAPHING UTILITY DISPLAY

As background for this discussion, you need to recall the definitions from intermediate algebra for expressions of the form $\sqrt[n]{x}$ and $x^{1/n}$, where n is a positive integer. See Appendices B.2 and B.3 if you need a reminder, or have someone in the group consult the appendix and present the definitions and relevant examples to the group at large.

Some graphing utilities produce incomplete graphs for root functions $y = \sqrt[n]{x}$ when n is an odd integer. To find out how your graphing utility operates in this respect, try the function $y = \sqrt[3]{x}$ on your graphing utility, and compare your result to Figures A and B. (You may need to enter the function in the form $y = x^{1/3}$.)

The reason for the incomplete graph in Figure A is that the graphing utility used requires nonnegative inputs in computing roots, even cube roots. Suppose for this discussion that you have this type of graphing utility. Find a way to use the techniques in this section to produce a correct graph, one similar to that shown in Figure B.

Figure A
An incomplete graph of $y = \sqrt[3]{x}$

Figure B
A complete graph of $y = \sqrt[3]{x}$

3.5 METHODS OF COMBINING FUNCTIONS. ITERATION

The notation ϕx to indicate a function of x was introduced by . . . [John Bernoulli] in 1718, . . . but the general adoption of symbols like f, F, ϕ, and ψ . . . to represent functions, seems to be mainly due to Euler and Lagrange. —W. W. Rouse Ball in *A Short Account of the History of Mathematics* (New York: Dover Publications, 1960)

The composition of functions is basic to the study of iteration and chaos. —Hartmut Jürgens et al., *Fractals for the Classroom: Strategic Activities,* Vol. 2 (New York: Springer-Verlag, 1992)

Two given numbers a and b can be combined in various ways to produce a third number. For instance, we can form the sum $a + b$ or the difference $a - b$ or the product ab. Also, if $b \neq 0$, we can form the quotient a/b. Similarly, two functions can be combined in various ways to produce a third function. Suppose, for example, that we start with the two functions $y = x^2$ and $y = x^3$. It seems natural to define their sum, difference, product, and quotient as follows:

$$\text{sum:} \quad y = x^2 + x^3$$
$$\text{difference:} \quad y = x^2 - x^3$$
$$\text{product:} \quad y = x^2 x^3 = x^5$$
$$\text{quotient:} \quad y = \frac{x^2}{x^3} = \frac{1}{x} \quad (\text{if } x \neq 0)$$

Indeed, this is the idea behind the formal definitions we now give.

DEFINITION | **Arithmetical Operations with Functions**

Let f and g be two functions. Then the **sum** $f + g$, the **difference** $f - g$, the **product** fg, and the **quotient** f/g are functions defined by the following equations:

$$(f + g)(x) = f(x) + g(x) \tag{1}$$
$$(f - g)(x) = f(x) - g(x) \tag{2}$$
$$(fg)(x) = f(x) \cdot g(x) \tag{3}$$
$$(f/g)(x) = f(x)/g(x) \qquad \text{provided } g(x) \neq 0 \tag{4}$$

For the functions defined by equations (1), (2), and (3), the domain is the set of all inputs x belonging to both the domain of f and the domain of g. For the quotient function in equation (4), we impose the additional restriction that the domain exclude all inputs x for which $g(x) = 0$.

EXAMPLE 1   Combining functions arithmetically

Let $f(x) = 3x + 1$ and $g(x) = x - 1$. Compute each of the following:
(a) $(f + g)(x)$;   **(b)** $(f - g)(x)$;   **(c)** $(fg)(x)$;   **(d)** $(f/g)(x)$.

SOLUTION
(a) $(f + g)(x) = f(x) + g(x) = (3x + 1) + (x - 1) = 4x$
(b) $(f - g)(x) = f(x) - g(x) = (3x + 1) - (x - 1) = 2x + 2$
(c) $(fg)(x) = [f(x)][g(x)] = (3x + 1)(x - 1) = 3x^2 - 2x - 1$
(d) $(f/g)(x) = \dfrac{f(x)}{g(x)} = \dfrac{3x + 1}{x - 1}$, where $x \neq 1$

For the remainder of this section, we are going to discuss a method of combining functions known as **composition of functions.** As you will see, this method is based on the familiar algebraic process of substitution. Suppose, for example, that f and g are two functions defined by

$$f(x) = x^2 \qquad g(x) = 3x + 1$$

Choose any number in the domain of g—say, $x = -2$. We can compute $g(-2)$:

$$g(-2) = 3(-2) + 1 = -5$$

Now let's use the output -5 that g has produced as an *input* for f. We obtain

$$f(-5) = (-5)^2 = 25$$

Consequently,

$$f[g(-2)] = 25$$

So, beginning with the input -2, we've successively applied g and then f to obtain the output 25. Similarly, we could carry out this same procedure for any other number in the domain of g. Here is a summary of the procedure:

1. Start with an input x and calculate $g(x)$.
2. Use $g(x)$ as an input for f; that is, calculate $f[g(x)]$.

We use the notation $f \circ g$ to denote the function, or rule, that tells us to assign the output $f[g(x)]$ to the initial input x. In other words, $f \circ g$ denotes the rule consisting of two steps: *First apply g; then apply f.* We read the notation $f \circ g$ as *f circle g* or *f composed with g.* In Figure 1, we summarize these ideas.

Figure 1
Diagram for the function $f \circ g$.

When we write $g(x)$, we assume that x is in the domain of the function g. Likewise, for the notation $f[g(x)]$ to make sense, the outputs $g(x)$ must themselves be acceptable inputs for the function f. Our formal definition, then, for the composite function $f \circ g$ is as follows.

DEFINITION	Composition of Functions: $f \circ g$

Given two functions f and g, the function $f \circ g$ is defined by

$$(f \circ g)(x) = f[g(x)]$$

The domain of $f \circ g$ consists of those inputs x (in the domain of g) for which $g(x)$ is in the domain of f.

EXAMPLE 2   Comparing $f \circ g$ to $g \circ f$

Let $f(x) = x^2$ and $g(x) = 3x + 1$. Compute $(f \circ g)(x)$ and $(g \circ f)(x)$.

SOLUTION

$$
\begin{aligned}
(f \circ g)(x) &= f[g(x)] && \text{definition of } f \circ g \\
&= f(3x + 1) && \text{definition of } g \\
&= (3x + 1)^2 && \text{definition of } f \\
&= 9x^2 + 6x + 1
\end{aligned}
$$

$$
\begin{aligned}
(g \circ f)(x) &= g[f(x)] && \text{definition of } g \circ f \\
&= g(x^2) && \text{definition of } f \\
&= 3(x^2) + 1 && \text{definition of } g \\
&= 3x^2 + 1
\end{aligned}
$$

Notice that the two results obtained in Example 2 are not the same. This shows that, in general, $f \circ g$ and $g \circ f$ represent different functions.

EXAMPLE 3   Two ways to compute a composite value

Let f and g be defined as in Example 2: $f(x) = x^2$ and $g(x) = 3x + 1$. Compute $(f \circ g)(-2)$.

SOLUTION
We will show two methods.

FIRST METHOD   Using the formula for $(f \circ g)(x)$ developed in Example 2, we have

$$(f \circ g)(x) = 9x^2 + 6x + 1$$

and, therefore,

$$
\begin{aligned}
(f \circ g)(-2) &= 9(-2)^2 + 6(-2) + 1 \\
&= 36 - 12 + 1 \\
&= 25
\end{aligned}
$$

ALTERNATIVE METHOD   Working directly from the definition of $f \circ g$, we have

$$
\begin{aligned}
(f \circ g)(-2) &= f[g(-2)] \\
&= f[3(-2) + 1] \\
&= f(-5) \\
&= (-5)^2 \\
&= 25
\end{aligned}
$$

In Examples 2 and 3, the domain of both f and g is the set of all real numbers. And as you can easily check, the domain of both $f \circ g$ and $g \circ f$ is also the set of all real numbers. In Example 4, however, some care needs to be taken in describing the domain of the composite function.

EXAMPLE 4   Finding the domain of a composite function

Let f and g be defined as follows:

$$f(x) = x^2 + 1 \qquad g(x) = \sqrt{x}$$

Compute $(f \circ g)(x)$. Find the domain of $f \circ g$ and sketch its graph.

SOLUTION

$$
\begin{aligned}
(f \circ g)(x) &= f[g(x)] \\
&= f(\sqrt{x}) = (\sqrt{x})^2 + 1 = x + 1
\end{aligned}
$$

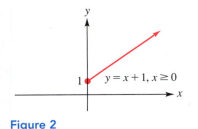

Figure 2
The graph of the function $f \circ g$ in Example 4.

So we have $(f \circ g)(x) = x + 1$. Now, what about the domain of $f \circ g$? Our first inclination may be to say (incorrectly!) that the domain is the set of all real numbers, since any real number can be used as an input in the expression $x + 1$. However, the definition of $f \circ g$ on page 211 tells us that the inputs for $f \circ g$ must first of all be acceptable inputs for g. Given the definition of g, then, we must require that x be nonnegative. On the other hand, for any nonnegative input x, the number $g(x)$ will be an acceptable input for f. (Why?) In summary, then, the domain of $f \circ g$ is the interval $[0, \infty)$. The graph of $f \circ g$ is shown in Figure 2.

EXAMPLE 5   An application of composition of functions

Suppose that an offshore oil rig is leaking and that the oil forms a circular region whose radius r increases over the first 12 hours according to the function

$$r = f(t) = \frac{t}{2t + 4} \qquad 0 \le t \le 12$$

where r is measured in miles, t is in hours, and $t = 0$ corresponds to the instant that the leak begins.

(a) Using function notation, the area of a circle of radius r is given by $A(r) = \pi r^2$. Find $(A \circ f)(t)$ and interpret the result.

(b) Compute the rate of change of the function $A \circ f$ over the interval $0 \le t \le 6$.

SOLUTION

(a) $(A \circ f)(t) = A[f(t)] = A\left(\dfrac{t}{2t + 4}\right)$

$$= \pi \left(\frac{t}{2t + 4}\right)^2$$

After squaring the fraction in this last expression, we obtain

$$(A \circ f)(t) = \frac{\pi t^2}{4t^2 + 16t + 16} \tag{1}$$

Since $f(t)$ expresses the radius in terms of time, our result in equation (1) gives the area in terms of time.

(b) By definition (from Section 3.3), the average rate of change of the function $A \circ f$ on the interval $[0, 6]$ is given by

$$\frac{\Delta(A \circ f)}{\Delta t} = \frac{(A \circ f)(6) - (A \circ f)(0)}{6 - 0} \qquad (2)$$

To evaluate the difference quotient on the right side of equation (2) we need to find $(A \circ f)(6)$ and $(A \circ f)(0)$. Exercise 33 asks you to use equation (1) to check that $(A \circ f)(6) = 9\pi/64$ and $(A \circ f)(0) = 0$. Assuming these results, equation (2) becomes

$$\begin{aligned}
\frac{\Delta(A \circ f)}{\Delta t} &= \frac{9\pi/64}{6} \\
&= 3\pi/128 \qquad \text{\color{red}Check the arithmetic.} \\
&\approx 0.07 \text{ (sq. miles)/hour}
\end{aligned}$$

That is, on average over the first six hours, the area of the oil spill is increasing at approximately 0.07 (sq. miles)/hour. (Exercise 34 asks you to compute the rate of change of the area over the next six-hour period and to compare it to this result for the first six hours.)

AFTERWORD ON THE SOLUTION TO EXAMPLE 5(b) A graphing utility can also be used for the computations. For instance, with a TI-83 graphing calculator, in the $\boxed{Y=}$ screen, one enters the three functions

$$Y_1 = \frac{x}{2x + 4} \qquad Y_2 = \pi x^2 \qquad Y_3 = Y_1(Y_2)$$

Then, in the home screen, the required rate of change is obtained by evaluating the expression $\dfrac{Y_3(6) - Y_3(0)}{6 - 0}$, or more simply, $\dfrac{Y_3(6)}{6}$, since in this example $Y_3(0) = 0$. For other types of graphing utilities, consult your owners manual.

One reason for studying the composition of functions is that it lets us express a given function in terms of simpler functions. This procedure is often useful in calculus. Suppose, for example, that we wish to express the function C defined by

$$C(x) = (2x^3 - 5)^2$$

as a composition of simpler functions. That is, we want to come up with two functions f and g so that the equation

$$C(x) = (f \circ g)(x)$$

holds for every x in the domain of C.

We begin by thinking what we would do to compute $(2x^3 - 5)^2$ for a given value of x. First, we would compute the quantity $2x^3 - 5$, then we would square the result. Therefore, recalling that the rule $f \circ g$ tells us to do g *first*, we let $g(x) = 2x^3 - 5$. Then, since the next step is squaring, we let $f(x) = x^2$. Now let's see whether these choices for f and g are correct; that is, let us calculate $(f \circ g)(x)$ and see whether it really is the same as $C(x)$.

Using $f(x) = x^2$ and $g(x) = 2x^3 - 5$, we have

$$(f \circ g)(x) = f[g(x)] = f(2x^3 - 5) = (2x^3 - 5)^2 = C(x)$$

This equation shows that our choices for f and g were indeed correct, and we have expressed C as a composition of two simpler functions.*

Note that in expressing C as $f \circ g$, we chose g to be the "inner" function—that is, the quantity inside the parentheses: $g(x) = 2x^3 - 5$. This observation is used in Example 6.

EXAMPLE 6   Expressing a function in terms of two simpler functions

Let $s(x) = \sqrt{1 + x^4}$. Express the function s as a composition of two simpler functions f and g.

SOLUTION
Let g be the "inner" function; that is, let $g(x)$ be the quantity inside the radical:

$$g(x) = 1 + x^4$$

and let's take f to be the square root function:

$$f(x) = \sqrt{x}$$

Now we need to verify that these are the appropriate choices for f and g; that is, we need to check that the equation $(f \circ g)(x) = s(x)$ is true for every x in the domain of s. We have

$$(f \circ g)(x) = f[g(x)] = f(1 + x^4) = \sqrt{1 + x^4} = s(x)$$

Thus, $(f \circ g)(x) = s(x)$, as required.

We conclude this section by describing the process of *iteration* for a function. This simple process is of fundamental importance in modern mathematics and science in the study of fractals and chaos.[†] The process of iteration is sequential; it proceeds step by step. We begin with a function f and an initial input x_0. In the first step, we compute the output $f(x_0)$. For the second step, we use the number $f(x_0)$ as an *input* for f and compute $f(f(x_0))$. The process then continues in this way: at each step, we use the output from the previous step as the new input.

For example, suppose that we start with the function $f(x) = x/2$ and the input $x_0 = 6$. Then, for the first three steps in the iteration process, we have

$$f(6) = \frac{6}{2} = 3$$

$$f(3) = \frac{3}{2} = 1.5$$

$$f(1.5) = \frac{1.5}{2} = 0.75$$

---

*Other answers are possible, too. For instance, if $F(x) = (x - 5)^2$ and $G(x) = 2x^3$, then (as you should verify for yourself $C(x) = F[G(x)]$.

[†] See, for example, pp. 314–363 in *The Nature and Power of Mathematics* by Donald M. Davis (Princeton, N.J.: Princeton University Press, 1993). For a wider view, see the book by James Gleick: *Chaos: Making a New Science* (New York: Penguin Books, 1988).

After this, a calculator becomes convenient. Check for yourself that the iterations run as follows:

$$6 \rightarrow 3 \rightarrow 1.5 \rightarrow 0.75 \rightarrow 0.375 \rightarrow 0.1875 \rightarrow 0.09375 \rightarrow 0.046875 \rightarrow \cdots$$

This list of numbers is called the *orbit* of 6 under the function *f*. In the list or orbit, the number 3 is the *first iterate* of 6, the number 1.5 is the *second iterate* of 6, and so on. In the box that follows, we summarize these ideas and introduce a useful notation for the iterates.

DEFINITION | Iterates

Given a function *f* and an input x_0, the **iterates** of x_0 are the numbers $f(x_0), f(f(x_0))$, $f(f(f(x_0)))$, and so on. The number $f(x_0)$ is called the **first iterate,** the number $f(f(x_0))$ is called the **second iterate,** and so on. Subscript notation is used to denote the iterates as follows.

$$\begin{aligned} x_1 &= f(x_0) & \text{the first iterate} \\ x_2 &= f(f(x_0)) & \text{the second iterate} \\ x_3 &= f(f(f(x_0))) & \text{the third iterate} \\ &\vdots \end{aligned}$$

The **orbit** of x_0 under the function *f* is the list of numbers $x_0, x_1, x_2, x_3, \ldots$.

Notice that the subscripts in this notation tell you how many times to apply the function *f*. For example, x_1 indicates that *f* is applied once to obtain $f(x_0)$, and x_2 indicates that the function is applied twice to obtain $f(f(x_0))$.

EXAMPLE **7** | **Computing iterates**

Compute the first four iterates in each case. Use a calculator as necessary. In part (b), round the final answers to three decimal places.

(a) $f(x) = x^2, x_0 = -2$ **(b)** $g(x) = \sqrt{x}, x_0 = 0.1$

SOLUTION

(a) $x_1 = (-2)^2 = 4$
$x_2 = (4)^2 = 16$
$x_3 = (16)^2 = 256$
$x_4 = (256)^2 = 65{,}536$

(b) We use the square root key (or keys) on the calculator repeatedly. As you should check for yourself, this yields

$$x_1 = \sqrt{0.1} = 0.316227\ldots \approx 0.316 \qquad \text{rounding to three} \\ \text{decimal places}$$

$$\begin{aligned} x_2 &= \sqrt{0.316227\ldots} = 0.562341\ldots \approx 0.562 \\ x_3 &= \sqrt{0.562341\ldots} = 0.749894\ldots \approx 0.750 \\ x_4 &= \sqrt{0.749989\ldots} = 0.865964\ldots \approx 0.866 \end{aligned}$$

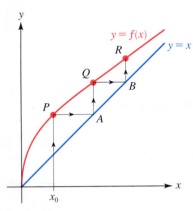

Figure 3

The y-coordinates of the points P, Q, and R are, respectively, the first three iterates of x_0 under the function f.

The process of iterating a function can be visualized in an x-y coordinate system. We sketch the function $y = f(x)$ along with the line $y = x$, as in Figure 3. Then, starting with the initial input x_0 on the x-axis, we follow the horizontal and vertical segments to determine points P, Q, and R on the graph of f. As the next paragraph explains, the y-coordinates of the points P, Q, and R are the first, second, and third iterates, respectively, of x_0. Additional iterates are obtained by continuing this pattern. In Figure 4 we show how this looks in a specific case. The y-coordinates of the five points P, Q, R, S, and T are the first five iterates of $x_0 = 0.1$ under the function $g(x) = -4x^2 + 4x$. Although Figure 4 at first appears more complicated than Figure 3, it is important to understand that the pattern is formed in exactly the same way. We start with the initial input on the x-axis and travel vertically to determine the point P on the curve. After that, each successive point on the curve is generated in the same way: Go horizontally to the line $y = x$ and then vertically to the curve.

Figure 4

Graphical iteration of $g(x) = -4x^2 + 4x$ with initial input $x_0 = 0.1$. The y-coordinates of the five points P, Q, R, S, and T are, respectively, the first five iterates of x_0.

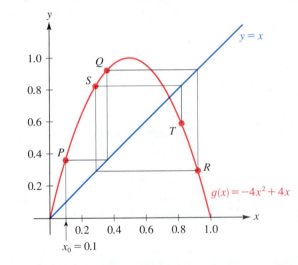

As we've said, this paragraph explains why the y-coordinates of the three points P, Q, and R in Figure 3 are, respectively, the first three iterates of x_0 under the function f. In Figure 3, the x-coordinate of P is given to be x_0. So (according to the definition of the graph of a function), the y-coordinate of P is $f(x_0)$. In other words, the y-coordinate of P is the first iterate of x_0. Next in Figure 3, we move horizontally from the point P to the point A. Because the movement is horizontal, the y-coordinate doesn't change; it's the y-coordinate of P, or $f(x_0)$. Now, what about the x-coordinate of A? Since the point A lies on the line $y = x$, the x- and y-coordinates of A must be identical. Consequently, the x-coordinate of A must be the number $f(x_0)$. Finally, we move vertically in Figure 3 from A to Q. The point Q has the same x-coordinate as the point A—namely, $f(x_0)$. Now, since Q lies on the graph of the function f, we conclude that the y-coordinate of Q is $f(f(x_0))$. That is, the y-coordinate of Q is indeed the second iterate of x_0. The reasoning to show that the y-coordinate of R is the third iterate of x_0 is entirely similar, so we omit giving the details here.

EXERCISE SET 3.5

A

In Exercises 1–8, compute each expression, given that the functions f, g, h, k, and m are defined as follows:

$$f(x) = 2x - 1 \qquad k(x) = 2, \quad \text{for all } x$$
$$g(x) = x^2 - 3x - 6 \qquad m(x) = x^2 - 9$$
$$h(x) = x^3$$

1. (a) $(f + g)(x)$
   (b) $(f - g)(x)$
   (c) $(f - g)(0)$

2. (a) $(fh)(x)$
   (b) $(h/f)(x)$
   (c) $(f/h)(1)$

3. (a) $(m - f)(x)$
   (b) $(f - m)(x)$

4. (a) $(fg)(x)$
   (b) $(fg)(1/2)$

5. (a) $(fk)(x)$
   (b) $(kf)(x)$
   (c) $(fk)(1) - (kf)(2)$

6. (a) $(g + m)(x)$
   (b) $(g + m)(x) - (g - m)(x)$

7. (a) $(f/m)(x) - (m/f)(x)$
   (b) $(f/m)(0) - (m/f)(0)$

8. (a) $[h \cdot (f + m)](x)$ *Note*: h and $(f + m)$ are two functions; the notation $h \cdot (f + m)$ denotes the product function.
   (b) $(hf)(x) + (hm)(x)$

9. Let $f(x) = 3x + 1$ and $g(x) = -2x - 5$. Compute the following.
   (a) $(f \circ g)(x)$
   (b) $(f \circ g)(10)$
   (c) $(g \circ f)(x)$
   (d) $(g \circ f)(10)$

10. Let $f(x) = 1 - 2x^2$ and $g(x) = x + 1$. Compute the following.
   (a) $(f \circ g)(x)$
   (b) $(f \circ g)(-1)$
   (c) $(g \circ f)(x)$
   (d) $(g \circ f)(-1)$
   (e) $(f \circ f)(x)$
   (f) $(g \circ g)(-1)$

11. Compute $(f \circ g)(x)$, $(f \circ g)(-2)$, $(g \circ f)(x)$, and $(g \circ f)(-2)$ for each pair of functions.
   (a) $f(x) = x^2 - 3x - 4$; $g(x) = 2 - 3x$
   (b) $f(x) = 2^x$; $g(x) = x^2 + 1$
   (c) $f(x) = x$; $g(x) = 3x^5 - 4x^2$
   (d) $f(x) = 3x - 4$; $g(x) = (x + 4)/3$

12. Let $h(x) = 4x^2 - 5x + 1$, $k(x) = x$, and $m(x) = 7$ for all x. Compute the following.
   (a) $h[k(x)]$
   (b) $k[h(x)]$
   (c) $h[m(x)]$
   (d) $m[h(x)]$
   (e) $k[m(x)]$
   (f) $m[k(x)]$

13. Let $F(x) = \dfrac{3x - 4}{3x + 3}$ and $G(x) = \dfrac{x + 1}{x - 1}$. Compute the following.
   (a) $(F \circ G)(x)$
   (b) $F[G(t)]$
   (c) $(F \circ G)(2)$
   (d) $(G \circ F)(x)$
   (e) $G[F(y)]$
   (f) $(G \circ F)(2)$

14. Let $f(x) = (1/x^2) + 1$ and $g(x) = 1/(x - 1)$
   (a) Compute $(f \circ g)(x)$.
   (b) What is the domain of $f \circ g$?
   (c) Graph the function $f \circ g$.

15. Let $M(x) = (2x - 1)/(x - 2)$.
   (a) Compute $M(7)$ and then $M[M(7)]$.
   (b) Compute $(M \circ M)(x)$.
   (c) Compute $(M \circ M)(7)$, using the formula you obtained in part (b). Check that your answer agrees with that obtained in part (a).

16. Let $F(x) = (x + 1)^5$, $f(x) = x^5$, and $g(x) = x + 1$. Which of the following is true for all x?

$$(f \circ g)(x) = F(x) \qquad \text{or} \qquad (g \circ f)(x) = F(x)$$

17. Refer to the graphs of the functions f, g, and h to compute the required quantities. Assume that all the axes are marked off in one-unit intervals

(a)   (b)

(c)

   (a) $f[g(3)]$
   (b) $g[f(3)]$
   (c) $f[h(3)]$
   (d) $(h \circ g)(2)$
   (e) $h\{f[g(3)]\}$
   (f) $(g \circ f \circ h \circ f)(2)$
   Note: This means first do f, then h, then f, then g.

18. (a) Let $T(x) = 4x^3 - 3x^2 + 6x - 1$ and $I(x) = x$. Find $(T \circ I)(x)$ and $(I \circ T)(x)$.
   (b) Let $G(x) = ax^2 + bx + c$ and $I(x) = x$. Find $(G \circ I)(x)$ and $(I \circ G)(x)$.
   (c) What general conclusion do you arrive at from the results of parts (a) and (b)?

19. The domain of a function f consists of the numbers $-1, 0,$ 1, 2, and 3. The following table shows the output that f assigns to each input.

x	1	0	1	2	3
f(x)	2	2	0	3	1

The domain of a function g consists of the numbers 0, 1, 2, 3, and 4. The following table shows the output that g assigns to each input.

x	0	1	2	3	4
g(x)	3	2	0	4	-1

Use this information to complete the following tables for $f \circ g$ and $g \circ f$. *Note*: Two of the entries will be undefined.

x	0	1	2	3	4
(f ∘ g)(x)					

x	-1	0	1	2	3	4
(g ∘ f)(x)						

20. The following two tables show certain pairs of inputs and outputs for functions f and g.

x	0	π/6	π/4	π/3	π/2
f(x)	0	1/2	√2/2	√3/2	1

y	0	1/4	√2/4	1/2	√2/2	3/4	√3/2	1
g(y)	π/2	π	0	π/3	π/4	0	π/6	0

Use this information to complete the following table of values for $(g \circ f)(x)$.

x	0	π/6	π/4	π/3	π/2
(g ∘ f)(x)					

21. Let $f(x) = x^3 - 2x$ and $g(x) = x + 4$.
(a) What is the relationship between the graphs of the two functions f and $f \circ g$? (The idea here is to answer without looking at the graphs; use a concept from Section 3.4.)
(G) (b) Use a graphing utility to check your answer in part (a).

22. Let $f(x) = \sqrt{x^3 + 2x + 17}$ and $g(x) = x + 6$.
(a) What is the relationship between the graphs of the two functions f and $f \circ g$? (As in Exercise 21, the idea here is to answer without looking at the graphs.)
(G) (b) Use a graphing utility to check your answer in part (a).

23. Let $g(x) = \sqrt{x} - 3$ and $f(x) = x - 1$.
(a) Sketch a graph of g. Specify the domain and range.
(b) Sketch a graph of f. Specify the domain and range.
(c) Compute $(f \circ g)(x)$. Graph the function $f \circ g$ and specify its domain and range.
(d) Find a formula for $g[f(x)]$. Which values of x are acceptable inputs here? That is, what is the domain of $g \circ f$?
(e) Use the results of part (d) to sketch a graph of the function $g \circ f$.

24. Let $F(x) = -x^2$ and $G(x) = \sqrt{x}$. Determine the domains of $F \circ G$ and $G \circ F$.

25. Suppose that an oil spill in a lake covers a circular area and that the radius of the circle is increasing according to the formula $r = f(t) = 15 + t^{1.65}$, where t represents the number of hours since the spill was first observed and the radius r is measured in meters. (Thus, when the spill was first discovered, $t = 0$ hr and the initial radius was $r = f(0) = 15 + 0^{1.65} = 15$ m.)
(a) Let $A(r) = \pi r^2$, as in Example 5. Compute a table of values for the composite function $A \circ f$ with t running from 0 to 5 in increments of 0.5. (Round each output to the nearest integer.) Then use the table to answer the questions that follow in parts (b) through (d).
(b) After one hour, what is the area of the spill (rounded to the nearest 10 m²)?
(c) Initially, what was the area of the spill (when $t = 0$)? Approximately how many hours does it take for this area to double?
(d) Compute the average rate of change of the area of the spill from $t = 0$ to $t = 2.5$ and from $t = 2.5$ to $t = 5$. Over which of the two intervals is the area increasing faster?

26. A spherical weather balloon is being inflated in such a way that the radius is given by

$$r = g(t) = \tfrac{1}{2}t + 2$$

Assume that r is in meters and t is in seconds, with $t = 0$ corresponding to the time that inflation begins. If the volume of a sphere of radius r is given by

$$V(r) = \tfrac{4}{3}\pi r^3$$

compute $V[g(t)]$ and use this to find the time at which the volume of the balloon is 36π m³.

27. Suppose that a manufacturer knows that the daily production cost to build x bicycles is given by the function C, where

$$C(x) = 100 + 90x - x^2 \qquad (0 \le x \le 40)$$

That is, $C(x)$ represents the cost in dollars of building x bicycles. Furthermore, suppose that the number of bicycles that can be built in t hr is given by the function f, where

$$x = f(t) = 5t \qquad (0 \le t \le 8)$$

(a) Compute $(C \circ f)(t)$.
(b) Compute the production cost on a day that the factory operates for $t = 3$ hr.
(c) If the factory runs for 6 hr instead of 3 hr, is the cost twice as much?

28. Suppose that in a certain biology lab experiment, the number of bacteria is related to the temperature T of the environment by the function

$$N(T) = -2T^2 + 240T - 5400 \qquad (40 \leq T \leq 90)$$

Here, $N(T)$ represents the number of bacteria present when the temperature is T degrees Fahrenheit. Also, suppose that t hr after the experiment begins, the temperature is given by

$$T(t) = 10t + 40 \qquad (0 \leq t \leq 5)$$

(a) Compute $N[T(t)]$.

(b) How many bacteria are present when $t = 0$ hr? When $t = 2$ hr? When $t = 5$ hr?

29. Express each function as a composition of two functions.

(a) $F(x) = \sqrt[3]{3x + 4}$ **(c)** $H(x) = (ax + b)^5$

(b) $G(x) = |2x - 3|$ **(d)** $T(x) = 1/\sqrt{x}$

30. Let $a(x) = x^2$, $b(x) = |x|$, and $c(x) = 3x - 1$. Express each of the following functions as a composition of two of the given functions.

(a) $f(x) = (3x - 1)^2$ **(c)** $h(x) = 3x^2 - 1$

(b) $g(x) = |3x - 1|$

31. Let $a(x) = 1/x$, $b(x) = \sqrt[3]{x}$, $c(x) = 2x + 1$, and $d(x) = x^2$. Express each of the following functions as a composition of two of the given functions.

(a) $f(x) = \sqrt[3]{2x + 1}$ **(d)** $K(x) = 2\sqrt[3]{x} + 1$

(b) $g(x) = 1/x^2$ **(e)** $l(x) = \dfrac{2}{x} + 1$

(c) $h(x) = 2x^2 + 1$ **(f)** $m(x) = \dfrac{1}{2x + 1}$

32. Express $n(x) = x^{2/3}$ as a composition of two of the functions given at the begining of Exercise 31. (If you don't recall the definition for fractional exponents, see Appendix B.3 at the back of this book.)

33. For this exercise, refer to equation (1) in Example 5.

(a) Show that $(A \circ f)(0) = 0$ and that $(A \circ f)(6) = 9\pi/64$.

Ⓖ **(b)** Use a graphing utility to graph the average-rate-of-change function $y = (A \circ f)(t)$. Note that the graph appears to pass through the point $(0, 0)$, which supports the first result in part (a).

Ⓖ **(c)** By zooming in, estimate the y-coordinate of the point on the graph corresponding to $t = 6$. What does this have to do with part (a) of this exercise?

34. Refer to Example 5(b).

(a) Compute the average rate of change of $A \circ f$ over the time interval from $t = 6$ to $t = 12$ hours. On average, was the area of the oil leak increasing faster over the first six hours or the second six hours?

Ⓖ **(b)** Graph the average-rate-of-change function $y = (A \circ f)(t)$ in the viewing rectangle $[0, 12]$ by $[0, 0.8]$. Then draw two line segments, one connecting the two points on the graph corresponding to $t = 0$ and $t = 6$, the other connecting the two points cor-

responding to $t = 6$ and $t = 12$. (If your graphing utility does not have a draw feature, use hard copy and draw using a ruler.) Explain in complete sentences how the picture supports your answer to the question in part (a).

In Exercises 35–40, use the given function and compute the first six iterates of each initial input x_0. In cases where a calculator answer contains four or more decimal places, round the final answer to three decimal places. (However, during the calculations, work with all of the decimal places that your calculator affords.)

35. $f(x) = 2x$ **36.** $f(x) = \frac{1}{4}x$

   (a) $x_0 = 1$ **(a)** $x_0 = 16$

   (b) $x_0 = 0$ **(b)** $x_0 = 0$

   (c) $x_0 = -1$ **(c)** $x_0 = -16$

37. $g(x) = 2x + 1$ **38.** $g(x) = \frac{1}{4}x + 3$

   (a) $x_0 = -2$ **(a)** $x_0 = 3$

   (b) $x_0 = -1$ **(b)** $x_0 = 4$

   (c) $x_0 = 1$ **(c)** $x_0 = 5$

39. $F(x) = x^2$ **40.** $G(x) = x^2 + 0.25$

   (a) $x_0 = 0.9$ **(a)** $x_0 = 0.4$

   (b) $x_0 = 1$ **(b)** $x_0 = 0.5$

   (c) $x_0 = 1.1$ **(c)** $x_0 = 0.6$

41. The accompanying figure shows a portion of the iteration process for $f(x) = \sqrt{x}$ with initial input $x_0 = 0.1$. Use the figure to estimate (to within 0.05) the first four iterates of $x_0 = 0.1$. Then, use a calculator to compute the four iterates. (Round to three decimal places, where appropriate.) Check that your calculator results are consistent with the estimates obtained from the graph.

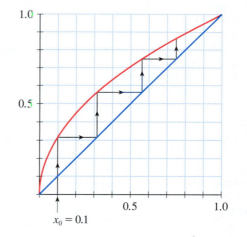

42. The following figure shows a portion of the iteration process for $f(x) = 3.6(x - x^2)$ with initial input $x_0 = 0.3$. Use the figure to estimate (to within 0.1, or closer if it seems appropriate) the first seven iterates of $x_0 = 0.3$. Then, use a calculator to compute the seven iterates. (Round the final answers to three decimal places.) Check that your calculator results are consistent with the estimates obtained from the graph.

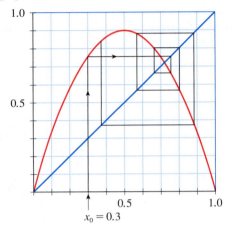

$x_0 = 0.3$

B

43. *The 3x + 1 conjecture* Define a function f, with domain the positive integers, as follows:

$$f(x) = \begin{cases} 3x + 1 & \text{if } x \text{ is odd} \\ x/2 & \text{if } x \text{ is even} \end{cases}$$

(a) Compute $f(1), f(2), f(3), f(4), f(5),$ and $f(6)$.
(b) Compute the first three iterates of $x_0 = 1$.
(c) Compute the iterates of $x_0 = 3$ until you obtain the value 1. [After this, the iterates will recycle through the simple pattern obtained in part (b).]
(d) *The 3x + 1 conjecture* asserts that for any positive integer x_0, the iterates eventually return to the value 1. Verify that this conjecture is valid for each of the following values of x_0: 2, 4, 5, 6, and 7.

Remark: At present, the 3x + 1 conjecture is indeed a conjecture, not a theorem. No one yet has found a proof that the assertion is valid for *every* positive integer. Computer checks, however, have verified the conjecture on a case-by-case basis for very large values of x_0. As of April 2000, the conjecture had been verified for all values of x_0 up to approximately 1.8×10^{16}.

44. *The 3x + 1 conjecture* (continued from Exercise 43) If you have access to the Internet, use Alfred Wassermann's 3x + 1 on-line calculator located at

http://did.mat.uni-bayreuth.de/personen/wassermann/fun/3np1_.html

to answer the following questions. For which n does x_n first reach 1 if $x_0 = 100$? If $x_0 = 1000$? If $x_0 = 10^4$? (The Web address above was accessible at the time of this writing, March 2002.)

45. Let $g(x) = 4x - 1$. Find $f(x)$, given that the equation $(g \circ f)(x) = x + 5$ is true for all values of x.
46. Let $g(x) = 2x + 1$. Find $f(x)$, given that $(g \circ f)(x) = 10x - 7$.
47. Let $f(x) = -2x + 1$ and $g(x) = ax + b$. Find a and b so the equation $f[g(x)] = x$ holds for all values of x.
48. Let $f(x) = (3x - 4)/(x - 3)$.
(a) Compute $(f \circ f)(x)$.
(b) Find $f[f(113/355)]$. (Try not to do it the hard way.)
49. Let $f(x) = x^2$ and $g(x) = 2x - 1$.
(a) Compute $\dfrac{f[g(x)] - f[g(a)]}{g(x) - g(a)}$.
(b) Compute $\dfrac{f[g(x)] - f[g(a)]}{x - a}$.

Exercises 50 and 51 indicate how iteration is used in finding roots of numbers and roots of equations. (The functions that are given in each exercise were determined using Newton's method, a process studied in calculus.)

50. Let $f(x) = 0.5\left(x + \dfrac{3}{x}\right)$.
(a) Compute the first ten iterates of $x_0 = 1$ under the function f. What do you observe?
(b) Use your calculator to evaluate $\sqrt{3}$ and compare the answer to your results in part (a). What do you observe?
(c) It can be shown that for any positive number x_0, the iterates of x_0 under the function $f(x) = 0.5(x + 3/x)$ always approach the number $\sqrt{3}$. (You'll see the reasons for this in Section 4.3.) Looking at your results in parts (a) and (b), which is the first iterate that agrees with $\sqrt{3}$ through the first three decimal places? Through the first eight decimal places?
(d) Compute the first ten iterates of $x_0 = 50$ under the function f, then answer the questions presented in part (c).

51. Let $f(x) = \dfrac{2x^3 + 7}{3x^2}$.
(a) Compute the first ten iterates of $x_0 = 1$ under the function f. What do you observe?
(b) Evaluate the expression $\sqrt[3]{7}$ and compare the answer to your results in part (a). What do you observe?
(c) It can be shown that for any positive number x_0, the iterates of x_0 under the function $f(x) = \dfrac{2x^3 + 7}{3x^2}$ always approach the number $\sqrt[3]{7}$. Looking at your results in parts (a) and (b), which is the first iterate that agrees with $\sqrt[3]{7}$ through the first three decimal places? Through the first eight decimal places?

C

52. Let $f(x) = \dfrac{x - 3}{x + c}$, where c denotes a constant.

(a) If $c = -1$, show that $f(f(x)) = x$.

(b) Use a graphing utility to support the result in part (a).

That is, enter the function $f(x) = \dfrac{x - 3}{x - 1}$, and then have the

machine graph the two functions $y = x$ and $y = f(f(x))$ in the same picture.

(c) What does the result in part (a) tell you about the iteration process for the function $y = (x - 3)/(x - 1)$?

That is, what pattern emerges in the iterates? Answer in complete sentences.

(d) Now assume $c = 1$, instead of -1. Show that $f(f(f(x))) = x$.

(e) Use a graphing utility to support the result in part (d).

(f) What does the result in part (d) tell you about the iteration process for the function $y = (x - 3)/(x + 1)$? That is, what pattern emerges in the iterates? Answer in complete sentences.

MINI PROJECT

A GRAPHICAL APPROACH TO COMPOSITION OF FUNCTIONS

As we saw in this section, the iteration process can be carried out either algebraically or graphically. Composition of functions can also be represented graphically, although Examples 2 through 4 demonstrate only the algebraic approach. The accompanying figure shows the graphs of two functions f and g, the line $y = x$, and an input a. Use the figure and the ideas of this section to discover and then explain how to find the point $(a, g[f(a)])$ on the graph of the composite function $g \circ f$.

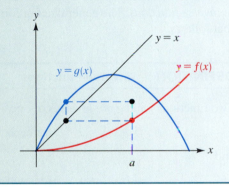

3.6 INVERSE FUNCTIONS

If anybody ever told me why the graph of $y = x^{1/2}$ is the reflection of the graph of $y = x^2$ in a 45° line, it didn't sink in. To this day, there are textbooks that expect students to think that it is so obvious as to need no explanation. This is a pity, if only because [in calculus] it is such a common practice to define the natural logarithm first and then define the exponential function as its inverse. —Professor Ralph P. Boas, recalling his student days in the article "Inverse Functions" in *The College Mathematics Journal,* vol. 16 (1985), p. 42.

PROBLEM
SOLUTION

© 1981 Scott Kim. From *Inversions* (W. H. Freeman, 1989).

We shall introduce the idea of inverse functions through an easy example. Consider the function f defined by $f(x) = 2x$. A short table of values for f and the graph of f are displayed in Figure 1 on the next page.

Now we ask this question: What happens if we take the table for f (in Figure 1) and interchange the entries in the x and y columns to obtain a new table that looks like the following?

x	y
-2	-4
-1	-2
0	0
1	2
2	4

$f(x) = 2x$

Figure 1

x	y
-4	-2
-2	-1
0	0
2	1
4	2

← New table formed by interchanging the inputs and outputs for f

Several observations are in order. Our new table does itself represent a function (in this case): for each input x there is exactly one output y. Let's call this new function g. Note that each output is one-half of the corresponding input. So the function g can be described by the formula $g(x) = x/2$. In summary, we began with the "doubling function," $f(x) = 2x$; then, by interchanging the inputs with the outputs, we obtained the "halving function," $g(x) = x/2$.

The doubling function f and the halving function g are in a certain sense opposites; each reverses, or undoes, the effect of the other. That is, if you begin with x and then calculate $g[f(x)]$, you get x again; similarly, $f[g(x)]$ is also equal to x. (Verify these last two statements for yourself.)

The two functions f and g that we've been discussing are an example of a pair of *inverse functions*. The general definition for inverse functions is given in the box that follows.

DEFINITION | **Inverse Functions**

Two functions f and g are **inverses** of one another provided that

$$f[g(x)] = x \qquad \text{for each } x \text{ in the domain of } g$$

and

$$g[f(x)] = x \qquad \text{for each } x \text{ in the domain of } f$$

EXAMPLE 1 **Applying the definition of inverse functions**

Suppose that f and g are a pair of inverse functions.
(a) If 13 is in the domain of g, find $f[g(13)]$.
(b) If $\sqrt{5}$ is in the domain of f, find $g[f(\sqrt{5})]$.

SOLUTION
(a) By the definition of inverse functions, the equation $f[g(x)] = x$ is true for each x in the domain of g. In particular then, using $x = 13$, we have $f[g(13)] = 13$.
(b) Since $\sqrt{5}$ is in the domain of f, the definition of inverse functions gives us $g[f(\sqrt{5})] = \sqrt{5}$.

EXAMPLE 2   **Applying the definition of inverse functions**

Verify that the functions f and g are inverses, where

$$f(x) = \frac{1}{3}x + 2 \quad \text{and} \quad g(x) = 3x - 6$$

SOLUTION
In view of the definition, we must check that $f[g(x)] = x$ and $g[f(x)] = x$. We have

$$f[g(x)] = f(3x - 6)$$
$$= \frac{1}{3}(3x - 6) + 2 = x \qquad \text{(Check the algebra.)}$$

Thus, $f[g(x)] = x$. Now we still need to check that $g[f(x)] = x$. We have

$$g[f(x)] = g\left(\frac{1}{3}x + 2\right)$$
$$= 3\left(\frac{1}{3}x + 2\right) - 6 = x \qquad \text{Again, check the algebra.}$$

Having shown that $f[g(x)] = x$ and $g[f(x)] = x$, we conclude that f and g are indeed inverse functions.

EXAMPLE 3   **Two perspectives on inverse functions**

Suppose that f and g are a pair of inverse functions. If $f(2) = 3$, what is $g(3)$?

SOLUTION
We can use either of two methods.

FIRST METHOD   The quantities 3 and $f(2)$ are declared to be equal. Thus, whether we use 3 or $f(2)$ as an input for g, the result must be the same. We then have

$$3 = f(2)$$
$$g(3) = g[f(2)]$$
$$g(3) = 2 \quad \text{using the fact that}$$
$$g[f(x)] = x \text{ for any}$$
$$x \text{ in the domain of } f$$

Thus $g(3) = 2$.

ALTERNATIVE METHOD   Think of a table of values for the function f. Since $f(2) = 3$, one entry in the table must look like this:

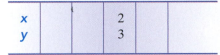

x		2	
y		3	

Now, g reverses the roles of x and y, so in the table for g, one entry must look like this:

x		3	
y		2	

Thus, $g(3) = 2$, as obtained previously.

It is customary to use the notation f^{-1} (read f *inverse*) for the function that is the inverse of f. So in Example 2, for instance, we have

$$f(x) = \frac{1}{3}x + 2 \qquad \text{and} \qquad f^{-1}(x) = 3x - 6 \tag{1}$$

CAUTION   In the context of functions, $f^{-1}(x)$ does not in general mean $\dfrac{1}{f(x)}$. For instance, regarding the pair of inverse functions in equations (1), it's certainly not true in general that $3x - 6 = \dfrac{1}{\frac{1}{3}x + 2}$. (For other examples, see Exercises 11 and 12 at the end of this section.) For reference now, in the box that follows, we rewrite the defining equations for inverse functions using the f^{-1} notation. Figure 2 provides a graphic summary of these equations.

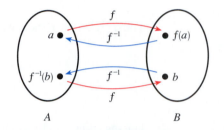

Figure 2

The action of inverse functions: The set A in the figure is both the domain of f and the range of f^{-1}. The set B is both the domain of f^{-1} and the range of f.

$$f[f^{-1}(x)] = x \qquad \text{for each } x \text{ in the domain of } f^{-1}$$

and

$$f^{-1}[f(x)] = x \qquad \text{for each } x \text{ in the domain of } f$$

EXAMPLE 4   Using the definition of inverse functions

Suppose $f(x) = 4x^3 + 7$. Find $f[f^{-1}(5)]$. (Assume that f^{-1} exists and that 5 is in its domain.)

SOLUTION
As in Example 1, this is just an application of the definition of inverse functions. We have, by definition, $f[f^{-1}(x)] = x$ for every x in the domain of f^{-1}. Replacing x with 5 throughout this last equation, we immediately obtain

$$f[f^{-1}(5)] = 5$$

Note that we did not need to use the equation $f(x) = 4x^3 + 7$, nor did we need to find a formula for $f^{-1}(x)$.

EXAMPLE 5   An equation to solve that involves inverse functions

Solve the following equation for x, given that the domain of both f and f^{-1} is $(-\infty, \infty)$ and that $f(1) = -2$:

$$3 + f^{-1}(x - 1) = 4$$

SOLUTION

$$3 + f^{-1}(x - 1) = 4$$
$$f^{-1}(x - 1) = 1$$
$$f[f^{-1}(x - 1)] = f(1) \qquad \text{applying } f \text{ to both sides}$$
$$x - 1 = f(1) \qquad \text{definition of inverse function}$$
$$x = 1 + f(1) = 1 + (-2) = -1$$

We'll postpone a discussion until the end of this section as to which functions have inverses and which do not. For functions that do have inverses, however, there's a simple method that can often be used to determine those inverses. This method is illustrated in Example 6.

EXAMPLE 6   Determining the inverse function

Let $f(x) = \dfrac{x - 1}{3x + 5}$. Find $f^{-1}(x)$.

SOLUTION

Step 1   We begin by writing the given equation as

$$y = \frac{x - 1}{3x + 5} \qquad (2)$$

We know that f^{-1} interchanges the inputs and outputs of f, so to determine f^{-1}, we need only switch the x's and y's in equation (2). That gives us

$$x = \frac{y - 1}{3y + 5} \qquad (3)$$

Step 2   Now we solve equation (3) for y.

$$3xy + 5x = y - 1 \qquad \text{multiplying both sides of equation (3) by } 3y + 5$$
$$3xy - y = -5x - 1 \qquad \text{rearranging}$$
$$y(3x - 1) = -5x - 1$$
$$y = \frac{-5x - 1}{3x - 1}$$

Thus, the inverse function is $y = (-5x - 1)/(3x - 1)$. Using function notation, we can write this result as

$$f^{-1}(x) = \frac{-5x - 1}{3x - 1}$$

REMARK   Actually, we should call this result only a candidate for f^{-1}, because certain technical matters remain to be discussed. However, if you compute $f[f^{-1}(x)]$ and $f^{-1}[f(x)]$ and find they both do equal x, then the matter is settled, by definition. Exercise 21 at the end of this section asks you to carry out those calculations.

For reference, in the box that follows we summarize our procedure for calculating $f^{-1}(x)$.

To Find $f^{-1}(x)$ for the Function $y = f(x)$

Step 1   Interchange x and y in the equation $y = f(x)$.

Step 2   Solve the resulting equation for y.

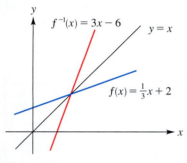

Figure 3

There is a certain type of symmetry that always occurs when we graph a function and its inverse. Consider, for example, the pair of inverse functions $f(x) = \frac{1}{3}x + 2$ and $f^{-1}(x) = 3x - 6$ that are graphed in Figure 3. (In Example 2 we verified that these were indeed inverse functions.) To bring out the symmetry, the figure also shows the line $y = x$. Note that the graphs of f and f^{-1} appear to be *symmetric about the line $y = x$*. That is, if the page were folded along the dashed line $y = x$ and held up to the light, then the graphs of f and f^{-1} would coincide. To say it differently, the graphs of f and f^{-1} are mirror images of one another about the line $y = x$. As background for the remainder of our work on inverse functions, we give a definition of symmetry about the line $y = x$ in the box that follows. (After you read the definition, note that the ideas involved are the same as those for symmetry about the x- or y-axis, as presented in Section 1.7.)

DEFINITION | **Symmetry about the Line $y = x$**

Refer to Figure 4. Two points P and Q are symmetric about the line $y = x$ provided that

1. $\overline{PQ}$ is perpendicular to the line $y = x$; and
2. the points P and Q are equidistant from the line $y = x$.

Figure 4

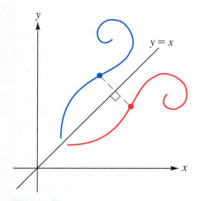

Figure 5

The two curves are symmetric about the line $y = x$.

This definition says that P and Q are symmetric about the line $y = x$ if $y = x$ is the perpendicular bisector of line segment $\overline{PQ}$. In Figure 4, we say that the two points P and Q are **reflections** of each other about the line $y = x$ and that $y = x$ is the **axis of symmetry.** In addition, we say that two curves are symmetric about $y = x$ if each point on one curve is the reflection of a corresponding point on the other curve, and vice versa; see Figure 5.

Figure 6

Figure 7
Two examples of the fact that (a, b) and (b, a) are symmetric about $y = x$.

EXAMPLE 7 **Showing two given points are symmetric about $y = x$**

Verify that the points $P(4, 1)$ and $Q(1, 4)$ are symmetric about the line $y = x$; see Figure 6.

SOLUTION
We have to show that the line $y = x$ is the perpendicular bisector of the line segment $\overline{PQ}$. Now, the slope of $\overline{PQ}$ is

$$m = \frac{4 - 1}{1 - 4} = -1$$

We know that the slope of the line $y = x$ is 1. Since these two slopes are negative reciprocals, we conclude that the line $y = x$ is perpendicular to $\overline{PQ}$. Next, we must show that the line $y = x$ passes through the midpoint of $\overline{PQ}$. However, (as you should check for yourself), the midpoint of $\overline{PQ}$ is $(\frac{5}{2}, \frac{5}{2})$, which does lie on the line $y = x$. (Why?) In summary, then, we've shown that the line $y = x$ passes through the midpoint of $\overline{PQ}$ and is perpendicular to $\overline{PQ}$. Thus P and Q are symmetric about the line $y = x$, as we wished to show.

■

Just as we've shown that $(4, 1)$ and $(1, 4)$ are symmetric about the line $y = x$, we can also show that in general, (a, b) and (b, a) are always symmetric about the line $y = x$. Figure 7 displays two examples of this, and Exercise 51 at the end of this section asks you to supply a proof for the general situation.

Now let's see why the graphs of f and f^{-1} are always mirror images of each other about the line $y = x$. First, recall that the function f^{-1} switches the inputs and outputs of f. Thus, (a, b) is on the graph of f if and only if (b, a) is on the graph of f^{-1}. But as we have just stated, the points (a, b) and (b, a) are symmetric about the line $y = x$. It follows therefore that the graphs of f and f^{-1} are symmetric about the line $y = x$. For reference, we restate this useful fact in the box that follows.

▌ PROPERTY SUMMARY   The Graphs of f and f^{-1}

The graphs of f and f^{-1} are symmetric about the line $y = x$.

EXAMPLE 8

The graph of a function f consists of the line segment joining the points $(-2, -3)$ and $(-1, 4)$, as shown in Figure 8. Sketch a graph of f^{-1}.

SOLUTION

The graph of f^{-1} is obtained by reflecting the graph of f about the line $y = x$. The reflection of $(-2, -3)$ is $(-3, -2)$, and the reflection of $(-1, 4)$ is $(4, -1)$. To graph f^{-1}, then, we plot the reflected points $(-3, -2)$ and $(4, -1)$ and connect them with a line segment, as shown in Figure 9. For reference, Figure 9 also shows the graphs of f and $y = x$.

Figure 8

Figure 9

EXAMPLE 9

Let $g(x) = x^3$. Find $g^{-1}(x)$, and then, on the same set of axes, sketch the graphs of g, g^{-1}, and $y = x$.

SOLUTION

We begin by writing $y = x^3$, then switching x and y and solving for y. We first have

$$x = y^3$$

To solve this equation for y, we take the cube root of both sides to obtain

$$\sqrt[3]{x} = y$$

So the inverse function is $g^{-1}(x) = \sqrt[3]{x}$. We could graph the function g^{-1} by plotting points, but the easier way is to reflect the graph of $g(x) = x^3$ about the line $y = x$; see Figure 10.

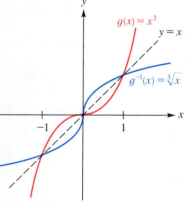

Figure 10

Earlier in this section, we mentioned that not every function has an inverse function. For instance, Tables 1(a) and (b) indicate what happens if we begin with the function defined by $F(x) = x^2$ and interchange the inputs and outputs. The

TABLE 1

(a) $F(x) = x^2$

x	y
1	1
2	4
3	9
−1	1
−2	4
−3	9

(b) Reversing the Inputs and Outputs

x	y
1	1
4	2
9	3
1	−1
4	−2
9	−3

resulting table, Table 1(b), does not define a function. For instance, the input 4 in Table 1(b) produces two distinct outputs, 2 and −2, but the definition of a function requires exactly one output for a given input.

Why didn't this difficulty arise when we looked for the inverse of $f(x) = 2x$ at the beginning of this section? The answer is that there is an essential difference between the functions defined by $f(x) = 2x$ and $F(x) = x^2$. With $f(x) = 2x$, different inputs never yield the same output. To put it another way, for the function $f(x) = 2x$, the only time it can happen that $f(a) = f(b)$ is when $a = b$. This condition guarantees that interchanging the inputs and outputs of f will yield a function. With $F(x) = x^2$, however, it *can* happen that $F(a) = F(b)$ and yet $a \neq b$. For instance,

$$F(2) = F(-2) \qquad \text{because } 2^2 = 4 = (-2)^2$$

but, of course, $2 \neq -2$.

It's useful to have a name for functions such as $f(x) = 2x$, in which distinct inputs always yield distinct outputs. We call these types of functions *one-to-one* functions.

DEFINITION | **One-to-One**

A function f is **one-to-one** provided that the following condition holds for all a and b in the domain of f:

If $f(a) = f(b)$ then $a = b$

EXAMPLES

$f(x) = 2x$ is one-to-one.

$F(x) = x^2$ is not one-to-one because $F(2) = F(-2)$, yet $2 \neq -2$.

Using graphs, there is an easy way to tell which functions are one-to-one:

Horizontal Line Test

A function f is one-to-one if and only if each horizontal line intersects the graph of $y = f(x)$ in at most one point.

Figure 11 will help you to see why the horizontal line test is valid. The figure shows a horizontal line intersecting the graph of a function f at two distinct points with x-coordinates a and b. From the graph, we see that $f(a) = f(b)$, even though $a \neq b$. Thus, the function represented in Figure 11 is not one-to-one.

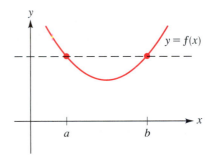

Figure 11

EXAMPLE 10

Use the horizontal line test to determine which functions in Figure 12 are one-to-one.

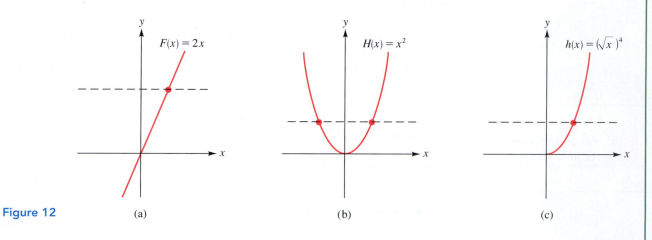

Figure 12   (a)   (b)   (c)

SOLUTION

The horizontal line test tells us that F and h are one-to-one, but H is not.

The following theorem tells us that the functions with inverses are precisely the one-to-one functions. (The proof of the theorem isn't difficult, but we'll omit it here.)

THEOREM

A function f has an inverse function f^{-1} if and only if f is one-to-one.

EXAMPLE 11

Which of the three functions in Figure 12 have inverse functions?

SOLUTION

According to our theorem, the functions with inverses are those that are one-to-one. In Example 10, we saw that F and h were one-to-one but H was not. So F and h each have inverses but H does not.

EXERCISE SET 3.6

A

1. If h and k are a pair of inverse functions, then (fill in the blanks).
   (a) $h[k(x)] =$ _____ for every x in the domain of _____.
   (b) $k[h(x)] =$ _____ for every x in the domain of _____.

2. Let $f(x) = x^2$ and $g(x) = 1/x$. Compute $f[g(x)]$ and $g[f(x)]$, and note that the results are identical. Then say why f and g do *not* qualify as a pair of inverse functions.

3. Verify that the given pairs of functions are inverse functions, as in Example 2.
   (a) $f(x) = 3x$; $g(x) = x/3$
   (b) $f(x) = 4x - 1$; $g(x) = \frac{1}{4}x + \frac{1}{4}$
   (c) $g(x) = \sqrt{x}$; $h(x) = x^2$ [Assume that the domain of both g and h is $[0, \infty)$.]

4. Which pairs of functions are inverses?
   (a) $f(x) = -3x + 2$; $g(x) = \frac{2}{3} - \frac{1}{3}x$
   (b) $F(x) = 2x + 1$; $G(x) = \frac{1}{2}x - 1$
   (c) $G(x) = x^3$; $H(x) = 1 - x^3$
   (d) $f(t) = t^3$; $g(t) = \sqrt[3]{t}$

Ⓖ 5. (a) Use a graphing utility to graph the following three functions on the same set of axes.

$$f(x) = \sqrt[3]{x^3 + 3} - 1$$
$$g(x) = \sqrt[3]{x^3 + 1} - 3$$
$$h(x) = \sqrt[3]{(x + 1)^3} - 3$$

   If necessary, adjust the picture so that it shows true proportions.
   (b) Which two graphs appear to be symmetric about the line $y = x$? Use algebra (as in Example 2) to verify that the two functions are indeed inverses.

Ⓖ 6. (a) Use a graphing utility to graph both $g(x) = \sqrt[3]{2 - x^3}$ and $y = x$ on the same set of axes.
   (b) What kind of symmetry do you observe? What does this tell you about the inverse for g? [If you are uncertain how to answer this last question, try coming back to it after working part (c).]
   (c) Starting with the equation $g(x) = \sqrt[3]{2 - x^3}$, follow the two-step procedure on page 226 to find $g^{-1}(x)$. What do you observe?

In Exercises 7 and 8, suppose that f and g are a pair of inverse functions.

7. If $f(7) = 12$, what is $g(12)$? (If you need a hint, reread Example 3.)

8. If $g(-2) = 0$, what is $f(0)$?

In Exercises 9 and 10, let $f(x) = x^3 + 2x + 1$, and assume that the domain of f^{-1} is $(-\infty, \infty)$. Simplify each expression (as in Example 4).

9. (a) $f[f^{-1}(4)]$
   (b) $f^{-1}[f(-1)]$
   (c) $(f \circ f^{-1})(\sqrt{2})$
   (d) $f[f^{-1}(t + 1)]$

10. (a) $f(0)$
    (b) $f^{-1}(1)$ *Hint*: Use the result in part (a).
    (c) $f(-1)$
    (d) $f^{-1}(-2)$

11. Let $f(x) = 2x + 1$.
    (a) Find $f^{-1}(x)$.
    (b) Calculate $f^{-1}(5)$ and $1/f(5)$. Are your answers the same?

12. Let $g(t) = \dfrac{1}{t} + 1$.
    (a) Find $g^{-1}(t)$.
    (b) Calculate $g^{-1}(2)$ and $1/g(2)$, and note that the answers are not the same.

13. Let $f(x) = 3x - 1$.
    (a) Compute $f^{-1}(x)$.
    (b) Verify that $f[f^{-1}(x)] = x$ and that $f^{-1}[f(x)] = x$.
    (c) On the same set of axes, sketch the graphs of f, f^{-1}, and the line $y = x$. Note that the graphs of f and f^{-1} are symmetric about the line $y = x$.

14. Follow Exercise 13, but use $f(x) = \frac{1}{3}x - 2$.

15. Follow Exercise 13, but use $f(x) = \sqrt{x - 1}$. [The domain of f^{-1} will be $[0, \infty)$.]

16. Follow Exercise 13, but use $f(x) = 1/x$.

17. Let $f(x) = (x + 2)/(x - 3)$.
    (a) Find the domain and range of the function f.
    (b) Find $f^{-1}(x)$.
    (c) Find the domain and range of the function f^{-1}. What do you observe?

18. Let $f(x) = (2x - 3)/(x + 4)$. Find $f^{-1}(x)$. Find the domain and range for f and f^{-1}. What do you observe?

19. Let $f(x) = 2x^3 + 1$. Find $f^{-1}(x)$.

20. In our discussion at the beginning of this section, we considered the functions $f(x) = 2x$ and $g(x) = x/2$. Verify that $f[g(x)] = x$ and that $g[f(x)] = x$. On the same set of axes, sketch the graphs of f, g, and $y = x$.

21. This exercise refers to the comments made at the end of Example 6. Compute $f[f^{-1}(x)]$ and $f^{-1}[f(x)]$ using the functions $f(x) = (x - 1)/(3x + 5)$ and $f^{-1}(x) = (-5x - 1)/(3x - 1)$. By actually carrying out the calculations, you'll find that the result in each case is x, which *proves* these are indeed inverse functions.

22. Let $f(x) = (x - 3)^3 - 1$.
    (a) Compute $f^{-1}(x)$.
    (b) On the same set of axes, sketch the graphs of f, f^{-1}, and the line $y = x$. *Hint*: Sketch f using the ideas of Section 3.4. Then reflect f to get f^{-1}.

For Exercises 23 and 24, refer to the following figure, which shows the graph of a function f. (The axes are marked off in one-unit intervals.) In each case, graph the indicated function.

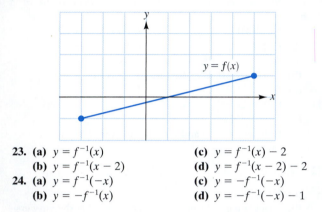

y = f(x)

23. (a) $y = f^{-1}(x)$
    (b) $y = f^{-1}(x - 2)$
    (c) $y = f^{-1}(x) - 2$
    (d) $y = f^{-1}(x - 2) - 2$

24. (a) $y = f^{-1}(-x)$
    (b) $y = -f^{-1}(x)$
    (c) $y = -f^{-1}(-x)$
    (d) $y = -f^{-1}(-x) - 1$

In Exercises 25–28, assume that the domain of f and f^{-1} is $(-\infty, \infty)$. Solve the equation for x or for t (whichever is appropriate) using the given information.

25. (a) $7 + f^{-1}(x - 1) = 9; f(2) = 6$
    (b) $4 + f(x + 3) = -3; f^{-1}(-7) = 0$

26. (a) $f^{-1}(2x + 3) = 5; f(5) = 13$
    (b) $f(1 - 2x) = -4; f^{-1}(-4) = -5$

27. $f^{-1}\left(\dfrac{t + 1}{t - 2}\right) = 12; f(12) = 13$

28. $f\left(\dfrac{1 - 2t}{1 + 2t}\right) = 7; f^{-1}(7) = -3$

For Exercises 29–33, use the horizontal line test to determine whether the function is one-to-one (and therefore has an inverse). (You should be able to sketch the graph of each function on your own, without using a graphing utility.)

29. $f(x) = -x^2 + 1$ 30. $g(x) = x^3 - 1$

31. $g(x) = 5$ (for all x)

32. $f(x) = \begin{cases} x^2 & \text{if } -1 \le x \le 0 \\ x^2 + 1 & \text{if } x > 0 \end{cases}$

33. $g(x) = \begin{cases} x^2 & \text{if } -1 \le x < 0 \\ x^2 + 1 & \text{if } x \ge 0 \end{cases}$

34. Which of the six basic functions graphed in Figure 7 in Section 3.2 are one-to-one?

Ⓖ *In Exercises 35–40, use a graphing utility to graph each function and then apply the horizontal line test to see whether the function is one-to-one.*

35. $y = x^2 + 2x$ 36. $y = x^3 + 2x$

37. $y = 2x^3 + x^2$ *Suggestion:* Begin with the standard viewing rectangle and then use a zoom-in view. (First looks can be deceiving.)

38. $y = 0.01x^4 - 1$ 39. $y = 2x^5 + x - 1$

40. (a) $f(x) = x^3 + x^2 + x$
    (b) $g(x) = x^3 - x^2 + x$
    (c) $h(x) = x^3 - x^2 - x$

B

Ⓖ 41. (a) In the standard viewing rectangle, graph two functions

$$f(x) = x^3 + 1.73x^2 + x + 2$$

and

$$g(x) = x^3 + 1.74x^2 + x + 2$$

(b) Use the zoom feature of your graphing utility to find out whether either function is one-to-one.

42. Let $f(x) = \sqrt{x}$.
    (a) Find $f^{-1}(x)$. What is the domain of f^{-1}? [The domain is not $(-\infty, \infty)$.]
    (b) In each case, determine whether the given point lies on the graph of f or f^{-1}.
        (i) $(4, 2)$
        (ii) $(2, 4)$
        (iii) $(5, \sqrt{5})$
        (iv) $(\sqrt{5}, 5)$
        (v) $(a, f(a))$, where $a \ge 0$
        (vi) $(f(a), a)$
        (vii) $(b, f^{-1}(b))$, where $b \ge 0$
        (viii) $(f^{-1}(b), b)$

43. In the following figure, determine the coordinates of the points A, B, C, and D. Express your answers in terms of the function f and the number a. (Each dashed line through A is parallel to an axis.)

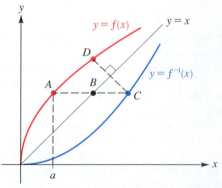

44. In the following figure, determine the coordinates of the points A, B, C, D, E, and F. Specify the answers in terms of the functions f and f^{-1} and the number b.

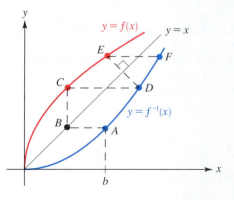

45. Let $f(x) = 3/(x - 1)$.
- **(a)** Find the average rate of change of f on the interval $[4, 9]$.
- **(b)** Find $f^{-1}(x)$ and then compute the average rate of change of f^{-1} on the interval $[f(4), f(9)]$. What do you observe?

46. The accompanying figure shows the graph of a function f on the interval $[a, b]$. If the average rate of change of f on the interval $[a, b]$ is k, show that the average rate of change of the function f^{-1} on the interval $[f(a), f(b)]$ is $1/k$, where $k \neq 0$.

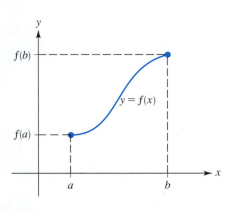

47. Suppose that (a, b) is a point on the graph of $y = f(x)$. Match the functions defined in the left-hand column with the points in the right-hand column. For example, the appropriate match for (a) in the left-hand column is determined as follows. The graph of $y = f(x) + 1$ is obtained by translating the graph of $y = f(x)$ up one unit. Thus, the point (a, b) moves up to $(a, b + 1)$ and, consequently, (E) is the appropriate match for (a).

- **(a)** $y = f(x) + 1$
- **(b)** $y = f(x + 1)$
- **(c)** $y = f(x - 1) + 1$
- **(d)** $y = f(-x)$
- **(e)** $y = -f(x)$
- **(f)** $y = -f(-x)$
- **(g)** $y = f^{-1}(x)$
- **(h)** $y = f^{-1}(x) + 1$
- **(i)** $y = f^{-1}(x - 1)$
- **(j)** $y = f^{-1}(-x) + 1$
- **(k)** $y = -f^{-1}(x)$
- **(l)** $y = -f^{-1}(-x) + 1$
- **(m)** $y = 1 - f^{-1}(x)$
- **(n)** $y = f(1 - x)$

- **(A)** $(-a, b)$
- **(B)** (b, a)
- **(C)** $(a - 1, b)$
- **(D)** $(-b, a + 1)$
- **(E)** $(a, b + 1)$
- **(F)** $(1 - a, b)$
- **(G)** $(-a, -b)$
- **(H)** $(-b, 1 - a)$
- **(I)** $(b, -a)$
- **(J)** $(a, -b)$
- **(K)** $(b + 1, a)$
- **(L)** $(a + 1, b + 1)$
- **(M)** $(b, a + 1)$
- **(N)** $(b, 1 - a)$

48. Let $f(x) = \dfrac{1}{1 - x}$. Show that the function $f \circ f$ is the inverse of f.

49. Let $f(x) = -\dfrac{2x + 2}{x}$.
- **(a)** Find $f[f(x)]$.
- **(b)** Use a graphing utility to graph $y = f[f(x)]$. Display the graph using true proportions. What type of symmetry does the graph appear to have?
- **(c)** The result in part (b) suggests that the inverse of the function $f \circ f$ is again $f \circ f$. Use algebra to show that this is indeed correct.

50. In this exercise you'll investigate the inverse of a composite function. In parts (b) and (c), which involve graphing, be sure to use the same size unit and scale on both axes so that symmetry about the line $y = x$ can be checked visually.
- **(a)** Let $f(x) = 2x + 1$ and $g(x) = \frac{1}{4}x - 3$. Compute each of the following:
  - **(i)** $f(g(x))$
  - **(ii)** $g(f(x))$
  - **(iii)** $f^{-1}(x)$
  - **(iv)** $g^{-1}(x)$
  - **(v)** $f^{-1}(g^{-1}(x))$
  - **(vi)** $g^{-1}(f^{-1}(x))$
- **(b)** On the same set of axes, graph the two answers that you obtained in (i) and (v) of part (a). Note that the graphs are *not* symmetric about $y = x$. The conclusion here is that the inverse function for $f(g(x))$ is not $f^{-1}(g^{-1}(x))$.
- **(c)** On the same set of axes, graph the two answers that you obtained in (i) and (vi) of part (a); also put the line $y = x$ into the picture. Note that the two graphs *are* symmetric about the line $y = x$. The conclusion here is that the inverse function for $f(g(x))$ is $g^{-1}(f^{-1}(x))$. In fact, it can be shown that this result is true in general. For reference, then, we summarize this fact about the inverse of a composite function in the box that follows.

> **Inverse of a Composition**
>
> $$(f \circ g)^{-1} = g^{-1} \circ f^{-1}$$

51. Use the method of Example 7 to show that the points (a, b) and (b, a) are symmetric about the line $y = x$.

MINI PROJECT | A FREQUENTLY ASKED QUESTION ABOUT INVERSES

> *Student*:   If $f[g(x)] = x$, isn't it automatically true then that $g[f(x)] = x$?
> *Professor*:   No; that's only when you have a pair of inverse functions.
> *Student*:   But in Example 2 and in similar homework problems, whenever
> I find that $f[g(x)] = x$, it *always* turns out that $g[f(x)] = x$.

This project provides an example of two functions f and g satisfying the following two conditions:

1. The equation $g[f(x)] = x$ holds for every real number x.
2. There are some real numbers x for which $f[g(x)] \neq x$.

Figures A and B show these two functions f and g; but you are not told which is which. Determine which function is f and which is g. In investigating $f[g(x)]$ and $g[f(x)]$ be sure you cover all cases.

$$y = \begin{cases} x & \text{if } x \leq 0 \\ x+1 & \text{if } x > 0 \end{cases}$$

$$y = \begin{cases} x & \text{if } x \leq 0 \\ 1 & \text{if } 0 < x \leq 1 \\ x-1 & \text{if } x > 0 \end{cases}$$

Figure A

Figure B

3.7   VARIATION

Perhaps the most dramatic moment in the development of physics during the 19th century occurred to J. C. Maxwell one day in the 1860's. . . . Maxwell noted that the equations or the laws [for electricity, magnetism, and light] . . . were mutually inconsistent . . . , and in order for the whole system to be consistent, he had to add another term to his equations. With this new term there came an amazing prediction, which was that a part of the electric and magnetic fields would fall off much more slowly with the distance than the inverse square, namely, inversely as the first power of the distance! And so he realized that electric currents in one place can affect other charges far away, and he predicted the basic effects with which we are familiar today—radio transmission, radar, and so on. —Richard Feynman in *The Feynman Lectures on Physics* (Reading, Mass.: Addison-Wesley, 1963)

Certain functions and relationships between variables occur so frequently in the applications of mathematics that there is a special terminology used to describe them. For a function defined by an equation of the form $y = kx$, where k is a nonzero constant, we say that y **varies directly** as x or that y is **directly proportional** to x. The number k in the equation $y = kx$ is called the **constant of proportionality.** In geometric terms, note that k is just the slope of the line $y = kx$, and that the graph of this line must pass through the origin. (How do we know it passes through the origin?)

EXAMPLE | 1 | **Working with direct variation**

Suppose that F varies directly as a. If $F = 60$ when $a = 5$, find F when $a = 6$.

SOLUTION
Since F varies directly as a, we write

$$F = ka$$

To find k, we substitute the given pair of values $F = 60$ and $a = 5$ in this equation. This yields

$$60 = k \cdot 5 \qquad \text{and therefore} \qquad k = 12$$

Now that we know $k = 12$, the equation $F = ka$ becomes

$$F = 12a$$

From this equation we can calculate F whenever a is given. In particular, then, if $a = 6$, we obtain $F = 12(6)$. Thus $F = 72$, as required. The formula $F = 12a$ is called the **law of variation** for this problem.

EXAMPLE | 2 | **Running at a constant speed**

At a constant speed, the distance traveled varies directly as the time. Suppose a runner covers 1 mile in 7 min.
(a) Find the law of variation.
(b) Using a graphing utility with a TRACE feature, graph the law of variation and estimate the distance the runner covers in 30 min.
(c) Use the law of variation to compute the exact distance that the runner covers in 30 min. Round the answer to two decimal places.

SOLUTION
(a) Let d and t denote distance and time, respectively. The statement that distance varies directly as time is written algebraically as

$$d = kt$$

As in Example 1, we use the given data to calculate k. Substituting $d = 1$ and $t = 7$ in the equation $d = kt$ yields

$$1 = k \cdot 7 \qquad \text{and therefore} \qquad k = \frac{1}{7}$$

Thus, the law of variation is

$$d = \frac{1}{7}t$$

(b) In Figure 1(a), we've graphed the function $d = \frac{1}{7}t$. We display only a first-quadrant portion of the line since, in general, time t and distance d are

GRAPHICAL PERSPECTIVE

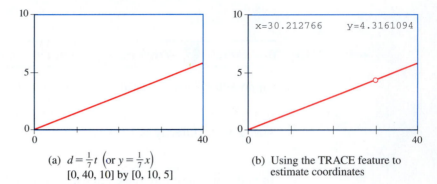

(a) $d = \frac{1}{7}t$ (or $y = \frac{1}{7}x$)
[0, 40, 10] by [0, 10, 5]

(b) Using the TRACE feature to estimate coordinates

Figure 1
$d = \frac{1}{7}t$ (or $y = \frac{1}{7}x$)

nonnegative. In Figure 1(b), we've used the TRACE feature to locate a point on the screen with a first coordinate as close as possible to 30. As indicated in the figure, the corresponding second coordinate is approximately 4.3. Thus, our estimate is that in 30 min. the runner covers about 4.3 miles.

(c) Substituting $t = 30$ in the equation $d = \frac{1}{7}t$ yields

$$d = \frac{1}{7}(30) = \frac{30}{7} \text{ miles}$$
$$\approx 4.29 \text{ miles}$$

Note that the result is consistent with the graphical estimate obtained in part (b).

Another type of relationship occurring frequently in applications involves functions of the form

$$y = k/x \quad \text{where } k \text{ is a nonzero constant}$$

In this case, we say that y **varies inversely** as x, or that y is **inversely proportional** to x. As in the case with direct variation, we refer to k as the *constant of proportionality*. A classic example of inverse variation occurs in chemistry and physics when we study a sample of gas held at a constant temperature. According to Boyle's law, the pressure p of the gas is inversely proportional to the volume v; that is, $p = k/v$, for some constant k. Figure 2 shows the general form for the graph of a function $p = k/v$, assuming p, v, and k are positive. Note that the graph has the same general shape as that of our familiar $y = 1/x$. Another example: For a stringed instrument such as a guitar or violin, the rate of vibration (or *frequency*) of a string is inversely proportional to the length of the string (assuming that the tension in the string is kept constant).

$p = k/v$

Figure 2
For a sample of gas held at a constant temperature, the pressure p varies inversely as the volume v.

EXAMPLE **3** **Working with inverse variation**

Given that y varies inversely as x and that y is 6 when x is -2, determine y when x is 4.

SOLUTION
The statement that y varies inversely as x translates algebraically to

$$y = \frac{k}{x}$$

To first find k, we substitute the given pair of values $x = -2$ and $y = 6$ in this equation. This yields

$$6 = k/(-2) \qquad \text{and therefore} \qquad k = -12$$

Now that we know $k = -12$, the equation $y = k/x$ becomes

$$y = -12/x$$

Using this law of variation, we can calculate y whenever x is given. In particular, then, if $x = 4$, we obtain

$$y = -12/4 = -3$$

Thus when x is 4, we have y is -3; this is the required answer.

The list that follows shows some common formulas that can be expressed in the language of variation. In this list, k denotes a constant in each case, the so-called **constant of proportionality.**

Formula or Law of Variation	Terminology
1. $y = kx^n$	y varies directly as x^n; y is directly proportional to x^n.
2. $y = \dfrac{k}{x^n}$	y varies inversely as x^n; y is inversely proportional to x^n.
3. $y = kxz$	y varies jointly as x and z; y varies directly as the product of x and z.
4. $y = \dfrac{kx}{z}$	y varies directly as x and inversely as z.

The rest of this section and the exercises that follow deal with this terminology. Notice that we have already discussed formulas 1 and 2 for the case $n = 1$.

EXAMPLE 4 Finding the law of variation

The volume of a sphere varies directly as the cube of the radius. When the radius is 2 cm, the volume is $32\pi/3$ cm^3. Find the constant of proportionality k. Write the law of variation.

SOLUTION
Let V denote the volume and r the radius. The statement that the volume varies directly as the cube of the radius translates algebraically to

$$V = kr^3$$

To find the constant of proportionality k, we substitute the given values for r and V in this equation to obtain

$$\frac{32\pi}{3} = k \cdot 2^3 = 8k$$

$$k = \frac{32\pi}{3 \cdot 8} = \frac{4\pi}{3}$$

Now that we have found the constant k, we can write the law of variation simply by substituting this value of k in the equation $V = kr^3$. This yields $V = (4\pi/3)r^3$.

EXAMPLE 5 Another example involving inverse variation

The quantity B varies inversely as the square of C. What happens to the value of B if C is tripled?

SOLUTION

The statement "B varies inversely as the square of C" is written algebraically as

$$B = \frac{k}{C^2}$$

To see what happens when C is tripled, replace C by the quantity $3C$ in our equation.

$$B = \frac{k}{(3C)^2} = \frac{k}{9C^2}$$

Now compare the original expression for B with the new expression.

$$\text{original expression for } B: \quad \frac{k}{C^2}$$

$$\text{new expression for } B \text{ after } C \text{ is tripled:} \quad \frac{1}{9} \cdot \frac{k}{C^2}$$

From this we can see that if C is tripled, B is one-ninth of its original value. This is the required solution.

Example 5 differs from Examples 1–4 in that we did not need to evaluate the constant of proportionality k specifically. Indeed, the problem did not supply enough information to calculate a value for k. Example 6 is similar in this respect.

EXAMPLE 6 Newton's law of gravitation

The force of attraction between two objects varies directly as the product of their masses and inversely as the square of the distance between them. (This is Newton's Law of Gravitation.) What happens to the force if both masses are tripled and the distance is doubled?

SOLUTION
We begin by assigning letters to denote the relevant quantities and then writing the law of variation. Let F, m, M, and d denote, respectively, the force, the two masses, and the distance. Then the given statement translates algebraically to

$$F = \frac{kmM}{d^2}$$

We want to find out what happens to F when both masses are tripled and the distance is doubled. To do this, we make the following substitutions in our law of variation:

replace m by $3m$
replace M by $3M$
replace d by $2d$

We then have

$$F = \frac{k(3m)(3M)}{(2d)^2} = \frac{9kmM}{4d^2}$$

Now we compare the original expression for F with the expression obtained after tripling the masses and doubling the distance:

original expression for F: $\dfrac{kmM}{d^2}$

new expression for F: $\dfrac{9}{4}\left(\dfrac{kmM}{d^2}\right)$

From this we can see that the new force is nine-fourths of the original force. In other words, the new force is $2\frac{1}{4}$ times as strong as the original. This is the required solution.

EXAMPLE **7**

Suppose that A varies jointly as X and Y. Also, suppose that X varies inversely as the square of Y. Show that A varies inversely as Y.

SOLUTION
Translating the two given statements into algebraic equations gives us

$$A = kXY \tag{1}$$

and

$$X = \frac{c}{Y^2} \tag{2}$$

Both k and c are constants here. We do need the two different letters since there is no reason to suppose that $k = c$. Now we are asked to show that A varies inversely as Y. To do this, we just substitute for X in equation (1), using the value given in equation (2). We have

$$A = kXY = k\left(\frac{c}{Y^2}\right)Y$$

and therefore

$$A = \frac{kc}{Y}$$

This last equation says that A equals a constant divided by Y; that is, A varies inversely as Y. If we want, we can denote kc by the single letter K. Then we have $A = K/Y$, which again says that A varies inversely as Y.

EXERCISE SET 3.7

A

In Exercises 1–5, translate the given statement into an equation.

1. **(a)** y varies directly as x.
   (b) A varies inversely as B.
2. **(a)** y varies directly as the cube of z.
   (b) s varies inversely as t^2.
3. **(a)** x varies jointly as u and v^2.
   (b) z varies jointly as A^2 and B^3.
4. **(a)** y varies directly as u and inversely as v.
   (b) y varies directly as the square root of x and inversely as the product of w and z.
5. **(a)** F is inversely proportional to r^2.
   (b) V^2 is directly proportional to the sum of U^2 and T^2.
6. If y varies directly as x, and y is 20 when $x = 4$, find y when $x = -6$.
7. Suppose A varies inversely as B, and $A = -1$ when $B = 2$.
   (a) Find the constant of proportionality.
   (b) Write the law of variation.
   (c) Find A when $B = 5/4$.
8. Suppose T varies directly as V^2, and $T = 8$ when $V = 4$.
   (a) Find the constant of proportionality and write the law of variation.
   (b) Find T when $V = 10$.
   (c) Find T when $V = \sqrt{10}$.
9. Assume x varies jointly as y and z. Also $x = 9$ when $y = 2$ and $z = -3$. Find x when both y and z are 4.
10. Suppose F varies directly as the product of q and Q and inversely as the square of d. When $q = 2$, $Q = 4$, and $d = 1/2$, the value of F is 8.
    (a) Find the constant of proportionality and write the law of variation.
    (b) Find F when $q = 3$, $Q = 6$, and $d = 2$.
11. If A varies jointly as B and C, what happens to A if B is tripled and C is doubled?
12. If S varies inversely as the square of T, what happens to S if T is cut in half?
13. If x varies jointly as B and C and inversely as the square root of A, what happens to x if A, B, and C are all quadrupled?

14. Suppose that the value of xy is always 3/4. Does y vary directly or inversely with x? What is the constant of proportionality in this case?
15. The surface area S of a sphere varies directly as the square of the radius r. If the radius is 2 cm, the surface area is 16π cm^2.
    (a) Find the law of variation.
    (G) (b) Using a graphing utility with a TRACE feature (as in Example 2), graph the law of variation and estimate the surface area when the radius is 5 cm.
    (c) Use the law of variation to compute the surface area corresponding to a radius of 5 cm. Give an exact answer in terms of π and also a calculator approximation rounded to two decimal places. Check that your result is consistent with the estimate in part (b).
16. For motion at a constant speed, the distance d varies directly as the time t. Suppose a car covers 10 miles in 15 min.
    (a) Find the law of variation.
    (G) (b) Using a graphing utility with a TRACE feature, graph the law of variation and estimate the distance covered in 50 min.
    (c) Use the law of variation to compute the distance covered in 50 min. Give an exact expression and also a calculator approximation rounded to two decimal places. Check that your result is consistent with the estimate in part (b).
17. Suppose that the rate at which oil flows through a pipeline is inversely proportional to the length of the pipeline. When the length L is 1000 m, the rate R is 5 m^3/min.
    (a) Find the law of variation.
    (G) (b) Using a graphing utility with a TRACE feature, graph the law of variation and estimate the rate of flow if $L = 1300$ m.
    (c) Use the law of variation to compute the rate of flow if $L = 1300$ m. Give an exact expression and also a calculator approximation rounded to two decimal places. Check that your result is consistent with the estimate in part (b).

18. The price p for a certain item varies inversely as the number n of items sold, provided $100 \leq n \leq 1000$. When the price is set at \$22 per item, 100 items can be sold.
  (a) Find the law of variation.
  (b) Using a graphing utility with a TRACE feature, graph the law of variation and estimate what the price should be to yield sales of $n = 650$ items.
  (c) Use the law of variation and some algebra to find the price p that will produce sales of $n = 650$ items. Check that your result is consistent with the estimate in part (b).

19. The force of attraction between two objects varies jointly as their masses and inversely as the square of the distance between them. What happens to the force if one mass is tripled, the other mass is quadrupled, and the distance is cut in half?

20. The magnitude of the force between two electrical charges Q_1 and Q_2 varies directly as the product of Q_1 and Q_2 and inversely as the square of the distance between the charges. (This is Coulomb's Law, named after the French physicist Charles Coulomb, who verified the law experimentally in 1785.) What happens to the magnitude of the force if Q_1 is doubled, Q_2 is quadrupled, and the distance is tripled?

21. Neglecting air resistance, the distance any object falls starting from rest is directly proportional to the square of the time. (This law was discovered experimentally by Galileo, sometime before the year 1590. In his first experiments, Galileo used his own pulse to measure time.) Given that an object starting from rest will fall 490 m in 10 sec, how far will an object fall in 5 sec, starting from rest?

22. The volume of a right circular cone varies jointly as the height and the square of the radius of the base. When the radius and height are both 3 cm, the volume is 9π cm^3. Find the volume of a cone having a height of 4 cm and a radius of $\frac{3}{2}$ cm.

23. (Boyle's Law)   In 1662, Robert Boyle observed that, for a sample of gas at a constant temperature, the volume is inversely proportional to the pressure.
  (a) Suppose that at a certain fixed temperature, the pressure of a 2-liter sample of gas is 1.025 atm. (atm is the abbreviation for atmospheres, a unit of pressure, where 1 atm = 14.7 lb/in^2., the air pressure at sea level.) Calculate the constant of proportionality and write the law of variation.
  (b) Suppose that, with the temperature still fixed, the pressure on our sample of gas is reduced to 1 atm. Calculate the new volume of the gas.

24. (Charles's Law)   In 1787, Jacques Charles stated that, for a sample of gas held at a constant pressure, the volume is directly proportional to the Kelvin temperature. Suppose that initially you have 300 ml of a gas at 300K. Then, holding the pressure constant, you heat the gas to 310K. What is the new volume?

25. **(a)** For a satellite in a circular orbit about the earth, the kinetic energy varies directly as the mass of the satellite and inversely as the radius of the orbit. What happens to the kinetic energy if we cut the radius in half and triple the mass?
  (b) The velocity of an earth satellite in a circular orbit varies inversely as the square root of the radius of the orbit. What happens to the velocity if we cut the radius in half?

26. (Hooke's Law)   In 1676, the English physicist Robert Hooke found that the distance a spring is stretched varies directly as the force pulling on the spring. Suppose that, for a certain spring, a pull of 5 lb results in a stretch of 3 in.
  (a) Find the constant of proportionality and write the law of variation.
  (b) How far will a pull of 12 lb stretch the spring?

27. The weight of an object above the earth's surface varies inversely as the square of its distance from the center of the earth. How much would an astronaut weigh 500 miles above the surface of the earth, assuming her weight at the surface is 140 lb? (Assume that the radius of the earth is 4000 miles.)

28. One of the ways in which the human body adjusts blood flow is through changes in the radii of small arteries called *arterioles*. According to Poiseuille's Law [named after the French physiologist J. L. Poiseuille (1779–1869)], the blood flow in a given arteriole (measured, for example, in cm^3/sec) varies directly as the fourth power of the radius. By what fraction is the blood flow through an arteriole multiplied if the radius is decreased by 20%?

B

29. The graphs of six functions are displayed here and on page 242. Which graph best represents the given statement?
  (a) y varies directly as x, where the constant of proportionality is negative.
  (b) y varies inversely as x.
  (c) y varies directly as x^2, with a positive constant of proportionality.
  (d) y varies directly as x, with a positive constant of proportionality.
  (e) y varies directly as x^2, with a negative constant of proportionality.

C   D

E   F

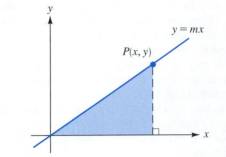

$y = mx$

$P(x, y)$

30. Three solid lead spheres of radii a, b, and c are melted down and recast into one large sphere of radius R. Show that $R = \sqrt[3]{a^3 + b^3 + c^3}$. Use the fact that the volume of a sphere varies directly as the cube of the radius.

31. The *period* of a pendulum is the time required for one complete swing, forward and back. If the period of a pendulum varies directly as the square root of the length of the pendulum, by how much should a pendulum of length L be shortened in order to cut the period in half?

32. At highway speeds, the wind resistance against a car varies directly as the square of the speed.
    (a) If the speed is increased from 40 mph to 60 mph, by what factor is the wind resistance multiplied?
    (b) At what speed is the wind resistance twice as much as that at 50 mph?

33. The intensity of illumination due to a small light source varies inversely as the square of the distance from the light. Suppose a book is 6 ft from a lamp. At what distance from the lamp will the illumination be twice as great?

34. The volume of a coin varies jointly as the thickness and the square of the radius. Suppose that two silver dollars are melted down (illegal!) and cast into one coin of the same thickness as a silver dollar. Show that the diameter of the new coin is $\sqrt{2}$ times the diameter of a silver dollar.

35. Show that the area of the shaded triangle in the figure varies jointly as m and the square of x.

36. The weight of an object on or above the earth's surface varies inversely as the square of the object's distance from the center of the earth. At what height x above the earth's surface will an object weigh half as much as it does on the surface? Assume that the radius of the earth is 4000 miles.
    Answer: $4000(\sqrt{2} - 1) \approx 1656.9$ miles.

37. In the year 1619, Johannes Kepler announced his Third Law of Planetary Motion. This law states that the square of the time it takes for a planet to complete one revolution about the sun varies directly as the cube of the planet's average distance from the sun. A table giving the average distance of each planet from the sun follows. Use a calculator to find the time it takes for each planet to complete one revolution about the sun. For Earth, assume the time is 365.2564 days.

Planet	Average distance from sun measured in astronomical units
Mercury	0.387099
Venus	0.723332
Earth	1.000000
Mars	1.523691
Jupiter	5.204377
Saturn	9.577971
Uranus	19.26020
Neptune	30.09421
Pluto	39.82984

38. According to *Hubble's Law,* named after the American astronomer Edwin Hubble (1889–1953), the universe is expanding in such a way that the galaxies are receding from one another at velocities that vary directly as their distances. Given that a galaxy in Virgo, which is 78,000,000 light years from us, is receding at 745 miles/sec, estimate the distance from us of the Hydra galaxy, which is receding from us at 37,881 miles/sec.

C

39. Suppose that x varies directly as z and that y also varies directly as z.
    (a) Does $x + y$ vary directly as z?
    (b) Does xy vary directly as z?

(c) Does $\sqrt{xy}$ vary directly as z? (Assume that x, y, and z are positive.)

40. Suppose that A varies directly as B. Show that AB varies directly as $A^2 + B^2$.

41. Suppose that $x + y$ varies directly as $x - y$. Also suppose that the constant of proportionality is not ± 1.
(a) Show that y varies directly as x.
(b) Show that $x^2 + y^2$ varies jointly as x and y.
(c) Show that $x^3 + y^3$ varies jointly as x, y, and $x + y$.

42. Suppose that $A^3 + \dfrac{1}{B^3}$ varies directly as $A^3 - \dfrac{1}{B^3}$. Also suppose that the constant of proportionality is not ± 1. Show that B varies inversely as A.

43. Show that the area of the shaded triangle in the figure varies directly as a^2.

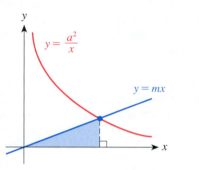

44. For this exercise, refer to the following figure.
(a) Show that the area of $\triangle OAB$ varies inversely as b.
(b) Show that the area of triangle I varies directly as a and inversely as the product of b and $a - b$.
(c) Show that the area of triangle II varies inversely as $a - b$.

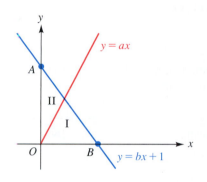

Chapter 3   Summary of Principal Terms and Formulas

Terms or notation	Page reference	Comments
1. Function	160	Given two nonempty sets A and B, a function from A to B is a rule of correspondence that assigns to each element of A exactly one element in B.
2. Domain	160	The domain of a function is the set of all inputs for that function. If f is a function from A to B, then the domain is the set A.
3. Range	160	The range of a function from A to B is the set of all elements in B that are actually used as outputs.
4. $f(x)$	163	Given a function f, the notation $f(x)$ denotes the output that results from the input x.
5. Graph of a function	173	The graph of a function f consists of those points (x, y) such that x is in the domain of f and $y = f(x)$.

Terms or notation	Page reference	Comments
6. Vertical line test	174	A graph in the x-y plane represents a function $y = f(x)$ provided that any vertical line intersects the graph in at most one point.
7. Turning point	186	A turning point on a graph is a point where the graph changes from rising to falling, or vice versa. See, for example, Figure 1 on page 186.
8. Maximum value and minimum value	186	An output $f(a)$ is a maximum value of the function f if $f(a) \geq f(x)$ for every x in the domain of f. An output $f(b)$ is a minimum value if $f(b) \leq f(x)$ for every x in the domain of f.
9. Increasing function	186	A function f is increasing on an interval if the following condition holds: If a and b are in the interval and $a < b$, then $f(a) < f(b)$. Geometrically, this means that the graph is rising as we move in the positive x-direction.
10. Decreasing function	186	A function f is decreasing on an interval if the following condition holds: If a and b are in the interval and $a < b$, then $f(a) > f(b)$. Geometrically, this means that the graph is falling as we move in the positive x-direction.
11. Average rate of change	189	The average rate of change, $\Delta f/\Delta x$, of a function f on an interval $[a, b]$ is the slope of the line segment through the two points $(a, f(a))$ and $(b, f(b))$: $$\frac{\Delta f}{\Delta x} = \frac{f(b) - f(a)}{b - a}$$
12. Techniques for graphing functions	200	The box on page 200 lists six techniques involving translation and reflection. For a comprehensive example, see Example 7 on page 205.
13. $f \circ g$	211	The composition of two functions f and g: $(f \circ g)(x) = f[g(x)]$
14. Iterates	215	Given a function f and an input x_0, the iterates of x_0 are the numbers $f(x_0), f(f(x_0)), f(f(f(x_0))), \ldots$. The number $f(x_0)$ is called the first iterate, the number $f(f(x_0))$ is called the second iterate, and so on. The orbit of x_0 under the function f is the list of numbers $$x_0, f(x_0), f(f(x_0)), \ldots$$
15. Inverse functions	222	Two functions f and g are inverses of one another provided that the following two conditions are met. First, $f[g(x)] = x$ for each x in the domain of g. Second, $g[f(x)] = x$ for each x in the domain of f.
16. f^{-1}	224	f^{-1} denotes the inverse function for f. *Note*: In this context, f^{-1} does not mean $1/f$.

Terms or notation	Page reference	Comments
17. One-to-one function	229	A function f is said to be one-to-one provided that distinct inputs always yield distinct outputs. Geometrically, this means that every horizontal line intersects the graph of $y = f(x)$ in at most one point. Algebraically, this means that f satisfies the following condition: If a and b are in the domain of f, and $f(a) = f(b)$, then $a = b$. The relationship between inverse functions and one-to-one functions is this: A function f has an inverse if and only if f is one-to-one.
18. y varies directly as x	234	$y = kx$, where k is a constant
19. y varies inversely as x	236	$y = k/x$, where k is a constant
20. y varies jointly as x and z	237	$y = kxz$, where k is a constant

Writing Mathematics

1. Section 3.6 on inverse functions begins with the following quotation from Professor R. P. Boas:

   > If anybody ever told me why the graph of $y = x^{1/2}$ is the reflection of the graph of $y = x^2$ [$x \geq 0$] in a 45° line, it didn't sink in. To this day, there are textbooks that expect students to think that it is so obvious as to need no explanation.

   Show your instructor that you are not in the dark about inverse functions. Write out an explanation of why the graphs of $y = x^{1/2}$ and $y = x^2$ ($x \geq 0$) are indeed reflections of one another in a 45° line. (As preparation for your writing, you'll probably need to look over Section 3.6 and make a few notes.)

2. Decide which of the following rules are functions. Write out your reasons in complete sentences. (In each case, assume that the domain is the set of students in your school.)
   (a) F is the rule that assigns to each person his or her brother.
   (b) G is the rule that assigns to each person his or her aunt.
   (c) H is the rule that assigns to each person his or her mother.
   (d) K is the rule that assigns to each person his or her mother or father.

Chapter 3   Review Exercises

1. **(a)** Find the domain of the function defined by
   $f(x) = \sqrt{15 - 5x}$.
   (b) Find the range of the function defined by
   $g(x) = (3 + x)/(2x - 5)$.

2. Let $f(x) = 3x^2 - 4x$ and $g(x) = 2x + 1$. Compute
   each of the following:
   (a) $(f - g)(x)$ **(c)** $f[g(-1)]$
   (b) $(f \circ g)(x)$

3. A **linear function** is a function defined by an equation of the form $F(x) = ax + b$, where a and b are real numbers. Suppose f and g are linear functions. Is the function $f \circ g$ a linear function?

4. A **quadratic function** is a function defined by an equation of the form $F(x) = ax^2 + bx + c$, where a, b, and c are real numbers and $a \neq 0$.
   (a) Give an example of quadratic functions f and g for which $f + g$ is also a quadratic function.
   (b) Give an example of quadratic functions f and g for which $f + g$ is not a quadratic function.
   (c) Suppose f and g are quadratic functions. Is the function $f \circ g$ a quadratic function?

5. **(a)** Compute $\dfrac{F(x) - F(a)}{x - a}$ given that $F(x) = 1/x$.
   (b) Compute $\dfrac{g(x + h) - g(x)}{h}$ for the function defined by $g(x) = x - 2x^2$.

6. The y-intercept for the graph of $y = f(x)$ is 4. What is the y-intercept for the graph of $y = -f(x) + 1$?

7. **(a)** Find $g^{-1}(x)$ given that $g(x) = \dfrac{1 - 5x}{3x}$.
   (b) The figure displays the graph of a function f. Sketch the graph of f^{-1}.

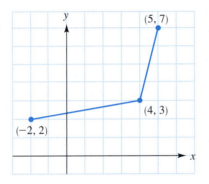

8. Refer to the function f in Exercise 7(b).
   (a) Compute $\Delta f/\Delta x$ on $[-2, 5]$.
   (b) Find the average rate of change of f^{-1} on $[2, 7]$.

9. Graph each function and specify the intercepts.
   (a) $y = |x + 2| - 3$ **(b)** $y = \dfrac{1}{x + 2} - 1$

10. Refer to the following graph of $y = g(x)$. The domain of g is $[-5, 2]$.
    (a) What are the coordinates of the turning point(s)?
    (b) What is the maximum value of g?
    (c) Which input yields a minimum value for g?
    (d) On which interval(s) is g increasing?
    (e) Compute $\Delta g/\Delta x$ for each of the following intervals: $[-5, -3]$; $[-3, -2]$; $[-5, 2]$.

11. Let $f(x) = 3x^2 - 2x$. Find   **(a)** $f(-1)$;
    (b) $f(1 - \sqrt{2})$.

12. Graph the function G defined by
    $$G(x) = \begin{cases} \sqrt{1 - x^2} & \text{if } -1 \leq x < 0 \\ \sqrt{x} & \text{if } x \geq 0 \end{cases}$$

13. Express the slope of a line passing through the points $(5, 25)$ and $(5 + h, (5 + h)^2)$ as a function of h.

14. Graph the function $y = |9 - x^2|$. Specify symmetry and intercepts.

15. The following figure shows the graph of a function $y = f(x)$. Sketch the graph of $y = f(-x)$.

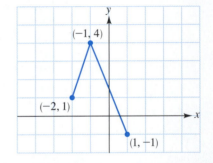

16. Given that the domains of f and f^{-1} are both $(-\infty, \infty)$ and that $f^{-1}(1) = -4$, solve the following equation for t: $2 + f(3t + 5) = 3$.

17. Let $f(x) = \frac{1}{2}x + 2$.

    (a) Compute the first three iterates of $x_0 = 1$.

    (b) Refer to the following figure. Explain why the y-coordinates of the points A, B, and C are the iterates that you computed in part (a).

    (c) Determine the y-coordinate of the point where the lines $y = x$ and $y = \frac{1}{2}x + 2$ intersect.

    (d) If we continue the pattern of dashed lines in the figure, it appears that the iterates of $x_0 = 1$ will approach 4. Verify this empirically by using your calculator to compute the fourth through tenth iterates. (Round your final answers to three decimal places.)

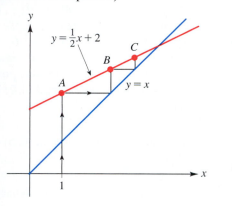

In Exercises 18–34, sketch the graph and specify any x- or y-intercepts.

18. $y = \dfrac{1}{x} + 1$

19. $f(x) = \dfrac{1}{x + 1}$

20. $y = \dfrac{1}{x + 1} + 1$

21. $y = |x + 3|$

22. $g(x) = -\sqrt{x - 4}$

23. $h(x) = \sqrt{1 - x^2}$

24. $y = 4 - \sqrt{-x}$

25. $y = (\sqrt{x})^2$

26. $f \circ g$, where $g(x) = x + 3$ and $f(x) = x^2$

27. $f \circ g$, where $g(x) = \sqrt{x - 1}$ and $f(x) = -x^2$

28. $f(x) = \begin{cases} \sqrt{1 - x^2} & \text{if } -1 \le x \le 0 \\ \sqrt{x} + 1 & \text{if } x > 0 \end{cases}$

29. $y = \begin{cases} 1/x & \text{if } 0 < x \le 1 \\ 1/(x - 1) & \text{if } 1 < x \le 2 \end{cases}$

30. $(y + |x| - 1)(y - |x| + 1) = 0$

31. f^{-1}, where $f(x) = \frac{1}{2}(x + 1)$

32. g^{-1}, where $g(x) = \sqrt[3]{x + 2}$

33. $f \circ f^{-1}$, where $f(x) = \sqrt{x - 2}$

34. $f^{-1} \circ f$, where $f(x) = \sqrt{x - 2}$

In Exercises 35–40, find the domain of the function.

35. $y = 1/(x^2 - 9)$

36. $y = x^3 - x^2$

37. $y = \sqrt{8 - 2x}$

38. $y = \dfrac{x}{6x^2 + 7x - 3}$

39. $y = \sqrt{x^2 - 2x - 3}$

40. $y = \sqrt{5 - x^2}$

In Exercises 41–44, determine the range of the function.

41. $y = \dfrac{x + 4}{3x - 1}$

42. $f \circ g$, where $f(x) = 1/x$ and $g(x) = 3x + 4$

43. $g \circ f$, where $f(x) = \dfrac{x + 2}{x - 1}$ and $g(x) = \dfrac{x + 1}{x + 4}$

44. f^{-1}, where $f(x) = \dfrac{x}{3x - 6}$

In Exercises 45–52, express the function as a composition of two or more of the following functions:

$$f(x) = 1/x \quad g(x) = x - 1 \quad F(x) = |x| \quad G(x) = \sqrt{x}$$

45. $a(x) = \dfrac{1}{x - 1}$

46. $b(x) = \dfrac{1}{x} - 1$

47. $c(x) = \sqrt{x - 1}$

48. $d(x) = \sqrt{x} - 1$

49. $A(x) = (1/\sqrt{x}) - 1$

50. $B(x) = |x - 2|$

51. $C(x) = \sqrt[4]{x} - 1$

52. $D(x) = 1/\sqrt{x - 3}$

For Exercises 53–80, compute the indicated quantity using the functions f, g, and F defined as follows:

$$f(x) = x^2 - x \quad g(x) = 1 - 2x \quad F(x) = \dfrac{x - 3}{x + 4}$$

53. $f(-3)$

54. $f(1 + \sqrt{2})$

55. $F(3/4)$

56. $f(t)$

57. $f(-t)$

58. $g(2x)$

59. $f(x - 2)$

60. $g(x + h)$

61. $f(x^2)$

62. $f(x)/x \quad (x \ne 0)$

63. $[f(x)][g(x)]$

64. $f[f(x)]$

65. $f[g(x)]$

66. $g[f(3)]$

67. $(g \circ f)(x)$

68. $(g \circ f)(x) - (f \circ g)(x)$

69. $(F \circ g)(x)$

70. $\dfrac{g(x + h) - g(x)}{h}$

71. $\dfrac{f(x + h) - f(x)}{h}$

72. $\dfrac{F(x) - F(a)}{x - a}$

73. $F^{-1}(x)$

74. $F[F^{-1}(x)]$

75. $(g \circ g^{-1})(x)$

76. $g^{-1}(x)$

77. $g^{-1}(-x)$

78. $\dfrac{g^{-1}(x + h) - g^{-1}(x)}{h}$

79. $F^{-1}[F(22/7)]$

80. $T^{-1}(x)$, where $T(x) = f(x)/x \quad (x \ne 0)$

In Exercises 81–94, refer to the graph of the function f in the figure. (The axes are marked off in one-unit intervals.)

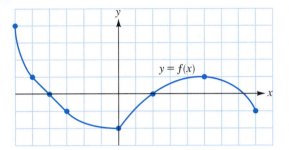

$y = f(x)$

81. Is $f(0)$ positive or negative?
82. Specify the domain and range of f.
83. Find $f(-3)$.
84. Which is larger, $f(-5/2)$ or $f(-1/2)$?
85. Compute $f(0) - f(8)$.
86. Compute $|f(0) - f(8)|$.
87. Specify the coordinates of the turning points.
88. What are the minimum and the maximum values of f?
89. On which interval(s) is f decreasing?
90. For which x-values is it true that $1 \leq f(x) \leq 4$?
91. What is the largest value of $f(x)$ when $|x| \leq 2$?
92. Is f a one-to-one function?
93. Does f possess an inverse function?
94. Compute $f[f(-4)]$.

For Exercises 95–110, refer to the graphs of the functions f and g in the following figure. Assume that the domain of each function is $[0, 10]$.

$y = g(x)$

$y = f(x)$

95. For which x-value is $f(x) = g(x)$?
96. For which x-values is it true that $g(x) \leq f(x)$?
97. (a) For which x-value is $f(x) = 0$?
    (b) For which x-value is $g(x) = 0$?
98. Compute $f(0) + g(0)$.

99. Compute each of the following.
    (a) $(f + g)(8)$ (c) $(fg)(8)$
    (b) $(f - g)(8)$ (d) $(f/g)(8)$
100. Compute each of the following.
    (a) $g[f(5)]$ (c) $(g \circ f)(5)$
    (b) $f[g(5)]$ (d) $(f \circ g)(5)$
101. Which is larger, $(f \circ f)(10)$ or $(g \circ g)(10)$?
102. Compute $g[f(10)] - f[g(10)]$.
103. For which x-values is it true that $f(x) \geq 3$?
104. For which x-values is it true that $|f(x) - 3| \leq 1$?
105. What is the largest number in the range of g?
106. Specify the coordinates of the highest point on the graph of each of the following equations.
    (a) $y = g(-x)$ (d) $y = f(-x)$
    (b) $y = -g(x)$ (e) $y = -f(x)$
    (c) $y = g(x - 1)$ (f) $y = -f(-x)$
107. On which intervals is the function g decreasing?
108. What are the coordinates of the turning points of g?
109. For which values of x in the interval $(4, 7)$ is the quantity $\dfrac{f(x) - f(5)}{x - 5}$ negative?
110. For which values of x in the interval $(0, 5)$ is the quantity $\dfrac{f(x) - f(2)}{x - 2}$ positive?
111. Suppose A varies jointly as x and y^2. What happens to the value of A when x is tripled and y is doubled?
112. Suppose B varies directly as x and inversely as y^3. When $x = 6$ and $y = 2$, the value of B is 9. Find B when $x = 3$ and $y = 4$.
113. The intensity of illumination from a small light source varies inversely as the square of the distance from the light. Suppose that a book is 10 ft from a lamp. At what distance from the lamp will the illumination be five times as great?
114. The velocity of an earth satellite in a circular orbit varies inversely as the square root of the radius of the orbit. What happens to the velocity if the radius is decreased by 10 percent?
115. The length of a side in an equilateral triangle varies directly as the square root of the area of the triangle. When the area is $16\sqrt{3}$ cm^2, the length of a side is 8 cm. How long is a side when the area is 20 cm^2?

Chapter 3   Test

1. Find the domain of the function defined by
$f(x) = \sqrt{x^2 - 5x - 6}$.

2. Find the range of the function defined by
$$g(x) = \frac{2x - 8}{3x + 5}$$

3. Let $f(x) = 2x^2 - 3x$ and $g(x) = 2 - x$. Compute each of the following:
   (a) $(f - g)(x)$; **(b)** $(f \circ g)(x)$; **(c)** $f[g(-4)]$.

4. Let $f(t) = 2/t$. Compute $\dfrac{f(t) - f(a)}{t - a}$.

5. Let $g(x) = 2x^2 - 5x$. Compute $\dfrac{g(x + h) - g(x)}{h}$.

6. Find $g^{-1}(x)$ given that $g(x) = \dfrac{-4x}{6x + 1}$.

7. The graph of a function f is a line segment joining the two points $(-3, 1)$ and $(5, 6)$. Determine the slope of the line segment that results from graphing the equation $y = -f^{-1}(x)$.

8. Graph each function and specify the intercepts:
   (a) $y = -|x - 3| + 1$;   **(b)** $y = \dfrac{1}{x + 3} - 2$.   Ⓖ

9. The figure shows the graph of a function g with domain $[-4, 2]$.

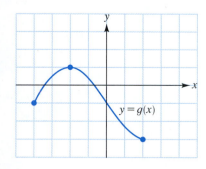

$y = g(x)$

   (a) Specify the range of g.
   (b) What are the coordinates of the turning point?
   (c) What is the minimum value of g?
   (d) Which input yields a maximum value for g? What is that maximum value?
   (e) On which interval is g decreasing?
   (f) Use the horizontal line test to say whether or not g is one-to-one.

10. Let $f(x) = x^2 - 3x - 1$. Compute each of the following:
    (a) $f(-3/2)$;   **(b)** $f(\sqrt{3} - 2)$.

11. Graph the function F defined by
$$F(x) = \begin{cases} |x + 1| & \text{if } x < -1 \\ -x^2 + 1 & \text{if } x > -1 \end{cases}$$
What is the domain of this function?

12. Suppose that z varies directly as x^2 and inversely as the square root of t. When $x = 6$ and $t = 9$, the value of z is 12. Find z when $x = 3$ and $t = 16$.

13. (Refer to the accompanying figure.) The rate r of vibration of a tuning fork is inversely proportional to the square of the length L. When the length $L = 0.05$ m, $r = 1020$ vibrations per second.

L

   (a) Find the law of variation.
   (b) Using a graphing utility with a TRACE feature, graph the law of variation and estimate the rate of vibration if $L = 0.08$ m. (Round the answer to the nearest integer.)
   (c) Use the law of variation to compute the rate of vibration when $L = 0.08$ m. Round your answer to one decimal place. [Check to see that your answer is consistent with the approximation obtained in part (b).]

14. In the figure below, the graph of $f(x) = 1 - |2x - 1|$ is shown in red ($y = x$ is included also).

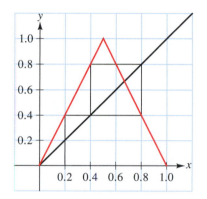

(a) Use the figure to specify the first six iterates of $x_0 = 0.2$.

(b) Use a calculator to compute the first six iterates of $x_0 = 0.2$. Are your results consistent with those in part (a)?

15. The graph of f is a line segment joining the points $(1, 3)$ and $(5, -2)$. Sketch the graph of $y = f(-x)$.

16. Given that the domains of f and f^{-1} are both $(-\infty, \infty)$ and that $f^{-1}(-3) = 1$, solve the equation $5 + f(4t - 3) = 2$ for t.

17. Let $f(x) = 1/x$. Find a value for b so that on the interval $[1, b]$, we have $\Delta f/\Delta x = -1/10$.

18. During the first five hours of a lab experiment, the temperature of a solution is given by

$$F(t) = 0.16t^2 - 1.6t + 35 \qquad (0 \le t \le 5)$$

where t is measured in hours, $t = 0$ corresponds to the time that the experiment begins, and $F(t)$ is the temperature (in degrees Celsius) at time t.

(a) Find (and simplify) a formula for $\Delta F/\Delta t$ over the interval $[a, 5]$.

(b) Use the result in part (a) to compute the average rate of change of temperature during the period from $t = 4$ hr to $t = 5$ hr. (Be sure to include the units as part of the answer.)

4 CHAPTER

Polynomial and Rational Functions. Applications to Optimization

POLYNOMIAL was used by François Viéta (1540–1603). The term RATIONAL FUNCTION was used by Joseph Louis Lagrange (1736–1813) in "Réflexions sur la résolution algébrique des équations," . . .—From Jeff Miller's Web site Earliest Known Uses of Some of the Words of Mathematics
http://members.aol.com/jeff570/mathword.html

In the previous chapter we studied some rather general rules for working with functions and graphs. With that as background, we are ready to focus our attention on a few specific types of functions and their applications. Some contexts in which we'll use the functions and ideas introduced in this chapter are:

- Modeling data relating smoking and lung cancer
  (Example 5 in Section 4.1)
- Summarizing global trends in airline passenger miles using a quadratic model
  (Exercise 1 in Exercise Set 4.2)
- Modeling the spread of the AIDS disease using linear and quadratic functions
  (Example 1 in Section 4.2)
- Finding the minimum value of a quadratic function to gain insight into how least-squares lines are defined and determined
  (Project at end of Section 4.5)
- Determining long-term behavior in a population model by computing the horizontal asymptote of a rational function
  (Example 7 in Section 4.7)

4.1 LINEAR FUNCTIONS

All decent functions are practically linear. —Professor Andrew Gleason

I had a moment of mixed joy and anguish, when my mind took over. It raced well ahead of my body and drew my body compellingly forward. I felt that the moment of a lifetime had come. — Dr. Roger Bannister, first person to run the mile in under four minutes, recalling the last lap of his record-breaking run at the Ilffley Road Track, Oxford, England, May 6, 1954. The quotation is from Roger Bannister, *The Four Minute Mile* (New York: Dodd, Mead & Co., 1955).

By a **linear function** we mean a function defined by an equation of the form

$$f(x) = Ax + B$$

where A and B are constants. In this chapter, the constants A and B will always be real numbers. From our work in Chapter 1, we know that the graph of $y = Ax + B$ is a straight line.

EXAMPLE **1** **Finding an equation defining a linear function**

Suppose that f is a linear function. If $f(1) = 0$ and $f(2) = 3$, find an equation defining f.

SOLUTION
From the statement of the problem, we know that the graph of f is a straight line passing through the points $(1, 0)$ and $(2, 3)$. Thus the slope of the line is

$$m = \frac{y_2 - y_1}{x_2 - x_1} = \frac{3 - 0}{2 - 1} = 3$$

Now we can use the point–slope formula using the point $(1, 0)$ to find the required equation. We have

$$y - y_1 = m(x - x_1)$$
$$y - 0 = 3(x - 1)$$
$$y = 3x - 3$$

This is the equation defining f. If we wish, we can rewrite it using function notation: $f(x) = 3x - 3$.

One basic application of linear functions that occurs in business and economics is *linear* or *straight-line depreciation*. In this situation we assume that the value $V(t)$ of an asset (such as a machine or an apartment building) decreases linearly over time t:

$$V(t) = mt + b$$

where $t = 0$ corresponds to the time when the asset is new (or when its value is first assessed), and the slope m is negative.

EXAMPLE **2** **Using a linear model for depreciation**

A factory owner buys a new machine for $8000. After ten years, the machine has a salvage value of $500.
(a) Assuming linear depreciation (as indicated in Figure 1), find a formula for the value $V(t)$ of the machine after t yr, where $0 \le t \le 10$.
(b) Use the depreciation function determined in part (a) to find the value of the machine after seven years.

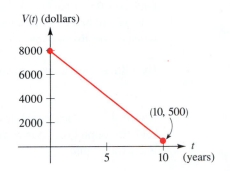

Figure 1

SOLUTION

(a) We need to determine m and b in the equation $V(t) = mt + b$. From Figure 1 we see that the V-intercept of the line segment is $b = 8000$. Furthermore, since the line segment passes through the two points $(10, 500)$ and $(0, 8000)$, the slope m is

$$\frac{8000 - 500}{0 - 10} = -750$$

So we have $m = -750$ and $b = 8000$, and the required equation is

$$V(t) = -750t + 8000 \qquad (0 \le t \le 10)$$

(b) Substituting $t = 7$ in the equation $V(t) = -750t + 8000$ yields

$$V(7) = -750(7) + 8000 = -5250 + 8000 = 2750$$

Thus the value of the machine after seven years is $2750.

It is instructive to keep track of the units associated with the slope in Example 2. Repeating the slope calculation and keeping track of the units, we have

$$m = \frac{\$8000 - \$500}{0 \text{ yr} - 10 \text{ yr}} = -\$750/\text{yr} \qquad (1)$$

Thus, slope is a *rate of change*. (We met this idea earlier, back in Section 1.6.) In this example, the slope represents the rate of change of the value of the machine. Notice that the calculations in (1) are exactly those that we would have used in Chapter 3 to compute the *average* rate of change $\Delta V/\Delta t$. In the context of linear functions, we often suppress the word "average" and refer simply to the *rate of change*, because no matter what two points we choose on the graph of a linear function, the slope will always be the same.

As a second example of slope as a rate of change, let us suppose that a small manufacturer of handmade running shoes knows that her total cost in dollars, $C(x)$, for producing x pairs of shoes each business day is given by the linear function

$$C(x) = 10x + 50$$

Figure 2(a) displays a table and the graph for this function. Actually, since x represents the daily number of pairs of shoes, x can assume only whole-number values. So, technically, the graph that should be given is the one in Figure 2(b). However,

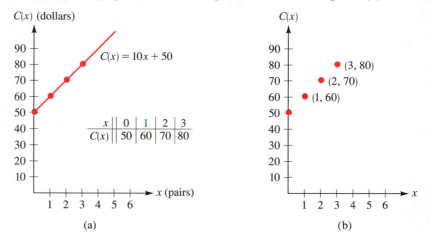

x	0	1	2	3
$C(x)$	50	60	70	80

Figure 2

(a)   (b)

the graph in Figure 2(a) turns out to be useful in practice, and we shall follow this convention.

The slope of the line $C(x) = 10x + 50$ in Figure 2(a) is 10, the coefficient of x. To understand the units involved, let's calculate the slope using two of the points in Figure 2(a). Using the points $(0, 50)$ and $(1, 60)$ and keeping track of the units, we have

$$m = \frac{\$60 - \$50}{1 \text{ pair} - 0 \text{ pairs}} = \frac{\$10}{1 \text{ pair}} = \$10/\text{pair}$$

Again the slope is a rate. In this case, the slope represents the rate of increase of cost; each additional pair of shoes produced costs the manufacturer $10.

In the study of economics, a function that gives the cost $C(x)$ for producing x units of a commodity is called a **cost function.** When the graph of a cost function is a line, we define the **marginal cost** as the additional cost to produce one more unit. Thus, in the preceding example, the marginal cost is $10 per pair, and we see that the slope of the line in Figure 2(a) represents the marginal cost.

EXAMPLE **3**   **A cost function that is linear**

Suppose that the cost $C(x)$ in dollars of producing x bicycles is given by the linear function

$$C(x) = 625 + 45x$$

(a) Find the cost of producing 10 bicycles.
(b) What is the marginal cost?
(c) Use the answers in parts (a) and (b) to find the cost of producing 11 bicycles. Then check the answer by evaluating $C(11)$.

SOLUTION
(a) Using $x = 10$ in the cost equation, we have

$$C(10) = 625 + 45(10) = 1075$$

Thus, the cost of producing ten bicycles is $1075.

(b) Since C is a linear function, the marginal cost is the slope and we have

$$\text{marginal cost} = \$45 \text{ per bicycle}$$

(c) According to the result in part (b), each additional bicycle costs $45. Therefore, we can compute the cost of 11 bicycles by adding $45 to the cost for 10 bicycles:

$$\text{cost of 11 bicycles} = \text{cost of 10 bicycles} + \text{marginal cost}$$
$$= \$1075 + \$45 = \$1120$$

So the cost of producing 11 bicycles is $1120. We can check this result by using the cost function $C(x) = 625 + 45x$ to compute $C(11)$ directly:

$$C(11) = 625 + 45(11)$$
$$= 625 + 495$$
$$= \$1120 \qquad \text{as obtained previously}$$

Interpreting slope as a rate of change is not restricted to applications in business or economics. Suppose, for example, that you are driving a car at a steady rate of 50 mph. Using the distance formula from elementary mathematics,

$$\text{distance} = \text{rate} \times \text{time}$$

or

$$d = rt$$

We have, in this case, $d = 50t$, where d represents the distance traveled (in miles) in t hours. In Figure 3, we show a graph of the linear function $d = 50t$. The slope of this line is 50, the coefficient of t. But 50 is also the given rate of speed, in miles per hour. So again, slope is a rate of change. In this case, slope is the **velocity,** or *rate of change of distance with respect to time.*

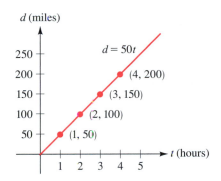

Figure 3

The slope of the line is the rate of change of distance with respect to time.

We conclude this section by indicating one way that linear functions are used in statistical applications. Tables 1 and 2 show the evolution of the world record for the one-mile run during the years 1911–1993. In case you're wondering why Table 1 begins with 1911, John Paul Jones was the first twentieth-century runner to break the

TABLE 1   Evolution of the Record for the Mile Run, 1911–1954

x (year)	y (time)	
1911	4:15.4	(John Paul Jones, United States)
1913	4:14.6	(John Paul Jones, United States)
1915	4:12.6	(Norman Taber, United States)
1923	4:10.4	(Paavo Nurmi, Finland)
1931	4:09.2	(Jules Ladoumegue, France)
1933	4:07.6	(Jack Lovelock, New Zealand)
1934	4:06.8	(Glen Cunningham, United States)
1937	4:06.4	(Sidney Wooderson, Great Britain)
1942	4:06.2	(Gunder Haegg, Sweden)
1942	4:06.2	(Arne Andersson, Sweden)
1942	4:04.6	(Gunder Haegg, Sweden)
1943	4:02.6	(Arne Andersson, Sweden)
1944	4:01.6	(Arne Andersson, Sweden)
1945	4:01.4	(Gunder Haegg, Sweden)
1954	3:59.4	(Roger Bannister, Great Britain)
1954	3:58.0	(John Landy, Australia)

TABLE 2   Evolution of the Record for the Mile Run, 1957–1993

x (year)	y (time)	
1957	3:57.2	(Derek Ibbotson, Great Britain)
1958	3:54.5	(Herb Elliott, Australia)
1962	3:54.4	(Peter Snell, New Zealand)
1964	3:54.1	(Peter Snell, New Zealand)
1965	3:53.6	(Michel Jazy, France)
1966	3:51.3	(Jim Ryun, United States)
1967	3:51.1	(Jim Ryun, United States)
1975	3:51.0	(Filbert Bayi, Tanzania)
1975	3:49.4	(John Walker, New Zealand)
1979	3:49.0	(Sebastian Coe, Great Britain)
1980	3:48.8	(Steve Ovett, Great Britain)
1981	3:48.53	(Sebastian Coe, Great Britain)
1981	3:48.40	(Steve Ovett, Great Britain)
1981	3:47.33	(Sebastian Coe, Great Britain)
1985	3:46.32	(Steve Cram, Great Britain)
1993	3:44.39	(Noureddine Morceli, Algeria)

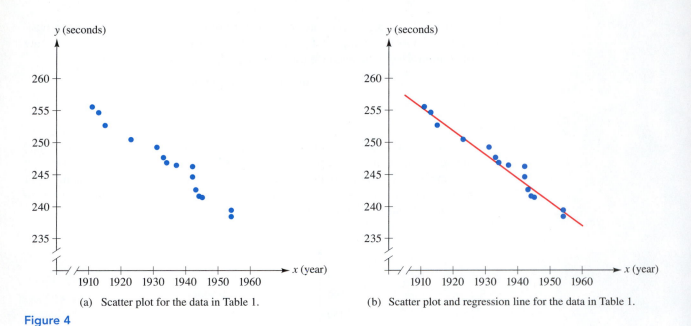

(a)   Scatter plot for the data in Table 1.

(b)   Scatter plot and regression line for the data in Table 1.

Figure 4

previous century's record of 4:15.6, set in 1895. Table 1 ends with the year 1954, the first year in which the "four-minute barrier" was broken. (See the quote by Roger Bannister at the beginning of this section.)

Let's begin by looking at the data in Table 1 graphically. In Figure 4(a), we've plotted the (x, y) pairs given in the table. The resulting plot is called a **scatter diagram** or **scatter plot.**

A striking feature of the scatter diagram in Figure 4(a) is that the records do not appear to be leveling off; rather, they seem to be decreasing in an approximately linear fashion. Using the *least-squares technique* from the field of statistics, it can be shown that the linear function that best fits the data in Table 1 is

$$f(x) = -0.370x + 962.041 \qquad (2)$$

The graph of this line is shown in Figure 4(b). The line itself is referred to as the **regression line,** or the **least-squares line.**

General formulas for determining the regression line are given in Exercise 45. (The derivation, however, is not given, because that requires calculus.) Many graphing utilities and spreadsheet applications have features for computing regression lines when two columns or lists of numbers are entered. We'll cite but one example here using the popular spreadsheet program Microsoft *Excel*™. For more details, or for other types of graphing utilities, see the owner's manual that came with your graphing utility.

Suppose, for example, that we want to obtain a scatter plot and regression line for the four data points

$$(0, 1) \quad (1, 2) \quad (3, 6) \quad (6, 7)$$

As indicated in Figure 5(a), we enter the x-y pairs using adjacent columns of the *Excel* spreadsheet. (In the figure, the x-values are in column A and the y-values in column B.) Next, of the many chart types that *Excel* provides for displaying data visually, we choose the "Scatter" option and obtain the result shown in Figure 5(b). Finally, for the required regression line, we go to the "Add Trendline" menu and

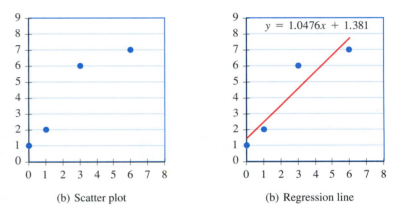

(a) Data

		A	B	
	1	0	1	
	2	1	2	
	3	3	6	
	4	6	7	

(b) Scatter plot   (b) Regression line

Figure 5
Using a spreadsheet application to obtain a regression line

choose "Linear" along with "Display equation on chart." The final result is shown in Figure 5(c).

EXAMPLE 4 **Using a regression line**

(a) The regression line given by equation (2) is based on records from 1911 through 1954. Use this regression line to estimate what the world record for the mile run might have been in 1999.

(b) Check the prediction against the actual 1999 record 3:43.13, set by Hicham El Guerrouj of Morocco. Compute the percentage error.

SOLUTION

(a) Substituting the value $x = 1999$ in equation (2) yields

$$f(1999) = -0.370(1999) + 962.041 \approx 222.41 \text{ sec}$$
$$= 3 \text{ min and } 42.41 \text{ sec}$$

(b) The prediction obtained in part (a) is 3:42.41 and the actual record is 3:43.13. Thus, the actual record is slightly higher than the prediction, but the difference is less than one second. For computing the percentage error, let's first express the actual record of 3:43.13 in seconds. [From part (a), we already know the predicted record expressed in seconds.] We have

$$3 \text{ min and } 43.13 \text{ sec} = 3(60) \text{ sec} + 43.13 \text{ sec}$$
$$= 223.13 \text{ sec}$$

Now we're ready to compute the percentage error in the prediction:

$$\text{percentage error} = \frac{|(\text{actual value}) - (\text{predicted value})|}{\text{actual value}} \times 100$$
$$= \frac{|223.13 - 222.41|}{223.13} \times 100 \approx 0.32\%$$

The percentage error in the prediction is 0.32%.

Remark: The prediction in this example has turned out to be extremely good. However, see part (c) in the next example for an important word of caution regarding predictions such as this.

EXAMPLE 5   A regression line in a medical application

The data in Table 3 are extracted from a 1954 paper on smoking and lung cancer that appeared in the journal *Danish Medical Bulletin*. In Table 3, the left-hand column of numbers shows annual cigarette consumption in 1930 for four countries. The right-hand column gives the mortality rates from lung cancer in those same countries 20 years later.

TABLE 3   Cigarette Consumption and Lung Cancer

	Cigarette consumption for 1930 (per person)	Lung cancer deaths in 1950 (per 100,000 males)
Sweden	320	11.1
Netherlands	444	28.3
Finland	1106	35.3
England-Wales	1200	53.0

Data extracted from "Bronchial Carcinoma—A Pandemic," by A. Nielsen and J. Clemmennsen, *Danish Medical Bulletin*, vol. 1 (1954), pp. 194–199. In some instances the indicated years 1930 and 1950 are only approximate. For instance, for Sweden, the mortality rate pertains to 1951, rather than 1950.

(a) Use a graphing utility to create a scatter plot and to find and graph the regression line. (Use the consumption data for the *x*-values and the mortality rates for the corresponding *y*-values.)

(b) Use the regression line to obtain estimates for the mortality rates in 1950 for Denmark and Norway, given that the 1930 cigarette consumption figures for Denmark and Norway were 353 and 257, respectively.

(c) Compute the percentage error in each estimate, given that the actual death rates (per 100,000) in Denmark and Norway were 18.4 and 9.2, respectively.

Smokescreen? The figure is from a 1948 magazine ad for Camel cigarettes. A 1929 ad for Lucky Strike cigarettes announced "Many prominent athletes smoke Luckies all day long with no harmful effects to wind or physical condition." A 1953 ad by Liggett & Meyers said "It's so satisfying to know that a doctor reports no adverse effects to the nose, throat and sinuses from smoking Chesterfield."

Source: The University of Alabama Center for the Study of Tobacco and Society (Dr. Alan Blum, M.D.)

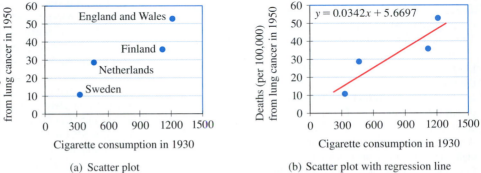

Figure 6

(a) Scatter plot

(b) Scatter plot with regression line

SOLUTION

(a) See Figure 6 for the scatter plot and regression line.

(b) As indicated in Figure 6(b), the regression line is $y = 0.0342x + 5.697$.

For Denmark with $x = 373$:
$$y = 0.0342(373) + 5.6697$$
$$\approx 18.4$$

For Norway with $x = 257$:
$$y = 0.0342(257) + 5.6697$$
$$\approx 14.5$$

According to the calculations, the projected mortality rate for Denmark in 1950 is 18.4 deaths per 100,000; for Norway the projection is 14.5 per 100,000.

(c) For Denmark, the projection is 18.4 and so is the actual death rate. So in this case, we happen to have 0% error. For Norway, we have

$$\text{percentage error} = \frac{|(\text{actual value}) - (\text{projected value})|}{\text{actual value}} \times 100$$
$$= \frac{|9.2 - 14.5|}{9.2} \times 100 \approx 58\%$$

In summary, for Denmark, the projected death rate from lung cancer agrees with the actual rate. For Norway, however, there is a relatively large discrepancy between the projection and actual value; the percentage error there is 58%. This highlights one of the shortcomings in using a regression line. Without further analysis and the use of more advanced statistical concepts (and perhaps more data) it's difficult or impossible to know ahead of time how much one should rely on these projections.

EXERCISE SET 4.1

A

In Exercises 1–8, find the linear functions satisfying the given conditions.

1. $f(-1) = 0$ and $f(5) = 4$
2. $f(3) = 2$ and $f(-3) = -4$
3. $g(0) = 0$ and $g(1) = \sqrt{2}$
4. The graph passes through the points $(2, 4)$ and $(3, 9)$.
5. $f(\frac{1}{2}) = -3$ and the graph of f is a line parallel to the line $x - y = 1$.
6. $g(2) = 1$ and the graph of g is perpendicular to the line $6x - 3y = 2$.
7. The graph of the inverse function passes through the points $(-1, 2)$ and $(0, 4)$.
8. The x- and y-intercepts of the inverse function are 5 and -1, respectively.
9. Let $f(x) = 3x - 4$ and $g(x) = 1 - 2x$. Determine whether the function $f \circ g$ is linear.
10. Explain why there is no linear function with a graph that passes through all three of the points $(-3, 2)$, $(1, 1)$, and $(5, 2)$.
11. A factory owner buys a new machine for $20,000. After eight years, the machine has a salvage value of $1000. Find a formula for the value of the machine after t yr, where $0 \leq t \leq 8$.
12. A manufacturer buys a new machine costing $120,000. It is estimated that the machine has a useful lifetime of ten years and a salvage value of $4000 at that time.
    (a) Find a formula for the value of the machine after t yr, where $0 \leq t \leq 10$.
    (b) Find the value of the machine after eight years.
13. A factory owner installs a new machine costing $60,000. Its expected lifetime is five years, and at the end of that time the machine has no salvage value.
    (a) Find a formula for the value of the machine after t yr, where $0 \leq t \leq 5$.
    (b) Complete the following depreciation schedule.

End of year	Yearly depreciation	Accumulated depreciation	Value V
0	0	0	60,000
1			
2			
3			
4			
5		60,000	0

14. Let x denote a temperature on the Celsius scale, and let y denote the corresponding temperature on the Fahrenheit scale.
    (a) Find a linear function relating x and y; use the facts that $32°F$ corresponds to $0°C$ and $212°F$ corresponds to $100°C$. Write the function in the form $y = Ax + B$.

    (b) What Celsius temperature corresponds to $98.6°F$?
    (c) Find a number z for which $z°F = z°C$.
15. Suppose that the cost $C(x)$, in dollars, of producing x electric fans is given by $C(x) = 450 + 8x$.
    (a) Find the cost to produce 10 fans.
    (b) Find the cost to produce 11 fans.
    (c) Use your answers in parts (a) and (b) to find the marginal cost. (Then check that your answer is the slope of the line.)
16. Suppose that the cost to a manufacturer of producing x units of a certain motorcycle is given by $C(x) = 220x + 4000$, where $C(x)$ is in dollars.
    (a) Find the marginal cost.
    (b) Find the cost of producing 500 motorcycles.
    (c) Use your answers in parts (a) and (b) to find the cost of producing 501 motorcycles.
17. Suppose that the cost $C(x)$, in dollars, of producing x compact discs is given by $C(x) = 0.5x + 500$.
    (a) Graph the given equation.
    (b) Compute $C(150)$.
    (c) If you add the marginal cost to the answer in part (b), you obtain a certain dollar amount. What does this amount represent?
18. The following graphs each relate distance and time for a moving object. Determine the velocity in each case.

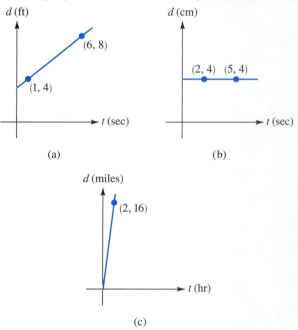

19. Two points A and B move along the x-axis. After t sec, their positions are given by the equations

$$A: \quad x = 3t + 100$$
$$B: \quad x = 20t - 36$$

(a) Which point is traveling faster, A or B?

(b) Which point is farther to the right when $t = 0$?

(c) At what time t do A and B have the same x-coordinate?

20. A point moves along the x-axis, and its x-coordinate after t sec is $x = 4t + 10$. (Assume that x is in centimeters.)

(a) What is the velocity?

(b) What is the x-coordinate when $t = 2$ sec?

(c) Use your answers in parts (a) and (b) to find the x-coordinate when $t = 3$ sec.   *Hint*: What are the units of the velocity in part (a)? Check your answer by letting $t = 3$ in the given equation.

21. The following table gives the population of California in 1995 and in 1997.

x (Year)	y (Population)
1995	31,493,525
1997	32,217,708

(a) Find the equation of the linear function whose graph passes through the two (x, y) points given in the table.

(b) Use the linear function determined in part (a) to make a projection for the population of California in 2000. (Round your answer to the nearest 1000.)

(c) The actual California population for 2000 was 33,871,648. Does the linear function yield a projection that is too high or too low? Compute the percentage error.

22. The following table gives the population of Florida in 1985 and 1990.

x (Year)	y (Population)
1985	11,351,118
1990	13,018,365

Follow the steps in Exercise 21 to make a projection for the population of Florida in 2000. (Use 1985 and 1990 in place of 1995 and 1997, respectively.) For part (c), the actual population of Florida in 2000 was 15,982,378.

23. The following table indicates total motion picture receipts (including video tape rentals) in the United States for the years 1994 and 1995.

Motion Picture Receipts

x (Year)	y (Receipts) (in millions of dollars)
1994	53,504
1995	57,184

Source: U.S. Census Bureau, *Statistical Abstract of the United States: 1999*

(a) Find the equation of the linear function whose graph passes through the two (x, y) points given in the table.

Ⓖ **(b)** Use a graphing utility to graph the line in part (a). Then use a TRACE or ZOOM feature to estimate what

motion picture receipts might have been for the year 1997.

(c) Compute the percentage error in the estimate in part (b), given that the actual figure for 1997 was $63,010 millions.

24. During the 1990s, the percentage of TV households viewing cable and satellite TV programs increased while the percentage viewing network affiliate shows (ABC, CBS, NBC, and FOX) generally decreased. The following table shows the primetime ratings for the network affiliates in the years 1993 and 1995. (In the table, the *rating percentage* is defined as the percentage of TV households viewing a TV program in an average minute.)

Primetime Ratings for Network Affiliates

x (Year)	y (Rating percentage)
1993	40.9
1995	37.3

Source: Nielsen Media Research, *1998 Report on Television*

(a) Find the equation of the linear function whose graph passes through the two (x, y) points given in the table.

Ⓖ **(b)** Using a graphing utility to graph the line in part (a). Then use a TRACE or ZOOM feature to estimate what the rating percentage might have been for the year 1997.

(c) Compute the percentage error in the estimate in part (b), given that the actual figure for 1997, according to Nielsen Media Research, was 33.1.

In general, the growth of plants or animals does not follow a linear pattern. For relatively short intervals of time, however, a linear function may provide a reasonable description of the growth. Exercises 25 and 26 provide examples of this.

25. (a) In an experiment with sunflower plants, H. S. Reed and R. H. Holland measured the height of the plants every seven days for several months. [The experiment is reported in *Proceedings of the National Academy of Sciences,* vol. 5 (1919), p. 140.] The following data are from this experiment.

x (Number of days)	21	49
y [Average height (cm) of plants after x days]	67.76	205.50

Find the linear function whose graph passes through the two points given in the table. (Round each number in the answer to two decimal places.)

(b) Use the linear function determined in part (a) to estimate the average height of the plants after 28 days. (Round your answer to two decimal places.)

(continues)

(c) In the experiment, Reed and Holland found that the average height after 28 days was 98.10 cm. Is your estimate in part (b) too high or too low? Compute the percentage error in your estimate.

(d) Follow parts (b) and (c) using $x = 14$ days. For the computation of percentage error, you need to know that Reed and Holland found that the average height after 14 days was 36.36 cm.

(e) As you've seen in parts (c) and (d), your estimates are quite close to the actual values obtained in the experiment. This indicates that for a relatively short interval, the growth function is nearly linear. Now repeat parts (b) and (c) using $x = 84$ days, which is a longer interval of time. You'll find that the linear function does a poor job in describing the growth of the sunflower plants. To compute the percentage error, you need to know that Reed and Holland determined the average height after 84 days to be 254.50 cm.

26. (a) The biologist R. Pearl measured the population of a colony of fruit flies (*Drosophila melanogaster*) over a period of 39 days. [The experiment is discussed in Pearl's book, *The Biology of Population Growth* (New York: Alfred Knopf, 1925).] Two of the measurements made by Pearl are given in the following table.

x (Number of days)	12	18
y (Population)	105	225

Find the linear function whose graph passes through the two points given in the table.

(b) Use the linear function determined in part (a) to estimate the population after 15 days. Then compute the percentage error in your estimate, given that the actual population after 15 days, as found by Pearl, was 152.

(c) Use the linear function determined in part (a) to estimate the population after 9 days. Then compute the percentage error in your estimate, given that after 9 days, Pearl found the actual population to be 39.

(d) As you've seen in parts (b) and (c), your estimates are quite close to the actual values obtained in the experiment. This indicates that for relatively short intervals of time, the growth is nearly linear. Now repeat part (b) using $x = 39$ days, which covers a longer interval of time. You'll find that the linear function does a poor job in describing the population growth. For the computation of percentage error, you need to know that Pearl determined the population after 39 days to be 938.

27. (a) On graph paper, plot the following points: $(1, 0)$, $(2, 3)$, $(3, 6)$, $(4, 7)$.

(b) In your scatter diagram from part (a), sketch a line that best seems to fit the data. Estimate the slope and the y-intercept of the line.

(c) The actual regression line in this case is $y = 2.4x - 2$. Add the graph of this line to your sketch from parts (a) and (b).

28. (a) On graph paper, plot the following points: $(1, 2)$, $(3, 2)$, $(5, 4)$, $(8, 5)$, $(9, 6)$.

(b) In your scatter diagram from part (a), sketch a line that best seems to fit the data.

(c) Using your sketch from part (b), estimate the y-intercept and the slope of the regression line.

(d) The actual regression line is $y = 0.518x + 1.107$. Check to see whether your estimates in part (c) are consistent with the actual y-intercept and slope. Graph this line along with the points given in part (a). (In sketching the line, use the approximation $y = 0.5x + 1.1$.)

29. The table in the following figure shows the population of Los Angeles over the years 1930–1990. The accompanying graph shows the corresponding scatter plot and regression line. The equation of the regression line is

$$f(x) = 37,546.068x - 71,238,863.429$$

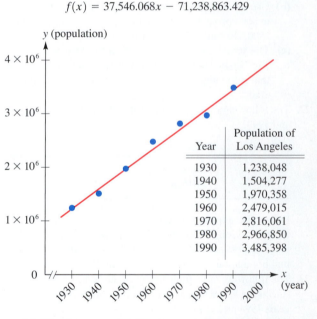

Year	Population of Los Angeles
1930	1,238,048
1940	1,504,277
1950	1,970,358
1960	2,479,015
1970	2,816,061
1980	2,966,850
1990	3,485,398

(a) Use the regression line to compute an estimate for what the population of Los Angeles might have been in 2000. (Round the answer to the nearest thousand.) Then compute the percentage error in the estimate, given that the actual figure for 2000 is 3.823 million.

(b) Find $f^{-1}(x)$.

(c) Use your answer in part (b) to estimate the year in which the population of Los Angeles might reach 4 million.   *Hint*: For the function f, the inputs are years and the outputs are populations; for f^{-1}, the inputs are populations and the outputs are years.

30. Perhaps foreshadowing the end of the Cold War in the early 1990s, the number of nuclear warheads worldwide began to decrease after the year 1986 (when the number was at an all-time high). The table in the following figure 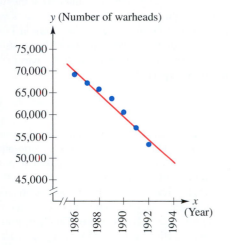 indicates the global number of nuclear warheads over the years 1986–1992. The graph shows the corresponding scatter plot and regression line. The equation for the regression line is

$$f(x) = -2810.96x + 5,653,063.25$$

Year	Number of nuclear warheads
1986	69,075
1987	67,302
1988	65,932
1989	63,645
1990	60,642
1991	57,017
1992	53,136

Source of data: R. S. Norris and W. M. Arkin, *Bulletin of Atomic Scientists,* Nov./Dec. 1997

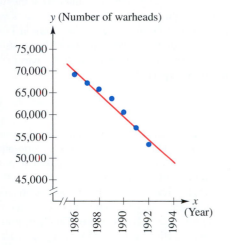

(a) Use the regression line to make a projection for the number of nuclear warheads in 1997. Then compute the percentage error, given that the actual number was 36,110.
(b) When might there be only (!) 10,000 nuclear warheads worldwide? Round the answer to the nearest integer. Also, if the trend were to continue, when would there be no warheads remaining?
31. The following table shows world grain production for selected years over the period 1950–1990.

Year	World grain production (million tons)
1950	631
1960	824
1970	1079
1980	1430
1990	1769

Source: Lester R. Brown et al. in *Vital Signs 1999* (New York: W. W. Norton & Co, 1999)

(a) Use a graphing utility or spreadsheet to determine the equation of the regression line. For the x-y data pairs, use x for the year and y for the grain production. Create a graph showing both the scatter plot and regression line.
(b) Use the equation of the regression line to make a projection for the world grain production in 1993. Then compute the percentage error in your projection, given that the actual grain production in 1993 was 1714 million tons. *Remark*: Your projection will turn out to be higher than the actual figure. One (among many) reasons for this: In 1993 there was a drop in world grain production due largely to the effects of poor weather on the U.S. corn crop.
(c) Use the equation of the regression line to make a projection for the world grain production in 1998. Then compute the percentage error in your projection, given that the actual world grain production in 1998 was 1844 million tons. *Remark*: Again, your projection will turn out to be too optimistic. According to Lester Brown's *Vital Signs: 1999,* world grain production dropped in 1998 "due largely to severe drought and heat in Russia on top of an overall deterioration of that country's economy."
32. The following table shows global natural gas production for the years 1990–1996. The abbreviation tcf stands for *trillion cubic feet.*

Year	Natural gas production (tcf)
1990	71.905
1991	73.037
1992	73.219
1993	74.570
1994	75.190
1995	76.614
1996	80.045

Source: *BP Amoco Statistical Review of World Energy* (49th ed.) http://www.bp.com/worldenergy

(a) Use a graphing utility or spreadsheet to determine the equation of the regression line. For the x-y data pairs, use x for the year and y for the natural gas production. Create a graph showing both the scatter plot and regression line.
(b) What are the units associated with the slope of the regression line in part (a)?
(c) In your graph of the regression line, use a TRACE or ZOOM feature to make an estimate for natural gas production in the year 1999.
(d) The actual production figure for 1999 is 83.549 tcf. Is your estimate high or low? Compute the percentage error.
(e) The table above shows that global production of natural gas increased over the period 1990–1996, and in fact it still continues to increase each year. However,

for purposes of making a very conservative estimate, let's assume for the moment that natural gas production levels off at its 1999 value of 83.549 tcf per year. According to the *BP Amoco Statistical Review of World Energy,* as of the end of 1999, proved world reserves of natural gas were 5171.8 tcf. Carry out the following calculation and interpret your answer. *Hint*: Keep track of the units.

$$\frac{5171.8 \text{ tcf}}{83.549 \text{ tcf/yr}} = \cdots$$

33. This exercise illustrates one of the pitfalls that can arise in using a regression line: large discrepancies can occur when the regression line is used to make long-term projections. The following table shows the population of California during the years 1860–1900. The accompanying graph displays the scatter plot of the data; the equation of the regression line is

$$f(x) = 28{,}632.69x - 52{,}928{,}780$$

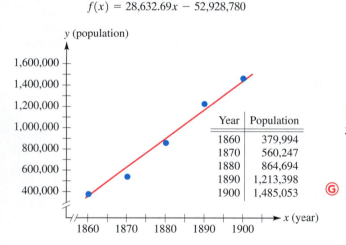

Year	Population
1860	379,994
1870	560,247
1880	864,694
1890	1,213,398
1900	1,485,053

As you can see from the scatter plot, the population growth over this period is very nearly linear.
(a) Use the regression line to estimate the population of California in 1990.
(b) According to the U.S. Bureau of the Census, the population of California in 1990 was 29,839,250. Is your estimate in part (a) close to this figure?

34. (a) The June 1976 issue of *Scientific American* magazine contained an article entitled "Future Performance in Footracing" by H. W. Ryder, H. J. Carr, and P. Herget. According to the authors of this article, "It appears likely that within 50 years the record [for the mile] will be down to 3:30." Use the regression line defined in equation (2) on page 256 to make a projection for the mile time in the year 2026 (which will be 50 years after the article appeared). Is your projection close to the one given in the *Scientific American* article?
(b) The regression line defined in equation (2) is based only on the data from Table 1. If we use all the data

from both Table 1 and Table 2, then the least-squares technique can be used to derive the following regression line:

$$f(x) = -0.400x + 1019.472$$

Use this equation to make a projection for the mile time in the year 2026. Which projection is closer to the value predicted in the *Scientific American* article, this one or the one calculated in part (a)?

35. The data in Table 2 of Section 4.1 covers the years 1957–1993. The equation of the regression line for the data is

$$y = -0.318x + 858.955$$

(a) Use this equation to estimate what the mile record might have been for the year 1954. Then check Table 1 to see the actual record for 1954, which was set by John Landy. Is your estimate too high or too low? Compute the percentage error in your estimate.
(b) Use the given equation to estimate what the mile record might have been back in 1911. Check your estimate against the actual record as shown in Table 1, and compute the percentage error. How does the percentage error here compare to that in part (a)? Why is this to be expected?

B

36. Linear functions can be used to approximate more complicated functions. This is one of the meanings or implications of the quotation by Professor Gleason on page 251. This exercise illustrates that idea.
(a) Using calculus, it can be shown that the equation of the line that is tangent to the curve $y = x^2$ at the point $(1, 1)$ is $y = 2x - 1$. Verify this visually by graphing the two functions $y = x^2$ and $y = 2x - 1$ on the same set of axes. (*Suggestion*: Use a viewing rectangle that extends from -2 to 3 in the x-direction and from -3 to 4 in the y-direction.) Note that the tangent line is virtually indistinguishable from the curve in the immediate vicinity of the point $(1, 1)$.
(b) For numerical rather than visual evidence of how well the linear function $y = 2x - 1$ approximates the function $y = x^2$ in the immediate vicinity of $(1, 1)$, complete the following tables.

x	0.9	0.99	0.999
x^2			
$2x - 1$			

x	1.1	1.01	1.001
x^2			
$2x - 1$			

In Exercises 37–40, let f and g be the linear functions defined by

$$f(x) = Ax + B \quad (A \neq 0) \qquad and \qquad g(x) = Cx + D \quad (C \neq 0)$$

In each case, compute the average rate of change of the given function on the interval [a, b].

37. (a) f
    (b) $f \circ f$
    (c) $g \circ f$
    (d) $f \circ g$

38. (a) f^{-1}
    (b) g^{-1}

39. (a) $(f \circ g)^{-1}$
    (b) $(g \circ f)^{-1}$

40. (a) $f^{-1} \circ g^{-1}$
                              (b) $g^{-1} \circ f^{-1}$

41. Show that the linear function $f(x) = mx$ satisfies the following identities:
    (a) $f(a + b) = f(a) + f(b)$;   **(b)** $f(ax) = af(x)$.

42. Let f be a linear function such that

$$f(a + b) = f(a) + f(b)$$

for all real numbers a and b. Show that the graph of f passes through the origin. *Hint:* Let $f(x) = Ax + B$ and show that $B = 0$.

43. Find a linear function $f(x) = mx + b$ such that m is positive and $(f \circ f)(x) = 9x + 4$.

44. (a) Let f be a linear function. Show that

$$f\left(\frac{x_1 + x_2}{2}\right) = \frac{f(x_1) + f(x_2)}{2}$$

    (In words: The output of the average is the average of the outputs.)

    (b) Show (by using an example) that the equation in part (a) does not hold for the function $f(x) = x^2$.

45. This exercise shows how to compute the slope and the y-intercept of the regression line. As an example, we'll work with the simple data set given in Exercise 27.

x	1	2	3	4
y	0	3	6	7

    (a) Let Σx denote the sum of the x-coordinates in the data set, and let Σy denote the sum of the y-coordinates. Check that $\Sigma x = 10$ and $\Sigma y = 16$.

    (b) Let Σx^2 denote the sum of the squares of the x-coordinates, and let Σxy denote the sum of the products of the corresponding x- and y-coordinates. Check that $\Sigma x^2 = 30$ and $\Sigma xy = 52$.

    (c) The slope m and the y-intercept b of the regression line satisfy the following pair of simultaneous equations [in the first equation, n denotes the number of points (x, y) in the data set]:

$$\begin{cases} nb + (\Sigma x)m = \Sigma y \\ (\Sigma x)b + (\Sigma x^2)m = \Sigma xy \end{cases}$$

    In the present example, these equations become

$$\begin{cases} 4b + 10m = 16 \\ 10b + 30m = 52 \end{cases}$$

Solve this pair of equations for m and b, and check that your answers agree with the values in Exercise 27(c).

In Exercises 46–49:

(a) *Use the method described in Exercise 45 to find the equation of the regression line for the given data set.*

(b) *For Exercises 48 and 49, if your graphing utility has a feature for computing regression lines, use it to check your answer in part (a).*

46.
x	2	4	8	10
y	−7	−5	−2	−1

47.
x	1	2	3	4	5
y	2	3	9	9	11

48.
x	1	2	3	4	5
y	16	13.1	10.5	7.5	2

49.
x	520	740	560	610	650
y	81	98	83	88	95

C

50. Suppose that f is a linear function satisfying the condition $f(kx) = kf(x)$ for all real numbers k. Prove that the graph of f passes through the origin.

51. (a) Find all linear functions f satisfying the identity $f(f(x)) = 2x + 1$. (For your answer, rationalize any denominators containing radicals.)

    (b) Find all linear functions f satisfying the identity

$$f(f(f(x))) = 2x + 1$$

    (Rationalize any denominators containing radicals.)

52. Let a and b be real numbers, and suppose that the inequality $ax + b \leq 0$ has no solutions. What can you say about the linear function $f(x) = ax + b$? Answer in complete sentences and justify what you say.

53. Are there any linear functions f satisfying the identity $f^{-1}(x) = f(f(x))$? If so, list them; if not, explain why not.

54. Let $f(x) = ax + b$, where a and b are positive numbers, and assume that the following equation holds for all values of x: $f(f(x)) = bx + a$. Show that $a + b = 1$.

MINI PROJECT | WHO ARE BETTER RUNNERS, MEN OR WOMEN?

An oversimplified answer here would be to say that men are better because in any given event, the men's world record time is less than the women's. For instance, as of the year 2000, the men's and women's world records for the 100-m dash were 9.79 sec and 10.49 sec, respectively. This reasoning, however, ignores the fact that, historically, women have been training and competing on the world stage for far less time than men. Likewise, it's only relatively recently that women have been supported or sponsored to a degree approaching that for men. Indeed, currently, women's records are falling at a faster rate than men's. Complete the following research and calculations, then discuss this issue.

The following table gives the names of athletes who set world records in the 100-m dash over the years 1968–1999. (The table is restricted to times that were recorded electronically, rather than manually; 1968 was the year in which electric timing made its major debut (in the 1968 Olympics in Mexico City).

Selected record-setting performances in 100-m dash, 1968–1999.*

Year	Men	Women
1968	James Hines, USA (10/14/68)	Wyomia Tyus, USA (10/15/68)
1972		Renate Stecher, DDR (East Germany) (9/2/72)
1976		Annegret Richter, FRG (West Germany) (7/25/76)
1977		Marlies Oelsner (Göhr) DDR (7/1/77)
1983	Calvin Smith, USA (7/3/83)	Evelyn Ashford, USA (7/3/83)
1984		Evelyn Ashford, USA (8/22/84)
1988	Carl Lewis, USA (9/24/88)	Florence Griffith-Joyner, USA (7/17/88)
1991	Carl Lewis, USA (8/25/91)	
1994	Leroy Burrell, USA (7/6/94)	
1996	Donovan Bailey, CAN (7/29/96)	
1999	Maurice Greene, USA (6/16/99)	

*In cases where the record was broken more than once in a given year, only the fastest performance is indicated.

(a) Using the library or the Internet, look up the 100-m times for each athlete in the table.

(b) For the men's records: Let x represent the year, with $x = 0$ corresponding to 1960, and let y represent the record time in that year. Find the regression line $y = f(x)$.

(c) For the women's records: Again, let x represent the year, with $x = 0$ corresponding to 1960, and let y represent the record time in that year. Find the regression line $y = g(x)$.

(d) Solve the equation $f(x) = g(x)$ to compute a projection for the year in which the men's and women's records might be equal. What would that common record be?

4.2 QUADRATIC FUNCTIONS

In the text and exercises for the previous section, we saw examples in which data sets and their scatter plots were modeled using a linear function. Clearly, however, linear functions with their straight line graphs cannot be appropriate models for every data set. Consider, for instance, the scatter plot in Figure 1, which shows the daily trading volumes for Amazon.com stock in a particular month. It appears that no line could adequately summarize the situation. Next, in Figure 2, look at the data and scatter plot relating to the spread of the disease AIDS. Here it appears that some sort of curve, rather than straight line, would be better for summarizing the trend in the data. As you'll see in a moment, one type of curve that does fit the data in Figure 2 quite well is obtained by graphing a *quadratic function*.

x Years after 1980	4	5	6	7	8	9	10
y AIDS cases (millions)	0.2	0.4	0.7	1.1	1.6	2.3	3.2

Figure 1
Daily number of shares of Amazon.com stock traded (i.e., sold) in March 2000. $x = 1$ corresponds to March 1.

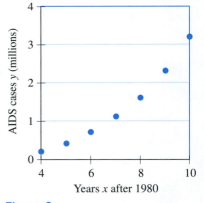

Figure 2
Estimated cumulative number of AIDS cases worldwide, 1984–1990.

Source: Joint United Nations Programme on HIV/AIDS

After the linear functions, the next simplest are the **quadratic functions,** which are defined by equations of the form

$$f(x) = ax^2 + bx + c \qquad (a \neq 0)$$

where a, b, and c are constants and a is not zero. (The word "quadratic" is derived from the Latin word for a square, *quadratus*.) We will see that the graph of any quadratic function is a curve called a **parabola** that is similar or identical in shape to the basic $y = x^2$ graph. Figure 3 displays the graphs of two typical quadratic functions. Subsequent examples will demonstrate that the parabola opens upward when $a > 0$ and downward when $a < 0$. As Figure 3 indicates, the turning point on the parabola is called the **vertex.** The **axis of symmetry** of the parabola $y = ax^2 + bx + c$ is the vertical line passing through the vertex. (Mini Project 1 at the end of this section indicates how to *prove* that the graph does indeed have this symmetry.)

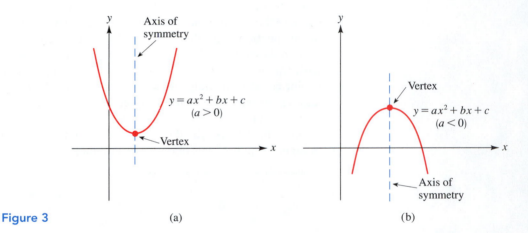

Figure 3   (a)   (b)

We mentioned in Section 4.1 that many graphing utilities have features for determining the regression line for a given set of data points. This is the line that best fits the data points. (We are using the phrase "best fits" in an intuitive sense here; the technical meaning is explained in a mini project at the end of Section 4.4.) In addition to fitting a linear function to a data set, many of these graphing utilities can fit other types of functions as well, including quadratics. In Example 1, we find and compare linear and quadratic models for the AIDS data in Figure 2. For details on using a graphing calculator to determine a quadratic function that best fits a data set, see the owner's manual that came with your graphing utility. [On a TI graphing calculator, go to the "STAT menu," choose "CALC," and then select "QuadReg" (which stands for *quadratic regression*). In Microsoft *Excel*™, go to the "Chart" menu, select "Add Trendline," then choose "Polynomial," and type in "2" for the order.]

EXAMPLE	**1**	**Using linear and quadratic functions to model AIDS data**

(a) Use a graphing utility to find and graph the linear and quadratic models for the AIDS data in Figure 2 on page 267.

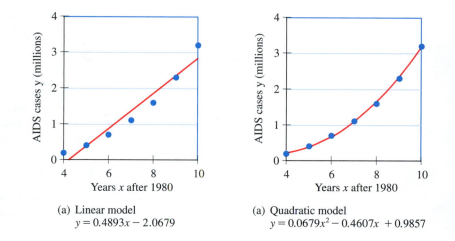

Figure 4
Linear and quadratic models for AIDS data from Figure 2

(a) Linear model
$$y = 0.4893x - 2.0679$$

(a) Quadratic model
$$y = 0.0679x^2 - 0.4607x + 0.9857$$

(b) Use each function to make projections for the years 1992 and 1997. Then use the following information to see which model is more accurate in each case. The cumulative numbers of AIDS cases in 1992 and 1997 were 5.5 million and 15.1 million, respectively.

SOLUTION

(a) See Figure 4.

(b) The results of the calculations (rounded to one decimal place) are shown in Table 1. As perspective in evaluating these models, keep in mind that they are obtained using the data in Figure 2, which covers the years 1984–1990. So we are using the models to project two years and seven years beyond the data—that is, beyond 1990. For 1992, both projections fall short of the actual figure, but the quadratic model is much closer. As you should compute for yourself, the percentage errors for the linear and quadratic projections for 1992 are approximately 31% and 5%, respectively. For 1997, both projections again fall short of the actual figure. The linear model is even farther off this time; it predicts less than half of the actual figure. The quadratic model is again closer to reality, but as you can check, the percentage error in this case is a relatively large 15%. In summary: The quadratic model works quite well in the short run (two years), but it misses the mark considerably for the longer term (seven years). The linear model misses badly in both cases. For a health professional or a United Nations administrator writing a budget proposal to deal with the AIDS epidemic in 1990, evidently even the quadratic model doesn't grow fast enough and so might mislead the

TABLE 1 Comparison of Linear and Quadratic Models for Cumulative Number of AIDS Cases Worldwide in 1992 and 1997

Year	Linear model $y = 0.4893x - 2.0679$ (millions)	Quadratic model $y = 0.0679x^2 - 0.4607x + 0.9857$ (millions)	Actual figures (millions)
1992 ($x = 12$)	3.8	5.2	5.5
1997 ($x = 17$)	6.3	12.8	15.1

person into underfunding the project. (In later work we look at types of functions that grow faster than any quadratic. See, for example, Exercise 64 in Section 4.6, or see Section 5.1.)

In the next set of examples, the main focus will be on graphing quadratic functions by hand, rather than with a graphing utility. The goal is to *understand* why the graphs look as they do and to see how they relate to the basic $y = x^2$. The techniques that we'll use in analyzing quadratic functions have already been developed in previous chapters. In particular, the following two topics are prerequisites for understanding the examples in this section:

*   Completing the square—for a review see either Section 1.7 or 2.1
*   Translations and reflections—for a review see Section 3.4

EXAMPLE **2** **Graphing a quadratic function using completing the square and translation**

Graph the function $y = x^2 - 2x + 3$.

SOLUTION
The idea here is to use the technique of completing the square; this will enable us to obtain the required graph simply by shifting the basic $y = x^2$ graph. We begin by writing the given equation

$$y = x^2 - 2x \qquad +3$$

To complete the square for the x-terms we want to add 1. (Check this.) Of course, to keep the equation in balance we have to account for this by writing

$$y = (x^2 - 2x + 1) + 3 - 1 \qquad \text{adding zero to the right side}$$

or

$$y = (x - 1)^2 + 2$$

Now, as we know from Section 3.4, the graph of this last equation is obtained by moving the parabola $y = x^2$ one unit in the positive x-direction and two units in the positive y-direction. This shifts the vertex from the origin to the point $(1, 2)$. See Figure 5.

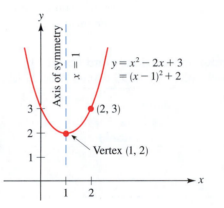

Figure 5

NOTE   As a guide to sketching the graph, you'll want to know the y-intercept. To find the y-intercept, substitute $x = 0$ in the given equation to obtain $y = 3$. Then, given the vertex $(1, 2)$ and the point $(0, 3)$, a reasonably accurate graph can be quickly sketched. [Actually, once you find that $(0, 3)$ is on the graph, you also know that the reflection of this point about the axis of symmetry is on the graph. This is why the point $(2, 3)$ is shown in Figure 5.]

Now we want to compare the graphs of $y = x^2$, $y = 2x^2$, and $y = \frac{1}{2}x^2$. The last two functions were not specifically discussed in the previous chapters, so for now you can graph them by first setting up tables. If you do this, rather than using a graphing utility, you'll gain some insight into why the graphs in Figure 6 look as they do. All three graphs are parabolas that open upward, but $y = 2x^2$ appears to be narrower than $y = x^2$, while $y = \frac{1}{2}x^2$ appears wider than $y = x^2$. Another way (besides setting up tables) to see why the graphs appear this way is to start from the inequalities $\frac{1}{2} < 1 < 2$. Then, assuming for the moment that $x \neq 0$, we can multiply through by the *positive* quantity x^2 to obtain

$$\frac{1}{2}x^2 < x^2 < 2x^2 \qquad \text{(provided } x \neq 0\text{)}$$

This tells us that, with the exception of the origin, the graph of $y = \frac{1}{2}x^2$ is *below* that of $y = x^2$, which in turn is *below* $y = 2x^2$. This is why, on any given interval (for example $-2 \le x \le 2$, as in Figure 6), the graph of $y = \frac{1}{2}x^2$ appears wider than $y = x^2$, which in turn appears wider than $y = 2x^2$.

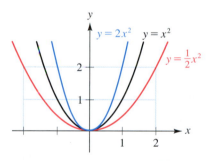

Figure 6

The observations that we've just made about the relative shapes of three parabolas also apply to $y = -x^2$, $y = -2x^2$, and $y = -\frac{1}{2}x^2$, except in these cases the parabolas open downward rather than upward. In the box that follows we summarize and generalize our observations up to this point.

PROPERTY SUMMARY   The Graph of $y = ax^2$

1. The graph of $y = ax^2$ is a parabola with vertex at the origin. It is similar in shape to $y = x^2$.
2. The parabola $y = ax^2$ opens upward if $a > 0$, downward if $a < 0$.
3. The parabola $y = ax^2$ is narrower than $y = x^2$ if $|a| > 1$, wider than $y = x^2$ if $|a| < 1$.

EXAMPLE	3	Graphing parabolas that appear narrower than the basic $y = x^2$

Sketch the graphs of the following quadratic functions:

(a) $y = 3(x - 1)^2$; **(b)** $y = -3(x - 1)^2$.

SOLUTION

(a) Because of the $x - 1$, we shift the basic parabola $y = x^2$ one unit to the right. The factor of 3 in the given equation tells us that we want to draw a parabola that is narrower than $y = x^2$ but that has the same vertex, $(1, 0)$. To see exactly how narrow to draw $y = 3(x - 1)^2$, we need to know another point on the graph other than the vertex, $(1, 0)$. An easy point to obtain is the y-intercept. Setting $x = 0$ in the equation yields $y = 3(0 - 1)^2 = 3$. Now that we know the vertex, $(1, 0)$, and the y-intercept, 3, we can sketch a reasonably accurate graph; see Figure 7(a).

(b) In part (a) we sketched the graph of $y = 3(x - 1)^2$. By reflecting that graph in the x-axis, we obtain the graph of $y = -3(x - 1)^2$, as shown in Figure 7(b).

Figure 7

(a) (b)

 | EXAMPLE | 4 | Analyzing a quadratic function |

Graph the function $f(x) = -2x^2 + 12x - 16$ and specify the vertex, axis of symmetry, maximum or minimum value of f, and x- and y-intercepts.

SOLUTION

The idea is to complete the square, as in Example 2. We have

$$
\begin{aligned}
y &= -2x^2 + 12x \quad\quad - 16 \\
&= -2(x^2 - 6x \quad\quad) - 16 \\
&= -2(x^2 - 6x + 9) - 16 + 18 \quad\quad \text{adding } (2)(9) = 18 \text{ to keep} \\
&\quad\quad\quad\quad\quad\quad\quad\quad\quad\quad\quad\quad\quad\quad\quad \text{the equation in balance} \\
&= -2(x - 3)^2 + 2 \quad\quad\quad\quad\quad\quad\quad\quad\quad\quad\quad\quad\quad\quad (1)
\end{aligned}
$$

From equation (1), we see that the required graph is obtained simply by shifting the graph of $y = -2x^2$ "right 3, up 2," so that the vertex is $(3, 2)$. As a guide to sketching the graph, we want to compute the intercepts. The y-intercept is -16. (Why?) For the x-intercepts, we replace y with 0 in equation (1) to obtain

$$-2(x - 3)^2 = -2$$
$$(x - 3)^2 = 1$$
$$x - 3 = \pm 1$$
$$x = 4 \quad \text{or} \quad 2$$

Thus the x-intercepts are $x = 4$ and $x = 2$. Knowing these intercepts and the vertex, we can sketch the graph as in Figure 8. You should check for yourself that the information accompanying Figure 8 is correct.

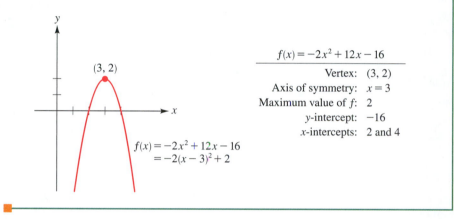

Figure 8

The next three examples involve maximum and minimum values of functions. As background for this, we first summarize our basic technique for graphing parabolas.

PROPERTY SUMMARY
The Graph of the Parabola $y = ax^2 + bx + c$

By completing the square, the equation of the parabola $y = ax^2 + bx + c$ can always be rewritten in the form

$$y = a(x - h)^2 + k$$

In this form, the vertex of the parabola is (h, k) and the axis of symmetry is the line $x = h$. The parabola opens upward if $a > 0$ and downward if $a < 0$.

EXAMPLE	5	Determining the input at which a quadratic function has a minimum value

The graph of the quadratic function $g(x) = x^2 - \frac{6}{7}x + 2$ is a parabola that opens upward because the coefficient of x^2 is positive. Thus, it makes sense to talk about a minimum rather than maximum value for this function. Let x_0 denote the input for which $g(x_0)$ is minimum.

(a) Use a TRACE and/or ZOOM feature on a graphing utility to estimate x_0 to the nearest one-tenth. What is the corresponding estimate for $g(x_0)$?

(b) Use algebra to determine the exact value of x_0 and the minimum value of the function. Check that the results are consistent with the visual evidence in part (a).

SOLUTION

(a) From the preliminary view obtained in Figure 9(a), we can see that the input x_0 at which the function has a minimum is roughly 0.5, a bit less actually. The zoom-in view in Figure 9(b) reveals that x_0 is closer to 0.4 than to 0.5. So we have $x_0 \approx 0.4$, to the nearest tenth. Figure 9(b) shows that the output $g(x_0)$ is between 1.81 and 1.82, so certainly to the nearest tenth we can say that $g(x_0) \approx 1.8$. *Suggestion*: Before working this example with your own graphing utility, reread the comments on page 188 regarding zooming in on turning points.

GRAPHICAL PERSPECTIVE

Figure 9

Zooming in on the turning point of $g(x) = x^2 - \frac{6}{7}x + 2$

(a) $[-1, 2, 0.5]$ by $[1, 4, 0.5]$ (b) $[0.30, 0.60, 0.05]$ by $[1.81, 1.84, 0.01]$

(b) By completing the square as in Example 4, we find that the given equation can be rewritten

$$g(x) = \left(x - \frac{3}{7} \right)^2 + \frac{89}{49}$$

(You should verify this for yourself, using Example 4 as a model.) This last equation tells us that the vertex of the parabola is the point $(\frac{3}{7}, \frac{89}{49})$. Thus, the minimum value of the function is $\frac{89}{49}$ and this minimum occurs at $x_0 = \frac{3}{7}$. If you pick up a calculator now, you'll find that $\frac{3}{7} \approx 0.429$ and $\frac{89}{49} \approx 1.816$. Both of these values are consistent with the graphical estimates in part (a).

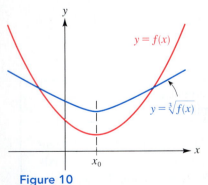

Figure 10

The minimum value occurs at x_0 for both of the functions $y = \sqrt[3]{f(x)}$ and $y = f(x)$.

For the previous two examples, keep in mind that we were able to find the maximum or minimum easily only because the functions were quadratics. In contrast, you cannot expect to find the minimum of $y = x^4 - 8x$ using the method of Examples 4 and 5 because it is not a quadratic function. In general, the techniques of calculus are required to find maxima and minima for functions other than quadratics. There are some cases, however, in which our present method can be adapted to functions that are closely related to quadratics. For instance, in the next example, we look for an input that minimizes a function of the form $y = \sqrt[3]{f(x)}$, where f is a quadratic function. As indicated in Figure 10, we need only find the input that minimizes the quadratic function $y = f(x)$, because this same input also minimizes $y = \sqrt[3]{f(x)}$.

(The outputs of the two functions are, of course, different, but the point here is that the same *input* does the job for both functions.)

| EXAMPLE | 6 | **Finding the minimum of a function that's related to a quadratic** |

Let $f(x) = \sqrt[3]{x^2 - x + 1}$. Which x-value yields the minimum value for the function f? What is this minimum value?

SOLUTION
According to the remarks just prior to this example, the x-value that minimizes the function $f(x) = \sqrt[3]{x^2 - x + 1}$ will be the same x-value that minimizes the function $y = x^2 - x + 1$. To find this x-value we complete the square, just as we've done previously in this section. We have

$$y = x^2 - x \qquad + 1$$
$$= x^2 - x + \frac{1}{4} + 1 - \frac{1}{4}$$
$$= \left(x - \frac{1}{2}\right)^2 + \frac{3}{4}$$

This shows that the vertex of the parabola $y = x^2 - x + 1$ is $(1/2, 3/4)$. Since this parabola opens upward, we conclude that the input $x = 1/2$ will produce the minimum value for $y = x^2 - x + 1$ and also for $f(x) = \sqrt[3]{x^2 - x + 1}$. For the minimum value of the function f, we compute

$$f(1/2) = \sqrt[3]{(1/2)^2 - (1/2) + 1}$$
$$= \sqrt[3]{3/4} \approx 0.91$$

In summary: The minimum value of the function f occurs when $x = 1/2$. This minimum value is $\sqrt[3]{3/4}$, which is approximately 0.91. See Figure 11 for a graphical perspective.

GRAPHICAL PERSPECTIVE

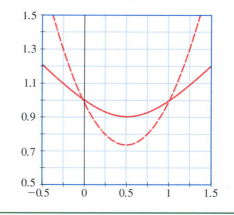

Figure 11
The dashed graph is $y = x^2 - x + 1$; the other is $f(x) = \sqrt[3]{x^2 - x + 1}$. The minimum value for both functions occurs when $x = 0.5$. The minimum value of f is $\sqrt[3]{3/4}$, which is approximately 0.9.

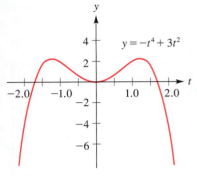

Figure 12

EXAMPLE 7 Finding inputs that maximize a function related to a quadratic

As indicated in Figure 12, the maximum value of the function $y = -t^4 + 3t^2$ occurs when t is between 1.0 and 1.5 and also when t is between -1.0 and -1.5. Determine these t-values exactly, then obtain calculator approximations rounded to three decimal places.

SOLUTION
The substitution $t^2 = x$ will reduce this question to one about a quadratic function. If $t^2 = x$, then $t^4 = x^2$ and we have

$$
\begin{aligned}
y &= -t^4 + 3t^2 \\
&= -x^2 + 3x \\
&= -\left(x - \frac{3}{2}\right)^2 + \frac{9}{4}
\end{aligned}
$$

completing the square, as we've done throughout this section (Check the algebra!)

The graph of this last equation is a parabola that opens downward, and the maximum occurs when x is $3/2$. So we have

$$ t^2 = x = 3/2 $$

and therefore

$$ t = \pm\sqrt{\frac{3}{2}} = \pm\frac{\sqrt{3}}{\sqrt{2}} = \pm\frac{\sqrt{6}}{2} \approx \pm 1.225 $$

Note that these calculator values are consistent with Figure 12.

In many real-life applications in science and business, data are collected that can be arranged in an x-y table of values. (We've seen examples in this and the previous section.) We conclude this section with a useful theorem that tells how to determine whether a given x-y table can be generated by a linear function or a quadratic function (or neither). As background for the theorem, we explain what is meant by *first differences* and *second differences*. Suppose we have a list of numbers—say, 2, 7, 8, 4, 15. Then the list of **first differences** is a new list formed by subtracting adjacent members of the given list as follows:

$$ 7 - 2(=5), \qquad 8 - 7(= 1), \qquad 4 - 8(= -4), \qquad 15 - 4 = 11 $$

So, for this example, the list of first differences is 5, 1, -4, 11. The process can be summarized conveniently as shown in Figure 13.

Once we have the list of first differences, we can follow the same procedure to form the list of **second differences** as shown in Figure 14. For the second differences, we are subtracting adjacent numbers in the list of first differences.

With the terminology we've now introduced, we are ready to state the theorem for determining whether or not a given table of x-y values can be generated by a lin-

Original list ⟶ 2   7   8   4   15
First differences ⟶ 5   1 -4 11

Figure 13

Original list ⟶ 2   7   8   4   15
First differences ⟶ 5   1 -4 11
Second differences ⟶ -4 -5 15

Figure 14

ear or quadratic function. In the statement of the theorem, reference is made to x values that are *equally spaced.* That means, for example, we could have $x = 1, 2, 3,$ and so on, or $x = 5, 10, 15,$ and so on, but not $x = 1, 2, 4, \ldots$.

THEOREM   How to Determine Whether a Data Set Can Be Generated by a Linear or Quadratic Function

Suppose we have a table of x-y values and the x-values are equally spaced. Then:

(a) The data can be generated by a linear function if and only if the first differences of the y-values are constant.
(b) The data can be generated by a quadratic function if and only if the second differences of the y-values are constant.

The proof for part (a) of this theorem is outlined in Exercises 63 and 64. We omit the proof of part (b) because of the lengthy algebra that is involved.

EXAMPLE   8

The x-y values in Figure 15 are generated by a linear function because the first differences of the y's are constant. The x-y values in Figure 16 are generated by neither a linear function nor a quadratic function because neither the first nor second differences are constant.

x	0	2	4	6	8
y	−1	5	11	17	23

First differences ⟶ 6   6   6   6

Figure 15

x	1	2	3	4	5
y	3	4	8	13	21

First differences ⟶ 1   4   5   8

Second differences ⟶ 3   1   3

Figure 16

EXERCISE SET 4.2

A

Exercises 1 and 2 require graphing utilities that can create scatter plots and compute linear and quadratic regression models.

1. The following table shows the number of air passenger miles flown worldwide for selected years from 1950 to 1985.

(a) Use a graphing utility to create a scatter plot for the data. Then determine the linear model and the quadratic model that fit the data as closely as possible. Add the graph of each model to your scatter plot.
(b) Use each model to make projections for the number of passenger miles for 1990 and for 1998.

(continues)

Year x ($x = 0$ is 1950)	0	5	10	15	20	25	30	35
Passenger miles y (billions)	16.94	36.90	65.94	119.78	278.57	421.83	658.88	827.19

Data computed from *Vital Signs 1999,* Lester R. Brown et al. (New York: W.W. Norton & Co., 1999)

(c) Use the following information to compute the percentage errors in the projections in part (b). For each case, which model produces the smaller percentage error? The number of passenger miles for 1990 and 1998 were 1145.94 billion and 1585.74 billion, respectively.

2. The following table shows the number of cellular phone subscribers worldwide over the years 1990–1995.

Year x (x = 0 is 1990)	0	1	2	3	4	5
Cell phone subscribers (millions)	11	16	23	34	55	91

Source: *Vital Signs, 2000,* Lester R. Brown et al. (New York: W.W. Norton & Co., 2000)

(a) Use a graphing utility to create a scatter plot for the data. Then determine the quadratic model that fit the data as closely as possible. Add the graph of the quadratic function to your scatter plot.

(b) Use the model to make projections for the number of cell phone subscribers for 1996 and for 1998. Which projection do you think might be more accurate? Now compute the percentage error in each projection using the following information. The number of cell phone subscribers in 1996 and 1998 were 142 million and 319 million, respectively.

For Exercises 3 and 4, refer to the following figures. Figure A shows a scatter plot and quadratic model for the U.S. population over the period 1790–1950. Figure B shows similar information for the world population over the period 1750–1950. For both figures, x represents the number of years after 1750. (Thus, x = 0 corresponds to the year 1750, and x = 200 corresponds to 1950.)

$$y = 0.00572x^2 - 0.44741x + 12.70012$$

Figure A
Scatter plot and quadratic model for U.S. population, 1790–1950

$$y = 0.000042x^2 - 0.000140x + 0.824000$$

Figure B
Scatter plot and quadratic model for world population, 1750–1950

3. Use the quadratic model in Figure A to complete the following table. Round each projection to two decimal places; round the percentage error to the nearest integer.

	Projected U.S. population (millions)	Actual U.S. population (millions)	Projection too high or too low?	% Error
1970		203.30		
2000		275.60		

4. Use the quadratic model in Figure B to complete the following table. Round each projection to two decimal places; round the percentage error to the nearest integer.

	Projected world population (billions)	Actual world population (billions)	Projection too high or too low?	% Error
1970		3.708		
2000		6.080		

Afterword on Exercises 3 and 4: For the time spans shown in Figures A and B, quadratic models describe both the U.S. and the world population quite well. That is, the graph of each quadratic function comes very close to the data points in the scatter plot. Exercise 3 shows that a quadratic model for the U.S. population does a reasonably good job of projecting the population figures 20 and even 50 years beyond the base data. Exercise 4, on the other hand, shows that the quadratic model is inappropriate for similar projections of the world population, because the quadratic model grows much too slowly. In Chapter 5 we study *exponential functions,* which can be much more appropriate for modeling very rapid growth.

In Exercises 5–20, graph the quadratic function. Specify the vertex, axis of symmetry, maximum or minimum value, and intercepts.

5. $y = (x + 2)^2$

6. $y = -(x + 2)^2$

7. $y = 2(x + 2)^2$

8. $y = 2(x + 2)^2 + 4$

9. $y = -2(x + 2)^2 + 4$

10. $y = x^2 + 6x - 1$

11. $f(x) = x^2 - 4x$

12. $F(x) = x^2 - 3x + 4$

13. $g(x) = 1 - x^2$

14. $y = 2x^2 + \sqrt{2}x$

15. $y = x^2 - 2x - 3$

16. $y = 2x^2 + 3x - 2$

17. $y = -x^2 + 6x + 2$

18. $y = -3x^2 + 12x$

19. $s = 2 + 3t - 9t^2$

20. $s = -\frac{1}{4}t^2 + t - 1$

For Exercises 21–26, determine the input that produces the largest or smallest output (whichever is appropriate). State whether the output is largest or smallest.

21. $y = 2x^2 - 4x + 11$

22. $f(x) = 8x^2 + x - 5$

23. $g(x) = -6x^2 + 18x$

24. $s = -16t^2 + 196t + 80$

25. $f(x) = x^2 - 10$

26. $h(x) = x^2 - 10x$

In Exercises 27–30, find the maximum or minimum value for each function (whichever is appropriate). State whether the value is a maximum or minimum.

27. $y = x^2 - 8x + 3$

28. $y = \frac{1}{2}x^2 + x + 1$

29. $y = -2x^2 - 3x + 2$

30. $y = -\frac{1}{3}x^2 - 2x$

In Exercises 31–34, you are given a quadratic function.

(a) By looking at the coefficient of the square term, state whether the function has a maximum or a minimum value.

Ⓖ **(b)** Use a TRACE and/or ZOOM feature on a graphing utility to estimate the input x_0 for which the function obtains its maximum or minimum value. (Estimate to the nearest one-tenth, as in Example 5.) What is the corresponding estimate for the maximum or minimum value?

(c) Use algebra to determine the exact value of x_0 and the corresponding maximum or minimum value of the function. Check to see that the results are consistent with the graphical estimates obtained in part (b).

31. $f(x) = 6x^2 - x - 4$

32. $g(t) = 40t - 7t^2$

33. $y = -9t^2 + 40t + 1$

34. $y = \sqrt{2}x^2 + 4x + 3$

35. How far from the origin is the vertex of the parabola $y = x^2 - 6x + 13$?

36. Find the distance between the vertices of the parabolas $y = -\frac{1}{2}x^2 + 4x$ and $y = 2x^2 - 8x - 1$.

For Exercises 37–42, the functions f, g, and h are defined as follows:

$$f(x) = 2x - 3 \qquad g(x) = x^2 + 4x + 1 \qquad h(x) = 1 - 2x^2$$

In each exercise, classify the function as linear, quadratic, or neither.

37. $f \circ g$

38. $g \circ f$

39. $g \circ h$

40. $h \circ g$

41. $f \circ f$

42. $h \circ h$

In Exercises 43 and 44, determine the inputs that yield the minimum values for each function. Compute the minimum value in each case.

43. **(a)** $f(x) = \sqrt{x^2 - 6x + 73}$

(b) $g(x) = \sqrt[3]{x^2 - 6x + 73}$

(c) $h(x) = x^4 - 6x^2 + 73$

44. **(a)** $F(x) = (4x^2 - 4x + 109)^{1/2}$

(b) $G(x) = (4x^2 - 4x + 109)^{1/3}$

(c) $H(x) = 4x^4 - 4x^2 + 109$

Ⓖ For Exercises 45 and 46, first, use a graphing utility to estimate to the nearest one-tenth the maximum value of the function. Then use algebra to determine the exact value, and check that your answer is consistent with the graphical estimate.

45. **(a)** $f(x) = \sqrt{-x^2 + 4x + 12}$

(b) $g(x) = \sqrt[3]{-x^2 + 4x + 12}$

(c) $h(x) = -x^4 + 4x^2 + 12$

46. **(a)** $F(x) = (27x - x^2)^{1/2}$

(b) $G(x) = (27x - x^2)^{1/3}$

(c) $H(x) = 27x^2 - x^4$

In Exercises 47–50, determine whether the x-y values are generated by a linear function, a quadratic function, or neither.

47.

x	−4	−2	0	2	4
y	25	3	−4	7	33

48.

x	1	2	3	4	5
y	−12	−8.5	−5	−1.5	2

49.

x	0	1	2	3	4
y	−21	−3	7	9	3

50.

x	0.25	0.50	0.75	1.00	1.25
y	−0.40	−0.16	0.08	0.32	0.62

Ⓖ **51.** On the same set of axes, graph the four parabolas $y = x^2$, $2x^2$, $3x^2$, and $8x^2$. Relative to your graphs, where do you think the graph of $y = 50x^2$ would fit in? After answering, check by adding the graph of $y = 50x^2$ to the picture.

Ⓖ **52.** Graph the four parabolas $y = x^2, 0.5x^2, 0.25x^2$, and $0.125x^2$. Relative to your graphs, where do you think the graph of $y = 0.02x^2$ would fit in? After answering, check by adding the graph of $y = 0.02x^2$ to the picture.

B

53. Let $f(x) = x^2$. Find the average rate of change $\Delta f/\Delta x$ on the interval $[a, x]$.

54. If $f(x) = ax^2 + bx + c$, show that
$$\frac{f(x + h) - f(x)}{h} = 2ax + ah + b.$$

55. Find the x-coordinate of the vertex of the parabola $y = (x - a)(x - b)$. (Your answer will be in terms of the constants a and b.) *Hint:* It's easier here to rely on symmetry than on completing the square.

Ⓖ **56. (a)** Graph the two functions $y = x^2 + 4x$ and $y = x^2 - 4x$. How are the two graphs related (in terms of symmetry)?
　　(b) Follow part (a) using the two functions $y = -2x^2 + 3x + 4$ and $y = -2x^2 - 3x + 4$.
　　(c) Which one of the graphing techniques from Section 3.4 relates to what you have observed in parts (a) and (b)?

In Exercises 57–60, find quadratic functions satisfying the given conditions.

57. The graph passes through the origin, and the vertex is the point $(2, 2)$. *Hint:* What do h and k stand for in the general equation $y = a(x - h)^2 + k$?

58. The graph is obtained by translating $y = x^2$ four units in the negative x-direction and three units in the positive y-direction.

59. The vertex is $(3, -1)$, and one x-intercept is 1.

60. The axis of symmetry is the line $x = 1$. The y-intercept is 1. There is only one x-intercept.

61. Let $g(x) = x^2 + bx$. Are there any values for b for which the minimum value of this function is -1? If so, what are they? If not, explain why.

62. For which value of c will the minimum value of the function $f(x) = x^2 + 2x + c$ be $\sqrt{2}$?

63. This exercise shows that if we have a table generated by a linear function and the x-values are equally spaced, then the first differences of the y-values are constant.
　　(a) In the following data table, the three x-entries are equally spaced. Compute the three entries in the $f(x)$ row assuming that f is the linear function given by $f(x) = mx + b$. (Don't worry about the fact that your answers contain all four of the letters m, b, a, and h.)

x	a	a + h	a + 2h
f(x)			

　　(b) Compute the first differences for the three quantities that you listed in the $f(x)$ row in part (a). (The two first differences that you obtain should turn out to be equal, as required.)

64. Suppose we have a table of x-y values with the x-values equally spaced. This exercise shows that if the first differences of the y-values are constant, then the table can be generated by a linear function. *Note:* To show that the data points are generated by a linear function, it is enough to show that the points all lie on one nonvertical line.

Consider the following table with the x-values equally spaced by an amount $h \neq 0$. (We are assuming that h is nonzero to guarantee that the three x-values are distinct.)

x	a	a + h	a + 2h
y	y_1	y_2	y_3

Assuming that the first differences of the y-values are constant, we have $y_2 - y_1 = y_3 - y_2 = k$, where k is a constant.
　(a) Check that the slope of the line joining the two data points (a, y_1) and $(a + h, y_2)$ is k/h.
　(b) Likewise, check that the slope of the line joining the two data points $(a + h, y_2)$ and $(a + 2h, y_3)$ is k/h.
From parts (a) and (b), we conclude that the three data points lie on a nonvertical line, as required. (The slope k/h is well defined because we assumed $h \neq 0$.)

65. The following table and scatter plot show global coal consumption for the years 1990–1995.

Global Coal Consumption 1990–1995

Year x x = 0 ↔ 1990	Coal consumption y (billion tons)
0	3.368
1	3.285
2	3.258
3	3.243
4	3.261
5	3.311

Source: World Resources Institute

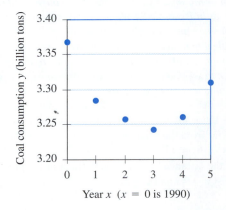

(a) Use a graphing utility to find a quadratic model for the data. Then use the model to make estimates for global coal consumption in 1989 and 1996.
(b) Use the following information to show that, in terms of percentage error, the 1996 estimate is better than the 1989 estimate, but in both cases the percentage error is less than 2%. The actual figures for coal consumption in 1989 and 1996 are 3.408 and 3.428 billion tons, respectively.

(c) Use the model to project worldwide coal consumption in 1998. Then show that the percentage error is more than 9%, given that the actual 1998 consumption was 3.329 billion tons.

66. *Driven by rising consumer demand and growing dissatisfaction with conventional farming practices, the organic agriculture industry is soaring.*—*Vital Signs 2000,* Lester R. Brown et al. (New York: W.W. Norton & Co., 2000)

The following table and scatter plot show the area devoted to organic farming in the European Union over the period 1988–1998. (One hectare equals 2.471 acres or 10,000 sq. meters.)

Year x	Area y (million hectares)
1988	0.17
1990	0.34
1992	0.69
1994	1.15
1996	1.90
1998	3.17

Data from *Vital Signs 2000*

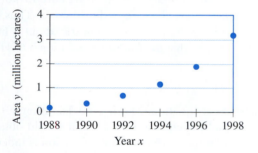

(a) In view of the scatter plot and the opening quotation, a quadratic model might be useful here, at least in making short-term predictions. Indeed, if you look at the scatter plot without knowing that it was generated from real-life data, in the context of this section you might think that the points all lie exactly on the graph of a parabola. Show that this is not the case: Compute the second differences of the y-values in the given table, and note that they are not equal. (In fact, no two are equal.)

(b) In the following table, fill in the missing entries so that all six data points are generated by one quadratic function. *Hint*: Work back from the second differences; you want them all to be equal.

1988	0.17
1990	0.34
1992	0.69
1994	?
1996	?
1998	?

C

67. By completing the square, show that the coordinates of the vertex of the parabola $y = ax^2 + bx + c$ are $(-b/2a, -D/4a)$, where $D = b^2 - 4ac$.

68. Compute the average of the two x-intercepts of the graph of $y = ax^2 + bx + c$. (Assume $b^2 - 4ac > 0$.) How does your answer relate to the result in Exercise 67?

69. Consider the quadratic function $y = px^2 + px + r$, where $p \neq 0$.
    (a) Show that if the vertex lies on the x-axis, then $p = 4r$.
    (b) Show that if $p = 4r$, then the vertex lies on the x-axis.

70. Let $f(x) = ax^2 + bx$, where $a \neq 0$ and $b \neq 0$. Find a value for b such that the equation $f(f(x)) = 0$ has exactly three real roots.

MINI PROJECT 1 | HOW DO YOU KNOW THAT THE GRAPH OF A QUADRATIC FUNCTION IS ALWAYS SYMMETRIC ABOUT A VERTICAL LINE?

(a) A short answer that sweeps everything under the rug: The graph of the basic parabola $y = x^2$ is symmetric about a vertical axis, and the graph of $f(x) = ax^2 + bx + c$ inherits this type of symmetry. In a group, discuss the reasons supporting this short answer. (The discussion won't be complete without taking into account the following three concepts: completing the square, translations, and reflection in the y-axis.) After the discussion, write a paragraph or two answering, as carefully as you can, the question raised in the title of this mini project.

(b) Here's a completely different way to go about answering the question in the title of this mini project. An advantage to this alternate approach is that, as a by-product, it gives you the equation of the axis of symmetry without the need for completing the square. We start with the definition of what it means for the graph of $y = f(x)$ to be symmetric about a vertical line—say, $x = x_0$.

DEFINITION | **Symmetry about the Line $x = x_0$**

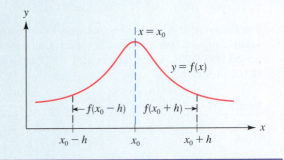

The graph of a function f is symmetric about the line $x = x_0$ provided the following condition holds. Whenever $x_0 + h$ is in the domain of f, then so is $x_0 - h$ and

$$f(x_0 + h) = f(x_0 - h)$$

Now suppose we have a quadratic function

$$f(x) = ax^2 + bx + c \qquad (a \neq 0) \tag{1}$$

Because the domain is the set of all real numbers, we don't have to worry about the part of the definition in the box above regarding domain. Thus, to show that the graph of the quadratic function is symmetric about a line $x = x_0$ we need to find a value for x_0 so that the following equation holds for all values of h:

$$f(x_0 + h) = f(x_0 - h) \tag{2}$$

Now your job: Evaluate each side of equation (2) using the definition of f in equation (1). Then solve the resulting equation for x_0. You should obtain $x_0 = -b/2a$. This shows that the graph of the quadratic function is indeed symmetric about a vertical line—namely, $x = -b/2a$.

MINI PROJECT 2 | WHAT'S LEFT IN THE TANK?

(a) The following table shows global oil consumption for the years 1990–1995, 1998, and 1999. Using only the data for the years 1990–1995, use a graphing utility to find a linear model and a quadratic model for global oil consumption.

(b) Provide the detailed calculations to back up the following statements regarding the use of these models in projecting global oil consumption for the years 1998 and 1999.

> For both 1998 and 1999, the linear model underestimates oil consumption, while the quadratic model overestimates. For both models, however, the percentage error is always less than 5%. For both years, the quadratic model gives the better projection (as measured by percentage error).

Global Oil Consumption, 1990–1995, 1998, 1999

Year x ($x = 0$ corresponds to 1990)	0	1	2	3	4	5	8	9
Global oil consumption y (billion barrels)	23.886	23.913	24.105	23.997	24.506	24.893	26.251	26.723

Source: U.S. Energy Information Administration

(c) Oil, as you know, is a nonrenewable energy resource; at some point in the future, it will be used up. When this will occur will depend on, among other things, consumption patterns and the amount of oil that exists on the planet. In this portion of the mini project you'll compute estimates for the *depletion date* for oil using three different consumption scenarios. Based on figures from the U.S. Geological Survey and the Energy Information Administration, we'll assume that in 1990 the total amount of oil remaining was 2886 billion barrels. [This estimate takes into account reserves and resources as of 1990, plus resources discovered in the 1990s, plus current estimates (at the end of the 1990s) for as yet undiscovered resources.]

A consequence of the linear yearly consumption model   Suppose that the consumption y (in billions of barrels) in year x is given by a linear function, $y = mx + b$. Then, using results from Chapter 9, it can be shown that the total amount A of oil consumed over the years $x = 0$ through $x = T$ is given by the quadratic model

$$A = \frac{m}{2}T^2 + \left(\frac{m}{2} + b \right)T + b$$

On the left side of this equation, replace A with 2886; on the right side, replace m and b with the values that you obtained for the linear model in part (a). Now, find the positive root of the resulting quadratic equation. (Use either the quadratic formula or a graphing utility.) The root represents the "life expectancy" of oil under this scenario. Round the answer to the nearest ten years and add to 1990 to get a depletion date for oil.

A consequence of the quadratic yearly consumption model   Suppose that the consumption y (in billions of barrels) in year x is given by a quadratic function, $y = ax^2 + bx + c$. Then, using results from Chapter 9, it can be shown that the total amount A of oil consumed over the years $x = 0$ through $x = T$ is given by the cubic model

$$A = \frac{a}{3}T^3 + \frac{a + b}{2}T^2 + \left(\frac{a + 3b + 6c}{6} \right)T + c$$

On the left side of this equation, replace A with 2886; on the right side replace a, b, and c with the values that you obtained for the quadratic model in part (a). Now, use a graphing utility to find the positive root of the resulting cubic equation. The root gives you the "life expectancy" of oil under this scenario. Round the answer to the nearest ten years and add to 1990 to get a depletion date for oil.

Constant consumption model If you've done the preceding calculations correctly, you will have found that the depletion dates under both models occur within the present century. Now try the following "what if." Go back to the linear consumption model and redo the calculations using for m the value 0 rather than the value from part (a). In terms of consumption, what's the interpretation of $m = 0$? What depletion date do you obtain?

(d) Write a paragraph or two summarizing the results from part (c). (Don't forget to include what the assumptions are in each case.)

4.3 USING ITERATION TO MODEL POPULATION GROWTH (Optional Section)

Although I shall henceforth adopt the habit of referring to the variable X as "the population," there are countless situations outside population biology where . . . [iteration of functions] applies. . . . Examples in economics include models for the relationship between commodity quantity and price, for the theory of business cycles, and for the temporal sequences generated by various other economic quantities.

. . . I would therefore urge that people be introduced to, say, [the iteration process for $f(x) = kx(1 - x)$] early in their mathematical education. This equation can be studied phenomenologically by iterating it on a calculator, or even by hand. Its study does not involve as much conceptual sophistication as does elementary calculus. Such study would greatly enrich the student's intuition about nonlinear systems. —Biologist Robert M. May, "Simple mathematical models with very complicated dynamics," *Nature,* vol. 261 (1976), pp. 459–467.

The size of a population or its genetic makeup may change from one generation to the next. In the study of *discrete dynamics,* we use functions and the iteration process (from Section 3.5) to investigate and analyze changes such as these that occur over discrete intervals of time. We begin by introducing the notion of a *fixed point* of a function.

If we start with a function f and an input x, it's usually not the case that $f(x)$ turns out to be the same as x itself. That is, usually, the output is not the same as the input. But sometimes this does happen. Take, for example, the function $f(x) = 3x - 2$ and the input $x = 1$. Then we have

$$f(1) = 3(1) - 2 = 1$$

So, for this particular function, the input $x = 1$ is an instance where "input = output." The input $x = 1$ in this case is called a *fixed point* of the function $f(x) = 3x - 2$. In the box that follows, we give the general definition of a fixed point.

DEFINITION | **Fixed Point of a Function**

A fixed point of a function f is an input x in the domain of f such that

$$f(x) = x$$

EXAMPLE

Both 0 and 1 are fixed points for $f(x) = x^2$ because $f(0) = 0^2 = 0$ and $f(1) = 1^2 = 1$.

| EXAMPLE | 1 | Finding fixed points |

Find the fixed points (if any) for each function:

(a) $f(x) = 1 - x$;   **(b)** $g(x) = 1 + x$;   **(c)** $h(x) = x^2 - x - 3$.

SOLUTION

(a) We're looking for a number x such that $f(x) = x$. In view of the definition of f, this last equation becomes

$$1 - x = x$$

or

$$2x = 1$$

and therefore

$$x = 1/2$$

This result shows that the function f has one fixed point; it is $x = 1/2$.

(b) If x is a fixed point of $g(x) = 1 + x$, we have

$$1 + x = x$$

But then subtracting x from both sides of this last equation yields $1 = 0$, which is impossible. We conclude from this that there is no fixed point for the function $g(x) = 1 + x$.

(c) The fixed points (if any) are the solutions of the quadratic equation $x^2 - x - 3 = x$. Subtracting x from both sides of this equation, we have

$$x^2 - 2x - 3 = 0$$
$$(x - 3)(x + 1) = 0$$

Looking at this last equation, we can see that there are two roots: 3 and -1. Each of these numbers is a fixed point for the given function $h(x) = x^2 - x - 3$. That is, $h(3) = 3$ and $h(-1) = -1$. [You should verify each of these last two statements for yourself by actually computing $h(3)$ and $h(-1)$.]

A fixed point of a function can be interpreted geometrically: it is the x-coordinate of a point where the given function intersects the line $y = x$. Figure 1 shows the fixed points for the functions in the example that we've just completed.

Figure 1

Fixed points for the functions in Example 1.

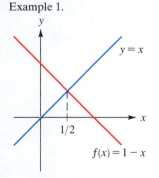

(a) The fixed point of f is 1/2.

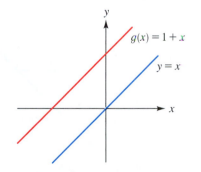

(b) The function g has no fixed points.

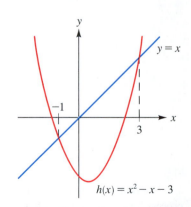

(c) The fixed points of h are -1 and 3.

Fixed points are related to the iteration process in several ways. Suppose that a number a is a fixed point of the function f. Then by definition we have $f(a) = a$, which says that the first iterate of a is equal to a itself. Similarly, all of the subsequent iterates of the fixed point a will be equal to a. For instance, for the second iterate, we have

$$f(f(a)) = f(a) \qquad \text{\textcolor{red}{substituting a for $f(a)$ on the left-hand side}}$$

$$ = a \qquad \text{\textcolor{red}{again because $f(a) = a$}}$$

This shows that the second iterate is equal to a. The same type of calculation will show that any subsequent iterate of the fixed point a is equal to a.

Another connection between fixed points and iteration is this: for some functions, a fixed point can be a "target value" for other iterates. We'll explain this using Figure 2.

Figure 2 shows the first four steps in the iteration process for $f(x) = \frac{1}{2}x + 2$ with $x_0 = 1$. (See Section 3.5 if you need to review graphical iteration.) As indicated in Figure 2, the input 4 is a fixed point for the function f, and the iteration process follows a staircase pattern that approaches the point $(4, 4)$. We say in this case that the iterates of $x_0 = 1$ **approach** the fixed point 4 and that this target value 4 is an **attracting fixed point** of the function f. Table 1 gives you a more numerical look at what is meant by saying that the iterates approach the target value 4.

In the table, notice, for example, that

$$x_5 \text{ differs from 4 by less than 0.1}$$

$$x_{10} \text{ differs from 4 by less than 0.01}$$

and $\qquad\qquad x_{15}$ differs from 4 by less than 0.0001

What's important here is that the differences between the iterates and 4 can be made as small as we please by carrying out the iteration process sufficiently far. The idea of a target value or *limit* is made more precise in calculus. But, for our purposes, Figure 2 and Table 1 will certainly give you an intuitive understanding of the idea and what we mean by saying that the iterates approach 4.

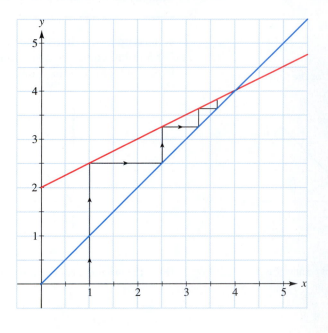

Figure 2

The first four steps in the iteration process for $f(x) = \frac{1}{2}x + 2$ with $x_0 = 1$.

TABLE 1	The Iterates Approach 4		
x_1	2.5		
x_2	3.25		
x_3	3.625		
x_4	3.8125	x_{11}	3.9985 ...
x_5	3.906 ...	x_{12}	3.99926 ...
x_6	3.953 ...	x_{13}	3.99963 ...
x_7	3.976 ...	x_{14}	3.99981 ...
x_8	3.988 ...	x_{15}	3.999908 ...
x_9	3.9941 ...	x_{20}	3.9999971 ...
x_{10}	3.9970 ...	x_{25}	3.999999910 ...

(a) x_1 through x_{10} **(b)** x_{11} through x_{15} along with x_{20} and x_{25}

Figures 3 and 4 show two more ways that the iteration process may relate to fixed points. In Figure 3, there is an attracting fixed point for the iterates of $x_0 = -0.1$, but this time the iteration process approaches the fixed point through a spiral pattern rather than a staircase pattern. To find the fixed point (and thereby determine the number that the iterates are approaching), we need to solve the quadratic equation $x^2 - 0.5 = x$. As you should check for yourself by means of the quadratic formula and then a calculator, the relevant root here is $(1 - \sqrt{3})/2 \approx -0.366$. Notice that this value is consistent with Figure 3. In Figure 4, the fixed point 1 is a **repelling fixed point** for the iterates of $x_0 = 1.25$. As Figure 4 indicates, the iterates of $x_0 = 1.25$ move farther and farther away from the value 1. Indeed, as you can check with a calculator, the first five iterates of 1.25 are as follows (we're rounding to two decimal places):

$$x_1 \approx 1.56 \qquad x_2 \approx 2.44 \qquad x_3 \approx 5.96 \qquad x_4 \approx 35.53 \qquad x_5 \approx 1262.18$$

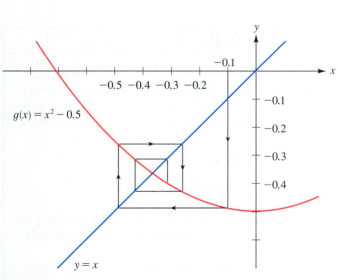

Figure 3
The iterates of $x_0 = -0.1$ under the function $g(x) = x^2 - 0.5$ approach the attracting fixed point $\frac{1}{2}(1 - \sqrt{3}) \approx -0.366$.

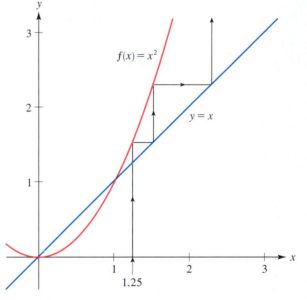

Figure 4
The iterates of $x_0 = 1.25$ under the function $f(x) = x^2$ move away from the repelling fixed point 1.

The iteration process for functions is often applied in the study of population growth. The word "population" here is used in a general sense. It needn't refer only to human populations. For instance, biological or ecological studies may involve animal, insect, or bacterial populations. (Also, see the quotation at the beginning of this section.) The following equation defines one type of quadratic function that has been studied extensively in this context:

$$f(x) = kx(1 - x) \tag{1}$$

In using this idealized model, we assume that the population size is measured by a number between 0 and 1, where 1 corresponds to the maximum possible population size in the given environment and 0 corresponds to the case where the population has become extinct. We start with a given input x_0 ($0 \le x_0 \le 1$) that represents the fraction of the maximum population size that is initially present. For instance, if the maximum possible population of catfish in a pond is 100 and initially there were 70 catfish, then we would have $x_0 = 70/100 = 0.7$.

The next basic assumption in using equation (1) to model population size is that the iterates of x_0 represent the fraction of the maximum possible population present after each successive time interval. That is,

x_1 is the (fraction of the maximum) population after the first time interval

x_2 is the (fraction of the maximum) population after the second time interval

and, in general,

x_n is the (fraction of the maximum) population after the nth time interval

In a given study, the time intervals might be measured, for example, in years, in months, or in breeding seasons. The constant k in equation (1) is the *growth parameter*; it is related to the rate of growth of the particular population being studied. Science writer James Gleick has described k this way: "In a pond, it might correspond to the fecundity of the fish, the propensity of the population not just to boom but also to bust. . . ." [*Chaos: Making a New Science* (New York: Viking Penguin, Inc., 1988)]

EXAMPLE 2   Using iteration in analyzing population size

In the Mississippi Delta region, many farmers have replaced unproductive cotton fields with catfish ponds. Suppose that a farmer has a catfish pond with a maximum population size of 500 and that initially the pond is stocked with 50 catfish. Also, assume that the growth parameter for this population is $k = 2.9$, so that equation (1) becomes

$$f(x) = 2.9x(1 - x) \tag{2}$$

Finally, assume that the time intervals here are breeding seasons.
(a) What is the value for x_0?
(b) Use Figure 5 to estimate the iterates x_1 through x_5. Then use a calculator to compute these values. Round the final answers to three decimal places. Interpret the results.
(c) As indicated in Figure 5, the iteration process is spiraling in on a fixed point of the function. (Figure 6 at the end of this section will demonstrate this in greater detail.) Find this fixed point and interpret the result.

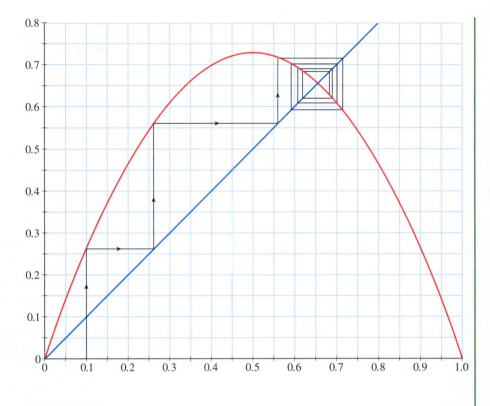

Figure 5
The first ten iterations of $x_0 = 0.1$ under $f(x) = 2.9x(1 - x)$.

SOLUTION

(a) $x_0 = \dfrac{\text{initial population}}{\text{maximum population}} = \dfrac{50}{500} = 0.1$

(b) Looking at Figure 5, it appears that x_1, the first iterate of x_0, is between 0.25 and 0.30, much closer to the former number than the latter. As an estimate, let's say that x_1 is about 0.26. This and the other estimates are given in Table 2. *Suggestion*: Make the estimates for yourself before looking at the estimates we give. Some slight discrepancies are okay. In the bottom row of Table 2 are the values of the iterates obtained using a calculator. You should verify these for yourself. (There should be no discrepancies here.)

   The results in Table 2 tell us what is happening to the population through the first five breeding seasons. From an initial population of 50 catfish, the population size steadily increases through the first three breeding seasons (the numbers in the table are getting bigger). The population size then drops after the fourth breeding season, and goes back up after the fifth season. (These facts can be deduced from Figure 5, as well as from Table 2.) To

TABLE 2
Iterates of $x_0 = 0.1$ under the Function $f(x) = 2.9x(1 - x)$ (Calculator values rounded to three decimal places)

	x_1	x_2	x_3	x_4	x_5
From graph	0.26	0.56	0.71	0.58	0.70
From calculator	0.261	0.559	0.715	0.591	0.701

TABLE 3 Catfish Population

n	0	1	2	3	4	5
Number of fish after n breeding seasons	50	131	280	358	296	351

compute the actual numbers of fish, we need to multiply each number in the bottom row of Table 2 by 500. (Remember that the iterates in Table 2 represent fractions of the maximum possible population size 500.) For example, to compute the actual number of fish at the end of the third breeding season, we multiply the initial population size of 500 by x_3:

$$x_3 \times 500 \approx 0.715 \times 500$$
$$\approx 358 \text{ catfish}$$

The number of catfish at the end of each of the other breeding seasons is obtained in the same manner. See Table 3; use your calculator to check each of the entries in the table.

(c) The fixed point we are looking for occurs when the parabola in Figure 5 intersects the line $y = x$. So, following the method in Example 1, we need to solve the equation $2.9x(1 - x) = x$. We have

$$2.9x(1 - x) = x$$
$$-2.9x^2 + 1.9x = 0$$
$$x(-2.9x + 1.9) = 0$$

Therefore, $x = 0$ or $-2.9x + 1.9 = 0$—that is,

$$x = \frac{-1.9}{-2.9} = 0.655 \qquad \text{using a calculator and rounding to three decimal places}$$

This shows that there are two fixed points for the function f—namely, 0 and 19/29. Looking at Figure 5, we know that 19/29 is the fixed point that we are interested in here, not 0. So the iterates of $x_0 = 0.1$ are approaching the value 19/29, which is approximately 0.655.

We now summarize and interpret the results. There were initially 50 catfish in a pond that could hold at most 500. As we saw in part (b), the population size increases over the first three breeding seasons. After this, as Figure 5 shows, the population oscillates up and down but draws closer and closer to an equilibrium population corresponding to the fixed point $x = 19/29$. This equilibrium population is

$$\frac{19}{29} \times 500 \approx 328 \text{ catfish}$$

We conclude this section with some pictures indicating only three of the many possibilities that can arise in the iteration of quadratic functions of the form $f(x) = kx(1 - x)$. Figure 6 concerns the function with growth parameter $k = 2.9$

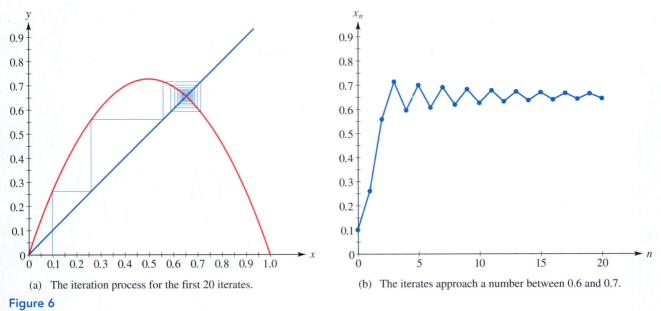

(a)   The iteration process for the first 20 iterates.

(b)   The iterates approach a number between 0.6 and 0.7.

Figure 6
The iteration of $f(x) = kx(1 - x)$ with growth parameter $k = 2.9$ and $x_0 = 0.1$.

that we used in the catfish example: $f(x) = 2.9x(1 - x)$. In Figure 6(a), we've carried out the iteration of $x_0 = 0.1$ through the twentieth iterate. (In the catfish example, Figure 5 goes only as far as the tenth iterate.) Figure 6(a) indicates quite clearly that the iterates are indeed approaching a fixed point of the function. Figure 6(b) presents another way to visualize the long-term behavior of the iterates. Values of n are marked on the horizontal axis, values of the iterates x_n are marked on the vertical axis, and the points with coordinates (n, x_n) are then plotted. For example, since $x_0 = 0.1$, we plot the point $(0, 0.1)$; and since $x_1 = 0.261$, we plot the point $(1, 0.261)$. The line segments in Figure 6(b) are drawn in only to help the eye see the pattern that is emerging. Three facts that can be inferred from Figure 6(b) are as follows: after the first few iterates, the iterates oscillate up and down; the magnitude of the oscillations is decreasing; and, in the long run, the iterates are approaching a number between 0.6 and 0.7. (In Example 2 we found this value to be approximately 0.655.)

Unlike the iteration pictured in Figure 6(a), Figure 7(a) shows a case in which the iteration process is spiraling away from, rather than toward, a fixed point. Again, we've used the initial input $x_0 = 0.1$, but this time the growth parameter is $k = 3.2$. As indicated in Figure 7(b), the long-term behavior of the iterates becomes quite predictable: they alternate between two values. Figure 7(b) shows that the smaller of these two values is between 0.5 and 0.6, while the larger is approximately 0.8. Exercise 36 gives you formulas for computing these two limiting values for the iterates. (They turn out to be, approximately, 0.513 and 0.799.)

In Figure 8, once again we take the initial input to be $x_0 = 0.1$, but this time the growth parameter is $k = 3.9$. Now the iterates seem to fluctuate widely with no apparent pattern, in sharp contrast to the previous two figures, in which there were clear patterns. Phenomena such as this are the subject of **chaos theory,** a new branch of twentieth-century mathematics with wide application. [For a readable and nontechnical introduction to this relatively new subject, see James Gleick's *Chaos: Making a New Science* (New York: Viking Penguin, Inc., 1988). For a little more detail on the mathematics, see the paperback by Donald M. Davis, *The Nature and Power of Mathematics* (Princeton, N.J.: Princeton University Press, 1993), pp. 314–363.]

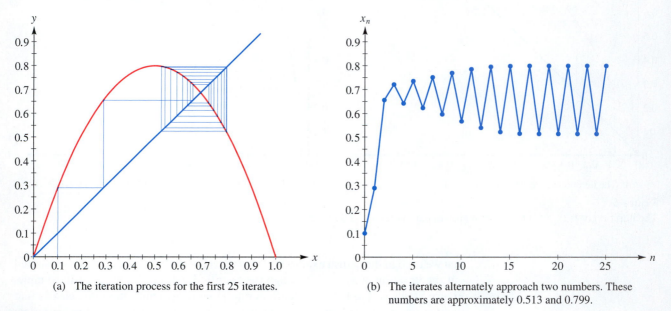

(a) The iteration process for the first 25 iterates.

(b) The iterates alternately approach two numbers. These numbers are approximately 0.513 and 0.799.

Figure 7
The iteration of $f(x) = kx(1 - x)$ with growth parameter $k = 3.2$ and $x_0 = 0.1$.

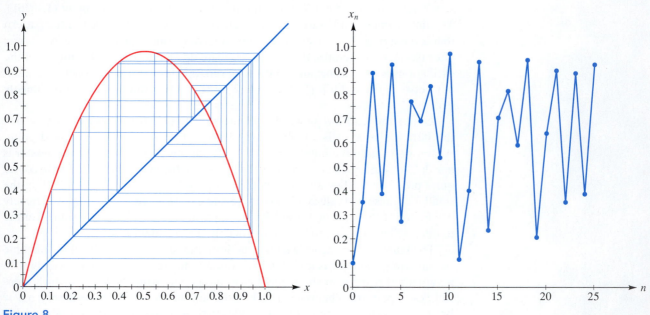

Figure 8
The iteration of $kx(1 - x)$ with growth parameter $k = 3.9$ and $x_0 = 0.1$.

EXERCISE SET 4.3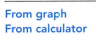

A

In Exercises 1–16, find all real numbers (if any) that are fixed points for the given functions.

1. $f(x) = -4x + 5$
2. $g(x) = 3x - 14$
3. $G(x) = \frac{1}{2} + x$
4. $F(x) = (7 - 2x)/8$
5. $h(x) = x^2 - 3x - 5$
6. $H(t) = 3t^2 + 18t - 6$
7. $f(t) = t^2 - t + 1$
8. $F(t) = t^2 - t - 1$
9. $k(t) = t^2 - 12$
10. $K(t) = t^2 + 12$
11. $T(x) = 1.8x(1 - x)$
12. $T(y) = 3.4y(1 - y)$
13. $g(u) = 2u^2 + 3u - 4$
14. $G(u) = 3u^2 - 4u - 2$
15. $f(x) = 7 + \sqrt{x - 1}$
16. $f(x) = \sqrt{10 + 3x} - 4$

Ⓖ *In Exercises 17–22:*

(a) *Graph each function along with the line $y = x$. Use the graph to determine how many (if any) fixed points there are for the given function.*

(b) *For those cases in which there are fixed points, use the zoom-in capability of the graphing utility to estimate the fixed point. (In each case, continue the zoom-in process until you are sure about the first three decimal places.)*

17. $f(x) = x^3 + 3x + 2$
18. $g(x) = x^3 - 3x + 2$
19. $h(x) = x^3 - 3x - 3.07$
20. $k(x) = x^3 - 3x - 3.08$
21. $s(t) = t^4 + 3t - 2$
22. $u(t) = t^4 + 3t + 2$

23. This exercise refers to the function $g(x) = x^2 - 0.5$ in Figure 3 in the text.

(a) Use the quadratic formula to verify that one of the fixed points of this function is $(1 - \sqrt{3})/2$, then use your calculator to check that this is approximately -0.366.

(b) According to the text, the iterates of -0.1 approach the value determined in part (a). Use your calculator: which is the first iterate to have the digit 3 in the first decimal place?

(c) Use your calculator: which is the first iterate to have the digit 6 in the second decimal place?

24. This exercise refers to the function $f(x) = x^2$ in Figure 4 in the text. According to the text, the iterates of 1.25 move farther and farther away from the fixed point 1. In this exercise you'll see that iterates of other points even closer to the fixed point 1 nevertheless are still "repelled" by 1.

(a) Let $x_0 = 1.1$. Use your calculator: which is the first iterate to exceed 10? Which is the first iterate to exceed one million?

(b) Let $x_0 = 1.001$. Use your calculator: which is the first iterate to exceed 2? Which is the first iterate to exceed one million?

(c) Let $x_0 = 0.99$. Compute x_1 through x_{10} to see that the iterates are indeed moving farther and farther away from the fixed point 1. What value are the iterates approaching? Is this value a fixed point of the function?

25. The accompanying figure shows the first eight steps in the iteration process for $f(x) = -0.7x + 2$, with $x_0 = 0.4$.

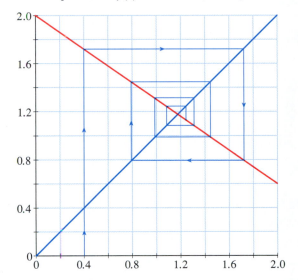

(a) Complete the following table. For the values obtained from the graph, estimate to the nearest one-tenth; for the calculator values, round the final answers to three decimal places.

	x_1	x_2	x_3	x_4	x_5	x_6	x_7	x_8
From graph								
From calculator								

(b) The figure shows that the iterates are approaching a fixed point of the function. Determine the exact value of this fixed point and then give a calculator approximation rounded to three decimal places.

(c) In the table for part (a), you used a calculator to compute the first eight iterates. Which of these iterates is the first to have the same digit in the first decimal place as the fixed point?

26. The figure on the next page shows the first six steps in the iteration process for $f(x) = 2.9x(1 - x)$, with $x_0 = 0.2$.

(a) Complete the following table. For the values obtained from the graph, estimate to the nearest 0.05, or closer if you can. For instance, to the nearest 0.05, the first iterate is 0.45. But, since the graph shows the iterate a bit above 0.45, the estimate 0.46 would be better. For the calculator work, round the final answers to three decimal places.

	x_1	x_2	x_3	x_4	x_5	x_6
From graph						
From calculator						

(continues)

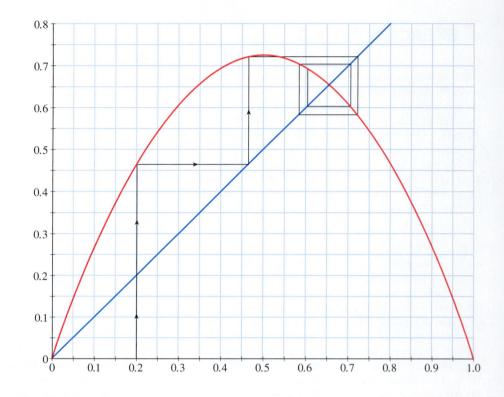

Figure for Exercise 26

(b) The figure shows that the iterates are approaching a fixed point of the function. Use the figure to estimate, to the nearest 0.05, a value for this fixed point. Then, determine the exact value of this fixed point, and also give a calculator approximation rounded to three decimal places.

(c) In the table for part (a), you used a calculator to compute the first six iterates. Which of these iterates is the first to have the same digit in the first decimal place as the fixed point? *Remark*: Agreement in the second decimal place doesn't occur until the 26th iterate.

For Exercises 27 and 28, refer to the figure on page 295, which shows the first nine steps in the iteration process for $f(x) = 4x(1 - x)$, with $x_0 = 0.9$. (Qualitatively, note that this figure is quite different from those in the previous two exercises or in Figures 2 through 7 in the text. Here, no clear pattern in the iterates seems to emerge.)

27. Complete the following table. For the values obtained from the graph, estimate to the nearest 0.05, or closer if you can. For instance, to the nearest 0.05, the first iterate is 0.35. But, since the graph shows the iterate a bit above 0.35, the estimate 0.36 would be better. For the calculator work, round the final answers to three decimal places.

	x_1	x_2	x_3	x_4	x_5	x_6	x_7	x_8	x_9
From graph									
From calculator									

28. As in Exercise 27, we work with the function $f(x) = 4x(1 - x)$. Furthermore, in part (a), we use an input that is very close to the one in Exercise 27. As you will see, however, the results will be remarkably different. This phenomenon, whereby a very small change in the initial input results in a completely different pattern in the iterates, is called **sensitivity to initial conditions.** Sensitivity to initial conditions is one of the characteristic behaviors studied in chaos theory.

(a) For the initial input, use $x_0 = \frac{1}{8}(5 + \sqrt{5}) \approx 0.9045$. Note that this differs from the input in Exercise 27 by less than 0.01. (Exercise 36 shows how this seemingly off-the-wall input was obtained.) Use algebra (not a calculator!) to compute exact expressions for x_1 and x_2. What do you observe? What are x_3 and x_4? What's the general pattern here? *Note*: If you were to use a calculator rather than algebra for all of this, due to rounding errors you could miss seeing the patterns.

(b) Take $x_0 = 0.905$. Use a calculator to compute x_1 through x_{10}. Round the final results to three decimal places. Is the behavior of the iterates more like that in Exercise 27 or in part (a) of this exercise?

As background for Exercises 29–32, you need to have read Example 2 in this section. As in Example 2, assume that there is a catfish pond with a maximum population size of 500 catfish. Also assume, unless stated otherwise, that the initial population size is 50 catfish (so that $x_0 = 0.1$).

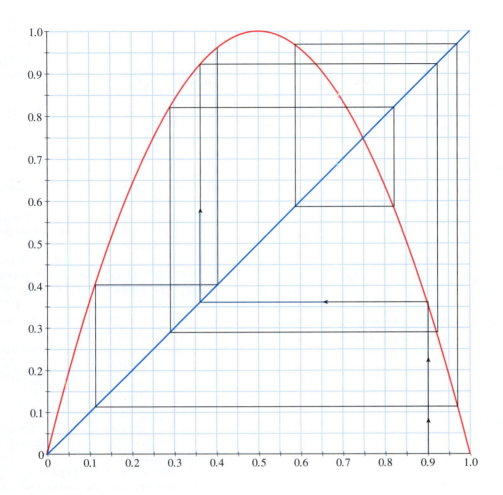

29. (a) In Example 2, we used a growth parameter of $k = 2.9$ and we computed the first five iterates of $x_0 = 0.1$. Now (under these same assumptions) compute x_{21} through x_{25}, given that $x_{20} \approx 0.64594182$. Round your final answers to four decimal places.

(b) Use the results in part (a) to complete the following table. [Compare your table to Table 3 in the text; note that the population size continues to oscillate up and down, but now the sizes of the oscillations are much smaller. This provides additional evidence that the population size is approaching an equilibrium value. (In Example 2 we determined this equilibrium size to be about 328 catfish.)]

n	20	21	22	23	24	25
Number of fish after n breeding seasons						

(c) In parts (a) and (b) and in Example 2, we worked with a growth parameter of 2.9, and we found that the iterates of $x_0 = 0.1$ were approaching a fixed point of the function. Now assume instead that the growth parameter is $k = 0.75$ (but, still, that $x_0 = 0.1$). Determine the iterates x_1 through x_5. (Round to five decimal places.) Are the iterates approaching a fixed point of the function $f(x) = 0.75x(1 - x)$? Interpret your results.

30. As in Example 2, take $x_0 = 0.1$, but now assume that the growth parameter is $k = 3$, so that equation (1) in the text becomes $f(x) = 3x(1 - x)$.

(a) Complete the following tables. (Round the final answers for the iterates to three decimal places.) Notice that after the third breeding season, the population oscillates up and down, as in Example 2.

n	0	1	2	3	4	5
x_n	0.1					
Number of fish after n breeding seasons	50					

n	6	7	8	9	10
x_n					
Number of fish after n breeding seasons					

(b) Complete the following table, given that $x_{100} \approx 0.643772529$. Round your final answers for the iterates to four decimal places. In your results, note that the population continues to oscillate up and down, but that the sizes of the oscillations are less than those observed in part (a).

n	101	102	103	104	105	106
x_n						
Number of fish after n breeding seasons						

(c) The iterates that you computed in parts (a) and (b) are approaching a fixed point of the function $f(x) = 3x(1 - x)$. Find this fixed point and the corresponding equilibrium population size.

31. (a) As in Example 2, take $x_0 = 0.1$, but now assume that the growth parameter is $k = 3.1$, so that equation (1) in the text becomes $f(x) = 3.1x(1 - x)$. Complete the following three tables. For the third table, use the fact that $x_{20} \approx 0.56140323$. Round your final answers for the iterates to four decimal places.

n	0	1	2	3	4	5
x_n	0.1					
Number of fish after n breeding seasons	50					

n	6	7	8	9	10
x_n					
Number of fish after n breeding seasons					

n	21	22	23	24	25	26
x_n						
Number of fish after n breeding seasons						

(b) Your results in part (a) will show that the population size is oscillating up and down, but that the iterates don't seem to be approaching a fixed point. Indeed, determine the (nonzero) fixed point of the function $f(x) = 3.1x(1 - x)$. Then note that the successive iterates in part (a) actually move farther and farther away from this fixed point.

(c) It can be shown that the long-term behavior of the iterates in this case resembles the pattern in Figure 7(b) on page 292; that is, the iterates alternately approach two values. Use the following formulas, with $k = 3.1$,

to determine these two values a and b that the iterates are alternately approaching. (The formulas are developed in Exercise 36.) Round the answers to four decimal places. Check to see that your answers are consistent with the results in part (a).

$$a = \frac{1 + k + \sqrt{(k - 3)(k + 1)}}{2k}$$

$$b = \frac{1 + k - \sqrt{(k - 3)(k + 1)}}{2k}$$

(d) What are the two populations corresponding to these two numbers a and b?

32. As in Example 2, assume that the maximum population size of the pond is 500 catfish, but now suppose that the growth parameter is $k = 0.6$.

(a) Suppose that the initial population is again 50 catfish, so that $x_0 = 0.1$. Compute the first ten iterates. Do they appear to be approaching a fixed point? Interpret the results.

(b) Follow part (a), but assume that the initial population is 450, so that $x_0 = 450/500 = 0.9$.

(c) What relationship do you see between the iterates in part (a) and in part (b)?

B

33. Suppose that c and d are inputs for a function g and that $g(c) = d$ and $g(d) = c$. Then we say the set $\{c, d\}$ is a **2-cycle** for the function g.

(a) Assuming that $\{c, d\}$ is a 2-cycle for the function g, list the first six iterates of c and the first six iterates of d. Describe in a complete sentence or two the pattern you see.

(b) The following figure shows the iteration process for the function $T(x) = 1 - |2x - 1|$, with initial input $x_0 = 0.4$. Use the figure to list the first six iterates of 0.4 under the function T.

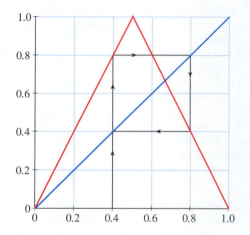

(c) In the following sentence, fill in the two blank spaces with numbers. The work in part (b) shows that $\{\_\_, \_\_\}$ is a 2-cycle for the function T.

(d) In part (b) you used a graph to list the first six iterates of $x_0 = 0.4$ under the function $T(x) = 1 - |2x - 1|$. Now, using a calculator (or just simple arithmetic), actually compute x_1 and x_2 and thereby check your results.

34. As background for this exercise, you need to have worked part (a) in the previous exercise so that you know the definition of a 2-cycle.

(a) The curve in the following figure is the graph of the function $Q(x) = x^2 - 7$. Use the figure to list the first six iterates of -3 under the function Q. Also, list the first six iterates of 2.

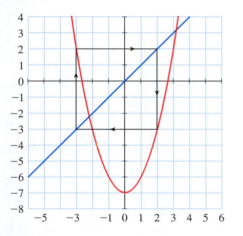

(b) In the following sentence, fill in the two blank spaces with numbers. The work in part (a) shows that $\{\_\_, \_\_\}$ is a 2-cycle for the function Q.

(c) In part (a) you used a graph to list the first six iterates of -3 under the function Q. Now, using the formula $Q(x) = x^2 - 7$, actually compute x_1 and x_2 and thereby check your results.

(d) Use your calculator to compute the first six iterates of $x_0 = -2.99$ under the function $Q(x) = x^2 - 7$. Note that the behavior of the iterates is vastly different than that observed in part (a), even though the initial inputs differ by only 0.01. (As pointed out in Exercise 28, this type of behavior is referred to as *sensitivity to initial conditions*.)

35. Let $f(x) = 4x(1 - x)$. In this exercise we find distinct inputs a and b such that $f(a) = b$ and $f(b) = a$. As indicated in Exercise 33, the set $\{a, b\}$ is called a *2-cycle for the function f*.

(a) From the equation $f(a) = b$ and the definition of f, we have

$$4a(1 - a) = b \qquad (1)$$

Likewise, from the equation $f(b) = a$ and the definition of f, we have

$$4b(1 - b) = a \qquad (2)$$

Subtract equation (2) from equation (1) and show that the resulting equation can be written

$$4(b - a)(b + a - 1) = b - a \qquad (3)$$

(b) Divide both sides of equation (3) by the quantity $b - a$. (The quantity $b - a$ is nonzero because we are assuming that a and b are distinct.) Then solve the resulting equation for b in terms of a. You should obtain

$$b = \frac{5}{4} - a \qquad (4)$$

(c) Use equation (4) to substitute for b in equation (1). After simplifying, you should obtain

$$16a^2 - 20a + 5 = 0 \qquad (5)$$

(d) Use the quadratic formula to solve equation (5) for a. You should obtain

$$a = (5 \pm \sqrt{5})/8$$

(e) Using the positive root for the moment, suppose $a = (5 + \sqrt{5})/8$. Use this expression to substitute for a in equation (4). Show that the result is $b = (5 - \sqrt{5})/8$. Now check that these values of a and b satisfy the conditions of the problem. That is, given that $f(x) = 4x(1 - x)$, show that

$$f\left[\frac{1}{8}(5 + \sqrt{5})\right] = \frac{1}{8}(5 - \sqrt{5})$$

and

$$f\left[\frac{1}{8}(5 - \sqrt{5})\right] = \frac{1}{8}(5 + \sqrt{5})$$

Note: If we'd begun part (e) by using the other root of the quadratic—namely, $a = (5 - \sqrt{5})/8$—then we would have found $b = (5 + \sqrt{5})/8$, so no new information would have been obtained. In summary, the 2-cycle for the function f consists of the two numbers that are the roots of equation (5).

36. Let $f(x) = kx(1 - x)$ and assume that $k > 0$. Follow the method of Exercise 35 to show that the values of a and b for which $f(a) = b$ and $f(b) = a$ are given by the formulas

$$a = \frac{1 + k + \sqrt{(k - 3)(k + 1)}}{2k}$$

and

$$b = \frac{1 + k - \sqrt{(k - 3)(k + 1)}}{2k}$$

Note that for $k > 3$, both of these expressions represent real numbers (because the quantity beneath the radical sign is then positive). In summary: For all values of k greater than 3, the function $f(x) = kx(1 - x)$ has a unique 2-cycle $\{a, b\}$, where a and b are given by the preceding formulas.

37. **(a)** Let $f(x) = 3.5x(1 - x)$. Use the formulas in Exercise 36 to find values for a and b such that $\{a, b\}$ is a 2-cycle for this function [that is, so that $f(a) = b$ and $f(b) = a$]. Exact answers are required, not calculator approximations.

    (b) The figure on the right shows the iteration process for the 2-cycle determined in part (a). Use the answers in part (a) to specify the coordinates of the four points P, Q, R, and S.

38. Let $f(x) = 3.2x(1 - x)$. Use the formulas in Exercise 36 to determine the values of a and b such that $\{a, b\}$ is a 2-cycle for this function. Use a calculator to evaluate the answers and round to three decimal places. [You'll find that these are the two numbers referred to in the caption for Figure 7(b) in this section.]

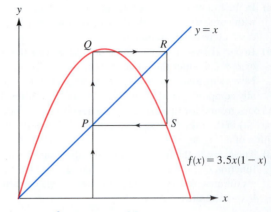

Figure for Exercise 37

▌ **4.4   SETTING UP EQUATIONS THAT DEFINE FUNCTIONS**

I hope that I shall shock a few people in asserting that the most important single task of mathematical instruction in the secondary schools is to teach the setting up of equations to solve word problems.
—George Polya (1887–1985)

Each problem that I solved became a rule which afterwards served to solve other problems. —René Descartes (1596–1650)

One of the first steps in problem solving often involves defining a function. The function then serves to describe or summarize a given situation in a way that is both concise and (one hopes) revealing. In this section we practice setting up equations that define such functions. In Section 4.5 we will use this skill in solving an important class of applied problems involving quadratic functions.

For many of the examples in this section, we'll rely on the following four-step procedure to set up the required equation. You may eventually want to modify this procedure to fit your own style. The important point, however, is that it *is* possible to approach these problems in a systematic manner. A word of advice: You're accustomed to working mathematics problems in which the answers are numbers. In this section, the answers are functions (or, more precisely, equations defining functions); you may need to get used to this.

Steps for Setting Up Equations

Step 1   After reading the problem carefully, draw a picture that conveys the given information.

Step 2   State in your own words, as specifically as you can, what the problem is asking for. (This usually requires rereading the problem.) Now, assuming that the problem asks you to find a particular quantity (or a formula for a particular quantity), assign a variable to denote that key quantity.

Step 3   Label any other quantities in your figure that appear relevant. Are there equations relating these quantities?

Step 4   Find an equation involving the key variable that you identified in Step 2. (Some people prefer to do this right after Step 2.) Now, as necessary, substitute in this equation using the auxiliary equations from Step 3 to obtain an equation involving only the required variables.

Perimeter = 100 cm

x

Figure 1

EXAMPLE 1 **The area of a rectangle as a function of its width**

The perimeter of a rectangle is 100 cm. Express the area of the rectangle in terms of the width x.

SOLUTION
Let's follow our four-step procedure.
Step 1 See Figure 1. The figure conveys the given information that the perimeter is 100 cm and that the width is x. (For the moment, ignore the label L at the base of the rectangle; it enters the picture in Step 3, but it's not part of the given information.)
Step 2 We want to express the area of the rectangle in terms of x, the width. Let A stand for the area of the rectangle.
Step 3 Call the length of the rectangle L, as indicated in Figure 1. Then, since the perimeter is given as 100 cm, we have

$$2x + 2L = 100$$
$$x + L = 50$$
$$L = 50 - x \tag{1}$$

Step 4 The area of a rectangle equals width times length:

$$A = x \cdot L$$
$$= x(50 - x) \qquad \text{substituting for } L \text{ using equation (1)}$$
$$= 50x - x^2 \tag{2}$$

This is the required equation expressing the area of the rectangle in terms of the width x.

To emphasize this dependence of A on x, we can use function notation to rewrite equation (2):

$$A(x) = 50x - x^2$$

The domain of this area function is the open interval $(0, 50)$. To see why this is so, first note that $x > 0$ because x denotes a width. Furthermore, in view of equation (1), we must have $x < 50$, otherwise L, the length, would be zero or negative.

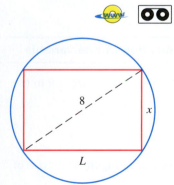

Figure 2

EXAMPLE 2 **The perimeter of a rectangle as a function of its width**

A rectangle is inscribed in a circle of diameter 8 cm. Express the perimeter of the rectangle as a function of its width x.

SOLUTION
We follow our four-step procedure.
Step 1 See Figure 2. Notice that the diagonal of the rectangle is a diameter of the circle.
Step 2 The problem asks us to come up with a formula or a function that gives us the perimeter of the rectangle in terms of x, the width. Let P denote the perimeter.

Step 3 Let L denote the length of the rectangle, as shown in Figure 2. Then, by the Pythagorean theorem, we have

$$L^2 + x^2 = 8^2 = 64$$

This equation relates the length L and the width x. Rather than leaving the equation in this form, however, we'll solve for L in terms of x (because the instructions for the problem mention x, not L):

$$L^2 = 64 - x^2$$
$$L = \sqrt{64 - x^2} \tag{3}$$

Step 4 The perimeter of the rectangle is the sum of the lengths of the four sides. Thus,

$$P = x + x + L + L = 2x + 2L \tag{4}$$

This equation expresses P in terms of x and L. However, the problem asks for P in terms of just x. Using equation (3) to substitute for L in equation (4), we have

$$P = 2x + 2\sqrt{64 - x^2} \tag{5}$$

This is the required equation. It expresses the perimeter P as a function of the width x, so if you know x, you can calculate P.

To emphasize this dependence of P on x, we can employ function notation to rewrite equation (5):

$$P(x) = 2x + 2\sqrt{64 - x^2} \tag{6}$$

Before leaving this example, we need to specify the domain of the perimeter function in equation (6). An easy way to do this is to look again at Figure 2. Since x represents a width, we certainly want $x > 0$. Furthermore, Figure 2 tells us that $x < 8$, because in any right triangle, a leg is always shorter than the hypotenuse. Putting these observations together, we conclude that the domain of the perimeter function in equation (6) is the open interval $(0, 8)$.

NOTE Although it would make sense *algebraically* to use an input such as $x = -1$ in equation (6), it does not make sense in our *geometric* context, where x denotes the width.

EXAMPLE **3** **Expressing a distance as a function of one variable**

Let $P(x, y)$ be a point on the curve $y = \sqrt{x}$. Express the distance from P to the point $(1, 0)$ as a function of x.

SOLUTION
Step 1 See Figure 3.
Step 2 We want to express the length of the dashed line in Figure 3 in terms of x. Call this length D.

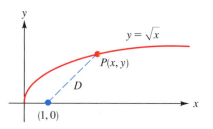

Figure 3

Step 3 There are no other quantities in Figure 3 that need labeling. But don't forget that we are given

$$y = \sqrt{x} \tag{7}$$

Step 4 By the distance formula, we have

$$D = \sqrt{(x-1)^2 + (y-0)^2}$$
$$= \sqrt{x^2 - 2x + 1 + y^2} \tag{8}$$

Now we can use equation (7) to eliminate y in equation (8):

$$D(x) = \sqrt{x^2 - 2x + 1 + (\sqrt{x})^2}$$
$$= \sqrt{x^2 - 2x + 1 + x}$$
$$= \sqrt{x^2 - x + 1} \tag{9}$$

Equation (9) expresses the distance as a function of x, as required. What about the domain of this distance function? Since the x-coordinate of a point on the curve $y = \sqrt{x}$ can be any nonnegative number, the domain of the distance function is $[0, \infty)$.

EXAMPLE 4 **Expressing the area of a triangle as a function of one variable**

A point $P(x, y)$ lies in the first quadrant on the parabola $y = 16 - x^2$, as indicated in Figure 4. Express the area of the triangular region in Figure 4 as a function of x.

SOLUTION
Step 1 See Figure 4.
Step 2 We want to express the area of the shaded triangle in terms of x. Let A denote the area of this triangle.
Step 3 Since the coordinates of P are (x, y), the base of our triangle is x and the height is y. Also, x and y are related by the given equation

$$y = 16 - x^2 \tag{10}$$

Step 4 The area of a triangle equals $\frac{1}{2}$(base)(height):

$$A = \tfrac{1}{2}(x)(y)$$

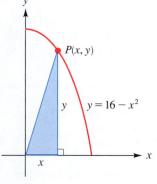

Figure 4

so

$$A(x) = \tfrac{1}{2}(x)(16 - x^2) \qquad \text{substituting for } y \text{ using equation (10)}$$
$$= 8x - \tfrac{1}{2}x^3$$

This last equation expresses the area of the triangle as a function of x, as required. (Exercise 52 at the end of this section will ask you to specify the domain of this area function.)

| EXAMPLE | 5 | **Expressing the area of a circle as a function of the circumference** |

A piece of wire x in. long is bent into the shape of a circle. Express the area of the circle in terms of x.

SOLUTION

Step 1 See Figure 5.

Step 2 We are supposed to express the area of the circle in terms of x, the circumference. Let A denote the area.

Step 3 The general formula for circumference in terms of radius is $C = 2\pi r$. Since in our case the circumference is given as x, our equation becomes

$$x = 2\pi r \qquad \text{or} \qquad r = \frac{x}{2\pi} \qquad (11)$$

Step 4 The general formula for the area of a circle in terms of the radius is

$$A = \pi r^2 \qquad (12)$$

This expresses A in terms of r, but we want A in terms of x. So we replace r in equation (12) by the quantity given in equation (11). This yields

$$A(x) = \pi\left(\frac{x}{2\pi}\right)^2 = \frac{\pi x^2}{4\pi^2} = \frac{x^2}{4\pi}$$

Thus, we have

$$A(x) = \frac{x^2}{4\pi}$$

This is the required equation. It expresses the area of the circle in terms of the circumference x. In other words, if we know the length of the piece of wire that is to be bent into a circle, we can use that length to calculate what the area of the circle will be.

x in.

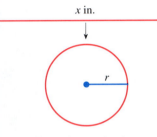

Circumference is *x* in.

Figure 5

| EXAMPLE | 6 | **Expressing the surface area of a cylinder as a function of the radius of the base** |

Figure 6 displays a right circular cylinder, along with the formulas for its volume V and total surface area S. Given that the volume is 10 cm^3, express the surface area S as a function of r, the radius of the base.

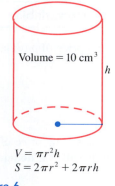

Volume $= 10 \text{ cm}^3$

h

$V = \pi r^2 h$
$S = 2\pi r^2 + 2\pi rh$

Figure 6

SOLUTION

Step 1   See Figure 6.

Step 2   We are given a formula that expresses the surface area S in terms of both r and h. We want to express S in terms of just r.

Step 3   We are given that $V = 10$ and also that $V = \pi r^2 h$. Thus,

$$\pi r^2 h = 10$$

and consequently, expressing h in terms of r,

$$h = \frac{10}{\pi r^2} \tag{13}$$

Step 4   We take the given formula for S—namely,

$$S = 2\pi r^2 + 2\pi rh$$

and replace h with the quantity given in equation (13). We obtain

$$S(r) = 2\pi r^2 + 2\pi r \left(\frac{10}{\pi r^2} \right)$$

$$= 2\pi r^2 + \frac{20}{r}$$

This is the required equation. It expresses the total surface area in terms of the radius r. Since the only restriction on r is that it be positive, the domain of the area function is $(0, \infty)$.

We have followed the same four-step procedure in Examples 1 through 6. Of course, no single method can cover all possible cases. As usual, common sense and experience are often necessary. Also, you should not feel compelled to follow this procedure at any cost. Keep this in mind as you study the last two examples in this section.

EXAMPLE   7   Expressing a certain product as a function of a single variable

Two numbers add up to 8. Express the product P of these two numbers in terms of a single variable.

SOLUTION

If we call the two numbers x and $8 - x$, then their product P is given by

$$P(x) = x(8 - x) = 8x - x^2$$

That's it. This last equation expresses the product as a function of the variable x. Since there are no restrictions on x (other than its being a real number), the domain of this function is $(-\infty, \infty)$.

EXAMPLE	8	Expressing revenue as a function of one variable

In economics, the revenue R generated by selling x units at a price of p dollars per unit is given by

$$R = x \cdot p$$

In Figure 7, we are given a hypothetical function relating the selling price of a certain item to the number of units sold. Such a function is called a **demand function.** Express the revenue as a function of x.

SOLUTION

More than anything else, this problem is an exercise in reading. After reading the problem several times, we find that it comes down to this:

given: $R = x \cdot p$ and $p = -\dfrac{1}{3}x + 40$ $(0 \le x \le 117)$

find: an equation expressing R in terms of x

In view of this, we write

$$R = x \cdot p = x\left(-\frac{1}{3}x + 40\right)$$

Thus we have

$$R(x) = -\frac{1}{3}x^2 + 40x \qquad (0 \le x \le 117)$$

This is the required function. It allows us to calculate the revenue when we know the number of units sold. Note that this revenue function is a quadratic function.

QUESTION FOR REVIEW AND ALSO PREVIEW   How would you find the maximum revenue in this case?

Figure to the left:

p (Price per unit)

$p = -\dfrac{1}{3}x + 40 \ (0 \le x \le 117)$

Number of units sold x

Figure 7

EXERCISE SET 4.4

A

1. (a) The perimeter of a rectangle is 16 cm. Express the area of the rectangle in terms of the width x.
     Suggestion: First reread Example 1.
   (b) The area of a rectangle is 85 cm^2. Express the perimeter as a function of the width x.

2. A rectangle is inscribed in a circle of diameter 12 in.
   (a) Express the perimeter of the rectangle as a function of its width x.   *Suggestion*: First reread Example 2.
   (b) Express the area of the rectangle as a function of its width x.

3. A point $P(x, y)$ is on the curve $y = x^2 + 1$.
   (a) Express the distance from P to the origin as a function of x.   *Suggestion*: First reread Example 3.

   (b) In part (a), you expressed the length of a certain line segment as a function of x. Now express the slope of that line segment in terms of x.

4. A point $P(x, y)$ lies on the curve $y = \sqrt{x}$, as shown in the figure.

(a) Express the area of the shaded triangle as a function of x. *Suggestion*: First reread Example 4.

(b) Express the perimeter of the shaded triangle in terms of x.

5. A piece of wire πy inches long is bent into a circle.

   (a) Express the area of the circle as a function of y. *Suggestion*: First reread Example 5.

   (b) If the original piece of wire were bent into a square instead of a circle, how would you express the area in terms of y?

6. The volume of a right circular cylinder is 20 in.^{3} Express the total surface area of the cylinder as a function of r, the radius of the base. *Suggestion*: First reread Example 6.

7. Two numbers add to 16.

   (a) Express the product of the two numbers in terms of a single variable. *Suggestion*: First reread Example 7.

   (b) Express the sum of the squares of the two numbers in terms of a single variable.

   (c) Express the difference of the cubes of the two numbers in terms of a single variable. (There are two answers.)

   (d) What happens when you try to express the average of the two numbers in terms of one variable?

8. The product of two numbers is 16. Express the sum of the squares of the two numbers as a function of a single variable.

9. Given a demand function $p = -\frac{1}{4}x + 8$, express the revenue as a function of x. *Suggestion*: First reread Example 8.

10. In Example 1, we considered a rectangle with perimeter 100 cm. We found that the area of such a rectangle is given by $A(x) = 50x - x^2$, where x is the width of the rectangle. Compute the numbers $A(1)$, $A(10)$, $A(20)$, $A(25)$, and $A(35)$. Which width x seems to yield the largest area $A(x)$?

11. In Example 2, we considered a rectangle inscribed in a circle of diameter 8 cm. We found that the perimeter of such a rectangle is given by

$$P(x) = 2x + 2\sqrt{64 - x^2}$$

where x is the width of the rectangle.

   (a) Use a calculator to complete the following table, rounding each answer to two decimal places.

x	1	2	3	4	5	6	7
P(x)							

   (b) In your table, what is the largest value for $P(x)$? What is the width x in this case?

   (c) Using calculus, it can be shown that among all possible widths x, the width $x = 4\sqrt{2}$ cm yields the largest possible perimeter. Use a calculator to compute the perimeter in this case, and check to see that the value you obtain is indeed larger than all of the values obtained in part (a).

12. In economics, the demand function for a given commodity tells us how the unit price p is related to the number

of units x that are sold. Suppose we are given a demand function

$$p = 5 - \frac{x}{4} \qquad (p \text{ in dollars})$$

   (a) Graph this demand function.

   (b) How many units can be sold when the unit price is $3? Locate the point on the graph of the demand function that conveys this information.

   (c) To sell 12 items, how should the unit price be set? Locate the point on the graph of the demand function that conveys this information.

   (d) Find the revenue function corresponding to the given demand function. (Use the formula $R = x \cdot p$.) Graph the revenue function.

   (e) Find the revenue when $x = 2$, when $x = 8$, and when $x = 14$.

   (f) According to your graph in part (d), which x-value yields the greatest revenue? What is that revenue? What is the corresponding unit price?

13. Let $2s$ denote the length of the side of an equilateral triangle.

   (a) Express the height of the triangle as a function of s.

   (b) Express the area of the triangle as a function of s.

   (c) Use the function you found in part (a) to determine the height of an equilateral triangle, each side of which is 8 cm long.

   (d) Use the function you found in part (b) to determine the area of an equilateral triangle, each side of which is 5 in. long.

14. If x denotes the length of a side of an equilateral triangle, express the area of the triangle as a function of x.

15. The height of a right circular cylinder is twice the radius. Express the volume as a function of the radius.

16. Using the information given in Exercise 15, express the radius as a function of the volume.

17. The volume of a right circular cylinder is 12π in.^{3}

   (a) Express the height as a function of the radius.

   (b) Express the total surface area as a function of the radius.

18. The total surface area of a right circular cylinder is 14 in.^{2} Express the volume as a function of the radius.

19. The volume V and the surface area S of a sphere of radius r are given by the formulas $V = \frac{4}{3}\pi r^3$ and $S = 4\pi r^2$. Express V as a function of S.

20. The base of a rectangle lies on the x-axis, while the upper two vertices lie on the parabola $y = 10 - x^2$. Suppose that the coordinates of the upper right vertex of the rectangle are (x, y). Express the area of the rectangle as a function of x.

21. The hypotenuse of a right triangle is 20 cm. Express the area of the triangle as a function of the length x of one of the legs.

22. (a) Express the area of the shaded triangle in the accompanying figure as a function of x.
(b) Express the perimeter of the shaded triangle as a function of x.

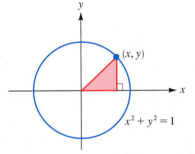

$x^2 + y^2 = 1$

23. For the following figure, express the length AB as a function of x.   *Hint:* Note the similar triangles.

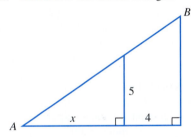

24. Five hundred feet of fencing is available to enclose a rectangular pasture alongside a river, which serves as one side of the rectangle (so only three sides require fencing—see the figure). Express the area of the rectangular pasture as a function of x.

ⓖ *For Exercises 25–32:*
(a) *Is this a quadratic function? Use a graphing utility to draw the graph.*
(b) *How many turning points are there within the given interval?*
(c) *On the given interval, does the function have a maximum value? A minimum value?*

25. $A(x) = 50x - x^2$, $0 < x < 50$ (This is the function that we obtained in Example 1.)
26. $P(x) = 2x + 2\sqrt{64 - x^2}$, $0 < x < 8$ (from Example 2)
27. $D(x) = \sqrt{x^2 - x + 1}$, $x \geq 0$ (from Example 3)
28. $A(x) = 8x - \frac{1}{2}x^3$, $0 < x < 4$ (from Example 4)
29. $S(r) = 2\pi r^2 + 20/r$, $r > 0$ (from Example 6)
30. $R(x) = x(-\frac{1}{3}x + 40)$, $x \geq 0$ (from Example 8)

31. (a) $f(x) = x + \dfrac{1}{x}$, $x > 0$ **(b)** $g(x) = x - \dfrac{1}{x}$, $x > 0$

32. (a) $F(x) = \sqrt{x} - \dfrac{1}{x}$ **(b)** $G(x) = \sqrt{x} + \dfrac{1}{x}$

ⓖ 33. (a) Suppose that the product of two positive numbers is $\sqrt{11}$. Express the sum of the two numbers as a function of a single variable, and then use a graphing utility to draw the graph. Based on the graph, does the sum have a minimum value or maximum value?
(b) Suppose that the sum of two positive numbers is $\sqrt{11}$. Express the product of the two numbers as a function of a single variable. Without drawing a graph, explain why the product has a maximum value.

B

34. A piece of wire 4 m long is cut into two pieces, then each piece is bent into a square. Express the combined area of the two squares in terms of one variable.

35. A piece of wire 3 m long is cut into two pieces. Let x denote the length of the first piece and $3 - x$ the length of the second. The first piece is bent into a square and the second into a rectangle in which the width is half the length. Express the combined area of the square and the rectangle as a function of x. Is the resulting function a quadratic function?

In Exercises 36–39, refer to the following figure, which displays a right circular cone along with the formulas for the volume V and the lateral surface area S.

$$V = \frac{1}{3}\pi r^2 h$$
$$S = \pi r \sqrt{r^2 + h^2}$$

36. The volume of a right circular cone is 12π cm^3.
(a) Express the height as a function of the radius.
(b) Express the radius as a function of the height.

37. Suppose that the height and radius of a right circular cone are related by the equation $h = \sqrt{3}r$.
(a) Express the volume as a function of r.
(b) Express the lateral surface area as a function of r.

38. The volume of a right circular cone is 2 ft.^{3} Show that the lateral surface area as a function of r is given by

$$S = \frac{\sqrt{\pi^2 r^6 + 36}}{r}$$

39. In a certain right circular cone, the volume is numerically equal to the lateral surface area.
(a) Express the radius as a function of the height.
(b) Express the height as a function of the radius.

40. A line is drawn from the origin O to a point $P(x, y)$ in the first quadrant on the graph of $y = 1/x$. From point P, a line is drawn perpendicular to the x-axis, meeting the x-axis at B.

(a) Draw a figure of the situation described.

(b) Express the perimeter of $\triangle OPB$ as a function of x.

(c) Try to express the area of $\triangle OPB$ as a function of x. What happens?

41. A piece of wire 14 in. long is cut into two pieces. The first piece is bent into a circle, the second into a square. Express the combined total area of the circle and the square as a function of x, where x denotes the length of the wire that is used for the circle.

42. A wire of length L is cut into two pieces. The first piece is bent into a square, the second into an equilateral triangle. Express the combined total area of the square and the triangle as a function of x, where x denotes the length of wire used for the triangle. (Here, L is a constant, not another variable.)

43. An athletic field with a perimeter of $\frac{1}{4}$ mile consists of a rectangle with a semicircle at each end, as shown in the figure. Express the area of the field as a function of r, the radius of the semicircle.

44. A square of side x is inscribed in a circle. Express the area of the circle as a function of x.

45. An equilateral triangle of side x is inscribed in a circle. Express the area of the circle as a function of x.

46. (a) Refer to the accompanying figure. An offshore oil rig is located at point A, which is 10 miles out to sea. An oil pipeline is to be constructed from A to a point C on the shore and then to an oil refinery at point D, farther up the coast. If it costs \$8000 per mile to lay the pipeline in the sea and \$2000 per mile on land, express the cost of laying the pipeline in terms of x, where x is the distance from B to C.

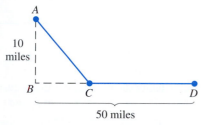

(b) Use a calculator and your result in part (a) to complete the following table, rounding the answers to the nearest 100 dollars.

x (miles)	0	10	20	30	40	50
Cost (dollars)						

(c) Based on the results in part (b), does it appear that the cost function is increasing, decreasing, or neither

on the interval $0 \le x \le 50$? Which of the x-values in the table yields the lowest cost?

(d) Use a calculator and your result in part (a) to complete the following table, rounding the answers to the nearest 100 dollars.

x (miles)	0	4	8	12	16	20
Cost (dollars)						

(e) Again, answer the questions in part (c), but now take into account the results in part (d) also.

47. (a) An open-top box is constructed from a 6-by-8-inch rectangular sheet of tin by cutting out equal squares at each corner and then folding up the flaps, as shown in the figure. Express the volume of the box as a function of x, the length of the side of each cutout square.

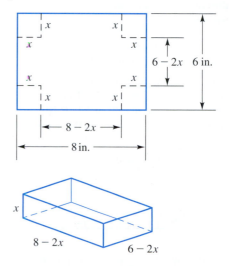

(b) Use a calculator and your result in part (a) to complete the following table. Round the answers to the nearest 0.5 in.[3]

x (in.)	0	0.5	1.0	1.5	2.0	2.5	3.0
Volume (in.[3])							

(c) Based on the results in part (b), which x-value in the interval $0 \le x \le 3$ appears to yield the largest volume for the box?

(d) Use a calculator and your result in part (a) to complete the following table. Round off the answers to the nearest 0.1 in.[3]

x (in.)	0.8	0.9	1.0	1.1	1.2	1.3	1.4
Volume (in.[3])							

(e) Again, answer the questions in part (c), but now take into account the results in part (d), too.

48. Follow Exercise 47(a), but assume that the original piece of tin is a square, 12 in. on each side.

49. A Norman window is in the shape of a rectangle surmounted by a semicircle, as shown in the figure. Assume that the perimeter of the window is 32 ft.

(a) Express the area of the window as a function of r, the radius of the semicircle.

(b) The function you were asked to find in part (a) is a quadratic function, so its graph is a parabola. Does the parabola open upward or downward? Does it pass through the origin? Show that the vertex of the parabola is

$$\left(\frac{32}{\pi + 4}, \frac{512}{\pi + 4}\right)$$

50. Refer to the following figure. Express the lengths CB, CD, BD, and AB in terms of x.   *Hint*: Recall the theorem from geometry stating that in a 30°-60°-90° right triangle, the side opposite the 30° angle is half the hypotenuse.

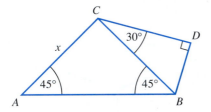

51. Refer to the following figure; let s denote the ratio of y to z.

(a) Express y as a function of s.

(b) Express s as a function of y.

(c) Express z as a function of s.

(d) Express s as a function of z.

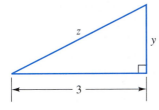

52. Refer to Example 4.

(a) What is the x-intercept of the curve $y = 16 - x^2$ in Figure 4?

(b) What is the domain of the area function in Example 4, assuming that the point P does not lie on the x- or y-axis?

53. The following figure shows the parabola $y = x^2$ and a line segment $\overline{AP}$ drawn from the point $A(0, -1)$ to the point $P(a, a^2)$ on the parabola.

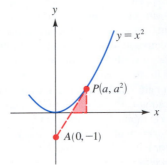

(a) Express the slope of $\overline{AP}$ in terms of a.

(b) Show that the area of the shaded triangle in the figure is given by

$$\text{area} = \frac{a^5}{2(a^2 + 1)}$$

54. A rancher who wishes to fence off a rectangular area finds that the fencing in the east-west direction will require extra reinforcement due to strong prevailing winds. Fencing in the east-west direction will therefore cost $12 per (linear) yard, as opposed to a cost of $8 per yard for fencing in the north-south direction. Given that the rancher wants to spend $4800 on fencing, express the area of the rectangle as a function of x, its width. The required function is in fact a quadratic, so its graph is a parabola. Does the parabola open upward or downward? By considering this graph, find which width x yields the rectangle of largest area. What is this maximum area?

C

55. The following figure shows two concentric squares. Express the area of the shaded triangle as a function of x.

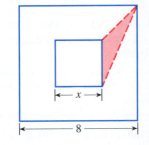

56. The following figure shows two concentric circles of radii r and R. Let A denote the area within the larger circle but outside the smaller one. Express A as a function of x (where x is defined in the figure).

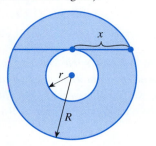

57. A straight line with slope m ($m < 0$) passes through the point $(1, 2)$ and intersects the line $y = 4x$ at a point in the first quadrant. Let A denote the area of the triangle bounded by $y = 4x$, the x-axis, and the given line of slope m. Express A as a function of m.

58. A line with slope m ($m < 0$) passes through the point (a, b) in the first quadrant and intersects the line $y = Mx$ ($M > 0$) at another point in the first quadrant. Let A denote the area of the triangle bounded by $y = Mx$, the x-axis, and the given line with slope m. Show that A can be written as

$$A = \frac{M(am - b)^2}{2m(m - M)}$$

59. A line with slope m ($m < 0$) passes through the point (a, b) in the first quadrant. Express the area of the triangle bounded by this line and the axes in terms of m.

60. One corner of a sheet of paper of width a is folded over and just reaches the opposite side, as indicated in the figure. Express L, the length of the crease, in terms of x and a.

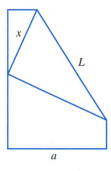

MINI PROJECT

GROUP WORK ON FUNCTIONS OF TIME

In each of the examples and exercises of this section, the goal was to express a certain quantity as a function of a single variable. In the three problems that follow, that single variable will be time t. For each problem, after the group finds a solution, each individual should write up the solution on his or her own. (The problems are in order of increasing difficulty.)

1. (a) This problem is based on Example 1. So first, have someone in your group present the example.
  (b) Under the conditions of Example 1, assume the following additional information. The width x is increasing at the rate of 3 cm/min, and at time $t = 0$ min, the width x is 2 cm. Express the area of the rectangle as a function of time t.
  (c) During the time interval from $t = 0$ min to $t = 16$ min, when is the area of the rectangle increasing? When is it decreasing? Why are we using the value $t = 16$ min?

2. Starting at time $t = 0$ hr, two cars move away from the same intersection. Car A goes north at 40 miles/hr and car B goes east at 30 miles/hr. Express the distance between the cars as a function of the time t. Is your result a linear function, a quadratic function, or neither?

3. Car C passes through an intersection, heading north at 40 miles/hr. One quarter hour later, car D goes through the intersection, heading east at 30 miles/hr. Express the distance between the two cars as a function of time. Is your result a linear function, a quadratic function, or neither?

4.5   MAXIMUM AND MINIMUM PROBLEMS

. . . problems on maxima and minima, although new features in an English textbook, stand so little in need of apology with the scientific public that I offer none. —G. Chrystal in his preface to *Textbook of Algebra* (1886)

In daily life it is constantly necessary to choose the best possible (optimal) solution. A tremendous number of such problems arise in economics and in technology. In such cases it is frequently useful to resort to mathematics. —V. K. Tikhomirov in *Stories about Maxima and Minima* (translated from the Russian by Abe Shenitzer) (Providence, R.I.: The American Mathematical Society, 1990)

You have already seen several examples of maximum and minimum problems in Section 4.2. Before reading further in the present section, you should first review Examples 5 through 7 on pages 273–276.

Actually, we begin this section's discussion of maximum and minimum problems at a more intuitive level. Consider, for example, the following question: If two numbers add to 9, what is the largest possible value of their product? To gain some insight here, we carry out a few preliminary calculations. See Table 1.

We have circled the 20 in the right-hand column of Table 1 because it *appears* to be the largest product. We say "appears" because our table is incomplete. For instance, what if we allowed x- and y-values that are not whole numbers? Might we get a product exceeding 20? Table 2 shows the results of some additional calculations along these lines.

As you can see from Table 2, there is a product exceeding 20—namely, 20.25. Now the question is, if we further expand our tables, can we find an even larger product, one exceeding 20.25? And here we have come about as far as we want to go using this approach involving arithmetic and tables. For no matter what candidate we come up with for the largest product, there will always be the question of whether we might do still better using a larger table.

Nevertheless, this approach was useful, for it showed us what is really at the heart of a typical maximum or minimum problem. Essentially, we are trying to sort through an infinite number of possible cases and pick out the required extreme case. In the example at hand, there are infinitely many pairs of numbers x and y adding to 9. We are asked to look at the products of all of these pairs and see which (if any!) is the largest. Example 1 shows how to apply our knowledge of quadratic functions to solve this problem in a definitive manner.

TABLE 1

x	y	$x + y$	xy
−2	11	9	−22
−1	10	9	−10
0	9	9	0
1	8	9	8
2	7	9	14
3	6	9	18
4	5	9	⟨20⟩

TABLE 2

x	y	$x + y$	xy
1	8	9	8
1.5	7.5	9	11.25
2	7	9	14
2.5	6.5	9	16.25
3	6	9	18
3.5	5.5	9	19.25
4	5	9	20
4.5	4.5	9	⟨20.25⟩

EXAMPLE   1   Maximizing a product

Two numbers add to 9. What is the largest possible value for their product?

SOLUTION
Call the two numbers x and $9 - x$. Then their product P is given by

$$P = x(9 - x) = 9x - x^2$$

The graph of this quadratic function is the parabola in Figure 1. Note the accompanying calculations for the vertex. As Figure 1 and the accompanying

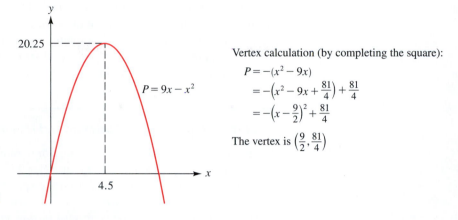

$P = 9x - x^2$

Vertex calculation (by completing the square):

$$P = -(x^2 - 9x)$$
$$= -\left(x^2 - 9x + \frac{81}{4}\right) + \frac{81}{4}$$
$$= -\left(x - \frac{9}{2}\right)^2 + \frac{81}{4}$$

The vertex is $\left(\frac{9}{2}, \frac{81}{4}\right)$

Figure 1

calculations show, the largest value of the product P is 20.25. This is the required solution. (Note from the graph that there is no smallest value of P.)

In the example just completed, we used the technique of completing the square to determine the vertex of a parabola. Another way to determine the vertex, approximately at least, is to use a graphing utility, as in Example 5(a) in Section 4.2. There is a third method and we shall make use of it for the remaining examples in this section. This third method uses the *vertex formula* given in the box that follows. You should check with your instructor to see which of the three methods he or she wants you to use in a given context. We're going to use the vertex formula in the exposition here so that you can better focus on the problem-solving aspects of the examples.

THEOREM   The Vertex Formula

The x-coordinate of the vertex of the parabola $y = ax^2 + bx + c$ $(a \neq 0)$ is given by

$$x = \frac{-b}{2a}$$

EXAMPLE

For the parabola $y = 2x^2 - 12x + 5$ the x-coordinate of the vertex is

$$x = \frac{-b}{2a} = \frac{-(-12)}{2(2)} = 3$$

The vertex formula can be derived by completing the square or by using the method outlined in the mini project at the end of Section 4.2. Here's another way, perhaps the simplest. The parabola $y = ax^2 + bx + c$ will have the same x-coordinate for its vertex as does $y = ax^2 + bx$ because the two graphs are just vertical translates of one another by a distance $|c|$. But due to symmetry, it's easy to find the x-coordinate of the vertex of $y = ax^2 + bx$. As Figure 2 indicates, the x-coordinate of the vertex is midway between the two x-intercepts. To find these x-intercepts, we have

$$ax^2 + bx = 0$$
$$x(ax + b) = 0$$
$$x = 0 \quad \bigg| \quad ax + b = 0$$
$$x = -b/a$$

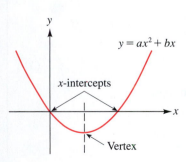

Figure 2

Now we want the x-value that is halfway between $x = 0$ and $x = -b/a$—namely, $\frac{1}{2}(-b/a)$. Thus, the x-coordinate of the vertex is $-b/2a$, as we wished to show.

Except for the fact that we completed the square rather than using the vertex formula, Example 1 indicates the general strategy that we're going to follow for solving the maximum and minimum problems in this section.

Strategy for Solving the Maximum and Minimum Problems in This Section

1. Express the quantity to be maximized or minimized in terms of a single variable. For instance, in Example 1, we found $P = 9x - x^2$. In setting up such functions, you'll want to keep in mind the four-step procedure used in Section 4.4.
2. Assuming that the function you have determined is a quadratic, note whether its graph, a parabola, opens upward or downward. Check whether this is consistent with the requirements of the problem. For instance, the parabola in Figure 1 opens downward; so it makes sense to look for a *largest*, not a smallest, value of P. Now use the vertex formula $x = -b/2a$ to locate the x-coordinate of the vertex. (If the function is not a quadratic but is closely related to a quadratic, these ideas may still apply. See, for instance, Examples 6 and 7 in Section 4.2.)
3. After you have determined the x-coordinate of the vertex, you must relate that information to the original question. In Example 1, for instance, we were asked for the product P, not for x.

EXAMPLE	2	Finding the dimensions of the rectangle with maximum area

Among all rectangles having a perimeter of 10 ft, find the dimensions (length and width) of the one with the greatest area.

SOLUTION

First we want to set up a function that expresses the area of the rectangle in terms of a single variable. In doing this, we'll be guided by the four-step procedure that we used in Section 4.4. Figure 3(a) displays the given information. Our problem is to determine the dimensions of the rectangle that has the greatest area. As Figure 3(b) indicates, we can label the dimensions x and y. Since the perimeter is given as 10 ft, we have

$$2x + 2y = 10$$
$$x + y = 5$$
$$y = 5 - x \tag{1}$$

Letting A denote the area of the rectangle, we can write

$$A = xy$$
$$A(x) = x(5 - x) \qquad \text{substituting for } y \text{ using equation (1)}$$
$$= 5x - x^2$$

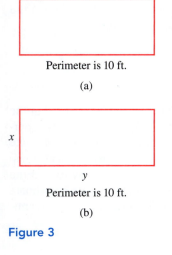

Perimeter is 10 ft.

(a)

Perimeter is 10 ft.

(b)

Figure 3

GRAPHICAL PERSPECTIVE

Figure 4
The graph of the area function,
$A(x) = 5x - x^2$, is a parabola that
opens downward. The maximum
area occurs when $x = 2.5$ ft.

This expresses the area of the rectangle in terms of the width x. Since the graph of this quadratic function is a parabola that opens downward, it does make sense to talk about the maximum. We calculate the x-coordinate of the vertex:

$$x = \frac{-b}{2a}$$

$$= \frac{-5}{2(-1)} = 2.5 \text{ ft}$$

This is the width of the rectangle with the greatest area. Now, the problem asks for the length and width of this rectangle, not its area. So we don't want to calculate $A(x)$; rather, we want y. Using equation (1), we have

$$y = 5 - x = 5 - 2.5 = 2.5 \text{ ft}$$

Thus, among all rectangles having a perimeter of 10 ft, the one with the greatest area is actually the square with dimensions of 2.5 ft by 2.5 ft. See Figure 4 for a graphical view of this result.

EXAMPLE 3 Finding the maximum height of a baseball

Suppose that a baseball is tossed straight up and that its height as a function of time is given by the function

$$h(t) = -16t^2 + 64t + 6$$

where $h(t)$ is measured in feet, t is in seconds, and $t = 0$ corresponds to the instant that the ball is released. What is the maximum height of the ball?

GRAPHICAL PERSPECTIVE

Figure 5
The graph of the height function
$h(t) = -16t^2 + 64t + 6$ is a
parabola that opens downward.
The maximum height is 70 ft, and
this occurs when $t = 2$ sec.

SOLUTION
The given function tells us how the height $h(t)$ of the ball depends upon the time t. We want to know the largest possible value for $h(t)$. Since the graph of the given height function is a parabola opening downward, we can determine the largest value of $h(t)$ just by finding the vertex of the parabola. We use the vertex formula $x = -b/2a$, letting t play the role of x. This yields

$$t = \frac{-b}{2a} = \frac{-64}{2(-16)} = 2$$

Thus, when $t = 2$ sec, the ball reaches its maximum height. To find that maximum height (and the problem does ask for height, not time), we substitute the value $t = 2$ in the equation $h(t) = -16t^2 + 64t + 6$ to obtain

$$h(2) = -16(2)^2 + 64(2) + 6$$
$$= 70 \text{ ft} \quad \text{(Check the arithmetic.)}$$

The maximum height of the ball is 70 ft. See Figure 5 for a graphical perspective. *Caution*: The path of the baseball is not the parabola. Why?

In the next example, we look for an input to minimize a function of the form $y = \sqrt{f(x)}$, where f is a quadratic function with values that are always nonnegative.

(There was an example similar to this in Section 4.2.) As indicated in Figure 6, we need only find the input that minimizes the quadratic function $y = f(x)$, because this same input also minimizes $y = \sqrt{f(x)}$. (The *outputs* of the two functions are different; but the same *input* serves to minimize both functions.)

| EXAMPLE | 4 | **Working with a function that is closely related to a quadratic** |

Which point on the curve $y = \sqrt{x}$ is closest to the point $(1, 0)$?

SOLUTION
In Figure 7, we let D denote the distance from a point (x, y) on the curve to the point $(1, 0)$. We are asked to find out exactly which point (x, y) will make the distance D as small as possible. Using the distance formula, we have

$$D = \sqrt{(x - 1)^2 + (y - 0)^2}$$
$$= \sqrt{x^2 - 2x + 1 + y^2} \tag{2}$$

This expresses D in terms of both x and y. To express D in terms of x alone, we use the given equation $y = \sqrt{x}$ to substitute for y in equation (3). This yields

$$D(x) = \sqrt{x^2 - 2x + 1 + (\sqrt{x})^2}$$
$$= \sqrt{x^2 - x + 1}$$

Now, according to the remarks just prior to this example, the x-value that minimizes the function $D(x) = \sqrt{x^2 - x + 1}$ is the same x-value that minimizes $y = x^2 - x + 1$. (Actually, to apply those remarks, we need to know that the expression $x^2 - x + 1$ is nonnegative for all inputs x. See Exercise 43 for this.) Since the graph of $y = x^2 - x + 1$ is a parabola that opens upward, the required x-value is the x-coordinate of the vertex, and we have

$$x = \frac{-b}{2a} = \frac{-(-1)}{2(1)} = \frac{1}{2}$$

Now that we have the x-coordinate of the vertex, we want to calculate the y-coordinate. [We don't want to calculate $D(x)$ because this particular problem asks us to specify a *point*, not a distance. Go back and reread the wording of the question if you are not clear on this.] Substituting $x = 1/2$ in the equation $y = \sqrt{x}$ yields

$$y = \sqrt{\frac{1}{2}} = \frac{\sqrt{1}}{\sqrt{2}} = \frac{1}{\sqrt{2}}$$
$$= \frac{1}{\sqrt{2}} \cdot \frac{\sqrt{2}}{\sqrt{2}} = \frac{\sqrt{2}}{2}$$

Thus, among all points on the curve $y = \sqrt{x}$, the point closest to $(1, 0)$ is $(1/2, \sqrt{2}/2)$. *Question*: Why would it not have made sense to ask instead for the point farthest from $(1, 0)$?

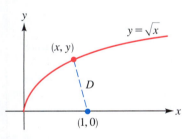

Figure 6
The same input x_0 minimizes both functions $y = f(x)$ and $y = \sqrt{f(x)}$ [provided $f(x)$ is nonnegative for all x].

Figure 7

| EXAMPLE | 5 | Finding the dimensions that maximize an area |

Suppose that you have 600 m of fencing with which to build two adjacent rectangular corrals. The two corrals are to share a common fence on one side, as shown in Figure 8. Find the dimensions x and y so that the total enclosed area is as large as possible.

SOLUTION

You have 600 m of fencing to be set up as shown in Figure 8. The question is how to choose x and y so that the total area is maximum. Since the total length of fencing is 600 m, we can relate x and y by writing

$$x + x + x + y + y = 600$$
$$3x + 2y = 600$$
$$2y = 600 - 3x \qquad (4)$$
$$y = 300 - \frac{3}{2}x$$

Letting A denote the total area, we have

$$A = xy$$
$$A(x) = x\left(300 - \frac{3}{2}x\right) \qquad \text{using equation (4) to substitute for } y$$
$$= 300x - \frac{3}{2}x^2$$

This last equation expresses the area as a function of x. Note that the graph of this function is a parabola opening downward, so it does make sense to talk about a maximum. The x-coordinate of the vertex is

$$x = \frac{-b}{2a} = \frac{-300}{2(-3/2)} = \frac{300}{3} = 100$$

Now that we have the x-value that maximizes the area $A(x)$, the remaining dimension y can be calculated using equation (4):

$$y = 300 - \frac{3}{2}x = 300 - \frac{3}{2}(100) = 150$$

Thus, by choosing x to be 100 m and y to be 150 m, the total area in Figure 8 will be as large as possible. Incidentally, note that the exact location of the fence dividing the two corrals does not influence our work or the final answer.

In the next example, a demand function is given and we want to maximize the revenue. As background, you should review Example 8 in Section 4.4.

| EXAMPLE | 6 | A maximization problem involving revenue |

Suppose that the following demand function relates the selling price, p, of an item to the quantity sold, x:

$$p = -\frac{1}{3}x + 40 \qquad (p \text{ in dollars})$$

x x x

y

Figure 8

For which value of x will the revenue be a maximum? Compute the corresponding unit price and maximum revenue.

SOLUTION

First, recall the formula for revenue given on page 304.

$$R = \text{numbers of units} \times \text{price per unit}$$

Using this, we have

$$R = x \cdot p$$

$$R(x) = x\left(-\frac{1}{3}x + 40\right) = -\frac{1}{3}x^2 + 40x \tag{5}$$

We want to know which value of x yields the largest revenue $R(x)$. Since the graph of the revenue function in equation (5) is a parabola opening downward, the required x-value is the x-coordinate of the vertex of the parabola. We have

$$x = \frac{-b}{2a} = \frac{-40}{2(-1/3)} = \frac{20}{1/3} = 60$$

This tells us that to earn the maximum revenue, 60 items should be sold. We calculate the corresponding unit price by substituting $x = 60$ in the given equation $p = -\frac{1}{3}x + 40$. This yields

$$p = -\frac{1}{3}(60) + 40$$

$$= -20 + 40 = 20$$

Thus, setting the price at \$20 per item will maximize the revenue. That maximum revenue can be calculated by using $x = 60$ in equation (5). More simply, since we now have both values $x = 60$ and $p = 20$, we can compute

$$R = x \cdot p = 60 \times 20 = 1200$$

The maximum revenue is \$1200.

GRAPHICAL PERSPECTIVE

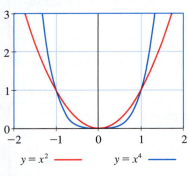

$y = x^2$ ——— $y = x^4$ ———

Figure 9

The next example, the last for this section, is similar to Example 3 in that we are looking for the minimum value of a function that, although not a quadratic function, is closely related to one. Also, in this example you'll see a translation of the function $y = x^4$. Since we have not yet studied this function, for background, we show its graph along with that of the familiar $y = x^2$ in Figure 9. In the next section we do study functions of the form $y = x^n$, where n is a positive integer.

EXAMPLE   7   Minimizing the length of a connecting road

Figure 10 shows portions of two train routes. One route follows the curve $y = x^4 + 10$ and the other follows $y = 4x^2$. Assume that distance along both axes is measured in miles. The railroad wants to construct a north-south maintenance road $\overline{PQ}$ between the two routes, as indicated in the figure. Where should

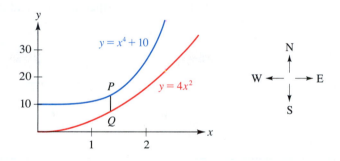

Figure 10

Portions of two train routes and a north-south maintenance road $\overline{PQ}$.

the road be located so that it is as short as possible? What is the minimum length?

SOLUTION

We want to determine points P and Q on the curves $y = x^4 + 10$ and $y = 4x^2$, respectively, so that the distance from P to Q is as small as possible. Since P and Q lie on a north-south line, they have the *same* x-coordinate, and we can write the coordinates of P and Q as

$$P(x, x^4 + 10) \quad \text{and} \quad Q(x, 4x^2)$$

The vertical distance PQ between these two points is just the difference in the y-coordinates:

$$PQ = (x^4 + 10) - 4x^2 = x^4 - 4x^2 + 10$$

So the function that we want to minimize is

$$y = x^4 - 4x^2 + 10 \tag{6}$$

The substitution $x^2 = t$ will transform this into a quadratic function. (You saw this substitution technique before, in Example 7 of Section 4.2.) Note that if $x^2 = t$, then $x^4 = t^2$, and equation (6) becomes

$$y = t^2 - 4t + 10 \tag{7}$$

Since the graph of equation (7) will be a parabola opening upward, the input t that yields a minimum value for this function is

$$t = \frac{-b}{2a} = \frac{-(-4)}{2(1)} = 2$$

That gives us t, but what about x? Substituting the value $t = 2$ in the equation $x^2 = t$ gives us $x^2 = 2$ and consequently $x = \sqrt{2}$. (We're using the positive square root because x is in the first quadrant.) With this value for x, we can calculate the y-coordinates of P and Q as follows:

y-coordinate of P	y-coordinate of Q
$(\sqrt{2})^4 + 10 = 4 + 10 = 14$	$4(\sqrt{2})^2 = 8$

Putting things together now, the road should connect the points P and Q given by

$$P: (\sqrt{2}, 14) \approx (1.41, 14) \quad \text{and} \quad Q: (\sqrt{2}, 8) \approx (1.41, 8)$$

The length of the road is $14 - 8 = 6$ mi.

▍EXERCISE SET 4.5

A

1. Two numbers add to 5. What is the largest possible value of their product?

2. Find two numbers adding to 20 such that the sum of their squares is as small as possible.

3. The difference of two numbers is 1. What is the smallest possible value for the sum of their squares?

4. For each quadratic function, state whether it would make sense to look for a highest or a lowest point on the graph. Then determine the coordinates of that point.
   (a) $y = 2x^2 - 8x + 1$
   (b) $y = -3x^2 - 4x - 9$
   (c) $h = -16t^2 + 256t$
   (d) $f(x) = 1 - (x + 1)^2$
   (e) $g(t) = t^2 + 1$
   (f) $f(x) = 1000x^2 - x + 100$

5. Among all rectangles having a perimeter of 25 m, find the dimensions of the one with the largest area.

6. What is the largest possible area for a rectangle with a perimeter of 80 cm?

7. What is the largest possible area for a right triangle in which the sum of the lengths of the two shorter sides is 100 in.?

8. The perimeter of a rectangle is 12 m. Find the dimensions for which the diagonal is as short as possible.

9. Two numbers add to 6.
   (a) Let T denote the sum of the squares of the two numbers. What is the smallest possible value for T?
   (b) Let S denote the sum of the first number and the square of the second. What is the smallest possible value for S?
   (c) Let U denote the sum of the first number and twice the square of the second number. What is the smallest possible value for U?
   (d) Let V denote the sum of the first number and the square of twice the second number. What is the smallest possible value for V?

10. Suppose that the height of an object shot straight up is given by $h = 512t - 16t^2$. (Here h is in feet and t is in seconds.) Find the maximum height and the time at which the object hits the ground.

11. A baseball is thrown straight up, and its height as a function of time is given by the formula $h = -16t^2 + 32t$ (where h is in feet and t is in seconds).
    (a) Find the height of the ball when $t = 1$ sec and when $t = 3/2$ sec.
    (b) Find the maximum height of the ball and the time at which that height is attained.
    (c) At what times is the height 7 ft?

12. Find the point on the curve $y = \sqrt{x}$ that is nearest to the point $(3, 0)$.

13. Which point on the curve $y = \sqrt{x - 2} + 1$ is closest to the point $(4, 1)$? What is this minimum distance?

14. Find the coordinates of the point on the line $y = 3x + 1$ closest to $(4, 0)$.

15. (a) What number exceeds its square by the greatest amount?
    (b) What number exceeds twice its square by the greatest amount?

16. Suppose that you have 1800 m of fencing with which to build three adjacent rectangular corrals, as shown in the figure. Find the dimensions so that the total enclosed area is as large as possible.

17. Five hundred feet of fencing is available for a rectangular pasture alongside a river, the river serving as one side of the rectangle (so only three sides require fencing). Find the dimensions yielding the greatest area.

18. Let $A = 3x^2 + 4x - 5$ and $B = x^2 - 4x - 1$. Find the minimum value of $A - B$.

19. Let $R = 0.4x^2 + 10x + 5$ and $C = 0.5x^2 + 2x + 101$. For which value of x is $R - C$ a maximum?

20. Suppose that the revenue generated by selling x units of a certain commodity is given by $R = -\frac{1}{5}x^2 + 200x$. Assume that R is in dollars. What is the maximum revenue possible in this situation?

21. Suppose that the function $p = -\frac{1}{4}x + 30$ relates the selling price p of an item to the number of units x that are sold. Assume that p is in dollars. For which value of x will the corresponding revenue be a maximum? What is this maximum revenue and what is the unit price?

22. The action of sunlight on automobile exhaust produces air pollutants known as *photochemical oxidants*. In a study of cross-country runners in Los Angeles, it was shown that running performances can be adversely affected when the oxidant level reaches 0.03 part per million. Suppose that on a given day, the oxidant level L is approximated by the formula

$$L = 0.059t^2 - 0.354t + 0.557 \qquad (0 \le t \le 7)$$

where t is measured in hours, with $t = 0$ corresponding to 12 noon, and L is in parts per million. At what time is the oxidant level L a minimum? At this time, is the oxidant level high enough to affect a runner's performance?

23. (a) Find the smallest possible value of the quantity $x^2 + y^2$ under the restriction that $2x + 3y = 6$.
    (b) Find the radius of the circle whose center is at the origin and that is tangent to the line $2x + 3y = 6$. How does this answer relate to your answer in part (a)?

24. (a) Find the coordinates of the vertex of the parabola $y = 2x^2 - 4x + 7$. Then, on the same set of axes, graph this parabola along with $y = x^2$.
    (b) Rework Example 7 using the two curves in part (a) rather than the two used in Example 7.

Ⓖ 25. (a) Using a graphing utility, graph the two functions $y = x^4 + 2$ and $y = x^2$ in the viewing rectangle $[0, 4, 1] \times [0, 16, 2]$. If your graphing utility allows, include the y-axis gridlines in the picture; they'll be helpful in part (b).
    Ⓖ (b) Suppose that we use the two functions in part (a) as the given functions in Example 7. Based on the graph in part (a), make a rough estimate for the minimum length of the road.
    (c) Use the technique of Example 7 to find the exact value for the minimum length.

B

26. Through a type of chemical reaction known as *autocatalysis,* the human body produces the enzyme trypsin from the enzyme trypsinogen. (Trypsin then breaks down proteins into amino acids, which the body needs for growth.) Let r denote the rate of the chemical reaction in which trypsin is formed from trypsinogen. It has been shown experimentally that $r = kx(a - x)$, where k is a positive constant, a is the initial amount of trypsinogen, and x is the amount of trypsin produced (so x increases as the reaction proceeds). Show that the reaction rate r is a maximum when $x = a/2$. In other words, the speed of the reaction is greatest when the amount of trypsin formed is half the original amount of trypsinogen.

27. (a) Let $x + y = 15$. Find the minimum value of the quantity $x^2 + y^2$.
    (b) Let C be a constant and $x + y = C$. Show that the minimum value of $x^2 + y^2$ is $C^2/2$. Then use this result to check your answer in part (a).

28. Suppose that A, B, and C are positive constants and that $x + y = C$. Show that the minimum value of $Ax^2 + By^2$ occurs when $x = BC/(A + B)$ and $y = AC/(A + B)$.

29. The following figure shows a square inscribed within a unit square. For which value of x is the area of the inner square a minimum? What is the minimum area? *Hint:* Denote the lengths of the two segments that make up the base of the unit square by t and $1 - t$. Now use the Pythagorean theorem and congruent triangles to express x in terms of t.

30. (a) The coordinates of the point on the line $y = mx + b$ that is closest to origin the are given by

$$\left(\frac{-mb}{1 + m^2}, \frac{b}{1 + m^2} \right)$$

    (b) Show that the perpendicular distance from the origin to the line $y = mx + b$ is $|b|/\sqrt{1 + m^2}$. *Suggestion:* Use the result in part (a).
    (c) Use part (b) to show that the perpendicular distance from the origin to the line $Ax + By + C = 0$ is $|C|/\sqrt{A^2 + B^2}$.

31. The point P lies in the first quadrant on the graph of the line $y = 7 - 3x$. From the point P, perpendiculars are drawn to both the x-axis and the y-axis. What is the largest possible area for the rectangle thus formed?

32. Show that the largest possible area for the shaded rectangle shown in the figure is $-b^2/4m$. Then use this to check your answer to Exercise 31.

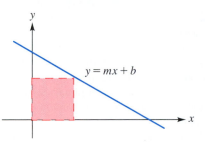

33. Show that the maximum possible area for a rectangle inscribed in a circle of radius R is $2R^2$. *Hint:* Maximize the square of the area.

34. An athletic field with a perimeter of $\frac{1}{4}$ mile consists of a rectangle with a semicircle at each end, as shown in the figure. Find the dimensions x and r that yield the greatest possible area for the rectangular region.

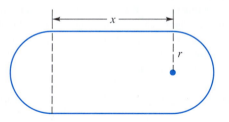

35. A rancher who wishes to fence off a rectangular area finds that the fencing in the east-west direction will require extra reinforcement due to strong prevailing winds. Because of this, the cost of fencing in the east-west direction will be $12 per (linear) yard, as opposed to a cost of $8 per yard for fencing in the north-south direction. Find the dimensions of the largest possible rectangular area that can be fenced for $4800.

36. Let $f(x) = (x - a)^2 + (x - b)^2 + (x - c)^2$, where a, b, and c are constants. Show that $f(x)$ will be a minimum when x is the average of a, b, and c.

37. Let $y = a_1(x - x_1)^2 + a_2(x - x_2)^2$, where a_1, a_2, x_1, and x_2 are all constants. In addition, suppose that a_1 and a_2 are both positive. Show that the minimum of this function occurs when

$$x = \frac{a_1 x_1 + a_2 x_2}{a_1 + a_2}$$

38. Among all rectangles with a given perimeter P, find the dimensions of the one with the shortest diagonal.

39. By analyzing sales figures, the economist for a stereo manufacturer knows that 150 units of a compact disc player can be sold each month when the price is set at $p = \$200$ per unit. The figures also show that for each $10 hike in price, five fewer units are sold each month.

(a) Let x denote the number of units sold per month and let p denote the price per unit. Find a linear function relating p and x. *Hint:* $\Delta p / \Delta x = 10/(-5) = -2$

(b) The revenue R is given by $R = xp$. What is the maximum revenue? At what level should the price be set to achieve this maximum revenue?

40. Let $f(x) = x^2 + px + q$, and suppose that the minimum value of this function is 0. Show that $q = p^2/4$.

41. Among all possible inputs for the function $f(t) = -t^4 + 6t^2 - 6$, which ones yield the largest output?

42. Let $f(x) = x - 3$ and $g(x) = x^2 - 4x + 1$.

(a) Find the minimum value of $g \circ f$.

(b) Find the minimum value of $f \circ g$.

(c) Are the results in parts (a) and (b) the same?

43. This exercise completes a detail mentioned in Example 4. In that example, we used the result that the input minimizing the function $D(x) = \sqrt{x^2 - x + 1}$ is the same input that minimizes the function $y = x^2 - x + 1$. For this result to be applicable, we need to know that the quantity $x^2 - x + 1$ is nonnegative for all x. Parts (a) and (b) each suggest a way to verify that this is indeed the case. [Actually, parts (a) and (b) both show that the quantity $x^2 - x + 1$ is positive for all x.]

(a) Use the fact that $x^2 - x + 1$ can be written as $(x - \frac{1}{2})^2 + \frac{3}{4}$ to explain (in complete sentences) why the quantity $x^2 - x + 1$ is positive for all inputs x.

(b) Compute the discriminant of the quadratic $x^2 - x + 1$ and explain why the graph of $y = x^2 - x + 1$ has no x-intercepts. Then, use this fact to explain why the quantity $x^2 - x + 1$ is always positive. *Hint:* Is the graph of the parabola $y = x^2 - x + 1$ U-shaped up or U-shaped down?

44. Let $f(x) = (x - 1)^2 - 4$.

(a) Sketch the graph of the function f and note that the minimum value occurs when x is 1.

(b) Although the minimum value of the function f occurs when $x = 1$, the minimum value of the function $y = \sqrt{f(x)}$ does not occur when $x = 1$. Explain why this does not contradict the statement in the caption to Figure 6 on page 314.

(c) Find the value(s) of x that minimize the function $y = \sqrt{f(x)}$. *Hint:* What is the domain of $y = \sqrt{f(x)}$?

45. A piece of wire 16 in. long is to be cut into two pieces. Let x denote the length of the first piece and $16 - x$ the length of the second. The first piece is to be bent into a circle and the second piece into a square.

(a) Express the total combined area A of the circle and the square as a function of x.

(b) For which value of x is the area A a minimum?

(c) Using the x-value that you found in part (b), find the ratio of the lengths of the shorter to the longer piece of wire. *Answer:* $\pi/4$

46. A 30-in. piece of string is to be cut into two pieces. The first piece will be formed into the shape of an equilateral triangle and the second piece into a square. Find the length of the first piece if the combined area of the triangle and the square is to be as small as possible.

Exercises 47–52 are maximum-minimum problems in which the function that you set up will not turn out to be a quadratic, nor will it be closely related to one. Thus the vertex formula will not be applicable.

(a) *Set up the appropriate function to be maximized or minimized just as you have practiced in this section and Section 4.4.*

Ⓖ **(b)** *Use a graphing utility to draw the graph of the function and determine, as least approximately, the number or numbers that the problem asks for. (General techniques for determining the exact values are developed in calculus.)*

Exercises 47–52 are quoted from the following calculus textbooks with the permission of the publishers.

EXERCISE   TEXT

47.   *Calculus: A New Horizon,* 6th ed., Howard Anton (New York: John Wiley & Sons, 1998)

48.   *Calculus,,* 6th ed., Roland E. Larson et al. (Boston: Houghton Mifflin, 1997)

49.   *Calculus and Analytic Geometry,* 5th ed., Sherman K. Stein and Anthony Barcellos (New York: McGraw-Hill, 1992)

50.   *Calculus with Analytic Geometry,* 5th ed., C. H. Edwards, Jr. and David E. Penney (Upper Saddle River, N.J.: Prentice-Hall, 1997)

51.   *Calculus: One and Several Variables,* 8th ed., Saturnino L. Salas and Garret J. Etgen (New York: John Wiley & Sons, 1998)

52.   *Calculus,* 4th ed., James Stewart (Pacific Grove, Calif.: Brooks/Cole, 1999)

47. A cylindrical can, open at the top, is to hold 500 cm³ of liquid. Find the height and radius that minimize the amount of material needed to manufacture the can.

48. A power station is on one side of a river that is 1/2 mi. wide, and a factory is 6 mi. downstream on the other side. It costs $6 per foot to run power lines overland and $8 per foot to run them underwater. Find the most economical path for the transmission line from the power station to the factory.

49. What point on the parabola $y = x^2$ is closest to the point $(3, 0)$?

50. A rectangle of fixed perimeter 36 is rotated about one of its sides, thus sweeping out a figure in the shape of a right circular cylinder. [Refer to the accompanying figure.] What is the maximum possible volume of that cylinder?

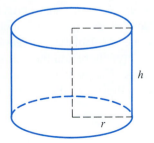

51. Let ABC be a triangle with vertices $A(-3, 0)$, $B(0, 6)$, $C(3, 0)$. Let P be a point on the line segment that joins B to the origin. Find the position of P that minimizes the sum of the distances between P and the vertices.

52. A fence 8 ft tall runs parallel to a tall building at a distance of 4 ft from the building. What is the length of the shortest ladder that will reach from the ground over the fence to the wall of the building?

C

Ⓖ **53. (a)** Use a graphing utility to graph the parabolas $y = x^2 - 2x$, $y = x^2 + 2x$, $y = x^2 - 3x$, and $y = x^2 + 3x$. Check visually that, in each case, the vertex of the parabola appears to lie on the curve $y = -x^2$.

   (b) Prove that for all real numbers k, the vertex of the parabola $y = x^2 + kx$ lies on the curve $y = -x^2$.

Ⓖ **54. (a)** Graph the four parabolas $y = x^2 + 2kx + 1$ corresponding to $k = 2, 3, 0.75, 1.5$. Among the four parabolas, which one appears to have the vertex closest to the origin?

   (b) Let $f(x) = x^2 + 2kx + 1$. Find a positive value for k so that the distance from the origin to the vertex of the parabola is as small as possible. Check that your answer is consistent with your observations in part (a).

55. The figure shows a rectangle inscribed in a given triangle of base b and height h. Find the ratio of the area of the triangle to the area of the rectangle when the area of the rectangle is maximum.

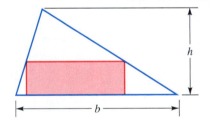

56. A Norman window is in the shape of a rectangle surmounted by a semicircle, as shown in the figure. Assume that the perimeter of the window is P, a constant. Show that the area of the window is a maximum when both x and r are equal to $P/(\pi + 4)$. Show that this maximum area is $\frac{1}{2}P^2/(\pi + 4)$.

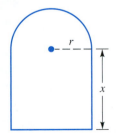

57. A triangle is inscribed in a semicircle of diameter $2R$, as shown in the figure. Show that the smallest possible value for the area of the shaded region is $(\pi - 2)R^2/2$.

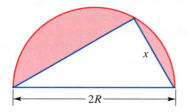

Hint: The area of the shaded region is a minimum when the area of the triangle is a maximum. Find the value of x that maximizes the *square* of the area of the triangle. This will be the same x that maximizes the area of the triangle.

58. (a) Complete the following table. Which x-y pair in the table yields the smallest sum $x + y$?

x	0.5	1	1.5	2	2.5	3	3.5
y							
xy	12	12	12	12	12	12	12
x + y							

(b) Find two positive numbers with a product of 12 and as small a sum as possible. *Hint:* The quantity that you need to minimize is $x + (12/x)$, where $x > 0$. But

$$x + \frac{12}{x} = \left(\sqrt{x} - \sqrt{\frac{12}{x}} \right)^2 + 2\sqrt{12}$$

This last expression is minimized when the quantity within parentheses is zero. Why?

(c) Use a calculator to verify that the two numbers obtained in part (b) produce a sum that is smaller than any of the sums obtained in part (a).

59. What is the smallest possible value for the sum of a positive number and its reciprocal? *Hint:* After setting up the appropriate function, adapt the hint given in Exercise 58(b).

60. Suppose that a and b are positive numbers whose sum is 1.
(a) Find the maximum possible value of the product ab.
(b) Prove that $\left(1 + \dfrac{1}{a} \right)\left(1 + \dfrac{1}{b} \right) \geq 9$.

PROJECT | THE LEAST-SQUARES LINE

The great advances in mathematical astronomy made during the early years of the nineteenth century were due in no small part to the development of the method of least squares. The same method is the foundation for the calculus of errors of observation now occupying a place of great importance in the scientific study of social, economic, biological, and psychological problems. — A Source Book in Mathematics, by David Eugene Smith (New York: Dover Publications, 1959)

Of all the principles which can be proposed . . . [for finding the line or curve that best fits a data set], *I think there is none more general, more exact, and more easy of application than . . . rendering the sum of the squares of the errors a minimum.* —Adrien Marie Legendre (1752–1833) [Legendre and Carl Friedrich Gauss (1777–1855) developed the least-squares method independently.]

When we worked with the least-squares (or regression) line in Section 4.1, we said that this is the line that "best fits" the given set of data points. In this project we explain the meaning of "best fits" and we use quadratic functions to gain some insight into how the least-squares line is calculated. *Suggestion:* Two or three people could "volunteer" to study the text here and then present the material to another group or to the class at large. The presentation will be deemed successful if the audience can then work on their own the two exercises at the end of this project.

Suppose that we have a data point (x_i, y_i) and a line $y = f(x) = mx + b$ (not necessarily the least-squares line). As indicated in Figure A, we define the *deviation* e_i of this line from the data point as

$$e_i = y_i - f(x_i)$$

Figure A

In Figure A, where the data point is above the line, the deviation e_i represents the vertical distance between the data point and the line. If the data point were below the line, then the deviation $y_i - f(x_1)$ would be the negative of that distance. Think of e_i as the *error* made if $f(x_i)$ were used as a prediction for the value of y_i.

Figure B and the accompanying calculations show two examples using this definition. For the data point $(x_1, y_1) = (4, 3)$, the result $e_1 = 2$ does represent the vertical distance between the data point and the line. For the data point $(8, 2)$, the deviation is $e_2 = -1$, and the absolute value of this gives the vertical distance between the data point and the line.

Figure B

For the data point $(x_1, y_1) = (4, 3)$:

$$e_1 = y_1 - f(x_1)$$
$$= 3 - \left(\tfrac{1}{2}(4) - 1\right) = 2$$

For the data point $(x_2, y_2) = (8, 2)$:

$$e_2 = y_2 - f(x_2)$$
$$= 2 - \left(\tfrac{1}{2}(8) - 1\right) = -1$$

Now suppose we have a given data set and a line. One way to measure how closely this line fits the data would be to simply add up all the deviations. The problem with this, however, is that positive and negative deviations would tend to cancel one another out. For this reason, we use the sum of the squares of the deviations as the measure of how closely a line fits a data set. (The sum of the absolute values of the deviations would be another alternative, but in calculus, squares are easier to work with than absolute values.)

As an example, suppose we have the following data set:

x	1	2	3
y	2	4	5

or equivalently

$(x_1, y_1) = (1, 2)$
$(x_2, y_2) = (2, 4)$
$(x_3, y_3) = (3, 5)$

DEFINITION | Line of Best Fit

Suppose we have a set of n data points (x_1, y_1), (x_2, y_2), (x_3, y_3), . . . , (x_n, y_n). For any line, let E denote the sum of the squares of the deviations:

$$E = e_1^2 + e_2^2 + e_3^2 + \cdots + e_n^2$$

Then the **line of best fit** is that line for which E is as small as possible. (It can be shown that there is only one such line.) The line of best fit is also known as the **least-squares line** or the **regression line.**

Then, for any line $y = mx + b$ we have

$$
\begin{aligned}
e_1 &= 2 - [m(1) + b] & e_2 &= 4 - [m(2) + b] & e_3 &= 5 - [m(3) + b] \\
&= 2 - m - b & &= 4 - 2m - b & &= 5 - 3m - b
\end{aligned}
$$

Now, by definition, for the least-squares line we want to find the values for m and b so that the quantity $E = e_1^2 + e_2^2 + e_3^2$ is as small as possible. Using the expressions we've just calculated for e_1, e_2, and e_3, we have

$$
\begin{aligned}
E &= e_1^2 + e_2^2 + e_3^2 \\
&= (2 - m - b)^2 + (4 - 2m - b)^2 + (5 - 3m - b)^2
\end{aligned}
\tag{1}
$$

Equation (1) expresses E as a function of *two* variables, m and b. In all the work we've done previously on maximum-minimum problems, we were always able to express one variable in terms of the other, and then apply our knowledge of quadratics. In this case though, m and b are not related and so that strategy doesn't work here. In fact, in general, the methods of calculus are required to handle a two variable maximum-minimum problem. Therefore, to make this problem accessible within our context, we'll assume the result (from calculus) that in this problem b turns out to be 2/3. Go ahead now and make the substitution $b = 2/3$ in equation (1). After simplifying, you'll have a quadratic function, and you know how to find the value of m that minimizes the function. Then, with both m and b known, you have determined the least-squares line $y = mx + b$. As a visual check, graph the data set along with the least-squares line. See whether things look reasonable. Finally, use a graphing utility to compute the least-squares line and check your answer.

EXERCISES

In each case you are given a data set and the value for b in the least-squares line $y = mx + b$. Find the equation of the least-squares line using the method indicated above. Check your answer using an appropriate graphing utility. Then graph the data set along with the least-squares line. In Exercise 2, in reporting the final answer, round both b and m to two decimal places.

1. (*Easy numbers*)

x	1	2	3
y	2	5	11

$b = -3$

2. (*"Actual" numbers*) The data in the following table are from a study relating air pollution and incidence of respiratory disease. The data were collected in five cities over the years 1955, 1957, and 1958. The subjects were women working in RCA factories. In the table, x denotes the average concentration of particulate sulfates ("sulfur dust") in the air of the given city, measured in micrograms per cubic meter ($\mu g/m^3$). The variable y denotes the number of absences due to respiratory disease per 1000 employees per year. (Only absences lasting more than seven days were counted.)

$b = -20.8117$

City	x Concentration of suspended particulate sulfates ($\mu g/m^3$)	y Number of respiratory disease related absences per 1000 employees per year
Cincinnati, Oh.	7.4	18.5
Indianapolis, Ind.	13.2	44.2
Woodbridge, N.J.	13.6	50.3
Camden, N.J.	17.1	58.3
Harrison, N.J.	19.8	85.7

Source: F. Curtis Dohan, "Air Pollutants and Incidence of Respiratory Disease," *Archives of Environmental Health*, 3(1961), pp. 387–395

4.6 POLYNOMIAL FUNCTIONS

After a few hitches or twitches, the polynomial functions settle down to behavior in which things simply get bigger or smaller. An example is $f(x) = x^2$. Just once, as the values of x approach 0, the graph of this function dips downward to alter its shape; thereafter it ascends solemnly like a helium-filled balloon. —David Berlinski, *A Tour of the Calculus* (New York: Vintage Books, 1995)

We can rephrase the definitions of linear and quadratic functions using the terminology for polynomials that was reviewed in Section 1.3. A function f is *linear* if $f(x)$ is a polynomial of degree 1:

$$f(x) = a_1 x + a_0 \tag{1}$$

A function f is *quadratic* if $f(x)$ is a polynomial of degree 2:

$$f(x) = a_2 x^2 + a_1 x + a_0 \tag{2}$$

In the box that follows, we give the general definition for a polynomial function of degree n.

DEFINITION | **Polynomial Function**

A **polynomial function** is a function defined by an equation of the form

$$f(x) = a_n x^n + a_{n-1} x^{n-1} + \cdots + a_1 x + a_0$$

where n is a nonnegative integer and the a_i's are constants. If $a_n \neq 0$, the **degree** of the polynomial function is n, the largest exponent on the input variable.

EXAMPLES

$f(x) = 2x^3 - 3x^2 + x - 4$
(degree 3)

$g(u) = 3u^7 - \sqrt{3}u^2 - \pi$
(degree 7)

$L(x) = 5x + 1$ (degree 1)

A nonzero constant function—for example, $y = 7$ for all x—is a polynomial function of *degree zero*. (Think of the defining equation $y = 7$ as $y = 7x^0$.) The one polynomial function for which degree is not defined is the *zero function*: $f(x) = 0$ for all x. (What does the graph of that function look like?)

A function of the form $y = x^n$, where n is any real-number constant, is called a **power function.** If we only consider exponents n that are nonnegative integers, then these power functions are the simplest polynomial functions. In a sense, they are the building blocks for all other polynomial functions. Figure 1 shows examples of power functions with which we are already familiar. Our focus here is on the power functions in Figures 1(a) through 1(d), where n is a nonnegative integer.

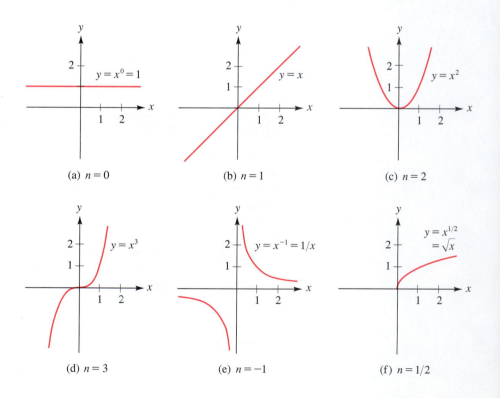

Figure 1
Some familiar power functions
$y = x^n$, where n is a real number

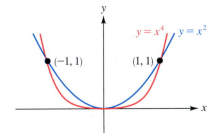

Figure 2

When n is an integer greater than 3, the graph of $y = x^n$ resembles the graph of $y = x^2$ or $y = x^3$, depending on whether n is even or odd. Consider, for instance, the graph of $y = x^4$, shown in Figure 2 along with the graph of $y = x^2$. Just as with $y = x^2$, the graph of $y = x^4$ is a symmetric, U-shaped curve passing through the three points $(-1, 1)$, $(1, 1)$, and $(0, 0)$. However, in the interval $-1 < x < 1$, the graph of $y = x^4$ is flatter than that of $y = x^2$. Similarly, the graph of $y = x^6$ in this interval would be flatter still. The data in Table 1 show why this is true. Figure 3 displays the graphs of $y = x^2$, $y = x^4$, and $y = x^6$ for the interval $0 \le x \le 1$.

■ TABLE 1

x	0.2	0.4	0.6	0.8	1.0
x^2	0.04	0.16	0.36	0.64	1.0
x^4	0.0016	0.0256	0.1296	0.4096	1.0
x^6	0.000064	0.004096	0.046656	0.262144	1.0

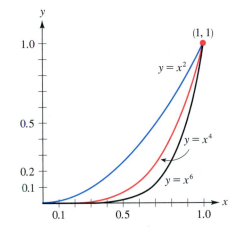

Figure 3

Incidentally, Figure 3 indicates one of the practical difficulties you may encounter in trying to draw an accurate graph of $y = x^n$. Suppose, for instance, that you want to graph $y = x^6$, and the lines you draw are 0.01 cm thick. Also suppose that you use the same scale on both axes, taking the common unit to be 1 cm. Then, in the first quadrant, your graph of $y = x^6$ will be indistinguishable from the x-axis when $x^6 < 0.01$, or $x < \sqrt[6]{0.01} \approx 0.46$ cm (using a calculator). This explains why sections of the graphs in Figure 3 appear horizontal.

For $|x| > 1$, the graph of $y = x^4$ rises more rapidly than that of $y = x^2$. Similarly, the graph of $y = x^6$ rises still more rapidly. This is shown in Figures 4(a) and (b). (Note the different scales used on the y-axes in the two figures.)

Figure 4 (a) (b)

EXAMPLE 1 Comparing average rates of change of two power functions

Let $f(x) = x^4$ and $g(x) = x^8$.
(a) Use Figure 5 to estimate and compare the averages rates of change of f and g on the interval $[0, 0.8]$ and on the interval $[0, 1.25]$.
(b) Use the given equations for f and g to compute the rates of change mentioned in part (a). Round the answers to three decimal places.

GRAPHICAL PERSPECTIVE

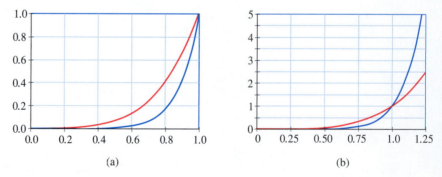

(a) (b)

SOLUTION
(a) Refer to Figure 5(a). On the interval $[0, 0.8]$ we have $g(0) = 0$, and from the graph, $g(0.8)$ is roughly 0.2. So on this interval we have

$$\frac{\Delta g}{\Delta x} = \frac{g(0.8) - g(0)}{0.8 - 0} \approx \frac{0.2 - 0}{0.8} = 0.25$$

Also on the interval $[0, 0.8]$, $f(0) = 0$ and $f(0.8) \approx 0.4$. Thus we have

$$\frac{\Delta f}{\Delta x} = \frac{f(0.8) - f(0)}{0.8 - 0} \approx \frac{0.4 - 0}{0.8} = 0.5$$

In summary, on the interval $[0, 0.8]$, the average rate of change of g is approximately 0.25, which is half of the value estimated for the average rate of change of f.

Exercise 45 asks you to follow this same procedure for the interval $[0, 1.25]$ shown in Figure 5(b). The exercise asks you to show that the estimates for this interval are $\Delta g/\Delta x \approx 4$ and $\Delta f/\Delta x \approx 2$. So, on this interval, it's g rather than f that has the larger average rate of change. The estimate for the average rate of change of g is twice the value estimated for the average rate of change of f.

(b) For the interval $[0, 0.8]$,

$$\frac{\Delta g}{\Delta x} = \frac{g(0.8) - g(0)}{0.8 - 0} = \frac{(0.8)^8 - 0^8}{0.8} \qquad \frac{\Delta f}{\Delta x} = \frac{f(0.8) - f(0)}{0.8 - 0} = \frac{(0.8)^4 - 0^4}{0.8}$$

$$\approx 0.210 \quad \text{rounding to three} \qquad\qquad = 0.512 \quad \text{(exactly)}$$
$$\text{decimal places}$$

Exercise 46(a) asks you to carry out similar calculations for the interval $[0, 1.25]$ to obtain the results $\Delta g/\Delta x \approx 4.768$ and $\Delta f/\Delta x \approx 1.953$.

Figure 6

As mentioned earlier, when n is odd, the graph of $y = x^n$ resembles that of $y = x^3$. In Figure 6 we compare the graphs of $y = x^3$ and $y = x^5$. Notice that both curves pass through $(0, 0)$, $(1, 1)$, and $(-1, -1)$. For reasons similar to those explained for even n, the graph of $y = x^5$ is flatter than that of $y = x^3$ in the interval $-1 < x < 1$, and the graph of $y = x^7$ is flatter still. For $|x| > 1$, the graph of $y = x^5$ is steeper than that of $y = x^3$, and $y = x^7$ is steeper still.

EXAMPLE 2 Translating the graph of a power function

Sketch the graph of $y = (x + 2)^5$ and specify the y-intercept.

SOLUTION
The graph of $y = (x + 2)^5$ is obtained by moving the graph of $y = x^5$ two units to the left. As a guide to drawing the curve, we recall that $y = x^5$ passes through the points $(1, 1)$ and $(-1, -1)$. Thus, $y = (x + 2)^5$ must pass through $(-1, 1)$ and $(-3, -1)$, as shown in Figure 7.

Although the curve rises and falls very sharply, it is important to realize that it is never really vertical. For instance, the curve eventually crosses the y-axis.

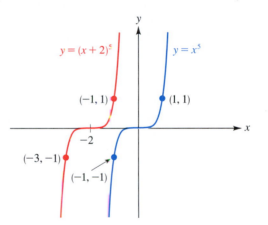

Figure 7

To find the y-intercept, we set $x = 0$ to obtain $y = 2^5 = 32$. Thus, the y-intercept is 32.

In Section 4.2, we observed the effect of the constant a on the graph of $y = ax^2$. Those same comments apply to the graph of $y = ax^n$. For instance, the graph of $y = \frac{1}{2}x^4$ is wider than that of $y = x^4$, whereas the graph of $y = -\frac{1}{2}x^4$ is obtained by reflecting $y = \frac{1}{2}x^4$ in the x-axis.

EXAMPLE 3   Translating a reflection of a power function

Graph the function $y = -2(x - 3)^4$.

SOLUTION
We begin with the graph of $y = -2x^4$, in Figure 8(a). The points $(1, -2)$ and $(-1, -2)$ are obtained by substituting $x = 1$ and $x = -1$, respectively, in the equation $y = -2x^4$. Now if we replace x with $x - 3$ in the equation $y = -2x^4$, we have $y = -2(x - 3)^4$, which we can graph by translating the graph in Figure 8(a) to the right 3 units; see Figure 8(b).

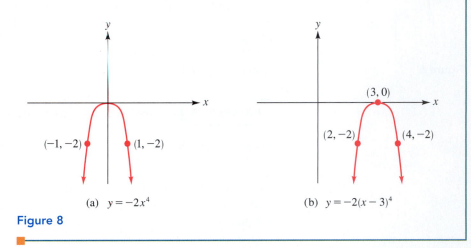

(a) $y = -2x^4$

(b) $y = -2(x - 3)^4$

Figure 8

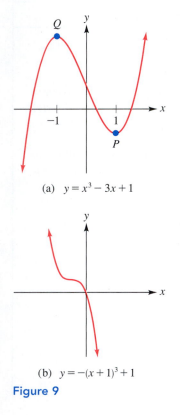

(a) $y = x^3 - 3x + 1$

(b) $y = -(x + 1)^3 + 1$

Figure 9

In principle, we can obtain the graph of any polynomial function by setting up a table and plotting a sufficient number of points. Indeed, this is just the way a graphing calculator or computer operates. However, to *understand* why the graphs look as they do, we want to discuss some additional methods for graphing polynomial functions.

There are three facts that we shall need. By way of example, look at the graphs of the polynomial functions in Figure 9. First, notice that both graphs are unbroken, smooth curves with no "corners." As in shown in calculus, this is true for the graph of every polynomial function. In contrast, the graphs in Figures 10 and 11 cannot represent polynomial functions. The graph in Figure 10 has a break in it, and the graph in Figure 11 has what's called a **cusp.**

Now look back at the graph in Figure 9(a). Recall (from Section 3.2) that points such as P and Q are called **turning points.** These are points where the graph changes from rising to falling, or vice versa. It is a fact (proved in calculus) that the graph of a polynomial function of degree n has *at most* $n - 1$ turning points. For instance, in

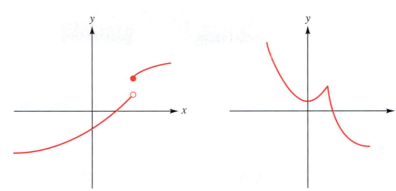

Figure 10
Since the graph has a break, it cannot represent a polynomial function.

Figure 11
Since the graph has a cusp, it cannot represent a polynomial function.

Figure 12
When $|x|$ is very large, the graph of $y = x^3 - 3x + 1$ appears indistinguishable from that of $y = x^3$.

Figure 9(a), there are two turning points, and the degree of the polynomial is 3. However, as Figure 9(b) indicates, we needn't have any turning points at all.

A third property of polynomial functions concerns their behavior when $|x|$ is very large. We'll illustrate this property using the function $y = x^3 - 3x + 1$, graphed in Figure 9(a). Now, in Figure 9(a), the x-values are relatively small; for instance, the x-coordinates of P and Q are 1 and -1, respectively. In Figure 12, however, we show the graph of this same function using units of 100 on the x-axis. On this scale, the graph appears indistinguishable from that of $y = x^3$. In particular, note that as $|x|$ gets very large, $|y|$ grows very large.

It's easy to see why the function $y = x^3 - 3x + 1$ resembles $y = x^3$ when $|x|$ is very large. First, let's rewrite the equation $y = x^3 - 3x + 1$ as

$$y = x^3\left(1 - \frac{3}{x^2} + \frac{1}{x^3}\right)$$

Now, when $|x|$ is very large, both $3/x^2$ and $1/x^3$ are close to zero. So we have

$$y \approx x^3(1 - 0 + 0)$$
$$\approx x^3 \qquad \text{when } |x| \text{ is very large}$$

The same technique that we've just used in analyzing $y = x^3 - 3x + 1$ can be applied to any (nonconstant) polynomial function. The result is summarized in item 3 in the following box.

▌PROPERTY SUMMARY   Graphs of Polynomial Functions

1. The graph of a polynomial function of degree 2 or greater is an unbroken smooth curve. (For degrees 1 and 0, the graph is a line.)
2. The graph of a polynomial function of degree n has at most $n - 1$ turning points.
3. For the graph of any polynomial function (other than a constant function), as $|x|$ gets very large, $|y|$ grows very large. If

$$f(x) = a_n x^n + a_{n-1} x^{n-1} + \cdots + a_1 x + a_n \qquad (a_n \neq 0)$$

then

$$f(x) \approx a_n x^n \qquad \text{when } |x| \text{ is very large}$$

EXAMPLE **4**

A function f is defined by

$$f(x) = -x^3 + x^2 + 9x + 9$$

Which of the graphs in Figure 13 might represent this function?

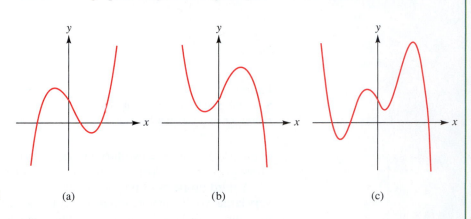

Figure 13   (a)   (b)   (c)

SOLUTION

When $|x|$ is very large, $f(x) \approx -x^3$. This rules out the graph in Figure 13(a). The graph in Figure 13(c) can also be ruled out, but for a different reason. That graph has four turning points, whereas the graph of the cubic function f can have at most two turning points. The graph in Figure 13(b), on the other hand, does have two turning points; furthermore, that graph does behave like $y = -x^3$ when $|x|$ is very large. The graph in Figure 13(b) might be (in fact, it is) the graph of the given function f.

EXAMPLE **5**   **Comparing polynomials of the same degree**

GRAPHICAL PERSPECTIVE

Figure 14
$f(x) = x^4$ (blue) and
$g(x) = x^4 - 3x^3 + 5x + 4$ (red) in
the viewing rectangle $[-3, 3, 1]$
by $[-3, 12, 3]$

Consider the two polynomial functions f and g defined by

$$f(x) = x^4 \qquad g(x) = x^4 - 3x^3 + 5x + 4$$

(a) With a graphing utility, find a viewing rectangle that highlights the differences between these two functions.

(b) Find a sequence of viewing rectangles demonstrating that as x gets larger and larger, the graph of g looks more and more like the graph of f.

SOLUTION

(a) The view in Figure 14 indicates that the graph of g has three turning points, while the graph of f has but one. The graph of f is symmetric about the y-axis; the graph of g is not.

(b) See the three views in Figure 15. For the scale and size used in Figure 15(c), the two graphs are virtually indistinguishable from one another.

GRAPHICAL PERSPECTIVE

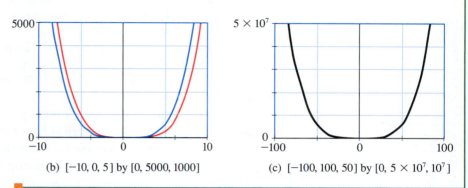

(a) $[-5, 5, 5]$ by $[0, 200, 50]$

(b) $[-10, 0, 5]$ by $[0, 5000, 1000]$

(c) $[-100, 100, 50]$ by $[0, 5 \times 10^7, 10^7]$

Figure 15

In views (a) and (b), the blue graph is $f(x) = x^4$ and the red is $g(x) = x^4 - 3x^2 + 5x + 4$. In view (c), both graphs are black; they are indistinguishable from one another at this scale and size.

▌ TABLE 2

Interval	$x(x + 1)(x - 3)$
$(-\infty, -1)$	negative
$(-1, 0)$	positive
$(0, 3)$	negative
$(3, \infty)$	positive

We can use our work on solving inequalities (in Section 2.6) to graph polynomial functions that are in factored form. Consider, for example, the function

$$f(x) = x(x + 1)(x - 3)$$

First of all, by inspection, we see that $f(x) = 0$ when $x = 0$, $x = -1$, or $x = 3$. These are the x-intercepts for the graph. Also, note that the y-intercept is 0. (Why?) Next, we want to know what the graph looks like in the intervals between the x-intercepts. To do this, we solve the two inequalities $f(x) > 0$ and $f(x) < 0$ using the technique in Section 2.6. Table 2 shows the results. (You should check these results for yourself; if you need a review, the details of this example are worked out on pages 141–142.)

Now we interpret the results in Table 2 graphically. When x is in either of the intervals $(-\infty, -1)$ or $(0, 3)$, the graph lies below the x-axis (because $f(x) < 0$); and when x is in either of the intervals $(-1, 0)$ or $(3, \infty)$, the graph lies above the x-axis (because $f(x) > 0$). This information is summarized in Figure 16(a). The three dots in the figure indicate the x-intercepts of the graph. The shaded regions are the **excluded regions** through which the graph cannot pass. The graph must pass only through the unshaded regions (and through the three x-intercepts). In Figure 16(b) we have drawn a rough sketch of a curve satisfying these conditions.

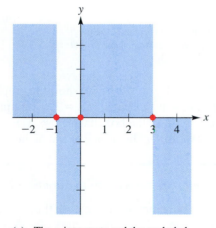

(a) The x-intercepts and the excluded regions for the graph of $f(x) = x(x + 1)(x - 3)$

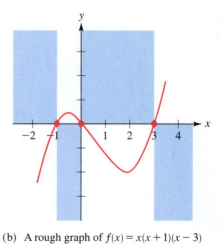

(b) A rough graph of $f(x) = x(x + 1)(x - 3)$

Figure 16

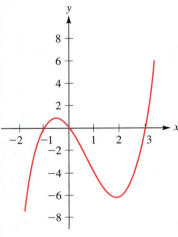

Figure 17
$f(x) = x(x + 1)(x - 3)$

Notice that to draw a smooth curve satisfying the conditions of Figure 16(a), we need at least two turning points: one between $x = -1$ and $x = 0$ and another between $x = 0$ and $x = 3$. On the other hand, since the degree of $f(x)$ is 3, there can be no more than two turning points. Thus Figure 16(b) has exactly two turning points. As another check on our rough sketch, we note that for large values of $|x|$, the graph indeed resembles that of $y = x^3$. (We are using the third property in the summary box on page 331.)

Although the precise location of the turning points is a matter for calculus, we can nevertheless improve upon the sketch in Figure 16(b) by computing $f(x)$ for some specific values of x. Some reasonable choices in this case are the inputs $-2, -\frac{1}{2}, 1, 2$, and 4. As you can check, the resulting points on the graph are $(-2, -10)$, $(-\frac{1}{2}, \frac{7}{8})$, $(1, -4)$, $(2, -6)$, and $(4, 20)$. We can now sketch the graph, as shown in Figure 17.

In the example just concluded, the polynomial $f(x) = x(x + 1)(x - 3)$ has no *repeated factors*. That is, none of the factors is squared, cubed, or raised to a higher power. Now let us look at two examples in which there are repeated factors. The observations we make will help us in determining the general shape of a graph without the need for plotting a large number of individual points. In Figure 18(a), we show the graph of $g(x) = x(x + 1)(x - 3)^2$. Notice that in the immediate vicinity of the intercept at $x = 3$, the graph has the same general shape as that of $y = A(x - 3)^2$.

(a) $g(x) = x(x + 1)(x - 3)^2$
Near the x-intercept at 3, the graph has the same general shape as that of $y = A(x - 3)^2$.

(b) $h(x) = x(x + 1)(x - 3)^3$
Near the x-intercept at 3, the graph has the same general shape as that of $y = A(x - 3)^3$.

Figure 18

Similarly, in Figure 18(b), we show the graph of $h(x) = x(x + 1)(x - 3)^3$; notice that in the immediate vicinity of $x = 3$, the graph has the same general shape as that of $y = A(x - 3)^3$. In the box that follows, we state the general principle underlying these observations. (The principle can be justified using calculus.)

The Behavior of a Polynomial Function Near an x-Intercept

Let $f(x)$ be a polynomial and suppose that $(x - a)^n$ is a factor of $f(x)$. [Furthermore, assume that none of the other factors of $f(x)$ contains $(x - a)$.] Then, in the immediate vicinity of the x-intercept at a, the graph of $y = f(x)$ closely resembles that of $y = A(x - a)^n$.

The principle that we have just stated is easy to apply because we already know how to graph functions of the form $y = A(x - a)^n$. The next example shows how this works.

| EXAMPLE 6 | **Determining the shape of the graph of a polynomial function near an x-intercept** |

Describe the behavior of each function in the immediate vicinity of the indicated x-intercept.

(a) $f(x) = \frac{1}{2}(x - 3)(x - 1)^3$; intercept: $x = 1$

(b) $g(x) = (x + 1)(x + 4)(x + 3)^2$; intercept: $x = -3$

SOLUTION

(a) We make the following observation:

$$f(x) = \frac{1}{2}\underbrace{(x - 3)}_{\uparrow}(x - 1)^3$$

When x is close to 1,
this factor is close
to $1 - 3$, or -2.

So if x is very close to 1, we have the approximation

$$f(x) \approx \frac{1}{2}(1 - 3)(x - 1)^3 = \frac{1}{2}(-2)(x - 1)^3 = -(x - 1)^3$$

Thus, in the immediate vicinity of $x = 1$, the graph of f closely resembles $y = -(x - 1)^3$. Notice the technique we used to obtain this result. We retained the factor corresponding to the intercept $x = 1$, and we approximated the remaining factor using the value $x = 1$. See Figure 19.

(b) We use the approximation technique shown in part (a).

$$g(x) = \underbrace{(x + 1)}\underbrace{(x + 4)}(x + 3)^2$$

When x is close to -3,
this factor is close
to $-3 + 1$, or -2.

When x is close to -3,
this factor is close
to $-3 + 4$, or 1.

Thus, when x is close to -3, we have the approximation

$$g(x) \approx (-3 + 1)(-3 + 4)(x + 3)^2 = -2(x + 3)^2$$

This tells us that in the immediate vicinity of $x = -3$, the graph of g resembles $y = -2(x + 3)^2$. See Figure 20.

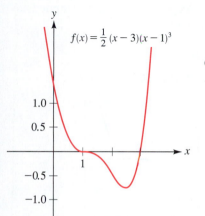

Figure 19

When x is close to 1, the graph of $f(x) = \frac{1}{2}(x - 3)(x - 1)^3$ closely resembles $y = -(x - 1)^3$.

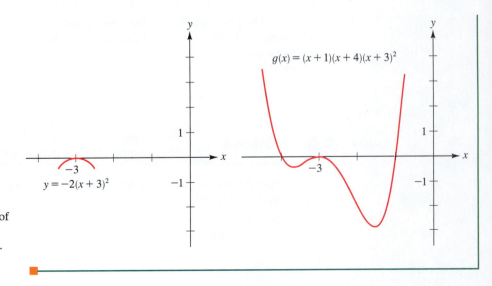

Figure 20
When x is close to -3, the graph of
$g(x) = (x + 1)(x + 4)(x + 3)^2$
closely resembles $y = -2(x + 3)^2$.

The technique introduced in Example 6, whereby we approximate the behavior of the factored polynomial near each x-intercept, allows us to draw by hand a reasonably accurate graph. (Of course, some judicious point plotting is necessary too, as mentioned previously.) It's interesting though, to use a graphing utility to confirm visually just how well that simple approximation technique does work. As an example, consider the polynomial function f defined by

$$f(x) = 2(x + 1)(x - 2)^3$$

Near the x-intercept $x = 2$, our approximation technique yields

$$f(x) \approx 2(2 + 1)(x - 2)^3$$
$$f(x) \approx 6(x - 2)^3 \qquad \text{near } x = 2$$

In Figure 21, we've used a graphing utility to plot both the graph of f and the graph of the approximating polynomial $y = 6(x - 2)^2$. As you can see, in the immediate vicinity of $x = 2$, the two graphs are remarkably close.

GRAPHICAL PERSPECTIVE

Figure 21
$f(x) = 2(x + 1)(x - 2)^3$
and $y = 6(x - 2)^3$

$\text{———}\ f(x) = 2(x + 1)(x - 2)^3$

$\text{———}\ y = 6(x - 2)^3$

EXERCISE SET 4.6

A

1. Which of the functions in Figure 1 on page 326 are polynomial functions?
2. (a) Give an example of a power function that is not a polynomial function.
   (b) Give an example of a polynomial function that is not a power function.
3. Let $f(x) = x^2$ and $g(x) = x^3$.
   (a) Either by hand or with a graphing utility, on the same set of axes draw the graphs of f and g on the interval $[0, 1]$.
   (b) Compute and compare the average rates of change of f and g on the interval $[0, 1]$ and on the interval $[0, 1/2]$.
4. Let $f(t) = t^4$ and $g(t) = t^5$.
   (a) Using a graphing utility, draw the graphs of f and g on the same set of axes.
   (b) Which of the three types of symmetry discussed on page 64 does the graph of each function possess?
   (c) Compute the average rate of change of each function on the interval $[-2, 2]$.
   (d) Explain why each answer in part (c) is predictable from the symmetry of the graph.

In Exercises 5–16, sketch the graph of each function and specify all x- and y-intercepts.

5. $y = (x - 2)^2 + 1$
6. $y = -3x^4$
7. $y = -(x - 1)^4$
8. $y = -(x + 2)^3$
9. $y = (x - 4)^3 - 2$
10. $y = -(x - 4)^3 - 2$
11. $y = -2(x + 5)^4$
12. $y = -2x^4 + 5$
13. $y = (x + 1)^5/2$
14. $y = \frac{1}{2}x^5 + 1$
15. $y = -(x - 1)^3 - 1$
16. $y = x^8$

In Exercises 17–20, give a reason (as in Example 1) why each graph cannot represent a polynomial function of degree 3.

17.
18.

19.
20.

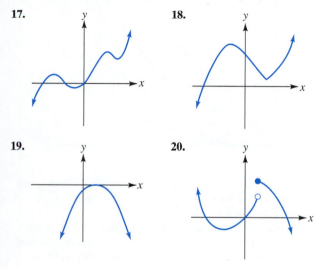

In Exercises 21–24, give a reason why each graph cannot represent a polynomial function that has the highest-degree term $2x^5$.

21.
22.

23.
24.

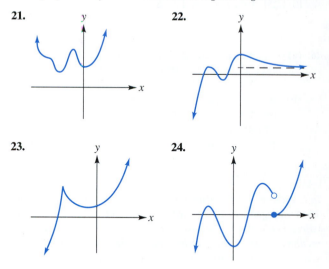

For Exercises 25 and 26:
(a) *With a graphing utility, find a viewing rectangle that highlights the differences between the two functions, as in Example 5;*
(b) *Find a sequence of viewing rectangles demonstrating that as x gets larger and larger, the graph of g looks more and more like the graph of f.*

25. $f(x) = 2x^2, g(x) = 2x^2 - 12x - 5$
26. $f(x) = 4x^3, g(x) = 4x^3 + 72x^2 + 420x + 805$

In Exercises 27–34:
(a) *determine the x- and y-intercepts and the excluded regions for the graph of the given function. Specify your results using a sketch similar to Figure 16(a). In Exercises 31–34, you will first need to factor the polynomial.*
(b) *Graph each function.*

27. $y = (x - 2)(x - 1)(x + 1)$
28. $y = (x - 3)(x + 2)(x + 1)$
29. $y = 2x(x - 2)(x - 1)$
30. $y = (x - 3)(x - 2)(x + 2)$
31. $y = x^3 - 4x^2 - 5x$
32. $y = x^3 - 9x$
33. $y = x^3 + 3x^2 - 4x - 12$
34. $y = x^3 - 5x^2 - x + 5$

In Exercises 35–44:
(a) *determine the x- and y-intercepts and the excluded regions for the graph of the given function. Specify your results using a sketch similar to Figure 16(a).*
(b) *Describe the behavior of the function at each x-intercept that corresponds to a repeated factor. Specify your results using a sketch similar to the left-hand portion of Figure 20.*
(c) *Graph each function.*

35. $y = x^3(x + 2)$
36. $y = (x - 1)(x - 4)^2$
37. $y = 2(x - 1)(x - 4)^3$
38. $y = (x - 1)^2(x - 4)^2$
39. $y = (x + 1)^2(x - 1)(x - 3)$
40. $y = x^2(x - 4)(x + 2)$

41. $y = -x^3(x - 4)(x + 2)$ **42.** $y = 4(x - 2)^2(x + 2)^3$

43. $y = -4x(x - 2)^2(x + 2)^3$ **44.** $y = -3x^3(x + 1)^4$

45. This exercise refers to Example 1(a) in the text. Use Figure 5(b) to obtain the following estimates: on the interval $[0, 1.25]$, $\Delta g/\Delta x \approx 4$, and $\Delta f/\Delta x \approx 2$.

B

46. (a) This exercise refers to Example 1(b), in which $f(x) = x^4$ and $g(x) = x^8$. Show that on the interval $[0, 1.25]$, we have $\Delta g/\Delta x \approx 4.768$ and $\Delta f/\Delta x \approx 1.953$.

(b) Let $f(x) = x^4$ and $g(x) = x^8$, as in part (a). Find a positive constant b so that on the interval $[0, b]$, the average rate of change of g is 100 times the average rate of change of f. Give two forms for the answer: an exact expression and a decimal approximation rounded to three decimal places.

In Exercises 47–50, first use the graph to estimate the x-intercepts. Then use algebra to determine each x-intercept. If an intercept involves a radical, give that answer as well as a calculator approximation rounded to three decimal places. Be sure to check that your results are consistent with the initial graphical estimates.

47.

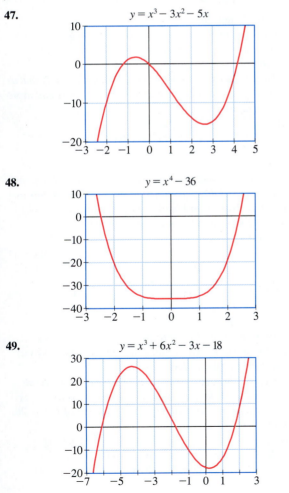

$y = x^3 - 3x^2 - 5x$

48.

$y = x^4 - 36$

49.

$y = x^3 + 6x^2 - 3x - 18$

50.

$y = x^4 - 2x^2 - 5$

For Exercises 51 and 52:

Ⓖ **(a)** *Use a graphing utility to draw a graph of each function.*

Ⓖ **(b)** *For each x-intercept, zoom in until you can estimate it accurately to the nearest one-tenth.*

(c) *Use algebra to determine each x-intercept. If an intercept involves a radical, give that answer as well as a calculator approximation rounded to three decimal places. Check to see that your results are consistent with the graphical estimates obtained in part (b).*

51. $N(t) = t^7 + 8t^4 + 16t$ **52.** $W(u) = 2u^4 - 17u^2 + 35$

In Exercises 53–58, six functions are defined as follows:

$$f(x) = x \qquad g(x) = x^2 \qquad h(x) = x^3$$
$$F(x) = x^4 \qquad G(x) = x^5 \qquad H(x) = x^6$$

Refer also to the following figure.

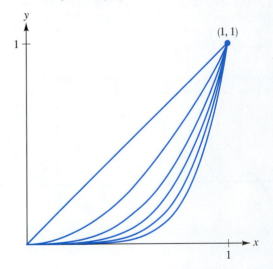

53. The six graphs in the figure are the graphs of the six given functions for the interval $[0, 1]$, but the graphs are not labeled. Which is which?

54. For which x-values in $[0, 1]$ will the graph of g lie strictly below the horizontal line $y = 0.1$? Use a calculator to evaluate your answer. Round off the result to two significant figures.

55. Follow Exercise 54, using the function H instead of g.

56. Find a number t in $[0, 1]$ such that the vertical distance between $f(t)$ and $g(t)$ is $1/4$.

57. Is there a number t in $[0, 1]$ such that the vertical distance between $g(t)$ and $F(t)$ is 0.26?

58. Find all numbers t in $[0, 1]$ such that $F(t) = G(t) + H(t)$.

59. The figure shows the graphs of $y = 4x$ and $y = x^2/100$ in a standard viewing rectangle.

   (a) Do the graphs intersect anywhere other than at the origin?

Ⓖ **(b)** Use a graphing utility to support your answer in part (a).

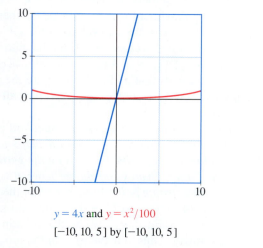

$y = 4x$ and $y = x^2/100$

$[-10, 10, 5]$ by $[-10, 10, 5]$

60. (a) Factor the expression $4x^2 - x^4$. Then use the techniques explained in this section to graph the function defined by $y = 4x^2 - x^4$.

   (b) Find the coordinates of the turning points. *Hint*: As in previous sections, use the substitution $x^2 = t$.

61. (a) Graph the function $D(x) = x^2 - x^4$.

   (b) Find the turning points of the graph. (See the hint in Exercise 60.)

   (c) On the same set of axes, sketch the graphs of $y = x^2$ and $y = x^4$ for $0 \le x \le 1$. What is the maximum vertical distance between the graphs?

62. (a) An open-top box is to be constructed from a 6-by-8-in. rectangular sheet of tin by cutting out equal squares at each corner and then folding up the resulting flaps. Let x denote the length of the side of each cutout square. Show that the volume $V(x)$ is

$$V(x) = x(6 - 2x)(8 - 2x)$$

   (b) What is the domain of the volume function in part (a)? [The answer is *not* $(-\infty, \infty)$.]

Ⓖ **(c)** Use a graphing utility to graph the volume function, taking into account your answer in part (b).

Ⓖ **(d)** By zooming in on the turning point, estimate to the nearest one-hundredth the maximum volume.

63. The point P is in the first quadrant on the graph of $y = 1 - x^4$. From P, perpendiculars are drawn to the x- and y-axes, thus forming a rectangle.

   (a) Express the area of the rectangle as a function of a single variable.

Ⓖ **(b)** Use a graphing utility to graph the area function. Then, using the ZOOM feature, estimate to the nearest hundredth the maximum possible area for the rectangle.

64. In Example 1 in Section 4.2, we used a graphing utility to determine linear and quadratic functions to model the AIDS data in Figure 2 on page 267. We then used these models to make projections for the years 1992 and 1997. For both models and both years, we found that the projections were too low. We also saw that the quadratic projections were more accurate than the linear ones, as measured by percentage error.

Ⓖ **(a)** Use a graphing utility to find a polynomial of degree 3 (a *cubic model*) that best fits the data in Figure 2 on page 267.

   (b) Use the cubic model to make projections for the cumulative number of AIDS cases in 1992 and 1997.

   (c) Are the projections higher or lower than the actual figures given in Example 1 in Section 4.2? Compute the percentage error for each projection. In each case, which is the better predictor, the quadratic or the cubic model?

PROJECT

FINDING SOME MAXIMUM VALUES WITHOUT USING CALCULUS

In Section 4.2 it was stated that, *in general,* the techniques of calculus are required to find maximum or minimum values for functions that are not quadratics. Nevertheless, there are a number of cases involving polynomials or quotients of polynomials where algebra, rather than calculus, will suffice. The following theorem can be used to find maximum values in certain types of problems. (One way to establish this theorem is indicated at the end of this project.)

THEOREM

If the sum of three positive quantities s, t, and u is constant, then their product stu is maximum if and only if $s = t = u$.

Figure A
$f(x) = 2x(4 - x)^2$

We'll give three examples that show how to use this theorem; then you can try your hand at the three problems that follow. First sample problem: *The sum of three positive numbers is 1. What is the maximum possible value for their product?* Letting s, t, and u denote the three numbers, we have $s + t + u = 1$. According to the theorem, the product is maximum when the three numbers are equal. So in that case, the equation $s + t + u = 1$ is equivalent to $s + s + s = 1$, which implies $s = 1/3$. Therefore t and u are also equal to $1/3$, and the maximum possible value for the product is $stu = (\frac{1}{3})(\frac{1}{3})(\frac{1}{3}) = \frac{1}{27}$. As a quick empirical check, pick any three positive fractions or decimals that add to 1. Compute the product, and check that it's less than $\frac{1}{27}$ (assuming, of course, you didn't pick all three numbers to be $1/3$).

Second sample problem: *Let $f(x) = 2x(4 - x)^2$. By using the standard graphing techniques discussed in Section 4.6, we find that the general shape for the graph of f is as shown in Figure A. One turning point, $(4, 0)$, is on the x-axis; the second turning point is in the first quadrant, with an x-coordinate between 0 and 4. Find the exact coordinates of this second turning point.* First, note that the three factors comprising $f(x)$—namely, $2x$, $(4 - x)$, and $(4 - x)$ again—do have a constant sum: $2x + (4 - x) + (4 - x) = 8$. Furthermore, on the open interval $0 < x < 4$, each of these factors is positive. Thus, our theorem in the box is applicable. The maximum value of the product $f(x) = 2x(4 - x)^2$ on the open interval $0 < x < 4$ occurs when the three factors are equal. Setting $2x = 4 - x$, we obtain $3x = 4$, and consequently $x = 4/3$. This is the required x-coordinate of the turning point. For the y-coordinate, we have $f(\frac{4}{3}) = 2(\frac{4}{3})(4 - \frac{4}{3})^2$. As you can verify, that works out to $\frac{512}{27}$. In summary, the coordinates of the turning point are $(\frac{4}{3}, \frac{512}{27})$.

GRAPHICAL PERSPECTIVE

Figure B
$g(x) = x^2(1 - x)$

Third sample problem: *According to Figure B, the graph of the polynomial function defined by $g(x) = x^2(1 - x)$ has a turning point in the first quadrant. Find the x-coordinate of that turning point.* First, note that on the open interval $0 < x < 1$, each of the three factors x, x again, and $1 - x$ is positive. However, as you can check, the sum of the three factors is not constant. We can work around this by writing

$$(x)(x)(1 - x) = 4\left[\left(\frac{x}{2}\right)\left(\frac{x}{2}\right)(1 - x)\right]$$

Now notice that the three factors in brackets do have a constant sum. For a maximum then, we require that $\frac{x}{2} = 1 - x$. Solving this last equation yields $x = 2/3$, the required x-coordinate.

EXERCISES

1. **(a)** First, sketch a graph of the function $f(x) = x^2(6 - 2x)$ using the standard techniques from Section 4.6. Then, use the ideas presented above to find the exact coordinates of the turning point in the first quadrant.
   (b) Follow part (a), but use $g(x) = x^2(6 - 3x)$.
2. An open-top box is to be constructed from a 16-by-24-in. square sheet of cardboard by cutting out congruent squares at each corner and then folding up the flaps. See the following figure. Express the volume of the box as a function of a single variable. Then find the maximum possible volume for the box.

3. When a person coughs, the radius r of the trachea (windpipe) decreases. In a paper "The Human Cough" (Lexington, Mass.: COMAP, Inc., 1979), Philip Tuchinsky developed the following model for the average velocity of air through the trachea during a cough:

$$v(r) = c(r_0 - r)r^2 \qquad (\tfrac{1}{2}r_0 \le r \le r_0)$$

In this formula, the variable r represents the radius of the trachea; the constant r_0 is the normal radius when the person is not coughing; c is a positive constant (which, among other things, depends upon the length of the person's trachea); and $v(r)$ is the average velocity of the air in the trachea. Show that $v(r)$ is a maximum when $r = \tfrac{2}{3}r_0$. In other words, according to this model, in coughing to clear the air passages, the most effective cough occurs when the radius contracts to two-thirds its normal size.

ESTABLISHING THE THEOREM

Exercise 49 in Section 2.5 outlines a proof of the following result, known as the *arithmetic-geometric mean inequality* for three numbers:

If a, b, and c are nonnegative real numbers, then $\sqrt[3]{abc} \le (a + b + c)/3$, with equality holding if and only if $a = b = c$.

On your own or through a group discussion, determine why our theorem in the box at the beginning of this project is a consequence of the arithmetic-geometric mean inequality. Then, on your own, write a paragraph carefully explaining this. (You are not being asked here to prove the arithmetic-geometric mean inequality.)

4.7 RATIONAL FUNCTIONS

We shall not attempt to explain the numerous situations in life sciences where the study of such graphs is important nor the chemical reactions which give rise to the rational functions r(x). . . . Let it suffice to mention that, to the experimental biochemist, theoretical results concerning the shapes of rational functions are of considerable interest.
—W. G. Bardsley and R. M. W. Wood in "Critical Points and Sigmoidicity of Positive Rational Functions," *The American Mathematical Monthly,* vol. 92 (1985), pp. 37–42.

After the polynomial functions, the next simplest functions are the **rational functions.** These are functions defined by equations of the form

$$y = \frac{f(x)}{g(x)}$$

where $f(x)$ and $g(x)$ are polynomials. In general, throughout this section, when we write a function such as $y = f(x)/g(x)$, we assume that $f(x)$ and $g(x)$ contain no common factors (other than constants). (Exercises 44 and 45 ask you to consider several cases in which $f(x)$ and $g(x)$ do contain common factors.) Also, for each of the examples that we discuss, the degree of the numerator $f(x)$ is less than or equal to the degree of the denominator $g(x)$. (A case in which the degree of $f(x)$ exceeds the degree of $g(x)$ is developed in the exercises.)

We'll use the graphs in Figure 1 to introduce some of the ideas and terminology involved in analyzing rational functions. Figure 1(a) shows the familiar rational function $y = 1/x$ that we graphed in Section 3.2. The graph of $y = 1/x$ differs from the graph of every polynomial function in two important aspects. First, the graph has a break in it; it is composed of two distinct pieces or **branches.** (Recall from Section 4.6 that the graph of a polynomial function never has a break or gap in it.) In general, the graph of a rational function has one more branch than the number of distinct real values for which the denominator is zero. In Figure 1(b), for example, there are two distinct numbers for which the denominator is zero (what are they?), and the graph has three branches.

The second way that the graph of $y = 1/x$ differs from that of a polynomial function is related to the *asymptotes.* A line is an **asymptote** for a curve if the distance between the line and the curve approaches zero as we move out farther and farther

GRAPHICAL PERSPECTIVE

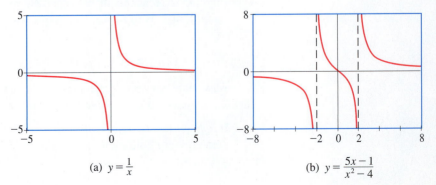

Figure 1
Graphs of two rational functions

(a) $y = \dfrac{1}{x}$

(b) $y = \dfrac{5x-1}{x^2-4}$

along the line. In Figure 1(a), the x-axis is a horizontal asymptote for the curve; the y-axis is a vertical asymptote. In Figure 1(b), the x-axis is a horizontal asymptote, and each of the lines $x = 2$ and $x = -2$ is a vertical asymptote. Note that 2 and -2 are also the x-values for which the denominator $x^2 - 4$ is zero. In general, this type of correspondence will always occur for a rational function $y = f(x)/g(x)$ as long as the two polynomials have no common factors (other than constants). Thus, to find the vertical asymptotes for the graph of $y = f(x)/g(x)$, just find the x-values for which the denominator is zero. What if there are no real-number values for x that make the denominator zero? Then the graph of the rational function has no vertical asymptotes. Figure 2 shows such an example.

GRAPHICAL PERSPECTIVE

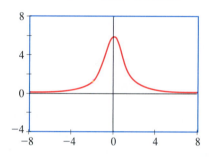

Figure 2

$y = 6/(1 + x^2)$

There are no real numbers x for which the denominator is zero. The graph has no vertical asymptotes, and there is no break in the graph.

We can use the functions in Figures 1(b) and 2 as examples in determining domain and intercepts. For the domain of the function $y = (5x - 1)/(x^2 - 4)$ in Figure 1(b), rewrite the equation with the denominator in factored form:

$$y = \frac{5x - 1}{(x - 2)(x + 2)}$$

In this form, we can see than any x-value other than -2 or 2 will yield an unambiguous output value for y. Thus, the domain consists of all real numbers except for -2 and 2. [Note that the graph in Figure 1(b) supports this conclusion.] Using interval notation, the domain is

$$(-\infty, -2) \cup (-2, 2) \cup (2, \infty)$$

For the function in Figure 2, $y = 6/(x^2 + 1)$, the denominator is never zero (why?). Thus, any real number x is an appropriate input and the domain is the set of all real numbers.

Now look back at Figure 1(b). Here we can see that the x- and the y-intercepts are each close to zero, if not zero. To calculate the y-intercept, set $x = 0$ in the given equation. That yields

$$y = \frac{5(0) - 1}{0^2 - 4} = \frac{1}{4}$$

Thus, the y-intercept is $1/4$. For the x-intercept in Figure 1(b), set $y = 0$ in the given equation:

$$0 = \frac{5x - 1}{x^2 - 4}$$

The fraction in this last equation, like any fraction, can only be zero if the numerator is zero (and the denominator isn't zero). Setting $5x - 1$ equal to zero gives us $x = 1/5$, which is the required x-intercept. By the way, this simple idea that a fraction can only be zero when the numerator is zero can tell you right away, even before seeing a graph, that some rational functions have no x-intercepts. For instance, the numerator of $y = 6/(x^2 + 1)$ is certainly never zero, and therefore the graph of $y = 6/(x^2 + 1)$ has no x-intercepts. (See Figure 2.)

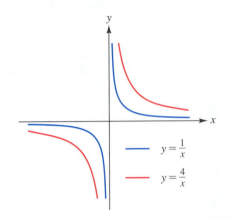

Figure 3

If k is a positive constant, the graph of $y = k/x$ resembles that of $y = 1/x$. Figure 3 shows the graphs of $y = 1/x$ and $y = 4/x$. Once we know about the graph of $y = k/x$, we can graph any rational function of the form

$$y = \frac{ax + b}{cx + d}$$

The next three examples show how this is done using the translation and reflection techniques explained in Section 3.4.

EXAMPLE **1** **Graphing a translate of y = 1/x**

Graph $y = 1/(x - 1)$.

SOLUTION
By translating the graph of $y = 1/x$ one unit to the right, we obtain the graph of $y = 1/(x - 1)$, shown in Figure 4. Notice that the horizontal asymptote is unchanged by the translation. However, the vertical asymptote moves one unit to the right. (The y-intercept is -1; why?)

Figure 4
The graph of $y = 1/(x - 1)$.

EXAMPLE **2** **Two graphs obtained from that of y = 2/x**

Graph $y = 2/(x - 1)$ and $y = -2/(x - 1)$.

SOLUTION
The graph of $y = 2/(x - 1)$ has the same basic shape and location as the graph of $y = 1/(x - 1)$ shown in Figure 4. As a further guide to sketching $y = 2/(x - 1)$, we can pick several convenient x-values on either side of the asymptote $x = 1$ and then compute the corresponding y-values. After doing this, we obtain the graph shown in Figure 5(a). By reflecting this graph about the x-axis, we obtain the graph of $y = -2/(x - 1)$, which is shown in Figure 5(b).

Figure 5

(a) $y = \dfrac{2}{x-1}$

(b) $y = \dfrac{-2}{x-1}$

EXAMPLE 3 **Long division as an aid in graphing**

Graph $y = \dfrac{4x - 2}{x - 1}$.

SOLUTION
First, as you can readily check, the x- and y-intercepts are 1/2 and 2, respectively. Next, using long division, we find that

$$\frac{4x - 2}{x - 1} = 4 + \frac{2}{x - 1}$$

(If you need a review of (or reintroduction to) the long division process from basic or intermediate algebra, see Section 7.2.) We conclude that the required graph can be obtained by moving the graph of $y = 2/(x - 1)$ up 4 units in the y-direction; see Figure 6. Notice that the vertical asymptote is still $x = 1$, but the horizontal asymptote is now $y = 4$ instead of the x-axis.

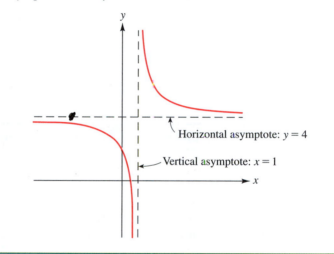

Figure 6

The graph of $y = \dfrac{4x - 2}{x - 1}$.

Now let's look at rational functions of the form $y = 1/x^n$. First we'll consider $y = 1/x^2$. As with $y = 1/x$, the domain consists of all real numbers except $x = 0$. For $x \neq 0$, the quantity $1/x^2$ is always positive. This means that the graph will always lie above the x-axis. Furthermore, the graph will be symmetric about the y-axis. (Why?) As $|x|$ becomes very large, the quantity $1/x^2$ approaches zero; this is true whether x itself is positive or negative. So when $|x|$ is very large, we expect the graph of $y = 1/x^2$ to look as shown in Figure 7(a). On the other hand, when x is a very small fraction, close to zero, either negative or positive, the quantity $1/x^2$ is very large. For instance, if $x = 1/10$, we find that $y = 100$, and if $x = 1/100$, we find that $y = 10,000$. Thus, as x approaches zero, from the right or from the left, the graph of $y = 1/x^2$ must look as shown in Figure 7(b).

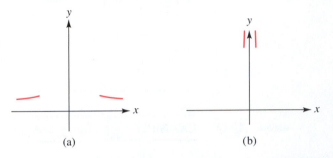

(a)   (b)

Figure 7

Now, by plotting several points and taking Figures 7(a) and 7(b) into account, we obtain the graph of $y = 1/x^2$, shown in Figure 8. Also, by following a similar line of reasoning, we find that the graph of $y = 1/x^3$ looks as shown in Figure 9. Note that $y = 1/x^3$ is symmetric about the origin, and $y = 1/x^2$ is symmetric about the y-axis.

In general, when n is an even integer greater than 2, the graph of $y = 1/x^n$ resembles that of $y = 1/x^2$. When n is an odd integer greater than 3, the graph of $y = 1/x^n$ resembles that of $y = 1/x^3$.

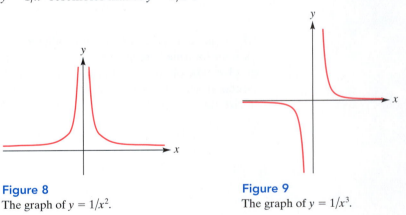

Figure 8
The graph of $y = 1/x^2$.

Figure 9
The graph of $y = 1/x^3$.

EXAMPLE **4** **A graph based on that of $y = 1/x^2$**

Graph $y = -1/(x + 3)^2$.

SOLUTION
Refer to Figure 10. Begin with the graph of $y = 1/x^2$. By moving the graph three units to the left, we obtain the graph of $y = 1/(x + 3)^2$. Then by re-

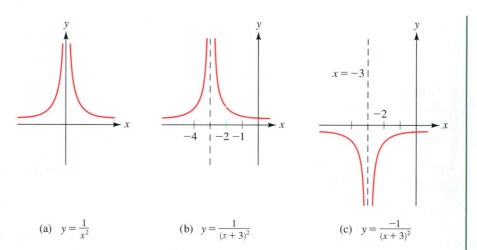

Figure 10

(a) $y = \dfrac{1}{x^2}$

(b) $y = \dfrac{1}{(x+3)^2}$

(c) $y = \dfrac{-1}{(x+3)^2}$

flecting the graph of $y = 1/(x + 3)^2$ about the x-axis, we get the graph of $y = -1/(x + 3)^2$.

QUESTION   What are the y-intercepts in Figures 10(b) and (c)?

EXAMPLE 5   A graph based on that of $y = 1/x^3$

Graph $y = 4/(x - 2)^3$.

SOLUTION
Moving the graph of $y = 1/x^3$ two units to the right gives us the graph of the function $y = 1/(x - 2)^3$. The graph of $y = 4/(x - 2)^3$ will have the same basic shape and location. As a further guide to sketching the required graph, we can pick several convenient x-values near the asymptote $x = 2$ and compute the corresponding y-values. Using $x = 0, x = 1$, and $x = 3$, we find that the points $(0, -\frac{1}{2})$, $(1, -4)$, and $(3, 4)$ are on the graph. With this information, the graph can be sketched as in Figure 11.

Figure 11
The graph of $y = 4/(x - 2)^3$.

As we have seen in Examples 1 through 5, the vertical asymptotes of $y = f(x)/g(x)$ are found by solving the equation $g(x) = 0$. In Example 6 we demonstrate a method for determining the horizontal asymptote. The method involves dividing both the numerator and the denominator by the highest power of x that appears.

EXAMPLE 6   Finding a horizontal asymptote

Determine the horizontal asymptote for the graph of $y = \dfrac{x^2 - 5x + 6}{2x^2 - 4x + 3}$.

SOLUTION
The strategy is to divide every term in the given fraction by the highest power of x that appears in the fraction, taking both the numerator and denominator into

account. In this particular example, it's a tie between the numerator and denominator: the highest power is x^2. We then have

$$y = \frac{x^2 - 5x + 6}{2x^2 - 4x + 3} = \frac{\dfrac{x^2}{x^2} - \dfrac{5x}{x^2} + \dfrac{6}{x^2}}{\dfrac{2x^2}{x^2} - \dfrac{4x}{x^2} + \dfrac{3}{x^2}} = \frac{1 - \dfrac{5}{x} + \dfrac{6}{x^2}}{2 - \dfrac{4}{x} + \dfrac{3}{x^2}}$$

Now, in that last expression, look at the four resulting fractions appearing within the numerator and denominator of the main fraction: $5/x$, $6/x^2$, $4/x$, and $3/x^2$. The point is that as $|x|$ gets larger and larger, all four of those fractions approach zero. We therefore have

$$y = \frac{1 - \dfrac{5}{x} + \dfrac{6}{x^2}}{2 - \dfrac{4}{x} + \dfrac{3}{x^2}} \approx \frac{1 - 0 + 0}{2 - 0 + 0} = \frac{1}{2} \quad \text{as } |x| \text{ gets larger and larger}$$

This tells us that the line $y = 1/2$ is a horizontal asymptote for the graph of the given function. See Table 1 and Figure 12 for numerical and graphical perspectives.

TABLE 1
As x Grows Larger and Larger, the Values of the Function Approach 0.5

x	$y = \dfrac{x^2 - 5x + 6}{2x^2 - 4x + 3}$
10	0.343558 ...
10^2	0.484925 ...
10^3	0.498499 ...
10^4	0.499849 ...
10^5	0.499984 ...
10^6	0.499998 ...

GRAPHICAL PERSPECTIVE

(a) $[-8, 8, 8]$ by $[-1, 3, 1]$ (b) $[-25, 25, 25]$ by $[-1, 3, 1]$

Figure 12
Two views of the rational function $y = \dfrac{x^2 - 5x + 6}{2x^2 - 4x + 3}$ (red) with horizontal asymptote $y = 1/2$ (blue)

EXAMPLE 7 Using rational functions to model bacterial growth

A group of agricultural scientists has been studying how the growth of a particular type of bacteria is affected by the acidity level of the soil. One colony of the bacteria is placed in a soil that is slightly acidic. A second colony of the same size is placed in a neutral soil. Suppose that after analyzing the data, the scientists determine that the size of each population over time can be modeled by the following functions.

$$\text{colony in neutral soil:} \quad y = (2t + 1)/(t + 1) \quad t \geq 0$$
$$\text{colony in acidic soil:} \quad y = (4t + 3)/(t^2 + 3) \quad t \geq 0$$

In both cases, y represents the population, in thousands, after t hours. Figure 13 shows the graphs of these two population functions over the first hour of the

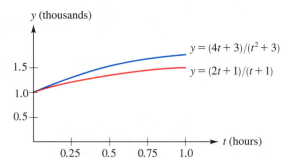

Figure 13

Population models for two colonies of bacteria

experiment. (Note from both the graphs and the equations that each colony begins with 1000 bacteria.) Find the horizontal asymptotes for each graph and thereby determine the long-term behavior of each colony.

SOLUTION:

Colony in neutral soil:

$$y = \frac{2t + 1}{t + 1}$$

$$= \frac{2 + 1/t}{1 + 1/t} \qquad \text{dividing each term by } t$$

$$\approx \frac{2 + 0}{1 + 0} = 2 \qquad \text{as } t \text{ gets larger and larger}$$

Colony in acidic soil:

$$y = \frac{4t + 3}{t^2 + 3}$$

$$= \frac{4/t + 3/t^2}{1 + 3/t^2} \qquad \text{dividing each term by } t^2$$

$$\approx \frac{0 + 0}{1 + 0} = 0 \qquad \text{as } t \text{ gets larger and larger}$$

For the colony in neutral soil, the horizontal asymptote for the growth function is $y = 2$. So in the long run, this colony approaches a population of 2000. For the colony in acidic soil, the horizontal asymptote is $y = 0$ and the population approaches extinction (or becomes extinct, depending on the interpretation). See Figure 14.

GRAPHICAL PERSPECTIVE

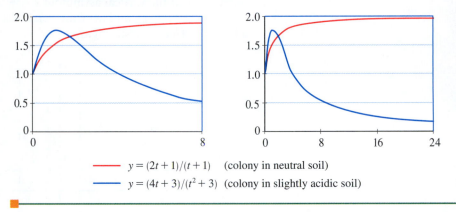

Figure 14

Snapshots of the population functions after $t = 8$ hr and $t = 24$ hr. The vertical axis is the population y, in thousands.

EXAMPLE | 8 | **Graphing a rational function**

Graph the function $y = \dfrac{x^2 - x - 6}{x^2 - x - 2} = \dfrac{(x - 3)(x + 2)}{(x + 1)(x - 2)}$.

SOLUTION
By looking at the factored form of the fraction, we see that the x-intercepts are 3 and -2. Also, the vertical asymptotes are the lines $x = -1$ and $x = 2$. For the y-intercept, it's a bit simpler to set $x = 0$ in the unfactored, rather than factored, expression. As you can check, the result in either case is that the y-intercept is 3. The horizontal asymptote works out to be the line $y = 1$; you should verify this for yourself using the method demonstrated in the previous two examples.

Now we want to see how the graph looks in the immediate vicinity of the x-intercepts and the vertical asymptotes. To do this, we use the approximation technique explained in the previous section for polynomial functions. Let's start with the x-intercept at $x = 3$. As in Section 4.6, we'll retain the factor $x - 3$ and approximate the remaining factors using $x = 3$. So, for x near 3, we have

$$y = \frac{(x - 3)(x + 2)}{(x + 1)(x - 2)} \approx \frac{(x - 3)(3 + 2)}{(3 + 1)(3 - 2)} = \frac{5}{4}(x - 3)$$

Thus, in the immediate vicinity of the x-intercept $x = 3$, the required graph will closely resemble the line $y = \frac{5}{4}(x - 3)$. The remaining calculations for approximating the graph near the other x-intercept and near the two vertical asymptotes are carried out in exactly the same manner. As Exercise 43 will ask you to verify, the results are:

x near 3: $\quad y \approx \dfrac{5}{4}(x - 3)$	x near -1: $\quad y \approx \dfrac{4/3}{x + 1}$
x near -2: $\quad y \approx -\dfrac{5}{4}(x + 2)$	x near 2: $\quad y \approx \dfrac{-4/3}{x - 2}$

We summarize these results as follows. As the graph passes through the points $(3, 0)$ and $(-2, 0)$, it resembles the lines $y = \frac{5}{4}(x - 3)$ and $y = -\frac{5}{4}(x + 2)$, respectively. Near the vertical asymptote $x = -1$, the graph has the same basic shape as $y = 1/(x + 1)$, and near the vertical asymptote $x = 2$, the graph resembles $y = -1/(x - 2)$; see Figure 15.

Figure 15

The graph of $y = \dfrac{(x - 3)(x + 2)}{(x + 1)(x - 2)}$ near its x-intercepts and vertical asymptotes.

Figure 16

The graph of $y = \dfrac{(x-3)(x+2)}{(x+1)(x-2)}$ near x-intercepts and its horizontal and vertical asymptotes.

Figure 17

The graph of $y = \dfrac{(x-3)(x+2)}{(x+1)(x-2)}$.

TABLE 2
When |x| is Large,
y is Less Than 1

x	y
10	0.95
100	0.99
−10	0.96
−100	0.9996

Finally, we want to find out how the graph approaches the horizontal asymptote $y = 1$. Perhaps the simplest way is to do some calculations. In Table 2 we have computed the outputs for some relatively large values of $|x|$. From the table, we see that when $|x|$ is large, y is less than 1. Graphically, this means that the curve approaches the asymptote $y = 1$ from below. In Figure 16 we summarize what we have discovered up to this point. Then, using Figure 16 as a guide, we can sketch the required graph, as shown in Figure 17.

NOTE In this particular case, however, the coordinates can be found by applying algebraic techniques. See Exercise 46 for the details.

EXERCISE SET 4.7

A

In Exercises 1–8:

(a) *Find the domain, x- and y-intercepts, vertical asymptotes, and horizontal asymptotes for each rational function.*

G **(b)** *Use a graphing utility to graph the function. Check to see that the graph is consistent with your results in part (a).*

1. $y = (3x + 15)/(4x - 12)$
2. $y = (x + 6)(x + 4)/(x - 1)^2$
3. $y = (6x^2 - 5x + 1)/(2x^2)$
4. $y = x/(x^2 + x + 1)$
5. $y = (x^2 - 9)/(4x^2 - 1)$
6. $y = (x^2 - 9)/(4x^2 + 1)$
7. $y = (3x^2 - 2x - 8)/(2x^3 + x^2 - 3x)$
8. $y = (x^3 - 27)/(x^4 - 2x^3 + 9x^2 - 18x)$

In Exercises 9–34, sketch the graph of each rational function. Specify the intercepts and the asymptotes.

9. $y = 1/(x + 4)$
10. $y = -1/(x + 4)$
11. $y = 3/(x + 2)$
12. $y = -3/(x + 2)$
13. $y = -2/(x - 3)$
14. $y = (x - 1)/(x + 1)$
15. $y = (x - 3)/(x - 1)$
16. $y = 2x/(x + 3)$
17. $y = (4x - 2)/(2x + 1)$
18. $y = (3x + 2)/(x - 3)$
19. $y = 1/(x - 2)^2$
20. $y = -1/(x - 2)^2$
21. $y = 3/(x + 1)^2$
22. $y = -3/(x + 1)^2$
23. $y = 1/(x + 2)^3$
24. $y = -1/(x + 2)^3$
25. $y = -4/(x + 5)^3$
26. $y = x/[(x + 1)(x - 3)]$
27. $y = -x/[(x + 2)(x - 2)]$
28. $y = 2x/(x + 1)^2$
29. **(a)** $y = 3x/[(x - 1)(x + 3)]$
    (b) $y = 3x^2/[(x - 1)(x + 3)]$

30. (a) $y = (4x^2 + 1)/(x^2 - 1)$
(b) $y = (4x^2 + 1)/(x^2 + 1)$
31. $y = (4x^2 + x - 5)/(2x^2 - 3x - 5)$
32. $y = (4x^2 - x - 3)/(2x^2 - 3x - 5)$
Ⓖ **33. (a)** $f(x) = (x - 2)(x - 4)/[x(x - 1)]$
(b) $g(x) = (x - 2)(x - 4)/[x(x - 3)]$
[Compare the graphs you obtain in parts (a) and (b). Notice how a change in only one constant can radically alter the nature of the graph.]
Ⓖ **34. (a)** $f(x) = (x - 1)(x + 2.75)/[(x + 1)(x + 3)]$
(b) $g(x) = (x - 1)(x + 3.25)/[(x + 1)(x + 3)]$
[Compare the graphs you obtain in parts (a) and (b). Notice how a relatively small change in one of the constants can radically alter the graph.]
35. The population y of a colony of bacteria after t hr is given by

$$y = (t + 12)/(0.0004t + 0.024), \quad \text{where } t \geq 0$$

(a) Find the initial population (i.e., the population when $t = 0$ hr).
(b) Determine the long-term behavior of the population (as in Example 7).
36. The population y (in thousands) of a colony of bacteria after t hr is given by

$$y = (6t + 12)/(2t + 1), \quad \text{where } t \geq 0$$

(a) Find the initial population and the long-term population. Which is larger?
Ⓖ **(b)** Use a graphing utility to graph the population function. Is the function increasing or decreasing? Check that your response here is consistent with your answers in part (a).

B

37. A desktop publisher designing a small rectangular poster decides to make the area 500 in.² The margins on the top and bottom of the poster are to be 3 in. and 4 in., respectively. The left and right margins are each to be 1.5 in.

(a) Express the area of the printed portion of the poster as a function of x, the width of the entire poster. (The gray region in the figure represents the printed area.)

Ⓖ **(b)** Use a graphing utility to graph the area function that you found in part (a).
Ⓖ **(c)** Use a ZOOM feature to estimate (to the nearest one-tenth) the width x for which the printed area is a maximum. What is the corresponding length of the poster in this case?
38. The accompanying figure gives formulas for the volume V and total surface area S of a circular cylinder with radius r and height h. For a cylinder of given volume, we are interested in finding the dimensions that minimize the surface area.

$$V = \pi r^2 h$$
$$S = 2\pi r^2 + 2\pi rh$$

(a) Assume that the volume of the cylinder is 1000 cm³. Express the surface area as a function of the radius. After combining terms in your answer, show that the resulting function can be written

$$S(r) = \frac{2\pi r^3 + 2000}{r} \quad (r > 0)$$

Ⓖ **(b)** Use a graphing utility to graph the surface area function obtained in part (a).
Ⓖ **(c)** Estimate, to the nearest one-hundredth, the radius r that minimizes the surface area. What is the corresponding value for h?

In Exercises 39–42, graph the functions. Note: In each case, the graph crosses its horizontal asymptote once. To find the point where the rational function $y = f(x)$ crosses its horizontal asymptote $y = k$, you'll need to solve the equation $f(x) = k$.

39. $y = \dfrac{(x - 4)(x + 2)}{(x - 1)(x - 3)}$
40. $y = \dfrac{(x - 1)(x - 3)}{(x + 1)^2}$

41. $y = \dfrac{(x + 1)^2}{(x - 1)(x - 3)}$
42. $y = \dfrac{2x^2 - 3x - 2}{x^2 - 3x - 4}$

43. (This exercise refers to Example 8.) Let

$$y = \frac{(x - 3)(x + 2)}{(x + 1)(x - 2)}$$

Verify each of the following approximations.
(a) When x is close to -2, then $y \approx -\frac{5}{4}(x + 2)$.
(b) When x is close to -1, then $y \approx \dfrac{4/3}{x + 1}$.
(c) When x is close to 2, then $y \approx \dfrac{-4/3}{x - 2}$.

In Exercises 44 and 45, graph the functions. Notice in each case that the numerator and denominator contain at least one common factor. Thus, you can simplify each quotient; but don't lose track of the domain of the function as it was initially defined.

44. (a) $y = \dfrac{x + 2}{x + 2}$ (c) $y = \dfrac{x - 1}{(x - 1)(x - 2)}$

   (b) $y = \dfrac{x^2 - 4}{x - 2}$

45. (a) $y = \dfrac{x^2 - 9}{x + 3}$

   (b) $y = \dfrac{x^2 - 5x + 6}{x^2 - 2x - 3}$

   (c) $y = \dfrac{(x - 1)(x - 2)(x - 3)}{(x - 1)(x - 2)(x - 3)(x - 4)}$

46. This exercise shows you how to determine the coordinates of the lowest point on the middle branch of the curve in Figure 17. The basic idea is as follows. Suppose that the required coordinates are (h, k). Then the horizontal line $y = k$ is the unique horizontal line intersecting the curve in one and only one point. (Any other horizontal line intersects the curve either in two points or not at all.) In steps (a) through (c) that follow, we use these observations to determine the point (h, k).

   (a) Given any horizontal line $y = k$, its intersection with the curve in Figure 17 is determined by solving the following pair of simultaneous equations:

$$\begin{cases} y = k \\ y = \dfrac{x^2 - x - 6}{x^2 - x - 2} \end{cases}$$

   In the second equation of the system, replace y with k and show that the resulting equation can be written

$$(k - 1)x^2 - (k - 1)x + (6 - 2k) = 0 \qquad (1)$$

   (b) If k is indeed the required y-coordinate, then equation (1) must have exactly one real solution. Set the discriminant of the quadratic equation equal to zero to obtain

$$(k - 1)^2 - 4(k - 1)(6 - 2k) = 0$$

   and deduce from this that $k = 1$ or $k = 25/9$. The solution $k = 1$ can be discarded. (To see why, look at Figure 17.)

   (c) Using the value $y = 25/9$, show that the corresponding x-coordinate is $1/2$. Thus, the required point is $\left(\frac{1}{2}, \frac{25}{9}\right)$.

47. Graph the function $y = x/(x - 3)^2$. Use the technique explained in Exercise 46 to find the coordinates of any turning points on the graph.

48. Graph the function $y = 2/(x - x^2)$. Use the technique explained in Exercise 46 to find the coordinates of any turning points on the graph.

An asymptote that is neither horizontal nor vertical is called a **slant** or **oblique asymptote**. For example, as indicated in the following figure, the line $y = x$ is a slant asymptote for the graph of $y = (x^2 + 1)/x$. To understand why the line $y = x$ is an asymptote, we carry out the indicated division and write the function in the form

$$y = x + \frac{1}{x} \qquad (2)$$

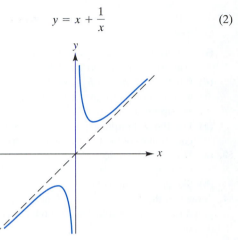

From equation (2), we see that if $|x|$ is very large then $y \approx x + 0$; that is, $y \approx x$, as we wished to show. Equation (2) actually tells us more than this. When $|x|$ is very close to zero, equation (2) yields $y \approx 0 + (1/x)$. In other words, as we approach the y-axis, the curve looks more and more like the graph of $y = 1/x$. In general, if we have a rational function $f(x)/g(x)$ in which the degree of $f(x)$ is 1 greater than the degree of $g(x)$, then the graph has a slant asymptote that is obtained as follows. Divide $f(x)$ by $g(x)$ to obtain an equation of the form

$$\frac{f(x)}{g(x)} = (mx + b) + \frac{h(x)}{g(x)}$$

where the degree of $h(x)$ is less than the degree of $g(x)$. Then the equation of the slant asymptote is $y = mx + b$. (For instance, using the previous example, we have $y = mx + b = x$ and $h(x)/g(x) = 1/x$.) In Exercises 49–51, you are asked to graph functions that have slant asymptotes.

49. Let $y = F(x) = \dfrac{x^2 + x - 6}{x - 3}$.

   (a) Use long division to show that

$$\frac{x^2 + x - 6}{x - 3} = (x + 4) + \frac{6}{x - 3}$$

   (b) The result in part (a) shows that the line $y = x + 4$ is a slant asymptote for the graph of the function F. Verify this fact empirically by completing the following two tables.

x	$x + 4$	$\dfrac{x^2 + x - 6}{x - 3}$
10		
100		
1000		

(continues)

x	$x + 4$	$\dfrac{x^2 + x - 6}{x - 3}$
-10		
-100		
-1000		

(c) Determine the vertical asymptote and the x- and y-intercepts of the graph of F.

(d) Graph the function F. (Use the techniques in this section along with the fact that $y = x + 4$ is a slant asymptote.)

(e) Use the technique explained in Exercise 46 to find the coordinates of the two turning points on the graph of F.

50. (a) Show that the line $y = x - 2$ is a slant asymptote for the graph of $F(x) = x^2/(x + 2)$.

(b) Sketch the graph of F.

51. Show that the line $y = -x$ is a slant asymptote for the graph of $y = (1 - x^2)/x$. Then sketch the graph of this function.

In Section 4.7 we've studied rational functions in which the degree of the numerator is less than or equal to the degree of the denominator. Exercises 52 and 53 examine cases in which the degree of the numerator is greater than that of the denominator.

52. Let $f(x) = (x^3 + 2x^2 + 1)/(x^2 + 2x)$.

(a) Graph the function f using a viewing rectangle extending from -5 to 5 in both the x- and y-directions.

(b) Add the graph of the line $y = x$ to your picture in part (a). Note that to the right of the origin, as x increases, the graph of f begins to look more and more like the line $y = x$. This also occurs to the left of the origin as x decreases.

(c) The results in part (b) suggest that the line $y = x$ may be an asymptote for the graph of f. Verify this visually by changing the viewing rectangle so that it extends from -20 to 20 in both the x- and the y-directions. What do you observe?

(d) Using algebra, verify the identity

$$\frac{x^3 + 2x^2 + 1}{x^2 + 2x} = \frac{1}{x^2 + 2x} + x$$

Then explain why, for large values of $|x|$, the graph of f looks more and more like the line $y = x$. *Hint:* Substitute some large numbers (such as 100 or 1000) into the expression $1/(x^2 + 2x)$. What happens?

53. Let $f(x) = (x^5 + 1)/x^2$.

(a) Graph the function f using a viewing rectangle that extends from -4 to 4 in the x-direction and from -8 to 8 in the y-direction.

(b) Add the graph of the curve $y = x^3$ to your picture in part (a). Note that as $|x|$ increases (that is, as x moves away from the origin), the graph of f looks more and more like the curve $y = x^3$. For additional perspective, first change the viewing rectangle so that y extends from -20 to 20. (Retain the x-settings for the moment.) Describe what you see. Next, adjust the viewing rectangle so that x extends from -10 to 10 and y extends from -100 to 100. Summarize your observations.

(c) In the text we said that a line is an asymptote for a curve if the distance between the line and the curve approaches zero as we move further and further out along the curve. The work in part (b) shows that a *curve* can be an asymptote for another curve. In particular, part (b) shows that the distance between the curve $y = x^3$ and the graph of the given function f approaches zero as we move further and further out along the graph of f. That is, the curve $y = x^3$ is an asymptote for the graph of the given function f. Complete the following two tables for a numerical perspective on this. In the tables, d denotes the vertical distance between the curve $y = x^3$ and the graph of f:

$$d = \left| \frac{x^5 + 1}{x^2} - x^3 \right|$$

x	5	10	50	100	500
d					

x	-5	-10	-50	-100	-500
d					

(d) Parts (b) and (c) have provided both a graphical and a numerical perspective. For an algebraic perspective that ties together the previous results, verify the following identity, and then use it to explain why the results in parts (b) and (c) were inevitable:

$$\frac{x^5 + 1}{x^2} = x^3 + \frac{1}{x^2}$$

PROJECT

FINDING SOME MINIMUM VALUES WITHOUT USING CALCULUS

In this project you'll solve some minimum problems that involve rational functions. The theorem that we'll rely on is given in the box that follows. Methods for establishing this theorem are indicated in Question 3 at the end of this project. (This theorem can be stated in a more general form for n positive quantities, rather than two or three, but we won't need that for the applications in this project.)

THEOREM   On Minimum Sums

(a) If the product of two positive quantities s and t is constant, then their sum $s + t$ is a minimum if and only if $s = t$.

(b) If the product of three positive quantities s, t, and u is constant, then their sum $s + t + u$ is a minimum if and only if $s = t = u$.

First, just to get a feeling for what part (a) of the theorem is saying, complete the following two tables, and in each case note which pair of numbers a and b yields the smallest sum.

a	b	ab	a + b	a	b	ab	a + b
1	36			1	12		
2	18			2	6		
3	12			3	4		
4	9			$3\frac{1}{2}$	$3\frac{3}{7}$		
6	6			$2\sqrt{3}$	$2\sqrt{3}$		

1. Use part (a) of the theorem to solve the following problems.
   (a) Find the smallest possible value for the sum of a positive number and its reciprocal.
   (b) The accompanying figure shows a line with slope m ($m < 0$) passing through the point $(2, 1)$. Find the smallest possible area for the triangle that is formed. *Hint:* First use the point–slope formula to write the equation of a line with slope m that passes through the point $(2, 1)$. Next, find the x- and y-intercepts of the line in terms of m. Then, express the area of the triangle in terms of m and find a way to apply part (a) of the theorem.

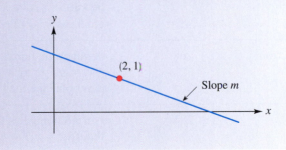

2. Use part (b) of the theorem to solve the following problems.

   (a) After using a graphing utility to estimate the minimum value of the function $f(x) = x^2 + \dfrac{1}{x}$ for $x > 0$, find the exact value for this minimum.

   Hint: Write $x^2 + \dfrac{1}{x}$ as $x^2 + \dfrac{1}{2x} + \dfrac{1}{2x}$.

   (b) Use a graphing utility to estimate the minimum value of the function $g(x) = (2x^3 + 3)/x$ for $x > 0$. Then find the exact minimum. (Adapt the hint in the previous exercise.)

   (c) A box with a square base and no top is to be constructed. The volume is to be 27 ft^3. Find the dimensions of the box so that the surface area is a minimum.

   (d) In Exercise 38 on page 352, a graphing utility is used to estimate the minimum possible surface area for a circular cylinder with a volume 1000 cm^3. Find the exact value for this minimum. Also confirm the following fact: For the values of r and h that yield the minimum surface area, we have $2r = h$.

3. (a) *Proof for part (a) of the theorem*   Verify that the following simple identity holds for all real numbers a and b.

   $$(a + b)^2 = 4ab + (a - b)^2$$

   Then use this identity to explain why part(a) of the theorem is valid. *Suggestion*: First discuss the reasoning within a group; then, on your own, write a paragraph carefully explaining the reasoning in your own words.

   (b) *Proof for part (b) of the theorem*   Use the result in Exercise 49 of Section 2.5 to explain why part (b) of the theorem is valid.

Chapter 4

Summary of Principal Terms

Terms	Page reference	Comments
1. Linear function	251	A linear function is a function of the form $f(x) = Ax + B$. The graph of a linear function is a straight line. An important idea that arose in several of the examples is that the slope of a line can be interpreted as a rate of change. Two instances of this are marginal cost and velocity.
2. Quadratic function	268	A quadratic function is a function of the form $f(x) = ax^2 + bx + c$. The graph of a quadratic function is a parabola. See the boxes on pages 271 and 273.
3. Fixed point	284	A fixed point of a function f is an input x in the domain of f such that $f(x) = x$. For example, $x = 1$ is a fixed point for $f(x) = \sqrt{x}$ because $f(1) = \sqrt{1} = 1$.

Terms	Page reference	Comments
4. Vertex formula	311	The x-coordinate of the vertex of the parabola $y = ax^2 + bx + c$ is given by $x = -b/2a$.
5. Polynomial function	326	A polynomial function is a function of the form $$f(x) = a_n x^n + a_{n-1} x^{n-1} + \cdots + a_1 x + a_0$$ where n is a nonnegative integer. Three basic properties of polynomial functions are summarized in the box on page 331. If $a_n \neq 0$, the degree of the polynomial function is n, the largest exponent on the input variable.
6. Rational function	342	A rational function is a function of the form $y = f(x)/g(x)$, where $f(x)$ and $g(x)$ are polynomials.
7. Asymptote	342	A line is said to be an asymptote for a curve if the distance between the line and the curve approaches zero as we move out farther and farther along the line.

Writing Mathematics

1. At the start of Section 4.6 is a quotation by David Berlinski describing polynomial functions. Reread the quotation, and then write a paragraph or two explaining its contents using the ideas and terminology from Section 4.6. With the aid of a graphing utility, use examples as appropriate to illustrate your points.

2. Consider the following two problems.

PROBLEM 1
The perimeter of a rectangle is 2 m.
(i) Express the area as a function of the width w.
(ii) Find the maximum possible area.

PROBLEM 2
The area of a rectangle is 2 m^2.
(i) Express the perimeter as a function of the width w.
(ii) Find the minimum possible perimeter.

(a) After working out Problem 1 for yourself, write out the solution in complete sentences, as if you were explaining it to a classmate or to your instructor. Be sure to let the reader know where you are headed and why each of the main steps is necessary.

(b) After working out part (i) of Problem 2 for yourself, write out the solution to part (i) in complete sentences.

(c) Explain why the methods of Section 4.5 are not applicable for solving part (ii) of Problem 2.

(d) The following result is known as the *arithmetic-geometric mean inequality*:

For all nonnegative real numbers a and b we have $a + b \geq 2\sqrt{ab}$, with equality holding if and only if $a = b$.

By working with classmates or your instructor, find a way to apply this result in part (ii) of Problem 2. Then write out your solution in complete sentences. How does your final answer here compare with that in Problem 1?

Chapter 4 | Review Exercises

1. Find $G(0)$ if G is a linear function such that $G(1) = -2$ and $G(-2) = -11$.
2. **(a)** Let $f(x) = 3x^2 + 6x - 10$. For which input x is the value of the function a minimum? What is that minimum value?
   (b) Let $g(t) = 6t^2 - t^4$. For which input t is the value of the function a maximum?
3. Suppose the function $p = -\frac{1}{8}x + 100 \quad (0 \leq x \leq 12)$ relates the selling price p of an item to the quantity x that is sold. Assume that p is in dollars. What is the maximum revenue possible in this situation?
4. Graph the function $y = -1(x + 1)^3$.
5. Graph the function $y = (x - 4)(x - 1)(x + 1)$.
6. Graph the function $f(x) = x^2 + 4x - 5$. Specify the vertex, the x- and y-intercepts, and the axis of symmetry.
7. A factory owner buys a new machine for $1000. After five years, the machine has a salvage value of $100. Assuming linear depreciation, find a formula for the value V of the machine after t yr, where $0 \leq t \leq 5$.
8. Graph the function $y = 2(x - 3)^4$. Does the graph cross the y-axis? If so, where?
9. Graph the function $y = (3x + 5)/(x + 2)$. Specify all intercepts and asymptotes.
10. What is the largest area possible for a right triangle in which the sum of the lengths of the two shorter sides is 12 cm?
11. Graph the function $y = x/[(x + 2)(x - 4)]$.
12. Let $P(x, y)$ be a point [other than $(-1, -1)$] on the graph of $f(x) = x^3$. Express the slope of the line passing through the points P and $(-1, -1)$ as a function of x. Simplify your answer as much as possible.

13. A rectangle is inscribed in a circle. The circumference of the circle is 12 cm. Express the perimeter of the rectangle as a function of its width w.
14. Give a reason why each of the following two graphs cannot represent a polynomial function with highest-degree term $-\frac{1}{3}x^3$.

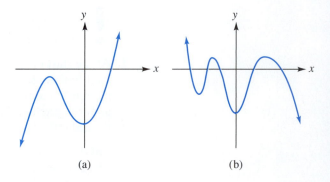

(a)   (b)

In Exercises 15–18, find equations for the linear functions satisfying the given conditions. Write each answer in the form $f(x) = mx + b$.

15. $f(4) = -1$ and the graph of f is parallel to the line $3x - 8y = 16$.
16. The graph passes through $(6, 1)$ and the x-intercept is twice the y-intercept.
17. $f(-3) = 5$, and the graph of the inverse function passes through $(2, 1)$.
18. The graph of f passes through the vertices of the two parabolas $y = x^2 + 4x + 1$ and $y = \frac{1}{2}x^2 + 9x + \frac{81}{2}$.

In Exercises 19–22, graph the quadratic functions. In each case, specify the vertex and the x- and y-intercepts.

19. $y = x^2 + 2x - 3$ **20.** $f(x) = x^2 - 2x - 15$

21. $f(x) = 2x^2 - 2x + 1$ **22.** $y = -3x^2 + 12x$

23. Find the distance between the vertices of the two parabolas $y = x^2 - 4x + 6$ and $y = -x^2 - 4x - 5$.

24. Find the value of a, given that the maximum value of the function $f(x) = ax^2 + 3x - 4$ is 5.

25. Suppose that an object is thrown vertically upward (from ground level) with an initial velocity of v_0 ft/sec. It can be shown that the height h (in feet) after t sec is given by the formula $h = v_0 t - 16t^2$.
  (a) At what time does the object reach its maximum height? What is that maximum height?
  (b) At what time does the object strike the ground?

26. Let $f(x) = 4x^2 - x + 1$ and $g(x) = (x - 3)/2$.
  (a) For which input will the value of the function $f \circ g$ be a minimum?
  (b) For which input will the value of $g \circ f$ be a minimum?

27. Find all values of b such that the minimum distance from the point $(2, 0)$ to the line $y = \frac{4}{3}x + b$ is 5.

28. What number exceeds one-half its square by the greatest amount?

29. Suppose that $x + y = \sqrt{2}$. Find the minimum value of the quantity $x^2 + y^2$.

30. For which numbers t will the value of $9t^2 - t^4$ be as large as possible?

31. Find the maximum area possible for a right triangle with a hypotenuse of 15 cm.   *Hint*: Let x denote the length of one leg. Show that the area is $A = x\sqrt{225 - x^2}/2$. Now work with A^2.

32. For which point (x, y) on the curve $y = 1 - x^2$ is the sum $x + y$ a maximum?

33. Let $f(x) = x^2 - (a^2 + 2a)x + 2a^3$, where $0 < a < 2$. For which value of a will the distance between the x-intercepts of f be a maximum?

34. Suppose that the function $p = 160 - \frac{1}{5}x$ relates the selling price p of an item to the quantity x that is sold. Assume that p is in dollars. For which value of x will the revenue R be a maximum? What is the selling price p in this case?

35. A piece of wire 16 cm long is cut into two pieces. Let x denote the length of the first piece and $16 - x$ the length of the second. The first piece is formed into a rectangle in which the length is twice the width. The second piece of wire is also formed into a rectangle, but with the length three times the width. For which value of x is the total area of the two rectangles a minimum?

36. Find all real numbers that are fixed points of the given functions.
  (a) $f(x) = x^2 - 8$ **(c)** $y = 4x - x^3$
  (b) $g(x) = x^2 + 8$ **(d)** $y = 8x^2 - x - 15$

37. In the following figure, $PQRS$ is a square with sides parallel to the coordinate axes. The coordinates of points A and B are $A(a, 0)$ and $B(b, 0)$. Show that $f(a) = b$ and $f(b) = a$.

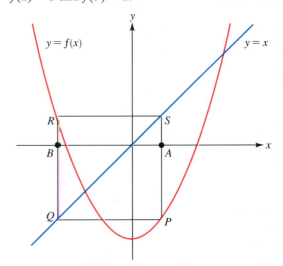

38. Let $f(x) = x^2 - 2$. Find two numbers a and b ($a \neq b$) such that $f(a) = b$ and $f(b) = a$.

In Exercises 39–50, graph each function and specify the x- and y-intercepts and asymptotes, if any.

39. $y = (x + 4)(x - 2)$ **40.** $y = (x + 4)(x - 2)^2$

41. $y = -x^2(x + 1)$ **42.** $y = -x^3(x + 1)$

43. $y = x(x - 2)(x + 2)$ **44.** $y = \dfrac{1 - 2x}{x}$

45. $y = \dfrac{-1}{(x - 1)^2}$ **46.** $y = \dfrac{x + 1}{x + 2}$

47. $y = \dfrac{x - 2}{x - 3}$ **48.** $y = \dfrac{x}{(x - 2)(x + 4)}$

49. $y = \dfrac{x^2 - 2x + 1}{x^2 - 4x + 4}$ **50.** $y = \dfrac{x(x - 2)}{(x - 4)(x + 4)}$

51. The range of the function $y = x^2 - 2x + k$ is the interval $[5, \infty)$. Find the value of k.

52. Find a value for b such that the range of the function $f(x) = x^2 + bx + b$ is the interval $[-15, \infty)$.

53. Find the range of the function $y = \dfrac{(x-1)(x-3)}{x-4}$.

Hint: Solve the equation for x in terms of y using the quadratic formula. If you're careful with the algebra, you will find that the expression under the resulting radical sign is $y^2 - 8y + 4$. The range of the given function can then be found by solving the inequality $y^2 - 8y + 4 \geq 0$.

54. In the following figure, triangle OAB is equilateral and $\overline{AB}$ is parallel to the x-axis. Find the length of a side and the area of the triangle OAB.

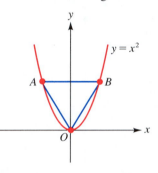

55. Let A denote the area of the right triangle in the first quadrant that is formed by the y-axis and the lines $y = mx$ and $y = m$. (Assume $m > 0$.) Express the area of the triangle as a function of m.

56. (a) The figure shows the parabola $y = x^2 - 6x$ and a circle that passes through the vertex and x-intercepts of the parabola. Find the center and radius of the circle.

(b) Suppose that a circle passes through the vertex and x-intercepts of a parabola $y = x^2 - 2bx$, where $b > 0$. Show that the circumference of the circle as a function of b is $c = \pi b^2 + \pi$.

57. In the accompanying figure, the radius of the circle is $OC = 1$. Express the area of $\triangle ABC$ as a function of x.

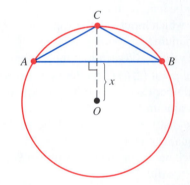

58. (a) Factor the expression $x^3 - 3x^2 + 4$.
Hint: Subtract and add 1, then factor by grouping.

(b) Use the factorization from part (a) to graph the function $y = x^3 - 3x^2 + 4$. Check your result using a graphing utility.

Chapter 4   Test

1. A linear function L satisfies the conditions $L(-2) = -4$ and $L(5) = 1$. Find $L(0)$.

2. (a) Let $F(x) = 4x - 2x^2$. What is the maximum value for this function? On which interval is this function increasing?

(b) Let $G(t) = 9t^4 + 6t^2 + 2$. For which value of t is this function a minimum?

3. Let $f(x) = (x - 3)(x + 4)^2$.
(a) Determine the intercepts and the excluded regions for the graph of f.
(b) Determine the behavior of f when x is very close to -4.
(c) Sketch the graph of f.

4. Graph each function:
   (a) $y = -1/(x - 3)^3$;
   (b) $y = -1/(x - 3)^2$.

5. Graph the function $y = -x^2 + 7x + 6$. Specify the intercepts, the axis of symmetry, and the coordinates of the turning point.

6. Suppose that the function $p = -\frac{1}{6}x + 80$
   $(0 \le x \le 400)$ relates the selling price of an item to the quantity x that is sold. Assume that p is in dollars. What is the maximum revenue possible in this situation? Which price p generates this maximum revenue?

7. The price of a new machine is $14,000. After ten years, the machine has a salvage value of $750. Assuming linear depreciation, find a formula for the value of the machine after t yr, where $0 \le t \le 10$.

8. Graph the function $y = -\frac{1}{2}(3 - x)^3$. Specify the intercepts. *Hint:* First graph $y = -\frac{1}{2}(x + 3)^3$.

9. Graph the function $y = (2x - 3)/(x + 1)$. Specify the intercepts and asymptotes.

10. (a) Suppose that $P(x, y)$ is a point on the line $y = 3x - 1$ and Q is the point $(-1, 3)$. Express the length PQ as a function of x.
    (b) For which value of x is the length PQ a minimum?

11. Let $f(x) = x(x - 2)/(x^2 - 9)$.
    (a) Find the vertical and the horizontal asymptotes.
    (b) Use a sketch to show the behavior of f near the x-intercepts.
    (c) Use a sketch to show the behavior of f near the asymptotes.
    (d) Sketch the graph of f.

12. A rectangle is inscribed in a semicircle of diameter 8 cm. (See the accompanying figure.) Express the area of the rectangle as a function of the width w of the rectangle.

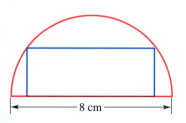

13. Explain why each of the following graphs cannot represent a polynomial function with highest-degree term $-x^3$.

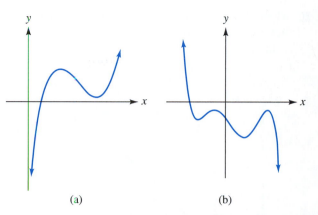

(a)   (b)

14. The following table shows U.S. expenditures for national defense for the years 1992–1995.

Federal Budget Outlays for National Defense, 1992–1995

Year t ($t = 0$ is 1990)	2	3	4	5
Expenditures y (billions of dollars)	298.4	291.1	281.6	272.1

Source: U.S. Office of Management and Budget

(a) Draw (by hand) a scatter plot of the data. (Show the portion of the y-axis from 270 to 300, in increments of 5.) You'll see that the data points appear to fall roughly on a straight line. Sketch in a line that seems to fit the data. Estimate the slope of the line. (Include the units for the slope.)

Ⓖ (b) Use a graphing utility to determine the equation of the line that best fits the data. Is your slope estimate in part (a) close to the slope obtained here?

(c) Use the least-squares equation obtained in part (b) to make estimates for what the expenditures might have been for 1996 and for 1999. Then compare the percentage errors in these two estimates, given that the actual expenditures (in billions of dollars) for 1996 and 1999 were 265.8 and 276.7, respectively.

Problems 15 and 16, which follow, are based on the work in Optional Section 4.3.

15. Find the real numbers (if any) that are fixed points of the given functions.
    (a) $f(x) = 2x(1 - x)$
    (b) $g(x) = (4x - 1)/(3x + 6)$

16. The following figure shows the first six steps in the iteration process for the function $f(x) = 0.6 - x^2$ with initial input $x_0 = 0.2$.

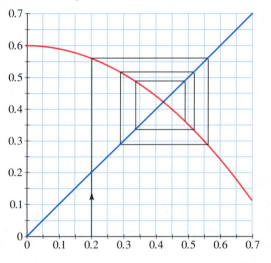

(a) Use the figure and your calculator to complete the following table. For your answers from the graph, try to estimate to the nearest 0.02 (or 0.01). For the calculator answers, round to three decimal places.

	x_1	x_2	x_3	x_4	x_5	x_6
From graph						
From calculator						

(b) Compute the two fixed points of the function $f(x) = 0.6 - x^2$. Give two forms for each answer: an exact expression involving a radical and a calculator approximation rounded to four decimal places. Which fixed point do the iterates in part (a) approach?

(c) Use your calculator to compute the first six iterates of $x_0 = 1$. Do the iterates seem to approach either of the fixed points determined in part (b)?

... in 1958 ... C. D. Keeling of The Scripps Institution of Oceanography ... began a series of painstaking measurements of CO_2 concentration on a remote site at what is now the Mauna Loa Observatory in Hawaii. His observations, continued to the present, show an exponential growth of atmospheric carbon dioxide. ... Keeling's observations have been duplicated at other stations in various parts of the world over shorter periods of time. In all sets of observations, the exponential increase is clear. ... —Gordon MacDonald in "Scientific Basis for the Greenhouse Effect," from *The Challenge of Global Warming*, Dean Edwin Abrahamson, ed. (Washington, D.C.: Island Press, 1989)

In this chapter we study two major types of functions with widespread applications: the *exponential functions* and their inverses, and the *logarithmic functions*. Exponential functions are functions of the form $y = ab^{cx}$ where a, b, and c are constants (with b positive and not equal to 1). As an introductory example, suppose that when you were born your parents opened a savings account in your name. Assume that the initial deposit was $100, with an annual interest rate of 8%, compounded twice a year. (If you are unfamiliar with compound interest, the details are in Section 5.6; but you won't need that to understand the point of the present discussion.) Under these conditions, the dollar amount y in the account after x years is given by the exponential function $y = 100 \times 1.04^{2x}$. In Table 1 we show the amount in the account at five-year intervals for the first 30 years. Figure 1(a) shows a graph of the function $y = 100 \times 1.04^{2x}$ and the data points from Table 1. For comparison, Figure 1(b) shows the same scatter plot along with a quadratic function fitted to the data (using the quadratic regression option on a graphing utility). Over this relatively short interval of time, it appears that there is not much of a difference between the two models. For longer time intervals, however, the situation changes drastically.

Table 2 shows additional values for the exponential function $y = 100 \times 1.04^{2x}$ at ten-year intervals. In Figure 2, we've plotted these data points along with the original data set, and we show the exponential and quadratic models. As you can see, over the longer time interval, the quadratic function is no longer a good model for the amount of money in your account; the quadratic grows too slowly to keep pace with the exponential. Indeed, this is one of the key features of exponential growth: Eventually it will outpace not only a quadratic, but any polynomial, no matter how high the degree.

Some of the applications you will see in this chapter for logarithmic or exponential functions involve:

- Modeling the temperature of a cup of coffee (Project in Section 5.2)
- Using the Richter scale to quantify the size of an earthquake (Example 8 in Section 5.3)

TABLE 1 $100 Invested at 8% per Year, Compounded Semiannually.						
x (years)	0	5	10	15	20	25	30
y (dollars)	100	148.02	219.11	324.34	480.10	710.67	1051.96

(a) Exponential function
$$y = 100 \times 1.04^{2x}$$

(b) Quadratic function fitted to the data in Table 1.
$$y = 0.894x^2 + 1.395x + 108.09$$

Figure 1

Initially, the exponential and quadratic models appear quite similar (but see Figure 2).

TABLE 2 Amount in the Account After 40, 50, and 60 Years.		
x (years)	40	50	60
y (dollars)	2304.98	5050.50	11066.26

Figure 2

The exponential function eventually grows much faster than the quadratic.

- Computing the annual yield (= effective rate) for money earning compound interest (Example 4 in 5.6)
- Modeling the producer price index over time (Exercise 33 in Section 5.6)
- Analyzing world population growth (Examples 2–4 in Section 5.7)
- Describing the increasing levels of carbon dioxide in the atmosphere (Example 6 and Exercises 15 and 16 in Section 5.7)

5.1 EXPONENTIAL FUNCTIONS

TABLE 1	
x (Day number)	**y** (Amount earned that day)
1	2¢ $(= 2^1)$
2	4¢ $(= 2^2)$
3	8¢ $(= 2^3)$
4	16¢ $(= 2^4)$
5	32¢ $(= 2^5)$
.	.
.	.
.	.
x	2^x

We begin with an example. Suppose that your mathematics instructor, in an effort to improve classroom attendance, offers to pay you each day for attending class! Suppose you are to receive 2¢ on the first day you attend class, 4¢ the second day, 8¢ the third day, and so on, as shown in Table 1. How much money will you receive for attending class on the 30th day?

As you can see by looking at Table 1, the amount y earned on day x is given by the rule, or *exponential function*,

$$y = 2^x$$

Thus on the 30th day (when $x = 30$), you will receive

$$y = 2^{30} \text{ cents}$$

If you use a calculator, you will find this amount to be well over 10 million dollars! The point here is simply this: Although we begin with a small amount, $y = 2$¢, repeated doubling quickly leads to a very large amount. In other words, the exponential function grows very rapidly.

Before leaving this example, we mention a simple method for quickly estimating numbers such as 2^{30} (or any power of two) in terms of the more familiar powers of ten. Begin by observing that

$$2^{10} \approx 10^3 \qquad \text{(a useful coincidence, worth remembering)}$$

Now just cube both sides to obtain

$$(2^{10})^3 \approx (10^3)^3 \qquad \text{or} \qquad 2^{30} \approx 10^9$$

Thus 2^{30} is about one billion. To convert this number of cents to dollars, we divide by 100 or 10^2 to obtain

$$\frac{10^9}{10^2} = 10^7 \text{dollars}$$

which is 10 million dollars, as mentioned before.

EXAMPLE 1 Estimating a power of 2

Estimate 2^{40} in terms of a power of 10.

SOLUTION
Take the basic approximation $2^{10} \approx 10^3$ and raise both sides to the fourth power. This yields

$$(2^{10})^4 \approx (10^3)^4$$

or

$$2^{40} \approx 10^{12}, \qquad \text{as required}$$

EXAMPLE **2** **Estimating an exponent**

Estimate the power to which 10 must be raised to yield 2.

SOLUTION
We begin with our approximation

$$10^3 \approx 2^{10}$$

Raising both sides to the power 1/10 yields

$$(10^3)^{1/10} \approx (2^{10})^{1/10}$$

and, consequently, $$10^{3/10} \approx 2$$

Thus the power to which 10 must be raised to yield 2 is approximately 3/10.

In Appendix B.3 we define the expression b^x, where x is a rational number. We also state that if x is irrational, then b^x can be defined so that the usual properties of exponents continue to hold. Although a rigorous definition of irrational exponents requires concepts from calculus, we can nevertheless convey the basic idea by means of an example. (We need to do this before we give the general definition for exponential functions.)

How shall we assign a meaning to $2^{\sqrt{2}}$, for example? The basic idea is to evaluate the expression 2^x successively by using rational numbers x that are closer and closer to $\sqrt{2}$. Table 2 displays the results of some calculations along these lines.

TABLE 2 **Values of 2^x for Rational Numbers x Approaching $\sqrt{2}$ (= 1.41421356 . . .)**

x	1.4	1.41	1.414	1.4142	1.41421	1.414213
2^x	2.6 . . .	2.65 . . .	2.664 . . .	2.6651 . . .	2.66514 . . .	2.665143 . . .

The data in the table suggest that as x approaches $\sqrt{2}$, the corresponding values of 2^x approach a unique real number, call it t, with a decimal expansion that begins as 2.665. Furthermore, by continuing this process we can obtain (in theory, at least) as many places in the decimal expansion of t as we wish. The value of the expression $2^{\sqrt{2}}$ is then defined to be this number t. The following results (stated here without proof) summarize this discussion and also pave the way for the definition we will give for exponential functions.

PROPERTY SUMMARY **Real Number Exponents**

Let b denote an arbitrary positive real number. Then:

1. For each real number x, the quantity b^x is a unique real number.
2. When x is irrational, we can approximate b^x as closely as we wish by evaluating b^r, where r is a rational number sufficiently close to the number x.
3. The properties of rational exponents continue to hold for irrational exponents.
4. If $b^x = b^y$ and $b \neq 1$, then $x = y$.

EXAMPLE 3 | **Simplifying expressions containing irrational exponents**

Use the properties of exponents to simplify each expression:
(a) $(3^{\sqrt{2}})^{\sqrt{2}}$;   **(b)** $(3^{\sqrt{2}})^2$.

SOLUTION
(a) $(3^{\sqrt{2}})^{\sqrt{2}} = 3^{\sqrt{2} \times \sqrt{2}} = 3^2 = 9$
(b) $(3^{\sqrt{2}})^2 = (3^2)^{\sqrt{2}}$ (Why?)
$\qquad\quad = 9^{\sqrt{2}}$

EXAMPLE 4 | **An equation with the unknown in the exponent**

Solve the equation $4^x = 8$.

SOLUTION
First let's estimate x, just to get a feeling for what kind of answer to expect. Since $4^1 = 4$, which is less than 8, and $4^2 = 16$, which is more than 8, we know that our final answer should be a number between 1 and 2. To obtain this answer, we take advantage of the fact that both 4 and 8 are powers of 2. Using this fact, we can write the given equation as

$$(2^2)^x = 2^3$$
$$2^{2x} = 2^3$$
$$2x = 3 \qquad \text{using Property 4 in the Property Summary box}$$
$$x = 3/2$$

Note that the answer $x = 3/2$ is indeed between 1 and 2, as we estimated.

For the remainder of this section, b denotes an arbitrary positive constant other than 1. In the box that follows, we define the exponential function with base b.

DEFINITION | **The Exponential Function with Base b ($b > 0$, $b \neq 1$)**

Let b denote an arbitrary positive constant other than 1. The **exponential function with base b** is defined by the equation

$$y = b^x$$

Note In many applications, functions of the form $y = ab^{kx}$, where a, b, and k are constants, are also referred to as exponential functions.

EXAMPLES
1. The equations $y = 2^x$ and $y = 3^x$ define the exponential functions with bases 2 and 3, respectively.
2. The equation $y = (1/2)^x$ defines the exponential functions with base $1/2$.
3. The equations $y = x^2$ and $y = x^3$ do not define exponential functions.

To help with our analysis of exponential functions, let's set up a table and use it to graph the exponential function $y = 2^x$. This is done in Figure 1. In drawing a

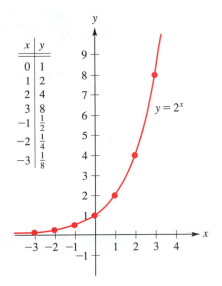

x	y
0	1
1	2
2	4
3	8
−1	$\frac{1}{2}$
−2	$\frac{1}{4}$
−3	$\frac{1}{8}$

Figure 1

smooth and unbroken curve, we are actually relying on the results in the Property Summary box on the previous page. The key features of the exponential function $y = 2^x$ and its graph are:

1. The domain of $y = 2^x$ is the set of all real numbers. The range is the set of all *positive* real numbers.
2. The y-intercept of the graph is 1. The graph has no x-intercept.
3. For $x > 0$, the function increases or grows very rapidly. For $x < 0$, the graph rapidly approaches the x-axis; the x-axis is a horizontal asymptote for the graph. (Recall from Section 4.7 that a line is an **asymptote** for a curve if the separation distance between the curve and the line approaches zero as we move farther and farther out along the line.)

You should memorize the basic shape and features of the graph of $y = 2^x$ so that you can sketch it as needed without first setting up a table. The next example shows why this is useful.

EXAMPLE **5** **Sketching graphs related to that of $y = 2^x$**

Graph each of the following functions. In each case specify the domain, the range, the intercept(s), and the asymptote.

(a) $y = -2^x$ **(b)** $y = 2^{-x}$ **(c)** $y = (1/2)^x$ **(d)** $y = 2^{-x} - 2$

SOLUTION

(a) Recall that -2^x means $-(2^x)$, not $(-2)^x$. The graph of $y = -2^x$ is obtained by reflecting the graph of $y = 2^x$ in the x-axis. See Figure 2(a).

(b) Similarly, the graph of $y = 2^{-x}$ is obtained from the graph of $y = 2^x$ by reflection in the y-axis. See Figure 2(b).

(c) Next, regarding $y = (1/2)^x$, observe that

$$\left(\frac{1}{2}\right)^x = \frac{1^x}{2^x} = \frac{1}{2^x} = 2^{-x}$$

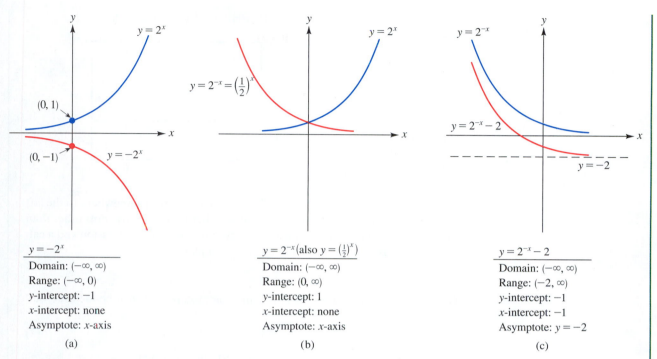

$y = -2^x$
Domain: $(-\infty, \infty)$
Range: $(-\infty, 0)$
y-intercept: -1
x-intercept: none
Asymptote: x-axis

(a)

$y = 2^{-x} \left(\text{also } y = \left(\tfrac{1}{2}\right)^x\right)$
Domain: $(-\infty, \infty)$
Range: $(0, \infty)$
y-intercept: 1
x-intercept: none
Asymptote: x-axis

(b)

$y = 2^{-x} - 2$
Domain: $(-\infty, \infty)$
Range: $(-2, \infty)$
y-intercept: -1
x-intercept: -1
Asymptote: $y = -2$

(c)

Figure 2

In other words, $y = \left(\tfrac{1}{2}\right)^x$ is really the same function as $y = 2^{-x}$, which we already graphed in Figure 2(b).

(d) Finally, to graph $y = 2^{-x} - 2$, take the graph of $y = 2^{-x}$ in Figure 2(b) and move it two units in the negative y-direction, as shown in Figure 2(c). Note that the asymptote and y-intercept will also move down two units. To find the x-intercept, we set $y = 0$ in the given equation to obtain

$$2^{-x} - 2 = 0$$
$$2^{-x} = 2^1$$
$$-x = 1 \qquad \text{\color{red}{using Property 4 on page 366}}$$
$$x = -1$$

Thus, the x-intercept is -1.

In the next example, we apply our knowledge about the graph of $y = 2^x$ to solve an equation. In particular, we use the fact that the graph of $y = 2^x$ always lies above the x-axis; for no value of x is 2^x ever zero.

EXAMPLE 6   **Using the fact that 2^x is always positive to solve an equation**

Consider the equation

$$x^3 2^x - 3(2^x) = 0 \qquad (1)$$

The graph in Figure 3 indicates that equation (1) has a positive root between 0 and 2. However, the viewing rectangle does not tell us whether equation (1) has

Figure 3
$y = x^3 2^x - 3(2^x)$
$[-8, 2, 2]$ by $[-5, 10, 5]$

other roots. (For instance, perhaps there is a negative root somewhere to the left of -8). By solving the equation, find out whether there are any roots other than the one between 0 and 2. For each root, give both an exact expression and a calculator approximation rouded to two decimal places.

SOLUTION

(a) On the left side of equation (1), the common term 2^x can be factored out. This yields

$$2^x(x^3 - 3) = 0 \qquad (2)$$

Now applying the zero-product property in equation (2), we have

$$2^x = 0 \qquad \text{or} \qquad x^3 - 3 = 0$$

The equation $2^x = 0$ has no solutions because 2^x is positive for all values of x. From the equation $x^3 - 3 = 0$ we conclude that $x^3 = 3$, and therefore $x = \sqrt[3]{3}$. With a calculator, we find that $\sqrt[3]{3} \approx 1.44$. This is the root between 0 and 2 in Figure 3. The algebra that we've carried out indicates that this is the only root of the given equation.

Figure 4
$y = 1.5^x \qquad y = 2^x \qquad y = 4^x$
$[-3, 3, 1]$ by $[0, 4, 1]$

Now what about exponential functions with bases other than 2? As Figure 4 indicates, the graphs are similar to $y = 2^x$. The graph of $y = 4^x$ rises more rapidly than $y = 2^x$ when x is positive. For negative x-values, the graph of $y = 4^x$ is below that of $y = 2^x$. You can see why this happens by taking $x = -1$, for example, and comparing the values of 4^x and 2^x. If $x = -1$, then

$$2^x = 2^{-1} = \frac{1}{2} \qquad \text{but} \qquad 4^x = 4^{-1} = \frac{1}{4}$$

Therefore $4^x < 2^x$ when $x = -1$. Notice also in Figure 4 that all three graphs have the same y-intercept of 1. This follows from the fact that $b^0 = 1$ for any positive number b.

The exponential functions in Figure 4 all have bases larger than 1. To see examples in which the bases are in the interval $0 < b < 1$, we need only reflect the graphs in Figure 4 in the y-axis. For instance, the reflection of $y = 4^x$ in the y-axis gives us the graph of $y = (1/4)^x$. (We discussed the idea behind this in Example 5; see Figure 2(b), for instance.)

In the box that follows, we summarize what we've learned up to this point regarding the exponential function $y = b^x$.

PROPERTY SUMMARY   The Exponential Function $y = b^x$

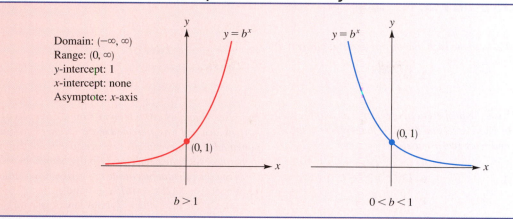

Domain: $(-\infty, \infty)$
Range: $(0, \infty)$
y-intercept: 1
x-intercept: none
Asymptote: x-axis

$y = b^x$

$(0, 1)$

$b > 1$

$y = b^x$

$(0, 1)$

$0 < b < 1$

  EXAMPLE **7** **Reflecting and translating the graph of an exponential function**

Graph the function $y = -3^{-x} + 1$.

SOLUTION
The required graph is obtained by reflecting and translating the graph of $y = 3^x$, as shown in Figure 5.

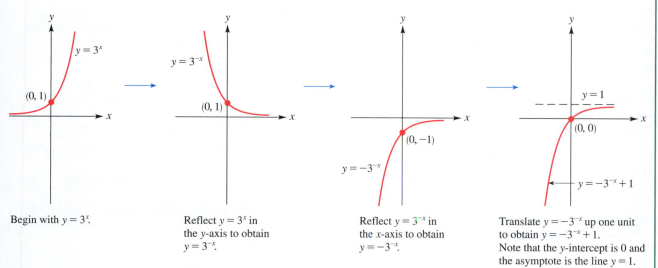

$y = 3^x$

$(0, 1)$

Begin with $y = 3^x$.

$y = 3^{-x}$

$(0, 1)$

Reflect $y = 3^x$ in the y-axis to obtain $y = 3^{-x}$.

$(0, -1)$

$y = -3^{-x}$

Reflect $y = 3^{-x}$ in the x-axis to obtain $y = -3^{-x}$.

$y = 1$

$(0, 0)$

$y = -3^{-x} + 1$

Translate $y = -3^{-x}$ up one unit to obtain $y = -3^{-x} + 1$. Note that the y-intercept is 0 and the asymptote is the line $y = 1$.

Figure 5

EXERCISE SET 5.1

A

In Exercises 1 and 2, estimate each quantity in terms of powers of ten, as in Example 1.

1. (a) 2^{30}
   (b) 2^{50}

2. (a) 2^{90}
   (b) 4^{50}

In Exercises 3–10, use the properties of exponents to simplify each expression. In Exercises 9 and 10, write the answers in the form b^n, where b and n are real numbers.

3. $(5^{\sqrt{3}})^{\sqrt{3}}$

4. $(\sqrt{2}^{\sqrt{2}})^{\sqrt{2}}$

5. $(4^{1+\sqrt{2}})(4^{1-\sqrt{2}})$

6. $(3^{2+\sqrt{5}})(3^{2-\sqrt{5}})$

7. $\dfrac{2^{4+\pi}}{2^{1+\pi}}$

8. $\dfrac{10^{\pi+2}}{10^{\pi-2}}$

9. $(\sqrt{5}^{\sqrt{2}})^2$

10. $[(\sqrt{3})^\pi]^4$

In Exercises 11 and 12, solve each equation, as in Example 4.

11. (a) $3^x = 27$
   (b) $9^t = 27$
   (c) $3^{1-2y} = \sqrt{3}$
   (d) $3^z = 9\sqrt{3}$

12. (a) $2^x = 32$
   (b) $2^t = 1/4$
   (c) $2^{3y+1} = \sqrt{2}$
   (d) $8^{z+1} = 32\sqrt{2}$

In Exercises 13–16, specify the domain of the function.

13. $y = 2^x$

14. $y = 1/2^x$

15. $y = 1/(2^{x-1})$

16. $y = 1/(2^x - 1)$

In Exercises 17–24, graph the pair of functions on the same set of axes.

17. $y = 2^x$; $y = 2^{-x}$

18. $y = 3^x$; $y = 3^{-x}$

19. $y = 3^x$; $y = -3^x$

20. $y = 4^x$; $y = -4^x$

21. $y = 2^x$; $y = 3^x$

22. $y = (1/3)^x$; $y = 3^x$

23. $y = (1/2)^x$; $y = (1/3)^x$

24. $y = (1/2)^{-x}$; $y = (1/3)^{-x}$

For Exercises 25–32, graph the function and specify the domain, range intercept(s), and asymptote.

25. $y = -2^x + 1$

26. $y = -3^x + 3$

27. $y = 3^{-x} + 1$

28. $y = 3^{-x} - 3$

29. $y = 2^{x-1}$

30. $y = 2^{x-1} - 1$

31. $y = 3^{x+1} + 1$

32. $y = 1 - 3^{x-1}$

For Exercises 33–38:

(a) *Use paper and pencil to determine the intercepts and asymptotes for the graph of each function.*

G (b) *Use a graphing utility to graph each function. Your results in part (a) will be helpful in choosing an appropriate viewing rectangle that shows the essential features of the graph.*

33. $y = -3^{x-2}$

34. $y = -3^{x-2} + 1$

35. $y = 4^{-x} - 4$

36. $y = 4 + (1/4)^x$

37. $y = 10^{x-1}$

38. $y = -10^{x-1} + 0.5$

For Exercises 39–42, solve the equation, as in Example 6.

39. $3x(10^x) + 10^x = 0$

40. $4x^2(2^x) - 9(2^x) = 0$

41. $3(3^x) - 5x(3^x) + 2x^2(3^x) = 0$

42. $\dfrac{(x+4)10^x}{x-3} = 2x(10^x)$

B

Exercises 43 and 44 provide some simple examples of an important idea that was presented in the chapter introduction: Exponential growth eventually outpaces polynomial growth (no matter how high the degree of the polynomial). In each case, carry out the calculations needed to verify the given statement.

43. Let $f(x) = 1.125x^2 - 0.75x + 1$ and $g(x) = 2^x$.

(a) On the interval $[0, 2]$, both functions have the same average rate of change. This is also true for the interval $[2, 4]$.

(b) On the interval $[4, 6]$, however, $\Delta g/\Delta x$ is more than twice $\Delta f/\Delta x$. And on the interval $[6, 8]$, $\Delta g/\Delta x$ is more than six times $\Delta f/\Delta x$.

(c) On the interval $[10, 12]$, the average rate of change of the exponential function is 64 times that of the quadratic function.

44. Let $g(x) = x^4$ and $h(x) = 3^x$.

(a) On the interval $[0, 5]$, the average rate of change of g is more than two and one-half times the average rate of change of h. On the interval $[5, 10]$, however, the average rate of change of h is more than six times that for g.

(b) On the interval $[10, 15]$, the average rate of change of the exponential function is more than 350 times that of the polynomial function.

45. Let $f(x) = 2^x$. Show that

$$\frac{f(x+h) - f(x)}{h} = 2^x\left(\frac{2^h - 1}{h}\right)$$

46. Let $\phi(t) = 1 + a^t$. Show that $\dfrac{1}{\phi(t)} + \dfrac{1}{\phi(-t)} = 1$.

47. Let $f(x) = 2^x$ and let g denote the function that is the inverse of f.

(a) On the same set of axes, sketch the graphs of f, g, and the line $y = x$.

(b) Using the graph you obtained in part (a), specify the domain, range, intercept, and asymptote for the function g.

48. Let $S(x) = (2^x - 2^{-x})/2$ and $C(x) = (2^x + 2^{-x})/2$. Compute $[C(x)]^2 - [S(x)]^2$.

For Exercises 49–56, refer to the following figure, which shows portions of the graphs of $y = 2^x$, $y = 3^x$, and $y = 5^x$. In each case, (a) use the figure to estimate the indicated quantity, and (b) use a calculator to compute the indicated quantity, rounding the result to two decimal places.

$y = 5^x$ $y = 3^x$ $y = 2^x$

$y = 2.0 \rightarrow$

$y = 1.5 \rightarrow$

$y = 1.0 \rightarrow$

$x = 0$ $x = 0.5$ $x = 1.0$

49. $\sqrt{2}$ *Hint:* $\sqrt{2} = 2^{1/2}$ **50.** $\sqrt[5]{2}$

51. $2^{3/5}$ **52.** $\sqrt[5]{8}$ *Hint:* $8^{1/5} = 2^?$

53. $\sqrt{3}$ **54.** $\sqrt[3]{3}$ **55.** $5^{3/10}$ **56.** $\sqrt[4]{5}$

In Exercises 57–60, refer to the following graph of the exponential function $y = 10^x$. Use the graph to estimate (to the nearest tenth) the solution of each equation. (After we've studied logarithmic functions later in the chapter, we will be able to obtain precise solutions.)

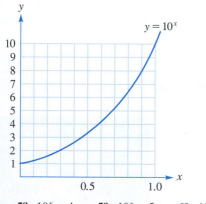

y

$y = 10^x$

10
9
8
7
6
5
4
3
2
1

0.5   1.0 x

57. $10^x = 2$ **58.** $10^x = 4$ **59.** $10^x = 5$ **60.** $10^x = 8$

Ⓖ **61.** If you add two quadratic functions, the result is again a quadratic function (assuming that the x^2-terms don't add to zero). For example, if $f(x) = x^2 + 2x$ and $g(x) = 2x^2 - 1$, then the sum is $(f + g)(x) = 3x^2 + 2x - 1$, another quadratic. But this is not the case for exponential functions. To explore a visual example, consider the function $y = 2^x + 2^{-x}$. This is the sum of the two exponentials $y = 2^x$ and $y = (1/2)^x$.

(a) Graph the function $y = 2^x + 2^{-x}$ in the standard viewing rectangle. As you can see, the resulting graph is U-shaped.

(b) For comparison, add the graph of $y = x^2$ to your picture. Note that (at this scale) the graphs are very similar for $|x| > 2$.

(c) Actually, the graph of $y = 2^x + 2^{-x}$ rises much more steeply than does $y = x^2$. To demonstrate this, change

the viewing rectangle so that x extends from -100 to 100 and y extends from 0 to 100,000.

Ⓖ **62.** Follow the general procedure in Exercise 61, but use the two functions $y = 2^x - 2^{-x}$ and $y = x^3$.

Ⓖ **63.** As background for this exercise, reread Example 2 in the text. (The example shows how to estimate the value of x for which $10^x = 2$.)

(a) In the same picture, graph the two functions $y = 10^x$ and $y = 2$. Use a viewing rectangle in which x runs from 0 to 1 and y runs from 0 to 10.

(b) Explain how your picture in part (a) supports the answer 3/10 that was obtained in Example 2.

C

64. This exercise serves as a preview for the work on logarithms in Section 5.3. Follow Steps (a)–(f) to complete the table. (Notice that one entry in the table is already filled in. Reread Example 2 in the text to see how that entry was obtained.)

(a) Fill in the entries in the right-hand column corresponding to $x = 1$ and $x = 10$.

(b) Note that 4 and 8 are powers of 2. Use this information along with the approximation $10^{0.3} \approx 2$ to find the entries in the table corresponding to $x = 4$ and $x = 8$.

(c) Find the entry corresponding to $x = 5$.
Hint: $5 = 10/2 \approx 10/10^{0.3}$.

(d) Find the entry corresponding to $x = 7$.
Hint: $7^2 \approx 50 = 5 \times 10$; now make use of your answer in part (c).

(e) Find the entry corresponding to $x = 3$.
Hint: $3^4 \approx 80 = 8 \times 10$.

(f) Find the entries corresponding to $x = 6$ and $x = 9$.
Hint: $6 = 3 \times 2$ and $9 = 3^2$.

Remark: This table is called a *table of logarithms to the base* 10. We say, for example, that the logarithm of 2 to the base 10 is (about) 0.3. We write this symbolically as $\log_{10} 2 \approx 0.3$.

x	Exponent to which 10 must be raised to yield x
1	
2	≈ 0.3
3	
4	
5	
6	
7	
8	
9	
10	

MINI PROJECT

USING DIFFERENCES TO COMPARE EXPONENTIAL AND POLYNOMIAL GROWTH

In the chapter introduction (on pages 363–364), you saw an example comparing and contrasting exponential and quadratic growth models. Another way to compare exponential growth and quadratic growth (or, more generally, polynomial growth) uses the idea of *differences,* as discussed at the end of Section 4.2.

(a) Appoint a member of the group in which you are working to summarize for the group the definitions and results on differences on page 277.
(b) Just as second differences are obtained by subtracting adjacent first differences, *third differences* are obtained by subtracting adjacent second differences. For the cubic polynomial function $y = x^3$, set up a table of x-y values using $x = 0, 1, 2, \ldots, 6$. Then compute the first, second, and third differences. What do you observe?
(c) For the exponential function $y = 2^x$, set up a table of x-y values using $x = 0, 1, 2, \ldots, 6$. Then compute the first, second, and third differences. What do you observe? Write a paragraph summarizing your observations and contrasting the results with what occurs in the case of linear and quadratic functions and the cubic function in part (b).

5.2 THE EXPONENTIAL FUNCTION $y = e^x$

Why did he [the Swiss mathematician Leonhard Euler (1707–1783)] *choose the letter* e? *There is no general consensus. According to one view, Euler chose it because it is the first letter of the word* exponential. *More likely, the choice came to him naturally as the first "unused" letter of the alphabet, since the letters* a, b, c, *and* d *frequently appear elsewhere in mathematics. It seems unlikely that Euler chose the letter because it is the initial of his own name, as occasionally has been suggested: he was an extremely modest man . . .*
—Eli Maor, e: *The Story of a Number* (Princeton, N.J.: Princeton University Press, 1994)

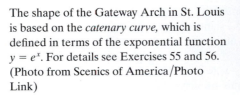

The shape of the Gateway Arch in St. Louis is based on the *catenary curve,* which is defined in terms of the exponential function $y = e^x$. For details see Exercises 55 and 56. (Photo from Scenics of America/Photo Link)

From the standpoint of calculus and scientific applications, one particular base for exponential functions is by far the most useful. This base is a certain irrational

TABLE 1

x	$f(x) = (1 + \frac{1}{x})^x$
10^1	2.59374246
10^2	2.70481382
10^3	2.71692393
10^4	2.71814592
10^5	2.71826823
10^6	2.71828046
10^7	2.71828169

Note: As x grows larger and larger, the values of $[1 + (1/x)]^x$ approach the number e. [Compare the digits in the right column with those in the expression for e given in equation (1).]

GRAPHICAL PERSPECTIVE

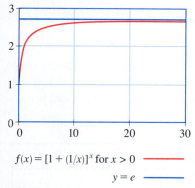

$f(x) = [1 + (1/x)]^x$ for $x > 0$ ———
$y = e$ ———

Figure 1
The line $y = e$ is an asymptote for the graph of $f(x) = [1 + (1/x)]^x$. As x gets larger and larger, the values of $[1 + (1/x)]^x$ approach the number e.

number that lies between 2 and 3 and is denoted by the letter e. To ten decimal places, without rounding off, the value of e is

$$e = 2.7182818284 \ldots \tag{1}$$

For most purposes of approximation and simple estimation, all you need to remember is that

$$e \approx 2.7$$

At the precalculus level, it's hard to escape the feeling that $y = 2^x$ or $y = 10^x$ is by far more simple and more natural than $y = e^x$. At the end of this section we'll explain why the function $y = e^x$ makes life simpler, not more complex.

There are several different (but equivalent) ways that the number e can be defined. One way involves investigating the values of the function $f(x) = [1 + (1/x)]^x$ as x becomes larger and larger. As indicated in Table 1, the values of this function get closer and closer to e as x gets larger and larger. Indeed, in some calculus books, the number e is *defined* as the target value or *limit* of the function $f(x) = [1 + (1/x)]^x$ as x grows ever larger. Now admittedly, we have not defined here the meaning of the terms *target value* or *limit*—that's a topic for calculus. Nevertheless, Table 1 will give you a reasonable, if intuitive, appreciation of the idea. In Figure 1, we give a graphical interpretation.

One more comment about Table 1 and the number e: In Section 5.6, you'll see that the data in Table 1 can be interpreted in terms of banking and compound interest. For this reason, the constant e is sometimes called "the banker's constant."

In calculus the number e is sometimes introduced in a way that involves slopes of lines. You know that the graph of each exponential function $y = b^x$ passes through the point $(0, 1)$. Figure 2(a) shows the exponential function $y = 2^x$ along with a line that is tangent to the curve at the point $(0, 1)$. By carefully measuring rise and run, it can be shown that the slope of this tangent line is about 0.7. Figure 2(b) shows a similar situation with the curve $y = 3^x$. Here the slope of the tangent line through $(0, 1)$ is approximately 1.1. Now, since the slope of the tangent to $y = 2^x$ is a bit less than 1, while that for $y = 3^x$ is a bit more than 1, it seems reasonable to suppose that there is a number between 2 and 3, call it e, with the property that the

Figure 2

(a)   (b)

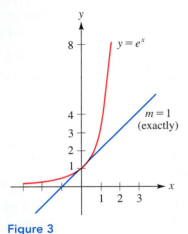

Figure 3
The slope of the tangent to the curve $y = e^x$ at the point $(0, 1)$ is $m = 1$.

slope of the tangent through $(0, 1)$ is exactly 1. See Figure 3 and the Property Summary box that follows.

PROPERTY SUMMARY   The Exponential Function $y = e^x$

Domain: $(-\infty, \infty)$
Range: $(0, \infty)$
y-intercept: 1
Asymptote: x-axis

$y = e^x$

Figure 4

REMARK   It's certainly not obvious that this definition for e is equivalent to the definition involving the expression $(1 + 1/x)^x$. The methods of calculus are required to demonstrate that the definitions are indeed equivalent. (For an informal justification, however, see Exercise 62.)

EXAMPLE **1** **Sketching graphs related to that of $y = e^x$**

Graph each of the following functions, specifying the domain, range, intercept, and asymptote: **(a)** $y = e^{x-1}$;   **(b)** $y = -e^{x-1}$;   **(c)** $y = -e^{x-1} + 1$.

SOLUTION
(a) We begin with the graph of $y = e^x$ (as in Figure 4). Moving the graph to the right one unit yields the graph of $y = e^{x-1}$, shown in Figure 5(a). The x-axis is still an asymptote for this translated graph, but the y-intercept will no longer be 1. To find the y-intercept, we replace x with 0 in the given equation to obtain

$$y = e^{-1} = \frac{1}{e} \approx 0.37 \qquad \text{using a calculator}$$

(b) Reflecting the graph from part (a) in the x-axis yields the graph of $y = -e^{x-1}$; see Figure 5(b). Note that under this reflection, the y-intercept moves from $1/e$ to $-1/e$.
(c) Translating the graph in Figure 5(b) up one unit produces the graph of $y = -e^{x-1} + 1$, shown in Figure 5(c). Under this translation, the asymptote moves from $y = 0$ (the x-axis) to $y = 1$. Also, the y-intercept moves from

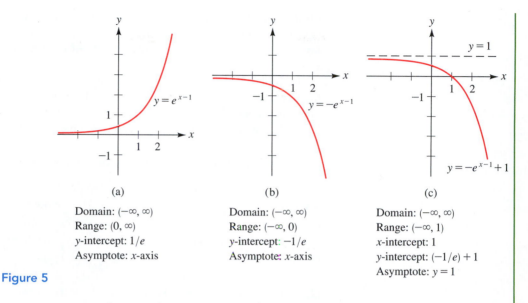

(a)

Domain: $(-\infty, \infty)$
Range: $(0, \infty)$
y-intercept: $1/e$
Asymptote: x-axis

(b)

Domain: $(-\infty, \infty)$
Range: $(-\infty, 0)$
y-intercept: $-1/e$
Asymptote: x-axis

(c)

Domain: $(-\infty, \infty)$
Range: $(-\infty, 1)$
x-intercept: 1
y-intercept: $(-1/e) + 1$
Asymptote: $y = 1$

Figure 5

$-1/e$ to $(-1/e) + 1$ (≈ 0.63). The x-intercept in Figure 5(c) is obtained by setting $y = 0$ in the equation $y = -e^{x-1} + 1$. This yields

$$0 = -e^{x-1} + 1$$
$$e^{x-1} = e^0$$
$$x - 1 = 0 \qquad \text{(Why?)}$$
$$x = 1$$

Near the beginning of this section we said that using e as the base for an exponential function makes things simpler. In order to explain the context in which this is true, we introduce the concept of the *instantaneous rate of change* of a function. We are going to rely on examples and numerical calculations. A more rigorous development properly belongs to calculus. Our first two examples will involve simple polynomial functions; then we'll move on to the exponential function $f(x) = e^x$. In Table 2(a) we've calculated the average rates of change of the function $f(x) = x^2$ on a sequence of increasingly short intervals, each interval having 1 as the left endpoint. Table 2(b) shows similar calculations with intervals that have 1 as the right endpoint.

TABLE 2 The Average Rates of Change Approach 2.

Interval	$\Delta f/\Delta x$ for $f(x) = x^2$	Interval	$\Delta f/\Delta x$ for $f(x) = x^2$
$[1, 1.1]$	2.1	$[0.9, 1]$	1.9
$[1, 1.01]$	2.01	$[0.99, 1]$	1.99
$[1, 1.001]$	2.001	$[0.999, 1]$	1.999
$[1, 1.0001]$	2.0001	$[0.9999, 1]$	1.9999
$[1, 1.00001]$	2.00001	$[0.99999, 1]$	1.99999
	(a)		(b)

From the tables, we make the following observation: As the lengths of the intervals get smaller and smaller, the average rates of change get closer and closer to 2. We say in this case that the **instantaneous rate of change** of the function $f(x) = x^2$ at $x = 1$ is 2.

In Tables 3(a) and (b), we are looking for the instantaneous rate of change of the function $g(x) = x^3$ at $x = 2$. From the tables, it appears that as the lengths of the intervals get smaller and smaller, the average rates of change get closer and closer to 12. We say that the instantaneous rate of change of $g(x) = x^3$ at $x = 2$ is 12.

▌ TABLE 3   The Average Rates of Change Approach 12.

Interval	$\Delta g / \Delta x$ for $g(x) = x^3$	Interval	$\Delta g / \Delta x$ for $g(x) = x^3$
[2, 2.1]	12.61	[1.9, 2]	11.41
[2, 2.01]	12.0601	[1.99, 2]	11.9401
[2, 2.001]	12.006001	[1.999, 2]	11.994001
[2, 2.0001]	12.00060001	[1.9999, 2]	11.99940001
[2, 2.00001]	12.0000600001	[1.99999, 2]	11.9999400001
	(a)		(b)

For a more intuitive example of instantaneous rate of change, something closer to your experience, suppose you are driving a car, and at a particular instant the speedometer reads 45 mi/hr. At that instant, the instantaneous rate of change of distance with respect to time is 45 mi/hr. We say that the *instantaneous velocity* at that instant is 45 mi/hr. (For a specific example relating average velocity and instantaneous velocity, go back to Section 3.3 and see Example 6 and the related Exercise 36.)

Now we are prepared to say what is so special about the exponential function $f(x) = e^x$. For reference, we state this result in the box that follows. (Remember, we are leaving the precise definitions and the proofs to calculus; we content ourselves here with examples.)

THEOREM   The Instantaneous Rate of Change of the Function $f(x) = e^x$

Let a be any real number. Then the instantaneous rate of change of the function $f(x) = e^x$ at $x = a$ is e^a. In other words, at each point on the graph of $f(x) = e^x$, the instantaneous rate of change is just the y-coordinate of the point. Furthermore, the *only* functions with this property are the exponential function $f(x) = e^x$ and constant multiples of this function (i.e., functions of the form $y = ce^x$, where c is a constant).

EXAMPLES

1. At $x = 1$, the instantaneous rate of change of
$$f(x) = e^x \quad \text{is} \quad e^1 = e$$

2. At $x = 2$, the instantaneous rate of change of
$$f(x) = e^x \quad \text{is} \quad e^2$$

Table 4 provides a numerical check on Example 2 in the box above. (Exercise 29 asks you to complete a similar table for intervals in which 2 is the right, rather than left, endpoint.)

> **TABLE 4**
> **The Average Rates of Change Approach $e^2 = 7.3890560. \ldots$**

Interval	$\Delta f / \Delta x$ for $f(x) = e^x$ (7 decimal places, no rounding)
[2, 2.1]	7.7711381
[2, 2.01]	7.4261248
[2, 2.001]	7.3927518
[2, 2.0001]	7.3894255
[2, 2.00001]	7.3890930
[2, 2.000001]	7.3890597

EXAMPLE 2   **Average and instantaneous rates of change of a population**

In a biology experiment, let $N(t)$ be the number of bacteria in a colony after t hours, where $t = 0$ corresponds to the time the experiment begins. Suppose that during the period from $t = 4$ hours to $t = 8$ hours the number of bacteria is modeled by the exponential function $N(t) = e^t$. (See Figure 6.)

GRAPHICAL PERSPECTIVE

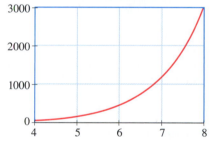

Figure 6
$N(t) = e^t, 4 \leq t \leq 8$

(a) Find the average rate of change of the population over the time period $5 \leq t \leq 7$.

(b) What is the instantaneous rate of change of the population at the instant $t = 7$ hours?

SOLUTION

(a) On the interval from $t = 5$ to $t = 7$ hours, we have

$$\frac{\Delta N}{\Delta t} = \frac{e^7 - e^5}{7 - 5} \quad \text{bacteria/hour}$$

$$\approx 474 \quad \text{bacteria/hour} \qquad \textcolor{red}{\text{using a calculator and rounding}}$$

Thus, on average over this 2-hour time period, the population is increasing at a rate of 474 bacteria/hour.

(b) Applying the theorem in the box on page 378, the instantaneous rate of change of the function $N(t) = e^t$ at $t = 7$ is e^7. Thus, at the instant when $t = 7$ hours, the population is changing at a rate of

$$e^7 \quad \text{bacteria/hour}$$

$$\approx 1097 \quad \text{bacteria/hour} \qquad \textcolor{red}{\text{using a calculator and rounding}}$$

EXERCISE SET 5.2

A

In Exercises 1–8, answer True or False. *You do not need a cal-culator for these exercises. Rather, use the fact that e is approxi-mately 2.7.*

1. $e < 1/2$ **2.** $e < 5/2$ **3.** $\sqrt{e} < 1$ **4.** $e^2 < 4$
5. $e^2 < 9$ **6.** $e^3 < 27$ **7.** $e^{-1} < 0$ **8.** $e^0 = 1$

For Exercises 9 and 10, you are given some simple rational ap-proximations for the irrational numbers π and e. In each case, use a calculator to find out how many decimal places of the ap-proximation agree with the irrational number. *Remark: Each of the approximations in Exercise 10 is the best rational approx-imation using numerators and denominators less than 1000.*

9. (a) $\pi \approx 22/7$ **10.** (a) $\pi \approx 355/113$
  (b) $e \approx 19/7$ (b) $e \approx 878/323$

In Exercises 11–20, graph the function and specify the domain, range, intercept(s), and asymptote.

11. $y = e^x$ **12.** $y = e^{-x}$
13. $y = -e^x$ **14.** $y = -e^{-x}$
15. $y = e^x + 1$ **16.** $y = e^{x+1}$
17. $y = e^{x+1} + 1$ **18.** $y = e^{x-1} - 1$
19. $y = e - e^x$ **20.** $y = e^{-x} - e$

In Exercises 21 and 22, first tell what translations or reflections are required to obtain the graph of the given function from that of $y = e^x$. Then graph the given function along with $y = e^x$ and check that the picture is consistent with what you've said.

21. (a) $y = e^{x+2}$ **22.** (a) $y = 1 - e^{-x}$
  (b) $y = e^{-x+2}$ (b) $y = 1 - e^{-x+1}$

In Exercises 23 and 24, let $f(x) = x^2$, $g(x) = 2^x$, and $h(x) = e^x$. In each case, carry out the calculations needed to verify the given statement.

23. On the interval $[0, 1]$: The average rate of change of h is about 1.7 times that of g. The functions f and g have the same average rate of change on the interval $[0, 1]$.

24. On the interval $[9, 10]$: The average rate of change of g is more than 25 times greater than that of f. The average rate of change of h is approximately 733 times that of f.

25. In this exercise we compare three functions:

$$y = 2^x \qquad y = e^x \qquad y = 3^x$$

  (a) Begin by graphing all three functions in the standard viewing rectangle. The picture confirms three facts that you know from the text: on the positive x-axis, the functions increase very rapidly; on the negative x-axis, the graphs approach the asymptote (which is the x-axis) as you move to the left; the y-intercept in each case is 1.

  (b) To compare the functions for positive values of x, use a viewing rectangle in which x extends from 0 to 3 and y extends from 0 to 10. Note that the graph of e^x is

bounded between the graphs of 2^x and 3^x, just as the number e is between 2 and 3. In particular, the picture that you obtain demonstrates the following fact: For positive values of x,

$$2^x < e^x < 3^x$$

  (c) To see the graphs more clearly when x is negative, change the viewing rectangle so that x extends from -3 to 0 and y extends from 0 to 1. Again, note that the graph of e^x is bounded between the graphs of 2^x and 3^x, but now the graph of 3^x is the bottom (rather than the top) curve in the picture. This demonstrates the following fact: For negative values of x,

$$3^x < e^x < 2^x$$

26. This exercise introduces the approximation

$$e^x \approx x + 1 \quad \text{provided } x \text{ is close to zero}$$

This approximation has an important consequence when we study population growth in Section 5.7. (The details are in Exercise 61 in Section 5.7.)

  (a) Use the information in Figure 3 to explain why the equation of the line tangent to the curve $y = e^x$ at the point $(0, 1)$ is $y = x + 1$.

  (b) Verify visually that the line $y = x + 1$ is tangent to the curve $y = e^x$ at the point $(0, 1)$ by using a graphing utility to display the graphs of both functions in the same standard viewing rectangle. Note that the curve and the line are virtually indistinguishable in the im-mediate vicinity of the point $(0, 1)$.

  (c) Zoom in on the point $(0, 1)$; use a viewing rectangle in which x extends from -0.05 to 0.05 and y extends from 0.95 to 1.05. Again, note that the line and the curve are virtually indistinguishable in the immediate vicin-ity of the point $(0, 1)$. For a numerical look at this, complete the following table. In the columns for e^x and $e^x - (x + 1)$, round each entry to four decimal places. When you are finished, observe that the closer x is to 0, the closer the agreement between the two quantities $x + 1$ and e^x.

x	$x + 1$	e^x	$e^x - (x + 1)$
-0.04			
-0.03			
-0.02			
-0.01			
0.00			
0.01			
0.02			
0.03			
0.04			

27. Complete the following two tables. Based on the results you obtain, what would you say is the instantaneous rate of change of the function $f(x) = x^2$ at $x = 3$?

Interval	$\Delta f/\Delta x$ for $f(x) = x^2$	Interval	$\Delta f/\Delta x$ for $f(x) = x^2$
[3, 3.1]		[2.9, 3]	
[3, 3.01]		[2.99, 3]	
[3, 3.001]		[2.999, 3]	
[3, 3.0001]		[2.9999, 3]	
[3, 3.00001]		[2.99999, 3]	

28. Tables 2(a) and (b) on page 377 were used to determine the instantaneous rate of change of $f(x) = x^2$ at $x = 1$. Set up and complete similar tables to find the instantaneous rate of change of $g(x) = x^3$ at $x = 1$.

29. Let $f(x) = e^x$. Complete the following table. In the right column, give the average rates of change to six decimal places without rounding. Note that the average rates of change seem to approach $e^2 = 7.3890560. \ldots$ This result along with Table 4 in the text supports a statement in the box on page 379: The instantaneous rate of change of $f(x) = e^x$ at $x = 2$ is e^2.

Interval	$\Delta f/\Delta x$ for $f(x) = e^x$
[1.9, 2]	
[1.99, 2]	
[1.999, 2]	
[1.9999, 2]	
[1.99999, 2]	
[1.999999, 2]	
[1.9999999, 2]	

30. Complete a table similar to the one shown in Exercise 29, but use the function $h(x) = 3^x$, rather than $f(x) = e^x$. Note that the average rates of change do *not* approach $3^2 = 9$. What does this have to do with the statements in the box on page 378? (Explain using complete sentences.)

Exercises 31 and 32 refer to Example 2 in the text. Round all answers to the nearest integer. (Don't forget to include units with your answers.)

31. (a) Find the average rate of change of the population over the time period $5 \le t \le 6$.
(b) What is the instantaneous rate of change of the population at $t = 5$ hr? At $t = 5.5$ hr?

32. (a) Find the average rate of change of the population over the time period $4 \le t \le 6$.
(b) What is the instantaneous rate of change of the population at $t = 6$ hr?

33. As indicated in the accompanying figure, the three points P, Q, and R lie on the graph of $y = e^x$, and the x-coordinates of these points are $-1, 0,$ and 0.5, respectively. Use

the theorem in the box on page 378 to specify the instantaneous rate of change of the function $y = e^x$ at each of these points. For one of the points, a numerical answer can be obtained without a calculator. For the other points, give two forms of each answer: an exact expression involving e and a calculator approximation rounded to two decimal places.

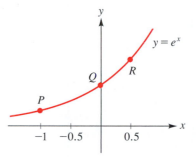

34. Suppose that during the first hour and 15 minutes of a physics experiment, the surface temperature of a small iron block is modeled by the exponential function $f(t) = 15e^t$, where $f(t)$ is the Celsius temperature t hours after the experiment begins.

(a) Use a graphing utility to graph the function $f(t) = 15e^t$ over the interval $0 \le t \le 1.25$.
(b) Compute the average rate of change of temperature over the second half hour of the experiment (i.e., over the interval $0.5 \le t \le 1$). Round the answer to one decimal place.
(c) Use one of the facts in the box on page 378 to determine the instantaneous rate of change of temperature: 30 minutes after the start of the experiment; 1 hour after the start of the experiment. Round each answer to one decimal place.

B

In Exercises 35–42, decide which of the following properties apply to each function. (More than one property may apply to a function.)

A. *The function is increasing for $-\infty < x < \infty$.*
B. *The function is decreasing for $-\infty < x < \infty$.*
C. *The function has a turning point.*
D. *The function is one-to-one.*
E. *The graph has an asymptote.*
F. *The function is a polynomial function.*
G. *The domain of the function is $(-\infty, \infty)$.*
H. *The range of the function is $(-\infty, \infty)$.*

35. $y = e^x$
36. $y = e^{-x}$
37. $y = -e^x$
38. $y = e^{|x|}$
39. $y = x + e$
40. $y = x^2 + e$
41. $y = x/e$
42. $y = e/x$

In Exercises 43–50, refer to the following graph of $y = e^x$. In each case, use the graph to estimate the indicated quantity to the nearest tenth (or closer, if it seems appropriate). Also, use a calculator to obtain a second estimate. Round the calculator values to three decimal places.

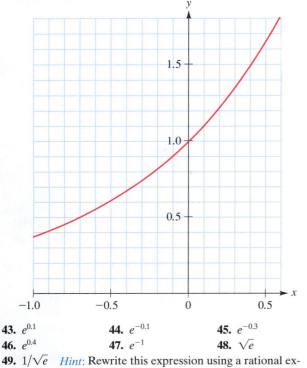

43. $e^{0.1}$ **44.** $e^{-0.1}$ **45.** $e^{-0.3}$

46. $e^{0.4}$ **47.** e^{-1} **48.** $\sqrt{e}$

49. $1/\sqrt{e}$ *Hint*: Rewrite this expression using a rational exponent—that is, $1/\sqrt{e} = e^?$.

50. $\sqrt[5]{e}$

51. (a) Use the graph preceding Exercise 43 to estimate the value of x for which $e^x = 1.5$.

    (b) One of the keys on your calculator will allow you to solve the equation in part (a) for x; it is the "ln" key. Use your calculator to compute $\ln(1.5)$. Round your answer to three decimal places and check to see that it is consistent with the value obtained graphically in part (a). *Remark*: The "ln" function, which is called the *natural logarithm* function, is discussed in detail in the next section.

52. Follow Exercise 51 using the equation $e^x = 0.6$ (rather than $e^x = 1.5$).

53. Follow Exercise 51 using the equation $e^x = 1.8$.

54. Follow Exercise 51 using the equation $e^x = 0.4$.

55. The **hyperbolic cosine function**, denoted by **cosh**, is defined by the equation

$$\cosh(x) = \frac{1}{2}(e^x + e^{-x})$$

    (a) Without a calculator, find cosh(0). Using a calculator, find cosh(1) and cosh(−1), rounding the answers to two decimal places.

    (b) What is the domain of the function cosh? *Hint*: Visualize the graphs of $y = e^x$ and $y = e^{-x}$.

    (c) Show that $\cosh(-x) = \cosh(x)$. What does this say about the graph of $y = \cosh(x)$?

(G) (d) Use a graphing utility to graph $y = \cosh(x)$. Check that the picture is consistent with your answer in part (c).

(G) 56. (Continuation of Exercise 55.) Suppose that a flexible cable of uniform density is suspended between two supports, as indicated in Figure A. (Example: a telephone line between two phone poles.) Using calculus and physics, it can be shown that the curve formed is a portion of the graph of $y = a \cosh(x/a)$, where a is a constant. See Figure B. (The actual value for a depends on the tension in the cable and the density of the cable.) The graph of an equation of the form $y = a \cosh(x/a)$ is called a **catenary** (from the Latin *catena*, meaning *chain*).

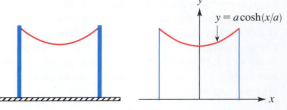

Figure A   **Figure B**

    (a) Use a graphing utility to graph the four catenaries $y = a \cosh(x/a)$ corresponding to $a = 0.2, 1, 3$, and 5. Describe (in complete sentences) how the shape and location of the curve changes as the value of a increases.

    (b) The Gateway Arch in St. Louis (shown on page 374) is built in the form of an inverted catenary. The equation for the curve is given in Figure C. In this equation, both x and y are measured in feet. Determine the width and the height of the Gateway Arch by using a graphing utility to graph the equation given in Figure C. *Remark*: The Gateway Arch is the tallest monument in the U.S. National Park System. Designed by architect Eero Saarinen and completed in 1965, the Gateway Arch commemorates the Louisiana Purchase, and more generally, the westward expansion of the United States over the period 1803–1890.

$$y = 757.71 - 127.71 \cosh(x/127.71)$$

Figure C

Gateway Arch design. Distance on axes is measured in feet.

57. The **hyperbolic sine function,** denoted by **sinh,** is defined by the equation

$$\sinh(x) = \frac{1}{2}(e^x - e^{-x})$$

Note: For speaking and reading purposes, *sinh* is pronounced as "cinch."

(a) Without a calculator, find sinh(0). Using a calculator, find sinh(1) and sinh(−1), rounding the answers to two decimal places.

(b) What is the domain of the function sinh?

(c) Show that $\sinh(-x) = -\sinh(x)$. What does this say about the graph of $y = \sinh(x)$?

Ⓖ (d) Use a graphing utility to graph $y = \sinh(x)$. Check that the picture is consistent with your answer in part (c).

Ⓖ (e) How many points are there where the two curves $y = \sinh(x)$ and $y = x^3$ intersect?

58. For this exercise you need to know the definitions of the functions cosh and sinh, which were given in Exercises 55 and 57, respectively.

Ⓖ (a) Use a graphing utility to graph the equation $y = \cosh(x) + \sinh(x)$. What do you observe?

(b) Use algebra to show why the graph in part (a) looks as it does. [That is, use the definitions of cosh(x) and sinh(x) to simplify the quantity cosh(x) + sinh(x).]

59. The following figure shows portions of the graphs of the exponential functions $y = e^x$ and $y = \pi^x$.

(a) Use the figure (not your calculator) to determine which number is larger, e^π or π^e.

(b) Use your calculator to check your answer in part (a). *Remark:* The number e^π is known to be irrational. This was proven by the Russian mathematician A. O. Gelfond in 1929. It is still not known, however, whether π^e is irrational.

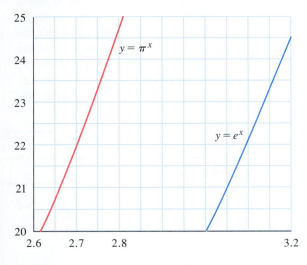

60. (a) Use your calculator to approximate the numbers $e\pi$ and $e + \pi$. *Remark:* It is not known whether these numbers are rational or irrational.

(b) Use your calculator to approximate $(e^{\pi\sqrt{163}})^{1/3}$. *Remark:* Contrary to the evidence on your calculator, it is known that this number is irrational.

61. Let $f(x) = e^x$. Let L denote the function that is the inverse of f.

(a) On the same set of axes, sketch the graphs of f and L. *Hint:* You do not need the equation for $L(x)$.

(b) Specify the domain, range, intercept, and asymptote for the function L and its graph.

(c) Graph each of the following functions. Specify the intercept and asymptote in each case.
  (i) $y = -L(x)$ **(ii)** $y = L(-x)$ **(iii)** $y = L(x - 1)$

62. The text showed two ways to define the number e. One involved the expression $(1 + 1/x)^x$ and the other involved a tangent line. In this exercise you'll see that these two approaches are, in fact, related. For convenience, we'll write the expression $(1 + 1/x)^x$ using the letter n rather than x (so that we can use x for something else in a moment.) Thus, we have

$$(1 + 1/n)^n \approx e \qquad \text{as } n \text{ becomes larger and} \qquad (1)$$
$$\text{larger without bound}$$

Working from this approximation, we'll obtain evidence that the slope of the tangent to the curve $y = e^x$ at $x = 0$ is 1. (A rigorous proof requires calculus.)

(a) Define x by the equation $x = 1/n$. As n becomes larger and larger without bound, what happens to the corresponding values of x? Complete the following table.

n	10^2	10^3	10^6	10^9
x				

(b) Substitute $x = 1/n$ in approximation (1) to obtain

$$(1 + x)^{1/x} \approx e \qquad \text{as } x \text{ approaches } 0$$

Next, raise both sides to the power x to obtain

$$(1 + x) \approx e^x \qquad \text{as } x \text{ approaches } 0 \qquad (2)$$

Approximation (2) says that the curve $y = e^x$ looks more and more like the line $y = 1 + x$ as x approaches 0. For numerical perspective on this, complete the following tables. These results suggest (but do not *prove*) that the line $y = 1 + x$ is tangent to the curve $y = e^x$ at $x = 0$. Thus, the results suggest that the slope of the tangent to $y = e^x$ at $x = 0$ is 1.

x	0.3	0.2	10^{-1}	10^{-2}	10^{-3}	10^{-4}
x + 1						
e^x						

x	−0.3	−0.2	-10^{-1}	-10^{-2}	-10^{-3}	-10^{-4}
x + 1						
e^x						

C

For Exercises 63 and 64, you need to know the definitions of the functions cosh and sinh given in Exercises 55 and 57, respectively.

63. Consider the following two equations:

$$[\cosh(x)]^2 - 2[\sinh(x)]^2 = 1 \qquad [\cosh(x)]^2 - [\sinh(x)]^2 = 1$$

Ⓖ

    (a) One of the given equations is an identity, and the other is not. Use a graphing utility to determine which equation is *not* an identity. For that equation, make a conjecture about the root(s). Then use the definitions of $\cosh(x)$ and $\sinh(x)$ to verify algebraically that your conjecture is valid.

    (b) For the other equation, use the definitions of $\cosh(x)$ and $\sinh(x)$ to prove that the equation is indeed an identity.

64. Consider the following three equations:

$$[\cosh(x)]^2 = \tfrac{1}{2}[\cosh(2x) + 1] \qquad [\sinh(x)]^2 = \tfrac{1}{2}[\cosh(2x) - 1]$$

$$[\sinh(x)]^2 = \tfrac{1}{2}[\sinh(2x) + 1]$$

    (a) Only one of the three equations is not an identity. Use a graphing utility to find out which. For that equation, are there any values of x for which the equation becomes a true statement? If so, use the graphing utility to estimate to the nearest hundredth each such value.

    (b) For each of the other two equations, use the definitions of $\cosh(x)$ and $\sinh(x)$ to prove that the equation is indeed an identity.

PROJECT

COFFEE TEMPERATURE

Suppose that a cup of hot coffee at a temperature of T_0 is set down to cool in a room where the temperature is kept at T_1. Then the temperature of the coffee as it cools can be modeled by the function

$$f(t) = (T_0 - T_1)e^{kt} + T_1 \qquad (1)$$

Here, $f(t)$ is the temperature of the coffee after t minutes; $t = 0$ corresponds to the initial instant when the temperature of the hot coffee is T_0; and k is a (negative) constant that depends, among other factors, on the dimensions of the cup and the material from which it is constructed. (This model is derived in calculus. It is based on *Newton's law of cooling*: The rate of change of temperature of a cooling object is proportional to the difference between the temperature of the object and the surrounding temperature.)

    After studying the following example, solve Problem A below by applying equation (1) and using the technique shown in the example. Then try your hand at Problem B. You won't be able to complete Problem B using the techniques developed in this text up to now. In complete sentences, explain exactly at which point you get stuck. What would you have to know how to do to complete the solution? (In the next section we discuss *logarithms,* which will allow us to complete Problem B.) For now, use a graphing utility to obtain an approximate answer for Problem B.

EXAMPLE | Coffee Temperature Project

Given that the graph of the function $y = ae^{kx}$ passes through the points $(0, 2)$ and $(3, 5)$, find y when $x = 7$.

SOLUTION

Substituting $x = 0$ and $y = 2$ in the given equation yields $2 = ae^0$, and therefore $a = 2$. Next, substituting $x = 3$ and $y = 5$ in the equation $y = 2e^{kx}$ yields $5 = 2e^{k(3)}$, or $e^{3k} = 2.5$. This last equation can be rewritten $(e^k)^3 = 2.5$. Taking the cube root of both sides gives us

$$e^k = 2.5^{1/3} \tag{2}$$

The original function can now be written as follows

$$y = 2e^{kx} = 2(e^k)^x$$
$$= 2(2.5^{1/3})^x \qquad \text{using equation (2)}$$

That is, $y = 2(2.5^{x/3})$. In this form, we can calculate y when x is given. In particular then, when $x = 7$, we obtain

$$y = 2(2.5^{7/3})$$
$$\approx 16.97 \qquad \text{using a calculator}$$

PROBLEM A

Suppose that a cup of hot coffee at a temperature of 185°F is set down to cool in a room where the temperature is kept at 70°F. What is the temperature of the coffee ten minutes later?

PROBLEM B

Suppose that a cup of hot coffee at a temperature of 185°F is set down to cool in a room where the temperature is kept at 70°F. How long will it take for the coffee to cool to 140°F?

5.3 LOGARITHMIC FUNCTIONS

But if logarithms have lost their role as the centerpiece of computational mathematics, the logarithmic function remains central to almost every branch of mathematics, pure or applied. It shows up in a host of applications, ranging from physics and chemistry to biology, psychology, art, and music. —Eli Maor in *e: The Story of a Number* (Princeton, N.J.: Princeton University Press, 1994)

[John Napier] hath set my head and hands a work with his new and admirable logarithms. I hope to see him this summer, if it please God, for I never saw book that pleased me better, or made me more wonder. —Henry Briggs, March 10, 1615

It has been thought that the earliest reference to the logarithmic curve was made by the Italian Evangelista Torricelli in a letter on the year 1644, but Paul Tannery made it practically certain that Descartes knew the curve in 1639. —Florian Cajori in *A History of Mathematics*, 4th ed. (New York: Chelsea Publishing Co., 1985)

For the number whose logarithm is unity, let e be written, . . . —Leonhard Euler in a letter written in 1727 or 1728

In the previous two sections we studied exponential functions. Now we consider functions that are inverses of exponential functions. These inverse functions are called *logarithmic functions*.

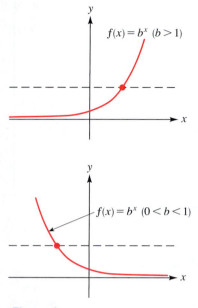

Figure 1
The exponential function $y = b^x$ is one-to-one.

Having said this, let's back up for a moment to review briefly some of the basic ideas behind inverse functions (as discussed in Section 3.6). We start with a given function F—say, $F(x) = 3x$ for example—that is one-to-one. (That is, for each output there is only one input.) Then, by interchanging the inputs and outputs, we obtain a new function, the so-called inverse function. In the case of $F(x) = 3x$, it's easy to find an equation defining the inverse function. We just interchange x and y in the equation $y = 3x$ to obtain $x = 3y$. Solving for y in this last equation then gives us $y = \frac{1}{3}x$, which defines the inverse function. Using function notation, we can summarize the situation by writing $F(x) = 3x$ and $F^{-1}(x) = \frac{1}{3}x$.

In the preceding paragraph we saw that a particular linear function had an inverse, and we found a formula for that inverse. Now let's repeat that same reasoning beginning with an exponential function. First, we must make sure the exponential function f defined by $f(x) = b^x$ is one-to-one. We can see this by applying the horizontal line test, as indicated in Figure 1. Next, since $f(x) = b^x$ is one-to-one, it has an inverse function. Let's study this inverse.

We begin by writing the exponential function $f(x) = b^x$ in the form

$$y = b^x \tag{1}$$

Then, in order to obtain an equation for f^{-1}, we interchange x and y in equation (1). This gives us

$$x = b^y \tag{2}$$

The crucial step now is to express equation (2) in words:

$$y \text{ is the exponent to which } b \text{ must be raised to yield } x. \tag{3}$$

Statement (3) defines the function that is the inverse of $y = b^x$. Now we introduce a notation that will allow us to write this statement in a more compact form.

DEFINITION | $\log_b x$

We define the expression $\mathbf{\log_b x}$ to mean "the exponent to which b must be raised to yield x." ($\log_b x$ is read *log base b of x* or *the logarithm of x to the base b.*)

EXAMPLES
(a) $\log_2 8 = 3$, since 3 is the exponent to which 2 must be raised to yield 8.
(b) $\log_{10}(1/10) = -1$, since -1 is the exponent to which 10 must be raised to yield 1/10.
(c) $\log_5 1 = 0$, since 0 is the exponent to which 5 must be raised to yield 1.

■ TABLE 1

Exponential form of equation	Logarithmic form of equation
$8 = 2^3$	$\log_2 8 = 3$
$\dfrac{1}{9} = 3^{-2}$	$\log_3 \dfrac{1}{9} = -2$
$1 = e^0$	$\log_e 1 = 0$
$a = b^c$	$\log_b a = c$

Using this notation, statement (3) becomes

$$y = \log_b x$$

Since (2) and (3) are equivalent, we have the following important relationship.

$$\boxed{y = \log_b x \quad \text{is equivalent to} \quad x = b^y}$$

We say that the equation $y = \log_b x$ is in **logarithmic form** and that the equivalent equation $x = b^y$ is in **exponential form.** Table 1 displays some examples.

Let us now summarize our discussion up to this point.

1. According to the horizontal line test, the function $f(x) = b^x$ is one-to-one and therefore possesses an inverse. This inverse function is written

$$f^{-1}(x) = \log_b x$$

2. $\log_b a = c$ means that $a = b^c$.

To graph the function $y = \log_b x$, we recall from Section 3.5 that the graph of a function and its inverse are reflections of one another about the line $y = x$. Thus, to graph $y = \log_b x$, we need only reflect the curve $y = b^x$ about the line $y = x$. This is shown in Figure 2.

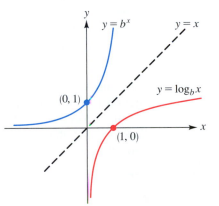

Figure 2

NOTE   For the rest of this chapter we assume that the base b is greater than 1 when we use the expression $\log_b x$.

With the aid of Figure 2, we can make the following observations about the function $y = \log_b x$.

PROPERTY SUMMARY
The Logarithmic Function $y = \log_b x$ $(b > 1)$

Domain: $(0, \infty)$
Range: $(-\infty, \infty)$
y-intercept: none
x-intercept: 1
Asymptote: y-axis

Figure 3

One aspect of the function $y = \log_b x$ may not be immediately apparent to you from Figures 2 and 3: The function grows or increases *very* slowly. Consider $y = \log_2 x$, for example. Let's ask how large x must be before the curve reaches the height $y = 10$ (see Figure 4).

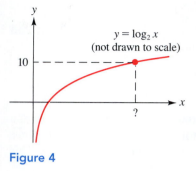

Figure 4

To answer this question, we substitute $y = 10$ in the equation $y = \log_2 x$:

$$10 = \log_2 x$$

Writing this equation in exponential form yields

$$x = 2^{10} = 1024$$

In other words, we must go out beyond 1000 on the x-axis before the curve $y = \log_2 x$ reaches a height of 10 units. Exercise 63 at the end of this section asks you to show that the graph of $y = \log_2 x$ doesn't reach a height of 100 until x is greater than 10^{30}. (Numbers as large as 10^{30} rarely occur in any of the sciences. For instance, the distance in inches to the Andromeda galaxy is less than 10^{24}.) The point we are emphasizing here is this: The graph of $y = \log_2 x$ (or $\log_b x$) is always rising, but very slowly.

Before going on to consider some numerical examples, let's pause for a moment to think about the notation we've been using. In the expression

$$\log_b x$$

the name of the function is $\log_b$ and the input is x. To emphasize this, we might be better off writing $\log_b x$ as $\log_b(x)$, so that the similarity to the familiar $f(x)$ notation is clear. However, for historical reasons,* the convention is to suppress the parentheses, and we follow that convention here. In the box that follows, we indicate some errors that can occur upon forgetting that $\log_b$ is the name of a function, not a number.

ERRORS TO AVOID

Error	Correction	Comment
$\dfrac{\log_2 8}{\log_2 4} \neq \dfrac{8}{4}$	$\dfrac{\log_2 8}{\log_2 4} = \dfrac{3}{2}$	$\log_2$ is the name of a function. It is not a factor that can be "cancelled" from the numerator and denominator.
$\dfrac{\log_2 16}{16} \neq \log_2$	$\dfrac{\log_2 16}{16} = \dfrac{4}{16} = \dfrac{1}{4}$	This "equation" is nonsense. On the left-hand side there is a quotient of two real numbers, which is a real number. On the right-hand side is the name of a function.

We conclude this section with a set of examples involving logarithms and logarithmic functions. In one way or another, every example makes use of the key fact that the equation $\log_b a = c$ is equivalent to $b^c = a$.

EXAMPLE 1   Comparing two logarithms by using the definition

Which quantity is larger: $\log_3 10$ or $\log_7 40$?

SOLUTION
First we estimate $\log_3 10$. This quantity represents the exponent to which 3 must be raised to yield 10. Since $3^2 = 9$ (less than 10), but $3^3 = 27$ (more than 10), we

---

*The notation *Log* was introduced in 1624 by the astronomer Johannes Kepler (1571–1630). In Leonhard Euler's text *Introduction to Analysis of the Infinite*, first published in 1748, appears the statement, "It has been customary to designate the logarithm of y by the symbol log y."

conclude that the quantity $\log_3 10$ lies between 2 and 3. In a similar way we can estimate $\log_7 40$; this quantity represents the exponent to which 7 must be raised to yield 40. Since $7^1 = 7$ (less than 40) while $7^2 = 49$ (more than 40), we conclude that the quantity $\log_7 40$ lies between 1 and 2. It now follows from these two estimates that $\log_3 10$ is larger than $\log_7 40$.

EXAMPLE 2 **Evaluating a logarithmic expression**

Evaluate $\log_4 32$.

SOLUTION

Let $y = \log_4 32$. The exponential form of this equation is $4^y = 32$. Now, since both 4 and 32 are powers of 2, we can rewrite the equation $4^y = 32$ using the same base on both sides:

$$(2^2)^y = 2^5$$
$$2^{2y} = 2^5$$
$$2y = 5 \qquad \text{using Property 4 on page 366}$$
$$y = 5/2 \qquad \text{as required}$$

EXAMPLE 3 **Sketching the graph of a logarithmic function and its reflection**

Graph the following equations:
(a) $y = \log_{10} x$; **(b)** $y = -\log_{10} x$.

SOLUTION

(a) The function $y = \log_{10} x$ is the inverse function for the exponential function $y = 10^x$. Thus we obtain the graph of $y = \log_{10} x$ by reflecting the graph of $y = 10^x$ in the line $y = x$; see Figure 5(a).
(b) To graph $y = -\log_{10} x$, we reflect $y = \log_{10} x$ in the x-axis; see Figure 5(b).

Figure 5

(a)   (b)

EXAMPLE **4** Finding the domain of a function defined by a logarithm

Find the domain of the function $f(x) = \log_2(12 - 4x)$.

SOLUTION
As you can see by looking back at Figure 3 on page 387, the inputs for the logarithmic function must be positive. So, in the case at hand, we require that the quantity $12 - 4x$ be positive. Consequently, we have

$$12 - 4x > 0$$
$$-4x > -12$$
$$x < 3$$

Therefore the domain of the function $f(x) = \log_2(12 - 4x)$ is the interval $(-\infty, 3)$.

The next example concerns the exponential function $y = e^x$ and its inverse function, $y = \log_e x$. Many books, as well as calculators, abbreviate the expression $\log_e x$ by $\ln x$, read *natural log of x*.* For reference and emphasis, we repeat this fact in the following box. (Incidentally, on most calculators, "log" is an abbreviation for $\log_{10}$.)

DEFINITION The "ln" Notation for Base *e* Logarithms

$\ln x$ means $\log_e x$

EXAMPLES
1. $\ln e = 1$ because $\ln e$ stands for $\log_e e$, which equals 1.
2. $\ln(e^2) = 2$ because $\ln(e^2)$ stands for $\log_e(e^2)$, which equals 2.
3. $\ln 1 = 0$ because $\ln 1$ stands for $\log_e 1$, which equals 0. (The exponential form of the equation $\ln 1 = 0$ is $e^0 = 1$.)

EXAMPLE **5** Sketching the graph of ln x and a translation

Graph the following functions.
(a) $y = \ln x$; **(b)** $y = \ln(x - 1)$.

SOLUTION
(a) The function $y = \ln x$ ($= \log_e x$) is the inverse of $y = e^x$. Thus its graph is obtained by reflecting $y = e^x$ in the line $y = x$, as in Figure 6(a).
(b) To graph $y = \ln(x - 1)$, we take the graph of $y = \ln x$ and move it one unit in the positive x-direction; see Figure 6(b).

---

* According to the historian Florian Cajori, the notation $\ln x$ was used by (and perhaps first introduced by) Irving Stringham in his text *Uniplanar Algebra* (San Francisco: University Press, 1893).

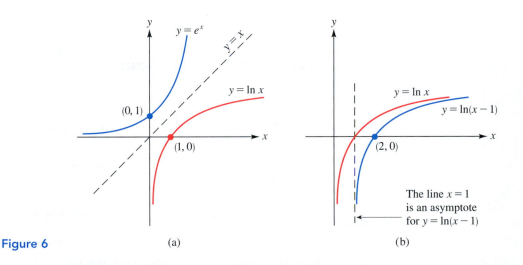

Figure 6   (a)   (b)

| EXAMPLE | 6 | **Solving an equation involving an exponential function** |

Consider the equation $10^{x/2} = 16$.

(a) Use a graphing utility to show that the equation has exactly one root. Then estimate the root to the nearest one-tenth (by finding an appropriate viewing rectangle).

(b) Solve the given equation algebraically by rewriting it in logarithmic form. Give two forms for the answer: an exact expression and a calculator approximation rounded to three decimal places. Check to see that the result is consistent with the graphical estimate obtained in part (a).

SOLUTION

(a) Figure 7(a) shows a viewing rectangle in which the graphs of $y = 10^{x/2}$ and $y = 16$ intersect. The x-coordinate of this intersection point is a root of the given equation. As you can check by zooming out from this view, there are no other intersection points, and thus no other roots of the given equation.

From Figure 7(a), we know that the required root is between 2 and 3, a bit closer to 2 than to 3. This suggests that we try a viewing rectangle with x running from 2 to 2.5, as indicated in Figure 7(b). This view shows that the

GRAPHICAL PERSPECTIVE

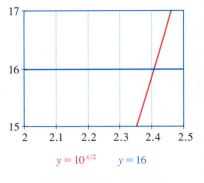

$y = 10^{x/2}$ $y = 16$ $y = 10^{x/2}$ $y = 16$

Figure 7   (a) $[-1, 3, 1]$ by $[0, 20, 10]$ (b) $[2, 2.5, 0.1]$ by $[15, 17, 1]$

root is closer to 2.4 than to either 2.3 or 2.5. Thus, to the nearest one-tenth, the required root is $x \approx 2.4$.

(b) Rewriting the equation $10^{x/2} = 16$ in the equivalent logarithmic form gives us $x/2 = \log_{10} 16$, and consequently

$$x = 2 \log_{10} 16 \qquad \text{multiplying both sides by 2}$$
$$\approx 2.408 \qquad \text{using a calculator}$$

Note that this value is consistent with the graphical estimate 2.4 that was obtained in part (a).

EXAMPLE 7   Using the natural logarithm function to solve an equation

Solve the equation $e^{2t-5} = 5000$. Give two forms for the answer: an exact expression and a calculator approximation rounded to three decimal places.

SOLUTION
Rewriting the equation in logarithmic form yields $2t - 5 = \ln 5000$. Consequently, we have

$$2t = 5 + \ln 5000$$
$$t = (5 + \ln 5000)/2 \qquad \text{dividing by 2}$$
$$\approx 6.759 \qquad \text{using a calculator}$$

Remark: This result says that, for the function $y = e^{2t-5}$, a relatively small input of about 6.8 produces a very large output of 5000. This is a reminder of the fact that exponential functions grow very rapidly (when the exponents are positive).

The next example indicates one of the many ways that logarithmic functions occur in applications. Geologists record the vibrations from an earthquake on a device known as a *seismograph*. Figure 8 shows a (simplified) output from a seismograph during an earthquake.

Figure 8

The maximum amplitude of the shock wave provides a measure of the intensity or strength of the particular earthquake. One of the problems faced by the American geologist Charles F. Richter (1900–1985) in devising a meaningful scale for comparing the sizes of earthquakes had to do with the fact that the maximum amplitudes for different earthquakes can vary greatly, by factors of thousands or more. For instance, an earthquake so small that most people would not feel it might register a maximum amplitude of 0.2 mm, whereas a destructive (yet not major) quake might register a maximum amplitude of 1000 mm. But a comparison scale for rating earthquakes that ranged from 0.2 to 1000, say, would not be especially useful; the wide range of numbers would make it nonintuitive. For instance, you might ask a friend to rate on a scale of 1 to 10 a movie he or she had seen, but you wouldn't want to use a scale of 1 to 1000. It would be nonintuitive and not especially informative in most cases. On the other hand, if we introduce a logarithm function in this example, we have $\log_{10} 0.2 \approx -0.7$ and $\log_{10} 1000 = 3$. So, while the original amplitudes range numerically from (approximately) 0 to 1000, the logs range from (approximately) -1 to 3, a much smaller, more manageable scale in this case. This is the observation that Dr. Richter used in defining his scale for comparing earthquakes. In his words:

> . . . *the range between the largest and smallest magnitudes seemed unmanageably large. Dr. Beno Gutenberg* [1889–1960] *then made the natural suggestion to plot the amplitudes logarithmically.*
> Charles F. Richter in an interview by Henry Spall of the U.S. Geological Survey, http://wwwneic.cr.usgs.gov/neis/seismology/people/int_richter.html

In its simplest form, the Richter magnitude M of an earthquake is defined by the equation

$$M = \log_{10}(A/A_0)$$

where A is the maximum amplitude recorded on a standard seismograph located 100 km from the epicenter of the earthquake, and the constant A_0 is the maximum amplitude of a certain standard intensity earthquake. In Example 8 we use this definition to compare the intensities of two earthquakes. *Remark*: Richter introduced this formula for the magnitude of an earthquake in 1935. Beginning in 1936, he worked with Gutenberg to modify the formula to account for different types of seismographs located at arbitrary distances from the epicenter of the quake. For more information about earthquakes and hands-on experience in computing the Richter magnitude of an earthquake, see the following *VirtualEarthquake* Web pages by Professor Gary Novak:

> http://vcourseware.calstatela.edu/VirtualEarthquake/VQuakeIntro.html

EXAMPLE	8	Using the Richter magnitude scale

The Richter magnitudes for two large earthquakes are given in Table 2. Use the formula $M = \log_{10}(A/A_0)$ to determine how many times stronger was the earthquake in India than the earthquake in California. Assume that A (the maximum amplitude on the seismograph) is a measure of the strength of an earthquake.

TABLE 2 Data for Two Earthquakes

Location of earthquake	Date	Richter magnitude M
Northridge, California	January 17, 1994	6.8
Gujarat Province, India	January 26, 2001	7.9

SOLUTION

We begin by solving the Richter magnitude equation $M = \log_{10}(A/A_0)$ for A. We want to do this so that we can compare the values of A for the two quakes. Rewriting the equation in its equivalent exponential form yields $A/A_0 = 10^M$, and therefore

$$A = A_0 10^M$$

Now, using this last equation and the values of M given in Table 2, we have

For Northridge quake	For Gujarat quake
$A = A_0 10^{6.8}$	$A = A_0 10^{7.9}$

Computing the ratio of these two A-values gives us

$$\frac{A\text{-value for Gujarat}}{A\text{-value for Northridge}} = \frac{A_0 10^{7.9}}{A_0 10^{6.8}} = 10^{7.9-6.8} = 10^{1.1}$$

$$\approx 12.6 \qquad \text{using a calculator}$$

In summary, the quake in India was more than 12 times stronger than the one in California, even though the Richter magnitudes differ by only 1.1. *Remark*: The Richter scale is an example of what we call a *logarithmic scale*. Exercise 45 asks you to show that if the Richter magnitudes of two quakes differ by an amount d, then the larger quake is 10^d (not just d) times stronger than the smaller quake. Other applications of logarithmic scales are given in Exercises 46–52.

EXERCISE SET 5.3

A

In Exercises 1–4, fill in the blanks.

1. If f and g are inverse functions, then:
   (a) $f[g(x)] = $ _____ for each x in the domain of g.
   (b) $g[f(x)] = $ _____ for each x in the domain of _____.

2. Suppose that f and g are inverse functions, 5 is in the domain of f, and 3 is in the domain of g. Then $f[g(3)] = $ _____, and $g[f(5)] = $ _____.

3. (a) The functions $f(x) = 2^x$ and $g(x) = $ _____ are inverse functions.
   (b) In view of part (a) we have $2^{\log_2 x} = $ _____ for each positive number x, and $\log_2(2^x) = $ _____ for all real numbers x.
   (c) So, for example, without calculating, I know that $2^{\log_2 99} = $ _____ and also that $\log_2(2^{-\pi}) = $ _____.

4. (a) The functions $f(x) = e^x$ and $g(x) = $ _____ are inverse functions.
   (b) In view of part (a) we have $e^{\ln x} = $ _____ for each positive number x, and $\ln(e^x) = $ _____ for all real numbers x.
   (c) So, for example, without calculating, I know that $e^{\ln 99} = $ _____ and also that $\ln(e^{-\pi}) = $ _____.

Exercises 5–8 provide additional review on the inverse function concept. The exercises do not explicitly involve logarithmic functions (but may be helpful to you because we've defined logarithmic functions in terms of inverse functions).

5. Which of the following functions are one-to-one and therefore possess an inverse?
   (a) $y = x + 1$
   (b) $y = (x + 1)^2$
   (c) $y = (x + 1)^3$

6. Let $f(x) = 4x - 5$.
   (a) Find $f^{-1}(x)$.
   (b) Find $f[f^{-1}(x)]$.
   (c) Find $[f(x)][f^{-1}(x)]$.
   (d) Note that your answers for parts (b) and (c) are not the same (or at least they shouldn't be). Of the two expressions given in parts (b) and (c), which one is involved in the definition of an inverse function?

7. Let $f(x) = x^3 + 5x + 1$. Given that 6 is in the domain of f^{-1}, evaluate $f[f^{-1}(6)]$.

8. The graph of $y = f(x)$ is a line segment joining the two points $(3, -2)$ and $(-1, 5)$. Specify the corresponding endpoints for the graph of:
   (a) $y = f^{-1}(x)$;   **(b)** $y = f^{-1}(x - 1)$.

In Exercises 9 and 10, write each equation in logarithmic form.

9. **(a)** $9 = 3^2$
   (b) $1000 = 10^3$
   (c) $7^3 = 343$
   (d) $\sqrt{2} = 2^{1/2}$

10. **(a)** $1/125 = 5^{-3}$
   (b) $e^0 = 1$
   (c) $5^x = 6$
   (d) $e^{3t} = 8$

In Exercises 11 and 12, write each equation in exponential form.

11. **(a)** $\log_2 32 = 5$
   (b) $\log_{10} 1 = 0$
   (c) $\log_e \sqrt{e} = 1/2$
   (d) $\ln(1/e) = -1$

12. **(a)** $\ln u = s$
   (b) $\log_a b = c$

In Exercises 13 and 14, complete the tables.

13.

x	1	10	10^2	10^3	10^{-1}	10^{-2}	10^{-3}
$\log_{10} x$							

14.

x	1	e	e^2	e^3	e^{-1}	e^{-2}	e^{-3}
$\ln x$							

In Exercises 15 and 16, rely on the definition of $\log_b x$ (in the box on page 386) to decide which of the two quantities is larger.

15. **(a)** $\log_5 30$ or $\log_8 60$
   (b) $\ln 17$ or $\log_{10} 17$

16. **(a)** $\log_9 80$ or $\ln(e^2)$
   (b) $\log_2 3$ or $\log_3 2$

In Exercises 17 and 18, evaluate each expression.

17. **(a)** $\log_9 27$
   (b) $\log_4(1/32)$
   (c) $\log_5 5\sqrt{5}$

18. **(a)** $\log_{25}(1/625)$
   (b) $\log_{16}(1/64)$
   (c) $\log_{10} 10$
   (d) $\log_2 8\sqrt{2}$

In Exercises 19 and 20, solve each equation for x by converting to exponential form. In Exercises 19(b) and 20, give two forms for each answer: one involving e and the other a calculator approximation rounded to two decimal places.

19. **(a)** $\log_4 x = -2$
   (b) $\ln x = -2$

20. **(a)** $\log_5 x = e$
   (b) $\ln x = -e$

In Exercises 21 and 22, find the domain of each function.

21. **(a)** $y = \log_4 5x$
   (b) $y = \log_{10}(3 - 4x)$
   (c) $y = \ln(x^2)$
   (d) $y = (\ln x)^2$
   (e) $y = \ln(x^2 - 25)$

22. **(a)** $y = \ln(2 - x - x^2)$
   (b) $y = \log_{10}\dfrac{2x + 3}{x - 5}$

23. In the accompanying figure, what are the coordinates of the four points A, B, C, and D?

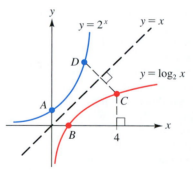

24. In the accompanying figure, what are the coordinates of the points A, B, C, and D?

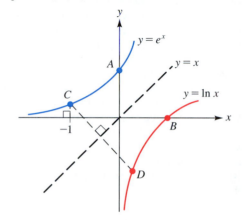

In Exercises 25–30, graph each function and specify the domain, range, intercept(s), and asymptote.

25. **(a)** $y = \log_2 x$
   (b) $y = -\log_2 x$
   (c) $y = \log_2(-x)$
   (d) $y = -\log_2(-x)$

26. **(a)** $y = \ln x$
   (b) $y = -\ln x$
   (c) $y = \ln(-x)$
   (d) $y = -\ln(-x)$

27. $y = -\log_3(x - 2) + 1$

28. $y = -\log_{10}(x + 1)$

29. $y = \ln(x + e)$

30. $y = \ln(-x) + e$

In Exercises 31 and 32, simplify each expression.

31. **(a)** $\ln e^4$
   (b) $\ln(1/e)$
   (c) $\ln \sqrt{e}$

32. **(a)** $\ln e$
   (b) $\ln e^{-2}$
   (c) $(\ln e)^{-2}$

For Exercises 33–40:

(a) *Use a graphing utility to estimate the root(s) of the equation to the nearest one-tenth (as in Example 6).*

(b) *Solve the given equation algebraically by first rewriting it in logarithmic form. Give two forms for each answer: an exact expression and a calculator approximation rounded to three decimal places. Check to see that each result is consistent with the graphical estimate obtained in part (a).*

33. $10^x = 25$

34. $10^{2x-1} = 145$

35. $10^{x^2} = 40$

36. $(10^x)^2 = 40$

37. $e^{2t+3} = 10$ **38.** $e^{t-1} = 16$

39. $e^{1-4t} = 12.405$ **40.** $e^{3x^2} = 112$

Ⓖ **41.** Use graphs to help answer the following questions.
  (a) For which x-values is $\ln x < \log_{10} x$?
  (b) For which x-values is $\ln x > \log_{10} x$?
  (c) For which x-values is $e^x < 10^x$?
  (d) For which x-values is $e^x > 10^x$?

Ⓖ **42.** This exercise demonstrates the very slow growth of the natural logarithm function $y = \ln x$. We consider the following question: How large must x be before the graph of $y = \ln x$ reaches a height of 10?
  (a) Graph the function $y = \ln x$ using a viewing rectangle that extends from 0 to 10 in the x-direction and 0 to 12 in the y-direction. Note how slowly the graph rises. Use the graphing utility to estimate the height of the curve (the y-coordinate) when $x = 10$.
  (b) Since we are trying to see when the graph of $y = \ln x$ reaches a height of 10, add the horizontal line $y = 10$ to your picture. Next, adjust the viewing rectangle so that x extends from 0 to 100. Now use the graphing utility to estimate the height of the curve when $x = 100$. [As both the picture and the y-coordinate indicate, we're still not even halfway to 10. Go on to part (c).]
  (c) Change the viewing rectangle so that x extends to 1000, then estimate the y-coordinate corresponding to $x = 1000$. (You'll find that the height of the curve is almost 7. We're getting closer.)
  (d) Repeat part (c) with x extending to 10,000. (You'll find that the height of the curve is over 9. We're almost there.)
  (e) The last step: change the viewing rectangle so that x extends to 100,000, then use the graphing utility to estimate the x-value for which $\ln x = 10$. As a check on your estimate, rewrite the equation $\ln x = 10$ in exponential form, and evaluate the expression that you obtain for x.

In Exercises 43 and 44, Richter magnitudes of earthquakes are given. Follow the method of Example 8 to determine how many times stronger one quake was than the other.

43.

Location of earthquake	Date	Richter magnitude M
Bombay, India	September 29, 1993	6.4
(15 mi. east of) San Salvador, El Salvador	February 13, 2001	6.6

44.

Location of earthquake	Date	Richter magnitude M
Napa, California	September 3, 2000	5.2
Sumatra, Indonesia	June 4, 2000	8.0

B

As background for Exercises 45 and 46, you need to have read Example 8.

45. Suppose that an earthquake has a Richter magnitude of M_0, and a second earthquake has a magnitude of $M_0 + d$ (where $d > 0$). Show that the second earthquake is 10^d times stronger than the first earthquake.

46. In 1956, the geologists Gutenberg and Richter developed the following formula for estimating the amount of energy E released in an earthquake: $\log_{10} E = 4.4 + (1.5)M$, where E is the energy in joules and M is the Richter magnitude. Now refer to the following table, and let E_1 denote the energy of the quake in the Philippines and E_2 the energy of the quake in Washington. Compute the ratio E_1/E_2 to compare the energies of the two quakes.

Location of earthquake	Date	Richter magnitude M
Mindanao, Philippines	January 1, 2001	7.5
(northeast of) Olympia, Washington	February 28, 2001	6.8

47. The intensity of the sounds that the human ear can detect varies over a very wide range of values. For instance, a whisper from 1 meter away has an intensity of approximately 10^{-10} watts per square meter (W/m^2), whereas, from a distance of 50 meters, the intensity of a launch of the Space Shuttle is approximately 10^8 W/m^2. For a sound with intensity I, the sound level β is defined by

$$\beta = 10 \log_{10}(I/I_0)$$

where the constant I_0 is the sound intensity of a barely audible sound at the threshold of hearing. The units for the sound level β are *decibels,* abbreviated dB.
  (a) Solve the equation $\beta = 10 \log_{10}(I/I_0)$ for I by first dividing by 10 and then converting to exponential form.
  (b) The sound level for a power lawnmower is $\beta = 100$ db and, for a cat purring, it is $\beta = 10$ db. Use your result in part (a) to determine how many times more intense is the power mower sound than the cat's purring.

48. A sound level of $\beta = 120$ db is at the threshold of pain. (Some loud rock concerts reach this level.) The sound intensity that corresponds to $\beta = 120$ db is 1 W/m². Use this information and the equation $\beta = 10 \log_{10}(I/I_0)$ to determine I_0, the intensity of a barely audible sound at the threshold of hearing. What is the corresponding decibel level β?

In Exercises 49–52, use the following information on pH. Chemists define pH by the formula $pH = -\log_{10}[H^+]$, where $[H^+]$ is the hydrogen ion concentration measured in moles per liter. For example, if $[H^+] = 10^{-5}$, then $pH = 5$. Solutions with a pH of 7 are said to be neutral; a pH below 7 indicates an acid; and a pH above 7 indicates a base.

49. (a) For some fruit juices, $[H^+] = 3 \times 10^{-4}$. Determine the pH and classify these juices as acid or base.
   (b) For sulfuric acid, $[H^+] = 1$. Find the pH.
50. An unknown substance has a hydrogen ion concentration of 3.5×10^{-9}. Classify the substance as acid or base.
51. What is the hydrogen ion concentration for black coffee if the pH is 5.9?
52. A chemist adds some acid to a solution changing the pH from 6 to 4. By what factor does the hydrogen ion concentration change?

In Exercises 53–60, decide which of the following properties apply to each function. (More than one property may apply to a function.)
A. *The function is increasing for $-\infty < x < \infty$.*
B. *The function is decreasing for $-\infty < x < \infty$.*
C. *The function has a turning point.*
D. *The function is one-to-one.*
E. *The graph has an asymptote.*
F. *The function is a polynomial function.*
G. *The domain of the function is $(-\infty, \infty)$.*
H. *The range of the function is $(-\infty, \infty)$.*
53. $y = \ln x$ **54.** $y = \ln(-x)$
55. $y = -\ln x$ **56.** $y = \ln|x|$
57. $y = \ln x + e$ **58.** $y = x + \ln e$
59. $y = -\ln(-x)$ **60.** $y = (\ln x)^2$
61. Let $f(x) = e^{x+1}$. Find $f^{-1}(x)$ and sketch its graph. Specify any intercept or asymptote.
62. Let $g(t) = \ln(t - 1)$. Find $g^{-1}(t)$ and draw its graph. Specify any intercept or asymptote.
63. Estimate a value for x such that $\log_2 x = 100$. Use the approximation $10^3 \approx 2^{10}$ to express your answer as a power of 10.   *Answer:* 10^{30}
64. (a) How large must x be before the graph of $y = \ln x$ reaches a height of $y = 100$?
   (b) How large must x be before the graph of $y = e^x$ reaches a height of **(i)** $y = 100$? **(ii)** $y = 10^6$?
G **65.** This exercise uses the natural logarithm and a regression line (i.e., a linear function) to find a simple formula relating the following two quantities:

   x: the average distance of a planet from the sun
   y: the period of a planet (i.e., the time required for a

planet to make one complete revolution around the sun)

(a) Complete the following table. (In the table, the abbreviation AU stands for *astronomical unit*, a unit of distance. By definition, one AU is the average distance from the earth to the sun.

Planet	x Average distance from the sun (AU)	y Period (years)	ln x	ln y
Mercury	0.387	0.241		
Venus	0.723	0.615		
Earth	1.000	1.000		
Mars	1.523	1.881		
Jupiter	5.202	11.820		

(b) Use a graphing utility to obtain a scatter plot and a regression equation $Y = AX + B$ for the pairs of numbers (ln x, ln y). After some rounding, you should obtain $A = 1.5$ and $B = 0$. Thus the quantities ln x and ln y are related as follows:

$$\ln y = 1.5 \ln x$$

(c) Show that the equation obtained in part (b) can be simplified to

$$y = x^{1.5} \qquad (1)$$

(d) Equation (1) was obtained using data for the first five planets from the sun. Complete the following table to see how well equation (1) fits the data for the remaining planets.

Planet	x (AU)	y (years) [calculated from eqn. (1)]	y (years) (from astronomical observations)
Saturn	9.555		29.46
Uranus	19.22		84.01
Neptune	30.11		164.79
Pluto	39.44		248.50

Remark: Equation (1) is *Kepler's third law of planetary motion.* According to George F. Simmons in *Calculus Gems* (New York: McGraw-Hill, Inc., 1992), "Kepler had struggled for more than twenty years to find this connection between a planet's *distance* from the sun and the *time* required to complete its orbit. He published his discovery in 1619 in a work entitled *Harmonices Mundi* (The Harmonies of the World)."

G **66.** This exercise uses the natural logarithm and a regression line to find a simple formula relating the following two quantities:

   x: the average distance of a planet from the sun
   y: the average orbital velocity of the planet

(a) Complete the following table.

Planet	x Average distance from the sun (millions of miles)	y Average orbital velocity (miles/sec)	ln x	ln y
Mercury	35.98	29.75		
Venus	67.08	21.76		
Earth	92.96	18.51		
Mars	141.64	14.99		

(b) Use a graphing utility to obtain a scatter plot and a regression equation $Y = AX + B$ for the pairs of numbers $(\ln x, \ln y)$. You will find that $A = -0.500$ and $B = 5.184$. Thus the quantities $\ln x$ and $\ln y$ are related as follows:

$$\ln y = -0.500 \ln x + 5.184$$

(c) Show that the equation obtained in part (b) can be rewritten in the form

$$y = \frac{e^{5.184}}{\sqrt{x}} \qquad (2)$$

Hint: Convert from logarithmic to exponential form.

(d) Equation (2) was obtained using data for the first four planets from the sun. Complete the following table to see how well equation (2) fits the data for the remaining planets. (Round the answers to two decimal places.)

Planet	x Average distance from the sun (millions of miles)	y Average orbital velocity (miles/sec) [calculated from eqn. (2)]	y Average orbital velocity (miles/sec) (from astronomical observations)
Jupiter	483.63		8.12
Saturn	888.22		5.99
Uranus	1786.55		4.23
Neptune	2799.06		3.38
Pluto	3700.75		2.95

Ⓖ **67.** *Logarithmic regression* In Sections 4.1 and 4.2, we looked at applications in which linear functions or quadratic functions were used to model data sets. In each case, we used a graphing utility to obtain the equation of the linear function or quadratic function that "best fitted" the data. In this exercise we use a function of the form $y = a + b \ln x$ to model a given data set, and we compare a projection made with this model to one based on a linear model.

(a) The following table gives the population of Los Angeles County at ten-year intervals over the period 1950–1990. Use a graphing utility to obtain a scatter plot for the data, and then use the *logarithmic regression* option on the graphing utility to find a function of the form $y = a + b \ln x$ that best fits the data. (Let $x = 50$ correspond to 1950. The reason we don't use the seemingly simpler choice $x = 0$ for 1950 is that $x = 0$ is not in the domain of the function $y = a + b \ln x$.) Display the graph of the function and the scatter plot on the same screen.

Population of Los Angeles County

Year	1950	1960	1970	1980	1990
Population (millions)	4.152	6.040	7.032	7.478	8.863

Source: U.S. Bureau of the Census

(b) Follow part (a), but use a linear model. Display the scatter plot and the graphs of the linear and logarithmic models on the same screen.

(c) Use each model to make a projection for what the population might have been in the year 2000.

(d) According to the U.S. Bureau of the Census, the population of Los Angeles County in 2000 was 9.519 million. Which answer in part (c) is closer to this? Compute the percentage error for each projection.

68. This exercise indicates one of the ways the natural logarithm function is used in the study of prime numbers. Recall that a prime number is a natural number greater than 1 with no factors other than itself and 1. For example, the first ten prime numbers are 2, 3, 5, 7, 11, 13, 17, 19, 23, and 29.

(a) Let $P(x)$ denote the number of prime numbers that do not exceed x. For instance, $P(6) = 3$, since there are three prime numbers (2, 3, and 5) that do not exceed 6. Compute $P(10)$, $P(18)$, and $P(19)$.

(b) According to the **prime number theorem**, $P(x)$ can be approximated by $x/(\ln x)$ when x is large and, in fact, the ratio $\dfrac{P(x)}{x/(\ln x)}$ approaches 1 as x grows larger and larger. Verify this empirically by completing the following table. Round your results to three decimal places. (These facts were discovered by Carl Friedrich Gauss in 1792, when he was 15 years old. It was not until 1896, more than 100 years later, that the prime number theorem was formally proved by the French mathematician J. Hadamard and also by the Belgian mathematician C. J. de la Vallée-Poussin.)

x	P(x)	$\dfrac{x}{\ln x}$	$\dfrac{P(x)}{x/\ln x}$
10^2	25		
10^4	1229		
10^6	78498		
10^8	5761455		
10^9	50847534		
10^{10}	455052512		

(c) In 1808, A. M. Legendre found that he could improve upon Gauss's approximation for $P(x)$ by using the expression $x/(\ln x - 1.08366)$ rather than $x/(\ln x)$. Complete the following table to see how well Legendre's expression approximates $P(x)$. Round your results to four decimal places.

x	$P(x)$	$\dfrac{x}{\ln x - 1.08366}$	$\dfrac{P(x)}{x/(\ln x - 1.08366)}$
10^2	25		
10^4	1229		
10^6	78498		
10^8	5761455		
10^9	50847534		
10^{10}	455052512		

C

69. (a) Find the domain of the function f defined by
$$f(x) = \ln(\ln x).$$
(b) Find $f^{-1}(x)$ for the function f in part (a).

70. Find the domain of the function g defined by
$$g(x) = \ln(\ln(\ln x)).$$

71. Let $g(x) = \ln(\ln(\ln x))$. ⓖ
(a) Using a graphing utility, display the graph of g, first in the standard viewing rectangle, then in the viewing rectangle $[0, 100, 10]$ by $[-5, 5, 1]$.
(b) Find the range of the function g. (See Example 4 in Section 3.1 if you need a hint on how to begin.)
(c) Could you have guessed the answer in part (b) from the graphs in part (a)?

MINI PROJECT | MORE COFFEE

As a prerequisite for this mini project, someone in your group or in the class at large needs to have completed the Section 5.2 project that dealt with the temperature of a cooling cup of coffee. This is a continuation (a completion, actually) of Problem B in that project. As a start here, have that person present a detailed summary of the questions, methods, and results from the Section 5.2 project. Then, for this mini project: Use the techniques from both Sections 5.2 and 5.3 to work out a complete solution to Problem B. Round the answer to the nearest one second. Check that your result is consistent with the approximation from the graphing utility obtained in the Section 5.2 project.

. . . *he* [John Napier] *invented the word "logarithms," using two Greek words,*

. . . *arithmos, "number" and logos, "ratio." It is impossible to say exactly what he had in mind when making up this word.* —Alfred Hooper in *Makers of Mathematics* (New York: Random House, Inc., 1948)

5.4 PROPERTIES OF LOGARITHMS

A few basic properties of logarithms are used repeatedly. Our procedure in this section will be to state these properties, discuss their proofs, and then look at examples.

Properties of Logarithms

1. (a) $\log_b b = 1$ **(b)** $\log_b 1 = 0$
2. $\log_b PQ = \log_b P + \log_b Q$
   The log of a product is the sum of the logs of the factors.
3. $\log_b(P/Q) = \log_b P - \log_b Q$
   The log of a quotient is the log of the numerator minus the log of the denominator.
   As a useful particular case, we have $\log_b(1/Q) = -\log_b Q$.
4. $\log_b P^n = n \log_b P$
5. $b^{\log_b P} = P$

NOTE P and Q are assumed to be positive in Properties 2–5.

In essence, each of these properties follows from the equivalence of the two equations $y = \log_b x$ and $b^y = x$. For instance, the equivalent exponential forms for Properties 1(a) and 1(b) are $b^1 = b$ and $b^0 = 1$, respectively, both of which are certainly valid.

To prove Property 2, we begin by letting $x = \log_b P$. The equivalent exponential form of this equation is

$$P = b^x \tag{1}$$

Similarly, we let $y = \log_b Q$. The exponential form of this equation is

$$Q = b^y \tag{2}$$

If we multiply equation (1) by equation (2), we get

$$PQ = b^x b^y$$

and, therefore,

$$PQ = b^{x+y} \tag{3}$$

Next we write equation (3) in its equivalent logarithmic form. This yields

$$\log_b PQ = x + y$$

But, using the definitions of x and y, this last equation is equivalent to

$$\log_b PQ = \log_b P + \log_b Q$$

That completes the proof of Property 2.

The proof of Property 3 is quite similar to the proof given for Property 2. Exercise 72(a) asks you to carry out the proof of Property 3.

We turn now to the proof of Property 4. We begin by letting $x = \log_b P$. In exponential form, this last equation becomes

$$b^x = P \tag{4}$$

Now we raise both sides of equation (4) to the power n. This yields

$$(b^x)^n = b^{nx} = P^n$$

The logarithmic form of this last equation is

$$\log_b P^n = nx$$

or (from the definition of x)

$$\log_b P^n = n \log_b P$$

The proof of Property 4 is now complete.

Property 5 is again just a restatement of the meaning of $\log_b P$. To derive this property, let $x = \log_b P$. Therefore,

$$b^x = P$$

Now, in this last equation, we simply replace x with $\log_b P$. The result is

$$b^{\log_b P} = P \qquad \text{as required}$$

Now let's see how these properties are used. To begin with, we display some simple numerical examples in the box that follows.

PROPERTY SUMMARY   Properties of Logarithms

Property	Example
$\log_b P + \log_b Q = \log_b PQ$	Simplify: $\log_{10} 50 + \log_{10} 2$ Solution: $\log_{10} 50 + \log_{10} 2 = \log_{10}(50 \cdot 2)$ $= \log_{10} 100$ $= 2$
$\log_b P - \log_b Q = \log_b \dfrac{P}{Q}$	Simplify: $\log_8 56 - \log_8 7$ Solution: $\log_8 56 - \log_8 7 = \log_8 \dfrac{56}{7}$ $= \log_8 8 = 1$
$\log_b P^n = n \log_b P$	Simplify: $\log_2 \sqrt[5]{16}$ Solution: $\log_2 \sqrt[5]{16} = \log_2(16^{1/5})$ $= \dfrac{1}{5} \log_2 16 = \dfrac{4}{5}$
$b^{\log_b P} = P$	Simplify: $3^{\log_3 7}$ Solution: $3^{\log_3 7} = 7$

The examples in the box showed how we can simplify or shorten certain expressions involving logarithms. The next example is also of this type.

EXAMPLE   1   Using log properties to shorten expressions

Express as a single logarithm with a coefficient of 1:

$$\frac{1}{2}\log_b x - \log_b(1 + x^2)$$

SOLUTION

$\dfrac{1}{2}\log_b x - \log_b(1 + x^2) = \log_b x^{1/2} - \log_b(1 + x^2)$ using Property 4 on page 399

$= \log_b \dfrac{x^{1/2}}{1 + x^2}$ using Property 3

This last expression is the required answer.

Property 2 says that the logarithm of a product of two factors is equal to the sum of the logarithms of the two factors. This can be generalized to any number of factors. For instance, with three factors we have

$\log_b(ABC) = \log_b[A(BC)]$

$= \log_b A + \log_b BC$ using Property 2

$= \log_b A + \log_b B + \log_b C$ using Property 2 again

The next example makes use of this idea.

EXAMPLE **2** **Using log properties to shorten expressions**

Express as a single logarithm with a coefficient of 1:

$$\ln(x^2 - 9) + 2 \ln \frac{1}{x + 3} + 4 \ln x \qquad (x > 3)$$

SOLUTION

$$\ln(x^2 - 9) + 2 \ln \frac{1}{x + 3} + 4 \ln x = \ln(x^2 - 9) + \ln\left[\left(\frac{1}{x + 3}\right)^2\right] + \ln x^4$$

$$= \ln(x^2 - 9) + \ln \frac{1}{(x + 3)^2} + \ln x^4$$

$$= \ln\left[(x^2 - 9) \cdot \frac{1}{(x + 3)^2} \cdot x^4\right]$$

$$= \ln \frac{(x^2 - 9)x^4}{(x + 3)^2}$$

This last expression can be simplified still further by factoring the quantity $x^2 - 9$ as $(x - 3)(x + 3)$. Then a factor of $x + 3$ can be divided out of the numerator and denominator of the fraction. The result is

$$\ln(x^2 - 9) + 2 \ln \frac{1}{x + 3} + 4 \ln x = \ln \frac{(x - 3)x^4}{x + 3}$$

$$= \ln \frac{x^5 - 3x^4}{x + 3}$$

In the examples we have considered so far, we've used the properties of logarithms to shorten given expressions. We can also use these properties to expand an expression. (This is useful in calculus.)

EXAMPLE **3** **Using log properties to expand expressions**

Write each of the following quantities as sums and differences of simpler logarithmic expressions. Express each answer in such a way that no logarithm of products, quotients, or powers appears.

(a) $\log_{10} \sqrt{3x}$ **(b)** $\log_{10} \sqrt[3]{\dfrac{2x}{3x^2 + 1}}$ **(c)** $\ln \dfrac{x^2\sqrt{2x - 1}}{(2x + 1)^{3/2}}$

SOLUTION

(a) $\log_{10} \sqrt{3x} = \log_{10}(3x)^{1/2} = \frac{1}{2}\log_{10}(3x)$
$\qquad = \frac{1}{2}(\log_{10} 3 + \log_{10} x)$

(b) $\log_{10} \sqrt[3]{\dfrac{2x}{3x^2 + 1}} = \log_{10}\left[\left(\dfrac{2x}{3x^2 + 1}\right)^{1/3}\right]$

$\qquad = \frac{1}{3}\log_{10}\dfrac{2x}{3x^2 + 1}$

$\qquad = \frac{1}{3}\left[\log_{10} 2x - \log_{10}(3x^2 + 1)\right]$

$\qquad = \frac{1}{3}\left[\log_{10} 2 + \log_{10} x - \log_{10}(3x^2 + 1)\right]$

(c) $\ln \dfrac{x^2\sqrt{2x-1}}{(2x+1)^{3/2}} = \ln \dfrac{x^2(2x-1)^{1/2}}{(2x+1)^{3/2}}$

$\qquad\qquad = \ln x^2 + \ln(2x-1)^{1/2} - \ln(2x+1)^{3/2}$

$\qquad\qquad = 2\ln x + \tfrac{1}{2}\ln(2x-1) - \tfrac{3}{2}\ln(2x+1)$

| EXAMPLE | 4 | Practice applying the log properties |

Given that $\log_b 2 = A$ and $\log_b 6 = B$, express each of the following in terms of A and/or B.

(a) $\log_b 8$ **(c)** $\log_b 12$ **(e)** $\log_{\sqrt{b}} 2$

(b) $\log_b \sqrt{6}$ **(d)** $\log_b 3$ **(f)** $\log_b(b/36)$

SOLUTION

(a) $\log_b 8 = \log_b 2^3 = 3\log_b 2 = 3A$

(b) $\log_b \sqrt{6} = \log_b 6^{1/2} = \tfrac{1}{2}\log_b 6 = \tfrac{1}{2}B$

(c) $\log_b 12 = \log_b(2 \times 6) = \log_b 2 + \log_b 6 = A + B$

(d) $\log_b 3 = \log_b(6/2) = \log_b 6 - \log_b 2 = B - A$

(e) Let $\log_{\sqrt{b}} 2 = x$. (So we want to express x in terms of A and/or B.) Writing this equation in exponential form, we have $(\sqrt{b})^x = 2$, and therefore

$$(b^{1/2})^x = 2$$
$$b^{x/2} = 2$$

Now we rewrite this last equation in logarithmic form to obtain $\log_b 2 = x/2$, and therefore

$$x = 2\log_b 2 = 2A$$

(f) $\log_b(b/36) = \log_b b - \log_b 36$

$\qquad\qquad = 1 - \log_b 6^2 = 1 - 2\log_b 6 = 1 - 2B$

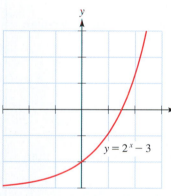

$y = 2^x - 3$

Figure 1

| EXAMPLE | 5 | Using the property $\log_b P^n = n \log_b P$ to find an x-intercept |

Figure 1 shows the graph of $y = 2^x - 3$. (The axes are marked off in one-unit intervals.) As indicated in the figure, the x-intercept of the curve is between 1 and 2. Determine both an exact expression for this x-intercept and a calculator approximation rounded to two decimal places.

SOLUTION

Setting y equal to zero in the given equation yields $2^x - 3 = 0$, or $2^x = 3$. Notice that the unknown, x, appears in the exponent. To solve for x, we take the logarithm of both sides of the equation. (Base e logarithms are used in the following computations; Exercise 71 asks you to check that the use of base 10 logarithms produces an equivalent answer.) We have

$$\ln 2^x = \ln 3$$
$$x \ln 2 = \ln 3 \qquad \text{using Property 4 on page 399}$$
$$x = \frac{\ln 3}{\ln 2} \approx 1.58 \qquad Caution: \frac{\ln 3}{\ln 2} \neq \ln 3 - \ln 2$$

| EXAMPLE | 6 | **Using the property** $\log_b P^n = n \log_b P$ **to solve an equation** |

Solve for x: $4^{5x+2} = 70$. Use a calculator to evaluate the final answer; round to three decimal places.

SOLUTION
The fact that the unknown appears in the exponent suggests that Property 4 (the property that "brings down" the exponent) may be useful. To put Property 4 into play, we'll take the logarithm of both sides of the given equation. Either base 10 or base e logarithms can be used here (because they're both on the calculator). Since we used base e logarithms in the previous example, let's use base 10 here just for the sake of variety. (You can check for yourself at the end that base e logarithms will yield the same numerical answer.)

Taking the base 10 logarithm of both sides of the given equation, we have

$$\log_{10} 4^{5x+2} = \log_{10} 70$$
$$(5x + 2)\log_{10} 4 = \log_{10} 70 \qquad \text{using Property 4}$$
$$5x + 2 = (\log_{10} 70)/(\log_{10} 4)$$
$$x = \frac{[(\log_{10} 70)/(\log_{10} 4)] - 2}{5}$$
$$x \approx 0.213$$

The required solution is approximately 0.213. Figure 2 shows a graphical interpretation of this result.

GRAPHICAL PERSPECTIVE

Figure 2
The curve $y = 4^{5x+2}$ intersects the line $y = 70$ at a point whose x-coordinate is slightly greater than 0.2. Example 6 shows that this x-coordinate is 0.213 (rounded to three decimal places).

It is sometimes necessary to convert logarithms in one base to logarithms in another base. After the next example, we will state a formula for this. However, as the next example indicates, it is easy to work this type of problem from the basics, without relying on a formula.

EXAMPLE 7 Changing bases

Express $\log_2 5$ in terms of base 10 logarithms.

SOLUTION

Let $z = \log_2 5$. The exponential form of this equation is

$$2^z = 5$$

We now take the base 10 logarithm of each side of this equation to obtain

$$\log_{10} 2^z = \log_{10} 5$$
$$z \log_{10} 2 = \log_{10} 5 \qquad \text{using Property 4 on page 399}$$
$$z = \frac{\log_{10} 5}{\log_{10} 2}$$

Given our definition of z, this last equation can be written

$$\log_2 5 = \frac{\log_{10} 5}{\log_{10} 2}$$

This is the required answer.

The method shown in Example 7 can be used to convert between any two bases. Exercise 72(b) at the end of this section asks you to follow this method, using letters rather than numbers, to arrive at the following general formula.

PROPERTY SUMMARY Change of Base for Logarithms

Formula	Examples
$\log_a x = \dfrac{\log_b x}{\log_b a}$	$\log_2 3 = \dfrac{\log_{10} 3}{\log_{10} 2} = \dfrac{\ln 3}{\ln 2}$ $\qquad$ $\log_{10} e = \dfrac{\ln e}{\ln 10} = \dfrac{1}{\ln 10}$

EXAMPLE 8 Changing bases to simplify an expression

Simplify the product $(\log_2 10)(\log_{10} 2)$.

SOLUTION

We'll use the change-of-base formula so that both factors in the given product involve logarithms with the same base. Expressing the first factor, $\log_2 10$, in terms of base 10 logarithms, we have

$$\log_2 10 = \frac{\log_{10} 10}{\log_{10} 2} = \frac{1}{\log_{10} 2}$$

and therefore $\quad (\log_2 10)(\log_{10} 2) = \dfrac{1}{\log_{10} 2}(\log_{10} 2) = 1$

So, the given product is equal to 1. [Using the same method, one can show that, in general, $(\log_a b)(\log_b a) = 1$.]

Examples 1–8 have dealt with applications of the five properties of logarithms. However, you also need to understand what the properties *don't* say. For instance, Property 3 does not apply to an expression such as $\dfrac{\log_{10} 5}{\log_{10} 2}$. [Property 3 *would* apply if the expression were $\log_{10}(5/2)$.] In the box that follows, we list some errors to avoid in working with logarithms.

ERRORS TO AVOID

Error	Correction	Comment
$\log_b(x + y) \neq \log_b x + \log_b y$	$\log_b(xy) = \log_b x + \log_b y$	In general, there is no simple identity involving $\log_b(x + y)$. This is similar to the situation for $\sqrt{x + y}$ (which is not equal to $\sqrt{x} + \sqrt{y}$).
$\dfrac{\log_b x}{\log_b y} \neq \log_b x - \log_b y$	$\log_b\left(\dfrac{x}{y}\right) = \log_b x - \log_b y$	In general, there is no simple identity involving the quotient $(\log_b x)/(\log_b y)$.
$(\ln x)^3 \neq 3 \ln x$	$(\ln x)^3 = (\ln x)(\ln x)(\ln x)$	$(\ln x)^3$ is not the same as $\ln(x^3)$. Regarding the latter, we do have $\ln(x^3) = 3 \ln x$, where $x > 0$.
$\ln \dfrac{x}{2} \neq \dfrac{\ln x}{2}$	$\ln \dfrac{x}{2} = \ln x - \ln 2 \quad$ for $x > 0$ and $\dfrac{\ln x}{2} = \dfrac{1}{2} \ln x = \ln \sqrt{x} \quad$ for $x > 0$	The confusion between $\ln \dfrac{x}{2}$ and $\dfrac{\ln x}{2}$ is sometimes due simply to careless handwriting.

EXERCISE SET 5.4

A

In Exercises 1–10, simplify the expression by using the definition and properties of logarithms.

1. $\log_{10} 70 - \log_{10} 7$
2. $\log_{10} 40 + \log_{10}(5/2)$
3. $\log_7 \sqrt{7}$
4. $\log_9 25 - \log_9 75$
5. $\log_3 108 + \log_3(3/4)$
6. $\ln e^3 - \ln e$
7. $-\frac{1}{2} + \ln \sqrt{e}$
8. $e^{\ln 3} + e^{\ln 2} - e^{\ln e}$
9. $2^{\log_2 5} - 3 \log_5 \sqrt[3]{5}$
10. $\log_b b^b$

In Exercises 11–19, write the expression as a single logarithm with a coefficient of 1.

11. $\log_{10} 30 + \log_{10} 2$
12. $2 \log_{10} x - 3 \log_{10} y$
13. $\log_5 6 + \log_5(1/3) + \log_5 10$
14. $p \log_b A - q \log_b B + r \log_b C$
15. (a) $\ln 3 - 2 \ln 4 + \ln 32$
    (b) $\ln 3 - 2(\ln 4 + \ln 32)$
16. (a) $\log_{10}(x^2 - 16) - 3 \log_{10}(x + 4) + 2 \log_{10} x$
    (b) $\log_{10}(x^2 - 16) - 3[\log_{10}(x + 4) + 2 \log_{10} x]$

17. $\log_b 4 + 3[\log_b(1 + x) - \frac{1}{2} \log_b(1 - x)]$
18. $\ln(x^3 - 1) - \ln(x^2 + x + 1)$
19. $4 \log_{10} 3 - 6 \log_{10}(x^2 + 1) + \frac{1}{2}[\log_{10}(x + 1) - 2 \log_{10} 3]$

In Exercises 20–26, write the quantity using sums and differences of simpler logarithmic expressions. Express the answer so that logarithms of products, quotients, and powers do not appear.

20. (a) $\log_{10} \sqrt{(x + 1)(x + 2)}$
    (b) $\ln \sqrt{\dfrac{(x + 1)(x + 2)}{(x - 1)(x - 2)}}$

21. (a) $\log_{10} \dfrac{x^2}{1 + x^2}$
    (b) $\ln \dfrac{x^2}{\sqrt{1 + x^2}}$

22. (a) $\log_b \dfrac{\sqrt{1 - x^2}}{x}$
    (b) $\ln \dfrac{x\sqrt[3]{4x + 1}}{\sqrt{2x - 1}}$

23. (a) $\log_{10} \sqrt{9 - x^2}$
    (b) $\ln \dfrac{\sqrt{4 - x^2}}{(x - 1)(x + 1)^{3/2}}$

24. (a) $\log_b \sqrt[3]{\dfrac{x + 3}{x}}$
    (b) $\ln \dfrac{1}{\sqrt{x^2 + x + 1}}$

25. (a) $\log_b \sqrt{x/b}$

  (b) $2 \ln \sqrt{(1 + x^2)(1 + x^4)(1 + x^6)}$

26. (a) $\log_b \sqrt[3]{\dfrac{(x - 1)^2(x - 2)}{(x + 2)^2(x + 1)}}$ **(b)** $\ln\left(\dfrac{e - 1}{e + 1}\right)^{3/2}$

In Exercises 27–36, suppose b is a positive constant greater than 1, and let A, B, and C be defined as follows:

$$\log_b 2 = A \qquad \log_b 3 = B \qquad \log_b 5 = C$$

In each case, use the properties of logarithms to evaluate the given expression in terms of A, B, and/or C. (In Exercises 31–36, use the change-of-base formula.)

27. (a) $\log_b 6$
  (b) $\log_b(1/6)$
  (c) $\log_b 27$
  (d) $\log_b(1/27)$

28. (a) $\log_b 10$
  (b) $\log_b 100$
  (c) $\log_b 0.01$
  (d) $\log_b 0.3$

29. (a) $\log_b(5/3)$
  (b) $\log_b 0.6$
  (c) $\log_b(5/9)$
  (d) $\log_b(5/16)$

30. (a) $\log_b \sqrt{5}$
  (b) $\log_b \sqrt{15}$
  (c) $\log_b \sqrt[3]{0.4}$
  (d) $\log_b \sqrt[4]{60}$

31. (a) $\log_3 b$
  (b) $\log_3(10b)$

32. (a) $\log_{b^2} 5$
  (b) $\log_{\sqrt{b}} 2$

33. (a) $\log_{3b} 2$
  (b) $\log_{3b} 15$

34. (a) $\log_{5b} 1.2$
  (b) $\log_{5b} 2.5$

35. (a) $(\log_b 5)(\log_5 b)$
  (b) $(\log_b 6)(\log_6 b)$

36. (a) $\log_{2b} 6 + \log_{2b}(1/6)$
  (b) $\log_{18}(1/b)$

In Exercises 37 and 38, suppose that $\log_{10} A = a$, $\log_{10} B = b$, and $\log_{10} C = c$. Express the following logarithms in terms of a, b, and c.

37. (a) $\log_{10} AB^2C^3$
  (b) $\log_{10} 10\sqrt{A}$
  (c) $\log_{10} \sqrt{10ABC}$
  (d) $\log_{10}(10A/\sqrt{BC})$

38. (a) $\log_{10} A + 2\log_{10}(1/A)$
  (b) $\log_{10}(A/10)$
  (c) $\log_{10} \dfrac{100 A^2}{B^4 \sqrt[3]{C}}$
  (d) $\log_{10} \dfrac{(4B)^5}{C}$

In Exercises 39 and 40, suppose that $\ln x = t$ and $\ln y = u$. Write each expression in terms of t and u.

39. (a) $\ln(ex)$
  (b) $\ln xy - \ln(x^2)$
  (c) $\ln\sqrt{xy} + \ln(x/e)$
  (d) $\ln(e^2 x\sqrt{y})$

40. (a) $\ln(e^{\ln x})$
  (b) $e^{\ln(\ln xy)}$
  (c) $\ln\left(\dfrac{ex}{y}\right) - \ln\left(\dfrac{y}{ex}\right)$
  (d) $\dfrac{(\ln x)^3 - \ln(x^4)}{\left(\ln \dfrac{x}{e^2}\right)\ln(xe^2)}$

In Exercises 41 and 42, graph the equations and determine the x-intercepts (as in Example 5).

41. (a) $y = 2^x - 5$
  (b) $y = 2^{x/2} - 5$

42. (a) $y = 3^{x-1} - 2$
  (b) $y = 3^{1-x} - 4$

In Exercises 43–47, solve the equations. Express the answers in terms of natural logarithms.

43. $5 = 2e^{2x-1}$ *Suggestion:* First divide by 2.

44. $3e^{1+t} = 2$ **45.** $2^x = 13$ **46.** $5^{3x-1} = 27$

47. $10^x = e$ **48.** $10^{2x+3} = 280$

In Exercises 49 and 50, solve the equations. Give two forms for each answer: one involving base 10 logarithms and the other a calculator approximation rounded to three decimal places.

49. $3^{x^2-1} = 12$ **50.** $2^{9-x^2} = 430.5389$

In Exercises 51–56, express the quantity in terms of base 10 logarithms.

51. $\log_2 5$ **52.** $\log_5 10$ **53.** $\ln 3$
54. $\ln 10$ **55.** $\log_b 2$ **56.** $\log_2 b$

In Exercises 57–61, express the quantity in terms of natural logarithms.

57. $\log_{10} 6$ **58.** $\log_2 10$ **59.** $\log_{10} e$
60. $\log_b 2$, where $b = e^2$ **61.** $\log_{10}(\log_{10} x)$

62. Give specific examples showing that each statement is false.
  (a) $\log(x + y) = \log x + \log y$
  (b) $(\log x)/(\log y) = \log x - \log y$
  (c) $(\log x)(\log y) = \log x + \log y$
  (d) $(\log x)^k = k \log x$

63. True or false?
  (a) $\log_{10} A + \log_{10} B - \frac{1}{2}\log_{10} C = \log_{10}(AB/\sqrt{C})$
  (b) $\log_e \sqrt{e} = 1/2$
  (c) $\ln \sqrt{e} = 1/2$
  (d) $\ln x^3 = \ln 3x$
  (e) $\ln x^3 = 3 \ln x$
  (f) $\ln 2x^3 = 3 \ln 2x$
  (g) $\log_a c = b$ means $a^b = c$.
  (h) $\log_5 24$ is between 5^1 and 5^2.
  (i) $\log_5 24$ is between 1 and 2.
  (j) $\log_5 24$ is closer to 1 than to 2.
  (k) The domain of $g(x) = \ln x$ is the set of all real numbers.
  (l) The range of $g(x) = \ln x$ is the set of all real numbers.
  (m) The function $g(x) = \ln x$ is one-to-one.

Use a calculator for Exercises 64–66.

64. (a) Check Property 2 using the values $b = 10$, $P = \pi$, and $Q = \sqrt{2}$.
  (b) Let $P = 3$ and $Q = 4$. Show that $\ln(P + Q) \neq \ln P + \ln Q$.
  (c) Check Property 3 using the values $b = 10$, $P = 2$, and $Q = 3$.
  (d) If $P = 10$ and $Q = 20$, show that $\ln(PQ) \neq (\ln P)(\ln Q)$.
  (e) Check Property 3 using natural logarithms and the values $P = 17$ and $Q = 76$.
  (f) Show that $(\log_{10} 17)/(\log_{10} 76) \neq \log_{10} 17 - \log_{10} 76$.

65. (a) Check Property 4 using the values $b = 10$, $P = \pi$, and $n = 7$.
  (b) Using the values given for b, P, and n in part (a), show that $\log_b P^n \neq (\log_b P)^n$.

(c) Verify Property 5 using the values $b = 10$ and $P = 1776$.

(d) Verify that $\ln 2 + \ln 3 + \ln 4 = \ln 24$.

(e) Verify that

$$\log_{10} A + \log_{10} B + \log_{10} C = \log_{10}(ABC)$$

using the values $A = 11$, $B = 12$, and $C = 13$.

(f) Let $f(x) = e^x$ and $g(x) = \ln x$. Compute $f[g(2345.6)]$.

(g) Let $f(x) = 10^x$ and $g(x) = \log_{10} x$. Compute $g[f(0.123456)]$.

66. Evaluate $[\frac{1}{\pi} \ln(640320^3 + 744)]^2$. *Remark*: Contrary to the empirical evidence from your calculator, it is known that this number is irrational.

B

67. *The approximation $\ln(1 + x) \approx x$*: As indicated in the accompanying graph, the values of $\ln(1 + x)$ and x are very close to one another for small positive values of x.

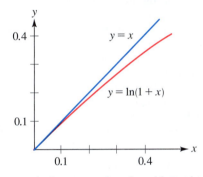

Using your calculator, complete the table to obtain numerical evidence of this. For your answers, report the first six decimal places of the calculator display (don't round off).

x	0.1	0.05	0.005	0.0005
ln(1 + x)				

68. As indicated in the figure, the graph of $y = ae^{bx}$ passes through the two points $(2, 10)$ and $(8, 80)$. Follow steps (a) through (c) to determine the values of the constants a and b.

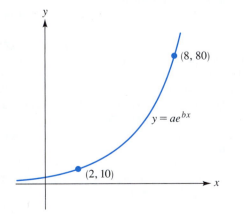

(a) Since the point $(8, 80)$ lies on the graph, the pair of values $x = 8$ and $y = 80$ satisfies the equation $y = ae^{bx}$; that is,

$$80 = ae^{8b} \tag{1}$$

Similarly, since the point $(2, 10)$ lies on the graph, we have

$$10 = ae^{2b} \tag{2}$$

Now use equations (1) and (2) to show that $b = (\ln 8)/6$.

(b) In equation (2), substitute for b using the expression obtained in part (a). Show that the resulting equation can be written $10 = a(e^{\ln 8})^{1/3}$.

(c) Use one of the properties of logarithms to simplify the right-hand side of the equation in part (b), then solve for a. (You should obtain $a = 5$.)

In Exercises 69 and 70, you are given the coordinates of two points on the graph of the curve $y = ae^{bx}$. In each case, determine the values of a and b. Hint: Use the method explained in Exercise 68.

69. $(-2, 324)$ and $(1/2, 4/3)$ **70.** $(1, 2)$ and $(4, 8)$

71. (a) Use base 10 logarithms to solve the equation $2^x - 3 = 0$.

(b) Use your calculator to evaluate the answer in part (a). Also use the calculator to evaluate $(\ln 3)/(\ln 2)$.

(c) Prove that $(\log_{10} 3)/(\log_{10} 2) = (\ln 3)/(\ln 2)$.

72. (a) Prove that $\log_b(P/Q) = \log_b P - \log_b Q$. *Hint*: Study the proof of Property 2 in the text.

(b) Prove the **change-of-base formula**:

$$\log_a x = \frac{\log_b x}{\log_b a}$$

Hint: Use the method of Example 7 in the text.

73. Show that $\log_b \dfrac{\sqrt{3} + \sqrt{2}}{\sqrt{3} - \sqrt{2}} = 2\log_b(\sqrt{3} + \sqrt{2})$.

74. (a) Show that $\log_b(P/Q) + \log_b(Q/P) = 0$.

(b) Simplify: $\log_a x + \log_{1/a} x$.

75. Simplify: $b^{3\log_b x}$.

76. Is there a constant k such that the equation $e^x = 2^{kx}$ holds for all values of x?

77. Prove that $\log_b a = 1/(\log_a b)$.

78. Simplify: $(\log_2 3)(\log_3 4)(\log_4 5)$.

79. (a) Without using your calculator, show that

$$\frac{1}{\log_2 \pi} + \frac{1}{\log_5 \pi} > 2. \qquad \text{*Hint*: Convert to base } \pi \text{ logarithms.}$$

(b) Without using your calculator, show that

$$\log_\pi 2 + \frac{1}{\log_\pi 2} > 2.$$

Hint: First explain, in complete sentences, how you know that the quantity $\log_\pi 2$ is positive. Then apply the inequality given in Exercise 45(b) in Section 2.5.

(c) Use your calculator (and the change-of-base formula) to find out which of the two quantities is larger:

$$\frac{1}{\log_2 \pi} + \frac{1}{\log_5 \pi} \quad \text{or} \quad \log_\pi 2 + \frac{1}{\log_\pi 2}.$$

80. A function f with domain $(1, \infty)$ is defined by the equation $f(x) = \log_x 2$.
   (a) Find a value for x such that $f(x) = 2$.
   (b) Is the number that you found in part (a) a fixed point of the function f?

C

81. Prove that $(\log_a x)/(\log_{ab} x) = 1 + \log_a b$.
82. Simplify a^x when $x = \log_b(\log_b a)/\log_b a$.
83. If $a^2 + b^2 = 7ab$, where a and b are positive, show that

$$\log\left[\frac{1}{3}(a + b)\right] = \frac{1}{2}(\log a + \log b)$$

no matter which base is used for the logarithms (but it is understood that the same base is used throughout).

84. Let $f(x) = \ln\left(1 - \frac{1}{x^2}\right)$.
   (a) Use the properties of logarithms (and some algebra) to show that

$$f(2) + f(3) + f(4) = \ln(5/8)$$

   (b) Use a calculator to check the result in part (a).

5.5   EQUATIONS AND INEQUALITIES WITH LOGS AND EXPONENTS

Initially, [John] Napier [1550–1617] called logarithms "artificial numbers" but later coined the term logarithm, meaning "number of the ratio." —Ronald Calinger, ed., *Classics of Mathematics* (Englewood Cliffs, N.J.: Prentice Hall, 1995)

. . . we would like to give a value for z, such that $a^z = y$. This value of z, insofar as it is viewed as a function of y, it is called the LOGARITHM of y. —Leonhard Euler (1707–1783), *Introductio in analysin infinitorum* (Lausanne: 1748) [This classic text has been translated by John D. Blanton (Berlin: Springer-Verlag, 1988).]

In Examples 2 through 9 in this section we illustrate some of the more common approaches for solving equations and inequalities involving exponential and logarithmic expressions. Although there is no single technique that can be used to solve every equation or inequality of this type, the methods in this section do have much in common: they all rely on the definition and the basic properties of logarithms.

Actually, in the previous two sections, we've already solved some equations involving exponential expressions. So, as background for the work to follow, you might want to review the following three examples. *Study suggestion*: Try solving for yourself the following three equations before you turn back to the text's solution.

EXAMPLE	EQUATION
Example 6 in Section 5.3	$10^{x/2} = 16$
Example 7 in Section 5.3	$e^{2t-5} = 5000$
Example 5 in Section 5.4	$2^x - 3 = 0$
Example 6 in Section 5.4	$4^{5x+2} = 70$

EXAMPLE **1** **Reviewing some possibilities for roots**

Consider the following two equations:

$$(\ln x)^2 = 2 \ln x \tag{1}$$
$$\ln(x^2) = 2 \ln x \tag{2}$$

(a) Is either one of the values $x = 1$ or $x = 2$ a root of equation (1)?
(b) Is either one of the values $x = 1$ or $x = 2$ a root of equation (2)?

SOLUTION
(a) To see whether the value $x = 1$ satisfies equation (1), we have

$$(\ln 1)^2 = 2 \ln 1$$
$$0^2 = 2(0) \qquad \text{True}$$

Thus $x = 1$ is a root of equation (1). To see whether $x = 2$ is a root, we write

$$(\ln 2)^2 = 2 \ln 2$$
$$\ln 2 = 2 \qquad \text{dividing through by } \ln 2 \, (\neq 0)$$

This last equation is not valid. You can see this by using a calculator, or, more directly, by rewriting the equation in exponential form to obtain $e^2 = 2$, which is clearly false. (Why?) Consequently, the value $x = 2$ is not a solution of equation (1).

(b) You've seen equation (2) before, or at least one very much like it. It's an example of one of the basic properties of logarithms that we studied in the previous section. That is, we know from the previous section that the equation $\ln(x^2) = 2 \ln x$ holds for every value of x in the domain of the function $y = \ln x$. In particular, then, since both $x = 1$ and $x = 2$ are in the domain of the function $y = \ln x$, we can conclude immediately, without any work required, that both of the values 1 and 2 satisfy equation (2).

The example that we have just completed serves to remind us of the difference between a *conditional equation* and an *identity*. An identity is true for all values of the variable in its domain. For example, the equation $\ln(x^2) = 2 \ln x$ is an identity; it is true for every positive real number x. In contrast to this, a conditional equation is true only for some (or perhaps even none) of the values of the variable. Equation (1) in Example 1 is a conditional equation; we saw that it is true when $x = 1$ and false when $x = 2$. The equation $2^x = -1$ is an example of a conditional equation that has no real-number root. (Why?) Most of the equations that we solve in this section are conditional equations, but watch for a few identities to pop up in the examples and exercises. It can make your work much easier.

EXAMPLE 2   **An equation where unknown appears in exponent**

Find all real-number roots of the equation $4^x = 3^{2x+1}$.

SOLUTION
Taking the natural logarithm of both sides of the given equation, we have

$$\ln(4^x) = \ln(3^{2x+1})$$

and therefore

$$x \ln 4 = (2x + 1)\ln 3$$
$$x \ln 4 = 2x \ln 3 + \ln 3$$
$$x(\ln 4 - 2 \ln 3) = \ln 3$$
$$x = \frac{\ln 3}{\ln 4 - 2 \ln 3}$$

Using a calculator now, we find that x is approximately -1.355. In Figure 1 we show a graphical check of this solution. Exercise 85 asks you to check the solution algebraically.

GRAPHICAL PERSPECTIVE

Figure 1

The curves $y = 4^x$ and $y = 3^{2x+1}$ intersect at a point whose x-coordinate is between -1.4 and -1.3. In Example 2 we found this x-value to be -1.355 (rounded to three decimal places). (Question: Which of the two graphs in the figure represents $y = 4^x$?)

EXAMPLE **3** **Some equations involving the natural logarithm**

Find all real-number roots of the following equations:
(a) $\ln(\ln x) = 2$; **(b)** $e^{\ln x} = -2$; **(c)** $e^{\ln x} = 2$.

SOLUTION

(a) For convenience, we let t stand for $\ln x$. Then the given equation becomes

$$\ln t = 2$$

Rewriting this equation in exponential form, we have

$$t = e^2$$

or, in view of the definition of t,

$$\ln x = e^2$$

Now we rewrite this last equation in exponential form to obtain

$$x = e^{e^2}$$
$$\approx 1618.2 \qquad \text{using a calculator}$$

As Exercise 86 asks you to verify, the value $x = e^{e^2}$ indeed checks in the original equation.

(b) Think for a moment about the graph of the basic exponential function $y = e^x$. The graph is always above the x-axis. This tells us that e raised to any exponent is never negative. Consequently, the equation $e^{\ln x} = -2$ has no real-number solution.

(c) From Section 5.2 (as well as 5.3), we know that $e^{\ln x} = x$ for all positive numbers x, so the given equation $e^{\ln x} = 2$ becomes simply $x = 2$, and we are done.

In the next example we solve equations involving logarithms by converting them to exponential form. For part (b) of the example, you need to recall the quadratic formula.

EXAMPLE **4** **Equations involving base 10 logarithms**

Find all real-number roots of the following equations:
(a) $\log_{10}(x^2 - 2) = 3$; **(b)** $\log_{10}(x^2 - 2x) = 3$.

SOLUTION

(a) Converting to exponential form, we have

$$x^2 - 2 = 10^3$$

and therefore

$$x^2 = 1002$$

or

$$x = \pm\sqrt{1002} \approx \pm 31.65$$

As Exercise 87(a) asks you to check, both of the values $\pm\sqrt{1002}$ satisfy the given equation.

(b) Again we begin by converting to exponential form. This yields

$$x^2 - 2x = 1000$$

and therefore

$$x^2 - 2x - 1000 = 0$$

Now we apply the quadratic formula:

$$x = \frac{-b \pm \sqrt{b^2 - 4ac}}{2a} = \frac{2 \pm \sqrt{(-2)^2 - 4(1)(-1000)}}{2}$$

$$= \frac{2 \pm \sqrt{4004}}{2} = \frac{2 \pm \sqrt{4(1001)}}{2}$$

$$= \frac{2 \pm 2\sqrt{1001}}{2} = 1 \pm \sqrt{1001}$$

As Exercise 87(b) asks you to verify, both of the numbers $1 + \sqrt{1001}$ and $1 - \sqrt{1001}$ satisfy the given equation. Using a calculator, we find that these two values are approximately 32.6 and -30.6, respectively. In Figure 2 we show a graphical check of the solutions that we've determined in this example.

GRAPHICAL PERSPECTIVE

(a) $y = \log_{10}(x^2 - 2)$ (b) $y = \log_{10}(x^2 - 2x)$

Figure 2

In (a), the figure shows that the curve $y = \log_{10}(x^2 - 2)$ intersects the horizontal line $y = 3$ at two points whose x-coordinates are approximately 30 and -30. In part (a) of Example 4, we found these two numbers to be 31.65 and -31.65, respectively (rounded to two decimal places). Similarly, in (b) the figure shows that the curve $y = \log_{10}(x^2 - 2x)$ intersects the horizontal line $y = 3$ at two points, one with an x-coordinate slightly greater than 30, the other with an x-coordinate very close to -30. In part (b) of the example, we found these two x-coordinates to be 32.6 and -30.6, respectively (rounded to one decimal place).

If you look over both parts of the example just completed, you'll see that we used the same general procedure in both cases: convert from logarithmic to exponential form, then solve the resulting quadratic equation. The next example works in the other direction: first we solve a quadratic equation, then we isolate the variable by converting from exponential to logarithmic form.

| EXAMPLE | 5 | An equation in quadratic form that involves exponentials |

As indicated in Figure 3, the graph of the function $y = 2(3^{2x}) - 3^x - 3$ has an x-intercept between 0.25 and 0.50. Find the exact value of this intercept and also a calculator approximation rounded to three decimal places.

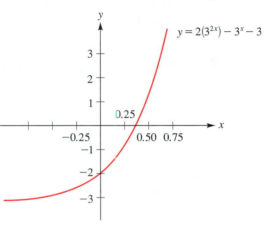

Figure 3

SOLUTION
We need to solve the equation $2(3^{2x}) - 3^x - 3 = 0$. The observation that helps here is that 3^{2x} can be written as $(3^x)^2$. This lets us rewrite the equation as

$$2(3^x)^2 - 3^x - 3 = 0$$

Now, for convenience, we let $t = 3^x$ so that the equation becomes

$$2t^2 - t - 3 = 0$$

or

$$(2t - 3)(t + 1) = 0 \qquad \text{Check the factoring.}$$

$$2t - 3 = 0 \quad \bigg| \quad t + 1 = 0$$
$$t = 3/2 \quad \bigg| \quad t = -1$$

Using the value $t = 3/2$ in the equation $t = 3^x$ gives us

$$3^x = 3/2$$

We can solve this last equation by taking the logarithm of both sides. Using base e logarithms, we have

$$\ln(3^x) = \ln 1.5$$

and therefore

$$x \ln 3 = \ln 1.5$$

Dividing both sides of this last equation by the quantity ln 3, we obtain $x = (\ln 1.5)/(\ln 3)$. As Exercise 88 asks you to verify, this value for x indeed checks in the original equation.

Now, what about the other value for t that we found? With $t = -1$, we have $3^x = -1$, but that is impossible. (Why?) Consequently, there is only one root of the given equation—namely, $x = (\ln 1.5)/(\ln 3)$. This is the required x-intercept for the graph shown in Figure 3. Using a calculator, we find that $\ln (1.5)/(\ln 3) \approx 0.369$. Note that this value is consistent with Figure 3, in which the x-intercept appears to be roughly halfway between 0.25 and 0.50.

When we solved equations in earlier chapters of this text, we learned that some techniques, such as squaring both sides of an equation, may introduce extraneous roots. The next example shows another type of situation in which an extraneous root may be introduced. We'll return to this point after the example.

EXAMPLE 6   Solving and then checking for extraneous roots

Solve for x: $\log_3 x + \log_3(x + 2) = 1$

SOLUTION
Using Property 2 (on page 399), we can write the given equation as

$$\log_3[x(x + 2)] = 1 \qquad \text{or} \qquad \log_3(x^2 + 2x) = 1$$

Writing this last equation in exponential form yields

$$x^2 + 2x = 3^1$$
$$x^2 + 2x - 3 = 0$$
$$(x + 3)(x - 1) = 0$$

Thus, we have

$$x + 3 = 0 \qquad \text{or} \qquad x - 1 = 0$$

and, consequently,

$$x = -3 \qquad \text{or} \qquad x = 1$$

Now let's check these values in the original equation to see whether they are indeed solutions.

If $x = -3$, the equation becomes

$$\log_3(-3) + \log_3(-1) \overset{?}{=} 1$$

Neither expression is defined, since the domain of the logarithm function does not contain negative numbers.

If $x = 1$, the equation becomes

$$\log_3 1 + \log_3 3 \overset{?}{=} 1$$
$$0 + 1 \overset{?}{=} 1 \qquad \text{True}$$

Thus, the value $x = 1$ is a solution of the original equation, but $x = -3$ is not.

In the example we just concluded, an extraneous solution ($x = -3$) was generated along with the correct solution, $x = 1$. How did this happen? It happened because in the second line of the solution, we used the property $\log_b PQ = \log_b P + \log_b Q$, which is valid only when both P and Q are positive. In the box that follows, we generalize this remark about extraneous solutions.

❚ PROPERTY SUMMARY
Extraneous Solutions for Logarithmic Equations

Using the properties of logarithms (on page 399) to solve logarithmic equations may introduce extraneous solutions that do not check in the original equation. (This occurs because the logarithm function requires positive inputs, but in solving an equation, we may not know ahead of time the sign of an input involving a variable.) Therefore, it is always necessary to check any candidates for solutions that are obtained in this manner.

Earlier in this book, in Sections 2.5 and 2.6, we learned how to solve inequalities involving polynomials. Those same techniques, along with the properties of logarithms, are often useful in solving inequalities involving logarithmic and exponential expressions. In the box that follows, we list some additional facts that are helpful in solving these types of inequalities.

❚ PROPERTY SUMMARY   Inequalities Involving Exponential and Logarithmic Functions

For each of the following, assume that b is a positive constant greater than 1.

1. (Refer to Figure 4.)
  (a) If $p < q$ then $b^p < b^q$.
  (b) Conversely, if $b^p < b^q$ then $p < q$.

2. (Refer to Figure 5.) Assume that p and q are positive.
  (a) If $p < q$ then $\log_b p < \log_b q$.
  (b) Conversely, if $\log_b p < \log_b q$ then $p < q$.

Figure 4

Figure 5

EXAMPLE   7   An inequality involving an exponential function

Solve the inequality $2(1 + 0.4^x) < 5$.

SOLUTION
Dividing both sides by 2, we have

$$1 + 0.4^x < 5/2$$
$$0.4^x < 3/2$$
$$\ln(0.4^x) < \ln(3/2) \qquad \text{using Property 2(a) above}$$
$$x \ln(0.4) < \ln(3/2)$$

Now, to isolate x, we want to divide both sides of this last inequality by the quantity $\ln(0.4)$, which is negative. (Without relying on a calculator, how do you know that this quantity is negative?) Since dividing both sides by a negative quantity reverses the inequality, we obtain

$$x > \frac{\ln(3/2)}{\ln(0.4)}$$

So the solution set consists of all real numbers greater than $\dfrac{\ln(3/2)}{\ln(0.4)}$.

Using a calculator to evaluate this last expression, we find that it is approximately -0.443. With this approximation, we can use interval notation to write the solution set as $(-0.443, \infty)$. See Figure 6 for a graphical view of this solution.

GRAPHICAL PERSPECTIVE

Figure 6
The graph of the equation
$y = 2(1 + 0.4^x)$ indicates that
the solution set of the inequality
$2(1 + 0.4^x) < 5$ is an interval of
the form (a, ∞), where a is approximately -0.5. In Example 7
we found that this value of a is
$(\ln 1.5)/(\ln 0.4) \approx -0.443$ (rounded
to three decimal places).

EXAMPLE **8** **Inequalities with logarithmic and exponential functions**

Solve each of the following inequalities:
(a) $\ln(2 - 3x) \leq 1$; **(b)** $e^{2-3x} \leq 1$.

SOLUTION
(a) As a preliminary but necessary first step, we need to determine the domain of the function $y = \ln(2 - 3x)$. Since the inputs for the natural logarithm function must be positive, we require $2 - 3x > 0$, and therefore

$$-3x > -2 \quad \text{or} \quad x < 2/3$$

Now we turn to the given inequality $\ln(2 - 3x) \leq 1$. Applying Property 1(a) on the previous page to this inequality, we can write

$$e^{\ln(2-3x)} \leq e^1$$

and consequently

$$2 - 3x \leq e$$
$$-3x \leq e - 2$$
$$x \geq \frac{e - 2}{-3} = \frac{2 - e}{3}$$

Putting things together now, we want x to be greater than or equal to $(2 - e)/3$ but less than $2/3$. Thus, the solution set is the interval $\left[\dfrac{2 - e}{3}, \dfrac{2}{3}\right)$.

NOTE For this interval notation to make sense, the fraction $(2 - e)/3$ must be smaller than $2/3$. Although you can verify this using a calculator, you can also explain it without relying on a calculator. Exercise 89 asks you to do this.

(b) Since the domain of the exponential function is the set of all real numbers, there is no preliminary restriction on x, as there was in part (a). Taking the natural logarithm of both sides of the given inequality [as in Property 2(a) on page 415], we obtain

$$\ln(e^{2-3x}) \leq \ln(1)$$

and therefore

$$2 - 3x \leq 0$$
$$-3x \leq -2$$
$$x \geq 2/3$$

The solution set of the given inequality is the interval $[2/3, \infty)$.

| EXAMPLE | 9 | Taking domain into account in determining the solution set |

(a) Determine the domain of the function $f(x) = \log_{10} x + \log_{10}(x - 2)$.
(b) Solve the inequality $\log_{10} x + \log_{10}(x - 2) \leq \log_{10} 24$.

SOLUTION
(a) As you know, the logarithm function requires positive inputs. So for the expression $\log_{10} x$ we require $x > 0$, whereas for the expression $\log_{10}(x - 2)$ we require

$$x - 2 > 0 \qquad \text{and therefore} \qquad x > 2$$

In summary, then, we want x to be greater than zero and, at the same time, we need to have x greater than 2. Notice now that if x is greater than 2, it's automatically true that x is greater than zero. Thus, the domain of the given function f consists of all real numbers greater than 2.

(b) On the left-hand side of the given inequality, we can use one of the basic properties of logarithms to write

$$\log_{10} x + \log_{10}(x - 2) = \log_{10}[x(x - 2)] = \log_{10}(x^2 - 2x)$$

With this result, the given inequality becomes

$$\log_{10}(x^2 - 2x) \leq \log_{10} 24$$

and therefore

$$x^2 - 2x \leq 24 \qquad \text{using Property 2(b) on page 415}$$

or

$$x^2 - 2x - 24 \leq 0 \qquad\qquad (1)$$

Inequality (1) is one of the types of inequalities that you learned to solve earlier in this text (in Section 2.6). As Exercise 90 at the end of this section asks you to show, the solution set for inequality (1) is the closed interval

$[-4, 6]$. This closed interval, however, is not the solution set for the original inequality; we need to take into account the result in part (a). Putting things together, then, we want only those numbers in the interval $[-4, 6]$ that are greater than 2. Consequently, the solution set for the given inequality is the interval $(2, 6]$. Figure 7 shows a graphical view of this result.

Figure 7
The curve $y = \log_{10} x + \log_{10}(x - 2)$ is below the horizontal line $y = \log_{10} 24$ when $2 < x < 6$. At $x = 6$ the curve and the line intersect. Thus, the solution set of the inequality $\log_{10} x + \log_{10}(x - 2) \leq \log_{10} 24$ is the interval $(2, 6]$.

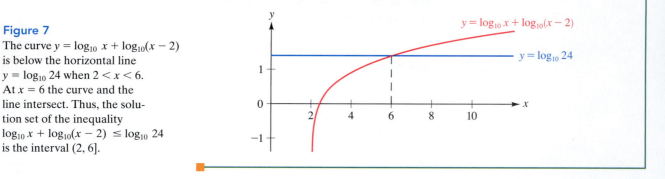

EXERCISE SET 5.5

A

To help you get started, Exercises 1–10 correlate directly with Examples 2–6 as shown in the chart. (Thus, if you need help in any of Exercises 1–10, first consult the indicated example.)

Exercise	1, 2	3, 4	5, 6	7, 8	9, 10
Based on Example No.	2	3(a)	4(a)	4(b)	5

In Exercises 1–41, find all the real-number roots of each equation. In each case, give an exact expression for the root and also (where appropriate) a calculator approximation rounded to three decimal places.

1. $5^x = 3^{2x-1}$
2. $7^{-4x} = 2^{1+3x}$
3. $\ln(\ln x) = 1.5$
4. $\log_3[\log_3(2x)] = -2$
5. $\log_{10}(x^2 + 36) = 2$
6. $\log_2(2x^2 - 4) = 5$
7. $\log_{10}(2x^2 - 3x) = 2$
8. $\log_9(x^2 + x) = 0.5$
9. $10^{2x} + 3(10^x) - 10 = 0$
10. $3(2^{2x}) - 11(2^x) - 4 = 0$
11. (a) $\ln(x^3) = 3 \ln x$
   (b) $(\ln x)^3 = 3 \ln x$
12. (a) $(\log_{10} x)^2 = 2 \log_{10} x$
   (b) $\log_{10}(x^2) = 2 \log_{10} x$
13. (a) $\log_3 6x = \log_3 6 + \log_3 x$
   (b) $\log_3 6x = 6 \log_3 x$
14. (a) $\ln x = (\log_{10} x)/(\log_{10} e)$
   (b) $\ln x = (\log_{10} e)/(\log_{10} x)$
15. $7^{\log_7 2x} = 2x$
16. $7^{\log_7 2x} = 7$
17. $\log_2(\log_3 x) = -1$
18. $\ln[\ln(\ln x)] = 1$
19. $\ln 4 - \ln x = (\ln 4)/(\ln x)$

20. $\log_5 x - \log_5 10 = \log_5(x/10)$
21. $\ln(3x^2) = 2 \ln(3x)$
22. $\ln(3x^2) = 2 \ln(\sqrt{3}x)$
23. $\log_{16} \dfrac{x+3}{x-1} = \dfrac{1}{2}$
24. $\log_{\sqrt{2}} \dfrac{1-4x}{1+4x} = 4$
25. (a) $e^{2x} + 2e^x + 1 = 0$
   (b) $e^{2x} - 2e^x + 1 = 0$
   (c) $e^{2x} - 2e^x - 3 = 0$
   (d) $e^{2x} - 2e^x - 4 = 0$
26. (a) $4e^{6x} - 12e^{3x} + 9 = 0$
   (b) $4e^{6x} + 12e^{3x} + 9 = 0$
   (c) $4e^{6x} - 16e^{3x} - 9 = 0$
   (d) $e^{6x} - 12e^{3x} - 9 = 0$
27. $e^x - e^{-x} = 1$ *Hint:* Multiply both side by e^x.
28. $e^x + e^{-x} = 2$
29. $2^{5x} = 3^x(5^{x+3})$
30. $e^{3x} = 10^{2x}(2^{1-x})$
31. $\log_6 x + \log_6(x + 1) = 1$
32. $\log_6 x + \log_6(x + 1) = 0$
33. $\log_9(x + 1) = \frac{1}{2} + \log_9 x$
34. $\log_2(x + 4) = 2 - \log_2(x + 1)$
35. $\log_{10}(2x + 4) + \log_{10}(x - 2) = 1$
36. $\ln x + \ln(x + 1) = \ln 12$
37. $\log_{10}(x + 3) - \log_{10}(x - 2) = 2$
38. $\ln(x + 1) = 2 + \ln(x - 1)$
39. $\log_{10}(x + 1) = 2 \log_{10}(x - 1)$
40. $\log_2(2x^2 + 4) = 5$
41. $\log_{10}(x - 6) + \log_{10}(x + 3) = 1$
42. Solve for x in terms of a:
   $\log_2(x + a) - \log_2(x - a) = 1$.
43. Solve for x in terms of y:
   (a) $\log_{10} x - y = \log_{10}(3x - 1)$;
   (b) $\log_{10}(x - y) = \log_{10}(3x - 1)$.
44. Solve for x in terms of b: $\log_b(1 - 3x) = 3 + \log_b x$.

In Exercises 45–50, you are given an equation of the form
$\ln x = f(x)$. *In each case, the equation cannot be solved using the algebraic techniques of this section.*

(a) *Without using a graphing utility, determine how many roots the equation has by sketching the graphs of $y = \ln x$ and $y = f(x)$ in the same coordinate system.*

Ⓖ **(b)** *Use a graphing utility to draw the graphs on the same screen. Then, by zooming in on each point where the graphs intersect, estimate to the nearest hundredth each root of the equation $\ln x = f(x)$.*

45. $\ln x = -x$
46. $\ln x = (x - 2)^2$
47. $\ln x = 1/(x + 1)$
48. $\ln x = e^{-x}$
49. $\ln x = |x - 2|$
50. $\ln x = -|x - 2|$

In Exercises 51–66, solve the inequalities.

51. $3(2 - 0.6^x) \le 1$
52. $6(5 - 1.6^x) \ge 13$
53. $4(10 - e^x) \le -3$
54. $\frac{2}{3}(1 - e^{-x}) \le -3$
55. $\ln(2 - 5x) > 2$
56. $3 \log_{10}(4x + 3) < 1$
57. $e^{2+x} \ge 100$
58. $4^{5-x} > 15$
59. $2^x > 0$
60. $\log_2 x \ge 0$
61. $\log_2 \dfrac{2x - 1}{x - 2} < 0$
62. $\ln \dfrac{3x - 2}{4x + 1} > \ln 4$
63. $e^{x^2 - 4x} \ge e^5$
64. $10^{-x^2} \le 10^{-12}$
65. $e^{(1/x)-1} > 1$
66. $e^{1/(x-1)} > 1$

67. (a) Specify the domain of the function $y = \ln x + \ln(x - 4)$.
  (b) Solve the inequality $\ln x + \ln(x - 4) \le \ln 21$.
68. (a) Specify the domain of the function $y = \ln x + \ln(x + 2)$.
  (b) Solve the inequality $\ln x + \ln(x + 2) \le \ln 35$.
69. Solve the inequality
  $\log_2 x + \log_2(x + 1) - \log_2(2x + 6) < 0$.
70. Solve the inequality $\log_{10}(x^2 - 6x - 6) > 0$.

B

71. Find all roots of the equation $\log_2 x = \log_x 2$, or explain why there are none.
72. Solve the equation $\log_2 x = \log_x 3$. For each root, give an exact expression and a calculator approximation rounded to two decimal places.
Ⓖ **73.** Use a graphing utility to graph the two functions $y = \ln(x^2)$ and $y = 2 \ln x$, first separately, then in the same viewing rectangle. What do you observe? What does this have to do with the statements made in the text in the solution of Example 1(b)?
Ⓖ **74. (a)** Graph the two functions $f(x) = (\ln x)/(\ln 3)$ and $g(x) = \ln x - \ln 3$. (Use a viewing rectangle in which x extends from 0 to 10 and y extends from -5 to 5.) Why aren't the two graphs identical? That is, doesn't one of the basic log identities say that $(\ln a)/(\ln b) = \ln a - \ln b$?
  (b) Your picture in part (a) indicates that
  $$\frac{\ln x}{\ln 3} > \ln x - \ln 3 \quad (0 < x \le 10).$$

Find a viewing rectangle in which
$(\ln x)/(\ln 3) \le \ln x - \ln 3$.
(c) Use the picture that you obtain in part (b) to estimate the value of x for which $(\ln x)/(\ln 3) = \ln x - \ln 3$.
(d) Solve the equation $(\ln x)/(\ln 3) = \ln x - \ln 3$ algebraically and use the result to check your estimate in part (c).

In Exercises 75–84, solve each equation. In Exercises 79–84, solve for x in terms of the other letters.

75. $3(\ln x)^2 - \ln(x^2) - 8 = 0$
76. $x^{1+\log_x 16} = 4x^2$
77. $\log_6 x = \dfrac{1}{\dfrac{1}{\log_2 x} + \dfrac{1}{\log_3 x}}$
78. $\dfrac{\ln(\sqrt{x + 4} + 2)}{\ln \sqrt{x}} = 2$
79. $\alpha \ln x + \ln \beta = 0$
80. $3 \ln x = \alpha + 3 \ln \beta$
81. $y = Ae^{kx}$
82. $\beta = 10 \log_{10}(x/x_0)$
83. $y = a/(1 + be^{-kx})$
84. $T = T_1 + (T_0 - T_1)e^{-kx}$

In Exercises 85–88, you are given an equation and a root that was obtained in an example in the text. In each case: (a) verify (algebraically) that the root indeed satisfies the equation; and (b) use a calculator to check that the root satisfies the equation.

85. *(From Example 2)* $4^x = 3^{2x+1}; x = (\ln 3)/(\ln 4 - 2 \ln 3)$
86. *[From Example 3(a)]* $\ln(\ln x) = 2; x = e^{e^2}$
87. (a) *[From Example 4(a)]* $\log_{10}(x^2 - 2) = 3; x = \pm\sqrt{1002}$
  (b) *[From Example 4(b)]* $\log_{10}(x^2 - 2x) = 3;$
    $x = 1 \pm \sqrt{1001}$
88. *(From Example 5)* $2(3^{2x}) - 3^x - 3 = 0; x = (\ln 1.5)/\ln 3$
89. *[From Example 8(a)]* Explain, in one or two complete sentences, how you know (without using a calculator) that the fraction $(2 - e)/3$ is less than $2/3$.
90. *[From Example 9(b)]* Solve the inequality $x^2 - 2x - 24 \le 0$. You should find that the solution set is the closed interval $[-4, 6]$.

C

In Exercises 91 and 92, solve the inequalities.

91. $\log_\pi[\log_4(x^2 - 5)] < 0$
92. $\dfrac{1}{\log_2 x} + \dfrac{1}{\log_3 x} + \dfrac{1}{\log_4 x} > 2 \quad (\text{for } x > 1)$

In Exercises 93 and 94, solve the equations.

93. $x^{(x^x)} = (x^x)^x$
94. $(\pi x)^{\log_{10} \pi} = (ex)^{\log_{10} e}$
95. Solve for x (assuming that $a > b > 0$):
  $$(a^4 - 2a^2b^2 + b^4)^{x-1} = (a - b)^{2x}(a + b)^{-2}$$
  Answer: $x = \dfrac{\ln(a - b)}{\ln(a + b)}$
96. Let $f(x) = \ln(x + \sqrt{x^2 + 1})$. Find $f^{-1}(x)$.
97. Suppose that $\log_{10} 2 = a$ and $\log_{10} 3 = b$. Solve for x in terms of a and b:
  $$6^x = \frac{10}{3} - 6^{-x}$$

$S = Pe^{rt}$. *This result is remarkable both because of its simplicity and the occurrence of e. (Who would expect that number to pop up in finance theory?)* —Philip Gillett in *Calculus and Analytic Geometry,* 3rd ed. (Lexington, Mass.: D.C. Heath, 1988)

5.6   COMPOUND INTEREST

We begin this section by considering how money accumulates in a savings account. Eventually, this will lead us back to the number e and to functions of the form $y = ae^{bx}$.

The following idea from arithmetic is a prerequisite for our discussion. To increase a given quantity by, say, 15%, we multiply the quantity by 1.15. For instance, suppose that we want to increase $100 by 15%. The calculations can be written

$$100 + 0.15(100) = 100(1 + 0.15) = 100(1.15)$$

Similarly, to increase a quantity by 30%, we would multiply by 1.30, and so on. The next example displays some calculations involving percentage increase. The results may surprise you unless you're already familiar with this topic.

EXAMPLE **1** **Computing percentage increase**

An amount of $100 is increased by 15% and then the new amount is increased by 15%. Is this the same as an overall increase of 30%?

SOLUTION

To increase $100 by 15%, we multiply by 1.15 to obtain $100(1.15). Now to increase this new amount by 15%, we multiply it by 1.15 to obtain

$$[(\$100)(1.15)](1.15) = \$100(1.15)^2 = \$132.25$$

Alternatively, if we increase the original $100 by 30%, we obtain

$$\$100(1.30) = \$130$$

Comparing, we see that the result of two successive 15% increases is greater than the result of a single 30% increase.

Now let's look at another example and use it to introduce some terminology. Suppose that you place $1000 in a savings account at 10% interest *compounded annually.* This means that at the end of each year, the bank contributes to your account 10% of the amount that is in the account at that time. Interest compounded in this manner is called **compound interest.** The original deposit of $1000 is called the **principal,** denoted P. The interest rate, expressed as a decimal, is denoted by r. Thus, $r = 0.10$ in this example. The variable A is used to denote the **amount** in the account at any given time. The calculations displayed in Table 1 show how the account grows.

We can learn several things from Table 1. First, consider how much interest is paid each year.

Interest paid for first year:	$1100 − $1000 = $100
Interest paid for second year:	$1210 − $1100 = $110
Interest paid for third year:	$1331 − $1210 = $121

Thus the interest earned each year is not a constant; it increases each year.

TABLE 1

Time period	Algebra	Arithmetic
After 1 year	$A = P(1 + r)$	$A = 1000(1.10)$ $= \$1100$
After 2 years	$A = [P(1 + r)](1 + r)$ $= P(1 + r)^2$	$A = 1100(1.10)$ $= \$1210$
After 3 years	$A = [P(1 + r)^2](1 + r)$ $= P(1 + r)^3$	$A = 1210(1.10)$ $= \$1331$

If you look at the algebra in Table 1, you can see what the general formula should be for the amount after t years.

Compound Interest Formula (Interest Compounded Annually)

Suppose that a principal of P dollars is invested at an annual rate r that is compounded annually. Then the amount A after t years is given by

$$A = P(1 + r)^t$$

EXAMPLE 2   Interest compounded annually

Suppose that \$2000 is invested at $7\frac{1}{2}\%$ interest compounded annually. How many years will it take for the money to double?

SOLUTION
In the formula $A = P(1 + r)^t$, we use the given values $P = \$2000$ and $r = 0.075$. We want to find how long it will take for the money to double; that is, we want to find t when $A = \$4000$. Making these substitutions in the formula, we obtain

$$4000 = 2000(1 + 0.075)^t$$

and, therefore,

$$2 = 1.075^t \qquad \text{dividing by 2000}$$

We can solve this exponential equation by taking the logarithm of both sides. We use base e logarithms. (Base 10 would also be convenient here.) This yields

$$\ln 2 = \ln 1.075^t$$

and, consequently,

$$\ln 2 = t \ln 1.075 \qquad \text{(Why?)}$$

To isolate t, we divide both sides of this last equation by $\ln 1.075$. This yields

$$t = \frac{\ln 2}{\ln 1.075} \approx 9.6 \, \text{years}$$

Now, assuming that the bank computes the compound interest only at the end of the year, we must round the preliminary answer of 9.6 years and say that when

TABLE 2
$A = 2000(1.075)^t$

t (years)	A (dollars)
9	3834.48
10	4122.06

$t = 10$ years, the initial \$2000 will have *more than* doubled. Table 2 adds some perspective to this. The table shows that after 9 years, something less than \$4000 is in the account; whereas after 10 years, the amount exceeds \$4000.

In Example 2, the interest was compounded annually. In practice, though, the interest is usually computed more often. For instance, a bank may advertise a rate of 8% per year *compounded semiannually*. This means that after half a year, the interest is compounded at 4%, and then after another half year, the interest is again compounded at 4%. (If you review Example 1, you'll see that two compoundings, each at a rate of $r/2$, is not the same as one compounding at the rate r. The former scheme yields more money overall than the latter.) In the case of a rate of 8% per year compounded semiannually, we say that the *periodic interest rate* is 4%. Similarly, if the interest rate of 8% per year were compounded four times per year, then the periodic interest rate would be 2 ($= 8/4$) percent. In Table 3, we show examples of this terminology assuming an annual rate of r%. Note that, in general, the **periodic interest rate** is equal to r/n, where r is the annual interest rate and n is the number of times per year that the interest is compounded.

TABLE 3
Periodic Interest Rates, Assuming an Annual Rate of R%.*

	n (number of times per year that interest is compounded)	Periodic interest rate (%)
Annually	1	r
Semiannually	2	$r/2$
Quarterly	4	$r/4$
Monthly	12	$r/12$
Daily (assuming 360-day year)	360	$r/360$
Daily (assuming 365-day year)	365	$r/365$

*Regarding the last two rows of the table: some banks base their calculations on a 360-day rather than 365-day year.

The compound interest formula

$$A = P(1 + r)^t \tag{1}$$

that we worked with in Example 2 can be generalized to cover the cases that we've just been discussing, where the interest is compounded more than once per year. The general formula is given in the box that follows.

Compound Interest Formula (Interest Compounded n Times per Year)

Suppose that a principal of P dollars is invested at an annual rate r that is compounded n times per year. Then the amount A after t years is given by

$$A = P\left(1 + \frac{r}{n}\right)^{nt} \tag{2}$$

Note that the quantity r/n in parentheses is the periodic interest rate, and the exponent nt represents the total number of compoundings.

Although we won't derive equation (2) from scratch, note that it is obtained from equation (1) as follows: In equation (1), replace the annual interest rate r by the periodic interest rate r/n; in equation (1), replace t (which not only represents time, but also the number of compoundings) with the quantity nt, which also represents the total number of compoundings.

EXAMPLE 3   Comparing annual to quarterly compounding

Suppose that $1000 is placed in a savings account at 10% per annum. How much is in the account at the end of 1 year if the interest is **(a)** compounded once each year ($n = 1$)? **(b)** compounded quarterly ($n = 4$)?

SOLUTION
We use the formula $A = P[1 + (r/n)]^{nt}$.
(a) For $n = 1$, we obtain   **(b)** For $n = 4$, we obtain

$$A = 1000\left(1 + \frac{0.10}{1}\right)^{(1)(1)} \qquad A = 1000\left(1 + \frac{0.10}{4}\right)^{4(1)}$$

$$= 1000(1.1) \qquad\qquad\quad = 1000(1.025)^4$$

$$= \$1100 \qquad\qquad\qquad\; = \$1103.81 \qquad \text{using a calculator}$$

Notice that compounding the interest quarterly rather than annually yields the greater amount. This is in agreement with our observations in Example 1.

The results in Example 3 will serve to illustrate some additional terminology used by financial institutions. In that example, the interest for the year under quarterly compounding was

$$\$1103.81 - \$1000 = \$103.81$$

Now, $103.81 is 10.381% of $1000. We say in this case that the **effective rate** of interest is 10.381%. The given rate of 10% per annum compounded once a year is called the **nominal rate.** (The nominal rate and the effective rate are also referred to as the *annual rate* and the *annual yield,* respectively.) The next example further illustrates these ideas.

EXAMPLE 4   Computing the effective rate

A bank offers a nominal interest rate of 6% per annum for certain accounts. Compute the effective rate if interest is compounded monthly.

SOLUTION
Let P denote the principal, which earns 6% ($r = 0.06$) compounded monthly ($n = 12$). Then with $t = 1$, our formula yields

$$A = P\left(1 + \frac{0.06}{12}\right)^{12(1)} \approx P(1.0617) \qquad \text{using a calculator}$$

This shows that the effective interest rate is about 6.17%.

> **TABLE 4**
> **Results of Compounding Interest More and More Frequently**
>
> | Number of compoundings, n | Amount, $\left(1 + \dfrac{1}{n}\right)^n$ |
> | --- | --- |
> | $n = 1$ (annually) | $[1 + (1/1)]^1 = 2$ |
> | $n = 2$ (semiannually) | $[1 + (1/2)]^2 = 2.25$ |
> | $n = 4$ (quarterly) | $[1 + (1/4)]^4 \approx 2.44$ |
> | $n = 12$ (monthly) | $[1 + (1/12)]^{12} \approx 2.61$ |
> | $n = 365$ (daily) | $[1 + (1/365)]^{365} \approx 2.7146$ |
> | $n = 8760$ (hourly) | $[1 + (1/8760)]^{8760} \approx 2.7181$ |
> | $n = 525{,}600$ (each minute) | $[1 + (1/525{,}600)]^{525{,}600} \approx 2.71827$ |
> | $n = 31{,}536{,}000$ (each second) | $[1 + (1/31{,}536{,}000)]^{31{,}536{,}000} \approx 2.71828$ |

There are two rather natural questions to ask on first encountering compound interest calculations:

Question 1 For a fixed period of time (say, one year), does more and more frequent compounding of interest continue to yield greater and greater amounts?

Question 2 Is there a limit on how much money can accumulate in a year when interest is compounded more and more frequently?

The answer to both of these questions is yes. If you look back over Example 1, you'll see evidence for the affirmative answer to Question 1. For additional evidence, and for the answer to Question 2, let's do some calculations. To keep things as simple as possible, suppose a principal of $1 is invested for 1 year at the nominal rate of 100% per annum. (More realistic figures could be used here, but the algebra becomes more cluttered.) With these data, our formula becomes

$$A = 1\left(1 + \frac{1}{n}\right)^{n(1)} \qquad \text{or} \qquad A = \left(1 + \frac{1}{n}\right)^n$$

Now, as we know from Section 5.2, the expression $\left(1 + \dfrac{1}{n}\right)^n$ approaches the value e as n grows larger and larger; see Table 4.

Table 4 shows that the amount does increase with the number of compoundings. But assuming that the bank rounds to the nearest penny, Table 4 also shows that there is no difference between compounding hourly, compounding each minute, and compounding each second. In each case, the rounded amount is $2.72.

Some banks advertise interest compounded not monthly, daily, or even hourly, but *continuously*—that is, at each instant. The formula (derived in calculus) for the amount earned under continuous compounding of interest is as follows.

Compound Interest Formula (Interest Compounded Continuously)

Suppose that a principal of P dollars is invested at an annual rate r that is compounded continuously. Then the amount A after t years is given by

$$A = Pe^{rt}$$

EXAMPLE 5 | Continuous compounding of interest

A principal of $1600 is placed in a savings account at 5% per annum compounded continuously. Assuming no subsequent withdrawals or deposits, when will the balance reach $2400?

SOLUTION
Substitute the values $A = 2400$, $P = 1600$, and $r = 0.05$ in the formula $A = Pe^{rt}$ to obtain $2400 = 1600e^{0.05t}$. To isolate the time t in this last equation, first divide both sides by 1600. This yields

$$1.5 = e^{0.05t}$$

and therefore

$$0.05t = \ln 1.5 \qquad \text{converting from exponential to logarithmic form}$$
$$t = (\ln 1.5)/0.05$$

Using a calculator now to evaluate this last expression, we find $t \approx 8.1$ years. In summary, it will take slightly longer that 8 years 1 month for the balance to reach $2400. *Suggestion*: You can use a graphing utility to check this result. Graph the function $A = 1600e^{0.05t}$ along with the horizontal line $A = 2400$ in an appropriate viewing rectangle. Is it then the first or second coordinate of the intersection point that you want to estimate?

In the next example we compare the results of continuous compounding and annual compounding of interest.

EXAMPLE 6 | Comparing continuous to annual compounding

A principal of $100 is deposited in Account #1 at 6% per annum, compounded continuously. At the same time, another principal of $100 is deposited in Account #2 at 6% per annum, compounded only once a year. Use a graphing utility to determine how long it will take until the amount in Account #1 exceeds that in Account #2 by (at least) ten dollars.

GRAPHICAL PERSPECTIVE

Figure 1
$f(t) = 100e^{0.06t} - 100(1.06^t)$
$[0, 25, 5]$ by $[0, 20, 5]$

SOLUTION
The amount in Account #1 after t years is given by $A = 100e^{0.06t}$; for Account #2 the amount is $A = 100(1 + 0.06)^t = 100(1.06)^t$. Define $f(t)$ to be the *difference* between the amounts in each account at time t:

$$f(t) = 100e^{0.06t} - 100(1.06)^t$$

We want to know when the difference $f(t)$ exceeds ten dollars. Using a graphing utility as in Figure 1, we find that when $t = 15$, the difference is less than 10, whereas when $t = 20$, the difference is more than 10.

At this point, we can either zoom in on the portion of the graph corresponding to $15 \le t \le 20$, or we can use the TABLE feature of the graphing utility. For this example, we'll use the TABLE feature to evaluate the amount in each account and the difference $f(t)$ corresponding to the years $t = 15$ to $t = 20$. The results

TABLE 5

t	Account #1 $A = 100e^{0.06t}$	Account #2 $A = 100(1.06)^t$	Difference $100e^{0.06t} - 100(1.06^t)$
15	245.96	239.66	6.30
16	261.17	254.04	7.13
17	277.32	269.28	8.04
18	294.47	285.43	9.03
19	312.68	302.56	10.12
20	332.01	320.71	11.30

are shown in Table 5. You should verify these results for yourself. (For help on the TABLE feature, consult the owner's manual for your graphing utility.)

As Table 5 indicates, after 18 years, the accounts differ by less than ten dollars, but after 19 years the difference is more than ten dollars. In summary, it takes 19 years (to the nearest year) before the accounts differ by at least ten dollars. *Remark*: An algebraic solution for this example would involve solving the inequality $100e^{0.06t} - 100(1.06^t) \geq 10$. This can't be done, however, using any of the algebraic techniques of this chapter. (Try it!)

In the next example, we compare the nominal rate with the effective rate under continuous compounding of interest.

EXAMPLE 7 **Nominal and effective rates under continuous compounding**

(a) Given a nominal rate of 8% per annum compounded continuously, compute the effective interest rate.
(b) Given an effective rate of 8% per annum, compute the nominal rate.

SOLUTION
(a) With the values $r = 0.08$ and $t = 1$, the formula $A = Pe^{rt}$ yields

$$A = Pe^{0.08(1)}$$
$$A \approx P(1.08329) \qquad \text{using a calculator}$$

This shows that the effective interest rate is approximately 8.33% per year.
(b) We now wish to compute the nominal rate r, given an effective rate of 8% per year. An effective rate of 8% means that the initial principal P grows to $P(1.08)$ by the end of the year. Thus, in the formula $A = Pe^{rt}$, we make the substitutions $A = P(1.08)$ and $t = 1$. This yields

$$P(1.08) = Pe^{r(1)}$$

Dividing both sides of this last equation by P, we have

$$1.08 = e^r$$

TABLE 6

Comparison of Nominal and Effective Rates in Example 7

Nominal rate (% per annum)	Effective rate (% per per annum)
8	8.33
7.70	8

To solve this equation for r, we rewrite it in its equivalent logarithmic form:

$$r = \ln(1.08)$$
$$r \approx 0.07696 \qquad \text{using a calculator}$$

Thus, a nominal rate of about 7.70% per annum yields an effective rate of 8%. Table 6 summarizes these results.

Now we come to one of the remarkable and characteristic features of compound interest and of growth governed by the formula $A = Pe^{rt}$. By the **doubling time** we mean, as the name implies, the amount of time required for a given principal to double. The surprising fact here is that the doubling time does not depend on the principal P. To see why this is so, we begin with the formula

$$A = Pe^{rt}$$

We are interested in the time t at which $A = 2P$. Replacing A by $2P$ in the formula yields

$$2P = Pe^{rt}$$
$$2 = e^{rt}$$
$$rt = \ln 2$$
$$t = \frac{\ln 2}{r}$$

Denoting the doubling time by T_2, we have the following formula.

Doubling Time under Continuous Compounding

$$\text{Doubling time} = T_2 = \frac{\ln 2}{r}$$

As you can see, the formula for the doubling time T_2 does not involve P, but only r. Thus, at a given rate under continuous compounding, \$2 and \$2000 would both take the same amount of time to double. (This idea takes some getting used to.)

EXAMPLE 8   Computing the doubling time

Compute the doubling time T_2 when a sum is invested at an interest rate of 4% per annum compounded continuously.

SOLUTION

$$T_2 = \frac{\ln 2}{r} = \frac{\ln 2}{0.04} \approx 17.3 \text{ years} \qquad \text{using a calculator}$$

There is a convenient approximation that allows us to estimate doubling times easily. As you can check with a calculator, we have $\ln 2 \approx 0.7$. Using this approximation in the doubling time formula, we have

$$T_2 \approx \frac{0.7}{r}$$

where r (as usual) is the annual interest rate expressed as a decimal. Now let R be the annual interest rate expressed as a percentage, so that $R = 100r$. Then we can write

$$T_2 \approx \frac{0.7}{r} = \frac{0.7}{r} \times \frac{100}{100}$$

$$= \frac{70}{100r} = \frac{70}{R}$$

That is,

$$T_2 \approx \frac{70}{R} \quad \text{where } R \text{ is the annual rate expressed as a percentage.}$$

This is the so-called *rule of 70* for estimating doubling times. For instance, for an annual percentage rate of 10%, we estimate the doubling time as $T_2 \approx 70/10 = 7$ years. As another example, let's use this rule of 70 to rework Example 8, in which the annual interest rate is 4%. We have

$$T_2 \approx \frac{70}{4} = \frac{35}{2} = 17.5 \text{ years}$$

This is quite close to the actual doubling time of 17.3 years computed in Example 8. *Remark*: A slightly less accurate approximation that's often used in business is the *rule of 72*, $T_2 \approx 72/R$. The number 72 is used instead of 70 simply because there are more numbers that divide into 72 evenly than into 70.

   The rule of 70 for estimating doubling times is useful when you want to sketch a graph of the function $A = Pe^{rt}$ without relying on a graphing utility. As an example, suppose that a principal of $1000 is invested at 10% per annum compounded continuously. Then the function that we wish to graph is

$$A = 1000e^{0.1t}$$

Approximating the doubling time using the rule of 70 gives us

$$T_2 \approx \frac{70}{R} = \frac{70}{10} = 7 \text{ years}$$

Now we just set up a table showing the results of doubling $1000 every 7 years, plot the points, and then join them with a smooth curve. See Figure 2.

t	A
0	1000
7	2000
14	4000
21	8000
28	16,000

Figure 2
Applying the rule of 70 to sketch a graph

EXERCISE SET 5.6

A

1. You invest $800 at 6% interest compounded annually. How much is in the account after 4 years, assuming that you make no subsequent withdrawal or deposit?

2. A sum of $1000 is invested at an interest rate of $5\frac{1}{2}$% compounded annually. How many years will it take until the sum exceeds $2500? (First find out when the amount equals $2500; then round off as in Example 2.)

3. At what interest rate (compounded annually) will a sum of $4000 grow to $6000 in 5 years?

4. A bank pays 7% interest compounded annually. What principal will grow to $10,000 in 10 years?

5. You place $500 in a savings account at 5% compounded annually. After 4 years you withdraw all your money and take it to a different bank, which advertises a rate of 6% compounded annually. What is the balance in this new account after 4 more years? (As usual, assume that no subsequent withdrawal or deposit is made.)

6. A sum of $3000 is placed in a savings account at 6% per annum. How much is in the account after 1 year if the interest is compounded **(a)** annually? **(b)** semiannually? **(c)** daily?

7. A sum of $1000 is placed in a savings account at 7% per annum. How much is in the account after 20 years if the interest is compounded **(a)** annually? **(b)** quarterly?

8. Your friend invests $2000 at $5\frac{1}{4}$% per annum compounded semiannually. You invest an equal amount at the same yearly rate, but compounded daily. How much larger is your account than your friend's after 8 years?

9. You invest $100 at 6% per annum compounded quarterly. How long will it take for your balance to exceed $120? (Round your answer up to the next quarter.)

10. A bank offers an interest rate of 7% per annum compounded daily. What is the effective rate?

11. What principal should you deposit at $5\frac{1}{2}$% per annum compounded semiannually so as to have $6000 after 10 years?

12. You place a sum of $800 in a savings account at 6% per annum compounded continuously. Assuming that you make no subsequent withdrawal or deposit, how much is in the account after 1 year? When will the balance reach $1000?

13. A principal of $600 dollars is placed in a savings account at 6.5% per annum compounded continuously.

    (G) **(a)** Use a graphing utility to estimate how long it will take for the balance to reach $800. (Adapt the suggestion at the end of Example 5.)

    (b) Use algebra (as in Example 5) to determine how long it will take for the balance to reach $800.

14. A principal of $4000 dollars is invested at 8% per annum compounded continuously.

    (G) **(a)** Use a graphing utility to estimate how long it will take for the balance to increase by 25%. (That is, you want

a balance of $4000 + $1000 = $5000.) (Adapt the suggestion at the end of Example 5.)

    (b) Use algebra, rather than a graphing utility, to solve the problem in part (a).

(G) 15. A principal of $3500 is deposited in Account #1 at 7% per annum, compounded continuously. At the same time, another principal of $3500 is deposited in Account #2 at 7% per annum, compounded semiannually. Use a graphing utility to determine how long it will take until the amount in Account #1 exceeds that in Account #2 by (at least) $200. (Follow the method of Example 6).

(G) 16. A principal of $1000 is deposited in Account #1 at 8% per annum, compounded continuously. At the same time, a principal of $1200 is deposited in Account #2 at 6% per annum, compounded once per year. Use a graphing utility to determine how long it will take until the amount in Account #1 exceeds that in Account #2. Give your answer to the nearest whole number of years, rounded upward.

17. A bank offers an interest rate of $6\frac{1}{2}$% per annum compounded continuously. What principal will grow to $5000 in 10 years under these conditions?

18. Given a nominal rate of 6% per annum, compute the effective rate under continuous compounding of interest.

19. Suppose that under continuous compounding of interest, the effective rate is 6% per annum. Compute the nominal rate.

20. You have two savings accounts, each with an initial principal of $1000. The nominal rate on both accounts is $5\frac{1}{4}$% per annum. In the first account, interest is compounded semiannually. In the second account, interest is compounded continuously. How much more is in the second account after 12 years?

21. You want to invest $10,000 for 5 years, and you have a choice between two accounts. The first pays 6% per annum compounded annually. The second pays 5% per annum compounded continuously. Which is the better investment?

22. Suppose that a certain principal is invested at 6% per annum compounded continuously.

    (a) Use the rule of 70, $T_2 \approx 70/R$, to estimate the doubling time.

    (b) Compute the doubling time using the formula $T_2 = (\ln 2)/r$.

    (c) Do your answers in (a) and (b) differ by more than 2 months?

23. A sum of $1500 is invested at 5% per annum compounded continuously.

    (a) Estimate the doubling time.

    (b) Compute the actual doubling time.

    (c) Let d_1 and d_2 denote the actual and estimated doubling times, respectively. Define d by $d = |d_1 - d_2|$. What percentage is d of the actual doubling time?

24. A sum of $5000 is invested at 10% per annum compounded continuously.
   (a) Estimate the doubling time.
   (b) Estimate the time required for the $5000 to grow to $40,000.

25. After carrying out the calculations in this problem, you'll see one of the reasons why some governments impose inheritance taxes and why laws are passed to prohibit savings accounts from being passed from generation to generation without restriction. Suppose that a family invests $1000 at 8% per annum compounded continuously. If this account were to remain intact, being passed from generation to generation, for 300 years, how much would be in the account at the end of those 300 years?

26. A principal of $500 is invested at 7% per annum compounded continuously.
   (a) Estimate the doubling time.
   (b) Sketch a graph similar to the one in Figure 2, showing how the amount increases with time.

27. A principal of $7000 is invested at 5% per annum compounded continuously.
   (a) Estimate the doubling time.
   (b) Sketch a graph showing how the amount increases with time.

28. In one savings account, a principal of $1000 is deposited at 5% per annum. In a second account, a principal of $500 is deposited at 10% per annum. Both accounts compound interest continuously.
   (a) Estimate the doubling time for each account.
   (b) On the same set of axes, sketch graphs showing the amount of money in each account over time. Give the (approximate) coordinates of the point where the two curves meet. In financial terms, what is the significance of this point? (In working this problem, assume that the initial deposits in each account were made at the same time.)
   (c) During what period of time does the first account have the larger balance?

Ⓖ *For Exercises 29–32, refer to the following tables for the **public debt** of the United States over the first five years of the 1980s and of the 1990s. (The public debt consists of the federal debt plus the debts owed by state and local governments.) The source for the data in the tables is the U.S. Bureau of the Public Debt.*

TABLE A   Public Debt 1980–1984

Year	1980	1981	1982	1983	1984
Public debt (trillion $)	0.930	1.029	1.197	1.411	1.663

TABLE B   Public Debt 1990–1994

Year	1990	1991	1992	1993	1994
Public debt (trillion $)	3.233	3.665	4.065	4.411	4.693

29. (a) Use a graphing utility to create a scatter plot for the data in Table A, letting $x = 0$ correspond to the year 1980.
   (b) Find the linear model that best fits the data. Add the graph of the linear model to the scatter plot.
   (c) Follow part (b) using an exponential model of the form $y = ab^x$ or $y = Ae^{Bx}$. (Most graphing calculators use the first form for exponential regressions. Some other types of graphing utilities, such as Microsoft's *Excel,* use the second form. Report your answer in the form that your graphing utility provides; you needn't give both forms.)

30. Use the models that you obtained in Exercise 29, along with the data in the following table, to provide the calculations that support each of the given statements.

Year	1985	1986	1989
Public debt (trillion $)	1.946	2.125	2.857

   (a) For 1985, the projections obtained from the linear model and from the exponential model both underestimate the actual public debt. However, the percentage error for the exponential model is less than 3%. The linear model is not as good in this case, with a percentage error more than three times larger than that for the exponential model.
   (b) For 1986, the linear model again underestimates the actual public debt, and the percentage error is more than 6%. The exponential model, on the other hand, now overestimates the public debt; the percentage error though is less than 4%.
   (c) For the projections for the end of the decade—that is, 1989—the linear model has grown much too slowly, the exponential much too rapidly. If an economist had used these models in mid-decade to make projections for 1989, it's possible that people in 1989 would claim both projections to have been misleading. The percentage error in the linear projection is roughly 10% and that for the exponential projection about twice as much.

31. (a) Use a graphing utility to create a scatter plot for the data in Table B, letting $x = 0$ correspond to the year 1980 (not 1990), so that $x = 10$ corresponds to 1990. [You'll see a reason for this at the end of part (c).]
   (b) Find the linear model that best fits the data. Add the graph of the linear model to the scatter plot.
   (c) Follow part (b) using a logarithmic model $y = a \ln x + b$. Note that $x = 0$ is not in the domain of this function. That's why we didn't let $x = 0$ correspond to the year 1990.

32. Use the models that you obtained in Exercise 31, along with the data in the following table, to provide the calculations that support each of the following statements.

Year	1995	1996	1997	2000
Public debt (trillion $)	4.974	5.225	5.413	5.674

(a) For 1995, the projections obtained from the linear model and from the logarithmic model both overestimate the actual public debt. This remains true for the years 1996, 1997, and 2000. In each of the four years, the logarithmic projection is closer to the actual value than the linear projection.

(b) In each of the three years 1995, 1996, and 1997, the percentage error in the logarithmic projection is less than 3%. For the linear projections, however, only the 1995 projection has a percentage error that is under 3%.

(c) Neither model appears to do a very good job in projecting the public debt for the year 2000: The percentage error in the logarithmic projection is roughly 10%, and for the linear projection it's more than twice that.

(G) 33. Economists use the **producer price index** to measure the average change in the price of wholesale goods. The following table lists the producer price index for certain years. In Figures A through D, we show scatter plots for the data along with regression functions that model the data. In each case, the years are on the horizontal axis, with $x = 0$ corresponding to 1970. In this exercise, you'll use a graphing utility to obtain the specific equation for each model and then compare projections.

TABLE A   Producer Price Index for Selected Years

Year x ($x = 0 \leftrightarrow 1970$)	1	5	10	15	20
Producer price index	38.1	58.4	89.8	103.2	116.3

Source: U.S. Bureau of Labor Statistics

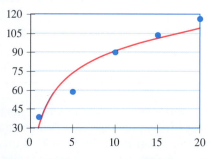

Figure A   Logarithmic model $y = a \ln x + b$

Figure B   Quadratic model $y = ax^2 + bx + c$

Figure C Power model $y = ax^b$

Figure D Linear model $y = ax + b$

(a) After looking at Figures A through D, say which model appears to fit the data points best. (That is, in your view, which curve comes closest to passing through all five of the given data points?)

(b) Using a graphing utility, create your own scatter plot of the data in Table A. Then, using the regression or trend line options on the graphing utility, find the specific equations for the models in Figures A through D.

(c) Use the functions that you determined in part (b) to complete Table B showing what each model projects for the producer price index in the indicated year. (Round each projection to one decimal place.) Then, in each row, circle the projection that is closest to the actual value given at the end of that row. Finally, compute the percentage error for each projection that you circled.

TABLE B   Producer Price Index (PPI) Projections Using Various Models

	PPI linear model	PPI logarithmic model	PPI quadratic model	PPI power model	PPI actual
1992					117.2
1995					124.8
2000					132.7

PROJECT

LOAN PAYMENTS

Amortize: Gradual paying off of a debt by making regular equal payments to cover interest and principal.
http://www.moneywords.com

Amortize: (obsolete) To deaden, render as if dead, destroy. —Oxford English Dictionary

The goode werkes that men don whil thay ben in good lif ben al amortised by synne folwyng.
—Chaucer (c 1386), as quoted in the Oxford English Dictionary

(a) In this first part of the project, we'll imagine a loan scenario and explain how the calculations work from month to month. Suppose that when you graduate college you are shopping for a new car. You decide that you'll need a loan of $15,000 toward the purchase of the particular model that you are interested in. The car dealer tells you that he can arrange this loan for you. It's a four-year loan at 6% per annum, compounded monthly. The dealer says that the monthly payments will each be $283.40.

In the following table, we will explain how the entries in the row for monthly payment #1 are obtained. Then, as a check on your understanding of the process, you should fill in the next two rows yourself. (Tables of this form are called *amortization schedules*.)

Monthly payment #	Monthly payment	Portion of payment toward interest	Portion of payment toward principal	Loan balance
0				$15,000.00
1	$283.40	$75.00	$208.40	$14,791.60
2	$283.40			
3	$283.40			

Portion of monthly payment #1 toward interest   Each month, a portion of your $283.40 payment is used to pay off the previous month's interest on the unpaid loan balance. Then, whatever is left from the $283.40 is used as a payment on the actual loan balance. We are assuming a rate of 6% per annum compounded monthly. So, the interest on an unpaid balance of P dollars for one month is

$$\text{interest} = P(1 + r/n)^{nt} - P$$
$$= P[(1 + 0.06/12)^{12(1/12)} - 1] = P[(1 + 0.06/12) - 1]$$
$$= 0.005P$$

For payment #1 then, we have

$$\text{interest} = 0.005P = 0.005(\$15,000) = \$75.00$$

Portion of monthly payment #1 toward principal:
$283.40 − $75.00 = $208.40

Loan balance: $15,000 − $208.40 = $14,791.60

The calculations corresponding to monthly payments #2 and #3 are now carried out in the same manner. In starting your work for monthly payment #2, remember that your unpaid balance now is $P = \$14{,}791.60$, rather than $P = \$15{,}000$. As a check on your work, after monthly payment #2 the loan balance should turn out to be $14,582.16.

(b) There are formulas (that we won't derive) from the mathematics of finance relating the loan amount, the monthly payments, and the interest rate. One such formula, which applies to our example (with monthly compounding and monthly payments) is

$$L = M\frac{(1 + p)^n - 1}{p} \tag{1}$$

where L is the loan amount, p is the periodic interest rate, and n is the total number of compoundings in the life of the loan (or, equivalently in our case, the length of the loan, in months). Use equation (1) to calculate what the monthly payments M should be for the example in part (a). Did the car dealer give you the correct figure?

(c) Solve equation (1) for n in terms of the other variables. Then use the result to compute an answer for the following problem: You are thinking of taking out a bank loan of $315,848 toward the purchase of a beachfront condominium. In your notes, you've written down that the interest rate is $7\frac{1}{2}\%$ per annum, compounded monthly, and that the monthly payments would be $953.90. You forgot, however, to write down the amount of time required to pay off the loan. Compute the required time.

Ⓖ **(d)** You graduate and the government reminds you that the time has come to begin repaying your student loan of $L = \$5800$. You have 4 years to repay this loan; the interest is compounded monthly; and the monthly payments are $M = \$113.88$. Find the periodic interest rate p and then the annual rate $r = 12p$. *Hint*: Substitute the given numerical values for L and M into equation (1) and then convince yourself that you cannot solve for p using the algebraic techniques you learned in this chapter. Now use a graphing utility and the ZOOM feature to find p. It may take some work to find an appropriate viewing rectangle.

5.7 EXPONENTIAL GROWTH AND DECAY

In newspapers and in everyday speech, the term *exponential growth* is used rather loosely to describe any situation involving rapid growth. In the sciences, however, **exponential growth** refers specifically to growth governed by functions of the form $y = ae^{bx}$, where a and b are positive constants. For example, since the function $A = Pe^{rt}$ (discussed in the previous section) has this general form, we say that money

Drawing by Professor Ann Jones,
University of Colorado, Boulder.
From the cover of *The Physics
Teacher,* vol. 14, no. 7 (October
1976).

grows exponentially under continuous compounding of interest. Similarly, in the sciences, **exponential decay** refers specifically to decrease or decay governed by functions of the form $y = ae^{bx}$, where a is positive and b is negative. As examples in this section, we shall consider population growth, global warming, and radioactive decay.

Under ideal conditions involving unlimited food and space, the size of a population of bacteria is modeled by a function of the form

$$\mathcal{N}(t) = \mathcal{N}_0 e^{kt}$$

In this *growth law,* $\mathcal{N}(t)$ is the population at time t, and k is a positive constant related to (but not equal to) the growth rate of the population. The constant k is referred to as the **growth constant.** The number $\mathcal{N}_0$ is also a constant; it represents the size of the population at time $t = 0$. You can see that this is the case by substituting $t = 0$ in the equation $\mathcal{N}(t) = \mathcal{N}_0 e^{kt}$. This yields

$$\mathcal{N}(0) = \mathcal{N}_0 e^0 = \mathcal{N}_0(1) = \mathcal{N}_0$$

That is, $\mathcal{N}(0) = \mathcal{N}_0$, which says that the population at time $t = 0$ is indeed $\mathcal{N}_0$. Figure 1 shows three examples. In each case, the initial population size is $\mathcal{N}_0 = 200$, and we've graphed the growth laws using three different values for the growth constant k.

A remark about notation: In science texts, the growth law is often written as

$$\mathcal{N} = \mathcal{N}_0 e^{kt} \quad \text{rather than} \quad \mathcal{N}(t) = \mathcal{N}_0 e^{kt}$$

In both cases, the quantity on the left side of the equation represents the population at time t. For ease in writing, we follow this convention too, in Examples 5 and 7. (For the record: The difference between the use of $\mathcal{N}$ in the two equations is this. On the left side of the equation $\mathcal{N}(t) = \mathcal{N}_0 e^{kt}$, $\mathcal{N}$ is the name of the function. On the left side of the equation $\mathcal{N} = \mathcal{N}_0 e^{kt}$, $\mathcal{N}$ serves as a dependent variable, just like the "y" in $y = x^2$.)

Figure 1
The graph of $\mathcal{N}(t) = 200e^{kt}$ for $k = 0.25$, $k = 0.5$, and $k = 1.5$

EXAMPLE | **1** | **Modeling bacterial growth**

Suppose that at the start of an experiment in a biology lab, 1200 bacteria are present in a colony. Two hours later, the size of the population is found to be 1440. Assume that the population size grows exponentially.

(a) Determine the growth constant k and the growth law for this population.

(b) How many bacteria were there 1.5 hours after the experiment began? Round the answer to the nearest 5 (i.e., nearest multiple of 5).

(c) When will the population size reach 4800?

SOLUTION

(a) The initial population size N_0 is 1200, so the equation $N(t) = N_0 e^{kt}$ becomes

$$N(t) = 1200 e^{kt}$$

We're also told that after 2 hours, the size of the population is 1440. Using this information in equation (1) gives us

$$N(2) = 1200 e^{k(2)}$$
$$1440 = 1200 e^{2k}$$

and therefore

$$e^{2k} = 1440/1200$$

To isolate k, we rewrite this last equation in logarithmic form (or take the natural logarithm of both sides) to obtain $2k = \ln(1440/1200)$, and consequently

$$k = \frac{\ln(1440/1200)}{2} \tag{2}$$
$$\approx 0.09116 \qquad \text{\color{red}{using a calculator and rounding to five decimal places}}$$

The growth law for the population is therefore

$$N(t) = 1200 e^{kt} \quad \text{where } k \text{ is given by equation (2)} \tag{3}$$

Notice that in our statement of the growth law we referenced the exact expression for k rather than the calculator approximation. In parts (b) and (c) of this problem, we don't round during the intermediate steps. This is standard procedure in scientific calculations because rounding errors can, in general, build up and adversely effect the accuracy of the final result.

(b) Substituting $t = 1.5$ in the growth law [that's equation (3)] yields

$$N(1.5) = 1200 e^{k(1.5)} \quad \text{where } k = \frac{\ln(1440/1200)}{2}$$

With a calculator now, you should verify that $N(1.5) \approx 1375.8$. In our context, $N(1.5)$ must be a whole number because it represents the number of bacteria. Furthermore, the problem asks us to round to the nearest 5. In summary then, after 1.5 hours the population is 1375.

(c) We want to find out when the population reaches 4800. Replacing $N(t)$ with 4800 in equation (3), we have $4800 = 1200 e^{kt}$, and therefore $4 = e^{kt}$. Rewriting this last equation in exponential form gives us

$$kt = \ln 4$$
$$t = (\ln 4)/k \approx 15.207 \qquad \text{\color{red}{using the expression for } k}$$
$$\text{\color{red}{in equation (2) and a calculator}}$$

Thus, it takes approximately 15.2 hours for the size of the population to reach 4800. As you can check with arithmetic, this is 15 hours 12 minutes. See Figure 2 for a graphical solution to this part of the example.

GRAPHICAL PERSPECTIVE

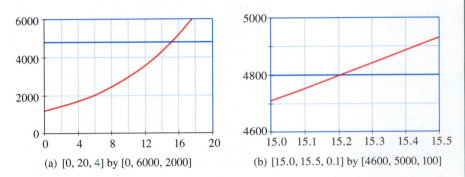

(a) [0, 20, 4] by [0, 6000, 2000]   (b) [15.0, 15.5, 0.1] by [4600, 5000, 100]

Figure 2
In both viewing rectangles, the red graph is the population growth function determined in part (a) of Example 1. The blue graph is the horizontal line $\mathcal{N}(t) = 4800$. The viewing rectangle in Figure 2(a) shows that the population reaches 4800 at a time that is roughly halfway between $t = 14$ and $t = 16$ hours. The zoom-in view in Figure 2(b) shows the time to be 15.2 hours, to the nearest tenth of an hour.

In Example 2 and in exercises, we are going to use the function $\mathcal{N}(t) = \mathcal{N}_0 e^{kt}$ as a model for the size of human populations. But in order to do that in a reasonably honest way (short of bringing in calculus), first we need to introduce the concept of a *relative growth rate*. Suppose that in a biology lab you obtain the following population figures in successive hours for a colony of bacteria:

3 P.M.:   12,000 bacteria   4 P.M.:   15,000 bacteria

Let us compute the percentage increase in the population over this 1-hour interval. We have

$$\text{percentage increase} = \frac{\text{change in population}}{\text{initial population}} \times 100$$

$$= \frac{15,000 - 12,000}{12,000} \times 100 = \frac{1}{4} \times 100 = 25\%$$

We say in this case that the **average relative growth rate** of the population is 25%/hour. In this section and in the exercises, we will shorten the phrase *average relative growth rate* to simply **relative growth rate.** Just as the interest rate in banking is sometimes expressed as a percentage and sometimes as a decimal, so too with the relative growth rate. In the example we are using then, the decimal form of the relative growth rate is 0.25/hour. Whether expressed as a percentage or a decimal, the relative growth rate over a unit of time is a measure of how the population increases (or decreases) compared to its starting value.

There is a relationship between the growth constant k in the equation $\mathcal{N}(t) = \mathcal{N}_0 e^{kt}$ and the relative growth rate. To demonstrate this, we'll use the growth

TABLE 1 Relative Growth Rates for the Function in Example 1 over Successive 1-Hour Intervals.*

t	Relative growth rate on interval $[t, t + 1]$	Calculator value (rounded to 3 places)
0	$\dfrac{\mathcal{N}(1) - \mathcal{N}(0)}{\mathcal{N}(0)}$	0.095
1	$\dfrac{\mathcal{N}(2) - \mathcal{N}(1)}{\mathcal{N}(1)}$	0.095
2	$\dfrac{\mathcal{N}(3) - \mathcal{N}(2)}{\mathcal{N}(2)}$	0.095
3	$\dfrac{\mathcal{N}(4) - \mathcal{N}(3)}{\mathcal{N}(3)}$	0.095

*The function is given by $\mathcal{N}(t) = 1200e^{kt}$, where $k = \dfrac{\ln(1440/1200)}{2} = 0.09116\ldots$

function obtained in Example 1. In Table 1, we've computed the relative growth rates (in decimal form) for this function over successive 1-hour time intervals.

We make two observations about the results in Table 1.

- The relative growth rates all are equal.
- The relative growth rate (0.095) is approximately equal to the growth constant k (0.091).

The observation that the relative growth rates all turn out to be the same seems surprising at first. This is, however, one of the characteristic features of exponential growth and not only a coincidence. To see why, we compute $[\mathcal{N}(t + 1) - \mathcal{N}(t)]/\mathcal{N}(t)$ for the function $\mathcal{N}(t) = \mathcal{N}_0 e^{kt}$. We have

$$\frac{\mathcal{N}(t + 1) - \mathcal{N}(t)}{\mathcal{N}(t)} = \frac{\mathcal{N}_0 e^{k(t+1)} - \mathcal{N}_0 e^{kt}}{\mathcal{N}_0 e^{kt}}$$

$$= \frac{e^{kt}e^k - e^{kt}}{e^{kt}} \qquad \text{dividing numerator and denominator} \quad (4)$$
$$\text{by } \mathcal{N}_0; \text{ and } e^{k(t+1)} = e^{kt+k} = e^{kt}e^k$$

$$= e^k - 1 \qquad \text{dividing numerator and}$$
$$\text{denominator by } e^{kt}$$

This shows that the relative growth rate over every interval $[t, t + 1]$ is the constant $e^k - 1$; the value does not depend upon t.

The second of the two observations we made was that the relative growth rate was nearly the same as the growth constant k. It turns out that this is true for every exponential growth function provided k is close to zero. (Exercise 61 will help you to see why this is so.) For this reason, in using the exponential growth law $\mathcal{N}(t) = \mathcal{N}_0 e^{kt}$ to model human populations, we'll simply use the relative growth rate for the value of k. This is convenient to do since most of the available data report growth in terms of a relative growth rate. Furthermore, since the census figures themselves are, in general, not precise, it wouldn't particularly meaningful to quibble over a fraction of a percent in this context. See Table 2.

TABLE 2 Comparison of Relative Growth Rate $e^k - 1$ [from Equation (4)] to growth constant k (computed assuming exponential growth)*

Country	Relative growth rate for year 2000		k
	as percentage	as decimal	
Liberia	3.2	0.032	0.03149...
United States	0.6	0.006	0.00598...
Sweden	−0.1	−0.001	−0.00099...

*When the numbers are close to zero, the relative growth rate (in decimal form) is a good approximation for k.
Source for relative growth rates: Population Reference Bureau, www.prb.org

EXAMPLE 2	Using an exponential growth model for world population growth

TABLE 3

Year	World population
1925	1.963
2000	6.067

Source: Population Reference Bureau

The Population Reference Bureau, www.prb.org, estimates that, in the year 1900, the size of the world's population was 1.628 billion, with a relative growth rate of about 0.7%/year.

(a) Use the exponential growth model to make population projections for the years 1925 and 2000.

(b) Use the data in Table 3 to compute the percentage errors in the projections, and summarize the results.

SOLUTION

(a) Let $t = 0$ correspond to 1900. The initial population is $\mathcal{N}_0 = 1.628$, in units of one billion. For the value of k, we convert the given relative growth rate 0.7%/year into its decimal form, which is 0.007/year. With these values, the growth law becomes $\mathcal{N}(t) = 1.628e^{0.007t}$. Now we're ready to compute.

For 1925, $t = 25$
$\mathcal{N}(25) = 1.628e^{0.007(25)}$
≈ 1.939 using a calculator

For 2000, $t = 100$
$\mathcal{N}(100) = 1.628e^{0.007(100)}$
≈ 3.278 using a calculator

The exponential model projects a world population of 1.939 billion in 1925 and 3.278 billion in 2000.

(b) For 1925, the projection is 1.939 billion, whereas the figure from Table 3 is 1.963 billion. The percentage error in the projection is then

$$\text{percentage error} = \frac{|\text{actual value} - \text{estimate}|}{\text{actual value}} \times 100$$

$$= \frac{|1.963 \text{ billion} - 1.939 \text{ billion}|}{1.963 \text{ billion}} \times 100$$

$$\approx 1.2\% \quad \text{using a calculator}$$

As you should check now for yourself using the year 2000 figures, the percentage error in the projection for 2000 is 46.0%. *Summary*: In both cases, the exponential projection is too low. However, the projection for 1925 *is* quite close, with a percentage error of only about 1%. In the case of the year 2000 projection, the model is way off; the percentage error is 46%.

| EXAMPLE | 3 | More on world population growth |

In the previous example we found that from 1900 to 2000, the world population grew faster than what was predicted by an exponential growth model with a relative growth rate of 0.7%/year. Actually, as you might well expect, the relative growth rate was not constant over the course of the entire century. But, assuming it had been, find the value of the growth constant k that would yield an accurate projection for the year 2000. (The year 2000 population is given in Table 3.)

SOLUTION
Again we let $t = 0$ correspond to 1900, and we have $\mathcal{N}_0 = 1.628$, in units of one billion. Our exponential growth law is $\mathcal{N}(t) = 1.628e^{kt}$. We want to find k so that the population in the year 2000 is 6.067 billion (according to Table 3). The year 2000 corresponds to $t = 100$, so we have

$$\mathcal{N}(100) = 1.628e^{k(100)}$$
$$6.067 = 1.628e^{100k}$$
$$\frac{6.067}{1.628} = e^{100k}$$

Writing this last equation in logarithmic form gives us

$$100k = \ln\frac{6.067}{1.628} \quad \text{and therefore} \quad k = \frac{\ln\dfrac{6.067}{1.628}}{100} \approx 0.0131\ldots \quad \textcolor{red}{\text{using a calculator}}$$

Thus, rounding to three decimal places, the required value of k is approximately 0.013. The corresponding relative growth rate in percent form is then 1.3%/year. *Note*: If you try checking this result using the approximation $k = 0.013$, you'll find that the population projection falls short of the required 6.067 billion. Don't round until the end. A good way to do this with your calculator is to first evaluate $k = (\ln\frac{6.067}{1.628})/100$, then *store* the result, and use the stored value in computing $1.628e^{k(100)}$. (If necessary, see your calculator owner's manual on how to store and then recall a constant.)

In the next example we compute the *doubling time* for a population that is growing exponentially. As the name indicates, the **doubling time** for a population is the amount of time required for the size of the population to double. Now, in the previous section we developed a formula for the doubling time T_2 of an amount of money under continuous compounding of interest. That formula is $T_2 = (\ln 2)/r$, where r is annual interest rate. The same formula is applicable now, because in both cases the underlying assumption is that of exponential growth. In the current context, the growth constant k plays the role of the annual interest rate r. Thus, for a population that grows exponentially, the doubling time is $T_2 = (\ln 2)/k$. Note that the doubling time depends only upon the growth constant k and not upon the initial population size $\mathcal{N}_0$. For reference, we state the formula for doubling time in the box that follows.

Doubling Time Formula

Suppose that a population grows exponentially. Then the doubling time T_2 is given by

$$T_2 = (\ln 2)/k$$

where k is the growth constant.

EXAMPLE 4 Computing a doubling time

According to the *Global Data Files* of the U.S. Bureau of the Census, the size of the world population in 1950 was approximately 2.555 billion, with an annual growth rate of 1.47%/year. Estimate the doubling time assuming exponential growth at 1.47%/year.

SOLUTION
Converting 1.47% to a decimal, we have $k = 0.0147$. The doubling time is therefore

$T_2 = (\ln 2)/k = (\ln 2)/0.0147$
≈ 47 years   using a calculator and rounding to the nearest year

The exponential model predicts a doubling time of approximately 47 years. That would be the year $1950 + 47 = 1997$. Note the fact that the 1950 population size was not needed in this computation. *Remark*: It's interesting to check this model against the actual data. The model predicts that the size of the population in 1997 is 2×2.555 billion $= 5.110$ billion. The actual figure from the U.S. Bureau of the Census is 5.847 billion. So the model projects too low a population for 1997. To put it another way, it actually took less than 47 years for the 1950 population to double in size. Indeed, a further check of the census data shows that the size of the population reached (our projected) 5.110 billion back in 1988.

In the next example we use the exponential growth model to describe the concentration of the so-called greenhouse gas carbon dioxide in the earth's atmosphere. As background, we quote from an article on the front page of the *New York Times,* June 7, 2001. The headline under which the article appears is "Panel Tells Bush Global Warming Getting Worse."

> A panel of top American scientists declared today that global warming was a real problem, and was getting worse . . . . In a much anticipated report from the National Academy of Sciences, 11 leading atmospheric scientists, including previous skeptics about global warming, reaffirmed the mainstream scientific view that the earth's atmosphere was getting warmer and that human activity was largely responsible. "Greenhouse gases are accumulating in the earth's atmosphere as a result of human activities, causing surface air temperatures and subsurface ocean temperature to rise," the report said.

New York Times article by Katharine Q. Seelye with Andrew C. Revkin, June 7, 2001

EXAMPLE	5	Using an exponential model for carbon dioxide levels in the atmosphere

The following statements about the increasing levels of carbon dioxide in the atmosphere appear in the book *The Challenge of Global Warming,* edited by Dean Edwin Abrahamson (Washington, D.C.: Island Press, 1989).

> Carbon dioxide, the single most important greenhouse gas, accounts for about half of the warming that has been experienced as a result of past emissions* and also for half of the projected future warming. The present [1988] concentration is now about 350 parts per million (ppm) and is increasing about 0.4% . . . per year. . . .

Assuming that the concentration of carbon dioxide in the atmosphere continues to increase exponentially at 0.4% per year, estimate when the concentration might reach 600 ppm. (This would be roughly twice the level estimated to exist prior to the Industrial Revolution.)

SOLUTION

Let $t = 0$ correspond to the year 1988. In the formula $\mathcal{N} = \mathcal{N}_0 e^{kt}$, we want to determine the time t when $\mathcal{N} = 600$. Using the values $\mathcal{N}_0 = 350$, $\mathcal{N} = 600$, and $k = 0.004$, we have

$$600 = 350\, e^{0.004t} \qquad \text{or} \qquad \frac{600}{350} = e^{0.004t}$$

and, consequently,

$$\ln(600/350) = 0.004t$$
$$t = \frac{\ln(600/350)}{0.004} \approx 135 \qquad \text{using a calculator and rounding to the nearest whole number}$$

Now, 135 years beyond the base year of 1988 is 2123. Rounding once more, we summarize our result this way: If the carbon dioxide levels continue to increase exponentially at 0.4% per year, then the concentration will reach 600 ppm by approximately 2125.

In Examples 1 through 4, we used the function $\mathcal{N}(t) = \mathcal{N}_0 e^{kt}$ to describe population growth. In Example 5, we used the function to model the increasing levels of carbon dioxide in the atmosphere. And in the previous section, a function of this form was used to describe how a sum of money grows under continuous compounding of interest. It is a remarkable fact that the same basic function, but now with $k < 0$, describes radioactive decay. How do scientists know that this is the appropriate model here? We mention two reasons—one empirical, one theoretical. As early as 1900–1903, the physicist Ernest Rutherford (Nobel prize, 1908) and the chemist Frederick Soddy (Nobel prize, 1921) carried out experiments measuring radioactive decay. They found that an equation of the form $\mathcal{N} = \mathcal{N}_0 e^{kt}$, with $k < 0$,

---

*The two principal sources of these emissions are the burning of fossil fuels and the burning of vegetation in the tropics.

indeed aptly described their data. Alternatively, on theoretical grounds, it can be argued that the decay rate at time t must be proportional to the amount $N(t)$ of the radioactive substance present. Under this condition, calculus can be used to show that a law of exponential decay applies. (Exercise 63 demonstrates the converse: Assuming that the law of exponential decay applies, then the rate of decay must be proportional to the amount of the substance present.)

In discussing radioactive decay, it is convenient to introduce the term *half-life*. As you'll see, this is analogous to the concept of *doubling time* for exponential growth.

DEFINITION   **Half-life**

The **half-life** of a radioactive substance is the time required for half of a given sample to disintegrate. The half-life is an intrinsic property of the substance; it does not depend on the given sample size.

EXAMPLE

Iodine-131 is a radioactive substance with a half-life of 8 days. Suppose that 2 g are present initially. Then:

at $t = 0$, 2 g are present;

at $t = 8$ days, 1 g is left;

at $t = 16$ days, $\frac{1}{2}$ g is left;

at $t = 24$ days, $\frac{1}{4}$ g is left;

at $t = 32$ days, $\frac{1}{8}$ g is left;

TABLE 4

t (days)	$N(t)$ (amount)
0	N_0
8	$N_0/2$
16	$N_0/4$
24	$N_0/8$
32	$N_0/16$

Just as we used the idea of doubling time to graph an exponential growth function in the previous section, we can use the half-life to graph an exponential decay function. Consider, for example, the radioactive substance iodine-131, which has a half-life of 8 days. Table 4 shows what fraction of an initial amount remains at 8-day intervals. Using the data in this table, we can draw the graph of the decay function $N(t) = N_0 e^{kt}$ for iodine-131 (see Figure 3). Notice that we are able to construct this graph without specifically evaluating the decay constant k. (The next example shows how to determine k.)

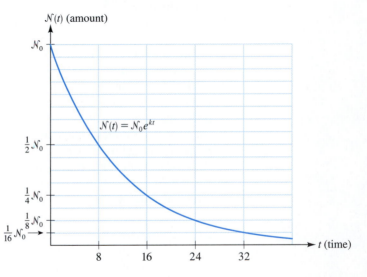

Figure 3
The exponential decay function for iodine-131.

| EXAMPLE | 6 | Using the exponential decay formula |

Hospitals utilize the radioactive substance iodine-131 in the diagnosis of the thyroid gland. The half-life of iodine-131 is 8 days.
(a) Determine the *decay constant k* for iodine-131.
(b) If a hospital acquires 2 g of iodine-131, how much of this sample will remain after 20 days?
(c) How long will it be until only 0.01 g remains?

SOLUTION
(a) We use the half-life information to find the value of the decay constant k. Substituting $t = 8$ in the decay law $N(t) = N_0 e^{kt}$ yields

$$N(8) = N_0 e^{k(8)}$$

$$\frac{1}{2}N_0 = N_0 e^{8k} \qquad \text{A half-life of 8 days means that } N(8) = \tfrac{1}{2}N_0$$

$$\frac{1}{2} = e^{8k} \qquad \text{dividing both sides by } N_0$$

In logarithmic form, this last equation becomes

$$8k = \ln\frac{1}{2} \qquad \text{and therefore} \qquad k = \left(\ln\frac{1}{2}\right)/8$$

$$\approx -0.08664 \qquad \text{rounding to five places}$$

(b) We are given $N_0 = 2$ g and we want to find $N(20)$, the amount remaining after 20 days. First, before using algebra and a calculator, let's estimate the answer to get a feeling for the situation. The half-life is 8 days. This means that after 8 days, 1 g remains; after 16 days, 0.5 g remains; and after 24 days, 0.25 g remains. Since 20 is between 16 and 24, it follows that after 20 days, the amount remaining will be between 0.5 g and 0.25 g. Now for the actual calculations: Substitute the values $N_0 = 2$ and $t = 20$ in the decay law $N(t) = N_0 e^{kt}$ to obtain

$$N(20) = 2e^{k(20)} \qquad \text{where } k = \left(\ln\frac{1}{2}\right)/8$$

$$\approx 0.354 \text{ g} \qquad \text{using a calculator}$$

Thus, after 20 days, approximately 0.35 g of the iodine-131 remains. Note that this amount is indeed between 0.5 g and 0.25 g as we first estimated.
(c) We want to find the time t for which $N(t) = 0.01$ g. Substituting 0.01 for $N(t)$ in the decay law gives us

$$0.01 = 2e^{kt} \qquad \text{and therefore} \qquad 0.005 = e^{kt}$$

The logarithmic form of this last equation is $kt = \ln(0.005)$, from which we conclude

$$t = \frac{\ln(0.005)}{k} \qquad \text{where } k = \left(\ln\frac{1}{2}\right)/8$$

$$\approx 61.2 \text{ days} \qquad \text{using a calculator}$$

In Example 6(a), we found that the decay constant for iodine-131 is given by $k = (\ln \frac{1}{2})/8$. Notice that the denominator in this expression is the half-life of iodine-131. By following the same reasoning used in Example 6(a), we find that the decay constant for any radioactive substance is always $\ln \frac{1}{2}$ divided by the half-life. For reference, this useful fact is restated in the box that follows.

PROPERTY SUMMARY   A Formula for the Decay Constant k

Formula

$$k = \frac{\ln \frac{1}{2}}{\text{half-life}}$$

Example

The half-life of strontium-90 is 28 years, and therefore $k = (\ln \frac{1}{2})/28$.

EXAMPLE 7   Calculations and estimates for radioactivity

An article on nuclear energy appeared in the January 1976 issue of *Scientific American*. The author of the article was Hans Bethe (1906–1992), a Nobel prize winner in physics. At one point in the article, Professor Bethe discussed the disposal (through burial) of radioactive waste material from a nuclear reactor. The particular waste product under discussion was plutonium-239.

> . . . Plutonium-239 has a half-life of nearly 25,000 years, and 10 half-lives are required to cut the radioactivity by a factor of 1000. Thus, the buried wastes must be kept out of the biosphere for 250,000 years.

(a) Supply the detailed calculations to support the statement that 10 half-lives are required before the radioactivity is reduced by a factor of 1000.
(b) Show how the figure of 10 half-lives can be obtained by estimation, as opposed to detailed calculation.

SOLUTION
(a) Let $\mathcal{N}_0$ denote the initial amount of plutonium-239 at time $t = 0$. Then the amount $\mathcal{N}$ present at time t is given by $\mathcal{N} = \mathcal{N}_0 e^{kt}$. We wish to determine t when $\mathcal{N} = \frac{1}{1000}\mathcal{N}_0$. First, since the half-life is 25,000 years, we have $k = (\ln \frac{1}{2})/25{,}000$. Now, substituting $\frac{1}{1000}\mathcal{N}_0$ for $\mathcal{N}$ in the decay law gives us

$$\frac{1}{1000}\mathcal{N}_0 = \mathcal{N}_0 e^{kt} \quad \text{where } k \text{ is } (\ln \tfrac{1}{2})/25{,}000$$

$$\frac{1}{1000} = e^{kt}$$

$$kt = \ln \frac{1}{1000} \quad \textcolor{red}{\text{converting from exponential to logarithmic form}}$$

$$t = \frac{\ln \frac{1}{1000}}{k} = \frac{\ln \frac{1}{1000}}{\left(\ln \frac{1}{2}\right)/25{,}000} = \frac{\left(\ln \frac{1}{1000}\right)(25{,}000)}{\ln \frac{1}{2}}$$

Using a calculator, we obtain the value 249,114.6 years; however, given the time scale involved, it would be ludicrous to announce the answer in this form. Instead, we round off the answer to the nearest thousand years and say that after 249,000 years, the radioactivity will have decreased by a factor of 1000. Notice that this result is consistent with Professor Bethe's ballpark estimate of 10 half-lives, or 250,000 years

(b) After 1 half-life: $\quad \mathcal{N} = \dfrac{\mathcal{N}_0}{2}$

After 2 half-lives: $\quad \mathcal{N} = \dfrac{1}{2}\left(\dfrac{\mathcal{N}_0}{2}\right) = \dfrac{\mathcal{N}_0}{2^2}$

After 3 half-lives: $\quad \mathcal{N} = \dfrac{1}{2}\left(\dfrac{\mathcal{N}_0}{2^2}\right) = \dfrac{\mathcal{N}_0}{2^3}$

Following this pattern, we see that after 10 half-lives, we should have

$$\mathcal{N} = \dfrac{\mathcal{N}_0}{2^{10}}$$

However, as we noted in the first section of this chapter, 2^{10} is approximately 1000. Therefore, we have

$$\mathcal{N} \approx \dfrac{\mathcal{N}_0}{1000} \qquad \text{after 10 half-lives}$$

This is in agreement with Professor Bethe's statement.

EXERCISE SET 5.7

A

In Exercises 1 and 2, assume that the populations grow exponentially—that is, according to the law $\mathcal{N}(t) = \mathcal{N}_0 e^{kt}$.

1. At the start of an experiment, 2000 bacteria are present in a colony. Two hours later, the population is 3800.
   (a) Determine the growth constant k.
   (b) Determine the population five hours after the start of the experiment.
   (c) When will the population reach 10,000?
2. At the start of an experiment, 2×10^4 bacteria are present in a colony. Eight hours later, the population is 3×10^4.
   (a) Determine the growth constant k.
   (b) What was the population two hours after the start of the experiment?
   (c) How long will it take for the population to triple?
3. The figure in the next column shows the graph of an exponential growth function $\mathcal{N}(t) = \mathcal{N}_0 e^{kt}$. Determine the values of $\mathcal{N}_0$ and k.

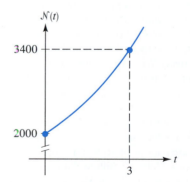

4. Suppose that you are helping a friend with his homework on the growth law, $\mathcal{N}(t) = \mathcal{N}_0 e^{kt}$, and on his paper you see the equation $e^k = -0.75$. How do you know at this point that your friend must have made an error?

TABLE A

Region	1995 Population (billions)	Percentage of population in 1995	Relative growth rate (%/year)	Year 2000 population (billions)	Percentage of world population in 2000
World	5.702	100	1.5	?	?
More developed regions	1.169	?	0.2	?	?
Less developed regions	4.533	?	1.9	?	?

TABLE B

Country	1995 Population (millions)	Relative growth rate (%/year)	Year 2000 population (millions)	Percentage increase in population
United States	263.2	0.7	?	?
People's Republic of China	1218.8	1.1	?	?
Mexico	93.7	2.2	?	?

In Exercises 5 and 6: **(a)** *Complete the table;* **(b)** *Use the information that is given to compute the percentage errors in the population projections in part (a). Round each answer to one decimal place. [The populations projections are in the second column from the right in each table in part (a).] The data in Exercises 5 and 6 is from The Population Reference Bureau,* www.prb.org.

5. (a) See Table A.
   (b) Year 2000 population in billions: world, 6.067; more developed, 1.184; less developed 4.883
6. (a) See Table B.
   (b) Year 2000 population in millions: United States, 275.6; People's Republic of China, 1264.5; Mexico, 99.6
7. As of the year 2000, the African nation of Chad had one of the highest population growth rates in the world, 3.3%/year. At the other extreme in 2000, the United Kingdom had one of the lowest (positive) growth rates, 0.1%/year. (There are countries with zero or with negative growth rates.)
   (a) In 2000, the sizes of the populations of Chad and the United Kingdom were 8.0 million and 59.8 million, respectively. Write the exponential growth law for each country, letting $t = 0$ correspond to 2000.
   (b) Assuming exponential growth, when would the two countries have populations of the same size? Round the answer to the nearest five years.   *Hint:* Equate the two expressions for $\mathcal{N}(t)$ obtained in part (a).
8. In 2000, the nations of Mali and Cuba had similar size populations: Mali 11.2 million, Cuba 11.1 million. However, the relative growth rate for Mali was 3.1%/year, whereas that for Cuba was 0.7%/year.
   (a) Assuming exponential growth at the given rates, make projections for each population in the year 2015.
   (b) When might the population of Mali reach 20 million? What would the population of Cuba be at that time?

9. In 2000, the nations of Niger and Portugal had similar size populations: Niger 10.1 million, Portugal 10.0 million. However, the relative growth rate for Niger was 3.0%/year, whereas that for Portugal was only 0.1%/year.
   (a) Assuming exponential growth at the given rates, make projections for each population in the year 2015.
   (b) When might the population of Niger reach 15 million? What would the population of Portugal be at that time?
10. The population of Guatemala in 2000 was 12.7 million.
   (a) Assuming exponential growth, what value for k would lead to a population of 20 million one quarter of a century later (i.e., in 2025)?   *Remark:* The answer you obtain is, in fact, less than the actual year 2000 growth rate, which was about 2.9%/year.
   (b) Again, assuming a population of 12.7 million in 2000, what value for k would lead to a population of 20 million, one century later (i.e., in 2100)?

As background for Exercises 11 and 12, here's a reminder of what we found in Example 2. Starting with data for the year 1900, we saw two cases where the world population had grown faster than what was predicted by an exponential model. In Exercises 11 and 12, you'll work with more recent data for the base year and find that now the population is, in fact, growing slower than the exponential model predicts.

11. According to the U.S. Bureau of the Census, the world population in 1975 was 4.088 billion, with a relative growth rate of 1.75%/year.
   (a) Use an exponential growth model to predict the year 2000 population.
   (b) According to Table 3 in the text, the world population in 2000 was 6.067 billion. Is your projection in part (a) higher or lower than this? Compute the percentage

error in the projection and round the answer to the nearest one percent.

12. Follow Exercise 11, but start from the following data. In 1990, the world population was 5.284 billion, with a relative growth rate of 1.56%/year.

13. In each case, you are given the relative growth rate of a country or region in the year 2000. Compute the doubling time for the population (assuming an exponential growth model). Round the answer to the nearest whole number of years.
  (a) United States: 0.6%/year
  (b) Tajikistan: 1.6%/year
  (c) Cambodia: 2.6%/year
  (d) Palestinian Territory: 3.7%/year

14. Refer to the following table, which gives global populations and relative growth rates for the year 2000. Assume an exponential growth model.

	2000 Population (billions)	Relative growth rate (%/year)
World	6.067	1.4
More developed regions	1.184	0.1
Less developed regions	4.883	1.7

Source: Population Reference Bureau

  (a) Compute the doubling time for the world population, and give the year in which that population would be reached. (Round each answer to the nearest year.)
  (b) Follow part (a) for the more developed regions.
  (c) Follow part (a) for the less developed regions.

15. In Example 5 in the text we modeled the concentration of carbon dioxide in the atmosphere using an exponential growth model with $k = 0.4$%/year and $N_0 = 350$ ppm, where $t = 0$ corresponds to 1988.
  (a) A more precise value for N_0 is 351.31 ppm. Using this value in the exponential growth model, complete the following table of projected values for the atmospheric carbon dioxide concentrations. Round the answers to one decimal place.

Year	1998 ($t = 10$)	2000 ($t = 12$)
Atmospheric concentration of carbon dioxide (ppm)		

Source for data: C. D. Keeling and T. P. Whorf, Scripps Institution of Oceanography.

  (b) The actual values for 1998 and 2000 were 366.7 ppm and 368.4 ppm, respectively. For each of these two years, say whether the exponential model projects too high or two low a value, and compute the percentage error.

16. For this exercise refer to the following table, which gives the atmospheric concentrations of carbon dioxide (CO_2) for selected years over the latter half of the 20th century.

Year	1960	1985	1990	1995	1999
Atmospheric concentration of carbon dioxide (ppm)	316.75	345.73	354.04	360.91	368.37

Source: C. D. Keeling and T. P. Whorf, Scripps Institution of Oceanography

  (a) Let $t = 0$ correspond to 1960, and assume an exponential growth model $N = 316.75e^{kt}$. Use the information for 1999 in the table to find the growth constant k.
  (b) Use the growth law determined in part (a) to make a projection for the CO_2 concentration in 1985. Then, referring to the table, say whether the exponential model predicts too high or low a figure, and compute the percentage error.
  (c) Follow part (b) for the year 1990 and 1995.
  (d) Why do you think that the year 1999 was not included in part (c)?

17. According to figures from the U.S. Bureau of the Census, in 2000, the size of the population of the state of New York was more than three times larger than that of Arizona. However, New York had one of the lower growth rates in the nation, and Arizona had the second highest.
  (a) Use the data in the following table to specify an exponential growth model for each state. (Let $t = 0$ correspond to the year 2000.)

State	Population in 2000 (millions)	Relative growth rate (%/year)
New York	18.976	0.6
Arizona	5.131	4.0

  (b) Assuming continued exponential growth, when would the two states have populations of the same size? Round the answer to the nearest five years. *Hint:* Equate the two expressions for $N(t)$ obtained in part (a).

18. The following table gives the population size and relative growth rates in 2000 for three states. Use exponential models to project the population of each state in the year 2010, listing the results from largest to smallest. Does the order remain the same as it was in 2000?

State	Population in 2000 (millions)	Relative growth rate (%/year)
Iowa	2.926	0.54
Arkansas	2.673	1.37
Nevada	1.998	6.63

19. Lester R. Brown, in his book *State of the World 1985* (New York: W. W. Norton and Co., 1985), makes the following statement:

> The projected [population] growth for North America, all of Europe, and the Soviet Union is less than the additions expected in either Bangladesh or Nigeria.

In this exercise, you are asked to carry out the type of calculations that could be used to support Lester Brown's projection for the period 1990–2025. The source of the data in the following table is the Population Division of the United Nations.

Region	1990 Population (millions)	Projected relative growth rate (%/year)	2025 Projected population (millions)
North America	275.2	0.7	?
(Former) Soviet Union	291.3	0.7	?
	291.3	0.7	?
Europe	499.5	0.2	?
Nigeria	113.3	3.1	?

(a) Complete the table, assuming that the populations grow exponentially and that the indicated growth rates are valid over the period 1990–2025.

(b) According to the projections in part (a), what will be the net increase in Nigeria's population over the period 1990–2025?

(c) According to the projections in part (a), what will be the net increase in the combined populations of North America, the (former) Soviet Union, and Europe over the period 1990–2025?

(d) Compare your answers in parts (b) and (c). Do your results support or contradict Lester Brown's projection?

20. The *2000 World Population Data Sheet* (published by the Population Reference Bureau, Washington, D.C.) lists ten nations with populations exceeding 100 million. (See Exercise 21 regarding a possible eleventh nation.) The populations, in order of decreasing size, along with relative growth rates, are listed in the accompanying table. Over a decade ago, in 1987, the U.S. Bureau of the Census made the following statement in its analysis of world population data: "The latest projections suggest that India's population may surpass China's in less than 60 years, or before today's youngsters in both countries reach old age." Using the data in the table, and assuming exponential growth, make projections for the populations of the ten countries in the year 2050 (which would be roughly 60 years after that statement by the Bureau of the Census). Do your results for China and India support the projection by the Bureau of the Census?

Country	2000 Population (millions)	Relative growth rate (%/year)
1. China	1264.5	0.9
2. India	1002.1	1.8
3. United States	275.6	0.6
4. Indonesia	212.2	1.6
5. Brazil	170.1	1.5
6. Pakistan	150.6	2.8
7. Russia	145.2	−0.6
8. Bangladesh	128.1	1.8
9. Japan	126.9	0.2
10. Nigeria	123.3	2.8

Source: Population Reference Bureau, www.prb.org

21. The population figures in Exercise 20 are actually for mid-2000. As of that time, the population of Mexico was 99.6 million, with a relative growth rate of 2.0%/year. Let $t = 0$ correspond to June 2000 and use the exponential growth model $\mathcal{N}(t) = 99.6e^{0.02t}$ to verify that $\mathcal{N}(0.5) > 100$. Interpret the result.

22. In 2000, the Philippines and Germany had similar size populations, but very different growth rates. The population of the Philippines was 80.3 million, with a relative growth rate of 2.0%/year. The population of Germany was 82.1 million, with a relative "growth" rate of −0.1%/year. Using exponential models, make a projection for the population of Germany in the year when the Philippine population has doubled.

In Exercises 23 and 24, use the half-life information to complete each table. (The formula $\mathcal{N} = \mathcal{N}_0 e^{kt}$ is not required.)

23. (a) Uranium-228: half-life = 550 seconds

t (seconds)	0	550	1100	1650	2200
$\mathcal{N}$ (grams)	8				

(b) Uranium-238: half-life = 4.9×10^9 years

t (years)	0				
$\mathcal{N}$ (grams)	10	5	2.5	1.25	0.625

24. (a) Polonium-210: half-life = 138.4 days.

t (years)	0	138.4	276.8	415.2	
$\mathcal{N}$ (grams)	0.4				0.025

(b) Polonium-214: half-life = 1.63×10^{-4} second

t (seconds)	0				6.52×10^{-4}
$\mathcal{N}$ (grams)	0.1	0.05	0.025	0.0125	

25. The half-life of iodine-131 is 8 days. How much of a one-gram sample will remain after 7 days?

26. The half-life of strontium-90 is 28 years. How much of a 10-g sample will remain after **(a)** 1 year? **(b)** 10 years?

27. The radioactive isotope sodium-24 is used as a tracer to measure the rate of flow in an artery or vein. The half-life of sodium-24 is 14.9 hours. Suppose that a hospital buys a 40-g sample of sodium-24.
  (a) How much of the sample will remain after 48 hours?
  (b) How long will it be until only 1 gram remains?

28. The radioactive isotope carbon-14 is used as a tracer in medical and biological research. Compute the half-life of carbon-14 given that the decay constant k is -1.2097×10^{-4}. (The units for k here are such that your half-life answer will be in years.)

29. (a) The half-life of radium-226 is 1620 years. Sketch a graph of the decay function for radium-226, similar to that shown in Figure 2.
  (b) The half-life of radium-A is 3 min. Sketch a graph of the decay function for radium-A.

30. (a) The half-life of thorium-232 is 1.4×10^{10} years. Sketch a graph of the decay function.
  (b) The half-life of thorium-A is 0.16 sec. Sketch a graph of the decay function.

31. The half-life of plutonium-241 is 13 years.
  (a) How much of an initial 2-g sample remains after 5 years?
  (b) Find the time required for 90% of the 2-g sample to decay. *Hint:* If 90% has decayed, then 10% remains.

32. The half-life of radium-226 is 1620 years.
  (a) How much of a 2-g sample remains after 100 years?
  (b) Find the time required for 80% of the 2-g sample to decay.

33. The half-life of thorium-229 is 7340 years.
  (a) Compute the time required for a given sample to be reduced by a factor of 1000. Show detailed calculations, as in Example 7(a).
  (b) Express your answer in part (a) in terms of half-lives.
  (c) As in Example 7(b), estimate the time required for a given sample for thorium-229 to be reduced by a factor of 1000. Compare your answer with that obtained in part (b).

34. The Chernobyl nuclear explosion (in the former Soviet Union, on April 26, 1986) released large amounts of radioactive substances into the atmosphere. These substances included cesium-137, iodine-131, and strontium-90. Although the radioactive material covered many countries, the actual amount and intensity of the fallout varied greatly from country to country, due to vagaries of the weather and the winds. One area that was particularly hard hit was Lapland, where heavy rainfall occurred just when the Chernobyl cloud was overhead.
  (a) Many of the pastures in Lapland were contaminated with cesium-137, a radioactive substance with a half-life of 33 years. If the amount of cesium-137 was found to be ten times the normal level, how long would it take until the level returned to normal? *Hint:* Let $\mathcal{N}_0$ be the amount that is ten times the normal level. Then you want to find the time when $\mathcal{N}(t) = \mathcal{N}_0/10$.
  (b) Follow part (a), but assume that the amount of cesium-137 was 100 times the normal level. *Remark:* Several days after the explosion, it was reported that the level of cesium-137 in the air over Sweden was 10,000 times the normal level. Fortunately there was little or no rainfall.

35. Strontium-90, with a half-life of 28 years, it is a radioactive waste product from nuclear fission reactors. One of the reasons great care is taken in the storage and disposal of this substance stems from the fact that strontium-90 is, in some chemical respects, similar to ordinary calcium. Thus, strontium-90 in the biosphere, entering the food chain via plants or animals, would eventually be absorbed into our bones.
  (a) Compute the decay constant k for strontium-90.
  (b) Compute the time required if a given quantity of strontium-90 is to be stored until the radioactivity is reduced by a factor of 1000.
  (c) Using half-lives, estimate the time required for a given sample to be reduced by a factor of 1000. Compare your answer with that obtained in (b).

36. (a) Suppose that a certain country violates the ban against above-ground nuclear testing and, as a result, an island is contaminated with debris containing the radioactive substance iodine-131. A team of scientists from the United Nations wants to visit the island to look for clues in determining which country was involved. However, the level of radioactivity from the iodine-131 is estimated to be 30,000 times the safe level. Approximately how long must the team wait before it is safe to visit the island? The half-life of iodine-131 is 8 days.
  (b) Rework part (a), assuming instead that the radioactive substance is strontium-90 rather than iodine-131. The half-life of strontium-90 is 28 years. Assume, as before, that the initial level of radioactivity is 30,000 times the safe level. (This exercise underscores the difference between a half-life of 8 days and one of 28 years.)

37. An article that appeared in the August 13, 1994, *New York Times* reported

German authorities have discovered . . . a tiny sample of weapons-grade nuclear material believed to have been smuggled out of Russia to interest foreign governments or terrorist groups that might want to build atomic bombs. . . . [the police] said they had seized the material, .028 ounces of highly enriched uranium-235, in June in . . . Bavaria . . . and have since arrested . . . [six] suspects. . . .

Suppose that the suspects, in an attempt to avoid arrest, had thrown the 0.028 ounces of uranium-235 into the Danube River, where it would sink to the bottom. How many ounces of the uranium-235 would still be in the river after 1000 years? The half-life of uranium-235 is 7.1×10^8 years.

38. In 1969 the United States National Academy of Sciences issued a report entitled *Resources and Man.* One conclusion in the report is that a world population of 10 billion "is close to (if not above) the maximum that an intensively managed world might hope to support with some degree of comfort and individual choice." (The figure "10 billion" is sometimes referred to as the *carrying capacity* of the Earth.)

(a) When the report was issued in 1969, the world population was about 3.6 billion, with a relative growth rate of 2% per year. Assuming continued exponential growth at this rate, estimate the year in which the Earth's carrying capacity of 10 billion might be reached.

(b) Repeat the calculations in part (a) using the following more recent data: In 2000, the world population was about 6.0 billion, with a relative growth rate of 1.4% per year. How does your answer compare with that in part (a)?

39. The following extract is from an article by Kim Murphy that appeared in the *Los Angeles Times* on September 14, 1994.

> CAIRO—Over a chorus of reservations from Latin America and Islamic countries still troubled about abortion and family issues, nearly 180 nations adopted a wide-ranging plan Tuesday on global population, the first in history to obtain partial endorsement from the Vatican.
>
> The plan, approved on the final day of the U.N. population conference here, for the first time tries to limit the growth of the world's population by preventing it from exceeding 7.2 billion people over the next two decades.

(a) In 1995, the world population was 5.7 billion, with a relative growth rate of 1.6%/year. Assuming continued exponential growth at this rate, make a projection for the world population in the year 2020. Round off the answer to one decimal place. How does your answer compare to the target value of 7.2 billion mentioned in the article?

(b) As in part (a), assume that in 1995 the world population was 5.7 billion. Determine a value for the growth constant k so that exponential growth throughout the years 1995–2020 leads to a world population of 7.2 billion in the year 2020.

G **40.** Economists define the **gross domestic product (GDP)** as the total market value of a nation's goods and services produced (within the borders of the nation) over a specified period of time. The GDP is one of the key measures of a nation's economic health. The following table gives the annual GDP for the United States for the period 1950–1990, at 10-year intervals. In Figures A and B, we show scatter plots for the data along with regression functions that model the data. Figure A uses a linear model, Figure B an exponential model. In each case, the years are on the horizontal axis, with $t = 0$ corresponding to 1900. In this exercise, you'll use a graphing utility to obtain the specific equations for these models and for two other models as well. You'll also compare projections using the various models.

TABLE A   Gross Domestic Product (GDP) in Constant 1996 Dollars

Year t $(t = 0 \leftrightarrow 1900)$	50	60	70	80	90
GDP (billions of dollars)	1686.6	2376.7	3578.0	4900.9	6707.9

Source: U.S. Bureau of Economic Analysis, http://www.bea.gov

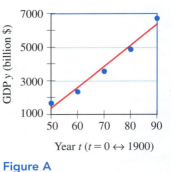

Figure A
Linear model $y = at + b$

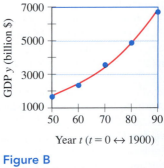

Figure B
Exponential model $y = ab^t$ or
$y = Ae^{Bt}$

(a) Use a graphing utility to create your own scatter plot of the data in Table A. Then, use the regression or trend line options of the graphing utility to obtain the specific equations for the linear model in Figure A and the exponential model in Figure B. (For the exponential model, you can report your answer in either one of the two forms indicated in the caption for Figure B, depending on what your graphing utility provides.) For comparison, graph both models in the same viewing rectangle.

(b) Use the regression or trend line options of the graphing utility to obtain the equation for a quadratic model $y = at^2 + bt + c$. Show the graph of the quadratic model with the scatter plot. Now repeat this process for a power model $y = at^b$.

(c) Use the functions that you determined in parts (a) and (b) to complete Table B, showing what each model projects for the gross domestic product in the indicated year. (Round each projection to one decimal place.) Then, in each row, circle the projection that is closest to the actual value given at the end of that row. Finally, compute the percentage error for each projection that you circled.

TABLE B   Gross Domestic Product (GDP) Projections Using Various Models

Year	GDP (billion $) linear model	GDP (billion $) exponential model	GDP (billion $) quadratic model	GDP (billion $) power model	GDP (billion $) actual
1995					7543.8
1999					8875.8
2001					9333.8

B

41. *Depletion of Nonrenewable Resources*: Suppose that the world population grows exponentially. Then, as a first approximation, it is reasonable to assume that the use of nonrenewable resources, such as petroleum and coal, also grows exponentially. Under these conditions, the following formula can be derived (using calculus):

$$A = \frac{A_0}{k}(e^{kT} - 1)$$

where A is the amount of the resource consumed from time $t = 0$ to $t = T$, the quantity A_0 is the amount of the resource consumed during the year $t = 0$, and k is the relative growth rate of annual consumption.

(a) Show that solving the formula for T yields

$$T = \frac{\ln[(Ak/A_0) + 1]}{k}$$

This formula gives the "life expectancy" T for a given resource. In the formula, A_0 and k are as previously defined, and A represents the total amount of the resource available.

(b) Over the years 1965–1972, world oil consumption grew exponentially. In this part of the exercise, you'll compute the life expectancy of oil, assuming continued exponential growth. Let $t = 0$ correspond to 1972. In that year, worldwide consumption of oil was approximately 18.7 billion barrels, with a relative growth rate of 7%/year. The global reserves of oil at that time were estimated to be 700 billion barrels. Compute the life expectancy T of oil under this exponential growth scenario and specify the *depletion year* ($= 1972 + T$). If the relative growth had been only 1%/year, what would the depletion year be?

(c) Beginning with the first Arab oil embargo of 1973–4, worldwide oil consumption ceased to grow exponentially. In fact, oil consumption actually decreased over the years 1973–1975, and it decreased again for 1979–1983, following the second Arab oil embargo. More recently, as indicated by the following data, scatter plot, and regression line, worldwide oil consumption has been increasing in a linear fashion. Use a graphing utility and the data in the table to determine the equation $y = mt + b$ for the regression line.

Year t ($t = 0 \leftrightarrow 1993$)	0	1	2	3
Global oil consumption y (billion barrels)	23.997	24.506	24.893	25.450

Year t ($t = 0 \leftrightarrow 1993$)	4	5	6
Global oil consumption y (billion barrels)	26.123	26.251	26.723

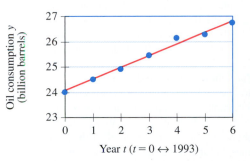

Year t ($t = 0 \leftrightarrow 1993$)

(d) If yearly consumption of a nonrenewable resource is given by a linear function $y = mt + b$, then it can be shown that the life expectancy T of the resource is given by the following formula,

$$T = \frac{-b + \sqrt{b^2 + 2mA}}{m}$$

where A represents the total amount of the resource available (i.e., the world reserves) in the year corresponding to $t = 0$. (The derivation does not require calculus, but we shall omit giving it here.) In 1993, world reserves of oil were estimated to be 1007 billion barrels. Using this value for A and the values of m and b determined in part (c), estimate the life expectancy for oil and the depletion date ($=1993 + T$). (Aside: The figure we've given here for the estimated world reserves in 1993 is greater than the one we gave for 1972. This reflects improved technology in locating oil fields and in extracting the oil.)

For Exercises 42 and 43, refer to the following table. (Data compiled from United States Geological Survey, World Resources Institute, and the World Almanac, 2000.)

Resource	1999 World consumption (million metric tons)	1999 World reserves (million metric tons)	Relative growth rate (%/year)
Aluminum (in bauxite)	23.1	34,000	3.0
Copper	12.6	650	2.4

42. (a) Suppose that the consumption of aluminum were growing exponentially. Use the formula for T in Exercise 41 to compute the life expectancy of aluminum and the depletion year.

(b) What if the world reserves for aluminum were actually twice that listed in the table, and the relative growth rate were half that in the table? Complete the depletion year under this scenario.

(c) Suppose from 1999 onward, consumption of aluminum stopped increasing exponentially and stabilized at the 1999 level of 23.1 million metric tons per year. How many years would it take to deplete the world reserves of 34,000 million metric tons?

43. Follow Exercise 42, using copper instead of aluminum.

Ⓖ **44.** The following table and scatter plots display data for the number of fixed-line phone connections worldwide and the number of cell phone subscribers over the years 1985–1998. As indicated by the scatter plots, the number of cell phone subscribers is far less than the number of phone lines, but it is increasing at a faster rate.

(a) Use the regression option on a graphing utility or spreadsheet to find a quadratic model $y = at^2 + bt + c$ describing the number of phone lines over the period 1985–1998 ($t = 0$ to $t = 13$). Similarly, find an exponential model of the form $y = ab^t$ or $y = Ae^{Bt}$ describing the number of cell phone subscribers. Then use the graphing utility to display the scatter plots and the two models in the same viewing rectangle.

(b) By finding an appropriate viewing rectangle, make a projection for the year in which there would be as many cell phone subscribers as phone lines.

Year t ($t = 0 \leftrightarrow 1985$)	Phone lines (millions)	Cell phone subscribers (millions)
0	407	1
1	426	1
2	446	2
3	469	4
4	493	7
5	520	11
6	546	16
7	574	23
8	606	34
9	645	55
10	691	91
11	738	142
12	788	215
13	844	319

Source: International Telecommunications Union data as reported in *Vital Signs 2000,* Lester Brown et al. (New York: W. W. Norton & Co., 2000)

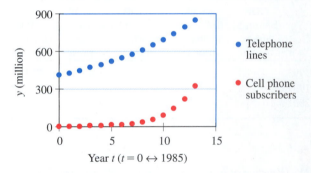

45. The age of some rocks can be estimated by measuring the ratio of the amounts of certain chemical elements within the rock. The method known as the *rubidium–strontium method* will be discussed here. This method has been used in dating the moon rocks brought back on the Apollo missions.

Rubidium-87 is a radioactive substance with a half-life of 4.7×10^{10} years. Rubidium-87 decays into the substance strontium-87, which is stable (nonradioactive). We are going to derive the following formula for the age of a rock:

$$T = \frac{\ln[(\mathcal{N}_s/\mathcal{N}_r) + 1]}{-k}$$

where T is the age of the rock, k is the decay constant for rubidium-87, $\mathcal{N}_s$ is the number of atoms of strontium-87 now present in the rock, and $\mathcal{N}_r$ is the number of atoms of rubidium-87 now present in the rock.

(a) Assume that initially, when the rock was formed, there were $\mathcal{N}_0$ atoms of rubidium-87 and none of strontium-87. Then, as time goes by, some of the rubidium atoms decay into strontium atoms, but the total number of atoms must still be N_0. Thus, after T years, we have $N_0 = N_r + N_s$ or, equivalently,

$$N_s = N_0 - N_r \qquad (1)$$

However, according to the law of exponential decay for the rubidium-87, we must have $N_r = N_0 e^{kT}$. Solve this equation for N_0 and then use the result to eliminate N_0 from equation (1). Show that the result can be written

$$N_s = N_r e^{-kT} - N_r \qquad (2)$$

(b) Solve equation (2) for T to obtain the formula given at the beginning of this exercise.

46. (Continuation of Exercise 45)
   (a) The half-life of rubidium-87 is 4.7×10^{10} years. Compute the decay constant k.
   (b) Analysis of lunar rock samples taken on the Apollo 11 mission showed the strontium–rubidium ratio to be

   $$\frac{N_s}{N_r} = 0.0588$$

   Estimate the age of these lunar rocks.

47. (Continuation of Exercise 45) Analysis of the so-called genesis rock sample taken on the Apollo 15 mission revealed a strontium–rubidium ratio of 0.0636. Estimate the age of this rock.

48. *Radiocarbon Dating*: Because rubidium-87 decays so slowly, the technique of rubidium–strontium dating is generally considered effective only for objects older than 10 million years. In contrast, archeologists and geologists rely on the *radiocarbon dating* method in assigning ages ranging from 500 to 50,000 years.

   Two types of carbon occur naturally in our environment: carbon-12, which is nonradioactive, and carbon-14, which has a half-life of 5730 years. All living plant and animal tissue contains both types of carbon, always in the same ratio. (The ratio is one part carbon-14 to 10^{12} parts carbon-12.) As long as the plant or animal is living, this ratio is maintained. When the organism dies, however, no new carbon-14 is absorbed, and the amount of carbon-14 begins to decrease exponentially. Since the amount of carbon-14 decreases exponentially, it follows that the level of radioactivity also must decrease exponentially. The formula describing this situation is

   $$N = N_0 e^{kT}$$

   where T is the age of the sample, N is the present level of radioactivity (in units of disintegrations per hour per gram of carbon), and N_0 is the level of radioactivity T years ago, when the organism was alive. Given that the half-life of carbon-14 is 5730 years and that $N_0 = 920$ disintegrations per hour per gram, show that the age T of a sample is given by

   $$T = \frac{5730 \ln(N/920)}{\ln(1/2)}$$

In Exercises 49–54, use the formula derived in Exercise 48 to estimate the age of each sample.   Note: Some technical complications arise in interpreting such results. Studies have shown that the ratio of carbon-12 to carbon-14 in the air (and therefore in living matter) has not in fact been constant over time. For instance, air pollution from factory smokestacks tends to increase the level of carbon-12. In the other direction, nuclear bomb testing increases the level of carbon-14.

49. Prehistoric cave paintings were discovered in the Lascaux cave in France. Charcoal from the site was analyzed and the level of radioactivity was found to be $N = 141$ disintegrations per hour per gram. Estimate the age of the paintings.

50. Before radiocarbon dating was used, historians estimated that the tomb of Vizier Hemaka, in Egypt, was constructed about 4900 years ago. After radiocarbon dating became available, wood samples from the tomb were analyzed, and it was determined that the radioactivity level was 510 disintegrations per hour per gram. Estimate the age of the tomb on the basis of this reading and compare your answer to the figure already mentioned.

51. Before radiocarbon dating, scholars believed that agriculture (farming, as opposed to the hunter-gatherer existence) in the Middle East began about 6500 years ago. However, when radiocarbon dating was used to study an ancient farming settlement at Jericho, the radioactivity level was found to be in the range $N = 348$ disintegrations per hour per gram carbon. Based on this evidence, estimate the age of the site, and compare your answer to the figure mentioned at the beginning of this exercise. Round your answer to the nearest 100 years. (Similar analyses in Iraq, Turkey, and other countries have since shown that agriculture was firmly established at least 9000 years ago.)

52. (a) Analyses of some ancient campsites in the Western Hemisphere reveal a carbon-14 radioactivity level of $N = 226$ disintegrations per hour per gram carbon. Show that this implies an age of 11,500 years, to the nearest 500 years.   *Remark: An age of 11,500 years corresponds to the last Ice Age, when sea level was significantly lower than it is today. According to the Clovis hypothesis, this was the time humans first entered the Western Hemisphere, across what would have been a land bridge extending over what is now the Bering Strait. (The name "Clovis" refers to Clovis, New Mexico. In 1933, a spearpoint and bones were discovered there that were subsequently found to be about 11,500 years old.)*
   (b) In the article "Coming to America" (*Time Magazine*, May 3, 1993), Michael D. Lemonick describes some of the evidence indicating that the Clovis hypothesis is not valid:

   > A team led by University of Kentucky archeologist Tom Dillehay discovered indisputable traces . . . [in Monte Verde, in southern Chile] of a human settlement that was inhabited between 12,800 and

12,300 years ago. Usually all scientists can find from that far back are stones and bones. In this case, thanks to a peat layer that formed during the late Pleistocene era, organic matter [to which radiocarbon dating can be applied] was mummified and preserved as well.

What range for the carbon-14 radioactivity level $\mathcal{N}$ corresponds to the ages mentioned in this article? (That is, compute the values of $\mathcal{N}$ corresponding to $T = 12,800$ and $T = 12,300$.)

53. The Dead Sea Scrolls are a collection of ancient manuscripts discovered in caves along the west bank of the Dead Sea. (The discovery occurred by accident when an Arab herdsman of the Taamireh tribe was searching for a stray goat.) When the linen wrappings on the scrolls were analyzed, the carbon-14 radioactivity level was found to be 723 disintegrations per hour per gram. Estimate the age of the scrolls using this information. Historical evidence suggests that some of the scrolls date back somewhere between 150 B.C. and A.D. 40. How do these dates compare with the estimate derived using radiocarbon dating?

54. **(a)** According to Exercise 48, the formula for the age T of a sample in terms of the radioactivity level $\mathcal{N}$ is

$$T = \frac{5730 \ln(\mathcal{N}/920)}{\ln(1/2)}.$$ Solve this equation for $\mathcal{N}$ and

show that the result can be written $\mathcal{N} = 920(2^{-T/5730})$.

(b) The famous prehistoric stone monument Stonehenge is located 8 miles north of Salisbury, England. Excavations and radiocarbon dating have led anthropologists and archeologists to distinguish three periods in the building of Stonehenge: period I, 2800 years ago; period II, 2100 years ago; and period III, 2000 years ago. Use the formula in part (a) to compute the carbon-14 radioactivity levels that would have led the scientists to assign these ages.

*Exercises 55–60 introduce a model for population growth that takes into account limitations on food and the environment. This is the **logistic growth model**, named and studied by the nineteenth century Belgian mathematician and sociologist Pierre Verhulst. (The word "logistic" has Latin and Greek origins meaning "calculation" and "skilled in calculation," respectively. However, that is not why Verhulst named the curve as he did. See Exercise 56 for more about this.)*

In the logistic model that we'll study, the initial population growth resembles exponential growth. But then, at some point, due perhaps to food or space limitations, the growth slows down and eventually levels off, and the population approaches an equilibrium level. The basic equation that we'll use for logistic growth is

$$\mathcal{N} = \frac{P}{1 + ae^{-bt}}$$

where $\mathcal{N}$ is the population at time t, P is the equilibrium population (or the upper limit for population), and a and b are positive constants.

55. The following figure shows the graph of the logistic function $\mathcal{N}(t) = 4/(1 + 8e^{-t})$. Note that in this equation the equilibrium population P is 4 and that this corresponds to the asymptote $\mathcal{N} = 4$ in the graph.

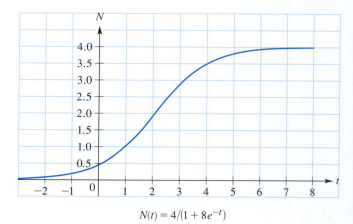

$$N(t) = 4/(1 + 8e^{-t})$$

(a) Use the graph and your calculator to complete the following table. For the values that you read from the graph, estimate to the nearest 0.25. For the calculator values, round to three decimal places.

	$\mathcal{N}(-1)$	$\mathcal{N}(0)$	$\mathcal{N}(1)$	$\mathcal{N}(4)$	$\mathcal{N}(5)$
From graph					
From calculator					

(b) As indicated in the graph, the line $\mathcal{N} = 4$ appears to be an asymptote for the curve. Confirm this empirically by computing $\mathcal{N}(10)$, $\mathcal{N}(15)$, and $\mathcal{N}(20)$. Round each answer to eight decimal places.

(c) Use the graph to estimate, to the nearest integer, the value of t for which $\mathcal{N}(t) = 3$.

(d) Find the exact value of t for which $\mathcal{N}(t) = 3$. Evaluate the answer using a calculator, and check that it is consistent with the result in part (c).

G 56. (Continuation of Exercise 55) The author's ideas for this exercise are based on Professor Bonnie Shulman's article "Math-Alive! Using Original Sources to Teach Mathematics in Social Context," *Primus*, vol. VIII (March 1998).

(a) The function $\mathcal{N}$ in Exercise 55 expresses population as a function of time. But as pointed out by Professor Shulman, in Verhulst's original work it was the other way around; he expressed time as a function of population. In terms of our notation, we would say that he was studying the function $\mathcal{N}^{-1}$. Given $\mathcal{N}(t) = 4/(1 + 8e^{-t})$, find $\mathcal{N}^{-1}(t)$.

(b) Use a graphing utility to draw the graphs of $\mathcal{N}$, $\mathcal{N}^{-1}$, and the line $y = x$ in the viewing rectangle $[-3, 8, 2]$ by $[-3, 8, 2]$. Use true portions. (Why?)

(c) In the viewing rectangle $[0, 5, 1]$ by $[-3, 2, 1]$, draw the graphs of $y = \mathcal{N}^{-1}(t)$ and $y = \ln t$. Note that the two graphs have the same general shape and characteristics. In other words, Verhulst's logistic function (our $\mathcal{N}^{-1}$) appears log-like, or *logistique*, as Verhulst actually named it in French. (For details, both historical and mathematical, see the paper by Professor Shulman cited above.)

57. The following figure shows the graph of a logistic function $\mathcal{N}(t) = P/(1 + ae^{-bt})$. As indicated in the figure, the graph passes through the two points $(0, 1)$ and $(1, 2)$, and the asymptote is $\mathcal{N} = 5$. In this exercise, we determine the values of the constants P, a, and b.

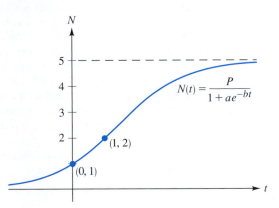

(a) As indicated in the figure, the logistic curve approaches an equilibrium population of 5. By definition, this is the value of the constant P, so the equation becomes

$$\mathcal{N}(t) = 5/(1 + ae^{-bt})$$

Now use the fact that the graph passes through the point $(0, 1)$ to obtain an equation that you can solve for a. Solve the equation; you should obtain $a = 4$.

(b) With $P = 5$ and $a = 4$, the logistic equation becomes $\mathcal{N}(t) = 5/(1 + 4e^{-bt})$. Use the fact that the graph of this equation passes through the point $(1, 2)$ to show that $b = \ln(8/3) \approx 0.9808$.

58. Biologist H. G. Thornton carried out an experiment in the 1920s to measure the growth of a colony of bacteria in a closed environment. As is common in the biology lab, Thornton measured the area of the colony, rather than count the number of individuals. (The reasoning is that the actual population size is directly proportional to the area, and the area is much easier to measure.) The following table and scatter plot summarize Thornton's measurements. Notice that the data points in the scatter plot suggest a logistic growth curve.

Days	0	1	2	3	4	5
Area of colony (cm²)	0.24	2.78	13.53	36.30	47.50	49.40

Source: H. G. Thornton, *Annals of Applied Biology*, 1922, p. 265

N [area of colony (cm²)]

(a) Use the technique shown in Exercise 57 to determine a logistic function $\mathcal{N}(t) = P/(1 + ae^{-bt})$ with a graph that passes through Thornton's two points $(0, 0.24)$ and $(2, 13.53)$. Assume that the equilibrium population is $P = 50$. Round the final values of a and b to two decimal places.

(b) Use the logistic function that you determined in part (a) to estimate the area of the colony after 1, 3, 4, and 5 days. Round your estimates to one decimal place. If you compare your estimates to the results that Thornton obtained in the laboratory, you'll see there is good agreement.

(c) Use the logistic function that you determined in part (a) to estimate the time at which the area of the colony was 10 cm². Express your answer in terms of days and hours, and round to the nearest half-hour.

59. The following data on the growth of *Lupinus albus*, a plant in the pea family, are taken from an experiment that was summarized in the classic text by Sir D'Arcy Wentworth Thompson, *On Growth and Form* (New York: Dover Publications, Inc., 1992).

Day	4	6	8	10	12
length (mm)	10.5	23.3	42.2	77.9	107.4

Day	14	16	18	20	21
length (mm)	132.3	149.7	158.1	161.4	161.6

(a) Follow Exercise 57 to determine an equation for a logistic function that models these data. Assume $P = 162$, $\mathcal{N}(0) = 1$, and $\mathcal{N}(10) = 77.9$.

(b) Using the model that you determined in part (a), compute $\mathcal{N}(4)$, $\mathcal{N}(8)$, $\mathcal{N}(12)$, and $\mathcal{N}(16)$. Round the answers to one decimal place. In each case, say whether your answer is higher or lower than the actual value.

(c) For the logistic model in general, it can be shown (using calculus) that the population is growing fastest at the instant when it reaches half the equilibrium level P. Use this fact to determine when the population in this experiment is growing fastest. Give your answer in the form d days, h hours, with the number of hours rounded to the nearest half-hour.

60. This exercise requires a graphing utility with a logistic regression feature. (This feature is included on the Texas Instruments *TI-83 Plus,* for example.)

(a) Use a graphing utility to create a scatter plot for the data in Exercise 59. Next, use the logistic regression option to find the equation of the logistic model that best fits the data.

(b) Based on the model that you determined in part (a), what is the equilibrium population P? (Remember that P has to be an integer.) Add the graph of the horizontal line $\mathcal{N} = P$ to the picture obtained in part (a).

(c) In the initial stages of logistic growth (when the population is small compared to the equilibrium level P), the growth resembles exponential growth. Demonstrate this as follows. Use a graphing utility to find an exponential growth model fitting only the first three data pairs in the table in Exercise 59. Add the graph of this exponential model to the picture that you obtained in part (b). Also include the graph of the horizontal line $\mathcal{N} = P/2$. Now (in complete sentences), summarize what you see.

61. (a) Let $\mathcal{N}(t) = \mathcal{N}_0 e^{kt}$. Show that $[\mathcal{N}(t+1) - \mathcal{N}(t)]/\mathcal{N}(t) = e^k - 1$. (This is actually done in detail in the text. So, ideally, you should look back only if you get stuck or want to check your answer.)

(b) Assume as given the following approximation, which was introduced in Exercise 26 of Section 5.2.

$$e^x \approx x + 1 \quad \text{provided } x \text{ is close to zero}$$

Use this approximation to explain why $e^k - 1 \approx k$,

provided k is close to zero.   *Remark*: Combining this result with that in part (a), we conclude that the relative growth rate for the function $\mathcal{N}(t) = \mathcal{N}_0 e^{kt}$ is approximately equal to the growth constant k. As explained in the text, this is one of the reasons why in applications we've not distinguished between the relative growth rate and the decay constant k.

62. In the text, we showed that the relative growth rate for the function $\mathcal{N}(t) = \mathcal{N}_0 e^{kt}$ is constant for all time intervals of unit length, $[t, t+1]$. Recall that we did this by computing the relative change $[\mathcal{N}(t+1) - \mathcal{N}(t)]/\mathcal{N}(t)$ and noting that the result was a constant, independent of t. (If you've completed the previous exercise, you've done this calculations for yourself.) Now, consider a time interval of arbitrary length, $[t, t+d]$. The relative change in the function $\mathcal{N}(t) = \mathcal{N}_0 e^{kt}$ over this time interval is $[\mathcal{N}(t+d) - \mathcal{N}(t)]/\mathcal{N}(t)$. Show that this quantity is a constant, independent of t. (The expression that you obtain for the constant will contain e and d, but not t. As a check on your work, replace d by 1 in the expression you obtain and make sure the result is the same as that in the text where we worked with intervals of length $d = 1$.)

63. Let $\mathcal{N} = \mathcal{N}_0 e^{kt}$. In this exercise we show that if Δt is very small, then $\Delta \mathcal{N}/\Delta t \approx k\mathcal{N}$. In other words, over very small intervals of time, the average rate of change of $\mathcal{N}$ is proportional to $\mathcal{N}$ itself.

(a) Show that the average rate of change of the function $\mathcal{N} = \mathcal{N}_0 e^{kt}$ on the interval $[t, t + \Delta t]$ is given by

$$\frac{\Delta \mathcal{N}}{\Delta t} = \frac{\mathcal{N}_0 e^{kt}(e^{k\Delta t} - 1)}{\Delta t} = \frac{\mathcal{N}(e^{k\Delta t} - 1)}{\Delta t}$$

(b) In Exercise 26 of Section 5.2, we saw that $e^x \approx x + 1$ when x is close to zero. Thus, if Δt is sufficiently small, we have $e^{k\Delta t} \approx k\,\Delta t + 1$. Use this approximation and the result in part (a) to show that $\Delta \mathcal{N}/\Delta t \approx kN$ when Δt is sufficiently close to zero.

PROJECT

A VARIABLE GROWTH CONSTANT?

Ⓖ As background for this project you need to have worked Exercise 39 in this section regarding the United Nation's proposed population limits. Or, if you are working in a group, have a group member present a summary of the problem and the results.

Over the past few decades, world population growth rates have, in general, been decreasing. (Over the period 1970–2000, for example, the rate fell from 2.0%/year to 1.4%/year.) In this project we consider a modification of the exponential model that takes this declining growth rate into account. In particular, we consider a growth equation of the form

$$\mathcal{N}(t) = \mathcal{N}_0 e^{[f(t)]t} \tag{1}$$

where the function f in the exponent describes the changing relative growth rate.

(a) The following table lists some relative growth rates for the world's population over the period 1985–2000. Set up a scatter plot for the information in the table, subject to the following specifications. For the horizontal time-axis, let $t = 0$ correspond to 1980. For the relative growth rates, use the decimal form, rather than the given percentage. Then use a graphing utility to determine the equation of the regression line $f(t) = mt + b$.

Year	1980	1985	1990	1995	2000
Relative growth rate (%/year)	1.8	1.8	1.7	1.6	1.4

Source: Population Reference Bureau

(b) In 1980, the world population was 4.448 billion. Use this value for $\mathcal{N}_0$ in equation (1). Also, in equation (1), replace $f(t)$ by the quantity $mt + b$ determined in part (a). Now, use a graphing utility to graph the resulting population growth model for the period 1980 to 2020 (i.e., $t = 0$ to $t = 40$). Use the graph to decide whether or not (according to this model) the projected population in 2020 exceeds United Nation's proposed limit of 7.2 billion.

(c) For purposes of comparison, add the graph of an exponential growth model to the viewing rectangle. For the model, use the value of $\mathcal{N}_0$ given in part (b) and refer to the table to obtain k. Does this model project a population excess of 7.2 billion in 2020?

(d) Switch to the viewing rectangle [0, 50, 200] by [0, 8, 4]. Give a description, in complete sentences, of the general behavior of the population after 2020 as projected by the model $\mathcal{N}(t) = \mathcal{N}_0 e^{[f(t)]t}$ in part (b). What do you think about this scenario?

Chapter 5

Summary of Principal Terms and Formulas

Terms or notation	Page reference	Comments
1. Exponential function with base b	367	The exponential function with base b is defined by the equation $y = b^x$. It is understood here that the base b is a positive number other than 1. More generally, functions of the form $y = ab^{cx}$ are also referred to as exponential functions.
2. The number e	375	The irrational number e is one of the basic constants in mathematics, as is the irrational number π. To five decimal places, the value of e is 2.71828. In calculus, e is the base most commonly used for exponential functions. The graph of the exponential function $y = e^x$ is shown in the box on page 376.
3. Instantaneous rate of change of the function $y = e^x$	378	The concept of the instantaneous rate of change of a function is introduced on pages 377–378. One of the distinguishing characteristics of the function $y = e^x$ is that the instantaneous rate of change at any point on the graph is just the y-coordinate of that point. See the examples in the box on page 378.
4. $\log_b x$	386	The expression $\log_b x$ denotes the exponent to which b must be raised to yield x. The equation $\log_b x = y$ is equivalent to $b^y = x$.
5. $\ln x$	390	The expression $\ln x$ means $\log_e x$. Logarithms to the base e are known as *natural logarithms*. For the graph of $y = \ln x$, see Figure 6(a) in Section 5.3.
6. $\log_a x = \dfrac{\log_b x}{\log_b a}$	405	This is the change-of-base formula for converting logarithms from one base to another.
7. $A = P\left(1 + \dfrac{r}{n}\right)^{nt}$	422	This formula gives the amount A that accumulates after t years when a principal of P dollars is invested at an annual rate r that is compounded n times per year.
8. $A = Pe^{rt}$	424	This formula gives the amount A that accumulates after t years when a principal of P dollars is invested at an annual rate r that is compounded continuously.
9. Doubling time; rule of 70	427	For money earning compound interest, the doubling time T_2 is the length of time required for an amount to double. As can be seen from the formula $T_2 = (\ln 2)/r$, the doubling time does not depend on the initial amount, but only on the interest rate r. (In the formula, the interest rate is expressed as a decimal.) The *rule of 70* gives an approximate value for the doubling time: $T_2 \approx 70/R$, where R is the interest rate expressed as a percentage.

Terms or notation	Page reference	Comments
10. $\mathcal{N} = \mathcal{N}_0 e^{kt}$ $(k > 0)$	434	This is the exponential growth model, where $\mathcal{N}_0$ is the size of the population at time $t = 0$, and k is a constant called the *growth constant*. In the applications we studied, we used the value of the relative growth rate for k. (As explained in the text, the two quantities are approximately equal when the values are near zero.)
11. $\mathcal{N} = \mathcal{N}_0 e^{kt}$ $(k < 0)$	441	In this exponential decay model for radioactive substances, $\mathcal{N}_0$ is the amount of the substance present at $t = 0$, and k is the decay constant.
12. Half-life	442	The half-life of a radioactive substance is the time required until only half of the initial amount remains. For a given radioactive substance, the half-life is independent of the amount initially present, as evidenced by the formula *half-life* $= (\ln \frac{1}{2})/k$.

Writing Mathematics

1. In Section 5.3, we defined logarithmic functions in terms of inverse functions.
   (a) Explain in general what is meant by a pair of inverse functions.
   (b) Assuming a knowledge of the function $f(x) = 2^x$, explain how to define and then graph the function $g(x) = \log_2 x$.

2. A student who was trying to simplify the expression $\dfrac{\ln(e^3)}{\ln(e^2)}$ wrote

$$\frac{\ln(e^3)}{\ln(e^2)} = \ln(e^3) - \ln(e^2) = 3 - 2 = 1$$

   (a) What is the error here? What do you think is the most probable reason for making this error?
   (b) Give the correct solution.

3. A student who wanted to simplify the expression $\dfrac{100}{\log_{10} 100}$ wrote

$$\frac{100}{\log_{10} 100} = \frac{\overset{1}{\cancel{100}}}{\log_{10} \underset{1}{\cancel{100}}} = \frac{1}{\log_{10}}$$

   Explain why this is nonsense, and then indicate the correct solution.

4. Look over the following method for solving the equation $\ln x^2 = 6$:

$$2 \ln x = 6$$
$$\ln x = 3$$
$$x = e^3$$

   As you can check, e^3 is a root of the given equation. The value $-e^3$, however, is also a root. Why is it that the demonstrated method fails to produce this root? Find a method that does produce both roots.

Chapter 5 | Review Exercises

1. Which is larger, $\log_5 126$ or $\log_{10} 999$?
2. Graph the function $y = 3^{-x} - 3$. Specify the domain, range, intercept(s), and asymptote.
3. Suppose that the population of a colony of bacteria increases exponentially. If the population at the start of an experiment is 8000, and 4 hours later it is 10,000, how long (from the start of the experiment) will it take for the population to reach 12,000? (Express the answer in terms of base e logarithms.)
4. Express $\log_{10} 2$ in terms of base e logarithms.
5. Let f be the function defined by

$$f(x) = \begin{cases} 2^{-x} & \text{if } x < 0 \\ x^2 & \text{if } x \geq 0 \end{cases}$$

   Sketch the graph of f and then use the horizontal line test to determine whether f is one-to-one.
6. Estimate 2^{60} in terms of an integral power of 10.
7. Solve for x: $\ln(x + 1) - 1 = \ln(x - 1)$.
8. Suppose that $5000 is invested at 8% interest compounded annually. How many years will it take for the money to double? Make use of the approximations $\ln 2 \approx 0.7$ and $\ln 1.08 \approx 0.08$ to obtain a numerical answer.
9. On the same set of axes, sketch the graphs of $y = e^x$ and $y = \ln x$. Specify the domain and range for each function.
10. Solve for x: $xe^x - 2e^x = 0$.
11. Simplify: $\log_9(1/27)$.
12. Given that $\ln A = a$, $\ln B = b$, and $\ln C = c$, express $\ln[(A^2\sqrt{B})/C^3]$ in terms of a, b, and c.
13. The half-life of plutonium-241 is 13 years. What is the decay constant? Use the approximation $\ln(1/2) \approx -0.7$ to obtain a numerical answer.
14. Express as a single logarithm with a coefficient of 1: $3\log_{10} x - \log_{10}(1 - x)$.
15. Solve for x, leaving your answer in terms of base e logarithms: $5e^{2-x} = 12$.
16. Let $f(x) = e^{x+1}$. Find a formula for $f^{-1}(x)$ and specify the domain of f^{-1}.
17. Suppose that in 1995 the population of a certain country was 2 million and increasing with a relative growth rate of 2%/year. Estimate the year in which the population will reach 3 million.
18. Simplify: $\ln e + \ln\sqrt{e} + \ln 1 + \ln e^{\ln 10}$.
19. A principal of $1000 is deposited at 10% per annum, compounded continuously. Estimate the doubling time and then sketch a graph that shows how the amount increases with time.
20. Simplify: $\ln(\log_8 56 - \log_8 7)$.

In Exercises 21–32, graph the function and specify the asymptote(s) and intercepts(s).

21. $y = e^x$
22. $y = -e^{-x}$
23. $y = \ln x$
24. $y = \ln(x + 2)$
25. $y = 2^{x+1} + 1$
26. $y = \log_{10}(-x)$
27. $y = (1/e)^x$
28. $y = (1/2)^{-x}$
29. $y = e^{x+1} + 1$
30. $y = -\log_2(x + 1)$
31. $y = \ln(e^x)$
32. $y = e^{\ln x}$

In Exercises 33–49, solve the equation for x. (When logarithms appear in your answer, leave the answer in that form, rather than using a calculator.)

33. $\log_4 x + \log_4(x - 3) = 1$
34. $\log_3 x + \log_3(2x + 5) = 1$
35. $\ln x + \ln(x + 2) = \ln 15$
36. $\log_6 \dfrac{x + 4}{x - 1} = 1$
37. $\log_2 x + \log_2(3x + 10) - 3 = 0$
38. $2\ln x - 1 = 0$
39. $3\log_9 x = 1/2$
40. $e^{2x} = 6$
41. $e^{1-5x} = 3\sqrt{e}$
42. $2^x = 100$
43. $\log_{10} x - 2 = \log_{10}(x - 2)$
44. $\log_{10}(x^2 - x - 10) = 1$
45. $\ln(x + 2) = \ln x + \ln 2$
46. $\ln(2x) = \ln 2 + \ln x$
47. $\ln x^4 = 4\ln x$
48. $(\ln x)^4 = 4\ln x$
49. $\log_{10} x = \ln x$
50. Solve for x: $(\ln x)/(\ln 3) = \ln x - \ln 3$. Use a calculator to evaluate your result and round to the nearest integer. *Answer:* 206,765

In Exercises 51–66, simplify the expression without using a calculator.

51. $\log_{10}\sqrt{10}$
52. $\log_7 1$
53. $\ln\sqrt[5]{e}$
54. $\log_3 54 - \log_3 2$
55. $\log_{10} \pi - \log_{10} 10\pi$
56. $\log_2 2$
57. 10^t, where $t = \log_{10} 16$
58. $e^{\ln 5}$
59. $\ln(e^4)$
60. $\log_{10}(10^{\sqrt{2}})$
61. $\log_{12} 2 + \log_{12} 18 + \log_{12} 4$
62. $(\log_{10} 8)/(\log_{10} 2)$
63. $(\ln 100)/(\ln 10)$
64. $\log_5 2 + \log_{1/5} 2$
65. $\log_2 \sqrt{16\sqrt[3]{2\sqrt{2}}}$
66. $\log_{1/8}(1/16) + \log_5 0.02$

In Exercises 67–70, express the quantity in terms of a, b, and c, where $a = \log_{10} A$, $b = \log_{10} B$, and $c = \log_{10} C$.

67. $\log_{10} A^2 B^3 \sqrt{C}$

68. $\log_{10} \sqrt[3]{AC/B}$

69. $16 \log_{10} \sqrt{A} \sqrt[4]{B}$

70. $6 \log_{10}[B^{1/3}/(A\sqrt{C})]$

In Exercises 71–76, find consecutive integers n and n + 1 such that the given expression lies between n and n + 1. Do not use a calculator.

71. $\log_{10} 209$

72. $\ln 2$

73. $\log_6 100$

74. $\log_{10}(1/12)$

75. $\log_{10} 0.003$

76. $\log_3 244$

77. (a) On the same set of axes, graph the curves $y = \ln(x + 2)$ and $y = \ln(-x) - 1$. According to your graph, in which quadrant do the two curves intersect?

(b) Find the x-coordinate of the intersection point.

78. The curve $y = ae^{bt}$ passes through the point $(3, 4)$ and has a y-intercept of 2. Find a and b.

79. A certain radioactive substance has a half-life of T years. Find the decay constant (in terms of T).

80. Find the half-life of a certain radioactive substance if it takes T years of one-third of a given sample to disintegrate. (Your answer will be in terms of T.)

81. A radioactive substance has a half-life of M minutes. What percentage of a given sample will remain after $4M$ minutes?

82. At the start of an experiment, a colony of bacteria has initial population a. After b hours, there are c bacteria present. Determine the population d hours after the start of the experiment.

83. The half-life of a radioactive substance is d days. If you begin with a sample weighing b grams, how long will it be until c grams remain?

84. Find the half-life of a radioactive substance if it takes D days for P percent of a given sample to disintegrate.

In Exercises 85–90, write the expression as a single logarithm with a coefficient of 1.

85. $\log_{10} 8 + \log_{10} 3 - \log_{10} 12$

86. $4 \log_{10} x - 2 \log_{10} y$

87. $\ln 5 - 3 \ln 2 + \ln 16$

88. $\ln(x^4 - 1) - \ln(x^2 - 1)$

89. $a \ln x + b \ln y$

90. $\ln(x^3 + 8) - \ln(x + 2) - 2 \ln(x^2 - 2x + 4)$

In Exercises 91–98, write the quantity using sums and differences of simpler logarithmic expressions. Express the answer so that logarithms of products, quotients, and powers do not appear.

91. $\ln \sqrt{(x - 3)(x + 4)}$

92. $\log_{10} \dfrac{x^2 - 4}{x + 3}$

93. $\log_{10} \dfrac{3}{\sqrt{1 + x}}$

94. $\ln \dfrac{x^2 \sqrt{2x + 1}}{2x - 1}$

95. $\ln\left(\dfrac{1 + 2e}{1 - 2e}\right)^3$

96. $\log_{10} \sqrt[3]{x\sqrt{1 + y^2}}$

97. Suppose that A dollars are invested at $R\%$ compounded annually. How many years will it take for the money to double?

98. A sum of \$2800 is placed in a savings account at 9% per annum. How much is in the account after 2 years if the interest is compounded quarterly?

99. A bank offers an interest rate of 9.5% per annum, compounded monthly. Compute the effective interest rate.

100. A sum of D dollars is placed in an account at $R\%$ per annum, compounded continuously. When will the balance reach E dollars?

101. Compute the doubling time for a sum of D dollars invested at $R\%$ per annum, compounded continuously.

102. Your friend invests D dollars at $R\%$ per annum, compounded semiannually. You invest an equal amount at the same yearly rate, but compounded daily. How much larger is your account than your friend's after T years?

103. (a) You invest \$660 at 5.5% per annum, compounded quarterly. How long will it take for your balance to reach \$1000? Round off your answer to the next quarter of a year.

(b) You invest D dollars at $R\%$ per annum, compounded quarterly. How long will it take for your balance to reach nD dollars? (Assume that $n > 1$.)

In Exercises 104–107, find the domain of each function.

104. (a) $y = \ln x$

(b) $y = e^x$

105. (a) $y = \log_{10} \sqrt{x}$

(b) $y = \sqrt{\log_{10} x}$

106. (a) $f \circ g$, where $f(x) = \ln x$ and $g(x) = e^x$

(b) $g \circ f$, where f and g are as in part (a)

107. $y = \dfrac{2 + \ln x}{2 - \ln x}$

108. Find the range of the function defined by $y = \dfrac{e^x + 1}{e^x - 1}$.

Chapter 5   Test

1. Graph the function $y = 2^{-x} - 3$. Specify the domain, range, intercepts, and asymptote.

2. Suppose that the population of a colony of bacteria increases exponentially. At the start of an experiment there are 6000 bacteria, and 1 hour later the population is 6200. How long (from the start of the experiment) will it take for the population to reach 10,000? Give two forms for the answer: one in terms of base e logarithms and the other a calculator approximation rounded to the nearest hour.

3. Which is larger, $\log_2 17$ or $\log_3 80$? Explain the reasons for your answer.

4. Express $\log_2 15$ in terms of base e logarithms.

5. Estimate 2^{40} in terms of an integral power of 10.

6. Suppose that $9500 is invested at 6% interest, compounded annually. How long will it take for the amount in the account to reach $12,000?

7. (a) For which values of x is the identity $e^{\ln x} = x$ valid?
   (b) On the same set of axes, graph $y = e^x$ and $y = \ln x$.

8. Given that $\log_{10} A = a$ and $\log_{10} B = b$, express $\log_{10}(A^3/\sqrt{B})$ in terms of a and b.

9. Simplify each expression.
   (a) $\log_5(1/\sqrt{5})$;
   (b) $\ln e^2 + \ln 1 - e^{\ln 3}$.

10. Solve for x: $x^3 e^x - 4xe^x = 0$.

11. The half-life of a radioactive substance is 4 days.
    (a) Find the decay constant.
    (b) How much of an initial 2-g sample will remain after 10 days?

12. Express as a single logarithm with a coefficient of 1: $2 \ln x - \ln \sqrt[3]{x^2 + 1}$.

13. Solve for x, leaving your answer in terms of base e logarithms: $-2e^{3x-1} = 9$.

14. Let $g(x) = \log_{10}(x - 1)$. Find a formula for $g^{-1}(x)$ and specify the range of g^{-1}.

15. A principal of $12,000 is deposited at 6% per annum, compounded continuously.
    (a) Use the rule of 70 to estimate the doubling time.
    (b) Sketch a graph showing how the amount increases with time.

16. (a) For which values of x is it true that $\ln(x^2) = 2 \ln x$?
    (b) For which values of x is it true that $(\ln x)^2 = 2 \ln x$?

17. Let $f(x) = e^x$.
    (a) Compute the average rate of change of the function f over the interval $[\ln 3, \ln 4]$. Give two forms for the answer: an expression simplified as much as possible and a calculator approximation rounded to two decimal places.
    (b) Specify the instantaneous rate of change of the function f at the point on the graph where $x = \ln 4$.

18. Solve for x: $\frac{1}{2} - \log_{16}(x - 3) = \log_{16} x$.

19. Solve for x: $6e^{2x} - 5e^x = 6$.

20. Solve each inequality:
    (a) $5(4 - 0.3^x) > 12$;
    (b) $\ln x + \ln(x - 3) \leq \ln 4$.

CHAPTER

6

Systems of Equations

Many problems in a variety of disciplines give rise to . . . [linear] systems. For example, in physics, in order to find the currents in an electrical circuit, . . . a system of linear equations must be solved. In chemistry, the balancing of chemical equations requires the solution of a system of linear equations. And in economics, the Leontief input-output model reduces problems concerning the production and consumption of goods to systems of linear equations. —Leslie Hogben in Elementary Linear Algebra (Pacific Grove, Calif.: Brooks/Cole, 1999)

In this chapter we consider systems of equations. Roughly speaking, a system of equations is just a collection of equations with a common set of unknowns. In solving such systems, we try to find values for the unknowns that simultaneously satisfy each of the equations in the system. Matrices are introduced as a tool for solving systems of equations. In this chapter you'll see systems of equations and/or matrices used to:

- Determine conditions for market equilibrium (Example 6 in Section 6.1)
- Solve an airline scheduling problem (Example 6 in Section 6.2)
- Determine production requirements in various sectors of an economy using input-output analysis (projects following Sections 6.2 and 6.4)
- Analyze a communications network (project following Section 6.3)
- Encode and decode messages (Examples 4 and 5 in Section 6.4)
- Find the most cost-efficient routes for transporting a product from warehouses to retail stores

6.1 SYSTEMS OF TWO LINEAR EQUATIONS IN TWO UNKNOWNS

Both in theory and in applications, it's often necessary to solve two equations in two unknowns. You may have been introduced to the idea of simultaneous equations in a previous course in algebra; however, to put matters on a firm foundation, we begin here with the basic definitions. By a **linear equation in two variables** we mean an equation of the form

$$ax + by = c$$

where the constants a and b are not both zero. The two variables needn't always be denoted by the letters x and y, of course; it is the *form* of the equation that matters. Table 1 displays some examples.

An ordered pair of numbers (x_0, y_0) is said to be a **solution of the linear equation** $ax + by = c$ provided we obtain a true statement when we replace x and y in the equation by x_0 and y_0, respectively. For example, the ordered pair $(3, 2)$ is a solution of the equation $x - y = 1$, since $3 - 2 = 1$. On the other hand, $(2, 3)$ is not a solution of $x - y = 1$, since $2 - 3 \neq 1$.

Now consider a **system** of two linear equations in two unknowns:

$$\begin{cases} ax + by = c \\ dx + ey = f \end{cases}$$

463

If we can find an ordered pair that is a solution to both equations, then we say that ordered pair is a **solution of the system.** Sometimes, to emphasize the fact that a solution must satisfy both equations, we refer to the system as a pair of **simultaneous equations.** A system that has at least one solution is said to be **consistent.** If there are no solutions, the system is **inconsistent.**

TABLE 1

Equations in two variables	Is it linear?	
	Yes	No
$3x - 8y = 12$	✓	
$-s + 4t = 0$	✓	
$2x - 3y^2 = 1$		✓
$y = 4 - 2x$	✓	
$\dfrac{4}{u} + \dfrac{5}{v} = 3$		✓

EXAMPLE 1 Recognizing a solution of a system of equations

Consider the system

$$\begin{cases} x + y = 2 \\ 2x - 3y = 9 \end{cases}$$

(a) Is $(1, 1)$ a solution of the system?
(b) Is $(3, -1)$ a solution of the system?

SOLUTION

(a) Although $(1, 1)$ is a solution of the first equation, it is not a solution of the system because it does not satisfy the second equation. (Check this for yourself.)

(b) $(3, -1)$ satisfies the first equation:

$$3 + (-1) \stackrel{?}{=} 2 \qquad \text{True}$$

$(3, -1)$ satisfies the second equation:

$$2(3) - 3(-1) \stackrel{?}{=} 9$$
$$6 + 3 \stackrel{?}{=} 9 \qquad \text{True}$$

Since $(3, -1)$ satisfies both equations, it is a solution of the system.

We can gain an important perspective on systems of linear equations by looking at Example 1 in graphical terms. Table 2 shows how each of the statements in that example can be rephrased using the geometric ideas with which we are already familiar.

TABLE 2

Algebraic idea	Corresponding geometric idea
1. The ordered pair $(1, 1)$ is a solution of the equation $x + y = 2$.	1. The point $(1, 1)$ lies on the line $x + y = 2$. See Figure 1.
2. The ordered pair $(1, 1)$ is not a solution of the equation $2x - 3y = 9$.	2. The point $(1, 1)$ does not lie on the line $2x - 3y = 9$. See Figure 1.
3. The ordered pair $(3, -1)$ is a solution of the system $$\begin{cases} x + y = 2 \\ 2x - 3y = 9 \end{cases}$$	3. The point $(3, -1)$ lies on both of the lines $x + y = 2$ and $2x - 3y = 9$. See Figure 1.

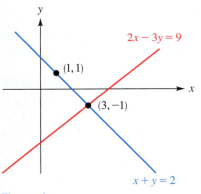

Figure 1

In Example 1, we verified that $(3, -1)$ is a solution of the system

$$\begin{cases} x + y = 2 \\ 2x - 3y = 9 \end{cases}$$

Are there any other solutions of this particular system? No: Figure 1 shows us that there are no other solutions, since $(3, -1)$ is clearly the only point common to both lines. In a moment we'll look at two important methods for solving systems of linear equations in two unknowns. But even before we consider these methods, we can say something about the solutions of linear systems.

▌PROPERTY SUMMARY   Possibilities for Solutions of Linear Systems

Given a system of two linear equation in two unknowns, exactly one of the following cases must occur.

CASE 1   The graphs of the two linear equations intersect in exactly one point. Thus there is exactly one solution to the system. See Figure 2.

CASE 2   The graphs of the two linear equations are parallel lines. Therefore the lines do not intersect, and the system has no solution. See Figure 3.

CASE 3   The two equations actually represent the same line. Thus, there are infinitely many points of intersection and correspondingly infinitely many solutions. See Figure 4.

Figure 2
A consistent system with exactly one solution.

Figure 3
An inconsistent system has no solution.

Figure 4
A consistent system with infinitely many solutions.

We are going to review two methods from intermediate algebra for solving systems of two linear equations in two unknowns. These methods are the **substitution method** and the **addition–subtraction method.** We'll begin by demonstrating the substitution method. Consider the system

$$\begin{cases} 3x + 2y = 17 & (1) \\ 4x - 5y = -8 & (2) \end{cases}$$

We first choose one of the two equations and then use it to express one of the variables in terms of the other. In the case at hand, neither equation appears simpler than the other, so let's just start with the first equation and solve for x in terms of y. We have

$$3x = 17 - 2y$$

$$x = \frac{1}{3}(17 - 2y) \qquad (3)$$

Now we use equation (3) to substitute for x in the equation that we have not yet used—namely, equation (2). This yields

$$4\left[\frac{1}{3}(17 - 2y)\right] - 5y = -8$$

$$4(17 - 2y) - 15y = -24 \qquad \textcolor{red}{\text{multiplying by 3}}$$

$$-23y = -92$$

$$y = 4$$

The value $y = 4$ that we have just obtained can now be used in equation (3) to find x. Replacing y with 4 in equation (3) yields

$$x = \frac{1}{3}[17 - 2(4)] = \frac{1}{3}(9) = 3$$

We have now found that $x = 3$ and $y = 4$. As you can easily check, this pair of values indeed satisfies both of the original equations. We write our solution as the ordered pair $(3, 4)$. Figure 5 summarizes the situation. It shows that the system is consistent and that $(3, 4)$ is the only solution.

Generally speaking, it is not necessary to graph the equations in a given system in order to decide whether the system is consistent. Rather, this information will emerge as you attempt to follow an algebraic method of solution. Examples 2 and 3 will illustrate this.

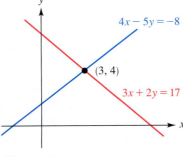

$4x - 5y = -8$

$(3, 4)$

$3x + 2y = 17$

Figure 5

EXAMPLE **2**   **Using the substitution method**

Solve the system

$$\begin{cases} \dfrac{3}{2}x - 3y = -9 \\ x - 2y = 4 \end{cases}$$

SOLUTION
We use the substitution method. Since it is easy to solve the second equation for x, we begin there:

$$x - 2y = 4$$

$$x = 4 + 2y$$

Now we substitute this result in the first equation of our system to obtain

$$\frac{3}{2}(4 + 2y) - 3y = -9$$

$$6 + 3y - 3y = -9$$

$$6 = -9 \qquad \text{False}$$

Since the substitution process leads us to this obviously false statement, we conclude that the given system has no solution; that is, the system is inconsistent.

QUESTION   What can you say about the graphs of the two given equations?

EXAMPLE 3   Another substitution example

Solve the system

$$\begin{cases} 3x + 4y = 12 \\ 2y = 6 - \dfrac{3}{2}x \end{cases}$$

SOLUTION
We use the method of substitution. Since it is easy to solve the second equation for y, we begin there:

$$2y = 6 - \frac{3}{2}x \qquad \text{and therefore} \qquad y = 3 - \frac{3}{4}x$$

Now we use this result to substitute for y in the first equation of the original system. The result is

$$3x + 4\left(3 - \frac{3}{4}x \right) = 12$$
$$3x + 12 - 3x = 12$$
$$3x - 3x = 12 - 12$$
$$0 = 0 \qquad \text{Always true}$$

This last identity imposes no restrictions on x. Graphically speaking, this says that our two lines intersect for every value of x. In other words, the two lines coincide. We could have foreseen this initially had we solved both equations for y. As you can verify, the result in both cases is

$$y = -\frac{3}{4}x + 3$$

Every point on this line yields a solution to our system of equations. In summary, then, our system is consistent and the solutions to the system have the form $(x, -\frac{3}{4}x + 3)$, where x can be any real number. For instance, when $x = 0$, we obtain the solution $(0, 3)$. When $x = 1$, we obtain the solution $(1, \frac{9}{4})$. The idea here is that for *each* value of x we obtain a solution; thus, there are infinitely many solutions.

Now let's turn to the addition–subtraction method of solving systems of equations. By way of example, consider the system

$$\begin{cases} 2x + 3y = 5 \\ 4x - 3y = -1 \end{cases}$$

Notice that if we add these two equations, the result is an equation involving only the unknown x:

$$6x = 4$$
$$x = 4/6 = 2/3$$

There are now several ways in which the corresponding value of y can be obtained. As you can easily check, substituting the value $x = 2/3$ in either of the original equations leads to the result $y = 11/9$.

Another way to find y is by multiplying both sides of the first equation by -2. (You'll see why in a moment.) We display the work this way:

$$2x + 3y = 5 \quad \xrightarrow{\text{Multiply by } -2} \quad -4x - 6y = -10$$

$$4x - 3y = -1 \quad \xrightarrow{\text{No change}} \quad 4x - 3y = -1$$

Adding the last two equations then gives us $-9y = -11$, and therefore $y = 11/9$. The required solution is then $(2/3, 11/9)$.

In the previous example, we were able to find x directly by adding the two equations. As the next example shows, it may be necessary first to multiply both sides of each equation by an appropriate constant.

EXAMPLE 4 Using the addition–subtraction method

Solve the system

$$\begin{cases} 5x - 3y = 4 \\ 2x + 4y = 1 \end{cases}$$

SOLUTION

To eliminate x, we could multiply the second equation by $5/2$ and then subtract the resulting equation from the first equation. However, to avoid working with fractions, we proceed as follows.

$$5x - 3y = 4 \quad \xrightarrow{\text{Multiply by 2}} \quad 10x - 6y = 8 \qquad (4)$$

$$2x + 4y = 1 \quad \xrightarrow{\text{Multiply by 5}} \quad 10x + 20y = 5 \qquad (5)$$

Subtracting equation (5) from equation (4) then yields

$$-26y = 3$$
$$y = -3/26$$

To find x, we return to the original system and work in a similar manner:

$$5x - 3y = 4 \quad \xrightarrow{\text{Multiply by 4}} \quad 20x - 12y = 16 \qquad (4)$$

$$2x + 4y = 1 \quad \xrightarrow{\text{Multiply by 3}} \quad 6x + 12y = 3 \qquad (5)$$

Upon adding the last two equations, we obtain

$$26x = 19$$
$$x = 19/26$$

The solution of the given system of equations is therefore $(19/26, -3/26)$.

We conclude this section with some problems that can be solved using simultaneous equations.

EXAMPLE 5 | Finding the equation of a parabola through given points

Determine the constants b and c so that the parabola $y = x^2 + bx + c$ passes through the points $(-3, 1)$ and $(1, -2)$.

SOLUTION
Since the point $(-3, 1)$ lies on the curve $y = x^2 + bx + c$, the coordinates must satisfy the equation. Thus we have

$$1 = (-3)^2 + b(-3) + c$$
$$-8 = -3b + c \qquad (6)$$

This gives us one equation in two unknowns. We need another equation involving b and c. Since the point $(1, -2)$ also lies on the graph of $y = x^2 + bx + c$, we must have

$$-2 = 1^2 + b(1) + c$$
$$-3 = b + c \qquad (7)$$

Rewriting equations (6) and (7), we have the system

$$\begin{cases} -3b + c = -8 & (8) \\ b + c = -3 & (9) \end{cases}$$

Subtracting equation (9) from (8) then yields

$$-4b = -5 \qquad \text{and therefore} \qquad b = 5/4$$

One way to obtain the corresponding value of c is to replace b by $5/4$ in equation (9). This yields

$$\frac{5}{4} + c = -3$$

$$c = -3 - \frac{5}{4} = -\frac{17}{4}$$

The required values of b and c are therefore

$$b = 5/4 \qquad c = -17/4$$

(Exercise 39(b) at the end of this section asks you to check that the parabola $y = x^2 + \frac{5}{4}x - \frac{17}{4}$ indeed passes through the given points.)

The next example involves an application in the economics of supply and demand. Figure 6 shows *supply* and *demand models* for a certain commodity. In the

Figure 6
Supply function:
$q = 30p + 100 \quad (p \geq 20)$

Demand function:
$q = -40p + 12{,}700 \quad (p \geq 20)$

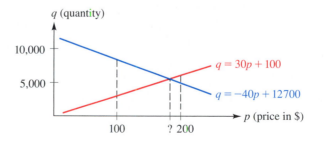

figure, the **supply function** $q = 30p + 100$ ($p \geq 20$) tells us the quantity q that the manufacturer will produce and supply (to the stores) when the selling price is p dollars per item. Note that this supply function is increasing; the higher the selling price, the more items the manufacturer will produce. The **demand function** in Figure 6, $q = -40p + 12,700$ ($p \geq 20$), gives us the quantity q that can be sold when the selling price is p dollars per item. Note that the demand function is decreasing; the higher the price, the fewer people there are who want to buy.

To see how these functions work, suppose that the selling price is $p = \$100$. Then we have

Supply: $\quad q = 30p + 100 = 30(100) + 100 = 3100$

Demand: $\quad q = -40p + 12,700 = -40(100) + 12,700 = 8700$

Thus, setting the market price at $100 per item will create a shortage; there are 8700 customers who will buy the item at this price, but only 3100 items are available. We can see this qualitatively in Figure 6. For $p = \$100$, the graph of the (blue) demand function is above that of the (red) supply function. In other words, demand exceeds supply in this case. Figure 6 also shows that if the price is set at $200, rather than $100, then the demand is less than the supply. So, there is a surplus in this case. The point where the graph of the supply function intersects the graph of the demand function is called the market **equilibrium point** for the given commodity. The values of p and q at this point are called the **equilibrium price** and the **equilibrium quantity,** respectively. At the equilibrium point, there is neither a surplus nor a shortage, and supply equals demand. In Figure 6, the equilibrium price is indicated with a question mark. We determine this price in Example 6.

EXAMPLE | 6 | **Using supply and demand functions**

Assume that the supply and demand functions for a commodity are as given in Figure 6.

$$\begin{cases} q = 30p + 100 \\ q = -40p + 12,700 \end{cases}$$

(As in Figure 6, we are assuming that p is in dollars and $p \geq 20$.) Find the equilibrium price and the corresponding equilibrium quantity.

SOLUTION

Use the first equation to substitute for q in the second equation. This yields

$$30p + 100 = -40p + 12,700$$
$$70p = 12,600$$
$$p = 12,600/70 = 1260/7 = 180$$

The equilibrium price is therefore $180. Notice that this is consistent with Figure 6, where the equilibrium price is between $100 and $200, closer to the latter than the former. For the equilibrium quantity, we can substitute $p = 180$ in either equation of the given system. Using the first equation, we obtain $q = 30(180) + 100 = 5500$. The equilibrium quantity is 5500. This too is consistent with Figure 6 (Why?).

In the next example, we solve a mixture problem using a system of two linear equations.

EXAMPLE	7	Solving a mixture problem

Suppose that a chemistry student can obtain two acid solutions from the stockroom. The first solution is 20% acid and the second solution is 45% acid. (The percentages are by volume.) How many milliliters of each solution should the student mix together to obtain 100 ml of a 30% acid solution?

SOLUTION
We begin by assigning letters to denote the required quantities.

Let x denote the number of milliliters of the 20% solution to be used.

Let y denote the number of milliliters of the 45% solution to be used.

We summarize the data in Table 3. Since the final mixture must total 100 ml, we have the equation

$$x + y = 100 \qquad (10)$$

TABLE 3

Type of solution	Number of ml	Percent of acid	Total acid (ml)
First solution (20% acid)	x	20	$(0.20)x$
Second solution (45% acid)	y	45	$(0.45)y$
Mixture	$x + y$	30	$(0.30)(x + y)$

This gives us one equation in two unknowns. However, we need a second equation. Looking at the data in the right-hand column of Table 3, we can write

$$\underbrace{0.20x}_{\substack{\text{Amount of acid} \\ \text{in } x \text{ ml of the} \\ \text{20\% solution}}} + \underbrace{0.45y}_{\substack{\text{Amount of acid} \\ \text{in } y \text{ ml of the} \\ \text{45\% solution}}} = \underbrace{(0.30)(x + y)}_{\substack{\text{Amount of acid in} \\ \text{the final mixture}}}$$

Thus,

$$\begin{aligned} 0.20x + 0.45y &= 0.30(x + y) \\ 20x + 45y &= 30(x + y) \\ 4x + 9y &= 6(x + y) = 6x + 6y \\ -2x + 3y &= 0 \qquad (11) \end{aligned}$$

Equations (10) and (11) can be solved by either the substitution method or the addition method. As Exercise 40 at the end of this section asks you to show, the results are

$$x = 60 \text{ ml} \quad \text{and} \quad y = 40 \text{ ml}$$

EXERCISE SET 6.1   *www*

A

1. Which of the following are linear equations in two variables?
   (a) $3x + 3y = 10$
   (c) $u - v = 1$
   (b) $2x + 4xy + 3y = 1$
   (d) $x = 2y + 6$

2. Which of the following are linear equations in two variables?
   (a) $y = x$
   (c) $\dfrac{4}{x} - \dfrac{3}{y} = -1$
   (b) $y = x^2$
   (d) $2w + 8z = -4w + 3$

3. Is $(5, 1)$ a solution of the following system?
$$\begin{cases} 2x - 8y = 2 \\ 3x + 7y = 22 \end{cases}$$

4. Is $(14, -2)$ a solution of the following system?
$$\begin{cases} x + y = 12 \\ x - y = 4 \end{cases}$$

5. Is $(0, -4)$ a solution of the following system?
$$\begin{cases} \frac{1}{6}x + \frac{1}{2}y = -2 \\ \frac{2}{3}x + \frac{3}{4}y = 2 \end{cases}$$

6. Is $(12, -8)$ a solution of the system in Exercise 5?

G *In Exercises 7–10, you are given a system of two linear equations. By graphing the pair of equations, determine which one of the three cases described in Figures 2 through 4 (on page 465) applies. (You're not being asked to solve the system.)*

7. $\begin{cases} 3x + 7y = 10 \\ 6x - 3y = 1 \end{cases}$
8. $\begin{cases} y = \sqrt{3}(1 - 3x)/3 \\ \sqrt{3}y + 3x - 1 = 0 \end{cases}$

9. $\begin{cases} 5y = 10.5x - 25.5 \\ 21x = 50 + 10y \end{cases}$
10. $\begin{cases} 2y = x - 18 \\ y = 0.4x + 1 \end{cases}$

In Exercises 11–18, use the substitution method to find all solutions of each system.

11. $\begin{cases} 4x - y = 7 \\ -2x + 3y = 9 \end{cases}$
12. $\begin{cases} 3x - 2y = -19 \\ x + 4y = -4 \end{cases}$

13. $\begin{cases} 6x - 2y = -3 \\ 5x + 3y = 4 \end{cases}$
14. $\begin{cases} 4x + 2y = 3 \\ 10x + 4y = 1 \end{cases}$

15. $\begin{cases} \frac{3}{2}x - 5y = 1 \\ x + \frac{3}{4}y = -1 \end{cases}$
16. $\begin{cases} 13x - 8y = -3 \\ -7x + 2y = 0 \end{cases}$

17. $\begin{cases} 4x + 6y = 3 \\ -6x - 9y = -\frac{9}{2} \end{cases}$
18. $\begin{cases} -\frac{2}{5}x + \frac{1}{4}y = 3 \\ \frac{1}{4}x - \frac{2}{5}y = -3 \end{cases}$

In Exercises 19 and 20:

G (a) *Graph the pair of equations, and by zooming in on the intersection point, estimate the solution of the system (each value to the nearest one-tenth).*

(b) *Use the substitution method to determine the solution. Check that your answer is consistent with the graphical estimate in part (a).*

19. $\begin{cases} 0.02x - 0.03y = 1.06 \\ 0.75x + 0.50y = -0.01 \end{cases}$
20. $\begin{cases} \sqrt{2}x - \sqrt{3}y = \sqrt{3} \\ \sqrt{3}x - \sqrt{8}y = \sqrt{2} \end{cases}$

In Exercises 21–28, use the addition–subtraction method to find all solutions of each system of equations.

21. $\begin{cases} 5x + 6y = 4 \\ 2x - 3y = -3 \end{cases}$
22. $\begin{cases} -8x + y = -2 \\ 4x - 3y = 1 \end{cases}$

23. $\begin{cases} 4x + 13y = -5 \\ 2x - 54y = -1 \end{cases}$
24. $\begin{cases} 16x - 3y = 100 \\ 16x + 10y = 10 \end{cases}$

25. $\begin{cases} \frac{1}{4}x - \frac{1}{3}y = 4 \\ \frac{2}{7}x - \frac{1}{7}y = \frac{1}{10} \end{cases}$ *Suggestion for Exercise 25*: First clear both equations of fractions.

26. $\begin{cases} 2.1x - 3.5y = 1.2 \\ 1.4x + 2.6y = 1.1 \end{cases}$
27. $\begin{cases} 8x + 16y = 5 \\ 2x + 5y = \frac{5}{4} \end{cases}$

28. $\begin{cases} \sqrt{6}x - \sqrt{3}y = 3\sqrt{2} - \sqrt{3} \\ \sqrt{2}x - \sqrt{5}y = \sqrt{6} + \sqrt{5} \end{cases}$

29. Find b and c, given that the parabola $y = x^2 + bx + c$ passes through $(0, 4)$ and $(2, 14)$.

30. Determine the constants a and b, given that the parabola $y = ax^2 + bx + 1$ passes through $(-1, 11)$ and $(3, 1)$.

31. Determine the constants A and B, given that the line $Ax + By = 2$ passes through the points $(-4, 5)$ and $(7, -9)$.

32. (a) Determine constants a and b so that the graph of $y = ax^3 + b$ passes through the two points $(2, 1)$ and $(-2, -7)$.

G (b) Using the values for a and b determined in part (a), graph the equation $y = ax^3 + b$ and see that it appears to pass through the two given points.

As background for Exercises 33 and 34, you need to have read the discussion on supply and demand preceding Example 6, as well as Example 6 itself. In each exercise, assume that the price p is in dollars.

33. Assume that the supply and demand functions for a commodity are as follows.
    Supply: $q = 200p$; Demand: $q = 9600 - 400p$.
    (a) If the price is set at $6, will there be a shortage or a surplus of the commodity? What if the price is doubled?
    (b) If the price is $20, will there be a shortage or a surplus?
    (c) Find the equilibrium price and the corresponding equilibrium quantity.

34. Assume that the supply and demand functions for a commodity are as follows.
    Supply: $q = 15p$; Demand: $q = 12{,}493 - 50p$.
    (a) If the price is set at $50, will there be a shortage or a surplus of the commodity? What if the price is tripled?
    (b) Show that if the price is $200, there will be a surplus. How many items will be left unsold?
    (c) Find the equilibrium price and the corresponding equilibrium quantity.

35. A student in a chemistry laboratory has access to two acid solutions. The first solution is 10% acid and the second is 35% acid. (The percentages are by volume.) How many cubic centimeters of each should she mix together to obtain 200 cm^3 of a 25% acid solution?

36. One salt solution is 15% salt and another is 20% salt. How many cubic centimeters of each solution must be mixed to obtain 50 cm³ of a 16% salt solution?

37. A shopkeeper has two types of coffee beans on hand. One type sells for $5.20/lb, the other for $5.80/lb. How many pounds of each type must be mixed to produce 16 lb of a blend that sells for $5.50/lb?

38. A certain alloy contains 10% tin and 30% copper. (The percentages are by weight.) How many pounds of tin and how many pounds of copper must be melted with 1000 lb of the given alloy to yield a new alloy containing 20% tin and 35% copper?

39. In this exercise you'll check the result of Example 5—first visually, then algebraically.
(a) Graph the parabola $y = x^2 + \frac{5}{4}x - \frac{17}{4}$ in an appropriate viewing rectangle to see that it appears to pass through the two points $(-3, 1)$ and $(1, -2)$.
(b) Verify algebraically that the two points $(-3, 1)$ and $(1, -2)$ indeed lie on the parabola $y = x^2 + \frac{5}{4}x - \frac{17}{4}$.

40. Consider the following system from Example 7:

$$\begin{cases} x + y = 100 \\ -2x + 3y = 0 \end{cases}$$

(a) Solve this system using the method of substitution. (As stated in Example 7, you should obtain $x = 60, y = 40$.)
(b) Solve the system using the addition–subtraction method.

B

41. Find constants a and b so that $(8, -7)$ is the solution of the system

$$\begin{cases} ax + by = 10 \\ bx + ay = -5 \end{cases}$$

42. (a) Sketch the triangular region in the first quadrant bounded by the lines $y = 5x$, $y = -3x + 6$, and the x-axis. One vertex of this triangle is the origin. Find the coordinates of the other two vertices. Then use your answers to compute the area of the triangle.
(b) More generally now, express the area of the shaded triangle in the following figure in terms of m, M, and b. Then use your result to check the answer you obtained in part (a) for that area.

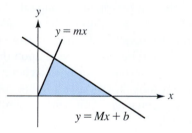

43. Find x and y in terms of a and b:

$$\begin{cases} \dfrac{x}{a} + \dfrac{y}{b} = 1 \\ \dfrac{x}{b} + \dfrac{y}{a} = 1 \end{cases}$$

Does your solution impose any conditions on a and b?

44. Solve the following system for x and y in terms of a and b, where $a \neq b$:

$$\begin{cases} ax + by = 1/a \\ b^2x + a^2y = 1 \end{cases}$$

45. Solve the following system for x and y in terms of a and b, where $a \neq b$:

$$\begin{cases} ax + a^2y = 1 \\ bx + b^2y = 1 \end{cases}$$

Does your solution impose any additional conditions on a and b?

46. Solve the following system for s and t:

$$\begin{cases} \dfrac{3}{s} - \dfrac{4}{t} = 2 \\ \dfrac{5}{s} + \dfrac{1}{t} = -3 \end{cases}$$

Hint: Make the substitutions $1/s = x$ and $1/t = y$ in order to obtain a system of two linear equations.

47. Solve the following system for s and t:

$$\begin{cases} \dfrac{1}{2s} - \dfrac{1}{2t} = -10 \\ \dfrac{2}{s} + \dfrac{3}{t} = 5 \end{cases}$$

(Use the hint in Exercise 46.)

48. Consider the following system: $\begin{cases} 2x^2 + 2y^2 = 55 \\ 4x^2 - 8y^2 = k \end{cases}$
(a) Assuming $k = 109$, solve the system.   *Hint:* The substitutions $u = x^2$ and $v = y^2$ will give you a linear system.
(b) Follow part (a) using: $k = 110$; $k = 111$.
(c) Use a graphing utility to shed light on why the number of solutions is different for each of the values of k considered in parts (a) and (b).

In Exercises 49–55, find all solutions of the given systems. For Exercises 51–55, use a calculator to round the final answers to two decimal places.

49.
$$\begin{cases} \dfrac{2w - 1}{3} + \dfrac{z + 2}{4} = 4 \\ \dfrac{w + 3}{2} - \dfrac{w - z}{3} = 3 \end{cases}$$

50. $\dfrac{x - y}{2} = \dfrac{x + y}{3} = 1$

51. $\begin{cases} 2\ln x - 5\ln y = 11 \\ \ln x + \ln y = -5 \end{cases}$

Hint: Let $u = \ln x$ and $v = \ln y$.

52. $\begin{cases} 3\ln x + \ln y = 3 \\ 4\ln x - 6\ln y = -7 \end{cases}$

53. $\begin{cases} e^x - 3e^y = 2 \\ 3e^x + e^y = 16 \end{cases}$

Hint: Let $u = e^x$ and $v = e^y$.

54. $\begin{cases} e^x + 2e^y = 4 \\ \frac{1}{2}e^x - e^y = 0 \end{cases}$

55. $\begin{cases} 4\sqrt{x^2 - 3x} - 3\sqrt{y^2 + 6y} = -4 \\ \frac{1}{2}(\sqrt{x^2 - 3x} + \sqrt{y^2 + 6y}) = 3 \end{cases}$

Hint: Let $u = \sqrt{x^2 - 3x}$ and $v = \sqrt{y^2 + 6y}$.

56. The sum of two numbers is 64. Twice the larger number plus five times the smaller number is 20. Find the two numbers. (Let x denote the larger number and let y denote the smaller number.)

57. In a two-digit number, the sum of the digits is 14. Twice the tens digit exceeds the units digit by one. Find the number.

58. You have two brands of dietary supplements on your shelf. Among other ingredients, both contain protein and carbohydrates. The amounts of protein and carbohydrates in one unit of each supplement are given in the following table as percentages of the *recommended daily amount* (RDA). How many units of each supplement do you need in a day to obtain the RDA for both protein and carbohydrates?

	Protein (% of RDA in one unit)	Carbohydrates (% of RDA in one unit)
Supplement #1	8	12
Supplement #2	16	4

59. Solve the following system for x and y in terms of a and b, where a and b are nonzero and $a \neq b$.

$$\begin{cases} \dfrac{a}{bx} + \dfrac{b}{ay} = a + b \\[2mm] \dfrac{b}{x} + \dfrac{a}{y} = a^2 + b^2 \end{cases}$$

60. Solve for x and y in terms of a, b, c, d, e, and f:

$$\begin{cases} ax + by = c \\ dx + ey = f \end{cases}$$

(Assume that $ae - bd \neq 0$.)

C

61. (a) Given that the lines $7x + 5y = 4$, $x + ky = 3$, and $5x + y + k = 0$ are concurrent (pass through a common point), what are the possible values for k?

(b) Check that your answers are reasonable: For each value of k that you find, use a graphing utility to draw the three lines. Do they appear to be concurrent?

62. Solve the following system for x and y in terms of a and b, where $ab \neq -1$:

$$\begin{cases} \dfrac{x + y - 1}{x - y + 1} = a \\[2mm] \dfrac{y - x + 1}{x - y + 1} = ab \end{cases}$$

Answer: $x = (a + 1)/(ab + 1)$
$y = a(b + 1)/(ab + 1)$

PROJECT | **GEOMETRY WORKBOOKS ON THE EULER LINE AND THE NINE-POINT CIRCLE**

The exercises in this project form a unit introducing two remarkable geometric results. Workbook I uses a specific example to introduce the *Euler line* and the *nine-point circle* of a triangle. Workbook II develops the Euler line for the general case. The exercises are straightforward in that they involve only basic coordinate geometry and systems of two linear equations. However, there is a good deal of algebraic calculation required in both workbooks, and just one error along the way will spoil the result at the end of that workbook. Thus, this is a good occasion for group work, assuming each person in a group takes an active role in checking results. For both workbooks, one group can be assigned to each exercise. Some exercises, however, require results from one or more previous exercises. For instance, in Workbook I, Exercises 4–7 depend upon the results of the previous exercises. So there is a need for some coordination here; the class as a whole needs to be certain that the results in Exercises 1–3 are correct before the groups assigned to Exercises 4–7 can begin their work. A similar situation occurs in several places in Workbook II.

WORKBOOK 1

For this workbook, you are given △ABC, with vertices as indicated in the following figure.

1. **(a)** A *median* of a triangle is a line segment drawn from a vertex to the midpoint of the opposite side. For △ABC, find the equation of the line containing the median to side $\overline{BC}$. Next, find the equation of the line containing the median to side $\overline{AC}$. Now, find the point where these two medians intersect.

   (b) A theorem from geometry states that, in any triangle, the three medians are concurrent (i.e., intersect in a single point). Use this to check your last answer in part (a) as follows. Find the equation of the line containing the median to $\overline{AB}$. Then check that this median passes through the intersection point found in part (a).

   Remark: For any triangle, the point where the three medians intersect is called the *centroid* of the triangle. It can be shown that the centroid is the center of gravity or balance point of the triangle.

   (c) Use a graphing utility to draw △ABC along with the three lines containing the medians. Check to see that the location of the centroid appears to be consistent with coordinates obtained in part (a).

2. An *altitude* of a triangle is a line segment drawn from a vertex perpendicular to the opposite side. A theorem from geometry states that, in any triangle, the three altitudes are concurrent. The point where they meet is called the *orthocenter* of the triangle. Follow Exercise 1 using the altitudes of △ABC, rather than the medians.

3. A theorem from geometry states that, in any triangle, the perpendicular bisectors of the three sides are concurrent. The point where they intersect is called the *circumcenter* of the triangle. The circumcenter is the center of the circle that passes through the three vertices of the triangle. Follow Exercise 1 using the perpendicular bisectors of the sides of △ABC, rather than the medians.

4. In Exercises 1–3 you found the coordinates of the centroid, the orthocenter, and the circumcenter of △ABC.

   (a) Show that these three points are collinear (i.e., all lie on a straight line).
   Hint: Use slopes.

   (b) Show that the distance from the orthocenter to the centroid is twice the distance from the centroid to the circumcenter.

   Remark: It can be shown (Workbook II) that the results in parts (a) and (b) are valid for all triangles. The line containing the centroid, orthocenter, and circumcenter is called the *Euler line* of the triangle.

5. **(a)** Find the center and radius of the circle that passes through the midpoints of the sides of $\triangle ABC$. *Hint:* Find the center as the intersection of two perpendicular bisectors, as in Exercise 3.
   (b) Show that the center lies on the Euler line of $\triangle ABC$, and that it is situated halfway between the orthocenter and circumcenter of $\triangle ABC$.
6. Let H denote the orthocenter of $\triangle ABC$ (determined in Exercise 2). Find the midpoints of the segments $\overline{HA}$, $\overline{HB}$, and $\overline{HC}$. Then check that each of these midpoints lies on the circle determined in Exercise 5.
7. In $\triangle ABC$, find the foot of each altitude (i.e., the point where the altitude intersects the opposite side.) Then check that each of these three points lies on the circle determined in Exercise 5.
   Remark: In view of Exercises 5–7, the following nine points all lie on the same circle: the feet of the three medians, the feet of the three altitudes, and the midpoints of the three segments drawn from the orthocenter to each vertex. It can be shown that this occurs no matter what triangle we begin with. The circle is called the *nine-point circle.*
Ⓖ 8. Use a graphing utility to display $\triangle ABC$, the nine-point circle, and the nine special points that all lie on the circle.

WORKBOOK 2

For this workbook, you are given an arbitrary triangle with vertices as indicated in the following figure. (All triangles can be represented in this fashion: the units on the y-axis are chosen so that the length of that altitude is 6; the constant a can be any negative real number, while b can be any positive real number. The factor 6 has been chosen so that the appearance of fractions is avoided in subsequent calculations.) For each exercise, supply the calculations to show that the given statements are correct.

1. The equations of the lines containing the medians of $\triangle ABC$ are
   $2x + (a + b)y = 6(a + b)$, $x - (a - 2b)y = 6b$, and $x - (b - 2a)y = 6a$.
2. The intersection point for each pair of medians of $\triangle ABC$ is $(2a + 2b, 2)$. This point is called the *centroid* of $\triangle ABC$, and it is usually denoted by the letter G.
3. The equations of the lines containing the altitudes of $\triangle ABC$ are $y = ax - 6ab$, $y = bx - 6ab$, and $x = 0$.
4. The intersection point for each pair of altitudes of $\triangle ABC$ is $(0, -6ab)$. This point is called the *orthocenter* of $\triangle ABC$, and it is usually denoted by the letter H.
5. The equations of the lines containing the perpendicular bisectors of the sides of $\triangle ABC$ are $x = 3a + 3b$, $bx - y = 3b^2 - 3$, and $ax - y = 3a^2 - 3$.

6. The intersection point for each pair of perpendicular bisectors is $(3a + 3b, 3ab + 3)$. This point is called the *circumcenter* of $\triangle ABC$, and it is usually denoted by the letter O.

7. The orthocenter H, the centroid G, and circumcenter O are collinear. The line through these three points is called the *Euler line* of $\triangle ABC$.

8. $HG = 2(GO)$. That is, the distance from the orthocenter to the centroid is twice the distance from the centroid to the circumcenter.

6.2 GAUSSIAN ELIMINATION

A method of solution is perfect if we can foresee from the start, and even prove, that following that method we shall obtain our aim. — Gottfried Wilhelm von Leibniz (1646–1716)

. . . the first electronic computer, the ABC, named after its designers Atanasoff and Berry, was built specifically to solve systems of 29 equations in 29 unknowns, a formidable task without the aid of a computer. —Angela B. Shiflet in *Discrete Mathematics for Computer Science* (St. Paul: West Publishing Co., 1987)

In the previous section, we solved systems of linear equations in two unknowns. In this section, we introduce the technique known as Gaussian elimination for solving systems of linear equations in which there are more than two unknowns.*

As a first example, consider the following system of three linear equations in the three unknowns x, y, and z:

$$\begin{cases} 3x + 2y - z = -3 \\ \quad\quad 5y - 2z = 2 \\ \quad\quad\quad\quad 5z = 20 \end{cases}$$

This system is easy to solve using the process of *back-substitution*. Dividing both sides of the third equation by 5 yields $z = 4$. Then, substituting $z = 4$ back into the second equation gives us

$$5y - 2(4) = 2$$
$$5y = 10$$
$$y = 2$$

Finally, substituting the values $z = 4$ and $y = 2$ back into the first equation yields

$$3x + 2(2) - 4 = -3$$
$$3x = -3$$
$$x = -1$$

We have now found that $x = -1$, $y = 2$, and $z = 4$. If you go back and check, you will find that these values indeed satisfy all three equations in the given system. Furthermore, the algebra we've just carried out shows that these are the only possible values for x, y, and z satisfying all three equations. We summarize by saying that the **ordered triple** $(-1, 2, 4)$ is the solution of the given system.

---

*The technique is named after Carl Friedrich Gauss (1777–1855). Early in the nineteenth century, Gauss used this technique (and introduced the method of least squares for minimizing errors) in analyzing the orbit of the asteroid Pallas. However, the essentials of Gaussian elimination were in existence long before Gauss's time. Indeed, a version of the method appears in the Chinese text *Chui-Chang Suan-Shu* ("Nine Chapters on the Mathematical Art"), written approximately two thousand years ago.

The system that we just considered was easy to solve (using back-substitution) because of the special form in which it was written. This form is called **upper-triangular form.** Although the following definition of upper-triangular form refers to systems with three unknowns, the same type of definition can be given for systems with any number of unknowns. Table 1 displays examples of systems in upper-triangular form.

Upper-Triangular Form (Three Variables)

A system of linear equations in x, y, and z is said to be in **upper-triangular form** provided x appears in no equation after the first and y appears in no equation after the second. (It is possible that y may not even appear in the second equation.)

TABLE 1	**Examples of Systems in Upper-Triangular Form**	
2 Unknowns: x, y	**3 Unknowns: x, y, z**	**4 Unknowns: x, y, z, t**

$$\begin{cases} 3x + 5y = 7 \\ 8y = 5 \end{cases}$$

$$\begin{cases} 4x - 3y + 2z = -5 \\ 7y + z = 9 \\ -4z = 3 \end{cases}$$

$$\begin{cases} x - y + z - 4t = 1 \\ 3y - 2z + t = -1 \\ 3z - 5t = 4 \\ 6t = 7 \end{cases}$$

$$\begin{cases} 15x - 2y + z = 1 \\ 3z = -8 \end{cases}$$

$$\begin{cases} 2x + y + 2z - t = -3 \\ 4z + 3t = 1 \\ 5t = 6 \end{cases}$$

$$\begin{cases} 8x + 3y - z + t = 2 \\ 2y + z - 4t = 1 \end{cases}$$

When we were solving linear systems of equations with two unknowns in the previous section, we observed that there were three possibilities: a unique solution, infinitely many solutions, and no solution. As the next three examples indicate, the situation is similar when dealing with larger systems.

EXAMPLE 1 Using back-substitution to solve a system

Find all solutions of the system

$$\begin{cases} x + y + 2z = 2 \\ 3y - 4z = -5 \\ 6z = 3 \end{cases}$$

SOLUTION

The system is in upper-triangular form, so we can use back-substitution. Dividing the third equation by 6 yields $z = 1/2$. Substituting this value for z back into the second equation then yields

$$3y - 4(1/2) = -5$$
$$3y = -3$$
$$y = -1$$

Now, substituting the values $z = 1/2$ and $y = -1$ back into the first equation, we obtain

$$x + (-1) + 2(1/2) = 2$$
$$x = 2$$

As you can easily check, the values $x = 2$, $y = -1$, and $z = 1/2$ indeed satisfy all three equations. Furthermore, the algebra we've just carried out shows that these are the only possible values for x, y, and z satisfying all three equations. We summarize by saying that the unique solution to our system is the ordered triple $(2, -1, 1/2)$.

EXAMPLE   2   A system with infinitely many solutions

Find all solutions of the system

$$\begin{cases} -2x + y + 3z = 6 \\ \qquad\qquad 2z = 10 \end{cases}$$

SOLUTION
Again, the system is in upper-triangular form, and we use back-substitution. Solving the second equation for z yields $z = 5$. Then replacing z by 5 in the first equation gives us

$$-2x + y + 3(5) = 6$$
$$-2x + y = -9$$
$$y = 2x - 9$$

At this point, we've made use of both equations in the given system. There is no third equation to provide additional restrictions on x, y, or z. We know from the previous section that the equation $y = 2x - 9$ has infinitely many solutions, all of the form

$$(x, 2x - 9) \quad \text{where } x \text{ is a real number}$$

It follows, then, that there are infinitely many solutions to the given system and they may be written

$$(x, 2x - 9, 5) \quad \text{where } x \text{ is a real number}$$

For instance, choosing in succession $x = 0$, $x = 1$, and $x = 2$ yields the solutions $(0, -9, 5)$, $(1, -7, 5)$, and $(2, -5, 5)$. (We remark in passing that any linear system in upper-triangular form in which the number of unknowns exceeds the number of equations will always have infinitely many solutions.)

EXAMPLE   3   A system with no solutions

Find all solutions of the system

$$\begin{cases} 4x - 7y + 3z = 1 \\ 3x + \ y - 2z = 4 \\ 4x - 7y + 3z = 6 \end{cases}$$

(Note that the system is not in upper-triangular form.)

SOLUTION

Look at the left-hand sides of the first and third equations: they are identical. Thus, if there were values for x, y, and z that satisfied both equations, it would follow that $1 = 6$, which is clearly impossible. We conclude that the given system has no solutions.

As Examples 1 and 2 have demonstrated, systems in upper-triangular form can be readily solved. In view of this, it would be useful to have a technique for converting a given system into an equivalent system in upper-triangular form. (The expression **equivalent system** means a system with exactly the same set of solutions as the original system.) **Gaussian elimination** is one such technique. We will demonstrate this technique in Examples 4, 5, 7, and 8. In using Gaussian elimination, we will rely on what are called the three **elementary operations,** listed in the box that follows. These are operations that, when performed on an equation in a system, produce an equivalent system.

The Elementary Operations

1. Multiply both sides of an equation by a nonzero constant.
2. Interchange the order in which two equations of a system are listed.
3. To one equation add a multiple of another equation in the system.

EXAMPLE **4** **Using Gaussian elimination to solve a system**

Find all solutions of the system

$$\begin{cases} x + 2y + z = 3 \\ 2x + y + z = 16 \\ x + y + 2z = 9 \end{cases}$$

SOLUTION

First we want to eliminate x from the second and third equations. To eliminate x from the second equation, we add to it -2 times the first equation. The result is the equivalent system

$$\begin{cases} x + 2y + z = 3 \\ -3y - z = 10 \\ x + y + 2z = 9 \end{cases}$$

To eliminate x from the third equation, we add to it -1 times the first equation. The result is the equivalent system

$$\begin{cases} x + 2y + z = 3 \\ -3y - z = 10 \\ -y + z = 6 \end{cases}$$

Now to bring the system into upper-triangular form, we need to eliminate y from the third equation. We could do this by adding $-1/3$ times the second equation to the third equation. However, to avoid working with fractions as long as possible, we proceed instead to interchange the second and third equations to obtain the equivalent system

$$\begin{cases} x + 2y + z = 3 \\ \quad -y + z = 6 \\ \quad -3y - z = 10 \end{cases}$$

Now we add -3 times the second equation to the last equation to obtain the equivalent system

$$\begin{cases} x + 2y + \ z = 3 \\ \quad -y + \ z = 6 \\ \quad\quad - 4z = -8 \end{cases}$$

The system is now in upper-triangular form, and back-substitution yields, in turn, $z = 2$, $y = -4$, and $x = 9$. (Check this for yourself.) The required solution is therefore $(9, -4, 2)$.

In Table 2 we list some convenient abbreviations used in describing the elementary operations. Plan on using these abbreviations yourself: they'll make it simpler for you (and your instructor) to check your work. In Table 2, the notation E_i stands for the ith equation in a system. For instance, for the initial system in Example 4, the symbol E_1 denotes the first equation: $x + 2y + z = 3$.

TABLE 2   Abbreviations for the Elementary Operations

Abbreviation	Explanation
1. cE_i	Multiply both sides of the ith equation by c.
2. $E_i \leftrightarrow E_j$	Interchange the ith and jth equations.
3. $cE_i + E_j$	Add c times the ith equation to the jth equation.

EXAMPLE 5 **Specifying the elementary operations in using Gaussian elimination**

Solve the system

$$\begin{cases} 4x - 3y + 2z = 40 \\ 5x + 9y - 7z = 47 \\ 9x + 8y - 3z = 97 \end{cases}$$

SOLUTION

$$\begin{cases} 4x - 3y + 2z = 40 \\ 5x + 9y - 7z = 47 \\ 9x + 8y - 3z = 97 \end{cases} \xrightarrow{(-1)E_2 + E_1} \begin{cases} -x - 12y + 9z = -7 \\ 5x + \ 9y - 7z = 47 \\ 9x + \ 8y - 3z = 97 \end{cases}$$
adding -1 times the second equation to the first equation

$$\xrightarrow[\ 9E_1 + E_3\ ]{5E_1 + E_2} \begin{cases} -x - 12y + \ 9z = -7 \\ \quad -51y + 38z = 12 \\ \quad -100y + 78z = 34 \end{cases}$$

$$\xrightarrow{\frac{1}{2}E_3} \begin{cases} -x - 12y + 9z = -7 \\ \quad\quad -51y + 38z = 12 \\ \quad\quad -50y + 39z = 17 \end{cases}$$

$$\xrightarrow{(-1)E_3 + E_2} \begin{cases} -x - 12y + 9z = -7 \\ \quad\quad -y - \quad z = -5 \\ \quad\quad -50y + 39z = 17 \end{cases}$$ to allow working with smaller but integral coefficients

$$\xrightarrow{-50E_2 + E_3} \begin{cases} -x - 12y + 9z = -7 \\ \quad\quad -y - \quad z = -5 \\ \quad\quad\quad 89z = 267 \end{cases}$$

The system is now in upper-triangular form. Solving the third equation, we obtain $z = 3$. Substituting this value back into the second equation yields $y = 2$. (Check this for yourself.) Finally, substituting $z = 3$ and $y = 2$ back into the first equation yields $x = 10$. (Again, check this for yourself.) The solution to the system is therefore $(10, 2, 3)$.

EXAMPLE	6	**Using Gaussian elimination to solve a scheduling problem**

A charter tour company specializing in Hawaiian vacations has a total of 1328 reservations for an August departure from Los Angeles to Honolulu. The company has three different kinds of jets available, each with three classes of seats: economy, economy-upgrade, and first class. The available seating on each type of jet is given in the following table.

	Jet type I	Jet type II	Jet type III
Economy	120	80	60
Economy-upgrade	60	35	26
First class	20	45	16

Of the 1328 reservations, 760 are in economy class, 354 are in economy-upgrade, and the remaining 214 are in first class. How many jets of each type are needed to accommodate these passengers, given that the tour company would like to have all seats occupied on each plane? Or is that not possible?

SOLUTION

Let x be the required number of type I jets, y the number of type II jets, and z the number of type III jets. We start by considering economy class. According to the table, on each type I jet there are 120 economy seats available. Thus, with x type I jets, we have $120x$ economy seats available. Similarly, for type II jets, there are $80y$ economy seats, and for type III jets, there are $60z$ economy seats. Therefore, we require that

$$\underbrace{120x + 80y + 60z}_{\substack{\text{\color{red}the number of economy}\\ \text{\color{red}seats available}}} = \overset{\substack{\color{red}\uparrow\\ \text{\color{red}the number of economy}\\ \text{\color{red}seats reserved}}}{760}$$

That gives us one equation involving the three unknowns x, y, and z. In a similar fashion, working with economy-upgrade, and then working with first class, we obtain the two additional equations $60x + 35y + 26z = 354$ and $20x + 45y + 16z = 214$. (Verify this for yourself.) In summary then, we have the following system of three linear equations:

$$\begin{cases} 120x + 80y + 60z = 760 \\ 60x + 35y + 26z = 354 \\ 20x + 45y + 16z = 214 \end{cases} \qquad (1)$$

This system can be solved using Gaussian elimination, just as in Example 5 (although the arithmetic is a bit messier). Exercise 30(a) asks you to solve the system. The result turns out to be $x = 3$, $y = 2$, and $z = 4$. This tells us that the tour company should use 3 type I jets, 2 type II jets, and 4 type III jets in order to accommodate all of the passengers, while filling all of the seats on each plane. If at least one of the values for x, y, or z in the solution had not been a whole number, then we would have needed to round up to the next whole number (why?), which would mean that at least one plane would be flying with one or more empty seats.

Each linear system that we considered in Examples 4–6 had exactly one solution. As the next two examples indicate, the method of Gaussian elimination also works in cases where the system has infinitely many solutions or no solution. Indeed, it's not necessary to know how many solutions there are (if any) at the start; rather, this information is revealed by the Gaussian elimination process.

EXAMPLE 7 A system with infinitely many solutions

Solve the system

$$\begin{cases} x + 2y + 4z = 0 \\ x + 3y + 9z = 0 \end{cases}$$

SOLUTION
This system is similar to the one in Example 2 in that there are fewer equations than there are unknowns. By subtracting the first equation from the second, we readily obtain an equivalent system in upper-triangular form:

$$\begin{cases} x + 2y + 4z = 0 \\ y + 5z = 0 \end{cases}$$

Although the system is now in upper-triangular form, notice that the second equation does not determine y or z uniquely; that is, there are infinitely many number pairs (y, z) satisfying the second equation. We can solve the second equation for y in terms of z; the result is $y = -5z$. Now we replace y with $-5z$ in the first equation to obtain

$$x + 2(-5z) + 4z = 0 \qquad \text{or} \qquad x = 6z$$

At this point we've used both of the equations in the system to express x and y in terms of z. Furthermore, there is no third equation in the system to provide additional restrictions on x, y, or z. We therefore conclude that the given system has infinitely many solutions. These solutions have the form

$$(6z, -5z, z) \quad \text{where } z \text{ is any real number}$$

EXAMPLE **8** **A system with no solutions**

Solve the system

$$\begin{cases} x - 4y + z = 3 \\ 3x + 5y - 2z = -1 \\ 7x + 6y - 3z = 2 \end{cases}$$

SOLUTION

$$\begin{cases} x - 4y + z = 3 \\ 3x + 5y - 2z = -1 \\ 7x + 6y - 3z = 2 \end{cases} \xrightarrow[-7E_1 + E_3]{-3E_1 + E_2} \begin{cases} x - 4y + z = 3 \\ 17y - 5z = -10 \\ 34y - 10z = -19 \end{cases}$$

$$\xrightarrow{-2E_2 + E_3} \begin{cases} x - 4y + z = 3 \\ 17y - 5z = -10 \\ 0 = 1 \end{cases}$$

From the third equation in this last system, we conclude that this system, and consequently the original system, has no solution. (Reason: If there *were* values for x, y, and z satisfying the original system, then it would follow that $0 = 1$, which is clearly impossible.)

Figure 1
Geometric interpretation for Example 7

In the examples in this section, we've seen the three possibilities regarding solutions of linear systems with more than two unknowns: a unique solution (as in Examples 1, 4, 5, and 6); infinitely many solutions (as in Examples 2 and 7); and no solution (as in Examples 3 and 8). Just as in the previous section, there are graphs to explain each case. But now, instead of lines in a two-dimensional coordinate system, the graphs involve *planes* in a three-dimensional coordinate system. We'll describe this in general, without going into detail.

In the familiar x-y coordinate system, an equation of the form $ax + by = c$ represents a line. And, as you know, a solution (x_0, y_0) of a linear system with two unknowns is a point that lies on both of the given lines. Similarly, in three dimensions, an equation of the form $ax + by + cz = d$ represents a plane. A solution (x_0, y_0, z_0) of a linear system involving three unknowns can be interpreted as a point that lies on all of the given planes. Thus, where we talked about intersecting lines in two dimensions, now we are talking about intersecting planes in three dimensions. Figure 1, for instance, shows a geometric interpretation for Example 7 where there were two equations in three unknowns. Each equation describes a plane. As indicated in Figure 1, these two planes intersect in a line. Each point on the line corresponds to a solution of the linear system. In Figure 2, we show some of the possibilities when there are three equations and three unknowns.

(a) The three planes intersect in a line. The corresponding system of three linear equations in three unknowns has infinitely many solutions.

(b) Although each pair of planes intersects in a line, the three planes have no point in common. The corresponding system of three equations has no solution.

(c) There is exactly one point common to all three planes. The corresponding system of three equations has exactly one solution.

(d) Two parallel planes cut a third plane. There is no point common to all three planes. The corresponding system of three equations has no solution.

Figure 2

Some of the ways that three planes can intersect (or fail to intersect)

EXERCISE SET 6.2

A

In Exercises 1–10, the systems of linear equations are in upper-triangular form. Find all solutions of each system.

1. $\begin{cases} 2x + y + z = -9 \\ 3y - 2z = -4 \\ 8z = -8 \end{cases}$

2. $\begin{cases} -3x + 7y + 2z = -19 \\ y + z = 1 \\ - 2z = -2 \end{cases}$

3. $\begin{cases} 8x + 5y + 3z = 1 \\ 3y + 4z = 2 \\ 5z = 3 \end{cases}$

4. $\begin{cases} 2x + 7z = -4 \\ 5y - 3z = 6 \\ 6z = 18 \end{cases}$

5. $\begin{cases} -4x + 5y = 0 \\ 3y + 2z = 1 \\ 3z = -1 \end{cases}$

6. $\begin{cases} 3x - 2y + z = 4 \\ 3z = 9 \end{cases}$

7. $\begin{cases} -x + 8y + 3z = 0 \\ 2z = 0 \end{cases}$

8. $\begin{cases} -x + y + z + w = 9 \\ 2y - z - w = 9 \\ 3z + 2w = 1 \\ 11w = 22 \end{cases}$

9. $\begin{cases} 2x + 3y + z + w = -6 \\ y + 3z - 4w = 23 \\ 6z - 5w = 31 \\ - 2w = 10 \end{cases}$

10. $\begin{cases} 7x - y - z + w = 3 \\ 2y - 3z - 4w = -2 \\ 3w = 6 \end{cases}$

In Exercises 11–30, find all solutions of each system.

11. $\begin{cases} x + y + z = 12 \\ 2x - y - z = -1 \\ 3x + 2y + z = 22 \end{cases}$

12. $\begin{cases} A + B - C = -1 \\ 3A - B + 2C = 9 \\ 5A + 3B + 3C = 1 \end{cases}$

13. $\begin{cases} 2x - 3y + 2z = 4 \\ 4x + 2y + 3z = 7 \\ 5x + 4y + 2z = 7 \end{cases}$

14. $\begin{cases} x \qquad\;\; + 2z = 5 \\ \qquad y - 30z = -16 \\ x - 2y + 4z = 8 \end{cases}$

15. $\begin{cases} 3x + 3y - 2z = 13 \\ 6x + 2y - 5z = 13 \\ 7x + 5y - 3z = 26 \end{cases}$

16. $\begin{cases} 2x + 5y - 3z = 4 \\ 4x - 3y + 2z = 9 \\ 5x + 6y - 2z = 18 \end{cases}$

17. $\begin{cases} x + y + z = 1 \\ -2x + y + z = -2 \\ 3x + 6y + 6z = 5 \end{cases}$

18. $\begin{cases} 7x + 5y - 7z = -10 \\ 2x + y + z = 7 \\ x + y - 3z = -8 \end{cases}$

19. $\begin{cases} 2x - y + z = -1 \\ x + 3y - 2z = 2 \\ -5x + 6y - 5z = 5 \end{cases}$

20. $\begin{cases} -2x + 2y - z = 0 \\ 3x - 4y + z = 1 \\ 5x - 8y + z = 4 \end{cases}$

21. $\begin{cases} 2x - y + z = 4 \\ x + 3y + 2z = -1 \\ 7x \qquad\;\; + 5z = 11 \end{cases}$

22. $\begin{cases} 3x + y - z = 10 \\ 8x - y - 6z = -3 \\ 5x - 2y - 5z = 1 \end{cases}$ (G)

23. $\begin{cases} x + y + z + w = 4 \\ x - 2y - z - w = 3 \\ 2x - y + z - w = 2 \\ x - y + 2z - 2w = -7 \end{cases}$

24. $\begin{cases} x + y - 3z + 2w = 0 \\ -2x - 2y + 6z + w = -5 \\ -x + 3y + 3z + 3w = -5 \\ 2x + y - 3z - w = 4 \end{cases}$

25. $\begin{cases} 2x + 3y + 2z = 5 \\ x + 4y - 3z = 1 \end{cases}$

26. $\begin{cases} 4x - y - 3z = 2 \\ 6x + 5y - z = 0 \end{cases}$

27. $\begin{cases} x - 2y - 2z + 2w = -10 \\ 3x + 4y - z - 3w = 11 \\ -4x - 3y - 3z + 8w = -21 \end{cases}$ (G)

28. $\begin{cases} 2x + y + z + w = 1 \\ x + 3y - 3z - 3w = 0 \\ -3x - 4y + 2z + 2w = -1 \end{cases}$

29. $\begin{cases} 4x - 2y + 3z = -2 \\ 6y - 4z = 6 \end{cases}$

30. This exercise concerns Example 6 in the text.
   (a) Solve system (1) on page 483. As stated in the text, you should obtain $x = 3$, $y = 2$, and $z = 4$.
   (b) There are Web sites on the Internet that will solve a (G) system of linear equations for you. They can be located by using any search engine to look for *equation solvers*. For instance, at the time of this writing, one such site is http://www.quickmath.com. Use one of these Web sites to solve system (1) on page 483.

31. Rework Example 6 in the text using the following new data. The number of reservations in economy, in economy-upgrade, and in first class are 830, 735, and 592, respec-

tively. The seating capacity on each type of jet is as follows.

	Jet type I	Jet type II	Jet type III
Economy	40	80	70
Economy-upgrade	35	75	60
First class	30	44	56

32. In Exercise 31, suppose that the company had made an error in counting the number of economy reservations, and that it should be 825 rather than 830. Solve the resulting system and interpret the result.

33. A parabola $y = ax^2 + bx + c$ passes through the three points $(1, -2)$, $(-1, 0)$, and $(2, 3)$.
   (a) Write down a system of three linear equations that must be satisfied by a, b, and c.
   (b) Solve the system in part (a).
   (c) With the values for a, b, and c that you found in part (b), use a graphing utility to draw the parabola $y = ax^2 + bx + c$. Find a viewing rectangle that seems to confirm that the parabola indeed passes through the three given points.
   (d) Another way to check your result in part (b): Apply the quadratic regression option on a graphing utility after entering the three given data points $(1, -2)$, $(-1, 0)$, and $(2, 3)$. (This is a valid check because, in general, three noncollinear points determine a unique parabola.)

34. A curve $y = x^3 + Ax^2 + Bx + C$ passes through the three points $(1, -2)$, $(2, 3)$, and $(3, 20)$.
   (a) Write down a system of three linear equations satisfied by A, B, and C.
   (b) Solve the system in part (a).
   (c) With the values for A, B, and C that you found in (G) part (b), use a graphing utility to draw the curve $y = x^3 + Ax^2 + Bx + C$, checking to see that it appears to pass through the three given points.

35. A manufacturer of office chairs makes three models: Utility, Secretarial, and Managerial. Three materials common to the manufacturing process for all of the models are cloth, steel, and plastic. The amounts required for one chair in each category are specified in the following table. The company wants to use up its inventory of these materials because of upcoming design changes. How many of each model should the manufacturer build to deplete its current inventory consisting of 476 units of cloth, 440 units of steel, and 826 units of plastic?

	Utility	Secretarial	Managerial
Cloth	3	4	2
Steel	2	5	8
Plastic	6	4	1

36. The U.S. Food and Drug Administration lists the following RDI's (reference daily intakes) for the antioxidants vitamin C, vitamin E, and zinc.

Vitamin C: 60 mg   Vitamin E: 30 mg   Zinc: 15 mg

Remark on terminology: The Food and Drug Administration defines *RDI* as a weighted average of the recommended daily allowances for all segments of the U.S. population.

Suppose that you have three brands of dietary supplements on your shelf. Among other ingredients, all three contain the antioxidants mentioned above. The amounts of these antioxidants in each supplement are indicated in the following table. How many ounces of each supplement should you combine to obtain the RDI's for vitamin C, vitamin E, and zinc?

	Vitamin C (mg/oz.)	Vitamin E (mg/oz.)	Zinc (mg/oz.)
Supplement I	12	4	1
Supplement II	5	1.25	2.5
Supplement III	2	3	0.5

B

37. After reviewing records from previous years, the owner of a small company notices some trends in the data. One of the trends is that monthly revenues from July through February, although different from year to year, seem to rise and fall in a pattern similar to the following. The (mathematically inclined) owner observes that a portion of a parabola may be a good model here.

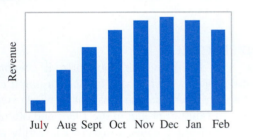

(a) This fiscal year, the monthly revenues for July, August, and September are as follows:

	Month		
	July	Aug.	Sept.
Revenue (Units of $100,000)	1	3.7	5.8

Let $t = 0$ correspond to July, $t = 1$ to August, and so on. Let $R(t)$ be the revenue (in units of $100,000) for the month t. Find the quadratic function $R(t) = at^2 + bt + c$ whose graph passes through the three given data points $(0, 1)$, $(1, 3.7)$, and $(2, 5.8)$.

(b) Use the vertex formula from Section 4.2 to determine the month for which the revenue is a maximum. Is your answer consistent with the bar graph shown at the start of this exercise? What is the corresponding maximum revenue?

38. Suppose that the height of an object as a function of time is given by $f(t) = at^2 + bt + c$, where t is time in seconds, $f(t)$ is the height in feet at time t, and a, b, and c are certain constants. If, after 1, 2, and 3 seconds, the corresponding heights are 184 ft, 136 ft, and 56 ft, respectively, find the time at which the object is at ground level (height $= 0$ ft).

39. Solve the following system for x, y, and z:

$$\begin{cases} e^x + \ e^y - 2e^z = 2a \\ e^x + 2e^y - 4e^z = 3a \\ \tfrac{1}{2}e^x - 3e^y + \ e^z = -5a \end{cases}$$

(Assume that $a > 0$.)   *Hint:* Let $A = e^x$, $B = e^y$, and $C = e^z$. Solve the resulting linear system for A, B, and C.

40. Solve the following system for α, β, and γ.

$$\begin{cases} \ln \alpha - \ \ln \beta - \ \ \ln \gamma = 2 \\ 3 \ln \alpha + 5 \ln \beta - 2 \ln \gamma = 1 \\ 2 \ln \alpha - 4 \ln \beta + \ \ \ln \gamma = 2 \end{cases}$$

41. The following figure shows a rectangular metal plate. The temperature at each red point on the boundary is maintained at the constant value shown (in Fahrenheit degrees). The temperature is not regulated at the interior points (indicated by the blue dots). Under these conditions, over time, it can be shown that the temperature at each interior point is the average of the temperatures of the four surrounding points (i.e., the points directly above, below, left, and right). For example, the temperature T_1 in the figure satisfies the condition $T_1 = (40 + 44 + T_2 + 41)/4$. Write down similar equations for T_2 and for T_3. Then solve the resulting system to determine the temperatures T_1, T_2, and T_3. *Remark:* Actually, the average-temperature condition that we've described is an approximation, and it holds only if the distances involved are sufficiently small.

42. The following figure displays three circles that are mutually tangent. The line segments joining the centers have lengths a, b, and c, as shown. Let r_1, r_2, and r_3 denote the radii of the circles, as indicated in the figure. Show that

$$r_1 = \frac{a + c - b}{2} \qquad r_2 = \frac{a + b - c}{2} \qquad r_3 = \frac{b + c - a}{2}$$

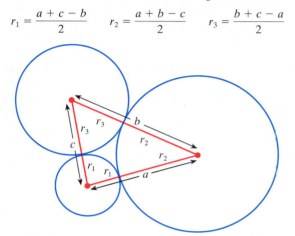

C

The following exercise appears in Algebra for Colleges and Schools *by H. S. Hall and S. R. Knight, revised by F. L. Sevenoak (New York: The Macmillan Co., 1906).*

43. A, B, and C are three towns forming a triangle. A man has to walk from one to the next, ride thence to the next, and drive thence to his starting point. He can walk, ride, and drive a mile in a, b, and c minutes, respectively. If he starts from B he takes $a + c - b$ hours, if he starts from C he takes $b + a - c$ hours, and if he starts from A he takes $c + b - a$ hours. Find the length of the circuit. [Assume that the circuit from A to B to C is counterclockwise.]

PROJECT

THE LEONTIEF INPUT-OUTPUT MODEL

Extract of press release from the Royal Swedish Academy of Sciences, October 18, 1973: The Royal Swedish Academy of Sciences has awarded the 1973 year's Prize in Economic Science in Memory of Alfred Nobel to Professor Wassily Leontief for the development of the input-output method and for its application to important economic problems. Professor Leontief is the sole and unchallenged creator of the input-output technique. This important innovation has given to economic sciences an empirically-useful method to highlight the general interdependence in the production system of a society. In particular, the method provides tools for a systematic analysis of the complicated interindustry transactions in an economy.

PART I EXPLANATION AND PRACTICE

Professor Wassily W. Leontief (1906–1999) won the Nobel prize in economics for his input-output model. In this project we explain the model in a simple case. As you'll see, the model yields a system of linear equations that we need to solve. Leontief began his work on the input-output model in the 1930s and often applied the model to the economy of the United States. You can see a readily available example of this at your local or college library in Leontief's article "The Structure of the U.S. Economy," which appears in the April 1965 issue of the magazine *Scientific American* (vol. 212, no. 4, pp. 25–35).

In the *Scientific American* article, Leontief divided the American economy into 81 different industries or *sectors*. (For instance, 5 of the 81 sectors were coal mining, glass and glass products, primary iron and steel manufacturing, apparel, and aircraft and parts.) To keep things simple, let's suppose now that we have an economy with only three sectors: steel, coal, and electricity. Further, suppose that the sectors are interrelated, as given by the following *input-output* table, Table 1. We'll explain how to interpret the table in the next paragraph.

TABLE 1 An Input-Output Table
for a Hypothetical Three-Sector Economy

	Outputs		
	Steel	Coal	Electricity
Inputs			
Steel	0.04	0.02	0.16
Coal	0.15	0	0.25
Electricity	0.14	0.10	0.04

The idea behind Table 1 is that each sector in the economy requires for its production process inputs from one or more of the other sectors. In the table, the first column of figures (beneath the output heading *Steel*) is interpreted as follows: The production of one unit of steel requires

0.04 unit of steel

0.15 unit of coal

0.14 unit of electricity

(While it's clear that coal and electricity would be required in the production of steel, you may be wondering why steel itself appears as one of the inputs. Think of the steel as being utilized in the building of equipment or factories for use by the steel industry.) The second and third columns of figures in the table are interpreted similarly. In particular, the second column of figures in Table 1 indicates that the production of 1 unit of coal requires 0.02 unit of steel, no units of coal, and 0.10 unit of electricity. The third column indicates that the production of 1 unit of electricity requires 0.16 unit of steel, 0.25 unit of coal, and 0.04 unit of electricity.

Table 1 can also be interpreted by reading across the rows, rather than down the columns. The first row of figures tells how much steel is required to produce 1 unit of output from each industry. In particular, to produce 1 unit of steel, 1 unit of coal, or 1 unit of electricity requires 0.04 unit of steel, 0.02 unit of steel, or 0.16 unit of steel, respectively. The second and third rows of figures are interpreted similarly.

Table 1 shows the demands that the three sectors place on one another for production. Now, outside of these three sectors, in other industries or in government, for example, there are, of course, additional demands for steel, coal, and electricity. These additional demands, from sources outside of the three given sectors, are referred to as *external demands*. By way of contrast, the demands in Table 1 are called *internal demands*. Let us suppose that the external demands are as given in Table 2. The problem to be solved then is as follows.

An Input-Output Problem

How many units should each of the three given sectors produce to satisfy both the internal and external demands on the economy?

TABLE 2

External demands*

Steel:	308 units
Coal:	275 units
Electricity:	830 units

*These are the demands for steel, coal, and electricity exclusive of the production requirements listed in Table 1.

To solve this input-output problem, let

x = the number of units of steel to be produced
y = the number of units of coal to be produced
z = the number of units of electricity to be produced

We'll use the information in Tables 1 and 2 to generate a system of three linear equations involving x, y, and z. First, since x represents the total number of units of steel to be produced, we have

(internal demands for steel) + (external demands for steel) = x

(internal demands for steel) + 308 = x using   (1)
                                                       Table 2

Regarding the internal demands for steel in equation (1), we'll make use of the first row of figures in Table 1 along with the following **proportionality assumption.** (For clarity, we state this for the case of steel output, but it's assumed for coal and electricity as well.) If the output of one unit of steel requires n units of a particular resource, then the output of x units steel requires $n \times x$ units of that resource. Now, we use this proportionality assumption, the first row of figures in Table 1, and the definitions of x, y, and z, to write

(internal demands for steel) = $0.04x + 0.02y + 0.16z$

Using this last equation to substitute in equation (1), we obtain

$$0.04x + 0.02y + 0.16z + 308 = x$$
$$-0.96x + 0.02y + 0.16z = -308 \qquad (2)$$

Equation (2) is our first of three equations in three unknowns. A second equation can be generated in the same manner, but using coal, rather than steel, as the output under consideration. Again, similarly, a third equation can be obtained by starting with electricity rather than steel.

Working in groups or individually, obtain the remaining two equations and then solve the resulting system. As a check, the required answers are $x = 500$ units of steel, $y = 600$ units of coal, and $z = 1000$ units of electricity. *Suggestion*: Work on this in groups. Then after your group has obtained the correct answers and you are clear on the details, work individually on the next problem.

PART II A PROBLEM TO DO AND THEN TO WRITE UP ON YOUR OWN

Consider a three-sector economy: plastics, energy, and transportation. The production of 1 unit of plastics requires 0.02 of unit plastics, 0.04 unit of energy, and 0.05 unit of transportation. The production of 1 unit of energy requires 0.03 unit of plastics, 0.08 unit of energy, and 0.01 unit of transportation. The production of 1 unit of transportation requires 0.01 unit of plastics, 0.02 unit of energy, and 0.06 unit of transportation. The external demands for plastics, energy, and transportation are 22.76 units, 36.68 units, and 43.45 units, respectively.

(a) Display the given data using tables similar to Tables 1 and 2.

(b) How many units should each sector produce in order to satisfy both the internal and external demands on the economy? *Hint for checking*: The final answers turn out to be integers.

6.3 MATRICES

Recall that in Gaussian Elimination, row operations are used to change the coefficient matrix to an upper triangular matrix. The solution is then found by back substitution, starting from the last equation in the reduced system. —Steven C. Althoen and Renate McLaughlin in "Gauss–Jordan Reduction: A Brief History," *American Mathematical Monthly,* vol. 94 (1987), pp. 130–142

Arthur Cayley (1821–1895) and James Joseph Sylvester (1814–1897), two English mathematicians, invented the matrix . . . in the 1850s. . . . The operations of addition and multiplication of matrices were later defined and the algebra of matrices was then developed. In 1925, Werner Heisenberg, a German physicist, used matrices in developing his theory of quantum mechanics, extending the role of matrices from algebra to the area of applied mathematics. —John K. Luedeman and Stanley M. Lukawecki in *Elementary Linear Algebra* (St. Paul: West Publishing Co., 1986)

As you saw in the previous section, there can be a good deal of bookkeeping involved in using Gaussian elimination to solve systems of equations. We can lighten the load somewhat by using a **matrix** (pl.: **matrices**), which is simply a rectangular array of numbers enclosed in parentheses or brackets. Here are three examples:

$$\begin{pmatrix} 2 & 3 \\ -5 & 4 \end{pmatrix} \quad \begin{pmatrix} -6 & 0 & 1 & \frac{1}{4} \\ \frac{2}{3} & 1 & 5 & 8 \end{pmatrix} \quad \begin{pmatrix} \pi & 0 & 0 \\ 0 & 1 & 9 \\ -1 & -2 & 3 \\ -4 & 8 & 6 \end{pmatrix}$$

The particular numbers constituting a matrix are called its **entries** or **elements.** For instance, the entries in the matrix $\begin{pmatrix} 2 & 3 \\ -5 & 4 \end{pmatrix}$ are the four numbers 2, 3, −5, and 4. In this section, the entries will always be real numbers. However, it is also possible to consider matrices in which some or all of the entries are nonreal complex numbers.

It is convenient to agree on a standard system for labeling the rows and columns of a matrix. The rows are numbered from top to bottom and the columns from left to right, as indicated in the following example:

$$\begin{array}{cc} & \text{column 1} \quad \text{column 2} \\ & \downarrow \qquad\quad \downarrow \\ \begin{array}{l} \text{row 1} \rightarrow \\ \text{row 2} \rightarrow \\ \text{row 3} \rightarrow \end{array} & \begin{pmatrix} 5 & -3 \\ 0 & 2 \\ -1 & 16 \end{pmatrix} \end{array}$$

We express the **size,** or **dimension,** of a matrix by specifying the number of rows and the number of columns, in that order. For instance, we would say that the matrix

$$\begin{pmatrix} 5 & -3 \\ 0 & 2 \\ -1 & 16 \end{pmatrix}$$

is a 3×2 (read "3 by 2") matrix, not 2×3. The following example will help fix in your mind the terminology that we have introduced.

EXAMPLE 1 Matrix terminology

Consider the matrix

$$\begin{pmatrix} 1 & 3 & 5 \\ 7 & 9 & 11 \end{pmatrix}$$

(a) List the entries.
(b) What is the size of the matrix?
(c) Which element is in the second row, third column?

SOLUTION

(a) The entries are 1, 3, 5, 7, 9, and 11.
(b) Since there are two rows and three columns, this is a 2×3 matrix.
(c) To locate the element in the second row, third column, we draw lines through the second row and the third column and see where they intersect:

$$\begin{pmatrix} 1 & 3 & 5 \\ 7 & 9 & 11 \end{pmatrix}$$

Thus, the entry in row 2, column 3 is 11.

There is a natural way to use matrices to describe and solve systems of linear equations. Consider, for example, the following system of linear equations in **standard form** (with the x-, y-, and z-terms lined up on the left-hand side and the constant terms on the right-hand side of each equation):

$$\begin{cases} x + 2y - 3z = 4 \\ 3x \quad\quad + z = 5 \\ -x - 3y + 4z = 0 \end{cases}$$

The **coefficient matrix** of this system is the matrix

$$\begin{pmatrix} 1 & 2 & -3 \\ 3 & 0 & 1 \\ -1 & -3 & 4 \end{pmatrix}$$

As the name implies, the coefficient matrix of the system is the matrix whose entries are the coefficients of x, y, and z, written in the same relative positions as they appear in the system. Notice the zero appearing in the second row, second column of the matrix. It is there because the coefficient of y in the second equation is in fact zero. The **augmented matrix** of the system of equations considered here is

$$\begin{pmatrix} 1 & 2 & -3 & 4 \\ 3 & 0 & 1 & 5 \\ -1 & -3 & 4 & 0 \end{pmatrix}$$

As you can see, the augmented matrix is formed by *augmenting* the coefficient matrix with the column of constant terms taken from the right-hand side of the given system of equations. To help relate the augmented matrix to the original system of equations, we will write the augmented matrix this way:

$$\begin{pmatrix} 1 & 2 & -3 & 4 \\ 3 & 0 & 1 & 5 \\ -1 & -3 & 4 & 0 \end{pmatrix}$$

EXAMPLE	2	Specifying the coefficient matrix and augmented matrix

Write the coefficient matrix and the augmented matrix for the system

$$\begin{cases} 8x - 2y + z = 1 \\ 3x - 4z + y = 2 \\ 12y - 3z - 6 = 0 \end{cases}$$

SOLUTION
First we write the system in standard form, with the x-, y-, and z-terms lined up and the constant terms on the right. This yields

$$\begin{cases} 8x - 2y + z = 1 \\ 3x + y - 4z = 2 \\ 12y - 3z = 6 \end{cases}$$

The coefficient matrix is then

$$\begin{pmatrix} 8 & -2 & 1 \\ 3 & 1 & -4 \\ 0 & 12 & -3 \end{pmatrix}$$

while the augmented matrix is

$$\left(\begin{array}{ccc|c} 8 & -2 & 1 & 1 \\ 3 & 1 & -4 & 2 \\ 0 & 12 & -3 & 6 \end{array} \right)$$

In the previous section we used the three elementary operations in solving systems of linear equations. In Table 1 we express these operations in the language of matrices. The matrix operations are called the **elementary row operations.**

TABLE 1

Elementary operations for a system of linear equations	Corresponding elementary row operations for a matrix
1. Multiply both sides of an equation by a nonzero constant.	1′. Multiply each entry in a given row by a nonzero constant.
2. Interchange two equations.	2′. Interchange two rows.
3. To one equation, add a multiple of another equation.	3′. To one row, add a multiple of another row.

Table 2 displays an example of each elementary row operation. In the table, notice that the notation for describing these operations is essentially the same as that introduced in the previous section. For example, in the previous section, $10E_1$ indicated that the first equation in a system was multiplied by 10. Now, $10R_1$ indicates that each entry in the first *row* of a matrix is multiplied by 10.

TABLE 2

Examples of the elementary row operations		Comments

$$\begin{pmatrix} 1 & 2 & 3 \\ 4 & 5 & 6 \\ 7 & 8 & 9 \end{pmatrix} \xrightarrow{\;10\,R_1\;} \begin{pmatrix} 10 & 20 & 30 \\ 4 & 5 & 6 \\ 7 & 8 & 9 \end{pmatrix}$$

Multiply each entry in row 1 by 10.

$$\begin{pmatrix} 1 & 2 & 3 \\ 4 & 5 & 6 \\ 7 & 8 & 9 \end{pmatrix} \xrightarrow{\;R_2 \leftrightarrow R_3\;} \begin{pmatrix} 1 & 2 & 3 \\ 7 & 8 & 9 \\ 4 & 5 & 6 \end{pmatrix}$$

Interchange rows 2 and 3.

$$\begin{pmatrix} 1 & 2 & 3 \\ 4 & 5 & 6 \\ 7 & 8 & 9 \end{pmatrix} \xrightarrow{\;-4R_1 + R_2\;} \begin{pmatrix} 1 & 2 & 3 \\ 0 & -3 & -6 \\ 7 & 8 & 9 \end{pmatrix}$$

To each entry in row 2, add -4 times the corresponding entry in row 1.

We are now ready to use matrices to solve systems of equations. In the example that follows, we'll use the same system of equations used in Example 5 of the previous section. *Suggestion*: After reading the next example, carefully compare each step with the corresponding one taken in Example 5 of the previous section.

EXAMPLE 3 **Using the elementary row operations to solve a system**

Solve the system

$$\begin{cases} 4x - 3y + 2z = 40 \\ 5x + 9y - 7z = 47 \\ 9x + 8y - 3z = 97 \end{cases}$$

SOLUTION

$$\begin{pmatrix} 4 & -3 & 2 & | & 40 \\ 5 & 9 & -7 & | & 47 \\ 9 & 8 & -3 & | & 97 \end{pmatrix} \xrightarrow{\;(-1)R_2 + R_1\;} \begin{pmatrix} -1 & -12 & 9 & | & -7 \\ 5 & 9 & -7 & | & 47 \\ 9 & 8 & -3 & | & 97 \end{pmatrix}$$

$$\xrightarrow[\;9R_1 + R_3\;]{\;5R_1 + R_2\;} \begin{pmatrix} -1 & -12 & 9 & | & -7 \\ 0 & -51 & 38 & | & 12 \\ 0 & -100 & 78 & | & 34 \end{pmatrix}$$

$$\xrightarrow{\;\frac{1}{2}R_3\;} \begin{pmatrix} -1 & -12 & 9 & | & -7 \\ 0 & -51 & 38 & | & 12 \\ 0 & -50 & 39 & | & 17 \end{pmatrix}$$

$$\xrightarrow{\;(-1)R_3 + R_2\;} \begin{pmatrix} -1 & -12 & 9 & | & -7 \\ 0 & -1 & -1 & | & -5 \\ 0 & -50 & 39 & | & 17 \end{pmatrix}$$

$$\xrightarrow{\;-50R_2 + R_3\;} \begin{pmatrix} -1 & -12 & 9 & | & -7 \\ 0 & -1 & -1 & | & -5 \\ 0 & 0 & 89 & | & 267 \end{pmatrix}$$

This last augmented matrix represents a system of equations in upper-triangular form:

$$\begin{cases} -x - 12y + 9z = -7 \\ \qquad -y - \;\; z = -5 \\ \qquad\qquad 89z = 267 \end{cases}$$

As you should now check for yourself, this yields the values $z = 3$, then $y = 2$, then $x = 10$. The solution of the original system is therefore $(10, 2, 3)$.

For the remainder of this section (and in part of the next) we will study matrices without referring to systems of equations. As motivation for this, we point out that matrices are essential tools in many fields of study. For example, a knowledge of matrices and their properties is needed for work in computer graphics. To begin, we need to say what it means for two matrices to be equal.

DEFINITION | **Equality of Matrices**

EXAMPLES

Two matrices are equal provided they have the same size (same number of rows, same number of columns) and the corresponding entries are equal.

$$\begin{pmatrix} 2 & 3 \\ 4 & 5 \end{pmatrix} = \begin{pmatrix} 2 & 3 \\ 4 & 5 \end{pmatrix}$$

$$\begin{pmatrix} 2 & 3 & 0 \\ 4 & 5 & 0 \end{pmatrix} \neq \begin{pmatrix} 2 & 3 \\ 4 & 5 \end{pmatrix}$$

$$\begin{pmatrix} 2 & 3 \\ 4 & 5 \end{pmatrix} \neq \begin{pmatrix} 2 & 3 \\ 5 & 4 \end{pmatrix}$$

Now we can define matrix addition and subtraction. These operations are defined only between matrices of the same size

---

DEFINITION | **Matrix Addition**

EXAMPLES

To add (or subtract) two matrices of the same size, add (or subtract) the corresponding entries.

$$\begin{pmatrix} 2 & 3 \\ -1 & 4 \end{pmatrix} + \begin{pmatrix} 6 & 1 \\ 0 & -4 \end{pmatrix} = \begin{pmatrix} 8 & 4 \\ -1 & 0 \end{pmatrix}$$

$$\begin{pmatrix} 2 & 3 \\ -1 & 4 \\ 9 & 10 \end{pmatrix} - \begin{pmatrix} 6 & 1 \\ 0 & -4 \\ 3 & 2 \end{pmatrix} = \begin{pmatrix} -4 & 2 \\ -1 & 8 \\ 6 & 8 \end{pmatrix}$$

Many properties of the real numbers also apply to matrices. For instance, matrix addition is *commutative*:

$$A + B = B + A \quad \text{where } A \text{ and } B \text{ are matrices of the same size}$$

Matrix addition is also *associative*:

$$A + (B + C) = (A + B) + C \quad \text{where } A, B, \text{ and } C \text{ are matrices of the same size}$$

In the next example, we verify these properties in two specific instances.

EXAMPLE **4** **Adding matrices**

Let $A = \begin{pmatrix} 1 & 2 \\ 3 & 4 \end{pmatrix}$, $B = \begin{pmatrix} 0 & -5 \\ 8 & -1 \end{pmatrix}$, and $C = \begin{pmatrix} 6 & 7 \\ 8 & 9 \end{pmatrix}$.

(a) Show that $A + C = C + A$.

(b) Show that $A + (B + C) = (A + B) + C$.

SOLUTION

(a) $A + C = \begin{pmatrix} 1 & 2 \\ 3 & 4 \end{pmatrix} + \begin{pmatrix} 6 & 7 \\ 8 & 9 \end{pmatrix} = \begin{pmatrix} 7 & 9 \\ 11 & 13 \end{pmatrix}$

$C + A = \begin{pmatrix} 6 & 7 \\ 8 & 9 \end{pmatrix} + \begin{pmatrix} 1 & 2 \\ 3 & 4 \end{pmatrix} = \begin{pmatrix} 7 & 9 \\ 11 & 13 \end{pmatrix}$

This shows that $A + C = C + A$, since both $A + C$ and $C + A$ represent the matrix $\begin{pmatrix} 7 & 9 \\ 11 & 13 \end{pmatrix}$.

(b) First we compute $A + (B + C)$:

$$A + (B + C) = \begin{pmatrix} 1 & 2 \\ 3 & 4 \end{pmatrix} + \left[\begin{pmatrix} 0 & -5 \\ 8 & -1 \end{pmatrix} + \begin{pmatrix} 6 & 7 \\ 8 & 9 \end{pmatrix} \right]$$

$$= \begin{pmatrix} 1 & 2 \\ 3 & 4 \end{pmatrix} + \begin{pmatrix} 6 & 2 \\ 16 & 8 \end{pmatrix}$$

$$= \begin{pmatrix} 7 & 4 \\ 19 & 12 \end{pmatrix}$$

Next we compute $(A + B) + C$:

$$(A + B) + C = \left[\begin{pmatrix} 1 & 2 \\ 3 & 4 \end{pmatrix} + \begin{pmatrix} 0 & -5 \\ 8 & -1 \end{pmatrix} \right] + \begin{pmatrix} 6 & 7 \\ 8 & 9 \end{pmatrix}$$

$$= \begin{pmatrix} 1 & -3 \\ 11 & 3 \end{pmatrix} + \begin{pmatrix} 6 & 7 \\ 8 & 9 \end{pmatrix}$$

$$= \begin{pmatrix} 7 & 4 \\ 19 & 12 \end{pmatrix}$$

We conclude from these calculations that $A + (B + C) = (A + B) + C$, since both sides of that equation represent the matrix $\begin{pmatrix} 7 & 4 \\ 19 & 12 \end{pmatrix}$.

We will now define an operation on matrices called **scalar multiplication.** First, the word *scalar* here just means *real number,* so we are talking about multiplying a matrix by a real number. (In more advanced work, nonreal complex scalars are also considered.)

DEFINITION | **Scalar Multiplication**

EXAMPLES

To multiply a matrix by a scalar, multiply each entry in the matrix by that scalar.

$2\begin{pmatrix} 5 & 9 & 0 \\ -1 & 2 & 3 \end{pmatrix} = \begin{pmatrix} 10 & 18 & 0 \\ -2 & 4 & 6 \end{pmatrix}$

$1\begin{pmatrix} 1 & 2 \\ 3 & 4 \end{pmatrix} = \begin{pmatrix} 1 & 2 \\ 3 & 4 \end{pmatrix}$

There are two simple but useful properties of scalar multiplication that are worth noting at this point. We'll omit the proofs of these two properties; however, Example 5 does ask us to verify them for a particular case.

> **PROPERTY SUMMARY** **Properties of Scalar Multiplication**
>
> 1. $c(kM) = (ck)M$, for all scalars c and k and any matrix M.
> 2. $c(M + N) = cM + cN$, where c is any scalar and M and N are any matrices of the same size.

EXAMPLE **5** **Demonstrating properties of scalar multiplication**

Let $c = 2$, $k = 3$, $M = \begin{pmatrix} 1 & 2 \\ 3 & 4 \end{pmatrix}$, and $N = \begin{pmatrix} 5 & 6 \\ 7 & 8 \end{pmatrix}$.

(a) Show that $c(kM) = (ck)M$.
(b) Show that $c(M + N) = cM + cN$.

SOLUTION

(a) $c(kM) = 2\left[3\begin{pmatrix} 1 & 2 \\ 3 & 4 \end{pmatrix} \right]$

$= 2\begin{pmatrix} 3 & 6 \\ 9 & 12 \end{pmatrix} = \begin{pmatrix} 6 & 12 \\ 18 & 24 \end{pmatrix}$

$(ck)M = (2 \cdot 3)\begin{pmatrix} 1 & 2 \\ 3 & 4 \end{pmatrix}$

$= 6\begin{pmatrix} 1 & 2 \\ 3 & 4 \end{pmatrix} = \begin{pmatrix} 6 & 12 \\ 18 & 24 \end{pmatrix}$

Thus, $c(kM) = (ck)M$, since in both cases the result is $\begin{pmatrix} 6 & 12 \\ 18 & 24 \end{pmatrix}$.

(b) $c(M + N) = 2\left[\begin{pmatrix} 1 & 2 \\ 3 & 4 \end{pmatrix} + \begin{pmatrix} 5 & 6 \\ 7 & 8 \end{pmatrix} \right]$

$= 2\begin{pmatrix} 6 & 8 \\ 10 & 12 \end{pmatrix} = \begin{pmatrix} 12 & 16 \\ 20 & 24 \end{pmatrix}$

$cM + cN = 2\begin{pmatrix} 1 & 2 \\ 3 & 4 \end{pmatrix} + 2\begin{pmatrix} 5 & 6 \\ 7 & 8 \end{pmatrix}$

$= \begin{pmatrix} 2 & 4 \\ 6 & 8 \end{pmatrix} + \begin{pmatrix} 10 & 12 \\ 14 & 16 \end{pmatrix} = \begin{pmatrix} 12 & 16 \\ 20 & 24 \end{pmatrix}$

Thus, $c(M + N) = cM + cN$, since both sides equal $\begin{pmatrix} 12 & 16 \\ 20 & 24 \end{pmatrix}$.

A matrix with zeros for all of its entries plays the same role in matrix addition as does the number zero in ordinary addition of real numbers. For instance, in the case of 2×2 matrices, we have

$$\begin{pmatrix} a & b \\ c & d \end{pmatrix} + \begin{pmatrix} 0 & 0 \\ 0 & 0 \end{pmatrix} = \begin{pmatrix} a & b \\ c & d \end{pmatrix}$$

and

$$\begin{pmatrix} 0 & 0 \\ 0 & 0 \end{pmatrix} + \begin{pmatrix} a & b \\ c & d \end{pmatrix} = \begin{pmatrix} a & b \\ c & d \end{pmatrix}$$

for all real numbers a, b, c, and d. The matrix $\begin{pmatrix} 0 & 0 \\ 0 & 0 \end{pmatrix}$ is called the **additive identity** for 2×2 matrices. Similarly, any matrix with all zero entries is the additive identity for matrices of that size. It is sometimes convenient to denote an additive identity matrix by a boldface zero: **0**. With this notation, we can write

$$A + \mathbf{0} = \mathbf{0} + A = A \qquad \text{for any matrix } A$$

For this matrix equation, it is understood that the size of the matrix **0** is the same as the size of A. With this notation, we also have

$$A - A = \mathbf{0} \qquad \text{for any matrix } A$$

Our last topic in this section is matrix multiplication. We will begin with the simplest case and then work up to the more general situation. By convention, a matrix with only one row is called a **row vector.** Examples of row vectors are

$$(2 \quad 13), \qquad (-1 \quad 4 \quad 3), \qquad \text{and} \qquad (0 \quad 0 \quad 0 \quad 1)$$

Similarly, a matrix with only one column is called a **column vector.** Examples are

$$\begin{pmatrix} 2 \\ 13 \end{pmatrix}, \qquad \begin{pmatrix} -1 \\ 4 \\ 3 \end{pmatrix}, \qquad \text{and} \qquad \begin{pmatrix} 0 \\ 0 \\ 0 \\ 1 \end{pmatrix}$$

The following definition tells us how to multiply a row vector and a column vector when they have the same number of entries.

DEFINITION | **The Inner Product of a Row Vector and a Column Vector**

EXAMPLES

Let A be a row vector and B a column vector, and assume that the number of columns in A is the same as the number of rows in B. Then the **inner product** $A \cdot B$ is defined to be the number obtained by multiplying the corresponding entries and then adding the products.

$$(1 \quad 2 \quad 3) \cdot \begin{pmatrix} 4 \\ 5 \\ 6 \end{pmatrix} = 1 \cdot 4 + 2 \cdot 5 + 3 \cdot 6 = 32$$

$$(1 \quad 2) \cdot \begin{pmatrix} 4 \\ 5 \\ 6 \end{pmatrix} \quad \text{is not defined}$$

$$(1 \quad 2 \quad 3) \cdot \begin{pmatrix} 4 \\ 5 \end{pmatrix} \quad \text{is not defined}$$

An important observation here is that the end result of taking the inner product is always just a number. The definition of matrix product that we now give depends on this observation.

DEFINITION | **The Product of Two Matrices**

Let A and B be two matrices, and assume that the number of columns in A is the same as the number of rows in B. Then the **product matrix** AB is computed according to the following rule:

The entry in the ith row and the jth column of AB is the inner product of the ith row of A with the jth column of B.

The matrix AB will have as many rows as A and as many columns as B.

As an example of matrix multiplication, we will compute the product AB, where $A = \begin{pmatrix} 1 & 2 \\ 3 & 4 \end{pmatrix}$ and $B = \begin{pmatrix} 5 & 6 & 0 \\ 7 & 8 & 1 \end{pmatrix}$. In other words, we will compute $\begin{pmatrix} 1 & 2 \\ 3 & 4 \end{pmatrix}\begin{pmatrix} 5 & 6 & 0 \\ 7 & 8 & 1 \end{pmatrix}$. Before we attempt to carry out the calculations of any matrix multiplication, however, we should check on two points.

1. Is the product defined? That is, does the number of columns in A equal the number of rows in B? In this case, yes; the common number is 2.
2. What is the size of the product? According to the definition, the product AB will have as many rows as A and as many columns as B. Thus, the size of AB will be 2×3.

Schematically, then, the situation looks like this:

$$\begin{pmatrix} 1 & 2 \\ 3 & 4 \end{pmatrix}\begin{pmatrix} 5 & 6 & 0 \\ 7 & 8 & 1 \end{pmatrix} = \begin{pmatrix} ? & ? & ? \\ ? & ? & ? \end{pmatrix}$$

We have six positions to fill. The computations are presented in Table 3. Reading from the table, we see that our result is

$$AB = \begin{pmatrix} 1 & 2 \\ 3 & 4 \end{pmatrix}\begin{pmatrix} 5 & 6 & 0 \\ 7 & 8 & 1 \end{pmatrix} = \begin{pmatrix} 19 & 22 & 2 \\ 43 & 50 & 4 \end{pmatrix}$$

TABLE 3

Position	How to compute	Computation
row 1, column 1	inner product of row 1 and column 1 $\begin{pmatrix} 1 & 2 \\ 3 & 4 \end{pmatrix}\begin{pmatrix} 5 & 6 & 0 \\ 7 & 8 & 1 \end{pmatrix}$	$1 \cdot 5 + 2 \cdot 7 = 19$
row 1, column 2	inner product of row 1 and column 2 $\begin{pmatrix} 1 & 2 \\ 3 & 4 \end{pmatrix}\begin{pmatrix} 5 & 6 & 0 \\ 7 & 8 & 1 \end{pmatrix}$	$1 \cdot 6 + 2 \cdot 8 = 22$
row 1, column 3	inner product of row 1 and column 3 $\begin{pmatrix} 1 & 2 \\ 3 & 4 \end{pmatrix}\begin{pmatrix} 5 & 6 & 0 \\ 7 & 8 & 1 \end{pmatrix}$	$1 \cdot 0 + 2 \cdot 1 = 2$
row 2, column 1	inner product of row 2 and column 1 $\begin{pmatrix} 1 & 2 \\ 3 & 4 \end{pmatrix}\begin{pmatrix} 5 & 6 & 0 \\ 7 & 8 & 1 \end{pmatrix}$	$3 \cdot 5 + 4 \cdot 7 = 43$
row 2, column 2	inner product of row 2 and column 2 $\begin{pmatrix} 1 & 2 \\ 3 & 4 \end{pmatrix}\begin{pmatrix} 5 & 6 & 0 \\ 7 & 8 & 1 \end{pmatrix}$	$3 \cdot 6 + 4 \cdot 8 = 50$
row 2, column 3	inner product of row 2 and column 3 $\begin{pmatrix} 1 & 2 \\ 3 & 4 \end{pmatrix}\begin{pmatrix} 5 & 6 & 0 \\ 7 & 8 & 1 \end{pmatrix}$	$3 \cdot 0 + 4 \cdot 1 = 4$

EXAMPLE 6 **Computing matrix products**

Let $A = \begin{pmatrix} 1 & 2 \\ 3 & 4 \end{pmatrix}$ and $B = \begin{pmatrix} 5 & 6 \\ 7 & 8 \end{pmatrix}$. By computing AB and then BA, show that $AB \neq BA$. This shows that, in general, matrix multiplication is not commutative.

SOLUTION

$$AB = \begin{pmatrix} 1 & 2 \\ 3 & 4 \end{pmatrix}\begin{pmatrix} 5 & 6 \\ 7 & 8 \end{pmatrix} = \begin{pmatrix} 1\cdot5 + 2\cdot7 & 1\cdot6 + 2\cdot8 \\ 3\cdot5 + 4\cdot7 & 3\cdot6 + 4\cdot8 \end{pmatrix} = \begin{pmatrix} 19 & 22 \\ 43 & 50 \end{pmatrix}$$

$$BA = \begin{pmatrix} 5 & 6 \\ 7 & 8 \end{pmatrix}\begin{pmatrix} 1 & 2 \\ 3 & 4 \end{pmatrix} = \begin{pmatrix} 5\cdot1 + 6\cdot3 & 5\cdot2 + 6\cdot4 \\ 7\cdot1 + 8\cdot3 & 7\cdot2 + 8\cdot4 \end{pmatrix} = \begin{pmatrix} 23 & 34 \\ 31 & 46 \end{pmatrix}$$

Comparing the two matrices AB and BA, we conclude that $AB \neq BA$.

Most graphing utilities allow for matrix multiplication. This can be useful when the matrices are large or contain decimal entries. We'll briefly outline two examples here using the simple two-by-two matrices from Example 6. (For the details on these and for other types of graphing utilities, see your owner's manual. Figure 1 shows the computation of the matrix product AB on a Texas Instruments TI-83+ graphing calculator. After specifying the dimensions of the two matrices and their elements in the matrix-edit menu, the matrices can be displayed as in Figure 1(a). The next screen, Figure 1(b), shows the matrix product AB.

```
[A]
        [[1 2]
         [3 4]]
[B]
        [[5 6]
         [7 8]]
```
(a)

```
[A][B]
        [[19 22]
         [43 50]]
```
(b)

Figure 1
Computing the matrix product AB using a Texas Instruments TI-83+ graphing calculator

Figure 2 shows the resulting calculation as carried out in a Microsoft *Excel* spreadsheet. The elements of the matrix A are typed into cells B1 through C2. Likewise, matrix B is entered in cells B4 through C5. The product matrix is then obtained by means of the function MMULT, which stands for "matrix multiplication." Some additional details are given in the figure caption.

Figure 2
Computing the matrix product AB using Microsoft *Excel*. After typing in the entries of the two matrices A and B, the so-called array formula {=MMULT(B1:C2,B4:C5)} is used to compute and display AB. (See the documentation or help menu that came with *Excel*.)

	A	B	C
1	A=	1	2
2		3	4
3			
4	B=	5	6
5		7	8
6			
7	AB=	19	22
8		43	50
9			

EXERCISE SET 6.3

A

In Exercises 1–4, specify the size of each matrix.

1. (a) $\begin{pmatrix} -4 & 0 & 5 \\ 2 & 8 & -1 \end{pmatrix}$ **2. (a)** $\begin{pmatrix} 1 & 0 \\ 0 & -1 \end{pmatrix}$

(b) $\begin{pmatrix} 7 & 1 \\ 4 & -3 \\ 0 & 0 \end{pmatrix}$ **(b)** $\begin{pmatrix} 1 \\ 6 \\ 8 \\ 1 \end{pmatrix}$

3. $\begin{pmatrix} 1 & a & b & c \\ a & 1 & 0 & a \\ b & 0 & 1 & b \\ c & a & b & 1 \\ 0 & 0 & 0 & 1 \end{pmatrix}$ **4.** $(-3 \quad 1 \quad 6 \quad 0)$

In Exercises 5–8, write the coefficient matrix and the augmented matrix for each system.

5. $\begin{cases} 2x + 3y + 4z = 10 \\ 5x + 6y + 7z = 9 \\ 8x + 9y + 10z = 8 \end{cases}$ **6.** $\begin{cases} 5x - y + z = 0 \\ 4y + 2z = 1 \\ 3x + y + z = -1 \end{cases}$

7. $\begin{cases} x + z + w = -1 \\ x + y + 2w = 0 \\ y + z + w = 1 \\ 2x - y - z = 2 \end{cases}$ **8.** $\begin{cases} 8x - 8y = 5 \\ x - y + z = 1 \end{cases}$

In Exercises 9–22, use matrices to solve each system of equations.

9. $\begin{cases} x - y + 2z = 7 \\ 3x + 2y - z = -10 \\ -x + 3y + z = -2 \end{cases}$ **10.** $\begin{cases} 2x - 3y + 4z = 14 \\ 3x - 2y + 2z = 12 \\ 4x + 5y - 5z = 16 \end{cases}$

11. $\begin{cases} x + z = -2 \\ -3x + 2y = 17 \\ x - y - z = -9 \end{cases}$ **12.** $\begin{cases} 5x + y + 10z = 23 \\ 4x + 2y - 10z = 76 \\ 3x - 4y = 18 \end{cases}$

13. $\begin{cases} x + y + z = -4 \\ 2x - 3y + z = -1 \\ 4x + 2y - 3z = 33 \end{cases}$ **14.** $\begin{cases} 2x + 3y - 4z = 7 \\ x - y + z = -\frac{3}{2} \\ 6x - 5y - 2z = -7 \end{cases}$

15. $\begin{cases} 3x - 2y + 6z = 0 \\ x + 3y + 20z = 15 \\ 10x - 11y - 10z = -9 \end{cases}$

16. $\begin{cases} 3A - 3B + C = 4 \\ 6A + 9B - 3C = -7 \\ A - 2B - 2C = -3 \end{cases}$

17. $\begin{cases} 4x - 3y + 3z = 2 \\ 5x + y - 4z = 1 \\ 9x - 2y - z = 3 \end{cases}$

18. $\begin{cases} 6x + y - z = -1 \\ -3x + 2y + 2z = 2 \\ 5y + 3z = 1 \end{cases}$

19. $\begin{cases} x - y + z + w = 6 \\ x + y - z + w = 4 \\ x + y + z - w = -2 \\ -x + y + z + w = 0 \end{cases}$

20. $\begin{cases} x + 2y - z - 2w = 5 \\ 2x + y + 2z + w = -7 \\ -2x - y - 3z - 2w = 10 \\ z + w = -3 \end{cases}$

21. $\begin{cases} 15A + 14B + 26C = 1 \\ 18A + 17B + 32C = -1 \\ 21A + 20B + 38C = 0 \end{cases}$

22. $\begin{cases} A + B + C + D + E = -1 \\ 3A - 2B - 2C + 3D + 2E = 13 \\ 3C + 4D - 4E = 7 \\ 5A - 4B + E = 30 \\ C - 2E = 3 \end{cases}$

In Exercises 23–50, the matrices A, B, C, D, E, F, and G are defined as follows:

$$A = \begin{pmatrix} 2 & 3 \\ -1 & 4 \end{pmatrix} \quad B = \begin{pmatrix} 1 & -1 \\ 3 & 0 \end{pmatrix} \quad C = \begin{pmatrix} 1 & 0 \\ 0 & 1 \end{pmatrix}$$

$$D = \begin{pmatrix} -1 & 2 & 3 \\ 4 & 0 & 5 \end{pmatrix} \quad E = \begin{pmatrix} 2 & 1 \\ 8 & -1 \\ 6 & 5 \end{pmatrix}$$

$$F = \begin{pmatrix} 5 & -1 \\ -4 & 0 \\ 2 & 3 \end{pmatrix} \quad G = \begin{pmatrix} 0 & 0 \\ 0 & 0 \\ 0 & 0 \end{pmatrix}$$

In each exercise, carry out the indicated matrix operations if they are defined. If an operation is not defined, say so.

23. $A + B$ **24.** $A - B$ **25.** $2A + 2B$ **26.** $2(A + B)$
27. AB **28.** BA **29.** AC **30.** CA
31. $3D + E$ **32.** $E + F$ **33.** $2F - 3G$ **34.** DE
35. ED **36.** DF **37.** FD **38.** $A + D$
39. $G + A$ **40.** DG **41.** GD
42. $(A + B) + C$ **43.** $A + (B + C)$ **44.** CD
45. DC **46.** $5E - 3F$ **47.** $A^2 \, (= AA)$
48. A^2A **49.** AA^2 **50.** C^2
51. Let

$$A = \begin{pmatrix} -1 & 3 & 4 \\ 3 & 2 & -3 \\ 9 & 1 & 6 \end{pmatrix} \quad B = \begin{pmatrix} 7 & 0 & 1 \\ 0 & 0 & 3 \\ -1 & 2 & 4 \end{pmatrix}$$

$$C = \begin{pmatrix} 4 & 6 & 1 \\ 2 & 1 & 3 \\ -1 & -1 & 2 \end{pmatrix}$$

(a) Compute $A(B + C)$.   **(c)** Compute $(AB)C$.
(b) Compute $AB + AC$.   **(d)** Compute $A(BC)$.

Ⓖ *In Exercises 52–55, use a graphing utility to compute the matrix products.*

52. $\begin{pmatrix} 1.03 & 2.1 \\ -0.45 & 3.09 \end{pmatrix} \begin{pmatrix} 2.33 & 4.17 \\ 0 & -1.24 \end{pmatrix}$

53. $\begin{pmatrix} -32 & 14 \\ 27 & 9 \end{pmatrix} \begin{pmatrix} 83 & -19 \\ 13 & 41 \end{pmatrix}$

54. $\begin{pmatrix} 12 & -10 & 13 \\ 5 & 7 & 25 \\ -8 & 9 & 28 \end{pmatrix} \begin{pmatrix} -11 & 31 & 6 \\ 0 & 1 & -14 \\ 41 & 12 & -17 \end{pmatrix}$

55. $\begin{pmatrix} -6 & 9 & -5 & 1 \\ 9 & -1 & -5 & 2 \\ -5 & -5 & 9 & -3 \\ 1 & 2 & -3 & 1 \end{pmatrix} \begin{pmatrix} 0.5 & 1 & 1.5 & 2 \\ 1 & 2 & 3.5 & 5.5 \\ 1.5 & 3.5 & 7 & 12.5 \\ 2 & 5.5 & 12.5 & 25 \end{pmatrix}$

B

56. Let $A = \begin{pmatrix} 1 & 2 \\ 3 & 4 \end{pmatrix}$ and $B = \begin{pmatrix} 5 & 6 \\ 7 & 8 \end{pmatrix}$. Let A^2 and B^2 denote the matrix products AA and BB, respectively. Compute each of the following.

(a) $(A + B)(A + B)$ (c) $A^2 + AB + BA + B^2$
(b) $A^2 + 2AB + B^2$

57. Let $A = \begin{pmatrix} 3 & 5 \\ 7 & 9 \end{pmatrix}$ and $B = \begin{pmatrix} 2 & 4 \\ 6 & 8 \end{pmatrix}$. Compute each of the following.

(a) $A^2 - B^2$ (c) $(A + B)(A - B)$
(b) $(A - B)(A + B)$ (d) $A^2 + AB - BA - B^2$

58. Let

$$A = \begin{pmatrix} 1 & 0 \\ 0 & 1 \end{pmatrix} \qquad B = \begin{pmatrix} 1 & 0 \\ 0 & -1 \end{pmatrix}$$

$$C = \begin{pmatrix} -1 & 0 \\ 0 & 1 \end{pmatrix} \qquad D = \begin{pmatrix} -1 & 0 \\ 0 & -1 \end{pmatrix}$$

Complete the following multiplication table.

	A	B	C	D
A				
B			D	
C				
D				

Hint: In the second row, third column, D is the proper entry because (as you can check) $BC = D$.

59. In this exercise, let's agree to write the coordinates (x, y) of a point in the plane as the 2×1 matrix $\begin{pmatrix} x \\ y \end{pmatrix}$.

(a) Let $A = \begin{pmatrix} 1 & 0 \\ 0 & -1 \end{pmatrix}$ and $Z = \begin{pmatrix} x \\ y \end{pmatrix}$. Compute the matrix AZ. After computing AZ, observe that it represents the point obtained by reflecting $\begin{pmatrix} x \\ y \end{pmatrix}$ about the x-axis.

(b) Let $B = \begin{pmatrix} -1 & 0 \\ 0 & 1 \end{pmatrix}$ and $Z = \begin{pmatrix} x \\ y \end{pmatrix}$. Compute the matrix BZ. After computing BZ, observe that it represents the point obtained by reflecting $\begin{pmatrix} x \\ y \end{pmatrix}$ about the y-axis.

(c) Let A, B, and Z represent the matrices defined in parts (a) and (b). Compute the matrix $(AB)Z$, and then interpret it in terms of reflection about the axes.

60. The **trace** of a 2×2 matrix $\begin{pmatrix} a & b \\ c & d \end{pmatrix}$ is defined by

$$\text{tr}\begin{pmatrix} a & b \\ c & d \end{pmatrix} = a + d$$

(a) If $A = \begin{pmatrix} 1 & 2 \\ 3 & 4 \end{pmatrix}$ and $B = \begin{pmatrix} 5 & 6 \\ 7 & 8 \end{pmatrix}$, verify that $\text{tr}(A + B) = \text{tr}\, A + \text{tr}\, B$.

(b) If $A = \begin{pmatrix} a & b \\ c & d \end{pmatrix}$ and $B = \begin{pmatrix} e & f \\ g & h \end{pmatrix}$, show that $\text{tr}(A + B) = \text{tr}\, A + \text{tr}\, B$.

61. Let $A = \begin{pmatrix} a & b \\ c & d \end{pmatrix}$. The **transpose** of A is the matrix denoted by A^T and defined by $A^T = \begin{pmatrix} a & c \\ b & d \end{pmatrix}$. In other words, A^T is obtained by switching the columns and rows of A. Show that the following equations hold for all 2×2 matrices A and B.

(a) $(A + B)^T = A^T + B^T$
(b) $(A^T)^T = A$
(c) $(AB)^T = B^T A^T$

62. Find an example of two 2×2 matrices A and B for which $AB = \mathbf{0}$ but neither A nor B is $\mathbf{0}$.

PROJECT

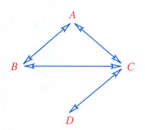

Figure 1
Communication links

$$
\begin{array}{c c c c c}
 & A & B & C & D \\
A & \begin{pmatrix} 0 & 1 & 1 & 0 \\ 1 & 0 & 1 & 0 \\ 1 & 1 & 0 & 1 \\ 0 & 0 & 1 & 0 \end{pmatrix} \\
B & \\
C & \\
D & \\
\end{array}
$$

Figure 2
Communication matrix $\mathcal{M}$

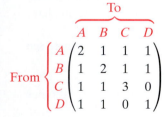

Figure 3
The one-intermediary matrix based on the spy network in Figure 1.

COMMUNICATIONS AND MATRICES

Matrices and matrix products are used in communication theory. Suppose that four spies—we'll call them A, B, C, and D—work for an intelligence agency under the following restrictions: Spies A, B, and C all can communicate directly with one another. Spy D, however, is super-secret and can only communicate directly (back and forth) with spy C. These communication links are summarized in Figure 1.

The communication links between the four spies can also be specified by means of the four-by-four *communication matrix* shown in Figure 2. We'll call this matrix $\mathcal{M}$.

In the communication matrix, the rows and columns are associated with the spies as indicated by the red letters outside the matrix. An entry of 1 means that the two spies associated with that matrix location can communicate directly. For example, in the location row one, column two, there is a 1 because the corresponding spies A and B can communicate directly (according to Figure 1). In cases where two spies cannot communicate directly, 0 is used in the matrix. For instance, there is a 0 in the location row two, column four because the two spies B and D cannot communicate directly. There are four locations in the matrix not covered by what we've said so far. These are the locations corresponding to only one spy (i.e., a spy and him/herself):

row 1, column 1: spy A	row 3, column 3: spy C
row 2, column 2: spy B	row 4, column 4: spy D

By definition, we use a 0 in those locations.

Looking at Figure 1, we see that spy D cannot send a message directly to spy A. However, spy D can get a message to A by using relays or *intermediaries*. For instance, using one intermediary, the message from D to A can be routed

$$D \to C \to A$$

Or, using two intermediaries, the message can be routed

$$D \to C \to B \to A$$

One could even have three intermediaries (perhaps to confuse the enemy) with the route

$$D \to C \to B \to C \to A$$

We now display and explain a four-by-four matrix that shows how many ways messages can be sent from one spy to another using only one intermediary. See Figure 3.

As an example of how to interpret the entries in the one-intermediary matrix in Figure 3, consider the 1 that appears in the second row, first column. This tells us that there is exactly one way to send a message from B to A using an intermediary. You can check that this is the case by looking at Figure 1; the route is $B \to C \to A$. As another example, consider the 2 in the first row, first column. This tells us that there are two ways that A can send a message to A through one intermediary. As Figure 1 shows, the two routes are $A \to B \to A$ and $A \to C \to A$. (One possible reason why spy A might want to send a message to him/herself through an intermediary would be to test the integrity of that intermediary.) As a final example in interpreting the matrix in Figure 3, consider the 0 in the bottom

Figure 4

To

$$\begin{array}{c} \\ \text{From} \begin{cases} A \\ B \\ C \\ D \end{cases} \end{array} \begin{array}{cccc} A & B & C & D \\ \begin{pmatrix} 2 & ? & ? & ? \\ ? & ? & ? & ? \\ ? & ? & ? & ? \\ ? & 1 & 3 & ? \end{pmatrix} \end{array}$$

Figure 5
Two-intermediary matrix for the spy network in Figure 1.

$$\begin{array}{c} A \\ B \\ C \\ D \\ E \end{array} \begin{array}{ccccc} A & B & C & D & E \\ \begin{pmatrix} 0 & 1 & 1 & 1 & 1 \\ 1 & 0 & 1 & 0 & 0 \\ 1 & 1 & 0 & 1 & 0 \\ 1 & 0 & 1 & 0 & 1 \\ 1 & 0 & 0 & 1 & 0 \end{pmatrix} \end{array}$$

Figure 6
Communication matrix M for a network with five spies.

row. This says that there is no way for D to send a message to C using exactly one intermediary. (Check this in Figure 1.)

There is a surprising connection between the communication matrix M in Figure 2 and the one-intermediary matrix in Figure 3. If you compute the matrix product of M with itself (this is abbreviated M^2), you get the one-intermediary matrix in Figure 3. You should verify this now for yourself. That is, either by hand or with a graphing utility, compute M^2 and check that the product is the same as the matrix in Figure 3. It can be shown that this situation holds in general.

EXERCISES

1. **(b)** Consider the communication links shown in Figure 4.
   (a) Specify the communication matrix M for this system.
   (b) By referring to Figure 4, work out the one-intermediary matrix for this system. Then, check your answer by computing M^2 and comparing the results.

2. Return to the spy network that we considered in Figure 1.
   (a) Work out the entries for the following *two-intermediary* matrix in Figure 5. As examples, three entries are already filled in. The 2 in the first row, first column indicates that there are two ways to send a message from A to A using two intermediaries. The routes are $A \to B \to C \to A$ and $A \to C \to B \to A$. As another example, the 1 in the fourth row, second column indicates that there is only one way to send a message from D to B using two intermediaries. As you can verify in Figure 1, the route is $D \to C \to A \to B$. Finally, consider the 3 in the fourth row, third column. This indicates that there are three ways to send a message from D to C using two intermediaries. The three routes are $D \to C \to A \to C$, $D \to C \to B \to C$, and $D \to C \to D \to C$. In these routes from D to C, note that D and C themselves can be used as intermediaries.
   (b) Let M be the communication matrix given in Figure 2. It can be shown that the matrix $M^3 [=M(M^2)]$ equals the two-intermediary matrix for this system. Compute M^3, either by hand or with a graphing utility, and then use the result to check the entries in your answer for part (a).

3. Consider a network of five spies, $A, B, C, D,$ and E with the communication matrix M shown in Figure 6.
   (a) Draw a diagram, of the type shown in Figure 1, showing the communication links.
   (b) Using the diagram from part (a): How many one-intermediary routes can you find for sending a message from D to A? How many two-intermediary routes can you find for sending a message from D to A?
   (c) Use a graphing utility to compute M^2 and M^3, and thereby check to see whether you missed any routes in part (b).

Concluding remarks:
1. As you can see after working this exercise, for larger networks, it becomes preferable to work from the matrices rather than a diagram. For instance, what if there were, say, 50 spies, rather than only 4 or 5?
2. The mathematics in this project can be applied in many other areas. For instance, instead of communication between spies, the subject could be air travel between cities. Then, for example, a one-intermediary route from New York to San Francisco would correspond to a flight between those two cities with one stopover—in Chicago, say.

6.4   THE INVERSE OF A SQUARE MATRIX

… when the chips are down we close the office door and compute with matrices like fury. —Paul Halmos (quoted here out of context), *Celebrating 50 Years of Mathematics,* J. H. Ewing et al. (ed.) (New York: Springer-Verlag, 1995)

We have noted Cayley's work in analytic geometry, especially in connection with the use of determinants; but [Arthur] *Cayley* [1821–1895] *also was one of the first men to study matrices, another instance of the British concern for form and structure in algebra.* —Carl B. Boyer in *A History of Mathematics,* 2nd ed., revised by Uta C. Merzback (New York: John Wiley and Sons, 1991)

A matrix that has the same number of rows and columns is called a **square matrix.** So, two examples of square matrices are

$$A = \begin{pmatrix} 1 & 2 \\ 3 & 4 \end{pmatrix} \quad \text{and} \quad B = \begin{pmatrix} -5 & 6 & 7 \\ \frac{1}{2} & 0 & 1 \\ 8 & 4 & -3 \end{pmatrix}$$

The matrix A is said to be a square matrix of **order two** (or, more simply, a 2×2 matrix); B is a square matrix of **order three** (that is, a 3×3 matrix). We will first present the concepts and techniques of this section in terms of square matrices of order two. After that, we'll show how the ideas carry over to larger square matrices.

We begin by defining a special matrix I_2:

$$I_2 = \begin{pmatrix} 1 & 0 \\ 0 & 1 \end{pmatrix}$$

The matrix I_2 plays the same role in the multiplication of 2×2 matrices as does the number 1 in the multiplication of real numbers. Specifically, I_2 has the following property: For every 2×2 matrix $A = \begin{pmatrix} a & b \\ c & d \end{pmatrix}$, we have (as you can easily verify)

$$\begin{pmatrix} a & b \\ c & d \end{pmatrix}\begin{pmatrix} 1 & 0 \\ 0 & 1 \end{pmatrix} = \begin{pmatrix} a & b \\ c & d \end{pmatrix} \quad \text{and} \quad \begin{pmatrix} 1 & 0 \\ 0 & 1 \end{pmatrix}\begin{pmatrix} a & b \\ c & d \end{pmatrix} = \begin{pmatrix} a & b \\ c & d \end{pmatrix}$$

In other words,

$$AI_2 = A \quad \text{and} \quad I_2 A = A$$

The matrix I_2 is called the **multiplicative identity** for square matrices of order two.

If we have two real numbers a and b such that $ab = 1$, then we say that a and b are (multiplicative) inverses. In the box that follows we apply this terminology to matrices. Note that in the case of inverse matrices, the multiplication *is* commutative. That is, AB and BA yield the same result (namely, I_2).

DEFINITION | **The Inverse of a 2 × 2 Matrix**

If A and B are 2×2 matrices such that

$$AB = I_2 \quad \text{and} \quad BA = I_2$$

then A and B are said to be **inverses** of one another.

EXAMPLE **1**   **Finding a matrix inverse from first principles**

Find the inverse of $A = \begin{pmatrix} 1 & 2 \\ 3 & 4 \end{pmatrix}$.

SOLUTION

We need to find numbers $a, b, c,$ and d such that the following two matrix equations are valid:

$$\begin{pmatrix} 1 & 2 \\ 3 & 4 \end{pmatrix}\begin{pmatrix} a & b \\ c & d \end{pmatrix} = \begin{pmatrix} 1 & 0 \\ 0 & 1 \end{pmatrix} \tag{1}$$

$$\begin{pmatrix} a & b \\ c & d \end{pmatrix}\begin{pmatrix} 1 & 2 \\ 3 & 4 \end{pmatrix} = \begin{pmatrix} 1 & 0 \\ 0 & 1 \end{pmatrix} \tag{2}$$

From equation (1) we have

$$\begin{pmatrix} a + 2c & b + 2d \\ 3a + 4c & 3b + 4d \end{pmatrix} = \begin{pmatrix} 1 & 0 \\ 0 & 1 \end{pmatrix}$$

For this last equation to be valid, the corresponding entries in the two matrices must be equal. (Why?) Consequently, we obtain four equations, two involving a and c, and two involving b and d:

$$\begin{cases} a + 2c = 1 \\ 3a + 4c = 0 \end{cases} \qquad \begin{cases} b + 2d = 0 \\ 3b + 4d = 1 \end{cases}$$

As Exercise 42(a) asks you to show, the solution to the first system is $a = -2$ and $c = 3/2$, and for the second system, $b = 1$ and $d = -1/2$. Furthermore (as you can check), these same values are obtained for a, b, c and d if we begin with matrix equation (2) rather than (1). Thus, the required inverse matrix is

$$\begin{pmatrix} -2 & 1 \\ \frac{3}{2} & -\frac{1}{2} \end{pmatrix}$$

Exercise 42(b) asks you to carry out the matrix multiplication to confirm that we indeed have

$$\begin{pmatrix} 1 & 2 \\ 3 & 4 \end{pmatrix}\begin{pmatrix} -2 & 1 \\ \frac{3}{2} & -\frac{1}{2} \end{pmatrix} = \begin{pmatrix} -2 & 1 \\ \frac{3}{2} & -\frac{1}{2} \end{pmatrix}\begin{pmatrix} 1 & 2 \\ 3 & 4 \end{pmatrix}$$
$$= \begin{pmatrix} 1 & 0 \\ 0 & 1 \end{pmatrix}$$

In Example 1 we found the inverse of a square matrix by solving systems of equations. However, as you've seen in previous sections, there are systems of equations that do not have solutions. In the present context, this implies that there are matrices that do not have inverses. For example (as Exercise 43 asks you to show), the matrix $\begin{pmatrix} 2 & 5 \\ 6 & 15 \end{pmatrix}$ does not have an inverse. A matrix that does not have an inverse is called a **singular matrix.** If a matrix does possess an inverse, we say that the matrix is **nonsingular.** So, according to the result in Example 1, the matrix $\begin{pmatrix} 1 & 2 \\ 3 & 4 \end{pmatrix}$ is nonsingular.

It can be proven that if a square matrix has an inverse, then that inverse is unique. (In other words, you can never find two different inverses for the same matrix.) The inverse of a nonsingular matrix A is denoted by A^{-1} (read A *inverse*).

There are several methods that can be used to compute the inverse of a matrix (or to show that there is no inverse). One method involves working with systems of equations, as in Example 1. A more efficient method, however, is the following. (We'll demonstrate the method with an example, but we won't give a formal proof of its validity.)

Suppose that we want to compute the inverse (if it exists) for the matrix

$$A = \begin{pmatrix} 6 & 2 \\ 13 & 4 \end{pmatrix}$$

We begin by writing a larger matrix, formed by merging A and I_2 as follows:

$$\left(\begin{array}{cc|cc} 6 & 2 & 1 & 0 \\ 13 & 4 & 0 & 1 \end{array} \right)$$

A convenient abbreviation for this matrix is $(A \mid I_2)$. Now we carry out elementary row operations on this matrix until one of the following two situations occurs:

1. Either the matrix takes on the form

$$\left(\begin{array}{cc|cc} 1 & 0 & a & b \\ 0 & 1 & c & d \end{array} \right)$$

In this case, $A^{-1} = \begin{pmatrix} a & b \\ c & d \end{pmatrix}$.

2. Or, one of the rows to the left of the dashed line consists entirely of zeros. In this case, A^{-1} does not exist.

The calculations for our example then run as follows.

$$\left(\begin{array}{cc|cc} 6 & 2 & 1 & 0 \\ 13 & 4 & 0 & 1 \end{array} \right) \xrightarrow{-2R_1 + R_2} \left(\begin{array}{cc|cc} 6 & 2 & 1 & 0 \\ 1 & 0 & -2 & 1 \end{array} \right)$$

$$\xrightarrow{R_1 \leftrightarrow R_2} \left(\begin{array}{cc|cc} 1 & 0 & -2 & 1 \\ 6 & 2 & 1 & 0 \end{array} \right)$$

$$\xrightarrow{-6R_1 + R_2} \left(\begin{array}{cc|cc} 1 & 0 & -2 & 1 \\ 0 & 2 & 13 & -6 \end{array} \right)$$

$$\xrightarrow{\frac{1}{2}R_2} \left(\begin{array}{cc|cc} 1 & 0 & -2 & 1 \\ 0 & 1 & \frac{13}{2} & -3 \end{array} \right)$$

The inverse matrix can now be read off:

$$A^{-1} = \begin{pmatrix} -2 & 1 \\ \frac{13}{2} & -3 \end{pmatrix}$$

(You should verify for yourself that we indeed have $AA^{-1} = I_2$ and $A^{-1}A = I_2$.)

Inverse matrices can be used to solve certain systems of equations in which the number of unknowns is the same as the number of equations. Before explaining how this works, we describe how a system of equations can be written in matrix form. Consider the system

$$\begin{cases} x + 2y = 8 \\ 3x + 4y = 6 \end{cases} \tag{3}$$

As defined in the previous section, the coefficient matrix A for this system is

$$A = \begin{pmatrix} 1 & 2 \\ 3 & 4 \end{pmatrix}$$

Now we define two matrices X and B:

$$X = \begin{pmatrix} x \\ y \end{pmatrix} \qquad B = \begin{pmatrix} 8 \\ 6 \end{pmatrix}$$

Then system (3) can be written as a single matrix equation:

$$AX = B \tag{4}$$

To see why this is so, we expand equation (4) to obtain

$$\begin{pmatrix} 1 & 2 \\ 3 & 4 \end{pmatrix}\begin{pmatrix} x \\ y \end{pmatrix} = \begin{pmatrix} 8 \\ 6 \end{pmatrix} \qquad \text{\color{red}using the definitions of } A, X, \text{ and } B$$

$$\begin{pmatrix} x + 2y \\ 3x + 4y \end{pmatrix} = \begin{pmatrix} 8 \\ 6 \end{pmatrix} \qquad \text{\color{red}carrying out the matrix multiplication}$$

By equating the corresponding entries of the matrices in this last equation, we obtain $x + 2y = 8$ and $3x + 4y = 6$, as given initially in system (3).

EXAMPLE 2 **Using matrix algebra to solve a system**

Use an inverse matrix to solve the system

$$\begin{cases} x + 2y = 8 \\ 3x + 4y = 6 \end{cases}$$

SOLUTION

As explained just prior to this example, the matrix form for this system is

$$AX = B \tag{5}$$

where

$$A = \begin{pmatrix} 1 & 2 \\ 3 & 4 \end{pmatrix} \qquad X = \begin{pmatrix} x \\ y \end{pmatrix} \qquad B = \begin{pmatrix} 8 \\ 6 \end{pmatrix}$$

From Example 1, we know that

$$A^{-1} = \begin{pmatrix} -2 & 1 \\ \frac{3}{2} & -\frac{1}{2} \end{pmatrix}$$

Now we multiply both sides of equation (5) by A^{-1} to obtain

$$A^{-1}(AX) = A^{-1}B$$
$$(A^{-1}A)X = A^{-1}B \qquad \text{\color{red}Matrix multiplication is associative.}$$
$$I_2 X = A^{-1}B$$
$$X = A^{-1}B \qquad \text{(Why?)}$$

Substituting the actual matrices into this last equation, we have

$$\begin{pmatrix} x \\ y \end{pmatrix} = \begin{pmatrix} -2 & 1 \\ \frac{3}{2} & -\frac{1}{2} \end{pmatrix} \begin{pmatrix} 8 \\ 6 \end{pmatrix} = \begin{pmatrix} -10 \\ 9 \end{pmatrix}$$

Therefore $x = -10$ and $y = 9$, as required.

All of the ideas we have discussed for square matrices of order two can be carried over directly to larger square matrices. It is easy to check that the following matrices, I_3 and I_4, are the multiplicative identities for 3×3 and 4×4 matrices, respectively:

$$I_3 = \begin{pmatrix} 1 & 0 & 0 \\ 0 & 1 & 0 \\ 0 & 0 & 1 \end{pmatrix} \qquad I_4 = \begin{pmatrix} 1 & 0 & 0 & 0 \\ 0 & 1 & 0 & 0 \\ 0 & 0 & 1 & 0 \\ 0 & 0 & 0 & 1 \end{pmatrix}$$

These identity matrices are described by saying that they have ones down the **main diagonal** and zeros everywhere else. (Larger identity matrices can be defined following this same pattern.) Sometimes, when it is clear from the context or as a matter of convenience, we'll omit the subscript and denote the appropriately sized identity matrix simply by I. (This is done in the box that follows.)

PROPERTY SUMMARY **The Inverse of a Square Matrix**

1. Suppose that A and B are square matrices of the same size, and let f denote the identity matrix of that size. Then A and B are said to be **inverses** of one another provided

$$AB = I \qquad \text{and} \qquad BA = I$$

2. It can be shown that every square matrix has at most one inverse.
3. A square matrix that has an inverse is said to be **nonsingular** (or **invertible**). A square matrix that does not have an inverse is called **singular.**
4. If the matrix A is nonsingular, then the inverse of A is denoted by A^{-1}. In this case, we have

$$AA^{-1} = I \qquad \text{and} \qquad A^{-1}A = I$$

EXAMPLE **3** **Using elementary row operations to compute a matrix inverse**

Let

$$A = \begin{pmatrix} 5 & 0 & 2 \\ 2 & 2 & 1 \\ -3 & 1 & -1 \end{pmatrix}$$

Use the elementary row operations to compute A^{-1}, if it exists.

SOLUTION

Following the method that we described for the 2×2 case, we first write down the matrix $(A \mid I_3)$ formed by merging A and I_3:

$$\begin{pmatrix} 5 & 0 & 2 & 1 & 0 & 0 \\ 2 & 2 & 1 & 0 & 1 & 0 \\ -3 & 1 & -1 & 0 & 0 & 1 \end{pmatrix}$$

Now we carry out the elementary row operations, trying to obtain I_3 to the left of the dashed line. We have

$$\begin{pmatrix} 5 & 0 & 2 & 1 & 0 & 0 \\ 2 & 2 & 1 & 0 & 1 & 0 \\ -3 & 1 & -1 & 0 & 0 & 1 \end{pmatrix} \xrightarrow{-2R_2+R_1} \begin{pmatrix} 1 & -4 & 0 & 1 & -2 & 0 \\ 2 & 2 & 1 & 0 & 1 & 0 \\ -3 & 1 & -1 & 0 & 0 & 1 \end{pmatrix}$$

$$\xrightarrow[3R_1+R_3]{-2R_1+R_2} \begin{pmatrix} 1 & -4 & 0 & 1 & -2 & 0 \\ 0 & 10 & 1 & -2 & 5 & 0 \\ 0 & -11 & -1 & 3 & -6 & 1 \end{pmatrix} \xrightarrow{1R_3+R_2} \begin{pmatrix} 1 & -4 & 0 & 1 & -2 & 0 \\ 0 & -1 & 0 & 1 & -1 & 1 \\ 0 & -11 & -1 & 3 & -6 & 1 \end{pmatrix}$$

$$\xrightarrow{(-1)R_2} \begin{pmatrix} 1 & -4 & 0 & 1 & -2 & 0 \\ 0 & 1 & 0 & -1 & 1 & -1 \\ 0 & -11 & -1 & 3 & -6 & 1 \end{pmatrix} \xrightarrow[11R_2+R_3]{4R_2+R_1} \begin{pmatrix} 1 & 0 & 0 & -3 & 2 & -4 \\ 0 & 1 & 0 & -1 & 1 & -1 \\ 0 & 0 & -1 & -8 & 5 & -10 \end{pmatrix}$$

$$\xrightarrow{(-1)R_3} \begin{pmatrix} 1 & 0 & 0 & -3 & 2 & -4 \\ 0 & 1 & 0 & -1 & 1 & -1 \\ 0 & 0 & 1 & 8 & -5 & 10 \end{pmatrix}$$

The required inverse is therefore

$$A^{-1} = \begin{pmatrix} -3 & 2 & -4 \\ -1 & 1 & -1 \\ 8 & -5 & 10 \end{pmatrix}$$

Just as graphing utilities can be used for matrix multiplication (as we saw in the previous section), they can also be used to compute matrix inverses. As in Section 6.3, we will briefly cite two examples, leaving the details for your owner's manual. We'll use the matrix A given in Example 3. Figure 1 shows the two stages in computing the inverse on a Texas Instruments 83+ graphing calculator. After specifying the dimension and elements of the matrix A, it can be displayed as in Figure 1(a). Figure 1(b) then shows the result of computing A^{-1}. As you can see, it indeed agrees with the result in Example 3.

Figure 2 shows the resulting calculation as carried in a Microsoft *Excel* spreadsheet. The elements of the matrix A have been entered in cells B1 through D3. The inverse matrix is obtained by means of the function MINVERSE.

Figure 1
Computing the inverse of the matrix A (from Example 3) on a Texas Instruments TI-83+ graphing calculator

(a)   (b)

Figure 2

Computing the inverse of the matrix A (from Example 3) using Microsoft *Excel*. After typing in the entries for A in cells B1 through D3, the array formula `{=MINVERSE(B1:D3)}` is used to compute and display the inverse in cells B5 through D7.

	A	B	C	D
1		5		2
2	A=	2	2	1
3		-3	1	-1
4				
5		-3	2	-4
6	A^{-1}=	-1	1	-1
7		8	-5	10
8				

We conclude this section with a description of how matrices can be used in sending and receiving coded messages. Any square matrix that has an inverse can be used to encode the message. Let's suppose that the *encoding matrix* is $\begin{pmatrix} 3 & 8 \\ 4 & 11 \end{pmatrix}$, and that the message to be encoded is *HIDE IT*. We begin by letting the letters of the alphabet correspond to the integers 1–26 in the natural way shown in Table 1. Additionally, as indicated in the table, the integer 0 is used either to indicate a space between words or as a placeholder at the end of the message.

TABLE 1

A	B	C	D	E	F	G	H	I	J	K	L	M
1	2	3	4	5	6	7	8	9	10	11	12	13

14	15	16	17	18	19	20	21	22	23	24	25	26
N	O	P	Q	R	S	T	U	V	W	X	Y	Z

Conventions regarding zero: The integer 0 is used either to indicate a space between words or as a placeholder at the end of the message.

Using the correspondences in Table 1, we have

$$\begin{array}{ccccccc} H & I & D & E & & I & T \\ 8 & 9 & 4 & 5 & 0 & 9 & 20 \end{array}$$

At this point, the message has been converted to a list of seven integers: 8, 9, 4, 5, 0, 9, 20. From this list, we will create some 2×1 matrices. (If the encoding matrix had been a square matrix of order three, rather than order two, then we would create 3×1 matrices; you'll see this in Example 4.) To create the 2×1 matrices, we first pair up the integers in our list:

$$8, 9 \quad 4, 5 \quad 0, 9 \quad 20, 0$$

The 0 at the end is inserted as a placeholder, because there was nothing to pair with 20. The four pairs are now written as one-column matrices:

$$\begin{pmatrix} 8 \\ 9 \end{pmatrix} \quad \begin{pmatrix} 4 \\ 5 \end{pmatrix} \quad \begin{pmatrix} 0 \\ 9 \end{pmatrix} \quad \begin{pmatrix} 20 \\ 0 \end{pmatrix}$$

The last step is to multiply each 2×1 matrix by the encoding matrix. As you should check for yourself, we obtain

$$\begin{pmatrix} 3 & 8 \\ 4 & 11 \end{pmatrix}\begin{pmatrix} 8 \\ 9 \end{pmatrix} = \begin{pmatrix} 96 \\ 131 \end{pmatrix}$$

$$\begin{pmatrix} 3 & 8 \\ 4 & 11 \end{pmatrix}\begin{pmatrix} 4 \\ 5 \end{pmatrix} = \begin{pmatrix} 52 \\ 71 \end{pmatrix}$$

$$\begin{pmatrix} 3 & 8 \\ 4 & 11 \end{pmatrix}\begin{pmatrix} 0 \\ 9 \end{pmatrix} = \begin{pmatrix} 72 \\ 99 \end{pmatrix}$$

$$\begin{pmatrix} 3 & 8 \\ 4 & 11 \end{pmatrix}\begin{pmatrix} 20 \\ 0 \end{pmatrix} = \begin{pmatrix} 60 \\ 80 \end{pmatrix}$$

The encoded message then consists of the eight integers that we have just calculated:

$$96, 131, 52, 71, 72, 99, 60, 80 \tag{6}$$

If you were to receive message (6) and you knew that the encoding matrix was $\begin{pmatrix} 3 & 8 \\ 4 & 11 \end{pmatrix}$, you could decode the message by undoing the previous steps as follows. Express message (6) as the four matrices

$$\begin{pmatrix} 96 \\ 131 \end{pmatrix} \qquad \begin{pmatrix} 52 \\ 71 \end{pmatrix} \qquad \begin{pmatrix} 72 \\ 99 \end{pmatrix} \qquad \begin{pmatrix} 60 \\ 80 \end{pmatrix} \tag{7}$$

Next, compute $\begin{pmatrix} 3 & 8 \\ 4 & 11 \end{pmatrix}^{-1}$. As you can check, this turns out to be $\begin{pmatrix} 11 & -8 \\ -4 & 3 \end{pmatrix}$.

Now multiply each of the matrices in (7) by $\begin{pmatrix} 11 & -8 \\ -4 & 3 \end{pmatrix}$. This yields

$$\begin{pmatrix} 11 & -8 \\ -4 & 3 \end{pmatrix}\begin{pmatrix} 96 \\ 131 \end{pmatrix} = \begin{pmatrix} 8 \\ 9 \end{pmatrix}$$

Use a calculator or graphing utility to check these results

$$\begin{pmatrix} 11 & -8 \\ -4 & 3 \end{pmatrix}\begin{pmatrix} 52 \\ 71 \end{pmatrix} = \begin{pmatrix} 4 \\ 5 \end{pmatrix}$$

$$\begin{pmatrix} 11 & -8 \\ -4 & 3 \end{pmatrix}\begin{pmatrix} 72 \\ 99 \end{pmatrix} = \begin{pmatrix} 0 \\ 9 \end{pmatrix}$$

$$\begin{pmatrix} 11 & -8 \\ -4 & 3 \end{pmatrix}\begin{pmatrix} 60 \\ 80 \end{pmatrix} = \begin{pmatrix} 20 \\ 0 \end{pmatrix}$$

This produces the list of eight integers 8, 9, 4, 5, 0, 9, 20, 0. Table 1 allows us to translate this list. As you can check, the result is the original message *HIDE IT*.

EXAMPLE 4 Encoding a message by means of a matrix

Encode the message *GO LATER* assuming that the encoding matrix is

$$\begin{pmatrix} 5 & 0 & 2 \\ 2 & 2 & 1 \\ -3 & 1 & -1 \end{pmatrix}.$$

SOLUTION

Step 1 Convert the message to a list of integers. Using Table 1 and the convention regarding zero, the message *GO LATER* corresponds to the list 7, 15, 0, 12, 1, 20, 5, 18.

Step 2 Put the integers into one-column matrices. In the previous discussion, where we were using a 2×2 encoding matrix, we created matrices by taking the integers *two* at a time. Now we have a 3×3 encoding matrix, and we take the integers *three* at a time to obtain

$$\begin{pmatrix} 7 \\ 15 \\ 0 \end{pmatrix} \qquad \begin{pmatrix} 12 \\ 1 \\ 20 \end{pmatrix} \qquad \begin{pmatrix} 5 \\ 18 \\ 0 \end{pmatrix}$$

The zero in the matrix on the right is inserted as a placeholder.

Step 3 Multiply each matrix in Step 2 by the encoding matrix:

$$\begin{pmatrix} 5 & 0 & 2 \\ 2 & 2 & 1 \\ -3 & 1 & -1 \end{pmatrix} \begin{pmatrix} 7 \\ 15 \\ 0 \end{pmatrix} = \begin{pmatrix} 35 \\ 44 \\ -6 \end{pmatrix}$$

$$\begin{pmatrix} 5 & 0 & 2 \\ 2 & 2 & 1 \\ -3 & 1 & -1 \end{pmatrix} \begin{pmatrix} 12 \\ 1 \\ 20 \end{pmatrix} = \begin{pmatrix} 100 \\ 46 \\ -55 \end{pmatrix}$$

$$\begin{pmatrix} 5 & 0 & 2 \\ 2 & 2 & 1 \\ -3 & 1 & -1 \end{pmatrix} \begin{pmatrix} 5 \\ 18 \\ 0 \end{pmatrix} = \begin{pmatrix} 25 \\ 46 \\ 3 \end{pmatrix}$$

The encoded message is therefore $35, 44, -6, 100, 46, -55, 25, 46, 3$.

EXAMPLE 5 Decoding a message using a matrix inverse

Check the result in Example 4 by decoding the message

$$35, 44, -6, 100, 46, -55, 25, 46, 3 \tag{8}$$

The encoding matrix is $\begin{pmatrix} 5 & 0 & 2 \\ 2 & 2 & 1 \\ -3 & 1 & -1 \end{pmatrix}$.

SOLUTION

Step 1 Put the integers in (8) into one-column matrices. Since the encoding matrix is 3×3, each one-column matrix will have three rows. This produces

$$\begin{pmatrix} 35 \\ 44 \\ -6 \end{pmatrix} \qquad \begin{pmatrix} 100 \\ 46 \\ -55 \end{pmatrix} \qquad \begin{pmatrix} 25 \\ 46 \\ 3 \end{pmatrix}$$

Step 2 Compute the inverse matrix for the encoding matrix $\begin{pmatrix} 5 & 0 & 2 \\ 2 & 2 & 1 \\ -3 & 1 & -1 \end{pmatrix}$.

As we saw in Example 3, the inverse of this matrix is given by

$$\begin{pmatrix} 5 & 0 & 2 \\ 2 & 2 & 1 \\ -3 & 1 & -1 \end{pmatrix}^{-1} = \begin{pmatrix} -3 & 2 & -4 \\ -1 & 1 & -1 \\ 8 & -5 & 10 \end{pmatrix}$$

Step 3 Multiply each one-column matrix in Step 1 by the inverse matrix obtained in Step 2.

$$\begin{pmatrix} -3 & 2 & -4 \\ -1 & 1 & -1 \\ 8 & -5 & 10 \end{pmatrix}\begin{pmatrix} 35 \\ 44 \\ -6 \end{pmatrix} = \begin{pmatrix} 7 \\ 15 \\ 0 \end{pmatrix} \quad \begin{pmatrix} -3 & 2 & -4 \\ -1 & 1 & -1 \\ 8 & -5 & 10 \end{pmatrix}\begin{pmatrix} 100 \\ 46 \\ -55 \end{pmatrix} = \begin{pmatrix} 12 \\ 1 \\ 20 \end{pmatrix} \quad \begin{pmatrix} -3 & 2 & -4 \\ -1 & 1 & -1 \\ 8 & -5 & 10 \end{pmatrix}\begin{pmatrix} 25 \\ 46 \\ 3 \end{pmatrix} = \begin{pmatrix} 5 \\ 18 \\ 0 \end{pmatrix}$$

This gives us the list of nine integers

$$7, 15, 0, 12, 1, 20, 5, 18, 0$$

Table 1 can now be used to translate this list. As you can verify, the result is *GO LATER*.

EXERCISE SET 6.4

A

In Exercises 1–4, the matrices A, B, C, and D are defined as follows.

$$A = \begin{pmatrix} 4 & -1 \\ -5 & 2 \end{pmatrix} \qquad B = \begin{pmatrix} \frac{1}{2} & 5 \\ 3 & 1 \end{pmatrix}$$

$$C = \begin{pmatrix} 3 & 0 & -2 \\ 0 & 5 & 6 \\ 1 & 4 & -7 \end{pmatrix} \qquad D = \begin{pmatrix} 1 & 2 & 3 \\ 4 & 5 & 6 \\ 7 & 8 & 9 \end{pmatrix}$$

1. Compute AI_2 and I_2A to verify that $AI_2 = I_2A = A$.
2. Compute BI_2 and I_2B to verify that $BI_2 = I_2B = B$.
3. Compute CI_3 and I_3C to verify that $CI_3 = I_3C = C$.
4. Compute DI_3 and I_3D to verify that $DI_3 = I_3D = D$.

In Exercises 5–12, compute A^{-1}, if it exists, using the method of Example 1.

5. $A = \begin{pmatrix} 7 & 9 \\ 4 & 5 \end{pmatrix}$ **6.** $A = \begin{pmatrix} 3 & -8 \\ 2 & -5 \end{pmatrix}$

7. $A = \begin{pmatrix} -3 & 1 \\ 5 & 6 \end{pmatrix}$ **8.** $A = \begin{pmatrix} -4 & 0 \\ 9 & 3 \end{pmatrix}$

9. $A = \begin{pmatrix} -2 & 3 \\ -4 & 6 \end{pmatrix}$ **10.** $A = \begin{pmatrix} \frac{5}{3} & -2 \\ -\frac{2}{3} & 1 \end{pmatrix}$

11. $A = \begin{pmatrix} \frac{1}{3} & \frac{1}{3} \\ -\frac{1}{9} & \frac{2}{9} \end{pmatrix}$ **12.** $\begin{pmatrix} -3 & 7 \\ 12 & -28 \end{pmatrix}$

In Exercises 13–26, compute the inverse matrix, if it exists, using elementary row operations (as shown in Example 3).

13. $\begin{pmatrix} 2 & 1 \\ 3 & 2 \end{pmatrix}$ **14.** $\begin{pmatrix} -6 & 5 \\ 18 & -15 \end{pmatrix}$

15. $\begin{pmatrix} 0 & -11 \\ 1 & 6 \end{pmatrix}$ **16.** $\begin{pmatrix} -2 & 13 \\ -4 & 25 \end{pmatrix}$

17. $\begin{pmatrix} \frac{2}{3} & -\frac{1}{4} \\ -8 & 3 \end{pmatrix}$ **18.** $\begin{pmatrix} -\frac{2}{5} & \frac{1}{3} \\ -6 & 5 \end{pmatrix}$

19. $\begin{pmatrix} -5 & 4 & -3 \\ 10 & -7 & 6 \\ 8 & -6 & 5 \end{pmatrix}$ **20.** $\begin{pmatrix} 1 & 0 & -2 \\ -3 & -1 & 6 \\ 2 & 1 & -5 \end{pmatrix}$

21. $\begin{pmatrix} 1 & 2 & -1 \\ 0 & 3 & 0 \\ -4 & 0 & 5 \end{pmatrix}$ **22.** $\begin{pmatrix} 1 & -4 & -8 \\ 1 & 2 & 5 \\ 1 & 1 & 3 \end{pmatrix}$

23. $\begin{pmatrix} -7 & 5 & 3 \\ 3 & -2 & -2 \\ 3 & -2 & -1 \end{pmatrix}$ **24.** $\begin{pmatrix} 2 & -1 & -1 \\ 1 & 0 & -1 \\ -2 & 1 & 2 \end{pmatrix}$

25. $\begin{pmatrix} 1 & 2 & 3 \\ 4 & 5 & 6 \\ 7 & 8 & 9 \end{pmatrix}$ **26.** $\begin{pmatrix} 2 & 1 & 3 \\ 4 & 5 & -7 \\ 2 & 1 & 3 \end{pmatrix}$

Ⓖ *For Exercises 27–32, the matrices A, B, C, D, and E are defined as follows. In each exercise, use a graphing utility to carry out the indicated matrix operations. If an operation is undefined, state this.*

$$A = \begin{pmatrix} 2 & 3 \\ 4 & 5 \end{pmatrix} \qquad B = \begin{pmatrix} 3 & 2 & 6 \\ 1 & 1 & 2 \\ 2 & 2 & 5 \end{pmatrix} \qquad C = \begin{pmatrix} 9 & 4 & 4 \\ 2 & 2 & 1 \\ -3 & 1 & -1 \end{pmatrix}$$

$$D = \begin{pmatrix} -2 & 0 \\ 3 & 1 \\ 5 & 2 \end{pmatrix} \qquad E = \begin{pmatrix} 3 & -2 & -1 \\ 7 & -4 & 0 \end{pmatrix}$$

27. A^{-1} **28.** B^{-1} **29.** D^{-1}
30. $(CD)^{-1}$ **31.** $(DE)^{-1}$
32. $(BC)^{-1} - C^{-1}B^{-1}$ *Hint:* The order for carrying out the matrix operations is similar to that for real numbers. Thus, the subtraction in this exercise is the last step.

33. If $A = \begin{pmatrix} 3 & 8 \\ 4 & 11 \end{pmatrix}$, then $A^{-1} = \begin{pmatrix} 11 & -8 \\ -4 & 3 \end{pmatrix}$. Use this fact and the method of Example 2 to solve the following systems.
(a) $\begin{cases} 3x + 8y = 5 \\ 4x + 11y = 7 \end{cases}$ (b) $\begin{cases} 3x + 8y = -12 \\ 4x + 11y = 0 \end{cases}$

34. If $A = \begin{pmatrix} 3 & -7 \\ 4 & -9 \end{pmatrix}$, then $A^{-1} = \begin{pmatrix} -9 & 7 \\ -4 & 3 \end{pmatrix}$. Use this fact and the method of Example 2 to solve the following systems.
(a) $\begin{cases} 3x - 7y = 30 \\ 4x - 9y = 39 \end{cases}$ (b) $\begin{cases} 3x - 7y = -45 \\ 4x - 9y = -71 \end{cases}$

35. The inverse of the matrix

$$A = \begin{pmatrix} 3 & 2 & 6 \\ 1 & 1 & 2 \\ 2 & 2 & 5 \end{pmatrix} \text{ is } A^{-1} = \begin{pmatrix} 1 & 2 & -2 \\ -1 & 3 & 0 \\ 0 & -2 & 1 \end{pmatrix}$$

Use this fact to solve the following system.

$$\begin{cases} 3x + 2y + 6z = 28 \\ x + y + 2z = 9 \\ 2x + 2y + 5z = 22 \end{cases}$$

36. The inverse of the matrix

$$A = \begin{pmatrix} 1 & -1 & 1 \\ 2 & -3 & 2 \\ -4 & 6 & 1 \end{pmatrix} \text{ is } A^{-1} = \begin{pmatrix} 3 & -\frac{7}{5} & -\frac{1}{5} \\ 2 & -1 & 0 \\ 0 & \frac{2}{5} & \frac{1}{5} \end{pmatrix}$$

Use this fact to solve the following system.

$$\begin{cases} x - y + z = 5 \\ 2x - 3y + 2z = -15 \\ -4x + 6y + z = 25 \end{cases}$$

Ⓖ *In Exercises 37–41:*
(a) *compute the inverse of the coefficient matrix for the system; and*
(b) *use matrix methods to solve the system. In cases where the final answer involves decimals, round to three decimal places.*

37. $\begin{cases} x + 4y = 7 \\ 2x + 7y = 12 \end{cases}$

38. $\begin{cases} 8x - 5y = -13 \\ 3x + 4y = 48 \end{cases}$

39. $\begin{cases} x + 2y + 2z = 3 \\ 3x + y = -1 \\ x + y + z = 12 \end{cases}$

40. $\begin{cases} 5x - 2y - 2z = 15 \\ 3x + y = 4 \\ x + y + z = -4 \end{cases}$

41. $\begin{cases} 2x + 3y + z + w = 3 \\ 6x + 6y - 5z - 2w = 15 \\ x - y + z + \frac{1}{6}w = -3 \\ 4x + 9y + 3z + 2w = -3 \end{cases}$

42. (a) Solve the following two systems and then check to see that your results agree with those given in Example 1.

$$\begin{cases} a + 2c = 1 \\ 3a + 4c = 0 \end{cases} \qquad \begin{cases} b + 2d = 0 \\ 3b + 4d = 1 \end{cases}$$

(b) At the end of Example 1, it is asserted that

$$\begin{pmatrix} 1 & 2 \\ 3 & 4 \end{pmatrix}\begin{pmatrix} -2 & 1 \\ \frac{3}{2} & -\frac{1}{2} \end{pmatrix} = \begin{pmatrix} -2 & 1 \\ \frac{3}{2} & -\frac{1}{2} \end{pmatrix}\begin{pmatrix} 1 & 2 \\ 3 & 4 \end{pmatrix}$$
$$= \begin{pmatrix} 1 & 0 \\ 0 & 1 \end{pmatrix}$$

Carry out the indicated matrix multiplications to verify that these equations are valid.

43. Use the technique in Example 1 to show that the matrix $\begin{pmatrix} 2 & 5 \\ 6 & 15 \end{pmatrix}$ does not have an inverse.

44. Follow Exercise 43 using $\begin{pmatrix} -5 & 2 \\ 20 & -8 \end{pmatrix}$.

In Exercises 45 and 46, encode each message using the matrix $\begin{pmatrix} 9 & 8 \\ 8 & 7 \end{pmatrix}$. *For Exercises 47 and 48, encode using the matrix* $\begin{pmatrix} 1 & 1 & 0 \\ 0 & -1 & 2 \\ 1 & 0 & 1 \end{pmatrix}$.

45. TRY IT
46. DROP OFF
47. TURN NOW
48. WAIT

Ⓖ *In Exercises 49 and 50, decode each message, assuming that the encoding matrix was* $\begin{pmatrix} 12 & 47 \\ -1 & -4 \end{pmatrix}$. *For Exercises 51 and 52, decode, assuming the encoding matrix was* $\begin{pmatrix} 2 & -1 & -1 \\ -2 & 1 & 2 \\ -1 & 1 & 1 \end{pmatrix}$.

49. $463, -39, 60, -5, 825, -70, 60, -5$
50. $1038, -88, 274, -23, 379, -32, 156, -13$
51. $-3, 24, 22, -17, 17, 21$
52. $-7, 27, 21, -27, 52, 27, -4, 5, 4, 36, -36, -18$

B

53. Use the elementary row operations (as in Example 3) to find the inverse of the following matrix.

$$\begin{pmatrix} 1 & 1 & 1 & 1 \\ 1 & 2 & 3 & 4 \\ 1 & 3 & 6 & 10 \\ 1 & 4 & 10 & 20 \end{pmatrix}$$

54. Let $A = \begin{pmatrix} 1 & -6 & 3 \\ 2 & -7 & 3 \\ 4 & -12 & 5 \end{pmatrix}$.

(a) Compute the matrix product AA. What do you observe?
(b) Use the result in part (a) to solve the following system.

$$\begin{cases} x - 6y + 3z = 19/2 \\ 2x - 7y + 3z = 11 \\ 4x - 12y + 5z = 19 \end{cases}$$

C

55. Let $A = \begin{pmatrix} a & b \\ c & d \end{pmatrix}$. Compute A^{-1}. (Assume that $ad - bc \neq 0$.)

56. (a) Use the result in Exercise 55 to find the inverse of the matrix $\begin{pmatrix} x & 1+x \\ 1-x & -x \end{pmatrix}$. What do you observe about the result?

(b) What's the inverse of the matrix $\begin{pmatrix} 11 & 12 \\ -10 & -11 \end{pmatrix}$? Of the matrix $\begin{pmatrix} \pi + 1 & \pi + 2 \\ -\pi & -1 - \pi \end{pmatrix}$?

PROJECT

THE LEONTIEF MODEL REVISITED

As background here, you need to be familiar with the project at the end of Section 6.2 on the Leontief input-output model. Additionally, you need to have studied Example 2 in the present section, where matrix methods are applied to solve a system of equations. If necessary, one or two groups can be assigned to work through or to review that material and then present it to the class at large. In this project, you'll use matrix methods to shorten the work in solving an input-output problem.

PART I HOW TO SOLVE THE MATRIX EQUATION $AX + D = X$

In Example 2 of Section 6.4, we solved the matrix equation $AX = B$. Along the same lines, now we want to solve the matrix equation

$$AX + D = X \tag{1}$$

(We assume that X is a square matrix with an inverse and that each matrix has the appropriate numbers of rows and columns for the calculations to make sense.) We start by subtracting X from both sides of equation (1) to obtain

$$AX + D - X = \mathbf{0} \qquad \text{Recall from Section 6.3 that } \mathbf{0}$$
$$\text{denotes the zero matrix, in which}$$
$$\text{every entry is the number 0.}$$

$$AX - X = -D \qquad \text{Subtracting } D \text{ from both sides and}$$
$$\text{using the property } \mathbf{0} - D = -D$$

$$(A - I)X = -D \qquad I \text{ denotes the identity matrix. We've}$$
$$\text{used the fact that the distributive}$$
$$\text{property is valid in matrix algebra}$$
$$\text{just as it is in ordinary algebra.}$$

To isolate X now, multiply both sides of this last equation by the matrix $(A - I)^{-1}$. This yields

$$(A - I)^{-1}(A - I)X = (A - I)^{-1}(-D)$$
$$IX = (A - I)^{-1}(-D)$$
$$X = (A - I)^{-1}(-D) \qquad \text{using } IX = X \tag{2}$$

Equation (2) is the required solution. To see how this works in a simple case, suppose we want to solve equation (1), given that $A = \begin{pmatrix} 3 & 5 \\ 1 & 4 \end{pmatrix}$, $D = \begin{pmatrix} 10 \\ 20 \end{pmatrix}$, and $X = \begin{pmatrix} x \\ y \end{pmatrix}$. That is, the equation to be solved is

$$\begin{pmatrix} 3 & 5 \\ 1 & 4 \end{pmatrix}\begin{pmatrix} x \\ y \end{pmatrix} + \begin{pmatrix} 10 \\ 20 \end{pmatrix} = \begin{pmatrix} x \\ y \end{pmatrix} \tag{3}$$

Using equation (2), the solution is

$$X = (A - I)^{-1}(-D)$$
$$= \left[\begin{pmatrix} 3 & 5 \\ 1 & 4 \end{pmatrix} - \begin{pmatrix} 1 & 0 \\ 0 & 1 \end{pmatrix} \right]^{-1} \begin{pmatrix} -10 \\ -20 \end{pmatrix}$$
$$= \begin{pmatrix} 2 & 5 \\ 1 & 3 \end{pmatrix}^{-1} \begin{pmatrix} -10 \\ -20 \end{pmatrix} \qquad \text{carrying out the matrix subtraction}$$

$$= \begin{pmatrix} 3 & -5 \\ -1 & 2 \end{pmatrix} \begin{pmatrix} -10 \\ -20 \end{pmatrix} \qquad \text{computing the inverse}$$

$$= \begin{pmatrix} 70 \\ -30 \end{pmatrix} \qquad \text{carrying out the matrix multiplication}$$

Thus, the solution of equation (3) is $X = \begin{pmatrix} 70 \\ -30 \end{pmatrix}$, and the values of the unknowns x and y are 70 and -30, respectively.

PRACTICE PROBLEMS FOR PART I

1. Solve $AX + D = X$, given that $A = \begin{pmatrix} 11 & 9 \\ 6 & 5 \end{pmatrix}$, $D = \begin{pmatrix} 25 \\ 15 \end{pmatrix}$, and $X = \begin{pmatrix} x \\ y \end{pmatrix}$.

2. Solve $AX + D = X$, given that $A = \begin{pmatrix} 1 & -3 & 2 \\ -2 & 7 & -7 \\ 1 & -6 & 12 \end{pmatrix}$, $D = \begin{pmatrix} 20 \\ 50 \\ 70 \end{pmatrix}$, $X = \begin{pmatrix} x \\ y \\ z \end{pmatrix}$.

PART II THE MATRIX EQUATION $AX + D = X$
APPLIED TO AN INPUT-OUTPUT PROBLEM

As our example, we'll use the same problem discussed in Part I of the project at the end of Section 6.2. For ease of reference, we restate the problem here. Given the input-output data in Table 1 and the external demands in Table 2, determine an appropriate production level for each sector so that both the internal and external demands on the economy are satisfied.

As in the solution in Section 6.2, we start by using x, y, and z to represent the required number of units of steel, coal, and electricity, respectively. Next, define three matrices, A, D, and X as follows.

$$A = \begin{pmatrix} 0.04 & 0.02 & 0.16 \\ 0.12 & 0 & 0.30 \\ 0.16 & 0.10 & 0.04 \end{pmatrix} \qquad D = \begin{pmatrix} 10 \\ 12 \\ 15 \end{pmatrix} \qquad X = \begin{pmatrix} x \\ y \\ z \end{pmatrix} \qquad (4)$$

The input-output matrix.   The demand   The matrix
based on Table 1   matrix based of unknowns
                          on Table 2

Your tasks now are to work through the following steps to solve this input-output problem using matrix methods.

1. Check that the matrix product AD is defined. What will be the dimension of the product AD?

TABLE 1   Input-Output Table (internal demands)

	Outputs		
	Steel	Coal	Electricity
Inputs			
Steel	0.04	0.02	0.16
Coal	0.12	0	0.30
Electricity	0.16	0.10	0.04

TABLE 2
External Demands

	External demands
Steel	10 units
Coal	12 units
Electricity	15 units

2. Explain why the matrix equation $AX + D = X$ holds. *Hint*: In the project for Section 6.2 look at the equation immediately preceding equation (2).

3. Use equation (2) from part I of this project to solve the equation $AX + D = X$. Make use of a graphing utility in carrying out the matrix operations, and round the answers for x, y, and z to two decimal places.

PART III A SEVEN-SECTOR MODEL OF THE U.S. ECONOMY

The input-output examples above (and in the project for Section 6.2) involve two- or three-sector economies. By way of contrast, input-output tables for the U.S. economy (published by the U.S. Bureau of Economic Analysis) involve as many as 486 sectors. Table 3, which follows, is an *aggregated version* of one of those tables, in which the 486 sectors have been grouped into only 7 general categories or sectors. The names of the 7 sectors, abbreviated in Table 3, are spelled out in Table 4, which contains the external demands. Use a graphing utility to solve the input-output problem for this 7-sector economy. Round the final answers to the nearest integer. Then, given that each unit represents one million dollars, express the answers in dollars.

TABLE 3
Seven-Sector Input-Output Table for the 1967 U.S. Economy *

	Outputs						
	Ag	**Min**	**Con**	**Mfg**	**T & T**	**Serv**	**Oth**
Inputs							
Ag	.2939	.0000	.0025	.0516	.0009	.0081	.0203
Min	.0022	.0504	.0090	.0284	.0002	.0099	.0075
Con	.0096	.0229	.0003	.0042	.0085	.0277	.0916
Mfg	.1376	.0940	.3637	.3815	.0634	.0896	.1003
T & T	.0657	.0296	.1049	.0509	.0530	.0404	.0775
Serv	.0878	.1708	.0765	.0734	.1546	.1676	.1382
Oth	.0001	.0054	.0008	.0055	.0183	.0250	.0012

TABLE 4 External Demands for 1967 U.S. Economy

External demands	
Agriculture	9300
Mining	1454
Construction	85,583
Manufacturing	278,797
Trade & transportation	141,468
Services	211,937
Others	2453

*Derived from more detailed tables published by the U.S. Bureau of Economic Analysis.
Source for Tables 3 and 4: *Input-Output Analysis: Foundations and Extensions,* Ronald E. Miller and Peter D. Blair (Englewood Cliffs, N.J.: Prentice Hall, 1985)

6.5 DETERMINANTS AND CRAMER'S RULE

The idea of the determinant . . . dates back essentially to Leibniz (1693), the Swiss mathematician Gabriel Cramer (1750), and Lagrange (1773); the name is due to Cauchy (1812).
Y. Mikami has pointed out that the Japanese mathematician Seki Kōwa had the idea of a determinant sometime before 1683. —Dirk J. Struik in *A Concise History of Mathematics,* 4th ed. (New York: Dover Publications, 1987)

As you saw in the previous section, square matrices and their inverses can be used to solve certain systems of equations. In this section, we will associate a number with each square matrix. This number is called the **determinant** of the matrix. As you'll see, this too has an application in solving systems of equations.

The determinant of a matrix A can be denoted by det A or $|A|$. Determinants are also denoted simply by replacing the parentheses of matrix notation with vertical lines. Thus, three examples of determinants are

$$\begin{vmatrix} 1 & 2 \\ 3 & 4 \end{vmatrix} \qquad \begin{vmatrix} 1 & 2 & 3 \\ 4 & 5 & 6 \\ 7 & 8 & 9 \end{vmatrix} \qquad \begin{vmatrix} 3 & 7 & 8 & 9 \\ 5 & 6 & 4 & 3 \\ -9 & 9 & 0 & 1 \\ 1 & 3 & -2 & 1 \end{vmatrix}$$

A determinant with n rows and n columns is said to be an **nth-order determinant.** Therefore, the determinants we've just written are, respectively, second-, third-, and fourth-order determinants. As with matrices, we speak of the numbers in a determinant as its **entries.** We also number the rows and the columns of a determinant as we do with matrices. However, unlike matrices, each determinant has a numerical value. The value of a second-order determinant is defined as follows.

DEFINITION	2 × 2 Determinant

$$\begin{vmatrix} a & b \\ c & d \end{vmatrix} = ad - bc$$

Table 1 illustrates how this definition is used to evaluate, or *expand*, a second-order determinant.

▌TABLE 1

Determinant	Value of determinant
$\begin{vmatrix} 3 & 7 \\ 5 & 10 \end{vmatrix}$	$\begin{vmatrix} 3 & 7 \\ 5 & 10 \end{vmatrix} = 3(10) - 7(5) = 30 - 35 = -5$
$\begin{vmatrix} 3 & -7 \\ 5 & 10 \end{vmatrix}$	$\begin{vmatrix} 3 & -7 \\ 5 & 10 \end{vmatrix} = 3(10) - (-7)(5) = 30 + 35 = 65$
$\begin{vmatrix} a & a^3 \\ 1 & a^2 \end{vmatrix}$	$\begin{vmatrix} a & a^3 \\ 1 & a^2 \end{vmatrix} = a(a^2) - a^3(1) = a^3 - a^3 = 0$

In general, the value of an nth-order determinant ($n > 2$) is defined in terms of certain determinants of order $n - 1$. For instance, the value of a third-order determinant is defined in terms of second-order determinants. We'll use the following example to introduce the necessary terminology here:

$$\begin{vmatrix} 8 & 3 & 5 \\ 2 & 4 & 6 \\ 9 & 1 & 7 \end{vmatrix}$$

Pick a given entry—say, 8—and imagine crossing out all entries occupying the same row and the same column as 8.

$$\begin{vmatrix} 8 & 3 & 5 \\ 2 & 4 & 6 \\ 9 & 1 & 7 \end{vmatrix}$$

Now we are left with the second-order determinant $\begin{vmatrix} 4 & 6 \\ 1 & 7 \end{vmatrix}$. This second-order determinant is called the **minor** of the entry 8. Similarly, to find the minor of the entry 6 in the original determinant, imagine crossing out all entries that occupy the same row and the same column as 6.

$$\begin{vmatrix} 8 & 3 & 5 \\ 2 & 4 & 6 \\ 9 & 1 & 7 \end{vmatrix}$$

We are left with the second-order determinant $\begin{vmatrix} 8 & 3 \\ 9 & 1 \end{vmatrix}$, which by definition is the minor of the entry 6. In the same manner, *the minor of any element is the determinant obtained by crossing out the entries occupying the same row and column as the given element.*

Closely related to the minor of an entry in a determinant is the cofactor of that entry. The **cofactor** of an entry is defined as the minor multiplied by $+1$ or -1, according to the scheme displayed in Figure 1.

After looking at an example, we'll give a more formal rule for computing cofactors, one that will not rely on a figure and that will also apply to larger determinants.

$$\begin{vmatrix} + & - & + \\ - & + & - \\ + & - & + \end{vmatrix}$$

Figure 1

---

www **OO** | **EXAMPLE** | **1** | **Computing a minor and a cofactor**

Consider the determinant

$$\begin{vmatrix} 1 & 2 & 3 \\ 4 & 5 & 6 \\ 7 & 8 & 9 \end{vmatrix}$$

Compute the minor and the cofactor of the entry 4.

SOLUTION
By definition, we have

$$\text{minor of } 4 = \begin{vmatrix} 2 & 3 \\ 8 & 9 \end{vmatrix} = 18 - 24 = -6$$

Thus, the minor of the entry 4 is -6. To compute the cofactor of 4, we first notice that 4 is located in the second row and first column of the given determinant. On checking the corresponding position in Figure 1, we see a negative sign. Therefore, we have by definition

$$\begin{aligned} \text{cofactor of } 4 &= (-1)(\text{minor of } 4) \\ &= (-1)(-6) = 6 \end{aligned}$$

The cofactor of 4 is therefore 6.

The following rule tells us how cofactors can be computed without relying on Figure 1. For reference, we also restate the definition of a minor. (You should verify for yourself that this rule yields results that are consistent with Figure 1.)

DEFINITION | **Minors and Cofactors**

The **minor** of an entry b in a determinant is the determinant formed by suppressing the entries in the row and in the column in which b appears.

Suppose that the entry b is in the ith row and the jth column. Then the **cofactor** of b is given by the expression

$$(-1)^{i+j}(\text{minor of } b)$$

We are now prepared to state the definition that tells us how to evaluate a third-order determinant. Actually, as you'll see later, the definition is quite general and may be applied to determinants of any size.

DEFINITION | **The Value of a Determinant**

Multiply each entry in the first row of the determinant by its cofactor and then add the results. The value of the determinant is defined to be this sum.

To see how this definition is used, let's evaluate the determinant

$$\begin{vmatrix} 1 & 2 & 3 \\ 4 & 5 & 6 \\ 7 & 8 & 9 \end{vmatrix}$$

The definition tells us to multiply each entry in the first row by its cofactor and then add the results. Carrying out this procedure, we have

$$\begin{vmatrix} 1 & 2 & 3 \\ 4 & 5 & 6 \\ 7 & 8 & 9 \end{vmatrix} = 1\begin{vmatrix} 5 & 6 \\ 8 & 9 \end{vmatrix} - 2\begin{vmatrix} 4 & 6 \\ 7 & 9 \end{vmatrix} + 3\begin{vmatrix} 4 & 5 \\ 7 & 8 \end{vmatrix}$$

$$= 1(45 - 48) - 2(36 - 42) + 3(32 - 35)$$

$$= 0 \quad (\text{Check the arithmetic!})$$

So the value of this particular determinant is zero. The procedure we've used here is referred to as **expanding the determinant along its first row.** The following theorem (stated here without proof) tells us that the value of a determinant can be obtained by expanding along any row or column; the results are the same in all cases.

THEOREM

Select any row or any column in a determinant and multiply each element in that row or column by its cofactor. Then add the results. The number obtained will be the value of the determinant. (In other words, the number obtained will be the same as that obtained by expanding the determinant along its first row.)

According to this theorem, we could have evaluated the determinant

$$\begin{vmatrix} 1 & 2 & 3 \\ 4 & 5 & 6 \\ 7 & 8 & 9 \end{vmatrix}$$

by expanding it along any row or any column. Let's expand it along the second column and check to see that the result agrees with the value we obtained earlier. Expanding along the second column, we have

$$\begin{vmatrix} 1 & 2 & 3 \\ 4 & 5 & 6 \\ 7 & 8 & 9 \end{vmatrix} = -2\begin{vmatrix} 4 & 6 \\ 7 & 9 \end{vmatrix} + 5\begin{vmatrix} 1 & 3 \\ 7 & 9 \end{vmatrix} - 8\begin{vmatrix} 1 & 3 \\ 4 & 6 \end{vmatrix}$$

$$= -2(36 - 42) + 5(9 - 21) - 8(6 - 12)$$

$$= -2(-6) + 5(-12) - 8(-6)$$

$$= 12 - 60 + 48 = 0 \qquad \text{as obtained previously}$$

We would have obtained the same result had we chosen to begin with any other row or column. (Exercise 13 at the end of this section asks you to verify this.)

There are three basic rules that make it easier to evaluate determinants. These are summarized in the box that follows. (Suggestions for proving these can be found in the exercises.)

PROPERTY SUMMARY   Rules for Manipulating Determinants

EXAMPLES

1. If each entry in a given row is multiplied by the constant k, then the value of the determinant is multiplied by k. This is also true for columns.

$$10\begin{vmatrix} 1 & 3 & 4 \\ 1 & 2 & 3 \\ 4 & 5 & 6 \end{vmatrix} = \begin{vmatrix} 10 & 30 & 40 \\ 1 & 2 & 3 \\ 4 & 5 & 6 \end{vmatrix}$$

$$k\begin{vmatrix} a & b & c \\ d & e & f \\ g & h & i \end{vmatrix} = \begin{vmatrix} a & kb & c \\ d & ke & f \\ g & kh & i \end{vmatrix}$$

2. If a multiple of one row is added to another row, the value of the determinant is not changed. This also applies to columns.

$$\begin{vmatrix} a & b & c \\ d & e & f \\ g & h & i \end{vmatrix} = \begin{vmatrix} a & b & c \\ d + ka & e + kb & f + kc \\ g & h & i \end{vmatrix}$$

3. If two rows are interchanged, then the value of the determinant is multiplied by -1. This also applies to columns.

$$\begin{vmatrix} 1 & 2 & 3 \\ 4 & 5 & 6 \\ a & b & c \end{vmatrix} = -\begin{vmatrix} 4 & 5 & 6 \\ 1 & 2 & 3 \\ a & b & c \end{vmatrix}$$

EXAMPLE **2** **Evaluating a determinant**

Evaluate the determinant $\begin{vmatrix} 15 & 14 & 26 \\ 18 & 17 & 32 \\ 21 & 20 & 42 \end{vmatrix}$.

SOLUTION

$$\begin{vmatrix} 15 & 14 & 26 \\ 18 & 17 & 32 \\ 21 & 20 & 42 \end{vmatrix} = (3 \times 2)\begin{vmatrix} 5 & 14 & 13 \\ 6 & 17 & 16 \\ 7 & 20 & 21 \end{vmatrix}$$

using Rule 1 to factor 3 from the first column and 2 from the third column

$$= 6\begin{vmatrix} 5 & 1 & 13 \\ 6 & 1 & 16 \\ 7 & -1 & 21 \end{vmatrix}$$

using Rule 2 to subtract the third column from the second column

$$= 6\begin{vmatrix} 12 & 0 & 34 \\ 13 & 0 & 37 \\ 7 & -1 & 21 \end{vmatrix}$$

using Rule 2 to add the third row to the first and second rows

$$= 6\left[-\left(-1\begin{vmatrix} 12 & 34 \\ 13 & 37 \end{vmatrix}\right)\right]$$

expanding the determinant along the second column

$$= 6\begin{vmatrix} 12 & 34 \\ 1 & 3 \end{vmatrix}$$

using Rule 2 to subtract the first row from the second row

$$= 6(36 - 34) = 12$$

The value of the given determinant is therefore 12. Notice the general strategy. We used Rules 1 and 2 until one column (or one row) contained two zeros. At that point, it is a simple matter to expand the determinant along that column (or row).

EXAMPLE   3   Applying the rules for manipulating a determinant

Show that

$$\begin{vmatrix} 1 & 1 & 1 \\ a & b & c \\ a^2 & b^2 & c^2 \end{vmatrix} = (b - a)(c - a)(c - b)$$

SOLUTION

$$\begin{vmatrix} 1 & 1 & 1 \\ a & b & c \\ a^2 & b^2 & c^2 \end{vmatrix} = \begin{vmatrix} 1 & 0 & 0 \\ a & b - a & c - a \\ a^2 & b^2 - a^2 & c^2 - a^2 \end{vmatrix}$$

using Rule 2 to subtract the first column from the second and third columns

$$= \begin{vmatrix} 1 & 0 & 0 \\ a & b - a & c - a \\ a^2 & (b - a)(b + a) & (c - a)(c + a) \end{vmatrix}$$

$$= (b - a)(c - a)\begin{vmatrix} 1 & 0 & 0 \\ a & 1 & 1 \\ a^2 & b + a & c + a \end{vmatrix}$$

using Rule 1 to factor $b - a$ from the second column and $c - a$ from the third column

$$= (b - a)(c - a)[(c + a) - (b + a)]$$

expanding along first row

$$= (b - a)(c - a)(c - b)$$

EXAMPLE **4**   **Two cases where the value of a determinant is zero**

Evaluate each determinant:

(a) $\begin{vmatrix} 9 & 4 & -3 \\ 13 & 17 & 5 \\ 9 & 4 & -3 \end{vmatrix}$ **(b)** $\begin{vmatrix} 2 & 7 & 3 \\ 15 & 40 & 79 \\ 20 & 70 & 30 \end{vmatrix}$

SOLUTION

(a) Notice that the first and third rows are identical. Thus, we have

$$\begin{vmatrix} 9 & 4 & -3 \\ 13 & 17 & 5 \\ 9 & 4 & -3 \end{vmatrix} = \begin{vmatrix} 9 & 4 & -3 \\ 13 & 17 & 5 \\ 0 & 0 & 0 \end{vmatrix} \qquad \text{\color{red}{subtracting row 1 from row 3}}$$

$$= 0 \qquad \text{\color{red}{expanding along the third row}}$$

(b) Notice that the third row is a multiple of the first row. (A more precise, but less convenient, way to state this is: Each entry in the third row is a constant multiple of the corresponding entry in the first row.) So, we have

$$\begin{vmatrix} 2 & 7 & 3 \\ 15 & 40 & 79 \\ 20 & 70 & 30 \end{vmatrix} = 10\begin{vmatrix} 2 & 7 & 3 \\ 15 & 40 & 79 \\ 2 & 7 & 3 \end{vmatrix} = 10 \times 0 \qquad \text{(Why?)}$$

$$= 0$$

The reasoning that we used in Example 4 can be generalized as indicated in the box that follows.

Two Instances When the Value of a Determinant Is Zero

1. If two rows are identical, the value of the determinant is zero. This also holds for two columns.
2. If one row is a multiple of another row, the value of the determinant is zero. This also holds for two columns.

Caution: The value of a determinant can still turn out to be zero even if neither of these conditions seems to apply. See, for instance, the example on page 521.

The definition that we gave (on page 521) for third-order determinants can be extended to apply to fourth-order (or larger) determinants. Consider, for example, the fourth-order determinant A given by

$$A = \begin{vmatrix} 2 & 7 & -1 & 9 \\ 4 & 0 & 3 & 6 \\ -8 & 5 & 1 & 3 \\ 11 & 2 & -6 & 1 \end{vmatrix}$$

By definition, we can evaluate this determinant by selecting the first row, multiplying each entry by its cofactor, and then adding the results. This yields

$$A = \begin{vmatrix} 2 & 7 & -1 & 9 \\ 4 & 0 & 3 & 6 \\ -8 & 5 & 1 & 3 \\ 11 & 2 & -6 & 1 \end{vmatrix}$$

$$= 2 \begin{vmatrix} 0 & 3 & 6 \\ 5 & 1 & 3 \\ 2 & -6 & 1 \end{vmatrix} - 7 \begin{vmatrix} 4 & 3 & 6 \\ -8 & 1 & 3 \\ 11 & -6 & 1 \end{vmatrix} + (-1) \begin{vmatrix} 4 & 0 & 6 \\ -8 & 5 & 3 \\ 11 & 2 & 1 \end{vmatrix} - 9 \begin{vmatrix} 4 & 0 & 3 \\ -8 & 5 & 1 \\ 11 & 2 & -6 \end{vmatrix}$$

The problem is now reduced to evaluating four third-order determinants. As Exercise 31(a) asks you to check, the final result is $A = 174$. If we had expanded across the second row, rather than the first, then there would be one less third-order determinant to evaluate because of the zero in the second row. (In doing this, we would be relying on the theorem on page 521, which is valid for all determinants, not only three-by-three determinants.) Exercise 31(b) asks you to expand across the second row and again verify that the result is $A = 174$.

As with matrix products and inverses, a graphing calculator can be used in computing determinants. We display two examples here using the four-by-four determinant A given above. (For the details on these and for other types of graphing utilities, see your owner's manual. Figure 2 shows the end results of the two steps in computing the determinant using a Texas Instruments TI-83+ graphing calculator. After the four-by-four matrix corresponding to A has been entered, the matrix can be displayed as in Figure 2(a). Figure 2(b) shows the viewing screen that then results after executing the command det([A]). In Figure 3, we've used Microsoft *Excel* to evaluate the determinant.

```
[A]
    [[2 7 -1 9]
     [4 0 3 6]
     [-8 5 1 3]
     [11 2 -6 1]]
```

(a)

```
[A]
    [[2 7 -1 9]
     [4 0 3 6]
     [-8 5 1 3]
     [11 2 -6 1]]
det([A])
                    174
```

(b)

Figure 2

Evaluating a four-by-four determinant using a Texas Instruments TI-83+ graphing calculator. (For details, see your owner's manual.)

	A	B	C	D	E
1					
2		2	7	−1	9
3	A =	4	0	3	6
4		−8	5	1	3
5		11	2	−6	1
6					
7	det(A) =	174			

Figure 3

Computing a determinant using Microsoft *Excel*. After typing in the entries in cells B2 through E5, the formula =MDETERM(B2:E5) is used to compute the determinant. (See your owner's manual.)

Determinants can be used to solve certain systems of linear equations in which there are as many unknowns as there are equations. In the box that follows, we state **Cramer's rule** for solving a system of three linear equations in three unknowns.*

----

*The rule is named after one of its discoverers, the Swiss mathematician Gabriel Cramer (1704–1752).

A more general but entirely similar version of Cramer's rule holds for n equations in n unknowns.

Cramer's Rule

Consider the system

$$\begin{cases} a_1x + b_1y + c_1z = d_1 \\ a_2x + b_2y + c_2z = d_2 \\ a_3x + b_3y + c_3z = d_3 \end{cases}$$

Let the four determinants D, D_x, D_y, and D_z be defined as follows:

$$D = \begin{vmatrix} a_1 & b_1 & c_1 \\ a_2 & b_2 & c_2 \\ a_3 & b_3 & c_3 \end{vmatrix} \qquad D_x = \begin{vmatrix} d_1 & b_1 & c_1 \\ d_2 & b_2 & c_2 \\ d_3 & b_3 & c_3 \end{vmatrix}$$

$$D_y = \begin{vmatrix} a_1 & d_1 & c_1 \\ a_2 & d_2 & c_2 \\ a_3 & d_3 & c_3 \end{vmatrix} \qquad D_z = \begin{vmatrix} a_1 & b_1 & d_1 \\ a_2 & b_2 & d_2 \\ a_3 & b_3 & d_3 \end{vmatrix}$$

Then if $D \neq 0$, the unique values of x, y, and z satisfying the system are given by

$$x = \frac{D_x}{D} \qquad y = \frac{D_y}{D} \qquad z = \frac{D_z}{D}$$

[If $D = 0$, the solutions (if any) can be found using Gaussian elimination.]

Notice that the determinant D in Cramer's rule is just the determinant of the coefficient matrix of the given system. If you replace the first column of D with the column of numbers on the right side of the given system, you obtain D_x. The determinants D_y and D_z are obtained similarly.

Before actually proving Cramer's rule, let's take a look at how it's applied.

EXAMPLE 5 Using Cramer's rule to solve a system

Use Cramer's rule to find all solutions of the following system of equations:

$$\begin{cases} 2x + 2y - 3z = -20 \\ x - 4y + z = 6 \\ 4x - y + 2z = -1 \end{cases}$$

SOLUTION
First we list the determinants D, D_x, D_y, and D_z:

$$D = \begin{vmatrix} 2 & 2 & -3 \\ 1 & -4 & 1 \\ 4 & -1 & 2 \end{vmatrix} \qquad D_x = \begin{vmatrix} -20 & 2 & -3 \\ 6 & -4 & 1 \\ -1 & -1 & 2 \end{vmatrix}$$

$$D_y = \begin{vmatrix} 2 & -20 & -3 \\ 1 & 6 & 1 \\ 4 & -1 & 2 \end{vmatrix} \qquad D_z = \begin{vmatrix} 2 & 2 & -20 \\ 1 & -4 & 6 \\ 4 & -1 & -1 \end{vmatrix}$$

The calculations for evaluating D begin as follows:

$$D = \begin{vmatrix} 2 & 2 & -3 \\ 1 & -4 & 1 \\ 4 & -1 & 2 \end{vmatrix} \uparrow = \begin{vmatrix} 0 & 10 & -5 \\ 1 & -4 & 1 \\ 0 & 15 & -2 \end{vmatrix}$$

Subtract twice the second row from the first.
Subtract 4 times the second row from the third.

Since we now have two zeros in the first column, it is an easy matter to expand D along that column to obtain

$$D = -1 \begin{vmatrix} 10 & -5 \\ 15 & -2 \end{vmatrix}$$
$$= -1[-20 - (-75)] = -1(55) = -55$$

The value of D is therefore -55. (Since this value is nonzero, Cramer's rule does apply.) As Exercise 37 at the end of this section asks you to verify, the values of the other three determinants are

$$D_x = 144 \qquad D_y = 61 \qquad D_z = -230$$

By Cramer's rule, then, the unique values of x, y, and z that satisfy the system are

$$x = \frac{D_x}{D} = \frac{144}{-55} = -\frac{144}{55} \qquad y = \frac{D_y}{D} = \frac{61}{-55} = -\frac{61}{55} \qquad z = \frac{D_z}{D} = \frac{-230}{-55} = \frac{46}{11}$$

One way we can prove Cramer's rule is to use Gaussian elimination to solve the system

$$\begin{cases} a_1 x + b_1 y + c_1 z = d_1 \\ a_2 x + b_2 y + c_2 z = d_2 \\ a_3 x + b_3 y + c_3 z = d_3 \end{cases} \qquad (1)$$

A much shorter and simpler proof, however, has been found by D. E. Whitford and M. S. Klamkin.* This is the proof we give here; it makes effective use of the rules employed in this section for manipulating determinants.

Consider the system of equations (1) and assume that $D \neq 0$. We will show that if x, y, and z satisfy the system, then in fact $x = D_x/D$, with similar equations giving y and z. (Exercise 68 at the end of this section then shows how to check that these values indeed satisfy the given system.) We have

$$D_x = \begin{vmatrix} d_1 & b_1 & c_1 \\ d_2 & b_2 & c_2 \\ d_3 & b_3 & c_3 \end{vmatrix} \qquad \text{by definition}$$

$$= \begin{vmatrix} (a_1 x + b_1 y + c_1 z) & b_1 & c_1 \\ (a_2 x + b_2 y + c_2 z) & b_2 & c_2 \\ (a_3 x + b_3 y + c_3 z) & b_3 & c_3 \end{vmatrix} \qquad \begin{array}{l} \text{using the equations} \\ \text{in (1) to substitute for} \\ d_1, d_2, \text{ and } d_3 \end{array}$$

---

*The proof was published in the *American Mathematical Monthly,* vol. 60 (1953); pp. 186–187.

$$= \begin{vmatrix} a_1x & b_1 & c_1 \\ a_2x & b_2 & c_2 \\ a_3x & b_3 & c_3 \end{vmatrix}$$

subtracting y times the second column as well as z times the third column from the first column

$$= x\begin{vmatrix} a_1 & b_1 & c_1 \\ a_2 & b_2 & c_2 \\ a_3 & b_3 & c_3 \end{vmatrix}$$

factoring x out of the first column

$$= xD$$

by definition

We now have $D_x = xD$, which is equivalent to $x = D_x/D$, as required. The formulas for y and z are obtained similarly.

EXERCISE SET 6.5

A

In Exercises 1–6, evaluate the determinants.

1. (a) $\begin{vmatrix} 2 & -17 \\ 1 & 6 \end{vmatrix}$ **2. (a)** $\begin{vmatrix} 1 & 0 \\ 0 & 1 \end{vmatrix}$

(b) $\begin{vmatrix} 1 & 6 \\ 2 & -17 \end{vmatrix}$ **(b)** $\begin{vmatrix} 0 & 1 \\ 0 & 1 \end{vmatrix}$

3. (a) $\begin{vmatrix} 5 & 7 \\ 500 & 700 \end{vmatrix}$ **4. (a)** $\begin{vmatrix} -8 & -3 \\ 4 & -5 \end{vmatrix}$

(b) $\begin{vmatrix} 5 & 500 \\ 7 & 700 \end{vmatrix}$ **(b)** $\begin{vmatrix} -3 & -8 \\ -5 & 4 \end{vmatrix}$

5. $\begin{vmatrix} \sqrt{2}-1 & \sqrt{2} \\ \sqrt{2} & \sqrt{2}+1 \end{vmatrix}$ **6.** $\begin{vmatrix} \sqrt{3}+\sqrt{2} & 1+\sqrt{5} \\ 1-\sqrt{5} & \sqrt{3}-\sqrt{2} \end{vmatrix}$

In Exercises 7–12, refer to the following determinant:

$$\begin{vmatrix} -6 & 3 & 8 \\ 5 & -4 & 1 \\ 10 & 9 & -10 \end{vmatrix}$$

7. Evaluate the minor of the entry 3.

8. Evaluate the cofactor of the entry 3.

9. Evaluate the minor of -10.

10. Evaluate the cofactor of -10.

11. (a) Multiply each entry in the first row by its minor and find the sum of the results.

(b) Multiply each entry in the first row by its cofactor and find the sum of the results.

(c) Which gives you the value of the determinant, part (a) or part (b)?

12. (a) Multiply each entry in the first column by its cofactor and find the sum of the results.

(b) Follow the same instructions as in part (a), but use the second column.

(c) Follow the same instructions as in part (a), but use the third column.

13. Let $A = \begin{vmatrix} 1 & 2 & 3 \\ 4 & 5 & 6 \\ 7 & 8 & 9 \end{vmatrix}$. Evaluate A by expanding it along

(a) the second row; **(b)** the third row; **(c)** the first column; **(d)** the third column.

In Exercises 14–24, evaluate the determinants.

14. $\begin{vmatrix} 1 & 2 & -1 \\ 2 & -1 & 1 \\ 4 & 0 & 2 \end{vmatrix}$ **15.** $\begin{vmatrix} 5 & 10 & 15 \\ 1 & 2 & 3 \\ -9 & 11 & 7 \end{vmatrix}$ **16.** $\begin{vmatrix} 8 & 4 & 2 \\ 3 & 9 & 3 \\ -2 & 8 & 6 \end{vmatrix}$

17. $\begin{vmatrix} 1 & 2 & -3 \\ 4 & 5 & -9 \\ 0 & 0 & 1 \end{vmatrix}$ **18.** $\begin{vmatrix} 9 & 9 & 12 \\ 4 & 4 & 6 \\ 7 & 7 & 5 \end{vmatrix}$ **19.** $\begin{vmatrix} 8 & 7 & 800 \\ 3 & 4 & 300 \\ 5 & 2 & 500 \end{vmatrix}$

20. $\begin{vmatrix} 12 & 21 & -4 \\ 0 & 0 & 0 \\ 73 & 82 & 14 \end{vmatrix}$ **21.** $\begin{vmatrix} 3 & 0 & 0 \\ 0 & 19 & 0 \\ 0 & 0 & 10 \end{vmatrix}$ **22.** $\begin{vmatrix} -6 & -8 & 18 \\ 25 & 12 & 15 \\ -9 & 4 & 13 \end{vmatrix}$

23. $\begin{vmatrix} 23 & 0 & 47 \\ -37 & 0 & 18 \\ 14 & 0 & 25 \end{vmatrix}$ **24.** $\begin{vmatrix} 16 & 0 & -64 \\ -8 & 15 & -12 \\ 30 & -20 & 10 \end{vmatrix}$

Ⓖ *In Exercises 25 and 26, use a graphing utility to evaluate the determinants. For Exercise 25, check your answer by referring to Example 2.*

25. $\begin{vmatrix} 15 & 14 & 26 \\ 18 & 17 & 32 \\ 21 & 20 & 42 \end{vmatrix}$ **26.** $\begin{vmatrix} 25 & 40 & 5 & 10 \\ 9 & 0 & 3 & 6 \\ -2 & 3 & 11 & -17 \\ -3 & 4 & 7 & 2 \end{vmatrix}$

27. Consider the two determinants

$$\begin{vmatrix} 1 & 2 & 3 \\ -7 & -4 & 5 \\ 9 & 2 & 6 \end{vmatrix} \quad \text{and} \quad \begin{vmatrix} 10 & 20 & 30 \\ -7 & -4 & 5 \\ 9 & 2 & 6 \end{vmatrix}$$

(a) According to Item 1 in the Property Summary box on page 522, how are the values of these determinants related?

(b) Evaluate each determinant to verify your answer in part (a).

28. Consider the two determinants

$$\begin{vmatrix} \sqrt{2} & \sqrt{5} & \sqrt{7} \\ -6 & 10 & 0 \\ 8 & -1 & -6 \end{vmatrix} \quad \text{and} \quad \begin{vmatrix} -6 & 10 & 0 \\ \sqrt{2} & \sqrt{5} & \sqrt{7} \\ 8 & -1 & -6 \end{vmatrix}$$

(a) According to Item 3 in the Property Summary box on page 522, how are the values of these determinants related?

(G) (b) Evaluate each determinant to verify your answer in part (a).

In Exercises 29 and 30:
(a) *Evaluate the determinant as in Example 2;*
(G) (b) *Use a graphing utility to evaluate the determinant.*

29. $\begin{vmatrix} 1 & -1 & 0 & 2 \\ 0 & 1 & -1 & 0 \\ 2 & 1 & 0 & -1 \\ -2 & 2 & 1 & 1 \end{vmatrix}$ **30.** $\begin{vmatrix} 3 & -2 & 3 & 4 \\ 1 & 4 & -3 & 2 \\ 6 & 3 & -6 & -3 \\ -1 & 0 & 1 & 5 \end{vmatrix}$

31. This exercise refers to the four-by-four determinant A on page 524.

(a) Evaluate A by expanding across the first row. As stated in the text, you should obtain $A = 174$.

(b) Evaluate A by expanding across the second row.

32. Use the method illustrated in Example 3 to show that

$$\begin{vmatrix} 1 & 1 & 1 \\ a & b & c \\ a^3 & b^3 & c^3 \end{vmatrix} = (b - a)(c - a)(c^2 - b^2 + ac - ab)$$

$$= (b - a)(c - a)(c - b)(a + b + c)$$

33. Use the method shown in Example 3 to express the determinant $\begin{vmatrix} 1 & x & x^2 \\ 1 & y & y^2 \\ 1 & z & z^2 \end{vmatrix}$ as a product.

34. Show that

$$\begin{vmatrix} 1 & 1 & 1 \\ a^2 & b^2 & c^2 \\ a^3 & b^3 & c^3 \end{vmatrix} = (b - a)(c - a)(bc^2 - b^2c + ac^2 - ab^2)$$

$$= (b - a)(c - a)(c - b)(bc + ac + ab)$$

35. Simplify the determinant $\begin{vmatrix} 1 & 1 & 1 \\ 1 & 1 + x & 1 \\ 1 & 1 & 1 + y \end{vmatrix}$.

36. Use the method shown in Example 3 to express the following determinant as a product of three factors:

$$\begin{vmatrix} 1 & 1 & 1 \\ a & b & c \\ bc & ca & ab \end{vmatrix}$$

37. Verify the following statements (from Example 5).

(a) $\begin{vmatrix} -20 & 2 & -3 \\ 6 & -4 & 1 \\ -1 & -1 & 2 \end{vmatrix} = 144$

(b) $\begin{vmatrix} 2 & -20 & -3 \\ 1 & 6 & 1 \\ 4 & -1 & 2 \end{vmatrix} = 61$

(c) $\begin{vmatrix} 2 & 2 & -20 \\ 1 & -4 & 6 \\ 4 & -1 & -1 \end{vmatrix} = -230$

38. Consider the following system:

$$\begin{cases} x + y + 3z = 0 \\ x + 2y + 5z = 0 \\ x - 4y - 8z = 9 \end{cases}$$

(a) Without doing any calculations, find one obvious solution of this system.

(b) Calculate the determinant D.

(c) List all solutions of this system.

In Exercises 39–46, use Cramer's rule to solve those systems for which $D \neq 0$. In cases where $D = 0$, use Gaussian elimination or matrix methods.

39. $\begin{cases} 3x + 4y - z = 5 \\ x - 3y + 2z = 2 \\ 5x - 6z = -7 \end{cases}$ **40.** $\begin{cases} 3A - B - 4C = 3 \\ A + 2B - 3C = 9 \\ 2A - B + 2C = -8 \end{cases}$

41. $\begin{cases} 3x + 2y - z = -6 \\ 2x - 3y - 4z = -11 \\ x + y + z = 5 \end{cases}$ **42.** $\begin{cases} 5x - 3y - z = 16 \\ 2x + y - 3z = 5 \\ 3x - 2y + 2z = 5 \end{cases}$

43. $\begin{cases} 2x + 5y + 2z = 0 \\ 3x - y - 4z = 0 \\ x + 2y - 3z = 0 \end{cases}$ **44.** $\begin{cases} 4u + 3v - 2w = 14 \\ u + 2v - 3w = 6 \\ 2u - v + 4w = 2 \end{cases}$

45. $\begin{cases} 12x - 11z = 13 \\ 6x + 6y - 4z = 26 \\ 6x + 2y - 5z = 13 \end{cases}$ **46.** $\begin{cases} 3x + 4y + 2z = 1 \\ 4x + 6y + 2z = 7 \\ 2x + 3y + z = 11 \end{cases}$

(G) *In Exercises 47 and 48, use Cramer's rule along with a graphing utility to solve the systems.*

47. $\begin{cases} x + y + z + w = -7 \\ x - y + z - w = -11 \\ 2x - 2y - 3z - 3w = 26 \\ 3x + 2y + z - w = -9 \end{cases}$

48. $\begin{cases} 2A - B - 3C + 2D = -2 \\ A - 2B + C - 3D = 4 \\ 3A - 4B + 2C - 4D = 12 \\ 2A + 3B - C - 2D = -4 \end{cases}$

B

49. Find all values of x for which

$$\begin{vmatrix} x - 4 & 0 & 0 \\ 0 & x + 4 & 0 \\ 0 & 0 & x + 1 \end{vmatrix} = 0$$

50. Find all values of x for which

$$\begin{vmatrix} 1 & x & x^2 \\ 1 & 1 & 1 \\ 4 & 5 & 0 \end{vmatrix} = 0$$

51. By expanding the determinant $\begin{vmatrix} a & b & c \\ a & b & c \\ d & e & f \end{vmatrix}$ down the first

column, show that its value is zero.

52. By expanding the determinant $\begin{vmatrix} ka & kb & kc \\ d & e & f \\ g & h & i \end{vmatrix}$ along its first

row, show that it is equal to $k\begin{vmatrix} a & b & c \\ d & e & f \\ g & h & i \end{vmatrix}$.

53. Show that

$$\begin{vmatrix} a_1 + A_1 & b_1 & c_1 \\ a_2 + A_2 & b_2 & c_2 \\ a_3 + A_3 & b_3 & c_3 \end{vmatrix} = \begin{vmatrix} a_1 & b_1 & c_1 \\ a_2 & b_2 & c_2 \\ a_3 & b_3 & c_3 \end{vmatrix} + \begin{vmatrix} A_1 & b_1 & c_1 \\ A_2 & b_2 & c_2 \\ A_3 & b_3 & c_3 \end{vmatrix}$$

54. Show that

$$\begin{vmatrix} a_1 & b_1 & c_1 \\ a_2 & b_2 & c_2 \\ a_3 & b_3 & c_3 \end{vmatrix} = \begin{vmatrix} a_1 + kb_1 & b_1 & c_1 \\ a_2 + kb_2 & b_2 & c_2 \\ a_3 + kb_3 & b_3 & c_3 \end{vmatrix}$$

55. By expanding each determinant along a row or column, show that

$$\begin{vmatrix} a_1 & b_1 & c_1 \\ a_2 & b_2 & c_2 \\ a_3 & b_3 & c_3 \end{vmatrix} = -\begin{vmatrix} a_2 & b_2 & c_2 \\ a_1 & b_1 & c_1 \\ a_3 & b_3 & c_3 \end{vmatrix}$$

56. Solve for x in terms of a, b, and c:

$$\begin{vmatrix} a & a & x \\ c & c & c \\ b & x & b \end{vmatrix} = 0 \qquad (c \neq 0)$$

57. Show that

$$\begin{vmatrix} 1+a & 1 & 1 & 1 \\ 1 & 1+b & 1 & 1 \\ 1 & 1 & 1+c & 1 \\ 1 & 1 & 1 & 1+d \end{vmatrix} = abcd\left(\frac{1}{a} + \frac{1}{b} + \frac{1}{c} + \frac{1}{d} + 1\right)$$

58. Show that

$$\begin{vmatrix} 1 & a & a^2 \\ a^2 & 1 & a \\ a & a^2 & 1 \end{vmatrix} = (a^3 - 1)^2$$

59. Show that

$$\begin{vmatrix} 1 & bc & b+c \\ 1 & ca & c+a \\ 1 & ab & a+b \end{vmatrix} = (b-c)(c-a)(a-b)$$

60. Evaluate the determinant

$$\begin{vmatrix} a & b & c \\ a & a+b & a+b+c \\ a & 2a+b & 3a+2b+c \end{vmatrix}$$

61. Show that

$$\begin{vmatrix} 1 & a & a & a \\ 1 & b & a & a \\ 1 & a & b & a \\ 1 & a & a & b \end{vmatrix} = (b-a)^3$$

62. Show that

$$\begin{vmatrix} a & 1 & 1 & 1 \\ 1 & a & 1 & 1 \\ 1 & 1 & a & 1 \\ 1 & 1 & 1 & a \end{vmatrix} = (a-1)^3(a+3)$$

63. Solve the following system for x, y, and z:

$$\begin{cases} ax + by + cz = k \\ a^2x + b^2y + c^2z = k^2 \\ a^3x + b^3y + c^3z = k^3 \end{cases}$$

(Assume that the values of a, b, and c are all distinct and all nonzero.)

64. Show that the equation

$$\begin{vmatrix} x & y & 1 \\ x_1 & y_1 & 1 \\ 1 & m & 0 \end{vmatrix} = 0$$

represents a line that has slope m and passes through the point (x_1, y_1).

For Exercise 65, use the fact that the equation of a line passing through (x_1, y_1) and (x_2, y_2) can be written

$$\begin{vmatrix} x & y & 1 \\ x_1 & y_1 & 1 \\ x_2 & y_2 & 1 \end{vmatrix} = 0$$

For Exercise 66, use the fact that the equation of a circle passing through (x_1, y_1), (x_2, y_2), and (x_3, y_3) can be written

$$\begin{vmatrix} x^2 + y^2 & x & y & 1 \\ x_1^2 + y_1^2 & x_1 & y_1 & 1 \\ x_2^2 + y_2^2 & x_2 & y_2 & 1 \\ x_3^2 + y_3^2 & x_3 & y_3 & 1 \end{vmatrix} = 0$$

65. Find the equation of the line passing through $(-3, -1)$ and $(2, 9)$. Write the answer in the form $y = mx + b$.

66. Find the equation of the circle passing through $(7, 0)$, $(5, -6)$, and $(-1, -4)$. Write the answer in the form $(x - h)^2 + (y - k)^2 = r^2$.

C

67. Show that the area of the triangle in Figure A is $\frac{1}{2}\begin{vmatrix} a & b \\ c & d \end{vmatrix}$.

Hint: Figure B indicates how the required area can be found using a rectangle and three right triangles.

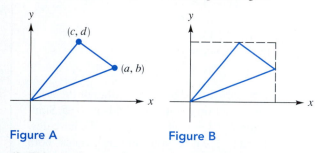

Figure A   **Figure B**

68. This exercise completes the derivation of Cramer's rule given in the text. Using the same notation, and assuming $D \neq 0$, we need to show that the values $x = D_x/D$, $y = D_y/D$, and $z = D_z/D$ satisfy the equations in (1) on page 527. We will show that these values satisfy the first equation in (1), the verification for the other equations being entirely similar.

   (a) Check that substituting the values $x = D_x/D$, $y = D_y/D$, and $z = D_z/D$ in the first equation of (1) yields an equation equivalent to

$$a_1 D_x + b_1 D_y + c_1 D_z - d_1 D = 0$$

(b) Show that the equation in part (a) can be written

$$a_1 \begin{vmatrix} b_1 & c_1 & d_1 \\ b_2 & c_2 & d_2 \\ b_3 & c_3 & d_3 \end{vmatrix} - b_1 \begin{vmatrix} a_1 & c_1 & d_1 \\ a_2 & c_2 & d_2 \\ a_3 & c_3 & d_3 \end{vmatrix}$$
$$+ c_1 \begin{vmatrix} a_1 & b_1 & d_1 \\ a_2 & b_2 & d_2 \\ a_3 & b_3 & d_3 \end{vmatrix} - d_1 \begin{vmatrix} a_1 & b_1 & c_1 \\ a_2 & b_2 & c_2 \\ a_3 & b_3 & c_3 \end{vmatrix} = 0$$

(c) Show that the equation in part (b) can be written

$$\begin{vmatrix} a_1 & b_1 & c_1 & d_1 \\ a_1 & b_1 & c_1 & d_1 \\ a_2 & b_2 & c_2 & d_2 \\ a_3 & b_3 & c_3 & d_3 \end{vmatrix} = 0$$

(d) Now explain why the equation in (c) indeed holds.

69. Let D denote the determinant of the matrix $\begin{pmatrix} a & b \\ c & d \end{pmatrix}$.

   (a) Show that the inverse of this matrix is $\frac{1}{D}\begin{pmatrix} d & -b \\ -c & a \end{pmatrix}$. Assume that $ad - bc \neq 0$.

   (b) Use the result in part (a) to find the inverse of the matrix $\begin{pmatrix} -6 & 7 \\ 1 & 9 \end{pmatrix}$.

70. Let $A = \begin{pmatrix} a & b \\ c & d \end{pmatrix}$ and $B = \begin{pmatrix} e & f \\ g & h \end{pmatrix}$. Is it true that $\det(AB) = (\det(A))(\det(B))$?

6.6 NONLINEAR SYSTEMS OF EQUATIONS

In the previous sections of this chapter, we looked at several techniques for solving systems of linear equations. In the present section, we consider **nonlinear systems of equations** — that is, systems in which at least one of the equations is not linear. There is no single technique that serves to solve all nonlinear systems. However, simple substitution will often suffice. The work in this section focuses on examples showing some of the more common approaches. In all of the examples and in the exercises, we will be concerned exclusively with solutions (x, y) in which both x and y are real numbers. In Example 1, the system consists of one linear equation and one quadratic equation. Such a system can always be solved by substitution.

EXAMPLE **1** **Solving a simple nonlinear system by substitution**

Find all solutions (x, y) of the following system, where x and y are real numbers:

$$\begin{cases} 2x + y = 1 \\ y = 4 - x^2 \end{cases}$$

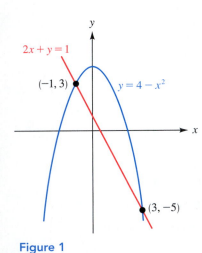

$2x + y = 1$

$(-1, 3)$

$y = 4 - x^2$

$(3, -5)$

Figure 1

SOLUTION

We use the second equation to substitute for y in the first equation. This will yield an equation with only one unknown:

$$2x + (4 - x^2) = 1$$
$$-x^2 + 2x + 3 = 0$$
$$x^2 - 2x - 3 = 0$$
$$(x - 3)(x + 1) = 0$$

$$x - 3 = 0 \quad | \quad x + 1 = 0$$
$$x = 3 \quad | \quad x = -1$$

The values $x = 3$ and $x = -1$ can now be substituted back into either of the original equations. Substituting $x = 3$ in the equation $y = 4 - x^2$ yields $y = -5$. Similarly, substituting $x = -1$ in the equation $y = 4 - x^2$ gives us $y = 3$. We have now obtained two ordered pairs, $(3, -5)$ and $(-1, 3)$. As you can easily check, both of these are solutions of the given system. Figure 1 displays the graphical interpretation of this result. The line $2x + y = 1$ intersects the parabola $y = 4 - x^2$ at the points $(-1, 3)$ and $(3, -5)$.

EXAMPLE **2** **Determining where two graphs intersect**

Where do the graphs of the parabola $y = x^2$ and the circle $x^2 + y^2 = 1$ intersect? See Figure 2.

SOLUTION

The system that we wish to solve is

$$\begin{cases} y = x^2 & (1) \\ x^2 + y^2 = 1 & (2) \end{cases}$$

In view of equation (1), we can replace the x^2-term of equation (2) by y. Doing this yields

$$y + y^2 = 1$$

or

$$y^2 + y - 1 = 0$$

This last equation can be solved by using the quadratic formula with $a = 1$, $b = 1$, and $c = -1$. As you can check, the results are

$$y = \frac{-1 + \sqrt{5}}{2} \quad \text{and} \quad y = \frac{-1 - \sqrt{5}}{2}$$

However, from Figure 2, it is clear that the y-coordinate at each intersection point is positive. Therefore, we discard the negative number $y = \frac{1}{2}(-1 - \sqrt{5})$ from further consideration in this context. Substituting the positive number $y = \frac{1}{2}(-1 + \sqrt{5})$ back in the equation $y = x^2$ then gives us

$$x^2 = (-1 + \sqrt{5})/2$$

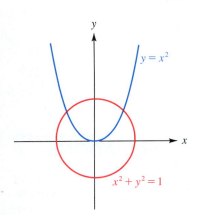

$y = x^2$

$x^2 + y^2 = 1$

Figure 2

Therefore,

$$x = \pm\sqrt{(-1 + \sqrt{5})/2}$$

By choosing the positive square root, we obtain the x-coordinate for the intersection point in the first quadrant. That point is therefore

$$\left(\sqrt{(-1 + \sqrt{5})/2}, (-1 + \sqrt{5})/2\right) \approx (0.79, 0.62) \qquad \text{using a calculator}$$

Similarly, the negative square root yields the x-coordinate for the intersection point in the second quadrant. That point is

$$\left(-\sqrt{(-1 + \sqrt{5})/2}, (-1 + \sqrt{5})/2\right) \approx (-0.79, 0.62) \qquad \text{using a calculator}$$

We have now found the two intersection points, as required.

EXAMPLE 3 | **Using substitution and the quadratic formula to solve a system**

Find all solutions (x, y) of the following system, where x and y are real numbers:

$$\begin{cases} xy = 1 & (3) \\ y = 3x + 1 & (4) \end{cases}$$

SOLUTION
Since these equations are easy to graph by hand, we do so, because that will tell us something about the required solutions. As Figure 3 indicates, there are two intersection points, one in the first quadrant, the other in the third quadrant. One way to begin now would be to solve equation (3) for one unknown in terms of the other. However, to avoid introducing fractions at the outset, let's use equation (4) to substitute for y in equation (3). This yields

$$x(3x + 1) = 1$$

or

$$3x^2 + x - 1 = 0$$

Figure 3

This last equation can be solved using the quadratic formula. As you should verify, the solutions are

$$x = (-1 + \sqrt{13})/6 \qquad \text{and} \qquad x = (-1 - \sqrt{13})/6$$

The corresponding y-values can now be obtained by substituting for x in either of the given equations. We will substitute in equation (4). (Exercise 35 at the end of this section asks you to substitute in equation (3) as well and then to show that the very different-looking answers obtained in that way are in fact equal to those found here.) Substituting $x = (-1 + \sqrt{13})/6$ into the equation $y = 3x + 1$ gives us

$$y = 3\left(\frac{-1 + \sqrt{13}}{6}\right) + 1$$

$$= \frac{-1 + \sqrt{13}}{2} + \frac{2}{2} = \frac{1 + \sqrt{13}}{2}$$

Thus, one of the intersection points is

$$((-1 + \sqrt{13})/16), (1 + \sqrt{13})/2)$$

Notice that this must be the first-quadrant point of intersection, since both co-ordinates are positive. The intersection point in the third quadrant is obtained in exactly the same manner. As you can check, substituting $x = -\frac{1}{6}(1 + \sqrt{13})$ in the equation $y = 3x + 1$ yields $y = \frac{1}{2}(1 - \sqrt{13})$. Thus, the other intersection point is

$$(-(1 + \sqrt{13})/6, (1 - \sqrt{13})/2)$$

As Figure 3 indicates, there are no other solutions.

EXAMPLE **4**   **A system with no solution**

Find all real numbers x and y that satisfy the system of equations

$$\begin{cases} y = \sqrt{x} \\ (x + 2)^2 + y^2 = 1 \end{cases}$$

SOLUTION
We use the first equation to substitute for y in the second equation. This yields

$$(x + 2)^2 + (\sqrt{x})^2 = 1$$
$$x^2 + 4x + 4 + x = 1$$
$$x^2 + 5x + 3 = 0$$

Using the quadratic formula to solve this last equation, we have

$$x = \frac{-5 \pm \sqrt{5^2 - 4(1)(3)}}{2(1)} = \frac{-5 \pm \sqrt{13}}{2}$$

The two values of x are thus

$$(-5 + \sqrt{13})/2 \qquad \text{and} \qquad (-5 - \sqrt{13})/2$$

However, notice that both of these quantities are negative. (The second is obviously negative; without using a calculator, can you explain why the first is negative?) Thus, neither of these quantities is an appropriate x-input in the equation $y = \sqrt{x}$, since we are looking for y-values that are real numbers. We conclude from this that there are no pairs of real numbers x and y satisfying the given system. In geometric terms, this means that the two graphs do not intersect; see Figure 4.

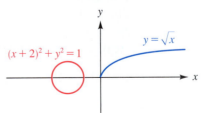

Figure 4

In the next example, we look at a system that can be reduced to a linear system through appropriate substitutions.

Solve the system

$$\begin{cases} \dfrac{2}{x^2} - \dfrac{3}{y^2} = -6 \\ \dfrac{3}{x^2} + \dfrac{4}{y^2} = 59 \end{cases}$$

SOLUTION

Let $u = 1/x^2$ and $v = 1/y^2$, so that the system becomes

$$\begin{cases} 2u - 3v = -6 \\ 3u + 4v = 59 \end{cases}$$

This is now a linear system. As you can verify using the methods of Section 6.1, the solution of this linear system is $u = 9$, $v = 8$. In view of the definitions of u and v, then, we have

$$\frac{1}{x^2} = 9 \qquad \qquad \frac{1}{y^2} = 8$$
$$x^2 = 1/9 \qquad \qquad y^2 = 1/8$$
$$x = \pm 1/3 \qquad \qquad y = \pm 1/\sqrt{8} = \pm 1/(2\sqrt{2}) = \pm\sqrt{2}/4$$

This gives us four possible solutions for the original system:

$$\left(\frac{1}{3}, \frac{\sqrt{2}}{4} \right) \qquad \left(\frac{1}{3}, -\frac{\sqrt{2}}{4} \right) \qquad \left(-\frac{1}{3}, \frac{\sqrt{2}}{4} \right) \qquad \left(-\frac{1}{3}, -\frac{\sqrt{2}}{4} \right)$$

As you can check, all four of these pairs satisfy the given system.

Determine all solutions (x, y) of the following system, where x and y are real numbers:

$$\begin{cases} y = 3^x & (5) \\ y = 3^{2x} - 2 & (6) \end{cases}$$

SOLUTION

We'll use the substitution method. First we rewrite equation (6) as

$$y = (3^x)^2 - 2 \qquad (7)$$

Now, in view of equation (5), we can replace 3^x with y in equation (7) to obtain

$$y = y^2 - 2$$
$$0 = y^2 - y - 2$$
$$0 = (y + 1)(y - 2)$$

From this last equation, we see that $y = -1$ or $y = 2$. With $y = -1$, equation (5) becomes $-1 = 3^x$, contrary to the fact that 3^x is positive for all real numbers x. Thus, we discard the case where $y = -1$. On the other hand, if $y = 2$, equation (5) becomes

$$2 = 3^x$$

We can solve this exponential equation by taking the logarithm of both sides. Using base e logarithms, we have

$$\ln 2 = \ln 3^x = x \ln 3$$

and, consequently,

$$x = \frac{\ln 2}{\ln 3} \qquad \text{CAUTION: } \frac{\ln 2}{\ln 3} \neq \ln 2 - \ln 3$$

We've now found that $x = (\ln 2)/(\ln 3)$ and $y = 2$. Figure 5 displays a graphical interpretation of this result. (Using a calculator for the x-coordinate, we find that $x \approx 0.6$, which is consistent with Figure 5.)

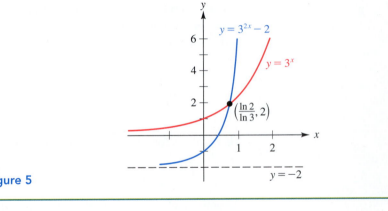

Figure 5

| **EXAMPLE** | **7** | **Using a graphing utility to approximate the solution of a nonlinear system** |

Use a graphing utility to approximate, to the nearest tenth, the solution(s) of the system

$$\begin{cases} y = 2 - x^2 \\ y = \ln x \end{cases}$$

SOLUTION
From work in previous chapters, we know how to quickly sketch by hand the graphs of these two equations. If you do that, either on paper or in your head, you'll see that the given system has exactly one solution. As indicated by the viewing rectangle in Figure 6(a), the required value of x is between 1 and 2, and y is between 0 and 1. After some experimenting, we obtain the viewing rectangle shown in Figure 6(b), which indicates that, to the nearest tenth, the solution of the given system is (1.3, 0.3).

GRAPHICAL PERSPECTIVE

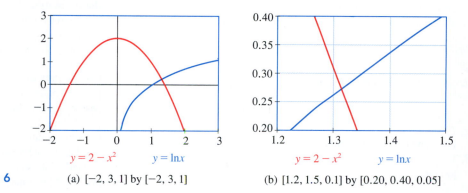

Figure 6   (a) $[-2, 3, 1]$ by $[-2, 3, 1]$ (b) $[1.2, 1.5, 0.1]$ by $[0.20, 0.40, 0.05]$

EXERCISE SET 6.6

A

In Exercises 1–22, find all solutions (x, y) of the given systems, where x and y are real numbers.

1. $\begin{cases} y = 3x \\ y = x^2 \end{cases}$

2. $\begin{cases} y = x + 3 \\ y = 9 - x^2 \end{cases}$

3. $\begin{cases} x^2 + y^2 = 25 \\ 24y = x^2 \end{cases}$

4. $\begin{cases} 3x + 4y = 12 \\ x^2 - y + 1 = 0 \end{cases}$

5. $\begin{cases} xy = 1 \\ y = -x^2 \end{cases}$

6. $\begin{cases} x + 2y = 0 \\ xy = -2 \end{cases}$

7. $\begin{cases} 2x^2 + y^2 = 17 \\ x^2 + 2y^2 = 22 \end{cases}$

8. $\begin{cases} x - 2y = 1 \\ y^2 - x^2 = 3 \end{cases}$

9. $\begin{cases} y = 1 - x^2 \\ y = x^2 - 1 \end{cases}$

10. $\begin{cases} xy = 4 \\ y = 4x \end{cases}$

11. $\begin{cases} xy = 4 \\ y = 4x + 1 \end{cases}$

12. $\begin{cases} \dfrac{2}{x^2} + \dfrac{5}{y^2} = 3 \\ \dfrac{3}{x^2} - \dfrac{2}{y^2} = 1 \end{cases}$

13. $\begin{cases} \dfrac{1}{x^2} - \dfrac{3}{y^2} = 14 \\ \dfrac{2}{x^2} + \dfrac{1}{y^2} = 35 \end{cases}$

14. $\begin{cases} y = -\sqrt{x} \\ (x - 3)^2 + y^2 = 4 \end{cases}$

15. $\begin{cases} y = -\sqrt{x - 1} \\ (x - 3)^2 + y^2 = 4 \end{cases}$

16. $\begin{cases} y = -\sqrt{x - 6} \\ (x - 3)^2 + y^2 = 4 \end{cases}$

17. $\begin{cases} y = 2^x \\ y = 2^{2x} - 12 \end{cases}$

18. $\begin{cases} y = e^{4x} \\ y = e^{2x} + 6 \end{cases}$

19. $\begin{cases} 2(\log_{10} x)^2 - (\log_{10} y)^2 = -1 \\ 4(\log_{10} x)^2 - 3(\log_{10} y)^2 = -11 \end{cases}$

20. $\begin{cases} y = \log_2(x + 1) \\ y = 5 - \log_2(x - 3) \end{cases}$

21. $\begin{cases} 2^x \cdot 3^y = 4 \\ x + y = 5 \end{cases}$

22. $\begin{cases} a^{2x} + a^{2y} = 10 \\ a^{x+y} = 4 \end{cases}$ $(a > 0)$

Hint: Use the substitutions $a^x = t$ and $a^y = u$.

In Exercises 23–34:

(a) *Use a graphing utility to approximate the solutions (x, y) of each system. Zoom in on the relevant intersection points until you are sure of the first two decimal places of each coordinate.*

(b) *In Exercises 23–28 only, also use an algebraic method of solution. Round the answers to three decimal places and check to see that your results are consistent with the graphical estimates obtained in part (a).*

23. $\begin{cases} y = x - 5 \\ y = -x^2 + 2 \end{cases}$

24. $\begin{cases} y = x^2 - 1 \\ y = -2x^4 + 3 \end{cases}$

25. $\begin{cases} y = \sqrt{x + 1} + 1 \\ 3x + 4y = 12 \end{cases}$

26. $\begin{cases} y = \sqrt{x} \\ y = 2x^3 \end{cases}$

27. $\begin{cases} y = 4^{2x} \\ y = 4^x + 3 \end{cases}$

28. $\begin{cases} y = \ln x \\ y = 1 + \ln(x - 5) \end{cases}$

29. $\begin{cases} y = e^{x/2} \\ y = x^2 \end{cases}$

30. $\begin{cases} x^2 - 2x + y = 0 \\ y = 3x^3 - x^2 - 10x \end{cases}$

31. $\begin{cases} y = \sqrt{x} - 1 \\ y = \ln x \end{cases}$

32. $\begin{cases} y = \sqrt[3]{x} \\ y = \ln x \end{cases}$

Hint: There are two solutions.

Hint: There are two solutions.

33. $\begin{cases} y = x^3 \\ y = (e^x + e^{-x})/2 \end{cases}$

34. $\begin{cases} y = x^3 \\ y = (e^x - e^{-x})/2 \end{cases}$

Hint: There are two solutions.

Hint: There are four solutions; use symmetry to cut the amount of work in half.

35. (a) Let $x = (-1 + \sqrt{13})/6$. Using this x-value, show that the equations $y = 3x + 1$ and $y = 1/x$ yield the same y-value. (This completes a detail mentioned in Example 3.)

(b) Solve the following system. (You should obtain $u = 9$ and $v = 8$, as stated in Example 5.)

$$\begin{cases} 2u - 3v = -6 \\ 3u + 4v = 59 \end{cases}$$

36. A sketch shows that the line $y = 100x$ intersects the parabola $y = x^2$ at the origin. Are there any other intersection points? If so, find them. If not, explain why not.

B

37. Solve the following system for x and y:

$$\begin{cases} ax + by = 2 \\ abxy = 1 \end{cases}$$

(Assume that neither a nor b is zero.)

38. Let a, b, and c be constants (with $a \neq 0$), and consider the system

$$\begin{cases} y = ax^2 + bx + c \\ y = k \end{cases}$$

For which value of k (in terms of a, b, and c) will the system have exactly one solution? What is that solution? What is the relationship between the solution you've found and the graph of $y = ax^2 + bx + c$?

39. Find all solutions of the system

$$\begin{cases} x^3 + y^3 = 3473 \\ x + y = 23 \end{cases}$$

40. Solve the following system for x, y, and z:

$$\begin{cases} yz = p^2 \\ zx = q^2 \\ xy = r^2 \end{cases}$$

(Assume that p, q, and r are positive constants.)

41. If the diagonal of a rectangle has length d and the perimeter of the rectangle is $2p$, express the lengths of the sides in terms of d and p.

42. Solve for x and y in terms of p, q, a, and b:

$$\begin{cases} q^{\ln x} = p^{\ln y} \\ (px)^{\ln a} = (qy)^{\ln b} \end{cases}$$

(Assume all constants and variables are positive.)

43. The accompanying figure shows the graph of a function $N = N_0 e^{kt}$. In each case, determine the constants N_0 and k so that the graph passes through
(a) (2, 3) and (8, 24); **(b)** (1/2, 1) and (4, 10).

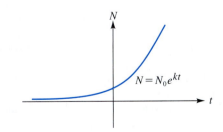

44. [As background for this exercise, first work Exercise 43(a).] Measurements (begun in 1958, by Charles D.

Keeling of the Scripps Institution of Oceanography) show that the concentration of carbon dioxide in the atmosphere has increased exponentially over the years 1958–2000. In 1958, the average carbon dioxide concentration was $N = 315$ ppm (parts per million), and in 2000 it was $N = 369$ ppm.

(a) Use the data to determine the constants k and N_0 in the growth law $N = N_0 e^{kt}$. *Hint*: The curve $N = N_0 e^{kt}$ passes through the two points (1958, 315) and (2000, 369). Use this to obtain a system of two equations in two unknowns.

(b) Stephen H. Schneider in his article "The Changing Climate" [*Scientific American*, vol. 261 (1989), pp. 70–79] estimates that, under certain circumstances, the concentration of carbon dioxide in the atmosphere could reach 600 ppm by the year 2080. Use your results in part (a) to make a projection for the year 2080. Is your projection higher or lower than Dr. Schneider's?

45. Solve the following system for x, y, and z in terms of p, q, and r.

$$\begin{cases} x(x + y + z) = p^2 \\ y(x + y + z) = q^2 \\ z(x + y + z) = r^2 \end{cases}$$

(Assume that p, q, and r are nonzero.) *Hint*: Denote $x + y + z$ by w; then add the three equations.

46. (a) Find the points where the line $y = -2x - 2$ intersects the parabola $y = \frac{1}{2}x^2$.

(b) On the same set of axes, sketch the line $y = -2x - 2$ and the parabola $y = \frac{1}{2}x^2$. Be certain that your sketch is consistent with the results obtained in part (a).

47. If a right triangle has area 180 cm^2 and hypotenuse 41 cm, find the lengths of the two legs.

48. The sum of two numbers is 8, while their product is -128. What are the two numbers?

49. If a rectangle has perimeter 46 cm and area 60 cm^2, find the length and the width.

50. Find all right triangles for which the perimeter is 24 units and the area is 24 square units.

51. Solve the following system for x and y using the substitution method:

$$\begin{cases} x^2 + y^2 = 5 & (1) \\ xy = 2 & (2) \end{cases}$$

52. The substitution method in Exercise 51 leads to a quadratic equation. Here is an alternative approach to solving that system; this approach leads to linear equations. Multiply equation (2) by 2 and add the resulting equation to equation (1). Now take square roots to conclude that $x + y = \pm 3$. Next, multiply equation (2) by 2 and subtract the resulting equation from equation (1). Take square roots to conclude that $x - y = \pm 1$. You now have the following four linear systems, each of which can be solved (with almost no work) by the addition–subtraction method.

$$\begin{cases} x + y = 3 \\ x - y = 1 \end{cases} \qquad \begin{cases} x + y = 3 \\ x - y = -1 \end{cases}$$

$$\begin{cases} x + y = -3 \\ x - y = 1 \end{cases} \qquad \begin{cases} x + y = -3 \\ x - y = -1 \end{cases}$$

Solve these systems and compare your results with those obtained in Exercise 51.

53. Solve the following system using the method explained in Exercise 52.

$$\begin{cases} x^2 + y^2 = 7 \\ xy = 3 \end{cases}$$

54. Solve the following system using the substitution method:

$$\begin{cases} 3xy - 4x^2 = 2 \\ -5x^2 + 3y^2 = 7 \end{cases}$$

(Begin by solving the first equation for y.)

55. Here is an alternative approach for solving the system in Exercise 54. Let $y = mx$, where m is a constant to be determined. Replace y with mx in both equations of the system to obtain the following pair of equations:

$$x^2(3m - 4) = 2 \qquad (1)$$
$$x^2(-5 + 3m^2) = 7 \qquad (2)$$

Now divide equation (2) by equation (1). After clearing fractions and simplifying, you can write the resulting equation as $2m^2 - 7m + 6 = 0$. Solve this last equation by factoring. The values of m can then be used in equation (1) to determine values for x. In each case, the corresponding y-values are determined by the equation $y = mx$.

C

56. Solve the following system for x and y:

$$\begin{cases} \dfrac{1}{x^2} + \dfrac{1}{xy} = \dfrac{1}{a^2} \\ \dfrac{1}{y^2} + \dfrac{1}{xy} = \dfrac{1}{b^2} \end{cases}$$

(Assume that a and b are positive.)

57. Solve the following system for x and y:

$$\begin{cases} x^4 = y^6 \\ \ln \dfrac{x}{y} = \dfrac{\ln x}{\ln y} \end{cases}$$

6.7 SYSTEMS OF INEQUALITIES. LINEAR PROGRAMMING

. . . the variety of problems in business, science, and industry that can be solved by using the linear programming model is impressive, to say the least. Problems in transportation, scheduling, capital budgeting, production planning, inventory control, advertising, and many more areas can be solved by the linear programming approach. —Stanley J. Farlow and Gary M. Haggard in *Finite Mathematics and Its Applications* (New York: Random House, 1988)

In the first section of this chapter, we solved systems of equations in two unknowns. Now we will consider systems of inequalities in two unknowns. The techniques we develop are used in calculus when discussing functions of two variables and in business and economics in the study of linear programming.

Let a, b, and c denote real numbers, and assume that a and b are not both zero. Then all of the following are called **linear inequalities:**

$$ax + by + c < 0 \qquad ax + by + c > 0$$
$$ax + by + c \leq 0 \qquad ax + by + c \geq 0$$

An ordered pair of numbers (x_0, y_0) is said to be a **solution** of a given inequality (linear or not) provided we obtain a true statement on substituting x_0 and y_0 for x and y, respectively. For instance, $(-1, 1)$ is a solution of the inequality $2x + 3y - 6 < 0$, since substitution yields

$$2(-1) + 3(1) - 6 < 0$$
$$-5 < 0 \qquad \text{True}$$

Figure 1

Figure 2
The graph of $y < 2x$.

Like a linear equation in two unknowns, a linear inequality has infinitely many solutions. For this reason, we often represent the solutions graphically. When we do this, we say that we are **graphing the inequality.**

For example, let's graph the inequality $y < 2x$. First we observe that the coordinates of the points on the line $y = 2x$ do not, by definition, satisfy this inequality. So it remains to consider points above the line and points below the line. We will in fact show that the required graph consists of all points that lie below the line. Take any point $P(x_0, y_0)$ not on the line $y = 2x$. Let $Q(x_0, 2x_0)$ be the point on $y = 2x$ with the same first coordinate as P. Then, as indicated in Figure 1, the point P lies below the line if and only if the y-coordinate of P is less than the y-coordinate of Q. In other words, (x_0, y_0) lies below the line if and only if $y_0 < 2x_0$. This last statement is equivalent to saying that (x_0, y_0) satisfies the inequality $y < 2x$. This shows that the graph of $y < 2x$ consists of all points below the line $y = 2x$ (see Figure 2). The broken line in Figure 2 indicates that the points on $y = 2x$ are not part of the required graph. If the original inequality had been $y \leq 2x$, then we would use a solid line rather than a broken one, indicating that the line is included in the graph. And if the original inequality had been $y > 2x$, the graph would be the region above the line.

Just as the graph of $y < 2x$ is the region below the line $y = 2x$, it is true in general that the graph of $y < f(x)$ is the region below the graph of the function f. For example, Figures 3 and 4 display the graphs of $y < x^2$ and $y \geq x^2$.

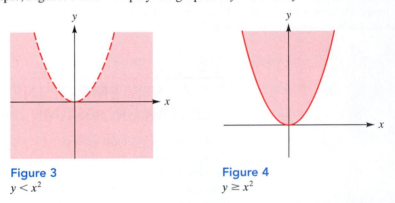

Figure 3
$y < x^2$

Figure 4
$y \geq x^2$

Example 1 summarizes the technique developed so far for graphing an inequality. Following this example, we will point out a useful alternative method.

EXAMPLE 1 **Graphing an inequality**

Graph the inequality $4x - 3y \leq 12$.

SOLUTION
The graph will include the line $4x - 3y = 12$ and either the region above the line or the region below it. To decide which region, we solve the inequality for y:

$$4x - 3y \leq 12$$
$$-3y \leq -4x + 12$$
$$y \geq \frac{4}{3}x - 4 \qquad \text{Multiplying or dividing by a negative number reverses an inequality.}$$

This last inequality tells us that we want the region above the line, as well as the line itself. Figure 5 displays the required graph.

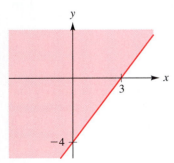

Figure 5
The graph of $4x - 3y \leq 12$.

There is another method that we can use in Example 1 to determine which side of the line we want. This method involves a **test point.** We pick any convenient point that is not on the line $4x - 3y = 12$. Then we test to see whether this point satisfies the given inequality. For example, let's pick the point $(0, 0)$. Substituting these coordinates in the inequality $4x - 3y \le 12$ yields

$$4(0) - 3(0) \overset{?}{\le} 12$$

$$0 \overset{?}{\le} 12 \qquad \text{True}$$

We conclude from this that the required side of the line includes the point $(0, 0)$. In agreement with Figure 5, we see this is the region above the line.

Next we discuss *systems* of inequalities in two unknowns. As with systems of equations, a **solution** of a system of inequalities is an ordered pair (x_0, y_0) that satisfies all of the inequalities in the system. As a first example, let's graph the points that satisfy the following nonlinear system.

$$\begin{cases} y - x^2 \ge 0 \\ x^2 + y^2 < 1 \end{cases}$$

By writing the first inequality as $y \ge x^2$, we see that it describes the set of points on or above the parabola $y = x^2$. For the second inequality, we must decide whether it describes the points inside or outside the circle $x^2 + y^2 = 1$. One way to do this is to choose $(0, 0)$ as a test point. Substituting the values $x = 0$ and $y = 0$ in the inequality $x^2 + y^2 < 1$ yields the true statement $0^2 + 0^2 < 1$. Since $(0, 0)$ lies within the circle and satisfies the inequality, we conclude that the inequality $x^2 + y^2 < 1$ describes the set of all points within the circle. Now we put our information together. We wish to graph the points that lie on or above the parabola $y = x^2$ but within the circle $x^2 + y^2 = 1$. This is the shaded region shown in Figure 6.

Figure 6

| **EXAMPLE** | **2** | **Graphing a system of inequalities** |

Graph the system

$$\begin{cases} -x + 3y \le 12 \\ x + y \le 8 \\ x \ge 0 \\ y \ge 0 \end{cases}$$

SOLUTION

Solving the first inequality for y, we have

$$3y \le x + 12$$
$$y \le \tfrac{1}{3}x + 4$$

Thus, the first inequality is satisfied by the points on or below the line $-x + 3y = 12$. Similarly, by solving the second inequality for y, we see that it describes the set of points on or below the line $x + y = 8$. The third inequality, $x \ge 0$, describes the points on or to the right of the y-axis. Similarly, the fourth inequality, $y \ge 0$, describes the points on or above the x-axis. We summarize these statements in Figure 7(a). The arrows indicate which side of each line we wish to consider.

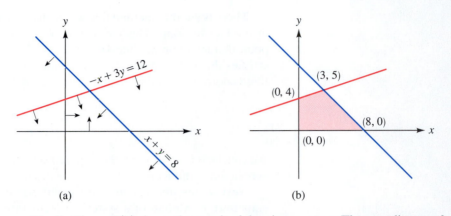

Figure 7   (a)   (b)

Finally, Figure 7(b) shows the graph of the given system. The coordinates of each point in the shaded region satisfy all four of the given inequalities. The coordinates (3, 5) in Figure 7(b) were found by solving the system of linear equations $-x + 3y = 12$ and $x + y = 8$. (You'll need to carry out this work as part of Exercise 25.)

We can use Figure 7(b) to introduce some terminology that is useful in describing sets of points in the plane. A **vertex** of a region is a corner, or point, where two adjacent bounding sides meet. Thus, the vertices of the shaded region in Figure 7(b) are the four points (0, 0), (8, 0), (3, 5), and (0, 4). The shaded region in Figure 7(b) is **convex.** This means that, given any two points in that region, the straight line segment joining these two points lies wholly within the region. Figure 8(a) displays another example of a convex set, while Figure 8(b) shows a set that is not convex. The shaded region in Figure 8(b) is also an example of a **bounded region.** By this we mean that the region can be wholly contained within some (sufficiently large) circle. Perhaps the simplest example of a region that is not bounded is the entire x-y plane itself. Figure 9 shows another example of an unbounded region.

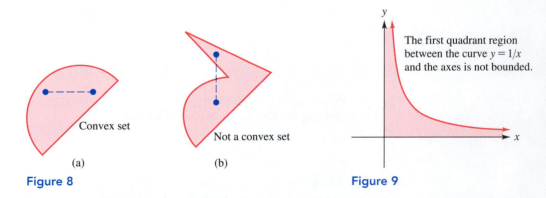

Convex set

Not a convex set

(a)   (b)

Figure 8

The first quadrant region between the curve $y = 1/x$ and the axes is not bounded.

Figure 9

We are prepared now to consider a typical problem in linear programming.

PROBLEM   Let $C = x + 3y$. Find the values of x and y that yield the largest possible value of C, given that x and y satisfy the following set of inequalities.

$$\begin{cases} -x + 3y \le 12 \\ x + y \le 8 \\ x \ge 0 \\ y \ge 0 \end{cases}$$

The function $C = x + 3y$ in this problem is known as the **objective function.** The given inequalities are called the **constraints.** In analyzing this problem, we will take advantage of the fact that we've already graphed this system of inequalities in Example 2. For convenience, we repeat the graph here (Figure 10). Since the coordinates of each point in the shaded region of Figure 10 satisfy the constraints of the present problem, we refer to the region as the set of **feasible solutions.** The problem then is to find that point among the feasible solutions that yields the largest possible value for C. A feasible solution (x_0, y_0) that yields the largest value for C is called an **optimal solution.** Some linear programming problems ask for a smallest rather than a largest value of the objective function. A feasible solution yielding the smallest value for C is also called an optimal solution.

We will see subsequently that there is a general method for solving this type of linear programming problem, but as a start, let's experiment. How large can the value of C be, subject to the given constraints? For instance, can C equal 30? With $C = 30$, we have

$$x + 3y = 30$$

Figure 11 shows the graph of this line. Since the line has no points in common with the set of feasible solutions, we conclude that C cannot equal 30 under the conditions of the problem.

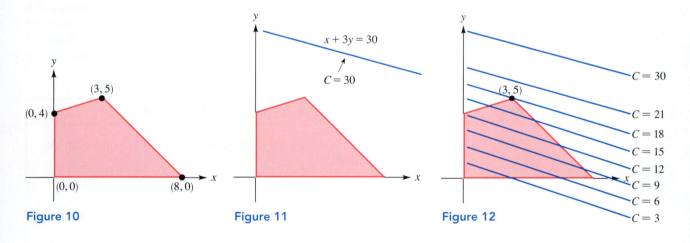

Figure 10

Figure 11

Figure 12

Still experimenting, in Figure 12 we have graphed the lines $x + 3y = C$ for various values of C. Observe that the lines are all parallel. To see why this is so, we solve the equation $x + 3y = C$ for y. This yields

$$y = -\frac{1}{3}x + \frac{1}{3}C$$

As you can see, the slope here is $m = -1/3$, independent of C. Since the lines all have the same slope, they are all parallel. On the other hand, the equation $y = -\frac{1}{3}x + \frac{1}{3}C$

shows how the value of C does influence the graph. The y-intercept of the line is $C/3$. So as C increases, the line moves *up,* but the slope is always $-1/3$.

Now, as you can see in Figure 12, when $C = 18$ the line $x + 3y = 18$ intersects the set of feasible solutions at the vertex $(3, 5)$. However, it is also evident from the figure and the remarks in the previous paragraph that when C is greater than 18, the line $x + 3y = C$ will not intersect the set of feasible solutions. We conclude therefore that the vertex $(3, 5)$ is the optimal solution. That is, subject to the given constraints, the value of C is largest when $x = 3$ and $y = 5$.

Notice that in the example just concluded, the optimal solution to the linear programming problem occurred at a vertex. The following theorem (stated here without proof) tells us that the optimal solution, when it exists, always occurs at a vertex. (In the statement of the theorem, the word "nonstrict" refers to inequalities that involve the symbols $\leq$ or $\geq$ rather than $<$ or $>$.)

Linear Programming Theorem

Consider the objective function $C = ax + by + c$, where the nonnegative variables x and y are subject to constraints specified by nonstrict linear inequalities. Then if C assumes a maximum (or minimum) value, this occurs at a vertex. Furthermore, if the set of feasible solutions is both convex and bounded, then a maximum and minimum always exist.

In view of this theorem, we can solve linear programming problems by first finding the vertices of the feasibility set and then computing and comparing the values of the objective function at those vertices. This is done in Example 3, which follows. Technically speaking, the theorem we've stated guarantees a maximum and a minimum only when the feasibility set is convex and bounded. However, this is not as great a restriction as it might at first seem. As it turns out, the graph of any set of linear inequalities forms a convex set. (For a look at a case in which the feasibility set is not bounded, see Exercise 44.)

 EXAMPLE 3 **Solving a linear programming problem**

Let $C = 3x - 2y$. Find the maximum and minimum values of C subject to the following set of constraints:

$$\begin{cases} 6y - x \geq 5 \\ 4x + 3y \leq 34 \\ 2y - x \leq 8 \\ 4x - y \geq 3 \end{cases}$$

What are the optimal solutions in each case?

SOLUTION
In Figure 13 we have graphed the given system of inequalities using the method discussed in Example 2. Now we need to determine the coordinates of the four vertices. To find the coordinates of E, we solve the system

$$\begin{cases} 6y - x = 5 & (1) \\ 4x - y = 3 & (2) \end{cases}$$

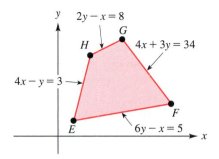

Figure 13

Solving equation (1) for x, we have

$$6y - 5 = x \qquad\qquad (3)$$

Now we can use equation (3) to substitute for x in equation (2). This yields

$$4(6y - 5) - y = 3$$
$$24y - 20 - y = 3$$
$$23y = 23$$
$$y = 1$$

Finally, substituting $y = 1$ in equation (3) gives us $x = 6(1) - 5 = 1$.

We have now found that the coordinates of the vertex E are $(1, 1)$. The coordinates of F, G, and H can be found in a similar fashion. The results are displayed in Figure 14. Notice that the set of feasible solutions forms a convex and bounded set. Thus the objective function attains a maximum and a minimum value at certain vertices. We compute the value of C at each vertex and tabulate the results as shown in Table 1.

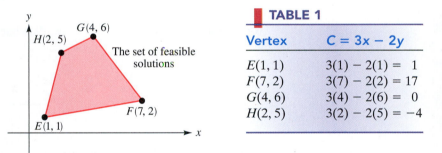

Figure 14

TABLE 1

Vertex	$C = 3x - 2y$
$E(1, 1)$	$3(1) - 2(1) = 1$
$F(7, 2)$	$3(7) - 2(2) = 17$
$G(4, 6)$	$3(4) - 2(6) = 0$
$H(2, 5)$	$3(2) - 2(5) = -4$

The largest value of C in this table is 17; this is the required maximum. The corresponding optimal solution is $(7, 2)$. Similarly, the smallest value of C is -4; this is the required minimum. The corresponding optimal solution is $(2, 5)$.

The next example deals with a type of linear programming problem known as the **transportation problem.** In this type of problem we are required to minimize the costs of shipping goods from warehouses to retail outlets.

EXAMPLE   4   A transportation problem in linear programming

A motorcycle manufacturer must fill orders from two dealers. The first dealer, D_1, has ordered 20 motorcycles, while the second dealer, D_2, has ordered 30 motorcycles. The manufacturer has the motorcycles stored in two warehouses, W_1 and

W_2. There are 40 motorcycles in W_1 and 15 in W_2. The shipping costs, per motorcycle, are as follows: \$15 from W_1 to D_1; \$13 from W_1 to D_2; \$14 from W_2 to D_1; and \$16 from W_2 to D_2. Under these conditions, find the number of motorcycles to be shipped from each warehouse to each dealer if the total shipping cost is to be a minimum. What is the minimum cost?

SOLUTION

The first step is to summarize the data. We do this as follows.

Shipping cost per motorcycle	Supply and demand	
	W_1	W_2
W_1 to D_1: \$15	(40 motorcycles in stock)	(15 motorcycles in stock)
W_1 to D_2: \$13		
W_2 to D_1: \$14	D_1	D_2
W_2 to D_2: \$16	(wants 20 motorcycles)	(wants 30 motorcycles)

The problem asks how many motorcycles should be shipped from each warehouse to each dealer.

1. Let x denote the number of motorcycles to be shipped from W_1 to D_1.
2. Then $20 - x$ denotes the number of motorcycles to be shipped from W_2 to D_1. (Why?)
3. Let y denote the number of motorcycles to be shipped from W_1 to D_2.
4. Then $30 - y$ denotes the number of motorcycles to be shipped from W_2 to D_2.

If C denotes the total shipping cost, we then have

$$C = 15x + 13y + 14(20 - x) + 16(30 - y)$$

As you can check, this equation simplifies to

$$C = x - 3y + 760$$

This is the objective function that we must minimize.

Now let's examine the constraints of the problem. The number of items shipped from the warehouse certainly must be a nonnegative number in each case. Thus we must have

$$x \geq 0$$
$$y \geq 0$$
$$20 - x \geq 0$$
$$30 - y \geq 0$$

Furthermore, since at most 40 motorcycles can be shipped from W_1, we must have

$$x + y \leq 40$$

Similarly, since at most 15 motorcycles can be shipped from W_2, we have

$$(20 - x) + (30 - y) \le 15$$

or

$$50 - x - y \le 15$$
$$-x - y \le -35$$
$$x + y \ge 35$$

In summary, then, we have the following set of constraints:

$$\begin{cases} x \ge 0 \\ y \ge 0 \\ 20 - x \ge 0 \\ 30 - y \ge 0 \\ x + y \le 40 \\ x + y \ge 35 \end{cases}$$

(5, 30) (10, 30)   The set of feasible solutions

(20, 20)

(20, 15)

Figure 15

Figure 15 shows the graph of the region described by this set of constraints. (Exercise 33 at the end of this section asks you to verify that the graph and the coordinates of the vertices are as indicated.) Since the region is both convex and bounded, we know that the objective function assumes its minimum value at one of the vertices. We compute the value of C at each vertex.

Vertex	$C = x - 3y + 760$
(20, 15)	$C = 20 - 3(15) + 760 = 735$
(20, 20)	$C = 20 - 3(20) + 760 = 720$
(10, 30)	$C = 10 - 3(30) + 760 = 680$
(5, 30)	$C = 5 - 3(30) + 760 = 675$

As the table shows, the minimum shipping cost is \$675. The corresponding vertex is (5, 30). Therefore, the shipping cost will be minimized if the deliveries are made as follows.

$$x = 5: \quad \text{ship 5 motorcycles from } W_1 \text{ to } D_1$$
$$20 - x = 15: \quad \text{ship 15 motorcycles from } W_2 \text{ to } D_1$$
$$y = 30: \quad \text{ship 30 motorcycles from } W_1 \text{ to } D_2$$
$$30 - y = 0: \quad \text{ship no motorcycles from } W_2 \text{ to } D_2$$

The linear programming problems that we have considered have involved only two unknowns, x and y. Because there were only two unknowns, we were able to utilize graphical methods in solving the problems. As you might guess, linear programming problems actually arising in business and industry may involve dozens or even thousands of unknowns. For such cases, there is a purely algebraic method of solution called the **simplex method,** which makes use of matrices. Furthermore, this method can be programmed for computer solution, and several commercial firms now market software for solving linear programming problems. [For an excellent discussion of the simplex method as well as methods of computer solution, see *An Introduction to Management Science,* 9th ed., by David R. Anderson et al. (Cincinnati, Ohio: South-Western Publishing, 1999).]

EXERCISE SET 6.7

A

In Exercises 1 and 2, decide whether or not the ordered pairs are solutions of the given inequality.

1. $4x - 6y + 3 \geq 0$

   (a) $(1, 2)$

   (b) $(0, \frac{1}{2})$

2. $5x + 2y < 1$

   (a) $(-1, 3)$

   (b) $(0, 0)$

In Exercises 3–16, graph the given inequalities.

3. $2x - 3y > 6$

4. $2x - 3y < 6$

5. $2x - 3y \geq 6$

6. $2x - 3y \leq 6$

7. $x - y < 0$

8. $y \leq \frac{1}{2}x - 1$

9. $x \geq 1$

10. $y < 0$

11. $x > 0$

12. $y \leq \sqrt{x}$

13. $y > x^3 + 1$

14. $y \leq |x - 2|$

15. $x^2 + y^2 \geq 25$

16. $y \geq e^x - 1$

In Exercises 17–22, graph the systems of inequalities.

17. $\begin{cases} y \leq x^2 \\ x^2 + y^2 \leq 1 \end{cases}$

18. $\begin{cases} y < x \\ x^2 + y^2 < 1 \end{cases}$

19. $\begin{cases} y \geq 1 \\ y \leq |x| \end{cases}$

20. $\begin{cases} x \geq 0 \\ y \geq 0 \\ y < \sqrt{x} \\ x \leq 4 \end{cases}$

21. $\begin{cases} x \geq 0 \\ y \geq 0 \\ y \leq 1 - x^2 \end{cases}$

22. $\begin{cases} y < 2x \\ y > \frac{1}{2}x \end{cases}$

In Exercises 23–34, graph the systems of linear inequalities. In each case specify the vertices. Is the region convex? Is the region bounded?

23. $\begin{cases} y \leq x + 5 \\ y \leq -2x + 14 \\ x \geq 0 \\ y \geq 0 \end{cases}$

24. $\begin{cases} y \geq x + 5 \\ y \geq -2x + 14 \\ x \geq 0 \\ y \geq 0 \end{cases}$

25. $\begin{cases} -x + 3y \leq 12 \\ x + y \leq 8 \\ x \geq 0 \\ y \geq 0 \end{cases}$

26. $\begin{cases} y \geq 2x \\ y \geq -x + 6 \end{cases}$

27. $\begin{cases} 0 \leq 2x - y + 3 \\ x + 3y \leq 23 \\ 5x + y \leq 45 \end{cases}$

28. $\begin{cases} 0 \leq 2x - y + 3 \\ x + 3y \leq 23 \\ 5x + y \leq 45 \\ x \geq 0 \\ y \geq 0 \end{cases}$

29. $\begin{cases} 5x + 6y < 30 \\ y > 0 \end{cases}$

30. $\begin{cases} 5x + 6y < 30 \\ x > 0 \end{cases}$

31. $\begin{cases} 5x + 6y < 30 \\ x > 0 \\ y > 0 \end{cases}$

32. $\begin{cases} 2x + 3y \geq 6 \\ 2x + 3y \leq 12 \end{cases}$

33. $\begin{cases} x \geq 0 \\ y \geq 0 \\ 20 - x \geq 0 \\ 30 - y \geq 0 \\ x + y \leq 40 \\ x + y \geq 35 \end{cases}$

34. $\begin{cases} x \geq 0 \\ y \geq 0 \\ 3x - y + 1 \geq 0 \\ 0 \leq x - y + 3 \\ y \leq 5 \\ x \leq \frac{1}{3}(17 - y) \\ x \leq \frac{1}{2}(y + 8) \end{cases}$

In Exercises 35–43, find the maximum and the minimum values of the objective function subject to the given constraints. Specify the corresponding optimal solutions in each case.

35. Objective function: $C = 3y - x$

   Constraints: $\begin{cases} y \leq x + 1 \\ y \leq -2x + 10 \\ x \geq 0 \\ y \geq 0 \end{cases}$

36. Objective function: $C = 2x - y + 8$

   Constraints: $\begin{cases} 4x + 3y \geq 12 \\ 6x + 5y \leq 30 \\ x \geq 0 \\ y \geq 0 \end{cases}$

37. Objective function: $C = 2x + y$

   Constraints: $\begin{cases} x \geq 0 \\ x \leq 1 \\ y \geq 0 \\ 2y - x \leq 3 \end{cases}$

38. Objective function: $C = 2x - y$

   Constraints: Use the same constraints given in Exercise 37.

39. Objective function: $C = 10y + 9x - 1$

   Constraints: $\begin{cases} x + y \geq 3 \\ y - x + 1 \geq 0 \\ y \leq -x + 12 \\ y \leq 6 \\ x \geq 1 \end{cases}$

40. Objective function: $C = 10y + 11x - 1$

   Constraints: Use the same constraints given in Exercise 39.

41. Objective function: $C = 10y + 3x + 100$

   Constraints: $\begin{cases} 0 \leq 4x - y + 4 \\ x + 5y \leq 41 \\ x + 3y \leq 27 \\ 2x + y \leq 24 \\ x \geq 0 \\ y \geq 0 \end{cases}$

42. Objective function: $C = 20y + 7x + 100$
Constraints: Use the same constraints given in Exercise 41.

43. (a) Objective function: $C = 19x + 100y$
Constraints: Use the same constraints given in Exercise 41.

(b) Objective function: $C = 21x + 100y$
Constraints: Use the same constraints given in Exercise 41.

(This exercise shows that a slight change in the objective function may yield a relatively large change in the optimal solution.)

44. Consider the objective function $C = 4y - x$ subject to the constraints

$$\begin{cases} y \geq -\tfrac{3}{2}x + 4 \\ 2y \geq x \\ y \geq 0 \\ x \geq 0 \end{cases}$$

(a) Given that there is an optimal solution yielding a minimum value of C, find that solution and the corresponding value of C.
(b) Find a feasible solution for which $C = 100$.
(c) Find a feasible solution for which $C = 1000$.
(d) Find a feasible solution for which $C = 10^6$.

45. An automobile manufacturer must fill orders from two dealers. The first dealer, D_1, has ordered 40 cars, and the second dealer, D_2, has ordered 25 cars. The manufacturer has the cars stored at two locations, W_1 and W_2. There are 30 cars at W_1 and 50 cars at W_2. The shipping costs per car are as follows: \$180 from W_1 to D_1; \$150 from W_1 to D_2; \$160 from W_2 to D_1; and \$170 from W_2 to D_2. Under these conditions, how many cars should be shipped from each storage location to each dealer so as to minimize the total shipping cost? What is this minimum cost?

46. A television manufacturer must fill orders from two retailers. The first retailer, R_1, has ordered 55 television sets, and the second retailer, R_2, has ordered 75 sets. The manufacturer has the television sets stored in two warehouses, W_1 and W_2. There are 100 sets in W_1 and 120 sets in W_2. The shipping costs per television set are as follows: \$8 from W_1 to R_1; \$12 from W_1 to R_2; \$13 from W_2 to R_1; and \$7 from W_2 to R_2. Under these conditions, find the number of television sets that should be shipped from each warehouse to each retailer if the total shipping cost is to be a minimum. What is this minimum cost?

47. A factory produces two products, A and B. The production time (in hours) for each unit is as follows:

Product	Assembly	Inspection and packing
A	0.40	0.20
B	0.50	0.40

Each day at most 500 worker-hours of assembly time and 300 worker-hours of inspection/packing time are available. The profit realized on each unit of A is \$0.60; on each unit of B it is \$0.80. Find the maximum daily profit under these conditions.   *Hint:* Let x denote the number of units of A produced each day and let y denote the number of units of B produced each day. After graphing the constraints you'll find that the coordinates of one vertex are not whole numbers. Round these off to the nearest whole number. (Why?)

48. Due to increased energy costs, the owner of the factory described in Exercise 47 decides to limit production. In particular, the owner decides that no more than 800 units of A and 500 units of B should be produced daily. Under these additional constraints, how many units of A and B should be produced daily to maximize profits? What is the maximum profit?

49. A farmer can allocate at most 600 acres for growing cherry tomatoes and regular tomatoes. The labor requirements and the profits are as follows.

	Cherry tomatoes	Regular tomatoes
Labor (hr/acre)	3	2
Profit (\$/acre)	50	36

What is the farmer's maximum profit if 1350 hr of labor are available? How many acres should be devoted to each crop to achieve the maximum profit?

50. Consider the situation described in Exercise 49. Now additionally suppose that due to market considerations, the farmer decides to devote at most 550 acres to regular tomatoes and at most 100 acres to cherry tomatoes. What is the farmer's maximum profit in this case?

51. Suppose that you inherit \$12,000 but that there are stipulations attached. Some (or all) of the money must be invested in two stocks, A and B. Stock A is low risk and B is medium risk. The stipulations require that at most \$6000 be invested in A and at least \$2000 be invested in B. Furthermore, the amount invested in B is not to exceed twice the amount invested in A. The expected returns on the stocks are \$0.06/dollar for A and \$0.08/dollar for B. Under these conditions, how much should you invest in each stock so as to maximize your expected returns?

52. You have \$10,000 to invest in three stocks, A, B, and C. At least \$1000 must be invested in each stock. The combined investment in A and B must be at least \$5000, and the investment in B cannot exceed five times that in A. The expected returns on the stocks are as follows: \$0.04/dollar for A; \$0.05/dollar for B; \$0.06/dollar for C. Under these conditions, how much should you invest in each stock so as to maximize the expected returns?   *Hint:* Let x and y denote the dollar amounts invested in A and B, respectively. Then $10{,}000 - x - y$ denotes the number of dollars invested in C.

53. In Exercise 52, suppose instead that the expected returns are \$0.05/dollar on A and \$0.04/dollar on B. The return on C remains at \$0.06/dollar. Now how much should you invest in each stock so as to maximize the expected returns?

54. In Exercise 52, suppose instead that the expected return on B is \$0.07/dollar. The expected returns on A and C remain the same. Now how much should you invest to maximize the expected returns?

55. A veterinarian wishes to prepare a supply of dog food by mixing two commercially available foods, A and B. The mixture is to contain at least 50 units of carbohydrates, at least 36 units of protein, and at least 40 units of fat. The specifications for the two commercially available products are listed in the following table.

Product	Units of carbohydrates (per pound)	Units of protein (per pound)	Units of fat (per pound)
A	3	2	2
B	5	4	8

The veterinarian can purchase food A for \$0.44/lb and B for \$0.80/lb. How many pounds of each food should be mixed to meet the dietary requirements at the least cost?

B

56. Graph the following system of inequalities and specify the vertices.

$$\begin{cases} x \ge 0 \\ y \ge e^x \\ y \le e^{-x} + 1 \end{cases}$$

A formula such as

$$f(x, y) = \sqrt{2x - y + 1} \qquad (1)$$

defines a **function of two variables.** *The inputs for such a function are ordered pairs (x, y) of real numbers. For example, using the ordered pair $(3, 5)$ as an input, we have*

$$f(3, 5) = \sqrt{2(3) - 5 + 1} = \sqrt{2}$$

So the input $(3, 5)$ yields an output of $\sqrt{2}$. We define the domain for this function just as we did in Chapter 3: The domain is the set of all inputs that yield real-number outputs. For instance, the ordered pair $(1, 4)$ is not in the domain of the function we have been discussing, because (as you should check for yourself) $f(1, 4) = \sqrt{-1}$, which is not a real number. We can determine the domain of the function in equation (1) by requiring that the quantity under the radical sign be nonnegative. Thus, we require that $2x - y + 1 \ge 0$ and, consequently, $y \le 2x + 1$. (Check this.) The following figure shows the graph of this inequality; the domain of our function is the set of ordered pairs making up the graph. In Exercises 57–62, follow a similar procedure and sketch the domain of the given function.

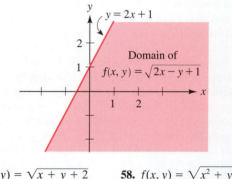

57. $f(x, y) = \sqrt{x + y + 2}$

58. $f(x, y) = \sqrt{x^2 + y^2 - 1}$

59. $g(x, y) = \sqrt{25 - x^2 - y^2}$

60. $g(x, y) = \ln(x^2 - y)$

61. $h(x, y) = \ln(xy)$

62. $h(x, y) = \sqrt{x} + \sqrt{y}$

Chapter 6

Summary of Principal Terms

Terms	Page reference	Comment
1. Linear equation in two variables	463	A linear equation in two variables is an equation of the form $ax + by = c$, where a, b, and c are constants and x and y are variables or unknowns. Similarly, a linear equation in three variables is an equation of the form $ax + by + cz = d$.
2. Solution of a linear equation	463	A solution of the linear equation $ax + by = c$ is an ordered pair of numbers (x_0, y_0) such that $ax_0 + by_0 = c$. Similarly, a solution of the linear equation $ax + by + cz = d$ is an ordered triple of numbers (x_0, y_0, z_0) such that $ax_0 + by_0 + cz_0 = d$.
3. Consistent system; inconsistent system	464	A system of equations is consistent if it has at least one solution; otherwise, the system is inconsistent. See Figures 2, 3, and 4 in Section 6.1 for a geometric interpretation of these terms.
4. Upper-triangular form (three variables)	478	A system of linear equations in x, y, and z is said to be in upper-triangular form if x appears in no equation after the first and y appears in no equation after the second. This definition can be extended to include systems with any number of unknowns. See Table 1 in Section 6.2 for examples of systems that are in upper-triangular form. When a system is in upper-triangular form, it is a simple matter to obtain the solutions; see, for instance, Examples 1 and 2 in Section 6.2.
5. Gaussian elimination	480, 481	This is a technique for converting a system of equations to upper-triangular form. See Examples 4 and 5 in Section 6.2.
6. Elementary operations	481	These are operations that can be performed on an equation in a system without altering the solution set. See the box on page 481 for a list of these operations.
7. Matrix	491	A matrix is a rectangular array of numbers, enclosed in parentheses or brackets. The numbers constituting the rectangular array are called the entries or the elements in the matrix. The size or dimension of a matrix is expressed by specifying the number of rows and the number of columns in that order. For examples of this terminology, see Example 1 in Section 6.3.
8. Matrix equality	495	Two matrices are said to be equal provided they are the same size and their corresponding entries are equal.
9. Matrix addition and subtraction	495	To add two matrices of the same size, add the corresponding entries. Similarly, to subtract two matrices of the same size, subtract the corresponding entries.

Terms	Page reference	Comment
10. Matrix multiplication	498	Let A and B be two matrices. The matrix product AB is defined only when the number of columns in A is the same as the number of rows in B. In this case, the matrix AB will have as many rows as A and as many columns as B. The entry in the ith row and jth column of AB is the number formed as follows: multiply the corresponding entries in the ith row of A and the jth column of B, then add the results.
11. Square matrix	505	A matrix that has the same number of rows and columns is called a square matrix. An $n \times n$ square matrix is said to be an nth-order square matrix.
12. Multiplicative identity matrix	505	For the set of $n \times n$ matrices, the multiplicative identity matrix I_n is the $n \times n$ matrix with ones down the main diagonal (upper left corner to bottom right corner) and zeros everywhere else. For example, $$I_2 = \begin{pmatrix} 1 & 0 \\ 0 & 1 \end{pmatrix} \quad \text{and} \quad I_3 = \begin{pmatrix} 1 & 0 & 0 \\ 0 & 1 & 0 \\ 0 & 0 & 1 \end{pmatrix}$$ For every $n \times n$ matrix A, we have $AI_n = A$ and $I_n A = A$.
13. Inverse matrix	505, 509	Given an $n \times n$ matrix A, the inverse matrix (if it exists) is denoted by A^{-1}. The defining equations for A^{-1} are $AA^{-1} = I$ and $A^{-1}A = I$. Here, I stands for the multiplicative identity matrix that is the same size as A. If a square matrix has an inverse, it is said to be nonsingular; if there is no inverse, then the matrix is singular.
14. Minor	520, 521	The minor of an entry b in a determinant is the determinant formed by suppressing the entries in the row and column in which b appears.
15. Cofactor	520, 521	If an entry b appears in the ith row and the jth column of a determinant, then the cofactor of b is computed by multiplying the minor of b times the number $(-1)^{i+j}$.
16. Determinant	521	The value of the 2×2 determinant $\begin{vmatrix} a & b \\ c & d \end{vmatrix}$ is defined to be $ad - bc$. For larger determinants, the value can be found as follows. Select any row or column and multiply each entry in that row or column by its cofactor, then add the results. The resulting sum is the value of the determinant. It can be shown that this value is independent of the particular row or column that is chosen.

Terms	Page reference	Comment
17. Cramer's rule	526	This rule yields the solutions of certain systems of linear equations in terms of determinants. For a statement of the rule, see page 526. Example 5 on pages 526–527 shows how the rule is applied. The proof of Cramer's rule begins on page 527.
18. Linear inequality in two variables	539	A linear inequality in two variables is any one of the four types of inequalities that result when the equal sign in the equation $ax + by = c$ is replaced by one of the four symbols $<, \leq, >, \geq$. For any of these linear inequalities, a solution is an ordered pair of numbers (x_0, y_0) with the property that a true statement is obtained when x and y (in the inequality) are replaced by x_0 and y_0, respectively.
19. Vertex	542	A vertex of a region in the x-y plane is a corner or point where two adjacent bounding sides meet.
20. Convex region	542	A region in the x-y plane is convex if, given any two points in the region, the line segment joining those points lies wholly within the region.
21. Bounded region	542	A region in the x-y plane is bounded if it can be completely contained within some (sufficiently large) circle.
22. Objective function and constraints	543	In a linear programming problem, the function to be maximized or minimized is called the objective function. The inequalities restricting the variables in the objective function are referred to as the constraints.
23. Set of feasible solutions	543	In a linear programming problem, the set of points obtained by graphing the constraints is called the set of feasible solutions.
24. Optimal solution	543	In a linear programming problem, a feasible solution that yields a largest or smallest value for the objective function is called an optimal solution.
25. Linear programming theorem	544	Consider an objective function $C = ax + by + c$, where the nonnegative variables x and y are subject to constraints specified by nonstrict inequalities. Then if C assumes a maximum (or a minimum) value, this occurs at a vertex. Furthermore, if the set of feasible solutions is both convex and bounded, then both a maximum and a minimum always exist.

Writing Mathematics

On your own or with a group of classmates, complete the following exercise. Then (strictly on your own) write out a detailed solution. This will involve a combination of English composition and algebra (much like the exposition in this textbook). At each stage, be sure to tell the reader (in complete sentences) where you are headed and why each of the main steps is necessary.

Three integers are said to form a **Pythagorean triple** if the square of the largest is equal to the sum of the squares of the other two. For example, 3, 4, and 5 form a Pythagorean triple because $3^2 + 4^2 = 5^2$. Do you know any other Pythagorean triples? This exercise shows how to develop expressions for generating an infinite number of Pythagorean triples.

(a) Refer to the figure. Suppose that the coordinates of the point P are $(a/c, b/c)$, where a, b, and c are integers. Explain why a, b, and c form a Pythagorean triple.

(b) Let m denote the slope of the line passing through P and $(0, -1)$. Using the methods of Section 6.6, show that the coordinates of the point P are

$$\left(\frac{2m}{m^2 + 1}, \frac{m^2 - 1}{m^2 + 1} \right)$$

(c) Use the results in parts (a) and (b) to explain why the three numbers $2m$, $m^2 - 1$, and $m^2 + 1$ form a Pythagorean triple. Then complete the following table to obtain examples of Pythagorean triples.

m	$2m$	$m^2 - 1$	$m^2 + 1$
2			
4			
6			
8			

$x^2 + y^2 = 1$

Chapter 6   Review Exercises

In Exercises 1–31, solve each system of equations. If there are no solutions in a particular case, say so. In cases where there are literal (rather than numerical) coefficients, specify any restrictions that your solutions impose on those coefficients.

1. $\begin{cases} x + y = -2 \\ x - y = 8 \end{cases}$

2. $\begin{cases} x - y = 1 \\ x + y = 5 \end{cases}$

3. $\begin{cases} 2x + y = 2 \\ x + 2y = 7 \end{cases}$

4. $\begin{cases} 3x + 2y = 6 \\ 5x + 4y = 4 \end{cases}$

5. $\begin{cases} 7x + 2y = 9 \\ 4x + 5y = 63 \end{cases}$

6. $\begin{cases} \dfrac{x}{2} + \dfrac{y}{3} = 9 \\ \dfrac{x}{5} - \dfrac{y}{2} = -4 \end{cases}$

7. $\begin{cases} 2x - \dfrac{y}{2} = -8 \\ \dfrac{x}{3} + \dfrac{y}{8} = -1 \end{cases}$

8. $\begin{cases} 3x - 14y - 1 = 0 \\ -6x + 28y - 3 = 0 \end{cases}$

9. $\begin{cases} 3x + 5y - 1 = 0 \\ 9x - 10y - 8 = 0 \end{cases}$ **10.** $\begin{cases} 9x + 15y - 1 = 0 \\ 6x + 10y + 1 = 0 \end{cases}$

11. $\begin{cases} \dfrac{2}{3}x = -\dfrac{1}{2}y - 12 \\ \dfrac{x}{2} = y + 2 \end{cases}$ **12.** $\begin{cases} 0.1x + 0.2y = -5 \\ -0.2x - 0.5y = 13 \end{cases}$

13. $\begin{cases} \dfrac{1}{x} + \dfrac{1}{y} = -1 \\ \dfrac{2}{x} + \dfrac{5}{y} = -14 \end{cases}$ **14.** $\begin{cases} \dfrac{2}{x} + \dfrac{15}{y} = -9 \\ \dfrac{1}{x} + \dfrac{10}{y} = -2 \end{cases}$

15. $\begin{cases} ax + (1 - a)y = 1 \\ (1 - a)x + y = 0 \end{cases}$

16. $\begin{cases} ax - by - 1 = 0 \\ (a - 1)x + by + 2 = 0 \end{cases}$

17. $\begin{cases} 2x - y = 3a^2 - 1 \\ 2y + x = 2 - a^2 \end{cases}$

18. $\begin{cases} 3ax + 2by = 3a^2 - ab + 2b^2 \\ 3bx + 2ay = 2a^2 + 5ab - 3b^2 \end{cases}$

19. $\begin{cases} 5x - y = 4a^2 - 6b^2 \\ 2x + 3y = 5a^2 + b^2 \end{cases}$ **20.** $\begin{cases} \dfrac{2b}{x} - \dfrac{3}{y} = 7ab \\ \dfrac{4a}{x} + \dfrac{5a}{by} = 3a^2 \end{cases}$

21. $\begin{cases} px - qy = q^2 \\ qx + py = p^2 \end{cases}$ **22.** $\begin{cases} x - y = \dfrac{a - b}{a + b} \\ x + y = 1 \end{cases}$

23. $\begin{cases} x + y + z = 9 \\ x - y - z = -5 \\ 2x + y - 2z = -1 \end{cases}$

24. $\begin{cases} x - 4y + 2z = 9 \\ 2x + y + z = 3 \\ 3x - 2y - 3z = -18 \end{cases}$

25. $\begin{cases} 4x - 4y + z = 4 \\ 2x + 3y + 3z = -8 \\ x + y + z = -3 \end{cases}$

26. $\begin{cases} x - 8y + z = 1 \\ 5x + 16y + 3z = 3 \\ 4x - 4y - 4z = -4 \end{cases}$

27. $\begin{cases} 4x + 2y - 3z = 15 \\ 2x + y + 3z = 3 \end{cases}$ **28.** $\begin{cases} 3x + 2y + 17z = 1 \\ x + 2y + 3z = 3 \end{cases}$

29. $\begin{cases} 9x + y + z = 0 \\ -3x + y - z = 0 \\ 3x - 5y + 3z = 0 \end{cases}$

30. $\begin{cases} ax + by - 2az = 4ab + 2b^2 \\ x + y + z = 4a + 2b \\ bx + ay + 4az = 5a^2 + b^2 \end{cases}$

31. $\begin{cases} x + y + z + w = 8 \\ 3x + 3y - z - w = 20 \\ 4x - y - z + 2w = 18 \\ 2x + 5y + 5z - 5w = 8 \end{cases}$

In Exercises 32–36, evaluate each of the determinants.

32. $\begin{vmatrix} \frac{1}{6} & 1 \\ 2 & 12 \end{vmatrix}$ **33.** $\begin{vmatrix} 2 & 6 & 4 \\ 6 & 18 & 24 \\ 15 & 5 & -10 \end{vmatrix}$

34. $\begin{vmatrix} 0 & 2 & 4 & 0 \\ 4 & 0 & 6 & 2 \\ 0 & 0 & 1 & 1 \\ 14 & 7 & 1 & 0 \end{vmatrix}$ **35.** $\begin{vmatrix} 1 & 0 & 0 & 0 \\ 0 & 2 & 0 & 0 \\ 0 & 0 & 3 & 0 \\ 0 & 0 & 0 & 4 \end{vmatrix}$

36. $\begin{vmatrix} 1 & a & b & c \\ 0 & 2 & d & e \\ 0 & 0 & 3 & f \\ 0 & 0 & 0 & 4 \end{vmatrix}$

37. Show that $\begin{vmatrix} a & b & c \\ b & c & a \\ c & a & b \end{vmatrix} = 3abc - a^3 - b^3 - c^3.$

38. Show that $\begin{vmatrix} 1 & 1 & 1 \\ 1 & 1 + x & 1 \\ 1 & 1 & 1 + x^2 \end{vmatrix} = x^3.$

39. Determine constants a and b so that the parabola $y = ax^2 + bx - 1$ passes through the points $(-2, 5)$ and $(2, 9)$.

40. Find two numbers whose sum and difference are 52 and 10, respectively.

41. This exercise appears in *Plane and Solid Analytic Geometry,* by W. F. Osgood and W. G. Graustein (New York: Macmillan, 1920): Let P be any point (a, a) of the line $x - y = 0$, other than the origin. Through P draw two lines, of arbitrary slopes m_1 and m_2, intersecting the x-axis in A_1 and A_2, and the y-axis in B_1 and B_2, respectively. Prove that the lines $\overline{A_1 B_2}$ and $\overline{A_2 B_1}$ will, in general, meet on the line $x + y = 0$.

42. Determine constants h, k, and r so that the circle $(x - h)^2 + (y - k)^2 = r^2$ passes through the three points $(0, 0)$, $(0, 1)$ and $(1, 0)$.

43. The vertices of a triangle are the points of intersection of the lines $y = x - 1$, $y = -x - 2$, and $y = 2x + 3$. Find the equation of the circle passing through these three intersection points.

44. The vertices of triangle ABC are $A(0, 0)$, $B(3, 0)$, and $C(0, 4)$.

(a) Find the center and the radius of the circle that passes though A, B, and C. This circle is called the **circumcircle** for triangle ABC.

(b) The figure shows the **inscribed circle** for triangle ABC; this is the circle that is tangent to all three sides of the triangle. Find the center and the radius of this circle using the following two facts.

(i) The center of the inscribed circle is the common intersection point of the three angle bisectors of the triangle.

(ii) The line that bisects angle B has slope $-1/2$.

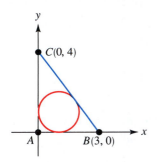

(c) Verify the following statement for triangle ABC. (The statement actually holds for any triangle; the result is known as *Euler's theorem*.) Let R and r denote the radii of the circles in parts (a) and (b), respectively. Then the distance d between the centers of the circles satisfies the equation

$$d^2 = R^2 - 2rR$$

(d) Verify the following statement for triangle ABC (which in fact holds for any triangle): The area of triangle ABC is equal to one-half the product of the perimeter and the radius of the inscribed circle.

(e) Verify the following statement for triangle ABC (which in fact holds for any triangle): The sum of the reciprocals of the lengths of the altitudes is equal to the reciprocal of the radius of the inscribed circle.

In Exercises 45–61, the matrices A, B, C, and D are defined as follows:

$$A = \begin{pmatrix} 3 & -2 \\ 1 & 5 \end{pmatrix} \qquad B = \begin{pmatrix} 2 & 1 \\ 1 & 8 \end{pmatrix}$$

$$C = \begin{pmatrix} -1 & 0 \\ 0 & -1 \end{pmatrix} \qquad D = \begin{pmatrix} 3 & -1 \\ 4 & 1 \\ -5 & 9 \end{pmatrix}$$

In each exercise, carry out the indicated matrix operations if they are defined. If an operation is not defined, say so.

45. $2A + 2B$ **46.** $2(A + B)$ **47.** $4B$

48. $B + 4$ **49.** AB **50.** BA

51. $AB - BA$ **52.** $B + D$ **53.** $B + C$

54. $A(B + C)$ **55.** $AB + AC$ **56.** $(B + C)A$

57. $BA + CA$ **58.** $(A + B) + C$ **59.** $A + (B + C)$

60. $(AB)C$ **61.** $A(BC)$

62. For a square matrix A, the notation A^2 means AA. Similarly, A^3 means AAA. If $A = \begin{pmatrix} 1 & 1 \\ 0 & 1 \end{pmatrix}$, verify that

$$A^2 = \begin{pmatrix} 1 & 2 \\ 0 & 1 \end{pmatrix} \qquad \text{and} \qquad A^3 = \begin{pmatrix} 1 & 3 \\ 0 & 1 \end{pmatrix}$$

63. Let A be the matrix $\begin{pmatrix} 0 & 0 & 0 \\ a & 0 & 0 \\ b & c & 0 \end{pmatrix}$. Compute A^2 and A^3.

For Exercises 64 and 65, in each case compute the inverse of the matrix in part (a), and then use that inverse to solve the system of equations in part (b).

64. (a) $\begin{pmatrix} 5 & -4 \\ 14 & -11 \end{pmatrix}$ **(b)** $\begin{cases} 5x - 4y = 2 \\ 14x - 11y = 5 \end{cases}$

65. (a) $\begin{pmatrix} 1 & -2 & 3 \\ 2 & -5 & 10 \\ -1 & 2 & -2 \end{pmatrix}$

(b) $\begin{cases} x - 2y + 3z = -2 \\ 2x - 5y + 10z = -3 \\ -x + 2y - 2z = 6 \end{cases}$

66. Find the inverse of the matrix $\begin{pmatrix} 1 & -1 & 0 & -3 \\ 5 & -2 & 3 & -11 \\ 2 & 3 & 2 & -3 \\ 4 & 5 & 5 & -5 \end{pmatrix}$.

In Exercises 67–74, compute D, D_x, D_y, D_z (and D_w where appropriate) for each system of equations. Use Cramer's rule to solve the systems in which $D \neq 0$. If $D = 0$, solve the system using Gaussian elimination or matrix methods.

67. $\begin{cases} 2x - y + z = 1 \\ 3x + 2y + 2z = 0 \\ x - 5y - 3z = -2 \end{cases}$ **68.** $\begin{cases} x + 2y - z = -1 \\ 2x - 3y + 3z = 3 \\ 2x + 3y + z = 1 \end{cases}$

69. $\begin{cases} x + 2y + 3z = -1 \\ 4x + 5y + 6z = 2 \\ 7x + 8y + 9z = -3 \end{cases}$ **70.** $\begin{cases} 3x + 2y - 2z = 0 \\ 2x + 3y - z = 0 \\ 8x + 7y - 5z = 0 \end{cases}$

71. $\begin{cases} 3x + 2y - 2z = 1 \\ 2x + 3y - z = -2 \\ 8x + 7y - 5z = 0 \end{cases}$

72. $\begin{cases} x + y + z + w = 5 \\ x - y - z + w = 3 \\ 2x + 3y + 3z + 2w = 21 \\ 4z - 3w = -7 \end{cases}$

73. $\begin{cases} 2x - y + z + 3w = 15 \\ x + 2y + 2w = 12 \\ 3y + 3z + 4w = 12 \\ -4x + y - 4z = -11 \end{cases}$

74. $\begin{cases} x + y + z = (a + b)^2 \\ \dfrac{bx}{a} + \dfrac{ay}{b} - z = 0 \\ x + y - z = (a - b)^2 \end{cases}$

In Exercises 75–86, find all solutions (x, y) for each system, where x and y are real numbers.

75. $\begin{cases} y = 6x \\ y = x^2 \end{cases}$ **76.** $\begin{cases} y = 4x \\ y = x^3 \end{cases}$

77. $\begin{cases} y = 9 - x^2 \\ y = x^2 - 9 \end{cases}$ **78.** $\begin{cases} x^3 - y = 0 \\ xy - 16 = 0 \end{cases}$

79. $\begin{cases} x^2 - y^2 = 9 \\ x^2 + y^2 = 16 \end{cases}$ **80.** $\begin{cases} 2x + 3y = 6 \\ y = \sqrt{x + 1} \end{cases}$

81. $\begin{cases} x^2 + y^2 = 1 \\ y = \sqrt{x} \end{cases}$ **82.** $\begin{cases} \dfrac{x}{11} + \dfrac{y}{12} = 2 \\ \dfrac{xy}{132} = 1 \end{cases}$

83. $\begin{cases} x^2 + 2xy + 3y^2 = 68 \\ 3x^2 - xy + y^2 = 18 \end{cases}$

84. $\begin{cases} \dfrac{x^2}{a^2} + \dfrac{y^2}{b^2} = 1 \\ \dfrac{x^2}{b^2} + \dfrac{y^2}{a^2} = 1 \end{cases}$ $(a > b > 0)$

85. $\begin{cases} 2(x - 3)^2 - (y + 1)^2 = -1 \\ -3(x - 3)^2 + 2(y + 1)^2 = 6 \end{cases}$

Hint: Let $u = x - 3$ and $v = y + 1$.

86. $\begin{cases} x^4 = y - 1 \\ y - 3x^2 + 1 = 0 \end{cases}$

*Exercises 87–92 appear (in German) in an algebra text by Leonhard Euler, first published in 1770. The English versions given here are taken from the translated version, Elements of Algebra, 5th ed., by Leonhard Euler (Lon-*don: Longman, Orme, and Co., 1840). [*This, in turn, has been reprinted by Springer-Verlag (New York, 1984).*]

87. Required two numbers, whose sum may be s, and their proportion as a to b.

Answer: $\dfrac{as}{a + b}$ and $\dfrac{bs}{a + b}$

88. The sum $2a$, and the sum of the squares $2b$, of two numbers being given; to find the numbers.

Answer: $a - \sqrt{b - a^2}$ and $a + \sqrt{b - a^2}$

89. To find three numbers, so that [the sum of] one-half of the first, one-third of the second, and one-quarter of the third, shall be equal to 62; one-third of the first, one-quarter of the second, and one-fifth of the third, equal to 47; and one-quarter of the first, one-fifth of the second, and one-sixth of the third, equal to 38.

Answer: 24, 60, 120

90. Required two numbers, whose product may be 105, and whose squares [when added] may together make 274.

91. Required two numbers, whose product may be m, and the sum of the squares n $(n \geq 2m)$.

92. Required two numbers such that their sum, their product, and the difference of their squares may all be equal.

In Exercises 93–96, graph each system of inequalities and specify whether the region is convex or bounded.

93. $\begin{cases} x^2 + y^2 \geq 1 \\ y - 4x \leq 0 \\ y - x \geq 0 \\ x \geq 0 \\ y \geq 0 \end{cases}$ **94.** $\begin{cases} x^2 + y^2 \leq 1 \\ x \geq 0 \\ y \geq 0 \end{cases}$

95. $\begin{cases} y - \sqrt{x} \leq 0 \\ y \geq 0 \\ x \geq 1 \\ x - 4 \leq 0 \end{cases}$ **96.** $\begin{cases} y - |x| \leq 0 \\ x + 1 \geq 0 \\ x - 1 \leq 0 \\ y + 1 \geq 0 \end{cases}$

In Exercises 97 and 98, find the maximum and the minimum values of the objective function subject to the given constraints. Specify the optimal solution in each case.

97.

Objective function: $C = 7x + 6y + 6$

Constraints: $2y - 5x \leq 10$, $3y - x \leq 28$

$2x + y \leq 32$, $y + 5x \leq 70$

$x \geq 0$, $y \geq 0$

98.

Objective function: $C = 7y - 5x + 80$

Constraints: $3x + y \geq 13$, $y - 2x \leq 3$

$7x - 4y \leq 5$, $2y - x \leq 15$

Chapter 6 ▮ Test 🌐

1. Determine all solutions of the system

$$\begin{cases} 3x + 4y = 12 \\ y = x^2 + 2x + 3 \end{cases}$$

2. Find all solutions of the system

$$\begin{cases} x - 2y = 13 \\ 3x + 5y = -16 \end{cases}$$

3. (a) Find all solutions of the following system using Gaussian elimination:

$$\begin{cases} x + 4y - z = 0 \\ 3x + y + z = -1 \\ 4x - 4y + 5z = -7 \end{cases}$$

(b) Compute D, D_x, D_y, and D_z for the system in part (a). Then check your answer in part (a) using Cramer's rule.

4. Suppose that the matrices A and B are defined as follows:

$$A = \begin{pmatrix} 1 & -3 \\ 2 & -1 \end{pmatrix} \qquad B = \begin{pmatrix} 0 & 4 \\ 1 & 3 \end{pmatrix}$$

(a) Compute $2A - B$.   **(b)** Compute BA.

5. Assume that the supply and the demand functions for a commodity are given by $q = 24p + 180$ and $q = -30p + 1314$, respectively, where p is in dollars. Find the equilibrium price and the corresponding equilibrium quantity.

6. Find all solutions of the system

$$\begin{cases} \dfrac{1}{2x} + \dfrac{1}{3y} = 10 \\ -\dfrac{5}{x} - \dfrac{4}{y} = -4 \end{cases}$$

7. Specify the coefficient matrix for the system

$$\begin{cases} x + y - z = -1 \\ 2x - y + 2z = 11 \\ x - 2y + z = 10 \end{cases}$$

Also specify the augmented matrix for this system.

8. Use matrix methods to find all solutions of the system displayed in the previous problem.

9. A student is taking four courses, Math, English, Chemistry, and Economics, and spends 40 hours per week studying. The following facts are known about how the student allocates study time each week. The combined amount of study time for English and Chemistry is the same as that for Math and Economics. The amount of time spent on Math is one-quarter of the amount spent on the other three courses combined. Finally, the combined amount of time for English and Economics is two-thirds as much as that for Math and Chemistry.

(a) Let w, x, y, and z denote the number of hours per week that the student studies Math, English, Chemistry, and Economics, respectively. Specify a system of four linear equations involving these four unknowns. Write the system in standard form.

(b) Write the coefficient matrix of the system in part (a).

Ⓖ **(c)** Use a graphing utility to find the inverse of the coefficient matrix. Use matrix methods to solve the system and thereby determine how many hours per week the student studies each subject.

10. Consider the determinant

$$\begin{vmatrix} 2 & 3 & -1 \\ 0 & 1 & 4 \\ 5 & -2 & 6 \end{vmatrix}$$

(a) What is the minor of the entry in the third row, second column?

(b) What is the cofactor of the entry in the third row, second column?

11. Evaluate the determinant

$$\begin{vmatrix} 4 & -5 & 0 \\ -8 & 10 & 7 \\ 16 & 20 & 14 \end{vmatrix}$$

12. Find all solutions of the system

$$\begin{cases} x^2 + y^2 = 15 \\ xy = 5 \end{cases}$$

13. Find the solutions of the system

$$\begin{cases} A + 2B + 3C = 1 \\ 2A - B - C = 2 \end{cases}$$

14. (a) Determine the inverse of the following matrix:

$$\begin{pmatrix} 10 & -2 & 5 \\ 6 & -1 & 4 \\ 1 & 0 & 1 \end{pmatrix}$$

(b) Use the inverse matrix determined in part (a) to solve the following system:

$$\begin{cases} 10u - 2v + 5w = -1 \\ 6u - v + 4w = -2 \\ u + w = 3 \end{cases}$$

15. Graph the inequality $5x - 6y \geq 30$.

16. Determine constants P and Q so that the parabola $y = Px^2 + Qx - 5$ passes through the two points $(-2, -1)$ and $(-1, -2)$.

17. Graph the inequality $(x - 2)^2 + y^2 > 1$. Is the solution set bounded? Is it convex?

18. Graph the following system of inequalities and specify the vertices:

$$\begin{cases} x \geq 0 \\ y \geq 0 \\ 2y - x \leq 14 \\ x + 3y \leq 36 \\ 9x + y \leq 99 \end{cases}$$

19. A factory assembles two types of motors, small and large. The cost of material is $15 for a small motor and $30 for a large motor. Two hours of labor are needed to assemble a small motor and 6 hours for a large motor. The cost of labor is $8/hr. The owner of the factory can allocate at most $1500/day for material and $2000/day for labor. Find the largest daily profit that the owner can make if the small motors are sold for $40 each and the large motors for $104 each. Assume that every motor produced is sold. *Hint:* Profit equals revenue minus cost.

20. Given that the inverse of the matrix

$$\begin{pmatrix} 5 & 0 & 2 \\ 2 & 2 & 1 \\ -3 & 1 & -1 \end{pmatrix} \text{ is the matrix } \begin{pmatrix} -3 & 2 & -4 \\ -1 & 1 & -1 \\ 8 & -5 & 10 \end{pmatrix},$$

solve the following system of equations:

$$\begin{cases} 5 \ln x + 2 \ln z = 3 \\ 2 \ln x + 2 \ln y + \ln z = -1 \\ -3 \ln x + \ln y - \ln z = 2 \end{cases}$$

Roots of Polynomial Equations

It is necessary that I make some general statements concerning the nature of equations. —René Descartes (1596–1650) in *La Géométrie* (1637)

The polynomial can be derived using Hückel's molecular orbital theory. The roots of the polynomial represent the allowed energy levels of the pi electrons. —Professor Jun-ichi Aihara in his article "Why Aromatic Compounds Are Stable," *Scientific American,* vol. 266 (March 1992), p. 65

A number of problems in computer graphics reduce to finding approximate real roots of quartic and cubic equations in one unknown. —Professor Don Herbison-Evans in his article "Solving Quartics and Cubics for Graphics," http://linus.socs.uts.edu.au/~don/pubs/solving.html

As indicated by the opening quotations, solving polynomial equations is an old subject with contemporary applications at the cutting edge of technology and science. This chapter continues the work we began in Sections 1.3 and 2.1 on solving polynomial equations. Here are some of the questions you should be able to answer after studying this chapter. See if you can answer any of them now, on the basis of your previous algebra courses. Come back to this list again when you have finished the chapter, as a way of evaluating your reading.

- What is synthetic division and how is it used to evaluate a polynomial and test for a root?
- In view of the quadratic formula, we know that every quadratic equation has two roots (although the two roots might be equal). In a similar way, does every cubic equation have three roots, and does every fourth-degree equation have four roots?
- Does every equation have a root?
- What does the fundamental theorem of algebra assert?
- The quadratic formula gives us the roots of a quadratic equation in terms of the coefficients. Is there a similar formula for cubic equations? What about a general formula for a polynomial equation of degree n?
- How can we determine whether an equation has rational roots?
- How can the table feature on a graphing utility be used to locate irrational roots? How can the ideas about iteration from Section 3.5 be used to locate irrational roots?
- Without a graphing utility, and without solving the equation, can we tell whether the roots will be positive, negative, or complex?
- Is it always possible to factor a polynomial?

7.1   THE COMPLEX NUMBER SYSTEM

In the following I shall denote the expression $\sqrt{-1}$ by the letter i so that ii = −1.—Leonhard Euler in a paper presented to the Saint Petersburg Academy in 1777.

In elementary algebra, complex numbers appear as expressions a + bi, where a and b are ordinary real numbers and $i^2 = -1$. . . . Complex numbers are manipulated by the usual rules of algebra, with the convention that i^2 is to be replaced by −1 whenever it occurs.—Ralph Boas in *Invitation to Complex Analysis* (New York: Random House, 1987)

The preceding quotation from Professor Boas summarizes the basic approach we will follow in this section. Near the end of the section, after you've become accustomed to working with complex numbers, we'll present a formal list of some of the basic definitions and properties that can be used to develop the subject more rigorously, as is required in more advanced courses.

When we solve equations in this chapter, you'll see instances in which the real-number system proves to be inadequate. In particular, since the square of a real number is never negative, there is no real number x such that $x^2 = -1$. To overcome this inconvenience, mathematicians define the symbol i by the equation

$$i^2 = -1$$

For reasons that are more historical than mathematical, i is referred to as the **imaginary unit.** This name is unfortunate in a sense, because to an engineer or a mathematician, i is neither less "real" nor less tangible than any real number. Having said this, however, we do have to admit that i does not belong to the real-number system.

Algebraically, we operate with the symbol i as if it were any letter in a polynomial expression. However, when we see i^2, we must remember to replace it by −1. Here are four sample calculations involving i.

1. $3i + 2i = 5i$
2. $-2i^2 + 6i = -2(-1) + 6i = 2 + 6i$
3. $(-i)^2 = i^2 = -1$
4. $0i = 0$

An expression of the form $a + bi$, where a and b are real numbers, is called a **complex number.*** Thus, four examples of complex numbers are:

$$2 + 3i$$
$$4 + (-5)i \qquad \text{(usually written } 4 - 5i)$$
$$1 - \sqrt{2}\,i \qquad \text{(also written } 1 - i\sqrt{2})$$
$$\frac{1}{2} + \frac{3}{2}i \qquad \left(\text{also written } \frac{1 + 3i}{2}\right)$$

Given a complex number $a + bi$, we say that a is the **real part** of $a + bi$, and b is the **imaginary part** of $a + bi$. For example, the real part of $3 - 4i$ is 3, and the imaginary part is −4. Observe that both the real part and the imaginary part of a complex number are themselves real numbers.

----

*The term *complex number* is attributed to Carl Friedrich Gauss (1777–1855). The term *imaginary number* originated with René Descartes (1596–1650).

We define the notion of *equality* for complex numbers in terms of their real and imaginary parts. Two complex numbers are said to be **equal** if their corresponding real and imaginary parts are equal. We can write this definition symbolically as follows:

$$a + bi = c + di \qquad \text{if and only if} \qquad a = c \qquad \text{and} \qquad b = d$$

EXAMPLE 1 Equality of complex numbers

Determine the real numbers c and d such that $10 + 4i = 2c + di$.

SOLUTION
Equating the real parts of the two complex numbers gives us $2c = 10$, and therefore $c = 5$. Similarly, equating the imaginary parts yields $d = 4$. These are the required values for c and d.

As the next example indicates, addition, subtraction, and multiplication of complex numbers are carried out using the usual rules of algebra, with the understanding (as mentioned before) that i^2 is always to be replaced with -1. (We'll discuss division of complex numbers subsequently.)

EXAMPLE 2 Operations with complex numbers

Let $z = 2 + 5i$ and $w = 3 - 4i$. Compute each of the following:
(a) $w + z$; **(b)** $3z$; **(c)** $w - 3z$; **(d)** zw; **(e)** wz.

SOLUTION
(a) $w + z = (3 - 4i) + (2 + 5i) = (3 + 2) + (-4i + 5i) = 5 + i$
(b) $3z = 3(2 + 5i) = 6 + 15i$
(c) $w - 3z = (3 - 4i) - (6 + 15i) = (3 - 6) + (-4i - 15i) = -3 - 19i$
(d) $zw = (2 + 5i)(3 - 4i) = 6 - 8i + 15i - 20i^2$
$$= 6 + 7i - 20(-1) = 26 + 7i$$
(e) $wz = (3 - 4i)(2 + 5i) = 6 + 15i - 8i - 20i^2$
$$= 6 + 7i - 20(-1) = 26 + 7i$$

If you look over the result of part (a) in Example 2, you can see that the sum is obtained simply by adding the corresponding real and imaginary parts of the given numbers. Likewise in part (c), the difference is obtained by subtracting the corresponding real and imaginary parts. In part (d), however, notice that the product is *not* obtained in a similar fashion. (Exercise 68 provides some perspective on this.) Finally, notice that the results in parts (d) and (e) are identical. In fact, it can be shown that for any two complex numbers z and w, we always have $zw = wz$. In other words, just as with real numbers, multiplication of complex numbers is *commutative*. Furthermore, along the same lines, it can be shown that all of the properties of real numbers listed in Appendix A.2 continue to hold for complex numbers. We'll return to this point at the end of this section and in the exercises.

As background for the discussion of division of complex numbers, we introduce the notion of a *complex conjugate,* or simply a *conjugate.**

DEFINITION | **Complex Conjugate**

Let $z = a + bi$. The **complex conjugate** of z, denoted by $\overline{z}$, is defined by

$$\overline{z} = a - bi$$

EXAMPLES
If $z = 3 + 4i$, then $\overline{z} = 3 - 4i$.
If $w = 9 - 2i$, then $\overline{w} = 9 + 2i$.

EXAMPLE 3   Computations involving conjugates

(a) If $z = 6 - 3i$, compute $z\overline{z}$.   **(b)** If $w = a + bi$, compute $w\overline{w}$.

SOLUTION

(a) $z\overline{z} = (6 - 3i)(6 + 3i)$
$\quad\quad = 36 + 18i - 18i - 9i^2$
$\quad\quad = 36 + 9 = 45$

(b) $w\overline{w} = (a + bi)(a - bi)$
$\quad\quad = a^2 - abi + abi - b^2i^2$
$\quad\quad = a^2 + b^2$

NOTE   The result in part (b) shows that the product of a complex number and its conjugate is always a real number.

Quotients of complex numbers are easy to compute using conjugates. Suppose, for example, that we wish to compute the quotient

$$\frac{5 - 2i}{3 + 4i}$$

To do this, we take the conjugate of the denominator—namely, $3 - 4i$—and then multiply the given fraction by $\dfrac{3 - 4i}{3 - 4i}$, which equals 1. This yields

$$\frac{5 - 2i}{3 + 4i} = \frac{5 - 2i}{3 + 4i} \cdot \frac{3 - 4i}{3 - 4i}$$

$$= \frac{15 - 26i + 8i^2}{9 - 16i^2}$$

$$= \frac{7 - 26i}{25} \quad\quad \text{since } 8i^2 = -8 \text{ and } -16i^2 = 16$$

$$= \frac{7}{25} - \frac{26}{25}i \quad\quad \text{as required}$$

*The term "conjugates" *(conjuguées)* was introduced by the nineteenth-century French mathematician A. L. Cauchy in his text *Cours d'Analyse Algébrique* (Paris: 1821).

In the box that follows, we summarize our procedure for computing quotients. The condition $w \neq 0$ means that w is any complex number other than $0 + 0i$.

Procedure for Computing Quotients

Let z and w be two complex numbers, $w \neq 0$. Then z/w is computed as follows:

$$\frac{z}{w} = \frac{z}{w} \cdot \frac{\overline{w}}{\overline{w}}$$

EXAMPLE	**4**	**Using conjugates to carry out division**

Let $z = 3 + 4i$ and $w = 1 - 2i$. Compute each of the following quotients:

(a) $\dfrac{1}{z}$;   **(b)** $\dfrac{z}{w}$.

SOLUTION

(a) $\dfrac{1}{z} = \dfrac{1}{z} \cdot \dfrac{\overline{z}}{\overline{z}} = \dfrac{1}{3 + 4i} \cdot \dfrac{3 - 4i}{3 - 4i}$

$\qquad = \dfrac{3 - 4i}{9 - 16i^2} = \dfrac{3 - 4i}{25}$

$\qquad = \dfrac{3}{25} - \dfrac{4}{25}i$

(b) $\dfrac{z}{w} = \dfrac{z}{w} \cdot \dfrac{\overline{w}}{\overline{w}} = \dfrac{3 + 4i}{1 - 2i} \cdot \dfrac{1 + 2i}{1 + 2i}$

$\qquad = \dfrac{3 + 10i + 8i^2}{1 - 4i^2} = \dfrac{-5 + 10i}{5}$

$\qquad = -1 + 2i$

We began this section by defining i with the equation $i^2 = -1$. This can be rewritten

$$\boxed{i = \sqrt{-1}}$$

provided that we agree to certain conventions regarding principal square roots and negative numbers. In dealing with the principal square root of a negative real number—say, $\sqrt{-5}$—we shall write

$$\sqrt{-5} = \sqrt{(-1)(5)} = \sqrt{-1}\sqrt{5} = i\sqrt{5}$$

In other words, we are allowing the use of the rule $\sqrt{ab} = \sqrt{a}\sqrt{b}$ when a is -1 and b is a positive real number. However, the rule $\sqrt{ab} = \sqrt{a}\sqrt{b}$ *cannot* be used when both a and b are negative. If that were allowed, we could write

$$1 = (-1)(-1)$$

and then $\sqrt{1} = \sqrt{(-1)(-1)} = \sqrt{-1}\sqrt{-1} = (i)(i)$

Consequently, $1 = i^2$ and therefore $1 = -1$

which is a contradiction. Again, the point here is that the rule $\sqrt{ab} = \sqrt{a}\sqrt{b}$ cannot be applied when both a and b are negative.

EXAMPLE 5 **Calculations involving the imaginary unit i**

Simplify: **(a)** i^4; **(b)** i^{101}; **(c)** $\sqrt{-12} + \sqrt{-27}$; **(d)** $\sqrt{-9}\sqrt{-4}$.

SOLUTION

(a) We make use of the defining equation for i, which is $i^2 = -1$. Thus we have

$$i^4 = (i^2)^2 = (-1)^2 = 1$$

(The result, $i^4 = 1$, is worth remembering.)

(b) $i^{101} = i^{100}i = (i^4)^{25}i = 1^{25}i = i$

(c) $\sqrt{-12} + \sqrt{-27} = \sqrt{12}\sqrt{-1} + \sqrt{27}\sqrt{-1}$
$$= \sqrt{4}\sqrt{3}i + \sqrt{9}\sqrt{3}i$$
$$= 2\sqrt{3}i + 3\sqrt{3}i = 5\sqrt{3}i$$

(d) $\sqrt{-9}\sqrt{-4} = (3i)(2i) = 6i^2 = -6$

NOTE $\sqrt{-9}\sqrt{-4} \neq \sqrt{36}$; why?

We have developed the complex-number system in this section for the same reason that complex numbers were developed historically: they're needed to solve polynomial equations. Consider, for example, the quadratic equation

$$ax^2 + bx + c = 0$$

where a, b, and c are real numbers and $a \neq 0$. In Section 2.1, we noted that if the discriminant $b^2 - 4ac$ is negative, then the equation has no real roots. As the next example indicates, such equations do have two complex-number roots. Furthermore (assuming the coefficients a, b, and c in the equation are real numbers) these roots always turn out to be complex conjugates.

EXAMPLE 6 **A quadratic equation with complex roots**

Solve the quadratic equation $x^2 - 4x + 6 = 0$.

SOLUTION

We use the quadratic formula with $a = 1$, $b = -4$, and $c = 6$:

$$x = \frac{-b \pm \sqrt{b^2 - 4ac}}{2a} = \frac{4 \pm \sqrt{16 - 4(1)(6)}}{2(1)}$$
$$= \frac{4 \pm \sqrt{-8}}{2} = \frac{4 \pm (2\sqrt{2})i}{2} \qquad \text{Check the algebra.}$$
$$= 2 \pm \sqrt{2}i$$

In summary, the two roots of the quadratic equation are the complex numbers $2 + \sqrt{2}i$ and $2 - \sqrt{2}i$. These numbers are complex conjugates.

NOTE When you write answers such as these on paper, be sure you make it clear that the symbol i is outside (not inside) the radical sign. To emphasize this distinction, sometimes it's helpful to write the roots in the form $2 + i\sqrt{2}$ and $2 - i\sqrt{2}$.

Near the beginning of this section, we mentioned that we would eventually present a formal list of some of the basic definitions and properties that can be used to develop the subject more rigorously. These are given in the two boxes that follow. As you'll see in some of the exercises, these definitions and properties are indeed consistent with our work in this section and with the properties of real numbers listed in Appendix A.2.

DEFINITION | **Addition, Subtraction, Multiplication, and Division for Complex Numbers**

Let $z = a + bi$ and $w = c + di$. Then $z + w$, $z - w$, and zw are defined as follows:

1. $z + w = (a + bi) + (c + di) = (a + c) + (b + d)i$
2. $z - w = (a + bi) - (c + di) = (a - c) + (b - d)i$
3. $zw = (a + bi)(c + di) = (ac - bd) + (ad + bc)i$

Furthermore, if $w \neq 0$, then z/w is defined as follows:

4. $\dfrac{z}{w} = \dfrac{a + bi}{c + di} \cdot \dfrac{c - di}{c - di} = \left(\dfrac{ac + bd}{c^2 + d^2} \right) + \left(\dfrac{bc - ad}{c^2 + d^2} \right)i$

PROPERTY SUMMARY **Properties of Complex Conjugates**

If $z = a + bi$, the complex conjugate, $\bar{z} = a - bi$, has the following properties.

1. $\bar{\bar{z}} = z$
2. $z = \bar{z}$ if and only if z is a real number
3. $\overline{z + w} = \bar{z} + \bar{w}$; $\overline{z - w} = \bar{z} - \bar{w}$
4. $\overline{zw} = \bar{z}\,\bar{w}$; $\dfrac{\bar{z}}{\bar{w}} = \overline{\left(\dfrac{z}{w} \right)}$
5. $(\bar{z})^n = \overline{z^n}$ for each natural number n

EXERCISE SET 7.1

A

1. Complete the table.

i^2	i^3	i^4	i^5	i^6	i^7	i^8
-1						

2. Simplify the following expression, and write the answer in the form $a + bi$.

$$1 + 3i - 5i^2 + 4 - 2i - i^3$$

For Exercises 3 and 4, specify the real and imaginary parts of each complex number.

3. (a) $4 + 5i$
   (b) $4 - 5i$
   (c) $\frac{1}{2} - i$
   (d) $16i$

4. (a) $-2 + \sqrt{7}i$
   (b) $1 + 5^{1/3}i$
   (c) $-3i$
   (d) 0

5. Determine the real numbers c and d such that

$$8 - 3i = 2c + di$$

6. Determine the real numbers a and b such that

$$27 - 64i = a^3 - b^3i$$

7. Simplify each of the following.
   (a) $(5 - 6i) + (9 + 2i)$ (b) $(5 - 6i) - (9 + 2i)$

8. If $z = 1 + 4i$, compute $z - 10i$.

9. Compute each of the following.
   (a) $(3 - 4i)(5 + i)$ (c) $\dfrac{3 - 4i}{5 + i}$
   (b) $(5 + i)(3 - 4i)$ (d) $\dfrac{5 + i}{3 - 4i}$

10. Compute each of the following.
   (a) $(2 + 7i)(2 - 7i)$ (b) $\dfrac{-1 + 3i}{2 + 7i}$

(c) $\dfrac{1}{2 + 7i}$

(d) $\dfrac{1}{2 + 7i} \cdot (-1 + 3i)$

In Exercises 11–36, evaluate each expression using the values
$z = 2 + 3i$, $w = 9 - 4i$, *and* $w_1 = -7 - i$.

11. (a) $z + w$

   (b) $\bar{z} + w$

   (c) $z + \bar{z}$

12. (a) $\bar{z} + \bar{w}$

   (b) $\overline{(z + w)}$

   (c) $w - \bar{w}$

13. $(z + w) + w_1$
14. $z + (w + w_1)$
15. zw

16. wz
17. $z\bar{z}$
18. $w\bar{w}$

19. $z(ww_1)$
20. $(zw)w_1$
21. $z(w + w_1)$

22. $zw + zw_1$
23. $z^2 - w^2$
24. $(z - w)(z + w)$

25. $(zw)^2$
26. $z^2 w^2$
27. z^3

28. z^4
29. z/w
30. w/z

31. $\bar{z}/\bar{w}$
32. $\overline{(z/w)}$
33. $z/\bar{z}$

34. $\bar{z}/z$
35. $(w - \bar{w})/(2i)$
36. $(w + \bar{w})/2$

For Exercises 37–40, compute each quotient.

37. $\dfrac{i}{5 + i}$
38. $\dfrac{1 - i\sqrt{3}}{1 + i\sqrt{3}}$
39. $\dfrac{1}{i}$
40. $\dfrac{i + i^2}{i^3 + i^4}$

In Exercises 41–44, use the fact that $i^4 = 1$ to simplify each expression [as in Example 5(b)].

41. i^{17}
42. i^{36}
43. i^{26}
44. i^{83}

In Exercises 45–52, simplify each expression.

45. $\sqrt{-49} + \sqrt{-9} + \sqrt{-4}$

46. $\sqrt{-25} + i$

47. $\sqrt{-20} - 3\sqrt{-45} + \sqrt{-80}$

48. $\sqrt{-4}\sqrt{-4}$

49. $1 + \sqrt{-36}\sqrt{-36}$
50. $i - \sqrt{-100}$

51. $3\sqrt{-128} - 4\sqrt{-18}$
52. $64 + \sqrt{-64}\sqrt{-64}$

In Exercises 53–60:

(a) *Compute the discriminant of the quadratic and note that it is negative (and therefore the equation has no real-number roots);*

(b) *Use the quadratic formula to obtain the two complex-conjugate roots of each equation.*

53. $x^2 - x + 1 = 0$
54. $x^2 - 6x + 12 = 0$

55. $5z^2 + 2z + 2 = 0$
56. $-10z^2 + 4z - 2 = 0$

57. $2z^2 + 3z + 4 = 0$
58. $3z^2 - 7z + 5 = 0$

59. $\frac{1}{6}z^2 - \frac{1}{4}z + 1 = 0$
60. $\frac{1}{2}z^2 + 2z + \frac{9}{4} = 0$

61. (a) Evaluate the expression $x^2 - 4x + 6$ when $x = 2 - i\sqrt{2}$.

   (b) How does your result in part (a) relate to Example 6 in this section?

62. (a) Find the roots of the quadratic equation $x^2 - 2x + 5 = 0$.

   (b) Compute the product of the two roots that you obtained in part (a).

   (c) Check your answer in part (b) by applying the theorem in the box on page 89.

63. Let $z = a + bi$ and $w = c + di$. Compute each of the following quantities, and then check that your results agree with the definitions in the first box on page 566.

   (a) $z + w$
   (b) $z - w$
   (c) zw
   (d) $\dfrac{z}{w}$

B

64. Show that $\left(\dfrac{-1 + i\sqrt{3}}{2}\right)^2 + \left(\dfrac{-1 - i\sqrt{3}}{2}\right)^2 = -1$.

65. Let $z = \dfrac{-1 + i\sqrt{3}}{2}$ and $w = \dfrac{-1 - i\sqrt{3}}{2}$. Verify the following statements.

   (a) $z^3 = 1$ and $w^3 = 1$

   (b) $zw = 1$

   (c) $z = w^2$ and $w = z^2$

   (d) $(1 - z + z^2)(1 + z - z^2) = 4$

66. Let $z = a + bi$ and $w = c + di$.

   (a) Show that $\bar{\bar{z}} = z$.

   (b) Show that $\overline{(z + w)} = \bar{z} + \bar{w}$.

67. Show that the complex number $0 \; (=0 + 0i)$ has the following properties.

   (a) $0 + z = z$ and $z + 0 = z$, for all complex numbers z.
     Hint: Let $z = a + bi$.

   (b) $0 \cdot z = 0$ and $z \cdot 0 = 0$, for all complex numbers z.

68. This exercise indicates one of the reasons why multiplication of complex numbers is not carried out simply by multiplying the corresponding real and imaginary parts of the numbers. (Recall that addition and subtraction *are* carried out in this manner.) Suppose for the moment that we were to define multiplication in this seemingly less complicated way:

$$(a + bi)(c + di) = ac + (bd)i \qquad (*)$$

   (a) Compute $(2 + 3i)(5 + 4i)$, assuming that multiplication is defined by $(*)$.

   (b) Still assuming that multiplication is defined by $(*)$, find two complex numbers z and w such that $z \neq 0$, $w \neq 0$, but $zw = 0$ (where 0 denotes the complex number $0 + 0i$).

Now notice that the result in part (b) is contrary to our expectation or desire that the product of two nonzero numbers be nonzero, as is the case for real numbers. On the other hand, it can be shown that when multiplication is carried out as described in the text, then the product of two complex numbers is nonzero if and only if both factors are nonzero.

69. (a) Show that addition of complex numbers is commutative. That is, show that $z + w = w + z$ for all complex numbers z and w. *Hint*: Let $z = a + bi$ and $w = c + di$.

   (b) Show that multiplication of complex numbers is commutative. That is, show that $zw = wz$ for all complex numbers z and w.

70. Let $z = a + bi$.

  (a) Show that $(\bar{z})^2 = \overline{z^2}$.   **(b)** Show that $(\bar{z})^3 = \overline{z^3}$.

In Exercises 71–74, find all roots of each equation.   *Hints:* First, factor by grouping. In Exercises 71 and 72, each equation has three roots; in Exercise 73, the equation has six roots; in Exercise 74, there are five roots.

71. $x^3 - 3x^2 + 4x - 12 = 0$ **72.** $2x^3 + 4x^2 + 3x + 6 = 0$

73. $x^6 - 9x^4 + 16x^2 - 144 = 0$ **74.** $x^5 + 4x^3 + 8x^2 + 32 = 0$

C

75. Let a and b be real numbers. Find the real and imaginary parts of the quantity

$$\frac{a + bi}{a - bi} + \frac{a - bi}{a + bi}$$

76. Find the real and imaginary parts of the quantity

$$\left(\frac{a + bi}{a - bi}\right)^2 - \left(\frac{a - bi}{a + bi}\right)^2$$

77. Find the real part of $\dfrac{(a + bi)^2}{a - bi} - \dfrac{(a - bi)^2}{a + bi}$.

78. (a) Let $\alpha = \left(\dfrac{\sqrt{a^2 + b^2} + a}{2}\right)^{1/2}$

  and let $\beta = \left(\dfrac{\sqrt{a^2 + b^2} - a}{2}\right)^{1/2}$

  Show that the square of the complex number $\alpha + \beta i$ is $a + bi$.

  (b) Use the result in part (a) to find a complex number z such that $z^2 = i$.

  (c) Use the result in part (a) to find a complex number z such that $z^2 = -7 + 24i$.

One of the basic properties of real numbers is that if a product is equal to zero, then at least one of the factors is zero. Exercises 79 and 80 show that this property also holds for complex numbers. For Exercises 79 and 80, assume that $zw = 0$, where $z = a + bi$ and $w = c + di$.

79. If $a \neq 0$, prove that $w = 0$. (That is, prove that $c = d = 0$.)

80. If $b \neq 0$, prove that $w = 0$. (That is, prove that $c = d = 0$.)

MINI PROJECT | A GEOMETRIC INTERPRETATION OF COMPLEX ROOTS

As you know, if a quadratic equation $f(x) = x^2 + Bx + C = 0$ has two real roots, there is a geometric interpretation: the roots are the x-intercepts for the graph of f. Less well known is the fact that there is a geometric interpretation when the quadratic equation has nonreal complex roots. If these roots are $a \pm bi$, where a and b are real numbers, then the coordinates of the vertex of the graph of f are (a, b^2).

(a) Check that this is true in the case of the quadratic equation
  $f(x) = x^2 + 6x + 25 = 0$.

(b) Show that the result holds in general. That is, assume the roots of
  $f(x) = x^2 + Bx + C = 0$ are $a \pm bi$, where a and b are real numbers with
  $b \neq 0$. Show that the vertex of the graph of f is (a, b^2).   *Hint:* Use the theorem in the box on page 89 to express B and C in terms of a and b.

7.2 DIVISION OF POLYNOMIALS

Although the process of long division for polynomials is often taught in elementary algebra courses, it usually does not receive sufficient emphasis there. As with ordinary long division for numbers, the process is best learned by first watching someone do examples and then practicing on your own. The terms *quotient, remainder, divisor,* and *dividend* will be used here in the same way they are used in ordinary division of numbers. For instance, when 7 is divided by 2, the quotient is 3 and the remainder is 1. We write this

$$\frac{7}{2} = 3 + \frac{1}{2}$$

or, equivalently,

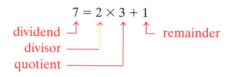

$$7 = 2 \times 3 + 1$$

dividend · divisor · quotient · remainder

The process of long division for polynomials follows the same four-step cycle used in ordinary long division of numbers: divide, multiply, subtract, bring down. As a first example, we divide $2x^2 - 7x + 8$ by $x - 2$. Notice that in setting up the division, we write both the dividend and the divisor in decreasing powers of x.

$$
\begin{array}{r}
2x \;-\; 3 \\
x - 2 \overline{\big)\, 2x^2 - 7x + 8} \\
\underline{2x^2 - 4x} \\
-3x + 8 \\
\underline{-3x + 6} \\
2
\end{array}
$$

1. Divide the first term of the dividend by the first term of the divisor: $\dfrac{2x^2}{x} = 2x$. The result becomes the first term of the quotient, as shown.

2. Multiply the divisor $x - 2$ by the term $2x$ obtained in the previous step. This yields the quantity $2x^2 - 4x$, which is written below the dividend, as shown.

3. From the quantity $2x^2 - 7x$ in the dividend, subtract the quantity $2x^2 - 4x$. This yields $-3x$.

4. Bring down the $+8$ in the dividend, as shown. The resulting quantity, $-3x + 8$, is now treated as the dividend and the entire process is repeated.

We've now found that when $2x^2 - 7x + 8$ is divided by $x - 2$, the quotient is $2x - 3$ and the remainder is 2. This is summarized by writing either

$$\frac{2x^2 - 7x + 8}{x - 2} = 2x - 3 + \frac{2}{x - 2} \tag{1}$$

or, after multiplying through by $x - 2$,

$$\underbrace{2x^2 - 7x + 8}_{\text{dividend}} = \underbrace{(x - 2)}_{\text{divisor}} \underbrace{(2x - 3)}_{\text{quotient}} + \underbrace{2}_{\text{remainder}} \tag{2}$$

There are two observations to be made here. First, notice that equation (2) is valid for all real numbers x, whereas equation (1) carries the implicit restriction that x may not equal 2. For this reason, we often prefer to write our results in the form of equation (2). Second, notice that the degree of the remainder is less than the degree of the divisor. This is very similar to the situation with ordinary division of positive integers, where the remainder is always less than the divisor.

As another example of the long division process, we divide $3x^4 - 2x^3 + 2$ by $x^2 - 1$. Notice in what follows that we have inserted the terms in the divisor and dividend that have coefficients of zero. These terms act as place holders.

$$
\begin{array}{r}
3x^2 - 2x\ + 3 \\
x^2 + 0x - 1\ \overline{\smash{\big)}\ 3x^4 - 2x^3 + 0x^2 + 0x + 2} \\
\underline{3x^4 + 0x^3 - 3x^2\ } \\
-2x^3 + 3x^2 + 0x\ \\
\underline{-2x^3 + 0x^2 + 2x\ } \\
3x^2 - 2x + 2 \\
\underline{3x^2 - 0x + 3} \\
-2x + 5
\end{array}
$$

We can write this result as

$$
\frac{3x^4 - 2x^3 + 2}{x^2 - 1} = 3x^2 - 2x + 3 + \frac{-2x + 5}{x^2 - 1}
$$

or, multiplying through by $x^2 - 1$,

$$
\underbrace{3x^4 - 2x^3 + 2}_{\text{dividend}} = \underbrace{(x^2 - 1)}_{\text{divisor}}\underbrace{(3x^2 - 2x + 3)}_{\text{quotient}} + \underbrace{(-2x + 5)}_{\text{remainder}}
$$

Notice that this last equation holds for all values of x, whereas the previous equation carries the restrictions that x may be neither 1 nor -1. Also, as in our previous example, observe that the degree of the remainder is less than that of the divisor.

There is a theorem, commonly referred to as the **division algorithm,** that summarizes rather nicely the key results of the long division process. We state the theorem here without proof.

The Division Algorithm

Let $p(x)$ and $d(x)$ be polynomials, and assume that $d(x)$ is not the zero polynomial. Then there are unique polynomials $q(x)$ and $R(x)$ such that

$$
p(x) = d(x) \cdot q(x) + R(x)
$$

where either $R(x)$ is the zero polynomial or the degree of $R(x)$ is less than the degree of $d(x)$.

The polynomials $p(x)$, $d(x)$, $q(x)$, and $R(x)$ are referred to, respectively, as the *dividend, divisor, quotient,* and *remainder.* When $R(x) = 0$, we have $p(x) = d(x) \cdot q(x)$ and we say that $d(x)$ and $q(x)$ are **factors** of $p(x)$. Also, since $d(x)$ is not the zero polynomial, notice that the equation $p(x) = d(x) \cdot q(x) + R(x)$ implies that the degree of $q(x)$ is less than or equal to the degree of $p(x)$. (Why?)

EXAMPLE **1** **Using long division to find a quotient and remainder**

Let $p(x) = x^3 + 2x^2 - 4$ and $d(x) = x - 3$. Use the long division process to find the polynomials $q(x)$ and $R(x)$ such that

$$
p(x) = d(x) \cdot q(x) + R(x)
$$

where either $R(x) = 0$ or the degree of $R(x)$ is less than the degree of $d(x)$.

SOLUTION
After inserting the term $0x$ in the dividend $p(x)$, we use long division to divide $p(x)$ by $d(x)$:

$$
\begin{array}{r}
x^2 + 5x \ + 15 \\
x - 3 \overline{\smash{\big)}\ x^3 + 2x^2 + \ 0x - \ \ 4} \\
\underline{x^3 - 3x^2} \\
5x^2 + \ \ 0x \\
\underline{5x^2 - 15x} \\
15x - \ \ 4 \\
\underline{15x - 45} \\
41
\end{array}
$$

We now have

$$
\underbrace{x^3 + 2x^2 - 4}_{p(x)} = \underbrace{(x - 3)}_{d(x)}\underbrace{(x^2 + 5x + 15)}_{q(x)} + \underbrace{41}_{R(x)}
$$

Thus, $q(x) = x^2 + 5x + 15$ and $R(x) = 41$. Notice that the degree of $R(x)$ is less than the degree of $d(x)$.

The long division procedure for polynomials can be streamlined when the divisor is of the form $x - r$. This shortened version, known as **synthetic division,** will be useful in subsequent sections when we are solving polynomial equations.

We can explain the idea behind synthetic division by using the long division carried out in Example 1. The basic idea is that in the long division process, it is the *coefficients* of the various polynomials that carry all the necessary information. In our example, for instance, the quotient and remainder can be abbreviated by writing down a sequence of four numbers:

$$1 \quad 5 \quad 15 \quad 41$$

By studying the long division process, you will find that these numbers are obtained through the following four steps:

Step 1 Write down the first coefficient of the dividend. This will be the first coefficient of the quotient.   *Result* $\boxed{1}$

Step 2 Multiply the 1 obtained in the previous step by the -3 in the divisor. Then subtract the result from the second coefficient of the dividend:

$$-3 \times 1 = -3 \qquad 2 - (-3) = 5 \qquad \textit{Result} \boxed{5}$$

Step 3 Multiply the 5 obtained in the previous step by the -3 in the divisor. Then subtract the result from the third coefficient of the dividend:

$$-3 \times 5 = -15 \qquad 0 - (-15) = 15 \qquad \textit{Result} \boxed{15}$$

Step 4 Multiply the 15 obtained in the previous step by the -3 in the divisor. Then subtract the result from the fourth coefficient of the dividend:

$$-3 \times 15 = -45 \qquad -4 - (-45) = 41 \qquad \textit{Result} \boxed{41}$$

A convenient format for setting up this process involves writing the constant term of the divisor and the coefficients of the dividend as follows:

$$
\begin{array}{r|rrrr}
-3 & 1 & 2 & 0 & -4 \\
\hline
\end{array}
$$

Now, using this format, let's again go through the four steps we have just described:

$$\underline{-3|}\ \ 1\ \ \ \ 2\ \ \ \ \ 0\ \ \ \ -4$$

Step 1 Bring down the 1.

$$1$$

Step 2 $-3 \times 1 = -3$
$2 - (-3) = 5$

$$\begin{array}{r} -3|\ \ 1\ \ \ \ \ 2\ \ \ \ \ 0\ \ \ \ -4 \\ -3 \\ \hline 1\ \ \ \ \ 5 \end{array}$$

Step 3 $-3 \times 5 = -15$
$0 - (-15) = 15$

$$\begin{array}{r} -3|\ \ 1\ \ \ \ \ 2\ \ \ \ \ \ 0\ \ \ \ -4 \\ -3\ \ \ -15 \\ \hline 1\ \ \ \ \ 5\ \ \ \ \ \ 15 \end{array}$$

Step 4 $-3 \times 15 = -45$
$-4 - (-45) = 41$

$$\begin{array}{r} -3|\ \ 1\ \ \ \ \ 2\ \ \ \ \ \ 0\ \ \ \ \ -4 \\ -3\ \ \ -15\ \ \ -45 \\ \hline 1\ \ \ \ \ 5\ \ \ \ \ \ 15\ \ \ \ \ \ 41 \end{array}$$

Although we have now obtained the required sequence of numbers, 1 5 15 41, there is one further simplification that can be made. In Steps 2 through 4, we can add instead of subtract if we use 3 instead of -3 in the initial format. (You will see the motivation for this in the next section when we discuss the remainder theorem.) With this change, let us now summarize the technique of synthetic division using the example with which we've been working. The method is applicable for any polynomial division where the divisor has the form $x - r$.

To Divide $x^3 + 2x^2 - 4$ by $x - 3$ Using Synthetic Division

COMMENTS

Format

$$\underline{3|}\ \ 1\ \ \ \ 2\ \ \ \ 0\ \ \ \ -4$$

Since the divisor is $x - 3$, the format begins with 3. The coefficients from the dividend are written in the order corresponding to decreasing powers of x. A zero coefficient is inserted as a place holder.

Procedure

$$\begin{array}{r} 3|\ \ 1\ \ \ \ 2\ \ \ \ \ 0\ \ \ \ -4 \\ 3\ \ \ \ 15\ \ \ \ \ 45 \\ \hline 1\ \ \ \ 5\ \ \ \ 15\ \ \ \ \ 41 \end{array}$$

Step 1 Bring down the 1.
Step 2 $3 \times 1 = 3$; $2 + 3 = 5$
Step 3 $3 \times 5 = 15$; $0 + 15 = 15$
Step 4 $3 \times 15 = 45$; $-4 + 45 = 41$

Answer

Quotient: $x^2 + 5x + 15$
Remainder: 41

The degree of the first term in the quotient is one less than the degree of the first term of the dividend.

EXAMPLE 2 **Synthetic division**

Use synthetic division to divide $x^3 - 6x + 4$ by $x - 2$.

SOLUTION

$$\begin{array}{r} 2|\ \ 1\ \ \ \ \ 0\ \ \ \ -6\ \ \ \ \ \ 4 \\ 2\ \ \ \ \ \ 4\ \ \ \ -4 \\ \hline 1\ \ \ \ \ 2\ \ \ \ -2\ \ \ \ \ \ 0 \end{array}$$

Looking at the third row of numbers in the synthetic division we've carried out, we see that the quotient is $x^2 + 2x - 2$ and the remainder is 0. In other words, both $x - 2$ and $x^2 + 2x - 2$ are factors of $x^3 - 6x + 4$, and we have

$$x^3 - 6x + 4 = (x - 2)(x^2 + 2x - 2)$$

EXAMPLE   3   Another example of synthetic division

Use synthetic division to divide $x^5 - a^5$ by $x - a$.

SOLUTION

$$
\begin{array}{r|rrrrrr}
\underline{a} & 1 & 0 & 0 & 0 & 0 & -a^5 \\
& & a & a^2 & a^3 & a^4 & a^5 \\
\hline
& 1 & a & a^2 & a^3 & a^4 & 0
\end{array}
$$

As before, we read off the quotient and remainder from the third row of numbers. The quotient is

$$x^4 + ax^3 + a^2x^2 + a^3x + a^4$$

and the remainder is zero, so we have

$$\underbrace{x^5 - a^5}_{\text{dividend}} = \underbrace{(x - a)}_{\text{divisor}}\underbrace{(x^4 + ax^3 + a^2x^2 + a^3x + a^4)}_{\text{quotient}} \underbrace{+\ 0}_{\text{remainder}}$$

or

$$x^5 - a^5 = (x - a)(x^4 + ax^3 + a^2x^2 + a^3x + a^4).$$

The last equation in Example 3 tells us how to factor $x^5 - a^5$. One factor is $x - a$. Notice the pattern in the second factor, $x^4 + ax^3 + a^2x^2 + a^3x + a^4$.

first term: a^0x^{5-1}

second term: a^1x^{5-2}

third term: a^2x^{5-3}

fourth term: a^3x^{5-4}

fifth term: a^4x^{5-5}

In the same way that we've found a factorization for $x^5 - a^5$, we can find a factorization for $x^n - a^n$ for any positive integer $n \geq 2$. In each case, the first factor is $x - a$, while the second factor follows the same pattern just described for $x^5 - a^5$. We state the general result in the box that follows. The result can be proved by using *mathematical induction* (discussed in a later chapter) or by using the *remainder theorem* (discussed in the next section).

Factorization of $x^n - a^n$

$$x^n - a^n = (x - a)(x^{n-1} + ax^{n-2} + a^2x^{n-3} + \cdots + a^{n-1})$$

In our development of synthetic division, we assumed that the form of the divisor was $x - r$. The next example shows what to do when the form of the divisor is $x + r$.

EXAMPLE 4 **Synthetic division when the divisor has the form $x + r$**

Use synthetic division to divide $x^4 - 2x^3 + 5x^2 - 4x + 3$ by $x + 1$.

SOLUTION
We first need to write the divisor $x + 1$ in the form $x - r$. We have

$$x + 1 = x - (-1)$$

In other words, r is -1, and this is the value we use to set up the synthetic division format. The format then is

$$\underline{-1|} \quad 1 \qquad -2 \qquad 5 \qquad -4 \qquad 3$$

Now we carry out the synthetic division procedure:

$$
\begin{array}{r|rrrrr}
-1 & 1 & -2 & 5 & -4 & 3 \\
   &   & -1 & 3 & -8 & 12 \\
\hline
   & 1 & -3 & 8 & -12 & 15
\end{array}
$$

The quotient is therefore $x^3 - 3x^2 + 8x - 12$, and the remainder is 15. We can summarize this result by writing

$$x^4 - 2x^3 + 5x^2 - 4x + 3 = (x + 1)(x^3 - 3x^2 + 8x - 12) + 15$$

(Notice that the degree of the remainder is less than the degree of the divisor, in agreement with the division algorithm.)

EXERCISE SET 7.2

A

In Exercises 1–20, use long division to find the quotients and the remainders. Also, write each answer in the form $p(x) = d(x) \cdot q(x) + R(x)$, as in equation (2) in the text.

1. $\dfrac{x^2 - 8x + 4}{x - 3}$

2. $\dfrac{x^3 - 4x^2 + x - 2}{x - 5}$

3. $\dfrac{x^2 - 6x - 2}{x + 5}$

4. $\dfrac{3x^2 + 4x - 1}{x - 1}$

5. $\dfrac{6x^3 - 2x + 3}{2x + 1}$

6. $\dfrac{x^4 - 4x^3 + 6x^2 - 4x + 1}{x - 1}$

7. $\dfrac{x^5 + 2}{x + 3}$

8. $\dfrac{4x^3 - x^2 + 8x - 1}{x^2 - x + 1}$

9. $\dfrac{x^6 - 64}{x - 2}$

10. $\dfrac{x^6 + 64}{x - 2}$

11. $\dfrac{5x^4 - 3x^2 + 2}{x^2 - 3x + 5}$

12. $\dfrac{8x^6 - 36x^4 + 54x^2 - 27}{2x^2 - 3}$

13. $\dfrac{3y^3 - 4y^2 - 3}{y^2 + 5y + 2}$

14. $\dfrac{4y^4 - y^3 + 2y - 1}{2y^2 - 3y - 4}$

15. $\dfrac{t^4 - 4t^3 + 4t^2 - 16}{t^2 - 2t + 4}$

16. $\dfrac{2t^5 - 6t^4 - t^2 + 2t + 3}{t^3 - 2}$

17. $\dfrac{z^5 - 1}{z - 1}$

18. $\dfrac{1 + z + z^2 + z^3}{1 + z + z^2}$

19. $\dfrac{ax^2 + bx + c}{x - r}$

20. $\dfrac{ax^3 + bx^2 + cx + d}{x - r}$

In Exercises 21–40, use synthetic division to find the quotients and remainders. Also, in each case, write the result of the division in the form $p(x) = d(x) \cdot q(x) + R(x)$, as in equation (2) in the text.

21. $\dfrac{x^2 - 6x - 2}{x - 5}$

22. $\dfrac{3x^2 + 4x - 1}{x - 1}$

23. $\dfrac{4x^2 - x - 5}{x + 1}$

24. $\dfrac{x^2 - 1}{x + 2}$

25. $\dfrac{6x^3 - 5x^2 + 2x + 1}{x - 4}$

26. $\dfrac{x^4 - 4x^3 + 6x^2 - 4x + 1}{x - 1}$

27. $\dfrac{x^3 - 1}{x - 2}$

28. $\dfrac{x^3 - 8}{x - 2}$

29. $\dfrac{x^5 - 1}{x + 2}$

30. $\dfrac{x^3 - 8x^2 - 1}{x + 3}$

31. $\dfrac{x^4 - 6x^3 + 2}{x + 4}$

32. $\dfrac{3x^3 - 2x^2 + x + 1}{x - \frac{1}{2}}$

33. $\dfrac{x^3 - 4x^2 - 3x + 6}{x - 10}$

34. $\dfrac{1 + 3x + 3x^2 + x^3}{x + 1}$

35. $\dfrac{x^3 - x^2}{x + 5}$

36. $\dfrac{5x^4 - 4x^3 + 3x^2 - 2x + 1}{x + \frac{1}{2}}$

37. $\dfrac{14 - 27x - 27x^2 + 54x^3}{x - \frac{2}{3}}$

38. $\dfrac{14 - 27x - 27x^2 + 54x^3}{x + \frac{2}{3}}$

39. $\dfrac{x^4 + 3x^2 + 12}{x - 3}$

40. (a) $\dfrac{x^4 - 16}{x - 2}$ **(b)** $\dfrac{x^4 + 16}{x + 2}$

In Exercises 41–44, each expression has the form $x^n - a^n$. Write each expression as a product of two factors (as in the box on page 573).

41. $x^5 - 32$ **42.** $y^6 - 1$ **43.** $z^4 - 81$ **44.** $x^7 - y^7$

B

In Exercises 45–48, use synthetic division to determine the quotient $q(x)$ and the remainder $R(x)$ in each case.

45. $\dfrac{6x^2 - 8x + 1}{3x - 4}$ *Hint:* Divide both numerator and denominator by 3. (Why?)

46. $\dfrac{4x^3 + 6x^2 - 6x - 5}{2x - 3}$

47. $\dfrac{6x^3 + 1}{2x + 1}$

48. $\dfrac{5x^3 - 3x^2 + 1}{3x + 1}$

49. When $x^3 + kx + 1$ is divided by $x + 1$, the remainder is -4. Find k.

50. (a) Show that when $x^3 + kx + 6$ is divided by $x + 3$, the remainder is $-21 - 3k$.

   (b) Determine a value of k such that $x + 3$ will be a factor of $x^3 + kx + 6$.

51. When $x^2 + 2px - 3q^2$ is divided by $x - p$, the remainder is zero. Show that $p^2 = q^2$.

52. Given that $x - 3$ is a factor of $x^3 - 2x^2 - 4x + 3$, solve the equation $x^3 - 2x^2 - 4x + 3 = 0$.

The process of synthetic division applies equally well when some or all of the coefficients are nonreal complex numbers. In Exercises 53–56, use synthetic division to determine the quotient $q(x)$ and the remainder $R(x)$ in each case.

53. $\dfrac{x^2 - 4x + 1}{x - i}$

54. $\dfrac{x^3 - 2x^2 - 4}{x - 3i}$

55. $\dfrac{x^2 - 2x + 2}{x - (1 + i)}$

56. $\dfrac{x^3 - x^2 + 4x - 4}{x + 2i}$

57. Given that the identity $f(x) = d(x) \cdot q(x) + R(t)$ holds for the following polynomials, evaluate $f(\sqrt{3})$.
   Hint (of sorts): There's an easy way and a tedious way.

$$f(x) = 2x^5 + 5x^4 - 8x^3 + 7x^2 - 9 \qquad d(x) = x^2 - 3$$
$$q(x) = 2x^3 + 5x^2 - 2x + 22 \qquad R(x) = -6x + 57$$

58. Given that the identity $f(t) = d(t) \cdot q(t) + R(x)$ holds for the following polynomials, evaluate $f(4)$.

$$f(t) = t^5 - 3t^4 + 2t^3 - 5t^2 + 6t - 7 \qquad d(t) = t - 4$$
$$q(t) = t^4 + t^3 + 6t^2 + 19t + 82 \qquad R(t) = 321$$

59. Find the remainder when $t^5 - 5a^4t + 4a^5$ is divided by $t - a$.

60. When $f(x)$ is divided by $(x - a)(x - b)$, the remainder is $Ax + B$. Apply the division algorithm to show that

$$A = \frac{f(a) - f(b)}{a - b} \qquad \text{and} \qquad B = \frac{bf(a) - af(b)}{b - a}$$

7.3 THE REMAINDER THEOREM AND THE FACTOR THEOREM

Descartes recommended [in his La Géométrie of 1637] that all terms [of an equation] should be taken to one side and equated with zero. Though he was not the first to suggest this, he was the earliest writer to realize the advantage to be gained. He pointed out that a polynomial f(x) was divisible by (x − a) if and only if a was a root of f(x).—David M. Burton in *The History of Mathematics, An Introduction,* 2d ed. (Dubuque, Iowa: Wm. C. Brown Publishers, 1991)

The techniques for solving polynomial equations of degree 2 were discussed in Sections 1.3 and 2.1. Now we want to extend those ideas. Our focus in this section and in the remainder of the chapter is on solving polynomial equations of any degree — that is, equations of the form

$$f(x) = a_n x^n + a_{n-1} x^{n-1} + \cdots + a_1 x + a_0 = 0 \qquad (1)$$

Here, as in Chapter 1, a **root** or **solution** of equation (1) is a number r that when substituted for x leads to a true statement. Thus, r is a root of equation (1) provided $f(r) = 0$. We also refer to the number r in this case as a **zero** of the function f.

EXAMPLE 1 **Checking for a zero or a root**

(a) Is -3 a zero of the function f defined by $f(x) = x^4 + x^2 - 6$?
(b) Is $\sqrt{2}$ a root of the equation $x^4 + x^2 - 6 = 0$?

SOLUTION

(a) By definition, -3 will be a zero of f if $f(-3) = 0$. We have

$$f(-3) = (-3)^4 + (-3)^2 - 6 = 81 + 9 - 6 = 84 \neq 0$$

Thus, -3 is not a zero of the function f.
(b) To check if $\sqrt{2}$ is a root of the given equation, we have

$$(\sqrt{2})^4 + (\sqrt{2})^2 - 6 \stackrel{?}{=} 0$$
$$4 + 2 - 6 \stackrel{?}{=} 0$$
$$0 \stackrel{?}{=} 0 \qquad \text{True}$$

Thus, $\sqrt{2}$ is a root of the equation $x^4 + x^2 - 6 = 0$.

There are cases in which a root of an equation is what we call a **repeated root.** Consider, for instance, the equation $x(x - 1)(x - 1) = 0$. We have

$$x(x - 1)(x - 1) = 0$$

$$x = 0 \quad \bigg| \quad \begin{matrix} x - 1 = 0 \\ x = 1 \end{matrix} \quad \bigg| \quad \begin{matrix} x - 1 = 0 \\ x = 1 \end{matrix}$$

The roots of the equation are therefore 0, 1, and 1. The repeated root here is $x = 1$. We say in this case that 1 is a **double root** or, equivalently, that 1 is a **root of multiplicity 2.** More generally, if a root is repeated k times, we call it a **root of multiplicity k.**

EXAMPLE **2**   **Specifying the multiplicity of a root**

State the multiplicity of each root of the equation

$$(x - 4)^2(x - 5)^3 = 0$$

SOLUTION
We have $(x - 4)(x - 4)(x - 5)(x - 5)(x - 5) = 0$. By setting each factor equal to zero, we obtain the roots 4, 4, 5, 5, and 5. From this we see that the root 4 has multiplicity 2, while the root 5 has multiplicity 3. It is not really necessary to write out all the factors as we did here; the exponents of the factors in the original equation give us the required multiplicities.

There is a connection between repeated roots and graphs. If a polynomial equation $f(x) = 0$ has a repeated root r, then the graph of the function f is tangent to the x-axis at $x = r$. We saw examples demonstrating this in Section 4.6, although we had not defined the term *multiple root* at that stage. Figure 1 shows two examples of this connection. (For the examples in Section 4.6, see Figures 18 through 21 on pages 334–336.)

GRAPHICAL PERSPECTIVE

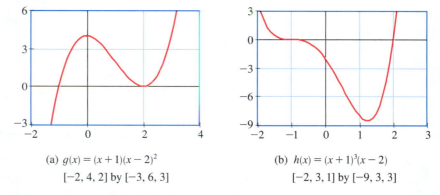

Figure 1
At a multiple root, the graph is tangent to the x-axis. In contrast, at a simple root [such as $x = 2$ in Figure 1(b)], the graph crosses the x-axis.

(a) $g(x) = (x + 1)(x - 2)^2$
$[-2, 4, 2]$ by $[-3, 6, 3]$

(b) $h(x) = (x + 1)^3(x - 2)$
$[-2, 3, 1]$ by $[-9, 3, 3]$

There are two simple but important theorems that will form the basis for much of our subsequent work with polynomials: the *remainder theorem* and the *factor theorem*. We begin with a statement of the remainder theorem.

The Remainder Theorem

When a polynomial $f(x)$ is divided by $x - r$, the remainder is $f(r)$.

Before turning to a proof of the remainder theorem, let's see what the theorem is saying in two particular cases. First, suppose that we divide the polynomial $f(x) = 2x^2 - 3x + 4$ by $x - 1$. Then according to the remainder theorem, the remainder in this case should be the number $f(1)$. Let's check:

$$
\begin{array}{r|rrr}
1\rfloor & 2 & -3 & 4 \\
 & & 2 & -1 \\
\hline
 & 2 & -1 & ③ \\
\end{array}
\qquad
\begin{aligned}
f(x) &= 2x^2 - 3x + 4 \\
f(1) &= 2(1)^2 - 3(1) + 4 \\
f(1) &= ③
\end{aligned}
$$

As the calculations show, the remainder is indeed equal to $f(1)$. As a second example, let us divide the polynomial $g(x) = ax^2 + bx + c$ by $x - r$. According to the remainder theorem, the remainder should be $g(r)$. Again, let us check:

$$
\begin{array}{r|llll}
r & a & b & c \\
& & ar & ar^2 + br \\
\hline
& a & ar + b & \boxed{ar^2 + br + c}
\end{array}
\qquad
\begin{aligned}
g(x) &= ax^2 + bx + c \\[1em]
g(r) &= \boxed{ar^2 + br + c}
\end{aligned}
$$

The calculations show that the remainder is equal to $g(r)$, as we wished to check.

In the example just concluded, we verified that the remainder theorem holds for any quadratic polynomial $g(x) = ax^2 + bx + c$. A general proof of the remainder theorem can easily be given along these same lines. The only drawback is that it becomes slightly cumbersome to carry out the synthetic division process when the dividend is

$$a_n x^n + a_{n-1} x^{n-1} + \cdots + a_1 x + a_0$$

For this reason, mathematicians often prefer to base the proof of the remainder theorem on the division algorithm given in the previous section. This is the path we will follow here.

To prove the remainder theorem, we must show that when the polynomial $f(x)$ is divided by $x - r$, the remainder is $f(r)$. Now, according to the division algorithm, we can write

$$f(x) = (x - r) \cdot q(x) + R(x) \tag{2}$$

In this identity, either $R(x)$ is the zero polynomial or the degree of $R(x)$ is less than that of $x - r$. But since the degree of $x - r$ is 1, we must have in this case that the degree of $R(x)$ is zero. Thus, in *either* case, the remainder $R(x)$ is a constant. Denoting this constant by c, we can rewrite equation (2) as

$$f(x) = (x - r) \cdot q(x) + c$$

Now if we set $x = r$ in this identity, we obtain

$$f(r) = (r - r) \cdot q(r) + c = c$$

We have now shown that $f(r) = c$. But by definition, c is the remainder $R(x)$. Thus, $f(r) = R(x)$. This proves the remainder theorem.

| **EXAMPLE** | **3** | **Using the remainder theorem to evaluate a function and check for a factor** |

Let $f(x) = 2x^3 - 5x^2 + x - 6$.
(a) Use the remainder theorem to evaluate $f(3)$.
(b) Is $x - 3$ a factor of $f(x) = 2x^3 - 5x^2 + x - 6$?

SOLUTION
(a) According to the remainder theorem, $f(3)$ is the remainder when $f(x)$ is divided by $x - 3$. Using synthetic division, we have

$$
\begin{array}{r|rrrr}
3 & 2 & -5 & 1 & -6 \\
& & 6 & 3 & 12 \\
\hline
& 2 & 1 & 4 & 6
\end{array}
$$

The remainder is 6, and therefore $f(3) = 6$.

(b) By definition, $x - 3$ is a factor of $f(x)$ if we obtain a zero remainder when $f(x)$ is divided by $x - 3$. But from our work in part (a), we know that the remainder is 6, not 0, so $x - 3$ is not a factor of $f(x)$.

EXAMPLE 4   Using the remainder theorem to check for a factor

In the previous section it was stated, but not proved, that $x - a$ is a factor of $f(x) = x^n - a^n$. Use the remainder theorem to prove this fact.

SOLUTION
We need to show that when $f(x) = x^n - a^n$ is divided by $x - a$, the remainder is zero. By the remainder theorem, the remainder is equal to $f(a)$, which is easy to find:

$$f(a) = a^n - a^n = 0$$

Thus, the remainder is zero, and $x - a$ is a factor of $x^n - a^n$, as required.

From our experience with quadratic equations, we know that there is a close connection between factoring a quadratic polynomial $f(x)$ and solving the polynomial equation $f(x) = 0$. The factor theorem states this relationship between roots and factors in a precise form. Furthermore, the factor theorem tells us that this relationship holds for polynomials of all degrees, not just quadratics.

The Factor Theorem

Let $f(x)$ be a polynomial. If $f(r) = 0$, then $x - r$ is a factor of $f(x)$. Conversely, if $x - r$ is a factor of $f(x)$, then $f(r) = 0$.

In terms of roots, we can summarize the factor theorem by saying that r is a root of the equation $f(x) = 0$ if and only if $x - r$ is a factor of $f(x)$. To prove the factor theorem, let us begin by assuming that $f(r) = 0$. We want to show that $x - r$ is a factor of $f(x)$. Now, according to the remainder theorem, if $f(x)$ is divided by $x - r$, the remainder is $f(r)$. So we can write

$$f(x) = (x - r) \cdot q(x) + f(r) \qquad \text{for some polynomial } q(x)$$

But since $f(r)$ is zero, this equation becomes

$$f(x) = (x - r) \cdot q(x)$$

This last equation tells us that $x - r$ is a factor of $f(x)$, as we wished to prove.

Now, conversely, let us assume that $x - r$ is a factor of $f(x)$. We want to show that $f(r) = 0$. Since $x - r$ is a factor of $f(x)$, we can write

$$f(x) = (x - r) \cdot q(x) \qquad \text{for some polynomial } q(x)$$

If we now let $x = r$ in this last equation, we obtain

$$f(r) = (r - r) \cdot q(r) = 0$$

as we wished to show.

The example that follows indicates how the factor theorem can be used to solve equations.

EXAMPLE **5**   Applying the factor theorem in solving an equation

Refer to Figure 2, which shows a graph of the function $y = x^3 - 5x + 2$. As indicated by the graph, the equation

$$x^3 - 5x + 2 = 0 \tag{3}$$

has (as least) three roots, one of which either is equal to 2 or is very close to 2. Confirm that $x = 2$ is indeed a root of equation (3), and then solve the equation.

GRAPHICAL PERSPECTIVE

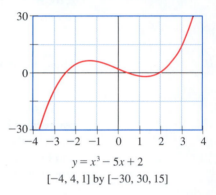

$y = x^3 - 5x + 2$

Figure 2 $[-4, 4, 1]$ by $[-30, 30, 15]$

SOLUTION

Replacing x by 2 in equation (3) yields $2^3 - 5(2) + 2 = 0$, which is true. Thus, 2 is a root of the given equation. Now, since $x = 2$ is a root, the factor theorem tells us that $x - 2$ is a factor of $x^3 - 5x + 2$. In other words,

$$x^3 - 5x + 2 = (x - 2) \cdot q(x) \qquad \text{for some polynomial } q(x) \tag{4}$$

To isolate $q(x)$, we divide $x^3 - 5x + 2$ by $x - 2$. As indicated in Figure 3, the result is $q(x) = x^2 + 2x - 1$. Thus, equation (4) becomes

$$x^3 - 5x + 2 = (x - 2)(x^2 + 2x - 1)$$

With this identity, the original equation becomes

$$(x - 2)(x^2 + 2x - 1) = 0$$

At this point, the problem is reduced to solving the linear equation $x - 2 = 0$ and the quadratic equation $x^2 + 2x - 1 = 0$. The linear equation yields $x = 2$, but we already know that 2 is a root. So, if we are to find any additional roots, they must come from the quadratic equation $x^2 + 2x - 1 = 0$. As you should now verify for yourself, the quadratic formula yields $x = -1 \pm \sqrt{2}$. Thus, we have three roots of the given cubic equation: 2, $-1 + \sqrt{2}$, and $-1 - \sqrt{2}$. As you'll see in the next section, a cubic equation can have at most three roots. So in this case, we've determined all the roots; that is, we have solved the equation. As you can check, the calculator approximations for the two roots containing radicals are $-1 + \sqrt{2} \approx 0.41$ and $-1 - \sqrt{2} \approx -2.41$. Note that these values are consistent with the graph shown in Figure 2.

$$\begin{array}{r|rrrr} 2| & 1 & 0 & -5 & 2 \\ & & 2 & 4 & -2 \\ \hline & 1 & 2 & -1 & 0 \end{array}$$

Figure 3

Using synthetic division to divide $x^3 - 5x + 2$ by $x - 2$. The quotient is $x^2 + 2x - 1$ with zero remainder.

Before going on to other examples, let's take a moment to summarize the technique we used in Example 5. We want to solve a polynomial equation $f(x) = 0$, given that one root is $x = r$. Since r is a root, the factor theorem tells us that $x - r$ is a factor of $f(x)$. Then, with the aid of synthetic division, we obtain a factorization

$$f(x) = (x - r) \cdot q(x) \qquad \text{for some polynomial } q(x)$$

This gives rise to the two equations $x - r = 0$ and $q(x) = 0$. Since the first of these only reasserts that r is a root, we try to solve the second equation, $q(x) = 0$. We refer to the equation $q(x) = 0$ as the **reduced equation.** Example 5 showed you the idea behind this terminology; the degree of $q(x)$ is one less than that of $f(x)$. If, as in Example 5, the reduced equation happens to be a quadratic equation, then we can always determine the remaining roots by factoring or by the quadratic formula. In subsequent sections, we will look at techniques that are helpful in cases where $q(x)$ is not quadratic.

| EXAMPLE | 6 | Determining a polynomial equation with prescribed roots |

Solve the equation $x^4 + 2x^3 - 7x^2 - 20x - 12 = 0$, given that $x = 3$ and $x = -2$ are roots.

SOLUTION
Since $x = 3$ is a root, the factor theorem tells us that $x - 3$ is a factor of the polynomial $x^4 + 2x^3 - 7x^2 - 20x - 12$. That is,

$$x^4 + 2x^3 - 7x^2 - 20x - 12 = (x - 3) \cdot q(x) \qquad \text{for some polynomial } q(x)$$

As in Example 5, we can find $q(x)$ by synthetic division. We have

$$
\begin{array}{r|rrrrr}
3 & 1 & 2 & -7 & -20 & -12 \\
  &   & 3 & 15 & 24 & 12 \\
\hline
  & 1 & 5 & 8 & 4 & 0
\end{array}
$$

Thus, $q(x) = x^3 + 5x^2 + 8x + 4$, and our original equation is equivalent to

$$(x - 3)(x^3 + 5x^2 + 8x + 4) = 0$$

Now, $x = -2$ is also a root of this equation. But $x = -2$ surely is not a root of the equation $x - 3 = 0$, so it must be a root of the reduced equation $x^3 + 5x^2 + 8x + 4 = 0$. Again, the factor theorem is applicable. Since $x = -2$ is a root of the reduced equation, $x + 2$ must be a factor of $x^3 + 5x^2 + 8x + 4$. That is,

$$x^3 + 5x^2 + 8x + 4 = (x + 2) \cdot p(x) \qquad \text{for some polynomial } p(x)$$

We can use synthetic division to determine $p(x)$:

$$
\begin{array}{r|rrrr}
-2 & 1 & 5 & 8 & 4 \\
   &   & -2 & -6 & -4 \\
\hline
   & 1 & 3 & 2 & 0
\end{array}
$$

Thus, $p(x) = x^2 + 3x + 2$, and our reduced equation can be written

$$(x + 2)(x^2 + 3x + 2) = 0$$

This gives rise to a second reduced equation:

$$x^2 + 3x + 2 = 0$$

In this case, the roots are readily obtained by factoring. We have

$$x^2 + 3x + 2 = 0$$
$$(x + 2)(x + 1) = 0$$

$$x + 2 = 0 \qquad | \qquad x + 1 = 0$$
$$x = -2 \qquad | \qquad x = -1$$

Now, of the two roots we've just found, -2 happens to be one of the roots that we were initially given in the statement of the problem. On the other hand, -1 is a distinct additional root. In summary, then, we have three distinct roots: 3, -2, and -1, where the root -2 has multiplicity 2. As you will see in the next section, a fourth-degree equation can have at most four (not necessarily distinct) roots. So in the case at hand, we have found all the roots of the given equation.

| **EXAMPLE** | **7** | **Finding polynomial equations satisfying given conditions** |

In each case, find a polynomial equation $f(x) = 0$ satisfying the given conditions. If there is no such equation, say so.
(a) The numbers -1, 4, and 5 are roots.
(b) A factor of $f(x)$ is $x - 3$, and -4 is a root of multiplicity 2.
(c) The degree of f is 4, the number -5 is a root of multiplicity 3, and 6 is a root of multiplicity 2.

SOLUTION
(a) If $f(x)$ is any polynomial containing the factors $x + 1$, $x - 4$, and $x - 5$, then the equation $f(x) = 0$ will certainly be satisfied with $x = -1$, $x = 4$, or $x = 5$. The simplest polynomial equation in this case is therefore

$$(x + 1)(x - 4)(x - 5) = 0$$

This is a polynomial equation with the required roots. If required, we can carry out the multiplication on the left-hand side of the equation. As you can check, this yields $x^3 - 8x^2 + 11x + 20 = 0$.
(b) According to the factor theorem, since -4 is a root, $x + 4$ must be a factor of $f(x)$. In fact, since -4 is a root of multiplicity 2, the quantity $(x + 4)^2$ must be a factor of $f(x)$. The following equation therefore satisfies the given conditions:

$$(x - 3)(x + 4)^2 = 0$$

(c) We are given two roots, one with multiplicity 3, the other with multiplicity 2. Thus the degree of $f(x)$ must be at least $3 + 2 = 5$. (Why?) But this then contradicts the given condition that the degree of f should be 4. Consequently, there is no polynomial equation that satisfies the given conditions.

EXERCISE SET 7.3

A

In Exercises 1–6, determine whether the given value for the variable is a root of the equation.

1. $12x - 8 = 112$; $x = 10$
2. $12x^2 - x - 20 = 0$; $x = 5/4$
3. $x^2 - 2x - 4 = 0$; $x = 1 - \sqrt{5}$
4. $1 - x + x^2 - x^3 = 0$; $x = -1$
5. $2x^2 - 3x + 1 = 0$; $x = 1/2$
6. $(x - 1)(x - 2)(x - 3) = 0$; $x = 4$

In Exercises 7–14, determine whether the given value is a zero of the function.

7. $f(x) = 3x - 2$; $x = 2/3$
8. $g(x) = 1 + x^2$; $x = -1$
9. $h(x) = 5x^3 - x^2 + 2x + 8$; $x = -1$
10. $F(x) = -2x^5 + 3x^4 + 8x^3$; $x = 0$
11. $f(t) = 1 + 2t + t^3 - t^5$; $t = 2$
12. $f(t) = 1 + 2t + t^3 - t^5$; $t = \sqrt{2}$
13. $f(x) = 2x^3 - 3x + 1$:
    (a) $x = (\sqrt{3} - 1)/2$ (b) $x = (\sqrt{3} + 1)/2$
14. $g(x) = x^4 + 8x^3 + 9x^2 - 8x - 10$:
    (a) $x = 1$ (c) $x = \sqrt{6} + 4$
    (b) $x = \sqrt{6} - 4$

In Exercises 15 and 16, list the distinct roots of each equation. In the case of a repeated root, specify its multiplicity.

15. $(x - 1)(x - 2)^3(x - 3) = 0$
16. $x(x + 5)^4 = 0$

Ⓖ *In Exercises 17 and 18, you are given a polynomial equation $f(x) = 0$. Specify the multiplicity of each repeated root. Then use a graphing utility to verify that the graph of $y = f(x)$ is tangent to the x-axis at each repeated root.*

17. (a) $(x + 1)^2(x + 2) = 0$ 18. (a) $x(x + 1)^2(x - 1) = 0$
    (b) $(x + 1)(x + 2)^3 = 0$ (b) $x^2(x + 1)^2(x - 1) = 0$
    (c) $(x + 1)^2(x + 2)^3 = 0$ (c) $x^3(x + 1)^2(x - 1) = 0$

In Exercises 19–24, use the remainder theorem (as in Example 3) to evaluate $f(x)$ for the given value of x.

19. $f(x) = 4x^3 - 6x^2 + x - 5$; $x = -3$
20. $f(x) = 2x^3 - x - 4$; $x = 4$
21. $f(x) = 6x^4 + 5x^3 - 8x^2 - 10x - 3$; $x = 1/2$
22. $f(x) = x^5 - x^4 - x^3 - x^2 - x - 1$; $x = -2$
23. $f(x) = x^2 + 3x - 4$; $x = -\sqrt{2}$
24. $f(x) = x^7 - 7x^6 + 5x^4 + 1$; $x = -3$
25. The graph in the accompanying figure indicates that the equation $2x^3 - x + 1 = 0$ has a root at or very near $x = -1$. Confirm that $x = -1$ is indeed a root, and then solve the equation, as in Example 5.

$$y = 2x^3 - x + 1$$

26. The graph in the accompanying figure indicates that the equation $x^3 - 14x - 8 = 0$ has a root at or very near $x = 4$. Confirm that $x = 4$ is indeed a root of the equation, and then solve the equation, as in Example 5.

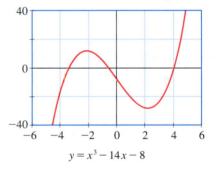

$$y = x^3 - 14x - 8$$

In Exercises 27–38, you are given a polynomial equation and one or more roots. Solve each equation using the method shown in Examples 5 and 6. To help you decide whether you have found all the roots in each case, you may rely on the following theorem, discussed in the next section: A polynomial equation of degree n has at most n (not necessarily distinct) roots.

27. $x^3 - 4x^2 - 9x + 36 = 0$; -3 is a root.
28. $x^3 + 7x^2 + 11x + 5 = 0$; -1 is a root.
29. $x^3 + x^2 - 7x + 5 = 0$; 1 is a root.
30. $x^3 + 8x^2 - 3x - 24 = 0$; -8 is a root.
31. $3x^3 - 5x^2 - 16x + 12 = 0$; -2 is a root.
32. $2x^3 - 5x^2 - 46x + 24 = 0$; 6 is a root.
33. $2x^3 + x^2 - 5x - 3 = 0$; $-3/2$ is a root.
34. $6x^4 - 19x^3 - 25x^2 + 18x + 8 = 0$; 4 and $-1/3$ are roots.
35. $x^4 - 15x^3 + 75x^2 - 125x = 0$; 5 is a root.
36. $2x^3 + 5x^2 - 8x - 20 = 0$; 2 is a root.
37. $x^4 + 2x^3 - 23x^2 - 24x + 144 = 0$; -4 and 3 are roots.
38. $6x^5 + 5x^4 - 29x^3 - 25x^2 - 5x = 0$; $\sqrt{5}$ and $-1/3$ are roots.

Ⓖ *In Exercises 39–42, each polynomial equation f(x) = 0 has a root r that is a rational number. Use a graphing utility to zoom in on the x-intercepts of the graph of y = f(x) until you think you know what r is. Then check algebraically that r is indeed a root. [You can check either by direct substitution or by using the remainder theorem and synthetic division to compute f(r).] After you know the rational root, follow the procedure shown in Example 5 to determine the remaining roots. For roots that contain radicals, give both the exact expression for the answer and a calculator approximation rounded to two decimal places. Finally, check to see that the decimal values are consistent with the x-intercepts that you looked at initially.*

39. $x^3 + x^2 - 18x + 10 = 0$ **40.** $-2x^3 + 7x^2 - 2x - 6 = 0$

41. $6x^3 - 28x^2 + 19x - 2 = 0$

42. $5x^5 - x^4 - 15x^3 + 3x^2 - 15x + 3 = 0$

Ⓖ *In Exercises 43 and 44, each polynomial equation f(x) = 0 has at least two rational roots. Use a graphing utility to zoom in on the x-intercepts of the graph of y = f(x) to see what the rational roots might be. For each suspected rational root, find out whether it is actually a root (either by using direct substitution or by using the remainder theorem and synthetic division). After you have determined the rational roots in this way, solve the equation as in Example 6.*

43. $4x^5 - 15x^4 + 8x^3 + 19x^2 - 12x - 4 = 0$

44. $x^4 - 10x^3 - 3.1x^2 + 28.9x + 21 = 0$

B

For Exercises 45 and 46, refer to the following tables for the functions f and g.

$f(t) = t^3 - 4t + 3$		$g(t) = t^5 + 2t^4 + t^3 + 2t^2 - t - 2$	
t	f(t)	t	g(t)
0.500	1.125000	−3.00	−89.0
0.750	0.421875	−2.50	−22.15625
1.000	0.000000	−2.25	−7.4228515625
1.250	−0.046875	−2.00	0.0
1.500	0.375000	−1.75	2.8603515625

45. (a) What is the remainder when $f(t)$ is divided by $t - \frac{1}{2}$?
  Hint: Don't calculate; use the remainder theorem.
  (b) What is the remainder when $f(t)$ is divided by $t - 1.25$?
  (c) Specify a linear factor of $f(t)$.
  (d) Solve the equation $f(t) = 0$.

46. (a) What is the remainder when $g(t)$ is divided by $t + 3$?
  (b) What is the remainder when $g(t)$ is divided by $t + \frac{5}{2}$?
  (c) Specify a linear factor of $g(t)$.
  (d) Solve the equation $g(t) = 0$.

In Exercises 47–52, find a polynomial equation f(x) = 0 satisfying the given conditions. If no such equation is possible, state this.

47. Degree 3; the coefficient of x^3 is 1; three roots are 3, −4, and 5

48. Degree 3; the coefficient are integers; $1/2, 2/5$, and $−3/4$ are roots

49. Degree 3; −1 is a root of multiplicity two; $x + 6$ is a factor of $f(x)$

50. Degree 3; 4 is a root of multiplicity two; −1 is a root of multiplicity two

51. Degree 4; $1/2$ is a root of multiplicity three; $x^2 - 3x - 4$ is a factor of $f(x)$

52. Degree 4; the coefficients are integers; $1/2$ is a root of multiplicity two; $2x^2 - 4x - 1$ is a factor of $f(x)$

53. In this exercise, we verify that the remainder theorem is valid for the cubic polynomial $g(x) = ax^3 + bx^2 + cx + d$.
  (a) Compute $g(r)$.
  (b) Using synthetic division, divide $g(x)$ by $x - r$. Check that the remainder you obtain is the same as the answer in part (a).

54. (a) In the standard viewing rectangle, graph the four equations $y = x^n - 2^n$, $(n = 2, 3, 4, 5)$.
  (b) Switch to a viewing rectangle that extends from −5 to 5 in the x-direction and from −40 to 20 in the y-direction. Name one property shared by all four graphs.
  (c) How does your answer in part (b) relate to the result in Example 4 of this section?

In Exercises 55 and 56, determine whether the given value is a zero of the function.

55. $Q(x) = ax^2 + bx + c; x = \dfrac{-b + \sqrt{b^2 - 4ac}}{2a}$
  Hint: Look before you leap!

56. $f(x) = x^3 - 3x^2 + 3x - 3$:
  (a) $x = \sqrt[3]{2} - 1$ **(b)** $x = \sqrt[3]{2} + 1$

57. Determine values for a and b such that $x - 1$ is a factor of both $x^3 + x^2 + ax + b$ and $x^3 - x^2 - ax + b$.

58. Determine a quadratic equation with the given roots:
  (a) $a/b, -b/a$
  (b) $-a + 2\sqrt{2b}, -a - 2\sqrt{2b}$

59. One root of the equation $x^2 + bx + 1 = 0$ is twice the other; find b. (There are two answers.)

60. Determine a value for a such that one root of the equation $ax^2 + x - 1 = 0$ is five times the other.

C

61. Solve the equation $x^3 - 12x + 16 = 0$, given that one of the roots has multiplicity two.

7.4   THE FUNDAMENTAL THEOREM OF ALGEBRA

One of the most intriguing problems was the question of the number of roots of an equation, which brought in negative and imaginary quantities, and led to the conclusion, in the work of Girard and Descartes, that an equation of degree n can have no more than n roots. The more precise statement, that an equation of degree n always has one root, and hence always has n roots (allowing for multiple roots), became known as the fundamental theorem of algebra. After several attempts by D'Alembert, Euler, and others, the proof was finally given by Gauss in 1799. —From *A Source Book in Mathematics, 1200–1800,* D. J. Struik, ed. (Princeton, N.J.: Princeton University Press, 1986)

Every equation of algebra has as many solutions as the exponent of the highest term indicates. —Albert Girard, 1629, as quoted by David M. Burton in *The History of Mathematics, An Introduction,* 2d ed. (Dubuque, Iowa: Wm. C. Brown Publishers, 1991)

. . . we arrive from all the investigations explained above at the rigorous proof of the theorem that every integral rational algebraic function of one variable can be decomposed into real factors of the first and second degree. —Carl Friedrich Gauss in his doctoral dissertation (1799)

Does every polynomial equation (of degree at least 1) have a root? Or, to put the question another way, is it possible to write a polynomial equation that has no solution? Certainly, if we consider only real roots (i.e., roots that are real numbers), then it is easy to specify an equation with no real roots:

$$x^2 = -1$$

This equation has no real roots because the square of a real number is never negative. On the other hand, if we expand our base of operations from the real-number system to the complex-number system, then $x^2 = -1$ does indeed have a root. In fact, both i and $-i$ are roots in this case.

As it turns out, the situation just described for the equation $x^2 = -1$ holds quite generally. That is, within the complex-number system, every polynomial equation of degree at least 1 has at least one root. This is the substance of a remarkable theorem that was first proved by the great mathematician Carl Friedrich Gauss in 1799. (Gauss was only 22 years old at the time.) Although there are many fundamental theorems in algebra, Gauss's result has come to be known as *the* **fundamental theorem of algebra.**

The Fundamental Theorem of Algebra

Every polynomial equation of the form

$$a_n x^n + a_{n-1} x^{n-1} + \cdots + a_1 x + a_0 = 0 \qquad (n \geq 1, a_n \neq 0)$$

has at least one root within the complex number system. (This root may be a real number.)

Although most of this section deals with polynomials with real coefficients, Gauss's theorem still applies in cases in which some or all of the coefficients are nonreal complex numbers. The proof of Gauss's theorem is usually given in the post-calculus course called complex variables.

The fundamental theorem of algebra asserts that every polynomial equation of degree at least 1 has a root. There are two initial observations to be made here. First, notice that the theorem says nothing about actually finding the root. Second, notice that the theorem deals only with polynomial equations. Indeed, it is easy to specify a nonpolynomial equation that does not have a root. Such an equation is $1/x = 0$. The expression on the left-hand side of this equation can never be zero because the numerator is 1.

EXAMPLE 1 **Understanding what the fundamental theorem says**

Which of the following equations has at least one root?
(a) $x^3 - 17x^2 + 6x - 1 = 0$ **(c)** $x^2 - 2ix + (3 + i) = 0$
(b) $\sqrt{2}x^{47} - \pi x^{25} + \sqrt{3} = 0$

SOLUTION
All three equations are polynomial equations, so, according to the fundamental theorem of algebra, each equation has at least one root.

◼

In Section 1.3 we used factoring as a tool for solving quadratic equations. The next theorem, a consequence of the fundamental theorem of algebra, tells us that (in principle, at least) any polynomial of degree n can be factored into a product of n linear factors. In proving the **linear factors theorem,** we will need to use the factor theorem. If you reread the proof of that theorem in the previous section, you will see that it makes no difference whether the number r appearing in the factor $x - r$ is a real number or a nonreal complex number. Thus, the factor theorem is valid in either case.

The Linear Factors Theorem

Let $f(x) = a_n x^n + a_{n-1} x^{n-1} + \cdots + a_1 x + a_0$, where $n \geq 1$ and $a_n \neq 0$. Then $f(x)$ can be expressed as a product of n linear factors:

$$f(x) = a_n(x - r_1)(x - r_2) \cdots (x - r_n)$$

(The complex numbers r_k appearing in these factors are not necessarily all distinct, and some or all of the r_k may be real numbers.)

Proof of the Linear Factors Theorem. According to the fundamental theorem of algebra, the equation $f(x) = 0$ has a root; let's call this root r_1. By the factor theorem, $x - r_1$ is a factor of $f(x)$, and we can write

$$f(x) = (x - r_1) \cdot Q_1(x)$$

for some polynomial $Q_1(x)$ that has degree $n - 1$ and leading coefficient a_n. If the degree of $Q_1(x)$ happens to be zero, we're done. On the other hand, if the degree of $Q_1(x)$ is at least 1, another application of the fundamental theorem of algebra followed by the factor theorem gives us

$$Q_1(x) = (x - r_2) \cdot Q_2(x)$$

where the degree of $Q_2(x)$ is $n - 2$ and the leading coefficient of $Q_2(x)$ is a_n. We now have

$$f(x) = (x - r_1)(x - r_2) \cdot Q_2(x)$$

We continue this process until the quotient is $Q_n(x) = a_n$. As a result, we obtain

$$f(x) = (x - r_1)(x - r_2) \cdots (x - r_n)a_n$$
$$= a_n(x - r_1)(x - r_2) \cdots (x - r_n)$$

as we wished to show.

The linear factors theorem tells us that any polynomial can be expressed as a product of linear factors. The theorem gives us no information, however, as to how those factors can actually be obtained. The next example demonstrates a case in which the factors are readily obtainable; this is always the case with quadratic polynomials.

> **EXAMPLE   2   Writing quadratics as the products of linear factors**
>
> Express each of the following second-degree polynomials in the form $a_n(x - r_1)(x - r_2)$:   **(a)** $3x^2 - 5x - 2$;   **(b)** $x^2 - 4x + 5$.
>
> **SOLUTION**
> **(a)** A factorization for $3x^2 - 5x - 2$ can be found by simple trial and error. We have
>
> $$3x^2 - 5x - 2 = (3x + 1)(x - 2)$$
>
> We now write the factor $3x + 1$ as $3(x + \frac{1}{3})$. This, in turn, can be written $3[x - (-\frac{1}{3})]$. The final factorization is then
>
> $$3x^2 - 5x - 2 = 3\left[x - \left(-\frac{1}{3}\right)\right](x - 2)$$
>
> **(b)** From the factor theorem, or from our more elementary work with quadratic equations, we know that if r_1 and r_2 are the roots of the equation $x^2 - 4x + 5 = 0$, then $x - r_1$ and $x - r_2$ are the factors of $x^2 - 4x + 5$. That is, $x^2 - 4x + 5 = (x - r_1)(x - r_2)$. The values for r_1 and r_2 in this case are readily obtained by using the quadratic formula. As you can check, the results are
>
> $$r_1 = 2 + i \qquad \text{and} \qquad r_2 = 2 - i$$
>
> The required factorization is therefore
>
> $$x^2 - 4x + 5 = [x - (2 + i)][x - (2 - i)]$$

Using the linear factors theorem, we can show that every polynomial equation of degree $n \geq 1$ has exactly n roots. To help you follow the reasoning, we make two preliminary comments. First, we agree that a root of multiplicity k will be counted as k roots. For example, although the third-degree equation $(x - 1)(x - 4)^2 = 0$ has only two distinct roots—namely, 1 and 4—it has three roots *if* we agree to count the repeated root 4 two times. The second preliminary comment concerns the *zero-product property,* which states that $pq = 0$ if and only if $p = 0$ or $q = 0$. When we stated this in Section 1.3, we were working within the real-number system. However, the property is also valid within the complex-number system. (See Exercises 79 and 80 in Section 7.1.) Now let's state and prove our theorem.

THEOREM

Every polynomial equation of degree $n \geq 1$ has exactly n roots, where a root of multiplicity k is counted k times.

Proof of the Theorem.   Using the linear factors theorem, we can write the nth-degree polynomial equation $f(x) = a_n x^n + a_{n-1} x^{n-1} + \cdots + a_1 x + a_0 = 0$ as

$$f(x) = a_n(x - r_1)(x - r_2) \cdots (x - r_n) = 0 \qquad (a_n \neq 0) \qquad (1)$$

By the factor theorem, each of the numbers $r_1, r_2, \ldots, r_n$ is a root. Some of these numbers may in fact be equal; in other words, we may have repeated roots in this list. In any case, if we agree to count a root of multiplicity k as k roots, then we obtain exactly n roots from the list $r_1, r_2, \ldots, r_n$. Furthermore, the equation $f(x) = 0$ can have no other roots, as we now show. Suppose that r is any number distinct from all the numbers $r_1, r_2, \ldots, r_n$. Replacing x with r in equation (1) yields

$$f(r) = a_n(r - r_1)(r - r_2) \cdots (r - r_n)$$

But the expression on the right-hand side of this last equation cannot be zero, because none of the factors is zero. Thus, $f(r)$ is not zero, and so r is not a root. This completes the proof of the theorem.

EXAMPLE	3	Finding $f(x)$ given the roots and multiplicities of $f(x) = 0$

TABLE 1

Root	Multiplicity
3	2
−2	1
0	2

Find a polynomial $f(x)$ with leading coefficient 1 such that the equation $f(x) = 0$ has only those roots specified in Table 1. What is the degree of this polynomial?

SOLUTION

The expressions $(x - 3)^2$, $x + 2$, and $(x - 0)^2$ all must appear as factors of $f(x)$ and, furthermore, no other linear factor can appear. The form of $f(x)$ is therefore

$$f(x) = a_n(x - 3)^2(x + 2)(x - 0)^2$$

Since the leading coefficient a_n is to be 1, we can rewrite this last equation as

$$f(x) = x^2(x - 3)^2(x + 2)$$

This is the required polynomial. The degree here is 5. This can be seen either by multiplying out the factors or by simply adding the multiplicities of the roots in Table 1. Figure 1 provides a graphical perspective on this example.

GRAPHICAL PERSPECTIVE

Figure 1
The roots of the equation
$f(x) = x^2(x - 3)^2(x + 2) = 0$ are
$-2, 0,$ and 3. The roots 0 and 3
each have multiplicity 2, and, correspondingly, the graph of f is tangent
to the x-axis at those x-intercepts.
The root -2 is a simple root, and
the graph of f is not tangent to the
x-axis at the intercept $x = -2$.

$$f(x) = x^2(x - 3)^2(x + 2)$$
$$[-3, 4, 1] \text{ by } [-40, 40, 20]$$

> **EXAMPLE 4**   **Using the linear factors theorem in determining a quadratic function**

Find a quadratic function f that has zeros of 3 and 5 and a graph that passes through the point $(2, -9)$.

SOLUTION

The general form of a quadratic function with 3 and 5 as zeros is $f(x) = a_n(x - 3)(x - 5)$. Since the graph passes through $(2, -9)$, we have

$$-9 = a_n(2 - 3)(2 - 5) = a_n(3)$$
$$-3 = a_n$$

The required function is therefore $f(x) = -3(x - 3)(x - 5)$. If we wish, we can carry out the multiplication and rewrite this as

$$f(x) = -3x^2 + 24x - 45$$

Figure 2 summarizes this result.

GRAPHICAL PERSPECTIVE

Figure 2
In Example 4 we determined that the function $f(x) = -3x^2 + 24x - 45$ has zeros of 3 and 5 and that the graph passes through the point $(2, -9)$.

$$f(x) = -3x^2 + 24x - 45$$
$$[1, 7, 1] \text{ by } [-12, 6, 3]$$

The next example shows how we can use the factored form of a polynomial $f(x)$ to determine the relationships between the coefficients of the polynomial and the roots of the equation $f(x) = 0$.

> **EXAMPLE 5**   **Applying the linear factors theorem to relate roots and coefficients**

Let r_1 and r_2 be the roots of the equation $x^2 + bx + c = 0$. Show that

$$r_1 r_2 = c \quad \text{and} \quad r_1 + r_2 = -b$$

REMARK   We actually established these two results previously, in Section 2.1. If you look back, you'll see we did this by means of the quadratic formula. The advantage to the following technique is that it does not require a formula for the roots; thus it can be applied to polynomial equations of any degree. (We'll return to this idea after the example.)

SOLUTION

Since r_1 and r_2 are the roots of the equation $x^2 + bx + c = 0$, we have the identity

$$x^2 + bx + c = (x - r_1)(x - r_2)$$

After multiplying out the right-hand side, we can rewrite this identity as

$$x^2 + bx + c = x^2 - (r_1 + r_2)x + r_1r_2$$

By equating coefficients, we readily obtain $r_1r_2 = c$ and $r_1 + r_2 = -b$, as required.

The technique used in Example 5 can be used to obtain similar relationships between the roots and the coefficients of polynomial equations of any given degree. In Table 2, for instance, we show the relationships obtained in Example 5, along with the corresponding relationships that can be derived for a cubic equation. (Exercise 47 at the end of this section asks you to verify the results for the cubic equation.)

TABLE 2

Equation	Roots	Relationships between roots and coefficients
$x^2 + bx + c = 0$	r_1, r_2	$r_1 + r_2 = -b$ $r_1r_2 = c$
$x^3 + bx^2 + cx + d = 0$	r_1, r_2, r_3	$r_1 + r_2 + r_3 = -b$ $r_1r_2 + r_2r_3 + r_3r_1 = c$ $r_1r_2r_3 = -d$

We conclude this section with some remarks concerning the solving of polynomial equations by formulas. You know that the roots of the quadratic equation $ax^2 + bx + c = 0$ are given by the formula

$$x = \frac{-b \pm \sqrt{b^2 - 4ac}}{2a}$$

The question is, are there similar formulas for the solutions of higher-degree equations? By "similar" we mean a formula involving the coefficients and radicals. To answer this question, we look at a bit of history. As early as 1700 B.C., Babylonian mathematicians were able to solve quadratic equations. This is clear from the study of the clay tablets with cuneiform numerals that archeologists have found. The ancient Greeks also were able to solve quadratic equations. Like the Babylonians, the Greeks worked without the aid of algebra as we know it. The mathematicians of ancient Greece used geometric constructions to solve equations. Of course, since all quantities were interpreted geometrically, negative roots were never considered.

The general quadratic formula was known to Islamic mathematicians sometime before A.D. 1000. For the next 500 years, mathematicians searched for, but did not discover, a formula to solve the general cubic equation. Indeed, in 1494, Luca Pacioli stated in his text *Summa di Arithmetica* that the general cubic equation could not be solved by the algebraic techniques then available. All of this was to change, however, within the next several decades.

Around 1515, the Italian mathematician Scipione del Ferro solved the cubic equation $x^3 + px + q = 0$ using algebraic techniques. This essentially constituted a solution of the seemingly more general equation $x^3 + bx^2 + cx + d = 0$. The reason for this is that if we make the substitution $x = y - b/3$ in the latter equation, the result is a cubic equation with no y^2-term. By 1540, the Italian mathematician Ludovico Ferrari had solved the general fourth-degree equation. Actually, at that time in Renaissance Italy, there was considerable controversy as to exactly who discovered the various formulas first. Details of the dispute can be found in any text on the history of mathematics. But for our purposes here, the point is simply that by the middle of the sixteenth century, all polynomial equations of degree 4 or less could be solved by the formulas that had been discovered. The common feature of these formulas was that they involved the coefficients, the four basic operations of arithmetic, and various radicals. For example, a formula for the solution of the equation $x^3 + px + q = 0$ is as follows:

$$ x = \sqrt[3]{\frac{-q}{2} + \sqrt{\frac{q^2}{4} + \frac{p^3}{27}}} + \sqrt[3]{\frac{-q}{2} - \sqrt{\frac{q^2}{4} + \frac{p^3}{27}}} $$

To get some idea of the practical difficulties inherent in computing with this formula, try using it to show that $x = -2$ is a root of the cubic equation $x^3 + 4x + 16 = 0$.

For more than 200 years after the cubic and quartic (fourth-degree) equations had been solved, mathematicians continued to search for a formula that would yield the solutions of the general fifth-degree equation. The first breakthrough, if it can be called that, occurred in 1770, when the French mathematician Joseph Louis Lagrange found a technique that served to unify and summarize all of the previous methods used for the equations of degrees 2, 3, and 4. However, Lagrange then showed that his technique could not work in the case of the general fifth-degree equation. Although we will not describe the details of Lagrange's work here, it is worth pointing out that he relied on the types of relationships between roots and coefficients that we looked at in Table 2 of this section.

Finally, in 1826, the Norwegian mathematician Niels Henrik Abel proved that for the general polynomial equation of degree 5 or higher, there could be no formula yielding the solutions in terms of the coefficients and radicals. This is not to say that such equations do not possess solutions. In fact they must, as we saw earlier in this section. It is just that we cannot in every case express the solutions in terms of the coefficients and radicals. For example, it can be shown that the equation $x^5 - 6x + 3 = 0$ has a real root between 0 and 1, but this number cannot be expressed in terms of the coefficients and radicals. (We can, however, compute the root to as many decimal places as we wish, as you will see in the next section.) In 1831, the French mathematician Evariste Galois completed matters by giving conditions for determining exactly which polynomial equations can be solved in terms of coefficients and radicals.

NOTE Historical background on polynomial equations and the fundamental theorem of algebra is available at the History of Mathematics Web site maintained by the University of St. Andrews, in Scotland. The home page is http://www-groups .dcs.st-and.ac.uk/~history. Also, there are a number of math history textbooks that discuss this material. To cite but three:

1. *A History of Mathematics: An Introduction,* 2nd ed., Victor. J. Katz (Boston: Addison-Wesley, 1998)
2. *History of Mathematics: An Introduction,* 3rd ed., David M. Burton (New York: McGraw Hill, 1997)
3. *Mathematics and Its History,* 2nd ed., John Stillwell (New York: Springer-Verlag, 2001)

EXERCISE SET 7.4

A

According to the fundamental theorem of algebra, which of the equations in Exercises 1 and 2 have at least one root?

1. (a) $x^5 - 14x^4 + 8x + 53 = 0$
   (b) $4.17x^3 + 2.06x^2 + 0.01x + 1.23 = 0$
   (c) $ix^2 + (2 + 3i)x - 17 = 0$
   (d) $x^{2.1} + 3x^{0.3} + 1 = 0$

2. (a) $\sqrt{3}x^{17} + \sqrt{2}x^{13} + \sqrt{5} = 0$
   (b) $17x^{\sqrt{3}} + 13x^{\sqrt{2}} + \sqrt{5} = 0$
   (c) $1/(x^2 + 1) = 0$
   (d) $2^{3x} - 2^x - 1 = 0$

G *In each of Exercises 3–10, you are given a polynomial equation $f(x) = 0$. According to the fundamental theorem of algebra, each of these equations has at least one root. However, the fundamental theorem does not tell you whether the equation has any real-number roots. Use a graph to determine whether the equation has at least one real root.* *Note: You are not being asked to solve the equation.*

3. $x^2 - 3x + 2.26 = 0$ **4.** $x^2 - 2x - 290 = 0$
5. $x^3 - 3x^2 + 3 = 0$ **6.** $x^4 - 3x^2 + 3 = 0$
7. $x^4 + x^3 + x^2 + x + 1 = 0$ **8.** $x^5 + x^3 + x^2 + x + 1 = 0$
9. $0.2x^3 + 4.4x^2 - 109x - 1 = 0$
10. $x^4 - \sqrt{35}x^2 + 2.79\pi = 0$

In Exercises 11–18, express each polynomial in the form $a_n(x - r_1)(x - r_2) \cdots (x - r_n)$.

11. $x^2 - 2x - 3$ **12.** $x^3 - 2x^2 - 3x$
13. $4x^2 + 23x - 6$ **14.** $6x^2 + x - 12$
15. $x^2 - 5$ **16.** $x^2 + 5$
17. $x^5 - 7x^3 - 18x$ **18.** $x^3 + 2x^2 - 3x - 6$

In Exercises 19–24, find a polynomial $f(x)$ with leading coefficient 1 such that the equation $f(x) = 0$ has the given roots and no others. If the degree of $f(x)$ is 7 or more, express $f(x)$ in factored form; otherwise, express $f(x)$ in the form $a_n x^n + a_{n-1} x^{n-1} + \cdots + a_1 x + a_0$.

19.

Root	1	-3
Multiplicity	2	1

20.

Root	0	4
Multiplicity	2	1

21.

Root	2	-2	2i	-2i
Multiplicity	1	1	1	1

22.

Root	2 + i	2 - i
Multiplicity	1	1

23.

Root	$\sqrt{3}$	$-\sqrt{3}$	$4i$	$-4i$
Multiplicity	2	2	1	1

24.

Root	5	1	$1 - i$	$1 + i$
Multiplicity	2	3	1	1

Exercises 25–30 are similar to Exercises 19–24, except that there is now a part (b) asking you to confirm that if r is a multiple root of $f(x) = 0$, then the graph of f is tangent to the x-axis at the corresponding x-intercept. For Exercises 25–30:

(a) *Find a polynomial $f(x)$ with leading coefficient 1 such that the equation $f(x) = 0$ has the given roots and no others. If the degree of $f(x)$ is more than 4, leave $f(x)$ in factored form, rather than multiplying it out.*

G (b) *Use a graphing utility to check the following fact, mentioned in this section: If $x = r$ is a multiple root of $f(x) = 0$, then the graph of the function f is tangent to the x-axis at $x = r$.*

25.

Root	0	1	3
Multiplicity	2	1	1

26.

Root	0	1	3
Multiplicity	1	2	1

27.

Root	0	1	3
Multiplicity	1	2	2

28.

Root	0	1	3
Multiplicity	2	1	2

29.

Root	0	1	3
Multiplicity	2	1	3

30.

Root	0	1	3
Multiplicity	3	2	1

In Exercises 31–34, express the polynomial $f(x)$ in the form $a_n x^n + a_{n-1} x^{n-1} + \cdots + a_1 x + a_0$.

31. Find a quadratic function that has zeros -4 and 9 and a graph that passes through the point $(3, 5)$.

32. (a) Find a quadratic function that has a maximum value of 2 and that has -2 and 4 as zeros.

G (b) Use a graphing utility to check that your answer in part (a) appears to be correct.

33. (a) Find a third-degree polynomial function that has zeros $-5, 2$, and 3 and a graph that passes through the point $(0, 1)$.

(G) **(b)** Use a graphing utility to check that your answer in part (a) appears to be correct.

34. Find a fourth-degree polynomial function that has zeros $\sqrt{2}, -\sqrt{2}, 1$, and -1 and a graph that passes through $(2, -20)$.

In Exercises 35–42, find a quadratic equation with the given roots. Write your answers in the form $Ax^2 + Bx + C = 0$. Suggestion: Make use of Table 2.

35. $r_1 = -i, r_2 = -\sqrt{3}$
36. $r_1 = 1 + i\sqrt{3}, r_2 = 1 - i\sqrt{3}$
37. $r_1 = 9, r_2 = -6$
38. $r_1 = 5, r_2 = 3/4$
39. $r_1 = 1 + \sqrt{5}, r_2 = 1 - \sqrt{5}$
40. $r_1 = 6 - 5i, r_2 = 6 + 5i$
41. $r_1 = a + \sqrt{b}, r_2 = a - \sqrt{b}$ $(b > 0)$
42. $r_1 = a + bi, r_2 = a - bi$

B

43. *Scipio Ferro of Bologna well-nigh thirty years ago discovered this rule and handed it on to Antonio Maria Fior of Venice, whose contest with Niccolò Tartaglia of Brescia gave Niccolò occasion to discover it. He [Tartaglia] gave it to me in response to my entreaties, though withholding the demonstration. Armed with this assistance, I sought out its demonstration in [various] forms.* —Girolamo Cardano, *Ars Magna* (Nuremberg, 1545)

The quotation is from the translation of *Ars Magna* by T. Richard Witmer (New York: Dover Publications, 1993).

In his book *Ars Magna* (*The Great Art*), the Renaissance mathematician Gerolamo Cardano (1501–1576) gave the following formula for a root of the equation $x^3 + ax = b$:

$$x = \sqrt[3]{\frac{b}{2} + \sqrt{\frac{b^2}{4} + \frac{a^3}{27}}} - \sqrt[3]{\frac{-b}{2} + \sqrt{\frac{b^2}{4} + \frac{a^3}{27}}}$$

(a) Use this formula and your calculator to compute a root of the cubic equation $x^3 + 3x = 76$.

(G) **(b)** Use a graph to check the answer in part (a). That is, graph the function $y = x^3 + 3x - 76$, and note the x-intercept. Also check the answer simply by substituting it in the equation $x^3 + 3x = 76$.

44. Consider the cubic equation $x^3 = 15x + 4$, which was solved by Rafael Bombelli in his text *L'Algebra parte maggiore del arithmetica* (Bologna: 1572). Bombelli was among the first to use complex numbers to obtain real-number solutions.

(a) Rewrite the equation in the form $x^3 - 15x = 4$, and apply Cardano's formula (given in the previous exercise). Show that this yields

$$x = \sqrt[3]{2 + 11i} - \sqrt[3]{-2 + 11i}$$

(b) Using paper and pencil (or a computer algebra system), show that

$$(2 + i)^3 = 2 + 11i \quad \text{and} \quad (-2 + i)^3 = -2 + 11i$$

(c) Use the results in part (b) to simplify the root given in part (a). You should obtain $x = 4$.

(G) **(d)** Use a graph to check visually that $x = 4$ is a root of $x^3 = 15x + 4$. (That is, graph the function $y = x^3 - 15x - 4$ and note that $x = 4$ appears to be one of the x-intercepts.) Also, check, algebraically that the value $x = 4$ satisfies the equation.

(e) Given that $x = 4$ is a root of the equation $x^3 = 15x + 4$, use the techniques from Section 7.3 to determine the remaining two roots. Check your answers by using a graphing utility to determine the x-intercepts of the graph in part (d).

45. Express the polynomial $x^4 + 64$ as a product of four linear factors. *Hint*: First add and subtract the quantity $16x^2$ so that you can use the difference-of-squares factoring formula.

46. Suppose that p and q are positive integers with $p > q$. Find a quadratic equation with integer coefficients and roots $\sqrt{p}/(\sqrt{p} \pm \sqrt{p - q})$.

47. Let r_1, r_2, and r_3 be the roots of the equation $x^3 + bx^2 + cx + d = 0$. Use the method shown in Example 5 to verify the following relationships:

$$r_1 + r_2 + r_3 = -b$$
$$r_1 r_2 + r_2 r_3 + r_3 r_1 = c$$
$$r_1 r_2 r_3 = -d$$

48. Let r_1, r_2, r_3, and r_4 be the roots of the equation $x^4 + bx^3 + cx^2 + dx + e = 0$. Use the method shown in Example 5 to prove the following facts:

$$r_1 + r_2 + r_3 + r_4 = -b$$
$$r_1 r_2 + r_2 r_3 + r_3 r_4 + r_4 r_1 + r_2 r_4 + r_3 r_1 = c$$
$$r_1 r_2 r_3 + r_2 r_3 r_4 + r_3 r_4 r_1 + r_4 r_1 r_2 = -d$$
$$r_1 r_2 r_3 r_4 = e$$

49. Solve the equation $x^3 - 75x + 250 = 0$, given that two of the roots are equal. *Suggestion*: Use Table 2.

C

50. (a) Let r_1, r_2, r_3, and r_4 be four real roots of the equation $x^4 + ax^2 + bx + c = 0$. Show that

$$r_1 + r_2 + r_3 + r_4 = 0$$

Hint: Use the first formula in Exercise 48.

(b) Suppose a circle intersects the parabola $y = x^2$ in the points $(x_1, y_1), \ldots, (x_4, y_4)$. Show that

$$x_1 + x_2 + x_3 + x_4 = 0$$

Hint: Use the results in part (a).

PROJECT

TWO METHODS FOR SOLVING CERTAIN CUBIC EQUATIONS

The first person known to have solved cubic equations algebraically was del Ferro but he told nobody of his achievement. On his deathbed, however, del Ferro passed on the secret to his (rather poor) student Fior. Fior began to boast that he was able to solve cubics and a challenge between him and Tartaglia was arranged in 1535. —From the MacTutor History of Mathematics Web pages: http://www-groups.dcs.st-and.ac.uk/~history. The site is maintained by the University of St. Andrews, Scotland.

Stated in modern language, one of the problems that Fior challenged Tartaglia to solve in 1535 was essentially as follows.

A Restatement of One of Fior's Challenge Problems for del Ferro

A tree is 12 units high. At what height should it be cut so that the length of the portion left standing equals the cube root of the length of the portion cut off?

In this project you'll learn two different methods that can be used to determine the root of the cubic equation that solves Fior's tree-cutting problem. As indicated in the figure above, we let x denote the length of the part of the tree left standing. The condition stated in the problem then gives us $x = \sqrt[3]{12 - x}$. After cubing both sides and rearranging, we have the cubic equation to be solved:

$$x^3 + x - 12 = 0 \tag{1}$$

Method 1. To find a root of an equation of the form $x^3 + ax + b = 0$, when $a > 0$, make the substitution $x = y - \dfrac{a}{3y}$. After simplifying, this will produce a sixth-degree equation that is of *quadratic type* and which therefore can be solved as in Section 2.2. Apply this strategy to solve equation (1). *Hints:* At the beginning, you have $a = 1$, and so the substitution to make in equation (1) is $x = y - \dfrac{1}{3y}$. After simplifying and then solving the resulting equation of quadratic type, you should obtain $y = (6 \pm \sqrt{36 + \frac{1}{27}})^{1/3}$. Then, choosing the positive square root for the moment, you'll obtain the required value for x:

$$x = y - \frac{1}{3y} = (6 + \sqrt{36 + \tfrac{1}{27}})^{1/3} - \frac{1}{3(6 + \sqrt{36 + \tfrac{1}{27}})^{1/3}}$$

$$\approx 2.14404 \qquad \text{using a calculator}$$

What value do you obtain for x if you use the negative square root—that is, $y = (6 - \sqrt{36 + \tfrac{1}{27}})^{1/3}$?

Method 2. This method uses function iteration. The explanation of iteration in Section 3.5 is a prerequisite for this discussion. It turns out that for certain functions f, a root of the equation $x = f(x)$ can be determined just by iterating the function f with an appropriate initial input. (Note that our tree-cutting equation was initially in the form $x = f(x)$; it was $x = \sqrt[3]{12 - x}$.)

We'll use Figures A and B to explain why iterating a function f can determine a root of the equation $x = f(x)$. Figure A shows the graph of a function f, along with the line $y = x$. Let (r, r) denote the point where the graph of $y = f(x)$ intersects the line $y = x$. (Why do the two coordinates of that point have to be the same?) The number r is then a root of the equation $x = f(x)$. We'd like to be able to compute the value of r to as many decimal places as might be required. Figure B shows the iteration process for f using an initial input x_0. From the pattern, you can see that the iterates approach the number r. Thus, we can, in principle at least, determine the root r to as many decimal places as we want by carrying out the iteration process sufficiently many times.

Figure A   Figure B

TABLE 1   The First Six Iterates of $x_0 = 5$ Under the Function $f(x) = \sqrt[3]{12 - x}$.

x_1	1.91293 . . .
x_2	2.16066 . . .
x_3	2.14283 . . .
x_4	2.14412 . . .
x_5	2.14403 . . .
x_6	2.14404 . . .

The numbers are not rounded. The digits in red show the agreement with the value obtained using Method 1.

Now let's see how this works in the case of Fior's tree-cutting equation $x = \sqrt[3]{12 - x}$. The function to be iterated is $f(x) = \sqrt[3]{12 - x}$. We need an initial input x_0. In general, one first makes an estimate of what the root might be and then uses that estimate or something close to it for x_0. But for the present function f, a remarkable fact is that any initial input x_0 will do. (A case where the choice of the initial input is important is given in Exercise 1(c) at the end of this project.) Selecting (somewhat at random) the initial input $x_0 = 5$ now, we obtain the results shown in Table 1. You should check the entries in this table yourself, and then set up similar tables using different initial inputs. In each case, you will find that the iterates approach the required root, as determined in Method 1.

EXERCISES

1. Consider the cubic equation $g(x) = x^3 + 6x - 2 = 0$.

   (a) Use a graphing utility to draw the graph of g. By zooming in on the x-intercept, approximate to the nearest hundredth the root of the given equation.

   (b) Use Method 1 to obtain an exact expression for the root. Then, use a calculator to check that the value is consistent with the result in part (a).

   (c) From the equation $x^3 + 6x - 2 = 0$, we have $6x = 2 - x^3$, and consequently $x = (2 - x^3)/6$. Use Method 2, with $f(x) = (2 - x^3)/6$, to determine a root of the equation $x = (2 - x^3)/6$. For the initial input x_0, use any number in the closed interval $[-2, 2]$. What happens if you take $x_0 = 3$?

(d) From the equation $x^3 + 6x - 2 = 0$, we have $x^3 = 2 - 6x$, and consequently $x = (2 - 6x)^{1/3}$. Try using Method 2, with $f(x) = (2 - 6x)^{1/3}$, to determine a root. Experiment with various values for x_0.

2. (a) Use Method 1 to find a real root of the cubic equation $x^3 + 3x - 36 = 0$.
   Answer: $x = (18 + 5\sqrt{13})^{1/3} - 1/(18 + 5\sqrt{13})^{1/3}$

(b) Use a calculator to evaluate the answer in part (a). Is the value shown on the calculator actually a root? Find all roots of the given equation.

3. (a) Method 1 will always yield a real root of the equation $x^3 + ax + b = 0$, when $a > 0$. In some cases, it also works for $a < 0$. As an example, use Method 1 to find a root of the equation $x^3 - 3x + 4 = 0$. For the answer, give both the exact expression containing radicals and a calculator approximation rounded to four decimal places.

(b) The equation $x^3 - 3x + 4 = 0$ can be rewritten $x = (x^3 + 4)/3$. Try using Method 2, with $f(x) = (x^3 + 4)/3$, to determine a root of the equation.

(c) Let $f(x) = (x^3 + 4)/3$ and take $x_0 = -2$. Graphically display the iteration process for $x_0 = -2$ under the function f. (Go at least as far as x_4.) Then do the same using $x_0 = -1$. What do you observe? Use your observations to explain the outcome in part (b).

7.5 RATIONAL AND IRRATIONAL ROOTS

Jacques Peletier (1517–1582), a French man of letters, poet, and mathematician, had observed as early as 1558, that the root of an equation is a divisor of the last term. —Florian Cajori in *A History of Mathematics,* 4th ed. (New York: Chelsea Publishing Co., 1985)

As we saw in the previous section, not every polynomial equation has a real root. Furthermore, even if a polynomial equation does possess a real root, that root isn't necessarily a rational number. (The equation $x^2 = 2$ provides a simple example.) If a polynomial equation with integer coefficients does have a rational root, however, we can find that root by applying the **rational roots theorem,** which we now state.

The Rational Roots Theorem

Consider the polynomial equation

$$a_n x^n + a_{n-1} x^{n-1} + \cdots + a_1 x + a_0 = 0 \qquad (n \geq 1, a_n \neq 0)$$

and suppose that all the coefficients are integers. Let p/q be a rational number, where p and q have no common factors other than ± 1. If p/q is a root of the equation, then p is a factor of a_0 and q is a factor of a_n.

A proof of the rational roots theorem is outlined in Exercise 41 at the end of this section. For the moment, though, let's just see why the theorem is plausible. Suppose that the two rational numbers a/b and c/d are the roots of a certain quadratic equation. Then, from our experience with quadratics (or by the linear factors theorem), we know that the equation can be written in the form

$$k\left(x - \frac{a}{b}\right)\left(x - \frac{c}{d}\right) = 0 \qquad (1)$$

where k is a constant. Now, as Exercise 30 will ask you to check, if we carry out the multiplication and clear of fractions, equation (1) becomes

$$(kbd)x^2 - (kad + kbc)x + kac = 0 \tag{2}$$

Observe that a and c (the numerators of the two roots) are factors of the constant term kac in equation (2), just as the rational roots theorem asserts. Furthermore, b and d (the denominators of the roots) are factors of the coefficient of the x^2-term in equation (2), again as the theorem asserts.

The following example shows how the rational roots theorem can be used to solve a polynomial equation.

| EXAMPLE | 1 | Applying the rational roots theorem |

Find the rational roots (if any) of the equation $2x^3 - x^2 - 9x - 4 = 0$. Then solve the equation.

SOLUTION
First we list the factors of a_0, the factors of a_n, and the possibilities for rational roots:

factors of $a_0 = -4$: $\pm 1, \pm 2, \pm 4$

factors of $a_3 = 2$: $\pm 1, \pm 2$

possible rational roots: $\pm\dfrac{1}{1}, \pm\dfrac{1}{2}, \pm\dfrac{2}{1}, \pm\dfrac{4}{1}$

Now we can use synthetic division to test whether or not any of these possibilities is a root. (A zero remainder will tell us that we have a root.) As you can check, the first three possibilities ($1, -1,$ and $1/2$) are not roots. However, using $-1/2$, we have

$$
\begin{array}{r|rrrr}
-1/2 & 2 & -1 & -9 & -4 \\
     &   & -1 & 1 & 4 \\
\hline
     & 2 & -2 & -8 & 0
\end{array}
$$

Thus, $x = -1/2$ is a root. We could now continue to check the remaining possibilities in this same manner. At this point, however, it is simpler to consider the reduced equation $2x^2 - 2x - 8 = 0$, or $x^2 - x - 4 = 0$. Since this is a quadratic equation, it can be solved directly. We have

$$x = \frac{-(-1) \pm \sqrt{(-1)^2 - 4(1)(-4)}}{2(1)} = \frac{1 \pm \sqrt{17}}{2}$$

We have now determined three distinct roots. Since the degree of the original equation is 3, there can be no other roots. We conclude that $x = -1/2$ is the only rational root. The three roots of the equation are $-1/2$ and $(1 \pm \sqrt{17})/2$.

As Example 1 indicates, the number of possibilities for rational roots can be relatively large, even for rather simple equations. The next theorem we develop allows us to reduce the number of possibilities. We say that a real number B is an **upper bound** for the roots of an equation if every real root is less than or equal to B.

Similarly, a real number b is a **lower bound** if every real root is greater than or equal to b. The following theorem tells us how synthetic division can be used in determining upper and lower bounds for roots.

The Upper and Lower Bound Theorem for Real Roots

Consider the polynomial equation

$$f(x) = a_n x^n + a_{n-1} x^{n-1} + \cdots + a_1 x + a_0 = 0$$

where all of the coefficients are real numbers and a_n is positive.

1. If we use synthetic division to divide $f(x)$ by $x - B$, where $B > 0$, and we obtain a third row containing no negative numbers, then B is an upper bound for the real roots of $f(x) = 0$.
2. If we use synthetic division to divide $f(x)$ by $x - b$, where $b < 0$, and we obtain a third row in which the numbers are alternately positive and negative, then b is a lower bound for the real roots of $f(x) = 0$. (In determining whether the signs alternate in the third row, zeros are counted as either positive or negative.)

CAUTION   A number may fail the lower bound test but still be a lower bound. For an example, see Exercise 55.

We will prove the first part of this theorem. A proof of the second part can be developed along similar lines. To prove the first part of the theorem, we use the division algorithm to write

$$f(x) = (x - B) \cdot Q(x) + R \tag{3}$$

The remainder R here is a constant that may be zero. To show that B is an upper bound, we must show that any number greater than B is not a root. Toward this end, let p be a number that is greater than B. Note that p must be positive, since B is positive. Then with $x = p$, equation (3) becomes

$$f(p) = (p - B) \cdot Q(p) + R \tag{4}$$

We will now show that the right-hand side of equation (4) is positive. This will tell us that p is not a root. First, look at the factor $(p - B)$. This is positive, since p is greater than B. Next consider $Q(p)$. By hypothesis, the coefficients of $Q(x)$ are all nonnegative. Furthermore, the leading coefficient of $Q(x)$ is a_n, which is positive. Since p is also positive, it follows that $Q(p)$ must be positive. Finally, the number R is nonnegative because, in the synthetic division of $f(x)$ by $x - B$, all the numbers in the third row are nonnegative. It now follows that the right-hand side of equation (4) is positive. Consequently, $f(p)$ is not zero and p is not a root of the equation $f(x) = 0$. This is what we wished to show.

EXAMPLE 2   **Solving an equation using the rational roots theorem and the upper and lower bound theorem**

Determine the rational roots, or show that none exist, for the equation

$$\frac{1}{4}x^4 - \frac{3}{4}x^3 + \frac{17}{4}x^2 + 4x + 5 = 0$$

SOLUTION

We will use the rational roots theorem along with the upper and lower bound theorem. First of all, if we are to apply the rational roots theorem, then our equation must have integer coefficients. In view of this, we multiply both sides of the given equation by 4 to obtain

$$x^4 - 3x^3 + 17x^2 + 16x + 20 = 0$$

As in the previous example, we list the factors of a_0, the factors of a_n, and the possibilities for rational roots:

factors of $a_0 = 20$: $\pm 1, \pm 2, \pm 4, \pm 5, \pm 10, \pm 20$

factors of $a_4 = 1$: ± 1

possible rational roots: $\pm 1, \pm 2, \pm 4, \pm 5, \pm 10, \pm 20$

Our strategy here will be to first check for positive roots, beginning with 1 and working upward. The checks for $x = 1$, $x = 2$, and $x = 4$ are as follows:

$\underline{1}$	1	−3	17	16	20
		1	−2	15	31
	1	−2	15	31	51

$\underline{2}$	1	−3	17	16	20
		2	−2	30	92
	1	−1	15	46	112

$\underline{4}$	1	−3	17	16	20
		4	4	84	400
	1	1	21	100	420

As you can see, none of the remainders here is zero. However, notice that in the division corresponding to $x = 4$, all the numbers appearing in the third row are nonnegative. It therefore follows that 4 is an upper bound for the roots of the given equation. In view of this, we needn't bother to check the remaining values $x = 5$, $x = 10$, and $x = 20$, since none of those can be roots. At this point we can conclude that the given equation has no positive rational roots.

Next we check for negative rational roots, beginning with -1 and working downward (if necessary). Checking $x = -1$, we have

$\underline{-1}$	1	−3	17	16	20
		−1	4	−21	5
	1	−4	21	−5	25

Two conclusions can be drawn from this synthetic division. First, $x = -1$ is not a root of the equation. Second, -1 is a lower bound for the roots because the signs in the third row of the synthetic division alternate. This means that we needn't bother to check any of the numbers $-2, -4, -5, -15$, and -20: none of them can be roots, since they are all less than -1.

Let's summarize our results. We have shown that the given equation has no positive rational roots and no negative rational roots. Furthermore, by inspection we see that zero is not a root of the equation. Thus, the given equation possesses no rational roots.

We conclude this section by demonstrating a method for approximating irrational roots. The method depends on the **location theorem.**

The Location Theorem

Let $f(x)$ be a polynomial, all of whose coefficients are real numbers. If a and b are real numbers such that $f(a)$ and $f(b)$ have opposite signs, then the equation $f(x) = 0$ has at least one real root between a and b.

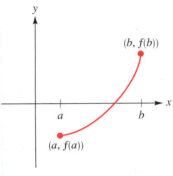

Figure 1

Figure 1 indicates why this theorem is plausible. If the point $(a, f(a))$ lies below the x-axis and $(b, f(b))$ lies above the x-axis, then it certainly seems that the graph of f must cross the x-axis at some point x_0 between a and b. At this intercept we have $f(x_0) = 0$; that is, x_0 is a root of the equation $f(x) = 0$. (The location theorem is a special case of the *intermediate value theorem;* the proof is usually discussed in calculus courses.)

Our technique for approximating (or "locating") irrational roots uses the **method of successive approximations.** We will demonstrate this method in Example 3.

EXAMPLE	3	Using the method of successive approximations to locate a root

As indicated by both Table 1 and Figure 2, the equation

$$f(x) = x^5 - 3x^2 + 4x + 2 = 0$$

has a real root between -1 and 0. Use the method of successive approximations to locate this root between successive hundredths.

TABLE 1 $f(x) = x^5 - 3x^2 + 4x + 2*$

x	-3	-2	-1	0	1	2
f(x)	-280	-50	-6	2	4	30

sign change

*The sign change is a signal that the equation $x^5 - 3x^2 + 4x + 2 = 0$ has a real root between -1 and 0.

GRAPHICAL PERSPECTIVE

$f(x) = x^5 - 3x^2 + 4x + 2$
$[-2, 2, 1]$ by $[-15, 15, 15]$

Figure 2

SOLUTION

The table or the graph gives us the location of the root between the successive integers -1 and 0. The strategy now is to locate the root between successive tenths, and then between successive hundredths. In Table 2, we've used a graphing utility to generate values of $f(x)$ for inputs x running from -1 up to 0 in increments of 0.1 (For details on using a graphing utility to create a table of function values, refer to the owner's manual for your graphing utility.) If a graphing utility is not available, the calculations can be carried out using synthetic division, the remainder theorem, and an ordinary calculator.

TABLE 2	Decimal values for $f(x)$ rounded to two places*									
x	−1.0	−0.9	−0.8	−0.7	−0.6	−0.5	−0.4	−0.3	−0.2	−0.1	0.0
f(x)	−6	−4.62	−3.45	−2.44	−1.56	−0.78	−0.09	0.53			

sign change

*Once a sign change is detected, it is not necessary to look at the remaining entries in the table
(unless one is looking for a possible second root).

As indicated in Table 2, there is a change in the sign of $f(x)$ as we go from
$x = -0.4$ to $x = -0.3$. Thus, according to the location theorem, the equation
$f(x) = 0$ has a root in the open interval $(-0.4, -0.3)$.

Next, we follow a similar procedure to locate the root between successive
hundredths. This time, the inputs begin with $x = -0.40$ and they increase in increments of 0.01. See Table 3.

TABLE 3	Decimal values for $f(x)$ rounded to two places*									
x	−0.40	−0.39	−0.38	−0.37	−0.36	−0.35	−0.34	−0.33	−0.32	−0.31	−0.30
f(x)	−0.09	−0.03	0.04								

sign change

*The sign change indicates the presence of a root in the open interval $(-0.39, -0.38)$.

In view of the location theorem, the sign change in Table 3 tells us that the
given equation has a root between $x = -0.39$ and $x = -0.38$. So, we've located
the root between successive hundredths, as required.

REMARK   An alternate procedure for this example would be to begin immediately with a table of values with the inputs running from $x = -1$ to 0, in increments of 0.01. Which method uses fewer calculations?

EXERCISE SET 7.5

A

1. (a) State the rational roots theorem.
   (b) List the possibilities for the rational roots of the equation $x^7 - 144x^2 - 8x - 11 = 0$.
2. Use the rational roots theorem to list the possibilities for
   the rational roots of each equation.
   (a) $x^4 - 32x^3 + 40x^2 + 12x - 3 = 0$
   (b) $3x^4 - 32x^3 + 40x^2 + 12x - 1 = 0$
3. (a) Use the rational roots theorem to explain why the
   equation $x^5 - 14x - 5 = 0$ cannot have a rational root
   that is larger than 5. Is 5 a root?
   (b) The graph in the following figure indicates that 2 may
   be a root of the equation $x^5 - 14x - 5 = 0$. Use the
   rational roots theorem to explain why 2 cannot be a
   root. Also, show directly that 2 is not a root by substituting 2 in the given equation.

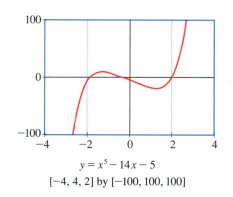

$y = x^5 - 14x - 5$

$[-4, 4, 2]$ by $[-100, 100, 100]$

(c) With a graphing utility, graph $y = x^5 - 14x - 5$; use a
viewing rectangle that shows 2 is not a root of the
equation $x^5 - 14x - 5 = 0$.

4. (a) The graph in the following figure indicates that 4 may be a root of the equation $x^6 - 4x^5 + 8x^2 - 3x - 101 = 0$. Use the rational roots theorem to explain why 4 cannot be a root.

$$y = x^6 - 4x^5 + 8x^2 - 3x - 101$$

$[-8, 8, 4]$ by $[-400, 2000, 400]$

(b) With a graphing utility, graph $y = x^6 - 4x^5 + 8x^2 - 3x - 101$; use a viewing rectangle that shows 4 is not a root of the equation $x^6 - 4x^5 + 8x^2 - 3x - 101 = 0$.

In Exercises 5–10, list the possibilities for rational roots.

5. $4x^3 - 9x^2 - 15x + 3 = 0$ **6.** $x^4 - x^3 + 10x^2 - 24 = 0$
7. $8x^5 - x^2 + 9 = 0$ **8.** $18x^4 - 10x^3 + x^2 - 4 = 0$
9. $\frac{2}{3}x^3 - x^2 - 5x + 2 = 0$
10. $\frac{1}{2}x^4 - 5x^3 + \frac{4}{3}x^2 + 8x - \frac{1}{3} = 0$

In Exercises 11–16, show that each equation has no rational roots.

11. $x^3 - 3x + 1 = 0$ **12.** $x^3 + 8x^2 - 1 = 0$
13. $x^3 + x^2 - x + 1 = 0$ **14.** $x^4 + 4x^3 + 4x^2 - 16 = 0$
15. $12x^4 - x^2 - 6 = 0$
16. $4x^5 - x^4 - x^3 - x^2 + x - 8 = 0$

For Exercises 17–27, find the rational roots of each equation and then solve the equation. (Use the rational roots theorem and the upper and lower bound theorem, as in Example 2.)

17. $x^3 + 3x^2 - x - 3 = 0$ **18.** $2x^3 - 5x^2 - 3x + 9 = 0$
19. $4x^3 + x^2 - 20x - 5 = 0$ **20.** $3x^3 - 16x^2 + 17x - 4 = 0$
21. $9x^3 + 18x^2 + 11x + 2 = 0$ **22.** $4x^3 - 10x^2 - 25x + 4 = 0$
23. $x^4 + x^3 - 25x^2 - x + 24 = 0$
24. $10x^4 + 107x^3 + 301x^2 + 171x + 23 = 0$
25. $x^4 - 4x^3 + 6x^2 - 4x + 1 = 0$
26. $x^3 - \frac{5}{2}x^2 - 23x + 12 = 0$
27. $x^3 - \frac{17}{3}x^2 - \frac{10}{3}x + 8 = 0$

In Exercises 28 and 29, determine integral upper and lower bounds for the real roots of the equations. (Follow the method used within the solution of Example 2.)

28. (a) $x^3 + 2x^2 - 5x + 20 = 0$
(b) $x^5 - 3x^2 + 100 = 0$

29. (a) $5x^4 - 10x - 12 = 0$
(b) $3x^4 - 4x^3 + 5x^2 - 2x - 4 = 0$
(c) $2x^4 - 7x^3 - 5x^2 + 28x - 12 = 0$

30. Referring to equation (1) in this section, multiply out the left-hand side, then clear the equation of fractions. Check that your result agrees with equation (2).

In Exercises 31–36, each equation has exactly one positive root. In each case, locate the root between successive hundredths. For Exercises 31 and 32, you are given the successive integer bounds for the root. For the other exercises, determine the successive integer bounds by computing $f(0), f(1), f(2)$, and so on, until you find a sign change.

31. $x^3 + x - 1 = 0$; root between 0 and 1
32. $x^3 - 2x - 5 = 0$; root between 2 and 3
33. $x^5 - 200 = 0$
34. $x^3 - 3x^2 + 3x - 26 = 0$
35. $x^3 - 8x^2 + 21x - 22 = 0$
36. $2x^4 - x^3 - 12x^2 - 16x - 8 = 0$

In Exercises 37–40, each polynomial equation has exactly one negative root.

(a) Use a graphing utility to determine successive integer bounds for the root.
(b) Use the method of successive approximations to locate the root between successive thousandths. (Make use of the graphing utility to generate the required tables.)

37. $x^3 + x^2 - 2x + 1 = 0$
38. $\dfrac{x^5}{10,000} - \dfrac{x^3}{50} + \dfrac{x}{1250} + \dfrac{1}{2000} = 0$
39. $x^3 + 2x^2 + 2x + 101 = 0$
40. $x^4 + 4x^3 - 6x^2 - 8x - 3 = 0$

B

41. This exercise outlines a proof of the rational roots theorem. At one point in the proof, we will need to rely on the following fact, which is proved in courses on number theory.

FACT FROM NUMBER THEORY   Suppose that A, B, and C are integers and that A is a factor of the number BC. If A has no factor in common with C (other than ± 1), then A must be a factor of B.

(a) Let $A = 2$, $B = 8$, and $C = 5$. Verify that the fact from number theory is correct here.
(b) Let $A = 20$, $B = 8$, and $C = 5$. Note that A is a factor of BC, but A is not a factor of B. Why doesn't this contradict the fact from number theory?
(c) Now we're ready to prove the rational roots theorem. We begin with a polynomial equation with integer coefficients:

$$a_n x^n + a_{n-1}x^{n-1} + \cdots + a_1 x + a_0 = 0 \qquad (n \geq 1, a_n \neq 0)$$

We assume that the rational number p/q is a root of the equation and that p and q have no common factors other than 1. Why is the following equation now true?

$$a_n\left(\frac{p}{q}\right)^n + a_{n-1}\left(\frac{p}{q}\right)^{n-1} + \cdots + a_1\left(\frac{p}{q}\right)a_0 = 0$$

(d) Show that the last equation in part (c) can be written

$$p(a_np^{n-1} + a_{n-1}qp^{n-2} + \cdots + a_1q^{n-1}) = -a_0q^n$$

Since p is a factor of the left-hand side of this last equation, p must also be a factor of the right-hand side. That is, p must be a factor of a_0q^n. But since p and q have no common factors, neither do p and q^n. Our fact from number theory now tells us that p must be a factor of a_0, as we wished to show. (The proof that q is a factor of a_n is carried out in a similar manner.)

42. The location theorem asserts that the polynomial equation $f(x) = 0$ has a root in the open interval (a, b) whenever $f(a)$ and $f(b)$ have unlike signs. If $f(a)$ and $f(b)$ have the same sign, can the equation $f(x) = 0$ have a root between a and b? *Hint*: Look at the graph of $f(x) = x^2 - 2x + 1$ with $a = 0$ and $b = 2$.

In Exercises 43–47, first graph the two functions. Then use the method of successive approximations to locate, between successive thousandths, the x-coordinate of the point where the graphs intersect. In Exercises 43 and 44, draw the graphs by hand. In Exercises 45–47, use a graphing utility to draw the graphs as well as to check your final answer. Finally, in Exercise 47, also check your answer by using an algebraic method to obtain the exact solution (as in Section 6.6).

43. $y = x^3 - 5$; $y = 2x - 3$

44. $y = x^3$; $y = 4 - x^2$

G 45. $y = e^{-x}$; $y = \ln x$ *Remark*: The method of successive approximations is not restricted to polynomial functions.

G 46. $y = 20x^2 + 25x + 9$; $y = x^3$

G 47. $y = x^5 + 100$; $y = x^5 + \frac{1}{2}x^3$

48. In a note that appeared in *The Two-Year College Mathematics Journal* [vol. 12 (1981), pp. 334–336], Professors Warren Page and Leo Chosid explain how the process of testing for rational roots can be shortened. In essence, their result is as follows. Suppose that we have a polynomial with integer coefficients and we are testing for a possible root p/q. Then, if a noninteger is generated at any point in the synthetic division process, p/q cannot be a root of the polynomial. For example, suppose we want to know whether $4/3$ is a root of $6x^4 - 10x^3 + 2x^2 - 9x + 8 = 0$. The first few steps of the synthetic division are as follows.

$$
\begin{array}{r|rrrrr}
\frac{4}{3} & 6 & -10 & 2 & -9 & 8 \\
 & & 8 & -\frac{8}{3} & & \\
\hline
 & 6 & -2 & & &
\end{array}
$$

Since the noninteger $-8/3$ has been generated in the synthetic division process, the process can be stopped;

$4/3$ is not a root of the polynomial. Use this idea to shorten your work in testing to see whether the numbers $3/4$, $1/8$, and $-3/2$ are roots of the equation

$$8x^5 - 5x^4 + 3x^2 - 2x - 6 = 0$$

49. In a note that appeared in *The College Mathematics Journal* [vol. 20 (1989), pp. 139–141], Professor Don Redmond proved the following interesting result.

Consider the polynomial equation $f(x) = a_nx^n + a_{n-1}x^{n-1} + \cdots + a_1x + a_0 = 0$, and suppose that the degree of $f(x)$ is at least 2 and that all of the coefficients are integers. If the three numbers a_0, a_n, and $f(1)$ are all odd, then the given equation has no rational roots.

Use this result to show that the following equations have no rational roots.

(a) $9x^5 - 8x^4 + 3x^2 - 2x + 27 = 0$

(b) $5x^5 + 5x^4 - 11x^2 - 3x - 25 = 0$

G 50. The following result is a particular case of a theorem proved by Professor David C. Kurtz in *The American Mathematical Monthly* [vol. 99 (1992), pp. 259–263].

Suppose we have a cubic equation $a_3x^3 + a_2x^2 + a_1x + a_0 = 0$ in which all of the coefficients are positive real numbers. Furthermore, suppose that the following two inequalities hold.

$$a_1^2 > 4a_0a_2 \quad \text{and} \quad a_2^2 > 4a_1a_3$$

Then the cubic equation has three distinct real roots.

(a) Check that these inequalities are valid in the case of the equation $2x^3 + 8x^2 + 7x + 1 = 0$. This implies that the equation has three distinct real roots. Use a graphing utility to verify this and to estimate each root to the nearest one hundredth.

(b) Follow part (a) for the equation $3x^3 + 40x^2 + 100x + 30 = 0$.

(c) Use a graphing utility to demonstrate that the graph of $y = 6x^3 + 15x^2 + 11x + 2$ has three distinct x-intercepts. Thus, the equation $6x^3 + 15x^2 + 11x + 2 = 0$ has three distinct real roots. Now check that the condition $a_2^2 > 4a_1a_3$ fails to hold in this case. Explain why this does not contradict the result from Professor Kurtz stated above.

C

*In Exercises 51–54, you need to know that a **prime number** is a positive integer greater than 1 with no factors other than itself and 1. Thus, the first seven prime numbers are 2, 3, 5, 7, 11, 13, and 17.*

51. Find all prime numbers p for which the equation $x^2 + x - p = 0$ has a rational root.

52. Find all prime numbers p for which the equation $x^3 + x^2 + x - p = 0$ has at least one rational root. For each value of p that you find, find the corresponding *real* roots of the equation.

53. Consider the equation $x^3 + px - q = 0$, where p and q are prime numbers. Observe that there are only four possible rational roots here: 1, -1, q, and $-q$.

    (a) Show that if $x = 1$ is a root, then we must have $q = 3$ and $p = 2$. What are the remaining roots in this case?

    (b) Show that none of the numbers -1, q, and $-q$ can be a root of the equation. *Hint:* For each case, assume the contrary and deduce a contradiction.

54. If p and q are prime numbers, show that the equation $x^3 + px - pq = 0$ has no rational roots.

55. Let $f(x) = x^3 + 3x^2 - x - 3$.

    (a) Factor $f(x)$ by using the basic factoring techniques in Appendix A.4.

    (b) Sketch the graph of $f(x) = x^3 + 3x^2 - x - 3$. Note that -3 is a lower bound for the roots.

    (c) Show that the number -3 fails the lower bound test. This shows that a number may fail the lower bound test and yet be a lower bound. (We say that the lower bound test provides a *sufficient* but not a *necessary* condition for a lower bound.)

7.6 CONJUGATE ROOTS AND DESCARTES'S RULE OF SIGNS

An equation can have as many true [positive] roots as it contains changes of sign, from plus to minus or from minus to plus; and as many false [negative] roots as the number of times two plus signs or two minus signs are found in succession. —René Descartes (1637)

We now proceed to investigate a remarkable theorem, implicit in the work of [Thomas] Harriot [1560–1621] but first used explicitly by Descartes (1637), which limits the number of positive or negative roots of an equation. . . .

Remarkable as the Harriot–Descartes's Rule of Signs is, it still leaves uncertainty as to the exact number of real roots in an equation: it only gives an upper limit to them. The problem of finding an exact test . . . was finally solved in 1829 by [Jacques Charles François] Sturm. —H. W. Turnbull in *Theory of Equations* (Edinburgh: Oliver and Boyd, 1939)

As you know from earlier work involving quadratic equations with real coefficients, when nonreal complex roots occur, they occur in conjugate pairs. For instance, as you can check by means of the quadratic formula, the roots of the equation $x^2 - 2x + 5 = 0$ are $1 + 2i$ and $1 - 2i$. The **conjugate roots theorem** tells us that the situation is the same for all polynomial equations with real coefficients.

The Conjugate Roots Theorem

Let $f(x)$ be a polynomial, all of whose coefficients are real numbers. Suppose that $a + bi$ is a root of the equation $f(x) = 0$, where a and b are real and $b \neq 0$. Then $a - bi$ is also a root of the equation.

To prove the conjugate roots theorem, we use four of the properties of complex conjugates listed in Section 7.1:

Property 1: $\overline{z_1 z_2} = \overline{z_1}\,\overline{z_2}$

Property 2: $(\overline{z})^m = \overline{z^m}$

Property 3: $\overline{r} = r$ for every real number r

Property 4: $\overline{z_1} + \overline{z_2} = \overline{z_1 + z_2}$

Using these properties, we can prove the theorem as follows. We begin with a polynomial with real coefficients:

$$f(x) = a_n x^n + a_{n-1}x^{n-1} + \cdots + a_1 x + a_0$$

We must show that if $z = a + bi$ is a root of $f(x) = 0$, then $\overline{z} = a - bi$ is also a root. We have

$$f(\bar{z}) = a_n(\bar{z})^n + a_{n-1}(\bar{z})^{n-1} + \cdots + a_1\bar{z} + a_0$$

$$= \overline{a_n}\,\overline{z^n} + \overline{a_{n-1}}\,\overline{z^{n-1}} + \cdots + \overline{a_1}\,\overline{z} + \overline{a_0} \qquad \text{Properties 3 and 2}$$

$$= \overline{a_n z^n} + \overline{a_{n-1}z^{n-1}} + \cdots + \overline{a_1 z} + \overline{a_0} \qquad \text{Property 1}$$

$$= \overline{a_n z^n + a_{n-1}z^{n-1} + \cdots + a_1 z + a_0} \qquad \text{Property 4}$$

$$= \overline{f(z)} = \bar{0} \qquad f(z) = 0, \text{ since } z \text{ is a root}$$

$$= 0 \qquad \text{Property 3}$$

We have now shown that $f(\bar{z}) = 0$, given that $f(z) = 0$. Thus, $\bar{z}$ is a root, as we wished to show.

Although the conjugate roots theorem concerns nonreal complex roots, it can nevertheless be used to obtain information about real roots, as the next two examples demonstrate.

EXAMPLE 1   **Using the conjugate roots theorem in solving an equation**

Solve the equation $f(x) = 2x^4 - 3x^3 + 12x^2 + 22x - 60 = 0$, given that one root is $1 + 3i$.

SOLUTION
Since all of the coefficients of $f(x)$ are real numbers, we know that the conjugate of $1 + 3i$ must also be a root. Thus, $1 + 3i$ and $1 - 3i$ are roots, from which it follows that $[x - (1 + 3i)]$ and $[x - (1 - 3i)]$ are factors of $f(x)$. As you can check, the product of these two factors is $x^2 - 2x - 10$. Thus we must have

$$f(x) = (x^2 - 2x + 10) \cdot Q(x)$$

for some polynomial $Q(x)$. We compute $Q(x)$ using long division:

$$
\begin{array}{r}
2x^2 + x - 6 \\
x^2 - 2x + 10 \,\overline{\big)\, 2x^4 - 3x^3 + 12x^2 + 22x - 60} \\
\underline{2x^4 - 4x^3 + 20x^2} \\
x^3 - 8x^2 + 22x \\
\underline{x^3 - 2x^2 + 10x} \\
-6x^2 + 12x - 60 \\
\underline{-6x^2 + 12x - 60} \\
0
\end{array}
$$

Thus, $Q(x) = 2x^2 + x - 6$, and the original equation becomes

$$f(x) = (x^2 - 2x + 10)(2x^2 + x - 6) = 0$$

We can now find any additional roots by solving the equation $2x^2 + x - 6 = 0$. We have

$$2x^2 + x - 6 = 0$$
$$(2x - 3)(x + 2) = 0$$

$$2x - 3 = 0 \qquad \Big| \qquad x + 2 = 0$$
$$x = 3/2 \qquad \Big| \qquad x = -2$$

We now have four distinct roots of the original equation: $1 + 3i$, $1 - 3i$, $3/2$, and -2. Since the degree of the equation is 4, there can be no other roots.

EXAMPLE	2	An analysis requiring the rational roots theorem and the conjugate roots theorem

Show that the equation $x^3 - 2x^2 + x - 1 = 0$ has at least one irrational root.

SOLUTION

Allowing for multiple roots, the equation has three roots. We will show that at least one of these three roots must be irrational. For convenience, we break the argument into two cases.

Case 1 *All three roots are real numbers.*
Applying the rational roots theorem to the given equation shows that there are no rational roots. (Verify this for yourself.) Thus, in this case, all three real roots must be irrational.

Case 2 *At least one root is a nonreal complex number.*
If we have one nonreal complex root $a + bi$, then we know that its conjugate, $a - bi$, is also a root. The third root then must be a real number. (Why?) As we pointed out in Case 1, however, the given equation has no rational roots. Thus, the real root must be irrational.

There is a theorem, similar to the conjugate roots theorem, that tells us about irrational roots of the form $a + b\sqrt{c}$. As background for this theorem, let's look at two preliminary examples. First, we consider the equation $x^2 - 2x - 5 = 0$. As you can check, the roots in this case are $1 + \sqrt{6}$ and $1 - \sqrt{6}$. However, it is not true in general that irrational roots such as these always occur in pairs. Consider as a second example the quadratic equation

$$(x + 2)(x - \sqrt{3}) = 0$$

or

$$x^2 + (2 - \sqrt{3})x - 2\sqrt{3} = 0$$

Here, one of the roots is $\sqrt{3}$, yet $-\sqrt{3}$ is not a root. This type of behavior can occur in polynomial equations where not all of the coefficients are rational. On the other hand, when the coefficients are all rational, we do have the following theorem. (See Exercise 45 at the end of this section for a proof.)

THEOREM

Let $f(x)$ be a polynomial in which all the coefficients are rational. Suppose that $a + b\sqrt{c}$ is a root of the equation $f(x) = 0$, where a, b, and c are rational and $\sqrt{c}$ is irrational. Then $a - b\sqrt{c}$ is also a root of the equation.

EXAMPLE	3	Finding a quadratic equation given one irrational root

Find a quadratic equation with rational coefficients and a leading coefficient of 1 such that one of the roots is $r_1 = 4 + 5\sqrt{3}$.

SOLUTION

If one root is $r_1 = 4 + 5\sqrt{3}$, then the other is $r_2 = 4 - 5\sqrt{3}$. We denote the required equation by $x^2 + bx + c = 0$. Thus, according to Table 2 in Section 7.4, we have

$$b = -(r_1 + r_2) = -[(4 + 5\sqrt{3}) + (4 - 5\sqrt{3})] = -8$$

and

$$c = r_1 r_2 = (4 + 5\sqrt{3})(4 - 5\sqrt{3}) = 16 - 75 = -59$$

The required equation is therefore $x^2 - 8x - 59 = 0$. This answer can also be obtained without using the table. Since the roots are $4 \pm 5\sqrt{3}$, we can write the required equation as $[x - (4 + 5\sqrt{3})][x - (4 - 5\sqrt{3})] = 0$. As you can now check by multiplying out the two factors, this equation is equivalent to $x^2 - 8x - 59 = 0$, as obtained previously.

◼

We conclude this section with a discussion of **Descartes's rule of signs.** This rule, published by Descartes in 1637, provides us with information about the types of roots an equation can have, even before we attempt to solve the equation. In order to state Descartes's rule of signs, we first explain what is meant by a variation in sign in a polynomial with real coefficients. Suppose that $f(x)$ is a polynomial with real coefficients, written in descending (or ascending) powers of x. For example, let $f(x) = 2x^3 - 4x^2 - 3x + 1$. Then we say that there is a **variation in sign** if two successive coefficients have opposite signs. In the case of $f(x) = 2x^3 - 4x^2 - 3x + 1$, there are two variations in sign, the first occurring as we go from 2 to -4 and the second occurring as we go from -3 to 1. In looking for variations in sign, we ignore terms with zero coefficients. Table 1 shows a few more examples of how we count variations in sign.

We now state Descartes's rule of signs and look at some examples. Although the proof of this theorem is not difficult, it is rather lengthy, and we shall omit it here.

TABLE 1

Polynomial	Number of variations in sign
$x^2 + 4x$	0
$-3x^5 + x^2 + 1$	1
$x^3 + 3x^2 - x + 6$	2

Descartes's Rule of Signs

Let $f(x)$ be a polynomial, all of whose coefficients are real numbers, and consider the equation $f(x) = 0$. Then:

(a) The number of positive roots either is equal to the number of variations in sign of $f(x)$ or is less than that by an even integer.

(b) The number of negative roots either is equal to the number of variations in sign of $f(-x)$ or is less than that by an even integer.

EXAMPLE **4** **Using Descartes's rule**

Use Descartes's rule of signs to obtain information regarding the roots of the equation $x^3 + 8x + 5 = 0$.

SOLUTION

Let $f(x) = x^3 + 8x + 5$. Then, since there are no variations in sign for $f(x)$, we see from part (a) of Descartes's rule that the given equation has no positive roots. Next we compute $f(-x)$ to learn about the possibilities for negative roots: we have $f(-x) = -x^3 - 8x + 5$. So $f(-x)$ has one sign change, and consequently [by part (b) of Descartes's rule] the original equation has one negative root. Furthermore, notice that zero is not a root of the equation. Thus, the equation has only one real root, a negative root. Since the equation has a total of three roots, we can conclude that we have one negative root and two nonreal complex roots. The two nonreal roots will be complex conjugates.

EXAMPLE 5 Another analysis using Descartes's rule

Use Descartes's rule to obtain information regarding the roots of the equation $x^4 + 3x^2 - 7x - 5 = 0$.

SOLUTION

Let $f(x) = x^4 + 3x^2 - 7x - 5$. Then $f(x)$ has one variation in sign, so according to part (a) of Descartes's rule, the equation has one positive root. That leaves three roots still to be accounted for, since the degree of the equation is 4. We have $f(-x) = x^4 + 3x^2 + 7x - 5$. Since $f(-x)$ has one sign change, we know from part (b) of Descartes's rule that the equation has one negative root. Noting now that zero is not a root, we conclude that the two remaining roots must be nonreal complex roots. In summary, then, the equation has one positive root, one negative root, and two nonreal complex (conjugate) roots.

EXAMPLE 6 A case where Descartes's rule doesn't completely determine the nature of the roots

Use Descartes's rule to obtain information regarding the roots of the equation $f(x) = x^3 - x^2 + 3x + 2 = 0$.

SOLUTION

Since $f(x)$ has two variations in sign, the given equation has either two positive roots or no positive roots. To see how many negative roots are possible, we compute

$$f(-x) = (-x)^3 - (-x)^2 + 3(-x) + 2$$
$$= -x^3 - x^2 - 3x + 2$$

Since $f(-x)$ has one variation in sign, we conclude from part (b) of Descartes's rule that the equation has exactly one negative root. In summary, then, there are two possibilities:

either: one negative root and two positive roots

or: one negative root and two nonreal complex roots

By using Descartes's rule in Examples 4 and 5, we were able to determine the exact numbers of positive and negative roots for the given equations. As Example

6 indicates, however, there are cases in which a direct application of Descartes's rule provides several distinct possibilities for the types of roots, rather than a single definitive result. (Exercises 55 and 56 in the Review Exercises for this chapter illustrate a technique that is sometimes useful in gaining additional information from Descartes's rule. In particular, Exercise 56 will show you that the equation in Example 6 has no positive roots.)

EXERCISE SET 7.6

A

In Exercises 1–16, an equation is given, followed by one or more roots of the equation. In each case, determine the remaining roots.

1. $x^2 - 14x + 53 = 0$; $x = 7 - 2i$
2. $x^2 - x - \frac{1535}{4} = 0$; $x = \frac{1}{2} + 8\sqrt{6}$
3. $x^3 - 13x^2 + 59x - 87 = 0$; $x = 5 + 2i$
4. $x^4 - 10x^3 + 30x^2 - 10x - 51 = 0$; $x = 4 + i$
5. $x^4 + 10x^3 + 38x^2 + 66x + 45 = 0$; $x = -2 + i$
6. $2x^3 + 11x^2 + 30x - 18 = 0$; $x = -3 - 3i$
7. $4x^3 - 47x^2 + 232x + 61 = 0$; $x = 6 - 5i$
8. $9x^4 + 18x^3 + 20x^2 - 32x - 64 = 0$; $x = -1 + \sqrt{3}i$
9. $4x^4 - 32x^3 + 81x^2 - 72x + 162 = 0$; $x = 4 + \sqrt{2}i$
10. $2x^4 - 17x^3 + 137x^2 - 57x - 65 = 0$; $x = 4 - 7i$
11. $x^4 - 22x^3 + 140x^2 - 128x - 416 = 0$; $x = 10 + 2i$
12. $4x^4 - 8x^3 + 24x^2 - 20x + 25 = 0$; $x = (1 + 3i)/2$
13. $15x^3 - 16x^2 + 9x - 2 = 0$; $x = (1 + \sqrt{2}i)/3$
14. $x^5 - 5x^4 + 30x^3 + 18x^2 + 92x - 136 = 0$; $x = -1 + i\sqrt{3}, x = 3 - 5i$
15. $x^7 - 3x^6 - 4x^5 + 30x^4 + 27x^3 - 13x^2 - 64x + 26 = 0$; $x = 3 - 2i, x = -1 + i, x = 1$
16. $x^6 - 2x^5 - 2x^4 + 2x^3 + 2x + 1 = 0$; $x = 1 + \sqrt{2}$

In Exercises 17–20, find a quadratic equation with rational coefficients, one of whose roots is the given number. Write your answer so that the coefficient of x^2 is 1. Use either of the methods shown in Example 3.

17. $r_1 = 1 + \sqrt{6}$ 18. $r_1 = 2 - \sqrt{3}$
19. $r_1 = (2 + \sqrt{10})/3$ 20. $r_1 = \frac{1}{2} + \frac{1}{4}\sqrt{5}$

21. Let $f(x) = 2x^4 - 3x^3 + 12x^2 + 22x - 60$.
    (a) Use Descartes's rule to verify that the equation $f(x) = 0$ has one negative root.
    (b) Use Descartes's rule to verify that the equation $f(x) = 0$ has either one or three positive roots.
    (c) Graph the equation $y = f(x)$. Use the graph to say which of the two cases in part (b) actually holds.
    (d) Use the graph to estimate the real roots of the equation $f(x) = 0$. Check that your answers are consistent with the values obtained in Example 1.

22. Let $f(x) = x^3 - 2x^2 + x - 1$.
    (a) Without using a graphing utility, explain (in complete sentences) why the equation $f(x) = 0$ must have either

three real roots or only one real root. (If you get stuck, see Example 2 in the text.)
    (b) Use a graph to demonstrate that the equation $f(x) = 0$ has, in fact, only one real root.
    (c) According to Example 2, the root (or x-intercept) that you observed in part (b) is an irrational number. Use the graphing utility to obtain an approximation for this root.

23. Let $f(x) = x^3 + 8x + 5$.
    (a) Use Descartes's rule to explain in complete sentences why the equation $f(x) = 0$ has no positive roots and exactly one negative root. (If you need help, see Example 4 in the text.)
    (b) Use a graph to confirm the results in part (a). That is, graph $y = f(x)$ and note that there is but one x-intercept and it is negative.
    (c) Use a graphing utility to compute the root of $x^3 + 8x + 5 = 0$. (Do this either by repeatedly zooming in on the x-intercept of the graph or by using a SOLVE key.)
    (d) Use the general formula given on page 591 to compute the root of $x^3 + 8x + 5 = 0$. Check that the answer agrees with the value you obtained in part (c).

24. Let $f(x) = x^3 - x^2 + 3x + 2$.
    (a) Use Descartes's rule to explain in complete sentences why the equation $f(x) = 0$ has
       either: one negative root and two positive roots;
       or: one negative root and two nonreal complex roots.
       (If you need help, review Example 6 in the text.)
    (b) Use a graph to determine which of the two possibilities in part (a) is actually the case.
    (c) Use a graphing utility to compute the real root(s) of the equation $f(x) = 0$.

In Exercises 25–40, use Descartes's rule of signs to obtain information regarding the roots of the equations.

25. $x^3 + 5 = 0$ 26. $x^4 + x^2 + 1 = 0$
27. $2x^5 + 3x + 4 = 0$ 28. $x^3 + 8x - 3 = 0$
29. $5x^4 + 2x - 7 = 0$ 30. $x^3 - 4x^2 + x - 1 = 0$
31. $x^3 - 4x^2 - x - 1 = 0$
32. $x^8 + 4x^6 + 3x^4 + 2x^2 + 5 = 0$
33. $3x^8 + x^6 - 2x^2 - 4 = 0$
34. $12x^4 - 5x^3 - 7x^2 - 4 = 0$

35. $x^9 - 2 = 0$

36. $x^9 + 2 = 0$

37. $x^8 - 2 = 0$

38. $x^8 + 2 = 0$

39. $x^6 + x^2 - x - 1 = 0$

40. $x^7 + x^2 - x - 1 = 0$

B

41. Consider the equation $x^4 + cx^2 + dx - e = 0$, where c, d, and e are positive. Show that the equation has one positive root, one negative root, and two nonreal complex roots.

42. Consider the equation $x^n - 1 = 0$.

(a) Show that the equation has $n - 2$ nonreal complex roots when n is even.

(b) How many nonreal complex roots are there when n is odd?

43. (a) Find the polynomial $f(x)$ of lowest degree, with integer coefficients and with leading coefficient 1, such that $\sqrt{3} + 2i$ is a root of the equation $f(x) = 0$.

(b) Find the other roots of the equation.

44. (a) Find a cubic polynomial $f(x)$ with integer coefficients and leading coefficient 1 such that $1 + \sqrt[3]{2}$ is a root of the equation $f(x) = 0$.

Ⓖ (b) Use a graphing utility to find out whether $1 - \sqrt[3]{2}$ is a root of the equation.

C

45. Let $f(x)$ be a polynomial, with rational coefficients. Suppose that $a + b\sqrt{c}$ is a root of $f(x) = 0$, where a, b, and c are rational and $\sqrt{c}$ is irrational. Complete the following steps to prove that $a - b\sqrt{c}$ is also a root of the equation $f(x) = 0$.

(a) If $b = 0$, we're done. Why?

(b) (From now on we'll assume that $b \neq 0$.) Let $d(x) = [x - (a + b\sqrt{c})][x - (a - b\sqrt{c})]$. Explain why $d(a + b\sqrt{c}) = 0$.

(c) Verify that $d(x) = (x - a)^2 - b^2c$. Thus, $d(x)$ is a quadratic polynomial with rational coefficients.

(d) Now suppose that we use the long division process to divide the polynomial $f(x)$ by the quadratic polynomial $d(x)$. We'll obtain a quotient $Q(x)$ and a remainder. Since the degree of $d(x)$ is 2, our remainder will be of degree 1 or less. In other words, the general form of this remainder will be $Cx + D$. Furthermore, C and D will have to be rational, because all of the coefficients in $f(x)$ and in $d(x)$ are rational. In summary, we have the identity

$$f(x) = d(x) \cdot Q(x) + (Cx + D)$$

Now make the substitution $x = a + b\sqrt{c}$ in this identity, and conclude that $C = D = 0$.

(e) Using the result in part (d), we have

$$f(x) = [x - (a + b\sqrt{c})][x - (a - b\sqrt{c})] \cdot Q(x)$$

Let $x = a - b\sqrt{c}$ in this last identity and conclude that $a - b\sqrt{c}$ is a root of $f(x) = 0$, as required.

46. (a) Find the polynomial $f(x)$ of lowest degree with integer coefficients and with leading coefficient 1, such that $\sqrt[3]{2} + \sqrt{2}$ is a root of the equation $f(x) = 0$.

Ⓖ (b) Use a graphing utility to find out whether any of the following three numbers seem to be a root of the equation that you determined in part (a):

$$-\sqrt[3]{2} + \sqrt{2} \qquad \sqrt[3]{2} - \sqrt{2} \qquad -\sqrt[3]{2} - \sqrt{2}$$

(c) For each number in part (b) that seems to be a root, carry out the necessary algebra to *prove* or *disprove* that it is a root.

7.7   INTRODUCTION TO PARTIAL FRACTIONS

In elementary Algebra, a group of fractions connected by the signs of addition and subtraction is reduced to a more simple form by being collected into one single fraction whose denominator is the lowest common denominator of the given fractions. But the converse process of separating a fraction into a group of simpler, or partial, *fractions is often required.*
—H. S. Hall and S. R. Knight, in *Higher Algebra* (London: Macmillan and Co., 1946). (This classic text was first published in 1887.)

In calculus, there are times when it is helpful to express a given function in terms of simpler functions. In Section 3.4 we learned that composition of functions provided one way to do this. For example, if $R(x) = 1/(x^2 - 1)$, then, as you can easily check, $R(x)$ can be expressed as a composition of two simpler functions as follows:

$$R(x) = f(g(x)) \qquad \text{where } g(x) = x^2 - 1 \quad \text{and} \quad f(x) = 1/x$$

In this section, we introduce another way that certain types of functions can be broken down into simpler functions. Instead of using composition of functions, now we'll be using the sum and difference of functions. We can again use as an example $R(x) = 1/(x^2 - 1)$. This time $R(x)$ can be expressed as a *difference* of two simpler expressions as follows:

$$\frac{1}{x^2 - 1} = \frac{1}{2(x - 1)} - \frac{1}{2(x + 1)}$$

You can check this result for yourself by using the least common denominator to combine the two fractions on the right-hand side of the equation.

Our basic goal in this and the next section is to be able to write a given fractional expression as a sum or difference of two or more simpler fractions. For instance, in Example 1, we will be given the fraction $\dfrac{2x - 3}{(x - 1)(x + 1)}$, and we will be asked to find constants A and B so that

$$\frac{2x - 3}{(x - 1)(x + 1)} = \frac{A}{x - 1} + \frac{B}{x + 1} \tag{1}$$

When A and B are determined, the right-hand side of equation (1) is called the **partial fraction decomposition** of the given fraction. The adjective *partial* is used because each denominator on the right-hand side of equation (1) is a *part* of the denominator on the other side of the equation.

One of the basic tools that can be used in finding partial fractions is supplied by the following theorem, which we state here without proof.

Equating-the-Coefficients Theorem

Suppose that $P(x)$ and $Q(x)$ are polynomials such that $P(x) = Q(x)$ for all x. Then the corresponding coefficients of the two polynomials are equal.

The next two statements supply examples of what this theorem is saying.

1. If $ax + b = 5x - 11$ for all x, then $a = 5$ and $b = -11$.
2. If $px^3 + qx^2 + rx + s = x^2 + 2$ for all x, then $p = 0$, $q = 1$, $r = 0$, and $s = 2$.

In Example 1, we apply the theorem to obtain a partial fraction decomposition.

EXAMPLE 1 A basic partial fractions example

Determine constants A and B so that the following equation is an identity:

$$\frac{2x - 3}{(x - 1)(x + 1)} = \frac{A}{x - 1} + \frac{B}{x + 1} \tag{2}$$

SOLUTION
First, to clear equation (2) of fractions, we multiply both sides by the least common denominator $(x - 1)(x + 1)$. This yields

$$\begin{aligned} 2x - 3 &= A(x + 1) + B(x - 1) \\ &= Ax + A + Bx - B \\ &= (A + B)x + (A - B) \end{aligned}$$

So we have

$$2x - 3 = (A + B)x + (A - B)$$

Now, since this last equation is supposed to be an identity, we can use our equating-the-coefficients theorem to obtain the two equations

$$\begin{cases} A + B = 2 \\ A - B = -3 \end{cases}$$

Adding these two equations gives us $2A = -1$ and therefore $A = -1/2$. Subtracting the two equations yields $2B = 5$, and consequently $B = 5/2$. We've now found that $A = -1/2$ and $B = 5/2$, as required. The partial fraction decomposition is therefore

$$\frac{2x - 3}{(x - 1)(x + 1)} = \frac{-1/2}{x - 1} + \frac{5/2}{x + 1}$$
$$= \frac{-1}{2(x - 1)} + \frac{5}{2(x + 1)}$$

You should check for yourself that combining the two fractions on the right-hand side of this last equation indeed yields $\dfrac{2x - 3}{(x - 1)(x + 1)}$.

In Example 1, we found the partial fraction decomposition by solving a system of two linear equations. There is a shortcut that is often useful in such problems. As before, we start by multiplying both sides of equation (2) by the common denominator $(x - 1)(x + 1)$ to obtain

$$2x - 3 = A(x + 1) + B(x - 1) \tag{3}$$

Now, equation (3) is an identity; in particular, it must hold when $x = 1$. (You're about to see why we've singled out $x = 1$.) Substituting $x = 1$ in equation (3) gives us

$$2(1) - 3 = A(1 + 1) + B(1 - 1)$$

or

$$-1 = 2A$$

and consequently

$$A = -1/2 \qquad \text{as obtained previously}$$

The value for B is obtained similarly. Go back to equation (3), and this time let $x = -1$. This gives us

$$2(-1) - 3 = A(-1 + 1) + B(-1 - 1)$$

or

$$-5 = -2B \qquad \text{and therefore} \qquad B = 5/2$$

Again, this agrees with the result obtained previously.

A graphing calculator can be used to provide a quick check on the result obtained in a partial fractions problem. For instance, to check Example 1, graph the two functions

$$y = \frac{2x - 3}{(x - 1)(x + 1)} \qquad \text{and} \qquad y = \frac{-1}{2(x - 1)} + \frac{5}{2(x + 1)}$$

in an appropriate viewing rectangle. If the graphs appear to be identical (as is the case in Figure 1), then it's unlikely that there is an error. (See, however, Exercise 25 and the mini project at the end of this section for cautionary examples.) On the other hand, if the graphs are not identical, there must be at least one algebraic error to find and correct in the partial fractions work.

GRAPHICAL PERSPECTIVE

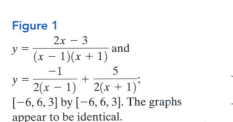

Figure 1

$$y = \frac{2x - 3}{(x - 1)(x + 1)} \text{ and}$$

$$y = \frac{-1}{2(x - 1)} + \frac{5}{2(x + 1)};$$

$[-6, 6, 3]$ by $[-6, 6, 3]$. The graphs appear to be identical.

EXAMPLE 2 **Solving a partial fractions problem in two ways**

Determine constants A and B so that the following equation is an identity:

$$\frac{x}{(x + 4)^2} = \frac{A}{x + 4} + \frac{B}{(x + 4)^2}$$

SOLUTION

We'll show two methods: first, the method that we used in Example 1, where we equated the coefficients; and second, the shortcut method explained after Example 1. For ease of reference, we'll call this shortcut method the *convenient-values method*.

FIRST METHOD *(equating coefficients)*   As in Example 1, we start by multiplying both sides of the given identity by the least common denominator. Here, the least common denominator is $(x + 4)^2$, and we obtain

$$x = A(x + 4) + B = Ax + 4A + B$$

That is,

$$x = Ax + (4A + B)$$

Now, by equating coefficients, we obtain the system of equations

$$\begin{cases} A = 1 \\ 4A + B = 0 \end{cases}$$

The first of these two equations gives us the value of A directly. Substituting $A = 1$ in the second equation then yields $4(1) + B = 0$, and therefore $B = -4$. In summary, $A = 1$, $B = -4$, and the partial fraction decomposition is

$$\frac{x}{(x + 4)^2} = \frac{1}{x + 4} - \frac{4}{(x + 4)^2}$$

You should check for yourself that combining the two fractions on the right-hand side of this last equation indeed produces the fraction on the left-hand side of the equation.

SECOND METHOD *(convenient values)*   As before, we start by multiplying both sides of the given identity by $(x + 4)^2$ to obtain

$$x = A(x + 4) + B \qquad (4)$$

Letting $x = -4$ in identity (4) gives us

$$-4 = A(-4 + 4) + B$$

or $\qquad\qquad\qquad\qquad B = -4 \qquad$ as obtained previously

With the value $B = -4$, identity (4) reads

$$x = A(x + 4) - 4 \qquad (5)$$

At this point, substituting any value for x (other than -4, which we've already exploited) will produce the required value for A. Using $x = 0$ then in equation (5) yields

$$0 = A(0 + 4) - 4 \qquad \text{or} \qquad -4A = -4$$

and therefore $\qquad\qquad A = 1 \qquad$ as obtained previously

In the examples up to this point, the convenient-values method appears to be somewhat more efficient (that is, shorter) than equating the coefficients. (In Example 2, perhaps it's a toss-up.) In the next example, we again show our two methods of solution. You can decide for yourself which method you prefer. In the convenient-values method, you'll see that one of the values we choose to substitute in the identity is a complex number. For purposes of completeness, we mention in passing the following theorem from post-calculus mathematics that justifies this technique.

If p and q are polynomial functions such that $p(x) = q(x)$ for all real numbers x, then $p(z) = q(z)$ for all complex numbers z.

As we've just said, we're stating this result only for the purposes of completeness; you certainly don't need to memorize this theorem or know any more about it to work the exercises.

EXAMPLE **3** | **Another demonstration of the two methods**

Determine real numbers A, B, and C so that the following equation is an identity:

$$\frac{7x^2 - 9x + 29}{(x - 2)(x^2 + 9)} = \frac{A}{x - 2} + \frac{Bx + C}{x^2 + 9}$$

SOLUTION

FIRST METHOD *(equating coefficients)*   The least common denominator is $(x - 2)(x^2 + 9)$. Multiplying both sides of the given identity by this common denominator yields

$$\begin{aligned}
7x^2 - 9x + 29 &= A(x^2 + 9) + (Bx + C)(x - 2) \\
&= Ax^2 + 9A + Bx^2 - 2Bx + Cx - 2C \\
&= (A + B)x^2 + (-2B + C)x + (9A - 2C)
\end{aligned}$$

Equating coefficients now gives us a system of three equations in three unknowns:

$$\begin{cases} A + B = 7 \\ -2B + C = -9 \\ 9A - 2C = 29 \end{cases}$$

Exercise 25(a) asks you to solve this system to obtain $A = 3$, $B = 4$, and $C = -1$. The partial fraction decomposition is therefore

$$\frac{7x^2 - 9x + 29}{(x - 2)(x^2 + 9)} = \frac{3}{x - 2} + \frac{4x - 1}{x^2 + 9}$$

SECOND METHOD *(convenient values)* Going back to the original identity and multiplying both sides by the least common denominator yields (as in the first method)

$$7x^2 - 9x + 29 = A(x^2 + 9) + (Bx + C)(x - 2) \tag{6}$$

Because of the factor $x - 2$ on the right-hand side of identity (6), we choose the convenient value $x = 2$. Substituting $x = 2$ in equation (6) gives us

$$7(2^2) - 9(2) + 29 = A(2^2 + 9) + 0$$

or

$$39 = 13A \qquad \text{and therefore} \qquad A = 3$$

Next, we want to choose another value for x, one that will cause the factor $x^2 + 9$ to be zero. Thus we choose $x = 3i$. (The other root of the equation $x^2 + 9 = 0$— namely, $x = -3i$—would also be an acceptable choice here.) Making the substitution $x = 3i$ (along with $x^2 = -9$) in identity (6) yields

$$7(-9) - 9(3i) + 29 = A(-9 + 9) + [B(3i) + C](3i - 2)$$
$$-34 - 27i = -9B - 6Bi + 3Ci - 2C$$
$$= (-9B - 2C) + (-6B + 3C)i$$

Equating the real parts and equating the imaginary parts from both sides of this last equation gives the equations

$$-9B - 2C = -34 \tag{7}$$

and

$$-6B + 3C = -27$$

or

$$-2B + C = -9 \qquad \text{dividing by 3} \tag{8}$$

As Exercise 25(b) asks you to check, the solution to the system consisting of equations (7) and (8) is $B = 4$ and $C = -1$. In summary, then, we have $A = 3$, $B = 4$, and $C = -1$, as obtained previously.

EXERCISE SET 7.7

A

In Exercises 1–22, determine the constants (denoted by capital letters) so that each equation is an identity. For Exercises 1–6, do each problem in two ways: **(a)** *use the equating-the-coefficients theorem, as in Example 1; and* **(b)** *use the convenient-values method that was explained after Example 1. For the remainder of the exercises, use either method (or a combination).*

1. $\dfrac{7x - 6}{(x - 2)(x + 2)} = \dfrac{A}{x - 2} + \dfrac{B}{x + 2}$

2. $\dfrac{5x + 27}{(x + 3)(x - 3)} = \dfrac{A}{x + 3} + \dfrac{B}{x - 3}$

3. $\dfrac{6x - 25}{(2x + 5)(2x - 5)} = \dfrac{A}{2x + 5} + \dfrac{B}{2x - 5}$

4. $\dfrac{x}{(4x + 3)(4x - 3)} = \dfrac{A}{4x + 3} + \dfrac{B}{4x - 3}$

5. $\dfrac{1}{(x + 1)(3x - 1)} = \dfrac{A}{x + 1} + \dfrac{B}{3x - 1}$

6. $\dfrac{1 - x}{(4x + 3)(2x - 1)} = \dfrac{A}{4x + 3} + \dfrac{B}{2x - 1}$

7. $\dfrac{8x + 3}{(x + 3)^2} = \dfrac{A}{x + 3} + \dfrac{B}{(x + 3)^2}$

8. $\dfrac{7x}{(x - 5)^2} = \dfrac{A}{x - 5} + \dfrac{B}{(x - 5)^2}$

9. $\dfrac{6 - x}{(5x + 4)^2} = \dfrac{A}{5x + 4} + \dfrac{B}{(5x + 4)^2}$

10. $\dfrac{30x - 17}{(6x - 1)^2} = \dfrac{A}{6x - 1} + \dfrac{B}{(6x - 1)^2}$

11. $\dfrac{3x^2 + 7x - 2}{(x - 1)(x^2 + 1)} = \dfrac{A}{x - 1} + \dfrac{Bx + C}{x^2 + 1}$

12. $\dfrac{15x^2 - 35x + 77}{(2x - 5)(x^2 + 3)} = \dfrac{A}{2x - 5} + \dfrac{Bx + C}{x^2 + 3}$

13. $\dfrac{x^2 + 1}{(x + 1)(x^2 + 4)} = \dfrac{A}{x + 1} + \dfrac{Bx + C}{x^2 + 4}$

14. $\dfrac{x - 7}{x(x^2 + 2)} = \dfrac{A}{x} + \dfrac{Bx + C}{x^2 + 2}$

15. $\dfrac{1}{x(x^2 - x + 1)} = \dfrac{A}{x} + \dfrac{Bx + C}{x^2 - x + 1}$

16. $\dfrac{2x^2 - 11x - 6}{(x + 2)(x^2 - 2x + 4)} = \dfrac{A}{x + 2} + \dfrac{Bx + C}{x^2 - 2x + 4}$

17. $\dfrac{3x^2 - 2}{(x - 2)(x + 1)(x - 1)} = \dfrac{A}{x - 2} + \dfrac{B}{x + 1} + \dfrac{C}{x - 1}$

18. $\dfrac{1}{(x - 1)(x - 2)(x - 3)} = \dfrac{A}{x - 1} + \dfrac{B}{x - 2} + \dfrac{C}{x - 3}$

19. $\dfrac{4x^2 - 47x + 133}{(x - 6)^3} = \dfrac{A}{x - 6} + \dfrac{B}{(x - 6)^2} + \dfrac{C}{(x - 6)^3}$

20. $\dfrac{x^2 + 2x}{(x + 1)^3} = \dfrac{A}{x + 1} + \dfrac{B}{(x + 1)^2} + \dfrac{C}{(x + 1)^3}$

21. $\dfrac{x^2 - 2}{(x^2 + 2)^2} = \dfrac{Ax + B}{x^2 + 2} + \dfrac{Cx + D}{(x^2 + 2)^2}$

22. $\dfrac{x^3 + 2x^2}{(x^2 + 3)^2} = \dfrac{Ax + B}{x^2 + 3} + \dfrac{Cx + D}{(x^2 + 3)^2}$

G 23. **(a)** Find an appropriate viewing rectangle to demonstrate that the following purported partial fraction decomposition is incorrect:
$$\frac{2x + 5}{(x - 4)(x + 3)} = \frac{13/7}{x - 4} + \frac{2/7}{x + 3}.$$

G **(b)** Follow part (a) using
$$\frac{2x + 5}{(x - 4)(x + 3)} = \frac{13/7}{x - 4} - \frac{1/7}{x + 3}.$$

(c) Determine the correct partial fraction decomposition, given that it has the general form
$$\frac{2x + 5}{(x - 4)(x + 3)} = \frac{A}{x - 4} + \frac{B}{x + 3}.$$

G 24. **(a)** Find an appropriate viewing rectangle to demonstrate that the following purported partial fraction decomposition is incorrect: $\dfrac{4}{x^2(x - 5)} = \dfrac{-4/5}{x^2} + \dfrac{4/25}{x - 5}.$

G **(b)** Follow part (a) using
$$\frac{4}{x^2(x - 5)} = \frac{-3/25}{x} + \frac{-2/5}{x^2} + \frac{6/25}{x - 5}.$$

(c) Determine the correct partial fraction decomposition, given that it has the general form
$$\frac{4}{x^2(x - 5)} = \frac{A}{x} + \frac{B}{x^2} + \frac{C}{x - 5}.$$

25. **(a)** Solve the following system of equations. (As indicated in Example 3, you should obtain $A = 3$, $B = 4$, and $C = -1$.)
$$\begin{cases} A + B = 7 \\ -2B + C = -9 \\ 9A - 2C = 29 \end{cases}$$

(b) Solve the following system of equations. (As indicated in the text, you should obtain $B = 4$ and $C = -1$.)
$$\begin{cases} -9B - 2C = -34 \\ -2B + C = -9 \end{cases}$$

Exercises 26 provides practice using the convenient-values method with complex numbers. As background, you should review Example 3.

26. In this exercise you'll use the convenient-values method to determine real numbers A, B, and C such that the following equation is an identity:

$$\frac{x+1}{x(x^2+4)} = \frac{A}{x} + \frac{Bx+C}{x^2+4}$$

(a) Multiplying both sides of the given identity by the least common denominator yields

$$x + 1 = A(x^2 + 4) + (Bx + C)x \qquad (1)$$

Determine A by substituting $x = 0$ in identity (1).

(b) In identity (1), substitute $x = 2i$ and show that the resulting equation can be written

$$2i + 1 = -4B + 2iC \qquad (2)$$

(c) Determine B and C by equating real and imaginary parts in equation (2).

In Exercises 27 and 28, the equating-the-coefficients theorem is used as a tool to factor polynomials and thereby solve equations.

27. Consider the polynomial equation

$$f(x) = x^4 - x^3 + x^2 - x + 1 = 0.$$

(a) Make use of the rational roots theorem in explaining why the equation has no rational roots.

(b) Use a graphing utility to graph the function $y = f(x)$, and conclude that the given polynomial equation has no real roots.

(c) Use the equating-the-coefficients theorem to find a factorization of $f(x)$ of the form

$$x^4 - x^3 + x^2 - x + 1 = (x^2 + bx + 1)(x^2 + cx + 1)$$

(d) Solve the equation $x^4 - x^3 + x^2 - x + 1 = 0$.

28. Let $f(x) = x^4 + 2x^3 + x^2 - 9$.

(a) Use a graphing utility to graph the function f. The graph indicates that the polynomial equation $f(x) = 0$ has two real roots. In this case, according to Section 7.4, how many complex roots must there be?

(b) Use the rational roots theorem to show that the equation $f(x) = 0$ had no rational roots. Why does this imply that the two real roots in part (a) must be irrational?

(c) Use the equating-the-coefficients theorem to find a factorization for $f(x)$ of the form $(x^2 + ax + b)(x^2 + ax - b)$.

(d) Use the factorization determined in part (c) to find the four roots of the equation $f(x) = 0$. For the two real roots, give both exact expressions and calculator approximations rounded to two decimal places. Check to see that these calculator values are consistent with the graph in part (a).

The following exercise provides an example in which an error in a partial fractions decomposition is not easily detected with a graphical approach. Indeed, this may be an example of a case in which, to check your partial fractions work, it's easier to repeat the algebra than to experiment with numerous viewing rectangles. Decide for yourself after completing the problem.

29. There is an error in the following partial fractions decomposition:

$$\frac{1}{(x+2)(x+5)(x-14)} = \frac{-1/48}{x+2} + \frac{1/57}{x+5} + \frac{1/305}{x-14}$$

(a) Let f and g denote the two functions defined by the expressions on the left side and the right side, respectively, in the above equation. Use a graphing utility to graph f and g, first in the standard viewing rectangle and then in the rectangle $[-15, 15, 5]$ by $[-0.02, 0.04, 0.02]$. In this latter rectangle, note that the graphs do *appear* to be identical. (People using a software graphing application and looking at the curves on a computer monitor may have a little advantage here over those drawing the graphs on a relatively small graphing calculator screen.)

(b) Find a viewing rectangle clearly demonstrating that the graphs of f and g are not identical.

(c) Find the correct partial fractions decomposition, given that the form is

$$\frac{1}{(x+2)(x+5)(x-14)} = \frac{A}{x+2} + \frac{B}{x+5} + \frac{C}{x-14}$$

MINI PROJECT

CHECKING A PARTIAL FRACTIONS DECOMPOSITION

The partial fractions decomposition for the expression $\dfrac{3}{x(x+17)(x^2+13)}$ has the form

$$\frac{A}{x} + \frac{B}{x+17} + \frac{Cx+D}{x^2+13}$$

Suppose that a classmate of yours worked out the following values for the constants:

$$A = \frac{3}{221}, \qquad B = -\frac{3}{5135}, \qquad C = -\frac{51}{3925}, \qquad D = -\frac{3}{302}$$

(a) Use a graphing utility to check this decomposition, as in the text, after Example 1. Unlike Example 1, however, not all of the constants given here are correct. Yet, initially, no difference in the two graphs is apparent. Can you find a viewing rectangle that clearly demonstrates that there must be an error in the decomposition? (It's certainly not the standard viewing rectangle.) Can you think of a simple nongraphical way to check for the presence of an error (short of rederiving the entire decomposition from scratch)?

(b) Determine correct values for the constants A, B, C, and D. If you have access to software or a graphing utility that computes partial fraction decompositions, use that. There are also interactive partial fraction calculators available on-line. Try using a search engine to locate "partial fraction expansion" or "partial fraction calculator." At the time of this writing, for example, one such calculator (developed by Professor Philip S. Crooke at Vanderbilt University) can be accessed from the Web page http://MSS.math .vanderbilt.edu. Finally, if neither of these options works out for you, use the methods of this section to find the decomposition.

7.8 MORE ABOUT PARTIAL FRACTIONS

If the denominator of a rational function has two relatively prime factors, then this rational function can be expressed as the sum of two fractions whose denominators are equal to the two factors. —Leonhard Euler (1707–1783) in *Introductio in analysis infinitorum* (1748)

In the examples and exercises in the previous section, you were told what the general form for each partial fraction should look like, and then the required constants were computed. In calculus, however, when a partial fraction decomposition is needed, you're usually not specifically told what the general form for the partial fractions should be. Rather, you're expected to know this. In this section we list the necessary guidelines. First, however, consider the following example indicating why these guidelines are needed.

Suppose, for instance, that we want to find the partial fraction decomposition for $1/[x(x^2+1)]$, and we guess that the general form is

$$\frac{1}{x(x^2+1)} = \frac{A}{x} + \frac{B}{x^2+1} \tag{1}$$

Multiplying both sides of this last equation by the least common denominator gives us

$$1 = A(x^2 + 1) + Bx$$
$$= Ax^2 + A + Bx$$
$$= Ax^2 + Bx + A$$

Equating coefficients now, as we did in the previous section, yields $A = 0$, $B = 0$, and $A = 1$. The two equations $A = 0$ and $A = 1$ are contradictory. The only way out is to conclude that there is no partial fraction decomposition of the form shown in equation (1). This example demonstrates the need for guidelines for setting up partial fraction decompositions.

We begin with a few definitions. Suppose that we have two polynomials $p(x)$ and $q(x)$. Following the terminology in Section 4.7, we call an expression of the form $p(x)/q(x)$ a **rational expression.** A rational expression is said to be **proper** if the degree of the numerator is less than the degree of the denominator. So, each of the following is a proper rational expression:

$$\frac{4x - 5}{2x^2 + 3x - 1} \qquad \frac{x^3 - 2x - 5}{x^5 - 9x^3 - 4} \qquad \frac{x^2 - 5}{(x + 1)(x + 2)(x + 3)}$$

An **improper** rational expression is one in which the degree of the numerator is greater than or equal to the degree of the denominator. Examples of improper rational expressions are $(x^2 + 1)/(x - 1)$ and $(x^3 - 4)/(x^3 + 2x^2 + 1)$. All of the partial fraction guidelines that we are about to discuss pertain to proper rational expressions. (Near the end of this section, we'll explain what to do for an improper rational expression.)

The final term that we need to define is *irreducible quadratic polynomial.* The definition is given in the box that follows.

DEFINITION | **Irreducible Quadratic Polynomial**

Let $f(x)$ be a quadratic polynomial, all of whose coefficients are real. Then $f(x)$ is said to be **irreducible** provided the equation $f(x) = 0$ has no real roots. This is equivalent to saying that $f(x)$ cannot be factored into the form $(ax + b)(cx + d)$, where a, b, c, and d are real numbers.

As examples, note that $x^2 + 5$ is irreducible, but $x^2 - 5$ is not irreducible.

The first step in determining a partial fraction decomposition is to factor the denominator. In the previous section, the denominators were already factored for you. If you glance back at the previous section, you'll see that the denominators contained one or more of the following types of factors, and no others:

$$\text{Linear factors:} \quad ax + b$$
$$\text{Powers of linear factors:} \quad (ax + b)^n, \quad n \geq 2$$
$$\text{Irreducible quadratic factors:} \quad ax^2 + bx + c$$
$$\text{Powers of irreducible quadratic factors:} \quad (ax^2 + bx + c)^n, \quad n \geq 2$$

A remarkable theorem tells us that, in fact, *every* polynomial with real coefficients can be factored (over the real numbers) using only the types of factors we've just listed. We state this theorem in the box that follows.

Linear and Quadratic Factors Theorem

Let $f(x)$ be a polynomial, all of whose coefficients are real numbers. Then $f(x)$ can be factored (over the real numbers) into a product of linear and/or irreducible quadratic factors.

To see a familiar example, consider the cubic polynomial $x^3 - 1$. From our work on factoring in Section 1.3, we know that

$$x^3 - 1 = (x - 1)(x^2 + x + 1)$$

So in this case, there is one linear factor and one irreducible quadratic factor. (You should check for yourself, using the discriminant, that the quadratic factor is indeed irreducible.) However, not every polynomial, of course, is so easily factored. For instance, although the theorem tells us that the polynomial $x^3 + x + 1$ *can* be factored into linear and/or irreducible quadratic factors, it doesn't tell us how to find those factors. Indeed, that can be a very difficult job, quite beyond this course. For the examples and exercises in this section (and in any calculus text), you'll be able to find the factors using factoring techniques from Appendix B.4 or from this chapter. At the end of this section we show how the linear and quadratic factors theorem follows from our earlier work in this chapter.

For all of the guidelines that we give, we assume that we're starting with a proper rational expression $p(x)/q(x)$ and that $q(x)$ has been factored into linear and/or irreducible quadratic factors. We also assume that $p(x)$ and $q(x)$ have no common factors; that is, the fraction $p(x)/q(x)$ has been reduced to lowest terms. The first two guidelines tell us what to do when $q(x)$ contains a linear factor $ax + b$ or a power of this factor.

Guidelines for the Partial Fractions Setup: Linear Factors

(a) If the denominator contains a linear factor $ax + b$ and no higher power of this factor, then the partial fraction setup must contain a term $A_1/(ax + b)$, where A_1 is a constant to be determined.

(b) [This encompasses part (a).] More generally, for each factor in the denominator of the form $(ax + b)^n$, the partial fraction setup must contain the following sum of n fractions:

$$\frac{A_1}{ax + b} + \frac{A_2}{(ax + b)^2} + \cdots + \frac{A_n}{(ax + b)^n}$$

where $A_1, A_2, \ldots, A_n$ are constants to be determined.

EXAMPLE 1 Specifying the setup for a decomposition

Determine the form of the partial fraction decomposition:

(a) $\dfrac{5x + 1}{4x^2 - 9}$;   **(b)** $\dfrac{x^2}{x^3 + 2x^2 - 5x - 10}$.

SOLUTION

(a) The denominator can be factored using difference-of-squares factoring:

$$4x^2 - 9 = (2x - 3)(2x + 3)$$

Thus, the denominator contains two distinct linear factors. So, according to the first guideline in the box preceding this example, the form of the partial fraction decomposition is

$$\frac{5x + 1}{4x^2 - 9} = \frac{A_1}{2x - 3} + \frac{A_2}{2x + 3}$$

(b) The denominator can be factored by grouping (as in Section B.4):

$$\begin{aligned} x^3 + 2x^2 - 5x - 10 &= (x^3 - 5x) + (2x^2 - 10) \\ &= x(x^2 - 5) + 2(x^2 - 5) \\ &= (x^2 - 5)(x + 2) \end{aligned} \qquad (2)$$

We're not finished with the factoring yet! If this had been a factoring problem from Section B.4, equation (2) would indeed be our final result. That's because in Section B.4 we were restricted to factors with integer (or possibly rational) coefficients. But here the restriction is removed, and we can factor $x^2 - 5$ as a difference of squares:

$$x^2 - 5 = (x - \sqrt{5})(x + \sqrt{5})$$

Combining this last result with equation (2), our final factorization looks this way:

$$x^3 + 2x^2 - 5x - 10 = (x - \sqrt{5})(x + \sqrt{5})(x + 2)$$

Thus, the denominator contains three distinct linear factors, and the form of the partial fraction decomposition is

$$\frac{x^2}{x^3 + 2x^2 - 5x - 10} = \frac{A_1}{x - \sqrt{5}} + \frac{A_2}{x + \sqrt{5}} + \frac{A_3}{x + 2}$$

EXAMPLE 2	**Another setup with distinct linear factors in the denominator**

Determine the form of the partial fraction decomposition for $\dfrac{6x - 1}{x^3 + 2x^2 - 5x - 6}$.

SOLUTION

In Example 1(b), we factored a denominator quite similar to this one by grouping the terms. As you can check, however, that won't work here. The next strategy then is to apply the rational roots theorem to the polynomial equation

$$x^3 + 2x^2 - 5x - 6 = 0 \qquad (3)$$

As Exercise 35 asks you to show, this leads to three rational roots: $-1, 2$, and -3. Consequently, we have a factorization into three distinct linear factors:

$$x^3 + 2x^2 - 5x - 6 = (x + 1)(x - 2)(x + 3)$$

The form of the partial fraction decomposition must therefore be

$$\frac{6x - 1}{x^3 + 2x^2 - 5x - 6} = \frac{A_1}{x + 1} + \frac{A_2}{x - 2} + \frac{A_3}{x + 3}$$

In Examples 1 and 2, none of the linear factors in the denominators are repeated. The next example shows cases in which there are repeated linear factors.

EXAMPLE 3   A setup with repeated linear factors in the denominator

Determine the form of the partial fraction decomposition for each of the following:

(a) $\dfrac{5x - 1}{x^2 - 6x + 9}$;   **(b)** $\dfrac{2x^4 + x + 1}{x^5 + 3x^4 + 3x^3 + x^2}$.

SOLUTION

(a) The denominator is a perfect square:

$$x^2 - 6x + 9 = (x - 3)^2$$

So, according to the second guideline in the box preceding Example 1, the form of the partial fraction decomposition is

$$\frac{5x - 1}{x^2 - 6x + 9} = \frac{A_1}{x - 3} + \frac{A_2}{(x - 3)^2}$$

(b) The first step in factoring is to look for a common term. (If you need to review factoring techniques, see Appendix B.4.) In this case there is one: it is x^2, and we have

$$x^5 + 3x^4 + 3x^3 + x^2 = x^2(x^3 + 3x^2 + 3x + 1)$$

Now what? Regarding the second factor on the right-hand side of this last equation, we could try to factor it by grouping. As you can check, that does work. It's more direct, however, to observe that this factor is actually a perfect cube: $x^3 + 3x^2 + 3x + 1 = (x + 1)^3$. Putting things together then, we have the final factorization

$$x^5 + 3x^4 + 3x^3 + x^2 = x^2(x + 1)^3$$

In summary, there are two repeated linear factors: the factor x occurs twice and the factor $x + 1$ occurs three times. The form of the partial fraction decomposition is therefore

$$\frac{2x^4 + x + 1}{x^5 + 3x^4 + 3x^3 + x^2} = \frac{A_1}{x} + \frac{A_2}{x^2} + \frac{B_1}{x + 1} + \frac{B_2}{(x + 1)^2} + \frac{B_3}{(x + 1)^3}$$

The next two guidelines apply in cases where the denominator contains one or more irreducible quadratic factors. After you read these guidelines, compare them to those that we listed previously for linear factors. You'll see that almost everything is the same. The only difference, in fact, is this: for the quadratic factors, the form of the numerator is $Ax + B$ rather than just A.

Guidelines for the Partial Fractions Setup: Irreducible Quadratic Factors

(a) If the denominator contains an irreducible quadratic factor $ax^2 + bx + c$ and no higher power of this factor, then the partial fractions setup must contain a term $\dfrac{A_1 x + B_1}{ax^2 + bx + c}$, where A_1 and B_1 are constants to be determined.

(b) [This encompasses part (a).] More generally, for each factor in the denominator of the form $(ax^2 + bx + c)^n$, where $ax^2 + bx + c$ is irreducible, the partial fractions setup must contain the following sum of n fractions:

$$\frac{A_1 x + B_1}{ax^2 + bx + c} + \frac{A_2 x + B_2}{(ax^2 + bx + c)^2} + \cdots + \frac{A_n x + B_n}{(ax^2 + bx + c)^n}$$

where the A_i and B_i are constants to be determined.

EXAMPLE 4   A denominator with linear and irreducible quadratic factors

Determine the partial fraction decomposition for $\dfrac{x^3 + 4x^2 + 1}{x^4 - 16}$.

SOLUTION
Our first job is to factor the denominator. Using difference-of-squares factoring, we have

$$x^4 - 16 = (x^2 - 4)(x^2 + 4)$$
$$= (x - 2)(x + 2)(x^2 + 4)$$

So there are two linear factors and one irreducible quadratic factor, none of which is repeated. The form of the partial fraction decomposition is therefore

$$\frac{x^3 + 4x^2 + 1}{(x - 2)(x + 2)(x^2 + 4)} = \frac{A}{x - 2} + \frac{B}{x + 2} + \frac{Cx + D}{x^2 + 4}$$

(For ease of reading and writing, we are denoting the constants by A, B, C, and D instead of, say, A_1, A_2, B_1, and C_1, respectively.) Multiplying both sides of this equation by the common denominator gives

$$x^3 + 4x^2 + 1 = A(x + 2)(x^2 + 4) + B(x - 2)(x^2 + 4) + (Cx + D)(x - 2)(x + 2) \quad (4)$$

Substituting $x = 2$ in identity (4) yields

$$2^3 + 4(2)^2 + 1 = A(2 + 2)(2^2 + 4) + 0 + 0$$
$$25 = 32A$$
$$A = 25/32$$

Similarly, letting $x = -2$ in identity (4) gives us

$$(-2)^3 + 4(-2)^2 + 1 = 0 + B(-4)(8) + 0$$
$$9 = -32B$$
$$B = -9/32$$

At this point, we've found A and B, but we still need C and D. If we substitute the values we've just found for A and B in equation (4), we have

$$x^3 + 4x^2 + 1 = \tfrac{25}{32}(x + 2)(x^2 + 4) - \tfrac{9}{32}(x - 2)(x^2 + 4) + (Cx + D)(x - 2)(x + 2) \quad (5)$$

Observe now that in identity (5), letting $x = 0$ will yield an equation involving D alone. Exercise 36(a) asks you to follow through with the arithmetic and algebra to obtain $D = 15/8$. Finally, in identity (5), we replace D by $15/8$ and make the substitution $x = 1$. As Exercise 36(b) asks you to verify, the end result is $C = 1/2$. Putting everything together now, the required partial fraction decomposition is

$$\frac{x^3 + 4x^2 + 1}{x^4 - 16} = \frac{\tfrac{25}{32}}{x - 2} + \frac{-\tfrac{9}{32}}{x + 2} + \frac{\tfrac{1}{2}x + \tfrac{15}{8}}{x^2 + 4}$$

$$= \frac{25}{32(x - 2)} - \frac{9}{32(x + 2)} + \frac{4x + 15}{8(x^2 + 4)}$$

EXAMPLE 5   **A denominator with a repeated irreducible quadratic factor**

Determine the partial fraction decomposition for the following expression:

$$\frac{3x^3 - x^2 + 7x - 3}{x^4 + 6x^2 + 9}$$

SOLUTION
The denominator factors as a perfect square:

$$x^4 + 6x^2 + 9 = (x^2 + 3)^2$$

Since the repeated factor $x^2 + 3$ is irreducible, our guidelines tell us that the form of the partial fraction decomposition is

$$\frac{3x^3 - x^2 + 7x - 3}{(x^2 + 3)^2} = \frac{Ax + B}{x^2 + 3} + \frac{Cx + D}{(x^2 + 3)^2}$$

Multiplying both sides of this identity by $(x^2 + 3)^2$ gives us

$$3x^3 - x^2 + 7x - 3 = (Ax + B)(x^2 + 3) + (Cx + D) \quad (6)$$

At this point we can use either of the two techniques discussed in the previous section, the equating-the-coefficients technique or the convenient-values technique. As Exercise 37 asks you to show, the results are $A = 3$, $B = -1$, $C = -2$, and $D = 0$. The partial fraction decomposition then is

$$\frac{3x^3 - x^2 + 7x - 3}{x^4 + 6x^2 + 9} = \frac{3x - 1}{x^2 + 3} - \frac{2x}{(x^2 + 3)^2}$$

Each of the examples we've considered in this and the previous section involved proper rational expressions. For improper rational expressions, we can first use long division to express the improper fraction in the general form

$$(\text{polynomial}) + (\text{proper rational expression})$$

Then the techniques we've developed can be applied to the second term in this sum. For instance, suppose we're given the improper rational expression $(2x^3 + 4x^2 - 15x - 36)/(x^2 - 9)$. As you can check for yourself by using long division, we obtain

$$\underbrace{\frac{2x^3 + 4x^2 - 15x - 36}{x^2 - 9} = 2x + 4}_{\text{polynomial}} + \underbrace{\frac{3x}{x^2 - 9}}_{\substack{\text{proper rational} \\ \text{expression}}}$$

At this stage, a partial fraction decomposition can be worked out for the proper rational expression $3x/(x^2 - 9)$. As Exercise 38(b) asks you to check, the result is

$$\frac{3x}{x^2 - 9} = \frac{3}{2(x - 3)} + \frac{3}{2(x + 3)}$$

So, our final decomposition of the given improper rational fraction is

$$\frac{2x^3 + 4x^2 - 15x - 36}{x^2 - 9} = 2x + 4 + \frac{3}{2(x - 3)} + \frac{3}{2(x + 3)}$$

OPTIONAL NOTE   The linear and quadratic factors theorem that we've discussed in this section is a direct consequence of two theorems that appeared earlier in the chapter: the linear factors theorem and the conjugate roots theorem. Here's why: If we start with a polynomial $f(x)$, the linear factors theorem says that we can decompose $f(x)$ into linear factors:

$$f(x) = (x - c_1)(x - c_2) \cdots (x - c_n) \tag{7}$$

Now, it's possible that some, or even all, of these c_i may be nonreal complex numbers. For instance, suppose that c_1 is a nonreal complex number, with $c_1 = a + bi$. Then, assuming that all of the coefficients of $f(x)$ are real numbers, the conjugate roots theorem tells us that in one of the other factors, call it $x - c_2$ for simplicity, we must have $c_2 = a - bi$. Now look what happens when we compute the product of these two factors:

$$\begin{aligned}
(x - c_1)(x - c_2) &= [x - (a + bi)][x - (a - bi)] \\
&= x^2 - (a - bi)x - (a + bi)x + (a + bi)(a - bi) \\
&= x^2 - 2ax + (a^2 + b^2) \quad \text{(Check the algebra.)}
\end{aligned}$$

As Exercise 45 asks you to check, this last quadratic polynomial is irreducible; that is, it has no real roots. In summary, the two linear factors $x - c_1$ and $x - c_2$ that contain nonreal complex numbers give rise to one irreducible quadratic factor with real coefficients. So, similarly, after pairing up any other linear factors in equation (7) that contain nonreal complex numbers, what are we left with? The right-hand side of equation (7) will contain only linear factors and/or irreducible quadratic factors, all with real coefficients, just as the linear and quadratic factors theorem asserts.

EXERCISE SET 7.8

A

In Exercises 1–4, determine whether the given quadratic polynomial is irreducible. [Recall from the text that a quadratic polynomial $f(x)$ is irreducible if the equation $f(x) = 0$ has no real roots.]

1. (a) $x^2 - 16$
   (b) $x^2 + 16$

2. (a) $x^2 + 17$
   (b) $x^2 - 17$

3. (a) $x^2 + 3x - 4$
   (b) $x^2 + 3x + 4$

4. (a) $24x^2 + x - 3$
   (b) $x^2 + 24x + 144$

In Exercises 5–16: **(a)** factor the denominator of the given rational expression; **(b)** determine the form of the partial fraction decomposition for the given rational expression; and **(c)** determine the values of the constants in the partial fraction decomposition that you gave in part (b). To help you in spotting errors, use the fact that in part (c), each of the required constants turns out to be an integer.

5. $\dfrac{11x + 30}{x^2 - 100}$

6. $\dfrac{x + 18}{x^2 - 36}$

7. $\dfrac{8x - 2\sqrt{5}}{x^2 - 5}$

8. $\dfrac{2\sqrt{11}}{x^2 - 11}$

9. $\dfrac{7x + 39}{x^2 - x - 6}$

10. $\dfrac{19x - 15}{4x^2 - 5x}$

11. $\dfrac{3x^2 + 17x - 38}{x^3 - 3x^2 - 4x + 12}$

12. $\dfrac{16x^2 + 9x - 2}{3x^3 + x^2 - 2x}$

13. $\dfrac{5x^2 + 2x + 5}{x^3 + x^2 + x}$

14. $\dfrac{x^2 - 3x - 1}{x^3 - x^2 + 2x - 2}$

15. $\dfrac{2x^3 + 5x - 4}{x^4 + 2x^2 + 1}$

16. $\dfrac{11x^3 + 35x - 7}{x^4 + 6x^2 + 9}$

In Exercises 17–34, determine the partial fraction decomposition for each of the given rational expressions. *Hint*: In Exercises 17 and 18, use the rational roots theorem to factor the denominator.

17. $\dfrac{x^2 + 2}{x^3 - 3x^2 - 16x - 12}$

18. $\dfrac{1}{x^3 + x^2 - 10x + 8}$

19. $\dfrac{5 - x}{6x^2 - 19x + 15}$

20. $\dfrac{2x}{32x^2 - 12x + 1}$

21. $\dfrac{2x + 1}{x^3 - 5x}$

22. $\dfrac{2x + 1}{x^3 + 5x}$

23. $\dfrac{x^3 + 2}{x^4 + 8x^2 + 16}$

24. $\dfrac{x^3 + 2}{x^4 - 8x^2 + 16}$

25. $\dfrac{x^3 + x - 3}{x^4 - 15x^3 + 75x^2 - 125x}$

26. $\dfrac{4x^2}{2x^3 - 5x^2 - 4x + 3}$

27. $\dfrac{1}{x^3 - 1}$

28. $\dfrac{x}{x^3 + 8}$

29. $\dfrac{7x^3 + 11x^2 - x - 2}{x^4 + 2x^3 + x^2}$

30. $\dfrac{4x - 5}{x^4 + 2x^3 + x^2 - 1}$
   Hint: Review the factoring in Exercise 29.

31. $\dfrac{x^3 - 5}{x^4 - 81}$

32. $\dfrac{x + 1}{x^4 - 16}$

33. $\dfrac{1}{x^4 + x^3 + 2x^2 + x + 1}$ *Hint*: To factor the denominator, replace $2x^2$ with $x^2 + x^2$ and group as follows: $(x^4 + x^3 + x^2) + (x^2 + x + 1)$.

34. $\dfrac{x^3 + 4x^2 + 16x}{x^4 + 16}$ *Hint*: To factor the denominator, add and subtract the term $16x^2$.

35. Use the rational roots theorem and the remainder theorem to determine the roots of the equation $x^3 + 2x^2 - 5x - 6 = 0$. (This is to verify a statement made in Example 2.)

36. This exercise completes two details mentioned in Example 4.
  (a) In identity (5) in the text, let $x = 0$ to obtain an equation involving D alone, then solve the equation. You should obtain $D = 15/8$.
  (b) In identity (5), we replaced D by $15/8$ and made the substitution $x = 1$. Check that this leads to the result $C = 1/2$.

37. This exercise completes calculations mentioned in Example 5.
  (a) Show that identity (6) in the text leads to the following system of equations:

$$\begin{cases} A = 3 \\ B = -1 \\ 3A + C = 7 \\ 3B + D = -3 \end{cases}$$

  (b) From the system in part (a), we have $A = 3$ and $B = -1$, which agrees with the values given in Example 5. Now determine C and D and check that your answers agree with the values given in Example 5.

38. This exercise completes the discussion of improper rational expressions in this section.
  (a) Use long division to obtain the following result:

$$\frac{2x^3 + 4x^2 - 15x - 36}{x^2 - 9} = (2x + 4) + \frac{3x}{x^2 - 9}$$

  (b) Find constants A and B such that $3x/(x^2 - 9) = A/(x - 3) + B/(x + 3)$. (According to the text, you should obtain $A = B = 3/2$.)

In Exercises 39–44, you are given an improper rational expression. First, use long division to rewrite the expression in the form

$$(polynomial) + (proper\ rational\ expression)$$

Next, obtain the partial fraction decomposition for the proper rational expression. Finally, rewrite the given improper rational expression in the form

$$(polynomial) + (partial\ fractions)$$

39. $\dfrac{6x^3 - 16x^2 - 13x + 25}{x^2 - 4x + 3}$

40. $\dfrac{2x^5 - 11x^4 - 4x^3 + 53x^2 - 24x - 5}{2x^3 + x^2 - 10x - 5}$

41. $\dfrac{x^5 - 10x^4 + 36x^3 - 55x^2 + 32x + 1}{x^4 - 6x^3 + 12x^2 - 8x}$

42. $\dfrac{x^6 + 3x^5 + 9x^3 + 26x^2 + 3x + 8}{x^3 + 8}$

43. $\dfrac{x^6 + 2x^5 + 5x^4 - x^2 - 2x - 4}{x^4 - 1}$

44. $\dfrac{2x^7 + 3x^6 - 2x^4 - 4x^3 + 2}{x^4 + 2x^3 + x^2 - 1}$

Hint: After the long division, you can factor the denominator by writing it as $x^2(x^2 + 2x + 1) - 1^2$ and then using the difference-of-squares technique.

45. This exercise completes a detail mentioned in the text in the derivation of the linear and quadratic factors theorem. Let a and b be real numbers with $b \neq 0$. Show that the quadratic polynomial $x^2 - 2ax + a^2 + b^2$ is irreducible.

B

In Exercises 46–52, determine the partial fraction decomposition for each of the given expressions.

46. $\dfrac{1}{(x - a)(x - b)}$ $(a \neq b)$

47. $\dfrac{px + q}{(x - a)(x - b)}$ $(a \neq b)$

48. $\dfrac{1}{(x - a)(x + a)}$ $(a \neq 0)$

49. $\dfrac{px + q}{(x - a)(x + a)}$ $(a \neq 0)$

50. $\dfrac{x^2 + px + q}{(x - a)(x - b)(x - c)}$ (Assume that a, b, and c are all unequal.)

51. $\dfrac{1}{(1 - ax)(1 - bx)(1 - cx)}$ (Assume that a, b, and c all are nonzero and all unequal.)

52. (a) $\dfrac{x^2 - 1}{(x^2 + 1)^3}$ **(b)** $\dfrac{x^5 - 1}{(x^2 + 1)^3}$

C

53. Find the partial fraction decomposition: $\dfrac{1}{x^4 + 1}$.

Hint: To factor the denominator, add and subtract a term.

54. (a) Determine the general form for the partial fraction decomposition of $1/(x^4 - x^3 + x^2 - x + 1)$. (Note that you are not required to find the numerical values for each constant.) *Hint*: See Exercise 29 in Section 7.7.

(b) Determine the general form for the partial fraction decomposition of $1/(x^5 - 1)$.

MINI PROJECT | AN UNUSUAL PARTIAL FRACTIONS PROBLEM

As background for this mini project, you need to have worked Exercise 27(c) in Exercise Set 7.7. That exercise indicates how the equating-the-coefficients theorem can be used to factor certain polynomials.

The guidelines in this section for setting up a partial fractions decomposition for a rational expression require us to first factor the denominator into linear and/or irreducible quadratic factors. Two students each used this approach in determining a partial fractions decomposition for

$$\frac{x}{x^4 - 7x^2 + 1}$$

Student #1 used the equating-the-coefficients theorem to find a factorization of $x^4 - 7x^2 + 1$ of the form

$$x^4 - 7x^2 + 1 = (x^2 + bx + 1)(x^2 - bx + 1)$$

After obtaining the factorization (and checking to see that each factor was irreducible), the student followed the methods of Sections 7.7 and 7.8 to obtain a partial fraction decomposition. Student #2 worked along these lines also, but looked for a factorization of $x^4 - 7x^2 + 1$ having the form

$$x^4 - 7x^2 + 1 = (x^2 + Bx - 1)(x^2 - Bx - 1)$$

(a) Retrace each student's work to see what results were obtained. Do the two students end up with the same partial fraction decomposition? If not, who is correct (either, both, or neither)?

(b) Factoring has also been a key element in previous sections where we solved polynomial equations. Consider the equation $x^4 - 7x^2 + 1 = 0$. Use a graphing utility to see that this equation has four distinct real roots. Then use the rational roots theorem to explain why none of the roots is a rational number.

(c) Solve the equation $x^4 - 7x^2 + 1 = 0$ using the factorization obtained by student #1. Leave the roots in radical form, rather than obtaining calculator approximations. Next, solve the equation again, but this time use the factorization obtained by student #2. Do you obtain the same four roots?

(d) Using the values for b and B that you determined in part (a), graph the following four quadratic functions in the viewing rectangle $[-4, 4]$ by $[-3, 3]$. What do you observe regarding the x-intercepts, and what does this have to do with the last question in part (c)?

$$y_1 = x^2 + bx + 1 \qquad y_3 = x^2 + Bx - 1$$
$$y_2 = x^2 - bx + 1 \qquad y_4 = x^2 - Bx - 1$$

Chapter 7   Summary of Principal Terms and Notation

Terms or formulas	Page reference	Comments
1. $i^2 = -1$	561	This is the defining equation for the symbol i. The quantity i is not a real number.
2. Complex numbers	561	A complex number is an expression of the form $a + bi$, where a and b are real numbers. With the exception of division, complex numbers are combined using the usual rules of algebra, along with the convention $i^2 = -1$. Division is carried out using complex conjugates (defined in Item 3), as indicated in the first box on page 564.
3. If $z = c + di$, then $\bar{z} = c - di$.	563	$\bar{z}$ is called the complex conjugate of z.
4. The division algorithm	570	This is a theorem that summarizes the results of the long division process for polynomials. Suppose that $p(x)$ and $d(x)$ are polynomials and $d(x)$ is not the zero polynomial. Then according to the division algorithm, there are unique polynomials $q(x)$ and $R(x)$ such that $$p(x) = d(x) \cdot q(x) + R(x)$$ where either $R(x)$ is the zero polynomial or the degree of $R(x)$ is less than the degree of $d(x)$.

Terms or formulas	Page reference	Comments
5. Root, solution, zero	576	A root, or solution, of a polynomial equation $f(x) = 0$ is a number r such that $f(r) = 0$. The root r is also called a zero of the function f.
6. The remainder theorem	577	The remainder theorem asserts that when a polynomial $f(x)$ is divided by $x - r$, the remainder is $f(r)$. Example 3 in Section 7.3 shows how this theorem can be used with synthetic division to evaluate a polynomial.
7. The factor theorem	579	The factor theorem makes two statements about a polynomial $f(x)$. First, if $f(r) = 0$, then $x - r$ is a factor of $f(x)$. Second, if $x - r$ is a factor of $f(x)$, then $f(r) = 0$.
8. The fundamental theorem of algebra	585	Let $f(x)$ be a polynomial of degree 1 or greater. The fundamental theorem of algebra asserts that the equation $f(x) = 0$ has at least one root among the complex numbers. (See Example 1 in Section 7.4.)
9. The linear factors theorem	586	Let $f(x) = a_n x^n + a_{n-1} x^{n-1} + \cdots + a_1 x + a_0$, where $n \geq 1$ and $a_n \neq 0$. Then this theorem asserts that $f(x)$ can be expressed as a product of n linear factors: $$f(x) = a_n(x - r_1)(x - r_2) \cdots (x - r_n)$$ (The complex numbers $r_1, r_2, \ldots, r_n$ are not necessarily all distinct, and some or all of them may be real numbers.) On page 587, the linear factors theorem is used to prove that every polynomial equation of degree $n \geq 1$ has exactly n roots, where a root of multiplicity k is counted k times.
10. The rational roots theorem	596	Given a polynomial equation, this theorem tells us which rational numbers are candidates for roots of the equation. For a statement of the theorem, see page 596. A proof of the theorem is outlined in Exercise 41, Exercise Set 7.5. For an example of how the theorem is applied, see Example 1 in Section 7.5.
11. Upper bound for roots; lower bound for roots	597	A real number B is an upper bound for the roots of an equation if every real root is less than or equal to B. Similarly, a real number b is a lower bound for the roots if every real root is greater than or equal to b.
12. The upper and lower bound theorem for real roots	598	This theorem tells how synthetic division can be used in determining upper and lower bounds for roots of equations. For the statement and proof of the theorem, see page 598. For a demonstration of how the theorem is applied, see Example 2 in Section 7.5.

Terms or formulas	Page reference	Comments
13. The location theorem	600	Let $f(x)$ be a polynomial with real coefficients. If a and b are real numbers such that $f(a)$ and $f(b)$ have opposite signs, then the equation $f(x) = 0$ has at least one root between a and b.
14. The conjugate roots theorem	604	Let $f(x)$ be a polynomial with real coefficients, and suppose that $a + bi$ is a root of the equation $f(x) = 0$, where a and b are real numbers and $b \neq 0$. Then this theorem asserts that $a - bi$ is also a root of the equation. (In other words, for polynomial equations in which all of the coefficients are real numbers, when complex nonreal roots occur, they occur in conjugate pairs.) For illustrations of how this theorem is applied, see Examples 1 and 2 in Section 7.6.
15. Variation in sign	607	Suppose that $f(x)$ is a polynomial with real coefficients, written in descending or ascending powers of x. Then a variation in sign occurs whenever two successive coefficients have opposite signs. For examples, see Table 1 in Section 7.6.
16. Descartes's rule of signs	607	Let $f(x)$ be a polynomial, all of whose coefficients are real numbers, and consider the equation $f(x) = 0$. Then, according to Descartes's rule: **a.** The number of positive roots either is equal to the number of variations in sign of $f(x)$ or is less than that by an even integer. **b.** The number of negative roots either is equal to the number of variations in sign of $f(-x)$ or is less than that by an even integer. Examples 4–6 in Section 7.6 show how Descartes's rule is applied.
17. Proper rational expression	619	Let $p(x)$ and $q(x)$ be polynomials such that the degree of $p(x)$ is less than the degree of $q(x)$. Then the fraction $p(x)/q(x)$ is called a proper rational expression.
18. Irreducible quadratic polynomial	619	Let $ax^2 + bx + c$ be a quadratic polynomial, all of whose coefficients are real numbers. The polynomial is irreducible provided the equation $ax^2 + bx + c = 0$ has no real-number roots.
19. The linear and quadratic factors theorem	620	Let $f(x)$ be a polynomial, all of whose coefficients are real numbers. Then $f(x)$ can be factored (over the real numbers) into a product of linear and/or irreducible quadratic factors.
20. Partial fraction decomposition	620	A proper rational expression $p(x)/q(x)$ can be decomposed into a sum of simpler partial fractions. The denominators of the partial fractions are built from the linear and/or irreducible quadratic factors of the original denominator $q(x)$. For details, see the boxes on pages 620 and 623.

Writing Mathematics

Discuss each of the following statements with a classmate and decide whether it is true or false. Then (on your own), write out the reason (or reasons) for each decision.

1. Every equation has a root.
2. Every polynomial equation of degree 4 has four distinct roots.
3. No cubic equation can have a root of multiplicity 4.
4. The degree of the polynomial $x(x - 1)(x - 2)(x - 3)$ is 3.
5. The degree of the polynomial $6(x + 1)^2(x - 5)^4$ is 6.
6. Every polynomial of degree n, where $n \geq 1$, can be written in the form $(x - r_1)(x - r_2) \cdots (x - r_n)$.
7. The sum of the roots of the polynomial equation $x^2 - px + q = 0$ is p.
8. The product of the roots of the polynomial equation $2x^3 - x^2 + 3x - 1 = 0$ is 1.
9. Although a polynomial equation of degree n may have n distinct roots, the fundamental theorem of algebra tells us how to find only one of the roots.
10. Every polynomial equation of degree $n \geq 1$ has at least one real root.
11. If all of the coefficients in a polynomial equation are real, then at least one of the roots must be real.
12. Every cubic equation with roots $\sqrt{5}$, $\sqrt{6}$, and $\sqrt{7}$ can be written in the form

$$a_n(x - \sqrt{5})(x - \sqrt{6})(x - \sqrt{7}) = 0$$

13. Every polynomial equation of degree ≥ 1 has at least one root.
14. According to the rational roots theorem, there are four possibilities for the rational roots of the equation $x^5 + 6x^2 - 2 = 0$.
15. According to the rational roots theorem, there are only two possibilities for the rational roots of the equation $\frac{1}{3}x^3 - 5x + 1 = 0$.
16. The sum of the roots of the equation $x^2 - 12x + 16 = 0$ is -12.
17. According to the location theorem, if $f(x)$ is a polynomial and $f(a)$ and $f(b)$ have the same sign, then the equation $f(x) = 0$ has at least one root between a and b.
18. According to Descartes's rule, the equation $x^7 - 4x + 3 = 0$ has two positive roots.

Chapter 7   Review Exercises

In Exercises 1 and 2, you are given polynomials $p(x)$ and $d(x)$. In each case, use synthetic division to determine polynomials $q(x)$ and $R(x)$ such that

$$p(x) = d(x) \cdot q(x) + R(x)$$

where either $R(x) = 0$ or the degree of $R(x)$ is less than the degree of $d(x)$.

1. $p(x) = x^4 + 3x^3 - x^2 - 5x + 1; d(x) = x + 2$
2. $p(x) = 4x^4 + 2x + 1; d(x) = x - 2$

In Exercises 3–8, use synthetic division to find the quotients and the remainders.

3. $\dfrac{x^4 - 2x^2 + 8}{x - 3}$

4. $\dfrac{x^3 - 1}{x - 2}$

5. $\dfrac{2x^3 - 5x^2 - 6x - 3}{x + 4}$

6. $\dfrac{x^2 + x - 3\sqrt{2}}{x - \sqrt{2}}$

7. $\dfrac{5x^2 - 19x - 4}{x + 0.2}$

8. $\dfrac{x^3 - 3a^2x^2 - 4a^4x + 9a^6}{x - a^2}$

In Exercises 9–16, use synthetic division and the remainder theorem to find the indicated values of the functions. Use a calculator for Exercises 15 and 16.

9. $f(x) = x^5 - 10x + 4; f(10)$
10. $f(x) = x^4 + 2x^3 - x; f(-2)$
11. $f(x) = x^3 - 10x^2 + x - 1; f(1/10)$
12. $f(x) = x^4 - 2a^2x^2 + 3a^3x - a^4; f(-a)$
13. $f(x) = x^3 + 3x^2 + 3x + 1; f(a - 1)$
14. $f(x) = x^3 - 1; f(1.1)$
15. $f(x) = x^4 + 4x^3 - 6x^2 - 8x - 2;$
    (a) $f(-0.3)$ (Round the result to 2 decimal places.)
    (b) $f(-0.39)$ (Round the result to 3 decimal places.)
    (c) $f(-0.394)$ (Round the result to 5 decimal places.)
16. $f(-4.907378)$, where f is the function in Exercise 15. (Round the result to three decimal places.)
17. Find a value for a such that 3 is a root of the equation $x^3 - 4x^2 - ax - 6 = 0$.
18. For which values of b will -1 be a root of the equation $x^3 + 2b^2x^2 + x - 48 = 0$?
19. For which values of a will $x - 1$ be a factor of the polynomial $a^2x^3 + 3ax^2 + 2$?
20. Use synthetic division to verify that $\sqrt{2} - 1$ is a root of the equation $x^6 + 14x^3 - 1 = 0$.
21. Let $f(x) = ax^3 + bx^2 + cx + d$ and suppose that r is a root of the equation $f(x) = 0$.
    (a) Show that $r - h$ is a root of $f(x + h) = 0$.
    (b) Show that $-r$ is a root of the equation $f(-x) = 0$.
    (c) Show that kr is a root of the equation $f(x/k) = 0$.
22. Suppose that r is a root of the equation $a_2x^2 + a_1x + a_0 = 0$. Show that mr is a root of the quadratic equation $a_2x^2 + ma_1x + m^2a_0 = 0$.

In Exercises 23–28, list the possibilities for the rational roots of the equations.

23. $x^5 - 12x^3 + x - 18 = 0$
24. $x^5 - 12x^3 + x - 17 = 0$
25. $2x^4 - 125x^3 + 3x^2 - 8 = 0$
26. $\frac{3}{5}x^3 - 8x^2 - \frac{1}{2}x + \frac{3}{2} = 0$
27. $x^3 + x - p = 0$, where p is a prime number
28. $x^3 + x - pq = 0$, where both p and q are prime numbers

In Exercises 29–36, each equation has at least one rational root. Solve the equations. Suggestion: Use the upper and lower bound theorem to eliminate some of the possibilities for rational roots.

29. $2x^3 + x^2 - 7x - 6 = 0$
30. $x^3 + 6x^2 - 8x - 7 = 0$
31. $2x^3 - x^2 - 14x + 10 = 0$
32. $2x^3 + 12x^2 + 13x + 15 = 0$
33. $\frac{3}{2}x^3 + \frac{1}{2}x^2 + \frac{1}{2}x - 1 = 0$
34. $x^4 - 2x^3 - 13x^2 + 38x - 24 = 0$
35. $x^5 + x^4 - 14x^3 - 14x^2 + 49x + 49 = 0$
36. $8x^5 + 12x^4 + 14x^3 + 13x^2 + 6x + 1 = 0$
37. Solve the equation $x^3 - 9x^2 + 24x - 20 = 0$, using the fact that one of the roots has multiplicity 2.
38. One root of the equation $x^2 + kx + 2k = 0 \; (k \neq 0)$ is twice the other. Find k and find the roots of the equation.
39. State each of the following theorems.
    (a) The division algorithm
    (b) The remainder theorem
    (c) The factor theorem
    (d) The fundamental theorem of algebra
40. Find a quadratic equation with roots $a - \sqrt{a^2 - 1}$ and $a + \sqrt{a^2 - 1}$, where $a > 1$.

In Exercises 41–44, write each polynomial in the form

$$a_n(x - r_1)(x - r_2) \cdots (x - r_n)$$

41. $6x^2 + 7x - 20$
42. $x^2 + x - 1$
43. $x^4 - 4x^3 + 5x - 20$
44. $x^4 - 4x^2 - 5$

Each of Exercises 45–48 gives an equation, followed by one or more roots. Solve the equation.

45. $x^3 - 7x^2 + 25x - 39 = 0; x = 2 - 3i$
46. $x^3 + 6x^2 - 24x + 160 = 0; x = 2 + 2i\sqrt{3}$
47. $x^4 - 2x^3 - 4x^2 + 14x - 21 = 0; x = 1 + i\sqrt{2}$
48. $x^5 + x^4 - x^3 + x^2 + x - 1 = 0; x = (1 + i\sqrt{3})/2,$
    $x = (-1 - \sqrt{5})/2$

In Exercises 49–54, use Descartes's rule of signs to obtain information regarding the roots of the equations.

49. $x^3 + 8x - 7 = 0$
50. $3x^4 + x^2 + 4x - 2 = 0$
51. $x^3 + 3x + 1 = 0$
52. $2x^6 + 3x^2 + 6 = 0$
53. $x^4 - 10 = 0$
54. $x^4 + 5x^2 - x + 2 = 0$
55. Consider the equation $x^3 + x^2 + x + 1 = 0$.
    (a) Use Descartes's rule to show that the equation has either one or three negative roots.
    (b) Now show that the equation cannot have three negative roots. *Hint:* Multiply both sides of the equation by $x - 1$. Then simplify the left-hand side and reapply Descartes's rule to the new equation.
    (c) Actually, the original equation can be solved using the basic factoring techniques discussed in Section 1.3. Solve the equation in this manner.
56. Use Descartes's rule to show that the equation $x^3 - x^2 + 3x + 2 = 0$ has no positive roots.

Hint: Multiply both sides of the equation by $x + 1$ and apply Descartes's rule to the resulting equation.

57. Let P be the point in the first quadrant where the curve $y = x^3$ intersects the circle $x^2 + y^2 = 1$. Locate the x-coordinate of P within successive hundredths.

58. Let P be the point in the first quadrant where the parabola $y = 4 - x^2$ intersects the curve $y = x^3$. Locate the x-coordinate of P within successive hundredths.

59. Consider the equation $x^3 - 36x - 84 = 0$.
   (a) Use Descartes's rule to check that this equation has exactly one positive root.
   (b) Use the upper and lower bound theorem to show that 7 is an upper bound for the positive root.
   (c) Using a calculator, locate the positive root within successive hundredths.

60. Consider the equation $x^3 - 3x + 1 = 0$.
   (a) Use Descartes's rule to check that this equation has exactly one negative root.
   (b) Use the upper and lower bound theorem to show that -2 is a lower bound for the negative root.
   (c) Using a calculator, locate the negative root within successive hundredths.

In Exercises 61–64, find polynomial equations that have integer coefficients and the given values as roots.

61. $4 - \sqrt{5}$

62. $a + b$ and $a - b$, where a and b are integers

63. $6 - 2i$ and $\sqrt{5}$

64. $\dfrac{5 + \sqrt{6}}{5 - \sqrt{6}}$ *Hint:* First rationalize the expression.

65. Find a fourth-degree polynomial equation with integer coefficients, such that $x = 1 + \sqrt{2} + \sqrt{3}$ is a root.   *Hint:* Begin by writing the given relationship as $x - 1 = \sqrt{2} + \sqrt{3}$; then square both sides.

66. Find a cubic equation with integer coefficients, such that $x = 1 + \sqrt[3]{2}$ is a root. Is $1 - \sqrt[3]{2}$ also a root of the equation?

In Exercises 67–70, first determine the zeros of each function; then sketch the graph.

67. $y = x^3 - 2x^2 - 3x$

68. $y = x^4 + 3x^3 + 3x^2 + x$

69. $y = x^4 - 4x^2$

70. $y = x^3 + 6x^2 + 5x - 12$

For Exercises 71–78, carry out the indicated operations and express your answer in the standard form $a + bi$.

71. $(3 - 2i)(3 + 2i) + (1 + 3i)^2$

72. $2i(1 + i)^2$

73. $(1 + i\sqrt{2})(1 - i\sqrt{2}) + (\sqrt{2} + i)(\sqrt{2} - i)$

74. $(2 + 3i)/(1 + i)$

75. $(3 - i\sqrt{3})/(3 + i\sqrt{3})$

76. $\dfrac{1 + i}{1 - i} + \dfrac{1 - i}{1 + i}$

77. $-\sqrt{-2}\sqrt{-9} + \sqrt{-8} - \sqrt{-72}$

78. $\dfrac{\sqrt{-4} - \sqrt{-3}\sqrt{-3}}{\sqrt{-100}}$

79. The **real part** of a complex number z is denoted by $\text{Re}(z)$. For instance, $\text{Re}(2 + 5i) = 2$. Show that for any complex number z, we have $\text{Re}(z) = \frac{1}{2}(z + \bar{z})$. *Hint:* Let $z = a + bi$.

80. The **imaginary part** of a complex number z is denoted by $\text{Im}(z)$. For instance, $\text{Im}(2 + 5i) = 5$. Show that for any complex number z, we have
$$\text{Im}(z) = \frac{1}{2i}(z - \bar{z}).$$

81. The **absolute value** of the complex number $a + bi$ is defined by $|a + bi| = \sqrt{a^2 + b^2}$.
   (a) Compute $|6 + 2i|$ and $|6 - 2i|$.
   (b) As you know, the absolute value of the real number -3 is 3. Now write -3 in the form $-3 + 0i$ and compute its absolute value using the new definition. (The point here is to observe that the two results agree.)
   (c) Let $z = a + bi$. Show that $z\bar{z} = |z|^2$.

In Exercises 82–86, verify that the formulas are correct by carrying out the operations indicated on the left-hand side of the equations. (This list of formulas appears in A Treatise on Algebra, *by George Peacock, published in 1845.)*

82. $\dfrac{1}{a + bi} + \dfrac{1}{a - bi} = \dfrac{2a}{a^2 + b^2}$

83. $\dfrac{1}{a - bi} - \dfrac{1}{a + bi} = \dfrac{2bi}{a^2 + b^2}$

84. $\dfrac{a + bi}{a - bi} + \dfrac{a - bi}{a + bi} = \dfrac{2(a^2 - b^2)}{a^2 + b^2}$

85. $\dfrac{a + bi}{a - bi} - \dfrac{a - bi}{a + bi} = \dfrac{4abi}{a^2 + b^2}$

86. $\dfrac{a + bi}{c + di} + \dfrac{a - bi}{c - di} = \dfrac{2(ac + bd)}{c^2 + d^2}$

In Exercises 87–92, determine the partial fraction decomposition for each expression.

87. $\dfrac{2x - 1}{100 - x^2}$

88. $\dfrac{x}{x^2 - 20}$

89. $\dfrac{1}{x^3 + 2x^2 + x}$

90. $\dfrac{1}{x^4 + x^2}$

91. $\dfrac{x^3 + 2}{x^4 + 6x^2 + 9}$

92. $\dfrac{4x^3 - 3x^2 + 24x - 15}{x^4 + 11x^2 + 30}$

93. (a) Let $f(x) = x^4 - 2x^3 + x^2 - 1$. Use the equating-the-coefficients theorem to find a factorization for $f(x)$ of the form
$$f(x) = (x^2 + bx + 1)(x^2 + cx - 1).$$

(b) Find the partial fraction decomposition for $x^3/(x^4 - 2x^3 + x^2 - 1)$.

94. (a) Compute the product $(x^2 + rx + c)(x^2 - rx + c)$. [This can be useful in part (b).]

(b) If r is a root of the equation $x^2 + bx + c = 0$, show that r^2 is a root of $x^2 + (2c - b^2)x + c^2 = 0$.

Chapter 7 Test

1. Let $f(x) = 6x^4 - 5x^3 + 7x^2 - 2x - 2$. Make use of the remainder theorem and synthetic division to compute $f(1/2)$.

2. Solve the equation $x^3 + x^2 - 11x - 15 = 0$, given that one of the roots is -3.

3. List the possibilities for the rational roots of the equation $2x^5 - 4x^3 + x - 6 = 0$.

4. Find a quadratic function with zeros 1 and -8 and with a y-intercept of -24.

5. Use synthetic division to divide $4x^3 + x^2 - 8x + 3$ by $x + 1$.

6. (a) State the factor theorem.

(b) State the fundamental theorem of algebra.

(c) State the linear and quadratic factors theorem.

7. (a) The equation $x^3 - 2x^2 - 1 = 0$ has just one positive root. Use the upper and lower bound theorem to determine the smallest integer that is an upper bound for the root.

(b) Locate the root between successive tenths. (Use a calculator.)

8. Solve the equation $x^5 - 6x^4 + 11x^3 + 16x^2 - 50x + 52 = 0$, given that two of the roots are $1 + i$ and $3 - 2i$.

9. Let $p(x) = x^4 + 2x^3 - x + 6$ and $d(x) = x^2 + 1$. Find polynomials $q(x)$ and $R(x)$ such that $p(x) = d(x) \cdot q(x) + R(x)$.

10. Express the polynomial $2x^2 - 6x + 5$ in the factored form $a_n(x - r_1)(x - r_2)$.

11. Consider the equation $x^4 - x^3 + 24 = 0$.

(a) List the possibilities for rational roots.

(b) Use the upper and lower bound theorem to show that 2 is an upper bound for the roots.

(c) In view of parts (a) and (b), what possibilities now remain for positive rational roots?

(d) Which (if any) of the possibilities in part (c) are actually roots?

12. (a) Find the rational roots of the cubic equation $2x^3 - x^2 - x - 3 = 0$.

(b) Find all solutions of the equation in part (a).

13. Use Descartes's rule of signs to obtain information regarding the roots of the following equation: $3x^4 + x^2 - 5x - 1 = 0$.

14. Find a cubic polynomial $f(x)$ with integer coefficients, such that $1 - 3i$ and -2 are roots of the equation $f(x) = 0$.

15. Find a polynomial $f(x)$ with leading coefficient 1, such that the equation $f(x) = 0$ has the following roots and no other:

Root	Multiplicity
2	1
$3i$	3
$1 + \sqrt{2}$	2

Write your answer in the form $a_n(x - r_1)(x - r_2) \cdots (x - r_n)$.

16. Simplify: $(3 + 2i)(5 - 3i) + \sqrt{-3}$

17. Write the expression $\dfrac{3 + i}{1 - 4i}$ in the form $a + bi$.

In Exercises 18–20, determine the partial fraction decomposition for each expression.

18. $\dfrac{3x - 1}{x^3 - 16x}$

19. $\dfrac{1}{x^3 - x^2 + 3x - 3}$

20. $\dfrac{4x^2 - 15x + 20}{x^3 - 4x^2 + 4x}$

CHAPTER 8

The Conic Sections

The Greeks knew the properties of the curves given by cutting a cone with a plane—the ellipse, the parabola and hyperbola. Kepler discovered by analysis of astronomical observations, and Newton proved mathematically . . . that the planets move in ellipses. The geometry of Ancient Greece thus became the cornerstone of modern astronomy. —John Lighton Synge (1897–1987)

In this chapter we study three curves: the *parabola,* the *ellipse,* and the *hyperbola.* As indicated in the accompanying figure, these curves (along with circles) are formed when planes intersect the surface of a right circular cone. Although this is the context in which these curves were first identified by the ancient Greeks, in this chapter we'll follow a more algebraic approach involving the familiar *x-y* coordinate system. It can be shown that the two approaches are equivalent. If, after studying this chapter, you want to see the details, do a search on the World Wide Web for "Dandelin spheres." (Germinal Dandelin was a nineteenth-century Belgian/French mathematician who found a relatively simple way to see the equivalence of the two approaches.)

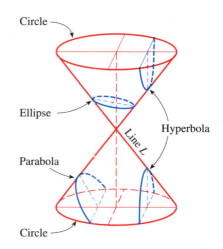

If not the first, certainly one of the first to discover and study the conics was the Greek mathematician Menaechmus (c. 380–320 B.C.), who was a tutor of Alexander the Great. However, until the seventeenth century, the conic sections were studied only as a portion of pure (as opposed to applied) mathematics. Then, in the seventeenth century, it was discovered that the conic sections were crucial in expressing some of the most important laws of nature. This is essentially the observation made in the opening quotation by the physicist J. L. Synge.

635

Only a few of the many applications of conics are included in this relatively brief chapter. You'll see examples where conics are used in:

- The design of a radio telescope (Example 6 in Section 8.1)
- The design of an arch (Exercise 41 in Section 8.1)
- Analyzing the orbit of a comet or planet (Example 7 in Section 8.3)
- Determining a location on earth without recourse to a global positioning system (Project for Section 8.5)

8.1   THE PARABOLA

In Section 4.2, we saw that the graph of a quadratic function $y = ax^2 + bx + c$ is a symmetric U-shaped curve called a *parabola*. In this section, we give a more general definition of the parabola, a definition that emphasizes the geometric properties of the curve.

DEFINITION | **The Parabola**

A **parabola** is the set of all points in the plane equally distant from a fixed line and a fixed point not on the line. The fixed line is called the **directrix;** the fixed point is called the **focus.**

Let us initially suppose that the focus of the parabola is the point $(0, p)$ and the directrix is the line $y = -p$. We will assume throughout this section that p is positive. To understand the geometric context of the definition of the parabola, the special graph paper displayed in Figure 1(a) is useful. The common center of the concentric circles in Figure 1(a) is the focus $(0, p)$. Thus, all the points on a given circle are at a fixed distance from the focus. The radii of the circles increase in increments of p units. Similarly, the broken horizontal lines in the figure are drawn at intervals that are multiples of p units from the directrix $y = -p$. By considering the points

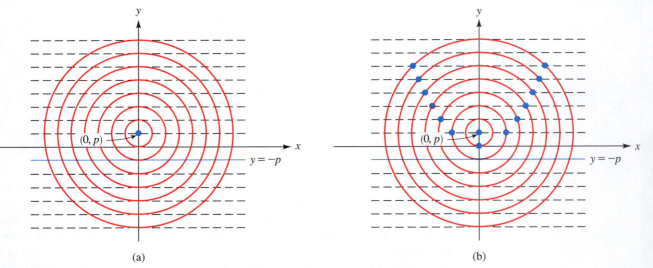

(a)   (b)

Figure 1

where the circles intersect the horizontal lines, we can find a number of points equally distant from the focus $(0, p)$ and the directrix $y = -p$; see Figure 1(b).

The graphical method we have just described lets us locate many points equally distant from the focus $(0, p)$ and the directrix $y = -p$. Figure 1(b) shows that the points on the parabola are symmetric about a line, in this case the y-axis. Also, by studying the figure, you should be able to convince yourself that in this case there can be no points below the x-axis that satisfy the stated condition. However, to describe the required set of points completely, and to show that our new definition is consistent with the old one, Figure 1(b) is inadequate. We need to bring algebraic methods to bear on the problem. Thus, let d_1 denote the distance from the point $P(x, y)$ to the focus $(0, p)$ and let d_2 denote the distance from $P(x, y)$ to the directrix $y = -p$, as shown in Figure 2. The distance d_1 is then

$$d_1 = \sqrt{(x - 0)^2 + (y - p)^2} = \sqrt{x^2 + y^2 - 2py + p^2}$$

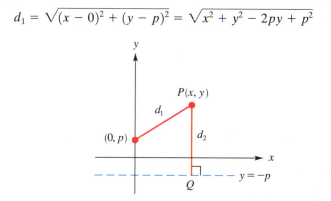

Figure 2

The distance d_2 in Figure 2 is just the vertical distance between the points P and Q. Thus,

$$d_2 = (y\text{-coordinate of } P) - (y\text{-coordinate of } Q)$$
$$= y - (-p) = y + p$$

(Absolute value signs are unnecessary here because, as noted earlier, P cannot lie below the x-axis.) The condition that P be equally distant from the focus $(0, p)$ and the directrix $y = -p$ can be expressed by the equation

$$d_1 = d_2$$

Using the expressions we've found for d_1 and d_2, this last equation becomes

$$\sqrt{x^2 + y^2 - 2py + p^2} = y + p$$

We can obtain a simpler but equivalent equation by squaring both sides. (Two non-negative quantities are equal if and only if their squares are equal.) Thus,

$$x^2 + y^2 - 2py + p^2 = y^2 + 2py + p^2$$

After combining like terms, this equation becomes

$$x^2 = 4py$$

This is the equation of a parabola with focus $(0, p)$ and directrix $y = -p$. In the box that follows, we summarize the properties of the parabola $x^2 = 4py$. As Figure 3(b) indicates, the terminology we've introduced applies equally well to an arbitrary parabola for which the axis of symmetry is not necessarily parallel to one of the co-ordinate axes and the vertex is not necessarily the origin.

▌ PROPERTY SUMMARY   The Parabola

1. The parabola is the set of points equidistant from a fixed line called the **directrix** and a fixed point, not on the line, called the **focus.**
2. The **axis** of a parabola is the line drawn through the focus and perpendicular to the directrix.
3. The **vertex** of a parabola is the point where the parabola intersects its axis. The vertex is located halfway between the focus and the directrix. See Figure 3.

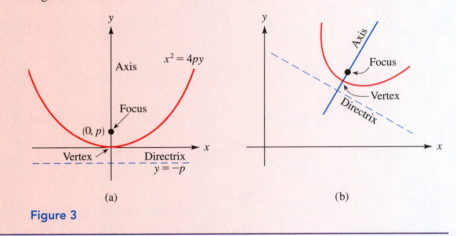

(a)   (b)

Figure 3

EXAMPLE   **1**   **Analyzing a parabola**

(a) Refer to Figure 4. Determine the focus and the directrix of the parabola $x^2 = 16y$.

(b) As indicated in Figure 4, the point Q is on the parabola $x^2 = 16y$ and the x-coordinate of Q is -12. Find the y-coordinate of Q.

(c) Verify that the point Q is equidistant from the focus and the directrix of the parabola.

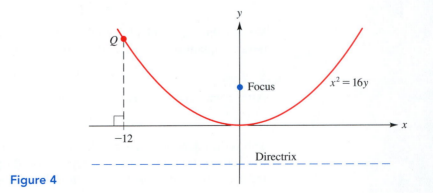

Figure 4

SOLUTION

(a) We know that for the basic parabola $x^2 = 4py$, the focus is the point $(0, p)$ and the directrix is the line $y = -p$. Comparing the given equation $x^2 = 16y$ with $x^2 = 4py$, we see that $4p = 16$ and therefore $p = 4$. Thus, the focus of the parabola $x^2 = 16y$ is $(0, 4)$ and the directrix is $y = -4$.

(b) We are given that the x-coordinate of Q is -12. Substituting $x = -12$ in the given equation $x^2 = 16y$ yields $(-12)^2 = 16y$. Therefore $y = 144/16 = 9$, and the coordinates of Q are $(-12, 9)$.

(c) In part (b) we found that the y-coordinate of Q is 9. From part (a) we know that the equation of the directrix is $y = -4$. Thus, the vertical distance from Q down to the directrix is $9 - (-4) = 13$. Next, to calculate the distance from $Q(-12, 9)$ to the focus $(0, 4)$, we can use the formula for the distance between two points. This yields

$$d = \sqrt{(x_2 - x_1)^2 + (y_2 - y_1)^2}$$
$$= \sqrt{(-12 - 0)^2 + (9 - 4)^2} = \sqrt{144 + 25} = \sqrt{169} = 13$$

We've now shown that the distance from Q to the focus is the same as the distance from Q to the directrix—namely, 13 units. In other words, Q is indeed equidistant from the focus and the directrix of the parabola.

EXAMPLE 2 **Finding the equation of a parabola**

Determine the equation of the parabola in Figure 5, given that the curve passes through the point $(3, 5)$. Specify the focus and the directrix.

SOLUTION

The general equation for a parabola in this position is $x^2 = 4py$. Since the point $(3, 5)$ lies on the curve, its coordinates must satisfy the equation $x^2 = 4py$. Thus,

$$3^2 = 4p(5)$$
$$9 = 20p$$
$$9/20 = p$$

With $p = 9/20$, the equation $x^2 = 4py$ becomes

$$x^2 = 4(9/20)y$$

or

$$x^2 = (9/5)y \qquad \text{as required}$$

Furthermore, since $p = 9/20$, the focus is $(0, 9/20)$ and the directrix is $y = -9/20$.

Figure 5

We have seen that the equation of a parabola with focus $(0, p)$ and directrix $y = -p$ is $x^2 = 4py$. By following the same method, we can obtain general equations for parabolas with other orientations. The basic results are summarized in Figure 6.

PROPERTY SUMMARY   Basic Equations for the Parabola

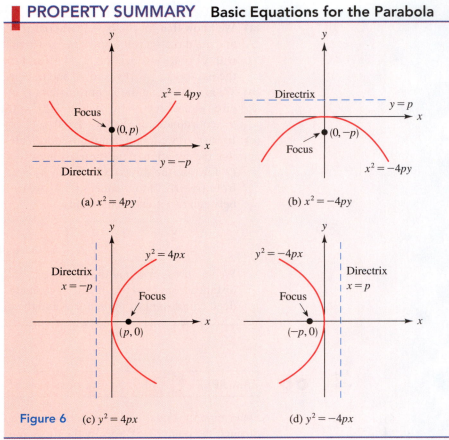

Figure 6   (a) $x^2 = 4py$ (b) $x^2 = -4py$ (c) $y^2 = 4px$ (d) $y^2 = -4px$

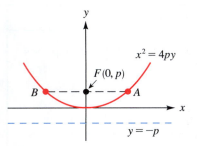

Figure 7
Focal width = $AB = 4p$

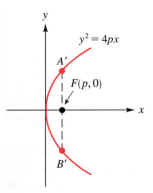

Figure 8
Focal width = $A'B' = 4p$

A **chord** of a parabola is a straight line segment joining any two points on the curve. If the chord passes through the focus, it is called a **focal chord.** For purposes of graphing, it is useful to know the length of the focal chord perpendicular to the axis of the parabola. This is the length of the horizontal line segment $\overline{AB}$ in Figure 7 and the vertical line segment $\overline{A'B'}$ in Figure 8. We will call this length the **focal width.**

In Figure 7, the distance from A to F is the same as the distance from A to the line $y = -p$. (Why?) But the distance from A to the line $y = -p$ is $2p$. Therefore, $AF = 2p$ and AB is twice this, or $4p$. We have shown that the focal width of the parabola is $4p$. In other words, given a parabola $x^2 = 4py$, the width at its focus is $4p$, the coefficient of y. In the same way, the length of the focal chord $\overline{A'B'}$ in Figure 8 is also $4p$, the coefficient of x in that case.

EXAMPLE | **3** | **Determining the focus and directrix of a parabola**

Find the focus and the directrix of the parabola $y^2 = -4x$, and sketch the graph.

SOLUTION
Comparing the basic equation $y^2 = -4px$ [in Figure 6(d)] with the equation at hand, we see that $4p = 4$ and thus $p = 1$. The focus is therefore $(-1, 0)$, and the

directrix is $x = 1$. The basic form of the graph will be as in Figure 6(d). For purposes of graphing, we note that the focal width is 4 (the absolute value of the coefficient of x). This, along with the fact that the vertex is $(0, 0)$, gives us enough information to draw the graph; see Figure 9.

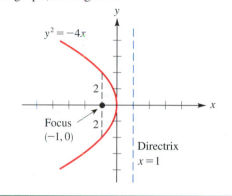

Figure 9

In Examples 1 through 3 and in Figure 6, the vertex of each parabola is located at the origin. Now we want to consider parabolas that are translated (shifted) from this standard position. As background for this, we review and generalize the results about translation in Section 3.4. Consider, as an example, the two equations

$$y = (x - 1)^2 \qquad \text{and} \qquad y = x^2 + 1$$

As we saw in Section 3.4, the graphs of these two equations can each be obtained by translating the graph of $y = x^2$. To graph $y = (x - 1)^2$, we translate the graph of $y = x^2$ to the right 1 unit; to graph $y = x^2 + 1$, we translate $y = x^2$ up 1 unit.

To see the underlying pattern here, let's rewrite the equation $y = x^2 + 1$ as $y - 1 = x^2$. Then the situation is this:

EQUATION	HOW GRAPH IS OBTAINED
$y = (x - 1)^2$	translation in positive x-direction
$y - 1 = x^2$	translation in positive y-direction

OBSERVATION   *In the equation $y = x^2$, the effect of replacing x with $x - 1$ is to translate the graph one unit in the positive x-direction. Similarly, replacing y with $y - 1$ translates the graph one unit in the positive y-direction.*

As a second example before we generalize, consider the equations

$$y = (x + 1)^2 \qquad \text{and} \qquad y = x^2 - 1$$

The first equation involves a translation in the negative x-direction; the second involves a translation in the negative y-direction. As before, to see the underlying pattern, we rewrite the second equation as $y + 1 = x^2$. Now we have the following situation:

EQUATION	HOW GRAPH IS OBTAINED
$y = (x + 1)^2$	translation in negative x-direction
$y + 1 = x^2$	translation in negative y-direction

OBSERVATION   *In the equation $y = x^2$, the effect of replacing x with $x + 1$ is to translate the graph one unit in the negative x-direction. Similarly, replacing y with $y + 1$ translates the graph one unit in the negative y-direction.*

Both of the examples that we have just considered are specific instances of the following basic result. (The result is valid whether or not the given equation and graph represent a function.)

PROPERTY SUMMARY   Translation and Coordinates

Suppose we have an equation that determines a graph in the x-y plane, and let h and k denote positive numbers. Then, replacing x with $x - h$ or $x + h$, or replacing y with $y - k$ or $y + k$, has the following effects on the graph of the original equation.

REPLACEMENT	RESULTING TRANSLATION
1. x replaced with $x - h$	h units in the positive x-direction
2. y replaced with $y - k$	k units in the positive y-direction
3. x replaced with $x + h$	h units in the negative x-direction
4. y replaced with $y + k$	k units in the negative y-direction

EXAMPLE 4   A parabola with a vertex other than the origin

Graph the parabola $(y + 1)^2 = -4(x - 2)$. Specify the vertex, the focus, the directrix, and the axis of symmetry.

SOLUTION
The given equation is obtained from $y^2 = -4x$ (which we graphed in the previous example) by replacing x and y with $x - 2$ and $y + 1$, respectively. So the required graph is obtained by translating the parabola in Figure 9 to the right 2 units and down 1 unit. In particular, this means that the vertex moves from $(0, 0)$ to $(2, -1)$; the focus moves from $(-1, 0)$ to $(1, -1)$; the directrix moves from $x = 1$ to $x = 3$; and the axis of symmetry moves from $y = 0$ (which is the x-axis) to $y = -1$. The required graph is shown in Figure 10.

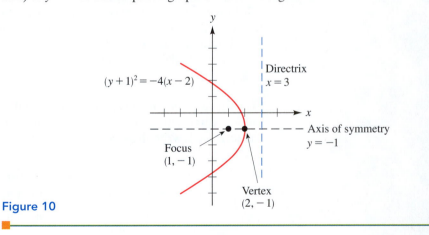

Figure 10

EXAMPLE 5 **Completing the square to analyze a parabola**

Graph the parabola $2y^2 - 4y - x + 5 = 0$, and specify each of the following: vertex, focus, directrix, axis of symmetry, and focal width.

SOLUTION
Just as we did in Section 4.2, we use the technique of completing the square:

$$2(y^2 - 2y) \qquad = x - 5$$
$$2(y^2 - 2y + 1) = x - 5 + 2 \qquad \text{adding 2 to both sides}$$
$$(y - 1)^2 = \frac{1}{2}(x - 3)$$

The graph of this last equation is obtained by translating the graph of $y^2 = \frac{1}{2}x$ "right 3, up 1." This moves the vertex from $(0, 0)$ to $(3, 1)$. Now, for $y^2 = \frac{1}{2}x$, the focus and directrix are determined by setting $4p = 1/2$. Therefore $p = 1/8$, and consequently the focus and directrix of $y^2 = \frac{1}{2}x$ are $(1/8, 0)$ and $x = -1/8$, respectively. Thus, the focus of the translated curve is $(3\frac{1}{8}, 1)$ and the directrix is $x = 2\frac{7}{8}$. Figure 11(a) shows the graph of $y^2 = \frac{1}{2}x$, and Figure 11(b) shows the translated graph. (You should check for yourself that the information accompanying Figure 11(b) is correct.)

$$y^2 = \frac{1}{2}x$$

Focus: $(1/8, 0)$
Directrix: $x = -1/8$
Focal width: $1/2$

(a)

$$2y^2 - 4y - x + 5 = 0$$
$$(y - 1)^2 = \frac{1}{2}(x - 3)$$

Vertex: $(3, 1)$
Focus: $(25/8, 1)$
Directrix: $x = 23/8$
Axis of symmetry: $y = 1$
Focal width: $1/2$

(b)

Figure 11

Incoming light rays

Focus

Parabolic mirror

Figure 12

There are numerous applications of the parabola in the sciences. Many of these involve parabolic reflectors. Figure 12 shows a cross section of a parabolic mirror in a telescope. As indicated in Figure 12, light rays coming in parallel to the axis of the parabola are reflected through the focus. Indeed, the word *focus* comes from a Latin word meaning "fireplace." [The mathematics (but not the physics) behind this focusing property is developed in Exercise 19 in Optional Section 8.2]. In addition to telescopes and radio telescopes, parabolic reflectors are used in communication systems such as satellite dishes for television, in surveillance systems, and in automobile headlights. To diagram the parabolic mirror of an automobile headlight, just reverse the directions of the arrows in Figure 12.

Geometrically, parabolic reflectors are described and designed as follows. We begin with a portion of a parabola and its axis of symmetry, as shown in Figure 13(a). By rotating the parabola about its axis, we obtain the bowl-shaped surface in Figure 13(b), known as a **paraboloid of revolution.** This is the surface used in a parabolic reflector.

Figure 13

Rotating a parabola about its axis of symmetry yields a bowl-shaped surface called a paraboloid of revolution.

(a) Parabola and axis of symmetry   (b) Paraboloid

As preparation for the next example, we define two terms that are used in describing optical systems: *focal length* and *focal ratio*. (Although we'll refer to a parabolic reflector in the definitions, the terms are often applied to other types of reflectors and lenses as well.)

DEFINITIONS | **Focal Length and Focal Ratio**

Consider a parabolic reflector with diameter d and vertex-to-focus distance p.

The **focal length** of the reflector is the distance p from the vertex to the focus. The **focal ratio** is the ratio of the focal length to the diameter; that is,

$$\text{focal ratio} = p/d$$

Thus, for example, if a parabolic reflector has a diameter of 2 ft and a focal length of 7 ft, then the focal ratio is $\dfrac{p}{d} = \dfrac{7\,\text{ft}}{2\,\text{ft}} = 3.5$.   *Remark*: In photography, the focal ratio is referred to as the *f-stop* or *f-number,* and a focal ratio of 3.5 is often written as f/3.5.

In the next example we determine the focal length and focal ratio of a parabolic reflector in a particular radio telescope. The following quote provides some general background about the field of radio astronomy.

Since its earliest beginnings in 1932 radio astronomy has developed into one of the most important means of investigating the universe. An impressive confirmation of this statement is given in the fact that all observing astronomers who received a Nobel prize in physics were working in the field of radio astronomy. This happened in 1974, in 1978, and for the last time in 1993. Electromagnetic radiation in the radio wavelength range from 0.35 mm up to about 15 m, which continuously hits the Earth from outer space, can be received using . . . [radio telescopes].
Max Planck Institute for Radio Astronomy, Bonn, Germany.
Web page http://www.mpifr-bonn.mpg.de/index_e.html, accessed November 2001.

EXAMPLE	6	**Finding the focal length and focal ratio of a parabolic reflector**

Figure 14(a) shows the parabolic reflector dish of the radio telescope at Parkes Observatory, which is located approximately 220 miles west of Sydney, Australia. Figure 14(b) gives the specifications for the parabolic cross section of the dish.
(a) Find the focal length of the dish. Round the answer to two decimal places.
(b) Find the focal ratio. Round the answer to two decimal places.

(a) Radio telescope at Parkes Observatory
(Photo copyright John Sarkissian)

(b) Dimensions of parabolic cross section

Figure 14
Photograph and specifications for the radio telescope dish at Parkes Observatory

SOLUTION
(a) According to Figure 14(b), the point $(32, 9.846)$ lies on the parabola $x^2 = 4py$. Thus we have $32^2 = 4p(9.846)$, and therefore

$$p = \frac{32^2}{4(9.846)}$$
$$\approx 26.00\,\text{m} \qquad \textcolor{red}{\text{using a calculator}}$$

Rounded to two decimal places, the focal length is 26.00 meters.
(b) From Figure 14(b), the diameter of the dish is 64 m ($= 32 + 32$). Now we can calculate the focal ratio using this value for the diameter along with the expression for the focal length p obtained in part (a). We have

$$\text{focal ratio} = \frac{p}{d} = \frac{32^2/(4 \times 9.846)}{64}$$
$$\approx 0.41 \qquad \textcolor{red}{\text{using a calculator}}$$

Rounded to two decimal places, the focal ratio is 0.41.

EXERCISE SET 8.1

A

In Exercises 1–22, graph the parabolas. In each case, specify the focus, the directrix, and the focal width. For Exercises 13–22, also specify the vertex.

1. $x^2 = 4y$ **2.** $x^2 = 16y$
3. $y^2 = -8x$ **4.** $y^2 = 12x$
5. $x^2 = -20y$ **6.** $x^2 - y = 0$
7. $y^2 + 28x = 0$ **8.** $4y^2 + x = 0$
9. $x^2 = 6y$ **10.** $y^2 = -10x$
11. $4x^2 = 7y$ **12.** $3y^2 = 4x$
13. $y^2 - 6y - 4x + 17 = 0$ **14.** $y^2 + 2y + 8x + 17 = 0$
15. $x^2 - 8x - y + 18 = 0$ **16.** $x^2 + 6y + 18 = 0$
17. $y^2 + 2y - x + 1 = 0$ **18.** $2y^2 - x + 1 = 0$
19. $2x^2 - 12x - y + 18 = 0$ **20.** $y + \sqrt{2} = (x - 2\sqrt{2})^2$
21. $2x^2 - 16x - y + 33 = 0$ **22.** $\frac{1}{4}y^2 - y - x + 1 = 0$

23. The accompanying figure shows the specifications for a cross section of the parabolic reflector dish for the Lovell radio telescope, located at the Jodrell Bank Observatory in England.
  (a) Find the focal length of the dish. Round the answer to one decimal place.
  (b) Find the focal ratio. Round the answer to one decimal place.

24. One of the largest radio telescopes in the world is the Effelsberg radio telescope, located 25 miles southwest of Bonn, Germany. Determine the diameter d of its parabolic reflector, given the specifications in the accompanying figure. (Assume that the units for x and y are meters.)

$$y^2 = 120x$$
Focal ratio: 0.3

25. The figure shows the specifications for a parabolic arch. (The blue curve is a portion of a parabola.)
  (a) Introduce a horizontal x-axis, passing through the vertex of the parabola, with the vertex corresponding to $x = 0$. Find an equation for the parabola of the form $x^2 = -4py$.

(b) Use the equation determined in part (a) to compute the height of the arch 5 feet away from the center (i.e., at $x = 5$ and $x = -5$).

26. The figure shows a diagram for the main section of a suspension bridge. The shape of the curved cable can be closely approximated by a parabola. As shown in the figure, there are 17 vertical support wires. These, along with the two 10.6-m end posts, are equally spaced at 5-m intervals. At the center of the bridge, the vertical support wire (from the vertex of the parabola to the roadway) has a length of 0.6 m.
  (a) In the figure, let the road correspond to the x-axis in an x-y coordinate system, with the vertex of the parabolic cable located at $(0, 0.6)$. Find an equation of the form $x^2 = 4p(y - k)$ for the parabola.
  (b) Find the lengths of the eight support wires on each side of the 0.6-m center wire.

B

For Exercises 27 and 28, refer to the figure below.

Figure A

27. Make a photocopy of Figure A. Then, in your copy of Figure A, indicate (by drawing dots) 11 points that are equidistant from the point $(-1, 1)$ and the line $x = 1$. What is the equation of the line of symmetry for the set of dots?

28. The 11 dots that you located in Exercise 27 are part of a parabola. Find the equation of that parabola and sketch its graph. Specify the vertex, the focus and directrix, and the focal width.

For Exercises 29–33, find the equation of the parabola satisfying the given conditions. In each case, assume that the vertex is at the origin.

29. The focus is $(0, 3)$.
30. The directrix is $y - 8 = 0$.
31. The directrix is $x + 32 = 0$.
32. The focus lies on the y-axis, and the parabola passes through the point $(7, -10)$.
33. The parabola is symmetric about the x-axis, the x-coordinate of the focus is negative, and the length of the focal chord perpendicular to the x-axis is 9.
34. Let P denote the point $(8, 8)$ on the parabola $x^2 = 8y$, and let $\overline{PQ}$ be a focal chord.
    (a) Find the equation of the line through the point $(8, 8)$ and the focus.
    (b) Find the coordinates of Q.
    (c) Find the length of $\overline{PQ}$.
    (d) Find the equation of the circle with this focal chord as a diameter.
    (e) Show that the circle determined in part (d) intersects the directrix of the parabola in only one point. Conclude from this that the directrix is tangent to the circle. Draw a sketch of the situation.
35. The segments $\overline{AA'}$ and $\overline{BB'}$ are focal chords of the parabola $x^2 = 2y$. The coordinates of A and B are $(4, 8)$ and $(-2, 2)$, respectively.
    (a) Find the equation of the line through A and B'.
    (b) Find the equation of the line through B and A'.
    (c) Show that the two lines you have found intersect at a point on the directrix.
36. Let $\overline{PQ}$ be the horizontal focal chord of the parabola $x^2 = 8y$. Let R denote the point where the directrix of the parabola meets the y-axis. Show that $\overline{PR}$ is perpendicular to $\overline{QR}$.
37. Suppose $\overline{PQ}$ is a focal chord of the parabola $y = x^2$ and that the coordinates of P are $(2, 4)$.
    (a) Find the coordinates of Q.
    (b) Find the coordinates of M, the midpoint of $\overline{PQ}$.
    (c) A perpendicular is drawn from M to the y-axis, meeting the y-axis at S. Also, a line perpendicular to the focal chord is drawn through M, meeting the y-axis at T. Find ST and verify that it is equal to one-half the focal width of the parabola.

38. Let F be the focus of the parabola $x^2 = 8y$, and let P denote the point on the parabola with coordinates $(8, 8)$. Let $\overline{PQ}$ be a focal chord. If V denotes the vertex of the parabola, verify that

$$PF \cdot FQ = VF \cdot PQ$$

39. In the following figure, $\triangle OAB$ is equilateral and $\overline{AB}$ is parallel to the x-axis. Find the length of a side and the area of triangle OAB.

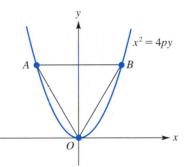

40. In designing an arch, architects and engineers sometimes use a parabolic arch rather than a semicircular arch. (One reason for this is that, in general, the parabolic arch can support more weight at the top than can the semicircular arch.) In the following figure, the blue arch is a semicircle of radius 1, centered at the origin. The red arch is a portion of a parabola. As indicated in the figure, the two arches have the same base and the same height. Assume that the unit of distance for each axis is the meter.

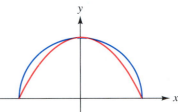

    (a) Find the equation of the parabola in the figure.
    (b) Using calculus, it can be shown that the area under this parabolic arch is $\frac{4}{3}$ m^2. Assuming this fact, show that the area beneath the parabolic arch is approximately 85% of the area beneath the semicircular arch.
    (c) Using calculus, it can be shown that the length of this parabolic arch is $\sqrt{5} + \frac{1}{2} \ln(2 + \sqrt{5})$ meters. Assuming this fact, show that the length of the parabolic arch is approximately 94% of the length of the semicircular arch.
41. (a) The span of the parabolic arch in Figure B (on page 648) is 8 m. At a distance of 2 m from the center, the vertical clearance is 4.5 m. Find the height of the arch. *Hint*: Choose a convenient coordinate system in which the equation of the parabola will have the form $x^2 = -4p(y - k)$.

(b) Figure C shows the same parabolic arch as in Figure B. Suppose that from a point P on the base of the arch, a laser beam is projected vertically upward. As indicated in the figure, the beam hits the arch at a point Q and is then reflected. Let R denote the point where the beam crosses the axis of the parabola. How far is it from R to the base of the arch? (That is, what is the perpendicular distance from R to the base?)

Figure B   **Figure C**

C

42. If $\overline{PQ}$ is a focal chord of the parabola $x^2 = 4py$ and the coordinates of P are (x_0, y_0), show that the coordinates of Q are

$$\left(\frac{-4p^2}{x_0}, \frac{p^2}{y_0} \right)$$

43. If $\overline{PQ}$ is a focal chord of the parabola $y^2 = 4px$ and the coordinates of P are (x_0, y_0), show that the coordinates of Q are

$$\left(\frac{p^2}{x_0}, \frac{-4p^2}{y_0} \right)$$

44. Let F and V denote the focus and the vertex, respectively, of the parabola $x^2 = 4py$. If $\overline{PQ}$ is a focal chord of the parabola, show that

$$PF \cdot FQ = VF \cdot PQ$$

45. Let $\overline{PQ}$ be a focal chord of the parabola $y^2 = 4px$, and let M be the midpoint of $\overline{PQ}$. A perpendicular is drawn from M to the x-axis, meeting the x-axis at S. Also from M, a line segment is drawn that is perpendicular to $\overline{PQ}$ and that meets the x-axis at T. Show that the length of $\overline{ST}$ is one-half the focal width of the parabola.

46. Let $\overline{AB}$ be a chord (not necessarily a focal chord) of the parabola $y^2 = 4px$, and suppose that $\overline{AB}$ subtends a right angle at the vertex. (In other words, $\angle AOB = 90°$, where O is the origin in this case.) Find the x-intercept of the segment $\overline{AB}$. What is surprising about this result? *Hint*: Begin by writing the coordinates of A and B as $A(a^2/4p, a)$ and $B(b^2/4p, b)$.

47. Let $\overline{PQ}$ be a focal chord of the parabola $x^2 = 4py$. Complete the following steps to prove that the circle with $\overline{PQ}$ as a diameter is tangent to the directrix of the parabola. Let the coordinates of P be (x_0, y_0).

(a) Show that the coordinates of Q are

$$\left(\frac{-4p^2}{x_0}, \frac{p^2}{y_0} \right)$$

(b) Show that the midpoint of $\overline{PQ}$ is

$$\left(\frac{x_0^2 - 4p^2}{2x_0}, \frac{y_0^2 + p^2}{2y_0} \right)$$

(c) Show that the length of $\overline{PQ}$ is $(y_0 + p)^2/y_0$. *Suggestion*: This can be done using the formula for the distance between two points, but the following is simpler. Let F be the focus. Then $PQ = PF + FQ$. Now, both PF and FQ can be determined by using the definition of the parabola rather than the distance formula.

(d) Show that the distance from the center of the circle to the directrix equals the radius of the circle. How does this complete the proof?

48. P and Q are two points on the parabola $y^2 = 4px$, the coordinates of P are (a, b), and the slope of $\overline{PQ}$ is m. Find the y-coordinate of the midpoint of $\overline{PQ}$. *Hint for checking*: The final answer is independent of both a and b.

MINI PROJECT 1   A BRIDGE WITH A PARABOLIC ARCH

Figure A shows a bridge across a river. The arch of the bridge is a parabola, and the six vertical cables that help support the road are equally spaced at 4-m intervals. Figure B shows the parabolic arch in an x-y coordinate system, with the left end of the arch at the origin. As indicated in Figure B, the length of the leftmost cable is 3.072 m. Determine an equation for the parabola of the form

$(x - h)^2 = -4p(y - k)$. Then use the equation to determine the lengths of the other cables and also the maximum height of the arch above the road.

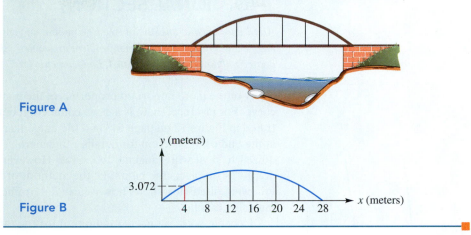

Figure A

Figure B

MINI PROJECT 2 | CONSTRUCTING A PARABOLA

This hands-on mini project shows you how to construct a parabola. Follow the instructions and then explain why the curve that you draw is indeed a parabola. In addition to a sheet of paper and a pen or marker, you'll need the following:

- A T-square and a drawing board or surface on which you can easily slide the T-square left and right in a straight horizontal path
- A piece of string that has the same length as the long bar of the T-square (see Figure C)
- Two thumb tacks

(Refer to Figure D.) Tack one end of the piece of string to a point on the drawing board. Call that point *F*. Tack the other end of the string to the top right corner of the T-square, as indicated in Figure D. Now, with the pen or marker, move the T-square horizontally subject to the following two constraints: The string must be kept taut by the pen, and the pen must stay against the edge of the T-square. The curve traced out by the pen will be a portion of a parabola. (In practice, the piece of string that you begin with should actually be just slightly longer than the length of the T-square, since a slight amount of the string is taken up in the tacking.)

Piece of string with same length as long bar of T-square

T-square

Figure C

Figure D

8.2   TANGENTS TO PARABOLAS (OPTIONAL SECTION)

And I dare say that this [the problem of finding tangents and normals to curves] *is not only the most useful and most general problem in geometry that I know, but even that I have ever desired to know.* —René Descartes, *La Géométrie* (1637).

Many of the more important properties of the parabola relate to the tangent to the curve. For instance, there is a close connection between the optical property illustrated in Figure 12 on page 643 and the tangents to the parabola. (See Exercise 19 at the end of this section for details.) In general, the techniques of calculus are required to deal with tangents to curves. However, for curves with equations and graphs as simple as the parabola, the methods of algebra are often adequate. The following discussion shows how we can determine the tangent to a parabola without using calculus.

To begin with, we need a definition for the tangent to a parabola. As motivation, recall from geometry that a tangent line to a circle is defined to be a line that intersects the circle in exactly one point (see Figure 1). However, this definition is not quite adequate for the parabola. For instance, the y-axis intersects the parabola $y = x^2$ in exactly one point, but surely it is not tangent to the curve. With this in mind, we adopt the definition in the following box for the tangent to a parabola. Along with this definition, we make the assumption that through each point P on a parabola, there is only one tangent line that can be drawn.

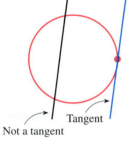

Tangent

Not a tangent

Figure 1

DEFINITION | **Tangent to a Parabola**

Let P be a point on a parabola. Then a line through P is said to be tangent to the parabola at P provided that the line intersects the parabola only at P and the line is not parallel to the axis of the parabola.

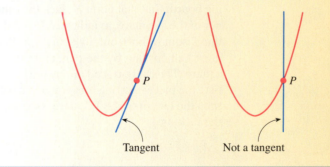

Tangent   Not a tangent

We'll demonstrate how to find tangents to parabolas by means of an example. The method used here can be used for any parabola. Suppose that we want to find the equation of the tangent to the parabola $x^2 = 4y$ at the point $(2, 1)$ on the curve (see Figure 2). Let m denote the slope of the tangent line. Since the line must pass through the point $(2, 1)$, its equation is:

$$y - 1 = m(x - 2)$$

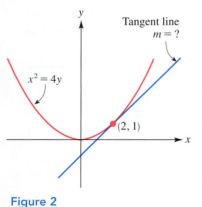

Figure 2

This is the tangent line, so it intersects the parabola in only one point—namely, (2, 1). Algebraically, this means that the ordered pair (2, 1) is the only solution of the following system of equations:

$$\begin{cases} y - 1 = m(x - 2) & (1) \\ x^2 = 4y & (2) \end{cases}$$

The strategy now is to solve this system of equations in terms of m. Then we'll reconcile the results with the fact that (2, 1) is known to be the unique solution. This will allow us to determine m. From equation (2), we have $y = x^2/4$. Then, substituting for y in equation (1), we obtain

$$\frac{x^2}{4} - 1 = m(x - 2)$$

$$x^2 - 4 = 4m(x - 2) \qquad \text{multiplying both sides by 4}$$

$$(x - 2)(x + 2) = 4m(x - 2) \qquad \text{factoring}$$

$$(x - 2)(x + 2) - 4m(x - 2) = 0 \qquad \begin{array}{l}\text{subtracting } 4m(x - 2)\\ \text{from both sides}\end{array}$$

Observe now that the quantity $x - 2$ appears twice on the left-hand side of the last equation, so we can factor it out to obtain

$$(x - 2)[(x + 2) - 4m] = 0$$

$$\begin{array}{c|c} x - 2 = 0 & x + 2 - 4m = 0 \\ x = 2 & x = 4m - 2 \end{array}$$

Now look at these two x-values that we've just determined. The first value, $x = 2$, yields no new information, since we knew from the start that this was the x-coordinate of the intersection point. However, the second value, $x = 4m - 2$, must also equal 2, since the system has but one solution. Thus we have

$$4m - 2 = 2$$

$$4m = 4$$

$$m = 1$$

Therefore, the slope of the tangent line is 1. Substituting this value of m in equation (1) yields

$$y - 1 = 1(x - 2)$$

or

$$y = x - 1 \qquad \text{as required}$$

EXERCISE SET 8.2

A

In Exercises 1–8, use the method shown in the text to find the equation of the tangent to the parabola at the given point. In each case, include a sketch with your answer.

1. $x^2 = y$; (2, 4)
2. $x^2 = -2y$; (2, -2)
3. $x^2 = 8y$; (4, 2)
4. $x^2 = 12y$; (6, 3)
5. $x^2 = -y$; (-3, -9)
6. $x^2 = -6y$; ($\sqrt{6}$, -1)
7. $y^2 = 4x$; (1, 2)
8. $y^2 = -8x$; (-8, -8)

B

In Exercises 9–11, find the slope of the tangent to the curve at the indicated point. (Use the method shown in the text for parabolas.)

9. $y = \sqrt{x}$; (4, 2)

   Hint: $x - 4 = (\sqrt{x} - 2)(\sqrt{x} + 2)$

10. $y = 1/x$; (3, 1/3)
11. $y = x^3$; (2, 8)

12. Consider the parabola $x^2 = 4y$, and let (x_0, y_0) denote a point on the parabola in the first quadrant.
(a) Find the y-intercept of the line tangent to the parabola at the point on the parabola where $y_0 = 1$.
(b) Repeat part (a) using $y_0 = 2$.
(c) Repeat part (a) using $y_0 = 3$.
(d) On the basis of your results in parts (a)–(c), make a conjecture about the y-intercept of the line that is tangent to the parabola at the point (x_0, y_0) on the curve. Verify your conjecture by computing this y-intercept.

C

13. Let (x_0, y_0) be a point on the parabola $x^2 = 4py$. Using the method explained in the text, show that the equation of the line tangent to the parabola at (x_0, y_0) is

$$y = \frac{x_0}{2p}x - y_0$$

Thus, the y-intercept of the line tangent to $x^2 = 4py$ at (x_0, y_0) is just $-y_0$.

14. Let (x_0, y_0) be a point on the parabola $y^2 = 4px$. Show that the equation of the line tangent to the parabola at (x_0, y_0) is

$$y = \frac{2p}{y_0}x + \frac{y_0}{2}$$

Show that the x-intercept of this line is $-x_0$.

Exercises 15–20 contain results about tangents to parabolas. In working these problems, you'll find it convenient to use the facts developed in Exercises 13 and 14. Also assume, as given, the results about focal chords in Exercises 42 and 43 of Exercise Set 8.1. In some of the problems, reference is made to the normal *line. The* **normal line** *or* **normal** *to a parabola at the point (x_0, y_0) on the parabola is defined as the line through (x_0, y_0) that is perpendicular to the tangent at (x_0, y_0).*

15. Let $\overline{PQ}$ be a focal chord of the parabola $x^2 = 4py$.
(a) Show that the tangents to the parabola at P and Q are perpendicular to each other.
(b) Show that the tangents to the parabola at P and Q intersect at a point on the directrix.
(c) Let D be the intersection point of the tangents at P and Q. Show that the line segment from D to the focus is perpendicular to $\overline{PQ}$.

16. Let $P(x_0, y_0)$ be a point [other than $(0, 0)$] on the parabola $y^2 = 4px$. Let A be the point where the normal line to the

parabola at P meets the axis of the parabola. Let B be the point where the line drawn from P perpendicular to the axis of the parabola meets the axis. Show that $AB = 2p$.

17. Let $P(x_0, y_0)$ be a point on the parabola $y^2 = 4px$. Let A be the point where the normal line to the parabola at P meets the axis of the parabola. Let F be the focus of the parabola. If a line is drawn from A perpendicular to $\overline{FP}$, meeting $\overline{FP}$ at Z, show that $ZP = 2p$.

18. Let $\overline{AB}$ be a chord (not necessarily a focal chord) of the parabola $x^2 = 4py$. Let M be the midpoint of the chord and let C be the point where the tangents at A and B intersect.
(a) Show that $\overline{MC}$ is parallel to the axis of the parabola.
(b) If D is the point where the parabola meets $\overline{MC}$, show that $CD = DM$.

19. In this exercise, we prove the *reflection property* of parabolas. The following figure shows a line tangent to the parabola $y^2 = 4px$ at $P(x_0, y_0)$. The dashed line through H and P is parallel to the axis of the parabola. We wish to prove that $\alpha = \beta$. That is, we wish to prove the reflection property: the angle of incidence equals the angle of reflection.

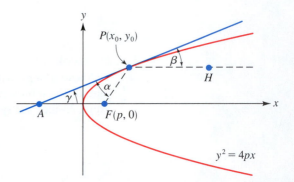

(a) Show that $FA = x_0 + p = FP$.   *Hint:* Regarding FP, it is easier to rely on the definition of a parabola than on the distance formula.
(b) Why does $\alpha = \gamma$? Why does $\gamma = \beta$?
(c) Conclude that $\alpha = \beta$, as required.

20. The segment $\overline{AB}$ is a focal chord of the parabola $y^2 = 4x$ and the coordinates of A are $(4, 4)$. Normals drawn through A and B meet the parabola again at A' and B', respectively. Prove that $\overline{A'B'}$ is three times as long as $\overline{AB}$.

8.3   THE ELLIPSE

We have here apparently [in the work of Anthemius of Tralles (a sixth-century Greek architect and mathematician)] *the first mention of the construction of an ellipse by means of a string stretched tight round the foci.*—Sir Thomas Heath in *A History of Greek Mathematics,* Vol. II (Oxford: The Clarendon Press, 1921)

The heavenly motions are nothing but a continuous song for several voices, to be perceived by the intellect, not by the ear.—Johannes Kepler (1571–1630)

In this section we discuss the symmetric, oval-shaped curve known as the *ellipse.* As Kepler discovered and Newton later proved, this is the curve described by the planets in their motions around the sun.

DEFINITION	The Ellipse

An **ellipse** is the set of all points in the plane, the sum of whose distances from two fixed points is constant. Each fixed point is called a **focus** (plural: **foci**).

Subsequently we will derive an equation describing the ellipse just as we found an equation for the parabola in Section 8.1. But first let's consider some rather immediate consequences of the definition. In fact, we can learn a great deal about the ellipse even before we derive its equation.

There is a simple mechanical method for constructing an ellipse that arises directly from the definition of the curve. Mark the given foci—say, F_1 and F_2—on a drawing board and insert thumbtacks at those points. Now take a piece of string that is longer than the distance from F_1 to F_2, and tie the ends of the string to the tacks. Next, pull the string taut with a pencil point and touch the pencil point to the drawing board. Then if you move the pencil while keeping the string taut, the curve traced out will be an ellipse, as indicated in Figure 1. The reason the curve is an ellipse is that for each point on the curve, the sum of the distances from the foci is constant, the constant being the length of the string. By actually carrying out this construction for yourself several times, each time varying the distance between the foci or the length of the string, you can learn a great deal about the ellipse. For instance, when the distance between the foci is small compared to the length of the string, the ellipse begins to resemble a circle, as in Figure 2. On the other hand, when the distance between the foci is nearly equal to the length of the string, the ellipse becomes relatively flat, as in Figure 3.

We now derive one of the standard forms for the equation of an ellipse. As indicated in Figure 4, we assume that the foci are $F_1(-c, 0)$ and $F_2(c, 0)$ and that the

Figure 1

Figure 2
When the distance between the foci is small compared to the length of the string, the ellipse resembles a circle.

Figure 3
When the distance between the foci is nearly equal to the length of the string, the ellipse is relatively flat.

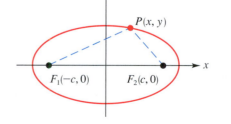

Figure 4
$F_1P + F_2P = 2a$

sum of the distances from the foci to a point $P(x, y)$ on the ellipse is $2a$. Since the point P lies on the ellipse, we have

$$F_1P + F_2P = 2a$$

and therefore

$$\sqrt{(x + c)^2 + (y - 0)^2} + \sqrt{(x - c)^2 + (y - 0)^2} = 2a$$

or

$$\sqrt{(x - c)^2 + y^2} = 2a - \sqrt{(x + c)^2 + y^2} \qquad (1)$$

Now, by following a straightforward but lengthy process of squaring and simplifying (as outlined in detail in Exercise 44), we find that equation (1) becomes

$$(a^2 - c^2)x^2 + a^2y^2 = a^2(a^2 - c^2) \qquad (2)$$

To write equation (2) in a more symmetric form, we define the positive number b by the equation

$$b^2 = a^2 - c^2 \qquad (3)$$

NOTE   For this definition to make sense, we need to know that the right-hand side of equation (3) is positive. (See Exercise 45 at the end of this section for details.)
    Finally, using equation (3) to substitute for $a^2 - c^2$ in equation (2), we obtain

$$b^2x^2 + a^2y^2 = a^2b^2$$

which can be written

$$\frac{x^2}{a^2} + \frac{y^2}{b^2} = 1 \qquad (a > b) \qquad (4)$$

We have now shown that the coordinates of each point on the ellipse satisfy equation (4). Conversely, it can be shown that if x and y satisfy equation (4), then the point (x, y) indeed lies on the ellipse. We refer to equation (4) as the **standard form** for the equation of an ellipse with foci $(-c, 0)$ and $(c, 0)$. For an ellipse in this form, it will always be the case that a is greater than b; this follows from equation (3).
    For purposes of graphing, we want to know the intercepts of the ellipse. To find the x-intercepts, we set $y = 0$ in equation (4) to obtain

$$\frac{x^2}{a^2} = 1$$

$$x^2 = a^2 \qquad \text{or} \qquad x = \pm a$$

The x-intercepts are therefore a and $-a$. In a similar fashion, you can check that the y-intercepts are b and $-b$. Also (according to the symmetry tests in Section 1.7), note that the graph of equation (4) must be symmetric about both coordinate axes. Figure 5 shows the graph of the ellipse $(x^2/a^2) + (y^2/b^2) = 1$. (Calculator exercises at the end of this section will help to convince you that the general shape of the curve in Figure 5 is correct.)
    In the box that follows, we define several terms that are useful in describing and analyzing the ellipse.

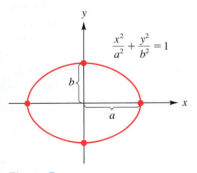

Figure 5
The intercepts of the ellipse $(x^2/a^2) + (y^2/b^2) = 1$ are $x = \pm a$ and $y = \pm b$.

DEFINITION | Terminology for the Ellipse

Figure 6

1. The **focal axis** is the line passing through the foci of the ellipse.
2. The **center** is the point midway between the foci. This is the point C in Figure 6.
3. The **vertices** (singular: **vertex**) are the two points where the focal axis meets the ellipse. These are the points V_1 and V_2 in Figure 6.
4. The **major axis** is the line segment joining the vertices. This is the line segment $\overline{V_1V_2}$ in Figure 6. Each of the (congruent) line segments $\overline{CV_1}$ and $\overline{CV_2}$ is referred to as a **semimajor axis.** The **minor axis** is the line segment through the center of the ellipse, perpendicular to the major axis, and with endpoints on the ellipse. In Figure 6, this is the line segment $\overline{AB}$. Each of the (congruent) line segments $\overline{CB}$ and $\overline{CA}$ is referred to as a **semiminor axis.**
5. The **eccentricity** is the ratio $\dfrac{c}{a} = \dfrac{\sqrt{a^2 - b^2}}{a}$.

The eccentricity (as defined in the box) provides a numerical measure of how much the ellipse deviates from being a circle. As Figure 7 indicates, the closer the eccentricity is to zero, the more the ellipse resembles a circle. In the other direction, as the eccentricity approaches 1, the ellipse becomes increasingly flat.

Figure 7
Eccentricity is a number between 0 and 1 that describes the shape of an ellipse. The narrowest ellipse in this figure has the same proportions as the orbit of Halley's comet. By way of contrast, the eccentricity of Earth's orbit is 0.0017; if an ellipse with this eccentricity were included in Figure 7, it would appear indistinguishable from a circle.

Eccentricity = 0.4   Eccentricity = 0.8   Eccentricity = 0.967

In the next box, we summarize our discussion up to this point. (We use the letter e to denote the eccentricity; this is the conventional choice, even though the same letter is used with a very different meaning in connection with exponential functions and logarithms.)

PROPERTY SUMMARY   The Ellipse $\dfrac{x^2}{a^2} + \dfrac{y^2}{b^2} = 1$ $(a > b)$

Foci: $(\pm c, 0)$, where $c^2 = a^2 - b^2$
Center: $(0, 0)$
Vertices: $(\pm a, 0)$
Length of major axis: $2a$
Length of semimajor axis: a
Length of minor axis: $2b$
Length of semiminor axis: b

Eccentricity: $e = \dfrac{c}{a} = \dfrac{\sqrt{a^2 - b^2}}{a}$

EXAMPLE 1 **Analyzing an ellipse**

Find the lengths of the major and minor axes of the ellipse $9x^2 + 16y^2 = 144$. Specify the coordinates of the foci and the eccentricity. Graph the ellipse.

SOLUTION
To convert the equation $9x^2 + 16y^2 = 144$ to standard form, we divide both sides by 144. This yields

$$\frac{9x^2}{144} + \frac{16y^2}{144} = \frac{144}{144}$$

$$\frac{x^2}{16} + \frac{y^2}{9} = 1$$

$$\frac{x^2}{4^2} + \frac{y^2}{3^2} = 1$$

This is the standard form. Comparing this equation with $(x^2/a^2) + (y^2/b^2) = 1$, we see that $a = 4$ and $b = 3$. Thus, the major and minor axes are 8 and 6 units, respectively. Next, to determine the foci, we use the equation $c^2 = a^2 - b^2$. We have

$$c^2 = 4^2 - 3^2 = 7 \qquad \text{and therefore} \qquad c = \sqrt{7}$$

(We choose the positive square root because $c > 0$.) It follows that the coordinates of the foci are $(-\sqrt{7}, 0)$ and $(\sqrt{7}, 0)$. We now calculate the eccentricity using the formula $e = c/a$. This yields $e = \sqrt{7}/4$. Figure 8 shows the graph along with the required information.

Figure 8

Major axis: 8
Minor axis: 6
Foci: $(\pm\sqrt{7}, 0)$
Eccentricity: $\dfrac{\sqrt{7}}{4} \approx 0.66$

EXAMPLE 2 **Finding the equation of an ellipse**

The foci of an ellipse are $(-1, 0)$ and $(1, 0)$, and the eccentricity is 1/3. Find the equation of the ellipse (in standard form), and specify the lengths of the major and minor axes.

SOLUTION
Since the foci are $(\pm 1, 0)$, we have $c = 1$. Using the equation $e = c/a$ with $e = 1/3$ and $c = 1$, we obtain

$$\frac{1}{3} = \frac{1}{a} \qquad \text{and therefore} \qquad a = 3$$

Recall now that b^2 is defined by the equation $b^2 = a^2 - c^2$. In view of this, we have

$$b^2 = 3^2 - 1^2 = 8$$
$$b = \sqrt{8} = 2\sqrt{2}$$

The equation of the ellipse in standard form is therefore

$$\frac{x^2}{3^2} + \frac{y^2}{(2\sqrt{2})^2} = 1$$

Furthermore, since $a = 3$ and $b = 2\sqrt{2}$, the lengths of the major and minor axes are 6 and $4\sqrt{2}$ units, respectively.

| EXAMPLE | 3 | **Finding the distance between the foci of an ellipse** |

The eccentricity of an ellipse is 4/5, and the sum of the distances from a point P on the ellipse to the foci is 10 units. Compute the distance between the foci F_1 and F_2. See Figure 9.

SOLUTION

We are required to find the distance between the foci F_1 and F_2. By definition, this is the quantity $2c$. Since the sum of the distances from a point on the ellipse to the foci is 10 units, we have

$$2a = 10 \qquad \text{by definition of } 2a$$
$$a = 5$$

Now we substitute the values $a = 5$ and $e = 4/5$ in the formula $e = c/a$:

$$\frac{4}{5} = \frac{c}{5} \qquad \text{and therefore} \qquad c = 4$$

It now follows that the required distance $2c$ is 8 units.

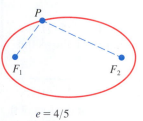

$e = 4/5$
$F_1P + F_2P = 10$

Figure 9

Suppose now that we translate the graph of $(x^2/a^2) + (y^2/b^2) = 1$ by h units in the x-direction and k units in the y-direction, as shown in Figure 10. Then the equation of the translated ellipse is

$$\frac{(x - h)^2}{a^2} + \frac{(y - k)^2}{b^2} = 1 \tag{5}$$

Equation (5) is another **standard form** for the equation of an ellipse. As indicated in Figure 10, the center of this ellipse is the point (h, k). In the next example, we use the technique of completing the square to convert an equation for an ellipse to standard form. Once the equation is in standard form, the graph is readily obtained.

Figure 10

| EXAMPLE | 4 | **Completing the square to analyze an ellipse** |

Determine the center, foci, and eccentricity of the ellipse

$$4x^2 + 9y^2 - 8x - 54y + 49 = 0$$

Graph the ellipse.

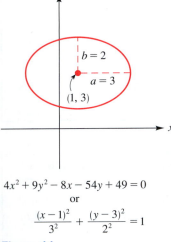

$$4x^2 + 9y^2 - 8x - 54y + 49 = 0$$

or

$$\frac{(x-1)^2}{3^2} + \frac{(y-3)^2}{2^2} = 1$$

Figure 11

SOLUTION

We will convert the given equation to standard form by using the technique of completing the square. We have

$$4x^2 - 8x + 9y^2 - 54y = -49$$
$$4(x^2 - 2x) + 9(y^2 - 6y) = -49$$
$$4(x^2 - 2x + 1) + 9(y^2 - 6y + 9) = -49 + 4(1) + 9(9)$$
$$4(x - 1)^2 + 9(y - 3)^2 = 36$$

Dividing this last equation by 36, we obtain

$$\frac{(x-1)^2}{9} + \frac{(y-3)^2}{4} = 1 \qquad \text{or} \qquad \frac{(x-1)^2}{3^2} + \frac{(y-3)^2}{2^2} = 1$$

This last equation represents an ellipse with center at $(1, 3)$ and with $a = 3$ and $b = 2$. We can calculate c using the formula $c^2 = a^2 - b^2$:

$$c^2 = 3^2 - 2^2 = 5 \qquad \text{and therefore} \qquad c = \sqrt{5}$$

Since the center of this ellipse is $(1, 3)$, the foci are therefore $(1 + \sqrt{5}, 3)$ and $(1 - \sqrt{5}, 3)$. Finally, the eccentricity is c/a, which is $\sqrt{5}/3$. Figure 11 shows the graph of this ellipse.

In developing the equation $(x^2/a^2) + (y^2/b^2) = 1$, we assumed that the foci of the ellipse were located on the x-axis at the points $(-c, 0)$ and $(c, 0)$. If instead the foci are located on the y-axis at the points $(0, c)$ and $(0, -c)$, then the same method we used in the previous case will show the equation of the ellipse to be

$$\frac{x^2}{b^2} + \frac{y^2}{a^2} = 1 \qquad (a > b)$$

We still assume that $2a$ represents the sum of the distances from a point on the ellipse to the foci. In the box that follows, we summarize the situation for the ellipse with foci $(0, c)$ and $(0, -c)$.

PROPERTY SUMMARY The Ellipse $\dfrac{x^2}{b^2} + \dfrac{y^2}{a^2} = 1$ $(a > b)$

Foci: $(0, \pm c)$, where $c^2 = a^2 - b^2$
Center: $(0, 0)$
Vertices: $(0, \pm a)$
Length of major axis: $2a$
Length of minor axis: $2b$

Eccentricity: $e = \dfrac{c}{a} = \dfrac{\sqrt{a^2 - b^2}}{a}$

We now have two standard forms for the ellipse:

Foci on the x-axis at $(\pm c, 0)$: $\dfrac{x^2}{a^2} + \dfrac{y^2}{b^2} = 1$ $(a > b)$

Foci on the y-axis at $(0, \pm c)$: $\dfrac{x^2}{b^2} + \dfrac{y^2}{a^2} = 1$ $(a > b)$

Because a is greater than b in both standard forms, it is always easy to determine by inspection whether the foci lie on the x-axis or the y-axis. Consider, for instance, the equation $(x^2/5^2) + (y^2/6^2) = 1$. In this case, since $6 > 5$, we have $a = 6$. And since 6^2 appears under y^2, we conclude that the foci lie on the y-axis.

EXAMPLE 5   An ellipse with vertices on the y-axis

The point $(5, 3)$ lies on an ellipse with vertices $(0, \pm 2\sqrt{21})$. Find the equation of the ellipse. Write the answer both in standard form and in the form $Ax^2 + By^2 = C$.

SOLUTION
Since the vertices are $(0, \pm 2\sqrt{21})$, the standard form for the equation in this case is $(x^2/b^2) + (y^2/a^2) = 1$. Furthermore, in view of the coordinates of the vertices, we have $a = 2\sqrt{21}$. Therefore,

$$\frac{x^2}{b^2} + \frac{y^2}{84} = 1$$

Now, since the point $(5, 3)$ lies on the ellipse, its coordinates must satisfy this last equation. We thus have

$$\frac{5^2}{b^2} + \frac{3^2}{84} = 1$$
$$\frac{25}{b^2} + \frac{3}{28} = 1$$
$$\frac{25}{b^2} = \frac{25}{28}$$

From this last equation, we see that $b^2 = 28$ and therefore

$$b = \sqrt{28} = 2\sqrt{7}$$

Now that we've determined a and b, we can write the equation of the ellipse in standard form. It is

$$\frac{x^2}{(2\sqrt{7})^2} + \frac{y^2}{(2\sqrt{21})^2} = 1$$

As you should verify for yourself, this equation can also be written in the equivalent form

$$3x^2 + y^2 = 84$$

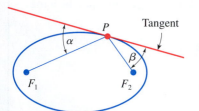

Reflection property
of the ellipse:
$\alpha = \beta$

Figure 12

As with the parabola, many of the interesting properties of the ellipse are related to tangent lines. We define a **tangent to an ellipse** as a line that intersects the ellipse in exactly one point. Figure 12 shows a line tangent to an ellipse at an arbitrary point P on the curve. Line segments $\overline{F_1P}$ and $\overline{F_2P}$ are drawn from the foci to the point of tangency. These two segments are called **focal radii.** One of the most basic properties of the ellipse is the **reflection property.** According to this property, the focal radii drawn to the point of tangency make equal angles with the tangent.

If $P(x_1, y_1)$ is a point on the ellipse $(x^2/a^2) + (y^2/b^2) = 1$, then the equation of the line tangent to the ellipse at $P(x_1, y_1)$ is

$$\frac{x_1 x}{a^2} + \frac{y_1 y}{b^2} = 1$$

This equation is easy to remember because it so closely resembles the equation of the ellipse itself. However, notice that the equation $(x_1 x/a^2) + (y_1 y/b^2) = 1$ is indeed linear, since $x_1, y_1, a,$ and b all denote constants. This equation for the tangent to an ellipse can be derived using the same technique we employed with the parabola in the previous section. (See Exercise 60 at the end of this section for the details.) The two examples that follow show how this equation is used.

EXAMPLE 6 **Determining the tangent to an ellipse**

Find the equation of the line that is tangent to the ellipse $x^2 + 3y^2 = 57$ at the point $(3, 4)$ on the ellipse; see Figure 13. Write the answer in the form $y = mx + b$.

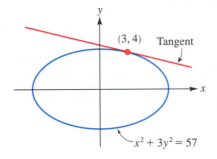

Figure 13

SOLUTION
We know that the equation of the line tangent to the ellipse at the point (x_1, y_1) is $(x_1 x/a^2) + (y_1 y/b^2) = 1$. We are given that $x_1 = 3$ and $y_1 = 4$. Thus we need to determine a^2 and b^2. To do that, we convert the equation $x^2 + 3y^2 = 57$ to standard form by dividing through by 57, which yields

$$\frac{x^2}{57} + \frac{y^2}{19} = 1$$

It follows that $a^2 = 57$ and $b^2 = 19$. The equation of the tangent line then becomes

$$\frac{3x}{57} + \frac{4y}{19} = 1$$
$$\frac{x}{19} + \frac{4y}{19} = 1$$
$$x + 4y = 19$$

Solving for y, we obtain

$$y = -\frac{1}{4}x + \frac{19}{4}$$

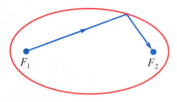

Figure 14

A consequence of the reflection property of the ellipse. A light ray or sound emitted from one focus is reflected through the other focus.

We conclude this section by mentioning a few of the applications of the ellipse. Some gears in machines are elliptical rather than circular. (In certain brands of racing bikes, this is true of one of the gears in front.) As with the parabola, the reflection property of the ellipse has applications in optics and acoustics. As indicated in Figure 14, a light ray or sound emitted from one focus of an ellipse is always reflected through the other focus. This property is used in the design of "whispering galleries." In these rooms (with elliptical cross sections), a person standing at focus F_2 can hear a whisper from focus F_1 while others closer to F_1 might hear nothing. Statuary Hall in the Capitol building in Washington, D.C., is a whispering gallery. This idea is used in a modern medical device known as the *lithotripter,* in which high-energy sound waves are used to break up kidney stones. The patient is positioned with the kidney stone at one focus in an elliptical water bath while sound waves are emitted from the other focus.

An important application of the ellipse occurs in astronomy: For a planet or comet revolving around the sun, the orbit is an ellipse with the sun at one focus. (This fact was discovered empirically by Johannes Kepler in the early 1600s and it was proved mathematically by Isaac Newton in the 1660s.) In this context, the vertices of the ellipse (that is, the endpoints of the major axis) have a special significance. We'll use Figure 15 to explain this. In Figure 15(a), the ellipse represents the orbit of a planet revolving about the sun located at focus F. The red circle in the figure is drawn with center F and radius FV_2, where V_2 is the vertex closer to F. The figure demonstrates that the planet comes closest to the sun (F) when it reaches the point V_2 in its orbit. (Why: Each point on the ellipse other than V_2 lies outside of the red circle and therefore is farther from F than is V_2.) The vertex V_2 in Figure 15(a) is referred to as the *perihelion* of the orbit. A similar argument, using Figure 15(b), shows that at the other vertex V_1, the planet reaches its farthest point from the sun. This point is the *aphelion* of the orbit. For ease of reference, we repeat this terminology in the box following Figure 15.

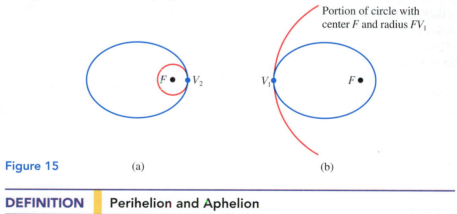

Figure 15   (a)   (b)

DEFINITION | **Perihelion and Aphelion**

(Refer to the figure below.) The **perihelion** is the point of the orbit at which a planet (or comet) is closest to the sun. The **aphelion** is the point farthest from the sun.

Figure 16
The distances from a planet or comet to the sun at perihelion and aphelion are $a - c$ and $a + c$, respectively.

As indicated in Figure 16, at perihelion, the distance from a planet or comet to the sun is given by the expression $a - c$, while at aphelion the distance is $a + c$. We use these expressions along with data for Halley's comet in Example 7.

EXAMPLE 7 Determining the distances to the sun at perihelion and aphelion

Halley's comet follows an elliptical orbit about the sun, reappearing in the skies above earth once every 76.6 years. The semimajor axis of the ellipse is $a = 17.8$ astronomical units (AU). (One AU is 92,956,495 miles, the average distance of the earth from the sun.) The eccentricity of the ellipse is $e = 0.967$. Find the distance of the comet from the sun at perihelion and at aphelion. That is, find the minimum and maximum distances of Halley's comet from the sun.

SOLUTION
According to Figure 16, the distances of the comet from the sun at perihelion and at aphelion are $a - c$ and $a + c$, respectively. Since we are given that $a = 17.8$ AU, we need to compute c. From the equation $e = c/a$, we have

$$\begin{aligned} c &= ae \\ &= (17.8)(0.967) \qquad \text{using the values given for } a \text{ for } e \\ &= 17.2 \,\text{AU} \qquad \text{using a calculator and rounding to one decimal place} \end{aligned}$$

Using this value for c, we compute

$$a - c = 17.8\,\text{AU} - 17.2\,\text{AU} = 0.6\,\text{AU}$$

and

$$a + c = 17.8\,\text{AU} + 17.2\,\text{AU} = 35.0\,\text{AU}$$

In summary: At perihelion, the distance from Halley's comet to the sun is approximately 0.6 AU. This is the closest the comet ever gets to the sun. At aphelion, the distance from the comet to the sun is approximately 35.0 AU. This is the farthest that the comet ever gets from the sun.

EXERCISE SET 8.3

A

In Exercises 1–24, graph the ellipses. In each case, specify the lengths of the major and minor axes, the foci, and the eccentricity. For Exercises 13–24, also specify the center of the ellipse.

1. $4x^2 + 9y^2 = 36$
2. $4x^2 + 25y^2 = 100$
3. $x^2 + 16y^2 = 16$
4. $9x^2 + 25y^2 = 225$
5. $x^2 + 2y^2 = 2$
6. $2x^2 + 3y^2 = 3$
7. $16x^2 + 9y^2 = 144$
8. $25x^2 + y^2 = 25$
9. $15x^2 + 3y^2 = 5$
10. $9x^2 + y^2 = 4$
11. $2x^2 + y^2 = 4$
12. $36x^2 + 25y^2 = 400$
13. $\dfrac{(x-5)^2}{5^2} + \dfrac{(y+1)^2}{3^2} = 1$
14. $\dfrac{(x-1)^2}{2^2} + \dfrac{(y+4)^2}{3^2} = 1$
15. $\dfrac{(x-1)^2}{1^2} + \dfrac{(y-2)^2}{2^2} = 1$
16. $\dfrac{x^2}{4^2} + \dfrac{(y-3)^2}{2^2} = 1$
17. $\dfrac{(x+3)^2}{3^2} + \dfrac{y^2}{1^2} = 1$
18. $\dfrac{(x-2)^2}{2^2} + \dfrac{(y-2)^2}{2^2} = 1$
19. $3x^2 + 4y^2 - 6x + 16y + 7 = 0$
20. $16x^2 + 64x + 9y^2 - 54y + 1 = 0$
21. $5x^2 + 3y^2 - 40x - 36y + 188 = 0$
22. $x^2 + 16y^2 - 160y + 384 = 0$
23. $16x^2 + 25y^2 - 64x - 100y + 564 = 0$
24. $4x^2 + 4y^2 - 32x + 32y + 127 = 0$

In Exercises 25–32, find the equation of the ellipse satisfying the given conditions. Write the answer both in standard form and in the form $Ax^2 + By^2 = C$.

25. Foci $(\pm 3, 0)$; vertices $(\pm 5, 0)$
26. Foci $(0, \pm 1)$; vertices $(0, \pm 4)$
27. Vertices $(\pm 4, 0)$; eccentricity 1/4
28. Foci $(0, \pm 2)$; endpoints of the minor axis $(\pm 5, 0)$
29. Foci $(0, \pm 2)$; endpoints of the major axis $(0, \pm 5)$
30. Endpoints of the major axis $(\pm 10, 0)$; endpoints of the minor axes $(0, \pm 4)$
31. Center at the origin; vertices on the x-axis; length of major axis twice the length of minor axis; $(1, \sqrt{2})$ lies on the ellipse
32. Eccentricity 3/5; one endpoint of the minor axis $(-8, 0)$; center at the origin
33. Find the equation of the tangent to the ellipse $x^2 + 3y^2 = 76$ at each of the given points. Write your answers in the form $y = mx + b$.
    (a) $(8, 2)$　　(b) $(-7, 3)$　　(c) $(1, -5)$
34. (a) Find the equation of the line tangent to the ellipse $x^2 + 3y^2 = 84$ at the point $(3, 5)$ on the ellipse. Write your answer in the form $y = mx + b$.
    (b) Repeat part (a), but at the point $(-3, -5)$ on the ellipse.
    (c) Are the lines determined in (a) and (b) parallel?
35. A line is drawn tangent to the ellipse $x^2 + 3y^2 = 52$ at the point $(2, 4)$ on the ellipse.
    (a) Find the equation of this tangent line.

(b) Find the area of the first-quadrant triangle bounded by the axes and this tangent line.
36. Tangent lines are drawn to the ellipse $x^2 + 3y^2 = 12$ at the points $(3, -1)$ and $(-3, -1)$ on the ellipse.
    (a) Find the equation of each tangent line. Write your answers in the form $y = mx + b$.
    (b) Find the point where the tangent lines intersect.

In Exercises 37 and 38, you are given the eccentricity e and the length a of the semimajor axis for the orbits of the planets Pluto and Mars. Compute the distance of each planet from the sun at perihelion and at aphelion (as in Example 7). For Pluto, round the final answers to two decimal places; for Mars, round to three decimal places.

37. Pluto: $e = 0.2484$; $a = 39.44$ AU
38. Mars: $e = 0.0934$; $a = 1.5237$ AU
39. As is the case for most asteroids in our solar system, the orbit of the asteroid Gaspra is located between the orbits of Mars and Jupiter. Given that its distances from the sun at perihelion and aphelion are 2.132 AU and 2.288 AU, respectively, compute the length of the semimajor axis of the orbit and the eccentricity. Round both answers to two decimal places. *Remark*: Within the past decade, the Galileo spacecraft (on a mission to Jupiter) took images of Gaspra, which can be viewed on the World Wide Web. For instance, at the time of this writing, one NASA Web site containing these images is http://www.jpl.nasa.gov/galileo. From this Web page, use the search feature there to find "Gaspra."
40. The small asteroid Icarus was first observed in 1949 at Palomar Observatory in California. Given that the distance from Icarus to the sun at aphelion is 1.969 AU and the eccentricity of the orbit is 0.8269, compute the semimajor axis of the orbit and the distance from Icarus to the sun at perihelion. Round the answers to two decimal places. *Remark*: One reason for interest in the orbit of Icarus is that it crosses earth's orbit. On June 14, 1968, there was a "close" approach in which Icarus came within approximately 4 million miles of earth. According to the Jet Propulsion Laboratory, the next close approach will be June 16, 2015, at which time Icarus will come within approximately 5 million miles of earth.

B

41. Consider the equation $(x^2/3^2) + (y^2/2^2) = 1$.
    (a) Use the symmetry tests from Section 1.7 to explain why the graph of this equation must be symmetric about both the x-axis and the y-axis.
    (b) Show that solving the equation for y yields
    $$y = \pm\tfrac{1}{3}\sqrt{36 - 4x^2}.$$

(c) Let $y = \frac{1}{3}\sqrt{36 - 4x^2}$. Use the techniques of Section 2.6 to find the domain of this function. (You need to solve the inequality $36 - 4x^2 \geq 0$.) You should find that the domain is the closed interval $[-3, 3]$.

(d) Use a calculator to complete the following table. Then plot the resulting points and connect them with a smooth curve. This gives you a sketch of the first quadrant portion of the graph of $(x^2/3^2) + (y^2/2^2) = 1$.

x	0	0.5	1	1.5	2	2.5	3
$y = \frac{1}{3}\sqrt{36 - 4x^2}$	2						0

(e) Use your graph in part (d) along with the symmetry results in part (a) to sketch a complete graph of $(x^2/3^2) + (y^2/2^2) = 1$.

42. In this exercise we consider how the eccentricity e influences the graph of an ellipse $(x^2/a^2) + (y^2/b^2) = 1$.

(a) For simplicity, we suppose that $a = 1$ so that the equation of the ellipse is $x^2 + (y^2/b^2) = 1$. Solve this equation for y to obtain

$$y = \pm b\sqrt{1 - x^2} \qquad (1)$$

(b) Assuming $a = 1$, show that b and e are related by the equation $b^2 = 1 - e^2$, from which it follows that $b = \pm\sqrt{1 - e^2}$. The positive root is appropriate here because $b > 0$. Thus, we have

$$b = \sqrt{1 - e^2} \qquad (2)$$

(c) Using equation (2) to substitute for b in equation (1) yields

$$y = \sqrt{1 - e^2}\sqrt{1 - x^2} \qquad \text{or} \qquad y = -\sqrt{1 - e^2}\sqrt{1 - x^2} \qquad (3)$$

This pair of equations represents an ellipse with semimajor axis 1 and eccentricity e. Using the value $e = 0.3$, graph equations (3) in the viewing rectangle $[-1, 1]$ by $[-1, 1]$. Use true proportions and, for comparison, add to your picture the circle with radius 1 and center $(0, 0)$. Note that the ellipse is nearly circular.

(d) Follow part (c) using $e = 0.017$. This is approximately the eccentricity for the earth's orbit about the sun. How does the ellipse compare to the circle in this case?

(e) Follow part (c) using, in turn, $e = 0.4$, $e = 0.6$, $e = 0.8$, $e = 0.9$, $e = 0.99$, and $e = 0.999$. Then, in complete sentences, summarize what you've observed.

43. In the accompanying figure, the center of the circle and one focus of the ellipse coincide with the focus of the parabola $y^2 = 4x$. All three curves are symmetric about the x-axis and pass through the origin. The eccentricity of

the ellipse is $3/4$. Find the equations of the circle and the ellipse.

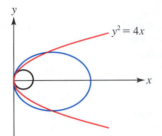

44. This exercise outlines the steps needed to complete the derivation of the equation $(x^2/a^2) + (y^2/b^2) = 1$.

(a) Square both sides of equation (1) on page 654. After simplifying, you should obtain

$$a\sqrt{(x + c)^2 + y^2} = a^2 + xc$$

(b) Square both sides of the equation in part (a). Show that the result can be written

$$a^2x^2 - x^2c^2 + a^2y^2 = a^4 - a^2c^2$$

(c) Verify that the equation in part (b) is equivalent to

$$(a^2 - c^2)x^2 + a^2y^2 = a^2(a^2 - c^2)$$

(d) Using the equation in part (c), replace the quantity $a^2 - c^2$ with b^2. Then show that the resulting equation can be rewritten $(x^2/a^2) + (y^2/b^2) = 1$, as required.

45. In the text we defined the positive number b^2 by the equation $b^2 = a^2 - c^2$. For this definition to make sense, we need to show that the quantity $a^2 - c^2$ is positive. This can be done as follows. First, recall that in any triangle, the sum of the lengths of two sides is always greater than the length of the third side. Now apply this fact to triangle F_1PF_2 in Figure 4 to show that $2a > 2c$. Conclude from this that $a^2 - c^2 > 0$, as required.

46. A line is drawn tangent to the ellipse $(x^2/a^2) + (y^2/b^2) = 1$ at the point (x_1, y_1) on the ellipse. Let P and Q denote the points where the tangent meets the y-and x-axes, respectively. Show that the midpoint of PQ is $(a^2/2x_1, b^2/2y_1)$.

47. A *normal* to an ellipse is a line drawn perpendicular to the tangent at the point of tangency. Show that the equation of the normal to the ellipse $(x^2/a^2) + (y^2/b^2) = 1$ at the point (x_1, y_1) can be written

$$a^2y_1x - b^2x_1y = (a^2 - b^2)x_1y_1$$

48. Let (x_1, y_1) be any point on the ellipse $(x^2/a^2) + (y^2/b^2) = 1$ $(a > b)$ other than one of the endpoints of the major or minor axis. Show that the normal at (x_1, y_1) does not pass through the origin. *Hint:* Find the y-intercept of the normal.

49. Find the points of intersection of the ellipses

$$\frac{x^2}{a^2} + \frac{y^2}{b^2} = 1 \quad \text{and} \quad \frac{x^2}{b^2} + \frac{y^2}{a^2} = 1 \quad (a > b)$$

Include a sketch with your answer.

50. Let F_1 and F_2 denote the foci of the ellipse $(x^2/a^2) + (y^2/b^2) = 1$ ($a > b$). Suppose that P is one of the endpoints of the minor axis and angle F_1PF_2 is a right angle. Compute the eccentricity of the ellipse.

51. Let $P(x_1, y_1)$ be a point on the ellipse $(x^2/a^2) + (y^2/b^2) = 1$. Let N be the point where the normal through P meets the x-axis, and let F be the focus $(-c, 0)$. Show that $FN/FP = e$, where e denotes the eccentricity.

52. Let $P(x_1, y_1)$ be a point on the ellipse $b^2x^2 + a^2y^2 = a^2b^2$. Suppose that the tangent to the ellipse at P meets the y-axis at A and the x-axis at B. If $AP = PB$, what are x_1 and y_1 (in terms of a and b)?

53. **(a)** Verify that the points $A(5, 1)$, $B(4, -2)$, and $C(-1, 3)$ all lie on the ellipse $x^2 + 3y^2 = 28$.

(b) Find a point D on the ellipse such that $\overline{CD}$ is parallel to $\overline{AB}$.

(c) If O denotes the center of the ellipse, show that the triangles OAC and OBD have equal areas. *Suggestion*: In computing the areas, the formula given in Section 1.4 at the end of Exercise 36 is useful.

54. Find the points of intersection of the parabola $y = x^2$ and the ellipse $b^2x^2 + a^2y^2 = a^2b^2$.

55. Recall that the two line segments joining a point on the ellipse to the foci are called *focal radii*. These are the segments $\overline{F_1P}$ and $\overline{F_2P}$ in the following figure.

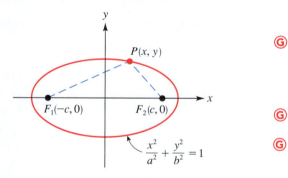

(a) Show that $F_1P = a + ex$. *Hint:* If you try to do this from scratch, it can involve a rather lengthy calculation. Begin instead with the equation $a\sqrt{(x + c)^2 + y^2} = a^2 + xc$ [from Exercise 44 (a)], and divide both sides by a.

(b) Show that $F_2P = a - ex$. *Hint:* Make use of the result in part (a), together with the fact that $F_1P + F_2P = 2a$, by definition.

56. Find the coordinates of a point P in the first quadrant on the ellipse $9x^2 + 25y^2 = 225$ such that $\angle F_2PF_1$ is a right angle.

57. The accompanying figure shows the two tangent lines drawn from the point (h, k) to the ellipse $b^2x^2 + a^2y^2 = a^2b^2$. Follow steps (a) through (d) to show that the equation of the line passing through the two points of tangency is $b^2hx + a^2ky = a^2b^2$.

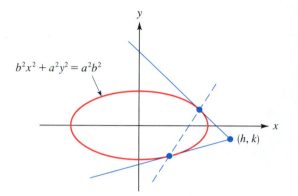

(a) Let (x_1, y_1) be one of the points of tangency. Check that the equation of the tangent line through this point is $b^2x_1x + a^2y_1y = a^2b^2$.

(b) Using the result in part (a), explain why $b^2x_1h + a^2y_1k = a^2b^2$.

(c) In a similar fashion, show that $b^2x_2h + a^2y_2k = a^2b^2$, where (x_2, y_2) is the other point of tangency.

(d) The equation $b^2hx + a^2ky = a^2b^2$ represents a line. Explain why this line must pass through the points (x_1, y_1) and (x_2, y_2).

58. **(a)** The *auxiliary circle* of an ellipse is defined to be the circle with diameter the same as the major axis of the ellipse. Determine the equation of the auxiliary circle for the ellipse $9x^2 + 25y^2 = 225$.

(b) Graph the ellipse $9x^2 + 25y^2 = 225$ along with its auxiliary circle. (Use true proportions.)

59. As background for this exercise, you need to have worked Exercise 58. This exercise illustrates an interesting geometric result concerning an ellipse and its auxiliary circle.

(a) Graph the ellipse $x^2 + 3y^2 = 12$ for $y \geq 0$.

(b) Find the values of a, b, and c for this ellipse.

(c) Find the equation of the auxiliary circle for this ellipse. Add the graph of the top half of this circle to your picture from part (a).

(d) Verify (algebraically) that the point $P(3, 1)$ lies on the ellipse. Then find the equation of the tangent line to the ellipse at P.

(e) Add the graph of the tangent from part (d) to the picture that you obtained in part (c).

(f) In part (b) you determined the value of c. Find the equation of the line passing through $(-c, 0)$ and perpendicular to the tangent in part (d). Then add the graph of this line to your picture. If you've done things correctly, you should obtain a figure similar to the following one. This figure provides an example of this general result: The line through the focus and perpendicular to the tangent meets the tangent at a point on the auxiliary circle.

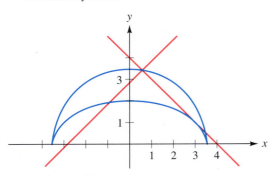

Figure for Problem 59
The line through the focus and perpendicular to a tangent meets the tangent at a point on the auxiliary circle.

C

60. This exercise outlines the steps required to show that the equation of the tangent to the ellipse $(x^2/a^2) + (y^2/b^2) = 1$ at the point (x_1, y_1) on the ellipse is $(x_1 x/a^2) + (y_1 y/b^2) = 1$.

(a) Show that the equation $(x^2/a^2) + (y^2/b^2) = 1$ is equivalent to

$$b^2 x^2 + a^2 y^2 = a^2 b^2 \quad (1)$$

Conclude that (x_1, y_1) lies on the ellipse if and only if

$$b^2 x_1^2 + a^2 y_1^2 = a^2 b^2 \quad (2)$$

(b) Subtract equation (2) from equation (1) to show that

$$b^2(x^2 - x_1^2) + a^2(y^2 - y_1^2) = 0 \quad (3)$$

Equation (3) is equivalent to equation (1) provided only that (x_1, y_1) lies on the ellipse. In the following steps, we will find the algebra much simpler if we use

equation (3) to represent the ellipse, rather than the equivalent and perhaps more familiar equation (1).

(c) Let the equation of the line tangent to the ellipse at (x_1, y_1) be

$$y - y_1 = m(x - x_1)$$

Explain why the following system of equations must have exactly one solution—namely, (x_1, y_1):

$$\begin{cases} b^2(x^2 - x_1^2) + a^2(y^2 - y_1^2) = 0 & (4) \\ y - y_1 = m(x - x_1) & (5) \end{cases}$$

(d) Solve equation (5) for y and then substitute for y in equation (4) to obtain

$$b^2(x^2 - x_1^2) + a^2 m^2(x - x_1)^2 + 2a^2 m y_1(x - x_1) = 0 \quad (6)$$

(e) Show that equation (6) can be written

$$(x - x_1)[b^2(x + x_1) + a^2 m^2(x - x_1) + 2a^2 m y_1] = 0 \quad (7)$$

(f) Equation (7) is a quadratic equation in x, but as pointed out earlier, $x = x_1$ must be the only solution. (That is, $x = x_1$ is a double root.) Thus, the factor in brackets must equal zero when x is replaced by x_1. Use this observation to show that

$$m = -\frac{b^2 x_1}{a^2 y_1}$$

This represents the slope of the line tangent to the ellipse at (x_1, y_1).

(g) Using this value for m, show that equation (5) becomes

$$b^2 x_1 x + a^2 y_1 y = b^2 x_1^2 + a^2 y_1^2 \quad (8)$$

(h) Now use equation (2) to show that equation (8) can be written

$$\frac{x_1 x}{a^2} + \frac{y_1 y}{b^2} = 1$$

which is what we set out to show.

MINI PROJECT | THE CIRCUMFERENCE OF AN ELLIPSE

As you know, there is a simple expression for the circumference of a circle of radius a—namely, $2\pi a$. However, there is no similar type of elementary expression for the circumference of an ellipse. (The circumference of an ellipse can be computed to as many decimal places as required using the methods of calculus.) Nevertheless, there are some interesting elementary formulas that allow us to approximate the circumference of an ellipse quite closely. Four such formulas follow, along with the names of their discoverers and approximate dates of discovery. Each formula yields an approximate value for the circumference of the ellipse $(x^2/a^2) + (x^2/b^2) = 1$.

Discoverer	Date	Formula
Giuseppe Peano	1887	$C_1 = \pi[a + b + \frac{1}{2}(\sqrt{a} - \sqrt{b})^2]$
Scrinivasa Ramanujan	1914	$C_2 = \pi[3(a + b) - \sqrt{(a + 3b)(3a + b)}]$
Roger A. Johnson	1930	$C_3 = \frac{\pi}{2}[a + b + \sqrt{2(a^2 + b^2)}]$
Roger Maertens	2000	$C_4 = 4(a^y + b^y)^{1/y}$, where $y = \dfrac{\ln 2}{\ln(\pi/2)}$

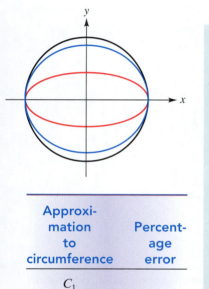

Approxi- mation to circumference	Percent- age error
C_1	
C_2	
C_3	
C_4	

(a) In the figure on the left, the red ellipse has an eccentricity of 0.9, and the blue ellipse an eccentricity of 0.5. The outer circle has a radius of 10, which is equal to the semimajor axis of each ellipse. As preparation for parts (b) and (c), determine the equation of each ellipse.

(b) For the red ellipse, use the approximation formulas given above to complete the table. Round the values of C_1, C_2, C_3, and C_4 to six decimal places. Round the percentage errors to two significant digits. In computing the percentage errors, use the fact that the actual circumference, rounded to six decimal places, is 23.433941. Which of the four approximations for the circumference of this ellipse is the best? Which is worst? How does the circumference of this ellipse compare to that of the black circle in the figure? This is, using the given six-place value for circumference, compute the ratio of the circumference of the ellipse to the circumference of the circle.

(c) Follow part (b) for the blue ellipse. The actual circumference here, rounded to six decimal places, is 29.349244. Compare your results for percentage errors to those in part (b), and summarize your observations (using complete sentences).

(d) What is the circumference of the circle in the given figure? What value does each approximation formula yield for the circumference of the circle?

8.4   THE HYPERBOLA

. . . Menaechmus [c. 350 B.C.] is reputed to have discovered the curves that were later known as the ellipse, the parabola, and the hyperbola. —Carl B. Boyer in *A History of Mathematics,* 2nd ed., revised by Uta C. Merzbach (New York: John Wiley and Sons, 1991)

The ellipse is the general shape of any closed orbit. . . . It is also possible to have orbits that are not closed, whose shapes are represented by open curves. Even if the two bodies are not bound together by their mutual gravitational attraction, their gravitational attraction for each other still affects their relative motion. Then the general shape of their orbit is a hyperbola. —Theodore P. Snow in *The Dynamic Universe: An Introduction to Astronomy,* 4th ed. (St. Paul, Minn.: West Publishing Co., 1990)

In the previous section, we defined an ellipse as the set of points P such that the sum of the distances from P to two fixed points is constant. By considering the difference instead of the sum, we are led to the definition of the hyperbola.

DEFINITION | **The Hyperbola**

A **hyperbola** is the set of all points in the plane, the difference of whose distances from two fixed points is a positive constant. Each fixed point is called a **focus.**

As with the ellipse, we label the foci F_1 and F_2. Before obtaining an equation for the hyperbola, we can see the general features of the curve by using two sets of concentric circles, with centers F_1 and F_2, to locate points satisfying the definition of a hyperbola. In Figure 1, we've plotted a number of points P such that either $F_1P - F_2P = 3$ or $F_2P - F_1P = 3$. By joining these points, we obtain the graph of the hyperbola shown in Figure 1.

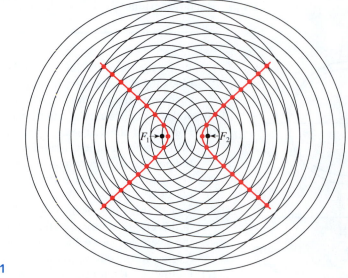

Figure 1

Unlike the parabola or the ellipse, the hyperbola is composed of two distinct parts, or **branches.** As you can check, the left branch in Figure 1 corresponds to the equation $F_2P - F_1P = 3$, while the right branch corresponds to the equation $F_1P - F_2P = 3$. Figure 1 also reveals that the hyperbola possesses two types of symmetry. First, it is symmetric about the line passing through the two foci F_1 and F_2; this line is referred to as the **focal axis** of the hyperbola. Second, the hyperbola is symmetric about the line that is the perpendicular bisector of the segment $\overline{F_1F_2}$.

To derive an equation for the hyperbola, let us initially assume that the foci are located at the points $F_1(-c, 0)$ and $F_2(c, 0)$, as indicated in Figure 2. We will use $2a$ to denote the positive constant referred to in the definition of the hyperbola. By definition, then, $P(x, y)$ lies on the hyperbola if and only if

$$|F_1P - F_2P| = 2a$$

or, equivalently,

$$F_1P - F_2P = \pm 2a$$

If we use the formula for the distance between two points, this last equation becomes

$$\sqrt{(x + c)^2 + (y - 0)^2} - \sqrt{(x - c)^2 + (y - 0)^2} = \pm 2a$$

or

$$\sqrt{(x + c)^2 + y^2} - \sqrt{(x - c)^2 + y^2} = \pm 2a$$

Figure 2

We can simplify this equation by carrying out the same procedure that we used for the ellipse in the previous section. As Exercise 44 asks you to verify, the resulting equation is

$$(c^2 - a^2)x^2 - a^2y^2 = a^2(c^2 - a^2) \tag{1}$$

Before further simplifying equation (1), we point out that the quantity $c^2 - a^2$ [which appears twice in equation (1)] is positive. To see why this is so, refer back to Figure 2. In triangle F_1F_2P (as in any triangle), the length of any side is less than the sum of the lengths of the other two sides. Thus

$$F_1P < F_1F_2 + F_2P \qquad \text{and} \qquad F_2P < F_1F_2 + F_1P$$

and therefore

$$F_1P - F_2P < F_1F_2 \qquad \text{and} \qquad F_2P - F_1P < F_1F_2$$

These last two equations tell us that

$$|F_1P - F_2P| < F_1F_2$$

Therefore, in view of the definitions of $2a$ and $2c$, we have

$$0 < 2a < 2c \qquad \text{or} \qquad 0 < a < c$$

This last inequality tells us that $c^2 - a^2$ is positive, as we wished to show.

Now, since $c^2 - a^2$ is positive, we can define the positive number b by the equation

$$b^2 = c^2 - a^2$$

With this notation, equation (1) becomes

$$b^2x^2 - a^2y^2 = a^2b^2$$

Dividing by a^2b^2, we obtain

$$\frac{x^2}{a^2} - \frac{y^2}{b^2} = 1 \tag{2}$$

We have now shown that the coordinates of every point on the hyperbola satisfy equation (2). Conversely, it can be shown that if the coordinates of a point satisfy equation (2), then the point satisfies the original definition of the hyperbola. Equation (2) is the **standard form** for the equation of a hyperbola with foci $F_1(-c, 0)$ and $F_2(c, 0)$.

The intercepts of the hyperbola are readily obtained from equation (2). To find the x-intercepts, we set y equal to zero to obtain

$$\frac{x^2}{a^2} = 1$$

$$x^2 = a^2 \qquad \text{or} \qquad x = \pm a$$

Thus, the hyperbola crosses the x-axis at the points $(-a, 0)$ and $(a, 0)$. On the other hand, the curve does not cross the y-axis, for if we set x equal to zero in equation (2), we obtain $-y^2/b^2 = 1$ or

$$y^2 = -b^2 \qquad (b > 0)$$

Since the square of any real number y is nonnegative, this last equation has no solution. Therefore, the graph does not cross the y-axis. Finally, let us note that (according to the symmetry tests in Section 1.7) the graph of equation (2) must be symmetric about both coordinate axes.

Before graphing the hyperbola $(x^2/a^2) - (y^2/b^2) = 1$, we point out the important fact that the two lines $y = (b/a)x$ and $y = -(b/a)x$ are asymptotes for the curve. We can see why as follows. First we solve equation (2) for y:

$$\frac{x^2}{a^2} - \frac{y^2}{b^2} = 1$$

$$b^2x^2 - a^2y^2 = a^2b^2 \qquad \text{multiplying both sides by } a^2b^2$$

$$-a^2y^2 = a^2b^2 - b^2x^2$$

$$y^2 = \frac{b^2x^2 - a^2b^2}{a^2} = \frac{b^2(x^2 - a^2)}{a^2}$$

$$y = \pm\frac{b}{a}\sqrt{x^2 - a^2} \tag{3}$$

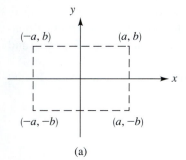

TABLE 1

x	$\sqrt{x^2 - 5^2}$
100	99.875
1000	999.987
10,000	9999.999

Now, as x grows arbitrarily large, the value of the quantity $\sqrt{x^2 - a^2}$ becomes closer and closer to x itself. Table 1 provides some empirical evidence for this statement in the case when $a = 5$. (A formal proof of the statement properly belongs to calculus.) In summary, then, we have the approximation $\sqrt{x^2 - a^2} \approx x$ as x grows arbitrarily large. So, in view of equation (3), we have

$$y = \pm\frac{b}{a}\sqrt{x^2 - a^2} \approx \pm\frac{b}{a}x \qquad \text{as } x \text{ grows arbitrarily large}$$

In other words, the two lines $y = \pm(b/a)x$ are asymptotes for the hyperbola.

A simple way to sketch the two asymptotes and then graph the hyperbola is as follows. First draw the rectangle with vertices (a, b), $(-a, b)$, $(-a, -b)$, and $(a, -b)$, as indicated in Figure 3(a). The slopes of the diagonals in this rectangle are b/a and $-b/a$. Thus by extending these diagonals as in Figure 3(b), we obtain the two asymptotes $y = \pm(b/a)x$. Now, since the x-intercepts of the hyperbola are a and $-a$, we can sketch the curve as shown in Figure 3(c).

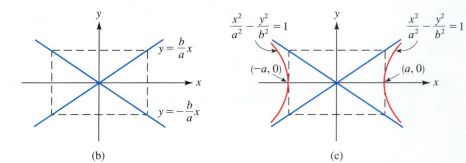

Figure 3
Steps in graphing the hyperbola $(x^2/a^2) - (y^2/b^2) = 1$ and its asymptotes.

For the hyperbola in Figure 3(c), the two points $(\pm a, 0)$, where the curve meets the x-axis, are referred to as **vertices.** The midpoint of the line segment joining the two vertices is called the **center** of the hyperbola. (Equivalently, we can define the center as the point of intersection of the two asymptotes.) For the hyperbola in Figure 3(c), the center coincides with the origin. The line segment joining the vertices of a hyperbola is the **transverse axis** of the hyperbola. For reference, in the box that follows we summarize our work on the hyperbola up to this point. Several new terms describing the hyperbola are also given in the box.

PROPERTY SUMMARY   The Hyperbola $\dfrac{x^2}{a^2} - \dfrac{y^2}{b^2} = 1$

1. The **foci** are the points $F_1(-c, 0)$ and $F_2(c, 0)$. The hyperbola is the set of points P such that $|F_1P - F_2P| = 2a$.
2. The **focal axis** is the line passing through the foci.
3. The **vertices** are the points at which the hyperbola intersects its focal axis. In Figure 4, these are the two points $V_1(-a, 0)$ and $V_2(a, 0)$.

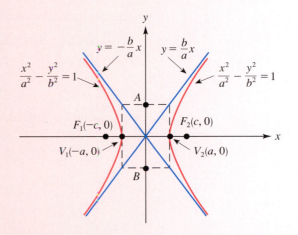

Figure 4

4. The **center** is the point on the focal axis midway between the foci. The center of the hyperbola in Figure 4 is the origin.
5. The **transverse axis** is the line segment joining the two vertices. In Figure 4, the length of the transverse axis $\overline{V_1V_2}$ is $2a$.
6. The **conjugate axis** is the line segment perpendicular to the transverse axis, passing through the center and extending a distance b on either side of the center. In Figure 4, this is the segment $\overline{AB}$.
7. The **eccentricity** e is defined by $e = c/a$, where $c^2 = a^2 + b^2$.

In general, if A, B, and C are positive numbers, then the graph of an equation of the form

$$Ax^2 - By^2 = C$$

will be a hyperbola of the type shown in Figure 4. Example 1 shows why this is so.

EXAMPLE | 1 | **Analyzing a hyperbola**

Graph the hyperbola $16x^2 - 9y^2 = 144$ after determining the following: vertices, foci, eccentricity, lengths of the transverse and conjugate axes, and asymptotes.

SOLUTION

First we convert the given equation to standard form by dividing both sides by 144. This yields

$$\frac{x^2}{9} - \frac{y^2}{16} = 1 \qquad \text{or} \qquad \frac{x^2}{3^2} - \frac{y^2}{4^2} = 1$$

By comparing this with the equation $(x^2/a^2) - (y^2/b^2) = 1$, we see that $a = 3$ and $b = 4$. The value of c can be determined by using the equation $c^2 = a^2 + b^2$. We have

$$c^2 = 3^2 + 4^2 = 25 \qquad \text{and therefore} \qquad c = 5$$

Now that we know the values of a, b, and c, we can list the required information:

Vertices: $(\pm 3, 0)$	Length of transverse axis $(= 2a)$: 6
Foci: $(\pm 5, 0)$	Length of conjugate axis $(= 2b)$: 8
Eccentricity: $\dfrac{5}{3}$	Asymptotes: $y = \pm\dfrac{4}{3}x$

The graph of the hyperbola is shown in Figure 5.

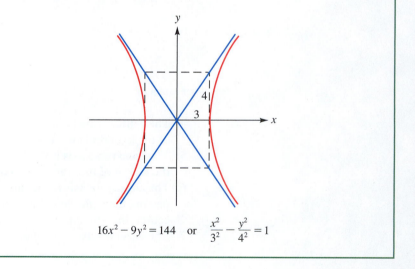

$$16x^2 - 9y^2 = 144 \quad \text{or} \quad \frac{x^2}{3^2} - \frac{y^2}{4^2} = 1$$

Figure 5

We can use the same method that we used to derive the equation for the hyperbola with foci $(\pm c, 0)$ when the foci are instead located on the y-axis at the points $(0, \pm c)$. The equation of the hyperbola in this case is

$$\frac{y^2}{a^2} - \frac{x^2}{b^2} = 1$$

We summarize the basic properties of this hyperbola in the following box.

PROPERTY SUMMARY The Hyperbola $\dfrac{y^2}{a^2} - \dfrac{x^2}{b^2} = 1$

Foci: $(0, \pm c)$, where $c^2 = a^2 + b^2$
Vertices: $(0, \pm a)$
Asymptotes: $y = \pm(a/b)x$
Length of transverse axis: $2a$
Length of conjugate axis: $2b$
Eccentricity: $e = c/a$

Figure 6

    EXAMPLE 2   **Completing the square to analyze and graph a hyperbola**

Use the technique of completing the square to show that the graph of the following equation is a hyperbola:

$$9y^2 - 54y - 25x^2 + 200x - 544 = 0$$

Graph the hyperbola and specify the center, the vertices, the foci, the length of the transverse axis, and the equations of the asymptotes.

SOLUTION

$$9(y^2 - 6y \quad\ ) - 25(x^2 - 8x \quad\ ) = 544 \qquad \textcolor{red}{\text{factoring}}$$
$$9(y^2 - 6y + 9) - 25(x^2 - 8x + 16) = 544 + 81 - 400 \qquad \textcolor{red}{\text{completing the squares}}$$

$$9(y - 3)^2 - 25(x - 4)^2 = 225$$
$$\frac{(y - 3)^2}{5^2} - \frac{(x - 4)^2}{3^2} = 1 \qquad \textcolor{red}{\text{dividing by 225}}$$

Now, the graph of this last equation is obtained by translating the graph of the equation

$$\frac{y^2}{5^2} - \frac{x^2}{3^2} = 1 \tag{4}$$

to the right 4 units and up 3 units. First we analyze the graph of equation (4). The general form of the graph is shown in Figure 6 in the preceding Property Summary box. In this case, we have $a = 5$, $b = 3$, and

$$c = \sqrt{a^2 + b^2} = \sqrt{5^2 + 3^2} = \sqrt{34} \ (\approx 5.8)$$

Consequently, the vertices [for equation (4)] are $(0, \pm 5)$, the foci are $(0, \pm\sqrt{34})$, and the asymptotes are $y = \pm\frac{5}{3}x$. Figure 7 shows the graph of this hyperbola.

Finally, by translating the graph in Figure 7 to the right 4 units and up 3 units, we obtain the graph of the original equation, as shown in Figure 8. You should verify for yourself that the information accompanying Figure 8 is correct.

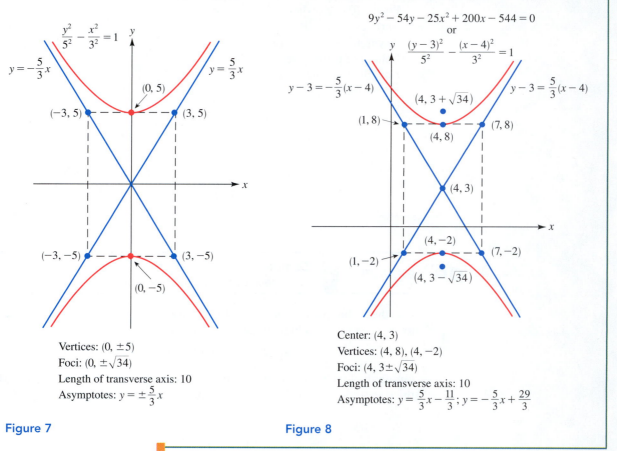

Vertices: $(0, \pm 5)$
Foci: $(0, \pm\sqrt{34})$
Length of transverse axis: 10
Asymptotes: $y = \pm\frac{5}{3}x$

Figure 7

Center: $(4, 3)$
Vertices: $(4, 8)$, $(4, -2)$
Foci: $(4, 3 \pm \sqrt{34})$
Length of transverse axis: 10
Asymptotes: $y = \frac{5}{3}x - \frac{11}{3}$; $y = -\frac{5}{3}x + \frac{29}{3}$

Figure 8

If you reread the example we have just completed, you'll see that it was not necessary to know in advance that the given equation represented a hyperbola. Rather, this fact emerged naturally after we completed the square. Indeed, completing the square is a useful technique for identifying the graph of any equation of the form

$$Ax^2 + Cy^2 + Dx + Ey + F = 0$$

EXAMPLE 3   Completing the square to identify a graph

Identify the graph of the equation

$$4x^2 - 32x - y^2 + 2y + 63 = 0$$

SOLUTION
As before, we complete the squares:

$$4(x^2 - 8x) - (y^2 - 2y) = -63$$
$$4(x^2 - 8x + 16) - (y^2 - 2y + 1) = -63 + 64 - 1$$
$$4(x - 4)^2 - (y - 1)^2 = 0$$

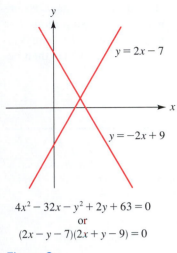

$$4x^2 - 32x - y^2 + 2y + 63 = 0$$
or
$$(2x - y - 7)(2x + y - 9) = 0$$

Figure 9

Since the right-hand side of this last equation is zero, dividing both sides by 4 will not bring the equation into one of the standard forms. Indeed, if we factor the left-hand side of the equation as a difference of two squares, we obtain

$$[2(x - 4) - (y - 1)][2(x - 4) + (y - 1)] = 0$$
$$(2x - y - 7)(2x + y - 9) = 0$$

$2x - y - 7 = 0$	$2x + y - 9 = 0$
$-y = -2x + 7$	$y = -2x + 9$
$y = 2x - 7$	

Thus the given equation is equivalent to the two linear equations $y = 2x - 7$ and $y = -2x + 9$. These two lines together constitute the graph. See Figure 9.

The two lines that we graphed in Figure 9 are actually the asymptotes for the hyperbola $4(x - 4)^2 - (y - 1)^2 = 1$. (Verify this for yourself.) For that reason, the graph in Figure 9 is referred to as a **degenerate hyperbola.** There are other cases similar to this that can arise in graphing equations of the form $Ax^2 + Cy^2 + Dx + Ey + F = 0$. For instance, as you can check for yourself by completing the squares, the graph of the equation

$$x^2 - 2x + 4y^2 - 16y + 17 = 0$$

consists of the single point $(1, 2)$. We refer to the graph in this case as a **degenerate ellipse.** Similarly, as you can check by completing the squares, the equation $x^2 - 2x + 4y^2 - 16y + 18 = 0$ has no graph; there are no points with coordinates that satisfy the equation.

We can obtain the equation of a tangent line to a hyperbola using the same ideas employed for the parabola and the ellipse in the previous sections. The result is this: The equation of the tangent to the hyperbola $(x^2/a^2) - (y^2/b^2) = 1$ at the point (x_1, y_1) on the curve is

$$\boxed{\frac{x_1 x}{a^2} - \frac{y_1 y}{b^2} = 1}$$

(See Exercise 53 at the end of this section for an outline of the derivation.) As with the parabola and the ellipse, many of the interesting properties of the hyperbola are related to the tangent lines. For instance, the hyperbola has a reflection property that is similar to the reflection properties of the parabola and the ellipse. To state this property, we first define a **focal radius** of a hyperbola as a line segment drawn from a focus to a point on the hyperbola. Then the **reflection property** of the hyperbola is that the tangent line bisects the angle formed by the focal radii drawn to the point of tangency, as indicated in Figure 10.

We conclude this section by listing several applications of the hyperbola. Some comets have hyperbolic orbits. Unlike Halley's comet, which has an elliptical orbit, these comets pass through the solar system once and never return. The Cassegrain telescope (invented by the Frenchman Sieur Cassegrain in 1672) uses both a hyperbolic mirror and a parabolic mirror. The Hubble Space Telescope, launched into orbit in April 1990, utilizes a Cassegrain telescope. The hyperbola is also used in some navigation systems. In the LORAN (LOng RAnge Navigation) system, an airplane or a ship at a point P receives radio signals that are transmitted simultaneously

Figure 10
Reflection property of the hyperbola: The tangent at P bisects the angle formed by the focal radii drawn to P, so $\alpha = \beta$.

from two locations, F_1 and F_2. The time difference between the two signals is converted to a difference in distances: $F_1P - F_2P$. This locates the ship along one branch of a hyperbola. Then, using data from a second set of signals, the ship is located along a second hyperbola. The intersection of the two hyperbolas then determines the location of the airplane or ship.

EXERCISE SET 8.4

A

For Exercises 1–24, graph the hyperbolas. In each case in which the hyperbola is nondegenerate, specify the following: vertices, foci, lengths of transverse and conjugate axes, eccentricity, and equations of the asymptotes. In Exercises 11–24, also specify the centers.

1. $x^2 - 4y^2 = 4$
2. $y^2 - x^2 = 1$
3. $y^2 - 4x^2 = 4$
4. $25x^2 - 9y^2 = 225$
5. $16x^2 - 25y^2 = 400$
6. $9x^2 - y^2 = 36$
7. $2y^2 - 3x^2 = 1$
8. $x^2 - y^2 = 9$
9. $4y^2 - 25x^2 = 100$
10. $x^2 - 3y^2 = 3$
11. $\dfrac{(x-5)^2}{5^2} - \dfrac{(y+1)^2}{3^2} = 1$
12. $\dfrac{(x-5)^2}{3^2} - \dfrac{(y+1)^2}{5^2} = 1$
13. $\dfrac{(y-2)^2}{2^2} - \dfrac{(x-1)^2}{1^2} = 1$
14. $\dfrac{(y-3)^2}{2^2} - \dfrac{x^2}{1^2} = 1$
15. $\dfrac{(x+3)^2}{4^2} - \dfrac{(y-4)^2}{4^2} = 1$
16. $\dfrac{(x+1)^2}{5^2} - \dfrac{(y-2)^2}{3^2} = 1$
17. $x^2 - y^2 + 2y - 5 = 0$
18. $16x^2 - 32x - 9y^2 + 90y - 353 = 0$
19. $x^2 - y^2 - 4x + 2y - 6 = 0$
20. $x^2 - 8x - y^2 + 8y - 25 = 0$
21. $y^2 - 25x^2 + 8y - 9 = 0$
22. $9y^2 - 18y - 4x^2 - 16x - 43 = 0$
23. $x^2 + 7x - y^2 - y + 12 = 0$
24. $9x^2 + 9x - 16y^2 + 4y + 2 = 0$
25. Let $P(x, y)$ be a point in the first quadrant on the hyperbola $(x^2/2^2) - (y^2/1^2) = 1$. Let Q be the point in the first quadrant with the same x-coordinate as P and lying on an asymptote to the hyperbola. Show that $PQ = (x - \sqrt{x^2 - 4})/2$.
26. The distance PQ in Exercise 25 represents the vertical distance between the hyperbola and the asymptote. Complete the following table to see numerical evidence that this separation distance approaches zero as x gets larger and larger. (Round each entry to one significant digit.)

x	10	50	100	500	1000	10,000
PQ						

In Exercises 27–36, determine the equation of the hyperbola satisfying the given conditions. Write each answer in the form $Ax^2 - By^2 = C$ or in the form $Ay^2 - Bx^2 = C$.

27. Foci $(\pm 4, 0)$; vertices $(\pm 1, 0)$
28. Foci $(0, \pm 5)$; vertices $(0, \pm 3)$
29. Asymptotes $y = \pm\frac{1}{2}x$; vertices $(\pm 2, 0)$
30. Asymptotes $y = \pm x$; foci $(0, \pm 1)$
31. Asymptotes $y = \pm\sqrt{10}\,x/5$; foci $(\pm\sqrt{7}, 0)$
32. Length of the transverse axis 6; eccentricity 4/3; center $(0, 0)$; focal axis horizontal
33. Vertices $(0, \pm 7)$; graph passes through the point $(1, 9)$
34. Eccentricity 2; foci $(\pm 1, 0)$
35. Length of the transverse axis 6; length of the conjugate axis 2; foci on the y-axis; center at the origin
36. Asymptotes $y = \pm 2x$; graph passes through $(1, \sqrt{3})$
37. Show that the two asymptotes of the hyperbola $x^2 - y^2 = 16$ are perpendicular to each other.
38. (a) Verify that the point $P(6, 4\sqrt{3})$ lies on the hyperbola $16x^2 - 9y^2 = 144$.
    (b) In Example 1, we found that the foci of this hyperbola were $F_1(-5, 0)$ and $F_2(5, 0)$. Compute the lengths F_1P and F_2P, where P is the point $(6, 4\sqrt{3})$.
    (c) Verify that $|F_1P - F_2P| = 2a$.
39. (a) Verify that the point $P(5, 6)$ lies on the hyperbola $5y^2 - 4x^2 = 80$.
    (b) Find the foci.
    (c) Compute the lengths of the line segments $\overline{F_1P}$ and $\overline{F_2P}$, where P is the point $(5, 6)$.
    (d) Verify that $|F_1P - F_2P| = 2a$.
Ⓖ 40. In this exercise you will look at the graph of the hyperbola $16x^2 - 9y^2 = 144$ from two perspectives.
    (a) Solve the given equation for y, then graph the two resulting functions in the standard viewing rectangle.
    (b) Determine the equations of the asymptotes. Add the graphs of the asymptotes to your picture from part (a).
    (c) Looking at your picture from part (b), you can see that the hyperbola seems to be moving closer and closer to its asymptotes as $|x|$ gets large. To see more dramatic evidence of this, change the viewing rectangle so that both x and y extend from -100 to 100. At this scale, the hyperbola is virtually indistinguishable from its asymptotes.

G **41.** In this exercise we graph the hyperbola

$$\frac{(y-3)^2}{5^2} - \frac{(x-4)^2}{3^2} = 1.$$

(a) Solve the equation for y to obtain

$$y = 3 \pm 5\sqrt{1 + (x-4)^2/9}$$

(b) In the standard viewing rectangle, graph the two equations that you obtained in part (a). Then, for a better view, adjust the viewing rectangle so that both x and y extend from -20 to 20.

B

42. **(a)** Let e_1 denote the eccentricity of the hyperbola $x^2/4^2 - y^2/3^2 = 1$, and let e_2 denote the eccentricity of the hyperbola $x^2/3^2 - y^2/4^2 = 1$. Verify that $e_1^2 e_2^2 = e_1^2 + e_2^2$.

(b) Let e_1 and e_2 denote the eccentricities of the two hyperbolas $x^2/A^2 - y^2/B^2 = 1$ and $y^2/B^2 - x^2/A^2 = 1$, respectively. Verify that $e_1^2 e_2^2 = e_1^2 + e_2^2$.

43. **(a)** If the hyperbola $x^2/a^2 - y^2/b^2 = 1$ has perpendicular asymptotes, show that $a = b$. What is the eccentricity in this case?

(b) Show that the asymptotes of the hyperbola $x^2/a^2 - y^2/a^2 = 1$ are perpendicular. What is the eccentricity of this hyperbola?

44. Derive equation (1) in this section from the equation that precedes it.

45. Let $P(x, y)$ be a point on the right-hand branch of the hyperbola $x^2/a^2 - y^2/b^2 = 1$. As usual, let F_2 denote the focus located at $(c, 0)$. The following steps outline a proof of the fact that the length of the line segment $\overline{F_2P}$ in this case is given by $F_2P = xe - a$.

(a) Explain why

$$\sqrt{(x+c)^2 + y^2} - \sqrt{(x-c)^2 + y^2} = 2a$$

(b) In the preceding equation, add the quantity $\sqrt{(x-c)^2 + y^2}$ to both sides and then square both sides. Show that the result can be written

$$xc - a^2 = a\sqrt{(x-c)^2 + y^2}$$

or $xc - a^2 = a(F_2P)$

(c) Divide both sides of the preceding equation by a to show that $xe - a = F_2P$, as required.

46. Let $P(x, y)$ be a point on the right-hand branch of the hyperbola $x^2/a^2 - y^2/b^2 = 1$. As usual, let F_1 denote the focus located at $(-c, 0)$. Show that $F_1P = xe + a$.
Hint: Use the result in Exercise 45 along with the fact that the right-hand branch is defined by the equation $F_1P - F_2P = 2a$.

47. Let P be a point on the right-hand branch of the hyperbola $x^2 - y^2 = k^2$. If d denotes the distance from P to the center of the hyperbola, show that

$$d^2 = (F_1P)(F_2P)$$

48. Let $P(x, y)$ be a point on the right-hand branch of the hyperbola $x^2/a^2 - y^2/b^2 = 1$. From P, a line segment is drawn perpendicular to the line $x = a/e$, meeting this line at D. If F_2 denotes (as usual) the focus located at $(c, 0)$, show that

$$\frac{F_2P}{PD} = e$$

The line $x = a/e$ is called the *directrix* of the hyperbola corresponding to the focus F_2. *Hint*: Use the expression for F_2P developed in Exercise 45.

49. By solving the system

$$\begin{cases} y = \frac{4}{3}x - 1 \\ 16x^2 - 9y^2 = 144 \end{cases}$$

show that the line and the hyperbola intersect in exactly one point. Draw a sketch of the situation. This demonstrates that a line that intersects a hyperbola in exactly one point does not have to be a tangent line.

In Exercises 50–52, find the equation of the line that is tangent to the hyperbola at the given point. Write your answer in the form $y = mx + b$.

50. $x^2 - 4y^2 = 16$; $(5, 3/2)$
51. $3x^2 - y^2 = 12$; $(4, 6)$
52. $16x^2 - 25y^2 = 400$; $(10, 4\sqrt{3})$

C

53. We define a *tangent line to a hyperbola* as a line that is not parallel to an asymptote and that intersects the hyperbola in exactly one point. Show that the equation of the line tangent to the hyperbola $x^2/a^2 - y^2/b^2 = 1$ at the point (x_1, y_1) on the curve is

$$\frac{x_1 x}{a^2} - \frac{y_1 y}{b^2} = 1$$

Hint: Allow for signs, but follow exactly the same steps as were supplied in Exercise 60 of Exercise Set 8.3, where we found the tangent to the ellipse. You should find that the slope in the present case is $m = (b^2 x_1/a^2 y_1)$. Explain why this slope cannot equal the slope of an asymptote as long as (x_1, y_1) is on the hyperbola.

54. The *normal line* to a hyperbola at a point P on the hyperbola is the line through P that is perpendicular to the tangent at P. If the coordinates of P are (x_1, y_1), show that the equation of the normal line is

$$a^2 y_1 x + b^2 x_1 y = x_1 y_1 (a^2 + b^2)$$

55. Suppose the point $P(x_1, y_1)$ is on the hyperbola $x^2/a^2 - y^2/b^2 = 1$. A tangent is drawn at P, meeting the x-axis at A and the y-axis at B. Also, perpendicular lines are drawn from P to the x- and y-axes, meeting these axes at C and D, respectively. If O denotes the origin, show that: **(a)** $OA \cdot OC = a^2$; and **(b)** $OB \cdot OD = b^2$.

56. At the point P on the hyperbola $x^2/a^2 - y^2/b^2 = 1$, a tangent line is drawn that meets the lines $x = a$ and $x = -a$ at S and T, respectively. Show that the circle with $\overline{ST}$ as a diameter passes through the two foci of the hyperbola.

PROJECT

USING HYPERBOLAS TO DETERMINE A LOCATION

An army base in the desert is located at the origin O of an x-y coordinate system. A soldier at the point P needs to determine his coordinates with respect to the army base, but he does not have a link to a global positioning system. He is, however, in radio contact with several small towns in the vicinity. As indicated in Figure 1, two of the towns are town A and town B. Town A is 6 miles due south of the army base; town B is 6 miles due north of the base. At an agreed upon instant, both towns send out identifying radio signals. The soldier receives the signal from B slightly before the signal from A, so he knows that he is closer to B than to A. Furthermore, by measuring the time difference between receiving the two signals, the soldier is able to compute that he is 8 miles closer to town B than to town A. Explain why the point P must lie on a certain hyperbola, and then find the equation of the hyperbola.

Figure 1

Figure 2

Next, the soldier follows a similar procedure using the towns C and D shown in Figure 2. Town C is 19 miles due west of the base; town D is 15 miles due east of the base. The soldier computes that he is 16 miles closer to town D than to town C. Using this information, explain why the point P must lie on a second hyperbola, and then determine its equation.

Use a graphing utility to draw the two hyperbolas and to estimate, to the nearest integers, the coordinates of the intersection point corresponding to the soldier's location P. Next, obtain more precise coordinates by using algebraic techniques (as in Section 6.6) to find the relevant intersection point of the two hyperbolas. At the end, use a calculator and round each coordinate to three decimal places.

8.5   THE FOCUS–DIRECTRIX PROPERTY OF CONICS

The most important contribution which Pappus [ca. 300] made to our knowledge of the conics was his publication of the focus, directrix, eccentricity theorem. —Julian Lowell Coolidge in *A History of the Conic Sections and Quadric Surfaces* (London: Oxford University Press, 1946)

We begin by recalling the focus–directrix property that we used in defining the parabola. A point P is on a parabola if and only if the distance from P to the focus is equal to the distance from P to the directrix. For the parabola in Figure 1, this means that for each point P on the parabola, we have $FP = PD$ or, equivalently,

$$\frac{FP}{PD} = 1$$

Parabola

Focus

Directrix

Figure 1

The ellipse and the hyperbola also can be characterized by focus–directrix properties. We'll begin with the ellipse. To help us in subsequent computations, we need to know the lengths of the line segments $\overline{F_1P}$ and $\overline{F_2P}$ in Figure 2.

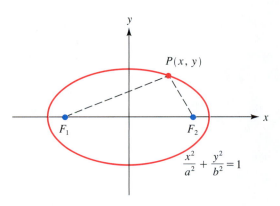

$$\frac{x^2}{a^2} + \frac{y^2}{b^2} = 1$$

Figure 2
The segments drawn from a point on the ellipse to the foci are called focal radii. Here $\overline{F_1P}$ and $\overline{F_2P}$ are the focal radii.

These line segments joining a point on the ellipse to the foci are called **focal radii.** In the box that follows, we give a formula for the length of each focal radius. In the formulas, e ($= c/a$) denotes the eccentricity of the ellipse. (For the derivation of these formulas, see Exercise 15 at the end of this section.)

The Focal Radii of the Ellipse $\dfrac{x^2}{a^2} + \dfrac{y^2}{b^2} = 1$

As indicated in Figure 2, let $P(x, y)$ be a point on the ellipse. Then the lengths of the focal radii are

$$F_1P = a + ex \quad \text{and} \quad F_2P = a - ex$$

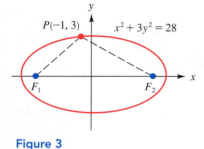

Figure 3

EXAMPLE **1** **Computing the lengths of the focal radii of an ellipse**

Figure 3 shows the ellipse $x^2 + 3y^2 = 28$. Compute the lengths of the focal radii drawn to the point $P(-1, 3)$.

SOLUTION

To apply the formulas in the box, we need to know the values of a and e for the given ellipse. (We already know that $x = -1$; that is given.) As you should verify for yourself using the techniques of Section 8.4, we have $a = 2\sqrt{7}$ and $e = \sqrt{6}/3$. Therefore

$$F_1P = a + ex$$
$$= 2\sqrt{7} + \left(\frac{\sqrt{6}}{3}\right)(-1) = \frac{6\sqrt{7} - \sqrt{6}}{3}$$

and

$$F_2P = a - ex$$
$$= 2\sqrt{7} - \left(\frac{\sqrt{6}}{3}\right)(-1) = \frac{6\sqrt{7} + \sqrt{6}}{3}$$

In Figure 4 we show the ellipse $(x^2/a^2) + (y^2/b^2) = 1$ and the vertical line $x = a/e$. The line $x = a/e$ is a **directrix** of the ellipse. In Figure 4, F_2P is the distance from the focus F_2 to a point P on the ellipse, and PD is the distance from P to the directrix $x = a/e$. We will prove the following remarkable *focus–directrix property of the ellipse:* For any point $P(x, y)$ on the ellipse, the ratio of F_2P to PD is equal to e, the eccentricity of the ellipse. To prove this, we proceed as follows:

$$\frac{F_2P}{PD} = \frac{a - ex}{PD} \qquad \text{using our formula for } F_2P$$

$$= \frac{a - ex}{\dfrac{a}{e} - x} \qquad \text{using Figure 4}$$

$$= \frac{e(a - ex)}{a - ex} \qquad \text{multiplying both numerator and denominator by } e$$

$$= e$$

So $F_2P/PD = e$, as we wished to show.

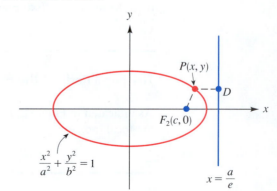

Figure 4

The directrix $x = a/e$ is associated with the focus $F_2(c, 0)$. What about the focus $F_2(-c, 0)$? From the symmetry of the ellipse, it follows that the line $x = -a/e$ is the directrix associated with this focus. More specifically, referring to Figure 5, not only do we have $F_2P/PD = e$, we also have $F_1P/PE = e$.

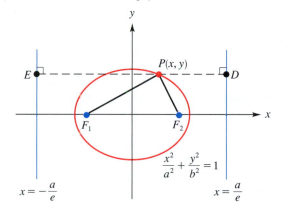

Figure 5
For every point P on the ellipse, we have $F_1P/PE = e$ and $F_2P/PD = e$.

We have seen that each point P on the ellipse in Figure 5 satisfies the following two equations:

$$\frac{F_1P}{PE} = e \tag{1}$$

$$\frac{F_2P}{PD} = e \tag{2}$$

Now, conversely, suppose that the point $P(x, y)$ in Figure 6 satisfies equations (1) and (2). We will show that $P(x, y)$ must, in fact, lie on the ellipse $(x^2/a^2) + (y^2/b^2) = 1$.*

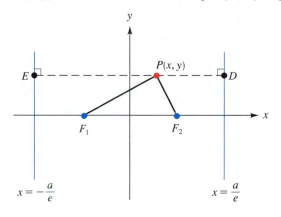

Figure 6

In Figure 6, the distance from E to D is $2a/e$. Therefore

$$PE + PD = 2a/e$$

and, consequently,

$$\frac{F_1P}{e} + \frac{F_2P}{e} = \frac{2a}{e} \qquad \text{\color{red}using equations (1) and (2) to substitute for } PE \text{ and } PD$$

---

*The short proof given here was communicated to the author by Professor Ray Redheffer. For a proof using equation (2) but not equation (1), see Exercise 16.

or

$$F_1P + F_2P = 2a \qquad \text{multiplying both sides by } e$$

Thus, by definition, P lies on the ellipse $(x^2/a^2) + (y^2/b^2) = 1$, as we wished to show.

In the box that follows, we summarize the focus–directrix properties of the ellipse $(x^2/a^2) + (y^2/b^2) = 1$. For reference, we also include the original defining property of the ellipse from Section 8.4.

Focus and Focus–Directrix Properties of the Ellipse $\dfrac{x^2}{a^2} + \dfrac{y^2}{b^2} = 1$

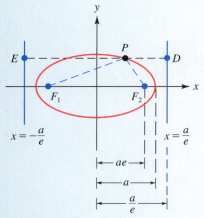

Figure 7
The ellipse $(x^2/a^2) + (y^2/b^2) = 1$ with foci $F_1(-c, 0)$ and $F_2(c, 0)$ and directrices $x = \pm a/e$.

Refer to Figure 7.

1. A point P is on the ellipse if and only if the sum of the distances from P to the foci $F_1(-c, 0)$ and $F_2(c, 0)$ is $2a$.
2. A point P is on the ellipse if and only if

$$\frac{F_2P}{PD} = e$$

where F_2P is the distance from P to the focus F_2, PD is the distance from P to the directrix $x = a/e$, and $e \,(= c/a)$ is the eccentricity of the ellipse.

3. A point P is on the ellipse if and only if

$$\frac{F_1P}{PE} = e$$

where F_1P is the distance from P to the focus F_1 and PE is the distance from P to the directrix $x = -a/e$.

EXAMPLE 2 **Using focus–directrix information to find the equation of an ellipse.**

The foci of an ellipse are $(\pm 3, 0)$ and the directrix corresponding to the focus $(3, 0)$ is $x = 5$. Find the equation of the ellipse. Write the answer in the form $Ax^2 + By^2 = C$.

SOLUTION
Using the given data (and referring to Figure 7), we have

$$ae = 3 \qquad \text{and} \qquad a/e = 5$$

From the first equation we obtain $e = 3/a$. Substituting this value of e in the second equation yields

$$\frac{a}{3/a} = 5 \qquad \text{and therefore} \qquad a^2 = 15$$

Now we can calculate b^2 by using the relation $b^2 = a^2 - c^2$ and the given information $c = 3$:

$$b^2 = a^2 - c^2 = 15 - 9 = 6$$

Substituting the values we've obtained for a^2 and b^2 in the equation $(x^2/a^2) + (y^2/b^2) = 1$ yields

$$\frac{x^2}{15} + \frac{y^2}{6} = 1$$

or

$$2x^2 + 5y^2 = 30 \qquad \text{multiplying both sides by 30}$$

The focus–directrix property can be developed for the hyperbola in the same way that we have proceeded for the ellipse. In fact, the algebra is so similar that we shall omit the details here and simply summarize the results. The hyperbola $(x^2/a^2) - (y^2/b^2) = 1$ has two directrices: they are the vertical lines $x = a/e$ in Figure 8(a) and $x = -a/e$ in Figure 8(b). Notice that these equations are identical to those for the directrices of the ellipse. In the box following Figure 8, we summarize the focus and the focus–directrix properties of the hyperbola.

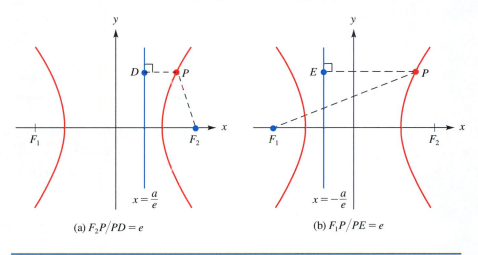

Figure 8
The hyperbola $(x^2/a^2) - (y^2/b^2) = 1$ has two directrices. The directrix $x = a/e$ corresponds to the focus $F_2(c, 0)$. The directrix $x = -a/e$ corresponds to the focus $F_1(-c, 0)$.

(a) $F_2P/PD = e$

(b) $F_1P/PE = e$

Focus and Focus–Directrix Properties of the Hyperbola $\dfrac{x^2}{a^2} - \dfrac{y^2}{b^2} = 1$

Refer to Figure 8.

1. A point P is on the hyperbola if and only if the difference of the distances from P to the foci $F_1(-c, 0)$ and $F_2(c, 0)$ is $2a$.
2. A point P is on the hyperbola if and only if

$$\frac{F_2P}{PD} = e$$

where F_2P is the distance from P to the focus F_2, PD is the distance from P to the directrix $x = a/e$, and $e \,(= c/a)$ is the eccentricity of the hyperbola.

3. A point P is on the hyperbola if and only if

$$\frac{F_1P}{PE} = e$$

where F_1P is the distance from P to the focus F_1 and PE is the distance from P to the directrix $x = -a/e$.

EXAMPLE **3**   **An eccentricity computation**

Determine the foci, the eccentricity, and the directrices for the hyperbola $9x^2 - 16y^2 = 144$.

SOLUTION

To convert the given equation to standard form, we divide both sides by 144. This yields

$$\frac{x^2}{4^2} - \frac{y^2}{3^2} = 1$$

So we have $a = 4$, $b = 3$, and

$$c = \sqrt{a^2 + b^2} = \sqrt{4^2 + 3^2} = 5$$

The foci are therefore $(\pm 5, 0)$, and the eccentricity is

$$e = \frac{c}{a} = \frac{5}{4}$$

The directrices are

$$x = \pm\frac{a}{e} = \pm\frac{4}{5/4} = \pm\frac{16}{5}$$

EXAMPLE **4**

Figure 9 shows the graph of the hyperbola $9x^2 - 16y^2 = 144$. (This is the hyperbola discussed in Example 3.)

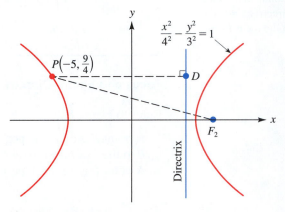

Figure 9

(a) Verify that the point $P(-5, 9/4)$ lies on this hyperbola.

(b) Compute F_2D and PD, and verify that

$$\frac{F_2P}{PD} = e$$

SOLUTION

(a) Substituting the values $x = -5$ and $y = 9/4$ in the equation of the hyperbola yields

$$9(-5)^2 - 16(9/4)^2 = 144$$

or

$$225 - 81 = 144$$

Since this last equation is correct, we conclude that the point $P(-5, 9/4)$ indeed lies on the hyperbola.

(b) From Example 3, we know that the coordinates of the focus F_2 are $(5, 0)$. So, using the distance formula, we have

$$F_2P = \sqrt{(5 - (-5))^2 + \left(0 - \frac{9}{4}\right)^2}$$
$$= \sqrt{100 + \frac{81}{16}} = \sqrt{\frac{1681}{16}} = \frac{41}{4}$$

Next, we use the fact (from Example 3) that the equation of the directrix in Figure 9 is $x = 16/5$. Thus, in Figure 9,

$$PD = |-5| + \frac{16}{5} = \frac{41}{5}$$

Finally, we compute the ratio F_2P/PD:

$$\frac{F_2P}{PD} = \frac{41/4}{41/5} = \frac{5}{4}$$

This is the same number that we obtained for the eccentricity in Example 3. So, we have verified in this case that the ratio of F_2P to PD is equal to the eccentricity.

The focus–directrix property provides a unified approach to the parabola, the ellipse, and the hyperbola. For both the ellipse and the hyperbola, we've seen that for any point P on the curve, we have

$$\frac{\text{distance from } P \text{ to a focus}}{\text{distance from } P \text{ to the corresponding directrix}} = e \qquad (3)$$

where e is a positive constant, the eccentricity of the curve. For the ellipse we have $e < 1$, and for the hyperbola we have $e > 1$. Equation (3) also holds for the parabola if we agree on the convention that the eccentricity of a parabola is 1. This is because with $e = 1$, equation (3) tells us that the distance from P to the focus equals the distance from P to the directrix, which is the defining condition for the parabola. The following theorem summarizes these remarks.

THEOREM The Focus–Directrix Property of Conics

Refer to Figure 10. Let $\mathcal{L}$ be a fixed line, F a fixed point, and e a positive constant. Consider the set of points P satisfying the condition $FP/PD = e$. (The point D is defined in the figure.) Then

(a) If $e = 1$, the set of points is a *parabola* with focus F and directrix $\mathcal{L}$.
(b) If $0 < e < 1$, the set of points is an *ellipse* with focus F, corresponding directrix $\mathcal{L}$, and eccentricity e.
(c) If $e > 1$, the set of points is a *hyperbola* with focus F, corresponding directrix $\mathcal{L}$, and eccentricity e.

Figure 10

NOTE In our development in this section, we've always considered cases in which the directrix is vertical, for simplicity. However, the preceding theorem is valid for any orientation of the directrix $\mathcal{L}$.

EXERCISE SET 8.5

A

In Exercises 1–4, you are given an ellipse and a point P on the ellipse. Find F_1P and F_2P, the lengths of the focal radii.

1. $x^2 + 3y^2 = 76$; $P(-8, 2)$
2. $x^2 + 3y^2 = 57$; $P(3, -4)$
3. $(x^2/15^2) + (y^2/5^2) = 1$; $P(9, 4)$
4. $2x^2 + 3y^2 = 14$; $P(-1, -2)$

In Exercises 5–10, determine the foci, the eccentricity, and the directrices for each ellipse and hyperbola.

5. (a) $(x^2/4^2) + (y^2/3^2) = 1$
   (b) $(x^2/4^2) - (y^2/3^2) = 1$
6. (a) $x^2 + 4y^2 = 1$
   (b) $x^2 - 4y^2 = 1$
7. (a) $12x^2 + 13y^2 = 156$
   (b) $12x^2 - 13y^2 = 156$
8. (a) $x^2 + 2y^2 = 2$
   (b) $x^2 - 2y^2 = 2$
9. (a) $25x^2 + 36y^2 = 900$
   (b) $25x^2 - 36y^2 = 900$
10. (a) $4x^2 + 25y^2 = 100$
    (b) $4x^2 - 25y^2 = 100$

In Exercises 11 and 12, use the given information to find the equation of the ellipse. Write the answer in the form $Ax^2 + By^2 = C$.

11. The foci are $(\pm 1, 0)$ and the directrices are $x = \pm 4$.
12. The foci are $(\pm\sqrt{3}, 0)$ and the eccentricity is $2/3$.

In Exercises 13 and 14, use the given information to find the equation of the hyperbola. Write the answer in the form $Ax^2 - By^2 = C$.

13. The foci are $(\pm 2, 0)$ and the directrices are $x = \pm 1$.
14. The foci are $(\pm 3, 0)$ and the eccentricity is 2.

B

15. In this exercise we show that the focal radii of the ellipse $x^2/a^2 + y^2/b^2 = 1$ are $F_1P = a + ex$ and $F_2P = a - ex$. The method used here, which avoids the use of radicals, appears in the eighteenth-century text *Traité analytique des sections coniques* (Paris: 1707) by the Marquis de l'Hospital (1661–1704). (For another method, one that does use radicals, see Exercise 55 in Section 8.3.) For convenience, let $d_1 = F_1P$ and $d_2 = F_2P$, as indicated in the accompanying figure.

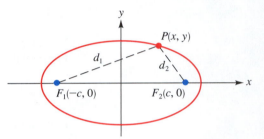

(a) Using the distance formula, verify that

$$d_1^2 = (x + c)^2 + y^2 \qquad \text{and} \qquad d_2^2 = (x - c)^2 + y^2$$

(b) Use the two equations in part (a) to show that
$$d_1^2 - d_2^2 = 4cx.$$

(c) Explain why $d_1 + d_2 = 2a$.

(d) Factor the left-hand side of the equation in part (b) and substitute for one of the factors using the equation in part (c). Show that the result can be written $d_1 - d_2 = 2cx/a$.

(e) Add the equations in parts (c) and (d). Show that the resulting equation can be written $d_1 = a + ex$, as required.

(f) Use the equation in part (c) and the result in part (e) to show that $d_2 = a - ex$.

16. In this exercise we show that if the point $P(x, y)$ in the accompanying figure satisfies the condition

$$\frac{F_2P}{PD} = e$$

then, in fact, the point $P(x, y)$ lies on the ellipse $x^2/a^2 + y^2/b^2 = 1$.

(a) From the given equation we have $(F_2P)^2 = e^2(PD)^2$. Use the distance formula and the figure to deduce from this equation that

$$(x - c)^2 + y^2 = e^2\left(\frac{a}{e} - x\right)^2$$

(b) In the equation in part (a), replace e with c/a. After carrying out the indicated operations and simplifying, show that the equation can be written

$$a^2x^2 - c^2x^2 + a^2y^2 = a^4 - a^2c^2$$

(c) The equation in part (b) is equivalent to $(a^2 - c^2)x^2 + a^2y^2 = a^2(a^2 - c^2)$. Now replace the quantity $a^2 - c^2$ by b^2 and show that the resulting equation can be written $x^2/a^2 + y^2/b^2 = 1$; thus, P lies on the ellipse, as we wished to show.

Chapter 8 | Summary of Principal Terms and Formulas

Terms and formulas	Page reference	Comments
1. Conic sections	635	As indicated by the figure in the chapter introduction, these are the curves formed when a plane intersects the surface of a right circular cone. These curves are the circle, the ellipse, the hyperbola, and the parabola.
2. Parabola	636	A parabola is the set of all points in the plane equally distant from a fixed line and a fixed point not on the line. The fixed line is called the *directrix,* and the fixed point is called the *focus.*
3. Axis (of a parabola)	638	This is the line that is drawn through the focus of the parabola, perpendicular to the directrix.
4. Vertex (of a parabola)	638	This is the point at which the parabola intersects its axis.
5. Focal chord (of a parabola)	640	A focal chord of a parabola is a line segment that passes through the focus and has endpoints on the parabola.
6. Focal width (of a parabola)	640	The focal width of a parabola is the length of the focal chord that is perpendicular to the axis of the parabola. For a given value of p, the two parabolas $x^2 = 4py$ and $y^2 = 4px$ have the same focal width; it is $4p$.
7. Tangent line (to a parabola)	650	A line that is not parallel to the axis of a parabola is tangent to the parabola provided that it intersects the parabola in exactly one point.
8. Focal length and focal ratio	644	Refer to the figure in the box on page 645, which shows a cross section of a parabolic reflector. The focal length of the reflector is the distance p from the vertex to the focus. The focal ratio is the ratio of the focal length p to the diameter d.

Terms and formulas	Page reference	Comments
9. Ellipse	653	An ellipse is the set of all points in the plane, the sum of whose distances from two fixed points is constant. Each fixed point is called a *focus* of the ellipse.
10. Eccentricity (of an ellipse)	655	The eccentricity is a number that measures how much the ellipse deviates from being a circle. See Figure 7 in Section 8.4. The eccentricity e is defined by the formula $e = c/a$, where c and a are defined by the following conventions. The distance between the foci is denoted by $2c$. The sum of the distances from a point on the ellipse to the two foci is denoted by $2a$.
11. Focal axis (of an ellipse)	655	This is the line passing through the two foci.
12. Center (of an ellipse)	655	This is the midpoint of the line segment joining the foci.
13. Vertices (of an ellipse)	655	The two points at which an ellipse meets its focal axis are called the vertices of the ellipse.
14. Major axis and semimajor axis (of an ellipse)	655	The major axis of an ellipse is the line segment joining the two vertices. The semimajor axis refers to either of the two (congruent) line segments from the center of the ellipse to a vertex.
15. Minor axis and semiminor axis (of an ellipse)	655	The minor axis of an ellipse is the line segment through the center, perpendicular to the major axis, and with endpoints on the ellipse. This is the segment $\overline{AB}$ in Figure 6 of Section 8.3. In that figure, each of the segments $\overline{CB}$ and $\overline{CA}$ is referred to as a semiminor axis.
16. $\dfrac{x^2}{a^2} + \dfrac{y^2}{b^2} = 1 \quad (a > b)$	655	This is the standard form for the equation of an ellipse with foci $(\pm c, 0)$.
$\dfrac{x^2}{b^2} + \dfrac{y^2}{a^2} = 1 \quad (a > b)$	658	This is the standard form for the equation of an ellipse with foci $(0, \pm c)$.
17. Tangent line to an ellipse	659	A tangent to an ellipse is a line that intersects the ellipse in exactly one point.
18. $\dfrac{x_1 x}{a^2} + \dfrac{y_1 y}{b^2} = 1$	660	This is the equation of the line that is tangent to the ellipse $x^2/a^2 + y^2/b^2 = 1$ at the point (x_1, y_1) on the ellipse.
19. Hyperbola	667	A hyperbola is the set of all points in the plane, the difference of whose distances from two fixed points is a positive constant. The two fixed points are the *foci,* and the line passing through the foci is the *focal axis.*

Terms and formulas	Page reference	Comments
20. $\dfrac{x^2}{a^2} - \dfrac{y^2}{b^2} = 1$	671	This is the standard form for the equation of a hyperbola with foci $(\pm c, 0)$.
$\dfrac{y^2}{a^2} - \dfrac{x^2}{b^2} = 1$	673	This is the standard form for the equation of a hyperbola with foci $(0, \pm c)$.
21. Asymptote	670	A line is said to be an asymptote for a curve if the distance between the line and the curve approaches zero as we move farther and farther out along the line. The asymptotes for the hyperbola $x^2/a^2 - y^2/b^2 = 1$ are the two lines $y = \pm(b/a)x$. For the hyperbola $y^2/a^2 - x^2/b^2 = 1$, the asymptotes are $y = \pm(a/b)x$.
22. Focal axis (of a hyperbola)	671	This is the line passing through the foci.
23. Vertices (of a hyperbola)	671	The two points at which the hyperbola intersects its focal axis are called vertices.
24. Center (of a hyperbola)	671	This is the point on the focal axis midway between the foci.
25. Transverse axis	671	This is the line segment joining the two vertices of a hyperbola.
26. Conjugate axis	671	This is the line segment perpendicular to the transverse axis of the hyperbola, passing through the center and extending a distance $b(=\sqrt{c^2 - a^2})$ on either side of the center.
27. Eccentricity (of a hyperbola)	671	For both of the standard forms for the hyperbola, the eccentricity e is defined by $e = c/a$.
28. $\dfrac{x_1 x}{a^2} - \dfrac{y_1 y}{b^2} = 1$	675	This is the equation of the tangent to the hyperbola $x^2/a^2 - y^2/b^2 = 1$ at the point (x_1, y_1) on the curve. (See Exercise 53 in Exercise Set 8.4 for the definition of a tangent to a hyperbola.)
29. Focus–directrix property of conics	685	Refer to Figure 10 in Section 8.5. Let $\mathcal{L}$ be a fixed line, F a fixed point, and e a positive constant. Consider the set of points P satisfying the condition $FP/PD = e$, where D is defined by Figure 10. Then: **(a)** If $e = 1$, the set of points is a *parabola* with focus F and directrix $\mathcal{L}$. **(b)** If $0 < e < 1$, the set of points is an *ellipse* with focus F, corresponding directrix $\mathcal{L}$, and eccentricity e. **(c)** If $e > 1$, the set of points is a *hyperbola* with focus F, corresponding directrix $\mathcal{L}$, and eccentricity e.

Writing Mathematics

Write out your answers to Questions 1 and 2 in complete sentences. If you draw a diagram to accompany your writing, or if you use equations, be sure that you clearly label any elements to which you refer.

1. Refer to Figure 1 on page 653. If the thumbtacks are 3 in. apart, what length of string should be used to produce an ellipse with eccentricity 2/3?
2. Refer to Figure 1 on page 653 to explain each of the following.
   (a) When the eccentricity of an ellipse is close to 1, the ellipse is very flat.
   (b) When the eccentricity of an ellipse is close to 0, the ellipse resembles a circle.
3. Investigate the geometric significance of the eccentricity of a hyperbola by completing the three steps that follow. Then, write a report telling what you have done, what patterns you have observed, and what relationship you have found between the eccentricity and the shape of the hyperbola.
   (a) Use the definition of the eccentricity e to show that $e = \sqrt{1 + (b/a)^2}$. (This shows that the eccentricity is always greater than 1.)
   (b) Compute the eccentricity for each of the following hyperbolas. (Use a calculator.)
      (i) $0.0201x^2 - y^2 = 0.0201$
      (ii) $3x^2 - y^2 = 3$
      (iii) $8x^2 - y^2 = 8$
      (iv) $15x^2 - y^2 = 15$
      (v) $99x^2 - y^2 = 99$
   (c) Use a graphing utility to compare the Quadrant I portions of the five hyperbolas in part (b).

Chapter 8   Review Exercises

In Exercises 1–4, find the equation of the parabola satisfying the given conditions. In each case, assume the vertex is $(0, 0)$.

1. (a) The focus is $(4, 0)$.   (b) The focus is $(0, 4)$.
2. The focus lies on the x-axis, and the curve passes through the point $(3, 1)$.
3. The parabola is symmetric about the y-axis, the y-coordinate of the focus is positive, and the length of the focal chord perpendicular to the y-axis is 12.
4. The focus of the parabola is the center of the circle $x^2 - 8x + y^2 + 15 = 0$.

In Exercises 5–8, find the equation of the ellipse satisfying the given conditions. Write your answers in the form $Ax^2 + By^2 = C$.

5. Foci $(\pm 2, 0)$; endpoints of the major axis $(\pm 8, 0)$
6. Foci $(0, \pm 1)$; endpoints of the minor axis $(\pm 4, 0)$
7. Eccentricity 4/5; one end of the minor axis at $(-6, 0)$; center at the origin
8. For any point P on a certain ellipse, the sum of the distances from $(1, 2)$ and $(-1, -2)$ is 12.   *Hint*: Use the distance formula and the definition of an ellipse.

In Exercises 9–12, find the equation of the hyperbola satisfying the given conditions. Write each answer in the form $Ax^2 - By^2 = C$ or in the form $Ay^2 - Bx^2 = C$.

9. Foci $(\pm 6, 0)$; vertices $(\pm 2, 0)$

10. Asymptotes $y = \pm 2x$; foci $(0, \pm 3)$

11. Eccentricity 4; foci $(\pm 3, 0)$

12. Length of the transverse axis 3; eccentricity 5/4; center $(0, 0)$; focal axis horizontal

In Exercises 13–17, graph the parabolas, and in each case specify the vertex, the focus, the directrix, and the focal width.

13. $x^2 = 10y$ **14.** $x^2 = 5y$

15. $x^2 = -12(y - 3)$ **16.** $x^2 = -8(y + 1)$

17. $(y - 1)^2 = -4(x - 1)$

In Exercises 18–23, graph the ellipses, and in each case specify the center, the foci, the lengths of the major and minor axes, and the eccentricity.

18. $x^2 + 2y^2 = 4$ **19.** $4x^2 + 9y^2 = 144$

20. $49x^2 + 9y^2 = 441$ **21.** $9x^2 + y^2 = 9$

22. $\dfrac{(x - 1)^2}{5^2} + \dfrac{(y + 2)^2}{3^2} = 1$

23. $\dfrac{(x + 3)^2}{3^2} + \dfrac{y^2}{3^2} = 1$

In Exercises 24–29, graph the hyperbolas. In each case specify the center, the vertices, the foci, the equations of the asymptotes, and the eccentricity.

24. $x^2 - 2y^2 = 4$ **25.** $4x^2 - 9y^2 = 144$

26. $49y^2 - 9x^2 = 441$ **27.** $9y^2 - x^2 = 9$

28. $\dfrac{(x - 1)^2}{5^2} - \dfrac{(y + 2)^2}{3^2} = 1$

29. $\dfrac{(y + 3)^2}{3^2} - \dfrac{x^2}{3^2} = 1$

In Exercises 30–43, use the technique of completing the square to graph the given equation. If the graph is a parabola, specify the vertex, axis, focus, and directrix. If the graph is an ellipse, specify the center, foci, and lengths of the major and minor axes. If the graph is a hyperbola, specify the center, vertices, foci, and equations of the asymptotes. Finally, if the equation has no graph, say so.

30. $3x^2 + 4y^2 - 6x + 16y + 7 = 0$

31. $y^2 - 16x - 8y + 80 = 0$

32. $y^2 + 4x + 2y - 15 = 0$

33. $16x^2 + 64x + 9y^2 - 54y + 1 = 0$

34. $16x^2 - 32x - 9y^2 + 90y - 353 = 0$

35. $x^2 + 6x - 12y + 33 = 0$

36. $5x^2 + 3y^2 - 40x - 36y + 188 = 0$

37. $x^2 - y^2 - 4x + 2y - 6 = 0$

38. $9x^2 - 90x - 16y^2 + 32y + 209 = 0$

39. $x^2 + 2y - 12 = 0$

40. $y^2 - 25x^2 + 8y - 9 = 0$

41. $x^2 + 16y^2 - 160y + 384 = 0$

42. $16x^2 + 25y^2 - 64x - 100y + 564 = 0$

43. $16x^2 - 25y^2 - 64x + 100y - 36 = 0$

44. Let F_1 and F_2 denote the foci of the hyperbola $5x^2 - 4y^2 = 80$.

(a) Verify that the point P with coordinates $(6, 5)$ lies on the hyperbola.

(b) Compute the quantity $(F_1P - F_2P)^2$.

45. Show that the coordinates of the vertex of the parabola $Ax^2 + Dx + Ey + F = 0$ are given by

$$x = -\frac{D}{2A} \quad \text{and} \quad y = \frac{D^2 - 4AF}{4AE}$$

46. If the equation $Ax^2 + Cy^2 + Dx + Ey + F = 0$ represents an ellipse or a hyperbola, show that the center is the point $(-D/2A, -E/2C)$.

47. The figure shows the parabola $x^2 = 4py$ and a circle with center at the origin and diameter $3p$. If V and F denote the vertex and focus of the parabola, respectively, show that the common chord of the circle and parabola bisects the line segment $\overline{VF}$.

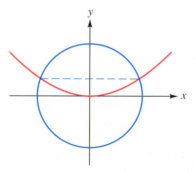

48. Verify that the point $P(5, 2)$ lies on the ellipse $2x^2 + 3y^2 = 62$, and compute the lengths F_1P and F_2P of the focal radii.

49. Verify that the point $P(4, -2)$ lies on the hyperbola $3x^2 - 5y^2 = 28$, and compute the lengths F_1P and F_2P of the focal radii.

50. Find the equation of the ellipse with foci $(\pm 2, 0)$ and directrices $x = \pm 5$.

51. Find the equation of the hyperbola with foci $(\pm 4, 0)$ and eccentricity 3/2.

52. In the following figure, V is the vertex of the parabola $y = ax^2 + bx + c$.

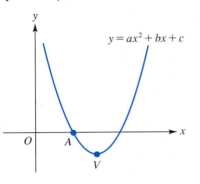

$y = ax^2 + bx + c$

If r_1 and r_2 are the roots of the equation $ax^2 + bx + c = 0$, show that

$$VO^2 - VA^2 = r_1 r_2$$

Chapter 8   Test

1. Find the focus and the directrix of the parabola $y^2 = -12x$, and sketch the graph.

2. Graph the hyperbola $x^2 - 4y^2 = 4$. Specify the foci and the asymptotes.

3. The distances from the planet Saturn to the sun at aphelion and at perihelion are 9.5447 AU and 9.5329 AU, respectively. Compute the eccentricity of the orbit and the length of the semimajor axis. Round each answer to three decimal places.

4. The foci of an ellipse are $(0, \pm 2)$ and the eccentricity is $1/2$. Determine the equation of the ellipse. Write your answer in standard form.

5. Determine the equation of the hyperbola with foci $(\pm 2, 0)$ and with asymptotes $y = \pm(1/\sqrt{3})x$. Write your answer in standard form.

6. Let F_1 and F_2 denote the foci of the hyperbola $5x^2 - 4y^2 = 80$.
   (a) Verify that the point $P(6, 5)$ lies on the hyperbola.
   (b) Compute the quantity $(F_1P - F_2P)^2$.

7. Graph the ellipse $4x^2 + 25y^2 = 100$. Specify the foci and the lengths of the major and minor axes.

8. Graph the equation $16x^2 + y^2 - 64x + 2y + 65 = 0$.

9. Graph the equation $\dfrac{(x + 4)^2}{3^2} - \dfrac{(y - 4)^2}{1^2} = 1$.

10. Graph the parabola $(x - 1)^2 = 8(y - 2)$. Specify the focal width and the vertex.

11. Consider the ellipse $(x^2/6^2) + (y^2/5^2) = 1$.
   (a) What are the equations of the directrices?
   (b) If P is a point on the ellipse in Quadrant I such that the x-coordinate of P is 3, compute F_1P and F_2P, the lengths of the focal radii.

12. Determine the equation of the line that is tangent to the ellipse $x^2 + 3y^2 = 52$ at the point $(-2, 4)$. Write your answer in the form $y = mx + b$.

13. Find the equation of the line that is tangent to the parabola $x^2 = 2y$ at the point $(4, 8)$. Write your answer in the form $y = mx + b$.

14. The following figure shows the specifications for a cross section of a parabolic reflector. Determine the focal length and the focal ratio of the reflector.

CHAPTER 9

Additional Topics in Algebra

A good background in algebra is an important prerequisite for calculus and for many courses in the fields of economics, business, probability, and statistics. In this final chapter we develop some additional topics that help provide a foundation in those areas of study. Some of the concepts and applications you'll see are as follows.

- We introduce the principle of mathematical induction for proving certain mathematical statements that involve infinitely many cases. (This is sometimes indicated by the phrase "and so on" in the statement of a problem.) An example of such a statement is the following:

  For each natural number n, the sum $1 + 2 + 3 + \cdots + n$ is equal to $n(n + 1)/2$.

  (Exercise 1 in Section 9.1 asks you to prove this statement using mathematical induction.)
- In Section 9.3, the size of a rabbit population is modeled using a recursive sequence. (See Example 5.) The mini project for this section lets you explore how changes in the model affect the size of the population.
- The United States Senate has 100 members. As an application of the work on combinations in Section 9.6, we calculate the number of four-person committees that can be formed in the Senate. (See Example 12 in Section 9.6.)
- As an example of the work on probability in Section 9.7, we find the probability that a given number of transistors in a shipment will all be non-faulty.

Mathematical induction is not a method of discovery but a technique of proving rigorously what has already been discovered.
—David M. Burton in *The History of Mathematics, an Introduction* (Boston: Allyn and Bacon, 1985)

9.1 MATHEMATICAL INDUCTION

Is mathematics an experimental science? The answer to this question is both yes and no, as the following example illustrates. Consider the problem of determining a formula for the sum of the first n odd natural numbers:

$$1 + 3 + 5 + \cdots + (2n - 1)$$

We begin by doing some calculations in the hope that this may shed some light on the problem. Table 1 shows the results of calculating the sum of the first n odd natural numbers for values of n ranging from 1 to 5. Upon inspecting the table, we observe that each sum in the right-hand column is the square of the corresponding entry in the left-hand column. For instance, for $n = 5$, we see that

$$\underbrace{1 + 3 + 5 + 7 + 9}_{\text{five terms}} = 5^2$$

TABLE 1

n	$1 + 3 + 5 + \cdots + (2n - 1)$	
1	1	$= 1$
2	$1 + 3$	$= 4$
3	$1 + 3 + 5$	$= 9$
4	$1 + 3 + 5 + 7$	$= 16$
5	$1 + 3 + 5 + 7 + 9 = 25$	

TABLE 2

n	2^n	$(n + 1)^2$
1	2	4
2	4	9
3	8	16
4	16	25
5	32	36

Now let's try the next case, where $n = 6$, and see if the pattern persists. That is, we want to know whether it is true that

$$\underbrace{1 + 3 + 5 + 7 + 9 + 11}_{\text{six terms}} = 6^2$$

As you can easily check, this last equation is true. Thus, based on the experimental (or empirical) evidence, we are led to the following conjecture:

Conjecture. The sum of the first n odd natural numbers is n^2. That is, $1 + 3 + 5 + \cdots + (2n - 1) = n^2$, for each natural number n.

At this point, the "law" we've discovered is indeed really only a conjecture. After all, we've checked it only for values of n ranging from 1 to 6. It is conceivable at this point (although we may feel it is unlikely) that the conjecture is false for certain values of n. For the conjecture to be useful, we must be able to prove that it holds without exception for *all* natural numbers n. In fact, we will subsequently prove that this conjecture is valid. But before explaining the method of proof to be used, let's look at one more example.

Again, let n denote a natural number. Then consider the following question: Which quantity is the larger, 2^n or $(n + 1)^2$? As before, we begin by doing some calculations. This is the experimental stage of our work. According to Table 2, the quantity $(n + 1)^2$ is larger than 2^n for each value of n up through $n = 5$. Thus, we make the following conjecture:

Conjecture. $(n + 1)^2 > 2^n$ for all natural numbers n.

Again, we note that this is only a conjecture at this point. Indeed, if we try the case where $n = 6$, we find that the pattern does not persist. That is, when $n = 6$, we find that 2^n is 64, while $(n + 1)^2$ is only 49. So in this example, the conjecture is not true in general; we have found a value of n for which it fails.

The preceding examples show that experimentation does have a place in mathematics, but we must be careful with the results. When experimentation leads to a conjecture, proof is required before the conjecture can be viewed as a valid law. For the remainder of this section, we shall discuss one such method of proof—*mathematical induction*.

In order to state the principle of mathematical induction, we first introduce some notation. Suppose that for each natural number n we have a statement P_n to be proved. Consider, for instance, the first conjecture we arrived at:

$$1 + 3 + 5 + \cdots + (2n - 1) = n^2$$

If we denote this statement by P_n, then we have, for example, that

P_1 is the statement that $1 = 1^2$

P_2 is the statement that $1 + 3 = 2^2$

P_3 is the statement that $1 + 3 + 5 = 3^2$

With this notation, we can now state the **principle of mathematical induction.**

The Principle of Mathematical Induction

Suppose that for each natural number n, we have a statement P_n for which the following two conditions hold:

1. P_1 is true.
2. For each natural number k, if P_k is true, then P_{k+1} is true.

Then all of the statements are true; that is, P_n is true for all natural numbers n.

The idea behind mathematical induction is a simple one. Think of each statement P_n as the rung of a ladder to be climbed. Then we make the analogy shown in Table 3.

▌ TABLE 3

Mathematical induction		Ladder analogy	
Hypotheses	1. P_1 is true. 2. If P_k is true, then P_{k+1} is true, for any k.	Hypotheses	1. You can reach the first rung. 2. If you are on the kth rung, you can reach the $(k+1)$st rung, for any k.
Conclusion	3. P_n is true for all n.	Conclusion	3. You can climb the entire ladder.

According to the principle of mathematical induction, we can prove that a statement or formula P_n is true for all n if we carry out the following two steps:

Step 1 Show that P_1 is true.
Step 2 Assume that P_k is true, and on the basis of this assumption, show that P_{k+1} is true.

In Step 2, the assumption that P_k is true is referred to as the **induction hypothesis.** (In computer science, Step 1 is sometimes referred to as the **initialization step.**) Now let's turn to some examples of proof by mathematical induction.

EXAMPLE 1 A basic induction proof

Use mathematical induction to prove that

$$1 + 3 + 5 + \cdots + (2n - 1) = n^2$$

for all natural numbers n.

SOLUTION
Let P_n denote the statement that $1 + 3 + 5 + \cdots + (2n - 1) = n^2$. Then we want to show that P_n is true for all natural numbers n.

Step 1 We must check that P_1 is true. But P_1 is just the statement that $1 = 1^2$, which is true.

Step 2 Assuming that P_k is true, we must show that P_{k+1} is true. Thus, we assume that

$$1 + 3 + 5 + \cdots + (2k - 1) = k^2 \qquad (1)$$

That is the induction hypothesis. We must now show that

$$1 + 3 + 5 + \cdots + (2k - 1) + [2(k + 1) - 1] = (k + 1)^2 \qquad (2)$$

In order to derive equation (2) from equation (1), we add the quantity $[2(k + 1) - 1]$ to both sides of equation (1). (The motivation for this stems from the observation that the left-hand sides of equations (1) and (2) differ only by the quantity $[2(k + 1) - 1]$.) We obtain

$$\begin{aligned}
1 + 3 + 5 + \cdots + (2k - 1) + [2(k + 1) - 1] &= k^2 + [2(k + 1) - 1] \\
&= k^2 + 2k + 1 \\
&= (k + 1)^2
\end{aligned}$$

This last equation is what we wanted to show. Having now carried out Steps 1 and 2, we conclude by the principle of mathematical induction that P_n is true for all natural numbers n.

EXAMPLE 2 **An induction proof requiring more algebra in Step 2**

Use mathematical induction to prove that

$$2^3 + 4^3 + 6^3 + \cdots + (2n)^3 = 2n^2(n + 1)^2$$

for all natural numbers n.

SOLUTION

Let P_n denote the statement that

$$2^3 + 4^3 + 6^3 + \cdots + (2n)^3 = 2n^2(n + 1)^2$$

Then we want to show that P_n is true for all natural numbers n.

Step 1 We must check that P_1 is true, where P_1 is the statement that

$$2^3 = 2(1^2)(1 + 1)^2 \qquad \text{or} \qquad 8 = 8$$

Thus, P_1 is true.

Step 2 Assuming that P_k is true, we must show that P_{k+1} is true. Thus, we assume that

$$2^3 + 4^3 + 6^3 + \cdots + (2k)^3 = 2k^2(k + 1)^2 \qquad (3)$$

We must now show that

$$2^3 + 4^3 + 6^3 + \cdots + (2k)^3 + [2(k + 1)]^3 = 2(k + 1)^2(k + 2)^2 \qquad (4)$$

Adding $[2(k + 1)]^3$ to both sides of equation (3) yields

$$\begin{aligned}
2^3 + 4^3 + 6^3 + \cdots + (2k)^3 + [2(k + 1)]^3 &= 2k^2(k + 1)^2 + [2(k + 1)]^3 \\
&= 2k^2(k + 1)^2 + 8(k + 1)^3 \\
&= 2(k + 1)^2[k^2 + 4(k + 1)] \\
&= 2(k + 1)^2(k^2 + 4k + 4) \\
&= 2(k + 1)^2(k + 2)^2
\end{aligned}$$

We have now derived equation (4) from equation (3), as we wished to do. Having carried out Steps 1 and 2, we conclude by the principle of mathematical induction that P_n is true for all natural numbers n.

EXAMPLE	3	A case where the statement to be proved is not a formula

TABLE 4

n	$2^{2n} - 1$
1	$3 \; (= 3 \cdot 1)$
2	$15 \; (= 3 \cdot 5)$
3	$63 \; (= 3 \cdot 21)$
4	$255 \; (= 3 \cdot 85)$

As indicated in Table 4, the number 3 is a factor of $2^{2n} - 1$ when $n = 1, 2, 3$, and 4. Use mathematical induction to show that 3 is a factor of $2^{2n} - 1$ for all natural numbers n.

SOLUTION

Let P_n denote the statement that 3 is a factor of $2^{2n} - 1$. We want to show that P_n is true for all natural numbers n.

Step 1 We must check that P_1 is true. But P_1 in this case is just the statement that 3 is a factor of $2^{2(1)} - 1$; that is, 3 is a factor of 3, which is surely true.

Step 2 Assuming that P_k is true, we must show that P_{k+1} is true. Thus, we assume that

$$3 \text{ is a factor of } 2^{2k} - 1 \qquad (5)$$

and we must show that

$$3 \text{ is a factor of } 2^{2(k+1)} - 1$$

The strategy here will be to rewrite the expression $2^{2(k+1)} - 1$ in such a way that the induction hypothesis, statement (5), can be applied. We have

$$2^{2(k+1)} - 1 = 2^{2k+2} - 1$$
$$= 2^2 \cdot 2^{2k} - 1$$
$$= 4 \cdot 2^{2k} - 4 + 3$$
$$2^{2(k+1)} - 1 = 4(2^{2k} - 1) + 3 \qquad (6)$$

Now look at the right-hand side of equation (6). By the induction hypothesis, 3 is a factor of $2^{2k} - 1$. Thus, 3 is a factor of $4(2^{2k} - 1)$, from which it certainly follows that 3 is a factor of $4(2^{2k} - 1) + 3$. In summary, then, 3 is a factor of the right-hand side of equation (6). Consequently, 3 must be a factor of the left-hand side of equation (6), which is what we wished to show.

Having now completed Steps 1 and 2, we conclude by the principle of mathematical induction that P_n is true for all natural numbers n. In other words, 3 is a factor of $2^{2n} - 1$ for all natural numbers n.

There are instances in which a given statement P_n is false for certain initial values of n but true thereafter. An example of this is provided by the statement

$$2^n > (n + 1)^2$$

As you can easily check, this statement is false for $n = 1, 2, 3, 4$, and 5. But, as Example 4 shows, the statement is true for $n \geq 6$. In Example 4, we adapt the principle of mathematical induction by beginning in Step 1 with a consideration of P_6 rather than P_1.

| | EXAMPLE 4 | A case where the statement to be proved is an inequality |

Use mathematical induction to prove that

$$2^n > (n + 1)^2 \qquad \text{for all natural numbers } n \geq 6$$

SOLUTION

Step 1 We must first check that P_6 is true. But P_6 is simply the assertion that

$$2^6 > (6 + 1)^2 \qquad \text{or} \qquad 64 > 49$$

Thus, P_6 is true.

Step 2 Assuming that P_k is true, where $k \geq 6$, we must show that P_{k+1} is true. Thus we assume that

$$2^k > (k + 1)^2 \qquad \text{where } k \geq 6 \tag{7}$$

We must show that

$$2^{k+1} > (k + 2)^2$$

Multiplying both sides of inequality (7) by 2 gives us

$$2(2^k) > 2(k + 1)^2 = 2k^2 + 4k + 2$$

This can be rewritten

$$2^{k+1} > k^2 + 4k + (k^2 + 2)$$

However, since $k \geq 6$, it is certainly true that

$$k^2 + 2 > 4$$

We therefore have

$$2^{k+1} > k^2 + 4k + 4 \qquad \text{or} \qquad 2^{k+1} > (k + 2)^2$$

as we wished to show.

Having now completed Steps 1 and 2, we conclude that P_n is true for all natural numbers $n \geq 6$.

EXERCISE SET 9.1

A

In Exercises 1–18, use the principle of mathematical induction to show that the statements are true for all natural numbers.

1. $1 + 2 + 3 + \cdots + n = n(n + 1)/2$
2. $2 + 4 + 6 + \cdots + 2n = n(n + 1)$
3. $1 + 4 + 7 + \cdots + (3n - 2) = n(3n - 1)/2$
4. $5 + 9 + 13 + \cdots + (4n + 1) = n(2n + 3)$
5. $1^2 + 2^2 + 3^2 + \cdots + n^2 = n(n + 1)(2n + 1)/6$
6. $2^2 + 4^2 + 6^2 + \cdots + (2n)^2 = 2n(n + 1)(2n + 1)/3$
7. $1^2 + 3^2 + 5^2 + \cdots + (2n - 1)^2 = n(2n - 1)(2n + 1)/3$
8. $2 + 2^2 + 2^3 + \cdots + 2^n = 2^{n+1} - 2$
9. $3 + 3^2 + 3^3 + \cdots + 3^n = (3^{n+1} - 3)/2$

10. $e^x + e^{2x} + e^{3x} + \cdots + e^{nx} = \dfrac{e^{(n+1)x} - e^x}{e^x - 1}$ $(x \neq 0)$

11. $1^3 + 2^3 + 3^3 + \cdots + n^3 = [n(n+1)/2]^2$

12. $2^3 + 4^3 + 6^3 + \cdots + (2n)^3 = 2n^2(n+1)^2$

13. $1^3 + 3^3 + 5^3 + \cdots + (2n-1)^3 = n^2(2n^2-1)$

14. $1 \cdot 2 + 3 \cdot 4 + 5 \cdot 6 + \cdots + (2n-1)(2n)$
$$= n(n+1)(4n-1)/3$$

15. $1 \cdot 3 + 3 \cdot 5 + 5 \cdot 7 + \cdots + (2n-1)(2n+1)$
$$= n(4n^2 + 6n - 1)/3$$

16. $\dfrac{1}{1 \times 3} + \dfrac{1}{2 \times 4} + \dfrac{1}{3 \times 5} + \cdots + \dfrac{1}{n(n+2)}$
$$= \dfrac{n(3n+5)}{4(n+1)(n+2)}$$

17. $1 + \dfrac{3}{2} + \dfrac{5}{2^2} + \dfrac{7}{2^3} + \cdots + \dfrac{2n-1}{2^{n-1}} = 6 - \dfrac{2n+3}{2^{n-1}}$

18. $1 + 2 \cdot 2 + 3 \cdot 2^2 + 4 \cdot 2^3 + \cdots + n \cdot 2^{n-1} = (n-1)2^n + 1$

19. Show that $n \leq 2^{n-1}$ for all natural numbers n.

20. Show that 3 is a factor of $n^3 + 2n$ for all natural numbers n.

21. Show that $n^2 + 4 < (n+1)^2$ for all natural numbers $n \geq 2$.

22. Show that $n^3 > (n+1)^2$ for all natural numbers $n \geq 3$.

In Exercises 23–26, prove that the statement is true for all natural numbers in the specified range. Use a calculator to carry out Step 1.

23. $(1.5)^n > 2n$; $n \geq 7$
24. $(1.25)^n > n$; $n \geq 11$
25. $(1.1)^n > n$; $n \geq 39$
26. $(1.1)^n > 5n$; $n \geq 60$

B

27. Let $f(n) = \dfrac{1}{1 \cdot 2} + \dfrac{1}{2 \cdot 3} + \dfrac{1}{3 \cdot 4} + \cdots + \dfrac{1}{n(n+1)}$.

(a) Complete the following table.

n	1	2	3	4	5
$f(n)$					

(b) On the basis of the results in the table, what would you guess to be the value of $f(6)$? Compute $f(6)$ to see if this is correct.

(c) Make a conjecture about the value of $f(n)$, and prove it using mathematical induction.

28. Let $f(n) = \dfrac{1}{1 \times 3} + \dfrac{1}{3 \times 5} + \dfrac{1}{5 \times 7} + \cdots$
$$+ \dfrac{1}{(2n-1)(2n+1)}.$$

(a) Complete the following table.

n	1	2	3	4
$f(n)$				

(b) On the basis of the results in the table, what would you guess to be the value of $f(5)$? Compute $f(5)$ to see if your guess is correct.

(c) Make a conjecture about the value of $f(n)$, and prove it using mathematical induction.

29. Suppose that a function f satisfies the following conditions:

$$f(1) = 1$$
$$f(n) = f(n-1) + 2\sqrt{f(n-1)} + 1 \quad (n \geq 2)$$

(a) Complete the table.

n	1	2	3	4	5
$f(n)$					

(b) On the basis of the results in the table, what would you guess to be the value of $f(6)$? Compute $f(6)$ to see if your guess is correct.

(c) Make a conjecture about the value of $f(n)$ when n is a natural number, and prove the conjecture using mathematical induction.

30. This exercise demonstrates the necessity of carrying out both Step 1 and Step 2 before considering an induction proof valid.

(a) Let P_n denote the statement that $n^2 + 1$ is even. Check that P_1 is true. Then give an example showing that P_n is not true for all n.

(b) Let Q_n denote the statement that $n^2 + n$ is odd. Show that Step 2 of an induction proof can be completed in this case, but not Step 1.

31. A *prime number* is a natural number that has no factors other than itself and 1. For technical reasons, 1 is not considered a prime. Thus, the list of the first seven primes looks like this: 2, 3, 5, 7, 11, 13, 17. Let P_n be the statement that $n^2 + n + 11$ is prime. Check that P_n is true for all values of n less than 10. Check that P_{10} is false.

32. Prove that if $x \neq 1$,

$$1 + 2x + 3x^2 + \cdots + nx^{n-1} = \dfrac{1 - x^n}{(1-x)^2} - \dfrac{nx^n}{1-x}$$

for all natural numbers n.

33. If $r \neq 1$, show that $1 + r + r^2 + \cdots + r^{n-1} = \dfrac{r^n - 1}{r - 1}$

for all natural numbers n.

34. Use mathematical induction to show that

$$x^n - 1 = (x - 1)(1 + x + x^2 + \cdots + x^{n-1})$$

for all natural numbers n.

35. Prove that 5 is a factor of $n^5 - n$ for all natural numbers $n \geq 2$.

36. Prove that 4 is a factor of $5^n + 3$ for all natural numbers n.

37. Prove that 5 is a factor of $2^{2n+1} + 3^{2n+1}$ for all nonnegative integers n.

38. Prove that 8 is a factor of $3^{2n} - 1$ for all natural numbers n.

39. Prove that 3 is a factor of $2^{n+1} + (-1)^n$ for all nonnegative integers n.

40. Prove that 6 is a factor of $n^3 + 3n^2 + 2n$ for all natural numbers n.

41. Use mathematical induction to show that $x - y$ is a factor of $x^n - y^n$ for all natural numbers n.   *Suggestion for Step 2:* Verify and then use the fact that

$$x^{k+1} - y^{k+1} = x^k(x - y) + (x^k - y^k)y.$$

In Exercises 42 and 43, use mathematical induction to prove that the formulas hold for all natural numbers n.

42. $\log_{10}(a_1 a_2 \ldots a_n) = \log_{10} a_1 + \log_{10} a_2 + \cdots + \log_{10} a_n$

43. $(1 + p)^n \geq 1 + np$, where $p > -1$

9.2   THE BINOMIAL THEOREM

A mathematician, like a painter or a poet, is a maker of patterns. —G. E. Hardy (1877–1947)

[Blaise Pascal] *made numerous discoveries relating to this array and set them forth in his* Traité du triangle arithmétique, *published posthumously in 1665, and among these was essentially our present Binomial Theorem for positive integral exponents.* —David Eugene Smith in *History of Mathematics,* Vol. II (New York: Ginn and Co., 1925)

Recall from basic algebra that an expression that is the sum or difference of two terms is referred to as a *binomial expression.* Three examples of binomial expressions are

$$u + v, \qquad 2x^2 - y^3, \qquad \text{and} \qquad 4m^3n + 5x^3y^4$$

Our goal in this section is to develop a general formula, known as the *binomial theorem,* for expanding any product of the form $(a + b)^n$, where n is a natural number.

We begin by looking for patterns in the expansion of $(a + b)^n$. To do this, let's list the expansions of $(a + b)^n$ for $n = 1, 2, 3, 4,$ and 5. (Exercises 1 and 2 ask you to verify these results simply by repeated multiplication.)

$$(a + b)^1 = a + b$$
$$(a + b)^2 = a^2 + 2ab + b^2$$
$$(a + b)^3 = a^3 + 3a^2b + 3ab^2 + b^3$$
$$(a + b)^4 = a^4 + 4a^3b + 6a^2b^2 + 4ab^3 + b^4$$
$$(a + b)^5 = a^5 + 5a^4b + 10a^3b^2 + 10a^2b^3 + 5ab^4 + b^5$$

After surveying these results, we note the following patterns.

PROPERTY SUMMARY   **Patterns Observed in $(a + b)^n$ for $n = 1, 2, 3, 4, 5$**

General Statement	Example
There are $n + 1$ terms.	There are 4 ($= 3 + 1$) terms in the expansion of $(a + b)^3$.
The expansion begins with a^n and ends with b^n.	$(a + b)^3$ begins with a^3 and ends with b^3.
The sum of the exponents in each term is n.	The sum of the exponents in each term of $(a + b)^3$ is 3.
The exponents of a decrease by 1 from term to term.	$(a + b)^3 = a^{③} + 3a^{②}b + 3a^{①}b^2 + a^{⓪}b^3$
The exponents of b increase by 1 from term to term.	$(a + b)^3 = a^3b^{⓪} + 3a^2b^{①} + 3ab^{②} + b^{③}$
When n is even, the coefficients are symmetric about the middle term.	The sequence of coefficients for $(a + b)^4$ is 1, 4, 6, 4, 1.
When n is odd, the coefficients are symmetric about the two middle terms.	The sequence of coefficients for $(a + b)^5$ is 1, 5, 10, 10, 5, 1.

Figure 1
"Pascal's" triangle, by Chu-Shi-Kie,
A.D. 1303.

The patterns we have just observed for $(a + b)^n$ persist for all natural numbers n. (This will follow from the binomial theorem, which is proved at the end of this section.) Thus, for example, the form of $(a + b)^6$ must be as follows. (Each question mark denotes a coefficient to be determined.)

$$(a + b)^6 = a^6 + ?a^5b + ?a^4b^2 + ?a^3b^3 + ?a^2b^4 + ?ab^5 + b^6$$

The problem now is to find the proper coefficient for each term. To do this, we need to discover additional patterns in the expansion of $(a + b)^n$.

We have already written out the expansions of $(a + b)^n$ for values of n ranging from 1 to 5. Let us now write only the coefficients appearing in those expansions. The resulting triangular array of numbers is known as **Pascal's triangle.**[*] For reasons of symmetry, we begin with $(a + b)^0$ rather than $(a + b)^1$.

$$(a + b)^0 \quad \ldots\ldots\ldots\ldots\ldots\ldots\ldots\ldots\ldots\ldots\ldots \quad 1$$
$$(a + b)^1 \quad \ldots\ldots\ldots\ldots\ldots\ldots\ldots\ldots\ldots \quad 1 \quad 1$$
$$(a + b)^2 \quad \ldots\ldots\ldots\ldots\ldots\ldots\ldots \quad 1 \quad 2 \quad 1$$
$$(a + b)^3 \quad \ldots\ldots\ldots\ldots\ldots\ldots \quad 1 \quad 3 \quad 3 \quad 1$$
$$(a + b)^4 \quad \ldots\ldots\ldots\ldots \quad 1 \quad 4 \quad 6 \quad 4 \quad 1$$
$$(a + b)^5 \quad \ldots\ldots\ldots \quad 1 \quad 5 \quad 10 \quad 10 \quad 5 \quad 1$$

The key observation regarding Pascal's triangle is this: *Each entry in the array* (other than the 1's along the sides) *is the sum of the two numbers diagonally above it.* For instance, the 6 that appears in the fifth row is the sum of the two 3's diagonally above it. Using this observation, we can form as many additional rows as we please. The coefficients for $(a + b)^n$ will then appear in the $(n + 1)$st row of the array.[†] For instance, to obtain the row corresponding to $(a + b)^6$, we have

sixth row, $(a + b)^5$:   1   5   10   10   5   1

seventh row, $(a + b)^6$:   1   6   15   20   15   6   1

Thus, the sequence of coefficients for $(a + b)^6$ is 1, 6, 15, 20, 15, 6, 1. This answers the question raised earlier about the expansion of $(a + b)^6$. We have

$$(a + b)^6 = a^6 + 6a^5b + 15a^4b^2 + 20a^3b^3 + 15a^2b^4 + 6ab^5 + b^6$$

For analytical work or for larger values of the exponent n, it is inefficient to rely on Pascal's triangle. For this reason, we point out another pattern in the expansions of $(a + b)^n$.

In the expansion of $(a + b)^n$, the coefficient of any term after the first can be generated as follows: From the *previous* term, multiply the coefficient by the exponent of a and then divide by the number of that previous term.

---

[*]The array is named after Blaise Pascal, a seventeenth-century French mathematician and philosopher. However, as Figure 1 indicates, the Pascal triangle was known to Chinese mathematicians centuries earlier.
[†]That these numbers actually are the appropriate coefficients follows from the binomial theorem, which is proved at the end of this section.

To see how this method is used, let's compute the second, third, and fourth coefficients in the expansion of $(a + b)^6$. To compute the coefficient of the second term, we go back to the first term, which is a^6. We have

coefficient of second term $= \dfrac{1 \cdot 6}{1} = 6$

coefficient of first term
exponent of a in first term
number of first term

Thus the second term is $6a^5b$ and, consequently, we have

coefficient of third term $= \dfrac{6 \cdot 5}{2} = 15$

coefficient of second term
exponent of a in second term
number of second term

Continuing now with this method, you should check for yourself that the coefficient of the fourth term in the expansion of $(a + b)^6$ is 20.

NOTE We now know that the first four coefficients are 1, 6, 15, and 20. By symmetry, it follows that the complete sequence of coefficients for this expansion is 1, 6, 15, 20, 15, 6, 1. No additional calculation of coefficients is necessary.

EXAMPLE 1 Expanding a binomial

Expand $(2x - y^2)^7$.

SOLUTION
First we write the expression of $(a + b)^7$ using the method explained just prior to this example, or using Pascal's triangle. As you should check for yourself, the expansion is

$$(a + b)^7 = a^7 + 7a^6b + 21a^5b^2 + 35a^4b^3 + 35a^3b^4 + 21a^2b^5 + 7ab^6 + b^7$$

Now we make the substitutions $a = 2x$ and $b = -y^2$. This yields

$$\begin{aligned}
[2x + (-y^2)]^7 &= (2x)^7 + 7(2x)^6(-y^2) + 21(2x)^5(-y^2)^2 \\
&\quad + 35(2x)^4(-y^2)^3 + 35(2x)^3(-y^2)^4 \\
&\quad + 21(2x)^2(-y^2)^5 + 7(2x)(-y^2)^6 + (-y^2)^7 \\
&= 128x^7 - 448x^6y^2 + 672x^5y^4 - 560x^4y^6 + 280x^3y^8 \\
&\quad - 84x^2y^{10} + 14xy^{12} - y^{14}
\end{aligned}$$

This is the required expansion. Notice how the signs alternate in the final answer; this is characteristic of all expressions of the form $(a - b)^n$.

In preparation for the binomial theorem, we introduce two notations that are used not only in connection with the binomial theorem, but also in many other areas of mathematics. The first of these notations is $n!$ (read "n factorial").

DEFINITION	**The Factorial Symbol**

EXAMPLES

$n! = 1 \cdot 2 \cdot 3 \cdots n$ where n is a natural number

$$3! = 1 \cdot 2 \cdot 3 = 6$$

$$\frac{6!}{4!} = \frac{6 \cdot 5 \cdot 4 \cdot 3 \cdot 2 \cdot 1}{4 \cdot 3 \cdot 2 \cdot 1}$$

$0! = 1$

$$= 6 \cdot 5 = 30$$

EXAMPLE 2   Practice with the factorial notation

Simplify the expression $\dfrac{(n + 1)!}{(n - 1)!}$.

SOLUTION

$$\frac{(n + 1)!}{(n - 1)!} = \frac{(n + 1) \cdot n \cdot (n - 1)!}{(n - 1)!} = (n + 1) \cdot n = n^2 + n$$

The second notation that we introduce in preparation for the binomial theorem is $\binom{n}{k}$. This notation is read "n choose k," because it can be shown that $\binom{n}{k}$ is equal to the number of ways of choosing a subset of k elements from a set with n elements.

DEFINITION	**The Binomial Coefficient** $\binom{n}{k}$

EXAMPLE

Let n and k be nonnegative integers, with $k \leq n$.

$$\binom{5}{2} = \frac{5!}{2!(5 - 2)!} = \frac{5!}{2!3!}$$

Then the binomial coefficient $\binom{n}{k}$ is defined by

$$= \frac{5 \cdot 4 \cdot 3 \cdot 2 \cdot 1}{(2 \cdot 1)(3 \cdot 2 \cdot 1)}$$

$$\binom{n}{k} = \frac{n!}{k!(n - k)!}$$

$$= \frac{5 \cdot 4}{2 \cdot 1} = 10$$

The binomial coefficients are so named because they are indeed the coefficients in the expansion of $(a + b)^n$. More precisely, the relationship is this: The coefficients in the expansion of $(a + b)^n$ are the $n + 1$ numbers

$$\binom{n}{0}, \binom{n}{1}, \binom{n}{2}, \ldots, \binom{n}{n}$$

Subsequently, we will see why this statement is true. For now, however, let's look at an example. Consider the binomial coefficients $\binom{3}{0}, \binom{3}{1}, \binom{3}{2},$ and $\binom{3}{3}$.

According to our statement, these four quantities should be the coefficients in the expansion of $(a + b)^3$. Let's check:

$$\binom{3}{0} = \frac{3!}{0!(3 - 0)!} = \frac{3!}{1(3!)} = 1$$

$$\binom{3}{1} = \frac{3!}{1!(3 - 1)!} = \frac{3 \cdot 2 \cdot 1}{1(2 \cdot 1)} = 3$$

$$\binom{3}{2} = \frac{3!}{2!(3 - 2)!} = \frac{3 \cdot 2 \cdot 1}{(2 \cdot 1)1} = 3$$

$$\binom{3}{3} = \frac{3!}{3!(3 - 3)!} = \frac{3!}{3!0!} = 1$$

The values of $\binom{3}{0}$, $\binom{3}{1}$, $\binom{3}{2}$, and $\binom{3}{3}$ are thus 1, 3, 3, and 1, respectively. But these last four numbers are indeed the coefficients in the expression of $(a + b)^3$, as we wished to check.

We are now in a position to state the binomial theorem, after which we will look at several applications. Finally, at the end of this section, we will use mathematical induction to prove the theorem. In the statement of the theorem that follows, we are assuming that the exponent n is a natural number.

The Binomial Theorem

$$(a + b)^n = \binom{n}{0}a^n + \binom{n}{1}a^{n-1}b + \binom{n}{2}a^{n-2}b^2 + \cdots + \binom{n}{n - 1}ab^{n-1} + \binom{n}{n}b^n$$

One of the uses of the binomial theorem is in identifying specific terms in an expansion without computing the entire expansion. This is particularly helpful when the exponent n is relatively large. Looking back at the statement of the binomial theorem, there are three observations we can make. First, the coefficient of the rth term is $\binom{n}{r - 1}$. For instance, the coefficient of the third term is $\binom{n}{3 - 1} = \binom{n}{2}$. The second observation is that the exponent for a in the rth term is $n - (r - 1)$. For instance, the exponent for a in the third term is $n - (3 - 1) = n - 2$. Finally, we observe that the exponent for b in the rth term is $r - 1$, the same quantity that appears in the lower position of the corresponding binomial coefficient. For instance, the exponent for b in the third term is $r - 1 = 3 - 1 = 2$. We summarize these three observations with the following statement.

The rth term in the expansion of $(a + b)^n$ is

$$\binom{n}{r - 1}a^{n-r+1}b^{r-1}$$

EXAMPLE   3   Finding a term in an expansion

Find the 15th term in the expansion of $\left(x^2 - \dfrac{1}{x} \right)^{18}$.

SOLUTION

Using the values $r = 15$, $n = 18$, $a = x^2$, and $b = -1/x$, we have

$$\binom{n}{r-1} a^{n-r+1} b^{r-1} = \binom{18}{15-1} (x^2)^{18-15+1} \left(\frac{-1}{x} \right)^{15-1}$$

$$= \binom{18}{14} x^8 \cdot \frac{1}{x^{14}}$$

$$= \frac{18 \cdot 17 \cdot 16 \cdot 15 \cdot (14!)}{14!(4 \cdot 3 \cdot 2 \cdot 1)} x^{-6}$$

$$= \frac{18 \cdot 17 \cdot 16 \cdot 15}{4 \cdot 3 \cdot 2 \cdot 1} x^{-6}$$

$$= 3060 x^{-6} \qquad \text{as required}$$

EXAMPLE   4   Finding the coefficient of a term in an expansion

Find the coefficient of the term containing x^4 in the expansion of $(x + y^2)^{30}$.

SOLUTION

Again we use the fact that the rth term in the expansion of $(a + b)^n$ is $\binom{n}{r-1} a^{n-r+1} b^{r-1}$. In this case, n is 30 and x plays the role of a. The exponent for x is then $n - r + 1$ or $30 - r + 1$. To see when this exponent is 4, we write

$$30 - r + 1 = 4 \qquad \text{and therefore} \qquad r = 27$$

The required coefficient is therefore $\binom{30}{27-1}$. We then have

$$\binom{30}{26} = \frac{30!}{26!(30-26)!} = \frac{30 \cdot 29 \cdot 28 \cdot 27}{4 \cdot 3 \cdot 2 \cdot 1}$$

After carrying out the indicated arithmetic, we find that $\binom{30}{26} = 27{,}405$. This is the required coefficient.

EXAMPLE   5   Finding the coefficient of a term in an expansion

Find the coefficient of the term containing a^9 in the expansion of $(a + 2\sqrt{a})^{10}$.

SOLUTION

The rth term in this expansion is

$$\binom{10}{r-1}a^{10-r+1}(2\sqrt{a})^{r-1}$$

We can rewrite this as

$$\binom{10}{r-1}a^{10-r+1}(2^{r-1})(a^{1/2})^{r-1}$$

or

$$2^{r-1}\binom{10}{r-1}a^{10-r+1+(r-1)/2}$$

From this we see that the general form of the coefficient we wish to find is $2^{r-1}\binom{10}{r-1}$. We now need to determine r when the exponent of a is 9. Thus, we require that

$$10 - r + 1 + \frac{r-1}{2} = 9$$

$$-r + \frac{r-1}{2} = -2$$

$$-2r + r - 1 = -4 \qquad \text{\textcolor{red}{multiplying both sides by 2}}$$

$$r = 3$$

The required coefficient is now obtained by substituting $r = 3$ in the expression $2^{r-1}\binom{10}{r-1}$. Thus, the required coefficient is

$$2^2\binom{10}{2} = \frac{4\cdot 10!}{2!(10-2)!} = 2\cdot 10\cdot 9 = 180$$

There are three simple identities involving the binomial coefficients that will simplify our proof of the binomial theorem.

IDENTITY 1 $\quad \binom{r}{0} = 1 \quad$ for all nonnegative integers r

IDENTITY 2 $\quad \binom{r}{r} = 1 \quad$ for all nonnegative integers r

IDENTITY 3 $\quad \binom{k}{r} + \binom{k}{r-1} = \binom{k+1}{r} \quad$ for all natural numbers k and r with $r \le k$

All three of these identities can be proved directly from the definitions of the binomial coefficients, without the need for mathematical induction. The proofs of the first two are straightforward, and we omit them here. The proof of the third identity runs as follows:

$$\binom{k}{r} + \binom{k}{r-1} = \frac{k!}{r!(k-r)!} + \frac{k!}{(r-1)!(k-r+1)!}$$

$$= \frac{k!}{r(r-1)!(k-r)!} + \frac{k!}{(r-1)!(k-r+1)(k-r)!}$$

The common denominator on the right-hand side of the last equation is $r(r-1)!(k-r+1)(k-r)!$ Thus, we have

$$\binom{k}{r} + \binom{k}{r-1} = \frac{k!(k-r+1)}{r(r-1)!(k-r+1)(k-r)!} + \frac{k!r}{r(r-1)!(k-r+1)(k-r)!}$$

$$= \frac{k!(k-r+1) + k!r}{r(r-1)!(k-r+1)(k-r)!}$$

$$= \frac{k!(k-r+1+r)}{r(r-1)!(k-r+1)(k-r)!} = \frac{k!(k+1)}{r!(k-r+1)!}$$

$$= \frac{(k+1)!}{r![(k+1)-r]!}$$

$$= \binom{k+1}{r} \qquad \text{as required}$$

Taken together, the three identities show why the $(n+1)$st row of Pascal's triangle consists of the numbers $\binom{n}{0}, \binom{n}{1}, \binom{n}{2}, \ldots, \binom{n}{n}$. Identities 1 and 2 tell us that this row of numbers begins and ends with 1. Identity 3 is just a statement of the fact that each entry in the row, other than the initial and final 1, is generated by adding the two entries diagonally above it.

We conclude this section by using mathematical induction to prove the binomial theorem. The statement P_n that we wish to prove for all natural numbers n is this:

$$(a+b)^n = \binom{n}{0}a^n + \binom{n}{1}a^{n-1}b + \cdots + \binom{n}{n-1}ab^{n-1} + \binom{n}{n}b^n$$

First, we check that P_1 is true. The statement P_1 asserts that

$$(a+b)^1 = \binom{1}{0}a^1 + \binom{1}{1}b$$

However, in view of Identities 1 and 2, this last equation becomes

$$(a+b)^1 = 1 \cdot a + 1 \cdot b$$

which is surely true. Now let's assume that P_k is true and, on the basis of this assumption, show that P_{k+1} is true. The statement P_k is

$$(a+b)^k = \binom{k}{0}a^k + \binom{k}{1}a^{k-1}b + \cdots + \binom{k}{k-1}ab^{k-1} + \binom{k}{k}b^k$$

Multiplying both sides of this equation by the quantity $(a+b)$ yields

$$(a + b)^{k+1} = (a + b)\left[\binom{k}{0}a^k + \binom{k}{1}a^{k-1}b + \cdots + \binom{k}{k-1}ab^{k-1} + \binom{k}{k}b^k\right]$$

$$= a\left[\binom{k}{0}a^k + \binom{k}{1}a^{k-1}b + \cdots + \binom{k}{k-1}ab^{k-1} + \binom{k}{k}b^k\right]$$

$$+ b\left[\binom{k}{0}a^k + \binom{k}{1}a^{k-1}b + \cdots + \binom{k}{k-1}ab^{k-1} + \binom{k}{k}b^k\right]$$

$$= \binom{k}{0}a^{k+1} + \binom{k}{1}a^k b + \cdots + \binom{k}{k-1}a^2 b^{k-1} + \binom{k}{k}ab^k$$

$$+ \binom{k}{0}a^k b + \binom{k}{1}a^{k-1}b^2 + \cdots + \binom{k}{k-1}ab^k + \binom{k}{k}b^{k+1}$$

$$= \binom{k}{0}a^{k+1} + \left[\binom{k}{1} + \binom{k}{0}\right]a^k b + \cdots + \left[\binom{k}{k} + \binom{k}{k-1}\right]ab^k + \binom{k}{k}b^{k+1}$$

We can now make some substitutions on the right-hand side of this last equation. The initial binomial coefficient $\binom{k}{0}$ can be replaced by $\binom{k+1}{0}$, because both are equal to 1, according to Identity 1. Similarly, the binomial coefficient $\binom{k}{k}$ appearing at the end of the equation can be replaced by $\binom{k+1}{k+1}$, since both are equal to 1 according to Identity 2. Finally, we can use Identity 3 to simplify each of the sums in the brackets. We obtain

$$(a + b)^{k+1} = \binom{k+1}{0}a^{k+1} + \binom{k+1}{1}a^k b + \cdots + \binom{k+1}{k}ab^k + \binom{k+1}{k+1}b^{k+1}$$

This last equation is just the statement P_{k+1}; that is, we have derived P_{k+1} from P_k, as we wished to do. The induction proof is now complete.

EXERCISE SET 9.2

A

In Exercises 1 and 2, verify each statement directly, without using the techniques developed in this section.

1. (a) $(a + b)^2 = a^2 + 2ab + b^2$

    (b) $(a + b)^3 = a^3 + 3a^2 b + 3ab^2 + b^3$

    Hint: $(a + b)^3 = (a + b)(a + b)^2$

2. (a) $(a + b)^4 = a^4 + 4a^3 b + 6a^2 b^2 + 4ab^3 + b^4$

    Hint: Use the result in Exercise 1(b) and the fact that $(a + b)^4 = (a + b)(a + b)^3$.

    (b) $(a + b)^5 = a^5 + 5a^4 b + 10a^3 b^2 + 10a^2 b^3 + 5ab^4 + b^5$

In Exercises 3–28, carry out the indicated expansions.

3. $(a + b)^9$
4. $(a - b)^9$
5. $(2A + B)^3$
6. $(1 + 2x)^6$
7. $(1 - 2x)^6$
8. $(3x^2 - y)^5$
9. $(\sqrt{x} + \sqrt{y})^4$
10. $(\sqrt{x} - \sqrt{y})^4$
11. $(x^2 + y^2)^5$
12. $(5A - B^2)^3$
13. $[1 - (1/x)]^6$
14. $(3x + y^2)^4$
15. $(x/2 - y/3)^3$
16. $(1 - z^2)^7$
17. $(ab^2 + c)^7$

18. $[x - (1/x)]^8$
19. $(x + \sqrt{2})^8$
20. $(4A - \frac{1}{2})^5$
21. $(\sqrt{2} - 1)^3$
22. $(1 + \sqrt{5})^4$
23. $(\sqrt{2} + \sqrt{3})^5$
24. $(\frac{1}{2} - 2a)^6$
25. $(2\sqrt[3]{2} - \sqrt[3]{4})^3$
26. $(x + y + 1)^4$

    Suggestion: Rewrite the expression as $[(x + y) + 1]^4$.

27. $(x^2 - 2x - 1)^5$

    Suggestion: Rewrite the expression as $[x^2 - (2x + 1)]^5$.

28. $[x^2 - 2x - (1/x)]^6$

In Exercises 29–38, evaluate or simplify each expression.

29. $5!$

30. (a) $3! + 2!$

    (b) $(3 + 2)!$

31. $\binom{7}{3}\binom{3}{2}$

32. $\dfrac{20!}{18!}$

33. (a) $\binom{5}{3}$

    (b) $\binom{5}{4}$

34. (a) $\binom{7}{7}$

    (b) $\binom{7}{0}$

35. $\dfrac{(n+2)!}{n!}$

36. $\dfrac{n[(n-2)!]}{(n+1)!}$

37. $\dbinom{6}{4} + \dbinom{6}{3} - \dbinom{7}{4}$

38. $(3!)! + (3!)^2$

39. Find the 15th term in the expansion of $(a+b)^{16}$.

40. Find the third term in the expansion of $(a-b)^{30}$.

41. Find the 100th term in the expansion of $(1+x)^{100}$.

42. Find the 23rd term in the expansion of $[x - (1/x^2)]^{25}$.

43. Find the coefficient of the term containing a^4 in the expansion of $(\sqrt{a} - \sqrt{x})^{10}$.

44. Find the coefficient of the term containing a^4 in the expansion of $(3a - 5x)^{12}$.

45. Find the coefficient of the term containing y^8 in the expansion of $[(x/2) - 4y]^9$.

46. Find the coefficient of the term containing x^6 in the expansion of $[x^2 - (1/x)]^{12}$.

47. Find the coefficient of the term containing x^3 in the expansion of $(1 - \sqrt{x})^8$.

48. Find the coefficient of the term containing a^8 in the expansion of $[a - (2/\sqrt{a})]^{14}$.

49. Find the term that does not contain A in the expansion of $[(1/A) + 3A^2]^{12}$.

50. Find the coefficient of B^{-10} in the expansion of $[(B^2/2) - (3/B^3)]^{10}$.

B

51. Show that the coefficient of x^n in the expansion of $(1+x)^{2n}$ is $(2n)!/(n!)^2$.

52. Find n so that the coefficients of the 11th and 13th terms in $(1+x)^n$ are the same.

53. **(a)** Complete the following table.

k	0	1	2	3	4	5	6	7	8
$\dbinom{8}{k}$									

(b) Use the results in part (a) to verify that

$$\binom{8}{0} + \binom{8}{1} + \binom{8}{2} + \cdots + \binom{8}{8} = 2^8.$$

(c) By taking $a = b = 1$ in the expansion of $(a+b)^n$, show that $\dbinom{n}{0} + \dbinom{n}{1} + \dbinom{n}{2} + \cdots + \dbinom{n}{n} = 2^n.$

C

54. Two real numbers A and B are defined by $A = \sqrt[99]{99!}$ and $B = \sqrt[100]{100!}$. Which number is larger, A or B?

Hint: Compare A^{9900} and B^{9900}.

55. This exercise outlines a proof of the identity

$$\binom{n}{0}^2 + \binom{n}{1}^2 + \binom{n}{2}^2 + \cdots + \binom{n}{n}^2 = \binom{2n}{n}.$$

(a) Verify that

$$(1+x)^n\left(1 + \frac{1}{x}\right)^n = \frac{(1+x)^{2n}}{x^n} \qquad (1)$$

(This requires only basic algebra, not the binomial theorem.)

(b) Show that the coefficient of the term independent of x on the right side of equation (1) is $\dbinom{2n}{n}$.

(c) Use the binomial theorem to expand $(1+x)^n$. Then show that the coefficient of the term independent of x on the left side of equation (1) is

$$\binom{n}{0}^2 + \binom{n}{1}^2 + \cdots + \binom{n}{n}^2.$$

9.3 **INTRODUCTION TO SEQUENCES AND SERIES**

Students of calculus do not always understand that infinite series are primarily tools for the study of functions. —George F. Simmons in *Calculus with Analytic Geometry* (New York: McGraw-Hill, 1985)

This section and the next two sections in this chapter deal with numerical sequences. We will begin with a somewhat informal definition of this concept. Then, after looking at some examples and terminology, we'll point out how the function concept is involved in defining a sequence. A **numerical sequence** is an ordered list of numbers. Here are four examples:

Example A: $1, \sqrt{2}, 10$

Example B: $2, 4, 6, 8, \ldots$

Example C: $1, \frac{1}{2}, \frac{1}{4}, \frac{1}{8}, \ldots$

Example D: $1, 1, 1, 1, \ldots$

The individual entries in a numerical sequence are called the **terms** of the sequence. In this chapter, the terms in each sequence will always be real numbers so, for convenience, we will drop the adjective *numerical* and refer simply to *sequences*. (It is worth pointing out, however, that in more advanced courses, sequences are studied in which the individual terms are functions.) Any sequence possessing only a finite number of terms is called a **finite sequence.** Thus the sequence in Example A is a finite sequence. On the other hand, Examples B, C, and D are examples of what we call **infinite sequences;** each contains infinitely many terms. As Example D indicates, it is not necessary that all the terms in a sequence be distinct. In this chapter, all the sequences we discuss will be infinite sequences.

In Examples B, C, and D, the three dots are read "and so on." In using this notation, we are assuming that it is clear what the subsequent terms of the sequence are. Toward this end, we often specify a formula for the nth term in a sequence. Example B in this case would appear this way:

$$2, 4, 6, 8, \ldots 2n, \ldots$$

A letter with subscripts is often used to denote the various terms in a sequence. For instance, if we denote the sequence in Example B by $a_1, a_2, a_3, \ldots$, then we have $a_1 = 2$, $a_2 = 4$, $a_3 = 6$, and, in general, $a_n = 2n$. Of course, there is nothing special about the letter a in this context; any other letter would do just as well.

EXAMPLE 1 Computing terms in a sequence

Consider the sequence $a_1, a_2, a_3, \ldots$, in which the nth term a_n is given by

$$a_n = \frac{n}{n + 1}$$

Compute the first three terms of the sequence, as well as the 1000th term.

SOLUTION
To obtain the first term, we replace n by 1 in the given formula. This yields

$$a_1 = \frac{1}{1 + 1} = \frac{1}{2}$$

The other terms are similarly obtained. We have

$$a_2 = \frac{2}{2 + 1} = \frac{2}{3}$$

$$a_3 = \frac{3}{3 + 1} = \frac{3}{4}$$

$$a_{1000} = \frac{1000}{1000 + 1} = \frac{1000}{1001}$$

A word about notation: In the example we just concluded, we denoted the sequence by $a_1, a_2, a_3, \ldots$. In this case the subscripts begin with 1 and run through the natural numbers. The next example shows a case where the subscripts start with 0 rather than 1. (There is no new concept here; as we've said, it's just a matter of notation.)

EXAMPLE 2 **Computing a sum of terms in a sequence**

Consider the sequence $b_0, b_1, b_2, \ldots$, in which the general term is given by

$$b_n = (-10)^n \quad \text{for } n \geq 0$$

Compute the sum of the first three terms of this sequence.

SOLUTION

We are asked to compute the sum $b_0 + b_1 + b_2$. Using the given formula for b_n, we have

$$b_0 = (-10)^0 = 1$$
$$b_1 = (-10)^1 = -10$$
$$b_2 = (-10)^2 = 100$$

Therefore, $b_0 + b_1 + b_2 = 1 + (-10) + 100 = 91$.

In Examples 1 and 2, we were given a formula for the general term in each sequence. The next example shows a different way of specifying a sequence: each term after the first (or after the first several) is defined in terms of what has come before. This is an example of a **recursive definition.** Recursive definitions are particularly useful in computer programming.

EXAMPLE 3 **Computing terms in a sequence that is defined recursively**

Compute the first three terms of the sequence $b_1, b_2, b_3, \ldots$, which is defined recursively by

$$b_1 = 4$$
$$b_n = 2(b_{n-1} - 1) \quad \text{for } n \geq 2$$

SOLUTION

We are given the first term: $b_1 = 4$. To find b_2, we replace n by 2 in the formula $b_n = 2(b_{n-1} - 1)$ to obtain

$$b_2 = 2(b_1 - 1) = 2(4 - 1) = 6$$

Thus, $b_2 = 6$. Next we use this value of b_2 in the formula $b_n = 2(b_{n-1} - 1)$ to obtain b_3. Replacing n by 3 in this formula yields

$$b_3 = 2(b_2 - 1) = 2(6 - 1) = 10$$

We have now found the first three terms of the sequence: $b_1 = 4$, $b_2 = 6$, and $b_3 = 10$.

If you think about the central idea behind the concept of a sequence, you can see that a function is involved: For each input n, we have an output a_n. As indicated in the previous examples, the inputs n may be the natural numbers $1, 2, 3, \ldots$, or they may be the nonnegative integers $0, 1, 2, 3, \ldots$. For these reasons, the formal definition of a sequence is stated as follows.

DEFINITION | **Sequence**

A sequence is a function whose domain is either the set of natural numbers or the set of nonnegative integers.

If we denote the function by f for the moment, then $f(1)$ denotes what we have been calling a_1. Similarly, $f(2) = a_2, f(3) = a_3$, and, in general, $f(n) = a_n$. So the sequence notation with subscripts is just another variety of function notation. Furthermore, since a sequence is a function, we can draw a graph, as indicated in the next example.

EXAMPLE 4 **Graphing a sequence**

Consider the sequence defined by

$$x_n = \frac{1}{2^n} \quad \text{for } n \geq 0$$

Graph this sequence for $n = 0, 1, 2,$ and 3.

SOLUTION
First we compute the required values of x_n, as shown in Table 1. Then, locating the inputs n on the horizontal axis and the outputs x_n on the vertical axis, we obtain the graph shown in Figure 1.

TABLE 1

n	0	1	2	3
x_n	1	$\frac{1}{2}$	$\frac{1}{4}$	$\frac{1}{8}$

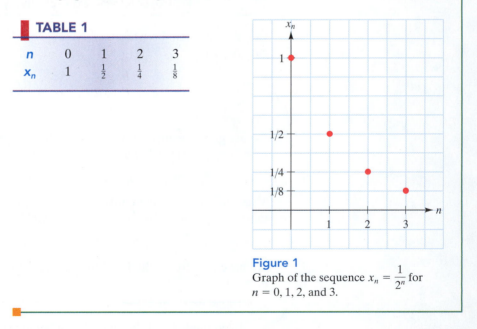

Figure 1
Graph of the sequence $x_n = \dfrac{1}{2^n}$ for $n = 0, 1, 2,$ and 3.

In part (b) of the next example, we use a graphing utility to compute the terms of a recursive sequence and to obtain the graph. For details on using a graphing utility in this fashion, consult the owner's manual for the graphing utility.

EXAMPLE	5	Using a recursive sequence to model population growth

Suppose that the following recursive sequence is a model for the size P_t of a population of rabbits on an island at the end of t years.

$$P_0 = 16$$
$$P_t = P_{t-1} + 0.0003P_{t-1}(3000 - P_{t-1}), \quad t \geq 1$$

(a) Find $P_0, P_1, \ldots, P_5$. After all the computations are complete, round each answer to the nearest integer (because the projected numbers of rabbits must be integers). Then use the rounded values to sketch the graph of the population sequence for $t = 0, 1, \ldots, 5$. Based on the graph, give a general description of the population trend over this period.

(b) Use a graphing utility to graph the population sequence through the end of the 15th year and describe the population trend.

SOLUTION

(a) We are given that $P_0 = 16$. This is the initial population. For P_1, we substitute $t = 1$ in the equation

$$P_t = P_{t-1} + 0.0003P_{t-1}(3000 - P_{t-1}) \tag{1}$$

This yields

$$\begin{aligned} P_1 &= P_0 + 0.0003P_0(3000 - P_0) \\ &= 16 + 0.0003(16)(3000 - 16) \qquad \text{using } P_0 = 16 \\ &= 30.3232 \qquad \text{using a calculator} \end{aligned}$$

Similarly, for P_2, we substitute $t = 2$ in equation (1) to obtain

$$\begin{aligned} P_2 &= P_1 + 0.0003P_1(3000 - P_1) \\ &= 30.3232 + 0.0003(30.3232)(3000 - 30.3232) \\ &\approx 57.33823106 \qquad \text{as displayed on calculator} \end{aligned}$$

The remaining calculations are carried out in the same way. The results are displayed in Table 2. The third row of the table shows the values rounded to the nearest integers. We interpret these rounded values as the population values as predicted by the given recursive model. In Figure 2 we've drawn a scatter plot showing the size of the rabbit population at the end of years $t = 0$ through $t = 5$.

TABLE 2

t	0	1	2	3	4	5
P_t	16	30.3232	57.33823106	107.9563372	201.6206695	370.8840037
P_t (to nearest integer)	16	30	57	108	202	371

Figure 2
The size P_t of the rabbit population
at the end of t years, $t = 0, 1, \ldots, 5$

Looking at Figure 2, it appears that the size of the population increases slowly at first, and then faster and faster. The picture suggests exponential growth; however, as you'll see in part (b), that model is inappropriate in the long run.

(b) Starting with the (unrounded) value of P_5 from Table 2 and applying equation (1), we use a graphing utility to compute P_6 through P_{15}. After rounding as in part (a), we obtain the population results displayed in Table 3. In Figure 3, we've used the graphing utility to draw a scatter plot using the data from both Tables 2 and 3.

TABLE 3 **The Size of the Rabbit Population at the End of Years $t = 6$ through $t = 15$**

t	6	7	8	9	10	11	12	13	14	15
P_t (to nearest integer)	663	1128	1762	2416	2839	2976	2997	3000	3000	3000

GRAPHICAL PERSPECTIVE

Figure 3
The size P_t of the rabbit population
at the end of t years, $t = 0, 1, \ldots, 15$

In Figure 2 we observed that the size of the population increases slowly at first, and then faster and faster (as in exponential growth). Now, with Figure 3, we see that the rapid growth continues only until the end of the ninth (or perhaps eighth) year. After the end of the ninth year, the size of the population begins to level off and approach 3000. Indeed, as indicated by both Table 3 and Figure 3, the size of the population reaches and remains at 3000 from $t = 13$ onward. The population size of 3000 is referred to as the *equilibrium population*. Actually, up to this point, we have only shown that the size of the population is 3000 for $t = 13, 14,$ and 15. We can use algebra to prove that once the population size reaches 3000, the recursive model projects that it will remain at that level for *all* subsequent years. In equation (1), assume that $P_{t-1} = 3000$. We then have

$$P_t = 3000 + 0.0003(3000)(3000 - 3000)$$
$$= 3000 + 0.0003(3000)(0) = 3000$$

This shows that if the size of the rabbit population is 3000 at the end of any year, then it will again be 3000 at the end of the next year.

We will often be interested in the sum of certain terms of a sequence. Consider, for example, the sequence

$$10, 20, 30, 40, \ldots$$

in which the *n*th term is $10n$. The sum of the first four terms in this sequence is

$$10 + 20 + 30 + 40 = 100$$

More generally, the sum of the first n terms in this sequence is indicated by

$$10 + 20 + 30 + \cdots + 10n$$

This expression is an example of a **finite series,** which simply means a sum of a finite number of terms.

We can indicate the sum of the first n terms of the sequence $a_1, a_2, a_3, \ldots$ by

$$a_1 + a_2 + a_3 + \cdots + a_n$$

Another way to indicate this sum uses what is called **sigma notation,** which we now introduce. The capital Greek letter sigma is written Σ. We define the notation $\sum_{k=1}^{n} a_k$ by the equation

$$\sum_{k=1}^{n} a_k = a_1 + a_2 + a_3 + \cdots + a_n$$

For example, $\sum_{k=1}^{3} a_k$ stands for the sum $a_1 + a_2 + a_3$, the idea in this case being to replace the subscript k successively by 1, 2, and 3 and then add the results.

For a more concrete example, let's evaluate $\sum_{k=1}^{4} k^2$. We have

$$\sum_{k=1}^{4} k^2 = 1^2 + 2^2 + 3^3 + 4^2$$
$$= 1 + 4 + 9 + 16 = 30$$

There is nothing special about the choice of the letter k in the expression $\sum_{k=1}^{4} k^2$. For instance, we could equally well write

$$\sum_{j=1}^{4} j^2 = 1^2 + 2^2 + 3^2 + 4^2 = 30$$

The letter k in the expression $\sum_{k=1}^{4} k^2$ is called the **index of summation.** Similarly, the letter j appearing in $\sum_{j=1}^{4} j^2$ is the index of summation in that case. As we have seen, the choice of the letter used for the index of summation has no effect on the value of the indicated sum. For this reason, the index of summation is referred to as a *dummy variable*. The next two examples provide further practice with sigma notation.

EXAMPLE 6 Understanding sigma notation

Express each of the following sums without sigma notation.

(a) $\displaystyle\sum_{k=1}^{3} (3k - 2)^2$ (b) $\displaystyle\sum_{i=1}^{4} ix^{i-1}$ (c) $\displaystyle\sum_{j=1}^{5} (a_{j+1} - a_j)$

SOLUTION

(a) The notation $\displaystyle\sum_{k=1}^{3} (3k - 2)^2$ directs us to replace k successively by 1, 2, and 3 in the expression $(3k - 2)^2$ and then to add the results. We thus obtain

$$\sum_{k=1}^{3} (3k - 2)^2 = 1^2 + 4^2 + 7^2 = 66$$

(b) The notation $\displaystyle\sum_{i=1}^{4} ix^{i-1}$ directs us to replace i successively by 1, 2, 3, and 4 in the expression ix^{i-1} and then to add the results. We have

$$\sum_{i=1}^{4} ix^{i-1} = 1x^0 + 2x^1 + 3x^2 + 4x^3$$
$$= 1 + 2x + 3x^2 + 4x^3$$

(c) To expand $\displaystyle\sum_{j=1}^{5} (a_{j+1} - a_j)$, we replace j successively by 1, 2, 3, 4, and 5 in the expression $a_{j+1} - a_j$ and then add. We obtain

$$\sum_{j=1}^{5} (a_{j+1} - a_j) = (a_2 - a_1) + (a_3 - a_2) + (a_4 - a_3) + (a_5 - a_4) + (a_6 - a_5)$$
$$= a_2 - a_1 + a_3 - a_2 + a_4 - a_3 + a_5 - a_4 + a_6 - a_5$$

Combining like terms, we have

$$\sum_{j=1}^{5} (a_{j+1} - a_j) = a_6 - a_1$$

Sums such as $\displaystyle\sum_{j=1}^{5} (a_{j+1} - a_j)$ are known as **collapsing** or **telescoping sums.**

EXAMPLE 7 Using sigma notation

Use sigma notation to rewrite each sum.

(a) $\dfrac{x}{1!} + \dfrac{x^2}{2!} + \dfrac{x^3}{3!} + \cdots + \dfrac{x^{12}}{12!}$ (b) $\dfrac{x}{2!} + \dfrac{x^2}{3!} + \dfrac{x^3}{4!} + \cdots + \dfrac{x^n}{(n+1)!}$

SOLUTION

(a) Since the exponents on x run from 1 to 12, we choose a dummy variable—say k, running from 1 to 12. Also, we notice that if x^k is the numerator of a given term in the sum, then $k!$ is the corresponding denominator. Consequently, the sum can be written

$$\frac{x}{1!} + \frac{x^2}{2!} + \frac{x^3}{3!} + \cdots + \frac{x^{12}}{12!} = \sum_{k=1}^{12} \frac{x^k}{k!}$$

(b) Since the exponents on x run from 1 to n, we choose a dummy variable—say k, running from 1 to n. (Note that both of the letters n and x would be inappropriate here as dummy variables.) Also, we notice that if the numerator of a given term in the sum is x^k, then the corresponding denominator is $(k + 1)!$. Thus, the given sum can be written

$$\frac{x}{2!} + \frac{x^2}{3!} + \frac{x^3}{4!} + \cdots + \frac{x^n}{(n + 1)!} = \sum_{k=1}^{n} \frac{x^k}{(k + 1)!}$$

EXERCISE SET 9.3

A

In Exercises 1–14, compute the first four terms in each sequence. For Exercises 1–10, assume that each sequence is defined for $n \geq 1$; in Exercises 11–14, assume that each sequence is defined for $n \geq 0$.

1. $a_n = n/(n + 1)$
2. $a_n = 1/n^2$
3. $b_n = (-1)^n$
4. $b_n = (n + 1)^2$
5. $c_n = 2^{-n}$
6. $c_n = (-1)^n(2^n)$
7. $x_n = 3n$
8. $x_n = (n - 1)^{n+1}$
9. $y_n = [1 + (1/n)]^n$
10. $y_n = [(-1)^n]\sqrt{n}$
11. $a_n = (n - 1)/(n + 1)$
12. $a_n = (-1)^n/(n + 2)$
13. $b_n = (-2)^{n+1}/(n + 1)^2$
14. $b_n = (n + 2)^{-n}$

In Exercises 15–22, the sequences are defined recursively. Compute the first five terms in each sequence.

15. $a_1 = 1; a_n = (1 + a_{n-1})^2, n \geq 2$
16. $a_1 = 2; a_n = \sqrt{a_{n-1}^2 + 1}, n \geq 2$
17. $a_1 = 2; a_2 = 2; a_n = a_{n-1}a_{n-2}, n \geq 3$
18. $F_1 = 1; F_2 = 1; F_n = F_{n-1} + F_{n-2}, n \geq 3$
19. $a_1 = 1; a_{n+1} = na_n, n \geq 1$
20. $a_1 = 1; a_2 = 2; a_n = a_{n-1}/a_{n-2}, n \geq 3$
21. $a_1 = 0; a_n = 2^{a_{n-1}}, n \geq 2$
22. $a_1 = 0; a_2 = 1; a_n = (a_{n-1} + a_{n-2})/2, n \geq 3$

In Exercises 23–32, sketch a graph showing the first five terms of the sequence. (A graphing utility is optional.)

23. $a_n = \dfrac{n}{2} + 1, n \geq 0$
24. $b_n = 16 - n^2, n \geq 0$
25. $c_n = 5/n, n \geq 1$
26. $d_n = (n - 1)!, n \geq 1$

27. $a_1 = 1; a_n = (a_{n-1})^2 - a_{n-1}, n \geq 2$
28. $a_0 = 3; a_n = (a_{n-1} + 1)/(a_{n-1} - 2), n \geq 1$
29. $b_0 = 2; b_n = (b_{n-1})^2 - 2b_{n-1} - 1, n \geq 1$
30. $b_1 = 0; b_n = b_{n-1} - n + 6, n \geq 2$
31. $A_0 = 0; A_n = \dfrac{A_{n-1} - 3}{A_{n-1} + 1}, n \geq 1$
32. $B_1 = 1; B_n = \dfrac{2B_{n-1} + 2}{-B_{n-1}}, n \geq 2$

In Exercises 33–38, find the sum of the first five terms of the sequence. (In Exercises 33–37, assume that the sequences are defined for $n \geq 1$.)

33. $a_n = 2^n$
34. $b_n = 2^{-n}$
35. $a_n = n^2 - n$
36. $b_n = (n - 1)!$
37. $a_n = (-1)^n/n!$
38. $a_1 = 1; a_n = \dfrac{1}{n - 1} - \dfrac{1}{n + 1}, n \geq 2$

39. Find the sum of the first five terms of the sequence that is defined recursively by $a_1 = 1; a_2 = 2; a_n = a_{n-1}^2 + a_{n-2}^2, n \geq 3$.

40. Find the sum of the first six terms of the sequence defined recursively by $a_1 = 1; a_2 = 2; a_{n+1} = a_n a_{n-1}, n \geq 2$.

In Exercises 41–52, express each of the sums without using sigma notation. Simplify your answers where possible.

41. $\displaystyle\sum_{k=1}^{3} (k - 1)$
42. $\displaystyle\sum_{k=1}^{5} k$
43. $\displaystyle\sum_{k=4}^{5} k^2$
44. $\displaystyle\sum_{k=2}^{6} (1 - 2k)$

45. $\sum_{n=1}^{3} x^n$

46. $\sum_{n=1}^{3} (n-1)x^{n-2}$

47. $\sum_{n=1}^{4} \frac{1}{n}$

48. $\sum_{n=0}^{4} 3^n$

49. $\sum_{j=1}^{9} \log_{10}\left(\frac{j}{j+1}\right)$

50. $\sum_{j=2}^{5} \log_{10} j$

51. $\sum_{j=1}^{6} \left(\frac{1}{j} - \frac{1}{j+1}\right)$

52. $\sum_{j=1}^{5} (x^{j+1} - x^j)$

In Exercises 53–62, rewrite the sums using sigma notation.

53. $5 + 5^2 + 5^3 + 5^4$

54. $5 + 5^2 + 5^3 + \cdots + 5^n$

55. $x + x^2 + x^3 + x^4 + x^5 + x^6$

56. $x + 2x^2 + 3x^3 + 4x^4 + 5x^5 + 6x^6$

57. $\frac{1}{1} + \frac{1}{2} + \frac{1}{3} + \cdots + \frac{1}{12}$

58. $\frac{1}{1} + \frac{1}{2} + \frac{1}{3} + \cdots + \frac{1}{n}$

59. $2 - 2^2 + 2^3 - 2^4 + 2^5$

60. $\binom{10}{3} + \binom{10}{4} + \binom{10}{5} + \cdots + \binom{10}{10}$

61. $1 - 2 + 3 - 4 + 5$

62. $\frac{1}{2} - \frac{1}{4} + \frac{1}{8} - \frac{1}{16} + \frac{1}{32} - \frac{1}{64} + \frac{1}{128}$

B

63. An important model that is used in population biology and ecology is the **Ricker model.** The Canadian biologist William E. Ricker introduced this model in his paper Stock and Recruitment (*Journal of the Fisheries Research Board of Canada,* **11** (1954) 559–623). For information on Ricker himself, see the Web page

http://www.science.ca/scientists/scientistprofile.php?pID=17

The general form of the Ricker model that we will use here is defined by a recursive sequence of the form

P_0 = initial population at time $t = 0$
$P_t = rP_{t-1}e^{-kP_{t-1}}$, for $t \geq 1$, and where r and k are
positive constants

(G) (a) Suppose that the initial size of a population is $P_0 = 300$ and that the size of the population at the end of year t is given by

$$P_t = 5P_{t-1}e^{-P_{t-1}/1000}, \quad t \geq 1$$

Use a graphing utility to compute the population sizes through the end of year $t = 5$. (As in Example 5, round the final answers to the nearest integers.) Then use the graphing utility to draw the population scatter plot for $t = 0, 1, \ldots, 5$. Describe in complete sentences how the size of the population changes over this period. Does the population seem to be approaching an equilibrium level?

(G) (b) Using a graphing utility, compute the sizes of the population in part (a) through the end of the year $t = 20$, and draw the corresponding scatter plot. Note that the population seems to be approaching an equilibrium level of 1609 (or 1610).

(c) Determine the equilibrium population algebraically by solving the following equation for P_{t-1}. For the final answer, use a calculator and round to the nearest integer.

$$P_{t-1} = 5P_{t-1}e^{-P_{t-1}/1000}$$

Hint: Divide both sides of the equation by P_{t-1}, and then use the techniques of Chapter 5.

(G) 64. (The Ricker model continued) Suppose that the initial size of a population is $P_0 = 300$ and that the size of the population at the end of year t is given by

$$P_t = 10P_{t-1}e^{-P_{t-1}/1000}, \quad t \geq 1$$

(a) Use a graphing utility to compute the population sizes through the end of year $t = 5$. (As in Example 5, round the final answers to the nearest integers.) Then use the graphing utility to draw the population scatter plot for $t = 0, 1, \ldots, 5$. Give a general description (in complete sentences) of how the size of the population changes over this period.

(b) Use a graphing utility to compute the population sizes through the end of year $t = 20$, and draw the scatter plot. To help you see the pattern, use the option on your graphing utility that connects adjacent dots in a scatter plot with line segments. Describe the population trend that emerges over the period $t = 15$ to $t = 20$.

(c) For a clearer view of the long-term population behavior, use a graphing utility to compute the population sizes for the period $t = 25$ to $t = 35$, and draw the scatter plot. As in part (b), use the option on your graphing utility that connects adjacent dots with line segments. Summarize (in complete sentences) what you observe.

*Exercises 65–73 involve the **Fibonacci sequence,** which is defined recursively as follows:*

$$F_1 = 1; \qquad F_2 = 1; \qquad F_{n+2} = F_n + F_{n+1} \quad \text{for } n \geq 1$$

This sequence was first studied by the Italian mathematician and merchant Leonardo of Pisa (c. 1170–1240), better known as Fibonacci ("son of Bonaccio").

65. (a) Complete the following table for the first ten terms of the Fibonacci sequence.

F_1	F_2	F_3	F_4	F_5	F_6	F_7	F_8	F_9	F_{10}
1	1								

(b) Given that $F_{20} = 6765$ and $F_{21} = 10946$, compute F_{22} and F_{19}.

(c) Given that $F_{29} = 514229$ and $F_{31} = 1346269$, compute F_{30}.

66. (a) Use the table in Exercise 65(a) to verify that $F_8 + F_9 + F_{10} = 2(F_8 + F_9)$.

(b) Show that $F_{100} + F_{101} + F_{102} = 2(F_{100} + F_{101})$.

(c) Show that $F_n + F_{n+1} + F_{n+2} = 2(F_n + F_{n+1})$ for all natural numbers $n \geq 3$.   *Hint:* Mathematical induction is not needed.

(d) Does the identity in part (c) hold for either of the values $n = 1$ or $n = 2$?

In Exercises 67–69: **(a)** Verify that the given equation holds for $n = 1$, $n = 2$, and $n = 3$; **(b)** Use mathematical induction to show that the equation holds for all natural numbers.

67. $F_1 + F_2 + F_3 + \cdots + F_n = F_{n+2} - 1$

68. $F_1^2 + F_2^2 + F_3^2 + \cdots + F_n^2 = F_n F_{n+1}$

69. $F_{n+1}^2 = F_n F_{n+2} + (-1)^n$ *Hint for part (b):* Add $F_{k+1}F_{k+2}$ to both sides of the equation in the induction hypothesis. Then factor F_{k+1} from the left-hand side and factor F_{k+2} from the first two terms on the right-hand side.

70. Use mathematical induction to prove that $F_n \geq n$ for all natural numbers $n \geq 5$.

71. We've seen that the Fibonacci sequence is defined recursively; each term after the second is the sum of the previous two terms. There is, in fact, an explicit formula for the nth term of the Fibonacci sequence:

$$F_n = \frac{(1 + \sqrt{5})^n - (1 - \sqrt{5})^n}{2^n \sqrt{5}}$$

This formula was discovered by the French-English mathematician Abraham deMoivre more than 500 years after Fibonacci first introduced the sequence in 1202 in his book *Liber Abaci*. (For a proof of this formula, see Exercises 72 and 73.)

(a) Using algebra (and not your calculator), check that this formula gives the right answers for F_1 and F_2.

(b) Use this formula and your calculator to computer F_{24} and F_{25}.

(c) Use the formula and your calculator to compute F_{26}. Then check your answer by using the results in part (b).

72. (This result will be used in Exercise 73.) Suppose that x is a real number such that $x^2 = x + 1$. Use mathematical induction to prove that

$$x^n = F_n x + F_{n-1} \quad \text{for } n \geq 2$$

Hint: You can carry out the induction proof without solving the given quadratic equation.

73. In this exercise, we use the result in Exercise 72 to derive the following formula for the nth Fibonacci number:

$$F_n = \frac{(1 + \sqrt{5})^n - (1 - \sqrt{5})^n}{2^n \sqrt{5}} \qquad (1)$$

The clever method used here was discovered by Erwin Just; it appeared in *Mathematics Magazine*, vol. 44 (1971), p. 199.

Let α and β denote the roots of the quadratic equation $x^2 = x + 1$. Then, according to Exercise 72, for $n \geq 2$ we have

$$\alpha^n = F_n \alpha + F_{n-1} \qquad (2)$$

and

$$\beta^n = F_n \beta + F_{n-1} \qquad (3)$$

(a) Subtract equation (3) from equation (2) to show that

$$F_n = (\alpha^n - \beta^n)/(\alpha - \beta) \quad \text{for } n \geq 2 \qquad (4)$$

(b) Use the quadratic formula to show that the roots of the equation $x^2 = x + 1$ are given by $\alpha = (1 + \sqrt{5})/2$ and $\beta = (1 - \sqrt{5})/2$.

(c) In equation (4), substitute for α and β using the values obtained in part (b). Show that this leads to equation (1).

(d) The work in parts (a) through (c) shows that equation (1) holds for $n \geq 2$. Now complete the derivation by checking that equation (1) also holds for $n = 1$.

74. This exercise requires a knowledge of matrix multiplication. Use mathematical induction to show that

$$\begin{pmatrix} 1 & 1 \\ 1 & 0 \end{pmatrix}^n = \begin{pmatrix} F_{n+1} & F_n \\ F_n & F_{n-1} \end{pmatrix} \quad \text{for } n \geq 2$$

MINI PROJECT | PERSPECTIVE AND ALTERNATIVE SCENARIOS FOR EXAMPLE 5

As background for this mini project, someone in your group or in the class at large should review and summarize Example 5 for everyone else.

1. Refer to Figure 2 on page 714, which shows a scatter plot for the size of the rabbit population in Example 5 at the end of years $t = 0, 1, \ldots, 5$. In the text it was mentioned that the picture suggests exponential growth. Another possibility is quadratic growth. Explore this as follows. Using a graphing utility, enter the six data points specified in Figure 2 and create a scatter plot similar to the one shown in Figure 2. Then use the exponential regression option on the graphing utility to find the exponential function that best fits the data.

Add the graph of the exponential function to your scatter plot. Next, add a quadratic regression curve to the picture. Which curve seems to fit the data points better?

2. In Example 5, suppose we make a change in the model so that the given recursive sequence is

$$P_0 = 16$$
$$P_t = 0.0003P_{t-1}(3000 - P_{t-1}), \quad t \geq 1$$

Use a graphing utility to investigate the population trend or trends. On paper, in complete sentences, compare and/or contrast your results with those in Example 5.

3. In Example 5, instead of beginning with 16 rabbits, suppose that we start with a much larger initial population size $P_0 = 3200$. (Assume, however, that the recursion equation from Example 5 remains the same.) Use a graphing utility to investigate the population trend or trends. On paper, in complete sentences, compare and/or contrast your results with those in Example 5.

4. (a) Follow Exercise 3, assuming that the size of the initial population is $P_0 = 6330$. (Compare and/or contrast your results with those in Exercise 3, rather than Example 5.)

(b) Find out what happens if $P_0 = 6334$. In terms of the rabbit population, can you suggest a possible interpretation of the results?

9.4 ARITHMETIC SEQUENCES AND SERIES

One of the most natural ways to generate a sequence is to begin with a fixed number a and then repeatedly add a fixed constant d. This yields the sequence

$$a, a + d, a + 2d, a + 3d, \ldots \tag{1}$$

Such a sequence is called an **arithmetic sequence** or **arithmetic progression.** Notice that the difference between any two consecutive terms is the constant d. We call d the **common difference.** Here are several examples of arithmetic sequences:

Example A: $1, 2, 3, \ldots$

Example B: $3, 7, 11, 15, \ldots$

Example C: $10, 5, 0, -5, \ldots$

In Example A, the first term is $a = 1$ and the common difference is $d = 1$. For Example B, we have $a = 3$. The value of d in this example is found by subtracting any two consecutive terms; thus $d = 4$. Finally, in Example C we have $a = 10$ and $d = -5$. Notice that when the common difference is negative, the terms of the sequence decrease.

There is a simple formula for the nth term in an arithmetic sequence. In arithmetic sequence (1), notice that

$$a_1 = a + 0d$$
$$a_2 = a + 1d$$
$$a_3 = a + 2d$$
$$a_4 = a + 3d$$

Following this pattern, it appears that the formula for a_n should be given by $a_n = a + (n-1)d$. Indeed, this is the correct formula, and Exercise 32 asks you to verify it using mathematical induction.

The nth Term of an Arithmetic Sequence

The nth term of an arithmetic sequence $a, a + d, a + 2d, \ldots$ is given by

$$a_n = a + (n-1)d$$

EXAMPLE **1** **Finding a term in an arithmetic sequence**

Determine the 100th term of the arithmetic sequence

$$7, 10, 13, 16, \ldots$$

SOLUTION
The first term is $a = 7$ and the common difference is $d = 3$. Substituting these values in the formula $a_n = a + (n-1)d$ yields

$$a_n = 7 + (n-1)3 = 3n + 4$$

To find the 100th term, we replace n by 100 in this last equation to obtain

$$a_{100} = 3(100) + 4 = 304 \qquad \text{as required}$$

EXAMPLE **2** **Determining an arithmetic sequence from given data**

Determine the arithmetic sequence in which the second term is -2 and the eighth term is 40.

SOLUTION
We are given that the second term is -2. Using this information in the formula $a_n = a + (n-1)d$, we have

$$-2 = a + (2-1)d = a + d$$

This gives us one equation in two unknowns. We are also given that the eighth term is 40. Therefore,

$$40 = a + (8-1)d = a + 7d$$

We now have a system of two equations in two unknowns:

$$\begin{cases} -2 = a + d \\ 40 = a + 7d \end{cases}$$

Subtracting the first equation from the second gives us

$$42 = 6d \qquad \text{or} \qquad 7 = d$$

To find a, we replace d by 7 in the first equation of the system. This yields

$$-2 = a + 7 \qquad \text{or} \qquad -9 = a$$

We have now determined the sequence, since we know that the first term is -9 and the common difference is 7. The first four terms of the sequence are $-9, -2, 5,$ and $12.$

Next we would like to derive a formula for the sum of the first n terms of an arithmetic sequence. Such a sum is referred to as an **arithmetic series.** If we use S_n to denote the required sum, we have

$$S_n = a + (a + d) + \cdots + [a + (n - 2)d] + [a + (n - 1)d] \qquad (1)$$

Of course, we must obtain the same sum if we add the terms from right to left rather than left to right. That is, we must have

$$S_n = [a + (n - 1)d] + [a + (n - 2)d] + \cdots + (a + d) + a \qquad (2)$$

Let's now add equations (1) and (2). Adding the left-hand sides is easy; we obtain $2S_n$. Now we add the corresponding terms on the right-hand sides. For the first term, we have

$$a \quad + \quad [a + (n - 1)d] = 2a + (n - 1)d$$

first term
in equation (1)   first term
in equation (2)

Next we add the second terms:

$$(a + d) \quad + \quad [a + (n - 2)d] = 2a + d + (n - 2)d = 2a + (n - 1)d$$

second term
in equation (1)   second term
in equation (2)

Notice that the sum of the second terms is again $2a + (n - 1)d$, the same quantity we arrived at with the first terms. As you can check, this pattern continues all the way through to the last terms. For instance,

$$[a + (n - 1)d] \quad + \quad a \quad = 2a + (n - 1)d$$

last term
in equation (1)   last term
in equation (2)

We conclude from these observations that by adding the right-hand sides of equations (1) and (2), the quantity $2a + (n - 1)d$ is added a total of n times. Therefore,

$$2S_n = n[2a + (n - 1)d]$$

$$S_n = \frac{n}{2}[2a + (n - 1)d]$$

This gives us the desired formula for the sum of the first n terms in an arithmetic sequence. There is an alternate form of this formula, which now follows rather quickly:

$$S_n = \frac{n}{2}[2a + (n - 1)d]$$

$$= \frac{n}{2}\{a + [a + (n - 1)d]\}$$

$$= \frac{n}{2}(a + a_n) = n\left(\frac{a + a_n}{2}\right)$$

This last equation is easy to remember. It says that the sum of an arithmetic series is obtained by averaging the first and last terms and then multiplying this average by n, the number of terms. For reference, we summarize both formulas as follows.

Formulas for the Sum of an Arithmetic Series

$$S_n = \frac{n}{2}[2a + (n-1)d]$$

$$S_n = n\left(\frac{a + a_n}{2}\right)$$

EXAMPLE 3 **Finding a sum of terms in an arithmetic sequence**

Find the sum of the first 30 terms of the arithmetic sequence $2, 6, 10, 14, \ldots$.

SOLUTION
We have $a = 2$, $d = 4$, and $n = 30$. Substituting these values in the formula $S_n = (n/2)[2a + (n-1)d]$ then yields

$$S_{30} = \frac{30}{2}[2(2) + (30-1)4]$$
$$= 15[4 + 29(4)]$$
$$= 1800 \quad \text{(Check the arithmetic.)}$$

EXAMPLE 4 **Finding a sum of terms and the common difference of an arithmetic sequence**

In a certain arithmetic sequence, the first term is 6 and the 40th term is 71. Find the sum of the first 40 terms and also the common difference for the sequence.

SOLUTION
We have $a = 6$ and $a_{40} = 71$. Using these values in the formula $S_n = n(a + a_n)/2$ yields

$$S_{40} = \frac{40}{2}(6 + 71) = 20(77) = 1540$$

The sum of the first 40 terms is thus 1540. The value of d can now be found by using the formula $a_n = a + (n-1)d$. We have $a_{40} = a + (40-1)d$, and therefore

$$71 = 6 + 39d$$
$$39d = 65$$
$$d = \frac{65}{39} = \frac{5}{3}$$

The required value of d is 5/3.

EXAMPLE 5 **An arithmetic series defined using sigma notation**

Show that $\sum_{k=1}^{50} (3k - 2)$ represents an arithmetic series, and compute the sum.

SOLUTION

There are two different ways to see that $\sum_{k=1}^{50} (3k - 2)$ is an arithmetic series. One way is simply to write out the first few terms and look at the pattern. We have

$$\sum_{k=1}^{50} (3k - 2) = 1 + 4 + 7 + 10 + \cdots + 148$$

From this it is clear that we are indeed summing the terms in an arithmetic sequence in which $d = 3$ and $a = 1$.

Another, more formal way to show that $\sum_{k=1}^{50} (3k - 2)$ represents an arithmetic series is to prove that the difference between successive terms in the indicated sum is a constant. Now, the form of a typical term in this sum is $3k - 2$. Thus, the form of the next term must be $[3(k + 1) - 2]$. The difference between these terms is then

$$[3(k + 1) - 2] - (3k - 2) = 3k + 3 - 2 - 3k + 2 = 3$$

The difference therefore is constant, as we wished to show.

To evaluate $\sum_{k=1}^{50} (3k - 2)$, we can use either of our two formulas for the sum of an arithmetic series. Using the formula $S_n = (n/2)[2a + (n - 1)d]$, we obtain

$$S_{50} = \frac{50}{2}[2(1) + 49(3)] = 25(149) = 3725$$

Thus, the required sum is 3725. You should check for yourself that the same value is obtained using the formula $S_n = n(a + a_n)/2$.

EXERCISE SET 9.4

A

1. Find the common difference d for each of the following arithmetic sequences.
   (a) $1, 3, 5, 7, \ldots$
   (b) $10, 6, 2, -2, \ldots$
   (c) $2/3, 1, 4/3, 5/3, \ldots$
   (d) $1, 1 + \sqrt{2}, 1 + 2\sqrt{2}, 1 + 3\sqrt{2}, \ldots$

2. Which of the following are arithmetic sequences?
   (a) $2, 4, 8, 16, \ldots$
   (b) $5, 9, 13, 17, \ldots$
   (c) $3, 11/5, 7/5, 3/5, \ldots$
   (d) $-1, -1, -1, -1, \ldots$
   (e) $-1, 1, -1, 1, \ldots$

In Exercises 3–8, find the indicated term in each sequence.

3. $10, 21, 32, 43, \ldots ; a_{12}$
4. $7, 2, -3, -8, \ldots ; a_{20}$
5. $6, 11, 16, 21, \ldots ; a_{100}$
6. $2/5, 4/5, 6/5, 8/5, \ldots ; a_{30}$
7. $-1, 0, 1, 2, \ldots ; a_{1000}$
8. $42, 1, -40, -81, \ldots ; a_{15}$
9. The fourth term in an arithmetic sequence is -6, and the 10th term is 5. Find the common difference and the first term.
10. The fifth term in an arithmetic sequence is $1/2$, and the 20th term is $7/8$. Find the first three terms of the sequence.
11. The 60th term in an arithmetic sequence is 105, and the common difference is 5. Find the first term.

12. Find the common difference in an arithmetic sequence in which $a_{10} - a_{20} = 70$.

13. Find the common difference in an arithmetic sequence in which $a_{15} - a_7 = -1$.

14. Find the sum of the first 16 terms in the sequence $2, 11, 20, 29, \ldots$.

15. Find the sum of the first 1000 terms in the sequence $1, 2, 3, 4, \ldots$.

16. Find the sum of the first 50 terms in an arithmetic series that has first term -8 and 50th term 139.

17. Find the sum: $\dfrac{\pi}{3} + \dfrac{2\pi}{3} + \pi + \dfrac{4\pi}{3} + \cdots + \dfrac{13\pi}{3}$.

18. Find the sum: $\dfrac{1}{e} + \dfrac{3}{e} + \dfrac{5}{e} + \cdots + \dfrac{21}{e}$.

19. Determine the first term of an arithmetic sequence in which the common difference is 5 and the sum of the first 38 terms is 3534.

20. The sum of the first 12 terms in an arithmetic sequence is 156. What is the sum of the first and 12th terms?

21. In a certain arithmetic sequence, the first term is 4 and the 16th term is -100. Find the sum of the first 16 terms and also the common difference for the sequence.

22. The fifth and 50th terms of an arithmetic sequence are 3 and 30, respectively. Find the sum of the first 10 terms.

23. The eighth term in an arithmetic sequence is 5, and the sum of the first 10 terms is 20. Find the common difference and the first term of the sequence.

In Exercises 24–26, find each sum.

24. $\displaystyle\sum_{i=1}^{10} (2i - 1) = 1 + 3 + 5 + \cdots + 19$

25. $\displaystyle\sum_{k=1}^{20} (4k + 3)$ **26.** $\displaystyle\sum_{n=5}^{100} (2n - 1)$

27. The sum of three consecutive terms in an arithmetic sequence is 30, and their product is 360. Find the three terms.   *Suggestion:* Let x denote the *middle* term and d the common difference.

28. The sum of three consecutive terms in an arithmetic sequence is 21, and the sum of their squares is 197. Find the three terms.

29. The sum of three consecutive terms in an arithmetic sequence is 6, and the sum of their cubes is 132. Find the three terms.

30. In a certain arithmetic sequence, $a = -4$ and $d = 6$. If $S_n = 570$, find n.

31. Let $a_1 = 1/(1 + \sqrt{2})$, $a_2 = -1$, and $a_3 = 1/(1 - \sqrt{2})$.
(a) Show that $a_2 - a_1 = a_3 - a_2$.
(b) Find the sum of the first six terms in the arithmetic sequence

$$\frac{1}{1 + \sqrt{2}}, -1, \frac{1}{1 - \sqrt{2}}, \ldots$$

32. Using mathematical induction, prove that the nth term of the sequence $a, a + d, a + 2d, \ldots$ is given by

$$a_n = a + (n - 1)d$$

B

33. Let b denote a positive constant. Find the sum of the first n terms in the sequence

$$\frac{1}{1 + \sqrt{b}}, \frac{1}{1 - b}, \frac{1}{1 - \sqrt{b}}, \ldots$$

34. The sum of the first n terms in a certain arithmetic sequence is given by $S_n = 3n^2 - n$. Show that the rth term is given by $a_r = 6r - 4$.

35. Let $a_1, a_2, a_3, \ldots$ be an arithmetic sequence, and let S_k denote the sum of the first k terms. If $S_n/S_m = n^2/m^2$, show that

$$\frac{a_n}{a_m} = \frac{2n - 1}{2m - 1}$$

36. If the common difference in an arithmetic sequence is twice the first term, show that

$$\frac{S_n}{S_m} = \frac{n^2}{m^2}$$

37. The lengths of the sides of a right triangle form three consecutive terms in an arithmetic sequence. Show that the triangle is similar to the 3-4-5 right triangle.

38. Suppose that $1/a$, $1/b$, and $1/c$ are three consecutive terms in an arithmetic sequence. Show that:
(a) $\dfrac{a}{c} = \dfrac{a - b}{b - c}$;   **(b)** $b = \dfrac{2ac}{a + c}$.

39. Suppose that a, b, and c are three positive numbers with $a > c > 2b$. If $1/a$, $1/b$, and $1/c$ are consecutive terms in an arithmetic sequence, show that

$$\ln(a + c) + \ln(a - 2b + c) = 2\ln(a - c)$$

9.5 GEOMETRIC SEQUENCES AND SERIES

A **geometric sequence** or **geometric progression** is a sequence of the form

$$a, ar, ar^2, ar^3, \ldots \quad \text{where } a \text{ and } r \text{ are constants}$$

As you can see, each term after the first in a geometric sequence is obtained by multiplying the previous term by r. The number r is called the **common ratio,** because the ratio of any term to the previous one is always r. For instance, the ratio of the fourth term to the third is $ar^3/ar^2 = r$. Here are two examples of geometric sequences:

$$1, \frac{1}{2}, \frac{1}{4}, \frac{1}{8}, \ldots$$

$$10, -100, 1000, -10,000, \ldots$$

In the first example, we have $a = 1$ and $r = 1/2$; in the second example, we have $a = 10$ and $r = -10$.

EXAMPLE 1 **Finding a term in a geometric sequence from given data**

In a certain geometric sequence, the first term is 2, the third term is 3, and the common ratio is negative. Find the second term.

SOLUTION
Let x denote the second term, so that the sequence begins

$$2, x, 3, \ldots$$

By definition, the ratios $3/x$ and $x/2$ must be equal. Thus, we have

$$\frac{3}{x} = \frac{x}{2}$$
$$x^2 = 6$$
$$x = \pm\sqrt{6}$$

Now, the second term must be negative, because the first term is positive and the common ratio is negative. Thus, the second term is $x = -\sqrt{6}$.

TABLE 1

n	a_n
1	ar^0
2	ar^1
3	ar^2
4	ar^3
$\vdots$	$\vdots$

The formula for the nth term of a geometric sequence is easily deduced by considering Table 1. The table indicates that the exponent on r is one less than the value of n in each case. On the basis of this observation, it appears that the nth term must be given by $a_n = ar^{n-1}$. Indeed, it can be shown by mathematical induction that this formula does hold for all natural numbers n. (Exercise 30 asks you to carry out the proof.) We summarize this result as follows:

nth Term of a Geometric Sequence

The nth term of the geometric sequence, $a, ar, ar^2, \ldots$ is given by

$$a_n = ar^{n-1}$$

EXAMPLE 2   Finding a term in a given geometric sequence

Find the seventh term in the geometric sequence $2, 6, 18, \ldots$.

SOLUTION

We can find the common ratio r by dividing the second term by the first. Thus, $r = 3$. Now, using $a = 2$, $r = 3$, and $n = 7$ in the formula $a_n = ar^{n-1}$, we have

$$a_7 = 2(3)^6 = 2(729) = 1458$$

The seventh term of the sequence is therefore 1458.

Suppose that we begin with a geometric sequence $a, ar, ar^2, \ldots$, in which $r \neq 1$. If we add the first n terms and denote the sum by S_n, we have

$$S_n = a + ar + ar^2 + \cdots + ar^{n-2} + ar^{n-1} \qquad (1)$$

This sum is called a **finite geometric series.** We would like to find a formula for S_n. To do this, we multiply equation (1) by r to obtain

$$rS_n = ar + ar^2 + ar^3 + \cdots + ar^{n-1} + ar^n \qquad (2)$$

We now subtract equation (2) from equation (1). This yields (after combining like terms)

$$S_n - rS_n = a - ar^n$$
$$S_n(1 - r) = a(1 - r^n)$$
$$S_n = \frac{a(1 - r^n)}{1 - r} \qquad (r \neq 1)$$

This is the formula for the sum of a finite geometric series. We summarize this result in the box that follows.

Formula for the Sum of a Geometric Series

Let S_n denote the sum $a + ar + ar^2 + \cdots + ar^{n-1}$, and assume that $r \neq 1$. Then

$$S_n = \frac{a(1 - r^n)}{1 - r}$$

EXAMPLE 3   Computing the sum of a finite geometric series

Evaluate the sum $\dfrac{1}{2^1} + \dfrac{1}{2^2} + \dfrac{1}{2^3} + \cdots + \dfrac{1}{2^{10}}$.

SOLUTION

This is a finite geometric series with $a = 1/2$, $r = 1/2$, and $n = 10$. Using these values in the formula for S_n yields

$$S_{10} = \frac{\frac{1}{2}\left[1 - \left(\frac{1}{2}\right)^{10}\right]}{1 - \frac{1}{2}} = \frac{\frac{1}{2}\left[1 - \frac{1}{1024}\right]}{\frac{1}{2}} = 1 - \frac{1}{1024} = \frac{1023}{1024}$$

We would now like to attach a meaning to certain expressions of the form

$$a + ar + ar^2 + \cdots$$

Such an expression is called an **infinite geometric series.** The three dots indicate (intuitively at least) that the additions are to be carried out indefinitely, without end. To see how to proceed here, let's start by looking at some examples involving finite geometric series. In particular, we'll consider the series

$$\frac{1}{2^1} + \frac{1}{2^2} + \frac{1}{2^3} + \cdots + \frac{1}{2^n}$$

for increasing values of n. The idea is to look for a pattern as n grows ever larger. Let $S_1 = \frac{1}{2}$, $S_2 = \frac{1}{2^1} + \frac{1}{2^2}$, $S_3 = \frac{1}{2^1} + \frac{1}{2^2} + \frac{1}{2^3}$, and, in general,

$$S_n = \frac{1}{2^1} + \frac{1}{2^2} + \cdots + \frac{1}{2^n}$$

| TABLE 2 |

n	$S_n = \dfrac{1}{2} + \dfrac{1}{2^2} + \cdots \dfrac{1}{2^n}$
1	0.5
2	0.75
5	0.96875
10	$0.999023437\ldots$
15	$0.999969482\ldots$
20	$0.999999046\ldots$
25	$0.999999970\ldots$

Then we can compute S_n for any given value of n by means of the formula for the sum of a finite geometric series. From Table 2, which displays the results of these calculations, it seems clear that as n grows larger and larger, the value of S_n grows ever closer to 1. More precisely (but leaving the details for calculus), it can be shown that the value of S_n can be made as close to 1 as we please, provided only that n is sufficiently large. For this reason, we say that the *sum of the infinite geometric series* $\dfrac{1}{2^1} + \dfrac{1}{2^2} + \dfrac{1}{2^3} + \cdots$ is 1. That is,

$$\frac{1}{2^1} + \frac{1}{2^2} + \frac{1}{2^3} + \cdots = 1$$

We can arrive at this last result another way. First we compute the sum of the *finite* geometric series $\dfrac{1}{2^1} + \dfrac{1}{2^2} + \cdots + \dfrac{1}{2^n}$. As you can check, the result is

$$S_n = 1 - (1/2)^n$$

Now, as n grows larger and larger, the value of $(1/2)^n$ gets closer and closer to zero. Thus, as n grows ever larger, the value of S_n will more and more resemble $1 - 0$, or 1.

Now let's repeat our reasoning to obtain a formula for the sum of the infinite geometric series

$$a + ar + ar^2 + \cdots \quad \text{where } |r| < 1$$

First we consider the finite geometric series

$$a + ar + ar^2 + \cdots + ar^{n-1}$$

The sum S_n in this case is

$$S_n = \frac{a(1 - r^n)}{1 - r}$$

We want to know how S_n behaves as n grows ever larger. This is where the assumption $|r| < 1$ is crucial. Just as $(1/2)^n$ approaches zero as n grows larger and larger, so will r^n approach zero as n grows larger and larger. Thus, as n grows ever larger, the sum S_n will more and more resemble

$$\frac{a(1-0)}{1-r} = \frac{a}{1-r}$$

For this reason, we say that the sum of the infinite geometric series is $a/(1-r)$. We will make free use of this result in the subsequent examples. However, a more rigorous development of infinite series properly belongs to calculus.

Formula for the Sum of an Infinite Geometric Series

Suppose that $|r| < 1$. Then the sum S of the infinite geometric series $a + ar + ar^2 + \cdots$ is given by

$$S = \frac{a}{1-r}$$

| EXAMPLE | 4 | Computing the sum of an infinite geometric series |

Find the sum of the infinite geometric series $1 + \dfrac{2}{3} + \dfrac{4}{9} + \cdots$.

SOLUTION
In this case, we have $a = 1$ and $r = 2/3$. Thus,

$$S = \frac{a}{1-r} = \frac{1}{1 - \frac{2}{3}} = \frac{1}{\frac{1}{3}} = 3$$

The sum of the series is 3.

| EXAMPLE | 5 | Expressing a repeating decimal as a fraction |

Find a fraction equivalent to the repeating decimal $0.2\overline{35}$.

SOLUTION
Let $S = 0.2\overline{35}$. Then we have

$$S = 0.2353535\ldots$$
$$= \frac{2}{10} + \frac{35}{1000} + \frac{35}{100,000} + \frac{35}{10,000,000} + \cdots$$

Now, the expression following $2/10$ on the right-hand side of this last equation is an infinite geometric series in which $a = 35/1000$ and $r = 1/100$. Thus,

$$S = \frac{2}{10} + \frac{a}{1-r}$$
$$= \frac{2}{10} + \frac{\frac{35}{1000}}{1 - \frac{1}{100}} = \frac{233}{990} \qquad \text{(Check the arithmetic!)}$$

The given decimal is therefore equivalent to $233/990$.

EXERCISE SET 9.5

A

1. Find the second term in a geometric sequence in which the first term is 9, the third term is 4, and the common ratio is positive.
2. Find the fifth term in a geometric sequence in which the fourth term is 4, the sixth term is 6, and the common ratio is negative.
3. The product of the first three terms in a geometric sequence is 8000. If the first term is 4, find the second and third terms.

In Exercises 4–8, find the indicated term of the given geometric sequence.

4. $9, 81, 729, \ldots; a_7$
5. $-1, 1, -1, 1, \ldots; a_{100}$
6. $1/2, 1/4, 1/8, \ldots; a_9$
7. $2/3, 4/9, 8/27, \ldots; a_8$
8. $1, -\sqrt{2}, 2, \ldots; a_6$
9. Find the common ratio in a geometric sequence in which the first term is 1 and the seventh term is 4096.
10. Find the first term in a geometric sequence in which the common ratio is 4/3 and the tenth term is 16/9.
11. Find the sum of the first ten terms of the sequence $7, 14, 28, \ldots$.
12. Find the sum of the first five terms of the sequence $-1/2, 3/10, -9/50, \ldots$.
13. Find the sum: $1 + \sqrt{2} + 2 + \cdots + 32$.
14. Find the sum of the first 12 terms in the sequence $-4, -2, -1, \ldots$.

In Exercises 15–17, evaluate each sum.

15. $\displaystyle\sum_{k=1}^{6} \left(\frac{3}{2}\right)^k$
16. $\displaystyle\sum_{k=1}^{6} \left(\frac{2}{3}\right)^{k+1}$
17. $\displaystyle\sum_{k=2}^{6} \left(\frac{1}{10}\right)^k$

In Exercises 18–22, determine the sum of each infinite geometric series.

18. $\dfrac{1}{4} + \dfrac{1}{4^2} + \dfrac{1}{4^3} + \cdots$
19. $\dfrac{2}{3} - \dfrac{4}{9} + \dfrac{8}{27} - \cdots$
20. $\dfrac{9}{10} + \dfrac{9}{100} + \dfrac{9}{1000} + \cdots$
21. $1 + \dfrac{1}{1.01} + \dfrac{1}{(1.01)^2} + \cdots$
22. $-1 - \dfrac{1}{\sqrt{2}} - \dfrac{1}{2} - \cdots$

In Exercises 23–27, express each repeating decimal as a fraction.

23. $0.555\ldots$
24. $0.\overline{47}$
25. $0.1\overline{23}$
26. $0.050505\ldots$
27. $0.\overline{432}$

B

28. The lengths of the sides in a right triangle form three consecutive terms of a geometric sequence. Find the common ratio of the sequence. (There are two distinct answers.)
29. The product of three consecutive terms in a geometric sequence is -1000, and their sum is 15. Find the common ratio. (There are two answers.)   *Suggestion:* Denote the terms by a/r, a, and ar.
30. Use mathematical induction to prove that the nth term of the geometric sequence $a, ar, ar^2, \ldots$ is ar^{n-1}.
31. Show that the sum of the following infinite geometric series is $3/2$:

$$\frac{\sqrt{3}}{\sqrt{3} + 1} + \frac{\sqrt{3}}{\sqrt{3} + 3} + \cdots$$

32. Let A_1 denote the area of an equilateral triangle, each side of which is 1 unit long. A second equilateral triangle is formed by joining the midpoints of the sides of the first triangle. Let A_2 denote the area of this second triangle. This process is then repeated to form a third triangle with area A_3; and so on. Find the sum of the areas: $A_1 + A_2 + A_3 + \cdots$.
33. Let $a_1, a_2, a_3, \ldots$ be a geometric sequence such that $r \neq 1$. Let $S = a_1 + a_2 + a_3 + \cdots + a_n$, and let
$$T = \frac{1}{a_1} + \frac{1}{a_2} + \cdots + \frac{1}{a_n}.$$ Show that $\dfrac{S}{T} = a_1 a_n$.
34. Suppose that a, b, and c are three consecutive terms in a geometric sequence. Show that $\dfrac{1}{a + b}$, $\dfrac{1}{2b}$, and $\dfrac{1}{c + b}$ are three consecutive terms in an arithmetic sequence.
35. A ball is dropped from a height of 6 ft. Assuming that on each bounce the ball rebounds to one-third of its previous height, find the total distance traveled by the ball.

9.6 PERMUTATIONS AND COMBINATIONS

I have all the words. What I am seeking is the perfect order of the words in the sentence. —James Joyce (1882–1941)

Suppose that you flip a coin twice, keeping track of the result each time. How many possible outcomes are there for this experiment? Using H and T to denote heads and tails, respectively, we can list the possible outcomes this way:

$$HH \quad HT \quad TH \quad TT$$

For instance, HH describes the experiment that results in two heads in succession. Looking at this list then, we can say that there are four possible outcomes for the experiment.

Now let's ask the same question for an experiment in which we flip a coin three times in succession. How many possible outcomes are there? One way to organize the counting procedure uses the *tree diagram* in Figure 1. A tree diagram shows us every possibility at each stage in the experiment. Each path through the tree (from left to right) corresponds to a possible outcome for the experiment. For instance, the topmost path corresponds to the case in which we obtain three heads in succession (HHH). As you can see in Figure 1, there are eight possible outcomes for our experiment, corresponding to the eight branches on the right side of the tree.

Figure 1

| EXAMPLE | 1 | **Using a tree diagram** |

There are four roads from town A to town B and three roads from town B to town C, as indicated schematically in Figure 2. Draw a tree diagram to find out how many routes there are from A to C via B.

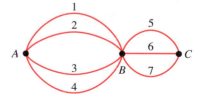

Figure 2

SOLUTION

Our tree diagram in Figure 3 has 12 branches on the right side (count them), so there are 12 routes from A to C via B.

Rather than actually counting all 12 branches in Figure 3, we could have reasoned as follows. There are four possible ways to go from A to B, and for each of these ways, there are then three ways to go from B to C. Thus there are 4 *times* 3, or 12, ways to go from A to C. The *multiplication principle*, which we now state, formalizes this idea.

The Multiplication Principle

Suppose that a task involves two consecutive operations. If the first operation can be performed in M ways, and the second operation can be performed in N ways, then the entire task can be performed in MN ways.

Figure 3

EXAMPLE 2 **An application of the multiplication principle**

How many different two-digit numbers can be formed using the digits 1, 3, 5, 7, and 9 if: **(a)** no digit is repeated (within a given number)? **(b)** repetitions are allowed?

SOLUTION

(a) We can think of the task of forming a two-digit number as composed of two consecutive operations: first choose a tens digit, then choose a units digit. There are five choices for the tens digit. Then, after choosing a tens digit (assuming that no repetitions are allowed), there will be four choices left for the units digit. It follows now from the multiplication principle that there are $5 \times 4 = 20$ possible choices. That is, there are 20 different two-digit numbers that can be formed.

(b) With repetitions allowed, there are five choices for the tens digit and five choices for the units digit. Therefore (using the multiplication principle again) there are $5 \times 5 = 25$ different two-digit numbers that can be formed.

The multiplication principle can be extended to tasks involving three or more consecutive operations. This idea is used in the next example.

EXAMPLE 3 **Another application of the multiplication principle**

You buy three books at a book sale: one by Hemingway, one by Joyce, and one by Fitzgerald. How many ways are there to arrange these on a (single) bookshelf?

SOLUTION

We have three positions to fill on the shelf: left, middle, and right. Starting with the left position, there are three choices. Then, once we make a choice for the left position, there are two choices remaining for the middle position. Finally, after choosing a book for that middle position, we'll have only one choice for the right position. Thus, according to the multiplication principle, there are $3 \times 2 \times 1 = 6$ ways to arrange the three books.

Let's list the six possible arrangements from Example 3. Using H for Hemingway, J for Joyce, and F for Fitzgerald, the six arrangements are:

$$HJF \quad JHF \quad FJH \quad HFJ \quad JFH \quad FHJ$$

Each of the six arrangements here is called a *permutation* of the set $\{H, J, F\}$. In general, by a **permutation** of a set of n distinct objects we just mean a listing or arrangement of those objects in a definite order. For instance, if we take the set $\{a, b, c, d\}$, then two (of the many possible) permutations of this set are *abcd* and *dcba*.

EXAMPLE **4** **Counting the number of permutations**

How many permutations are there of the set $\{a, b, c, d\}$?

SOLUTION
To form a permutation, we have to assign a letter to each of four positions:

$$\underline{\quad} \quad \underline{\quad} \quad \underline{\quad} \quad \underline{\quad}$$

We have four choices for the first position. Then, after filling that position, we'll have three letters left to choose from for the second position. After carrying that out, there will be only two letters to choose from for the third position. And finally, with the first three positions filled, there will be but one letter left for the fourth position. In summary, we have

$$\underline{\quad} \quad \underline{\quad} \quad \underline{\quad} \quad \underline{\quad}$$

4 choices   3 choices   2 choices   1 choice

Therefore (by the multiplication principle) the total number of permutations or arrangements is $4 \times 3 \times 2 \times 1 = 24$. Notice that this is 4!, where n! stands for the product of the first n natural numbers. (See Section 9.2.)

In the example we just concluded, we saw that there were 4! permutations of a set with four elements. By following the same kind of reasoning (and using mathematical induction), it can be shown that there are n! permutations of a set with n elements. For reference we state this fact in the box that follows.

The number of permutations of a set with n elements is n!

EXAMPLE **5** **Another example counting permutations**

A regular deck of playing cards contains 52 cards. How many arrangements (permutations) of the deck are there?

SOLUTION
The number of permutations of a set with 52 elements is 52!, and with the aid of a calculator we find that

$$52! \approx 8 \times 10^{67}$$

We've said that a permutation of a set with n elements is a listing or arrangement of all n of those elements in some definite order. Sometimes, though, we want to consider arrangements that use less than all n of the elements. For example, let's take the set $\{H, J, F\}$ that we used earlier and this time list the possible arrangements using only two elements at a time.

$$HJ \quad JH \quad FH \quad HF \quad JF \quad FJ$$

Each arrangement here is called a *permutation of three objects taken two at a time.* The number of such permutations is denoted by $P(3, 2)$. So we have $P(3, 2) = 6$, since there are six arrangements. More generally, we have the following definition.

DEFINITION | **Permutation of *n* Objects Taken *r* at a Time**

By a **permutation of *n* objects taken *r* at a time** we mean a listing or an arrangement of *r* of the objects in a definite order, where $r \leq n$. The number of such arrangements is denoted by

$$P(n, r)$$

EXAMPLE | **6** | **Making a list of permutations**

Using the set $\{a, b, c, d\}$, list all the permutations of these four letters taken two at a time. Thus, evaluate $P(4, 2)$.

SOLUTION

Using *a* and *b*:	*ab, ba*	Using *b* and *c*:	*bc, cb*
Using *a* and *c*:	*ac, ca*	Using *b* and *d*:	*bd, db*
Using *a* and *d*:	*ad, da*	Using *c* and *d*:	*cd, dc*

We've listed 12 permutations of the four given letters taken two at a time. (Convince yourself that we've listed all possibilities.) So $P(4, 2) = 12$.

We could have evaluated $P(4, 2)$ in Example 6 by reasoning as follows. To form a permutation of four letters taken two at a time, we have to fill two positions:

First letter   Second letter

There are four choices for the first letter. Then, after making a choice, we're left with three choices for the second letter. Therefore (according to the multiplication principle) there are $4 \times 3 = 12$ possible choices, in agreement with the result in Example 6. By using this type of reasoning, we can obtain the following general expression for the number of permutations of *n* objects taken *r* at a time. (Exercise 54 asks you to prove that the two expressions for $P(n, r)$ in the box are equal.)

$$P(n, r) = n(n - 1) \cdots (n - r + 1) = \frac{n!}{(n - r)!}$$

| EXAMPLE | 7 | **Finding the number of permutations without listing them** |

(a) Evaluate $P(8, 5)$, the number of permutations of 8 objects taken 5 at a time.
(b) Evaluate $P(8, 8)$.

SOLUTION
(a) In the first expression for $P(n, r)$ in the box, we let $n = 8$ and $r = 5$. Then

$$P(8, 5) = 8 \cdot 7 \cdots (8 - 5 + 1)$$
$$= 8 \cdot 7 \cdot 6 \cdot 5 \cdot 4 = 6720$$

So there are 6720 permutations of 8 objects taken 5 at a time.
(b) $P(8, 8) = 8 \cdot 7 \cdots (8 - 8 + 1)$
$$= 8 \cdot 7 \cdots 1 = 8! = 40{,}320$$

So there are 8! or 40,320 permutations of eight objects taken eight at a time. (Note that a permutation of eight objects taken eight at a time is just a permutation of those eight objects.)

| EXAMPLE | 8 | **Selecting and arranging pictures on a wall** |

From a set of ten pictures, five are to be selected and hung in a row. How many arrangements are possible?

SOLUTION
We want the number of permutations of ten objects taken five at a time. This number is

$$P(10, 5) = 10 \cdot 9 \cdots (10 - 5 + 1)$$
$$= 10 \cdot 9 \cdot 8 \cdot 7 \cdot 6 = 30{,}240$$

There are, therefore, 30,240 possible arrangements.

| EXAMPLE | 9 | **Counting arrangements that are subject to a side condition** |

How many possible arrangements are there of the letters $\{a, b, c, d, e\}$ in which a and b are next to each other?

SOLUTION
First let's think of a and b as a single object and count the number of arrangements of the *four* objects (ab), c, d, and e. This number is

$$P(4, 4) = 4! \qquad \text{(Why?)}$$
$$= 4 \cdot 3 \cdot 2 \cdot 1 = 24$$

Now, in each of these 24 arrangements, the letters a and b can appear in one of two ways, either ab or ba. Consequently (by the multiplication principle), the number of possible arrangements of $\{a, b, c, d, e\}$ in which a and b are adjacent is $24 \times 2 = 48$.

EXAMPLE 10 A consequence of Example 9

How many possible arrangements are there of the letters $\{a, b, c, d, e\}$ in which a and b are not next to each other?

SOLUTION
In this type of counting problem, the direct approach is not the simplest. Instead, we reason as follows. The number of arrangements in which a and b are not adjacent equals the total number of possible arrangements of the five letters *minus* the number of arrangements of the five letters in which a and b *are* adjacent. So, by the result in Example 9, the required number of arrangements is

$$P(5, 5) - 48 = 5 \cdot 4 \cdot 3 \cdot 2 \cdot 1 - 48$$
$$= 120 - 48 = 72$$

Therefore, there are 72 ways to arrange the five given letters such that a and b are not adjacent.

In working with permutations, we've seen that the order in which the objects are listed is an essential factor. Now we want to introduce the idea of a *combination*, in which the order of the objects is not a consideration. Our work will be based on the following definition.

DEFINITION Combination of *n* Objects Taken *r* at a Time

Suppose that r objects are selected from a set of n objects, where $r \leq n$. Then this selection or subset of r objects is called a **combination of the *n* objects taken *r* at a time.** The number of such combinations is denoted by

$$C(n, r)$$

To illustrate this definition, let's take the set $\{H, J, F\}$. As we saw earlier, there are six permutations of these letters taken two at a time.

Permutations involving H and J: HJ, JH

Permutations involving J and F: JF, FJ

Permutations involving H and F: HF, FH

Now, the two permutations HJ and JH use exactly the same letters. So, corresponding to these two permutations, we have the single combination or set of letters $\{H; J\}$. Also, whether we write this set as $\{H, J\}$ or $\{J, H\}$, it's still the same combination of letters. Similarly, corresponding to the two permutations JF and FJ, we have the single combination $\{J, F\}$. Finally, corresponding to the two permutations HF and FH, we have the combination $\{H, F\}$. Thus, although we have six permutations of the three letters taken two at a time, there are only three combinations of the letters taken two at a time. We can summarize these results by writing

$$P(3, 2) = 6 \quad \text{and} \quad C(3, 2) = 3$$

EXAMPLE **11** **Computing $P(n, r)$ and $C(n, r)$ by making lists**

Using the set $\{1, 2, 3, 4\}$, list both the combinations and the permutations of these four numbers taken three at a time. Evaluate $C(4, 3)$ and $P(4, 3)$.

SOLUTION

COMBINATION	PERMUTATIONS GENERATED FROM THE COMBINATION					
$\{1, 2, 3\}$	123	132	213	231	312	321
$\{1, 2, 4\}$	124	142	241	214	412	421
$\{1, 3, 4\}$	134	143	341	314	413	431
$\{2, 3, 4\}$	234	243	342	324	423	432

We've found four combinations. (Convince yourself that there are no other distinct combinations using three of the four given digits.) Thus $C(4, 3) = 4$. Next, let's evaluate $P(4, 3)$. Assuming that the listing we've made is complete, a direct count shows that $P(4, 3) = 24$. You can check this result for yourself using the formula for $P(n, r)$ that was given earlier. Actually, there is yet another way to calculate $P(4, 3)$, and we'll show it here because it will help you to follow the reasoning in the next paragraph. Each of our combinations contains three numbers; so corresponding to each combination there are 3! or six permutations. Therefore, according to the multiplication principle, we have $P(4, 3) = 6 \times C(4, 3) = 6 \times 4 = 24$.

We can derive a formula for $C(n, r)$ as follows. First, the number of permutations of n objects taken r at a time is

$$P(n, r) = n(n - 1) \cdots (n - r + 1)$$

We can think of the task of forming any one of these permutations as two consecutive operations:

1. Select r of the n objects.
2. Then arrange these r objects in some definite order.

Since Operation 1 can be performed in $C(n, r)$ ways, and then Operation 2 can be performed in $r!$ ways, the multiplication principle tells us that

$$P(n, r) = C(n, r) \cdot r!$$

Therefore,

$$C(n, r) = \frac{P(n, r)}{r!} = \frac{n(n - 1) \cdots (n - r + 1)}{r!}$$

We can simplify this last fraction. The technique is to multiply both the numerator and the denominator by the quantity $(n - r)!$; this gives us

$$C(n, r) = \frac{n(n - 1) \cdots (n - r + 1)[(n - r)!]}{r![(n - r)!]}$$

The numerator of this last fraction is just $n!$, so we have

$$C(n, r) = \frac{n!}{r!(n - r)!}$$

Do you recognize the fraction on the right-hand side of this last equation? It's the binomial coefficient $\binom{n}{r}$ that we defined in connection with our earlier work on the binomial theorem. In summary, then, we have the following two expressions for $C(n, r)$.

$$C(n, r) = \binom{n}{r} = \frac{n!}{r!(n - r)!}$$
$$= \frac{n(n - 1) \cdots (n - r + 1)}{r!}$$

Note that in the second expression, there are r factors in the numerator and r factors in the denominator.

EXAMPLE 12 An application using the formula for C(n, r)

There are 100 senators in the United States Senate. How many four-member committees are possible?

SOLUTION
Since we are interested in the members of a committee rather than the order in which they are chosen or seated, we use combinations rather than permutations. Thus we want the number of combinations of the 100 senators taken 4 at a time. This number is

$$C(100, 4) = \binom{100}{4} = \frac{100!}{4!(100 - 4)!}$$
$$= \frac{100(99)(98)(97)(96!)}{4!(96!)}$$
$$= \frac{100(99)(98)(97)}{4!}$$
$$= 3{,}921{,}225 \qquad \text{(using a calculator)}$$

So there are 3,921,225 possible four-member committees.

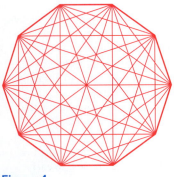

Figure 4

EXAMPLE 13   **Counting the number of diagonals in a polygon**

Figure 4 shows a **decagon** (ten-sided figure) with all its diagonals. How many diagonals are there?

SOLUTION
Each pair of vertices determines either a diagonal or a side of the decagon. So the total number of diagonals and sides is the same as the number of pairs of vertices. This latter number is $C(10, 2)$, the number of combinations of the ten vertices taken two at a time. Evaluating $C(10, 2)$, we have

$$C(10, 2) = \binom{10}{2} = \frac{10!}{2!(10-2)!} = \frac{10(9)(8!)}{2(8!)} = 45$$

Thus the total number of sides and diagonals is 45. Since 10 of these are sides, we're left with $45 - 10 = 35$ diagonals. (For a different approach to this problem, see Exercise 53.)

EXAMPLE 14   **An application of the multiplication principle and combinations**

On a certain committee there are eight Republicans and seven Democrats. How many possibilities are there for a five-member subcommittee that consists of three Republicans and two Democrats?

SOLUTION
Think of the task of forming a five-member subcommittee as composed of two consecutive operations: First choose three Republicans, then choose two Democrats. The number of ways to choose three Republicans is $C(8, 3)$, and the number of ways to choose two Democrats is $C(7, 2)$. Therefore, according to the multiplication principle, the total number of subcommittees here is

$$C(8, 3) \cdot C(7, 2) = \binom{8}{3} \cdot \binom{7}{2}$$
$$= \frac{8!}{3!(8-3)!} \cdot \frac{7!}{2!(7-2)!}$$
$$= \frac{8(7)(6)}{3(2)} \cdot \frac{7(6)}{2} = 1176 \qquad \text{(Check the arithmetic.)}$$

So there are 1176 possible subcommittees consisting of three Republicans and two Democrats.

EXERCISE SET 9.6

A

1. There are three roads from town A to town B and two roads from town B to town C. By drawing a tree diagram, find out how many routes there are from A to C via B. (Check your answer using the multiplication principle.)

2. There are five roads from town A to town B and three roads from town B to town C. Draw a tree diagram to find out how many routes there are from A to C via B. (Check your answer using the multiplication principle.)

3. There are three roads from town A to town B, two roads from town B to town C, and two roads from town C to town D. Draw a tree diagram to find out how many routes there are from town A to town D via B and C.

4. At the beginning of this section we noted that if a coin is flipped twice, there are four possible outcomes. Draw a tree diagram illustrating this.

5. How many different two-digit numbers can be formed using the digits 2, 4, 6, and 8 if no digit is repeated (within a given number)?

6. Follow Exercise 5, but allow repetitions.

7. How many three-digit numbers can be formed using the digits 1, 2, 3, 4, and 5 if no digit can be repeated (within a given number)?

8. How many even two-digit numbers can be formed using the digits 2, 4, 5, 7, and 9 if no digit is repeated (within a given number)?

9. How many three-digit numbers can be formed using the digits 3, 4, 5, 6, 7, and 8 if the number is to be less than 600 and repetitions are allowed?

10. In a certain state, the license plates contain three digits followed by three letters. (Repetitions of digits or letters on a given plate are allowed.) How many different license plates are possible?

11. Follow Exercise 10, but assume that repetitions are not allowed.

12. In how many ways can five books be arranged on a bookshelf?

13. In how many ways can ten books be arranged on a bookshelf?

14. How many permutations are there of the set $\{a, e, i, o, u\}$?

15. If the number of permutations of a certain set of letters is 24, how many letters must there be in the set?

16. Five framed photographs are going to be hung on the wall in one row. If one particular photograph must always be in the middle, how many arrangements are possible?

For Exercises 17–20, evaluate each expression.

17. $P(8, 1)$ 18. $P(8, 2)$
19. $P(6, 3)$ 20. $P(11, 3)$
21. Show that $P(n, n - 1) = n!$.
22. Show that $P(n, 1) = n$.
23. Which is larger, $P(100, 50)$ or $P(101, 51)$?

24. Which is larger, $P(100, 50)$ or $P(101, 52)$?

25. How many possible arrangements are there of the six letters $\{a, b, c, d, e, f\}$ in which b and c are next to each other? (Use the method shown in Example 9.)

26. How many arrangements are there of the six letters $\{a, b, c, d, e, f\}$ in which b and c are *not* next to each other? (Use the method shown in Example 10.)

27. How many four-digit numbers can be formed using the digits $\{2, 7, 1, 8\}$ such that 7 and 1 are next to each other? (No repetitions.)

28. How many four-digit numbers can be formed using the digits $\{2, 7, 1, 8\}$ such that 7 and 1 are *not* next to each other? (No repetitions.)

29. How many three-digit numbers can be formed from the ten digits $\{0, 1, 2, \ldots, 9\}$ if: **(a)** repetitions are allowed? (Don't forget, the first digit cannot be 0. **(b)** repetitions are not allowed?

30. How many three-digit numbers can be formed using the digits $\{1, 2, 3, 4, 5, 6\}$ if: **(a)** repetitions are not allowed and each number ends with 32? **(b)** repetitions are not allowed and the number formed in each case is odd? **(c)** repetitions are not allowed and each number formed is divisible by 5?

31. How many three-letter code words can be formed using the letters in the word "equations" if: **(a)** repetitions are allowed? **(b)** repetitions are not allowed? **(c)** repetitions are not allowed and each code word begins with q? **(d)** repetitions are not allowed and each code word begins with a vowel and ends with a consonant?

32. Using the set $\{a, b, c, d\}$, list both the combinations and the permutations of these four letters taken two at a time. Evaluate $C(4, 2)$ and $P(4, 2)$.

For Exercises 33–38, evaluate each expression.

33. **(a)** $C(12, 11)$ 34. **(a)** $C(15, 14)$
    (b) $C(12, 1)$ **(b)** $C(15, 1)$
35. **(a)** $C(8, 3)$ 36. **(a)** $C(10, 8)$
    (b) $C(8, 5)$ **(b)** $C(10, 2)$
37. **(a)** $C(12, 5)$ 38. **(a)** $C(15, 13)$
    (b) $C(12, 7)$ **(b)** $C(15, 2)$
39. Simplify $C(n, n - 2)$.   40. Simplify $C(n, 1)$.
41. There are eight people on a certain committee. How many three-member subcommittees are possible?

42. How many five-member subcommittees are possible from a committee with ten members?

43. There are ten people in a certain club.
    (a) How many different six-member committees are possible?
    (b) Of the ten people in the club, suppose that four are men and six are women. How many six-member committees are possible if each committee must be composed of three men and three women?

44. Suppose you are one of 12 people in a certain club.
  (a) How many five-member committees can be formed that include you?
  (b) How many five-member committees can be formed that exclude you?

45. Use the method shown in Example 13 to find out how many diagonals can be drawn in a hexagon (a six-sided polygon).

46. Use the method shown in Example 13 to count the number of diagonals that can be drawn in a regular 14-sided polygon.

47. Suppose that we are given eight points in the plane, and no three of the points lie on the same straight line. How many different triangles can be formed in which the vertices are chosen from these eight points?

48. Suppose that we are given 12 points lying on a certain circle. How many triangles can be formed if the vertices are chosen from the 12 points?

B

49. You have one penny, one nickel, one dime, and one quarter. How many different sums of money can you form using one or more of these coins? *Hint:* You need to compute $C(4, 1) + C(4, 2) + C(4, 3) + C(4, 4)$.

50. You have a $1 bill, a $5 bill, a $10 bill, a $20 bill, and a $100 bill. How many different sums of money can you form using one or more of these bills? (Study the hint in the previous exercise.)

51. A group consists of nine Republicans and eight Democrats. How many different eight-member committees can be formed that contain **(a)** six Republicans and two Democrats? **(b)** at least six Republicans?

52. (a) From a group of 12 people, a committee with eight members will be formed. Then, from those eight, a subcommittee with three people will be chosen. How many ways are there to carry out these two consecutive operations?
  (b) From a group of 12 people, a committee of three will be chosen. Later, five additional people are picked to increase the committee size to eight. How many ways are there to carry out these two consecutive operations?
  (c) Use parts (a) and (b) to explain why
  $$\binom{12}{8}\binom{8}{3} = \binom{12}{3}\binom{9}{5}$$
  (d) Verify the equality in part (c) by direct calculation (involving ratios of factorials).
  (e) Show that $\binom{n}{k}\binom{k}{r} = \binom{n}{r}\binom{n-r}{k-r}$, where $n \geq k \geq r$.
  Suggestion: You can do this either by direct calculation or by repeating the reasoning used in parts (a), (b), and (c) with the quantities 12, 8, and 3 replaced by n, k, and r, respectively.

53. In Example 13 we computed the number of diagonals in a decagon by considering combinations. Here is a different way to approach the problem. First observe that there are seven diagonals emanating from each of the ten vertices. Why does this imply that the total number of diagonals is $\frac{7 \cdot 10}{2} = 35$?

54. Show that $n(n-1)\cdots(n-r+1) = \dfrac{n!}{(n-r)!}$.

9.7 INTRODUCTION TO PROBABILITY

In order to define the mathematical notion of probability, we first introduce the idea of a *sample space* for an experiment. As a particularly simple first example, we can think of flipping a coin as an experiment in which there are two possible *outcomes*: heads (H) or tails (T). The set $\{H, T\}$ consisting of all possible outcomes for this experiment is called a **sample space.** Intuitively, at least, we believe that each outcome in this sample space is equally likely to occur.

As another example of this terminology, let's suppose that an experiment consists of tossing a single die and then noting the number of dots showing on the top face. Then the sample space for this experiment is the set

$$\{1, 2, 3, 4, 5, 6\}$$

Here, as in the previous example, our sample space consists of all possible outcomes for the experiment. Again, intuitively at least, we believe that each outcome is equally likely to occur. In this section, all the sample spaces that we consider are finite; that is, they contain finitely many outcomes.

Sometimes we wish to focus on a particular outcome, or a particular set of outcomes, in an experiment. For instance, in tossing the die we might be interested in whether the die shows an even number of dots. In this context, we refer to the set $\{2, 4, 6\}$ as an **event** in the sample space $\{1, 2, 3, 4, 5, 6\}$. More generally, any subset of a given sample space is referred to as an **event.** We shall use the notation $n(E)$ to denote the number of outcomes in an event E. For example, if E is the event "obtaining an even number" that we've just discussed, then $n(E) = 3$, because there are three outcomes in the set $E = \{2, 4, 6\}$.

EXAMPLE 1 Specifying events in a sample space

Use set notation to specify each of the following events in the die-tossing experiment using the sample space $\{1, 2, 3, 4, 5, 6\}$.

Event A: obtaining an odd number

Event B: obtaining a number less than 3

Event C: obtaining a 6

Event D: obtaining a number less than 7

Event E: obtaining a 7

SOLUTION

Event A: $\{1, 3, 5\}$ Event D: $\{1, 2, 3, 4, 5, 6\}$

Event B: $\{1, 2\}$ Event E: $\varnothing$

Event C: $\{6\}$

Two of the events we've listed here deserve special mention. First, notice that event D is the entire sample space, simply because every outcome in the sample space is a number less than 7. At the other extreme we have event E, "obtaining a 7." There are no outcomes in the sample space corresponding to this event, so (remembering that an event is a subset of the sample space) we say that event E is the **empty set**—that is, the set with no elements. By convention, the empty set is denoted by the symbol $\varnothing$.

EXAMPLE 2 Specifying the number of outcomes in an event

Using the events A, B, C, D, and E in Example 1, compute $n(A)$, $n(B)$, $n(C)$, $n(D)$, and $n(E)$.

SOLUTION

We need only count the number of outcomes in each set. The results are

$$n(A) = 3 \qquad n(D) = 6$$
$$n(B) = 2 \qquad n(E) = 0$$
$$n(C) = 1$$

For each event E in a finite sample space, we are going to define a number $P(E)$ called the *probability* of the event. Informally speaking, the number $P(E)$ is a rating of how likely it is that an event E will occur in any trial of the experiment. As you'll see, the number $P(E)$ always satisfies the condition $0 \leq P(E) \leq 1$.

DEFINITION	The Probability of an Event

Let S be a finite sample space in which each outcome is equally likely to occur. If E is an event in S, then the **probability** of E, denoted by $P(E)$, is defined by

$$P(E) = \frac{n(E)}{n(S)} = \frac{\text{the number of outcomes in } E}{\text{the number of outcomes in } S}$$

EXAMPLE	3	Computing the probability of an event

Find the probability for each of the events A, B, C, D, and E discussed in Examples 1 and 2.

SOLUTION

$$P(A) = P(\text{obtaining an odd number}) = \frac{n(A)}{n(S)} = \frac{3}{6} = \frac{1}{2}$$

$$P(B) = P(\text{obtaining a number less than 3}) = \frac{n(B)}{n(S)} = \frac{2}{6} = \frac{1}{3}$$

$$P(C) = P(\text{obtaining a 6}) = \frac{n(C)}{n(S)} = \frac{1}{6}$$

$$P(D) = P(\text{obtaining a number less than 7}) = \frac{n(D)}{n(S)} = \frac{6}{6} = 1$$

$$P(E) = P(\text{obtaining a 7}) = \frac{n(E)}{n(S)} = \frac{0}{6} = 0$$

Notice that each of the probabilities that we computed in Example 3 is a number between 0 and 1, as mentioned previously. Also note that a probability of 0 is assigned to the "impossible" event, whereas a probability of 1 is assigned to the "certain" event.

EXAMPLE	4	Using the formula for *C(n, r)* in computing a probability

Two (distinct) letters are selected at random from the set $\{a, b, c, d, e\}$. What is the probability that one of the letters is c?

SOLUTION
Let the sample space S consist of all combinations of the five given letters taken two at a time. For instance, ab, ac, ad, and ae are four of the outcomes in this

sample space. We needn't list every outcome in S in order to find $n(S)$; from our work in the previous section, we know that

$$n(S) = C(5, 2) = \frac{5!}{2!3!} = 10$$

Now we let E be the event "c is selected." We can specify E as follows:

$$E = \{ac, bc, dc, ec\}$$

Thus $n(E) = 4$ and, consequently,

$$P(E) = \frac{n(E)}{n(S)} = \frac{4}{10} = \frac{2}{5}$$

So the probability that c is one of the two letters selected is $2/5$. We can also express this by saying that there is a 40% chance that one of the two letters is c.

EXAMPLE 5 Computing the probability of choosing reliable transistors

Suppose that five faulty transistors are accidentally packaged with ten reliable transistors. What is the probability that three transistors chosen at random from this package will all be reliable?

SOLUTION
Let the sample space S consist of the combinations of the 15 transistors taken three at a time. Then we have

$$n(S) = C(15, 3) = \frac{15!}{3!12!} = 455$$

Next let E denote the event "choose three reliable transistors." Since these three must be chosen from the ten reliable ones, we have

$$n(E) = C(10, 3) = \frac{10!}{3!7!} = 120$$

Thus

$$P(E) = \frac{n(E)}{n(S)} = \frac{120}{455} = \frac{24}{91} \qquad \text{as required}$$

To continue our development of probability a bit further, it will be convenient to recall two basic ideas from set theory. Suppose that A and B are two sets. Then the **union** of these two sets, written $A \cup B$, is defined to be the set of elements that are in A *or* in B (or in both). For example, if

$$A = \{2, 3, 4, 5\} \qquad \text{and} \qquad B = \{4, 5, 6, 7\}$$

then

$$A \cup B = \{2, 3, 4, 5, 6, 7\}$$

The **intersection** of two sets A and B, written as $A \cap B$, is defined as the set of elements that belong to both A and B. For example, using A and B as just defined, we have

$$A \cap B = \{4, 5\}$$

because 4 and 5 are the only elements common to both sets. As another example, if $E = \{1, 2\}$ and $F = \{3, 4\}$, then $E \cap F = \emptyset$, because there are no elements common to the two sets. Two sets that have no elements in common are said to be **disjoint.**

If A and B are events (remember, that means subsets of a sample space), then $A \cup B$ is the event "A or B," whereas $A \cap B$ is the event "A and B." Here's an example. Let S be the sample space $\{1, 2, 3, 4, 5, 6\}$ that we've used for the die-tossing experiment, and let the events A and B be as follows:

A: obtaining an even number

B: obtaining a number greater than 4

Then we have

$$
\begin{aligned}
A \cup B &= \text{"obtaining an even number"} \; or \\
&\quad \text{"obtaining a number greater than 4"} \\
&= \{2, 4, 6\} \cup \{5, 6\} \\
&= \{2, 4, 5, 6\}
\end{aligned}
$$

whereas

$$
\begin{aligned}
A \cap B &= \text{"obtaining an even number"} \; and \\
&\quad \text{"obtaining a number greater than 4"} \\
&= \{2, 4, 6\} \cap \{5, 6\} \\
&= \{6\}
\end{aligned}
$$

EXAMPLE 6 **Computing the probability of the union of two events**

A coin is flipped three times and the result is recorded each time. Let A, B, and C denote the following events.

A: obtaining exactly two heads

B: obtaining no heads

C: obtaining more than one tail

Compute the following probabilities:
(a) $P(A)$, $P(B)$, and $P(C)$; **(b)** $P(A \text{ or } B)$; **(c)** $P(B \text{ or } C)$.

SOLUTION
Our sample space S will be as follows. (See Figure 1 in the previous section if you're not certain how we've obtained the eight outcomes in S.)

$$S = \{HHH, HHT, HTH, HTT, THH, THT, TTH, TTT\}$$

(a) Using the definitions given for A, B, and C, we have

$$A = \{HHT, HTH, THH\}$$
$$B = \{TTT\}$$
$$C = \{HTT, THT, TTH, TTT\}$$

Thus

$$P(A) = \frac{n(A)}{n(S)} = \frac{3}{8} \qquad P(B) = \frac{n(B)}{n(S)} = \frac{1}{8} \qquad P(C) = \frac{n(C)}{n(S)} = \frac{4}{8} = \frac{1}{2}$$

(b) The event "A or B" is $A \cup B$, so we have

$$A \cup B = \{HHT, HTH, THH, TTT\}$$

and, consequently,

$$P(A \cup B) = \frac{n(A \cup B)}{n(S)} = \frac{4}{8} = \frac{1}{2}$$

(c) Likewise, the event "B or C" is $B \cup C$, so we have

$$B \cup C = \{TTT, HTT, THT, TTH\}$$

and, consequently,

$$P(B \cup C) = \frac{n(B \cup C)}{n(S)} = \frac{4}{8} = \frac{1}{2} \qquad \text{as required}$$

We can use the results in Example 6 to introduce a rule that simplifies certain probability computations. If you review the results in Example 6, you'll see that $P(A \cup B) = 1/2$, $P(A) = 3/8$, and $P(B) = 1/8$. Thus we have

$$P(A \cup B) = P(A) + P(B) \qquad \text{because } \frac{1}{2} = \frac{3}{8} + \frac{1}{8}$$

That is,

$$P(A \text{ or } B) = P(A) + P(B) \tag{1}$$

So (in this case at least), we find that the probability of the event "A or B" is the sum of the individual probabilities of A and B. Furthermore, it can be shown that equation (1) remains valid for any two events (in a given sample space) that have no outcomes in common. *Terminology*: Two events with no outcomes in common are said to be **mutually exclusive.** For instance, the events A and B in Example 6 are mutually exclusive. (Why?)

The remarks in the previous paragraph can be summarized this way: If E and F are mutually exclusive, then $P(E \text{ or } F) = P(E) + P(F)$. This formula is *not* valid, however, in cases in which the two events fail to be mutually exclusive. To see an example of this, let's take the events B and C in Example 6. In that example we found that $P(B \cup C) = 1/2$, $P(B) = 1/8$, and $P(C) = 1/2$. Consequently we have

$$P(B \cup C) \neq P(B) + P(C) \qquad \text{because } \frac{1}{2} \neq \frac{1}{8} + \frac{1}{2}$$

The difference here, as we've said, has to do with the fact that B and C are not mutually exclusive. As you can check, B and C do have an outcome in common—namely, TTT. Because of this, when we compute $P(B \cup C)$, the outcome TTT is counted only once, but when we compute $P(B) + P(C)$, the outcome TTT is counted twice (once with B and once again with C). This is essentially the reason why we obtain different values for the quantities $P(B \cup C) =$ and $P(B) + P(C)$. By following this same type of reasoning (in which we keep track of how many times an outcome is counted) the following general formula can be derived. Notice that the formula agrees with equation (1) in the case when $A \cap B = \varnothing$ (that is, when A and B are mutually exclusive).

The Addition Theorem for Probabilities

If A and B are events in a sample space S, then

$$P(A \cup B) = P(A) + P(B) - P(A \cap B)$$

EXAMPLE 7   Using the addition theorem to compute a probability

Three balls are drawn at random from a bag containing eight red and seven yellow balls. Compute the probability for each of the following events:
(a) drawing three red balls;   **(b)** drawing three yellow balls;
(c) drawing either three red or three yellow balls.

SOLUTION
(a) The total number of possible outcomes in the experiment is $C(15, 3)$, the number of ways of choosing three balls from a set of 15. We have

$$C(15, 3) = \frac{15!}{3!12!} = \frac{15 \cdot 14 \cdot 13}{3 \cdot 2} = 455$$

Next, we need to know how many outcomes are in the event "drawing three red balls." This number is

$$C(8, 3) = \frac{8!}{3!5!} = \frac{8 \cdot 7 \cdot 6}{3 \cdot 2} = 56$$

Consequently, we have

$$P(\text{drawing 3 red balls}) = \frac{56}{455} = \frac{8}{65}$$

(b) Following the reasoning we used in part (a), we find that

$$P(\text{drawing 3 yellow balls}) = \frac{C(7, 3)}{C(15, 3)} = \frac{35}{455} \quad \text{(Check the arithmetic.)}$$
$$= \frac{1}{13}$$

(c) Let A and B denote the events described in parts (a) and (b), respectively. Then $A \cup B$ is the event "drawing either three red or three yellow balls," and we have

$$P(A \cup B) = P(A) + P(B) - P(A \cap B)$$
$$= \frac{8}{65} + \frac{1}{13} - P(\varnothing)$$
$$= \frac{8}{65} + \frac{5}{65} - 0 = \frac{13}{65} = \frac{1}{5}$$

So the probability of drawing either three red or three yellow balls is $1/5$.

EXAMPLE 8 | **Selecting a card at random from a deck of 52 cards**

A card is selected from a well-shuffled deck of 52 cards. What is the probability that it is a jack or a heart?

SOLUTION

In this experiment there are 52 possible outcomes. Let A denote the event "selecting a jack," and let B denote the event "selecting a heart." Then $P(A \cup B)$ is the probability of the event "selecting a jack or a heart," and we have

$$P(A \cup B) = P(A) + P(B) - P(A \cap B)$$
$$= \frac{4}{52} + \frac{13}{52} - \frac{1}{52} \quad \text{(Why?)}$$
$$= \frac{16}{52} = \frac{4}{13} \quad \text{as required}$$

EXERCISE SET 9.7

A

Exercises 1–20 refer to an experiment that consists of tossing a single die and then noting the number of dots showing on the top face. As in the text, assume that the sample space for this experiment is

$$S = \{1, 2, 3, 4, 5, 6\}$$

Let A, B, C, D, E, F, and G be the following events:

  A: obtaining an even number

  B: obtaining a number greater than 3

  C: obtaining a 1

  D: obtaining a number greater than 1

  E: obtaining an odd number

  F: obtaining a number less than 8

  G: obtaining an 8

Use set notation to specify the following events.

1. (a) A **2. (a)** C
  (b) B **(b)** D
  (c) $A \cup B$ **(c)** $C \cup D$
  (d) $A \cap B$ **(d)** $C \cap D$
3. (a) E **4. (a)** G
  (b) F **(b)** $G \cup B$
  (c) $E \cup A$ **(c)** $G \cap B$
  (d) $E \cap A$ **(d)** $G \cup F$

Are the following pairs of events mutually exclusive?

5. A and B **6.** A and C **7.** C and D **8.** B and E

For Exercises 9–20, compute each of the indicated quantities.

9. (a) $n(A)$ **10. (a)** $n(C)$ **11. (a)** $n(E)$
  (b) $n(B)$ **(b)** $n(D)$ **(b)** $n(F)$
  (c) $n(A \cup B)$ **(c)** $n(C \cup D)$ **(c)** $n(G)$
12. $n(B \cap E)$ **13. (a)** $P(a)$ **14. (a)** $P(C)$
                   (b) $P(B)$ **(b)** $P(D)$
15. (a) $P(E)$ **16.** $P(G)$ **17. (a)** $P(A \cup B)$
  (b) $P(F)$ **(b)** $P(A) + P(B)$
18. (a) $P(B \cup C)$ **19. (a)** $P(B \cup E)$
  (b) $P(B) + P(C)$ **(b)** $P(B) + P(E)$
20. (a) $P(D) + P(E) - P(D \cap E)$
  (b) $P(D \cup E)$

Exercises 21–40 refer to the experiment described in Example 6, in which a coin is flipped three times and the result is recorded each time. As in Example 6, let the sample space S be

$$S = \{HHH, HHT, HTH, HTT, THH, THT, TTH, TTT\}$$

Let A, B, C, D, E, and F be the following events:

  A: obtaining exactly two tails

  B: obtaining two heads on the first two tosses

  C: obtaining heads on the first toss

  D: obtaining no heads

  E: obtaining tails on both the second and third tosses

  F: obtaining tails on the first toss

Compute each of the following probabilities.

21. $P(A)$ **22.** $P(B)$ **23.** $P(C)$
24. $P(D)$ **25.** $P(E)$ **26.** $P(F)$

27. $P(A \cup B)$ **28.** $P(A \cap B)$ **29.** $P(A \cup C)$
30. $P(A \cap C)$ **31.** $P(C \cup F)$ **32.** $P(C \cap F)$
33. $P(E \cup F)$ **34.** $P(E \cap F)$ **35.** $P(B \cup C)$
36. $P(B \cap C)$ **37.** $P(D \cap A)$ **38.** $P(D \cup A)$
39. $P(D \cup E)$ **40.** $P(D \cap E)$

41. Two letters are selected at random (and without replacement) from the set $\{a, b, c, d, e, f\}$. What is the probability that one of the letters is d?

42. The numbers 1, 2, 3, and 4 are arranged in random order to form a four-digit number. What is the probability that the number is greater than 4000?

43. Suppose that two faulty transistors are accidentally packaged with a dozen reliable ones.
  (a) If one transistor is chosen at random from the package, what is the probability that it will be reliable?
  (b) If five transistors are chosen at random from the package, what is the probability that they are all reliable?

44. Suppose that five cartons of spoiled milk are inadvertently placed on a grocery shelf along with 15 cartons of fresh milk.
  (a) If three cartons are selected at random, what is the probability that they all contain unspoiled milk?
  (b) What is the probability that three cartons selected at random all contain spoiled milk?

In Exercises 45–48, use the addition theorem to compute the required probabilities.

45. If $P(A) = 1/3$, $P(B) = 1/4$, and $A \cap B = \varnothing$, compute $P(A \cup B)$.

46. If $P(A) = 2/5$, $P(B) = 3/10$, and $P(A \cap B) = 1/5$, compute $P(A \cup B)$.

47. If $P(A \cup B) = 7/9$, $P(A) = 5/9$, and $P(B) = 4/9$, compute $P(A \cap B)$.

48. If $P(A) = 1/3$, $P(A \cup B) = 2/3$, and $P(A \cap B) = 1/6$, compute $P(B)$.

49. Two balls are drawn at random from a bag containing six red and ten black balls. Compute the probability for each of the following events: **(a)** drawing two red balls; **(b)** drawing two black balls; **(c)** drawing either two red or two black balls.

50. A card is selected at random from a deck of 52 cards. Compute the probability of each of the following events: **(a)** selecting a 4 or 2; **(b)** selecting a red card or an ace; **(c)** selecting an ace or a picture card.

51. Your mathematics instructor claims that the probabilities of your getting A's on the first exam, on the second exam, and on both exams are 1/2, 3/10, and 1/10, respectively. According to the figures, what is the probability that you

will get an A on at least one exam?   *Hint:* Let G and H denote the events "A on first exam" and "A on second exam," respectively. Then you want to compute $P(G$ or $H)$.

52. Two cards are drawn at random from a regular 52-card deck. Find the probability that either both are black or both are 2s.

53. Five cards are drawn at random from a regular deck of 52 cards. Compute the probability that all five cards are clubs.

B

54. An experiment consists of selecting at random one of the 100 points in the x-y plane shown in the accompanying figure. Compute the probability for each of the following events.
  (a) The point lies on the line $y = x$.
  (b) The point lies on the line $y = 2x$.
  (c) The point lies on the circle $x^2 + y^2 = 25$.
  (d) The point lies within (but not on) the circle $x^2 + y^2 = 25$.
  (e) The point lies on or within the circle $x^2 + y^2 = 25$.
  (f) The point lies outside the circle $x^2 + y^2 = 25$.
  (g) The point lies inside the circle $x^2 + y^2 = 25$ but not on the graph of $y = |x - 3|$.
  (h) The point lies either on the circle $x^2 + y^2 = 25$ or on the line $y = x$.
  (i) The point lies on the x-axis or on the circle $x^2 + y^2 = 25$.
  (j) The sum of the x- and y-coordinates of the point is 18.
  (k) The sum of the x- and y-coordinates of the point is 21.
  (l) The y-coordinate of the point is greater than the square of the x-coordinate.

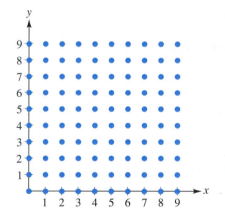

Chapter 9 Summary of Principal Terms and Formulas

Terms or formulas	Page reference	Comments
1. The principle of mathematical induction	695	This principle can be stated as follows. Suppose that for each natural number n we have a statement P_n. Suppose P_1 is true. Also suppose that P_{k+1} is true whenever P_k is true. Then according to the principle of mathematical induction, all of the statements are true; that is, P_n is true for all natural numbers n. The principle of mathematical induction has the status of an axiom; that is, we accept its validity without proof.
2. Pascal's triangle	701	Pascal's triangle refers to the triangular array of numbers displayed on page 701. Additional rows can be added to the triangle according to the following rule: Each entry in the array (other than the 1's along the sides) is the sum of the two numbers diagonally above it. The numbers in the nth row of Pascal's triangle are the coefficients of the terms in the expansion of $(a + b)^{n-1}$.
3. $n!$ (read: n factorial)	703	This denotes the product of the first n natural numbers. For example, $4! = (4)(3)(2)(1) = 24$.
4. $\dbinom{n}{k}$ (read: n choose k)	703	Let n and k be nonnegative integers with $k \leq n$. Then the *binomial coefficient* $\dbinom{n}{k}$ is defined by $$\binom{n}{k} = \frac{n!}{k!(n-k)!}$$
5. The binomial theorem	704	The binomial theorem is a formula that allows us to analyze and expand expressions of the form $(a + b)^n$. If we use *sigma notation,* then the statement of the binomial theorem that appears on page 704 can be abbreviated to read $$(a + b)^n = \sum_{k=0}^{n} \binom{n}{k} a^{n-k} b^k$$
6. Sequence	709	In the context of the present chapter, a sequence is an ordered list of real numbers: $a_1, a_2, a_3, \ldots$. The numbers a_k are called the *terms* of the sequence. A sequence can also be defined as a *function* whose domain is the set of natural numbers or the set of nonnegative integers.
7. $\displaystyle\sum_{k=1}^{n} a_k$	715	This expression stands for the sum $a_1 + a_2 + a_3 + \cdots + a_n$. The letter k in the expression is referred to as the *index of summation.*

Terms or formulas	Page reference	Comments
8. Arithmetic sequence	720	A sequence in which the successive terms differ by a constant is called an arithmetic sequence. The general form of an arithmetic sequence is $$a, a + d, a + 2d, a + 3d, \ldots$$ In this sequence, d is referred to as the *common difference*.
9. $a_n = a + (n - 1)d$	721	This is the formula for the nth term of an arithmetic sequence.
10. $S_n = \dfrac{n}{2}[2a + (n - 1)d]$ $S_n = n\left(\dfrac{a + a_n}{2}\right)$	723	These are the formulas for the sum of the first n terms of an arithmetic sequence.
11. Geometric sequence	726	A sequence in which the ratio of successive terms is constant is called a geometric sequence. The general form of a geometric sequence is $$a, ar, ar^2, \ldots$$ The number r is referred to as the *common ratio*.
12. $a_n = ar^{n-1}$	726	This is the formula for the nth term of a geometric sequence.
13. $S_n = \dfrac{a(1 - r^n)}{1 - r}$	727	This is the formula for the sum of the first n terms of a geometric sequence.
14. $S = \dfrac{a}{1 - r}$	729	This is the formula for the sum of the infinite geometric series $a + ar + ar^2 + \cdots$, where $|r| < 1$.
15. The multiplication principle	731	Suppose that a task involves two consecutive operations. If the first operation can be performed in M ways, and the second operation can be performed in N ways, then the entire task can be performed in MN ways.
16. Permutation	732	A permutation of a set of n objects is a listing or arrangement of those objects in a definite order.
17. $P(n, r) = \dfrac{n!}{(n - r)!}$ $= n(n - 1) \cdots (n - r + 1)$	734	$P(n, r)$ denotes the number of permutations of n objects taken r at a time, where $r \leq n$.
18. $C(n, r) = \dfrac{n!}{r!(n - r)!}$ $= \dfrac{n(n - 1) \cdots (n - r + 1)}{r!}$	738	$C(n, r)$ denotes the number of ways of selecting r objects from a set of n objects, where $r \leq n$. Each selection of subset of r objects is called a *combination* of the n objects taken r at a time.

Terms or formulas	Page reference	Comments
19. Sample space; event	741	The sample space for an experiment is the set of all possible outcomes of the experiment. A subset of the sample space is called an event. The notation $n(E)$ is used to denote the number of outcomes in an event E.
20. $P(E)$	743	Let S be a finite sample space in which each outcome is equally likely to occur. If E is an event in S, then the probability of E, denoted by $P(E)$, is defined by $$P(E) = \frac{n(E)}{n(S)} = \frac{\text{the number of outcomes in } E}{\text{the number of outcomes in } S}$$
21. The addition theorem for probabilities	747	If A and B are events in a sample space S, then $$P(A \cup B) = P(A) + P(B) - P(A \cap B)$$

Writing Mathematics

1. A sequence is defined recursively as follows: $a_0 = \pi$; $a_{n+1} = \dfrac{a_n}{a_n + 1}$ for $n \geq 0$.

   Compute a_0, a_1, and a_2. Describe (in a complete sentence or two) what pattern you observe. Make a conjecture about a general formula for a_n. Then use mathematical induction to prove that the formula is valid. (As in the examples in the text, use complete sentences in writing and explaining the proof.)

2. A sequence is defined recursively as follows.

$$a_1 = 3/2$$
$$a_{n+1} = a_n^2 - 2a_n + 2 \quad \text{for } n \geq 1$$

   Use mathematical induction to show that $1 < a_n < 2$ for all natural numbers n.

3. Study the procedure that was used on page 727 to obtain the formula for the sum of a finite geometric series. Then, work with classmates or your instructor to apply the procedure to find the sum of each of the following series. Finally, on your own, write a complete summary of what you have done.

   (a) $1 + 2 \ln x + 3(\ln x)^2 + 4(\ln x)^3 + \cdots + n(\ln x)^{n-1}$

   (b) $1 + 2e^x + 3e^{2x} + 4e^{3x} + \cdots + ne^{(n-1)x}$

4. Study the following purported "proof" by mathematical induction, and explain the fallacy.

   Let P_n denote the following statement: In any set of n numbers, the numbers are all equal.

   Step 1 P_1 is clearly true.

   Step 2 Assume that P_k is true. We must show that P_{k+1} is true. That is, given a set of $k + 1$ numbers $a_1, a_2, a_3, \ldots, a_{k+1}$, we must show that $a_1 = a_2 = a_3 = \cdots = a_{k+1}$. By the inductive hypothesis, however, we have $a_1 = a_2 = a_3 = \cdots = a_k$ and also $a_2 = a_3 = a_4 = \cdots = a_{k+1}$. These last two sets of equations imply that $a_1 = a_2 = a_3 = \cdots = a_k = a_{k+1}$, as we wished to show. Having now completed Steps 1 and 2, we conclude that P_n is true for every n.

5. Look up the word *anagram* in a dictionary, then explain which concept from this chapter is illustrated by the following special anagram:

    TELEGRAPH $\leftrightarrow$ GREAT HELP

6. (a) Write down a formula for the number of permutations of five objects taken r at a time.

    (b) Does the formula define a function? If your answer is *no,* explain your reasons. If your answer is *yes,* explain your reasons and describe the domain, the range, and the graph.

7. The following exercise appears in the text *Essentials of Finite Mathematics: Matrices, Linear Programming, Probability, Markov Chains* by Robert F. Brown and Brenda W. Brown (New York: Ardsley House Publishers, 1990). Working with a classmate or your instructor, solve the problem. Then, on your own, write out the complete solution. Be sure to let the reader know what you're doing and which principles you are applying.

    In a chemical experiment, six different chemicals will be mixed by adding one at a time to the mixture. There are two chemicals, out of the six, that cannot be added one after the other (in either order) without causing an undesirable reaction. In how many ways can the experiment be performed? *Hint:* In how many ways can the experiment be performed by adding the two reacting chemicals one after the other?

Chapter 9 Review Exercises

In Exercises 1–10, use the principle of mathematical induction to show that the statements are true for all natural numbers.

1. $5 + 10 + 15 + \cdots + 5n = \frac{5}{2}n(n + 1)$

2. $10 + 10^2 + 10^3 + \cdots + 10^n = \frac{10}{9}(10^n - 1)$

3. $1 \cdot 2 + 2 \cdot 3 + 3 \cdot 4 + \cdots$
$$+ \, n(n + 1) = \tfrac{1}{3}n(n + 1)(n + 2)$$

4. $\dfrac{1}{2} + \dfrac{2}{2^2} + \dfrac{3}{2^3} + \cdots + \dfrac{n}{2^n} = 2 - \dfrac{2 + n}{2^n}$

5. $1 + 3 \cdot 2 + 5 \cdot 2^2 + 7 \cdot 2^3 + \cdots$
$$+ \, (2n - 1) \cdot 2^{n-1} = 3 + (2n - 3) \cdot 2^n$$

6. $\dfrac{1}{1 \cdot 4} + \dfrac{1}{4 \cdot 7} + \dfrac{1}{7 \cdot 10} + \cdots$
$$+ \, \dfrac{1}{(3n - 2)(3n + 1)} = \dfrac{n}{3n + 1}$$

7. $1 + 2^2 \cdot 2 + 3^2 \cdot 2^2 + 4^2 \cdot 2^3 + \cdots + n^2 \cdot 2^{n-1}$
$$= (n^2 - 2n + 3)2^n - 3$$

8. 9 is a factor of $n^3 + (n + 1)^3 + (n + 2)^3$.

9. 3 is a factor of $7^n - 1$.

10. 8 is a factor of $9^n - 1$.

In Exercises 11–20, expand the given expressions.

11. $(3a + b^2)^4$

12. $(5a - 2b)^3$

13. $(x + \sqrt{x})^4$

14. $(1 - \sqrt{3})^6$

15. $(x^2 - 2y^2)^5$

16. $\left(\dfrac{1}{a} + \dfrac{2}{b}\right)^3$

17. $\left(1 + \dfrac{1}{x}\right)^5$

18. $\left(x^3 + \dfrac{1}{x^2}\right)^6$

19. $(a\sqrt{b} - b\sqrt{a})^4$

20. $(x^{-2} + y^{5/2})^8$

21. Find the fifth term in the expansion of $(3x + y^2)^5$.

22. Find the eighth term in the expansion of $(2x - y)^9$.

23. Find the coefficient of the term containing a^5 in the expansion of $(a - 2b)^7$.

24. Find the coefficient of the term containing b^8 in the expansion of $\left(2a - \dfrac{b}{3}\right)^{10}$.

25. Find the coefficient of the term containing x^3 in the expansion of $(1 + \sqrt{x})^8$.

26. Expand $(1 + \sqrt{x} + x)^6$. *Suggestion*: Rewrite the expression as $[(1 + \sqrt{x}) + x]^6$.

In Exercises 27–32, verify each assertion by computing the indicated binomial coefficients.

27. $\dbinom{2}{0}^2 + \dbinom{2}{1}^2 + \dbinom{2}{2}^2 = \dbinom{4}{2}$

28. $\dbinom{3}{0}^2 + \dbinom{3}{1}^2 + \dbinom{3}{2}^2 + \dbinom{3}{3}^2 = \dbinom{6}{3}$

29. $\dbinom{4}{0}^2 + \dbinom{4}{1}^2 + \dbinom{4}{2}^2 + \dbinom{4}{3}^2 + \dbinom{4}{4}^2 = \dbinom{8}{4}$

30. $\dbinom{2}{0} + \dbinom{2}{1} + \dbinom{2}{2} = 2^2$

31. $\dbinom{3}{0} + \dbinom{3}{1} + \dbinom{3}{2} + \dbinom{3}{3} = 2^3$

32. $\dbinom{4}{0} + \dbinom{4}{1} + \dbinom{4}{2} + \dbinom{4}{3} + \dbinom{4}{4} = 2^4$

In Exercises 33–38, compute the first four terms in each sequence. Also, in Exercises 33–35, graph the sequences for $n = 1, 2, 3,$ and 4.

33. $a_n = \dfrac{2n}{n+1}$

34. $a_n = \dfrac{3n-2}{3n+2}$

35. $a_n = (-1)^n \left(1 - \dfrac{1}{n+1}\right)$

36. $a_0 = 4$; $a_n = 2a_{n-1}, n \geq 1$

37. $a_0 = -3$; $a_n = 4a_{n-1}, n \geq 1$

38. $a_0 = 1$; $a_1 = 2$; $a_n = 3a_{n-1} + 2a_{n-2}, n \geq 2$

In Exercises 39 and 40, evaluate each sum.

39. (a) $\displaystyle\sum_{k=1}^{3} (-1)^k (2k+1)$

(b) $\displaystyle\sum_{k=0}^{8} \left(\dfrac{1}{k+1} - \dfrac{1}{k+2}\right)$

40. (a) $\displaystyle\sum_{j=1}^{4} \dfrac{(-1)^j}{j}$

(b) $\displaystyle\sum_{n=1}^{5} \left(\dfrac{1}{n} - \dfrac{1}{n+1}\right)$

In Exercises 41 and 42, rewrite each sum using sigma notation.

41. $\dfrac{5}{3} + \dfrac{5}{3^2} + \dfrac{5}{3^3} + \dfrac{5}{3^4} + \dfrac{5}{3^5}$

42. $\dfrac{1}{2} - \dfrac{2}{2^2} + \dfrac{3}{2^3} - \dfrac{4}{2^4} + \dfrac{5}{2^5} - \dfrac{6}{2^6}$

In Exercises 43–46, find the indicated term in each sequence.

43. $5, 9, 13, 17, \ldots ; a_{18}$

44. $5, 9/2, 4, 7/2, \ldots ; a_{20}$

45. $10, 5, 5/2, 5/4, \ldots ; a_{12}$

46. $\sqrt{2} + 1, 1, \sqrt{2} - 1, 3 - 2\sqrt{2}, \ldots ; a_{10}$

47. Determine the sum of the first 12 terms of an arithmetic sequence in which the first term is 8 and the 12th term is 43/2.

48. Find the sum of the first 45 terms in the sequence $10, 29/3, 28/3, 9, \ldots .$

49. Find the sum of the first 10 terms in the sequence $7, 70, 700, \ldots .$

50. Find the sum of the first 12 terms in the sequence $1/3, -2/9, 4/27, -8/81, \ldots .$

51. In a certain geometric sequence, the third term is 4 and the fifth term is 10. Find the sixth term, given that the common ratio is negative.

52. For a certain infinite geometric series, the first term is 2 and the sum is 18/11. Find the common ratio r.

In Exercises 53–56, find the sum of each infinite geometric series.

53. $\dfrac{3}{5} + \dfrac{3}{25} + \dfrac{3}{125} + \cdots$

54. $\dfrac{7}{10} + \dfrac{7}{100} + \dfrac{7}{1000} + \cdots$

55. $\dfrac{1}{9} - \dfrac{1}{81} + \dfrac{1}{729} - \cdots$

56. $1 + \dfrac{1}{1 + \sqrt{2}} + \dfrac{1}{(1 + \sqrt{2})^2} + \cdots$

57. Find a fraction equivalent of $0.\overline{45}$.

58. Find a fraction equivalent to $0.2\overline{13}$.

In Exercises 59–62, verify each equation using the formula for the sum of an arithmetic series:

$$S_n = \dfrac{n}{2}[2a + (n-1)d]$$

(The formulas given in Exercises 59–62 appear in Elements of Algebra by Leonhard Euler, first published in 1770.)

59. $1 + 2 + 3 + \cdots + n = n + n(n-1)/2$

60. $1 + 3 + 5 + \cdots \text{(to } n \text{ terms)} = n + 2n(n-1)/2$

61. $1 + 4 + 7 + \cdots \text{(to } n \text{ terms)} = n + 3n(n-1)/2$

62. $1 + 5 + 9 + \cdots \text{(to } n \text{ terms)} = n + 4n(n-1)/2$

63. In this exercise, you will use the following (remarkably simple) formula for approximating sums of powers of integers:

$$1^k + 2^k + 3^k + \cdots + n^k \approx \frac{(n + \frac{1}{2})^{k+1}}{k + 1} \qquad (1)$$

[This formula appears in the article, "Sums of Powers of Integers" by B. L. Burrows and R. F. Talbot, published in the *American Mathematical Monthly*, vol. 91 (1984), p. 394.]

(a) Use formula (1) to estimate the sum $1^2 + 2^2 + 3^2 + \cdots + 50^2$. Round your answer to the nearest integer.

(b) Compute the exact value of the sum in part (a) using the formula $\sum_{k=1}^{n} k^2 = \frac{1}{6}n(n + 1)(2n + 1)$. (This formula can be proved using mathematical induction.) Then compute the percentage error for the approximation obtained in part (a). The percentage error is given by

$$\frac{|\text{actual value} - \text{approximate value}|}{\text{actual value}} \times 100$$

(c) Use formula (1) to estimate the sum $1^4 + 2^4 + 3^4 + \cdots + 200^4$. Round your answer to six significant digits.

(d) The following formula for the sum $1^4 + 2^4 + \cdots + n^4$ can be proved using mathematical induction:

$$\sum_{k=1}^{n} k^4 = \frac{n(n + 1)(2n + 1)(3n^2 + 3n - 1)}{30}$$

Use this formula to compute the sum in part (c). Round your answer to six significant digits. Then use this result to compute the percentage error for the approximation in part (c).

64. According to *Stirling's formula* [named after James Stirling (1692–1770)], the quantity $n!$ can be approximated as follows:

$$n! \approx \sqrt{2\pi n}\left(\frac{n}{e}\right)^n$$

In this formula, e is the constant 2.718 . . . (discussed in Section 5.2). Use a calculator to complete the following table. Round your answers to five significant digits. As you will see, the numbers in the right-hand column approach 1 as n increases. This shows that,

in a certain sense, the approximation improves as n increases.

n	$n!$	$\sqrt{2\pi n}(n/e)^n$	$\dfrac{n!}{\sqrt{2\pi n}(n/e)^n}$
10			
20			
30			
40			
50			
60			
65			

65. How many different two-digit numbers can be formed using the digits 2, 4, 6, and 8? (Repetitions are allowed.)

66. How many permutations are there of the set $\{A, B, C, D, E\}$?

67. Using the set $\{a, b, c, d\}$, list the permutations of these four letters taken three at a time. Evaluate $P(4, 3)$.

68. Follow Exercise 67, using combinations rather than permutations, and evaluate $C(4, 3)$.

69. From a collection of 12 pictures, eight are to be selected and hung in a row. How many arrangements are possible?

70. How many possible arrangements are there of the letters a, b, c, d, e, f in which a, b, and c appear consecutively (but not necessarily in that order)?

71. How many possible arrangements are there of the letters a, b, c, d, e, f in which e and f are not next to each other?

72. How many four-member committees are possible in a club with 15 members?

73. On a certain committee there are nine males and ten females. How many possibilities are there for a four-member subcommittee consisting of two females and two males?

74. Given ten points in a plane, no three of which lie on the same line, find the number of different triangles that can be formed when the vertices are selected from the given points.

75. Suppose that three faulty batteries are accidentally shipped along with nine reliable batteries. What is the probability that two batteries selected at random from this shipment will both be reliable?

76. The accompanying figure shows a 12-sided figure with all its diagonals. How many diagonals are there?

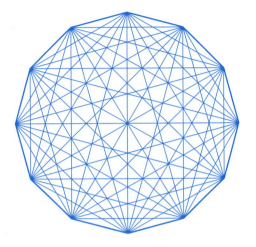

77. Two balls are selected at random from a bag containing five red and seven green balls. Compute the probability for each of the following events: **(a)** drawing two red balls; **(b)** drawing two green balls; **(c)** drawing either two red or two green balls.

For Exercises 78–88, consider an experiment in which two dice are tossed. Assume that one die is red and the other is white. Then the outcomes can be recorded as ordered pairs. Let us agree that the first element in any ordered pair corresponds to the number of dots showing on the red die and the second element corresponds to the number of dots showing on the white die. For example, (5, 4) denotes the outcome "5 on red and 4 on white." Compute the probabilities for the following events.

78. Obtaining a sum of 2
79. Obtaining a sum of 3
80. Obtaining a sum of 5
81. Obtaining a sum of 7
82. Obtaining a sum of 8
83. Obtaining a sum of either 7 or 11
84. Obtaining "doubles" (i.e., the same number of dots on both dice)
85. Obtaining a sum of at least 10
86. Obtaining a sum that is either even or less than 5
87. Obtaining any sum other than 7 and 8
88. Obtaining either doubles or a sum greater than 9
89. An integer between 1 and 15, inclusive, is selected at random. Compute the probabilities for the following events: **(a)** selecting an even number; **(b)** selecting an odd number; **(c)** selecting either a number greater than 10 or an even number; **(d)** selecting either a multiple of 2 or a multiple of 3.

90. (a) How many three-digit numbers are there with no repeated digits? (The first digit cannot be zero.)
(b) A three-digit number is selected at random. What is the probability that it has no repeated digits?

91. (a) How many three-digit numbers are there with one or more repeated digits?
(b) What is the probability that a three-digit number selected at random will contain one or more repeated digits?

92. The sum of three consecutive terms in a geometric sequence is 13, and the sum of the reciprocals is 13/9. What are the possible values for the common ratio? (There are four answers.)

93. The nonzero numbers a, b, and c are consecutive terms in a geometric sequence, and $a + b + c = 70$. Furthermore, $4a$, $5b$, and $4c$ are consecutive terms in an arithmetic sequence. Find a, b, and c.

94. The nonzero numbers a, b, and c are consecutive terms in a geometric sequence, and a, $2b$, and c are consecutive terms in an arithmetic sequence. Show that the common ratio in the geometric sequence must be either $2 + \sqrt{3}$ or $2 - \sqrt{3}$.

95. If the numbers $\dfrac{1}{b + c}, \dfrac{1}{c + a}$, and $\dfrac{1}{a + b}$ are consecutive terms in an arithmetic sequence, show that a^2, b^2, and c^2 are also consecutive terms in an arithmetic sequence.

96. If a, b, and c are consecutive terms in a geometric sequence, prove that $\dfrac{1}{a + b}, \dfrac{1}{2b}$, and $\dfrac{1}{c + b}$ are consecutive terms in an arithmetic sequence.

97. (a) Find a value for x such that $3 + x$, $4 + x$, and $5 + x$ are consecutive terms in a geometric sequence.
(b) Given three numbers a, b, and c, find a value for x (in terms of a, b, and c) such that $a + x$, $b + x$, and $c + x$ are consecutive terms in a geometric sequence.

98. If $\ln(A + C) + \ln(A + C - 2B) = 2 \ln(A - C)$, show that $1/A$, $1/B$, and $1/C$ are consecutive terms in an arithmetic sequence.

Chapter 9   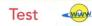 Test

1. Use the principle of mathematical induction to show that the following formula is valid for all natural numbers n:

$$1^2 + 2^2 + 3^2 + \cdots + n^2 = \frac{n(n+1)(2n+1)}{6}.$$

2. Express each of the following sums without using sigma notation, and then evaluate each sum.

   (a) $\displaystyle\sum_{k=0}^{2}(10k - 1)$ **(b)** $\displaystyle\sum_{k=1}^{3}(-1)^k k^2$

3. **(a)** Write the formula for the sum S_n of a finite geometric series.

   (b) Evaluate the sum $\dfrac{3}{2} + \dfrac{3^2}{2^2} + \dfrac{3^3}{2^3} + \cdots + \dfrac{3^{10}}{2^{10}}$.

4. **(a)** Determine the coefficient of the term containing a^3 in the expansion of $(a - 2b^3)^{11}$.

   (b) Find the fifth term of the expansion in part (a).

5. Expand the expression $(3x^2 + y^3)^5$.

6. Determine the sum of the first 12 terms of an arithmetic sequence in which the first term is 8 and the 12th term is 43/2.

7. Find the sum of the following infinite geometric series:

$$\frac{7}{10} + \frac{7}{100} + \frac{7}{1000} + \cdots.$$

8. A sequence is defined recursively as follows: $a_1 = 1$; $a_2 = 1$; and $a_n = (a_{n-1})^2 + a_{n-2}$ for $n \geq 3$. Determine the fourth and fifth terms in this sequence.

9. In a certain geometric sequence, the third term is 4 and the fifth term is 10. Find the sixth term, given that the common ratio is negative.

10. What is the 20th term in the arithmetic sequence $-61, -46, -31, \ldots$?

11. Graph the first four terms of each sequence.

    (a) $a_n = \frac{1}{2}(n^2 - n)$, $n \geq 0$

    (b) $b_1 = 1$; $b_n = (b_{n-1})^2 - nb_{n-1}$, $n \geq 2$

12. A code word consists of a sequence of three terms that can be letters (of the English alphabet) or digits, subject to the following constraints. The first term can either be a digit or a consonant. The second term must either be a vowel or a digit that is greater than 5. The third term can be any letter or one of the digits 0 or 1. How many code words are possible?

13. From a collection of 14 pictures, eight are to be selected and hung in a row. How many arrangements are possible?

14. Suppose that three faulty batteries are accidentally shipped along with 12 reliable batteries. What is the probability that two batteries selected at random from this shipment will both be reliable?

15. Two balls are selected at random from a bag containing three red and seven green balls. Compute the probability for each of the following events:
    (a) drawing two red balls; **(b)** drawing two green balls; **(c)** drawing either two red or two green balls.

Appendix

A.1 SIGNIFICANT DIGITS

Many of the numbers that we use in scientific work and in daily life are approximations. In some cases the approximations arise because the numbers are obtained through measurements or experiments. Consider, for example, the following statement from an astronomy textbook:

The diameter of the Moon is 3476 km.

We interpret this statement as meaning that the actual diameter D is closer to 3476 km than it is to either 3475 km or 3477 km. In other words,

$$3475.5\,\text{km} \leq D \leq 3476.5\,\text{km}$$

The interval [3475.5, 3476.5] in this example provides information about the accuracy of the measurement. Another way to indicate accuracy in an approximation is by specifying the number of *significant digits* it contains. The measurement 3476 km has four significant digits. In general, the number of significant digits in a given number is found as follows.

Significant Digits

The number of significant digits in a given number is determined by counting the digits from left to right, beginning with the leftmost nonzero digit.

EXAMPLES	
Number	Number of Significant Digits
1.43	3
0.52	2
0.05	1
4837	4
4837.0	5

Numbers obtained through measurements are not the only source of approximations in scientific work. For example, to five significant digits, we have the following approximation for the irrational number π:

$$\pi \approx 3.1416$$

This statement tells us that π is closer to 3.1416 than it is to either 3.1415 or 3.1417. In other words,

$$3.14155 \leq \pi \leq 3.14165$$

Table 1 provides some additional examples of the ideas we've introduced.

TABLE 1

Number	Number of significant digits	Range of measurement
37	2	[36.5, 37.5]
37.0	3	[36.95, 37.05]
268.1	4	[268.05, 268.15]
1.036	4	[1.0355, 1.0365]
0.036	2	[0.0355, 0.0365]

There is an ambiguity involving zero that can arise in counting significant digits. Suppose that someone measures the width w of a rectangle and reports the result as 30 cm. How many significant digits are there? If the value 30 cm was obtained by measuring to the nearest 10 cm, then only the digit 3 is significant, and we can conclude only that the width w lies in the range 25 cm $\leq w \leq$ 35 cm. On the other hand, if the 30 cm was obtained by measuring to the nearest 1 cm, then both the digits 3 and 0 are significant and we have 29.5 cm $\leq w \leq$ 30.5 cm.

By using **scientific notation** we can avoid the type of ambiguity discussed in the previous paragraph. A number written in the form

$$b \times 10^n \qquad \text{where } 1 \leq b < 10 \text{ and } n \text{ is an integer}$$

is said to be expressed in scientific notation. For the example in the previous paragraph, then, we would write

$$w = 3 \times 10^1 \text{ cm} \qquad \text{if the measurement is to the nearest 10 cm}$$

and

$$w = 3.0 \times 10^1 \text{ cm} \qquad \text{if the measurement is to the nearest 1 cm}$$

As the figures in Table 2 indicate, for a number $b \times 10^n$ in scientific notation, the number of significant digits is just the number of digits in b. (This is one of the advantages in using scientific notation; the number of significant digits, and hence the accuracy of the measurement, is readily apparent.)

TABLE 2

Measurement	Number of significant digits	Range of measurement
Mass of the earth:		
6×10^{27} g	1	[5.5×10^{27} g, 6.5×10^{27} g]
6.0×10^{27} g	2	[5.95×10^{27} g, 6.05×10^{27} g]
5.974×10^{27} g	4	[5.9735×10^{27} g, 5.9745×10^{27} g]
Mass of a proton:		
1.67×10^{-24} g	3	[1.665×10^{-24} g, 1.675×10^{-24} g]

Many of the numerical exercises in the text ask that you round the answers to a specified number of decimal places. Our rules for rounding are as follows.

Rules for Rounding a Number (with more than n Decimal Places) to n Decimal Places

1. If the digit in the $(n + 1)$st decimal place is greater than 5, increase the digit in the nth place by 1. If the digit in the $(n + 1)$st place is less than 5, leave the nth digit unchanged.
2. If the digit in the $(n + 1)$st decimal place is 5 and there is at least one nonzero digit to the right of this 5, increase the digit in the nth decimal place by 1.
3. If the digit in the $(n + 1)$st decimal place is 5 and there are no nonzero digits to the right of this 5, then increase the digit in the nth decimal place by 1 only if this results in an even digit.

The examples in Table 3 illustrate the use of these rules.

▌ TABLE 3

Number	Rounded to one decimal place	Rounded to three decimal places
4.3742	4.4	4.374
2.0515	2.1	2.052
2.9925	3.0	2.992

These same rules can be adapted for rounding a result to a specified number of significant digits. As examples of this, we have

2347	rounded to two significant digits is $2300 = 2.3 \times 10^3$
2347	rounded to three significant digits is $2350 = 2.35 \times 10^3$
975	rounded to two significant digits is $980 = 9.8 \times 10^2$
0.985	rounded to two significant digits is $0.98 = 9.8 \times 10^{-1}$

In calculator exercises that ask you to round your answers, it's important that you postpone rounding until the final calculation is carried out. For example, suppose that you are required to determine the hypotenuse x of the right triangle in Figure 1 to two significant digits. Using the Pythagorean theorem, we have

$$x = \sqrt{(1.36)^2 + (2.46)^2}$$

$$= 2.8 \qquad \text{using a calculator and rounding the final result to two significant digits}$$

Figure 1

On the other hand, if we first round each of the given lengths to two significant digits, we obtain

$$x = \sqrt{(1.4)^2 + (2.5)^2}$$

$$= 2.9 \qquad \text{to two significant digits}$$

This last result is inappropriate, and we can see why as follows. As Table 4 shows, the maximum possible values for the sides are 1.365 and 2.465, respectively.

▌ TABLE 4

Number	Range of measurement
1.36	[1.355, 1.365]
2.46	[2.455, 2.465]

Thus, the maximum possible value for the hypotenuse must be

$$\sqrt{(1.365)^2 + (2.465)^2} = 2.817\ldots \qquad \text{calculator display}$$
$$= 2.8 \qquad \text{to two significant digits}$$

This shows that the value 2.9 is indeed inappropriate, as we stated previously.

An error often made by people working with calculators and approximations is to report a final answer with a greater degree of accuracy than the data warrant. Consider, for example, the right triangle in Figure 2. Using the Pythagorean theorem and a calculator with an eight-digit display, we obtain

$$h = 3.6055513 \text{ cm}$$

This value for h is inappropriate, because common sense tells us that the answer should be no more accurate than the data used to obtain that answer. In particular, since the given sides of the triangle apparently were measured only to the nearest tenth of a centimeter, we certainly should not expect any improvement in accuracy for the resulting value of the hypotenuse. An appropriate form for the value of h here would be $h = 3.6$ cm. In general, for calculator exercises in this text that do not specify a required number of decimal places or significant digits in the final results, you should use the following guidelines.

Figure 2

Guidelines for Computing with Approximations

1. *For adding and subtracting:* Round the final result so that it contains only as many decimal places as there are in the data with the fewest decimal places.
2. *For multiplying and dividing:* Round the final result so that it contains only as many significant digits as there are in the data with the fewest significant digits.
3. *For powers and roots:* In computing a power or a root of a real number b, round the result so that it contains as many significant digits as there are in b.

A.2   PROPERTIES OF THE REAL NUMBERS

Today's familiar plus and minus signs were first used in 15th-century Germany as warehouse marks. They indicated when a container held something that weighed over or under a certain standard weight.
—Martin Gardner in "Mathematical Games" (*Scientific American,* June 1977)

I do not like $\times$ as a symbol for multiplication, as it is easily confounded with x; . . . often I simply relate two quantities by an interposed dot. . . . —G. W. Leibniz in a letter dated July 29, 1698

In this section we will first list the basic properties for the real number system. Then we will use those properties to prove the familiar rules of algebra for working with signed numbers.

The set of real numbers is **closed** with respect to the operations of addition and multiplication. This means that when we add or multiply two real numbers, the result (that is, the **sum** or the **product**) is again a real number. Some of the other most basic properties and definitions for the real-number system are listed in the following box. In the box, the lowercase letters a, b, and c denote arbitrary real numbers.

▍PROPERTY SUMMARY   Some Fundamental Properties of the Real Numbers

Commutative properties	$a + b = b + a$ $ab = ba$
Associative properties	$a + (b + c) = (a + b) + c$ $a(bc) = (ab)c$
Identity properties	1. There is a unique real number 0 (called **zero** or the **additive identity**) such that $$a + 0 = a \quad \text{and} \quad 0 + a = a$$ 2. There is a unique real number 1 (called **one** or the **multiplicative identity**) such that $$a \cdot 1 = a \quad \text{and} \quad 1 \cdot a = a$$
Inverse properties	1. For each real number a there is a real number $-a$ (called the **additive inverse** of a or the **opposite** of a) such that $$a + (-a) = 0 \quad \text{and} \quad (-a) + a = 0$$ 2. For each real number $a \neq 0$, there is a real number denoted by $\dfrac{1}{a}$ (or $1/a$ or a^{-1}) and called the **multiplicative inverse** or **reciprocal** of a, such that $$a \cdot \frac{1}{a} = 1 \quad \text{and} \quad \frac{1}{a} \cdot a = 1$$
Distributive properties	$a(b + c) = ab + ac$ $(b + c)a = ba + ca$

On reading this list of properties for the first time, many students ask the natural question, "Why do we even bother to list such obvious properties?" One reason is that all the other laws of arithmetic and algebra (including the "not-so-obvious" ones) can be derived from our rather short list. For example, the rule $0 \cdot a = 0$ can be proved using the distributive property, as can the rule that the product of two negative numbers is a positive number. We now state and prove those properties and several others.

THEOREM

Let a and b be real numbers. Then

(a) $a \cdot 0 = 0$ (c) $-(-a) = a$ (e) $(-a)(-b) = ab$

(b) $-a = (-1)a$ (d) $a(-b) = -ab$

Proof of Part (a)

$$a \cdot 0 = a \cdot (0 + 0) \qquad \text{additive identity property}$$
$$a \cdot 0 = a \cdot 0 + a \cdot 0 \qquad \text{distributive property}$$

Now since $a \cdot 0$ is a real number, it has an additive inverse, $-(a \cdot 0)$. Adding this to both sides of the last equation, we obtain

$$a \cdot 0 + [-(a \cdot 0)] = (a \cdot 0 + a \cdot 0) + [-(a \cdot 0)]$$
$$a \cdot 0 + [-(a \cdot 0)] = a \cdot 0 + \{a \cdot 0 + [-(a \cdot 0)]\} \qquad \text{associative property of addition}$$
$$0 = a \cdot 0 + 0 \qquad \text{additive inverse property}$$
$$0 = a \cdot 0 \qquad \text{additive identity property}$$

Thus $a \cdot 0 = 0$, as we wished to show.

Proof of Part (b)

$$0 = 0 \cdot a \qquad \text{using part (a) and the commutative of multiplication}$$

$$= [1 + (-1)]a \qquad \text{additive inverse property}$$

$$= 1 \cdot a + (-1)a \qquad \text{distributive property}$$

$$= a + (-1)a \qquad \text{multiplicative identity property}$$

Now, by adding $-a$ to both sides of this last equation, we obtain

$$-a + 0 = -a + [a + (-1)a]$$

$$-a = (-a + a) + (-1)a \qquad \text{additive identity property and associative property of addition}$$

$$-a = 0 + (-1)a \qquad \text{additive inverse property}$$

$$-a = (-1)a \qquad \text{additive identity property}$$

This last equation asserts that $-a = (-1)a$, as we wished to show.

Proof of Part (c)

$$-(-a) + (-a) = 0 \qquad \text{additive inverse property}$$

By adding a to both sides of this last equation, we obtain

$$[-(-a) + (-a)] + a = 0 + a$$

$$-(-a) + (-a + a) = a \qquad \text{associative property of addition and additive identity property}$$

$$-(-a) + 0 = a \qquad \text{additive inverse property}$$

$$-(-a) = a \qquad \text{additive identity property}$$

This last equation states that $-(-a) = a$, as we wished to show.

Proof of Part (d)

$$a(-b) = a[(-1)b] \qquad \text{using part (b)}$$

$$= [a(-1)]b \qquad \text{associative property of multiplication}$$

$$= [(-1)a]b \qquad \text{commutative property of multiplication}$$

$$= (-1)(ab) \qquad \text{associative property of multiplication}$$

$$= -(ab) \qquad \text{using part (b)}$$

Thus $a(-b) = -ab$, as we wished to show.

Proof of Part (e)

$$(-a)(-b) = -[(-a)b] \qquad \text{using part (d)}$$

$$= -[b(-a)] \qquad \text{commutative property of multiplication}$$

$$= -[-(ba)] \qquad \text{using part (d)}$$

$$= ba \qquad \text{using part (c)}$$

$$= ab \qquad \text{commutative property of multiplication}$$

We've now shown that $(-a)(-b) = ab$, as required.

A.3 $\sqrt{2}$ IS IRRATIONAL

The proof is by reductio ad absurdum, *and* reductio ad absurdum, *which Euclid loved so much, is one of a mathematician's finest weapons.*
—G. H. Hardy (1877–1947)

We will use an *indirect proof* to show that the square root of 2 is an irrational number. The strategy is as follows:

1. We suppose that $\sqrt{2}$ is a rational number.
2. Using (1) and the usual rules of logic and algebra, we derive a contradiction.
3. On the basis of the contradiction in (2), we conclude that the supposition in (1) is untenable; that is, we conclude that $\sqrt{2}$ is irrational.

In carrying out the proof, we'll assume that the following three statements are known:

If x is an even natural number, then $x = 2k$ for some natural number k.

Any rational number can be written in the form a/b, where the integers a and b have no common integral factors other than ± 1. (In other words, any fraction can be reduced to lowest terms.)

If x is a natural number and x^2 is even, then x is even.

Our indirect proof now proceeds as follows. Suppose that $\sqrt{2}$ is a rational number. Then we can write

$$\sqrt{2} = \frac{a}{b} \qquad \text{where } a \text{ and } b \text{ are natural numbers with no common factor other than 1} \qquad (1)$$

Since both sides of equation (1) are positive, we can square both sides to obtain the equivalent equation

$$2 = \frac{a^2}{b^2}$$

or

$$2b^2 = a^2 \qquad (2)$$

Since the left-hand side of equation (2) is an even number, the right-hand side must also be even. But if a^2 is even then a is even, and so

$$a = 2k \quad \text{for some natural number } k$$

Using this last equation to substitute for a in equation (2), we have

$$2b^2 = (2k)^2 = 4k^2$$

or

$$b^2 = 2k^2$$

Hence (reasoning as before) b^2 is even, and therefore b is even. But then we have that both b and a are even, contrary to our hypothesis that b and a have no common factor other than 1. We conclude from this that equation (1) cannot hold; that is, there is no rational number a/b such that $\sqrt{2} = a/b$. Thus $\sqrt{2}$ is irrational, as we wished to prove.

B   Appendix

B.1   REVIEW OF INTEGER EXPONENTS

I write a^{-1}, a^{-2}, a^{-3}, etc., for $\dfrac{1}{a}$, $\dfrac{1}{aa}$, $\dfrac{1}{aaa}$, etc. —Isaac Newton (June 13, 1676)

In basic algebra you learned the exponential notation a^n, where a is a real number and n is a natural number. Because that definition is basic to all that follows in this and the next two sections of this appendix, we repeat it here.

DEFINITION 1 │ Base and Exponent

Given a real number a and a natural number n, we define a^n by

$$a^n = \underbrace{a \cdot a \cdot a \cdots a}_{n \text{ factors}}$$

In the expression a^n, the number a is the **base** and n is the **exponent** or **power** to which the base is raised.

EXAMPLE 1   Using algebraic notation

Rewrite each expression using algebraic notation:
(a) x to the fourth power, plus five;
(b) the fourth power of the quantity x plus five;
(c) x plus y, to the fourth power;
(d) x plus y to the fourth power.

SOLUTION
(a) $x^4 + 5$ (b) $(x + 5)^4$ (c) $(x + y)^4$ (d) $x + y^4$

CAUTION   Note that parts (c) and (d) of Example 1 differ only in the use of the comma. So, for spoken purposes, the idea in (c) would be more clearly conveyed by saying "the fourth power of the quantity x plus y." In the case of (d), if you read it aloud to another student or your instructor, chances are you'll be asked, "Do you mean $x + y^4$ or do you mean $(x + y)^4$?"

MORAL   Algebra is a precise language; use it carefully. Learn to ask yourself whether what you've written will be interpreted in the manner you intended.

A-8

In basic algebra, the four properties in the following box are developed for working with exponents that are natural numbers. Each of these properties is a direct consequence of the definition of a^n. For instance, according to the first property, we have $a^2a^3 = a^5$. To verify that this is indeed correct, we note that

$$a^2a^3 = (aa)(aaa) = a^5$$

PROPERTY SUMMARY   Properties of Exponents

PROPERTY

1. $a^m a^n = a^{m+n}$
2. $(a^m)^n = a^{mn}$
3. $\dfrac{a^m}{a^n} = \begin{cases} a^{m-n} & \text{if } m > n \\ \dfrac{1}{a^{n-m}} & \text{if } m < n \\ 1 & \text{if } m = n \end{cases}$
4. $(ab)^m = a^m b^m;\quad \left(\dfrac{a}{b}\right)^m = \dfrac{a^m}{b^m}$

EXAMPLES

$a^5 a^6 = a^{11};\ (x+1)(x+1)^2 = (x+1)^3$

$(2^3)^4 = 2^{12};\ [(x+1)^2]^3 = (x+1)^6$

$\dfrac{a^6}{a^2} = a^4;\ \dfrac{a^2}{a^6} = \dfrac{1}{a^4};\ \dfrac{a^5}{a^5} = 1$

$(2x^2)^3 = 2^3 \cdot (x^2)^3 = 8x^6;\ \left(\dfrac{x^2}{y^3}\right)^4 = \dfrac{x^8}{y^{12}}$

Now we want to extend our definition of a^n to allow for exponents that are integers but not necessarily natural numbers. We begin by defining a^0.

DEFINITION 2 | Zero Exponent

For any nonzero real number a,

$$a^0 = 1$$

(0^0 is not defined.)

EXAMPLES

(a) $2^0 = 1$

(b) $(-\pi)^0 = 1$

(c) $\left(\dfrac{3}{1+a^2+b^2}\right)^0 = 1$

It's easy to see the motivation for defining a^0 to be 1. Assuming that the exponent zero is to have the same properties as exponents that are natural numbers, we can write

$$a^0 a^n = a^{0+n}$$

That is,

$$a^0 a^n = a^n$$

Now we divide both sides of this last equation by a^n to obtain $a^0 = 1$, which agrees with our definition.

Our next definition (see the box that follows) assigns a meaning to the expression a^{-n} when n is a natural number. Again, it's easy to see the motivation for this definition. We have

$$a^n a^{-n} = a^{n+(-n)} = a^0 = 1$$

That is,

$$a^n a^{-n} = 1$$

Now we divide both sides of this last equation by a^n to obtain $a^{-n} = 1/a^n$, in agreement with Definition 3.

DEFINITION 3 | **Negative Exponent**

For any nonzero real number a and natural number n,

$$a^{-n} = \frac{1}{a^n}$$

EXAMPLES

(a) $2^{-1} = \dfrac{1}{2^1} = \dfrac{1}{2}$

(b) $\left(\dfrac{1}{10}\right)^{-1} = \dfrac{1}{\left(\frac{1}{10}\right)^1} = 10$

(c) $x^{-2} = \dfrac{1}{x^2}$

(d) $(a^2b)^{-3} = \dfrac{1}{(a^2b)^3} = \dfrac{1}{a^6b^3}$

(e) $\dfrac{1}{2^{-3}} = \dfrac{1}{1/2^3} = 2^3 = 8$

It can be shown that the four properties of exponents that we listed earlier continue to hold now for all integer exponents. We make use of this fact in the next three examples.

EXAMPLE 2 | **Using properties of exponents**

Simplify the following expression. Write the answer in such a way that only positive exponents appear.

$$(a^2b^3)^2(a^5b)^{-1}$$

SOLUTION

FIRST METHOD

$$(a^2b^3)^2(a^5b)^{-1} = (a^2b^3)^2 \cdot \frac{1}{a^5b}$$

$$= \frac{a^4b^6}{a^5b} = \frac{b^{6-1}}{a^{5-4}} = \frac{b^5}{a}$$

ALTERNATE METHOD

$$(a^2b^3)^2(a^5b)^{-1} = (a^4b^6)(a^{-5}b^{-1})$$

$$= a^{4-5}b^{6-1} = a^{-1}b^5$$

$$= \frac{b^5}{a} \quad \text{as obtained} \atop \text{previously}$$

EXAMPLE 3 | **Using properties of exponents**

Simplify the following expression, writing the answer so that negative exponents are not used.

$$\left(\frac{a^{-5}b^2c^0}{a^3b^{-1}}\right)^3$$

SOLUTION

We show two solutions. The first makes immediate use of the property $(a^m)^n = a^{mn}$. In the second solution we begin by working within the parentheses.

FIRST SOLUTION

$$\left(\frac{a^{-5}b^2c^0}{a^3b^{-1}}\right)^3 = \frac{a^{-15}b^6}{a^9b^{-3}}$$

$$= \frac{b^{6-(-3)}}{a^{9-(-15)}} = \frac{b^9}{a^{24}}$$

ALTERNATE SOLUTION

$$\left(\frac{a^{-5}b^2c^0}{a^3b^{-1}}\right)^3 = \left(\frac{b^3}{a^8}\right)^3$$

$$= \frac{b^9}{a^{24}}$$

EXAMPLE   4   Simplifying with variable exponents

Simplify the following expressions, writing the answers so that negative exponents are not used. (Assume that p and q are natural numbers.)

(a) $\dfrac{a^{4p+q}}{a^{p-q}}$ **(b)** $(a^p b^q)^2 (a^{3p} b^{-q})^{-1}$

SOLUTION

(a) $\dfrac{a^{4p+q}}{a^{p-q}} = a^{4p+q-(p-q)}$

$$= a^{4p+q-p+q} = a^{3p+2q}$$

(b) $(a^p b^q)^2 (a^{3p} b^{-q})^{-1} = (a^{2p} b^{2q})(a^{-3p} b^q)$

$$= a^{-p} b^{3q}$$

$$= \frac{b^{3q}}{a^p}$$

EXAMPLE   5   Simplifying numerical quantities involving exponents

Use the properties of exponents to compute the quantity $\dfrac{2^{10} \cdot 3^{13}}{27 \cdot 6^{12}}$.

SOLUTION

$$\frac{2^{10} \cdot 3^{13}}{27 \cdot 6^{12}} = \frac{2^{10} \cdot 3^{13}}{(3^3)(3 \cdot 2)^{12}}$$

$$= \frac{2^{10} \cdot 3^{13}}{3^3 \cdot 3^{12} \cdot 2^{12}} = \frac{2^{10} \cdot 3^{13}}{3^{15} \cdot 2^{12}}$$

$$= \frac{1}{(3^{15-13})(2^{12-10})} = \frac{1}{3^2 \cdot 2^2} = \frac{1}{36}$$

As an application of some of the ideas in this section, we briefly discuss **scientific notation,** which is a convenient form for writing very large or very small numbers. Such numbers occur often in the sciences. For instance, the speed of light in a vacuum is

$$29{,}979{,}000{,}000 \, \text{cm/sec}$$

As written, this number would be awkward to work with in calculating. In fact, the number as written cannot even be displayed on some hand-held calculators, because there are too many digits. To write the number 29,979,000,000 (or any positive number) in scientific notation, we express it as a number between 1 and 10, multiplied by an appropriate power of 10. That is, we write it in the form

$$a \times 10^n \quad \text{where } 1 \leq a < 10 \text{ and } n \text{ is an integer}$$

According to this convention, the number 4.03×10^6 is in scientific notation, but the same quantity written as 40.3×10^5 is not in scientific notation. In order to convert a given number into scientific notation, we'll rely on the following two-step procedure.

To Express a Number Using Scientific Notation

1. First move the decimal point until it is to the immediate right of the first nonzero digit.
2. Then multiply by 10^n or 10^{-n}, depending on whether the decimal point was moved n places to the left or to the right, respectively.

For example, to express the number 29,979,000,000 in scientific notation, first we move the decimal point 10 places to the left so that it's located between the 2 and the 9, then we multiply by 10^{10}. The result is

$$29{,}979{,}000{,}000 = 2.9979000000 \times 10^{10}$$

or, more simply,

$$29{,}979{,}000{,}000 = 2.9979 \times 10^{10}$$

As additional examples we list the following numbers expressed in both ordinary and scientific notation. For practice, you should verify each conversion for yourself using our two-step procedure.

$$55{,}708 = 5.5708 \times 10^4$$
$$0.000099 = 9.9 \times 10^{-5}$$
$$0.0000002 = 2 \times 10^{-7}$$

All scientific calculators have keys for entering numbers in scientific notation. (Indeed, many calculators will convert numbers into scientific notation for you.) If you have questions about how your own calculator operates with respect to scientific notation, you should consult the owner's manual. We'll indicate only one example here, using a Texas Instruments graphing calculator. The keystrokes on other brands are quite similar. Consider the following number expressed in scientific notation: 3.159×10^{20}. (This gargantuan number is the diameter, in miles, of a galaxy at the center of a distant star cluster known as Abell 2029.) The sequence of keystrokes for entering this number on the Texas Instruments calculator is

3.519   20

EXERCISE SET B.1

A

In Exercises 1–4, evaluate each expression using the given value of x.

1. $2x^3 - x + 4; x = -2$

2. $1 - x + 2x^2 - 3x^3; x = -1$

3. $\dfrac{1 - 2x^2}{1 + 2x^3}; x = -\dfrac{1}{2}$

4. $\dfrac{1 - (x - 1)^2}{1 + (x - 1)^2}; x = -1$

In Exercises 5–16, use the properties of exponents to simplify each expression.

5. (a) $a^3 a^{12}$
   (b) $(a + 1)^3 (a + 1)^{12}$
   (c) $(a + 1)^{12}(a + 1)^3$

6. (a) $(3^2)^3 - (2^3)^2$
   (b) $(x^2)^3 - (x^3)^2$
   (c) $(x^2)^a - (x^a)^2$

7. (a) $yy^2 y^8$
   (b) $(y + 1)(y + 1)^2(y + 1)^8$
   (c) $[(y + 1)(y + 1)^8]^2$

8. (a) $\dfrac{x^{12}}{x^{10}}$
   (b) $\dfrac{x^{10}}{x^{12}}$
   (c) $\dfrac{2^{10}}{2^{12}}$

9. (a) $\dfrac{(x^2 + 3)^{10}}{(x^2 + 3)^9}$
   (b) $\dfrac{(x^2 + 3)^9}{(x^2 + 3)^{10}}$
   (c) $\dfrac{12^{10}}{12^9}$

10. (a) $(y^2 y^3)^2$
   (b) $y^2 (y^3)^2$
   (c) $2^m (2^n)^2$

11. (a) $\dfrac{t^{15}}{t^9}$
   (b) $\dfrac{t^9}{t^{15}}$
   (c) $\dfrac{(t^2 + 3)^{15}}{(t^2 + 3)^9}$

12. (a) $\dfrac{x^6}{x}$
   (b) $\dfrac{x}{x^6}$
   (c) $\dfrac{(x^2 + 3x - 2)^6}{x^2 + 3x - 2}$

13. (a) $\dfrac{x^6 y^{15}}{x^2 y^{20}}$
   (b) $\dfrac{x^2 y^{20}}{x^6 y^{15}}$
   (c) $\left(\dfrac{x^2 y^{20}}{x^6 y^{15}}\right)^2$

14. (a) $(x^2 y^3 z)^4$
   (b) $2(x^2 y^3 z)^4$
   (c) $(2x^2 y^3 z)^4$

15. (a) $4(x^3)^2$
   (b) $(4x^3)^2$
   (c) $\dfrac{(4x^2)^3}{(4x^3)^2}$

16. (a) $2(x - 1)^7 - (x - 1)^7$
   (b) $[2(x - 1)]^7 - (x - 1)^7$
   (c) $2(x - 1)^7 - [2(x - 1)]^7$

For Exercises 17–38, simplify each expression. Write the answers in such a way that negative exponents do not appear. In Exercises 35–38, assume that the letters p and q represent natural numbers.

17. (a) 64^0
   (b) $(64^3)^0$
   (c) $(64^0)^3$

18. (a) $2^0 + 3^0$
   (b) $(2^0 + 3^0)^3$
   (c) $(2^3 + 3^3)^0$

19. (a) $10^{-1} + 10^{-2}$
   (b) $(10^{-1} + 10^{-2})^{-1}$
   (c) $[(10^{-1})(10^{-2})]^{-1}$

20. (a) $4^{-2} + 4^{-1}$
   (b) $[(\frac{1}{4})^{-1} + (\frac{1}{4})^{-2}]^{-1}$
   (c) $[(\frac{1}{4})^{-1}(\frac{1}{4})^{-2}]^{-1}$

21. $(\frac{1}{3})^{-1} + (\frac{1}{4})^{-1}$

22. $(\frac{1}{3} + \frac{1}{4})^{-1}$

23. (a) $5^{-2} + 10^{-2}$
   (b) $(5 + 10)^{-2}$

24. (a) $(\frac{1}{4})^{-2} + (\frac{3}{4})^{-2}$
   (b) $(\frac{1}{4} + \frac{3}{4})^{-2}$

25. $(a^2 b c^0)^{-3}$

26. $(a^3 b)^3 (a^2 b^4)^{-1}$

27. $(a^{-2} b^{-1} c^3)^{-2}$

28. $(2^{-2} + 2^{-1} + 2^0)^{-2}$

29. $\left(\dfrac{x^3 y^{-2} z}{x y^2 z^{-3}}\right)^{-3}$

30. $\left(\dfrac{x^4 y^{-8} z^2}{x y^2 z^{-6}}\right)^2$

31. $\left(\dfrac{x^4 y^{-8} z^2}{x y^2 z^{-6}}\right)^{-2}$

32. $\left(\dfrac{a^3 b^{-9} c^2}{a^5 b^2 c^{-4}}\right)^0$

33. $\dfrac{x^2}{y^{-3}} \div \dfrac{x^2}{y^3}$

34. $(2x^2)^{-3} - 2(x^2)^{-3}$

35. $\dfrac{b^{p+1}}{b^p}$

36. $\dfrac{b^{p+2q}}{(b^2)^q}$

37. $(x^p)(x^p)$

38. $(2^{p-q})(2^{q-p+1})$

In Exercises 39–42, use the properties of exponents in computing each quantity, as in Example 5. (The point here is to do as little arithmetic as possible.)

39. $\dfrac{2^8 \cdot 3^{15}}{9 \cdot 3^{10} \cdot 12}$

40. $\dfrac{2^{12} \cdot 5^{13}}{10^{12}}$

41. $\dfrac{24^5}{32 \cdot 12^4}$

42. $\left(\dfrac{144 \cdot 125}{2^3 \cdot 3^2}\right)^{-1}$

For Exercises 43–50, express each number in scientific notation.

43. The average distance (in miles) from Earth to the Sun:

92,900,000

44. The average distance (in miles) from the planet Pluto to the Sun:

3,666,000,000

45. The average orbital speed (in miles per hour) of Earth:

66,800

46. The average orbital speed (in miles per hour) of the planet Mercury:

107,300

47. The average distance (in miles) from the Sun to the nearest star:

25,000,000,000,000,000,000

48. The equatorial diameter (in miles) of
   (a) Mercury: 3031
   (b) Earth: 7927
   (c) Jupiter: 88,733
   (d) the Sun: 865,000

49. The time (in seconds) for light to travel
   (a) one foot:

   $$0.000000001$$

   (b) across an atom:

   $$0.000000000000000001$$

   (c) across the nucleus of an atom:

   $$0.0000000000000000000000001$$

50. The mass (in grams) of
   (a) a proton:

   $$0.00000000000000000000000167$$

   (b) an electron:

   $$0.000000000000000000000000000911$$

B.2   REVIEW OF nTH ROOTS

Hindu mathematicians first recognized negative roots, and the two square roots of a positive number, . . . though they were suspicious also. —David Wells in *The Penguin Dictionary of Curious and Interesting Numbers* (Harmondsworth, Middlesex, England: Penguin Books, Ltd., 1986)

In this appendix we generalize the notion of square root that you studied in elementary algebra. The new idea is that of an *n*th root; the definition and some basic examples are given in the box that follows.

DEFINITION 1 | **nth Roots**

Let n be a natural number. If a and b are real numbers and

$$a^n = b$$

then we say that a is an **nth root** of b. When $n = 2$ and when $n = 3$, we refer to the roots as **square roots** and **cube roots,** respectively.

EXAMPLES

Both 3 and -3 are square roots of 9 because $3^2 = 9$ and $(-3)^2 = 9$.

Both 2 and -2 are fourth roots of 16 because $2^4 = 16$ and $(-2)^4 = 16$.

2 is a cube root of 8 because $2^3 = 8$.

-3 is a fifth root of -243 because $(-3)^5 = -243$.

As the examples in the box suggest, square roots, fourth roots, and all *even* roots of positive numbers occur in pairs, one positive and one negative. In these cases, we use the notation $\sqrt[n]{b}$ to denote the positive, or **principal,** *n*th root of b. As examples of this notation, we can write

$$\sqrt[4]{81} = 3 \qquad \text{(The principal fourth root of 81 is 3.)}$$

$$\sqrt[4]{81} \neq -3 \qquad \text{(The principal fourth root of 81 is not } -3.)$$

$$-\sqrt[4]{81} = -3 \qquad \text{(The negative of the principal fourth root of 81 is } -3.)$$

The symbol $\sqrt{}$ is called a **radical sign,** and the number within the radical sign is the **radicand.** The natural number n used in the notation $\sqrt[n]{}$ is called the **index** of the radical. For square roots, as you know from basic algebra, we suppress the index and simply write $\sqrt{}$ rather than $\sqrt[2]{}$. So, for example, $\sqrt{25} = 5$.

On the other hand, as we saw in the examples, cube roots, fifth roots, and in fact, all *odd* roots occur singly, not in pairs. In these cases, we again use the notation $\sqrt[n]{b}$ for the *n*th root. The definition and examples in the following box summarize our discussion up to this point.

DEFINITION 2	Principal *n*th Root

EXAMPLES

$\sqrt{25} = 5; \sqrt{25} \neq -5; -\sqrt{25} = -5$

1. Let n be a natural number. If a and b are nonnegative real numbers, then

$$\sqrt[n]{b} = a \quad \text{if and only if} \quad b = a^n$$

$$\sqrt[4]{\frac{1}{16}} = \frac{1}{2}; \sqrt[3]{8} = 2$$

The number a is the **principal *n*th root** of b.

2. If a and b are negative and n is an odd natural number, then

$$\sqrt[n]{b} = a \quad \text{if and only if} \quad b = a^n$$

$$\sqrt[5]{-\frac{1}{32}} = -\frac{1}{2}$$

There are five properties of *n*th roots that are frequently used in simplifying certain expressions. The first four are similar to the properties of square roots that are developed in elementary algebra. For reference, we list these properties side by side in the following box. Property 5 is listed here only for the sake of completeness; we'll postpone discussing it until appendix section B.5.

PROPERTY SUMMARY   Properties of *n*th Roots

Suppose that x and y are real numbers and that m and n are natural numbers. Then each of the following properties holds, provided only that the expressions on both sides of the equation are defined (and so represent real numbers)

CORRESPONDING PROPERTIES FOR SQUARE ROOTS

1. $(\sqrt[n]{x})^n = x$ $(\sqrt{x})^2 = x$

2. $\sqrt[n]{xy} = \sqrt[n]{x}\sqrt[n]{y}$ $\sqrt{xy} = \sqrt{x}\sqrt{y}$

3. $\sqrt[n]{\dfrac{x}{y}} = \dfrac{\sqrt[n]{x}}{\sqrt[n]{y}}$ $\sqrt{\dfrac{x}{y}} = \dfrac{\sqrt{x}}{\sqrt{y}}$

4. n even: $\sqrt[n]{x^n} = |x|$ $\sqrt{x^2} = |x|$

n odd: $\sqrt[n]{x^n} = x$

5. $\sqrt[m]{\sqrt[n]{x}} = \sqrt[mn]{x}$

Our immediate use for these properties will be in simplifying expressions involving *n*th roots. In general, we try to factor the expression under the radical so that one factor is the largest perfect *n*th power that we can find. Then we apply Property 2 or 3. For instance, the expression $\sqrt{72}$ is simplified as follows:

$$\sqrt{72} = \sqrt{(36)(2)} = \sqrt{36}\sqrt{2} = 6\sqrt{2}$$

In this procedure, we began by factoring 72 as $(36)(2)$. Note that 36 is the largest factor of 72 that is a perfect square. If we were to begin instead with a different factorization—say, $72 = (9)(8)$—we can still arrive at the same answer, but it would take longer. (Check this for yourself.) As another example, let's simplify $\sqrt[3]{40}$. First, what (if any) is the largest perfect-cube factor of 40? The first few perfect cubes are

$$1^3 = 1 \qquad 2^3 = 8 \qquad 3^3 = 27 \qquad 4^3 = 64$$

so we see that 8 is a perfect-cube factor of 40, and we write

$$\sqrt[3]{40} = \sqrt[3]{(8)(5)} = \sqrt[3]{8}\sqrt[3]{5} = 2\sqrt[3]{5}$$

EXAMPLE 1   Simplifying square roots

Simplify: **(a)** $\sqrt{12} + \sqrt{75}$;   **(b)** $\sqrt{\dfrac{162}{49}}$.

SOLUTION

(a) $\sqrt{12} + \sqrt{75} = \sqrt{(4)(3)} + \sqrt{(25)(3)}$
$= \sqrt{4}\sqrt{3} + \sqrt{25}\sqrt{3}$
$= 2\sqrt{3} + 5\sqrt{3} = (2 + 5)\sqrt{3}$
$= 7\sqrt{3}$

(b) $\sqrt{\dfrac{162}{49}} = \dfrac{\sqrt{162}}{\sqrt{49}} = \dfrac{\sqrt{81}\sqrt{2}}{7} = \dfrac{9\sqrt{2}}{7}$

EXAMPLE 2   Simplifying *n*th roots

Simplify: $\sqrt[3]{16} + \sqrt[3]{250} - \sqrt[3]{128}$.

SOLUTION

$\sqrt[3]{16} + \sqrt[3]{250} - \sqrt[3]{128} = \sqrt[3]{8}\sqrt[3]{2} + \sqrt[3]{125}\sqrt[3]{2} - \sqrt[3]{64}\sqrt[3]{2}$
$= 2\sqrt[3]{2} + 5\sqrt[3]{2} - 4\sqrt[3]{2}$
$= 3\sqrt[3]{2}$

EXAMPLE 3   Simplifying radicals containing variables

Simplify each of the following expressions by removing the largest possible perfect-square or perfect-cube factor from within the radical:

(a) $\sqrt{8x^2}$;

(b) $\sqrt{8x^2}$, where $x \geq 0$;

(c) $\sqrt{18a^7}$, where $a \geq 0$;

(d) $\sqrt[3]{16y^5}$.

SOLUTION

(a) $\sqrt{8x^2} = \sqrt{(4)(2)(x^2)} = \sqrt{4}\sqrt{2}\sqrt{x^2} = 2\sqrt{2}|x|$

(b) $\sqrt{8x^2} = \sqrt{4}\sqrt{2}\sqrt{x^2}$
$= 2\sqrt{2}x \qquad \textcolor{red}{\sqrt{x^2} = x \text{ because } x \geq 0}$

(c) $\sqrt{18a^7} = \sqrt{(9a^6)(2a)} = \sqrt{9a^6}\sqrt{2a}$
$= 3a^3\sqrt{2a}$

(d) $\sqrt[3]{16y^5} = \sqrt[3]{8y^3}\sqrt[3]{2y^2} = 2y\sqrt[3]{2y^2}$

EXAMPLE 4 Simplifying *n*th roots containing variables

Simplify each of the following, assuming that a, b, and c are positive:

(a) $\sqrt{8ab^2c^5}$; (b) $\sqrt[4]{\dfrac{32a^6b^5}{c^8}}$.

SOLUTION

(a) $\sqrt{8ab^2c^5} = \sqrt{(4b^2c^4)(2ac)} = \sqrt{4b^2c^4}\sqrt{2ac}$
$$= 2bc^2\sqrt{2ac}$$

(b) $\sqrt[4]{\dfrac{32a^6b^5}{c^8}} = \dfrac{\sqrt[4]{32a^6b^5}}{\sqrt[4]{c^8}}$

$$= \dfrac{\sqrt[4]{16a^4b^4}\sqrt[4]{2a^2b}}{c^2} = \dfrac{2ab\sqrt[4]{2a^2b}}{c^2}$$

In Examples 1 through 4, we used the definitions and properties of *n*th roots to simplify certain expressions. The box that follows shows some common errors to avoid in working with roots. Use the error box to test yourself: cover up the columns labeled "Correction" and "Comment" and try to decide for yourself where the errors lie.

ERRORS TO AVOID

Error	Correction	Comment
$\sqrt[4]{16} \neq \pm 2$	$\sqrt[4]{16} = 2$	Although -2 is one of the fourth roots of 16, it is not the *principal* fourth root. The notation $\sqrt[4]{16}$ is reserved for the principal fourth root.
$\sqrt{25} \neq -5$	$\sqrt{25} = 5$	Although -5 is one of the square roots of 25, it is not the *principal* square root.
$\sqrt{a+b} \neq \sqrt{a} + \sqrt{b}$	The expression $\sqrt{a+b}$ cannot be simplified.	The properties of roots differ with respect to addition versus multiplication. For $\sqrt{ab}$, we do have the simplification $\sqrt{ab} = \sqrt{a}\sqrt{b}$ (assuming a and b are nonnegative).
$\sqrt[3]{a+b} \neq \sqrt[3]{a} + \sqrt[3]{b}$	The expression $\sqrt[3]{a+b}$ cannot be simplified.	For $\sqrt[3]{ab}$, we do have $\sqrt[3]{ab} = \sqrt[3]{a}\sqrt[3]{b}$.

There are times when it is convenient to rewrite fractions involving radicals in alternate forms. Suppose, for example, that we want to rewrite the fraction $5/\sqrt{3}$ in an equivalent form not involving a radical in the denominator. This is called **rationalizing the denominator.** The procedure here is to multiply by 1 in this way:

$$\frac{5}{\sqrt{3}} = \frac{5}{\sqrt{3}} \cdot 1 = \frac{5}{\sqrt{3}} \cdot \frac{\sqrt{3}}{\sqrt{3}} = \frac{5\sqrt{3}}{3}$$

That is, $\dfrac{5}{\sqrt{3}} = \dfrac{5\sqrt{3}}{3}$, as required.

To rationalize a denominator of the form $a + \sqrt{b}$, we need to multiply the fraction not by $\sqrt{b}/\sqrt{b}$, but rather by $(a - \sqrt{b})/(a - \sqrt{b})\,(= 1)$. To see why this is necessary, notice that

$$(a + \sqrt{b})\sqrt{b} = a\sqrt{b} + b$$

which still contains a radical, whereas

$$(a + \sqrt{b})(a - \sqrt{b}) = a^2 - a\sqrt{b} + a\sqrt{b} - b$$
$$= a^2 - b$$

which is free of radicals. Similarly, to rationalize a denominator of the form $a - \sqrt{b}$, we multiply the fraction by $(a + \sqrt{b})/(a + \sqrt{b})$. (The quantities $a + \sqrt{b}$ and $a - \sqrt{b}$ are said to be **conjugates** of each other.) The next two examples make use of these ideas.

EXAMPLE 5   Rationalizing a square root denominator

Simplify: $\dfrac{1}{\sqrt{2}} - 3\sqrt{50}$.

SOLUTION
First, we rationalize the denominator in the fraction $1/\sqrt{2}$:

$$\frac{1}{\sqrt{2}} = \frac{1}{\sqrt{2}} \cdot 1 = \frac{1}{\sqrt{2}} \cdot \frac{\sqrt{2}}{\sqrt{2}} = \frac{\sqrt{2}}{2}$$

Next, we simplify the expression $3\sqrt{50}$:

$$3\sqrt{50} = 3\sqrt{(25)(2)} = 3\sqrt{25}\sqrt{2} = (3)(5)\sqrt{2} = 15\sqrt{2}$$

Now, putting things together, we have

$$\frac{1}{\sqrt{2}} - 3\sqrt{50} = \frac{\sqrt{2}}{2} - 15\sqrt{2} = \frac{\sqrt{2}}{2} - \frac{30\sqrt{2}}{2}$$
$$= \frac{\sqrt{2} - 30\sqrt{2}}{2} = \frac{(1 - 30)\sqrt{2}}{2}$$
$$= \frac{-29\sqrt{2}}{2}$$

EXAMPLE 6   Using the conjugate to rationalize a denominator

Rationalize the denominator in the expression $\dfrac{4}{2 + \sqrt{3}}$.

SOLUTION
We multiply by 1, writing it as $\dfrac{2 - \sqrt{3}}{2 - \sqrt{3}}$.

$$\frac{4}{2 + \sqrt{3}} \cdot 1 = \frac{4}{2 + \sqrt{3}} \cdot \frac{2 - \sqrt{3}}{2 - \sqrt{3}}$$

$$= \frac{4(2 - \sqrt{3})}{4 - (\sqrt{3})^2} = \frac{4(2 - \sqrt{3})}{4 - 3} = \frac{8 - 4\sqrt{3}}{1}$$

$$= 8 - 4\sqrt{3}$$

NOTE   Check for yourself that multiplying the original fraction $\frac{2 + \sqrt{3}}{2 + \sqrt{3}}$ does *not* eliminate radicals in the denominator.

In the next example, we are asked to rationalize the numerator rather than the denominator. This is useful at times in calculus.

EXAMPLE **7** **Rationalizing a numerator using the conjugate**

Rationalize the *numerator*: $\dfrac{\sqrt{x} - \sqrt{3}}{x - 3}$, where $x \geq 0$, $x \neq 3$.

SOLUTION

$$\frac{\sqrt{x} - \sqrt{3}}{x - 3} \cdot 1 = \frac{\sqrt{x} - \sqrt{3}}{x - 3} \cdot \frac{\sqrt{x} + \sqrt{3}}{\sqrt{x} + \sqrt{3}}$$

$$= \frac{(\sqrt{x})^2 - (\sqrt{3})^2}{(x - 3)(\sqrt{x} + \sqrt{3})} = \frac{x - 3}{(x - 3)(\sqrt{x} + \sqrt{3})}$$

$$= \frac{1}{\sqrt{x} + \sqrt{3}}$$

The strategy for rationalizing numerators or denominators involving *n*th roots is similar to that used for square roots. To rationalize a numerator or a denominator involving an *n*th root, we multiply the numerator or denominator by a factor that yields a product that itself is a perfect *n*th power. The next example displays two instances of this.

EXAMPLE **8** **Rationalizing denominators containing *n*th roots**

(a) Rationalize the denominator: $\dfrac{6}{\sqrt[3]{7}}$.

(b) Rationalize the denominator: $\dfrac{ab}{\sqrt[4]{a^2 b^3}}$, where $a > 0$, $b > 0$.

SOLUTION

(a) $\dfrac{6}{\sqrt[3]{7}} \cdot 1 = \dfrac{6}{\sqrt[3]{7}} \cdot \dfrac{\sqrt[3]{7^2}}{\sqrt[3]{7^2}}$

$$= \frac{6\sqrt[3]{7^2}}{\sqrt[3]{7^3}}$$

$$= \frac{6\sqrt[3]{49}}{7} \quad \text{as required}$$

(b) $\dfrac{ab}{\sqrt[4]{a^2b^3}} \cdot 1 = \dfrac{ab}{\sqrt[4]{a^2b^3}} \cdot \dfrac{\sqrt[4]{a^2b}}{\sqrt[4]{a^2b}}$

$= \dfrac{ab\sqrt[4]{a^2b}}{\sqrt[4]{a^4b^4}} = \dfrac{ab\sqrt[4]{a^2b}}{ab} = \sqrt[4]{a^2b}$

EXERCISE SET B.2

A

In Exercises 1–8, determine whether each statement is TRUE *or* FALSE.

1. $\sqrt{81} = -9$ **2.** $\sqrt{256} = 16$

3. $-\sqrt{100} = -10$ **4.** $\sqrt{49} = -7$

5. $\sqrt{4/3} = 2/\sqrt{3}$ **6.** $\sqrt{10 + 6} = \sqrt{10} + \sqrt{6}$

7. $(\sqrt[3]{10})^3 = 10$ **8.** $\sqrt{(-5)^2} = -5$

For Exercises 9–18, evaluate each expression. If the expression is undefined (i.e., does not represent a real number), say so.

9. (a) $\sqrt[3]{-64}$ **10. (a)** $\sqrt[5]{32}$

  (b) $\sqrt[4]{-64}$ **(b)** $\sqrt[5]{-32}$

11. (a) $\sqrt[3]{8/125}$ **12. (a)** $\sqrt[3]{-1/1000}$

  (b) $\sqrt[3]{-8/125}$ **(b)** $\sqrt[6]{-1/1000}$

13. (a) $\sqrt{-16}$ **14. (a)** $-\sqrt[4]{16}$

  (b) $\sqrt[4]{-16}$ **(b)** $-\sqrt[4]{-16}$

15. (a) $\sqrt[4]{256/81}$ **16. (a)** $\sqrt[6]{64}$

  (b) $\sqrt[3]{-27/125}$ **(b)** $\sqrt[6]{-64}$

17. (a) $\sqrt[5]{-32}$ **18. (a)** $\sqrt[4]{(-10)^4}$

  (b) $-\sqrt[5]{-32}$ **(b)** $\sqrt[3]{(-10)^3}$

In Exercises 19–44, simplify each expression. Unless otherwise specified, assume that all letters in Exercises 35–44 represent positive numbers.

19. (a) $\sqrt{18}$ **20. (a)** $\sqrt{150}$

  (b) $\sqrt[3]{54}$ **(b)** $\sqrt[3]{375}$

21. (a) $\sqrt{98}$ **22. (a)** $\sqrt{27}$

  (b) $\sqrt[5]{-64}$ **(b)** $\sqrt[3]{-108}$

23. (a) $\sqrt{25/4}$ **24. (a)** $\sqrt{225/49}$

  (b) $\sqrt[4]{16/625}$ **(b)** $\sqrt[5]{-256/243}$

25. (a) $\sqrt{2} + \sqrt{8}$ **26. (a)** $4\sqrt{3} - 2\sqrt{27}$

  (b) $\sqrt[3]{2} + \sqrt[3]{16}$ **(b)** $2\sqrt[3]{81} + 3\sqrt[3]{24}$

27. (a) $4\sqrt{50} - 3\sqrt{128}$

  (b) $\sqrt[4]{32} + \sqrt[4]{162}$

28. (a) $\sqrt{3} - \sqrt{12} + \sqrt{48}$

  (b) $\sqrt[5]{-2} + \sqrt[5]{-64} - \sqrt[5]{486}$

29. (a) $\sqrt{0.09}$

  (b) $\sqrt[3]{0.008}$

30. (a) $\sqrt[3]{-2} + \sqrt[3]{2}$

  (b) $\sqrt{81/121} - \sqrt[3]{-8/1331}$

31. $4\sqrt{24} - 8\sqrt{54} + 2\sqrt{6}$ **32.** $\sqrt[3]{192} + \sqrt[3]{-81} + \sqrt{\sqrt[3]{9}}$

33. $\sqrt{\sqrt[3]{64}}$ **34.** $\sqrt[3]{\sqrt{4096}}$

35. (a) $\sqrt{36x^2}$ **36. (a)** $\sqrt{225x^4y^3}$

  (b) $\sqrt{36y^2}$, where $y < 0$ **(b)** $\sqrt[4]{16a^4}$, where $a < 0$

37. (a) $\sqrt{ab^2}\sqrt{a^2b}$ **38. (a)** $\sqrt[3]{125x^6}$

  (b) $\sqrt{ab^3}\sqrt{a^3b}$ **(b)** $\sqrt[4]{64y^4}$, where $y < 0$

39. $\sqrt{72a^3b^4c^5}$ **40.** $\sqrt{(a + b)^5/(16a^2b^2)}$

41. $\sqrt[4]{16a^4b^5}$ **42.** $\sqrt[3]{8a^4b^6}$

43. $\sqrt[3]{16a^{12}b^2/c^9}$ **44.** $\sqrt[4]{ab^3}\sqrt[4]{a^3b}$

In Exercises 45–68, rationalize the denominators and simplify where possible. (Assume that all letters represent positive quantities.)

45. $4/\sqrt{7}$ **46.** $3/\sqrt{3}$

47. $1/\sqrt{8}$ **48.** $\sqrt{2}/\sqrt{3}$

49. $1/(1 + \sqrt{5})$ **50.** $\sqrt{2}/(1 - \sqrt{2})$

51. $(1 + \sqrt{3})/(1 - \sqrt{3})$ **52.** $(\sqrt{a} + \sqrt{b})/(\sqrt{a} - \sqrt{b})$

53. $(1/\sqrt{5}) + 4\sqrt{45}$ **54.** $(3/\sqrt{8}) - \sqrt{450}$

55. $1/\sqrt[3]{25}$ **56.** $4/\sqrt[3]{16}$

57. $3/\sqrt[4]{3}$ **58.** $\sqrt[3]{5}/\sqrt[3]{6}$

59. $3/\sqrt[5]{16a^4b^9}$ **60.** $3/\sqrt[4]{27a^5b^{11}}$

61. (a) $2/(\sqrt{3} + 1)$ **62. (a)** $1/(2\sqrt{5} + 1)$

  (b) $2/(\sqrt{3} - 1)$ **(b)** $1/(2\sqrt{5} - 1)$

63. (a) $x/(\sqrt{x} - 2)$ **64. (a)** $\sqrt{x}/(5 + \sqrt{x})$

  (b) $x/(\sqrt{x} - y)$ **(b)** $\sqrt{x}/(\sqrt{a} + \sqrt{x})$

65. (a) $(\sqrt{x} + \sqrt{2})/(\sqrt{x} - \sqrt{2})$

  (b) $(\sqrt{x} + \sqrt{a})/(\sqrt{x} - \sqrt{a})$

66. (a) $(\sqrt{3} + 2\sqrt{y})/(\sqrt{3} - 2\sqrt{y})$

  (b) $(\sqrt{x} - 2\sqrt{y})/(\sqrt{x} + 2\sqrt{y})$

67. $-2/(\sqrt{x + h} - \sqrt{x})$

68. $1/(\sqrt{x + h} - \sqrt{x - h})$

In Exercises 69–74, rationalize the numerator.

69. $(\sqrt{x} - \sqrt{5})/(x - 5)$ **70.** $(2\sqrt{y} + 3)/(4y - 9)$

71. $(\sqrt{2 + h} + \sqrt{2})/h$ **72.** $(\sqrt{1 + 2h} - 1)/(2h)$

73. $(\sqrt{x + h} - \sqrt{x})/h$

74. $(\sqrt{x + 2h} - \sqrt{x - 2h})/h$

In Exercises 75–78, refer to the following table. The left-hand column of the table lists four errors to avoid in working with expressions containing radicals. In each case, give a numerical example showing that the expressions on each side of the equation are, in general, not equal. Use a calculator as necessary.

Errors to avoid	Numerical example showing that the formula is not, in general, valid
75. $\sqrt{a+b} \neq \sqrt{a} + \sqrt{b}$	
76. $\sqrt{x^2 + y^2} \neq x + y$	
77. $\sqrt[3]{u+v} \neq \sqrt[3]{u} + \sqrt[3]{v}$	
78. $\sqrt[3]{p^3 + q^3} \neq p + q$	

B

79. (a) Use a calculator to evaluate $\sqrt{8 - 2\sqrt{7}}$ and $\sqrt{7} - 1$. What do you observe?

(b) Prove that $\sqrt{8 - 2\sqrt{7}} = \sqrt{7} - 1$.

Hint: In view of the definition of a principal square root, you need to check that $(\sqrt{7} - 1)^2 = 8 - 2\sqrt{7}$.

80. Use a calculator to provide empirical evidence indicating that both of the following equations may be correct:

$$\frac{2 - \sqrt{3}}{\sqrt{2} - \sqrt{2 - \sqrt{3}}} + \frac{2 + \sqrt{3}}{\sqrt{2} + \sqrt{2 + \sqrt{3}}} = \sqrt{2}$$

$$\sqrt{\sqrt{97.5 - (1/11)}} = \pi$$

The point of this exercise is to remind you that as useful as calculators may be, there is still the need for proofs in mathematics. In fact, it can be shown that the first equation is indeed correct, but the second is not.

B.3   REVIEW OF RATIONAL EXPONENTS

... Nicole Oresme, a bishop in Normandy (about 1323–1382), first conceived the notion of fractional powers, afterwards rediscovered by [Simon] Stevin [1548–1620] ...
—Florian Cajori in A History of Elementary Mathematics (New York: Macmillan Co., 1917)

If Descartes ... had discarded the radical sign altogether and had introduced the notation for fractional as well as integral exponents, ... it is conceivable that generations upon generations of pupils would have been saved the necessity of mastering the operations with two ... notations when one alone (the exponential) would have answered all purposes. —Florian Cajori, A History of Mathematical Notations, Vol. 1 (La Salle, Ill.: The Open Court Publishing Co., 1928)

We can use the concept of an nth root to give a meaning to fractional exponents that is useful and, at the same time, consistent with our earlier work. First, by way of motivation, suppose that we want to assign a value to $5^{1/3}$. Assuming that the usual properties of exponents continue to apply here, we can write

$$(5^{1/3})^3 = 5^1 = 5$$

That is,

$$(5^{1/3})^3 = 5$$

or

$$5^{1/3} = \sqrt[3]{5}$$

By replacing 5 and 3 with b and n, respectively, we can see that we want to define $b^{1/n}$ to mean $\sqrt[n]{b}$. Also, by thinking of $b^{m/n}$ as $(b^{1/n})^m$, we see that the definition for $b^{m/n}$ ought to be $(\sqrt[n]{b})^m$. These definitions are formalized in the box that follows.

DEFINITION | **Rational Exponents**

1. Let b denote a real number and n a natural number. We define $b^{1/n}$ by

$$b^{1/n} = \sqrt[n]{b}$$

(If n is even, we require that $b \geq 0$.)

2. Let m/n be a rational number reduced to lowest terms. Assume that n is positive and that $\sqrt[n]{b}$ exists. Then,

$$b^{m/n} = (\sqrt[n]{b})^m$$

or equivalently,

$$b^{m/n} = \sqrt[n]{b^m}$$

EXAMPLES

$4^{1/2} = \sqrt{4} = 2$

$(-8)^{1/3} = \sqrt[3]{-8} = -2$

$8^{2/3} = (\sqrt[3]{8})^2 = 2^2 = 4$

or, equivalently,

$8^{2/3} = \sqrt[3]{8^2} = \sqrt[3]{64} = 4$

It can be shown that the four properties of exponents that we listed in Appendix B.1 (on page A-9) continue to hold for rational exponents in general. In fact, we'll take this for granted rather than follow the lengthy argument needed for its verification. We will also assume that these properties apply to irrational exponents. For instance, we have

$$(2^{\sqrt{5}})^{\sqrt{5}} = 2^5 = 32$$

(The definition of irrational exponents is discussed in Section 5.1.) In the next three examples, we display the basic techniques for working with rational exponents.

EXAMPLE 1 Evaluating rational exponents

Simplify each of the following quantities. Express the answers using positive exponents. If an expression does not represent a real number, say so.
(a) $49^{1/2}$ **(b)** $-49^{1/2}$ **(c)** $(-49)^{1/2}$ **(d)** $49^{-1/2}$

SOLUTION
(a) $49^{1/2} = \sqrt{49} = 7$
(b) $-49^{1/2} = -(49^{1/2}) = -\sqrt{49} = -7$
(c) The quantity $(-49)^{1/2}$ does not represent a real number because there is no real number x such that $x^2 = -49$.
(d) $49^{-1/2} = \sqrt{49^{-1}} = \sqrt{1/49} = \sqrt{1}/\sqrt{49} = 1/7$
Alternatively, we have
$49^{-1/2} = (49^{1/2})^{-1} = 7^{-1} = 1/7$

EXAMPLE 2 Simplifying expressions containing rational exponents

Simplify each of the following. Write the answers using positive exponents. (Assume that $a > 0$.)
(a) $(5a^{2/3})(4a^{3/4})$

(b) $\sqrt[5]{\dfrac{16a^{1/3}}{a^{1/4}}}$

(c) $(x^2 + 1)^{1/5}(x^2 + 1)^{4/5}$

SOLUTION
(a) $(5a^{2/3})(4a^{3/4}) = 20a^{(2/3)+(3/4)}$

$\qquad\qquad = 20a^{17/12}$ because $\dfrac{2}{3} + \dfrac{3}{4} = \dfrac{17}{12}$

(b) $\sqrt[5]{\dfrac{16a^{1/3}}{a^{1/4}}} = \left(\dfrac{16a^{1/3}}{a^{1/4}}\right)^{1/5}$

$\qquad\qquad = (16a^{1/12})^{1/5}$ because $\dfrac{1}{3} - \dfrac{1}{4} = \dfrac{1}{12}$

$\qquad\qquad = 16^{1/5}a^{1/60}$

(c) $(x^2 + 1)^{1/5}(x^2 + 1)^{4/5} = (x^2 + 1)^1 = x^2 + 1$

EXAMPLE 3   Evaluating rational exponents

Simplify: **(a)** $32^{-2/5}$;   **(b)** $(-8)^{4/3}$.

SOLUTION

(a) $32^{-2/5} = (\sqrt[5]{32})^{-2} = 2^{-2} = \dfrac{1}{2^2} = \dfrac{1}{4}$

Alternatively, we have

$32^{-2/5} = (2^5)^{-2/5} = 2^{-2} = \dfrac{1}{2^2} = \dfrac{1}{4}$

(b) $(-8)^{4/3} = (\sqrt[3]{-8})^4 = (-2)^4 = 16$

Alternatively, we can write
$(-8)^{4/3} = [(-2)^3]^{4/3} = (-2)^4 = 16$

Rational exponents can be used to simplify certain expressions containing radicals. For example, one of the properties of nth roots that we listed but did not discuss in the previous appendix section is $\sqrt[m]{\sqrt[n]{x}} = \sqrt[mn]{x}$. Using exponents, it is easy to verify this property. We have

$$\sqrt[m]{\sqrt[n]{x}} = (x^{1/n})^{1/m}$$

$$= x^{1/mm} = \sqrt[mn]{x}   \quad \text{as we wished to show}$$

EXAMPLE 4   Using rational exponents to combine radicals

Consider the expression $\sqrt{x}\sqrt[3]{y^2}$, where x and y are positive.
(a) Rewrite the expression using rational exponents.
(b) Rewrite the expression using only one radical sign.

SOLUTION

(a) $\sqrt{x}\sqrt[3]{y^2} = x^{1/2}y^{2/3}$

(b) $\sqrt{x}\sqrt[3]{y^2} = x^{1/2}y^{2/3}$ rewriting the fractions using

$\qquad\qquad = x^{3/6}y^{4/6}$ a common denominator

$\qquad\qquad = \sqrt[6]{x^3}\sqrt[6]{y^4} = \sqrt[6]{x^3 y^4}$

EXAMPLE 5   Using rational exponents to simplify a radical expression

Rewrite the following expression using rational exponents (assume that x, y, and z are positive):

$$\sqrt{\dfrac{\sqrt[3]{x}\sqrt[4]{y^3}}{\sqrt[5]{z^4}}}$$

SOLUTION

$$\sqrt{\frac{\sqrt[3]{x}\sqrt[4]{y^3}}{\sqrt[5]{z^4}}} = \left(\frac{\sqrt[3]{x}\sqrt[4]{y^3}}{\sqrt[5]{z^4}}\right)^{1/2} = \left(\frac{x^{1/3}y^{3/4}}{z^{4/5}}\right)^{1/2}$$

$$= \frac{x^{1/6}y^{3/8}}{z^{2/5}} = x^{1/6}y^{3/8}z^{-2/5}$$

EXERCISE SET B.3

A

Exercises 1–14 are warm-up exercises to help you become famil-iar with the definition $b^{m/n} = (\sqrt[n]{b})^m = \sqrt[n]{b^m}$. In Exercises 1–8, write the expression in the two equivalent forms $(\sqrt[n]{b})^m$ and $\sqrt[n]{b^m}$. In Exercises 9–14, write the expression in the form $b^{m/n}$.

1. $a^{3/5}$ **2.** $x^{3/7}$ **3.** $5^{2/3}$

4. $10^{4/5}$ **5.** $(x^2 + 1)^{3/4}$ **6.** $(a + b)^{7/10}$

7. $2^{xy/3}$ **8.** $2^{(a+1)/b}$ **9.** $\sqrt[3]{p^2}$

10. $\sqrt[5]{R^3}$ **11.** $\sqrt[7]{(1 + u)^4}$ **12.** $\sqrt[6]{(1 + u)^5}$

13. $\sqrt[p]{(a^2 + b^2)^3}$ **14.** $\sqrt[3p]{(a^2 + b^2)^{2p}}$

In Exercises 15–50, evaluate or simplify each expression. Ex-press the answers using positive exponents. If an expression is undefined (i.e., does not represent a real number), say so. As-sume that all letters represent positive numbers.

15. $16^{1/2}$ **16.** $100^{1/2}$ **17.** $(1/36)^{1/2}$

18. $0.09^{1/2}$ **19.** $(-16)^{1/2}$ **20.** $(-1)^{1/2}$

21. $625^{1/4}$ **22.** $(1/81)^{1/4}$ **23.** $8^{1/3}$

24. $0.001^{1/3}$ **25.** $8^{2/3}$ **26.** $64^{2/3}$

27. $(-32)^{1/5}$ **28.** $(-1/125)^{1/3}$ **29.** $(-1000)^{1/3}$

30. $(-243)^{1/5}$ **31.** $49^{-1/2}$ **32.** $121^{-1/2}$

33. $(-49)^{-1/2}$ **34.** $(-64)^{-1/3}$ **35.** $36^{-3/2}$

36. $(-0.001)^{-2/3}$ **37.** $125^{2/3}$ **38.** $125^{-2/3}$

39. $(-1)^{3/5}$ **40.** $27^{4/3} + 27^{-4/3} + 27^0$

41. $32^{4/5} - 32^{-4/5}$ **42.** $64^{1/2} + 64^{-1/2} - 64^{4/3}$

43. $(9/16)^{-5/2} - (1000/27)^{4/3}$ **44.** $(256^{-3/4})^{4/3}$

45. $(2a^{1/3})(3a^{1/4})$ **46.** $\sqrt[5]{3a^4}$

47. $\sqrt[4]{64a^{2/3}/a^{1/3}}$ **48.** $(x^2 + 1)^{2/3}(x^2 + 1)^{4/3}$

49. $\dfrac{(x^2 + 1)^{3/4}}{(x^2 + 1)^{-1/4}}$

50. $\dfrac{(2x^2 + 1)^{-6/5}(2x^2 + 1)^{6/5}(x^2 + 1)^{-1/5}}{(x^2 + 1)^{9/5}}$

For Exercises 51–58, follow Example 4 in the text to rewrite each expression in two ways: **(a)** *using rational exponents; and*

(b) *using only one radical sign. (Assume that x, y, and z are positive.)*

51. $\sqrt{3}\sqrt[3]{6}$ **52.** $\sqrt{5}\sqrt[3]{7}$

53. $\sqrt[3]{6}\sqrt[4]{2}$ **54.** $\sqrt[3]{2}\cdot\sqrt[5]{2}$

55. $\sqrt[3]{x^2}\sqrt[5]{y^4}$ **56.** $\sqrt{x}\sqrt[3]{y}\sqrt[4]{z}$

57. $\sqrt[4]{x^a}\sqrt[3]{x^b}\sqrt{x^{a/b}}$ **58.** $\sqrt[3]{27\sqrt{64x}}$

In Exercises 59–66, rewrite each expression using rational ex-ponents rather than radicals. Assume that x, y, and z are positive.

59. $\sqrt[3]{(x + 1)^2}$ **60.** $(1/\sqrt{x}) + \sqrt{x}$

61. $(\sqrt[5]{x} + y)^2$ **62.** $\sqrt{\sqrt{x}}$

63. $\sqrt[3]{\sqrt{x}} + \sqrt{\sqrt[3]{x}}$ **64.** $\sqrt[3]{\sqrt{2}}$

65. $\sqrt{\sqrt[3]{x}\sqrt[4]{y}}$ **66.** $\sqrt[5]{\sqrt{x}\sqrt[3]{y}/\sqrt[4]{z^2}}$

67. Use a calculator to determine which is larger, $9^{10/9}$ or $10^{9/10}$.

68. Use a calculator to determine which number is closer to 3: $13^{3/7}$ or $6560^{1/8}$.

In Exercises 69–74, refer to the following table. The left-hand column of the table lists six errors to avoid in working with ex-pressions containing fractional exponents. In each case, give a numerical example showing that the expressions on each side of the equation are not equal. Use a calculator as necessary.

Errors to avoid	Numerical example showing that the formula is not, in general, valid
69. $(a + b)^{1/2} \neq a^{1/2} + b^{1/2}$	
70. $(x^2 + y^2)^{1/2} \neq x + y$	
71. $(u + v)^{1/3} \neq u^{1/3} + v^{1/3}$	
72. $(p^3 + q^3)^{1/3} \neq p + q$	
73. $x^{1/m} \neq \dfrac{1}{x^m}$	
74. $x^{-1/2} \neq \dfrac{1}{x^2}$	

B.4 REVIEW OF FACTORING

There are many cases in algebra in which the process of *factoring* simplifies the work at hand. To **factor** a polynomial means to write it as a product of two or more nonconstant polynomials. For instance, a factorization of $x^2 - 9$ is given by

$$x^2 - 9 = (x - 3)(x + 3)$$

In this case, $x - 3$ and $x + 3$ are the **factors** of $x^2 - 9$.

There is one convention that we need to agree on at the outset. If the polynomial or expression that we wish to factor contains only integer coefficients, then the factors (if any) should involve only integer coefficients. For example, according to this convention, we will not consider the following type of factorization in this section:

$$x^2 - 2 = (x - \sqrt{2})(x + \sqrt{2})$$

because it involves coefficients that are irrational numbers. (We should point out, however, that factorizations such as this are useful at times, particularly in calculus.) As it happens, $x^2 - 2$ is an example of a polynomial that cannot be factored using integer coefficients. In such a case we say that the polynomial is **irreducible over the integers.**

If the coefficients of a polynomial are rational numbers, then we do allow factors with rational coefficients. For instance, the factorization of $y^2 - \frac{1}{4}$ over the rational numbers is given by

$$y^2 - \tfrac{1}{4} = (y - \tfrac{1}{2})(y + \tfrac{1}{2})$$

We'll consider five techniques for factoring in this section. These techniques will be applied in the next two sections and throughout the text. In Table 1 we summarize these techniques. Notice that three of the formulas in the table are just

TABLE 1 Basic Factoring Techniques

Technique	Example or formula	Remark
Common factor	$3x^4 + 6x^3 - 12x^2 = 3x^2(x^2 + 2x - 4)$ $4(x^2 + 1) - x(x^2 + 1) = (x^2 + 1)(4 - x)$	In any factoring problem, the first step always is to look for the common factor of highest degree.
Difference of squares	$x^2 - a^2 = (x - a)(x + a)$	There is no corresponding formula for a sum of squares; $x^2 + a^2$ is irreducible over the integers.
Trial and error	$x^2 + 2x - 3 = (x + 3)(x - 1)$	In this example, the only possibilities, or trials, are: **(a)** $(x - 3)(x - 1)$ **(c)** $(x + 3)(x - 1)$ **(b)** $(x - 3)(x + 1)$ **(d)** $(x + 3)(x + 1)$ By inspection or by carrying out the indicated multiplications, we find that only case (c) checks.
Difference of cubes Sum of cubes	$x^3 - a^3 = (x - a)(x^2 + ax + a^2)$ $x^3 + a^3 = (x + a)(x^2 - ax + a^2)$	Verify these formulas for yourself by carrying out the multiplications. Then memorize the formulas.
Grouping	$x^3 - x^2 + x - 1 = (x^3 - x^2) + (x - 1)$ $ = x^2(x - 1) + (x - 1) \cdot 1$ $ = (x - 1)(x^2 + 1)$	This is actually an application of the common factor technique.

restatements of Special Products formulas. Remember, it is the form, or pattern, in the formula that is important—not the specific choice of letters.

The idea in factoring is to use one or more of these techniques until each of the factors obtained is irreducible. The examples that follow show how this works in practice.

EXAMPLE 1 **Factoring using the difference-of-squares technique**

Factor: **(a)** $x^2 - 49$;   **(b)** $(2a - 3b)^2 - 49$.

SOLUTION

(a) $x^2 - 49 = x^2 - 7^2$
$$= (x - 7)(x + 7) \qquad \text{difference of squares}$$

(b) Notice that the pattern or form is the same as in part (a): it's a difference of squares. We have

$(2a - 3b)^2 - 49 = (2a - 3b)^2 - 7^2$
$$= [(2a - 3b) - 7][(2a - 3b) + 7] \qquad \text{difference of squares}$$
$$= (2a - 3b - 7)(2a - 3b + 7)$$

EXAMPLE 2 **Factoring using common factor and difference of squares**

Factor: **(a)** $2x^3 - 50x$;   **(b)** $3x^5 - 3x$.

SOLUTION

(a) $2x^3 - 50x = 2x(x^2 - 25) \qquad \text{common factor}$
$$= 2x(x - 5)(x + 5) \qquad \text{difference of squares}$$

(b) $3x^5 - 3x = 3x(x^4 - 1) \qquad \text{common factor}$
$$= 3x[(x^2)^2 - 1^2]$$
$$= 3x(x^2 - 1)(x^2 + 1) \qquad \text{difference of squares}$$
$$= 3x(x - 1)(x + 1)(x^2 + 1) \qquad \text{difference of squares, again}$$

EXAMPLE 3 **Factoring by trial and error**

Factor: **(a)** $x^2 - 4x - 5$;   **(b)** $(a + b)^2 - 4(a + b) - 5$.

SOLUTION

(a) $x^2 - 4x - 5 = (x - 5)(x + 1) \qquad \text{trial and error}$

(b) Note that the form of this expression is

$$(\quad)^2 - 4(\quad) - 5$$

This is the same *form* as the expression in part (a), so we need only replace x with the quantity $a + b$ in the solution for part (a). This yields

$$(a + b)^2 - 4(a + b) - 5 = [(a + b) - 5][(a + b) + 1]$$
$$= (a + b - 5)(a + b + 1)$$

EXAMPLE 4 **Factoring by trial and error**

Factor: $2z^4 + 9z^2 + 4$.

SOLUTION
We use trial and error to look for a factorization of the form $(2z^2 + ?)(z^2 + ?)$. There are three possibilities:
(i) $(2z^2 + 2)(z^2 + 2)$ **(ii)** $(2z^2 + 1)(z^2 + 4)$ **(iii)** $(2z^2 + 4)(z^2 + 1)$
Each of these yields the appropriate first term and last term, but (after checking) only possibility (ii) yields $9z^2$ for the middle term. The required factorization is then $2z^4 + 9z^2 + 4 = (2z^2 + 1)(z^2 + 4)$.

QUESTION   Why didn't we consider any possibilities with subtraction signs in place of addition signs?

EXAMPLE 5 **Factoring expressions that are irreducible**

Factor: **(a)** $x^2 + 9$;   **(b)** $x^2 + 2x + 3$.

SOLUTION
(a) The expression $x^2 + 9$ is irreducible over the integers. (This can be discovered by trial and error.) If the given expression had instead been $x^2 - 9$, then it could have been factored as a difference of squares. Sums of squares, however, cannot in general be factored over the integers.
(b) The expression $x^2 + 2x + 3$ is irreducible over the integers. (Check this for yourself by trial and error.)

EXAMPLE 6 **Factoring using grouping and a special product**

Factor: $x^2 - y^2 + 10x + 25$.

SOLUTION
Familiarity with the special product $(a + b)^2 = a^2 + 2ab + b^2$ suggests that we try grouping the terms this way:

$$(x^2 + 10x + 25) - y^2$$

Then we have

$$\begin{aligned}
x^2 - y^2 + 10x + 25 &= (x^2 + 10x + 25) - y^2 \\
&= (x + 5)^2 - y^2 \\
&= [(x + 5) - y][(x + 5) + y] \qquad \text{difference of squares} \\
&= (x - y + 5)(x + y + 5)
\end{aligned}$$

EXAMPLE 7 Factoring using grouping

Factor: $ax + ay^2 + bx + by^2$.

SOLUTION
We factor a from the first two terms and b from the second two, to obtain

$$ax + ay^2 + bx + by^2 = a(x + y^2) + b(x + y^2)$$

We now recognize the quantity $x + y^2$ as a common expression that can be factored out. We then have

$$a(x + y^2) + b(x + y^2) = (x + y^2)(a + b)$$

The required factorization is therefore

$$ax + ay^2 + bx + by^2 = (x + y^2)(a + b)$$

As you may wish to check for yourself, this factorization can also be obtained by trial and error.

EXAMPLE 8 Factoring sum and difference of cubes

Factor: **(a)** $t^3 - 125$; **(b)** $8 + (a - 2)^3$.

SOLUTION
(a) $t^3 - 125 = t^3 - 5^3$
$= (t - 5)(t^2 + 5t + 25)$ difference of cubes
(b) $8 + (a - 2)^3 = 2^3 + (a - 2)^3$
$= [2 + (a - 2)][2^2 - 2(a - 2) + (a - 2)^2]$ sum of cubes
$= a(4 - 2a + 4 + a^2 - 4a + 4)$
$= a(a^2 - 6a + 12)$

As you can check now, the expression $a^2 - 6a + 12$ is irreducible over the integers. Therefore the required factorization is $a(a^2 - 6a + 12)$.

For some calculations (particularly in calculus), it's helpful to be able to factor an expression involving fractional exponents. For instance, suppose (as in the next example) that we want to factor the expression

$$x(2x - 1)^{-1/2} + (2x - 1)^{3/2}$$

The common expression to factor out here is $(2x - 1)^{-1/2}$; the technique is to *choose the expression with the smaller exponent.*

EXAMPLE 9   Factoring with rational exponents

Factor: $x(2x - 1)^{-1/2} + (2x - 1)^{3/2}$.

SOLUTION

$$x(2x - 1)^{-1/2} + (2x - 1)^{3/2} = (2x - 1)^{-1/2}[x + (2x - 1)^{3/2+1/2}]$$
$$= (2x - 1)^{-1/2}[x + (2x - 1)^2]$$
$$= (2x - 1)^{-1/2}(4x^2 - 3x + 1)$$

We conclude this section with examples of four common errors to avoid in factoring. The first two errors (in the box that follows) are easy to detect; simply multiplying out the supposed factorizations shows that they do not check. The third error may result from a lack of familiarity with the basic factoring techniques listed on page A-25. The fourth error indicates a misunderstanding of what is required in factoring; the final quantity in a factorization must be expressed as a *product*, not a sum, of terms or expressions.

ERRORS TO AVOID

Error	Correction	Comment
$x^2 + 6x + 9 \neq (x - 3)^2$	$x^2 + 6x + 9 = (x + 3)^2$	Check the middle term.
$x^2 + 64 \neq (x + 8)(x + 8)$	$x^2 + 64$ is irreducible over the integers.	A sum of squares is, in general, irreducible over the integers. (A difference of squares, however, can always be factored.)
$x^3 + 64$ is irreducible over the integers.	$x^3 + 64 = (x + 4)(x^2 - 4x + 16)$	Although a sum of squares is irreducible, a sum of cubes can be factored.
$x^2 - 2x + 3$ factors as $x(x - 2) + 3$	$x^2 - 2x + 3$ is irreducible over the integers.	The polynomial $x^2 - 2x + 3$ is the *sum* of the expression $x(x - 2)$ and the constant 3. By definition, however, the factored form of a polynomial must be a *product* (of two or more nonconstant polynomials).

EXERCISE SET B.4

A

In Exercises 1–66, factor each polynomial or expression. If a polynomial is irreducible, state this. (In Exercises 1–6, the factoring techniques are specified.)

1. (Common factor and difference of squares)
   (a) $x^2 - 64$
   (b) $7x^4 + 14x^2$
   (c) $121z - z^3$
   (d) $a^2b^2 - c^2$

2. (Common factor and difference of squares)
   (a) $1 - t^4$
   (b) $x^6 + x^5 + x^4$
   (c) $u^2v^2 - 225$
   (d) $81x^4 - x^2$

3. (Trial and error)
   (a) $x^2 + 2x - 3$
   (b) $x^2 - 2x - 3$
   (c) $x^2 - 2x + 3$
   (d) $-x^2 + 2x + 3$

4. (Trial and error)
  (a) $2x^2 - 7x - 4$
  (b) $2x^2 + 7x - 4$
  (c) $2x^2 + 7x + 4$
  (d) $-2x^2 - 7x + 4$

5. (Sum and difference of cubes)
  (a) $x^3 + 1$
  (b) $x^3 + 216$
  (c) $1000 - 8x^6$
  (d) $64a^3x^3 - 125$

6. (Grouping)
  (a) $x^4 - 2x^3 + 3x - 6$
  (b) $a^2x + bx - a^2z - bz$

7. (a) $144 - x^2$
  (b) $144 + x^2$
  (c) $144 - (y - 3)^2$

8. (a) $4a^2b^2 + 9c^2$
  (b) $4a^2b^2 - 9c^2$
  (c) $4a^2b^2 - 9(ab + c)^2$

9. (a) $h^3 - h^5$
  (b) $100h^3 - h^5$
  (c) $100(h + 1)^3 - (h + 1)^5$

10. (a) $x^4 - x^2$
  (b) $3x^4 - 48x^2$
  (c) $3(x + h)^4 - 48(x + h)^2$

11. (a) $x^2 - 13x + 40$
  (b) $x^2 - 13x - 40$

12. (a) $x^2 + 10x + 16$
  (b) $x^2 - 10x + 16$

13. (a) $x^2 + 5x - 36$
  (b) $x^2 - 13x + 36$

14. (a) $x^2 - x + 6$
  (b) $x^2 + x - 6$

15. (a) $3x^2 - 22x - 16$
  (b) $3x^2 - x - 16$

16. (a) $x^2 - 4x + 1$
  (b) $x^2 - 4x - 3$

17. (a) $6x^2 + 13x - 5$
  (b) $6x^2 - x - 5$

18. (a) $16x^2 + 18x - 9$
  (b) $16x^2 - 143x - 9$

19. (a) $t^4 + 2t^2 + 1$
  (b) $t^4 - 2t^2 + 1$
  (c) $t^4 - 2t^2 - 1$

20. (a) $t^4 - 9t^2 + 20$
  (b) $t^4 - 19t^2 + 20$
  (c) $t^4 - 19t^2 - 20$

21. (a) $4x^3 - 20x^2 - 25x$
  (b) $4x^3 - 20x^2 + 25x$

22. (a) $x^3 + x^2 + x$
  (b) $x^3 + 2x^2 + x$

23. (a) $ab - bc + a^2 - ac$
  (b) $(u + v)x - xy + (u + v)^2 - (u + v)y$

24. (a) $3(x + 5)^3 + 2(x + 5)^2$
  (b) $a(x + 5)^3 + b(x + 5)^2$

25. $x^2z^2 + xzt + xyz + yt$
26. $a^2t^2 + b^2t^2 - cb^2 - ca^2$
27. $a^4 - 4a^2b^2c^2 + 4b^4c^4$
28. $A^2 - B^2 + 16A + 64$
29. $A^2 + B^2$
30. $x^2 + 64$
31. $x^3 + 64$
32. $27 - (a - b)^3$
33. $(x + y)^3 - y^3$
34. $(a + b)^3 - 8c^3$
35. $x^3 - y^3 + x - y$
36. $8a^3 + 27b^3 + 2a + 3b$
37. (a) $p^4 - 1$
  (b) $p^8 - 1$
38. (a) $p^4 + 4$
  (b) $p^4 - 4$
39. $x^3 + 3x^2 + 3x + 1$
40. $-1 + 6x - 12x^2 + 8x^3$
41. $x^2 + 16y^2$
42. $4u^2 + 25v^2$

43. $\dfrac{25}{16} - c^2$

44. $\dfrac{81}{4} - y^2$

45. $z^4 - \dfrac{81}{16}$

46. $\dfrac{(a + b)^2}{4} - \dfrac{a^2b^2}{9}$

47. $\dfrac{125}{m^3n^3} - 1$

48. $\dfrac{x^3}{8} - \dfrac{512}{x^3}$

49. $\frac{1}{4}x^2 + xy + y^2$
50. $x^2 + x + 1$
51. $64(x - a)^3 - x + a$
52. $64(x - a)^4 - x + a$
53. $x^2 - a^2 + y^2 - 2xy$
54. $a^4 - (b + c)^4$
55. $21x^3 + 82x^2 - 39x$
56. $x^3a^2 - 8y^3a^2 - 4x^3b^2 + 32y^3b^2$
57. $12xy + 25 - 4x^2 - 9y^2$
58. $ax^2 + (1 + ab)xy + by^2$
59. $ax^2 + (a + b)x + b$
60. $(5a^2 - 11a + 10)^2 - (4a^2 - 15a + 6)^2$
61. $(x + 1)^{1/2} - (x + 1)^{3/2}$
62. $(x^2 + 1)^{3/2} + (x^2 + 1)^{7/2}$
63. $(x + 1)^{-1/2} - (x + 1)^{-3/2}$
64. $(x^2 + 1)^{-2/3} + (x^2 + 1)^{-5/3}$
65. $(2x + 3)^{1/2} - \frac{1}{3}(2x + 3)^{3/2}$
66. $(ax + b)^{-1/2} - \sqrt{ax + b}/b$

In Exercises 67 and 68, evaluate the given expressions using factoring techniques. (The point here is to do as little actual arithmetic as possible.)

67. (a) $100^2 - 99^2$
  (b) $8^3 - 6^3$
  (c) $1000^2 - 999^2$

68. (a) $10^3 - 9^3$
  (b) $50^2 - 49^2$
  (c) $\dfrac{15^3 - 10^3}{15^2 - 10^2}$

B

In Exercises 69–73, factor each expression.

69. $A^3 + B^3 + 3AB(A + B)$
70. $p^3 - q^3 - p(p^2 - q^2) + q(p - q)^2$
71. $2x(a^2 + x^2)^{-1/2} - x^3(a^2 + x^2)^{-3/2}$
72. $\frac{1}{2}(x - a)^{-1/2}(x + a)^{-1/2} - \frac{1}{2}(x + a)^{1/2}(x - a)^{-3/2}$
73. $y^4 - (p + q)y^3 + (p^2q + pq^2)y - p^2q^2$
74. (a) Factor: $x^4 + 2x^2y^2 + y^4$.
  (b) Factor: $x^4 + x^2y^2 + y^4$.   *Hint*: Add and subtract a term. [Keep part (a) in mind.]
  (c) Factor: $x^6 - y^6$ as a difference of squares.
  (d) Factor: $x^6 - y^6$ as a difference of cubes. [Use the result in part (b) to obtain the same answer as in part (c).]

B.5   REVIEW OF FRACTIONAL EXPRESSIONS

But if you do use a rule involving mechanical calculation, be patient, accurate, and systematically neat in the working. It is well known to mathematical teachers that quite half the failures in algebraic exercises arise from arithmetical inaccuracy and slovenly arrangement. — George Chrystal (1893)

The rules of basic arithmetic that you learned for working with simple fractions are also used for fractions involving algebraic expressions. In algebra, as in arithmetic, we say that a fraction is *reduced to lowest terms* or *simplified* when the numerator and denominator contain no common factors (other than 1 and -1). The factoring techniques reviewed in appendix section B.4 are used to reduce fractions. For example, to reduce the fraction $\dfrac{x^2 - 9}{x^2 + 3x}$, we write

$$\frac{x^2 - 9}{x^2 + 3x} = \frac{(x - 3)\cancel{(x + 3)}}{x\cancel{(x + 3)}} = \frac{x - 3}{x}$$

In the box that follows we review two properties of negatives and fractions that will be useful in this section. Notice that Property 2 follows from Property 1 because

$$\frac{a - b}{b - a} = \frac{a - b}{-(a - b)} = -1$$

PROPERTY SUMMARY   Negatives and Fractions

Property	Example
1. $b - a = -(a - b)$	$2 - x = -(x - 2)$
2. $\dfrac{a - b}{b - a} = -1$	$\dfrac{2x - 5}{5 - 2x} = -1$

EXAMPLE   1   Using Property 2 to simplify a fractional expression

Simplify: $\dfrac{4 - 3x}{15x - 20}$.

SOLUTION

$$\begin{aligned}
\frac{4 - 3x}{15x - 20} &= \frac{4 - 3x}{5(3x - 4)} \\
&= \frac{1}{5}\left(\frac{4 - 3x}{3x - 4}\right) = -\frac{1}{5} \qquad \text{using Property 2}
\end{aligned}$$

In Example 2 we display two more instances in which factoring is used to reduce a fraction. After that, Example 3 indicates how these skills are used to multiply and divide fractional expressions.

EXAMPLE   2   Using factoring to simplify a fractional expression

Simplify: **(a)** $\dfrac{x^3 - 8}{x^2 - 2x}$;   **(b)** $\dfrac{x^2 - 6x + 8}{a(x - 2) + b(x - 2)}$.

SOLUTION

(a) $\dfrac{x^3 - 8}{x^2 - 2x} = \dfrac{\cancel{(x-2)}(x^2 + 2x + 4)}{x\cancel{(x-2)}} = \dfrac{x^2 + 2x + 4}{x}$

(b) $\dfrac{x^2 - 6x + 8}{a(x-2) + b(x-2)} = \dfrac{\cancel{(x-2)}(x-4)}{\cancel{(x-2)}(a+b)} = \dfrac{x-4}{a+b}$

EXAMPLE 3   Using factoring to simplify products and quotients

Carry out the indicated operations and simplify:

(a) $\dfrac{x^3}{2x^2 + 3x} \cdot \dfrac{12x + 18}{4x^2 - 6x}$;

(c) $\dfrac{x^2 - 144}{x^2 - 4} \div \dfrac{x + 12}{x + 2}$.

(b) $\dfrac{2x^2 - x - 6}{x^2 + x + 1} \cdot \dfrac{x^3 - 1}{4 - x^2}$;

SOLUTION

(a) $\dfrac{x^3}{2x^2 + 3x} \cdot \dfrac{12x + 18}{4x^2 - 6x} = \dfrac{x^3}{x\cancel{(2x+3)}} \cdot \dfrac{6\cancel{(2x+3)}}{2x(2x-3)}$

$= \dfrac{6x^3}{2x^2(2x-3)} = \dfrac{3x}{2x-3}$

(b) $\dfrac{2x^2 - x - 6}{x^2 + x + 1} \cdot \dfrac{x^3 - 1}{4 - x^2} = \dfrac{(2x+3)(x-2)}{(x^2 + x + 1)} \cdot \dfrac{(x-1)(x^2 + x + 1)}{(2-x)(2+x)}$

$= \dfrac{-(2x+3)(x-1)}{2+x}$ using the fact

$= \dfrac{-2x^2 - x + 3}{2+x}$ that $\dfrac{x-2}{2-x} = -1$

(c) $\dfrac{x^2 - 144}{x^2 - 4} \div \dfrac{x + 12}{x + 2} = \dfrac{x^2 - 144}{x^2 - 4} \cdot \dfrac{x + 2}{x + 12}$

$= \dfrac{(x-12)(x+12)}{(x-2)(x+2)} \cdot \dfrac{x + 2}{x + 12} = \dfrac{x - 12}{x - 2}$

As in arithmetic, to add or subtract two fractions, the denominators must be the same. The rules in this case are

$$\frac{a}{b} + \frac{c}{b} = \frac{a+c}{b} \qquad \text{and} \qquad \frac{a}{b} - \frac{c}{b} = \frac{a-c}{b}$$

For example, we have

$$\frac{4x - 1}{x + 1} - \frac{2x - 1}{x + 1} = \frac{4x - 1 - (2x - 1)}{x + 1} = \frac{4x - 1 - 2x + 1}{x + 1} = \frac{2x}{x + 1}$$

Fractions with unlike denominators are added or subtracted by first converting to a common denominator. For instance, to add $9/a$ and $10/a^2$ we write

$$\frac{9}{a} + \frac{10}{a^2} = \frac{9}{a} \cdot \frac{a}{a} + \frac{10}{a^2}$$

$$= \frac{9a}{a^2} + \frac{10}{a^2} = \frac{9a + 10}{a^2}$$

Notice that the common denominator used was a^2. This is the **least common denominator.** In fact, other common denominators (such as a^3 or a^4) could be used here, but that would be less efficient. In general, the least common denominator for a given group of fractions is chosen as follows. Write down a product involving the irreducible factors from each denominator. The power of each factor should be equal to (but not greater than) the highest power of that factor appearing in any of the individual denominators. For example, the least common denominator for the two fractions $\dfrac{1}{(x + 1)^2}$ and $\dfrac{1}{(x + 1)(x + 2)}$ is $(x + 1)^2(x + 2)$. In the example that follows, notice that the denominators must be in factored form before the least common denominator can be chosen.

EXAMPLE 4 **Using the least common denominator**

Combine into a single fraction and simplify:

(a) $\dfrac{3}{4x} + \dfrac{7x}{10y^2} - 2$; (b) $\dfrac{x}{x^2 - 9} - \dfrac{1}{x + 3}$; (c) $\dfrac{15}{x^2 + x - 6} + \dfrac{x + 1}{2 - x}$.

SOLUTION

(a) Denominators: $2^2 \cdot x;\ 2 \cdot 5 \cdot y^2;\ 1$

Least common denominator; $2^2 \cdot 5xy^2 = 20xy^2$

$$\frac{3}{4x} + \frac{7x}{10y^2} - \frac{2}{1} = \frac{3}{4x} \cdot \frac{5y^2}{5y^2} + \frac{7x}{10y^2} \cdot \frac{2x}{2x} - \frac{2}{1} \cdot \frac{20xy^2}{20xy^2}$$

$$= \frac{15y^2 + 14x^2 - 40xy^2}{20xy^2}$$

(b) Denominators: $x^2 - 9 = (x - 3)(x + 3);\ x + 3$

Least common denominator: $(x - 3)(x + 3)$

$$\frac{x}{x^2 - 9} - \frac{1}{x + 3} = \frac{x}{(x - 3)(x + 3)} - \frac{1}{x + 3}$$

$$= \frac{x}{(x - 3)(x + 3)} - \frac{1}{x + 3} \cdot \frac{x - 3}{x - 3}$$

$$= \frac{x - (x - 3)}{(x - 3)(x + 3)} = \frac{3}{(x - 3)(x + 3)}$$

(c) $\dfrac{15}{x^2 + x - 6} + \dfrac{x + 1}{2 - x} = \dfrac{15}{(x - 2)(x + 3)} - \dfrac{x + 1}{x - 2}$ using the fact that $2 - x = -(x - 2)$

$$= \frac{15}{(x - 2)(x + 3)} - \frac{x + 1}{x - 2} \cdot \frac{x + 3}{x + 3}$$

$$= \frac{15 - (x^2 + 4x + 3)}{(x - 2)(x + 3)} = \frac{-x^2 - 4x + 12}{(x - 2)(x + 3)}$$

$$= \frac{(x - 2)(-x - 6)}{(x - 2)(x + 3)} = \frac{-x - 6}{x + 3}$$

Notice in the last line that we were able to simplify the answer by factoring the numerator and reducing the fraction. (This is why we prefer to leave the least common denominator in factored form, rather than multiply it out, in this type of problem.)

The next four examples illustrate using the least common denominator to simplify compound fractions, which are fractions "within" fractions.

EXAMPLE 5 Simplifying a compound fraction

Simplify: $\dfrac{\dfrac{1}{3a} - \dfrac{1}{4b}}{\dfrac{5}{6a^2} + \dfrac{1}{b}}$.

SOLUTION
The least common denominator for the four individual factors is $12a^2b$. Multiplying the given expression by $12a^2b/12a^2b$, which equals 1, yields

$$\frac{12a^2b}{12a^2b} \cdot \frac{\dfrac{1}{3a} - \dfrac{1}{4b}}{\dfrac{5}{6a^2} + \dfrac{1}{b}} = \frac{4ab - 3a^2}{10b + 12a^2}$$

EXAMPLE 6 Simplifying a fraction containing negative exponents

Simplify: $(x^{-1} + y^{-1})^{-1}$. (The answer is *not* $x + y$.)

SOLUTION
After applying the definition of negative exponent to rewrite the given expression, we'll use the method shown in Example 5.

$$(x^{-1} + y^{-1})^{-1} = \left(\frac{1}{x} + \frac{1}{y}\right)^{-1} = \frac{1}{\dfrac{1}{x} + \dfrac{1}{y}}$$

$$= \frac{xy}{xy} \cdot \frac{1}{\dfrac{1}{x} + \dfrac{1}{y}} = \frac{xy}{y + x}$$

EXAMPLE 7 Simplifying a compound fraction

Simplify: $\dfrac{\dfrac{1}{x + h} - \dfrac{1}{x}}{h}$. (This type of expression occurs in calculus.)

SOLUTION

$$\frac{\dfrac{1}{x + h} - \dfrac{1}{x}}{h} = \frac{(x + h)x}{(x + h)x} \cdot \frac{\dfrac{1}{x + h} - \dfrac{1}{x}}{h} \qquad \text{multiplying by } 1 = \frac{(x\text{th})x}{(x\text{th})x}$$

$$= \frac{x - (x + h)}{(x + h)xh} = \frac{-h}{(x + h)xh} = -\frac{1}{x(x + h)}$$

EXAMPLE 8   Simplifying a compound fraction

Simplify: $\dfrac{x - \dfrac{1}{x^2}}{\dfrac{1}{x^2} - 1}$.

SOLUTION

$$\frac{x - \dfrac{1}{x^2}}{\dfrac{1}{x^2} - 1} = \frac{x^2}{x^2} \cdot \frac{x - \dfrac{1}{x^2}}{\dfrac{1}{x^2} - 1} \qquad \text{multiplying by } 1 = \frac{x^2}{x^2}$$

$$= \frac{x^3 - 1}{1 - x^2} = -\frac{x^3 - 1}{x^2 - 1}$$

$$= -\frac{(x - 1)(x^2 + x + 1)}{(x - 1)(x + 1)} = -\frac{x^2 + x + 1}{x + 1}$$

EXERCISE SET B.5

A

In Exercises 1–12, reduce the fractions to lowest terms.

1. $\dfrac{x^2 - 9}{x + 3}$

2. $\dfrac{25 - x^2}{x - 5}$

3. $\dfrac{x + 2}{x^4 - 16}$

4. $\dfrac{x^2 - x - 20}{2x^2 + 7x - 4}$

5. $\dfrac{x^2 + 2x + 4}{x^3 - 8}$

6. $\dfrac{a + b}{ax^2 + bx^2}$

7. $\dfrac{9ab - 12b^2}{6a^2 - 8ab}$

8. $\dfrac{a^3b^2 - 27b^5}{(ab - 3b^2)^2}$

9. $\dfrac{a^3 + a^2 + a + 1}{a^2 - 1}$

10. $\dfrac{(x - y)^2(a + b)}{(x^2 - y^2)(a^2 + 2ab + b^2)}$

11. $\dfrac{x^3 - y^3}{(x - y)^3}$

12. $\dfrac{x^4 - y^4}{(x^4y + x^2y^3 + x^3y^2 + xy^4)(x - y)^2}$

In Exercises 13–55, carry out the indicated operations and simplify where possible.

13. $\dfrac{2}{x - 2} \cdot \dfrac{x^2 - 4}{x + 2}$

14. $\dfrac{ax + 3}{2a + 1} \div \dfrac{a^2x^2 + 3ax}{4a^2 - 1}$

15. $\dfrac{x^2 - x - 2}{x^2 + x - 12} \cdot \dfrac{x^2 - 3x}{x^2 - 4x + 4}$

16. $(3t^2 + 4tx + x^2) \div \dfrac{3t^2 - 2tx - x^2}{t^2 - x^2}$

17. $\dfrac{x^3 + y^3}{x^2 - 4xy + 3y^2} \div \dfrac{(x + y)^3}{x^2 - 2xy - 3y^2}$

18. $\dfrac{a^2 - a - 42}{a^4 + 216a} \div \dfrac{a^2 - 49}{a^3 - 6a^2 + 36a}$

19. $\dfrac{4}{x} - \dfrac{2}{x^2}$

20. $\dfrac{1}{3x} + \dfrac{1}{5x^2} - \dfrac{1}{30x^3}$

21. $\dfrac{6}{a} - \dfrac{a}{6}$

22. $\dfrac{1}{a} + \dfrac{1}{b} + \dfrac{1}{c}$

23. $\dfrac{1}{x + 3} + \dfrac{3}{x + 2}$

24. $\dfrac{4}{x - 4} - \dfrac{4}{x + 1}$

25. $\dfrac{3x}{x - 2} - \dfrac{6}{x^2 - 4}$

26. $1 + \dfrac{1}{x} - \dfrac{1}{x^2}$

27. $\dfrac{a}{x - 1} + \dfrac{2ax}{(x - 1)^2} + \dfrac{3ax^2}{(x - 1)^3}$

28. $\dfrac{a^2 + 5a - 4}{a^2 - 16} - \dfrac{2a}{2a^2 + 8a}$

29. $\dfrac{4}{x - 5} - \dfrac{4}{5 - x}$

30. $\dfrac{x}{x + a} + \dfrac{a}{a - x}$

31. $\dfrac{a^2 + b^2}{a^2 - b^2} + \dfrac{a}{a + b} + \dfrac{b}{b - a}$

32. $\dfrac{3}{2x + 2} - \dfrac{5}{x^2 - 1} + \dfrac{1}{x + 1}$

33. $\dfrac{1}{x^2 + x - 20} - \dfrac{1}{x^2 - 8x + 16}$

34. $\dfrac{4}{6x^2 + 5x - 4} + \dfrac{1}{3x^2 + 4x} - \dfrac{1}{2x - 1}$

35. $\dfrac{2q + p}{2p^2 - 9pq - 5q^2} - \dfrac{p + q}{p^2 - 5pq}$

36. $\dfrac{1}{x - 1} + \dfrac{1}{x^2 - 1} + \dfrac{1}{x^3 - 1}$

37. $\dfrac{\dfrac{1}{x} + 1}{\dfrac{1}{x} - 1}$

38. $\dfrac{\dfrac{4}{a} - a}{\dfrac{2}{a} + 1}$

39. $\dfrac{\dfrac{1}{x} - \dfrac{1}{a}}{x - a}$

40. $\dfrac{\dfrac{1}{a} + \dfrac{1}{b}}{\dfrac{1}{a} - \dfrac{1}{b}}$

41. $\dfrac{a - \dfrac{1}{a}}{1 + \dfrac{1}{a}}$

42. $\dfrac{\dfrac{1}{x^2} - \dfrac{1}{y^2}}{\dfrac{1}{x} + \dfrac{1}{y}}$

43. $\dfrac{\dfrac{1}{2 + h} - \dfrac{1}{2}}{h}$

44. $\dfrac{\dfrac{3}{x^2 + h} - \dfrac{3}{x^2}}{h}$

45. $\dfrac{\dfrac{a}{x^2} + \dfrac{x}{a^2}}{a^2 - ax + x^2}$

46. $\dfrac{x + \dfrac{xy}{y - x}}{\dfrac{y^2}{x^2 - y^2} + 1}$

47. $(x^{-1} + 2)^{-1}$

48. $\dfrac{(x^{-2} + 2x)^{-1}}{x^2}$

49. $\left(\dfrac{1}{a^{-1}} + \dfrac{1}{a^{-2}} \right)^{-1}$

50. $(a^{-1} + a^{-2})^{-1}$

51. $x(x + y)^{-1} + y(x - y)^{-1}$

52. $x(2x - 2y)^{-1} + y(2y - 2x)^{-1}$

B

53. $\dfrac{\dfrac{a + b}{a - b} + \dfrac{a - b}{a + b}}{\dfrac{a - b}{a + b} - \dfrac{a + b}{a - b}} \cdot \dfrac{ab^3 - a^3b}{a^2 + b^2}$

54. $\dfrac{\sqrt{x + a}}{\sqrt{x - a}} - \dfrac{\sqrt{x - a}}{\sqrt{x + a}}$

C

55. $\dfrac{\left(a + \dfrac{1}{b} \right)^a \left(a - \dfrac{1}{b} \right)^b}{\left(b + \dfrac{1}{a} \right)^a \left(b - \dfrac{1}{a} \right)^b}$

56. Consider the three fractions

$$\frac{b - c}{1 + bc}, \qquad \frac{c - a}{1 + ca}, \qquad \text{and} \qquad \frac{a - b}{1 + ab}$$

(a) If $a = 1$, $b = 2$, and $c = 3$, find the sum of the three fractions. Also compute their product. What do you observe?

(b) Show that the sum and the product of the three given fractions are, in fact, always equal.

Answers to Selected Exercises

CHAPTER 1

Exercise Set 1.1

1. (a) natural number, integer, rational number **(b)** integer, rational number
3. (a) rational number **(b)** irrational number **5. (a)** natural number, integer, rational number **(b)** rational number **7. (a)** rational number **(b)** rational number **9.** irrational number **11.** irrational number

13.

11/4

15.

$1 + \sqrt{2}$

17.

$\sqrt{2} - 1$

19.

$\sqrt{2} + \sqrt{3}$

21.

$\dfrac{1 + \sqrt{2}}{2}$

23.

$\pi/2$

25.

$\pi/6$

27.

-1

29.

$\pi/3$

31.

$2\pi + 1$

33.

$\dfrac{\sqrt{139} - 5}{3}$

35. False **37.** True **39.** False
41. False **43.** True

45.

2   5

47.

1   4

49.

0   3

51.

-3

53.

-1

55.

1

57.

π

59. (a) Both sides are 3.863703.
(b) Both sides are 3.162277.
(c) Both sides are 1.847759.
61. (a) one decimal place **(b)** two decimal places **(c)** six decimal places **(d)** nine decimal places
63. (a) $a = \sqrt{2}, b = \sqrt{8}$ **(b)** $a = \sqrt{2}$, $b = \sqrt{3}$ **65. (a)** $2^{1/2} = \sqrt{2}$
(b) $(\sqrt{2})^2 = 2$

Exercise Set 1.2

1. 3 **3.** 6 **5.** 2 **7.** 0 **9.** 0 **11.** 17
13. 1 **15.** 0 **17.** 25 **19.** -1
21. 0 **23.** 3 **25.** $\sqrt{2} - 2$ **27.** $x - 3$
29. $t^2 + 1$ **31.** $\sqrt{3} + 4$
33. $-2x + 7$ **35.** 1 **37.** $3x + 11$
39. $|x - 4| = 8$ **41.** $|x - 1| = \frac{1}{2}$
43. $|x - 1| \geq \frac{1}{2}$ **45.** $|y + 4| < 1$
47. $|y| < 3$ **49.** $|x^2 - a^2| < M$

51.

-4 4

53.

-1 1

55.

2   8

57.

-1 7

59.

$-11/6$ 7/6

61.

63. (a)

(b)

(c) The interval in part **(b)** does not include 2.   **65. (a)** $\min(6, 1) = 1$
(b) $\min(1, -6) = -6$
(c) $\min(-6, -6) = -6$

Exercise Set 1.3

1. is a solution   **3.** not a solution
5. is a solution   **7.** $x = -1$
9. $x = -8$ **11.** $m = 12$
13. $x = 3$ **15.** $x = -3$
17. $y = -15$ **19.** $x = -\frac{13}{7}$
21. $x = 8$ **23.** no solution
25. $x = -\frac{3}{5}$ **27.** no solution
29. $x = 3$ **31. (a)** no solution
(b) $x = \frac{2}{7}$ **(c)** $x = -\frac{7}{3}$
33. $x = 6$ or $x = -1$
35. $z = -\frac{1}{5}$ or $z = \frac{3}{2}$
37. $x = 1$ or $x = -3$
39. $x = \frac{1}{2}$ or $x = 6$

41. $x = -13$ or $x = 12$

43. $x = \dfrac{1 \pm \sqrt{21}}{2} \approx -1.79, 2.79$

45. $x = \dfrac{-3 \pm \sqrt{41}}{4} \approx -2.35, 0.85$

47. $x = -3 \pm \sqrt{7} \approx -5.65, -0.35$

49.

$x = \dfrac{-\sqrt{2} \pm \sqrt{82}}{4} \approx -2.62, 1.91$

51. $x = \frac{3}{4}, \frac{4}{3}$ **53.** $x = \pm 2\sqrt{6}$

55. $t = \pm \dfrac{\sqrt{2}}{4}$ **57. (a)** $u = -9$

(b) $u = -9 \pm 9\sqrt{2}$

59. $x = \dfrac{b + 1}{a}$ **61.** $x = 1$

63. $x = \dfrac{1}{a + b}$ **65.** $x = \dfrac{2ab}{b + a}$

67. $x = \dfrac{1}{2y}$ or $x = \dfrac{1}{y}$

69. $x = 0$ or $x = p + q$

71. $x = \dfrac{a^2 + ab + b^2}{a - b}$ **73.** $x = b$

75. $x = \dfrac{a + b}{2}$ **77.** $r = \dfrac{d}{1 - dt}$

79. $h = \dfrac{S - 2\pi r^2}{2\pi r}$

81. $x = \frac{5}{2}$ or $x = -4$

83. $x = \frac{1}{3}$ or $x = \frac{1}{2}$ **85.** no

solution   **87. (a)** $x = 0, 6, 7$

(b) $x = 3 \pm 2\sqrt{2}$

Exercise Set 1.4
1.

3. (a)

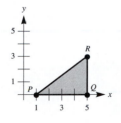

(b) 6 square units   **5. (a)** 5
(b) 13   **7. (a)** 10   **(b)** 9   **9.** 4
11. (a) $\left(4, \frac{1}{2}\right)$ is farther from the
origin   **(b)** $(-6, 7)$ is farther from
the origin   **13. (a)** is a right
triangle   **(b)** is a right triangle
(c) not a right triangle
15. area $= 0$; the three points
are collinear   **17. (a)** $(6, 5)$
(b) $\left(\frac{1}{2}, -\frac{3}{2}\right)$ **(c)** $(1, -4)$
19. (a) $(1, 1)$ **(b)** $\sqrt{34}$

21. (a)

(b) 1990: $(0, 25.8)$; 1992: $(2, 23.6)$
(c) 24.7%   **(d)** 0.80% error
23. (a)

(b) $1.8 million   **(c)** 45.5% error

25. (a)

t	1995	1997
n	14	30

22 million host computers

(b)

t	1985	1987
n	2	28

15 thousand host computers
(c) 1996: 0.83% error; 1986:
194.75% error
27. (a)

(b) both are $\left(0, \frac{7}{2}\right)$ **(c)** The diagonals of a parallelogram bisect each
other.   **29. (a)** $a = \sqrt{2}$; $b = \sqrt{3}$;
$c = 2$; $d = \sqrt{5}$; $e = \sqrt{6}$; $f = \sqrt{7}$;
$g = 2\sqrt{2}$

31. (b) both are $\left(\frac{a+b}{2}, \frac{c}{2}\right)$

Exercise Set 1.5

1. does not lie on the graph
3. does not lie on the graph
5. lies on the graph
7. (a) $y = \frac{2}{3}x + 1$;

x	-6	-3	0	3	6
y	-3	-1	1	3	5

(b)

9. x-intercept: 4; y-intercept: 3

11. x-intercept: 2; y-intercept: -4

13. x-intercept: 1; y-intercept: 1

15. (a) Xmin $= -3$, Xmax $= 3$,
Xscl $= 1$ and Ymin $= -200$,
Ymax $= 200$, Yscl $= 100$;
$[-3, 3, 1]$ by $[-200, 200, 100]$
(b) Xmin $= -1$, Xmax $= 1.25$,
Xscl $= 0.25$ and Ymin $= -0.2$,
Ymax $= 0.1$, Yscl $= 0.1$;
$[-1, 1.25, 0.25]$ by $[-0.2, 0.1, 0.1]$
17. (a) x-intercepts: $-2, -1$;
y-intercept: 2   **(b)** no x-intercepts;
y-intercept: 3   **19. (a)** x-intercepts:
$\frac{-1 \pm \sqrt{5}}{2}$; y-intercept: -1
(b) no x-intercepts; y-intercept: 1
21. x-intercepts:
$-1 \pm 2\sqrt{3} \approx -4.46, 2.46$;
y-intercept: 0   **23.** x-intercept: $-\frac{8}{3}$;
y-intercepts: $-2, 6$
25. (a)

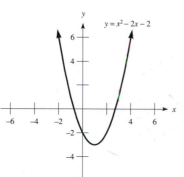

(b) $-0.7, 2.7$
(c) $1 \pm \sqrt{3} \approx -0.7, 2.7$
27. (a)

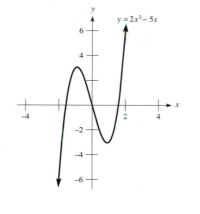

(b) $-1.6, 0, 1.6$
(c) $0, \pm\sqrt{\frac{5}{2}} \approx \pm 1.6$

29. (a)

(b) -1.7 **(c)** $\sqrt[3]{-5} \approx -1.7$
31. x-intercepts: $-1.879, 0.347, 1.532$

33. x-intercepts: $-0.815, 0.875, 5.998$

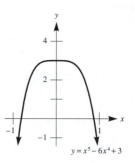

35. (a) The x-intercept is between -1 and 0.

(c) $[0, 10]$ **(d)** Start with a large viewing rectangle then zoom in near the x-intercepts.
37. $-20°$; This is the C-intercept of the graph.   **39. (a)** $\sqrt{2} \approx 1.4$
(b) $\sqrt{3} \approx 1.7$ **(c)** $\sqrt{6} \approx 2.4$
41. (a) 500 bacteria   **(b)** 1.5 hours
(c) $t = 3.5$ hours   **(d)** between $t = 3$ and $t = 4$
43. $A: x = \frac{2}{5}\sqrt{5} \approx 0.894$;
$B: x = \frac{1}{5}\sqrt{30} \approx 1.095$

Exercise Set 1.6
1. (a) -2 **(b)** 3 **(c)** $-\frac{7}{3}$
(d) 3 **3. (a)** slope $= 1$

(b) slope $= 0$

5. (a) $\Delta x = 5 \sec$; $\Delta y = 15 \text{ft}$;
$$\frac{\Delta y}{\Delta x} = 3 \text{ ft/sec}$$
(b) $\Delta x = 15 \sec$; $\Delta y = 45 \text{ft}$;
$$\frac{\Delta y}{\Delta x} = 3 \text{ ft/sec}$$
(c) $\Delta x = 25 \sec$; $\Delta y = 75 \text{ft}$;
$$\frac{\Delta y}{\Delta x} = 3 \text{ ft/sec}$$
7. (a) $\Delta x = 10$ years;
$\Delta y = 25{,}000{,}000$ people
(b) $\Delta x = 20$ years;
$\Delta y = 50{,}000{,}000$ people
(c) $\Delta x = 10$ years;
$\Delta y = 25{,}000{,}000$ people
9. $\Delta t = 3$ years;
$\Delta N = -0.119$ trillion cigarettes
11. $m_3 < m_2 < m_4 < m_1$
13. not collinear
15. (a) $y = -5x - 9$
(b) $y = \frac{1}{3}x + \frac{4}{3}$ **17. (a)** $y = 2x$
(b) $y = -2x - 4$ **(c)** $y = \frac{1}{7}x - \frac{11}{7}$
19. (a) $x = -3$ **(b)** $y = 4$
21. $x = 0$ **23. (a)** $y = -4x + 7$
(b) $y = 2x + \frac{3}{2}$
25. (a) $y = 4x + 11$

(b) $y = \frac{1}{2}x - \frac{5}{4}$

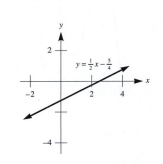

(c) $y = -\frac{5}{6}x + 5$

(d) $y = \frac{3}{4}x + \frac{3}{2}$

(e) $y = 4x - 2$

27. $y - 4 = 0$

29. (a) x-intercept: 5; y-intercept: 3; area $= \frac{15}{2}$; perimeter $= 8 + \sqrt{34}$
(b) x-intercept: 5; y-intercept: -3; area $= \frac{15}{2}$; perimeter $= 8 + \sqrt{34}$
31. (a) neither   **(b)** parallel
(c) perpendicular   **(d)** perpendicular
33. $y = \frac{2}{5}x + \frac{12}{5}$; $2x - 5y + 12 = 0$
35. $y = -\frac{4}{3}x + \frac{16}{3}$; $4x + 3y - 16 = 0$

37. They appear parallel as long as the viewing rectangle is square.

39. (a) $y = \dfrac{4}{3}x$

(b)

41. (a) $y = -\dfrac{2}{3}p + 410$

(b) 208 units   **(c)** $183
47. 44.1 square units
49. (a)

(b) $y = -2x + 5$ **(c)** $(4, -3)$
(d) $3\sqrt{5}$
51. (a)

(b) point: $(-1, 2)$

(c) $y = \frac{3}{5}x + \frac{12}{5}$; $y = -\frac{3}{4}x + \frac{3}{2}$; $\left(-\frac{2}{3}, 2\right)$

Exercise Set 1.7

1. (a)

(b)

(c)

3. (a)

(b)

(c)

5. (a)

(b)

(c)

7. x-intercepts: 2 and -2; y-intercept: 4; symmetric about the y-axis

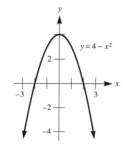

9. no x- or y-intercepts; symmetric about the origin

11. x- and y-intercepts: 0; symmetric about the y-axis

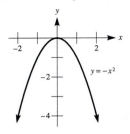

13. no x- or y-intercepts; symmetric about the origin

15. x- and y-intercepts: 0; symmetric about the y-axis

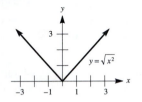

17. x- and y-intercepts: 1; no symmetry

19. x-intercept: 2; no y-intercept; symmetric about the x-axis

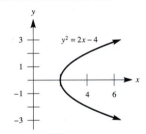

21. x-intercepts:
$$\frac{-1 \pm \sqrt{33}}{4} \approx -1.69, 1.19;$$
y-intercept: -4; no symmetry

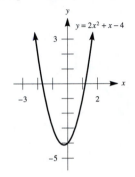

23. (a) x- and y-intercepts: 2; no symmetry

(b) x-intercepts: ± 2; y-intercept: 2; symmetric about the y-axis

25. (a)

(b) no symmetry

(c)

(d) no symmetry

27. (a)

(b) no symmetry

(c)

(d) no symmetry

29. (a)

(b) no symmetry

(c)

$y = \dfrac{1}{x^2 - x}$

$x = 1$

(d) no symmetry

31. (a)

$y = x^2 - 0.2x - 15$

(b) *y*-axis symmetry

(c)

$y = x^2 - 0.2x - 15$

(d) no symmetry

33. (a)

$y = \sqrt{|x|}$

(b) *y*-axis symmetry

(c)

$y = \sqrt{|x|}$

(d) *y*-axis symmetry

35. (a)

$y = 2x - x^3 - x^5 + x^7$

(b) origin symmetry

(c)

$y = 2x - x^3 - x^5 + x^7$

(d) origin symmetry

37. (a)

$y = x^4 - 10x^2 + \tfrac{1}{4}x$

(b) *y*-axis symmetry

(c)

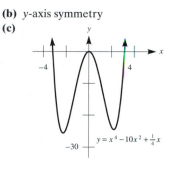

$y = x^4 - 10x^2 + \tfrac{1}{4}x$

(d) no symmetry   **39.** center:
(1, 5); radius: 13; (6, −7) lies on the
circle   **41.** center: (−8, 5); radius:
$\sqrt{13}$; (−5, 2) does not lie on the circle
43. center: (0, 0); radius: $\sqrt[4]{2}$;
y-intercepts: $\pm\sqrt[4]{2}$

45. center: (−4, 3); radius: 1; no
y-intercepts   **47.** center: $(-3, \tfrac{1}{3})$;
radius: $\sqrt{2}$; no *y*-intercepts

49. (a) center: $\left(2, -\dfrac{3}{2}\right)$; radius: $\tfrac{13}{4}$

(b) *x*-intercepts: −0.88 and 4.88

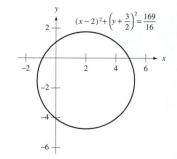

$(x - 2)^2 + \left(y + \dfrac{3}{2}\right)^2 = \dfrac{169}{16}$

(c) $x = \dfrac{8 \pm \sqrt{133}}{4} \approx -0.88, 4.88$

51.

$y = -\dfrac{4}{x}$

$y = -\dfrac{4}{x}$

53. $(x - 3)^2 + (y - 2)^2 = 169$
55. $(x - 3)^2 + (y - 5)^2 = 9$
57. $y = 2 \pm \sqrt{19}$
59. (a) *x*-intercepts for each graph: $\tfrac{8}{3}$;
y-intercept for $y = \tfrac{3}{4}x - 2$: −2;
y-intercept for $y = |\tfrac{3}{4}x - 2|$: 2
(b) on the interval $[\tfrac{8}{3}, \infty)$
(c) The graph of $y = |\tfrac{3}{4}x - 2|$ can
be obtained by reflecting $y = \tfrac{3}{4}x - 2$
about the *x*-axis for the interval
$(-\infty, \tfrac{8}{3})$. For the interval $[\tfrac{8}{3}, \infty)$,

no reflection is necessary.
61. (b)

(c)

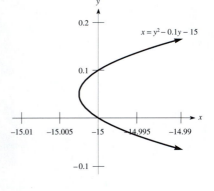

Chapter 1 Review Exercises

1. $|x - 6| = 2$ **3.** $|a - b| = 3$
5. $|x| > 10$ **7.** $\sqrt{6} - 2$
9. $x^4 + x^2 + 1$
11. (a) $-2x + 5$
(b) 1 **(c)** $2x - 5$
13.

15.

17.

19.

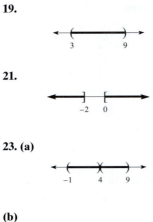

21.

23. (a)

(b)

25. $x = \frac{1}{3}$ **27.** $t = -\frac{37}{11}$
29. $t = -\frac{11}{3}$ **31.** $y = 8$
33. $x = \frac{1}{3}, -\frac{1}{2}$ **35.** $x = -6, 4$

37. $x = 10, -1$ **39.** $t = \dfrac{-1 \pm \sqrt{3}}{2}$
41. $y = -2x - 6$
43. $y = \frac{1}{4}x - \frac{5}{2}$ **45.** $y = 2x + 8$
47. $y = -2$ **49.** $y = x + 1$
51. $9x - 4y + 14 = 0$
53. $2x + y = 0$ **55.** $3x - 4y = 0$
57. $x + y - 1 = 0$ and
$x - 2y - 4 = 0$ **59.** y-axis
symmetry   **61.** y-axis symmetry
63. x-axis symmetry   **65.** x-axis,
y-axis, and origin symmetry
67. x-axis symmetry
69.

71.

73.

75. (a)

(b) intersect at $(-3, 0)$ and $(3, 0)$

77. (a)

(b)

(c) no intersection points

79.

81. (ab)

(c)

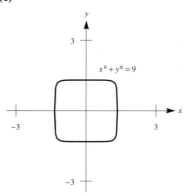

83. $MA = MB = MC = \sqrt{c^2 + b^2}$; all the same   **85. (a)** $\sqrt{130}$ **(b)** $-\frac{11}{3}$
(c) $\left(\frac{7}{2}, -\frac{1}{2}\right)$ **87.** $\frac{49}{6}$ square units
89. (a) $\left(\frac{5}{3}, \frac{11}{3}\right)$; They all intersect at the same point.
(b) $\left(\dfrac{2a}{3} + \dfrac{2b}{3}, \dfrac{2c}{3}\right)$;

They all intersect at the same point.
93. (a)

(b)

(c)

95. $\dfrac{13\sqrt{5}}{5}$ **97.** $\dfrac{17\sqrt{13}}{13}$

99. (a) $(x - 1)^2 + (y - 1)^2 = 1$
(b) $\overline{AT} : y = \frac{9}{8}x$; $\overline{BU} : y = -\frac{1}{4}x + 1$;
$\overline{CS} : y = -3x + 3$ **(c)** $(8/11, 9/11)$;
They intersect in the same point.

Chapter 1 Test

1. 1 **2. (a)** $\left(\dfrac{39}{10}, \dfrac{41}{10}\right)$ **(b)** $[2, \infty)$

3. $x = 5$ **4.** $x = \frac{14}{11}, -1$
5. (a) $x = -5, 1$

(b) $x = -2 \pm \sqrt{5}$ **6.** $x = \dfrac{de - b}{a - ce}$
7. (a) $\Delta N = 777$
(b) $\dfrac{\Delta N}{\Delta t} \approx 194$ stations per year

(c) $11{,}759$ stations
8. (a)

(b) 42.8 cm   **(c)** 18% error
9. $y = 5x - 7$ **10.** $y = \frac{6}{5}x - \frac{17}{5}$
11. $(3, 9)$ **12. (a)** origin symmetry
(b) y-axis symmetry
(c) no symmetry   **13.** x-intercept: 5;
y-intercept: -3

14. x-intercepts: $1 \pm \sqrt{5}$;
y-intercepts: $-2 \pm 2\sqrt{2}$

15. (a)

$y = x^3 - 2x^2 - 9x$

(b) *x*-intercepts: $-2.2, 0, 4.2$
(c) *x*-intercepts: 0,
$1 \pm \sqrt{10} \approx -2.2, 4.2$ **16.** yes
17. (a)

Reflection of $\overline{AB}$
$\overline{AB}$

(b)

$\overline{AB}$ Reflection of $\overline{AB}$

(c)

$\overline{AB}$
Reflection of $\overline{AB}$

18. $\frac{121}{5}$ square units

CHAPTER 2
Exercise Set 2.1

1. $x = -4 \pm 3\sqrt{2} \approx -8.24, 0.24$
3. $x = -2 \pm \sqrt{3} \approx -3.73, -0.27$

5. $y = \dfrac{5 \pm \sqrt{41}}{4} \approx -0.35, 2.85$

7. no real roots   **9.** $s = \frac{5}{2}$
11. $x = 4 \pm \sqrt{10} \approx 0.84, 7.16$

13. $x = \dfrac{1 \pm \sqrt{37}}{6} \approx -0.85, 1.18$
15.
$y = -4 \pm \sqrt{15} \approx -7.87, -0.13$
17. no real roots   **19. (a)** year 1917
(b) it is reasonable   **21. (a)** year
1990; 3 years off   **(b)** year 1997;
0 years off   **23. (a)** *x*-intercepts:
0.268 and 3.732

$y = x^2 - 4x + 1$

(b) $x = 2 \pm \sqrt{3} \approx 0.2679, 3.7321$
25. (a) *x*-intercepts: -16.367 and
0.367

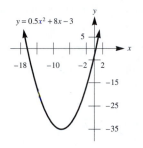

$y = 0.5x^2 + 8x - 3$

(b)
$x = -8 \pm \sqrt{70} \approx -16.3666, 0.3666$
27. (a) *x*-intercept: -2.550

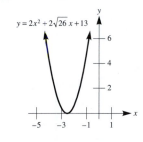

$y = 2x^2 + 2\sqrt{26}\,x + 13$

(b) $x = \dfrac{-\sqrt{26}}{2} \approx -2.5495$

29. sum: -8; product: -20
31. sum: 7; product: $\frac{9}{4}$
33. $x^2 - 14x + 33 = 0$
35. $x^2 - 2x - 1 = 0$
37. $4x^2 - 8x - 1 = 0$
39. $\sqrt{2}$ and $\sqrt{5}$ **41. (a)** 6 seconds
(b) 1 second (rising) 5 seconds
(falling)   **43.** two real roots
45. two real roots   **47.** one real

root   **49.** two real roots
51. $k = 36$ **53.** $k = \pm 2\sqrt{5}$

55. $r = \dfrac{-h \pm \sqrt{h^2 + 40}}{2}$

57. $t = 0$ or $t = \dfrac{v_0}{16}$

59. (a)

$y = x^2 + 8x + 16$ $y = x^2 - 8x + 16$

(b) -4 and 4; they are opposites
(c) $x = -4, 4$
65. $k = -\frac{10}{9}$ or $k = 2$
69. $x = 1$ or $x = \frac{1}{2}$
71. (a)

$c = 10$
$c = 8$ $c = 9$

(b) $c = 9$ **(c)** one root: $x = 3$
(d) $c > 9$

Exercise Set 2.2

1. $x = 4, 6$ **3.** $x = -\frac{11}{2}, -\frac{13}{2}$
5. $x = -\frac{10}{3}, 5$ **7.** $x = 5$
9. $x = 0, \frac{4}{5}$
11. (a)

$y = x^2 - 2.2x$
$y = 0.8$

(b) $x \approx 2.518$

(c) $x = \dfrac{2.2 \pm \sqrt{8.04}}{2} \approx 2.5177$

13. (a)

(b) $x \approx \pm 1.414$ **15.** $x = 0, 16$
17. $t = 5$ **19.** $x = 0, 2, -2$
21. $t = 0, -3, 1$ **23.** $x = 0, \frac{1}{4}, -6$
25. $x = \pm \sqrt{3}$ **27.** $y = -1, 1$
29. $t = \pm \frac{\sqrt{6}}{3}$
31. $x = 2 + \sqrt[3]{5} \approx 3.71$
33. $x = -4 - \sqrt[5]{16} \approx -5.74$
35. (a) $x = 3 \pm \sqrt[4]{30} \approx 5.34, 0.66$
(b) no solution **37.** $y = 1, -1$
39. $t = \pm \frac{\sqrt{6}}{3}$
41. $x = 0, \pm \sqrt{6}, \pm 2$
43. $x = \pm \sqrt{\dfrac{-1 + \sqrt{5}}{2}}$
45. $x = \pm \sqrt{\dfrac{-3 + \sqrt{17}}{2}}$
47. $x = -2, 1$ **49.** $t = \frac{1}{3}, \frac{1}{4}$
51. $y = \frac{3}{5}, 4$ **53.** $x = \pm \frac{1}{3}$
55. $t = \pm 27$ **57.** $x = \pm 3$
59. $y = 1 + \sqrt[3]{7}$ **61.** $t = -4$
63. $x = \pm \frac{1}{27}, \pm 1$
65. $x = \pm \sqrt{1 + \sqrt{2}} \approx \pm 1.55$
67. $x = -1$ **69.** $x = 4$
71. $x = -6$ **73.** $x = -\frac{1}{2}$
75. $x = \pm 2, \pm 3$ **77.** $x = -4, -\frac{20}{9}$
79. no solutions **81.** $y = 2$

83. (a) (a) $x \approx -1.65, -1.51,$
1.51, 1.65

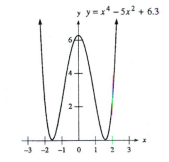

(b) $x = -\sqrt{\dfrac{5 - \sqrt{0.2}}{2}} \approx -1.51,$

$\sqrt{\dfrac{5 - \sqrt{0.2}}{2}} \approx 1.51,$

$-\sqrt{\dfrac{5 + \sqrt{0.2}}{2}} \approx -1.65,$

$\sqrt{\dfrac{5 + \sqrt{0.2}}{2}} \approx 1.65$

(b) (a) no real roots

(b) no real roots **85. (a)** $x \approx 0.49$
and 18.51

(b)
$x = \dfrac{19 \pm 5\sqrt{13}}{2} \approx 0.4861, 18.5139$

87. (a) $x \approx 19.94$

(b) $x = 11 + 4\sqrt{5} \approx 19.9443$
89. (a) $x \approx 2.78$

(b) $x = \frac{25}{9}$ **91. (a)** $x \approx -0.24$ and
4.24

(b) $x = \left(\dfrac{1 \pm \sqrt{5}}{2} \right)^3 \approx -0.2361, 4.2361$
95. $x = a + b$ **97.** $x = -\dfrac{a}{15}$

Exercise Set 2.3
1. (a) Let x and y denote the two
numbers. **(b)**

In English	In algebraic notation
Find two numbers.	x and y
Their sum is -9.	$x + y = -9$
The sum of their squares is 153.	$x^2 + y^2 = 153$

(c) $x^2 + 9x - 36 = 0$
(d) $x = -12, 3$ **(e)** The two
numbers are -12 and 3.

3. (a) Let x and y denote the two numbers.   **(b)**

In English	In algebraic notation
Find two numbers.	x and y
Their sum is 17.	$x + y = 17$
Their product is 52.	$xy = 52$

(c) $x^2 - 17x + 52 = 0$
(d) $x = 4, 13$ **(e)** The two numbers are 4 and 13.   **5. (a)** Let x and $x + 1$ denote these consecutive integers.   **(b)**

In English	In algebraic notation
Find two consecutive integers.	x and $x + 1$
The sum of their squares is 365.	$x^2 + (x + 1)^2 = 365$

(c) $x^2 + x - 182 = 0$ **(d)** $x = 13$
(e) The two numbers are 13 and 14.   **7. (a)** Let x and y denote the two numbers.   **(b)**

In English	In algebraic notation
Find two numbers.	x and y
Their average is 4.	$\frac{1}{2}(x + y) = 4$
The average of their squares is 80.	$\frac{1}{2}(x^2 + y^2) = 80$

(c) $x^2 - 8x - 48 = 0$
(d) $x = -4, 12$ **(e)** The two numbers are -4 and 12.
9. (a) Label the length l and the width w.

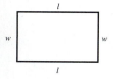

(b)

In English	In algebraic notation
Find the width and length of a rectangle.	w and l
The perimeter is 40 cm.	$2w + 2l = 40$
The area is 64 cm².	$lw = 64$

(c) $w^2 - 20w + 64 = 0$
(d) $w = 4$ **(e)** The width is 4 cm and the length is 16 cm.
11. (a) Let t denote the required number of years.

(b)
$(3000)(0.07)t + (5000)(0.08)t + (10000)(0.085)t = 8760$ **(c)** $t = 6$
(d) After 6 years the total interest will amount to $8,760.
13. (a) $7400(1 + \frac{r}{10}) +$
$12600(1 + \frac{1 + r}{20}) = 31260$
(b) $r = 5$ **(c)** The interest rate is 5%.   **15. (a)** The problem is asking you to find a and b.
(b)

In English	In algebraic notation
Find two numbers.	a and b
(a, b) is in the first quadrant.	$a > 0, b > 0$
(a, b) is on the line $y = 2x + 3$.	$b = 2a + 3$
The distance from (c, b) to $(0, 0)$ is 5.	$\sqrt{a^2 + b^2} = 5$

(c) $\sqrt{5a^2 + 12a + 9} = 5$
(d) $a = \dfrac{-6 + 2\sqrt{29}}{5} \approx 0.954$
(e) $\left(\dfrac{-6 + 2\sqrt{29}}{5}, \dfrac{3 + 4\sqrt{29}}{5} \right) \approx$
$(0.954, 4.908)$

17. (a) The problem is asking you to find a and b.
(b)

In English	In algebraic notation
Find two numbers.	a and b
(a, b) is in the third quadrant.	$a < 0, b < 0$
(a, b) is on the graph of $y = \frac{1}{x}$.	$b = \frac{1}{a}$
The distance from (a, b) to $(0, 0)$ is $\frac{\sqrt{257}}{4}$.	$\sqrt{a^2 + b^2} = \frac{\sqrt{257}}{4}$

(c) $\sqrt{a^2 + \dfrac{1}{a^2}} = \dfrac{\sqrt{257}}{4}$
(d) $a = -\frac{1}{4}, -4$ **(e)** $(-\frac{1}{4}, -4)$ or $(-4, -\frac{1}{4})$
19. (a)

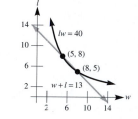

(b) $(5, 8)$ and $(8, 5)$
(c) w-intercepts: 5, 8; same as the w-coordinates of the intersection points

21. (a) intersection point: $(-2.5, -8.6)$

(b) $\left(\dfrac{3 - \sqrt{809}}{10}, \dfrac{-1 - 3\sqrt{809}}{10} \right) \approx$
$(-2.544, -8.633)$
23. (a) intersection points: $x \approx 1.2, 6.8$

(b) $x = 4 \pm 2\sqrt{2} \approx 1.17, 6.83$
27. (b) $a = -12, 11.2$

Exercise Set 2.4
1. 12 and -5 **3.** 13 and 4   **5.** 21
7. 8   **9.** 5   **11.** 12 and 6   **13.** 78
15. 44   **17.** 5 cm, 12 cm, 13 cm
19. 2.5 cm   **21.** 20 cm   **23.** 4.8 in., 5.2 in.   **25.** 10 cm, 30 cm, 30 cm
27. 8 years   **29.** 10 years
31. 6.6%   **33.** $8000 account:

7.5%; $9000 account: 8.5%
35. 80 cm³ of the 10% solution, 120 cm³ of the 35% solution
37. 15 pounds   **39.** 8 tons
41. 17 cm   **43. (a)** 6 seconds
(b) 1 second (rising) and 5 seconds (falling)   **45.** $12\frac{3}{8}$ minutes = 12 minutes, 22 seconds   **47.** 55 mph
49. 350 mph   **51.** 15 miles
53. $5117   **55.** $\frac{1}{24}$ mile/minute
57. $\dfrac{2v_1v_2}{v_1 + v_2}$ mph
59. $\dfrac{100(d + D)}{dr + DR}$ years
61. $\dfrac{a^2 - 2ab + 2b^2}{2(a - b)}$ inches
63. $\dfrac{b}{\sqrt{a^2 - 1}}$ units = $\dfrac{b\sqrt{a^2 - 1}}{a^2 - 1}$ units
65. $8 - 2\sqrt{6\pi - 16} \approx 4.62$ cm

Exercise Set 2.5

1. $(-\infty, -1)$ **3.** $[\frac{1}{3}, \infty)$
5. $(-\infty, -\frac{9}{2})$ **7.** $[-6, \infty)$
9. $(-\infty, \frac{5}{2})$ **11.** $[-\frac{17}{23}, \infty)$
13. $[4, 6]$ **15.** $[-\frac{1}{2}, 1]$
17. $(3.98, 3.998)$ **19. (a)** $[-\frac{1}{2}, \frac{1}{2}]$
(b) $(-\infty, -\frac{1}{2}) \cup (\frac{1}{2}, \infty)$
21. (a) $(-\infty, 0) \cup (0, \infty)$
(b) no solution   **23. (a)** $(-\infty, 3)$
(b) $(1, 3)$ **(c)** $(-\infty, 1) \cup (3, \infty)$
25. (a) $[-4, \infty)$ **(b)** $[-4, 6]$
(c) $(-\infty, -4) \cup (6, \infty)$
27. (a) $(a - c, \infty)$
(b) $(a - c, a + c)$
(c) $(-\infty, a - c] \cup [a + c, \infty)$
29. $(-10, 14)$ **31.** $(-11, 1)$
33. (a) $(-2h, h)$ **(b)** $(h, -2h)$
35. (a) $[a, \infty)$ **(b)** $a \approx 0.3$

(c) $[\frac{2}{7}, \infty)$ **37. (a)** $[a, b]$
(b) $a \approx 0.1$, $b \approx 0.6$

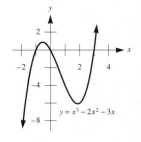

(c) $[\frac{1}{8}, \frac{5}{8}]$ **39.** $[-297°, 234°]$
41. $9200° \le F \le 11700°$
43. $(0, 3)$ **45. (a)**

a	b	$\sqrt{ab}$ (G.M.)	$\frac{a+b}{2}$ (A.M.)	Which is larger, G.M. or A.M.?
1	2	1.4142	1.5	A.M.
1	3	1.7320	2.0	A.M.
1	4	2.0000	2.5	A.M.
2	3	2.4495	2.5	A.M.
3	4	3.4641	3.5	A.M.
5	10	7.0711	7.5	A.M.
9	10	9.4868	9.5	A.M.
99	100	99.4987	99.5	A.M.
999	1000	999.4999	999.5	A.M.

47. (a) The negative slope tells us the sulfur dioxide emissions are decreasing.

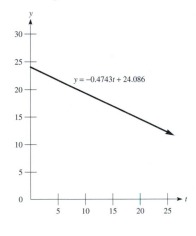

(b) year 2010

Exercise Set 2.6

1. (a) $(-\infty, -1] \cup [4, \infty)$
(b) $[-1, 4]$ **3. (a)** no solution
(b) $(-\infty, \infty)$
5. (a) $[-1, 1] \cup [3, \infty)$
(b) $(-\infty, -1) \cup (1, 3)$
7. The curve is positive on the set $(-1, 0) \cup (3, \infty)$, which represents the solution set for the inequality.

9. $(-3, 2)$ **11.** $(-\infty, 2) \cup (9, \infty)$
13. $(-\infty, 4] \cup [5, \infty)$
15. $(-\infty, -4] \cup [4, \infty)$
17. no solution
19. $(-7, -6) \cup (0, \infty)$
21. $(-\infty, \infty)$
23. $(-\infty, -\frac{3}{4}) \cup (-\frac{2}{3}, 0)$
25. $(-\infty, \frac{-1 - \sqrt{5}}{2}) \cup (\frac{-1 + \sqrt{5}}{2}, \infty)$
27. $[4 - \sqrt{14}, 4 + \sqrt{14}]$
29. $[-4, -3] \cup [1, \infty)$
31. $(-\infty, -6) \cup (-5, -4)$
33. $(-\infty, -\frac{1}{3}) \cup (\frac{1}{3}, 2) \cup (2, \infty)$
35. $(-\infty, -\frac{2}{3}] \cup \{3, -\frac{1}{3}\}$
37. $(-\infty, -\sqrt{5}] \cup [-2, 2] \cup [\sqrt{5}, \infty)$ **39.** $(-3, 3) \cup (4, \infty)$
41. $(-\sqrt{14}, -2\sqrt{2}) \cup (2\sqrt{2}, \sqrt{14})$
43. $(-1, 1]$ **45.** $(-\infty, \frac{3}{2}) \cup [2, \infty)$
47. $(-\infty, -1) \cup (0, 9)$
49. $(-\frac{7}{2}, -\frac{4}{3}) \cup (-1, 0) \cup (1, \infty)$
51. $(-\infty, -1)$ **53.** $(-1, 0)$
55. $[-1, 1) \cup (2, 4]$
57. $(-1, 2 - \sqrt{5}) \cup (1, 2 + \sqrt{5})$
59. (a) $[0.697, 4.303]$

(b) $x = \dfrac{5 \pm \sqrt{13}}{2} \approx 0.697, 4.303$

61. (a) $(-0.329, 24.329)$

(b) $x = 12 \pm 2\sqrt{38} \approx$
$-0.329, 24.329$

63. (a) $(-\infty, -1.554) \cup (1.554, \infty)$

(b) $x = \pm\sqrt{1 + \sqrt{2}} \approx \pm 1.554$

65. (a) $[-2.236, 2.236]$

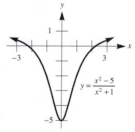

(b) $x = \pm\sqrt{5} \approx \pm 2.236$

67. (a) $(-2.236, 2.236)$

(b) $x = \pm\sqrt{5} \approx \pm 2.236$

69. $[-0.453, \infty)$

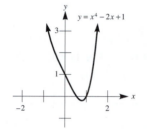

71. $(-\infty, 0.544) \cup (1, \infty)$

73. $(-3.079, 0) \cup (3.079, \infty)$

75. (a) April through September
(b) January, October, November, December
77. $(-\infty, -2] \cup [2, \infty)$
79. $(-\infty, 0) \cup (\frac{1}{2}, \infty)$ **81.** $(\frac{1}{5}, \frac{4}{5})$
83. $c = 2$

Chapter 2 Review Exercises

1. True   **3.** False   **5.** False
7. True   **9.** False   **11.** False
13. $x = \frac{1}{3}, -\frac{1}{2}$ **15.** $x = -6, 4$
17. $x = -1, 10$ **19.** $x = \frac{2}{3}, 3$
21. $t = \dfrac{-1 \pm \sqrt{3}}{2}$
23. $x = -6, -8$
25. $x = 256, 6561$ **27.** $x = \frac{1}{4}$
29. $x = 10$ **31.** $x = 2$

33. $x = \dfrac{1}{2y}$ **35.** $x = \dfrac{a + b}{ab}, -\dfrac{2}{a}$
37. $x = -a$ or $x = -b$
39. $(-2, 1)$ **41.** $[-\frac{1}{2}, \frac{1}{2}]$
43. $(-\infty, -2] \cup [3, \infty)$
45. $(-8, 5)$
47. $(3 - \sqrt{10}, 3 + \sqrt{10})$
49. $(-5, -3) \cup (3, 5)$
51. $(-2, \infty)$ **53.** $(-\frac{5}{2}, 4)$

55. $\left(-\infty, -\sqrt{\dfrac{3 + \sqrt{5}}{2}}\right) \cup$
$\left(-\sqrt{\dfrac{3 - \sqrt{5}}{2}}, 0\right) \cup \left(0, \sqrt{\dfrac{3 - \sqrt{5}}{2}}\right) \cup$
$\left(\sqrt{\dfrac{3 + \sqrt{5}}{2}}, \infty\right)$

57. $(-\infty, 3 - 2\sqrt{2}] \cup [3 + 2\sqrt{2}, \infty)$
59. $x^2 + 7x + 12 = 0$ **61.** 7 and
12   **63.** 47   **65.** $-1 + \sqrt{2}$ cm
67. $(16\,\text{cm}, \infty)$ **69.** 21, 22, 23
71. 7.5 mph; 1 hour 17 minutes
73. $\dfrac{-(a + b) + \sqrt{a^2 + 6ab + b^2}}{4}$
75. The ball thrown from 100 ft.
77. $x = \frac{\sqrt{2}}{2}$ **79.** $\dfrac{r}{1 + \sqrt{2}}$

Chapter 2 Test

1. (a) $x = -5, 1$
(b) $x = -2 \pm \sqrt{5}$
2. $x = \pm 2, \pm\sqrt{3}$ **3.** $x = -2, 2$
4. $x = -\frac{1}{3}, 1$ **5.** -4
6. $x^2 - 4x - 59 = 0$
7. $x = \pm\sqrt{\dfrac{3 + \sqrt{13}}{2}} \approx \pm 1.817$
8. (a) $x \approx -2.83, 2.83$

(b) $x = \pm 2\sqrt{2} \approx \pm 2.83$
9. $(-\infty, 3]$

10. $\left(\frac{27}{10}, \frac{31}{10}\right)$

11. $\left[\frac{7}{3}, 3\right]$ **12.** $[-8, \infty)$

13. $\left(-2, \frac{-3 - \sqrt{3}}{3}\right]$

$\cup \left(-1, \frac{-3 + \sqrt{3}}{3}\right] \cup (0, \infty)$

14. 168 pounds of Kona beans, 112 pounds of Columbian beans

15. 12:20 P.M.   **16.** $\left[-\frac{3}{2}, \frac{3}{2}\right]$

17. $(-4, -7)$ **18.** $x = 6$ is a solution to the original equation. However, $x = -1$ results in the statement $\sqrt{4} = -2$, which is false. Thus $x = -1$ is an extraneous solution of the original equation, even though it is a solution to the subsequent equation $x^2 - 5x - 6 = 0$.

CHAPTER 3
Exercise Set 3.1

1. (a) $g(1975) = \$2.00$
(b) $g(1995) - g(1975) = \$2.25$; The minimum wage increased $2.25 from 1975 to 1995.   **3. (a)** range
(b) $h(\text{Mars}) = 2$ **(c)** $h(\text{Neptune})$; Neptune has more moons than does Pluto.   **5. (a)** f and g
(b) range of f: $\{1, 2, 3\}$; range of g: $\{2, 3\}$ **7. (a)** g **(b)** range of g: $\{i, j\}$ **9. (a)** $(-\infty, \infty)$
(b) $\left(-\infty, \frac{1}{5}\right) \cup \left(\frac{1}{5}, \infty\right)$ **(c)** $\left(-\infty, \frac{1}{5}\right]$
(d) $(-\infty, \infty)$ **11. (a)** $(-\infty, \infty)$
(b) $(-\infty, -3) \cup (-3, 3) \cup (3, \infty)$
(c) $(-\infty, -3] \cup [3, \infty)$
(d) $(-\infty, \infty)$ **13. (a)** $(-\infty, \infty)$
(b) $(-\infty, 3) \cup (3, 5) \cup (5, \infty)$
(c) $(-\infty, 3] \cup [5, \infty)$
(d) $(-\infty, \infty)$
15. (a) $(-\infty, -3) \cup (-3, \infty)$
(b) $(-\infty, -3) \cup [2, \infty)$
(c) $(-\infty, -3) \cup (-3, \infty)$
17. domain: $(-\infty, \infty)$; range: $(-\infty, \infty)$ **19.** domain: $(-\infty, \infty)$; range: $(-\infty, \infty)$ **21.** domain: $(-\infty, 6) \cup (6, \infty)$; range: $\left(-\infty, \frac{4}{3}\right) \cup \left(\frac{4}{3}, \infty\right)$ **23. (a)** domain:

$(-\infty, 5) \cup (5, \infty)$; range: $(-\infty, 1) \cup (1, \infty)$ **(b)** domain: $(-\infty, \sqrt[3]{5}) \cup (\sqrt[3]{5}, \infty)$; range: $(-\infty, 1) \cup (1, \infty)$ **25.** domain: $(-\infty, \infty)$; range: $[4, \infty)$
27. (a) $y = (x - 3)^2$
(b) $y = x^2 - 3$ **(c)** $y = (3x)^2$
(d) $y = 3x^2$ **29. (a)** -1 **(b)** 1
(c) 5 **(d)** $-\frac{5}{4}$ **(e)** $z^2 - 3z + 1$
(f) $x^2 - x - 1$ **(g)** $a^2 - a - 1$
(h) $x^2 + 3x + 1$ **(i)** 1
(j) $4 - 3\sqrt{3}$ **(k)** $1 - \sqrt{2}$ **(l)** 2
31. (a) $12x^2$ **(b)** $6x^2$ **(c)** $3x^4$
(d) $9x^4$ **(e)** $\frac{3}{4}x^2$ **(f)** $\frac{3}{2}x^2$
33. (a) -3 **(b)** $-\frac{7}{18}$
(c) $-2x^2 - 4x - 1$
(d) $1 - 2x^2 - 4xh - 2h^2$
35. (a) 2 **(b)** 2 **(c)** 2
37. (a) $x = -2, 8$ **(b)** no real solutions   **(c)** $x = 3$ **39.** 1041 tee shirts per month   **41.** first model: 124 sales; second model: 100 sales
43. $9 - 6x$ **45. (a)** $2x^2 + 5x + 2$
(b) $2x^2 - 11x + 14$
(c) $16x - 12$ **47.** $a = 1$ and $b = 2$ **49.** $a = 2$ **51. (a)**
$f(a) = 0, f(2a) = \frac{1}{3}, f(3a) = \frac{1}{2}$; No
53. $a = \frac{1}{2}$ and $b = -\frac{3}{2}$ **59.** $k = -\frac{1}{3}$
61. (a) 0 **(b)** 1 **(c)** 2 **(d)** 6
(e) -1 **(f)** -2 **(g)** -6 **(h)** $\frac{1}{2}$
63. (a) $x_0 = -2$ **(b)** $x_0 = 4, -1$
(c) $x_0 = 0, 1$ **(d)** $x_0 = 1 \pm \sqrt{2}$
65. G is not a function, since it assigns more than one value to x. Let G assign the closest prime number greater than or equal to x.
67. (a) $P(1) = 17, P(2) = 19,$
$P(3) = 23, P(4) = 29; x = 17$
(b) $Q(1) = 41, Q(2) = 43,$
$Q(3) = 47, Q(4) = 53; x = 41$

Exercise Set 3.2

1. $y = \sqrt{3} \approx 1.732$

3. $y = \dfrac{\sqrt{5}}{5} \approx 0.447$ **5. (a)** yes
(b) no   **(c)** no   **(d)** yes
7. domain: $[-4, 2]$; range: $[-3, 3]$
9. domain: $[-3, 4]$; range: $[-2, 2]$
11. domain: $[-4, -1) \cup (-1, 4]$; range: $[-2, 3)$ **13.** domain:

$[-4, 3]$; range: $\{2\}$ **15. (a)** 1
(b) -3 **(c)** no   **(d)** $x = 2$ **(e)** -1
17. (a) positive   **(b)** $f(-2) = 4$;
$f(1) = 1; f(2) = 2; f(3) = 0$ **(c)** $f(2)$
(d) -3 **(e)** 3 **(f)** domain: $[-2, 4]$;
range: $[-2, 4]$ **19. (a)** $g(-2)$ **(b)** 5
(c) $f(2) - g(2)$ **(d)** $x = -2$ or $x = 3$
(e) range of f
21. (a)

(b)

(c)

23.

25.

27. (a)

(b)

29. (a)

(b)

31. (a)

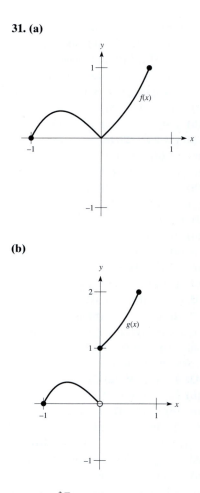

(b)

33. $P(4, \sqrt[3]{4}) \approx P(4, 1.587)$,
$Q(\sqrt[3]{4}, \sqrt[3]{4}) \approx Q(1.587, 1.587)$,
$R(\sqrt[3]{4}, \sqrt[9]{4}) \approx R(1.587, 1.167)$
35. $P(\sqrt{2}, -\sqrt{2}) \approx P(1.414, -1.414)$,
$Q(-\sqrt{2}, -\sqrt{2}) \approx Q(-1.414, -1.414)$,
$R(-\sqrt{2}, \sqrt{2}) \approx R(-1.414, 1.414)$
37. (a)

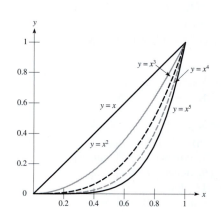

(b) As the power on x increases, the curve "hugs" the x-axis closer, before rising to the point $(1, 1)$.

Exercise Set 3.3

1. (a) A turning point is a point where the function changes from increasing to decreasing, or from decreasing to increasing. For example, $y = |x|$, $y = x^2$, and $y = \sqrt{1 - x^2}$ all have turning points at $x = 0$.   **(b)** A maximum value is a y-value corresponding to the highest point on the graph. For example, $y = 1$ is a maximum value for the function $y = \sqrt{1 - x^2}$ occurring at $x = 0$.   **(c)** A minimum value is a y-value corresponding to the lowest point on the graph. For example, $y = 0$ is a minimum value for the functions $y = |x|$ and $y = x^2$.   **(d)** A function f is increasing on an interval if $f(x_1) < f(x_2)$ whenever $x_1 < x_2$ on the interval. For example, $y = |x|$ and $y = x^2$ are increasing on $[0, \infty)$, $y = x^3$ is increasing on $(-\infty, \infty)$, $y = \sqrt{x}$ is increasing on $[0, \infty)$, and $y = \sqrt{1 - x^2}$ is increasing on $[-1, 0]$.   **(e)** A function f is decreasing on an interval if $f(x_1) > f(x_2)$ whenever $x_1 < x_2$ on the interval. For

example, $y = |x|$ and $y = x^2$ are decreasing on $(-\infty, 0]$, $y = \dfrac{1}{x}$ is decreasing on either $(-\infty, 0)$ or $(0, \infty)$, and $y = \sqrt{1 - x^2}$ is decreasing on $[0, 1]$.
3. (a) $[-1, 1]$　**(b)** 1 (occurring at $x = 1$)　**(c)** -1 (occurring at $x = 3$)　**(d)** $[0, 1]$ and $[3, 4]$　**(e)** $[1, 3]$　**5. (a)** $[-3, 0]$　**(b)** 0 (occurring at $x = 0$ and $x = 4$)　**(c)** -3 (occurring at $x = 2$)　**(d)** $[2, 4]$　**(e)** $[0, 2]$
7. turning points: $(-1.15, 1.08)$ and $(1.15, -5.08)$; increasing: $(-\infty, -1.15] \cup [1.15, \infty)$; decreasing: $[-1.15, 1.15]$　**9.** 10　**11.** 0
13. 2　**15. (a)** $\frac{1}{3}$ °C/min　**(b)** $\frac{4}{3}$ °C/min　**(c)** $\frac{1}{2}$ °C/min
17. (a) 1984–1990; 1978–1984: 1.8%/year; 1984–1990: 9.7%/year　**(b)** 1978–1984: 1.7%/year; 1984–1990: 9.7%/year
19. (a) $\dfrac{1}{16} \approx 0.06$ million tons/year
(b) 0.10 million tons/year
21. (a) $2x + 6$　**(b)** $2x + 2a$
23. (a) $4 + h$　**(b)** $2x + h$
25. (a) 8　**(b)** 8
27. (a) $x + a - 2$
(b) $2x + h - 2$　**29. (a)** $-\dfrac{1}{ax}$
(b) $-\dfrac{1}{x(x + h)}$
31. (a) $2x^2 + 2ax + 2a^2$
(b) $6x^2 + 6hx + 2h^2$
33. 48 feet/second
35. (a) $(64 + 16h)$ feet/second
(b)

h(seconds)	0.1	0.01	0.001	0.0001	0.00001
Average velocity $\dfrac{\Delta s}{\Delta t}$ on interval $[2, 2 + h]$	65.6	64.16	64.016	64.0016	64.00016

(c) 64 feet/second
37. $0 \le x \le 100$: $-\dfrac{3}{25}$ dollars/item;

$300 \le x \le 400$: $-\dfrac{3}{200}$ dollars/item

As more items are produced, the price should decrease.

39.

Function	$|x|$	x^2	x^3
Domain	$(-\infty, \infty)$	$(-\infty, \infty)$	$(-\infty, \infty)$
Range	$[0, \infty)$	$[0, \infty)$	$(-\infty, \infty)$
Turning Point	$(0, 0)$	$(0, 0)$	none
Maximum Value	none	none	none
Minimum Value	0	0	none
Interval(s) where Increasing	$(0, \infty)$	$(0, \infty)$	$(-\infty, \infty)$
Interval(s) where Decreasing	$(-\infty, 0)$	$(-\infty, 0)$	none

41. $b = 5$　**45.** turning point: $(-0.1875, -0.0016)$; increase: $[-0.1875, \infty)$; decrease: $(-\infty, -0.1875]$

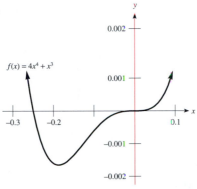

47. (a) They have the same graph.

Exercise Set 3.4
1. (a) C　**(b)** F　**(c)** I
(d) A　**(e)** J　**(f)** K

(g) D　**(h)** B　**(i)** E　**(j)** H
(k) G　**3.**

5.

7.

9.

11.

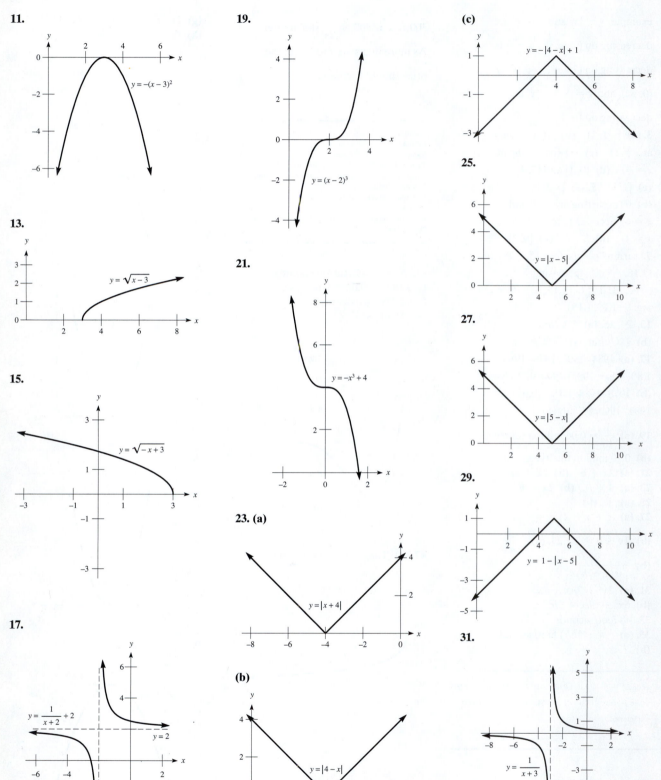

$y = -(x-3)^2$

13.

$y = \sqrt{x-3}$

15.

$y = \sqrt{-x+3}$

17.

$y = \dfrac{1}{x+2} + 2$

$y = 2$

$x = -2$

19.

$y = (x-2)^3$

21.

$y = -x^3 + 4$

23. (a)

$y = |x+4|$

(b)

$y = |4-x|$

(c)

$y = -|4-x| + 1$

25.

$y = |x-5|$

27.

$y = |5-x|$

29.

$y = 1 - |x-5|$

31.

$y = \dfrac{1}{x+3}$

$x = -3$

33.

$y = -\dfrac{1}{x+3}$

$x = -3$

(b) $f(x)$ reflected across the x-axis

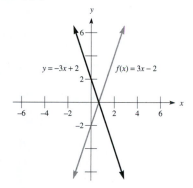

$y = -3x + 2$ $f(x) = 3x - 2$

(b) $f(x)$ reflected across the x-axis

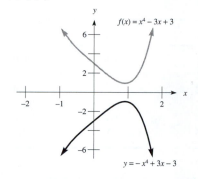

$f(x) = x^4 - 3x + 3$

$y = -x^4 + 3x - 3$

35.

$y = \sqrt{1 - (x-2)^2}$

43. (a) $f(x)$ reflected across the y-axis

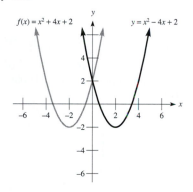

$f(x) = x^2 + 4x + 2$ $y = x^2 - 4x + 2$

(c) $f(x)$ reflected across the x-axis, then displaced up 3 units

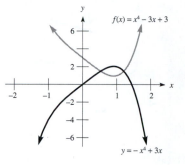

$f(x) = x^4 - 3x + 3$

$y = -x^4 + 3x$

37.

$y = 1 - \sqrt{1 - (x-2)^2}$

(b) $f(x)$ displaced down 2 units

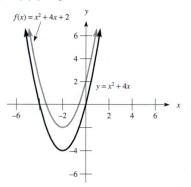

$f(x) = x^2 + 4x + 2$

$y = x^2 + 4x$

47. (a)

x	x^2	$x^2 - 1$	$x^2 + 1$
0	0	-1	1
± 1	1	0	2
± 2	4	3	5
± 3	9	8	10

(b) $y = x^2 - 1$ is a vertical displacement down one unit from $y = x^2$, while $y = x^2 + 1$ is a vertical displacement up one unit (from $y = x^2$)

39.

$y = \sqrt{1 - (2-x)^2}$

41. (a) $f(x)$ reflected across the y-axis

$y = -3x - 2$ $f(x) = 3x - 2$

45. (a) $f(x)$ reflected across the y-axis

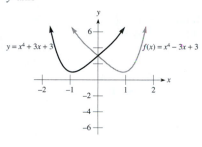

$y = x^4 + 3x + 3$ $f(x) = x^4 - 3x + 3$

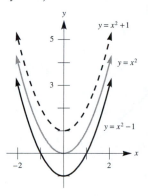

$y = x^2 + 1$

$y = x^2$

$y = x^2 - 1$

49. (a)

x	$\sqrt{x}$	$-\sqrt{x}$
0	0.0	0.0
1	1.0	−1.0
2	1.4	−1.4
3	1.7	−1.7
4	2.0	−2.0
5	2.2	−2.2

(b) $y = -\sqrt{x}$ is a reflection of $y = \sqrt{x}$ across the x-axis

51. (a)

(b)

53. (a)

(b)

55. (a) $R(a + c, b + d)$
(b) $T(a + c, b + d)$ **(c)** The order of translations do not affect the final point.

57. (a)

(c) Translate to the right 1 unit and 1 unit up. **59. (a)** $(-a, b + 2)$
(b) $(-a, -b + 2)$
(c) $(a + 3, -b)$
(d) $(a - 1, 1 - b)$
(e) $(-a + 1, b)$
(f) $(-a + 1, -b + 1)$

Exercise Set 3.5
1. (a) $x^2 - x - 7$
(b) $-x^2 + 5x + 5$ **(c)** 5
3. (a) $x^2 - 2x - 8$
(b) $-x^2 + 2x + 8$ **5. (a)** $4x - 2$
(b) $4x - 2$ **(c)** -4
7. (a) $\dfrac{-x^4 + 22x^2 - 4x - 80}{2x^3 - x^2 - 18x + 9}$
(b) $-\frac{80}{9}$

9. (a) $-6x - 14$ **(b)** -74
(c) $-6x - 7$ **(d)** -67
11. (a) $(f \circ g)(x) = 9x^2 - 3x - 6$,
$(f \circ g)(-2) = 36$,
$(g \circ f)(x) = -3x^2 + 9x + 14$,
$(g \circ f)(-2) = -16$
(b) $(f \circ g)(x) = 2^{x^2+1}$,
$(f \circ g)(-2) = 32$,
$(g \circ f)(x) = 2^{2x} + 1$,
$(g \circ f)(-2) = \frac{17}{16}$
(c) $(f \circ g)(x) = 3x^5 - 4x^2$,
$(f \circ g)(-2) = -112$,
$(g \circ f)(x) = 3x^5 - 4x^2$,
$(g \circ f)(-2) = -112$
(d) $(f \circ g)(x) = x$,
$(f \circ g)(-2) = -2, (g \circ f)(x) = x$,
$(g \circ f)(-2) = -2$

13. (a) $\dfrac{-x + 7}{6x}$ **(b)** $\dfrac{-t + 7}{6t}$ **(c)** $\frac{5}{12}$

(d) $\dfrac{1 - 6x}{7}$ **(e)** $\dfrac{1 - 6y}{7}$ **(f)** $-\frac{11}{7}$

15. (a) $M(7) = \frac{13}{5}$, $M[M(7)] = 7$
(b) $(M \circ M)(x) = x$
(c) $(M \circ M)(7) = 7$
17. (a) $f[g(3)] = 1$
(b) $g[f(3)] = -3$ **(c)** $f[h(3)] = -1$
(d) $(h \circ g)(2) = 2$
(e) $h\{f[g(3)]\} = 2$
(f) $(g \circ f \circ h \circ f)(2) = -3$
19.

x	0	1	2	3	4
$(f \circ g)(x)$	1	3	2	undef.	2

x	−1	0	1	2	3	4
$(g \circ f)(x)$	0	0	3	4	2	undef.

21. (a) $f \circ g$ is the graph of f displaced 4 units to the left
(b)

23. (a) domain: $[0, \infty)$; range: $[-3, \infty)$

(b) domain: $(-\infty, \infty)$; range: $(-\infty, \infty)$

(c) $(f \circ g)(x) = \sqrt{x} - 4$; domain: $[0, \infty)$; range: $[-4, \infty)$

(d) $g[f(x)] = \sqrt{x-1} - 3$; domain: $[1, \infty)$
(e)

25. (a)

t	A	t	A
0	707	3	1402
0.5	737	3.5	1648
1	804	4	1940
1.5	903	4.5	2284
2	1034	5	2685
2.5	1199		

(b) 800 m² **(c)** initially: 707 m²; after three hours this area has doubled **(d)** 0 to 2.5: 196.8 m²/hr; 2.5 to 5: 594.4 m²/hr; faster over the interval from $t = 2.5$ to $t = 5$ **27.**
(a) $(C \circ f)(t) = 100 + 450t - 25t^2$
(b) \$1225 **(c)** \$1900; no

29. (a) $f(x) = \sqrt[3]{x}, g(x) = 3x + 4$; $F(x) = (f \circ g)(x)$ **(b)** $f(x) = |x|$, $g(x) = 2x - 3$; $G(x) = (f \circ g)(x)$
(c) $f(x) = x^5, g(x) = ax + b$; $H(x) = (f \circ g)(x)$ **(d)** $f(x) = \frac{1}{x}$, $g(x) = \sqrt{x}; T(x) = (f \circ g)(x)$
31. (a) $f(x) = (b \circ c)(x)$
(b) $g(x) = (a \circ d)(x)$
(c) $h(x) = (c \circ d)(x)$
(d) $K(x) = (c \circ b)(x)$
(e) $l(x) = (c \circ a)(x)$
(f) $m(x) = (a \circ c)(x)$
33. (b)

(c) 0.44; $(A \circ f)(6) = \dfrac{9\pi}{64} \approx 0.44$

35. (a) $x_1 = 2, x_2 = 4, x_3 = 8$, $x_4 = 16, x_5 = 32, x_6 = 64$
(b) all six iterates are 0
(c) $x_1 = -2, x_2 = -4, x_3 = -8$, $x_4 = -16, x_5 = -32, x_6 = -64$
37. (a) $x_1 = -3, x_2 = -5, x_3 = -9$, $x_4 = -17, x_5 = -33, x_6 = -65$
(b) all six iterates are -1
(c) $x_1 = 3, x_2 = 7, x_3 = 15$, $x_4 = 31, x_5 = 63, x_6 = 127$
39. (a) $x_1 = 0.81, x_2 \approx 0.656$, $x_3 \approx 0.430, x_4 \approx 0.185, x_5 \approx 0.034$, $x_6 \approx 0.001$ **(b)** all six iterates are 1 **(c)** $x_1 = 1.21, x_2 \approx 1.464$, $x_3 \approx 2.144, x_4 \approx 4.595, x_5 \approx 21.114$, $x_6 \approx 445.792$ **41.** $x_1 \approx 0.316$, $x_2 \approx 0.562, x_3 \approx 0.750, x_4 \approx 0.866$
43. (a) $f(1) = 4, f(2) = 1$, $f(3) = 10, f(4) = 2, f(5) = 16$, $f(6) = 3$
(b) $f(1) = 4, f(4) = 2, f(2) = 1$
(c) $f(3) = 10, f(10) = 5$, $f(5) = 16, f(16) = 8, f(8) = 4$, $f(4) = 2, f(2) = 1$ **(d)** $x_0 = 2: 2, 1$ $x_0 = 4: 4, 2, 1$ $x_0 = 5: 16, 8, 4, 2, 1$ $x_0 = 6: 6, 3, 10, 5, 16, 8, 4, 2, 1$ $x_0 = 7: 22, 11, 34, 17, 52, 26, 13, 40,$ 20, 10, 5, 16, 8, 4, 2, 1

45. $f(x) = \dfrac{x + 6}{4}$

47. $a = -\frac{1}{2}$ and $b = \frac{1}{2}$
49. (a) $2x + 2a - 2$
(b) $4x + 4a - 4$ **51. (a)** $x_1 = 3$, $x_2 \approx 2.259259259$, $x_3 \approx 1.963308018$, $x_4 \approx 1.914212754$, $x_5 \approx 1.912932041$, $x_6 = \ldots = x_{10} \approx 1.912931183$
The iterates converge to 1.912931183.
(b) $\sqrt[3]{7} \approx 1.912931183$; they are the same **(c)** fifth iterate; sixth iterate

Exercise Set 3.6
1. (a) $h[k(x)] = x$ for every x in the domain of k. **(b)** $k[h(x)] = x$ for every x in the domain of h.
5. (a)

(b) f and h **7.** 7 **9. (a)** 4 **(b)** -1 **(c)** $\sqrt{2}$ **(d)** $t + 1$

11. (a) $f^{-1}(x) = \dfrac{x - 1}{2}$

(b) $f^{-1}(5) = 2, \dfrac{1}{f(5)} = \frac{1}{11}$; no

13. (a) $f^{-1}(x) = \dfrac{x + 1}{3}$
(c)

15. (a) $f^{-1}(x) = x^2 + 1$ for $x \geq 0$
(c)

17. (a) domain: $(-\infty, 3) \cup (3, \infty)$;
range: $(-\infty, 1) \cup (1, \infty)$

(b) $f^{-1}(x) = \dfrac{3x + 2}{x - 1}$

(c) domain: $(-\infty, 1) \cup (1, \infty)$;
range: $(-\infty, 3) \cup (3, \infty)$; domain
of f = range of f^{-1}, range of
f = domain of f^{-1}

19. $f^{-1}(x) = \sqrt[3]{\dfrac{x - 1}{2}}$

21. $f[f^{-1}(x)] = f^{-1}[f(x)] = x$

23. (a)

(b)

(c)

(d)

25. (a) $x = 7$ **(b)** $x = -3$
27. $t = \frac{9}{4}$ **29.** not one-to-one
31. not one-to-one   **33.** not
one-to-one   **35.** not one-to-one

37. not one-to-one

39. is one-to-one

41. (a)

(b) f is one-to-one, g is not one-to-
one   **43.** $A: (a, f(a)); B: (f(a), f(a)); C: (f(f(a)), f(a)); D: (f(a), f(f(a)))$ **45. (a)** $-\frac{1}{8}$

(b) $f^{-1}(x) = \dfrac{x + 3}{x}$; -8; they are

reciprocals
47. (a) E   **(b)** C   **(c)** L   **(d)** A
(e) J   **(f)** G   **(g)** B   **(h)** M
(i) K   **(j)** D   **(k)** I   **(l)** H
(m) N   **(n)** F   **49. (b)** symmetry
about the line $y = x$

Exercise Set 3.7

1. (a) $y = kx$ **(b)** $A = \dfrac{k}{B}$

3. (a) $x = kuv^2$ **(b)** $z = kA^2B^3$

5. (a) $F = \dfrac{k}{r^2}$

(b) $V^2 = k(U^2 + T^2)$

7. (a) $k = -2$ **(b)** $A = \dfrac{-2}{B}$

(c) $A = -\frac{8}{5}$ **9.** $x = -24$

11. A is six times its original value

13. x is eight times its original value **15. (a)** $S = 4\pi r^2$

(b) $S \approx 314.2$ cm^2

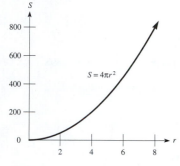

(c) $S = 100\pi \approx 314.16$ cm^2

17. (a) $R = \dfrac{5000}{L}$

(b) $R \approx 3.8$ m^3/min

(c) $R = \dfrac{50}{13} \approx 3.85$ m^3/min

19. the force is 48 times its original value **21.** 122.5 m

23. (a) $V = \dfrac{2.05}{P}$

(b) $V = 2.05$ liters **25. (a)** the kinetic energy is six times its original value **(b)** the velocity is $\sqrt{2}$ times its original value **27.** 110.6 lbs

29. (a) F **(b)** C **(c)** D **(d)** A **(e)** B **31.** shortened by $\frac{3}{4}L$ **33.** 4.24 ft **37.** Mercury:

87.9693 days; Venus: 224.7007 days; Earth: 365.2564 days; Mars: 686.9786 days Jupiter: 4,336.6159 days; Saturn: 10,826.9994 days; Uranus: 30,873.7244 days; Neptune: 60,300.6863 days Pluto: 91,814.3739 days **39. (a)** yes **(b)** no **(c)** yes

Chapter 3 Review Exercises

1. (a) $(-\infty, 3]$ **(b)** $(-\infty, \frac{1}{2}) \cup (\frac{1}{2}, \infty)$ **3.** yes

5. (a) $-\dfrac{1}{ax}$ **(b)** $1 - 4x - 2h$

7. (a) $g^{-1}(x) = \dfrac{1}{3x + 5}$

(b)

9. (a) x-intercepts: -5 and 1; y-intercept: -1

(b) x-intercept: -1; y-intercept: $-\frac{1}{2}$

11. (a) 5 **(b)** $7 - 4\sqrt{2}$
13. $m(h) = 10 + h$

15.

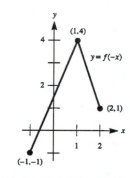

17. (a) 2.5, 3.25, 3.625 **(b)** A: $f(1) = 2.5$; B: $f(2.5) = 3.25$; C: $f(3.25) = 3.625$ **(c)** $y = 4$ **(d)** 3.813, 3.906, 3.953, 3.977, 3.988, 3.994, 3.997 **19.** no x-intercepts; y-intercept: 1

21. x-intercept: -3; y-intercept: 3

23. x-intercepts: ± 1; y-intercept: 1

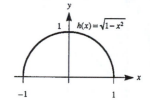

25. x-intercept: 0; y-intercept: 0

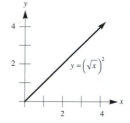

27. x-intercept: 1; no y-intercept

$(f \circ g)(x) = -\left(\sqrt{x-1}\right)^2$

29. no x- or y-intercepts

$x = 1$

31. x-intercept: $\frac{1}{2}$; y-intercept: -1

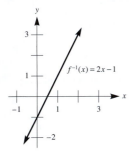

$f^{-1}(x) = 2x - 1$

33. x-intercept: 0; y-intercept: 0

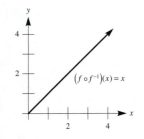

$(f \circ f^{-1})(x) = x$

35. $(-\infty, -3) \cup (-3, 3) \cup (3, \infty)$
37. $(-\infty, 4]$
39. $(-\infty, -1] \cup [3, \infty)$
41. $(-\infty, \frac{1}{3}) \cup (\frac{1}{3}, \infty)$
43. $(-\infty, \frac{2}{5}) \cup (\frac{2}{5}, \infty)$
45. $a(x) = (f \circ g)(x)$
47. $c(x) = (G \circ g)(x)$
49. $A(x) = (g \circ f \circ G)(x)$

51. $C(x) = (g \circ G \circ G)(x)$ **53.** 12
55. $-\frac{9}{19}$ **57.** $t^2 + t$
59. $x^2 - 5x + 6$ **61.** $x^4 - x^2$
63. $-2x^3 + 3x^2 - x$ **65.** $4x^2 - 2x$
67. $-2x^2 + 2x + 1$
69. $\dfrac{2x + 2}{2x - 5}$ **71.** $2x + h - 1$

73. $F^{-1}(x) = \dfrac{4x + 3}{1 - x}$ **75.** x

77. $\dfrac{1 + x}{2}$ **79.** $\frac{22}{7}$ **81.** negative

83. -1 **85.** -1 **87.** $(0, -2)$ and
$(5, 1)$ **89.** $[-6, 0]$ and $[5, 8]$
91. 0 at $x = 2$ **93.** no   **95.** $x = 4$
97. (a) $x = 10$ **(b)** $x = 0$
99. (a) 5   **(b)** -3 **(c)** 4   **(d)** $\frac{1}{4}$
101. $(f \circ f)(10)$ **103.** $[0, 4]$
105. 5   **107.** $(1, 3) \cup (6, 10)$
109. $(4, 7)$ **111.** A is 12 times its
original value   **113.** $2\sqrt{5} \approx 4.47$
feet   **115.** $\frac{4}{3}\sqrt{5}\sqrt[4]{27} \approx 6.80$ cm

Chapter 3 Test
1. $(-\infty, -1] \cup [6, \infty)$
2. $(-\infty, \frac{2}{3}) \cup (\frac{2}{3}, \infty)$
3. (a) $2x^2 - 2x - 2$
(b) $2x^2 - 5x + 2$ **(c)** 54

4. $-\dfrac{2}{at}$ **5.** $4x + 2h - 5$

6. $g^{-1}(x) = -\dfrac{x}{6x + 4}$ **7.** $-\frac{8}{5}$

8. (a) x-intercepts: 4 and 2;
y-intercept: -2

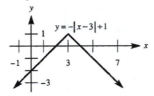

$y = -|x - 3| + 1$

(b) x-intercept: $-\frac{5}{2}$; y-intercept: $-\frac{5}{3}$

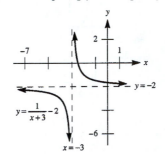

$y = \dfrac{1}{x+3} - 2$

$x = -3$

9. (a) $[-3, 1]$ **(b)** $(-2, 1)$
(c) -3 at $x = 2$ **(d)** 1 at $x = -2$
(e) $[-2, 2]$ **(f)** no   **10. (a)** $\frac{23}{4}$
(b) $12 - 7\sqrt{3}$ **11.** domain:
$(-\infty, -1) \cup (-1, \infty)$

$y = F(x)$

12. $z = \frac{9}{4}$

13. (a) $r = \dfrac{2.55}{L^2}$ **(b)** $r = 398$

vibrations per second

$r = \dfrac{2.55}{L^2}$

(c) $r \approx 398.4$ vibrations per second
14. (a) 0.4, 0.8, 0.4, 0.8, 0.4, 0.8
(b) 0.4, 0.8, 0.4, 0.8, 0.4, 0.8
15.

$(-1, 3)$

$(-5, -2)$

16. $t = 1$ **17.** $b = 10$

18. (a) $\dfrac{\Delta F}{\Delta t} = 0.16a - 0.8$

(b) $-0.16 \,°\text{C/hour}$

CHAPTER 4
Exercise Set 4.1

1. $f(x) = \frac{2}{3}x + \frac{2}{3}$ **3.** $g(x) = \sqrt{2}x$
5. $f(x) = x - \frac{7}{2}$ **7.** $f(x) = \frac{1}{2}x - 2$
9. yes **11.** $V(t) = -2375t + 20000$
13. (a) $V(t) = -12000t + 60000$
(b)

End of Year	Yearly Depreciation	Accumulated Depreciation	Value V
0	0	0	60,000
1	12,000	12,000	48,000
2	12,000	24,000	36,000
3	12,000	36,000	24,000
4	12,000	48,000	12,000
5	12,000	60,000	0

15. (a) $530 **(b)** $538
(c) $8/fan
17. (a)

(b) $575 **(c)** $575.50; the cost of producing 151 units
19. (a) B is traveling faster
(b) A is farther to the right
(c) $t = 8$ sec
21. (a) $y = 362091.5x - 690,879,017.5$
(b) 33,304,000 **(c)** too low; 1.7% error
23. (a) $y = 3680x - 7,284,416$
(b) $64,544 million

(c) 2.4% error
25. (a) $y = 4.92x - 35.55$
(b) 102.21 cm **(c)** too high;
4.2% error **(d)** 33.33 cm; too low;
8.3% error **(e)** 377.73 cm; too
high; 48.4% error
27. (a)

(b) slope: 2.5; y-intercept: -2

(c)

29. (a) 3,853,000; 0.8% error
(b) $f^{-1}(x) = \dfrac{x + 71238863.429}{37546.068}$
(c) year 2004
31. (a)

(b) 1809 million tons; 5.5% error
(c) 1954 million tons; 6.0% error
33. (a) 4,050,273 **(b)** no
35. (a) 3:57.6; too low; 0.17% error
(b) 4:11.3; too low; 1.61% error;
This percent error is much larger,
since 1911 is not in the data range.
37. (a) A **(b)** A^2 **(c)** AC
(d) AC **39. (a)** $\dfrac{1}{AC}$ **(b)** $\dfrac{1}{AC}$
43. $f(x) = 3x + 1$
45. $y = 2.4x - 2$ **47.** $y = 2.4x - 0.4$
49. $y = 0.084x + 37.241$
51. (a) $f(x) = \sqrt{2}x + (-1 + \sqrt{2})$
and $f(x) = -\sqrt{2}x + (-1 - \sqrt{2})$
(b) $f(x) = \sqrt[3]{2}x + (-1 + \sqrt[3]{2})$
53. $f(x) = x$

Exercise Set 4.2
1. (a)

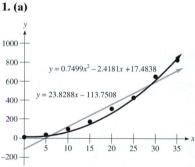

(b) 1990: linear $\approx$ 839.40 billion;
quadratic $\approx$ 1314.05 billion
1998: linear $\approx$ 1030.03 billion;
quadratic $\approx$ 1861.32 billion
(c) 1990: linear $\approx$ 26.8% error;
quadratic $\approx$ 14.7% error   1998:

linear ≈ 35.0% error; quadratic ≈ 17.4% error   The quadratic model produces the smaller percent error.

3.

Projected U.S. population (millions)	Actual U.S. population (millions)	Projection too high or too low?	% Error
1970	191.12	203.30	too low	5.99%
2000	258.35	275.60	too low	6.26%

5. vertex: $(-2, 0)$; axis of symmetry: $x = -2$; minimum value: 0; x-intercept: -2; y-intercept: 4

$y = (x + 2)^2$

7. vertex: $(-2, 0)$; axis of symmetry: $x = -2$; minimum value: 0; x-intercept: -2; y-intercept: 8

$y = 2(x + 2)^2$

9. vertex: $(-2, 4)$; axis of symmetry: $x = -2$; maximum value: 4; x-intercepts: $-2 \pm \sqrt{2}$; y-intercept: -4

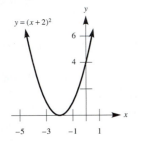

$y = -2(x + 2)^2 + 4$

11. vertex: $(2, -4)$; axis of symmetry: $x = 2$; minimum value: -4; x-intercepts: 0 and 4; y-intercept: 0

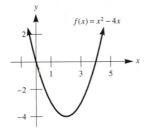

$f(x) = x^2 - 4x$

13. vertex: $(0, 1)$; axis of symmetry: $x = 0$; maximum value: 1; x-intercepts: ± 1; y-intercept: 1

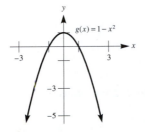

$g(x) = 1 - x^2$

15. vertex: $(1, -4)$; axis of symmetry: $x = 1$; minimum value: -4; x-intercepts: 3 and -1; y-intercept: -3

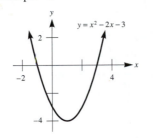

$y = x^2 - 2x - 3$

17. vertex: $(3, 11)$; axis of symmetry: $x = 3$; maximum value: 11; x-intercepts: $3 \pm \sqrt{11}$; y-intercept: 2

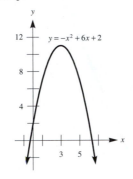

$y = -x^2 + 6x + 2$

19. vertex: $(\frac{1}{6}, \frac{9}{4})$; axis of symmetry: $t = \frac{1}{6}$; maximum value: $\frac{9}{4}$; t-intercepts: $-\frac{1}{3}$ and $\frac{2}{3}$; s-intercept: 2

$s = 2 + 3t - 9t^2$

21. $x = 1$; minimum output
23. $x = \frac{3}{2}$; maximum output
25. $x = 0$; minimum output
27. minimum value: -13
29. maximum value: $\frac{25}{8}$
31. (a) minimum value
(b) minimum value: -4.0; $x_0 \approx 0.2$
(c) minimum value: $-\frac{97}{24}$; $x_0 = \frac{1}{12}$
33. (a) maximum value
(b) maximum value: -43.4; $t_0 \approx 2.2$ **(c)** maximum value: $-\frac{391}{9}$; $t_0 = \frac{20}{9}$ **35.** 5 units
37. quadratic   **39.** neither
41. linear   **43. (a)** minimum value = 8 at $x = 3$ **(b)** minimum value = 4 at $x = 3$ **(c)** minimum value = 64 at $x = \pm\sqrt{3}$
45. (a) maximum value = 4
(b) maximum value = $2\sqrt[3]{2} \approx 2.5$
(c) maximum value = 16
47. neither   **49.** quadratic
51. The graph of $y = 50x^2$ should be "narrower".

$y = 8x^2$
$y = 3x^2$
$y = 2x^2$
$y = x^2$

53. $x + a$

55. $x = \dfrac{a + b}{2}$

57. $y = -\frac{1}{2}(x - 2)^2 + 2$

59. $y = \frac{1}{4}(x - 3)^2 - 1$

61. $b = \pm 2$ **63. (a)**

x	a	$a + h$	$a + 2h$
$f(x)$	$ma + b$	$ma + mh + b$	$ma + 2mh + b$

(b) mh **65. (a)**

$y = 0.015089x^2 - 0.086075x +$
3.364536; 1989: 3.466 billion tons;
1996: 3.391 billion tons   **(b)** 1989:
1.70% error; 1996: 1.08% error
(c) 3.642 billion tons; 9.40% error

Exercise Set 4.3

1. $x = 1$ **3.** no fixed points
5. $x = -1$ and $x = 5$ **7.** $t = 1$
9. $t = -3$ and $t = 4$ **11.** $x = 0$ and
$x = \frac{4}{9}$ **13.** $u = -2$ and $u = 1$
15. $x = 10$ **17. (a)** one fixed point

(b) $x \approx -0.771$
19. (a) three fixed points

(b) $x \approx -1.206, -1.103, 2.309$

---

21. (a) two fixed points

(b) $x \approx -1.495, 0.798$
23. (b) fourth iterate   **(c)** twelfth
iterate   **25. (a)**

	x_1	x_2	x_3	x_4	x_5	x_6	x_7	x_8
From graph	1.7	0.8	1.4	1.0	1.3	1.1	1.2	1.1
From calculator	1.72	0.796	1.443	0.990	1.307	1.085	1.240	1.132

(b) $x = \frac{20}{7} \approx 1.176$ **(c)** eighth
iterate   **27.**

	From graph	From calculator
x_1	0.36	0.36
x_2	0.92	0.922
x_3	0.29	0.289
x_4	0.82	0.822
x_5	0.58	0.585
x_6	0.97	0.971
x_7	0.11	0.113
x_8	0.40	0.402
x_9	0.96	0.962

29. (a) $x_{21} \approx 0.6632$, $x_{22} \approx 0.6477$,
$x_{23} \approx 0.6617$, $x_{24} \approx 0.6492$,
$x_{25} \approx 0.6605$ **(b)**

n	20	21	22	23	24	25
Number of fish after n breeding seasons	323	332	324	331	325	330

(c) $x_1 = 0.0675$, $x_2 \approx 0.04721$,
$x_3 \approx 0.03373$, $x_4 \approx 0.02445$,
$x_5 \approx 0.01789$.   The iterates are
approaching 0, which is a fixed point.
31. (a)

n	x_n	Number of fish after n breeding seasons
0	0.1	50
1	0.279	140
2	0.6236	312
3	0.7276	364
4	0.6143	307
5	0.7345	367
6	0.6046	302
7	0.7411	371
8	0.5948	297
9	0.7471	374
10	0.5857	293

---

n	x_n	Number of fish after n breeding seasons
21	0.7633	382
22	0.5601	280
23	0.7638	382
24	0.5592	280
25	0.7641	382
26	0.5587	279

(b) $x = \frac{21}{31} \approx 0.6774$
(c) $a \approx 0.7646, b \approx 0.5580$
(d) 382 fish, 279 fish   **33. (a)** $x_0 = c$:
$x_1 = d, x_2 = c, x_3 = d, x_4 = c,$
$x_5 = d, x_6 = c$; $x_0 = d$: $x_1 = c,$
$x_2 = d, x_3 = c, x_4 = d, x_5 = c,$
$x_6 = d$;   With an initial input of
either c or d, the sequence of
iterates alternates between c and d.
(b) 0.8, 0.4, 0.8, 0.4, 0.8, 0.4
(c) $\{0.4, 0.8\}$ **(d)** $x_1 = 0.8, x_2 = 0.4$
37. (a) $a = \frac{6}{7}, b = \frac{3}{7}$ **(b)** $P(\frac{3}{7}, \frac{3}{7})$,
$Q(\frac{3}{7}, \frac{6}{7})$, $R(\frac{6}{7}, \frac{6}{7})$, $S(\frac{6}{7}, \frac{3}{7})$

Exercise Set 4.4

1. (a) $A(x) = 8x - x^2$

(b) $P(x) = 2x + \dfrac{170}{x}$

3. (a) $D(x) = \sqrt{x^4 + 3x^2 + 1}$

(b) $m(x) = \dfrac{x^2 + 1}{x}$

5. (a) $A(y) = \dfrac{\pi y^2}{4}$

(b) $A(y) = \dfrac{\pi^2 y^2}{16}$

7. (a) $P(x) = 16x - x^2$
(b) $S(x) = 2x^2 - 32x + 256$
(c) $D(x) = x^3 - (16 - x)^3$ or
$D(x) = (16 - x)^3 - x^3$
(d) $A(x) = 8$; The average does not
depend on what the two numbers
are.   **9.** $R(x) = -\frac{1}{4}x^2 + 8x$
11. (a)

x	1	2	3	4	5	6	7
$P(x)$	17.88	19.49	20.83	21.86	22.49	22.58	21.75

(b) largest value: 22.58; $x = 6$
(c) 22.63   **13. (a)** $h(s) = \sqrt{3}s$
(b) $A(s) = \sqrt{3}s^2$ **(c)** $4\sqrt{3}$ cm

(d) $\dfrac{25\sqrt{3}}{4}$ in.^{2} **15.** $V(r) = 2\pi r^3$

17. (a) $h(r) = \dfrac{12}{r^2}$

(b) $S(r) = 2\pi r^2 + \dfrac{24\pi}{r}$

19. $V(S) = \dfrac{S\sqrt{S\pi}}{6\pi}$

21. $A(x) = \frac{1}{2}x\sqrt{400 - x^2}$

23. $AB(x) = \dfrac{(x + 4)\sqrt{x^2 + 25}}{x}$

25. (a) yes

(b) one turning point
(c) maximum value = 625; no
minimum value   **27. (a)** no

(b) one turning point   **(c)** no
maximum value; minimum
value ≈ 0.87 **29. (a)** no

(b) one turning point   **(c)** no
maximum value; minimum
value ≈ 25.69 **31. (a) (a)** no

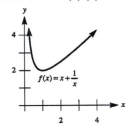

(b) one turning point   **(c)** no
maximum value; minimum
value = 2   **(b) (a)** no

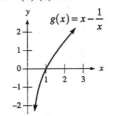

(b) no turning points   **(c)** no
maximum or minimum values

33. (a) $S(x) = x + \dfrac{\sqrt{11}}{x}$; minimum

value ≈ 3.642

(b) $P(x) = -x^2 + \sqrt{11}x$;
maximum value = 2.75

Graph labeled $P(x)=-x^2+\sqrt{11}\,x$

35. $A(x) = \frac{17}{144}x^2 - \frac{1}{3}x + \frac{1}{2}$; yes

37. (a) $V(r) = \dfrac{\sqrt{3}}{3}\pi r^3$

(b) $S(r) = 2\pi r^2$

39. (a) $r(h) = \dfrac{3h}{\sqrt{h^2 - 9}}$

(b) $h(r) = \dfrac{3r}{\sqrt{r^2 - 9}}$

41. $A(x) = \dfrac{4x^2 + \pi(14 - x)^2}{16\pi}$

43. $A(r) = \dfrac{r(1 - 4\pi r)}{4}$

45. $A(x) = \frac{\pi}{3}x^2$ **47. (a)** $V(x) =$

$4x^3 - 28x^2 + 48x$ **(b)**

x (in.)	0	0.5	1.0	1.5	2.0	2.5	3.0
volume (in.^{3})	0	17.5	24	22.5	16	7.5	0

(c) $x = 1.0$ **(d)**

x (in.)	0.8	0.9	1.0	1.1	1.2	1.3	1.4
volume (in.^{3})	22.5	23.4	24	24.2	24.2	23.9	23.3

(e) $x = 1.1$

49. (a) $A(r) = 32r - 2r^2 - \dfrac{\pi r^2}{2}$

(b) downward; yes

51. (a) $y(s) = \dfrac{3s}{\sqrt{1 - s^2}}$

(b) $s(y) = \dfrac{y}{\sqrt{y^2 + 9}}$

(c) $z(s) = \dfrac{3}{\sqrt{1 - s^2}}$

(d) $s(z) = \dfrac{\sqrt{z^2 - 9}}{z}$

53. (a) $m(a) = \dfrac{a^2 + 1}{a}$

55. $A(x) = \dfrac{8x - x^2}{4}$

57. $A(m) = \dfrac{2m^2 - 8m + 8}{m^2 - 4m}$

59. $A = \dfrac{(ma - b)^2}{-2m}$

Exercise Set 4.5

1. $\frac{25}{4}$ **3.** $\frac{1}{2}$ **5.** $\frac{25}{4}$ m by $\frac{25}{4}$ m

7. 1250 in.^{2} **9. (a)** 18   **(b)** $\frac{23}{4}$

(c) $\frac{47}{8}$ **(d)** $\frac{95}{16}$ **11. (a)** 16 feet; 12

feet   **(b)** 16 feet; 1 second

(c) $t = \frac{7}{4}$ sec or $t = \frac{1}{4}$ sec   **13.**

$\left(\frac{7}{2}, \frac{2 + \sqrt{6}}{2}\right)$; distance $= \frac{\sqrt{7}}{2}$ **15. (a)** $\frac{1}{2}$

(b) $\frac{1}{4}$ **17.** 125 ft by 250 ft

19. $x = 40$ **21.** $x = 60$; maximum

revenue = \$900; $p = 15

23. (a) $\frac{36}{13}$ **(b)** $\frac{6\sqrt{13}}{13}$; This is the square root.

25. (a)

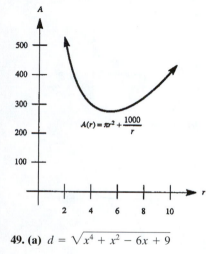

$y = x^4 + 2$

$y = x^2$

(b) 1.5 miles **(c)** $\frac{7}{4}$ miles

27. (a) $\frac{225}{2}$ **29.** $x = \frac{\sqrt{2}}{2}$; area $= \frac{1}{2}$

31. $\frac{49}{12}$ **35.** 100 yd by 150 yd

39. (a) $p(x) = -2x + 500$

(b) maximum revenue = $31,250; price = $250 **41.** $t = \sqrt{3}$ or $t = -\sqrt{3}$ **43. (a)** $(x - \frac{1}{2})^2 \geq 0$, thus $(x - \frac{1}{2})^2 + \frac{3}{4} \geq \frac{3}{4} > 0$

(b) Since $D = -3 < 0$, there are no x-intercepts. Since the graph is a parabola opening up, and there are no x-intercepts, the y-values must be positive for all values of x.

45. (a) $A(x) = \frac{4 + \pi}{16\pi}x^2 - 2x + 16$

(b) $x = \dfrac{16\pi}{4 + \pi}$ **(c)** $\dfrac{\pi}{4}$

47. (a) $A = \pi r^2 + \dfrac{1000}{r}$

(b) height ≈ 5.42 cm, radius ≈ 5.42 cm

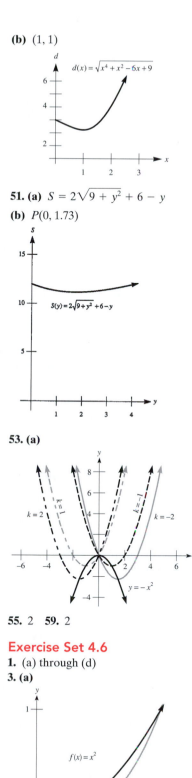

$A(r) = \pi r^2 + \dfrac{1000}{r}$

49. (a) $d = \sqrt{x^4 + x^2 - 6x + 9}$

(b) $(1, 1)$

$d(x) = \sqrt{x^4 + x^2 - 6x + 9}$

51. (a) $S = 2\sqrt{9 + y^2} + 6 - y$

(b) $P(0, 1.73)$

$S(y) = 2\sqrt{9 + y^2} + 6 - y$

53. (a)

$k = 2$ $k = 1$ $k = -1$ $k = -2$

$y = -x^2$

55. 2 **59.** 2

Exercise Set 4.6

1. (a) through (d)
3. (a)

$f(x) = x^2$

$g(x) = x^3$

(b) $[0, 1]$: $\dfrac{\Delta f}{\Delta x} = 1, \dfrac{\Delta g}{\Delta x} = 1$; $[0, \frac{1}{2}]$:

$\dfrac{\Delta f}{\Delta x} = \frac{1}{2}, \dfrac{\Delta g}{\Delta x} = \frac{1}{4}$ **5.** no x-intercepts; y-intercept: 5

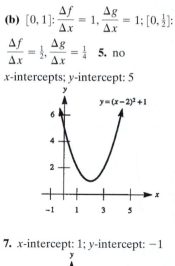

$y = (x - 2)^2 + 1$

7. x-intercept: 1; y-intercept: -1

$y = -(x - 1)^4$

9. x-intercept: $4 + \sqrt[3]{2}$; y-intercept: -66

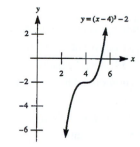

$y = (x - 4)^3 - 2$

11. x-intercept: -5; y-intercept: -1250

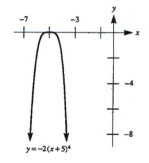

$y = -2(x + 5)^4$

13. x-intercept: -1; y-intercept: $\frac{1}{2}$

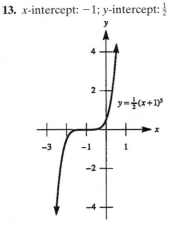

$y = \frac{1}{2}(x+1)^5$

15. x- and y-intercepts: 0

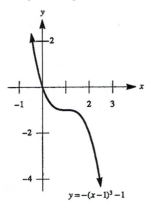

$y = -(x-1)^3 - 1$

17. A polynomial function of degree 3 can have at most 2 turning points. **19.** As $|x|$ gets very large, our function should be similar to $f(x) = a_3 x^3$. **21.** As $|x|$ gets very large with x negative, then the graph should resemble $2x^5$.
23. This graph has a corner.
25. (a)

$f(x)$ $g(x)$

(b)

$f(x)$ $g(x)$

27. (a) x-intercepts: 2, 1, -1;
y-intercept: 2

(b)

$y = (x-2)(x-1)(x+1)$

29. (a) x-intercepts: 0, 2, 1;
y-intercept: 0

(b)

$y = 2x(x-2)(x-1)$

31. (a) x-intercepts: 0, 5, -1;
y-intercept: 0

(b)

$y = x^3 - 4x^2 - 5x$

33. (a) x-intercepts: -3, -2, 2;
y-intercept: -12

(b)

$y = x^3 + 3x^2 - 4x - 12$

35. (a) x-intercepts: 0 and -2;
y-intercept: 0

(b)

$y = 2x^3$

(c)

$y = x^3(x + 2)$

37. (a) x-intercepts: 1, 4;
y-intercept: 128

(b)

$y = 6(x - 4)^3$

(c)

$y = 2(x - 1)(x - 4)^3$

39. (a) x-intercepts: $-1, 1, 3$;
y-intercept: 3

(b)

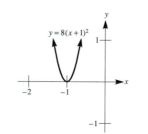

$y = 8(x + 1)^2$

(c)

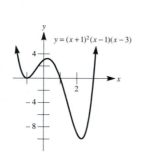

$y = (x + 1)^2(x - 1)(x - 3)$

41. (a) x-intercepts: $0, 4, -2$;
y-intercept: 0

(b)

$y = 8x^3$

(c)

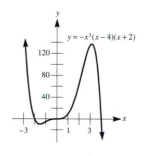

$y = -x^3(x - 4)(x + 2)$

43. (a) x-intercepts: $0, 2, -2$;
y-intercept: 0

(b)

$y = -512(x - 2)^2$

$y = 128(x + 2)^3$

(c)

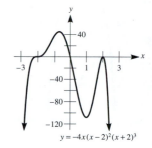

$y = -4x(x - 2)^2(x + 2)^3$

47.

$$x = 0, \frac{3 \pm \sqrt{29}}{2} \approx -1.193, 4.193$$

49. $x = -6, \pm\sqrt{3} \approx \pm1.732$

51. (a)

$N(t) = t^7 + 8t^4 + 16t$

(b) t-intercepts: $0.0, -1.6$
(c) $t = 0, -\sqrt[3]{4} \approx -1.587$
53. From left to right: $f(x) = x$,
$g(x) = x^2, h(x) = x^3, F(x) = x^4$,
$G(x) = x^5, H(x) = x^6$ **55.** $[0, 0.68]$
57. no such value
59. (a) $(400, 1600)$ **(b)**

61. (a)

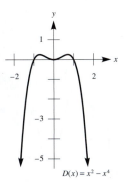

(b) $\left(\pm\frac{\sqrt{2}}{2}, \frac{1}{4}\right), (0, 0)$
(c) maximum vertical distance: $\frac{1}{4}$

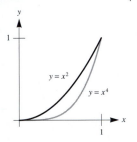

63. (a) $A(x) = x(1 - x^4) = x - x^5$

(b) maximum area: 0.53 square units

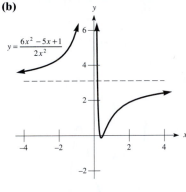

Exercise Set 4.7

1. (a) domain: $(-\infty, 3) \cup (3, \infty)$;
x-intercept: -5; y-intercept: $-\frac{5}{4}$;
vertical asymptote: $x = 3$;
horizontal asymptote: $y = \frac{3}{4}$
(b)

3. (a) domain: $(-\infty, 0) \cup (0, \infty)$;
x-intercepts: $\frac{1}{3}, \frac{1}{2}$; no y-intercept;
vertical asymptote: $x = 0$;
horizontal asymptote: $y = 3$
(b)

5. (a) domain:
$\left(-\infty, -\frac{1}{2}\right) \cup \left(-\frac{1}{2}, \frac{1}{2}\right) \cup \left(\frac{1}{2}, \infty\right)$;

x-intercepts: ± 3; y-intercept: 9;
vertical asymptotes: $x = \frac{1}{2}$ and
$x = -\frac{1}{2}$; horizontal asymptote:
$y = \frac{1}{4}$
(b)

7. (a) domain:
$\left(-\infty, -\frac{3}{2}\right) \cup \left(-\frac{3}{2}, 0\right) \cup (0, 1) \cup$
$(1, \infty)$; x-intercepts: $-\frac{4}{3}, 2$; no
y-intercept; vertical asymptotes:
$x = -\frac{3}{2}, x = 0, x = 1$; horizontal
asymptote: $y = 0$
(b)

9. no x-intercept; y-intercept: $\frac{1}{4}$;
horizontal asymptote: $y = 0$;
vertical asymptote: $x = -4$

11. no x-intercept; y-intercept: $\frac{3}{2}$; horizontal asymptote: $y = 0$; vertical asymptote: $x = -2$

13. no x-intercept; y-intercept: $\frac{2}{3}$; horizontal asymptote: $y = 0$; vertical asymptote: $x = 3$

15. x-intercept: 3; y-intercept: 3; horizontal asymptote: $y = 1$; vertical asymptote: $x = 1$

17. x-intercept: $\frac{1}{2}$; y-intercept: -2; horizontal asymptote: $y = 2$; vertical asymptote: $x = -\frac{1}{2}$

19. no x-intercept; y-intercept: $\frac{1}{4}$; horizontal asymptote: $y = 0$; vertical asymptote: $x = 2$

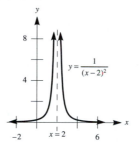

21. no x-intercept; y-intercept: 3; horizontal asymptote: $y = 0$; vertical asymptote: $x = -1$

23. no x-intercept; y-intercept: $\frac{1}{8}$; horizontal asymptote: $y = 0$; vertical asymptote: $x = -2$

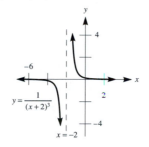

25. no x-intercept; y-intercept: $-\frac{4}{125}$; horizontal asymptote: $y = 0$; vertical asymptote: $x = -5$

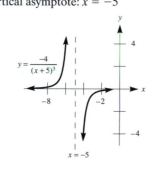

27. x- and y-intercepts: 0; horizontal asymptote: $y = 0$; vertical asymptotes: $x = -2$, $x = 2$

29. (a) x-intercept: 0; y-intercept: 0; horizontal asymptote: $y = 0$; vertical asymptotes: $x = 1$, $x = -3$

(b) x-intercept: 0; y-intercept: 0; horizontal asymptote: $y = 3$; vertical asymptotes: $x = 1$, $x = -3$

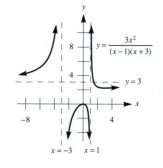

31. x-intercepts: $-\frac{5}{4}$, 1; y-intercept: 1; horizontal

asymptote: $y = 2$; vertical
asymptotes: $x = \frac{5}{2}$, $x = -1$

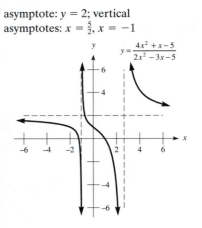

$$y = \frac{4x^2 + x - 5}{2x^2 - 3x - 5}$$

33. (a) x-intercepts: 2, 4; no y-intercept; horizontal asymptote: $y = 1$; vertical asymptotes: $x = 0$, $x = 1$

$$f(x) = \frac{(x-2)(x-4)}{x(x-1)}$$

(b) x-intercepts: 2, 4; no y-intercept; horizontal asymptote: $y = 1$; vertical asymptotes: $x = 0$, $x = 3$

$$g(x) = \frac{(x-2)(x-4)}{x(x-3)}$$

35. (a) 500 bacteria   **(b)** 2500 bacteria

37. (a) $A(x) = 521 - \dfrac{1500}{x} - 7x$

(b)

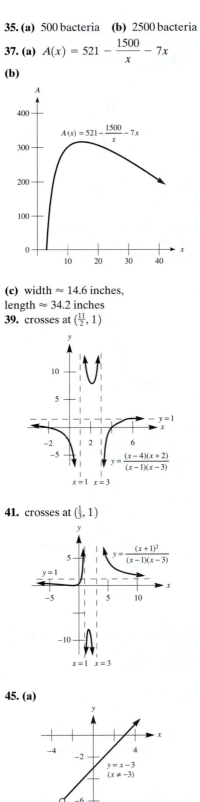

$$A(x) = 521 - \frac{1500}{x} - 7x$$

(c) width ≈ 14.6 inches,
length ≈ 34.2 inches

39. crosses at $\left(\frac{11}{2}, 1\right)$

$$y = \frac{(x-4)(x+2)}{(x-1)(x-3)}$$

41. crosses at $\left(\frac{1}{3}, 1\right)$

$$y = \frac{(x+1)^2}{(x-1)(x-3)}$$

45. (a)

$$y = x - 3 \quad (x \neq -3)$$

(b)

$$y = \frac{x-2}{x+1} \quad (x \neq 3)$$

(c)

$$y = \frac{1}{x-4} \quad (x \neq 1, 2, 3)$$

47. low point: $\left(-3, -\frac{1}{12}\right)$

$$y = \frac{x}{(x-3)^2}$$

49. (b)

x	$x + 4$	$\dfrac{x^2 + x - 6}{x - 3}$
10	14	14.8571
100	104	104.0619
1000	1004	1004.0060
-10	-6	-6.4615
-100	-96	-96.0583
-1000	-996	-996.0600

(c) vertical asymptote: $x = 3$;
x-intercepts: -3, 2; y-intercept: 2

(d)

(e) $(3 + \sqrt{6}, 7 + 2\sqrt{6})$,
$(3 - \sqrt{6}, 7 - 2\sqrt{6})$

51.

53. (a)

(b)

Note that as $|x|$ increases, the curve $f(x)$ approaches the curve $y = x^3$.

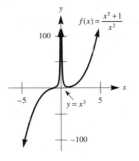

(c)

x	5	10	50	100	500
d	0.04	0.01	0.0004	0.0001	0.000004

x	-5	-10	-50	-100	-500
d	0.04	0.01	0.0004	0.0001	0.000004

(d) As $|x|$ increases, the quantity $\dfrac{1}{x^2}$ approaches 0, and thus $f(x) \approx x^3$ when $|x|$ gets very large.

Chapter 4 Review Exercises

1. $G(0) = -5$ **3.** \$20,000

5.

7. $V(t) = -180t + 1000$
9. x-intercept: $-\frac{5}{3}$; y-intercept: $\frac{5}{2}$; horizontal asymptote: $y = 3$; vertical asymptote: $x = -2$

11.

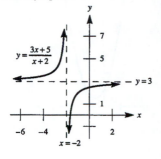

13. $P(w) = 2w + \dfrac{2\sqrt{144 - \pi^2 w^2}}{\pi}$

15. $f(x) = \frac{3}{8}x - \frac{5}{2}$

17. $f(x) = -\frac{3}{4}x + \frac{11}{4}$

19. vertex: $(-1, -4)$; x-intercepts: 1, -3; y-intercept: -3

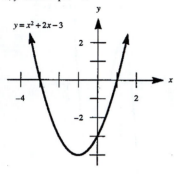

21. vertex: $\left(\frac{1}{2}, \frac{1}{2}\right)$; no x-intercept; y-intercept: 1

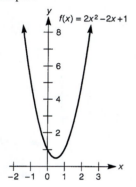

23. 5   **25. (a)** maximum height: $\frac{v_0^2}{32}$ ft; time: $t = \frac{v_0}{32}$ sec   **(b)** $t = \frac{v_0}{16}$ sec

27. $b = \frac{17}{3}, -11$ **29.** 1

31. $\frac{225}{4}$ cm^2 **33.** $a = 1$

35. $x = \frac{432}{59}$ cm ≈ 7.32 cm

39. x-intercepts: $-4, 2$;

y-intercept: -8

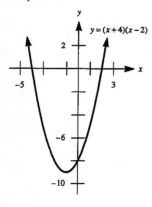

41. x-intercepts: $-1, 0$; y-intercept: 0

43. x-intercepts: $0, 2, -2$; y-intercept: 0

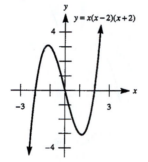

45. no x-intercept; y-intercept: -1; horizontal asymptote: $y = 0$; vertical asymptote: $x = 1$

47. x-intercept: 2; y-intercept: $\frac{2}{3}$; horizontal asymptote: $y = 1$; vertical asymptote: $x = 3$

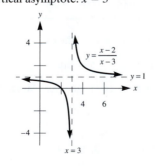

49. x-intercept: 1; y-intercept: $\frac{1}{4}$; horizontal asymptote: $y = 1$; vertical asymptote: $x = 2$

51. $k = 6$ **53.**

$(-\infty, 4 - 2\sqrt{3}] \cup [4 + 2\sqrt{3}, \infty)$

55. $A(m) = \dfrac{m}{2}$

57. $A(x) = (1 - x)\sqrt{1 - x^2}$

Chapter 4 Test

1. $L(0) = -\frac{18}{7}$ **2. (a)** maximum value: 2; increasing: $(-\infty, 1)$ **(b)** $t = 0$ **3. (a)** x-intercepts: $3, -4$; y-intercept: -48

(b)

(c)

4. (a)

$$y = \frac{-1}{(x-3)^3}$$

(b)

$$y = \frac{-1}{(x-3)^2}$$

5. turning point: $\left(\frac{7}{2}, \frac{73}{4}\right)$;

x-intercepts: $\dfrac{7 \pm \sqrt{73}}{2}$; y-intercept:

6; axis of symmetry: $x = \frac{7}{2}$

$$y = -x^2 + 7x + 6$$

6. maximum revenue $= \$9600$;
price $= \$40/\text{unit}$ **7.** $V(t) =$
$-1325t + 14000$ **8.** x-intercept: 3;
y-intercept: $-\frac{27}{2}$

$$y = -\tfrac{1}{2}(3 - x)^3$$

9. x-intercept: $\frac{3}{2}$; y-intercept: -3;
vertical asymptote: $x = -1$;
horizontal asymptote: $y = 2$

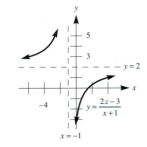

$$y = \frac{2x - 3}{x + 1}$$

10. (a)
$PQ(x) = \sqrt{10x^2 - 22x + 17}$
(b) $x = 1.1$ **11. (a)** vertical
asymptotes: $x = 3, x = -3$;
horizontal asymptote: $y = 1$
(b)

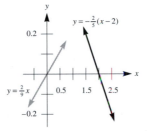

$$y = -\tfrac{2}{5}(x - 2)$$
$$y = \tfrac{2}{9}x$$

(c)

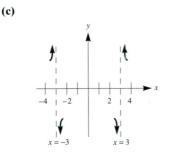

$$x = -3 \qquad x = 3$$

(d)

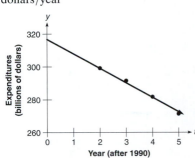

$$f(x) = \frac{x(x - 2)}{x^2 - 9}$$
$$y = 1$$
$$x = -3 \qquad x = 3$$

12. $A(w) = \frac{1}{2}w\sqrt{64 - w^2}$

13. (a) As $|x|$ increases in size for
x positive, $-x^3$ increases in the
negative direction, which the
pictured function does not.
(b) This graph has four turning
points, and a polynomial function
with highest degree term $-x^3$ can
have at most two turning points.
14. (a) slope ≈ -9 billion
dollars/year

Expenditures (billions of dollars)

Year (after 1990)

(b) $y = -8.84x + 316.74$; slope is
close   **(c)** 1996: 263.7 billion
dollars; 1999: 237.18 billion dollars;
1996: 0.8% error: 1999: 14.3% error
15. (a) $x = 0, \frac{1}{2}$ **(b)** no fixed
points   **16. (a)**

	x_1	x_2	x_3	x_4	x_5	x_6
From graph	0.56	0.28	0.52	0.33	0.49	0.36
From calculator	0.56	0.286	0.518	0.332	0.490	0.360

(b) $x = \dfrac{-5 - \sqrt{85}}{10} \approx -1.4220,$

$x = \dfrac{-5 + \sqrt{85}}{10} \approx 0.4220;$

The iterates are approaching
0.4220.   **(c)** $x_1 = -0.4$, $x_2 = 0.44$,
$x_3 = 0.4064$, $x_4 \approx 0.4348$,
$x_5 \approx 0.4109$, $x_6 \approx 0.4311$. Yes, they
are approaching the fixed point
0.4220.

CHAPTER 5
Exercise Set 5.1
1. (a) 10^9 **(b)** 10^{15} **3.** 125
5. 16   **7.** 8   **9.** $5^{\sqrt{2}}$ **11. (a)** $x = 3$
(b) $t = \frac{3}{2}$ **(c)** $y = \frac{1}{4}$ **(d)** $z = \frac{5}{2}$
13. $(-\infty, \infty)$ **15.** $(-\infty, \infty)$

17.

19.

21.

23.

25. domain: $(-\infty, \infty)$; range: $(-\infty, 1)$; x- and y-intercepts: 0; asymptote: $y = 1$

27. domain: $(-\infty, \infty)$; range: $(1, \infty)$; no x-intercept; y-intercept: 2; asymptote: $y = 1$

29. domain: $(-\infty, \infty)$; range: $(0, \infty)$; no x-intercept; y-intercept: $\frac{1}{2}$; asymptote: $y = 0$

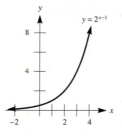

31. domain: $(-\infty, \infty)$; range: $(1, \infty)$; no x-intercept; y-intercept: 4; asymptote: $y = 1$

33. (a) no x-intercept; y-intercept: $-\frac{1}{9}$; asymptote: $y = 0$

(b)

35. (a) x-intercept: -1; y-intercept: -3; asymptote: $y = -4$

(b)

37. (a) no x-intercept; y-intercept: $\frac{1}{10}$; asymptote: $y = 0$

(b)

39. $x = -\frac{1}{3}$ **41.** $x = \frac{3}{2}, 1$

43. (a) $[0, 2]$: $\dfrac{\Delta f}{\Delta x} = 1.5$, $\dfrac{\Delta g}{\Delta x} = 1.5$;

$[2, 4]$: $\dfrac{\Delta f}{\Delta x} = 6$, $\dfrac{\Delta g}{\Delta x} = 6$

(b) $[4, 6]$: $\dfrac{\Delta f}{\Delta x} = 10.5$, $\dfrac{\Delta g}{\Delta x} = 24$;

$[6, 8]$: $\dfrac{\Delta f}{\Delta x} = 15$, $\dfrac{\Delta g}{\Delta x} = 96$

(c) $\dfrac{\Delta f}{\Delta x} = 24$, $\dfrac{\Delta g}{\Delta x} = 1536$

47. (a)

(b) domain: $(0, \infty)$; range: $(-\infty, \infty)$; x-intercept: 1; no y-intercept; asymptote: $x = 0$ **49. (a)** 1.4

(b) 1.41   **51. (a)** 1.5   **(b)** 1.52
53. (a) 1.7   **(b)** 1.73   **55. (a)** 1.6
(b) 1.62   **57.** $x \approx 0.3$
59. $x \approx 0.7$
61. (a)

(b)

(c)

63. (a)

(b) $(0.3, 2)$ is the point of intersection

Exercise Set 5.2

1. False   **3.** False   **5.** True
7. False   **9. (a)** two   **(b)** two
11. domain: $(-\infty, \infty)$; range: $(0, \infty)$; no x-intercept; y-intercept: 1; asymptote: $y = 0$

13. domain: $(-\infty, \infty)$; range: $(-\infty, 0)$; no x-intercept; y-intercept: -1; asymptote: $y = 0$

15. domain: $(-\infty, \infty)$; range: $(1, \infty)$; no x-intercept; y-intercept: 2; asymptote: $y = 1$

17. domain: $(-\infty, \infty)$; range: $(1, \infty)$; no x-intercept; y-intercept: $e + 1$; asymptote: $y = 1$

19. domain: $(-\infty, \infty)$; range: $(-\infty, e)$; x-intercept: 1; y-intercept: $e - 1$; asymptote: $y = e$

21. (a) $y = e^x$ displaced to the left 2 units

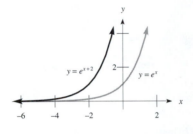

(b) $y = e^x$ displaced to the left 2 units, then reflected across the y-axis

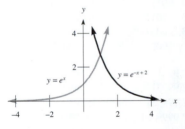

23. $\dfrac{\Delta f}{\Delta x} = 1, \dfrac{\Delta g}{\Delta x} = 1, \dfrac{\Delta h}{\Delta x} \approx 1.7$

25. (a)

(b)

(c)

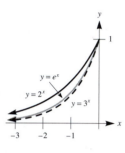

27. instantaneous rate of change = 6

Interval	$\Delta f/\Delta x$ for $f(x) = x^2$
[3, 3.1]	6.1
[3, 3.01]	6.01
[3, 3.001]	6.001
[3, 3.0001]	6.0001
[3, 3.00001]	6.00001
[2.9, 3]	5.9
[2.99, 3]	5.99
[2.999, 3]	5.999
[2.9999, 3]	5.9999
[2.99999, 3]	5.99999

29.

Interval	$\Delta f/\Delta x$ for $f(x) = e^x$
[1.9, 2]	7.031617
[1.99, 2]	7.352234
[1.999, 2]	7.385363
[1.9999, 2]	7.388687
[1.99999, 2]	7.389019
[1.999999, 2]	7.389053
[1.9999999, 2]	7.389056

31. (a) 255 bacteria per hour

(b) $t = 5$ hours: 148 bacteria per hour; $t = 5.5$ hours: 245 bacteria per hour

33. P: $e^{-1} \approx 0.37$; Q: 1; R: $e^{0.5} \approx 1.65$

35. A, D, E, G   **37.** B, D, E, G

39. A, D, F, G, H   **41.** A, D, F, G, H   **43.** graph: $e^{0.1} \approx 1.1$; calculator: $e^{0.1} \approx 1.105$

45. graph: $e^{-0.3} \approx 0.75$; calculator: $e^{-0.3} \approx 0.741$

47. graph: $e^{-1} \approx 0.35$; calculator: $e^{-1} \approx 0.368$

49. graph: $\dfrac{1}{\sqrt{e}} = e^{-0.5} \approx 0.6$; calculator: $\dfrac{1}{\sqrt{e}} = e^{-0.5} \approx 0.607$

51. (a) $x \approx 0.4$ **(b)** $\ln 1.5 \approx 0.405$

53. (a) $x \approx 0.6$ **(b)** $\ln 1.8 \approx 0.588$

55. (a) $\cosh(0) = 1$, $\cosh(1) \approx 1.54$, $\cosh(-1) \approx 1.54$

(b) $(-\infty, \infty)$ **(c)** The graph is symmetric about the y-axis.

(d)

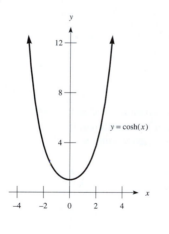

57. (a) $\sinh(0) = 0$, $\sinh(1) \approx 1.18$, $\sinh(-1) \approx -1.18$ **(b)** $(-\infty, \infty)$

(c) The graph is symmetric about the origin.

(d)

(e) three points

59. (a) e^{π} is larger   **(b)** e^{π} is larger
61. (a)

(b) domain: $(0, \infty)$; range: $(-\infty, \infty)$; x-intercept: 1; asymptote: $x = 0$ **(c) (i)** x-intercept: 1; asymptote: $x = 0$

(ii) x-intercept: -1; asymptote: $x = 0$

(iii) x-intercept: 2; asymptote: $x = 1$

63. (a)
$[\cosh(x)]^2 - 2[\sinh(x)]^2 = 1$ is not an identity; solution: $x = 0$

Exercise Set 5.3

1. (a) $f[g(x)] = x$

(b) $g[f(x)] = x$

3. (a) $g(x) = \log_2 x$ **(b)** $2^{\log_2 x} = x$;

$\log_2(2^x) = x$ **(c)** $2^{\log_2 99} = 99$;

$\log_2(2^{-\pi}) = -\pi$ **5. (a)** is one-to-one

(b) not one-to-one   **(c)** is one-to-one

7. $f[f^{-1}(6)] = 6$ **9. (a)** $\log_3 9 = 2$

(b) $\log_{10} 1000 = 3$ **(c)** $\log_7 343 = 3$

(d) $\log_2 \sqrt{2} = \frac{1}{2}$ **11. (a)** $2^5 = 32$

(b) $10^0 = 1$ **(c)** $e^{1/2} = \sqrt{e}$

(d) $e^{-1} = \dfrac{1}{e}$

13.

x	1	10	10^2	10^3	10^{-1}	10^{-2}	10^{-3}
$\log_{10} x$	0	1	2	3	-1	-2	-3

15. (a) $\log_5 30$ is larger   **(b)** $\ln 17$ is larger   **17. (a)** $\frac{3}{2}$ **(b)** $-\frac{5}{2}$ **(c)** $\frac{3}{2}$

19. (a) $x = \frac{1}{16}$ **(b)** $x = e^{-2} \approx 0.14$

21. (a) $(0, \infty)$ **(b)** $(-\infty, \frac{3}{4})$ **(c)** $(-\infty, 0) \cup (0, \infty)$ **(d)** $(0, \infty)$

(e) $(-\infty, -5) \cup (5, \infty)$ **23.** A: $(0, 1)$; B: $(1, 0)$; C: $(4, 2)$; D: $(2, 4)$

25. (a) domain: $(0, \infty)$; range: $(-\infty, \infty)$; x-intercept: 1; no y-intercept; asymptote: $x = 0$

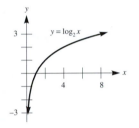

(b) domain: $(0, \infty)$; range: $(-\infty, \infty)$; x-intercept: 1; no y-intercept; asymptote: $x = 0$

(c) domain: $(-\infty, 0)$; range: $(-\infty, \infty)$; x-intercept: -1; no y-intercept; asymptote: $x = 0$

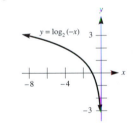

(d) domain: $(-\infty, 0)$; range: $(-\infty, \infty)$; x-intercept: -1; no y-intercept; asymptote: $x = 0$

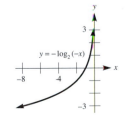

27. domain: $(2, \infty)$; range: $(-\infty, \infty)$; x-intercept: 5; no y-intercept; asymptote: $x = 2$

29. domain: $(-e, \infty)$; range: $(-\infty, \infty)$; x-intercept: $-e + 1$; y-intercept: 1; asymptote: $x = -e$

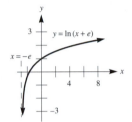

31. (a) 4   **(b)** -1 **(c)** $\frac{1}{2}$

33. (a) $x \approx 1.4$ **(b)** $x \approx 1.398$

35. (a) $x \approx \pm 1.3$

(b) $x \approx \pm 1.266$ **37. (a)** $t \approx -0.3$

(b) $t \approx -0.349$ **39. (a)** $t \approx -0.4$

(b) $t \approx -0.380$ **41. (a)** $(0, 1)$

(b) $(1, \infty)$

(c) $(0, \infty)$

(d) $(-\infty, 0)$

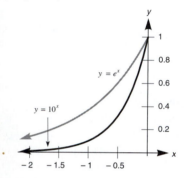

43. The San Salvador quake was 1.6 times stronger. **45.** The second quake is 10^d times stronger.
47. (a) $I = 10^{\beta/10}I_0$ **(b)** The power mower is 10 times more intense. **49. (a)** pH ≈ 3.5; acid
(b) pH $= 0$ **51.** $[H^+] = 10^{-5.9}$
53. A, D, E, H **55.** B, D, E, H
57. A, D, E, H **59.** A, D, E, H
61. $f^{-1}(x) = -1 + \ln x$;
x-intercept: e; asymptote: $x = 0$

63. $x \approx 10^{30}$

65. (a)

Planet	x	y	$\ln x$	$\ln y$
Mercury	0.387	0.241	-0.95	-1.42
Venus	0.723	0.615	-0.32	-0.49
Earth	1.000	1.000	0.00	0.00
Mars	1.523	1.881	0.42	0.63
Jupiter	5.202	11.820	1.65	2.47

(b) $\ln y = 1.50 \ln x$ **(d)**

Planet	x	y (calculated)	y (observed)
Saturn	9.555	29.54	29.46
Uranus	19.22	84.26	84.01
Neptune	30.11	165.22	164.79
Pluto	39.44	247.69	248.5

67. (a)
$y = -24.795459 + 7.453496 \ln x$

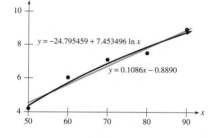

(b) $y = 0.1086x - 0.8890$
(c) logarithmic: 9.529 million; linear: 9.971 million
(d) logarithmic model is closer; logarithmic: 0.11% error; linear: 4.75% error **69. (a)** $(1, \infty)$
(b) $f^{-1}(x) = e^{e^x}$
71. (a)

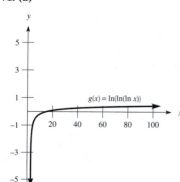

(b) $(-\infty, \infty)$ **(c)** no

Exercise Set 5.4
1. 1 **3.** $\frac{1}{2}$ **5.** 4 **7.** 0 **9.** 4
11. $\log_{10} 60$ **13.** $\log_5 20$
15. (a) $\ln 6$ **(b)** $\ln(\frac{3}{16384})$
17. $\log_b \left[\dfrac{4(1 + x)^3}{(1 - x)^{3/2}} \right]$
19. $\log_{10} \left[\dfrac{27\sqrt{x + 1}}{(x^2 + 1)^6} \right]$
21. (a) $2\log_{10} x - \log_{10}(1 + x^2)$
(b) $2\ln x - \frac{1}{2}\ln(1 + x^2)$
23. (a)
$\frac{1}{2}\log_{10}(3 + x) + \frac{1}{2}\log_{10}(3 - x)$
(b) $\frac{1}{2}\ln(2 + x) + \frac{1}{2}\ln(2 - x)$
$\quad - \ln(x - 1) - \frac{3}{2}\ln(x + 1)$
25. (a) $\frac{1}{2}\log_b x - \frac{1}{2}$
(b) $\ln(1 + x^2) + \ln(1 + x^4)$
$\quad + \ln(1 + x^6)$ **27. (a)** $A + B$
(b) $-A - B$ **(c)** $3B$ **(d)** $-3B$
29. (a) $C - B$ **(b)** $B - C$
(c) $C - 2B$ **(d)** $C - 4A$
31. (a) $\dfrac{1}{B}$ **(b)** $\dfrac{A + C + 1}{B}$
33. (a) $\dfrac{A}{B + 1}$ **(b)** $\dfrac{B + C}{B + 1}$
35. (a) 1 **(b)** 1 **37. (a)**
$a + 2b + 3c$ **(b)** $1 + \frac{1}{2}a$
(c) $\frac{1}{2}(1 + a + b + c)$
(d) $1 + a - \frac{1}{2}b - \frac{1}{2}c$ **39. (a)** $1 + t$
(b) $u - t$ **(c)** $\frac{3}{2}t + \frac{1}{2}u - 1$
(d) $2 + t + \frac{1}{2}u$
41. (a) x-intercept: $x \approx 2.32$

(b) *x*-intercept: $x \approx 4.64$

43. $x = \dfrac{\ln 5 - \ln 2 + 1}{2}$

45. $x = \dfrac{\ln 13}{\ln 2}$

47. $x = \dfrac{1}{\ln 10}$

49. $x = \pm\sqrt{1 + \dfrac{\log 12}{\log 3}} \approx \pm 1.806$

51. $\dfrac{\log_{10} 5}{\log_{10} 2}$

53. $\dfrac{\log_{10} 3}{\log_{10} e}$

55. $\dfrac{\log_{10} 2}{\log_{10} b}$

57. $\dfrac{\ln 6}{\ln 10}$

59. $\dfrac{1}{\ln 10}$

61. $\dfrac{\ln(\ln x) - \ln(\ln 10)}{\ln 10}$

63. (a) true **(b)** true **(c)** true
(d) false **(e)** true **(f)** false
(g) true **(h)** false **(i)** true
(j) false **(k)** false **(l)** true
(m) true **65. (f)** 2345.6
(g) 0.123456 **67.**

x	0.1	0.05	0.005	0.0005
$\ln(1+x)$	0.095310	0.048790	0.004987	0.000499

69. $a = 4, b = -\dfrac{\ln 243}{2.5}$

71. (a) $x = \dfrac{\log_{10} 3}{\log_{10} 2}$

(b) $\dfrac{\log_{10} 3}{\log_{10} 2} \approx 1.585, \dfrac{\ln 3}{\ln 2} \approx 1.585$

75. x^3

79. (c) $\log_\pi 2 + \dfrac{1}{\log_\pi 2}$ is larger

Exercise Set 5.5

1. $x = \dfrac{\ln 3}{2\ln 3 - \ln 5} \approx 1.869$

3. $x = e^{e^{1.5}} \approx 88.384$ **5.** $x = \pm 8$

7. $x = \dfrac{3 - \sqrt{809}}{4} \approx -6.361,$

$x = \dfrac{3 + \sqrt{809}}{4} \approx 7.861$

9. $x = \log 2 \approx 0.301$ **11. (a)** true
for all $x > 0$
(b) $x = e^{-\sqrt{3}} \approx 0.177,$
$x = e^{\sqrt{3}} \approx 5.652$ **13. (a)** true for
all $x > 0$ **(b)** $x = 6^{1/5} \approx 1.431$
15. true for all $x > 0$
17. $x = \sqrt{3} \approx 1.732$ **19.** no real
solutions **21.** no real solutions
23. $x = \frac{7}{3}$ **25. (a)** no real
solutions **(b)** $x = 0$
(c) $x = \ln 3 \approx 1.099$
(d) $x = \ln(1 + \sqrt{5}) \approx 1.174$

27. $x = \ln\left(\dfrac{1 + \sqrt{5}}{2}\right) \approx 0.481$

29. $x = \dfrac{3\ln 5}{5\ln 2 - \ln 3 - \ln 5} \approx 6.372$

31. $x = 2$ **33.** $x = \frac{1}{2}$ **35.** $x = 3$

37. $x = \frac{203}{99}$ **39.** $x = 3$ **41.** $x = 7$

43. (a) $x = \dfrac{10^y}{3(10^y) - 1}$

(b) $x = \dfrac{1 - y}{2}$ **45. (a)** one

solution **(b)** $x \approx 0.57$

47. (a) one solution **(b)** $x \approx 1.49$

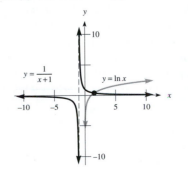

49. (a) two solutions
(b) $x \approx 1.56, x \approx 3.15$

51. $x \le -1$ **53.** $x \ge 2.375$
55. $x < -1.078$ **57.** $x \ge 2.605$
59. all real numbers
61. $-1 < x < \frac{1}{2}$ **63.** $x \le -1$ or
$x \ge 5$ **65.** $0 < x < 1$
67. (a) $(4, \infty)$ **(b)** $4 < x \le 7$
69. $0 < x < 3$ **71.** $x = \frac{1}{2}, 2$
73. The graphs are identical

75. $x = e^2 \approx 7.389$ **77.** all real

$x > 0$ **79.** $x = \beta^{-1/\alpha}$

81. $x = \dfrac{1}{k}\ln\dfrac{y}{A}$

83. $x = -\dfrac{1}{k}\ln\left(\dfrac{a - y}{by}\right)$

85. (b) $x \approx -1.355$ **87. (a)(b)**
$x \approx \pm 31.654$ **(b)(b)** $x \approx -30.639,$
32.639 **89.** Since $e > 0, 2 - e < 2,$

thus $\dfrac{2 - e}{3} < \frac{2}{3}$.

91. $(-3, -\sqrt{6}) \cup (\sqrt{6}, 3)$
93. $x = 1, x = 2$

95. $x = \dfrac{\ln(a - b)}{\ln(a + b)}$ **97.** $x = \dfrac{\pm b}{a + b}$

Exercise Set 5.6

1. $1009.98   **3.** 8.45%   **5.** $767.27
7. (a) $3869.68   **(b)** $4006.39
9. 13 quarters   **11.** $3487.50
13. (a) 4.4 years   **(b)** 4.43 years
15. 16 years   **17.** $2610.23
19. 5.83%   **21.** 6% investment
23. (a) 14 years   **(b)** 13.86 years
(c) 1.01%   **25.** $26.5 trillion
27. (a) 14 years
(b)

29. (b) $y = 0.1848x + 0.8764$

(c) $y = 0.906768(1.159292)^x$ or
$y = 0.906768e^{0.147809x}$

31. (b) $y = 0.3666x - 0.3858$

(c) $y = -6.81755 + 4.371057\ln x$
33. (a) quadratic model
(b) logarithmic:
$y = 31.216186 + 25.969647\ln x$;
quadratic: $y = -0.134362x^2 + 6.990301x + 30.04016$
power: $y = 36.171279 \cdot x^{0.380273}$;
linear: $y = 4.172184x + 38.603726$
(c) the power model is closest;
1992: 0% error; 1995: 1.4% error;
2000: 0.7% error

Year	PPI Linear	PPI Logarithmic	PPI Quadratic	PPI Power	PPI Actual
1992	130.4	111.5	118.8	117.2	117.2
1995	142.9	114.8	120.8	123.0	124.8
2000	163.8	119.5	118.8	131.8	132.7

Exercise Set 5.7

1. (a) $k \approx 0.3209$ **(b)** 9951
bacteria   **(c)** 5.0 hours
3. $N_0 = 2000; k \approx 0.1769$ **5. (a)**

Region	World	More dev.	Less dev.
1995 Population (billions)	5.702	1.169	4.533
Precent of Population in 1995	100	20.5	79.5
Relative Growth Rate (percent per year)	1.5	0.2	1.9
Year 2000 Population (billions)	6.146	1.181	4.985
Percent of World Population in 2000	100	19.22	81.11

(b) world: 1.3% error; more
developed: 0.3% error; less
developed: 2.1% error
7. (a) Chad: $N(t) = 8.0e^{0.033t}$;
United Kingdom: $N(t) = 59.8e^{0.001t}$
(b) 60 years   **9. (a)** Niger: 15.8
million; Portugal: 10.2 million
(b) 13 years; Portugal: 10.1 million
11. (a) 6.332 billion   **(b)** higher;
4% error   **13. (a)** 116 years
(b) 43 years   **(c)** 27 years
(d) 19 years   **15. (a)**

Year	1998 ($t = 10$)	2000 ($t = 12$)
Concentration of Carbon dioxide (ppm)	365.6	368.6

(b) 1998: too low, 0.3% error; 2000:
too high, 0.1% error   **17. (a)** New
York: $N(t) = 18.976e^{0.006t}$;
Arizona: $N(t) = 5.131e^{0.04t}$
(b) year 2040   **19. (a)**

Region	1990 Population (millions)	Growth Rate (%)	2025 Population
North America	275.2	0.7	351.6
Soviet Union	291.3	0.7	372.2
Europe	499.5	0.2	535.7
Nigeria	113.3	3.1	335.3

(b) 222.0 million   **(c)** North
America: 76.4 mil; Soviet Union:
80.9 mil; Europe: 36.2 mil;
combined: 193.5 mil   **(d)** Our
results support this projection.
21. By the end of the year 2000,
the population of Mexico exceeded
100 million people.   **23. (a)**

t(seconds)	0	550	1100	1650	2200
N(grams)	8	4	2	1	0.5

(b)

t(years)	0	4.9×10^9	9.8×10^9	14.7×10^9	19.6×10^9
N(grams)	10	5	2.5	1.25	0.625

25. 0.55 gram   **27. (a)** 4.29 grams
(b) 79 hours
29. (a)

(b)

31. (a) 1.53 grams   **(b)** 43 years
33. (a) 73,000 yrs   **(b)** 10 half-lives
(c) 10 half-lives   **35. (a)**
$k \approx -0.0248$ **(b)** 279 years
(c) 280 years   **37.** 0.02799997
ounces   **39. (a)** 8.5 billion; it is
greater   **(b)** 0.934% per year
41. (b) $T \approx 32$ yrs; year 2004
(c) $y = 0.4606x + 24.0385$
(d) $T \approx 32$ yrs; year 2025
43. (a) $T \approx 34$ yrs; year 2033
(b) $T \approx 67$ yrs; year 2066
(c) $T \approx 52$ yrs; year 2051
47. 4.181 billion years old
49. 15,505 years old   **51.** 8,000
years old; the site is older
53. years 10–5 B.C., which fit in
the historical range   **55.(a)**

	N(−1)	N(0)	N(1)	N(4)	N(5)
From graph	0.25	0.5	1.0	3.5	3.75
From calculator	0.176	0.444	1.014	3.489	3.795

(b) $N(10) \approx 3.99854773$,
$N(15) \approx 3.99999021$,
$N(20) \approx 3.99999993$ **(c)** $t \approx 3$
(d) $t = \ln 24 \approx 3.178$
57. (a) $a = 4$
(b) $b = \ln \frac{8}{3} \approx 0.9808$
59. (a) $N(t) = \dfrac{162}{1 + 161e^{-0.50t}}$
(b) $N(4) \approx 7.1$, $N(8) \approx 41.0$,
$N(12) \approx 115.8$, $N(16) \approx 153.7$;
$N(4)$ and $N(8)$ are lower; $N(12)$ and
$N(16)$ **(c)** 10 days 4 hours
61. (b) If k is close to 0, then
$e^k - 1 \approx (k + 1) - 1 = k$.

Chapter 5 Review Exercises

1. $\log_5 126$ **3.** $\dfrac{4\ln 1.5}{\ln 1.25}$ hours

5. f is not one-to-one

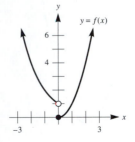

7. $x = \dfrac{e + 1}{e - 1}$ **9.** $y = e^x$: domain:
$(-\infty, \infty)$, range: $(0, \infty)$; $y = \ln x$:
domain: $(0, \infty)$, range: $(-\infty, \infty)$

11. $-\frac{3}{2}$ **13.** $k \approx -0.05$
15. $x = 2 - \ln \frac{12}{5}$ **17.** year 2015
19. doubling time: 7 years

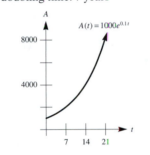

21. horizontal asymptote: $y = 0$; no
vertical asymptote; no x-intercept;
y-intercept: 1

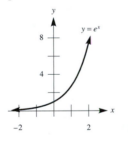

23. no horizontal asymptote;
vertical asymptote: $x = 0$;
x-intercept: 1; no y-intercept

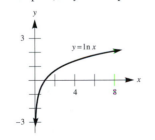

25. horizontal asymptote: $y = 1$; no
vertical asymptote; no x-intercept;
y-intercept: 3

27. horizontal asymptote: $y = 0$; no
vertical asymptote; no x-intercept;
y-intercept: 1

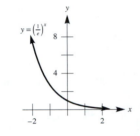

29. horizontal asymptote: $y = 1$; no
vertical asymptote; no x-intercept;
y-intercept: $e + 1$

31. no asymptotes; x- and
y-intercepts: 0

33. $x = 4$ **35.** $x = 3$ **37.** $x = \frac{2}{3}$
39. $x = \sqrt[3]{3}$ **41.** $x = \dfrac{1 - 2\ln 3}{10}$

43. $x = \frac{200}{99}$ **45.** $x = 2$ **47.** all real numbers $x > 0$ **49.** $x = 1$ **51.** $\frac{1}{2}$
53. $\frac{1}{5}$ **55.** -1 **57.** 16 **59.** 4 **61.** 2
63. 2 **65.** $\frac{9}{14}$ **67.** $2a + 3b + \frac{c}{2}$
69. $8a + 4b$ **71.** between 2 and 3
73. between 2 and 3 **75.** -3 and -2
77. (a) third quadrant

(b) $x = \dfrac{-2e}{e + 1} \approx -1.46$

79. $k = \dfrac{\ln\frac{1}{2}}{T}$ **81.** 6.25% remaining

83. $\dfrac{d\ln\frac{c}{b}}{\ln\frac{1}{2}}$ days **85.** $\log_{10} 2$

87. $\ln 10$ **89** $\ln x^a y^b$

91. $\frac{1}{2}\ln(x - 3) + \frac{1}{2}\ln(x + 4)$

93. $3\log_{10} x - \frac{1}{2}\log_{10}(1 + x)$

95. $3\ln(1 + 2e) - 3\ln(1 - 2e)$

97. $\dfrac{\ln 2}{\ln\left(1 + \frac{R}{100}\right)}$ years

99. 9.92% **101.** $\dfrac{100\ln 2}{R}$ years

103. (a) $7\frac{3}{4}$ years

(b) $\dfrac{\ln n}{\ln\left(1 + \frac{R}{400}\right)}$ years

105. (a) $(0, \infty)$ **(b)** $[1, \infty)$
107. $(0, e^2) \cup (e^2, \infty)$

Chapter 5 Test

1. domain: $(-\infty, \infty)$; range:
$(-3, \infty)$; x-intercept: $-\frac{\ln 3}{\ln 2}$;
y-intercept: -2; asymptote: $y = -3$

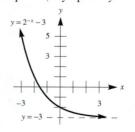

2. $\dfrac{\ln\frac{5}{3}}{\ln 1.033} \approx 16$ hrs **3.** $\log_2 17$

4. $\dfrac{\ln 15}{\ln 2}$ **5.** 10^{12} **6.** 4 years

7. (a) $(0, \infty)$ **(b)**

8. $3a - \frac{1}{2}b$ **9. (a)** $-\frac{1}{2}$ **(b)** -1
10. $x = 0, -2, 2$
11. (a) $k \approx -0.1733$ **(b)** 0.35 grams
12. $\ln\dfrac{x^2}{(x^2 + 1)^{1/3}}$ **13.** no solution
14. $g^{-1}(x) = 10^x + 1$; range: $(1, \infty)$
15. (a) 12 years
(b)

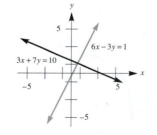

16. (a) $(0, \infty)$ **(b)** $x = 1, e^2$
17. (a) $\dfrac{1}{\ln\frac{4}{3}} \approx 3.48$ **(b)** 4
18. $x = 4$ **19.** $x = \ln\frac{3}{2} \approx 0.405$
20. (a) $x > \dfrac{\ln 1.6}{\ln 0.3} \approx -0.39$
(b) $(3, 4]$

CHAPTER 6
Exercise Set 6.1

1. (a) yes **(b)** no **(c)** yes
(d) yes **3.** yes **5.** no
7. consistent (one solution)

9. inconsistent (no solution)

11. $(3, 5)$ **13.** $\left(-\frac{1}{28}, \frac{39}{28}\right)$
15. $\left(-\frac{34}{49}, -\frac{20}{49}\right)$ **17.** $\left(x, \frac{3 - 4x}{6}\right)$,
$x =$ any real number
19. (a) $(16.30, -24.5)$

(b) $\left(\frac{5297}{325}, -\frac{7952}{325}\right) \approx (16.30, -24.47)$
21. $\left(-\frac{2}{9}, \frac{23}{27}\right)$ **23.** $\left(-\frac{283}{242}, -\frac{3}{121}\right)$
25. $\left(-\frac{226}{25}, -\frac{939}{50}\right)$ **27.** $\left(\frac{5}{8}, 0\right)$
29. $y = x^2 + 3x + 4$
31. $A = -28, B = -22$
33. (a) $p = \$6$: shortage of 6000 items;
$p = \$12$: shortage of 2400 items
(b) surplus of 2400 items **(c)** equilibrium price: \$16; equilibrium quantity:
3200 items **35.** 80 cc of 10% solution,
120 cc of 35% solution **37.** 8 pounds
of \$5.20 coffee, 8 pounds of \$5.80 coffee
39. (a)

41. $a = 3, b = 2$

43. $\left(\dfrac{ab}{a + b}, \dfrac{ab}{a + b}\right); a \neq \pm b$

45. $\left(\dfrac{a + b}{ab}, \dfrac{-1}{ab}\right); a \neq 0$ and $b \neq 0$

47. $\left(-\frac{1}{11}, \frac{1}{9}\right)$ **49.** $(5, 2)$

51. $(e^{-2}, e^{-3}) \approx (0.14, 0.05)$

53. $(\ln 5, 0) \approx (1.61, 0)$

55. $(4, -8), (4, 2), (-1, -8), (-1, 2)$

57. 59

59. $\left(\dfrac{1}{b}, \dfrac{1}{a}\right)$ **61. (a)** $k = -\frac{29}{7}$ and 2

(b) $k = -\frac{29}{7}$:

$k = 2$:

Exercise Set 6.2

1. $(-3, -2, -1)$ **3.** $\left(-\frac{1}{60}, -\frac{2}{15}, \frac{3}{5}\right)$

5. $\left(\frac{25}{36}, \frac{5}{9}, -\frac{1}{3}\right)$ **7.** $\left(x, \frac{x}{8}, 0\right), x =$ any

real number **9.** $(-1, 0, 1, -5)$

11. $\left(\frac{11}{3}, \frac{8}{3}, \frac{17}{3}\right)$ **13.** $(1, 0, 1)$

15. $(2, 3, 1)$ **17.** inconsistent

(no solution) **19.** $\left(\dfrac{z + 1}{-7}, \dfrac{5(z + 1)}{7}, z\right)$,

$z =$ any real number

21. $\left(\dfrac{11 - 5z}{7}, \dfrac{-3z - 6}{7}, z\right), z =$ any real

number **23.** $(4, 1, -3, 2)$

25. $\left(\dfrac{17 - 17z}{5}, \dfrac{8z - 3}{5}, z\right), z =$ any real

number

27. $\left(\dfrac{12 + 10w}{11}, \dfrac{146 + 19w}{55}, \dfrac{61w + 159}{55}, w\right)$,

$w =$ any real number

29. $\left(-\frac{5z}{12}, \frac{2z + 3}{3}, z\right), z =$ any real

number **31.** 6 Type I jets, 3 Type

II jets, 5 Type III jets

33. (a) $a + b + c = -2$

$a - b + c = 0$

$4a + 2b + c = 3$ **(b)** $a = 2$,

$b = -1, c = -3$

(c)

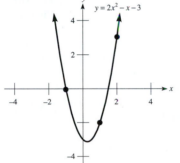

(d) $y = 2x^2 - x - 3$ **35.** 120 util-

ity chairs, 24 secretarial chairs,

10 managerial chairs

37. (a) $R(t) = -0.3t^2 + 3t + 1$

(b) December; $850,000

39. $x = \ln a, y = \ln 2a, z = \ln\frac{a}{2}$

41. $T_1 = 42, T_2 = 43, T_3 = 40$

43. 60 miles

Exercise Set 6.3

1. (a) two by three (2×3)

(b) three by two (3×2)

3. five by four (5×4)

5. coefficient matrix is

$\begin{pmatrix} 2 & 3 & 4 \\ 5 & 6 & 7 \\ 8 & 9 & 10 \end{pmatrix}$; augmented matrix:

$\begin{pmatrix} 2 & 3 & 4 & 10 \\ 5 & 6 & 7 & 9 \\ 8 & 9 & 10 & 8 \end{pmatrix}$

7. coefficient matrix:

$\begin{pmatrix} 1 & 0 & 1 & 1 \\ 1 & 1 & 0 & 2 \\ 0 & 1 & 1 & 1 \\ 2 & -1 & -1 & 0 \end{pmatrix}$;

augmented matrix:

$\begin{pmatrix} 1 & 0 & 1 & 1 & -1 \\ 1 & 1 & 0 & 2 & 0 \\ 0 & 1 & 1 & 1 & 1 \\ 2 & -1 & -1 & 0 & 2 \end{pmatrix}$

9. $(-1, -2, 3)$ **11.** $(-5, 1, 3)$

13. $(3, 0, -7)$ **15.** $(8, 9, -1)$

17. $\left(\dfrac{9z + 5}{19}, \dfrac{31z - 6}{19}, z\right), z =$ any real

number **19.** $(2, -1, 0, 3)$

21. inconsistent (no solution)

23. $\begin{pmatrix} 3 & 2 \\ 2 & 4 \end{pmatrix}$ **25.** $\begin{pmatrix} 6 & 4 \\ 4 & 8 \end{pmatrix}$

27. $\begin{pmatrix} 11 & -2 \\ 11 & 1 \end{pmatrix}$ **29.** $\begin{pmatrix} 2 & 3 \\ -1 & 4 \end{pmatrix}$

31. not defined

33. $\begin{pmatrix} 10 & -2 \\ -8 & 0 \\ 4 & 6 \end{pmatrix}$

35. $\begin{pmatrix} 2 & 4 & 11 \\ -12 & 16 & 19 \\ 14 & 12 & 43 \end{pmatrix}$

37. $\begin{pmatrix} -9 & 10 & 10 \\ 4 & -8 & -12 \\ 10 & 4 & 21 \end{pmatrix}$

39. not defined

41. $\begin{pmatrix} 0 & 0 & 0 \\ 0 & 0 & 0 \\ 0 & 0 & 0 \end{pmatrix}$ **43.** $\begin{pmatrix} 4 & 2 \\ 2 & 5 \end{pmatrix}$

45. not defined

47. $\begin{pmatrix} 1 & 18 \\ -6 & 13 \end{pmatrix}$ **49.** $\begin{pmatrix} -16 & 75 \\ -25 & 34 \end{pmatrix}$

51. (a) $\begin{pmatrix} -13 & 1 & 40 \\ 43 & 17 & 0 \\ 89 & 61 & 60 \end{pmatrix}$

(b) $\begin{pmatrix} -13 & 1 & 40 \\ 43 & 17 & 0 \\ 89 & 61 & 60 \end{pmatrix}$

(c) $\begin{pmatrix} -52 & -82 & 61 \\ 87 & 141 & 0 \\ 216 & 318 & 165 \end{pmatrix}$

(d) $\begin{pmatrix} -52 & -82 & 61 \\ 87 & 141 & 0 \\ 216 & 318 & 165 \end{pmatrix}$

53. $\begin{pmatrix} -2474 & 1182 \\ 2358 & -144 \end{pmatrix}$

55. $\begin{pmatrix} 0.5 & 0 & 0 & 0 \\ 0 & 0.5 & 0 & 0 \\ 0 & 0 & 0.5 & 0 \\ 0 & 0 & 0 & 0.5 \end{pmatrix}$

57. (a) $\begin{pmatrix} 16 & 20 \\ 24 & 28 \end{pmatrix}$ **(b)** $\begin{pmatrix} 18 & 26 \\ 18 & 26 \end{pmatrix}$

(c) $\begin{pmatrix} 14 & 14 \\ 30 & 30 \end{pmatrix}$ **(d)** $\begin{pmatrix} 18 & 26 \\ 18 & 26 \end{pmatrix}$

59. (a) $\begin{pmatrix} x \\ -y \end{pmatrix}$ **(b)** $\begin{pmatrix} -x \\ y \end{pmatrix}$

(c) $\begin{pmatrix} -x \\ -y \end{pmatrix}$; This would represent a reflection about the origin.

Exercise Set 6.4

5. $\begin{pmatrix} -5 & 9 \\ 4 & -7 \end{pmatrix}$ **7.** $\begin{pmatrix} -\frac{6}{23} & \frac{1}{23} \\ \frac{5}{23} & \frac{3}{23} \end{pmatrix}$

9. no A^{-1} exists **11.** $\begin{pmatrix} 2 & -3 \\ 1 & 3 \end{pmatrix}$

13. $\begin{pmatrix} 2 & -1 \\ -3 & 2 \end{pmatrix}$ **15.** $\begin{pmatrix} \frac{6}{11} & 1 \\ -\frac{1}{11} & 0 \end{pmatrix}$

17. does not exist

19. $\begin{pmatrix} -1 & 2 & -3 \\ 2 & 1 & 0 \\ 4 & -2 & 5 \end{pmatrix}$

21. $\begin{pmatrix} 5 & -\frac{10}{3} & 1 \\ 0 & \frac{1}{3} & 0 \\ 4 & -\frac{8}{3} & 1 \end{pmatrix}$

23. $\begin{pmatrix} 2 & 1 & 4 \\ 3 & 2 & 5 \\ 0 & -1 & 1 \end{pmatrix}$

25. does not exist

27. $A^{-1} = \begin{bmatrix} -2.5 & 1.5 \\ 2 & -1 \end{bmatrix}$

29. D^{-1} does not exist
31. $(DE)^{-1}$ **33. (a)** $x = -1$ and $y = 1$ **(b)** $x = -132$ and $y = 48$
35. $x = 2, y = -1, z = 4$

37. (a) $\begin{pmatrix} -7 & 4 \\ 2 & -1 \end{pmatrix}$ **(b)** $(-1, 2)$

39. (a) $\begin{pmatrix} -1 & 0 & 2 \\ 3 & 1 & -6 \\ -2 & -1 & 5 \end{pmatrix}$

(b) $(21, -64, 55)$

41. (a) $\begin{pmatrix} \frac{31}{108} & \frac{2}{27} & \frac{7}{18} & -\frac{11}{108} \\ -\frac{53}{162} & \frac{1}{81} & -\frac{5}{27} & \frac{31}{162} \\ -\frac{55}{54} & -\frac{1}{27} & \frac{5}{9} & \frac{23}{54} \\ \frac{131}{54} & -\frac{4}{27} & -\frac{7}{9} & -\frac{43}{54} \end{pmatrix}$

(b) $(\frac{10}{9}, -\frac{22}{27}, -\frac{59}{9}, \frac{88}{9})$

45. 324, 286, 225, 200, 241, 212
47. 41, 15, 38, 14, 28, 28, 38, −23, 15 **49.** SEE JOE **51.** STUDY

53. $\begin{pmatrix} 4 & -6 & 4 & -1 \\ -6 & 14 & -11 & 3 \\ 4 & -11 & 10 & -3 \\ -1 & 3 & -3 & 1 \end{pmatrix}$

55. $A^{-1} = \begin{pmatrix} \frac{d}{ad-bc} & \frac{-b}{ad-bc} \\ \frac{-c}{ad-bc} & \frac{a}{ad-bc} \end{pmatrix}$

Exercise Set 6.5

1. (a) 29 **(b)** −29 **3. (a)** 0
(b) 0 **5.** −1 **7.** −60 **9.** 9
11. (a) 314 **(b)** 674 **(c)** part (b)
13. (a) 0 **(b)** 0 **(c)** 0 **(d)** 0
15. 0 **17.** −3 **19.** 0 **21.** 570
23. 0 **25.** 12 **27. (a)** The right-hand determinant should be 10 times the left-hand determinant.
(b) det $A = 206$ and det $B = 2060$
29. (a) 20 **(b)** 20 **31. (a)** 174
(b) 174
33. $(y - x)(z - x)(z - y)$ **35.** xy
39. $(1, 1, 2)$ **41.** $(2, -3, -3)$
43. $(0, 0, 0)$
45. $(13 - \frac{11}{3}y, y, 13 - 4y)$, $y =$ any real number **47.** $(1, 0, -10, 2)$
49. $x = 4, -4, -1$

63. $\left(\frac{k(k-b)(k-c)}{a(a-b)(a-c)}, \right.$

$\left. \frac{k(k-a)(k-c)}{b(b-a)(b-c)}, \frac{k(k-a)(k-b)}{c(c-a)(c-b)} \right)$

65. $y = 2x + 5$

69. (b) $\begin{pmatrix} -\frac{9}{61} & \frac{7}{61} \\ \frac{1}{61} & \frac{6}{61} \end{pmatrix}$

Exercise Set 6.6

1. $(0, 0), (3, 9)$
3. $(2\sqrt{6}, 1), (-2\sqrt{6}, 1)$
5. $(-1, -1)$ **7.** $(2, 3), (2, -3),$ $(-2, 3), (-2, -3)$ **9.** $(1, 0), (-1, 0)$
11. $(\frac{-1 + \sqrt{65}}{8}, \frac{1 + \sqrt{65}}{2}),$ $(\frac{-1 - \sqrt{65}}{8}, \frac{1 - \sqrt{65}}{2})$ **13.** $(\frac{\sqrt{17}}{17}, 1),$ $(\frac{\sqrt{17}}{17}, -1), (\frac{-\sqrt{17}}{17}, 1), (\frac{-\sqrt{17}}{17}, -1)$
15. $(1, 0), (4, -\sqrt{3})$ **17.** $(2, 4)$
19. $(100, 1000), (100, \frac{1}{1000}),$ $(\frac{1}{100}, 1000), (\frac{1}{100}, \frac{1}{1000})$
21. $\left(\frac{2\ln 2 - 5\ln 3}{\ln 2 - \ln 3}, \frac{3\ln 2}{\ln 2 - \ln 3} \right)$

23. (a) $(2.19, -2.81),$ $(-3.19, -8.19)$

(b) $\left(\frac{-1 + \sqrt{29}}{2}, \frac{-11 + \sqrt{29}}{2} \right)$ $\approx (2.193, -2.807),$ $\left(\frac{-1 - \sqrt{29}}{2}, \frac{-11 - \sqrt{29}}{2} \right)$ $\approx (-3.193, -8.193)$
25. (a) $(0.85, 2.36)$

(b) $\left(\frac{32 - 4\sqrt{37}}{9}, \frac{1 + \sqrt{37}}{3} \right)$ $\approx (0.852, 2.361)$

27. (a) $(0.60, 5.30)$

(b) $\left(\frac{\ln\left(\frac{1 + \sqrt{13}}{2} \right)}{\ln 4}, \frac{7 + \sqrt{13}}{2} \right)$ $\approx (0.602, 5.303)$

29. $(-0.82, 0.67)$, $(1.43, 2.04)$

31. $(1, 0)$, $(12.34, 2.51)$

33. $(1.23, 1.86)$, $(6.14, 230.95)$

35. (b) $u = 9, v = 8$ **37.** $\left(\frac{1}{a}, \frac{1}{b}\right)$
39. $(9, 14)$, $(14, 9)$

41. $\dfrac{p - \sqrt{2d^2 - p^2}}{2}$ by

$\dfrac{p - \sqrt{2d^2 - p^2}}{2}$ **43. (a)** $N_0 = \frac{3}{2}$,

$k = \dfrac{\ln 8}{6}$ **(b)** $N_0 = 10^{-1/7}$,

$k = \frac{2}{7}\ln 10$ **45.** $\left(\frac{p^2}{A}, \frac{q^2}{A}, \frac{r^2}{A}\right)$ and

$\left(\frac{-p^2}{A}, \frac{-q^2}{A}, \frac{-r^2}{A}\right)$, where

$A = \sqrt{p^2 + q^2 + r^2}$ **47.** 9 cm and

40 cm   **49.** 3 cm by 20 cm

51. $(1, 2)$, $(-1, -2)$, $(2, 1)$,

$(-2, -1)$ **53.** $\left(\frac{1 + \sqrt{13}}{2}, \frac{-1 + \sqrt{13}}{2}\right)$,

$\left(\frac{-1 + \sqrt{13}}{2}, \frac{1 + \sqrt{13}}{2}\right)$, $\left(\frac{1 - \sqrt{13}}{2}, \frac{-1 - \sqrt{13}}{2}\right)$,

$\left(\frac{-1 - \sqrt{13}}{2}, \frac{1 - \sqrt{13}}{2}\right)$ **55.** $(2, 3)$, $(-2,$

$-3)$, $(1, 2)$, $(-1, -2)$ **57.** $(e^{9/2}, e^3)$

Exercise Set 6.7

1. (a) no **(b)** yes

3.

5.

7.

9.

11.

13.

15.

17.

19.

21.

23. convex and bounded; vertices:
$(0, 0), (7, 0), (3, 8), (0, 5)$

25. convex and bounded; vertices:
$(0, 0), (0, 4), (3, 5), (8, 0)$

27. convex but not bounded;
vertices: $(2, 7), (8, 5)$

29. convex but not bounded;
vertex: $(6, 0)$

31. convex and bounded; vertices:
$(0, 0), (0, 5). (6, 0)$

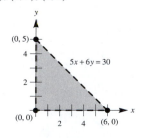

33. convex and bounded; vertices:
$(5, 30), (10, 30), (20, 15), (20, 20)$

35. maximum = 9 at $(3, 4)$; minimum = -5 at $(5, 0)$ **37.** maximum = 4 at $(1, 2)$; minimum = 0 at
$(0, 0)$ **39.** maximum = 113 at
$(6, 6)$; minimum = 27 at $(2, 1)$
41. maximum = 188 at $(6, 7)$; minimum = 100 at $(0, 0)$ **43. (a)** maximum = 819 at $(1, 8)$; minimum = 0
at $(0, 0)$ **(b)** maximum = 826 at
$(6, 7)$; minimum = 0 at $(0, 0)$
45. minimum cost = \$10,150;
W_1 to D_1: 0 cars; W_2 to D_1: 40 cars;
W_1 to D_2: 25 cars; W_2 to D_2: 0 cars

47. maximum profit = \$766.20; 833
units of A, 333 units of B
49. maximum profit = \$23,700;
150 acres of cherry, 450 acres of
regular   **51.** \$4000 in A, \$8000 in B
53. \$4000 in A, \$1000 in B, \$5000 in C
55. 10 pounds of A, 4 pounds of B

57.

59.

61.

Chapter 6 Review Exercises

1. $(3, -5)$ **3.** $(-1, 4)$
5. $(-3, 15)$ **7.** $\left(-\frac{18}{5}, \frac{8}{5}\right)$ **9.** $\left(\frac{2}{3}, -\frac{1}{5}\right)$
11. $(-12, -8)$ **13.** $\left(\frac{1}{3}, -\frac{1}{4}\right)$
15. $\left(\dfrac{-1}{a^2 - 3a + 1}, \dfrac{1 - a}{a^2 - 3a + 1}\right)$;
$a \neq \dfrac{3 \pm \sqrt{5}}{2}$
17. $(a^2, 1 - a^2)$
19. $(a^2 - b^2, a^2 + b^2)$
21. $\left(\dfrac{pq(p + q)}{p^2 + q^2}, \dfrac{p^3 - q^3}{p^2 + q^2}\right)$; $p \neq 0$
and $q \neq 0$ **23.** $(2, 3, 4)$
25. $(-1, -2, 0)$ **27.** $(x, 6 - 2x,$
$-1), x = $ any real number
29. $\left(-\frac{1}{6}z, \frac{1}{2}z, z\right)$, $z = $ any real number   **31.** $(4, 3, -1, 2)$ **33.** 960
35. 24   **39.** $a = 2$ and $b = 1$
43. $\left(x + \frac{17}{6}\right)^2 + \left(y + \frac{8}{3}\right)^2 = \frac{245}{36}$

45. $\begin{pmatrix} 10 & -2 \\ 4 & 26 \end{pmatrix}$ **47.** $\begin{pmatrix} 8 & 4 \\ 4 & 32 \end{pmatrix}$

49. $\begin{pmatrix} 4 & -13 \\ 7 & 41 \end{pmatrix}$

51. $\begin{pmatrix} -3 & -14 \\ -4 & 3 \end{pmatrix}$

53. $\begin{pmatrix} 1 & 1 \\ 1 & 7 \end{pmatrix}$ **55.** $\begin{pmatrix} 1 & -11 \\ 6 & 36 \end{pmatrix}$

57. $\begin{pmatrix} 4 & 3 \\ 10 & 33 \end{pmatrix}$ **59.** $\begin{pmatrix} 4 & -1 \\ 2 & 12 \end{pmatrix}$

61. $\begin{pmatrix} -4 & 13 \\ -7 & -41 \end{pmatrix}$

63. $A^2 = \begin{pmatrix} 0 & 0 & 0 \\ 0 & 0 & 0 \\ ac & 0 & 0 \end{pmatrix}$;

$A^3 = \begin{pmatrix} 0 & 0 & 0 \\ 0 & 0 & 0 \\ 0 & 0 & 0 \end{pmatrix}$

65. (a) $\begin{pmatrix} 10 & -2 & 5 \\ 6 & -1 & 4 \\ 1 & 0 & 1 \end{pmatrix}$

(b) $(16, 15, 4)$ **67.** $\left(-\frac{3}{5}, -\frac{13}{20}, \frac{31}{20}\right)$

69. inconsistent (no solution)

71. $(-5 - 4y, y, -8 - 5y), y =$ any real number   **73.** $(4, 1, -1, 3)$

75. $(0, 0)$ and $(6, 36)$ **77.** $(3, 0)$ and $(-3, 0)$ **79.** $\left(\frac{5\sqrt{2}}{2}, \frac{\sqrt{14}}{2}\right)$, $\left(-\frac{5\sqrt{2}}{2}, \frac{\sqrt{14}}{2}\right), \left(\frac{5\sqrt{2}}{2}, -\frac{\sqrt{14}}{2}\right)$, $\left(-\frac{5\sqrt{2}}{2}, -\frac{\sqrt{14}}{2}\right)$

81. $\left(\frac{-1 + \sqrt{5}}{2}, \frac{\sqrt{-2 + 2\sqrt{5}}}{2}\right)$

83. $(\sqrt{2}, 3\sqrt{2}), (-\sqrt{2}, -3\sqrt{2})$, $\left(\frac{7\sqrt{22}}{33}, \frac{31\sqrt{22}}{33}\right), \left(\frac{-7\sqrt{22}}{33}, \frac{-31\sqrt{22}}{33}\right)$

85. $(5, 2), (5, -4), (1, 2), (1, -4)$

87. $\dfrac{as}{a + b}$ and $\dfrac{bs}{a + b}$

89. 24, 60, 120

91. $\dfrac{\sqrt{n + 2m} + \sqrt{n - 2m}}{2}$ and $\dfrac{\sqrt{n + 2m} - \sqrt{n - 2m}}{2}$, $\dfrac{\sqrt{n - 2m} - \sqrt{n + 2m}}{2}$, and $\dfrac{-\sqrt{n - 2m} - \sqrt{n + 2m}}{2}$

93. neither convex nor bounded

95. convex and bounded

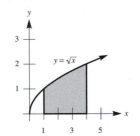

97. maximum value $= \frac{1046}{7}$ at $\left(\frac{68}{7}, \frac{88}{7}\right)$; minimum value $= 6$ at $(0, 0)$

Chapter 6 Test

1. $(0, 3)$ and $\left(-\frac{11}{4}, \frac{81}{16}\right)$ **2.** $(3, -5)$

3. (a) $(1, -1, -3)$ **(b)** $(1, -1, -3)$

4. (a) $\begin{pmatrix} 2 & -10 \\ 3 & -5 \end{pmatrix}$ **(b)** $\begin{pmatrix} 8 & -4 \\ 7 & -6 \end{pmatrix}$

5. equilibrium price $= \$21$; equilibrium quantity $= 684$ units

6. $\left(\frac{1}{116}, -\frac{1}{144}\right)$

7. coefficient matrix:

$\begin{pmatrix} 1 & 1 & -1 \\ 2 & -1 & 2 \\ 1 & -2 & 1 \end{pmatrix}$; augmented

matrix $= \begin{pmatrix} 1 & 1 & -1 & -1 \\ 2 & -1 & 2 & 11 \\ 1 & -2 & 1 & 10 \end{pmatrix}$

8. $(3, -3, 1)$

9. (a)
$$w + x + y + z = 40$$
$$-w + x + y - z = 0$$
$$4w - x - y - z = 0$$
$$-2w + 3x - 2y + 3z = 0$$

(b) $\begin{pmatrix} 1 & 1 & 1 & 1 \\ -1 & 1 & 1 & -1 \\ 4 & -1 & -1 & -1 \\ -2 & 3 & -2 & 3 \end{pmatrix}$

(c) inverse matrix:

$\begin{pmatrix} 0.2 & 0 & 0.2 & 0 \\ 0.1 & 0.5 & 0.2 & 0.2 \\ 0.4 & 0 & -0.2 & -0.2 \\ 0.3 & -0.5 & -0.2 & 0 \end{pmatrix}$

8 hours on math, 4 hours on English, 16 hours on chemistry, 12 hours on economics   **10. (a)** 8

(b) -8 **11.** -1120

12. $\left(\frac{5 + \sqrt{5}}{2}, \frac{5 - \sqrt{5}}{2}\right), \left(\frac{5 - \sqrt{5}}{2}, \frac{5 + \sqrt{5}}{2}\right)$, $\left(\frac{-5 + \sqrt{5}}{2}, \frac{-5 - \sqrt{5}}{2}\right), \left(\frac{-5 - \sqrt{5}}{2}, \frac{-5 + \sqrt{5}}{2}\right)$

13. $\left(1 - \frac{1}{5}C, -\frac{7}{5}C, C\right), C =$ any real number

14. (a) $\begin{pmatrix} 1 & -2 & 3 \\ 2 & -5 & 10 \\ -1 & 2 & -2 \end{pmatrix}$

(b) $(12, 38, -9)$

15.

16. $P = -1$ and $Q = -4$

17. neither bounded nor convex

18. vertices: $(0, 0), (0, 7), (6, 10)$, $\left(\frac{261}{26}, \frac{225}{26}\right), (11, 0)$

19. maximum profit $= \$1100$; 50 small motors, 25 large motors

20. $(e^{-19}, e^{-6}, e^{49})$

CHAPTER 7
Exercise Set 7.1

1.

i^2	i^3	i^4	i^5	i^6	i^7	i^8
-1	$-i$	1	i	-1	$-i$	1

3. (a) real part: 4; imaginary part: 5
(b) real part: 4; imaginary part: -5
(c) real part: $\frac{1}{2}$; imaginary part: -1
(d) real part: 0; imaginary part: 16
5. $c = 4, d = -3$ **7. (a)** $14 - 4i$
(b) $-4 - 8i$ **9. (a)** $19 - 17i$
(b) $19 - 17i$ **(c)** $\frac{11}{26} - \frac{23}{26}i$
(d) $\frac{11}{25} + \frac{23}{25}i$ **11. (a)** $11 - i$
(b) $11 - 7i$ **(c)** 4 **13.** $4 - 2i$
15. $30 + 19i$ **17.** 13
19. $-191 - 163i$ **21.** $19 - 4i$
23. $-70 + 84i$ **25.** $539 + 1140i$
27. $-46 + 9i$ **29.** $\frac{6}{97} + \frac{35}{97}i$
31. $\frac{6}{97} - \frac{35}{97}i$ **33.** $-\frac{5}{13} + \frac{12}{13}i$
35. -4 **37.** $\frac{1}{26} + \frac{5}{26}i$ **39.** $-i$
41. i **43.** -1 **45.** $12i$
47. $-3\sqrt{5}i$ **49.** -35 **51.** $12\sqrt{2}i$
53. (a) -3 **(b)** $x = \dfrac{1}{2} \pm \dfrac{\sqrt{3}}{2}i$
55. (a) -36 **(b)** $z = -\frac{1}{5} \pm \frac{3}{5}i$
57. (a) -23
(b) $z = -\dfrac{3}{4} \pm \dfrac{\sqrt{23}}{4}i$
59. (a) $-\frac{29}{48}$
(b) $z = \dfrac{3}{4} \pm \dfrac{\sqrt{87}}{4}i$
61. (a) 0 **(b)** This verifies the solution.
63. (a) $z + w = (a + c) + (b + d)i$
(b) $z - w = (a - c) + (b - d)i$
(c) $zw = (ac - bd) + (bc + ad)i$
(d) $\dfrac{z}{w} = \dfrac{ac + bd}{c^2 + d^2} + \dfrac{bc - ad}{c^2 + d^2}i$
71. $x = 3, \pm 2i$ **73.** $x = \pm 3,$
$\pm 2\sqrt{i}, \pm 2i\sqrt{i}$ **75.** real part:
$\dfrac{2a^2 - 2b^2}{a^2 + b^2}$; imaginary part: 0
77. real part: 0

Exercise Set 7.2

1. quotient: $x - 5$; remainder:
-11; $x^2 - 8x + 4 = (x - 3)$
$(x - 5) - 11$ **3.** quotient: $x - 11$;

remainder: 53; $x^2 - 6x - 2 =$
$(x + 5)(x - 11) + 53$
5. quotient: $3x^2 - \frac{3}{2}x - \frac{1}{4}$;
remainder: $\frac{13}{4}$; $6x^3 - 2x + 3 =$
$(2x + 1)(3x^2 - \frac{3}{2}x - \frac{1}{4}) + \frac{13}{4}$
7. quotient: $x^4 - 3x^3 + 9x^2 -$
$27x + 81$; remainder: -241;
$x^5 + 2 = (x + 3)(x^4 - 3x^3 +$
$9x^2 - 27x + 81) - 241$
9. quotient: $x^5 + 2x^4 + 4x^3 +$
$8x^2 + 16x + 32$; remainder: 0;
$x^6 - 64 = (x - 2)(x^5 + 2x^4 +$
$4x^3 + 8x^2 + 16x + 32) + 0$
11. quotient: $5x^2 + 15x + 17$;
remainder: $-24x - 83$;
$5x^4 - 3x^2 + 2 = (x^2 - 3x + 5)$
$(5x^2 + 15x + 17) + (-24x - 83)$
13. quotient: $3y - 19$;
remainder: $89y + 35$;
$3y^3 - 4y^2 - 3 = (y^2 + 5y + 2) \times$
$(3y - 19) + (89y + 35)$
15. quotient: $t^2 - 2t - 4$;
remainder: 0;
$t^4 - 4t^3 + 4t^2 - 16 = (t^2 - 2t + 4)$
$(t^2 - 2t - 4) + 0$ **17.** quotient:
$z^4 + z^3 + z^2 + z + 1$; remainder: 0;
$z^5 - 1 = (z - 1)(z^4 + z^3 +$
$z^2 + z + 1) + 0$ **19.** quotient:
$ax + (b + ar)$; remainder:
$c + r(b + ar) = ar^2 + br + c$;
$ax^2 + bx + c = (x - r)$
$(ax + (b + ar)) + (ar^2 + br + c)$
21. quotient: $x - 1$; remainder: -7;
$x^2 - 6x - 2 = (x - 5)(x - 1) - 7$
23. quotient: $4x - 5$; remainder: 0;
$4x^2 - x - 5 = (x + 1)(4x - 5) + 0$
25. quotient: $6x^2 + 19x + 78$;
remainder: 313; $6x^3 - 5x^2 + 2x +$
$1 = (x - 4)(6x^2 + 19x + 78) + 313$
27. quotient: $x^2 + 2x + 4$;
remainder: 7; $x^3 - 1 = (x - 2)$
$(x^2 + 2x + 4) + 7$ **29.** quotient:
$x^4 - 2x^3 + 4x^2 - 8x + 16$;
remainder: -33; $x^5 - 1 = (x + 2)$
$(x^4 - 2x^3 + 4x^2 - 8x + 16) - 33$
31. quotient: $x^3 - 10x^2 + 40x - 160$;

remainder: 642; $x^4 - 6x^3 +$
$2 = (x + 4)(x^3 - 10x^2 + 40x -$
$160) + 642$ **33.** quotient:
$x^2 + 6x + 57$; remainder: 576;
$x^3 - 4x^2 - 3x + 6 = (x - 10)$
$(x^2 + 6x + 57) + 576$
35. quotient: $x^2 - 6x + 30$;
remainder: -150; $x^3 - x^2 =$
$(x + 5)(x^2 - 6x + 30) - 150$
37. quotient: $54x^2 + 9x - 21$;
remainder: 0;
$54x^3 - 27x^2 - 27x + 14 =$
$(x - \frac{2}{3})(54x^2 + 9x - 21) + 0$
39. quotient: $x^3 + 3x^2 + 12x + 36$;
remainder: 120; $x^4 + 3x^2 +$
$12 = (x - 3)(x^3 + 3x^2 + 12x + 36)$
$+ 120$ **41.** $(x - 2)(x^4 + 2x^3 +$
$4x^2 + 8x + 16)$
43. $(z - 3)(z^3 + 3z^2 + 9z + 27)$
45. quotient: $2x$; remainder: 1
47. quotient: $3x^2 - \frac{3}{2}x + \frac{3}{4}$;
remainder: $\frac{1}{4}$ **49.** $k = 4$
53. quotient: $x + (-4 + i)$;
remainder: $-4i$ **55.** quotient:
$x + (-1 + i)$; remainder: 0
57. $-6\sqrt{3} + 57$ **59.** 0

Exercise Set 7.3

1. yes **3.** yes **5.** yes **7.** is a
zero **9.** is a zero **11.** not a zero
13. (a) is a zero **(b)** not a zero
15. $1, 2$ (multiplicity 3), 3
17. (a) multiplicity at -1: 2

$y = (x+1)^2(x+2)$

(b) multiplicity at -2: 3

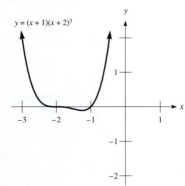

$y = (x + 1)(x + 2)^3$

41. $x = 2/3, \dfrac{4 - \sqrt{14}}{2} \approx 0.13,$

$\dfrac{4 + \sqrt{14}}{2} \approx 3.87$

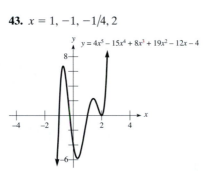

$y = 6x^3 - 28x^2 + 19x - 2$

43. $x = 1, -1, -1/4, 2$

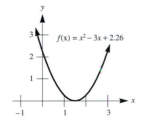

$y = 4x^5 - 15x^4 + 8x^3 + 19x^2 - 12x - 4$

(c) multiplicity at -1: 2;
multiplicity at -2: 3

$y = (x + 1)^2(x + 2)^3$

19. $f(-3) = -170$

21. $f\!\left(\tfrac{1}{2}\right) = -9$

23. $f(-\sqrt{2}) = -3\sqrt{2} - 2$

25. $x = -1, \dfrac{1 \pm i}{2}$ **27.** $x = \pm 3, 4$

29. $x = 1, -1 \pm \sqrt{6}$

31. $x = -2, \tfrac{2}{3}, 3$

33. $x = -\dfrac{3}{2}, \dfrac{1 \pm \sqrt{5}}{2}$

35. $x = 0, 5$ **37.** $x = -4, 3$

39. $x = -5, 2 - \sqrt{2} \approx 0.59,$
$2 + \sqrt{2} \approx 3.41$

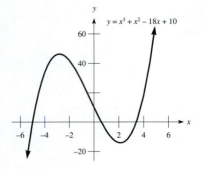

$y = x^3 + x^2 - 18x + 10$

45. (a) 1.125 **(b)** -0.046875

(c) $t - 1$ **(d)** $t = 1, \dfrac{-1 \pm \sqrt{13}}{2}$

47. $x^3 - 4x^2 - 17x + 60 = 0$

49. $x^3 + 8x^2 + 13x + 6 = 0$

51. no such polynomial exists

53. (a) $g(r) = ar^3 + br^2 + cr + d$

(b) $g(r) = ar^3 + br^2 + cr + d$

55. is a zero   **57.** $a = -1, b = -1$

59. $b = \pm\dfrac{3\sqrt{2}}{2}$ **61.** $x = 2$

(multiplicity 2), -4

Exercise Set 7.4

1. (a) yes   **(b)** yes   **(c)** yes
(d) no   **3.** no real roots

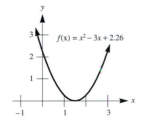

$f(x) = x^2 - 3x + 2.26$

5. three real roots

$f(x) = x^3 - 3x^2 + 3$

7. no real roots

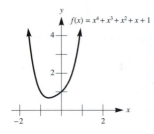

$f(x) = x^4 + x^3 + x^2 + x + 1$

9. one real root

$f(x) = 0.2x^3 + 4.4x^2 - 109x - 1$

11. $[x - (-1)](x - 3)$

13. $4\left(x - \tfrac{1}{4}\right)[x - (-6)]$

15. $[x - (-\sqrt{5})](x - \sqrt{5})$

17. $(x - 0)(x - (-3))(x - 3)$
$(x - (-\sqrt{2}i))(x - \sqrt{2}i)$

19. $f(x) = x^3 + x^2 - 5x + 3$

21. $f(x) = x^4 - 16$ **23.**
$f(x) = x^6 + 10x^4 - 87x^2 + 144$

25. (a) $f(x) = x^4 - 4x^3 + 3x^2$

(b)

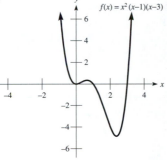

$f(x) = x^2(x-1)(x-3)$

27. (a) $f(x) = x(x - 1)^2(x - 3)^2$
(b)

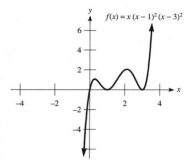

29. (a) $f(x) = x^2(x - 1)(x - 3)^3$
(b)

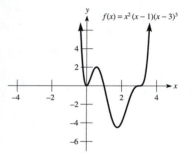

31. $f(x) = -\frac{5}{42}x^2 + \frac{25}{42}x + \frac{30}{7}$
33. (a) $f(x) = \frac{1}{30}x^3 - \frac{19}{30}x + 1$
(b)

35. $x^2 + (i + \sqrt{3})x + i\sqrt{3} = 0$
37. $x^2 - 3x - 54 = 0$ **39.** $x^2 - 2x - 4 = 0$ **41.** $x^2 - 2ax + a^2 - b = 0$ **43. (a)** $x = 4$
(b)

45. $(x + 2 - 2i)(x + 2 + 2i)$
$(x - 2 - 2i)(x - 2 + 2i)$
49. $x = 5$ (multiplicity 2), -10

Exercise Set 7.5

1. (a) see text   **(b)** $\pm 1, \pm 11$
3. (a) The possible rational roots are $\pm 1, \pm 5$, so there cannot be a rational root larger than 5. No, 5 is not a root.
(b) The possible rational roots are $\pm 1, \pm 5$, so 2 cannot be a root.
(c)

5. $\pm 1, \pm\frac{1}{2}, \pm\frac{1}{4}, \pm 3, \pm\frac{3}{2}, \pm\frac{3}{4}$ **7.** ± 1,
$\pm\frac{1}{2}, \pm\frac{1}{4}, \pm\frac{1}{8}, \pm 3, \pm\frac{3}{2}, \pm\frac{3}{4}, \pm\frac{3}{8}, \pm 9, \pm\frac{9}{2}$,
$\pm\frac{9}{4}, \pm\frac{9}{8}$
9. $\pm 1, \pm\frac{1}{2}, \pm 2, \pm 3, \pm\frac{3}{2}, \pm 6$
17. $x = 1, -1, -3$
19. $x = -\frac{1}{4}, \pm\sqrt{5}$
21. $x = -1, -\frac{2}{3}, -\frac{1}{3}$
23. $x = 1, -1, \dfrac{-1 \pm \sqrt{97}}{2}$
25. $x = 1$ (multiplicity 4)
27. $x = 1, -\frac{4}{3}, 6$ **29. (a)** 2 is an upper bound, -1 is a lower bound
(b) 2 is an upper bound, -1 is a lower bound   **(c)** 6 is an upper bound, -2 is a lower bound
31. between 0.68 and 0.69
33. between 2.88 and 2.89
35. between 4.31 and 4.32
37. (a) between -3 and -2
(b) between -2.148 and -2.147
39. (a) between -6 and -5
(b) between -5.265 and -5.264
41. (b) A and C have a common factor of 5, so the result does not apply.   **(c)** $x = \dfrac{p}{q}$ is a root of the equation

43. $x \approx 1.769$

45. $x \approx 1.310$

47. $x = \sqrt[3]{200} \approx 5.848$

51. $p = 2$
55. (a) $f(x) = (x + 3)(x + 1)(x - 1)$
(b)

Exercise Set 7.6

1. $7 + 2i$ **3.** $5 - 2i, 3$ **5.** $-2 - i$, -3 (multiplicity 2)   **7.** $6 + 5i, -\frac{1}{4}$
9. $4 - \sqrt{2}i, \pm\frac{3i}{2}$ **11.** $10 - 2i$, $1 \pm \sqrt{5}$ **13.** $\frac{1 - i\sqrt{2}}{3}, \frac{2}{5}$ **15.** $3 + 2i$, $-1 - i, -1 \pm \sqrt{2}$ **17.** $x^2 - 2x - 5 = 0$ **19.** $x^2 - \frac{4}{3}x - \frac{2}{3} = 0$
21. (c) one positive real root

$$f(x) = 2x^4 - 3x^3 + 12x^2 + 22x - 60$$

(d) $-2, \frac{3}{2}$ **23. (b)** one negative real root

$$f(x) = x^3 + 8x + 5$$

(c) -0.5982 **(d)**
$$x = \sqrt[3]{-2.5 + \sqrt{\tfrac{2723}{108}}} - \sqrt[3]{2.5 + \sqrt{\tfrac{2723}{108}}} \approx -0.5982$$
25. 2 complex roots and 1 negative real root   **27.** 4 complex roots and 1 negative real root   **29.** 2 complex roots, 1 positive real root and 1 negative real root   **31.** either 1 positive real root and 2 negative real roots, or 1 positive real root and 2 complex roots   **33.** 1 positive real root, 1 negative real root and 6 complex roots   **35.** 1 positive real root and 8 complex roots   **37.** 1 positive real root, 1 negative real root and 6 complex roots   **39.** 1 positive real root, 1 negative real root and 4 complex roots

41. 1 positive real root, 1 negative real root and 2 complex roots
43. (a) $f(x) = x^4 + 2x^2 + 49$
(b) $-\sqrt{3} \pm 2i$

Exercise Set 7.7

1. (a) $A = 2, B = 5$ **(b)** $A = 2$, $B = 5$ **3. (a)** $A = 4, B = -1$
(b) $A = 4, B = -1$
5. (a) $A = -\frac{1}{4}, B = \frac{3}{4}$
(b) $A = -\frac{1}{4}, B = \frac{3}{4}$
7. $A = 8, B = -21$
9. $A = -\frac{1}{5}, B = \frac{34}{5}$
11. $A = 4, B = -1, C = 6$
13. $A = \frac{2}{5}, B = \frac{3}{5}, C = -\frac{3}{5}$
15. $A = 1, B = -1, C = 1$
17. $A = \frac{10}{3}, B = \frac{1}{6}, C = -\frac{1}{2}$
19. $A = 4, B = 1, C = -5$
21. $A = 0, B = 1, C = 0, D = -4$
23. (a)

(b)

(c) $\dfrac{2x + 5}{(x - 4)(x + 3)} =$

$\dfrac{13/7}{x - 4} + \dfrac{1/7}{x + 3}$ **25. (a)** $A = 3$, $B = 4, C = -1$ **(b)** $B = 4, C = -1$
27. (a) The possible rational roots are -1 and 1, neither of which are roots.
(b)

$$f(x) = x^4 - x^3 + x^2 - x + 1$$

(c) $\left(x^2 + \dfrac{-1 + \sqrt{5}}{2}x + 1 \right) \times$
$\left(x^2 + \dfrac{-1 - \sqrt{5}}{2}x + 1 \right)$
(d) $x = \dfrac{1 - \sqrt{5} \pm i\sqrt{10 + 2\sqrt{5}}}{4}$, $\dfrac{1 + \sqrt{5} \pm i\sqrt{10 - 2\sqrt{5}}}{4}$
29. (a)

(b)

(c) $\dfrac{-1/48}{x + 2} + \dfrac{1/57}{x + 5} + \dfrac{1/304}{x - 14}$

Exercise Set 7.8

1. (a) no   **(b)** yes   **3. (a)** no
(b) yes   **5. (a)** $(x + 10)(x - 10)$

(b) $\dfrac{A}{x + 10} + \dfrac{B}{x - 10}$

(c) $\dfrac{4}{x + 10} + \dfrac{7}{x - 10}$

7. (a) $(x + \sqrt{5})(x - \sqrt{5})$

(b) $\dfrac{A}{x + \sqrt{5}} + \dfrac{B}{x - \sqrt{5}}$

(c) $\dfrac{5}{x + \sqrt{5}} + \dfrac{3}{x - \sqrt{5}}$

9. (a) $(x - 3)(x + 2)$

(b) $\dfrac{A}{x + 2} + \dfrac{B}{x - 3}$

(c) $\dfrac{-5}{x + 2} + \dfrac{12}{x - 3}$

11. (a) $(x - 3)(x - 2)(x + 2)$

(b) $\dfrac{A}{x - 3} + \dfrac{B}{x - 2} + \dfrac{C}{x + 2}$

(c) $\dfrac{8}{x - 3} + \dfrac{-2}{x - 2} + \dfrac{-3}{x + 2}$

13. (a) $x(x^2 + x + 1)$

(b) $\dfrac{A}{x} + \dfrac{Bx + C}{x^2 + x + 1}$

(c) $\dfrac{5}{x} + \dfrac{-3}{x^2 + x + 1}$

15. (a) $(x^2 + 1)^2$

(b) $\dfrac{Ax + B}{x^2 + 1} + \dfrac{Cx + D}{(x^2 + 1)^2}$

(c) $\dfrac{2x}{x^2 + 1} + \dfrac{3x - 4}{(x^2 + 1)^2}$

17. $\dfrac{\frac{19}{28}}{x - 6} + \dfrac{-\frac{3}{7}}{x + 1} + \dfrac{\frac{3}{4}}{x + 2}$

19. $\dfrac{10}{3x - 5} + \dfrac{-7}{2x - 3}$

21. $\dfrac{-1/5}{x} + \dfrac{(1 - 2\sqrt{5})/10}{x + \sqrt{5}} +$
$\dfrac{(1 + 2\sqrt{5})/10}{x - \sqrt{5}}$

23. $\dfrac{x}{x^2 + 4} + \dfrac{-4x + 2}{(x^2 + 4)^2}$ **25.**

$\dfrac{\frac{3}{125}}{x} + \dfrac{\frac{122}{125}}{x - 5} + \dfrac{\frac{253}{25}}{(x - 5)^2} + \dfrac{\frac{127}{5}}{(x - 5)^3}$

27. $\dfrac{\frac{1}{3}}{x - 1} + \dfrac{-\frac{1}{3}x - \frac{2}{3}}{x^2 + x + 1}$

29. $\dfrac{3}{x} + \dfrac{-2}{x^2} + \dfrac{4}{x + 1} + \dfrac{3}{(x + 1)^2}$

31. $\dfrac{\frac{8}{27}}{x + 3} - \dfrac{\frac{11}{54}}{x - 3} + \dfrac{\frac{1}{2}x + \frac{5}{18}}{x^2 + 9}$

33. $\dfrac{-x}{x^2 + 1} + \dfrac{x + 1}{x^2 + x + 1}$

35. $x = -3, -1, 2$

37. (b) $C = -2, D = 0$

39. $6x + 8 + \dfrac{2}{x - 3} + \dfrac{-1}{x - 1}$

41. $x - 4 + \dfrac{-\frac{1}{8}}{x} + \dfrac{\frac{1}{8}}{x - 2} +$
$\dfrac{\frac{3}{4}}{(x - 2)^2} + \dfrac{\frac{5}{2}}{(x - 2)^3}$ **43.**

$x^2 + 2x + 5 + \dfrac{-\frac{1}{4}}{x + 1} + \dfrac{\frac{1}{4}}{x - 1} +$
$\dfrac{-\frac{1}{2}}{x^2 + 1}$

47. $\dfrac{(pa + q)/(a - b)}{x - a} +$
$\dfrac{(pb + q)/(b - a)}{x - b}$

49. $\dfrac{(pa + q)/2a}{x - a} + \dfrac{(pa - q)/2a}{x + a}$

51. $\dfrac{\frac{a^2}{(a - b)(a - c)}}{1 - ax} + \dfrac{\frac{b^2}{(b - a)(b - c)}}{1 - bx} +$
$\dfrac{\frac{c^2}{(c - a)(c - b)}}{1 - cx}$ **53.**

$\dfrac{\frac{\sqrt{2}}{4}x + \frac{1}{2}}{x^2 + \sqrt{2}x - 1} + \dfrac{-\frac{\sqrt{2}}{4}x + \frac{1}{2}}{x^2 - \sqrt{2}x + 1}$

Chapter 7 Review Exercises

1. $q(x) = x^3 + x^2 - 3x + 1$,
$R(x) = -1$ **3.** quotient:
$x^3 + 3x^2 + 7x + 21$; remainder: 71
5. quotient: $2x^2 - 13x + 46$;
remainder: -187 **7.** quotient:
$5x - 20$; remainder: 0
9. $f(10) = 99{,}904$
11. $f(\frac{1}{10}) = -\frac{999}{1000}$
13. $f(a - 1) = a^3$
15. (a) $f(-0.3) \approx -0.24$
(b) $f(-0.39) \approx -0.007$
(c) $f(-0.394) \approx 0.00003$
17. $a = -5$ **19.** $a = -1, -2$

23. $\pm 1, \pm 2, \pm 3, \pm 6, \pm 9, \pm 18$
25. $\pm 1, \pm \frac{1}{2}, \pm 2, \pm 4, \pm 8$ **27.** $\pm p$,
± 1 **29.** $2, -\frac{3}{2}, -1$
31. $\frac{5}{2}, -1 \pm \sqrt{3}$ **33.** $\frac{2}{3}, \frac{-1 \pm i\sqrt{3}}{2}$
35. $-1, -\sqrt{7}$ (multiplicity 2),
$\sqrt{7}$ (multiplicity 2)
37. 2 (multiplicity 2), 5
39. see text
41. $6(x - \frac{4}{3})[x - (-\frac{5}{2})]$
43. $(x - 4)(x - (-\sqrt[3]{5})) \times$
$\left(x - \dfrac{\sqrt[3]{5} + i\sqrt{3}\sqrt[3]{25}}{2}\right) \times$
$\left(x - \dfrac{\sqrt[3]{5} - i\sqrt{3}\sqrt[3]{25}}{2}\right)$
45. $2 \pm 3i, 3$ **47.** $1 \pm i\sqrt{2}, \pm \sqrt{7}$
49. 1 positive real root and 2 complex roots   **51.** 1 negative real root and 2 complex roots   **53.** 1 positive real root, 1 negative real root and 2 complex roots   **55. (c)** -1,
$\pm i$ **57.** between 0.82 and 0.83
59. (c) between 6.93 and 6.94
61. $x^2 - 8x + 11 = 0$ **63.**
$x^4 - 12x^3 + 35x^2 + 60x - 200 = 0$
65. $x^4 - 4x^3 - 4x^2 + 16x - 8 = 0$
67. zeros: 0, 3, -1

69. zeros: 0, -2, 2

71. $5 + 6i$ **73.** 6   **75.** $\frac{1}{2} - \frac{\sqrt{3}}{2}i$

77. $3\sqrt{2} - 4\sqrt{2}i$ **81. (a)**

$|6 + 2i| = 2\sqrt{10}, |6 - 2i| = 2\sqrt{10}$

(b) $|-3| = 3$

87. $\dfrac{-\frac{21}{20}}{10 + x} + \dfrac{\frac{19}{20}}{10 - x}$

89. $\dfrac{1}{x} + \dfrac{-1}{x + 1} + \dfrac{-1}{(x + 1)^2}$

91. $\dfrac{x}{x^2 + 3} + \dfrac{-3x + 2}{(x^2 + 3)^2}$

93. (a) $(x^2 - x + 1)(x^2 - x - 1)$

(b) $\dfrac{\frac{1}{2}}{x^2 - x + 1} + \dfrac{x + \frac{1}{2}}{x^2 - x - 1}$

Chapter 7 Test

1. $f(\frac{1}{2}) = -\frac{3}{2}$ **2.** $-3, 1 \pm \sqrt{6}$

3. $\pm 1, \pm\frac{1}{2}, \pm 2, \pm 3, \pm\frac{3}{2}, \pm 6$

4. $f(x) = 3x^2 + 21x - 24$

5. quotient: $4x^2 - 3x - 5$;

remainder: 8   **6.** see text

7. (a) 3 is an upper bound

(b) between 2.2 and 2.3   **8.** $1 \pm i$,

$3 \pm 2i, -2$ **9.** $q(x) = x^2 + 2x - 1$,

$R(x) = -3x + 7$

10. $2[x - (\frac{3}{2} + \frac{1}{2}i)][x - (\frac{3}{2} - \frac{1}{2}i)]$

11. (a) $\pm 1, \pm 2, \pm 3, \pm 4, \pm 6, \pm 8,$

$\pm 12, \pm 24$ **(c)** $x = 1$ **(d)** none

12. (a) $\frac{3}{2}$ **(b)** $\frac{3}{2}, \frac{-1 \pm i\sqrt{3}}{2}$

13. 1 positive real root, 1 negative

real root, and 2 complex roots

14. $x^3 + 6x + 20 = 0$

15. $f(x) = (x - 2)(x - 3i)^3$

$[x - (1 + \sqrt{2})]^2$

16. $21 + (1 + \sqrt{3})i$ **17.** $-\frac{1}{17} + \frac{13}{17}i$

18. $\dfrac{\frac{1}{16}}{x} + \dfrac{-\frac{13}{32}}{x + 4} + \dfrac{\frac{11}{32}}{x - 4}$

19. $\dfrac{\frac{1}{4}}{x - 1} + \dfrac{-\frac{1}{4}x - \frac{1}{4}}{x^2 + 3}$

20. $\dfrac{5}{x} + \dfrac{-1}{x - 2} + \dfrac{3}{(x - 2)^2}$

CHAPTER 8
Exercise Set 8.1

1. focus: $(0, 1)$; directrix: $y = -1$;
focal width: 4

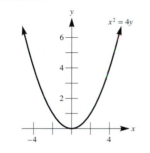

3. focus: $(-2, 0)$; directrix: $x = 2$;
focal width: 8

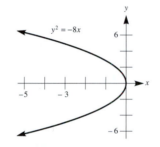

5. focus: $(0, -5)$; directrix: $y = 5$;
focal width: 20

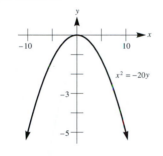

7. focus: $(-7, 0)$; directrix: $x = 7$;
focal width: 28

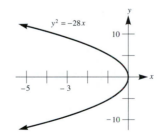

9. focus: $(0, \frac{3}{2})$; directrix:
$y = -\frac{3}{2}$; focal width: 6

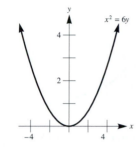

11. focus: $(0, \frac{7}{16})$; directrix: $y = -\frac{7}{16}$;
focal width: $\frac{7}{4}$

13. vertex: $(2, 3)$; focus: $(3, 3)$;
directrix: $x = 1$; focal width: 4

15. vertex: $(4, 2)$; focus: $(4, \frac{9}{4})$;
directrix: $y = \frac{7}{4}$; focal width: 1

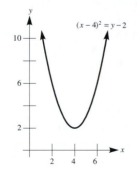

17. vertex: $(0, -1)$; focus: $\left(\frac{1}{4}, -1\right)$; directrix: $x = -\frac{1}{4}$; focal width: 1

19. vertex: $(3, 0)$; focus: $\left(3, \frac{1}{8}\right)$; directrix: $y = -\frac{1}{8}$; focal width: $\frac{1}{2}$

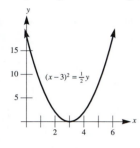

21. vertex: $(4, 1)$; focus: $\left(4, \frac{9}{8}\right)$; directrix: $y = \frac{7}{8}$; focal width: $\frac{1}{2}$

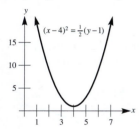

23. (a) 22.9 meters   **(b)** 0.3
25. (a) $x^2 = -\frac{80}{3}y$ **(b)** $14\frac{1}{16}$ feet
27. line of symmetry: $y = 1$

29. $x^2 = 12y$ **31.** $y^2 = 128x$
33. $y^2 = -9x$ **35. (a)** $y = \frac{9}{4}x - 1$
(b) $y = -\frac{9}{8}x - \frac{1}{4}$ **(c)** intersect at $\left(\frac{2}{9}, -\frac{1}{2}\right)$ **37. (a)** $\left(-\frac{1}{8}, \frac{1}{64}\right)$

(b) $\left(\frac{15}{16}, \frac{257}{128}\right)$ **(c)** $\overline{ST} = \frac{1}{2}$
39. side = $8\sqrt{3}p$ units, area = $48\sqrt{3}p^2$ sq. units
41. (a) 6 m   **(b)** $5\frac{1}{3}$ m

Exercise Set 8.2
1. $y = 4x - 4$

3. $y = x - 2$

5. $y = 6x + 9$

7. $y = x + 1$

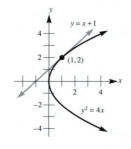

9. $m = \frac{1}{4}$ **11.** $m = 12$

Exercise Set 8.3
1. length of major axis: 6; length of minor axis: 4; foci: $(\pm\sqrt{5}, 0)$; eccentricity: $\dfrac{\sqrt{5}}{3}$

3. length of major axis: 8; length of minor axis: 2; foci: $(\pm\sqrt{15}, 0)$; eccentricity: $\dfrac{\sqrt{15}}{4}$

5. length of major axis: $2\sqrt{2}$; length of minor axis: 2; foci: $(\pm 1, 0)$; eccentricity: $\dfrac{\sqrt{2}}{2}$

7. length of major axis: 8; length of minor axis: 6; foci: $(0, \pm\sqrt{7})$; eccentricity: $\dfrac{\sqrt{7}}{4}$

9. length of major axis: $\dfrac{2\sqrt{15}}{3}$;

length of minor axis: $\dfrac{2\sqrt{3}}{3}$;

foci: $\left(0, \pm\dfrac{2\sqrt{3}}{3}\right)$; eccentricity:

$\dfrac{2\sqrt{5}}{5}$

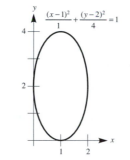

$$\dfrac{x^2}{1/3} + \dfrac{y^2}{5/3} = 1$$

11. length of major axis: 4;

length of minor axis:

$2\sqrt{2}$; foci: $(0, \pm\sqrt{2})$;

eccentricity: $\dfrac{\sqrt{2}}{2}$

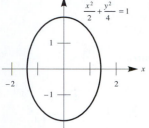

$$\dfrac{x^2}{2} + \dfrac{y^2}{4} = 1$$

13. center: $(5, -1)$; length of
major axis: 10; length of minor axis:
6; foci: $(9, -1)$ and $(1, -1)$;
eccentricity: $\frac{4}{5}$

$$\dfrac{(x-5)^2}{25} + \dfrac{(y+1)^2}{9} = 1$$

15. center: $(1, 2)$; length of
major axis: 4; length of minor

axis: 2; foci: $(1, 2 \pm \sqrt{3})$;

eccentricity: $\dfrac{\sqrt{3}}{2}$

$$\dfrac{(x-1)^2}{1} + \dfrac{(y-2)^2}{4} = 1$$

17. center: $(-3, 0)$; length of major
axis: 6; length of minor axis: 2; foci:

$(-3 \pm 2\sqrt{2}, 0)$; eccentricity: $\dfrac{2\sqrt{2}}{3}$

$$\dfrac{(x+3)^2}{9} + \dfrac{y^2}{1} = 1$$

19. center: $(1, -2)$; length of major
axis: 4; length of minor axis: $2\sqrt{3}$;
foci: $(2, -2)$ and $(0, -2)$;
eccentricity: $\frac{1}{2}$

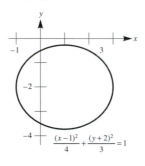

$$\dfrac{(x-1)^2}{4} + \dfrac{(y+2)^2}{3} = 1$$

21. center: $(4, 6)$; degenerate
ellipse

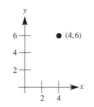

$\bullet$ $(4,6)$

23. no graph

25. $\dfrac{x^2}{25} + \dfrac{y^2}{16} = 1$ or

$16x^2 + 25y^2 = 400$

27. $\dfrac{x^2}{16} + \dfrac{y^2}{15} = 1$ or

$15x^2 + 16y^2 = 240$

29. $\dfrac{x^2}{21} + \dfrac{y^2}{25} = 1$ or

$25x^2 + 21y^2 = 525$

31. $\dfrac{x^2}{9} + \dfrac{y^2}{9/4} = 1$ or $x^2 + 4y^2 = 9$

33. (a) $y = -\frac{4}{3}x + \frac{38}{3}$

(b) $y = \frac{7}{9}x + \frac{76}{9}$ **(c)** $y = \frac{1}{15}x - \frac{76}{15}$

35. (a) $y = -6x + 26$ **(b)** $\frac{169}{3}$

37. perihelion: 29.64 AU; aphelion:
49.24 AU **39.** semimajor axis:
4.42 AU; eccentricity: 0.04

41. (c) $[-3, 3]$ **(d)**

x	0	0.5	1.0	1.5	2.0	2.5	3.0
y	± 2	± 1.97	± 1.89	± 1.73	± 1.49	± 1.11	0

(e)

$$\dfrac{x^2}{9} + \dfrac{y^2}{4} = 1$$

43. circle: $(x - 1)^2 + y^2 = 1$;

ellipse: $\dfrac{(x - 4)^2}{16} + \dfrac{y^2}{7} = 1$

49. intersection points: $\left(\dfrac{ab}{A}, \dfrac{ab}{A}\right)$,

$\left(\dfrac{ab}{A}, -\dfrac{ab}{A}\right)$, $\left(-\dfrac{ab}{A}, \dfrac{ab}{A}\right)$,

$\left(-\dfrac{ab}{A}, -\dfrac{ab}{A}\right)$,

where $A = \sqrt{a^2 + b^2}$

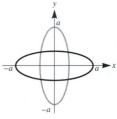

53. (b) $\left(-\frac{20}{7}, -\frac{18}{7}\right)$

59. (a)

(b) $a = 2\sqrt{3}, b = 2, c = 2\sqrt{2}$
(c) auxiliary circle: $x^2 + y^2 = 12$

(d) $x + y = 4$
(e)

(f) $y = x + 2\sqrt{2}$

Exercise Set 8.4

1. vertices: $(\pm 2, 0)$; length of transverse axis: 4; length of conjugate axis: 2; asymptotes: $y = \pm\frac{1}{2}x$ foci: $(\pm\sqrt{5}, 0)$; eccentricity: $\frac{\sqrt{5}}{2}$

3. vertices: $(0, \pm 2)$; length of transverse axis: 4; length of conjugate axis: 2; asymptotes: $y = \pm 2x$ foci: $(0, \pm\sqrt{5})$; eccentricity: $\frac{\sqrt{5}}{2}$

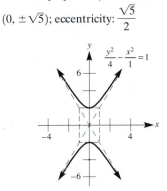

5. vertices: $(\pm 5, 0)$; length of transverse axis: 10; length of conjugate axis: 8; asymptotes: $y = \pm\frac{4}{5}x$ foci: $(\pm\sqrt{41}, 0)$; eccentricity: $\frac{\sqrt{41}}{5}$

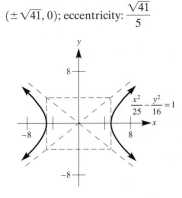

7. vertices: $\left(0, \pm\frac{\sqrt{2}}{2}\right)$; length of transverse axis: $\sqrt{2}$; length of conjugate axis: $\frac{2\sqrt{3}}{3}$; asymptotes: $y = \pm\frac{\sqrt{6}}{2}x$ foci: $\left(0, \pm\frac{\sqrt{30}}{6}\right)$; eccentricity: $\frac{\sqrt{15}}{3}$

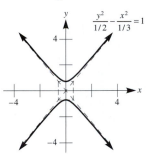

9. vertices: $(0, \pm 5)$; length of transverse axis: 10; length of conjugate axis: 4; asymptotes: $y = \pm\frac{5}{2}x$ foci: $(0, \pm\sqrt{29})$; eccentricity: $\frac{\sqrt{29}}{5}$

11. center: $(5, -1)$; vertices: $(10, -1)$ and $(0, -1)$; length of transverse axis: 10; length of conjugate axis: 6 asymptotes: $y = \frac{3}{5}x - 4$ and $y = -\frac{3}{5}x + 2$; foci: $(5 \pm \sqrt{34}, -1)$; eccentricity: $\frac{\sqrt{34}}{5}$

13. center: $(1, 2)$; vertices: $(1, 4)$ and $(1, 0)$; length of transverse axis: 4; length of conjugate axis: 2 asymptotes: $y = 2x$ and $y = -2x + 4$; foci: $(1, 2 \pm \sqrt{5})$;

eccentricity: $\dfrac{\sqrt{5}}{2}$

$$\dfrac{(y-2)^2}{4} - \dfrac{(x-1)^2}{1} = 1$$

15. center: $(-3, 4)$; vertices: $(1, 4)$ and $(-7, 4)$; length of transverse axis: 8; length of conjugate axis: 8 asymptotes: $y = x + 7$ and $y = -x + 1$; foci: $(-3 \pm 4\sqrt{2}, 4)$; eccentricity: $\sqrt{2}$

$$\dfrac{(x+3)^2}{16} - \dfrac{(y-4)^2}{16} = 1$$

17. center: $(0, 1)$; vertices: $(2, 1)$ and $(-2, 1)$; length of transverse axis: 4; length of conjugate axis: 4 asymptotes: $y = x + 1$ and $y = -x + 1$; foci: $(\pm 2\sqrt{2}, 1)$; eccentricity: $\sqrt{2}$

$$\dfrac{x^2}{4} - \dfrac{(y-1)^2}{4} = 1$$

19. center: $(2, 1)$; vertices: $(5, 1)$ and $(-1, 1)$; length of transverse axis: 6; length of conjugate axis: 6 asymptotes: $y = x - 1$ and

$y = -x + 3$; foci: $(2 \pm 3\sqrt{2}, 1)$; eccentricity: $\sqrt{2}$

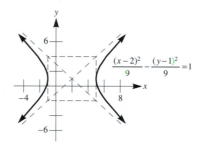

$$\dfrac{(x-2)^2}{9} - \dfrac{(y-1)^2}{9} = 1$$

21. center: $(0, -4)$; vertices: $(0, 1)$ and $(0, -9)$; length of transverse axis: 10; length of conjugate axis: 2; asymptotes: $y = 5x - 4$ and $y = -5x - 4$; foci: $(0, -4 \pm \sqrt{26})$; eccentricity: $\dfrac{\sqrt{26}}{5}$

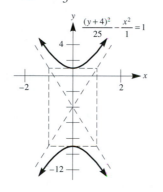

$$\dfrac{(y+4)^2}{25} - \dfrac{x^2}{1} = 1$$

23. degenerate hyperbola

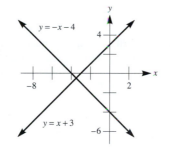

$y = -x - 4$

$y = x + 3$

27. $15x^2 - y^2 = 15$
29. $x^2 - 4y^2 = 4$
31. $2x^2 - 5y^2 = 10$
33. $y^2 - 32x^2 = 49$
35. $y^2 - 9x^2 = 9$ **39. (b)** $(0, \pm 6)$ **(c)** $F_1P = 5, F_2P = 13$

41. (b)

$$\dfrac{(y-3)^2}{25} - \dfrac{(x-4)^2}{9} = 1$$

(c) asymptotes: $y = \pm\dfrac{5}{3}(x - 4) + 3$

$$\dfrac{(y-3)^2}{25} - \dfrac{(x-4)^2}{9} = 1$$

49. intersection point: $\left(\dfrac{51}{8}, \dfrac{15}{2}\right)$

$y = \dfrac{4}{3}x - 1$

$\left(\dfrac{51}{8}, \dfrac{15}{2}\right)$

$16x^2 - 9y^2 = 144$

51. $y = 2x - 2$

Exercise Set 8.5

1. $F_1P = \dfrac{6\sqrt{19} - 8\sqrt{6}}{3}$,

$F_2P = \dfrac{6\sqrt{19} + 8\sqrt{6}}{3}$

3. $F_1P = 15 + 6\sqrt{2}$,

$F_2P = 15 - 6\sqrt{2}$ **5. (a)** foci:

$(\pm\sqrt{7}, 0)$; eccentricity: $\dfrac{\sqrt{7}}{4}$;

directrices: $x = \pm\dfrac{16\sqrt{7}}{7}$ **(b)** foci:

$(\pm 5, 0)$; eccentricity: $\frac{5}{4}$; directrices:

$x = \pm\frac{16}{5}$ **7. (a)** foci: $(\pm 1, 0)$;

eccentricity: $\dfrac{\sqrt{13}}{13}$; directrices:

$x = \pm 13$ **(b)** foci: $(\pm 5, 0)$;

eccentricity: $\dfrac{5\sqrt{13}}{13}$; directrices:

$x = \pm\frac{13}{5}$ **9. (a)** foci: $(\pm\sqrt{11}, 0)$;

eccentricity: $\dfrac{\sqrt{11}}{6}$; directrices:

$x = \pm\dfrac{36\sqrt{11}}{11}$ **(b)** foci: $(\pm\sqrt{61}, 0)$;

eccentricity: $\dfrac{\sqrt{61}}{6}$; directrices:

$x = \pm\dfrac{36\sqrt{61}}{61}$ **11.** $3x^2 + 4y^2 = 12$

13. $x^2 - y^2 = 2$

Chapter 8 Review Exercises

1. (a) $y^2 = 16x$ **(b)** $x^2 = 16y$
3. $x^2 = 12y$ **5.** $15x^2 + 16y^2 = 960$
7. $25x^2 + 9y^2 = 900$
9. $8x^2 - y^2 = 32$
11. $240x^2 - 16y^2 = 135$
13. vertex: $(0, 0)$; focus: $(0, \frac{5}{2})$;
directrix: $y = -\frac{5}{2}$; focal width: 10

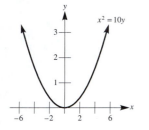

15. vertex: $(0, 3)$; focus: $(0, 0)$;
directrix: $y = 6$; focal width: 12

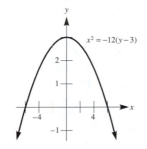

17. vertex: $(1, 1)$; focus: $(0, 1)$;
directrix: $x = 2$; focal width: 4

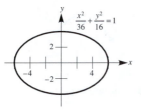

19. center: $(0, 0)$; length of major
axis: 12; length of minor axis: 8;
foci: $(\pm 2\sqrt{5}, 0)$; eccentricity: $\dfrac{\sqrt{5}}{3}$

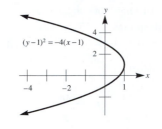

21. center: $(0, 0)$; length of major
axis: 6; length of minor axis: 2; foci:
$(0, \pm 2\sqrt{2})$; eccentricity: $\dfrac{2\sqrt{2}}{3}$

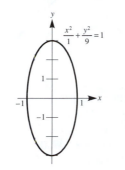

23. center: $(-3, 0)$; lengths of
major and minor axes: 6; focus:
$(-3, 0)$; eccentricity: 0

25. center: $(0, 0)$; vertices: $(\pm 6, 0)$;
asymptotes: $y = \pm\frac{2}{3}x$; foci:
$(\pm 2\sqrt{13}, 0)$; eccentricity: $\dfrac{\sqrt{13}}{3}$

27. center: $(0, 0)$; vertices: $(0, \pm 1)$;
asymptotes: $y = \pm\frac{1}{3}x$; foci:
$(0, \pm\sqrt{10})$; eccentricity: $\sqrt{10}$

29. center: $(0, -3)$; vertices: $(0, 0)$
and $(0, -6)$; asymptotes: $y = x - 3$
and $y = -x - 3$; foci:
$(0, -3 \pm 3\sqrt{2})$; eccentricity: $\sqrt{2}$

31. vertex: $(4, 4)$; axis of symmetry: $y = 4$; focus: $(8, 4)$; directrix: $x = 0$

33. center: $(-2, 3)$; length of major axis: 8; length of the minor axis: 6; foci: $(-2, 3 \pm \sqrt{7})$

35. vertex: $(-3, 2)$; axis of symmetry: $x = -3$; focus: $(-3, 5)$; directrix: $y = -1$

37. center: $(2, 1)$; vertices: $(5, 1)$ and $(-1, 1)$; asymptotes: $y = x - 1$ and $y = -x + 3$; foci: $(2 \pm 3\sqrt{2}, 1)$

39. vertex: $(0, 6)$; axis of symmetry: $x = 0$; focus: $(0, \frac{11}{2})$; directrix: $y = \frac{13}{2}$

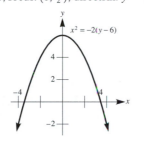

41. center: $(0, 5)$; length of major axis: 8; length of minor axis: 2; foci: $(\pm\sqrt{15}, 5)$

43. two lines: $4x - 5y = -2$ and $4x + 5y = 18$

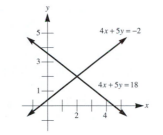

49. $F_1P = \sqrt{\frac{524 + 8\sqrt{3360}}{15}}$, $F_2P = \sqrt{\frac{524 - 8\sqrt{3360}}{15}}$

51. $45x^2 - 36y^2 = 320$

Chapter 8 Test

1. focus: $(-3, 0)$; directrix: $x = 3$

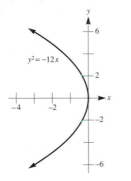

2. asymptotes: $y = \pm\frac{1}{2}x$; foci: $(\pm\sqrt{5}, 0)$

3. semimajor axis: 19.078 AU; eccentricity: 0.001

4. $\frac{x^2}{12} + \frac{y^2}{16} = 1$ **5.** $\frac{x^2}{3} - \frac{y^2}{1} = 1$

6. (b) $(F_1P - F_2P)^2 = 64$

7. length of major axis: 10; length of minor axis: 4; foci: $(\pm\sqrt{21}, 0)$

8.

9.

10. focal width: 8; vertex: (1, 2)

11. (a) $x = \pm \dfrac{36\sqrt{11}}{11}$

(b) $F_1P = \dfrac{12 + \sqrt{11}}{2}$,

$F_2P = \dfrac{12 - \sqrt{11}}{2}$ **12.** $y = \frac{1}{6}x + \frac{13}{3}$

13. $y = 4x - 8$ **14.** focal length: 12.8 m; focal ratio: 0.4

CHAPTER 9

Exercise Set 9.1

27. (a)

n	1	2	3	4	5
$f(n)$	$\frac{1}{2}$	$\frac{2}{3}$	$\frac{3}{4}$	$\frac{4}{5}$	$\frac{5}{6}$

(b) $\frac{6}{7}$ **(c)** $f(n) = \dfrac{1}{1 \times 2} + \dfrac{1}{2 \times 3} + \cdots + \dfrac{1}{n(n + 1)} = \dfrac{n}{n + 1}$

29. (a)

n	1	2	3	4	5
$f(n)$	1	4	9	16	25

(b) 36   **(c)** $f(n) = n^2$

Exercise Set 9.2

3. $a^9 + 9a^8b + 36a^7b^2 + 84a^6b^3 + 126a^5b^4 + 126a^4b^5 + 84a^3b^6 + 36a^2b^7 + 9ab^8 + b^9$

5. $8A^3 + 12A^2B + 6AB^2 + B^3$

7. $1 - 12x + 60x^2 - 160x^3 + 240x^4 - 192x^5 + 64x^6$

9. $x^2 + 4x\sqrt{xy} + 6xy + 4y\sqrt{xy} + y^2$

11. $x^{10} + 5x^8y^2 + 10x^6y^4 + 10x^4y^6 + 5x^2y^8 + y^{10}$

13. $1 - \dfrac{6}{x} + \dfrac{15}{x^2} - \dfrac{20}{x^3} + \dfrac{15}{x^4} - \dfrac{6}{x^5} + \dfrac{1}{x^6}$

15. $\dfrac{x^3}{8} - \dfrac{x^2y}{4} + \dfrac{xy^2}{6} - \dfrac{y^3}{27}$

17. $a^7b^{14} + 7a^6b^{12}c + 21a^5b^{10}c^2 + 35a^4b^8c^3 + 35a^3b^6c^4 + 21a^2b^4c^5 + 7ab^2c^6 + c^7$

19. $x^8 + 8\sqrt{2}x^7 + 56x^6 + 112\sqrt{2}x^5 + 280x^4 + 224\sqrt{2}x^3 + 224x^2 + 64\sqrt{2}x + 16$

21. $5\sqrt{2} - 7$ **23.** $89\sqrt{3} + 109\sqrt{2}$

25. $12 - 24\sqrt[3]{2} + 12\sqrt[3]{4}$

27. $x^{10} - 10x^9 + 35x^8 - 40x^7 - 30x^6 + 68x^5 + 30x^4 - 40x^3 - 35x^2 - 10x - 1$ **29.** 120

31. 105   **33. (a)** 10   **(b)** 5   **35.** $n^2 + 3n + 2$ **37.** 0   **39.** $120a^2b^{14}$

41. $100x^{99}$ **43.** 45   **45.** 294912

47. 28   **49.** 40095   **53. (a)**

k	0	1	2	3	4	5	6	7	8
$\binom{8}{k}$	1	8	28	56	70	56	28	8	1

Exercise Set 9.3

1. $\frac{1}{2}, \frac{2}{3}, \frac{3}{4}, \frac{4}{5}$ **3.** $-1, 1, -1, 1$ **5.** $\frac{1}{2}, \frac{1}{4}, \frac{1}{8}, \frac{1}{16}$ **7.** $3, 6, 9, 12$ **9.** $2, \frac{9}{4}, \frac{64}{27}, \frac{625}{256}$

11. $-1, 0, \frac{1}{3}, \frac{1}{2}$ **13.** $-2, 1, -\frac{8}{9}, 1$

15. $1, 4, 25, 676, 458329$ **17.** $2, 2, 4, 8, 32$ **19.** $1, 1, 2, 6, 24$ **21.** $0, 1, 2, 4, 16$ **23.**

25.

27.

29.

31.

33. 62   **35.** 40   **37.** $-\frac{19}{30}$ **39.** 903

41. 3   **43.** 41   **45.** $x + x^2 + x^3$

47. $\frac{25}{12}$ **49.** -1 **51.** $\frac{6}{7}$

53. $\displaystyle\sum_{j=1}^{4} 5^j$ **55.** $\displaystyle\sum_{j=1}^{6} x^j$

57. $\displaystyle\sum_{k=1}^{12} \frac{1}{k}$ **59.** $\displaystyle\sum_{j=1}^{5} (-1)^{j+1}2^j$

61. $\displaystyle\sum_{j=1}^{5} (-1)^{j+1}j$

63. (a)

t	0	1	2	3	4	5
P_t	300	1111	1829	1468	1691	1559

The population seems to be oscil-lating closer to a value near 1600.
(b)

t	P_t	t	P_t
0	300	11	1607
1	1111	12	1611
2	1829	13	1608
3	1468	14	1610
4	1691	15	1609
5	1559	16	1610
6	1640	17	1609
7	1591	18	1610
8	1621	19	1609
9	1602	20	1610
10	1614		

(c) $P_{t-1} \approx 1609$
65. (a)

F_1	F_2	F_3	F_4	F_5	F_6	F_7	F_8	F_9	F_{10}
1	1	2	3	5	8	13	21	34	55

(b) $F_{22} = 17{,}711$; $F_{19} = 4181$
(c) $F_{30} = 832{,}040$ **71. (a)** $F_1 = 1$;

$F_2 = 1$ **(b)** $F_{24} = 46{,}368$;
$F_{25} = 75{,}025$ **(c)** $F_{26} = 121{,}393$

Exercise Set 9.4
1. (a) 2   **(b)** -4 **(c)** $\frac{1}{3}$ **(d)** $\sqrt{2}$
3. $a_{12} = 131$ **5.** $a_{100} = 501$
7. $a_{1000} = 998$ **9.** $d = \frac{11}{6}$; $a = -\frac{23}{2}$
11. $a = -190$ **13.** $d = -\frac{1}{8}$
15. $S_{1000} = 500{,}500$
17. $S_{13} = \dfrac{91\pi}{3}$ **19.** $a = \frac{1}{2}$
21. $S_{16} = -768$; $d = -\frac{104}{15}$
23. $d = \frac{6}{5}$; $a = -\frac{17}{5}$ **25.** 900
27. 2, 10, 18 or 18, 10, 2   **29.** $-1, 2,$
5 or 5, 2, -1
31. (b) $S_6 = -6 - 9\sqrt{2}$
33. $S_n = \dfrac{n}{2(1-b)}[2 + (n-3)\sqrt{b}]$

Exercise Set 9.5
1. 6   **3.** 20; 100   **5.** $a_{100} = 1$
7. $a_8 = \frac{256}{6561}$ **9.** $r = \pm 4$
11. 7161   **13.** $63 + 31\sqrt{2}$ **15.** $\frac{1995}{64}$
17. $\frac{11111}{1000000} = 0.011111$ **19.** $\frac{2}{5}$
21. 101   **23.** $\frac{5}{9}$ **25.** $\frac{61}{495}$ **27.** $\frac{16}{37}$
29. $r = -\frac{1}{2}, -2$ **35.** 12 ft

Exercise Set 9.6
1. 6 routes

3. 12 routes

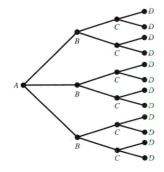

5. 12   **7.** 60   **9.** 108
11. 11,232,000   **13.** 3,628,800
15. 4 letters   **17.** 8   **19.** 120
23. $P(101, 51)$ **25.** 240   **27.** 12
29. (a) 900   **(b)** 648   **31. (a)** 729
(b) 504   **(c)** 56   **(d)** 140
33. (a) 12   **(b)** 12   **35. (a)** 56
(b) 56   **37. (a)** 792   **(b)** 792
39. $\dfrac{n(n-1)}{2}$ **41.** 56   **43. (a)** 210
(b) 80   **45.** 9 diagonals
47. 56 triangles   **49.** 15 sums
51. (a) 2352 committees
(b) 2649 committees
53. There are 7 diagonals from each vertex, which should only be counted once.

Exercise Set 9.7
1. (a) $\{2, 4, 6\}$ **(b)** $\{4, 5, 6\}$ **(c)**
$\{2, 4, 5, 6\}$ **(d)** $\{4, 6\}$ **3. (a)** $\{1, 3, 5\}$
(b) $\{1, 2, 3, 4, 5, 6\}$ **(c)** $\{1, 2, 3, 4,$
$5, 6\}$ **(d)** $\varnothing$ **5.** No   **7.** Yes
9. (a) 3   **(b)** 3   **(c)** 4   **11. (a)** 3
(b) 6   **(c)** 0   **13. (a)** $\frac{1}{2}$ **(b)** $\frac{1}{2}$
15. (a) $\frac{1}{2}$ **(b)** 1   **17. (a)** $\frac{2}{3}$ **(b)** 1
19. (a) $\frac{5}{6}$ **(b)** 1   **21.** $\frac{3}{8}$ **23.** $\frac{1}{2}$
25. $\frac{1}{4}$ **27.** $\frac{5}{8}$ **29.** $\frac{3}{4}$ **31.** 1   **33.** $\frac{5}{8}$
35. $\frac{1}{2}$ **37.** 0   **39.** $\frac{1}{4}$ **41.** $\frac{1}{3}$
43. (a) $\frac{6}{7}$ **(b)** $\frac{36}{91}$ **45.** $\frac{7}{12}$ **47.** $\frac{2}{9}$
49. (a) $\frac{1}{8}$ **(b)** $\frac{3}{8}$ **(c)** $\frac{1}{2}$ **51.** $\frac{7}{10}$
53. $\frac{33}{66640}$

Chapter 9 Review Exercises
11. $81a^4 + 108a^3b^2 +$
$54a^2b^4 + 12ab^6 + b^8$
13. $x^4 + 4x^3\sqrt{x} +$
$6x^3 + 4x^2\sqrt{x} + x^2$
15. $x^{10} - 10x^8y^2 +$
$40x^6y^4 - 80x^4y^6 +$
$80x^2y^8 - 32y^{10}$
17. $1 + \dfrac{5}{x} + \dfrac{10}{x^2} + \dfrac{10}{x^3} + \dfrac{5}{x^4} + \dfrac{1}{x^5}$
19. $a^4b^2 - 4a^3b^2\sqrt{ab} + 6a^3b^3 -$
$4a^2b^3\sqrt{ab} + a^2b^4$ **21.** $15xy^8$
23. 84   **25.** 28   **27.** Each side is 6.
29. Each side is 70.   **31.** Each side

is 8.   **33.** $1, \frac{4}{3}, \frac{3}{2}, \frac{8}{5}$

35. $-\frac{1}{2}, \frac{2}{3}, -\frac{3}{4}, \frac{4}{5}$

37. $-3, -12, -48, -192$

39. (a) -5 **(b)** $\frac{9}{10}$ **41.** $\displaystyle\sum_{k=1}^{5} \frac{5}{3k}$

43. $a_{18} = 73$ **45.** $a_{12} = \frac{5}{1024}$

47. $S_{12} = 177$

49. $S_{10} = 7{,}777{,}777{,}777$

51. $a_6 = -5\sqrt{10}$ **53.** $\frac{3}{4}$ **55.** $\frac{1}{10}$

57. $\frac{5}{11}$ **63. (a)** 42929 **(b)** 42925;
percent error: 0.00932%

(c) 6.48040×10^{10}

(d) 6.48027×10^{10}; percent error:
$2 \times 10^{-3}\%$ **65.** 16 **67.** $\{abc, acb,$
$bac, bca, cab, cba, abd, adb, bad,$
$bda, dab, dba, acd, adc, cad, cda,$
$dac, dca, bcd, bdc, cbd, cdb, dbc,$
$dcb\}$ $P(4,3) = 24$ **69.** $19{,}958{,}400$

71. 480 **73.** 1620 subcommittees

75. $\frac{6}{11}$ **77. (a)** $\frac{5}{33}$ **(b)** $\frac{7}{22}$ **(c)** $\frac{31}{66}$

79. $\frac{1}{18}$ **81.** $\frac{1}{6}$ **83.** $\frac{2}{9}$ **85.** $\frac{1}{6}$ **87.** $\frac{25}{36}$

89. (a) $\frac{7}{15}$ **(b)** $\frac{8}{15}$ **(c)** $\frac{2}{3}$ **(d)** $\frac{2}{3}$

91. (a) 252 **(b)** $\frac{7}{25}$ **93.** $40, 20, 10$
or $10, 20, 40$ **97. (a)** no such
value for x exists

(b) $x = \dfrac{b^2 - ac}{a + c - 2b}$

Chapter 9 Test

2. (a) 27 **(b)** -6

3. (a) $S_n = \dfrac{a(1 - r^n)}{1 - r}$ **(b)** $\frac{174075}{1024}$

4. (a) 42240 **(b)** $5280a^7b^{12}$

5. $243x^{10} + 405x^8y^3 +$
$270x^6y^6 + 90x^4y^9 + 15x^2y^{12} + y^{15}$

6. 177 **7.** $\frac{7}{9}$ **8.** 5 and 27

9. $a_6 = -5\sqrt{10}$ **10.** 224

11. (a)

(b)

12. 8680 words **13.** $121{,}080{,}960$
possible arrangements **14.** $\frac{22}{35}$

15. (a) $\frac{1}{15}$ **(b)** $\frac{7}{15}$ **(c)** $\frac{8}{15}$

APPENDIX B
Exercise Set B.1

1. -10 **3.** $\frac{2}{3}$ **5. (a)** a^{15}

(b) $(a + 1)^{15}$ **(c)** $(a + 1)^{15}$

7. (a) y^{11} **(b)** $(y + 1)^{11}$

(c) $(y + 1)^{18}$ **9. (a)** $x^2 + 3$

(b) $\dfrac{1}{x^2 + 3}$ **(c)** 12

11. (a) t^6 **(b)** $\dfrac{1}{t^6}$ **(c)** $(t^2 + 3)^6$

13. (a) $\dfrac{x^4}{y^5}$ **(b)** $\dfrac{y^5}{x^4}$ **(c)** $\dfrac{y^{10}}{x^8}$

15. (a) $4x^6$ **(b)** $16x^6$ **(c)** 4

17. (a) 1 **(b)** 1 **(c)** 1

19. (a) $\frac{11}{100}$ **(b)** $\frac{100}{11}$ **(c)** 1000

21. 7 **23. (a)** $\frac{1}{20}$ **(b)** $\frac{1}{225}$

25. $\dfrac{1}{a^6b^3}$ **27.** $\dfrac{a^4b^2}{c^6}$ **29.** $\dfrac{y^{12}}{x^6z^{12}}$

31. $\dfrac{y^{20}}{x^6z^{16}}$ **33.** y^6 **35.** b **37.** x^{2p}

39. 576 **41.** 12 **43.** 9.29×10^7
miles **45.** 6.68×10^4 mph

47. 2.5×10^{19} miles

49. (a) 1.0×10^{-9} seconds

(b) 1.0×10^{-18} seconds

(c) 1.0×10^{-24} seconds

Exercise Set B.2

1. false **3.** true **5.** true **7.** true

9. (a) -4 **(b)** not a real number

11. (a) $\frac{2}{5}$ **(b)** $-\frac{2}{5}$ **13. (a)** not a real
number **(b)** not a real number

15. (a) $\frac{4}{3}$ **(b)** $-\frac{3}{5}$ **17. (a)** -2

(b) 2 **19. (a)** $3\sqrt{2}$ **(b)** $3\sqrt[3]{2}$

21. (a) $7\sqrt{2}$ **(b)** $-2\sqrt[5]{2}$ **23. (a)** $\frac{5}{2}$

(b) $\frac{2}{5}$ **25. (a)** $3\sqrt{2}$ **(b)** $3\sqrt[3]{2}$

27. (a) $-4\sqrt{2}$ **(b)** $5\sqrt[4]{2}$

29. (a) 0.3 **(b)** 0.2 **31.** $-14\sqrt{6}$

33. $2\sqrt{2}$ **35. (a)** $6x$ **(b)** $-6y$

37. (a) $ab\sqrt{ab}$ **(b)** a^2b^2

39. $6ab^2c^2\sqrt{2ac}$ **41.** $2ab\sqrt[4]{b}$

43. $\dfrac{2a^4\sqrt[3]{2b^2}}{c^3}$ **45.** $\dfrac{4\sqrt{7}}{7}$ **47.** $\dfrac{\sqrt{2}}{4}$

49. $\dfrac{\sqrt{5} - 1}{4}$ **51.** $-2 - \sqrt{3}$

53. $\dfrac{61\sqrt{5}}{5}$ **55.** $\dfrac{\sqrt[3]{5}}{5}$ **57.** $\sqrt[4]{27}$

59. $\dfrac{3\sqrt[5]{2ab}}{2ab^2}$ **61. (a)** $\sqrt{3} - 1$

(b) $\sqrt{3} + 1$ **63. (a)** $\dfrac{x(\sqrt{x} + 2)}{x - 4}$

(b) $\dfrac{x(\sqrt{x} + y)}{x - y^2}$

65. (a) $\dfrac{x + 2\sqrt{2x} + 2}{x - 2}$

(b) $\dfrac{x + 2\sqrt{ax} + a}{x - a}$

67. $\dfrac{-2(\sqrt{x + h} + \sqrt{x})}{h}$

69. $\dfrac{1}{\sqrt{x} + \sqrt{5}}$

71. $\dfrac{1}{\sqrt{2 + h} - \sqrt{2}}$

73. $\dfrac{1}{\sqrt{x + h} + \sqrt{x}}$

75. $a = 9, b = 16$ **77.** $u = 1, v = 8$

79. (a) 1.645751

Exercise Set B.3

1. $\sqrt[5]{a^3} = (\sqrt[5]{a})^3$ **3.** $\sqrt[3]{5^2} = (\sqrt[3]{5})^2$

5. $\sqrt[4]{(x^2 + 1)^3} = \sqrt[4]{(x^2 + 1)^3}$

7. $\sqrt[3]{2^{xy}} = (\sqrt[3]{2})^{xy}$ **9.** $p^{2/3}$
11. $(1 + u)^{4/7}$ **13.** $(a^2 + b^2)^{3/p}$
15. 4   **17.** $\frac{1}{6}$ **19.** not a real number   **21.** 5   **23.** 2   **25.** 4   **27.** -2
29. -10 **31.** $\frac{1}{7}$ **33.** not a real number   **35.** $\frac{1}{216}$ **37.** 25   **39.** -1
41. $\frac{255}{16}$ **43.** $-\frac{28976}{243}$ **45.** $6a^{7/12}$
47. $2^{3/2}\,a^{1/12}$ **49.** $x^2 + 1$
51. (a) $2^{1/3}\,3^{5/6}$ **(b)** $\sqrt[6]{972}$
53. (a) $2^{7/12}\,3^{1/3}$ **(b)** $\sqrt[3]{10368}$
55. (a) $x^{2/3}\,y^{4/5}$ **(b)** $\sqrt[12]{x^{10}y^{12}}$
57. (a) $x^{(a+b)/3}$ **(b)** $\sqrt[15]{x^{a+b}}$
59. $(x + 1)^{2/3}$ **61.** $(x + y)^{2/5}$
63. $2x^{1/6}$ **65.** $x^{1/6}\,y^{1/8}$ **67.** $9^{10/9}$
69. $a = 9, b = 16$ **71.** $u = 1, v = 8$
73. $x = 4, m = 2$

Exercise Set B.4

1. (a) $(x + 8)(x - 8)$
(b) $7x^2(x^2 + 2)$
(c) $z(11 + z)(11 - z)$
(d) $(ab + c)(ab - c)$
3. (a) $(x + 3)(x - 1)$
(b) $(x - 3)(x + 1)$
(c) irreducible
(d) $(-x + 3)(x + 1)$ or $-(x - 3)(x + 1)$
5. (a) $(x + 1)(x^2 - x + 1)$
(b) $(x + 6)(x^2 - 6x + 36)$
(c) $8(5 - x^2)(25 + 5x^2 + x^4)$
(d) $(4ax - 5)(16a^2x^2 + 20ax + 25)$
7. (a) $(12 + x)(12 - x)$
(b) irreducible
(c) $(9 + y)(15 - y)$
9. (a) $h^3(1 + h)(1 - h)$
(b) $h^3(10 + h)(10 - h)$
(c) $(h + 1)^3(11 + h)(9 - h)$

11. (a) $(x - 8)(x - 5)$
(b) irreducible
13. (a) $(x + 9)(x - 4)$
(b) $(x - 9)(x - 4)$
15. (a) $(3x + 2)(x - 8)$
(b) irreducible
17. (a) $(3x - 1)(2x + 5)$
(b) $(6x + 5)(x - 1)$
19. (a) $(t^2 + 1)^2$
(b) $(t + 1)^2(t - 1)^2$
(c) irreducible
21. (a) $x(4x^2 - 20x - 25)$
(b) $x(2x - 5)^2$
23. (a) $(a - c)(b + a)$
(b) $(u + v - y)(x + u + v)$
25. $(xz + t)(xz + y)$
27. $(a^2 - 2b^2c^2)^2$ **29.** irreducible
31. $(x + 4)(x^2 - 4x + 16)$
33. $x(x^2 + 3xy + 3y^2)$
35. $(x - y)(x^2 + xy + y^2 + 1)$
37. (a) $(p^2 + 1)(p + 1)(p - 1)$
(b)
$(p^4 + 1)(p^2 + 1)(p + 1)(p - 1)$
39. $(x + 1)^3$ **41.** irreducible
43. $(\frac{5}{4} + c)(\frac{5}{4} - c)$
45. $(z^2 + \frac{9}{4})(z + \frac{3}{2})(z - \frac{3}{2})$
47. $\left(\dfrac{5}{mn} - 1\right)\left(\dfrac{25}{m^2n^2} + \dfrac{5}{mn} + 1\right)$
49. $(\frac{1}{2}x + y)^2$
51. $(x - a)(8x - 8a + 1) \times$ $(8x - 8a - 1)$
53. $(x - y - a)(x - y + a)$
55. $x(7x - 3)(3x + 13)$
57. $(5 - 2x + 3y)(5 + 2x - 3y)$
59. $(ax + b)(x + 1)$
61. $-x(x + 1)^{1/2}$
63. $x(x + 1)^{-3/2}$
65. $-\frac{2}{3}x(2x + 3)^{1/2}$ **67. (a)** 199
(b) 296   **(c)** 1999   **69.** $(A + B)^3$

71. $x(a^2 + x^2)^{-3/2}(2a^2 + x^2)$
73. $(y^2 - pq)(y - p)(y - q)$

Exercise Set B.5

1. $x - 3$ **3.** $\dfrac{1}{(x - 2)(x^2 + 4)}$
5. $\dfrac{1}{x - 2}$ **7.** $\dfrac{3b}{2a}$ **9.** $\dfrac{a^2 + 1}{a - 1}$
11. $\dfrac{x^2 + xy + y^2}{(x - y)^2}$ **13.** 2
15. $\dfrac{x^2 + x}{(x + 4)(x - 2)}$
17. $\dfrac{x^2 - xy + y^2}{(x - y)(x + y)}$
19. $\dfrac{4x - 2}{x^2}$ **21.** $\dfrac{36 - a^2}{6a}$
23. $\dfrac{4x + 11}{(x + 3)(x + 2)}$
25. $\dfrac{3x^2 + 6x - 6}{(x - 2)(x + 2)}$
27. $\dfrac{6ax^2 - 4ax + a}{(x - 1)^3}$ **29.** $\dfrac{8}{x - 5}$
31. $\dfrac{2a}{a + b}$ **33.** $\dfrac{-9}{(x + 5)(x - 4)^2}$
35. $\dfrac{-p^2 - pq - q^2}{p(2p + q)(p - 5q)}$
37. $\dfrac{1 + x}{1 - x}$ **39.** $-\dfrac{1}{ax}$ **41.** $a - 1$
43. $-\dfrac{1}{4 + 2h}$ **45.** $\dfrac{a + x}{a^2x^2}$
47. $\dfrac{x}{1 + 2x}$ **49.** $\dfrac{1}{a + a^2}$
51. $\dfrac{x^2 + y^2}{x^2 - y^2}$ **53.** $\dfrac{a^2 - b^2}{2}$
55. $\left(\dfrac{a}{b}\right)^{a+b}$

Index